12th International Seminar on Ocean and Coastal Engineering, Environmental and Natural Disaster Management 2024

IOP Conference Series: Earth and Environmental Science
Volume 1473

Surabaya, Indonesia
27 September 2024

ISBN: 979-8-3313-2008-9
ISSN: 1755-1307

Printed from e-media with permission by:

Curran Associates, Inc.
57 Morehouse Lane
Red Hook, NY 12571

Some format issues inherent in the e-media version may also appear in this print version.

Institute of Physics
Dirac House, Temple Back
Bristol BS1 6BE UK

Phone: 44 1 17 929 7481
Fax: 44 1 17 920 0979

techtracking@iop.org

Additional copies of this publication are available from:

Curran Associates, Inc.
57 Morehouse Lane
Red Hook, NY 12571 USA
Phone: 845-758-0400
Fax: 845-758-2633
Email: curran@proceedings.com
Web: www.proceedings.com

TABLE OF CONTENTS

 doi:10.1088/1755-1315/1473/1/011001

IOP Conf. Series: Earth and Environmental Science **1473** (2025) 011001 doi:10.1088/1755-1315/1473/1/011001

Preface ISOCEEN

Dear Colleagues,

It is an honor and pleasure to welcome you to the Joint Conference of 12th International Seminar on Ocean and Coastal Engineering, Environmental and Natural Disaster Management (ISOCEEN) with the 1st International Conference on Medicine, Marine Technology, and Social Science (ICoMMeS) on 27th September 2024.

On behalf of the organizing committee, I would like to warmly thank all the authors who, with their presentations, generously contributed to the lively exchange of scientific information that is so vital to the endurance of scientific conferences of this kind. Special thanks are also due to all keynote speakers and all committees for their efforts in preparing the manuscripts and managing the sessions, respectively. The theme of Seminar is "Development of Marine Medicine, Social Science, and Environmental Technology for a Sustainable Blue Economy". It will be interesting and useful for many engineers and scientists whose activities are related to research and solving problems in the marine, maritime, hydrodynamics, oil and gas, as well as Ocean and coastal area development also on marine medicine. Again, we welcome you to the Joint Conference. Hopefully you can take part and be able to enjoy the seminar hosted by the Department of Ocean Engineering, Institut Teknologi Sepuluh Nopember and Hang Tuah University.

Best regards,
Prof. Suntoyo, S.T., M.Eng., Ph.D.
Chairman ISOCEEN

IOP Conf. Series: Earth and Environmental Science **1473** (2025) 011001 doi:10.1088/1755-1315/1473/1/011001

Preface ICoMMeS

Dear Colleagues,

Prof. Dr.
Dian Mulawarmanti,
drg. MS.PBO.
University of Hang Tuah - Indonesia

International Conference on Medicine, Marine Technology and Social Science (ICoMMeS) 2024 is the 1st international hybrid seminar, collaboration with International Seminar on Ocean and Coastal Engineering, Environmental and Natural Disaster Management (ISOCEEN) dedicated to the advancement of Medicine, Marine Technology and Social Science has nurtured strong networks of researchers and professionals of related fields. With the growing number of collaborative activities between several countries, this seminar gives the opportunity to promote research collaboration among universities, industries, businesses, and ocean-coastal communities.

The theme of Seminar is Development of Marine Medicine, Social Science, and Environmental Technology for Sustainable Blue Economy. It will be interesting and useful for many engineers, doctors, practitioners and scientists whose activities are related to research and solving problems in medicine, pharmacies, social science, marine, maritime, hydrodynamics, oil and gas, fisheries as well as ocean area development which are solved in an integrated manner. Again, we welcome you to the ICoMMeS 2024. Hopefully you can take part and be able to enjoy the seminar hosted by Hang Tuah University collaboration with Department of Ocean Engineering Institut Teknologi Sepuluh Nopember.

Best regards,
Prof. Dr. Ir. Dian Mulawarmanti, drg. MS. PBO.
Chairman ICoMMeS

IOP Conf. Series: Earth and Environmental Science **1473** (2025) 011001 doi:10.1088/1755-1315/1473/1/011001

Editorial Team

IOP Conf. Series: Earth and Environmental Science **1473** (2025) 011001 doi:10.1088/1755-1315/1473/1/011001

Committee

Advisory Committee:

1. Rector of Institut Teknologi Sepuluh Nopember, Indonesia
2. Rector of Hang Tuah University, Indonesia
3. Vice Rector of Hang Tuah University
4. Dean of Marine Technology Faculty – ITS, Indonesia
5. Head of Ocean Engineering Department, FTK-ITS, Indonesia

Organizing Committee:

1. Prof. Suntoyo, Ph.D. (Chairman)
2. Silvianita, Ph.D. (Secretary)
3. Dr. Dendy Satrio (Paper and Proceeding Chair)
4. Wimala L. Dhanista, M.T. (Finance Chair)
5. Prof. Dr. Ir. Ninis T., M.P. (Chairman)
6. Assoc. Prof. Dr. Nurul Rosana, S.Pi., M.T. (Vice Chairman)
7. Drs. S. Teguh Wiyono, M.AP.(Secretary)
8. Akbar Rusmangera, S.T.(Secretary)
9. Assoc. Prof. Dr. Dewi Setyowati, S.H., M.H. (Finance Chair)
10. Asst. Prof. Erik Sugianto, S.T., M.T., Ph.D. (Paper and Proceeding Chair)
11. Asst. Prof. Dr. Apt. Giftania Wardani Sudjarwo, S.Farm., M.S. (Paper and Proceeding Vice Chair)
12. Dr. Wahyudi (ITS, Indonesia)
13. Yoyok Setyo Hadiwidodo, Ph.D. (ITS, Indonesia)
14. Dr. Hasan Ikhwani (ITS, Indonesia)
15. Dr.Eng. Rudi Walujo Prastianto (ITS, Indonesia)
16. Prof. Dr. Eng. Muhammad Zikra (ITS, Indonesia)
17. Dr.Eng. Shade Rahmawati (ITS, Indonesia)
18. Dr. Sujantoko (ITS, Indonesia)
19. Santi Frestiqauli, S.T., M.T. (ITS, Indonesia)
20. Ferdita Syalsabila, S.T., M.T., (ITS, Indonesia)
21. Anita Kusuma Wardhani, S.Kel., M.T. (ITS, Indonesia)
22. Sidharta Gautama, S.E (ITS, Indonesia)
23. Rizky Dwi Putra, S.A. (ITS, Indonesia)
24. Zuhud Ubaidillah, S.SI. (ITS, Indonesia)
25. Zahra Fitria Rosyadah (ITS, Indonesia)
26. Aris Pratomo (ITS, Indonesia)
27. Sutrisno (ITS, Indonesia)
28. Nuryadi (ITS, Indonesia)

29. Catur Wahyudiono (ITS, Indonesia)
30. M. Rizky Syarifudin, S.T. (ITS, Indonesia)

International Scientific Committee:

1. Prof. Hitoshi Tanaka, Dr. Eng. (Tohoku University, Japan)
2. Dr. Ir. Irwandi Idris, MSc. (Chairman, Indonesian Coastal Management Expert Association)
3. Prof. Widi Agoes Pratikto, PhD (ITS, Indonesia)
4. Prof. Daniel M. Rosyid, PhD (ITS, Indonesia)
5. Prof. Eko Budi Djatmiko, PhD (ITS, Indonesia)
6. Prof. Mukhtasor, Ph.D. (ITS, Indonesia)
7. Prof. Suntoyo, Ph.D (ITS, Indonesia)
8. Assoc. Prof. Dr. Alvaro Semedo (IHE-Delft, Netherlands)
9. Makoto Umeda, Dr.Eng (Tohoku University, Japan)
10. Dr. Joost Stronkhorst (HZ University of Applied Sciences, Netherlands)
11. Ir. Liliane Geerling R.T.D. (HZ Universityof Applied Sciences, Netherlands)
12. Assoc. Prof. Ahmad Sana, Ph.D. (Sultan Qaboos University, Oman)
13. Priyantha Ranjan, PhD (Curtin University, Australia)
14. Prof. Dr.Eng. Nguyen Trung Viet (Thuyloi University, Vietnam)
15. Agro Wisudawan, M.T. (University College Cork, Ireland)
16. Dr.Eng Totok Suprijo (ITB, Indonesia)
17. Prof. Dr. Muhammad Arsyad Thaha (Hasanuddin University, Indonesia)
18. Dr. Taufiqur Rachman (Hasanuddin University, Indonesia)
19. Dr.Eng Sitang Pilailar (Kasetsart University, Thailand)
20. Assoc. Prof Ts Dr Mohd Faris Khamidi (Qatar University, Qatar)
21. Assoc. Prof. Ir. Dr. Eng. Ahmad Fitriadhy (Universiti Malaysia Terengganu, Malaysia)
22. Dr.Eng. Purwanto Bekti Santoso (Unsoed, Indonesia)
23. Assoc. Prof. Dr. Herin Setianingsih, dr., M.Kes. (Hang Tuah University, Indonesia)
24. Assoc. Prof. Wienta Diarsvitri, dr., M.Sc., MHPE., Ph.D,FISPH.,FISCM. (Hang Tuah University, Indonesia)
25. Prof. Dr. Rima Parwati Sari, drg., M.Kes., PBO. (Hang Tuah University, Indonesia)
26. Asst. Prof. Kamarudin, S.HI., LL.M. (Hang Tuah University, Indonesia)
27. Asst. Prof. Husni Tamrin, S.AP., M.KP. (Hang Tuah University, Indonesia)
28. Prof. Dr. Drs.Viv Djanat Prasita, M.App.Sc. (Hang Tuah University, Indonesia)

Peer Review Statement

All papers published in this volume have been reviewed through processes administered by the Editors. Reviews were conducted by expert referees to the professional and scientific standards expected of a proceedings journal published by IOP Publishing.

- **Type of peer review:** Single Anonymous
- **Conference submission management system:** Morressier
- **Number of submissions received:** 152
- **Number of submissions sent for review:** 114
- **Number of submissions accepted:** 72
- **Acceptance Rate (Submissions Accepted / Submissions Received × 100):** 47%
- **Average number of reviews per paper:** 1
- **Total number of reviewers involved:** 18
- **Contact person for queries:**
  Name: Dr. Dendy Satrio, S.ST.
  Email: dendy.satrio@its.ac.id
  Affiliation: Institut Teknologi Sepuluh Nopember (ITS)

IOP Conf. Series: Earth and Environmental Science **1473** (2025) 012001 doi:10.1088/1755-1315/1473/1/012001

Planning the Coastal Area of Jangkaran Village with the Concept of Eco-Marine Tourism and Disaster Resilience

Aisya Nazifa[1] and Atrida Hadianti [1*]

[1] Department of Architecture and Planning, Gadjah Mada University, Yogyakarta, Indonesia

*E-mail: atrida.h@ugm.ac.id

Abstract. The coastal area of Jangkaran Village in the southern part of Kulon Progo Regency has abundant natural resources, including mangroves, natural habitats, and beaches, valuable assets to develop an attractive tourism village. In contrast, there are issues regarding poor accessibility, threat to mangrove conservation land due to tourism activities and risk of tsunami. Therefore, coastal area planning is needed to optimize the potential of natural resources by focusing on improving conservation, tourist facilities and disaster resiliency. Using the eco-marine tourism and disaster resilience concept, this study encourage the development of the tourism, fisheries, environmental conservation and disaster sectors in a sustainable and integrated manner. This concept is reflected in the coastal zoning and building layout; coastal facilities; signage; green space; circulation paths; and institutions. To produce the masterplan of Jangkaran Village coastal area, the study use Soft System Methodology (SSM) method with a comprehensive planning. This study shows the importance of in-depth understanding of the beneficiaries towards the 'rich picture' of the site. The deeper the study is towards stakeholders benefits the more directed the area plan will be developed, and the formulation of interventions carried out will be appropriate in achieving planning objectives.

1. Introduction

Jangkaran Village is a coastal village with a lot of natural and artificial resources, such as beaches, rivers, ponds, and mangrove forests making the village as of the potential attractions in Kulon Progo Regency. Some og its are, Pasir Mendit Beach, panoramic views of mangrove forests and ponds, namely Mangrove Pasir Kadilangu, Mangrove Jembatan Api-Api, and Eduwisata Wana Tirta. All of these attractions create a unique natural beauty with a combination of mangrove forest, beach and pond panoramas in one landscape. Pasir Mendit Tourism has an attraction that can be compete with Congot Beach tourism which is the fourth most popular tourist attraction in Kulon Progo Regency in 2020 [1]. This is possible with the adequate supporting facilities of the three mangrove forest tourist destinations and a diverse beach ranging from viewing posts, canoes, wooden trails, and a variety of photo spots. However, the three tourist destinations in Jangkaran Village do not yet have their own uniqueness because the attractions offered tend to be similar.

Aquaculture is the leading land use while vaname shrimp is the leading commodity [2]. The local community farm, harvest and export of Vaname shrimp farming is managed by local community groups and sold to the export market [3]. The Kulon Progo Regency Government established Jangkaran Village as a center for shrimp farming to advance the coastal area of

Jangkaran Village. The development of shrimp farming is supported by the placement of the pond area ±100 m from the coastline to avoid high sea waves. Despite having great potential in the field of shrimp farming, local residents who own ponds choose to develop natural attractions that are more profitable because of the increasing interest in tourist visits in recent years [4]. Nevertheless, both sectors provide benefits that are worth developing.

The development of Jangkaran Village aligns with the development direction of the Duchy Strategic Area (KSK) of the South Coast of Kulon Progo contained in the Regional Spatial Plan of Yogyakarta Special Region (RTRW) 2023-2043, which is planned as an agricultural zone, marine conservation reserve, and tourism. However, it is necessary to pay attention to the impact of development, especially the influence of the tourism sector on marine conservation zones and the threat of natural disasters. Tourism in mangrove forest can cause negative impacts on coastal ecosystems. A visible impacts in this Village, ncludes damage and inhibition of mangrove growth and decreased aesthetic value due to littering of tourism visitors [4]. In addition, the conversion of mangrove land indirectly increases the risk of tsunami [5]. However, the development of mangrove forest tourism at Pasir Mendit Beach and Pasir Kadilangu Beach has shown positive impacts on the economic and social development of the village, such as increased employment and community income [4]. Another positive impact is the enlargement the Pasir Kadilangu Beach mangrove forest, with an increase of 1,442.55 m^2, from 3,979.20 m^2 in 2014 to 5,421.75 m^2.

Based on the problems and potentials in Jangkaran Village, the development of the concept of eco-marine tourism community-based is the suitable solution in developing the area, integrating tsunami disaster resilience planning. The eco-marine tourism approach can increase the quantity of mangrove forests and biodiversity, contributing to the overall economic growth of the community. Simultaneously tsunami disaster resilient planning will stengthen strengthen used to increase the capacity of coastal areas and minimize environmental impacts due to disasters that can hamper the community's economy.

2. Literature Review

2.1 Planning Theories and Concepts

This planning uses several theories, namely the basic theory in the form of coastal areas and tourist destinations, as well as the main theory in the form of eco-marine tourism and tsunami disaster resilience.

a. *Coastal areas*

Coastal areas have special characteristics that affect both sea and land [6]. Therefore, coastal area planning needs a comprehensive approach through the application of the Integrated Coastal Zone Management (ICZM) [7] and regional planning component of the Building and Neighborhood Planning (RTBL).

b. *Tourist destinations*

Tourism destinations have tourist attractions, public facilities, tourism, accessibility, and mutual communities [8]. In supporting comprehensive tourism planning, the elaboration of 10A tourist destination attributes in the form of Attractions, Accessibility, Amenity, Accommodation [9][10], Activities, Awareness, Appearance, Appreciation, Action and Accountability [11], as well as 3 criteria in the form of something to see, something to do, and something to buy [12].

c. *Eco-marine tourism*

Ecotourism is tourism that emphasizes the value of nature conservation, the value of education or social empowerment, culture and the economy of local communities [13].

Marine tourism is the integrated management of marine tourism and conservation-based fisheries [14].

d. *Tsunami disaster resilience*
Disaster resilient villages have the ability to recognize threats and organize human resources to reduce vulnerability and increase disaster mitigation capacity [15].

Several planning concepts or theories that have been obtained from various previous literature reviews will be elaborated and used as a variable in conducting the planning process. The components and concepts used to plan Jangkaran Village are shown in Figure 1.

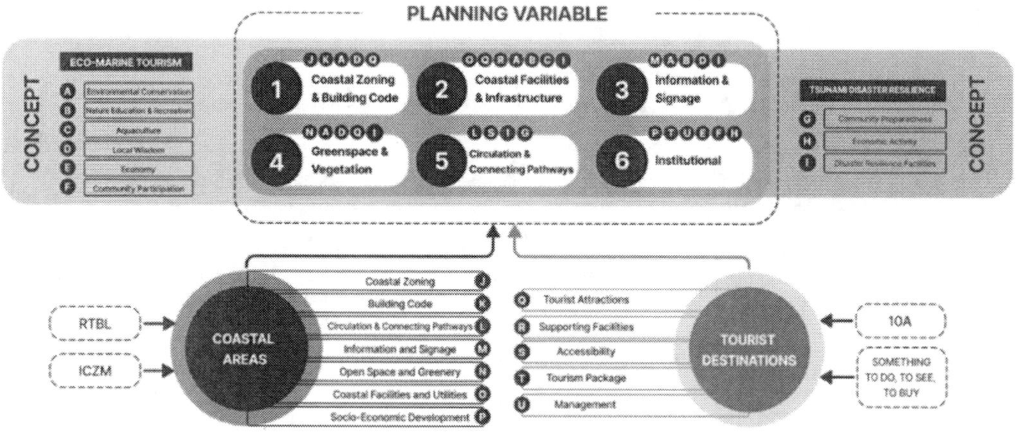

Figure 1. Elaboration diagram of the Planning Concept.

2.2 Ideal Concept in Area Planning

The study use an Ideal Concept to analyze the existing conditions of the coastal area of Jangkaran Village conduct gap analysis between ideal conditions and existing conditions. The indicators and criteria used in this planning are described in Table 1.

Table 1. Ideal concept

Variable	Indicator	Criteria	Foundation
Coastal Zoning & Building Code	Land use	Zoning is clearly divided according to the land allotments specified in the regulations	Permen PU No. 06/PRT/M/2007 [16]
	Coastal Zoning	High-density land use is restricted in the shoreline zone and buffer zone (within 100 meters from the shoreline), which are designated for protected ecosystem conservation	Kementrian ATR/BPN in Daulat (2021) [17]
	Environmental Conservation	Land use is oriented towards conservation, tourism, and fisheries areas, with a focus on controlling landscape changes	Sekartjakrarini (2020) [13]
	Building Intensity	The Basic Building Coefficient (KDB) is less than 80%, with a maximum of 1-2 floors allowed	Permen PU No. 06/PRT/M/2007 [16] & Yoeti (2002) [12]
	Building Condition	Buildings must be in good condition	
	Building Architecture	The design and shape of the building should align with the local character and enhance the image of the coastal tourism area	Troung (2009) [10], Islam and Savitri (2023) [11], Morrison (2014) [9], & Permen PU 06/PRT/M/2007 [16]
Coastal Facilities & Infrastructure	Availability of Facilities and Infrastructure	1. Supporting facilities for tourism activities include a souvenir center, an information center, and a musholla 2. Infrastructure and utilities are complete and evenly distributed, including electricity, clean water, drainage, sanitation, water treatment plant, sewage treatment plant, and disaster evacuation routes	
	Disaster Resilience Facilities	The existence of facilities to enhance tsunami preparedness, such as evacuation assembly point, search and rescue posts, green belts, and embankments	Oktari (2019) [18], Akhirianto et al. (2023) [19], & BNPB (2012) [20]
	Educational and Nature Tourism Activities	Available supporting facilities for attractive and educational activities for tourists include aquaculture education and environmental conservation programs focused on mangroves and turtles	Sekartjakrarini, 2020 [13], Cobbinah in Aziz (2023) [21]
	Leading Product Processing Activities	The presence of supporting facilities for the processing of regional superior products	Sekartjakrarini, 2020 [13], Jeami & Fahmi (2023) [22]
Information & Signage	Navigation and information system	1. There are clear signs indicating disaster threat levels, especially in coastal areas 2. There is an interactive and informative information board 3. There is a landmark serving as a village icon that reflects the character of the coastal area	Hainess-Young and Potcjin (2011) [6] Permen PU No. 06/PRT/M/2007 [16], & Cobbinah in Aziz (2023) [21]
	Tourist education	There are educational information boards providing details on nature conservation and aquaculture	Sekartjakrarini, (2020) [13]
Greenspace & Vegetation	Public Activity Center	1. There are attractive open spaces 2. Open spaces that accommodate local and tourist cultural activities, such as children's playgrounds, sports facilities, and cultural parks.	Permen PU No. 06/PRT/M/2007 [16] & Sekartjakrarini, (2020) [13]
	Quality of Open Space and Vegetation	1. Use of green belts and mangroves for tsunami protection and shading pathways 2. Use of rain gardens and porous pavement in public open space.	Oktari (2019) [18], Akhirianto et al. (2023) [19], & BNPB (2012) [20]

continued

continued

Variable	Indicator	Criteria	Foundation
Circulation & Connecting Pathways	Parking Lot	Parking spaces are organized by vehicle type and activity type, and are integrated with the activity center	Permen PU No. 06/PRT/M/2007 [16]
	Safety and Comfort	1. Pedestrian paths are well-maintained, connected throughout the area, and supported by complete street furniture, including lights, bench, bollards, and zebra crossings 2. There is a designated path for disabled pedestrians. 3. There is a map of disaster evacuation routes	Permen PU No. 06/PRT/M/2007 [16], Oktari (2019) [18], Akhirianto et al. (2023) [19], & BNPB (2012) [20]
	Road Circulation	The road is in good condition, has no potholes, and is accessible for mini tour buses	Troung (2009) [10], Islam and Savitri (2023) [11], & Permen PU No. 06/PRT/M/2007 [16]
	Private & Public Vehicle Circulation	1. Road circulation is one-way or two-way and accessible for motorcycles and cars. 2. Mini tour buses are available for travel to the Village Tourism	
Institutional	Community Participation	Tourism and fisheries development involve local communities	Sekartjakrarini, 2020 [13], Cobbinah in Aziz (2023) [21], Oktari (2019) [18], Akhirianto et al. (2023) [19], & BNPB (2012) [20]
	Capacity building of human resources	1. Environmental conservation activities conducted by the community 2. Training and guidance on conservation and aquaculture are provided to local communities 3. Procurement of tsunami preparedness training for local communities	
	Institutional System	1. There is an organization focused on tsunami disaster preparedness 2. There is a tourism management organization managed by the community and NGOs	BNPB (2012) [20] & Morrison (2014) [9]

3. Planning Methodology

3.1 Planning Approach

This plan aims to design the coastal area of Jangkaran Village as a community-based conservation and aquaculture educational tourism area to preserve coastal ecosystems and boost tsunami disaster response. It chooses comprehensive planning approach, which refers to a planning process that is carried out thoroughly, deeply and complexly so as to be able to formulate the main problems and find the right solutions in overcoming them.

3.2 Planning Location

The site is located in the coastal area of Jangkaran Village, Temon Sub-district, Kulon Progo Regency, Yogyakarta Special Region with an area of 144 ha or 39.4% of the total area of Jangkaran Village (365.64 ha). The location is selected based on the following considerations:

1. The coastal potential of Jangkaran Village has not been optimally utilized, such as shrimp farming centers, turtle and mangrove conservation, and the development of natural and educational tourism.
2. Based on the Regional Spatial Plan of Yogyakarta Special Region (RTRW) 2023-2043, the coastal area of Temon is included in the Strategic Area of the Duchy of the South Coast of

Kulon Progo and is geographically strategic because it is located near YIA Airport, which is 7 km away.

3. The evident signs of mangrove damage, the threat of tsunami & abrasion disasters and inadequate supporting facilities have hindered the development of village tourism.

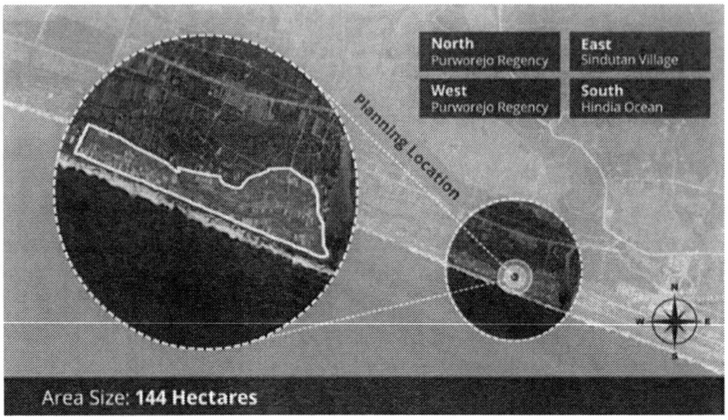

Figure 2. Site Plan.

3.3 Data Collection Method

The data collection methods in this planning are as follows:

1. Observation
   This planning uses anecdotal records (collection of all data in the field) and checklist notation (data collection according to a list) to record activities and necessary data, such as physical spatial, socio-cultural, and environmental, through direct observation of everything that happens incidentally. Observation data will be processed in GIS software to obtain digital data and information.
2. Documentation Studies
   Documentation is data derived from the results of recording on an object of observation using a camera to produce data in the form of images, sound or video. Documentation is used to describe the actual conditions in the field.
3. Interviews
   This study uses interviews to verify or recheck data or information obtained during observation and literature study.
4. Literature Review
   This method of data collection uses documentary sources in the form of agency documents, literature, similar research reports, seminar materials and journals.

3.4 Data Analysis Method

The data analysis method in this plan is as follows:

1. Soft System Methodology (SSM)
   A holistic approach to understanding the problem, forming a conceptual model, assessing feasibility and implementation.
2. Gap Analysis
   Differences or gaps from ideal conditions to actual conditions to analyze and formulate recommendations for plans or alternative plans.
3. Descriptive Analysis

This analysis method provides a detailed explanation of the actual conditions and the process of preparing the coastal area plan for Jangkaran Village.

3.5 Planning Method

The planning method used is as follows:

1. Substitute, Combine, Adapt, Modify, Put To Other Uses, Eliminate/Elaborate, Reverse (SCAMPER)
   The method collaborate findings from existing conditions with plans from ideal conditions to create a new concept from the modified results that solve the problem by applying the concept of eco-tourism and tsunami disaster resilience.
2. Analytical Hierarchy Process (AHP)
   The method will determine the most suitable alternative plan based on development priorities uses the 5E criteria (Efficacy, Efficiency, Effectiveness, Ethicality, Elegance)
3. Cost Benefit Analysis (CBA)
   Conducting CBA aims measure and interpret the feasibility of planning programs based on various calculations of the benefits that will be received from planning these activities.
4. Strength, Weakness, Opportunity, and Threat (SWOT)
   This method helps to identify challenges and opportunities in plan implementation.

3.6 Planning Framework

The planning stages from the planning site selection process to the planning evaluation are outlined in Figure 3.

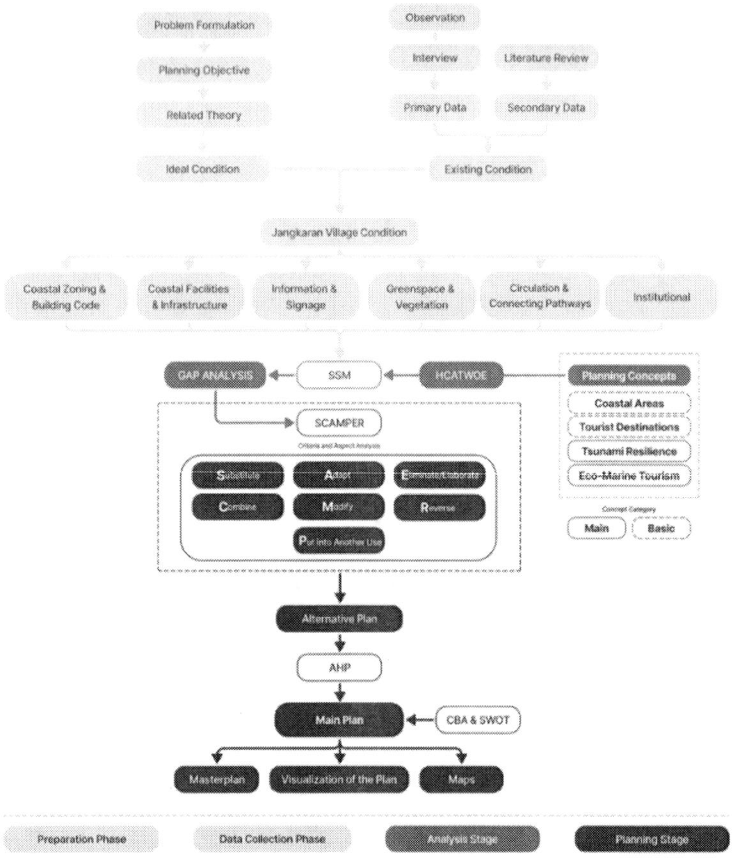

Figure 3. Planning Framework.

4. Results and Discussion

4.1 Overview of the Planning Site

Jangkaran Village is located at the mouth of the Bogowonto River which is famous for its coastal beauty with panoramic views of Kadilangu Sand Beach and Mangrove Forest. The coastal area of Jangkaran Village has slightly different characteristics from coastal areas in Kulon Progo Regency, namely having a fairly large mangrove forest conservation area of 7.07 hectares.

In general, Jangkaran Village is dominated by ponds. Majority of the villagers work as fishermen and cattle farmers. In addition, Jangkaran Village has been designated as a Center for Aquaculture, specializing in the superior commodity of Vaname Shrimp, and also designated as a Disaster Resilient Village in 2012 due to the threat of tsunami and high abrasion.

Jangkaran village has 3 potential tourist destinations. Geographically, the coastal area of Jangkaran Village is located ± 7 km from Yogyakarta International Airport (YIA) and adjacent to several tours, namely Congot Beach, Glagah Beach and Mangrove Demang Gedi. Thus encouraging the development of existing tourism should be directed into educational tourism and natural recreation that is interesting and unique.

4.2 Potential and Problem Analysis

The potential and problems of the area were obtained from the results of the plan gap analysis using planning variables (Table 2.). The results of the analysis are then used as a consideration for the formulation of alternative plans as a solution to solve existing problems and optimize the potential.

Table 2. Potential and Problem Analysis

Variable	Potential	Problem
Coastal Zoning & Building Code	• Buffer areas of mangroves located near settlements and pond areas create a distinctive coastal landscape.	• The land use zone has not yet been clearly defined. • There are buildings in the restricted zone (within 100 meters of the coastline). • Some buildings do not meet the applicable standards. • There are still buildings in poor condition and rundown. • Buildings do not reflect the value of coastal culture
Coastal Facilities & Infrastructure	• The area has a well-distributed electricity network, clean water supply, and a good evacuation system.	• Existing facilities are insufficient to support tourism, fisheries, and disaster resilience activities, such as lacking a tourist information center and temporary evacuation sites. • Existing infrastructure is incomplete and uneven, particularly waste and drainage facilities.
Information & Signage	• There are clear boundaries between residential and tourist areas, marked by gates and monuments.	• There is no disaster information board in the tourism and beach area. • The information boards currently available are not informative or interactive. • Existing landmarks and tourist monuments are not yet in harmony with the image of the coastal tourism area. • There are no educational boards about conservation and aquaculture.
Greenspace & Vegetation	• There is vacant land that can be developed into green open space.	• There is a lack of green open space and greenery in residential areas and along roads. • Rain gardens and porous pavements have not yet been implemented.
Circulation & Connecting Pathways	• There are already clear evacuation signs and circulation routes	• Parking facilities and roads are in poor condition and are not connected to pedestrian paths.
Institutional	• There is active community involvement in managing village tourism	• There is no integrated organization, such as Pokdarwis, for managing the area.

4.3 Alternative Development

Based on the analysis in Table 2. and the SCAMPER analysis, the planning of this area will focus on the development of tourism, conservation and fisheries sectors. This planning adapts Eco-marine tourism concept which focuses on providing supporting facilities for tourism, fisheries and institutional activities, and the concept of disaster resilience. The concept focuses on providing infrastructure, zoning regulations and improving the quality of human resources. Based on these planning objectives, two alternative plans were formulated, as detailed in Table 3.

Table 3. Alternative Plan

Alternative Plan	Descriptive
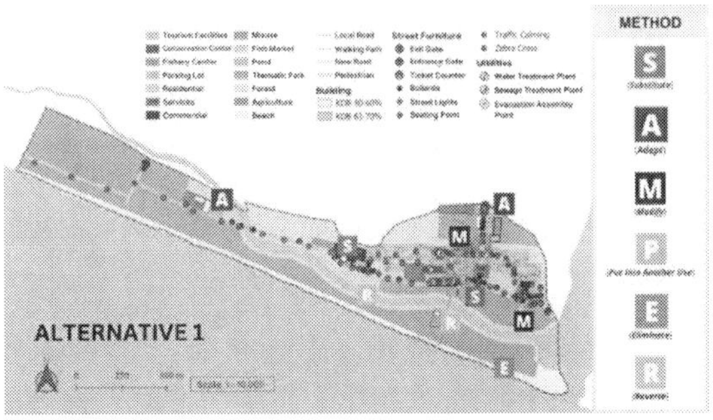	Alternative 1: Integrated Blue-Green Space Focus on optimizing the use of pond and mangrove land by enhancing the integration of tourism and fisheries sectors through marine tourism and special interest activities, improving regional connectivity, and supporting the equitable distribution of facilities.
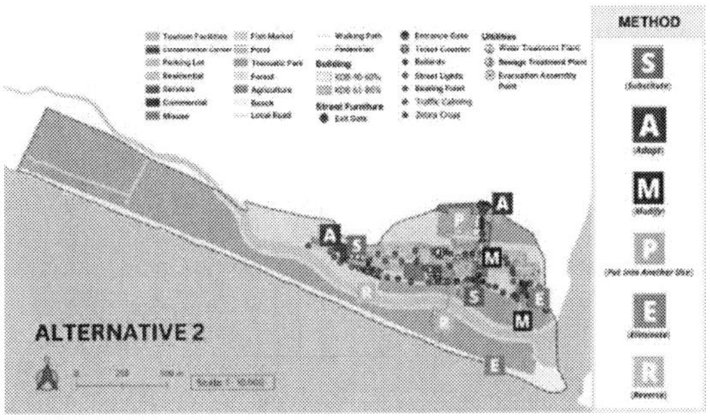	Alternative 2: Integrated Ecology Zone Focus on social and environmental aspects to preserve ecosystem services, prevent the conversion of mangroves and ponds, and enhance ecological functions by expanding areas designated for mangrove conservation and increasing the variety of fishery commodities.

Based on the results of the selection of alternative plans using the Analytical Hierarchy Process method in Table 4., it is known that the first alternative has a value of 0.561 or 56.1%. The first alternative has a superior value when compared to the second alternative value. The alternative plan used in this planning is the first alternative, namely Integrated Blue-Green Space.

Table 4. Analytical Hierarchy Process (AHP) Results

Alternative	Efficacy	Efficiency	Effectiveness	Ethicality	Elegance	Total	Percentage
Integrated Blue-Green Space	0,377	0,105	0,058	0,012	0,008	0,561	56,1 %
Integrated Ecology Zone	0,126	0,052	0,175	0,062	0,024	0,439	43,9 %

4.4 Masterplan

The objective of Jangkaran Village's coastal area masterplan is to create a coastal area as an educational tourism area for aquaculture & biological conservation that is disaster responsive and preserves coastal ecosystems. In terms of its physical spatial aspects, the highligt is mangrove's role as a conservation and disaster protection area as well as tourism, whereas the fisheries area remains as the leading sector and a source of community income. The planning will include the provision of activity centers, river boundary arrangements, provision of green spaces, road space planning, and also the provision of other supporting facilities. The overall intervention is described in Figure 4.

Figure 4. Masterplan Coastal Area of Jangkaran Village

4.5 Institutional Plan

The Jangkaran Village Coastal Area does not yet have an integrated institution. Currently each tourist destination has its own management institution. Therefore, it calls for the formation of a tourism awareness group (Pokdarwis) to mamage tourism activities in the coastal area in a directed, structured, and integrated manner.

The Jangkaran Village Tourism Awareness Group (Pokdarwis) is planned to have membership from trader groups, farmer groups, fishermen groups, tourism destination NGOs, and the local community. The planned organizational structure and role division within Pokdarwis are described in Table 5.

Table 5. Pokdarwis Plan in the Coastal Area of Jangkaran Village

Section	Actor	Roles
Tourism Activity Attraction	Fishermen and fish farmers group	• Tour guide for the cultivation and processing of fishery products. • Selling products at the fish market.
	NGO Wana Tirta	Mangrove and turtle conservation tour guide.
	NGO Pasir Kadilangu	Mangrove walk tour guide.
	NGO Mangrove Jembatan Api-Api	Mangrove recreation tour guide.
	Empowerment and family welfare group	Organizing events and performing arts stages.
Business Development	Empowerment and family welfare group	• Selling at the souvenir center. • Selling in commercial areas and food courts. • Managing the restaurant.
Order and Security	Youth organization and tourism NGOs	• Maintain the tourism information center. • Maintain evacuation assembly point. • Maintain the parking lot. • Guard the ticket counter.
Public Relations and Human Resource Development	Youth organization, empowerment and family welfare group, tourism NGOs, fishermen and farmers group	• Organizing human resource development training. • Promote and publicize of tourism activities.
Cleanliness and Aesthetics	Youth organization	Conduct regular monitoring and maintenance.

4.6 Development and Financing Plan

The phasing plan in the development of the coastal area of Jangkaran Village is carried out for ten years, starting from 2025 to 2034 with an expenditure of Rp 23.661.000.000. During the implementation phase, unlimited cycle of readjustment is allowed, as long as of the master plan maintains a maximum of 30% change and still follows the zoning plan and land use provisions in the current master plan. The phasing of the plan is divided into four stages:

1. Preparatory phase, focusing on deepening the study of the implementation of the regional master plan
2. The construction stage, focusing on meeting basic needs and improving existing tourist destinations which generates revenue for the next stage of development capital.
3. The development stage, focusing on the fulfillment of supporting facilities for tourist convenience.
4. Operational and maintenance stage, focusing on management, monitoring and evaluation of the entire development process.

4.7 Plan Evaluation

Planning evaluation is carried out to find out whether a plan can achieve its target and whether a planned project is feasible (profitable) or or likely to result in a loss. Therefore, the evaluation of the Jangkaran Village Coastal Area plan uses the Cost Benefit Analysis (CBA) and SWOT analysis.

An expenditure of Rp 23.661.000.000 and expected revenue of Rp 45.388.367.601 serves as the baseline data for Cost Benefit Analysis. The discount rate used is 6%, following the Indonesian interest rate per year 2024 [22], with the aim of anticipating inflation. Meanwhile, a discount rate of 10% was used to test whether the planned program would remain feasible if there was an increase in the discount rate.

At a discount rate of 6%, the plan has an Net Present Value (NPV) value of Rp 9,957,555,173 (NPV>0) and a Benefit Cost Ratio (BCR) value of 1.48 (BCR>1). Meanwhile, at discount rate value of 10% the NPV value is Rp 5,121,268,882 (NPV>0) and a BCR value is 1.27 (BCR>1). This proves eco-tourism and disaster resilience development plan for Jangkaran Village's coastal area is feasible for implementation. In addition, the Internal Rate of Return (IRR) value between the discount rate of 6% and 10% is 13.09% (IRR), greater than the discount rate (IRR>discount rate), indicating that the plan can be classified as a good project.

Based on the SWOT analysis, several planning strategies were formulated to increase strengths, reduce weaknesses, make good use of opportunities, and avoid threats that may occur in the future. The following SWOT analysis results are shown in Table 6.

5. Conclusion

The comprehensive planning of Jangkaran Village's coastal area involves various integrated aspects, including physical and non-physical aspects. The main focus of this plan is to maintain the role of coastal ecosystems as tsunami disaster protection areas and conservation reserves while ensuring active involvement of local communities in development and management of the site.

There are several challenges in the implementation of this plan including the limited capability of funding sources, the existence of similar tourism competitors and the threat of tsunami and abrasion during the area development process. Therefore, the strategy offered an implementation plan, financing scheme, as well as a partnership scheme with several scenarios to minimize losses in the development process.

Table 6. SWOT Analysis

SWOT	**Strength (S)** • Has abundant natural resources, including mangrove forests, ponds, beaches, and turtles. • Prioritizes utilizing village potential through the development of special interest and nature tourism. • The local community plays an active role in the area's development.	**Weakness (W)** • Inadequate infrastructure to support tourism activities. • Tourism facilities are in poor condition, which decreases their attractiveness. • There are no tour packages and integrated tourism management. • There is no tourism mobilizing institution, such as Pokdarwis. • Limited funding sources for the development of the area.
Opportunity (O) • Potential to establish partnerships with various parties, including government, local communities, the private sector, and investors. • Opportunity to attract both local and international tourists. • Lack of mangrove and sea turtle conservation educational tourism in Kulon Progo Regency.	**S-O Strategy** (Strategy of Using Strengths to Obtain Benefit) • Focusing on developing the coastal area into an eco-marine tourism village. • There is active participation from local communities.	**W-T Strategy** (Strategy to Reduce Weaknesses to Utilize Opportunities) • Revitalize and expand supporting infrastructure for tourism activities. • Develop tour packages and empower local communities to become tour guides. • Establish pentahelix cooperation with various stakeholders for plan implementation and financing.
Threat (T) • Environmental damage from waste generated by tourism activities. • Risk of natural disasters, such as tsunamis and coastal erosion • Potential competition from similar tourism destinations nearby, such as Demang Gedi Mangrove Tourism.	**S-T Strategy** (Strategy of Using Strengths to Overcome Threats) • Establish CHSE (Cleanliness, Health, Safety, and Environmental Sustainability) certification. • Enhance disaster mitigation efforts in facility development and improve local community capacity. • Strengthen tourism branding through various platforms	**W-T Strategy** (Strategy of Reducing Weaknesses to Overcome Threats) • Develop a plan implementation and financing scheme with scenarios: business as usual, low ambition (pessimistic), moderate ambition, and high ambition (optimistic). • Establish cooperation with neighboring villages that offer similar attractions to provide a diverse range of tour package options

References

[1] Asiya. Socioeconomic Conditions of Merchant Households in Congot Beach and Pasir Mendit Beach Tourism Objects, Jangkaran, Temon, Kulon Progo. 2017.

[2] Zakariya MI, Anna Z, Dhahiyat Y. Contribution of Marine Tourism to Fishermen's Income in Tidung Island, Thousand Islands Regency, DKI Jakarta Province. Journal of Fisheries and Marine Affairs. 2017;8(1):105–121.

[3] Kulon Progo Regency Government. Kulon Progo Regency Government. 2012. Regent Harvests Vaname Shrimp at Pasir Mendit Temon.

[4] Putri FM, Prasetiyowati SH. The Impact of Tourism Activities on the Mangrove Forest of Pasir Kadilangu Beach, Kulon Progo Regency, Yogyakarta Special Region. Journal of Empowerment Community and Education. 2021;1(3):187–94.

[5] Putri DAM, Handziko RC. Capacity Building for Disaster Resilient Villages against the Impacts of Climate Change in Jangkaran Village, Kulon Progo Regency. In: Proceedings of the UMS National Geography Seminar. 2016.

[6] Haines-Young R, Potschin M. Integrated Coastal Zone Management and the Ecosystem Approach. PEGASO. 2011;7:4–16.

[7] Post JC, Lundin CG. Guidelines for integrated coastal zone management. Vol. 9. Washington, DC: Wold Bank; 1996.

[8] Ministry of Tourism and Creative Economy (Permenparekraf). Regulation of the Minister of Tourism and Creative Economy of the Republic of Indonesia Number 9 of 2021 concerning Guidelines for Sustainable Tourism Destinations. Jakarta: Ministry of Tourism and Creative Economy; 2021.

IOP Conf. Series: Earth and Environmental Science **1473** (2025) 012001 doi:10.1088/1755-1315/1473/1/012001

[9] Truong TH, King B. An evaluation of satisfaction levels among Chinese tourists in Vietnam. International Journal of Tourism Research. 2009;11(6):521–35.
[10] Islam PI, Savitri FM. The Effect of Attraction, Accessibility, Amenity, Ancillary on Pilgrims' Revisit Interest in the Tomb of Sheikh Abu Bakar Jepara. Journal of Islamic Tourism, Halal Food, Islamic Traveling, and Creative Economy. 2023;3(1):76–81.
[11] Morrison AM. The 10 As of Successful Tourism Destinations. 2014.
[12] Yoeti OA. Marketing Strategy Planning for Tourist Destinations. Jakarta: Pradnya Paramita; 2002.
[13] Sekartjakrarini S. Criteria and Indicators of Indonesian Ecotourism. Bogor: IdeA; 2020.
[14] Millaty PK. Eco-tourism Business Empowerment Center in Munengan Godean with Bamboo Tectonics Approach. 2019.
[15] National Disaster Management Agency (BNPB). National Disaster Management Agency. 2016. National Disaster Management Agency (BNPB) Regulation Number 1 of 2012 on General Guidelines for Disaster Resilient Villages.
[16] Minister of Public Works. Regulation of the Minister of Public Works (Permen PU) No.06/PRT/M/2007 on General Guidelines for Building and Environmental Planning. 2007.
[17] Daulat A, Salim HL. Coastal spatial pattern based on disaster mitigation. In Disaster Management and Sustainable Development in the Sunda Strait Coast. Bogor: PT Penerbit IPB Press; 2021.
[18] Oktari RS. Capacity Building for Disaster Resilient Villages. Indonesian Journal of Community Engagement. 2019 Mar 31;4(2):189.
[19] Akhirianto NA, Giyarsih SR, Mardiatno D. Community preparedness of disaster resilient villages against tsunami threat in Cilacap Regency. Indonesian Geography Papers. 2023 Sep 12;37(2):158.
[20] National Disaster Management Agency (BNPB). Tsunami Disaster Risk Reduction Masterplan. 2012.
[21] Aziz MRA. Development of Kamulyan Beach Area of Cilacap Regency based on Disaster Mitigation Ecotourism. 2023.
[22] Bank Indonesia. Bank Indonesia. 2024. Bi-Rate Rises 25 Bps to 6.25%: Strengthening Stability and Safeguarding Growth from the Impact of Global Slowdown.

Analysis of the Suitability for Placing Interlocking Modular Hexareef Structures as an Environmentally Friendly Coastal Protection in the Gili Ketapang Probolinggo Marine Conservation Area

Suwardi[1*], H D Armono[1], D Satrio[1], E Jadmiko[2], A P F L Putro[2], and G J Ananta[3]

[1] Department of Ocean Engineering, Institut Teknologi Sepuluh Nopember, Surabaya 60111, Indonesia
[2] Department of Marine System Engineering, Institut Teknologi Sepuluh Nopember, Surabaya 60111, Indonesia
[3] Department of Oceanography, Hang Tuah University, Surabaya 60111, Indonesia

[*]E-mail: suwardicakwar@gmail.com

Abstract. Coral reefs are well-documented as natural coastal protection structures, alongside mangroves and seagrass beds. Reports indicate that these artificial reefs (AR), following deployment, often suffer damage due to factors such as improper sinking techniques due to heavy structures, wave impacts, displacement, sediment burial, or current drag. To reduce physical damage and facilitate easier deployment and seabed arrangement, will design an interlocking modular hexagonal artificial reef (IMHAR). This study aims to assess the suitability of the deployment site for this structure of interlocking modular hexagonal artificial reef (SIMHAR) based on oceanographic parameters, including depth, slope, substrate, wave height, current speed, and water activity. The methodology involves an overlay analysis using Geographic Information Systems (GIS) in the Gili Ketapang Marine Conservation Area, Probolinggo. The suitability assessment indicates that the northeastern part of the island is highly recommended for SIMHAR placement due to its relatively gentle slope (<5°), depth (>2 m), sandy coral substrate, extreme wave height of 0.69 m, highest current speed is 0.52 m/s, and minimal water activity such as boat mooring or aquaculture.

1. Introduction

Coral reef ecosystems are vulnerable to climate change impacts such as ocean acidification and rising sea temperatures, along with sedimentation from land runoff, hydrodynamic forces, destructive fishing practices, and marine debris. Rehabilitating coral reefs using artificial reefs is one strategy to address these issues.

An extensive review of coral reef rehabilitation programs in Indonesia [1] documented coral reef rehabilitation activities from 1990 to 2020, identifying 120,483 artificial reef units deployed throughout Indonesia using transplantation and artificial reefs. Artificial reefs are generally made from concrete in various shapes, such as pipes, domes, pyramids, bars, cubes, and cylinders.

The success rate of artificial reef rehabilitation varies in both ecological and hydrodynamic aspects. While many studies focus on ecological aspects, research on hydrodynamics is relatively limited. Most post-deployment studies of artificial reefs monitor ecological aspects, with only a few examining the physical condition of artificial reef [1]. Ecological monitoring includes coral community aspects with a focus on survival rates and/or growth of transplanted corals, fish community abundance, density, or benthic population monitoring.

Physical monitoring of artificial reefs only reports concrete structure conditions, showing that more than 45% of structures monitored after one to five years were damaged, buried in substrates, dispersed, or tilted from their original condition, as illustrated in Figure 1 [2]. This may be due to inadequate placement considerations of hydro-oceanographic aspects such as bathymetry and substrate conditions, or excessive wave impact. The use of artificial coral reefs is as important as the use of Ocean Renewable Energy [3–6] to support Sustainable Development Goals (SDGs) programs.

Figure 1. Condition of artificial reefs in several monitored locations: (a) Sabang-NAD, (b) Menjangan Besar Island-Central Java, (c) Sembilan Island-Sinjai South Sulawesi, and (d) Nusa Lembongan Island-Bali [2]

Oceanographic physical parameters at artificial reef deployment sites are crucial to ensure the growth and development of surrounding marine life. These parameters include depth, brightness, temperature, turbidity, substrate type, current speed, average wave height, and total suspended solids (TSS) [7]. Physical parameters for coral reefs can refer to seawater quality standards including brightness, TSS, temperature, and salinity [8]. Suitability for oceanographic parameters such as depth, slope, substrate, wave height, current speed, and water activity is obtained from primary/secondary sources, including:

1. Depth/bathymetry and slope from primary data collected in December 2022.
2. Waves from Copernicus Marine Service Physical Ocean Data.

3. Current speed from primary data using Acoustic Doppler Current Profiler in the northeastern part of Gili Ketapang Island in August 2024.
4. Substrate data from Allen Coral Atlas.
5. Water activity from ESRI World Imagery satellite base map imagery.

2. Design and Methodology

2.1 Design of Artificial Reef Modules

The innovation of interlocking and/or small modular artificial reefs has started to be applied in Indonesia. The Tanjungjati B Power Plant, in collaboration with Diponegoro University, conducted preliminary studies on the creation of artificial patch reefs (APR) and fish apartments using a concrete mixture from FABA tested in the waters of Panjang Island, Jepara Regency, Central Java. The application of APR modules has fostered over 3,000 transplanted coral colonies, with coral fragments growing up to 500% by the initial deployment (Figure 2) [9]. To form one APR unit, 107 components are required, with each part weighing between 22 kg and 80 kg and measuring approximately 20 cm x 25 cm x 135 cm.

Figure 2. Artificial Patch Reef (APR) model as a media for coral reef transplantation [10]

The Bali Coastal and Marine Resource Management Office (BPSPL) Denpasar, in collaboration with SORCE (Sustainable Oceanic Research, Conservation, and Education) Lombok Barat, is also testing coral transplantation media using MARS (Modular Artificial Reef Structure) blocks. Each block measures approximately 50 cm x 30 cm x 30 cm and weighs about 50 kg, with interlocking parts at the ends to form a strong and long structure (Figure 3). Each block has two intersecting holes for iron stakes to enhance attachment to the seabed and 36 small holes (0.5-inch diameter, 3 cm deep) for coral seedlings [11].

Figure 3. MARS block structure as a coral reef transplant media developed by BPSPL Denpasar and SORCE [11]

Reefy Company designs hydrodynamic blocks as core components for building artificial reefs (Reefy blocks), interconnected using "lego" techniques, as shown in Figure 4. The design focuses on hydrodynamic performance with internal holes and rounded angles to reduce resistance. Some block configurations leave gaps between blocks to create sheltered areas for fish. High porosity and complexity also allow nutrients to circulate within the structure [12]. Experiments show that wave transmission inversely correlates with the width-to-height ratio of incoming waves passing through Reefy blocks, while block roughness does not significantly affect wave reflection.

Figure 4. Interlocking reefy blocks by adopting the "lego" technique [12]

In this study, the choice of hexagonal artificial reef shapes (hexareef) is based on previous research indicating that hexagonal-shaped artificial reefs perform better than cylindrical and cubic reefs due to their smaller KT values for the same wave steepness [13]. The novelty of this research is in creating of interlocking of modular hexareef.

IMHAR is believed to make it easier to deployment and arrange on the bottom of the waters, does not require heavy equipment to sink it to the bottom of the waters, because each unit is lighter, compared to the previous AR that must be heavy to be stable from wave action. With the same arrangement,

IOP Conf. Series: Earth and Environmental Science **1473** (2025) 012002 doi:10.1088/1755-1315/1473/1/012002

SIMHAR will be more stable in the face of strong current and wave dynamics, because the IMHAR units are interlocking, so they are not easily shifted or fallen, compared to previous AR that are arranged without interconnected between their unit.

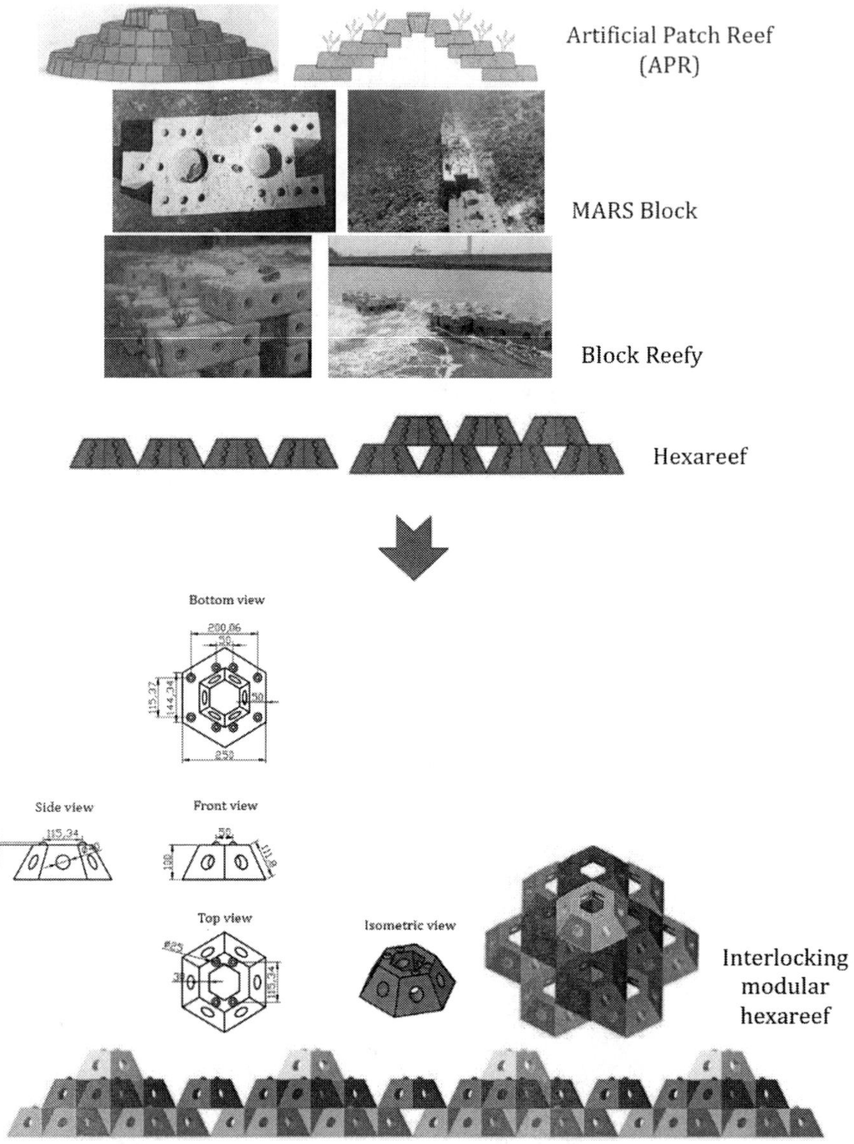

Figure 5. State of the art and novelty of research

The IMHAR is designed to stack and interlock like lego blocks. The IMHAR will be designed with the same dimensions as non-interlocking hexareef models from previous research [13], with a height of 100 mm, width of 250 mm, weight of 4.6 kg per module, and volume of 2,782 cm^3. The IMHAR will be assembled into a SIMHAR arranged in a tiered pyramid configuration with a 3-tier model consisting of 11 IMHAR units. The SIMHAR placement will be arranged in 1 row and 2 rows with spacing and zigzag combinations. An additional IMHAR will be added between SIMHAR units as connectors within the same row and between rows.

For field implementation, the height of the SIMHAR is planned to be 2/3 of the water depth, with a minimum depth of 2 meters, so the SIMHAR height, scaled up by a factor of 5, will be 1.5 meters as shown in Figure 6.

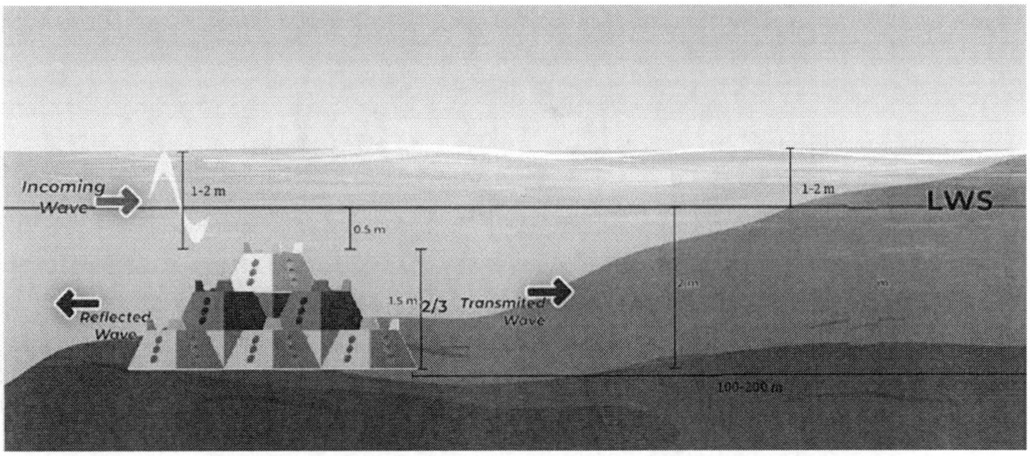

(a)

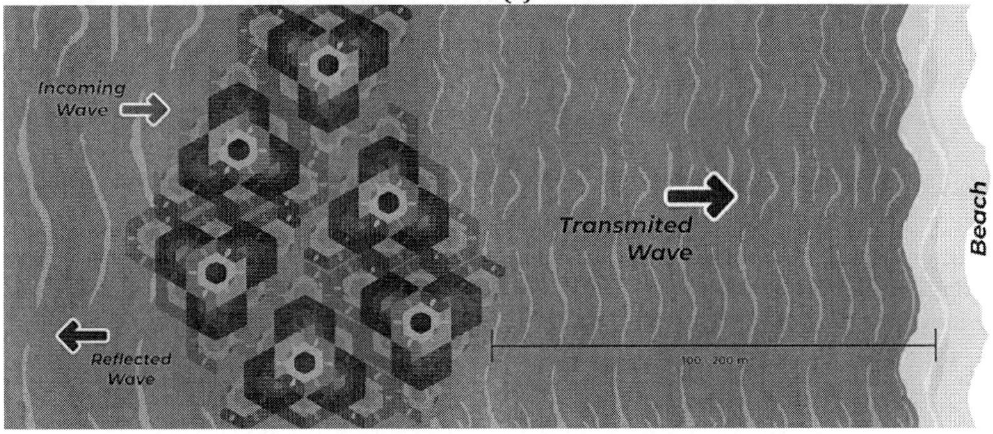

(b)

Figure 6. Design of the location for placing the SIMHAR in Gili Ketapang waters; (a) front view, (b) top view

2.2 Methodology

The study location for placing the SIMHAR is in the Gili Ketapang Marine Conservation Area, Probolinggo Regency, with the conservation goal of preserving and enhancing coral reef ecosystems within the conservation area. The waters of Gili Ketapang, located in the Madura Strait, Probolinggo Regency, East Java, are designated as a marine conservation area through Ministerial Decree No. 64/KEPMEN-KP/2020 concerning the Gili Ketapang Marine Conservation Area and Surrounding Waters in East Java Province. The area is divided into five zones: fishing, aquaculture, tourism, rehabilitation, and core zones, as shown in Figure 7.

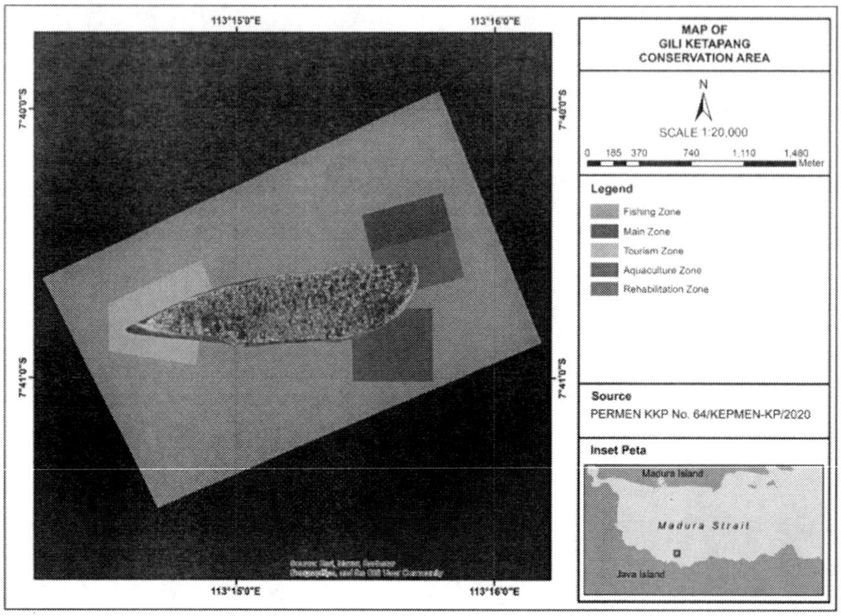

Figure 7. Map of Gili Ketapang Conservation Area

The suitability of the SIMHAR placement location is determined based on oceanographic parameters using overlay analysis based on Geographic Information System (GIS) with ArcGIS software. Suitability is categorized into three levels: unsuitable, suitable, and highly suitable. Suitability categories are determined by assigning values that affect the performance of the SIMHAR, including physical stability and natural coral recruitment. Each parameter is weighted 1, 3, or 5, considering the strength of the parameter's influence as a limiting factor for SIMHAR placement. For example, the stability of SIMHAR arrangements is strongly influenced by gentler slopes, and natural coral recruitment is highly influenced by sandy coral substrates. The suitability and parameter weights are listed in Table 1.

Table 1. Oceanographic parameters for SIMHAR placement

Parameter	Unit	Value			Weight
		Not Suitable	Suitable	Highly Suitable	
Depth	meter	0 – (-1)	(-1) – (-2)	<(-2)	3
Slope	degree	>15	5-15	<5	5
Substrate	type	mud	sand	sandy coral	5
Wave Height	meter	0-0,5	0,5-1	>1	1
Current Speed	m/s	0-0,5	0,5-1	>1	1
Water Activity	type	closed (high activity)	slightly closed (low activity)	open (no activity)	3

2.3 GIS Analysis

The GIS analysis involves:

1. Data Collection: Gathering primary and secondary data sources including bathymetry, wave height, current speed, substrate type, and water activity.
2. Data Processing: Inputting data into GIS software to create spatial layers for each parameter.
3. Overlay Analysis: Combining parameter layers to identify areas that meet suitability criteria using weighted scores.
4. Suitability Mapping: Generating a suitability map that indicates highly suitable, suitable, and not-suitable areas for SIMHAR placement.

3. Results and Discussion

3.1 Depth analysis

Bathymetric data reveal depths ranging from 0.5 meters to 25 meters, with areas over 2 meters deemed most suitable, as shown in Figure 8. The depth suitability map indicates that the majority of the study area meets the criteria for SIMHAR deployment.

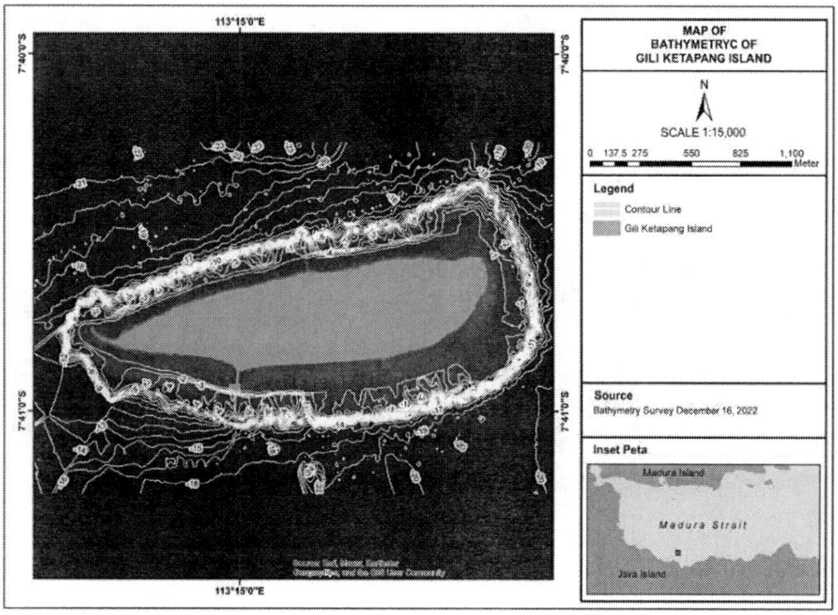

Figure 8. Map of bathymetric of Gili Ketapang Island

3.2 Slope analysis

To assess the suitability of a location for placing SIMHAR, it is important to evaluate the slope of each section within the quadrants of the study area. The slope is a key parameter that affects the stability and effectiveness of the SIMHAR. The assessment should provide details on the slope gradients in different quadrants to ensure the chosen location meets the required criteria for successful installation and performance of the SIMHAR.

Slope calculations for target depths up to -5 meters were performed using the rise over run formula:

$$Slope\ (\%) = \frac{rise}{run} \times 100 = \frac{depth}{distance} \times 100$$

$$Slope\ (degree) = \tan^{-1}\left(\frac{rise}{run}\right) = \tan^{-1}\frac{depth}{distance}$$

To calculate the slope in more detail, the research location was divided into four quadrants, and in each quadrant several sections were taken which were approximately 100 meters – 300 meters apart as shown in Figure 9 below.

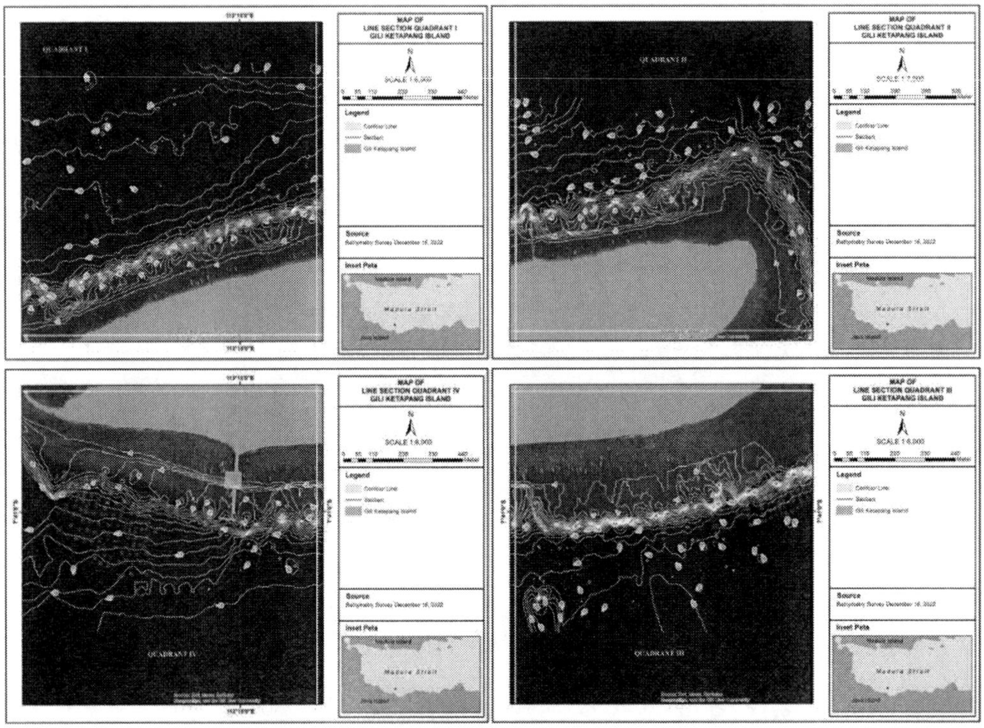

Figure 9. Map of profile section for each quadrant of Gili Ketapang Island

The results of slope calculations for each section and quadrant at a depth of 0 to -5 meters are presented in Table 2 below. Results indicate slopes around the northeastern part of Gili Ketapang are below 50°, making them suitable for SIMHAR placement.

Table 2. Slope of each quadrant section

Quadrant	Section	Distance section	Slope	
		(m)	(%)	(^{0})
I	1	152,73	3,95	2,266
	2	140,26	3,75	2,148
	3	123,27	4,27	2,447
II	1	111,06	4,38	2,513
	2	149,31	3,32	1,905
	3	301,56	1,66	0,955
	4	191,22	2,57	1,476
	5	214,34	2,33	1,337
	6	280,98	1,80	1,036
III	1	271,70	1,79	1,027
	2	282,28	1,77	1,019
	3	285,39	1,73	0,992
IV	1	234,49	2,12	1,219
	2	236,19	2,10	1,208
	3	192,58	2,62	1,505
	4	165,91	2,95	1,694

3.3 Wave height

Maximum wave height data show an extreme wave height of 0.69 meters with a period of 3.60 seconds. Wave height data indicate that the northeastern area experiences suitable wave conditions for SIMHAR placement. Areas with higher or extreme waves are preferred for determining SIMHAR stability, and the northeastern part of the study area meets this criterion.

3.4 Current Speed

Observations of current speed show that the highest average current speed occurs during the approach to high tide compared to during the approach to low tide, high tide, or low tide. During the approach to high tide, current speeds vary, with a minimum value of 0.15 meters per second and a maximum value of 0.52 meters per second. The dominant current direction during this period is towards the northwest. Areas with higher current speeds are preferred to determine of SIMHAR stability, and the northeastern part of the study area meets this criterion.

3.5 Substrate analysis

The substrate around Gili Ketapang shows sand, rubble and sandy coral as the predominant type in the study area, as shown in Figure 10. The substrate mapping indicates the presence of a sandy coral reef, which is the most suitable substrate for placing SIMHAR.

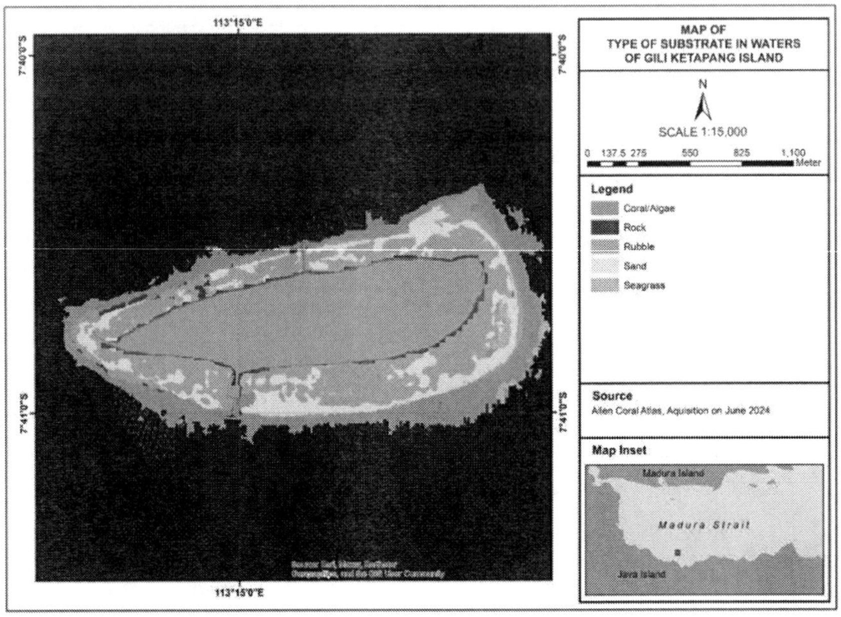

Figure 10. Map of substrate type in the waters of Gili Ketapang Island

3.6 Water activity

The level of activity or congestion in the waters at the proposed site for the placement of SIMHAR will impact their performance. This is in addition to the effects of water movement that can exert additional wave pressure. Conversely, the presence of the SIMHAR might interfere with existing community activities, such as docking areas or fish farming facilities.

The community activities around Gili Ketapang Island are illustrated in Figure 11: the northern side features docking areas and piers for fishing and inter-island boats; the western side is designated for snorkelling activities; and the southern side is used for boat docking and floating fish farming installations. Based on this mapping, the research area with minimal activity is located in the northeastern part of the island, making it more suitable for the placement of the SIMHAR.

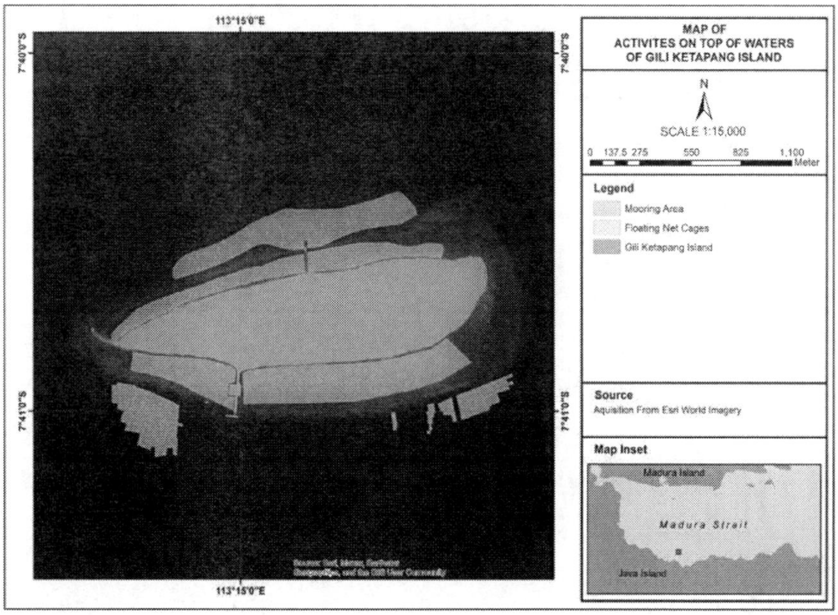

Figure 11. Map of activities on top of the waters in Gili Ketapang Island

3.7 Suitability map

By integrating all the oceanographic parameters and overlaying them with the weights as determined in Table 1, the suitability map, as shown in Figure 12, indicates that there are areas around the island that are highly suitable for the placement of the SIMHAR. However, there are areas in the northern and southern parts of the island that are highly unsuitable due to high levels of water activity, such as docking areas and fish farming facilities. The western part of the island is designated as a snorkelling tourism area based on available information.

Overall, the northeastern part of Gili Ketapang Island is recommended as the most suitable location for placing the SIMHAR. This area offers relatively flat seafloor conditions with a slope of less than 50 degrees, a depth greater than 2 meters, a sandy coral substrate, extreme wave heights of 0.69 meters, a maximum current speed of 0.52 meters per second, and minimal water activities.

IOP Conf. Series: Earth and Environmental Science **1473** (2025) 012002 doi:10.1088/1755-1315/1473/1/012002

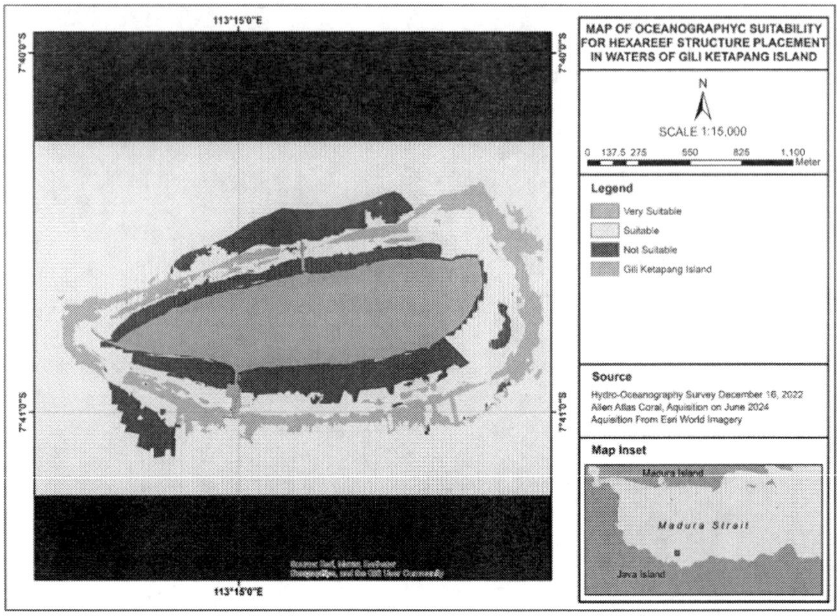

Figure 12. Map of suitability for SIMHAR placement in the waters of Gili Ketapang Island

Additionally, according to the zoning regulations of the Gili Ketapang Marine Conservation Area, this area falls within a rehabilitation zone, making it highly appropriate for placing SIMHAR as a measure to improve coral ecosystem conditions, support conservation targets, and reinforce the nearby protection zone.

4. Conclusion

The suitability assessment indicates that the northeastern part of Gili Ketapang Island is the most suitable location for placing SIMHAR based on oceanographic conditions, including depth, slope, substrate, wave height, current speed, and minimal water activity. This area offers optimal conditions for the placement of SIMHAR, thereby potentially enhancing their stability and effectiveness in coral rehabilitation and coastal protection. SIMHAR is believed to make it easier to deployment and arrange on the bottom of the waters, and will be more stable in the face of strong current and wave dynamics, compared to previous AR.

Acknowledgements

The authors thank the Department of Ocean Engineering at Institut Teknologi Sepuluh Nopember and Hang Tuah University for their support. Special thanks to the Ministry of Marine Affairs and Fisheries, East Java Province Maritime Affairs and Fisheries Service, and Gili Ketapang Village Head and for their ongoing commitment to marine conservation. This research was funded by the BIMA with Doctoral Research program from the Ministry of Education and Culture, Research and Technology with contract number 1716/PKS/ITS/2024.

IOP Conf. Series: Earth and Environmental Science **1473** (2025) 012002 doi:10.1088/1755-1315/1473/1/012002

References

[1] Razak T B, Boström-Einarsson L, Alisa C A G, Vida R T and Lamont T A C 2022 Coral reef restoration in Indonesia: A review of policies and projects *Mar. Policy* **137**

[2] Munasik 2008 Kondisi Terumbu Buatan Berbahan Beton pada Beberapa Perairan di Indonesia *Pros. Musyawarah Nas. Terumbu Karang* 11

[3] Satrio D, Suntoyo, Widyawan H B, Ramadhani M A and Muharja M 2024 Effects of the distance ratio of circular flow disturbance on vertical-axis turbine performance: An experimental and numerical study *Sustain. Energy Technol. Assessments* **63** 103635

[4] Erwandi E, Kasharjanto A, Satrio D, Rahuna D, Suyanto E M, Mintarso C S J, Irawanto Z and Ramadhani M A 2024 Numerical Analysis of Resistance and Motions on Trimaran Floating Platform for Tidal Current Power Plant *Int. Rev. Model. Simulations* **17** 6–16

[5] Fakhri M H, Satrio D and Suntoyo 2023 Numerical analysis of drag coefficient for damping plate on the submerged body of two-body point absorber device as wave energy converter *International Conference on Green Energy, Computing and Intelligent Technology (GEn-CITy 2023)* (IEEE Xplorer)

[6] Satrio D, Putra D J, Dhanistha W L, Utama I K A P, Putranto T, Hayati N, Muharja M and Madi M 2024 A Numerical Study of the Effect of Depth Immersion and Rotation Direction on the Performance of Cross-Flow Savonius Turbines *Pomorstov* **38** 250–62

[7] Fikri M, Isdianto A and Luthfi O M 2021 Physics Oseanography Around Artificial Reef On The Pantai Of Damas, Trenggalek District, East Java **10**

[8] Setneg 2021 *Peraturan Pemerintah Nomor 5 Tahun 2021 Tentang Penyelenggaraan Perizinan Berbasis Risiko*

[9] PLN P T J B 2023 *Pemanfaatan FABA PLTU Tanjung Jati B*

[10] Munasik M, Sabdono A, Hutapea E D, Sugiyanto S and Sugianto D N 2021 Coral Recruitment on Artificial Patch Reefs Deployed in The Marginal Reefs: Effect of Multilevel Substrate on Density of Coral Recruit *J. Segara* **17** 33

[11] Denpasar B 2023 *Rehabilitasi Habitat Kritis Jenis Ikan Dilindungi/Terancam Punah di Pantai Elak-Elak TWP Gita Nada*

[12] van de Brekel E 2021 Hydrodynamic and ecological performance of a new modular unit for living breakwaters Wave flume experiments and results

[13] Armono H D, Winarto A, Sujantoko and Ketut Suastika I 2021 A laboratory study on wave transmission over hexagonal artificial reef *IOP Conference Series: Earth and Environmental Science* vol 799 (IOP Publishing Ltd)

IOP Conf. Series: Earth and Environmental Science **1473** (2025) 012003 doi:10.1088/1755-1315/1473/1/012003

Longitudinal strength analysis of a crane barge during heavy lifting operation due to variations of trim

S P Selvia[1], R W Prastianto[1*] and Murdjito[1]

[1] Ocean Engineering Department, Institut Teknologi Sepuluh Nopember, Surabaya, Indonesia

*E-mail: rudiwp@oe.its.ac.id

Abstract. The increasing demand for oil and gas requires the continuous development and revitalization of offshore platforms. One of the common methods for the installation of these platforms involves the utilization of crane barges for lifting operations. However, one of the main concerns is the longitudinal strength of these barges under various loading conditions, particularly during heavy lifting operations which can induce changes in trim. This research intends to perform a comprehensive analysis of the longitudinal strength of a crane barge during heavy lifting operations under different trim conditions. This study will employ advanced modelling and simulation techniques using software to accurately simulate the variations in loads and their subsequent impact on the structural integrity of the barge. The results will show the changes in load distribution, which will be systematically compared across different trim conditions. This analysis included calculations of shear forces and bending moments to compare the conditions of the even keel up to 1-meter trim by stern. Furthermore, this study will compare the ultimate shear force and bending moment obtained with its allowable as defined by the applicable standards and regulations to verify the structural integrity of the crane barge under the specified conditions.

1. Introduction

The high levels of exploration and exploitation in Indonesia requires the developing of offshore infrastructure and revitalize existing offshore platforms. Fixed offshore platform installations generally employ the lifting method, which is a conventional approach that involves lifting structures using a crane barge [1]. One of the considerations that is a problem in choosing a barge is the ability of the barge to support the load it carries. The larger the dimensions of the barge will cause more problems [2]. Capacity of crane barge during operation is affected by the weight of the structure being lifted. As the structure's weight increases, the barge's capacity must also increase, impacting the crane's ability to lift the load [3].

When the crane barge is operating, there is a risk of accidents due to structural failure in the hull due to wave and cargo loads. One of external loads is buoyant force, which causes it to respond with stress and bending [4]. One of the barge's operational conditions is trim, which refers to its inclination caused by the forces exerted on it. Trim condition is divided into bow and stern. Due to the environmental load, a barge design must be considered specifically and fulfill operational criteria [5]. Heavy lifting typically refers to operations involving load weights between 60 and 60,000 metric tons, primarily for transportation purposes [6]. During such heavy lifting operations, the process of transporting substantial structural loads using a crane barge can have a considerable effect on the longitudinal strength of the barge. This impact is chiefly caused by the uneven load distribution that occurs during heavy lifting operations, leading to significant stress and strain on the barge's structure.. Consequently, it is imperative to conduct a comprehensive strength analysis to understand how the load of the structure affects the barge, ensuring that the minimum force required for safe and efficient operation is accurately determined [7].

This study will conduct an analysis of how changes in the barge's load, particularly as it is influenced by the crane's rotation angle, affect the structural integrity of the barge. The analysis will include detailed calculations of shear forces and bending moments, which will be used to compare the barge's structural conditions, from even keel up to 1-meter trim by stern. By thoroughly examining these factors, the study aims to provide a clearer understanding of the potential risks and necessary precautions involved in such heavy lifting operations, thereby contributing to safer and more reliable maritime engineering practices.

2. Research methods

2.1. Research data

The data required for this study includes detailed specifications of both the barge and the crane utilized in the installation process. This information is crucial for accurately modeling and analyzing the operational parameters during the installation. The crane is capable of operating in a radius of 6 meters to 42 meters, allowing for flexibility in various installation scenarios. However, for the purposes of this study, a fixed radius of 22 meters is selected for analysis, as it represents a typical operational scenario. At this radius, the crane's Safe Working Load (SWL) is rated at 97.1 tons, indicating its maximum load-carrying capacity under safe operating conditions. The data along with additional relevant specifications, are outlined in detail in Table 1.

Table 1. Barge and crane data.

Parameter	Value	Units
Length Overall	100.58	m
Length Between Perpendicular	100.58	m
Breadth	33.53	m
Depth	6.1	m
Operating Draft	4.7	m
Displacement	14345.41	tone
Crane Weight	424	tone
Lifting Load	87.59	tone
Crane Operating Radius	22	m
SWL Crane	97.1	tone

2.2. Barge modelling

The barge data presented in Table 1 is used for barge modelling in full load condition by using software. This model includes not only the barge's overall structure but also the compartments, such as the ballast tanks, which are incorporated specifically to simulate and generate changes in trim during operations. These ballast tanks are essential for adjusting the barge's stability and balance by altering the distribution of weight, particularly in scenarios that require changes in trim to ensure optimal operational performance.

To ensure the model's accuracy for further analysis, it is validated by comparing its hydrostatic parameters to those in the barge's stability booklet. The allowable margin of error is set at 2% between the software model and the booklet values, as specified in reference [8]. The validation results, shown in Table 2, confirm that all parameters meet this criterion, with errors falling below 2%, ensuring the model's reliability for subsequent analyses.

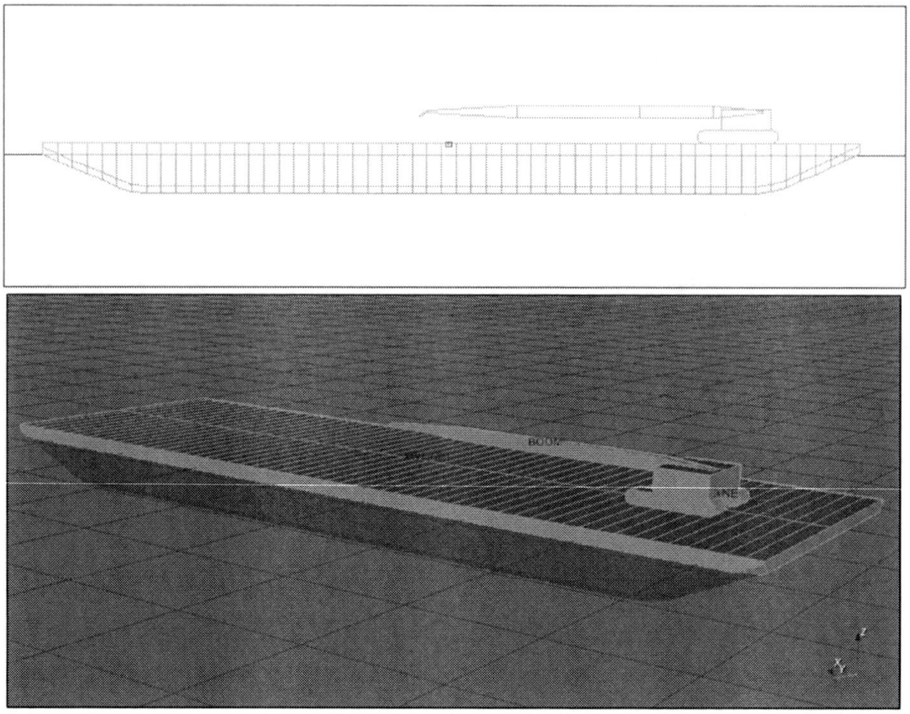

Figure 1. Crane barge modelling.

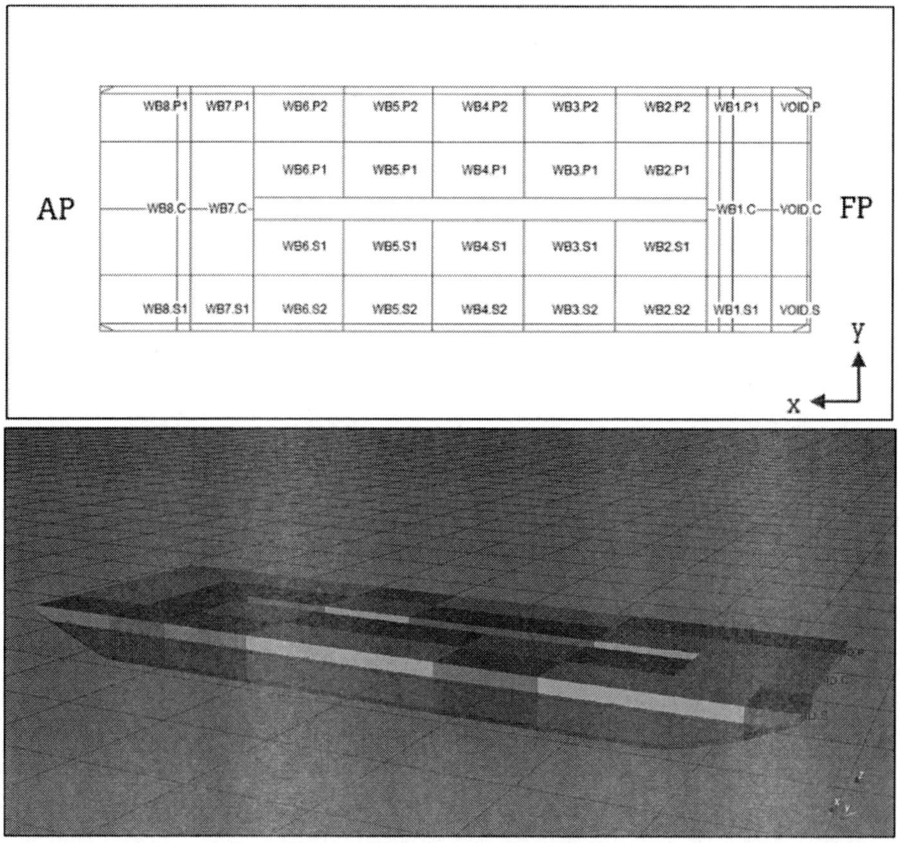

Figure 2. Ballast tank modelling.

Table 2. Barge modelling validation.

Parameter	Stability Booklet	Software	Error	Tolerance
Displacement (ton)	14345.41	14369.66	0.17%	2%
LCB (m)	50.07	50.46	0.77%	1%
VCB(m)	2.45	2.46	0.45%	1%
LCF (m)	50.08	50.51	0.86%	1%
KML (m)	202.46	202.08	0.19%	1%
KMT(m)	24.86	24.83	0.12%	1%

Table 3. Tank modelling validation.

Tank	Weight (ton)		Error
	Software	Stability Booklet	
WB1. C	891.2	897.96	0.753%
WB1. P1	366.7	370.34	0.983%
WB1. S1	366.7	370.34	0.983%
WB2. P1	597.4	597.26	0.023%
WB2. S1	592	597.26	0.881%
WB2. P2	597.4	591.93	0.924%
WB2. S2	592	591.93	0.012%
WB3. P1	597.4	597.35	0.008%
WB3. S1	592	597.35	0.896%
WB3. P2	597.4	592.02	0.909%
WB3. S2	592	592.02	0.003%
WB4. P1	597.4	597.3	0.017%
WB4. S1	592	597.3	0.887%
WB4. P2	597.4	591.98	0.916%
WB4. S2	592	591.98	0.003%
WB5. P1	597.4	597.3	0.017%
WB5. S1	592	597.3	0.887%
WB5. P2	597.4	591.98	0.916%
WB5. S2	592	591.98	0.003%
WB6. P1	597.4	597.3	0.017%
WB6. S1	592	597.3	0.887%
WB6. P2	597.4	591.98	0.916%
WB6. S2	592	591.98	0.003%
WB7. C	1024	1023.96	0.004%
WB7. P1	422.8	422.84	0.009%
WB7. S1	422.8	422.84	0.009%
WB8. C	963.1	972.1	0.926%
WB8. P1	395.9	399.65	0.938%
WB8. S1	395.9	399.65	0.938%

2.3. Tank capacity determination

To ensure that the crane barge's condition meets the required draft and trim for each scenario, ballast adjustments are necessary. Determination of condition limits trim refers to the recommended rules, which are trim 1 meter for a vessel length of 100 m [9]. Ballast calculation obtained through simulation using software by specifying draft mark readings for each variation trim. The following is a water-laden plan for each variation of conditions trim.

Table 4. Determination of draft mark readings.

Trim by Stern (m)	Draft Mark Readings (m)	
	AFT	FWD
Even Keel	4.69	4.69
0.25	4.71	4.46
0.5	4.72	4.22
0.75	4.72	3.97
1	4.73	3.73

2.4. Determination of operating scenarios

Longitudinal strength analysis is simulated by using software. Analysis is performed with the variation of crane swing rotation and the change of trim. The trim changes are carried out of the condition of evenkeel until trim by stern by 1 m [9]. This variation is based on the recommended trim is 1 m for a vessel length of 100 m. Condition scenarios trim illustrated as in Figure 3. Meanwhile, the crane swing rotation is varied by any 30^0 difference from the position of the load lifted on the bow and rotate in the direction of starboard up to 180^0. The illustration of the crane swing rotation during operation lifting shown in Figure 4.

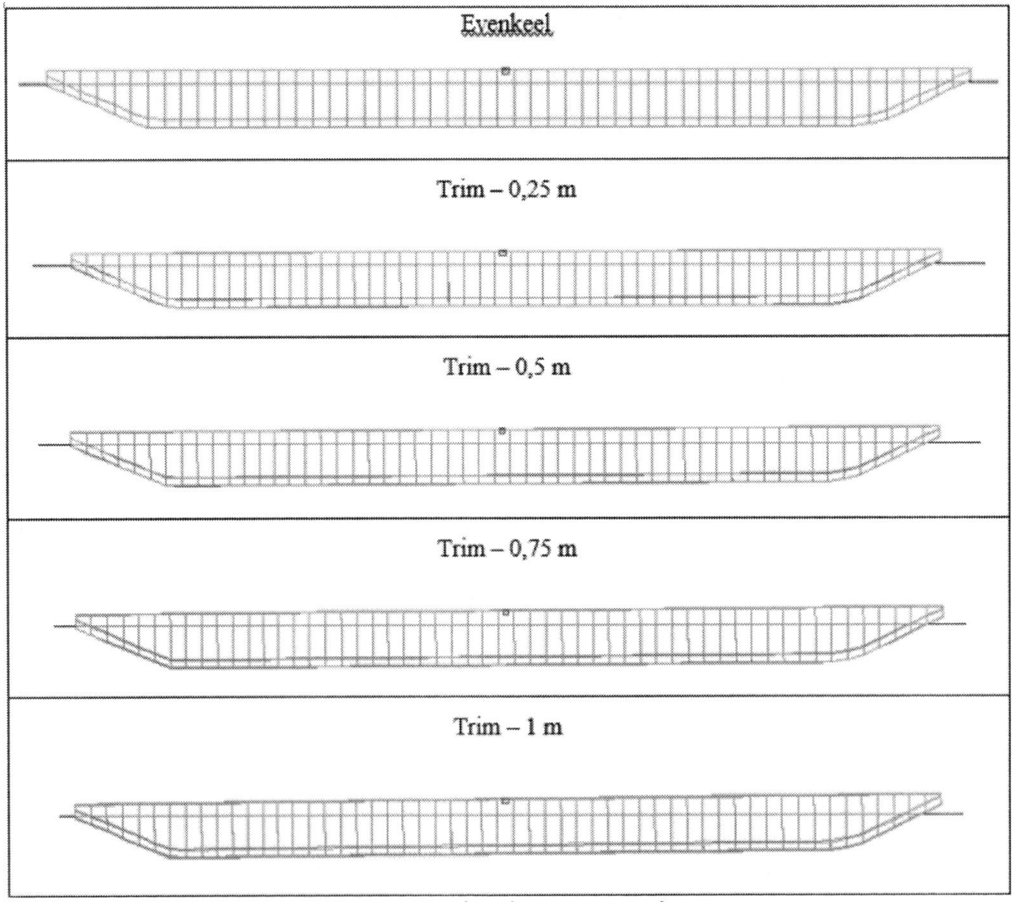

Figure 3. Trim changes scenario.

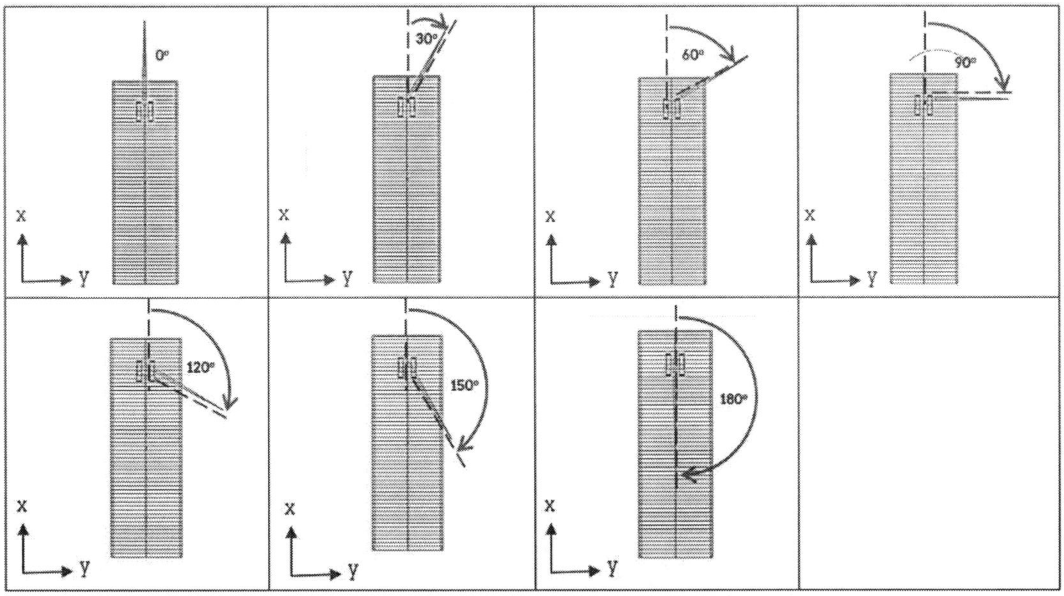

Figure 4. Crane swing rotation scenario.

2.5. Determination of combined COG

In modeling of the crane structure at the barge, make sure the clearance between the lifted structure and the end barge nearest to the FP. Clearance between the transported load and the end of the vessel at least 3 meters either horizontally or vertically [10].

After the topside structure is modeled in the lifting operation, a combined COG (Centre of Gravity) point between the crane and the topside is obtained in each configuration of the crane swing rotation. The initial COG of the crane barge lightship condition based on the stability booklet has been known to be 50.29 (X), 0 (Y), and 6.1 (Z). Table 5 is the combined COG of the crane + topside and the change of the COG of the barge crane after modeling the lifting conditions.

Table 5. Combined COG.

Swing Rotation (deg)	COG (Topside + crane)		
	x	y	z
0	3.19	-0.02	36.02
30	1.77	-5.28	36.02
60	-2.08	-9.11	36.02
90	-7.34	-10.51	36.02
120	-12.59	-9.09	36.02
150	-16.43	-5.23	36.02
180	-17.8	0.02	36.02
Crane Only	2.22	0.07	38.11

2.6. Determination of allowable shear force and bending moment

To determine the minimum amount of ship's longitudinal strength in cross-section along the ship, the shear force and bending moment must be calculated based on the total load of the combined still water and the consequences of the waves [11].

- Still Water
  - Shear force

$$Q_{sw-pos-min} = +\frac{5f_{qs}M_{sw}}{L} \quad (1)$$

$$Q_{sw-neg-min} = -\frac{5f_{qs}M_{sw}}{L} \quad (2)$$

  - Bending Moment

$$M_{sw-h-min} = +0{,}1f_{sw}[C_wL^2B(1{,}197 - 0{,}19C_B)] \quad (3)$$

$$M_{sw-s-min} = -0{,}05186f_{sw}[C_wL^2B(C_B + 0{,}7)] \quad (4)$$

- Wave load
  - Shear force

$$Q_{wv-pos} = 0{,}52f_{q-pos}f_pC_wLBC_B \quad (5)$$

$$Q_{wv-neg} = -(0{,}52f_{q-pos}f_pC_wLBC_B) \quad (6)$$

  - Bending Moment

$$M_{wv-h} = +0{,}19f_mf_pC_wL^2BC_B \quad (7)$$

$$M_{wv-s} = -0{,}10999f_mf_pC_wL^2B(C_B + 0{,}7) \quad (8)$$

with:
f_{qs} = distribution factor along the ship
M_{sw} = absolute maximum of M_{sw-h} and M_{sw-s}
f_{sw} = 1
L = length overall
C_w = wave coefficient
B = breadth
C_B = block coefficient

The results of the allowable calculation are shown in Table 6.

Table 6. Allowable shear force and bending moment.

No	Item	Still Water	Hogging	Sagging	Unit
1	Shear Force	1365	-1855.93	1855.93	tone
2	Bending Moment	27469	-48190	47452	mt.m

3. Result and discussion

3.1. Load distribution

The load distribution that occurs in the barge crane is the difference between the buoyancy force and the weight force. In other words, the load distribution is the net load, which is the resultant of both styles. Figure 5 shows the load distribution for all variations.

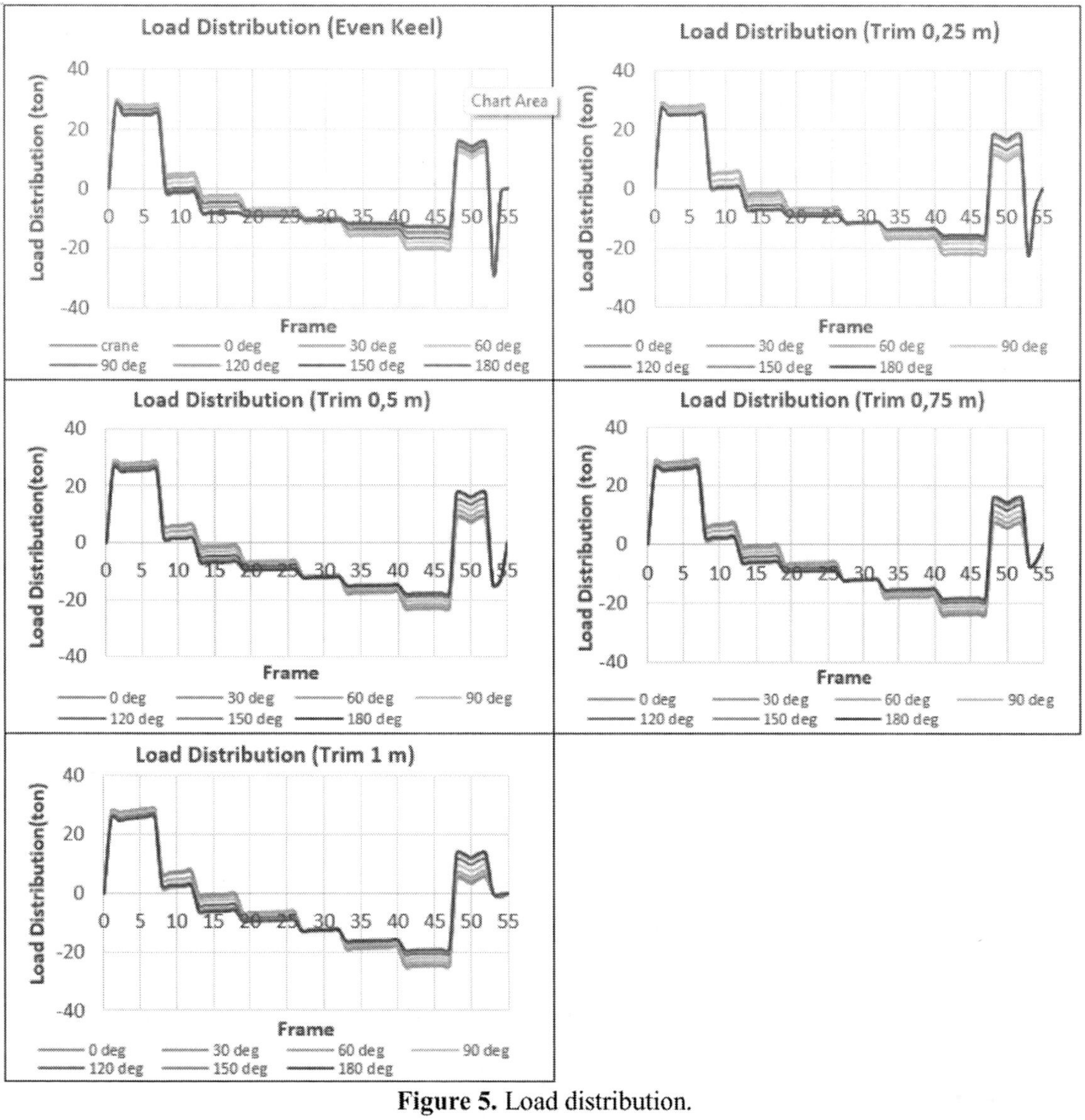

Figure 5. Load distribution.

Figure 5 shows a similar trendline to illustrate the load distribution of the crane barge at even keel conditions up to 1 meter trim. Based on the load distribution that occurs, it can be seen that there is a significant change in the distribution along the ship with the ballast tank, crane, and lifting load load that occurs on several frames depending on the placement of each load. The significant increase is located in frame 8 – frame 47 which is the location of the ballast tank and frame 50 – frame 55 which is the location of the pedestal crane.

3.2. Shear force and bending moment

The shear force and bending moment are the primary response of The result of how loads are distributed across the barge where the load has actually been counteracted by the buoyancy of the barge. In hogging and sagging conditions, wavelength is assumed to be the most critical condition, i.e. along the ship. Figure 6 to Figure 8 is the results of the shear force and bending moment at the maximum condition which is trim 1 m.

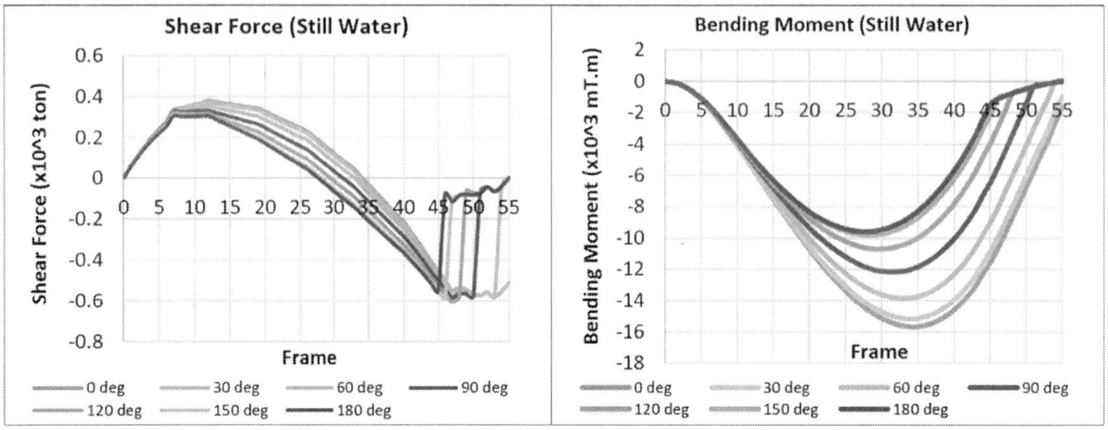

Figure 6. Shear force and bending moment in still water, trim 1 m condition.

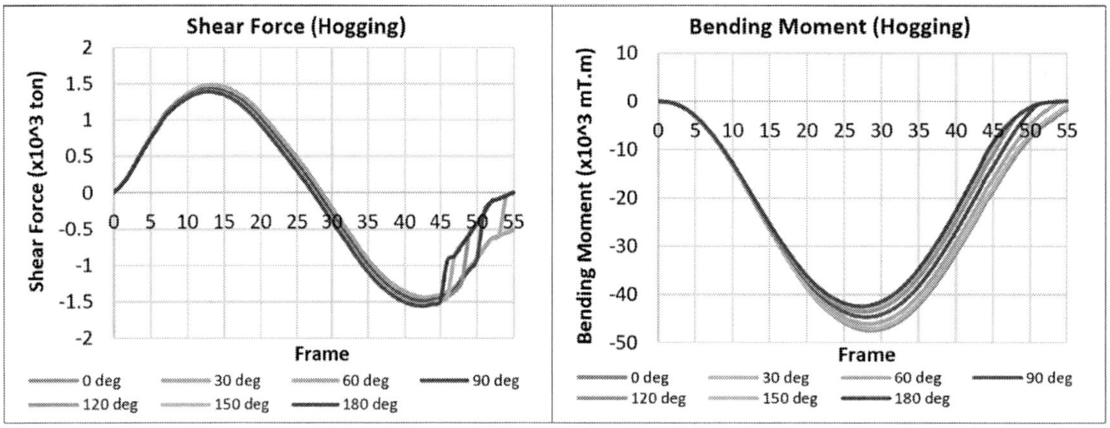

Figure 7. Shear force and bending moment in hogging, trim 1 m condition.

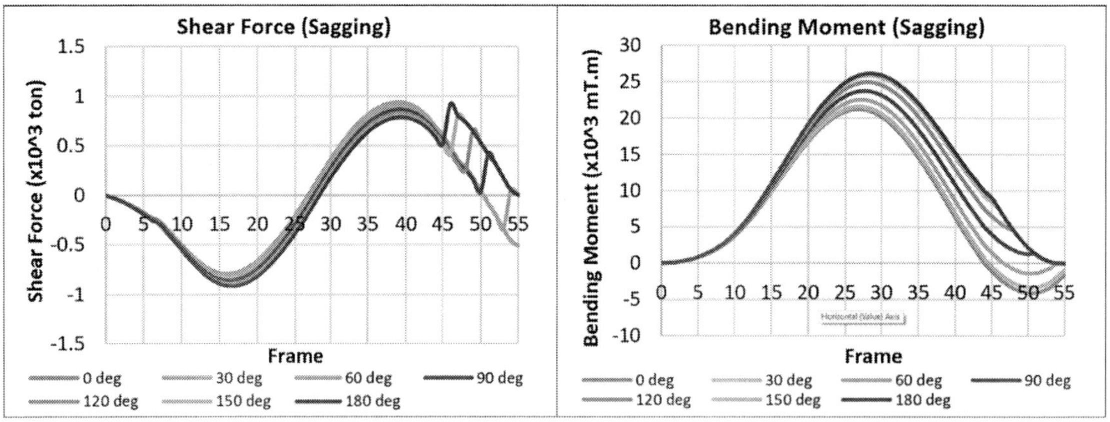

Figure 8. Shear force and bending moment in sagging trim 1 m condition.

The maximum shear force in the evenkeel condition, up to a trim of 1 meter, is observed at frame 13, with an upward trend in shear force across all crane swing rotation between frames 46 and 55. Similarly, the peak bending moment in the same condition, up to a 1-meter trim, is detected around frame 29, accompanied by an increase in bending moment values corresponding to each crane rotation direction.

Table 7. Summary of shear force and bending moment.

No.	Description	Still Water	Hogging	Sagging	Frame	Swing Rotation
A.	**Even Keel**					
1	SF (tone)	-540	1392	923	13	0 deg
2	BM (mt.m)	-14228	-43638	25472	29	0 deg
B.	**Trim 0,25 m**					
1	SF (tone)	-547	1419	935	13	0 deg
2	BM (mt.m)	-14776	-44857	25757	29	0 deg
C.	**Trim 0,5 m**					
1	SF (tone)	-551	1444	943	13	0 deg
2	BM (mt.m)	-15127	-45893	26108	29	0 deg
D.	**Trim 0,75 m**					
1	SF (tone)	-565	1465	944	13	0 deg
2	BM (mt.m)	-15426	-46765	26325	29	0 deg
E.	**Trim 1 m**					
1	SF (tone)	-577	1481	941	13	0 deg
2	BM (mt.m)	-15681	-47450	26084	29	0 deg

Referring to Table 7, the maximum bending moment is observed during crane barge operations under hogging conditions, with the crane positioned at 0 degrees. This peak bending moment occurs at a trim of 1 meter in hogging conditions, specifically at frame 29, with a magnitude of 47,450 ton.m. The area around frame 29 is identified as a critical area due to the significant bending moment, warranting further local analysis to ensure the structural stability of the barge.

3.3. Unity check

Checking the value of the shear force and bending moment that occurs uses the ratio between the shear force and the bending moment that occurs with its allowable or commonly known as a unity check (UC). The maximum combined UC value should not exceed 1 during installation. When the Maximum Combined UC value is below 1, the crane barge is declared safe during the operation process.

Table 8. UC checking on crane barge.

No.	Description	Max. Value	Allowable	Max. UC	Remarks
1	Bending Moment	47450	48190.63	0.98	PASS
2	Shear Force	1481	1855.93	0.80	PASS

4. Conclusion

Based on the results of the analysis, the load distribution on the barge crane during heavy lifting operations for all trim variations shows a similar trend. The significant load increase lies in frames 8–47, which are the location of the ballast tanks, and frames 50–55, where the pedestal loading of the crane is located. The ultimate shear force and bending moment occurs when the crane barge is trimmed by 1 meter in hogging conditions, with a maximum shear force of 1481 tons in frame 13 and a maximum bending moment of 47450 mt.m in frame 29. Both values meet the allowable criteria.

5. Refrences

[1] F. Ardhiansyah, R. W. Prastianto, Murdjito, K. Suastika, and S. Nugroho, *Dynamic Response of Floating Crane During Lifting Operation: A Parametric Study*, **vol. 455**. Springer Nature Singapore, 2024. doi: 10.1007/978-981-97-2136-8_15.

[2] C. Wang, J. Wu, and D. Wang, "Experimental and numerical investigations on the ultimate longitudinal strength of an ultra large container ship," *Ocean Eng.*, **vol. 192**, no. September, p. 106546, 2019, doi: 10.1016/j.oceaneng.2019.106546.

[3] Y. B. Hadasa, Murdjito, R. W. Prastianto, E. B. Djatmiko, Wahyudi, and M. Faiz, "Study on the effects of lifting loads on the longitudinal strength of a crane barge during heavy lifting operations due to various angle of crane rotation," *IOP Conf. Ser. Earth Environ. Sci.*, **vol. 1298**, no. 1, 2024, doi: 10.1088/1755-1315/1298/1/012021.

[4] T. Ragnar, "Studi Pengaruh Trim Pada Perhitungan Kekuatan Memanjang Kapal," Sepuluh Nopember Institute of Technology, **2021**.

[5] S. K. Prabu Chelladurai, A. K. Dash, V. Nagarajan, and O. P. Sha, "Longitudinal strength of high block coefficient merchant ships in irregular waves," *Ocean Eng.*, **vol. 283**, no. June 2022, p. 115066, 2023, doi: 10.1016/j.oceaneng.2023.115066.

[6] M. Faiz, Murdjito, R. W. Prastianto, E. B. Djatmiko, S. Nugroho, and Y. B. Hadasa, "Intact stability analysis of crane barge due to loading orientation effect during heavy lifting operation," *IOP Conf. Ser. Earth Environ. Sci.*, **vol. 1298**, no. 1, 2024, doi: 10.1088/1755-1315/1298/1/012020.

[7] B. Utomo and B. Santoso, "Strength Analysis of Vessel Ship Type PBL Conversion From Barge," no. January, **2019**, doi: 10.2991/icoma-18.2019.77.

[8] American Bureau of Shipping, *ABS Rules for Building and Classing Part 3: Hull Construction and Equipment*. New York: American Bureau of Shipping, **2014**.

[9] DNV GL Noble Denton, *DNVGL-ST-N001 Marine Operations and Marine Services*. DNV GL Noble Denton, **2016**.

[10] Noble Denton International Ltd, *GUIDELINES FOR LIFTING OPERATIONS BY FLOATING VESSELS*, 0027/NDI R. Noble Denton International Ltd, **2006**.

[11] DNV GL, *RULES FOR CLASSIFICATION Ships Edition October 2015 Part 3 Hull Chapter 14 Rudders and steering*, no. October. DNV GL, **2015**.

Sedimentation Analysis in Determining the Optimal Breakwater Layout for Grajagan Beach Fishing Port, Banyuwangi Regency

D T Laksono[1], V D Prasita[2,3], E A Kisnarti[2,3], R S Bintoro[2,3]

[1] Master student of Marine Engineering, Faculty of Engineering and Marine Science, Hang Tuah University, Surabaya.
[2]Master of Marine Engineering, Faculty of Engineering and Ocean Sciences, Hang Tuah University, Surabaya.
[3]Program of Oceanography, Faculty of Engineering and Marine Science, Hang Tuah University, Surabaya.

E-mail: joelene.laksono@gmail.com

Abstract. Sedimentation that occurs in Grajagan Fishing Harbor, based on Google Earth images for the 2017-2023 time span tends to gather in the southern and eastern breakwater areas. The existence of sedimentation around the harbor can hinder fishing boat traffic. Sedimentation that occurs in the Grajagan Fishing Port area is thought to be less than optimal in the existing breakwater layout so that an optimal breakwater layout is needed. This study aims to determine the optimal breakwater layout based on hydro-oceanographic parameters using quantitative descriptive methods. The data used, namely: tidal, wind, wave, and sediment data. The data is used to analyze tidal patterns, wind and wave characteristics, current distribution patterns, and sediments. The results of the analysis are for determining the optimal breakwater layout. In Layout breakwater plan C sedimentation only occurs on the west side of the breakwater up to 0.6 m, while in the shipping channel and the berthing pond neither sedimentation nor erosion occurs. The simulation results of Layout breakwater plan A, B, and C, for Layout breakwater plan C is the most optimal and efficient in maintaining the berth pond for Grajagan Beach Fishing Port, Banyuwangi Regency.

1. Introduction

Sedimentation around the estuary is the result of the confluence of two water masses, namely sea water and fresh water [12]. Sediment deposition in estuaries occurs due to the interaction between tides, ocean waves, and current velocities that carry sediments [18]. Tides cause tidal currents that can move water masses, thus affecting suspense sediment transport and sediment concentration [10]. Ocean waves are a form of energy that can affect sediment transport in a body of water [2].

Sediment buildup results in sedimentation that can interfere with human activities on the coast, such as: Harbors and coastal defense buildings [6],[7]. One of them is port activities in the estuary to the shipping channel, due to siltation around the port. Activities in the port, such as being used as a place to lean ships with wide and deep river estuary conditions [17]. Shallow estuary conditions can cause obstruction to fishing boat traffic. In addition, the shallow condition of the estuary has an impact on the maximum size of the ship anchored in the port [8]. One of the ports located in the estuary is Grajagan Fishing Port, Banyuwangi Regency. Judging from google earth images, in the 2017-2023 timeframe sedimentation tends to gather in the area south and east of the breakwater.

Efforts to overcome the size limitations of anchored ships, one of which is by preventing siltation due to sedimentation around the port. Sedimentation prevention can be done by making sedimentation control buildings, such as breakwater. Breakwater is a structure that can reduce wave energy and function as a sediment barrier caused by ocean currents [9], [13]. Breakwater is effective in reducing sedimentation that occurs at the port location, because the supply of sediment from outside the breakwater can no longer enter the inner location of the port [1].

Meanwhile, breakwater construction has been carried out in Puger Waters. However, there is a problem of less than optimal layout of breakwater buildings in reducing high waves and restraining sediment transport. In this case, a re-planning of the breakwater layout was carried out [5], [19]. In addition, in the Fishing Port at IPP (Installation of Fishing Port) Pancer, Banyuwangi Regency, a breakwater evaluation was carried out due to the increasing sea wave pressure. This causes instability in the breakwater layout. In research [15]. In Grajagan Fishing Port, Banyuwangi Regency, the development of breakwater buildings in supporting the arrangement and development of the Port against the influence of waves. The influence factors of water conditions used, namely: waves and wind only [15].

Based on the above review, the author considers it necessary to conduct further research on the effectiveness of breakwater in reducing sedimentation. This research is in the Port area in the estuary, especially Grajagan Port, Banyuwangi Regency, which is still limited in research. This research was conducted to optimize the existing breakwater layout. In addition, hydro-oceanographic parameters as supporting parameters in the model simulation process. The results of this study can be used in an effort to consider decisions for the development of breakwater buildings in Grajagan Fishing Port, Banyuwangi Regency.

2. Methodology

Data collection was conducted in November 2023. The research site was situated within the Grajagan Coastal Fishing Port area in Banyuwangi Regency. The study area encompasses approximately 1,479,000 square meters surrounding the Grajagan PPP (Figure 1). A quantitative descriptive methodology was employed for this research.

2.1 Data

The data in this study consisted of primary and secondary data. Primary data used included sediment (sampling on November 20, 2023 which was divided into two, namely sediment load floating in the water and seabed samples), bathymetry (sounding on November 19, 2023 with an area of approximately 1,479,000 m2 using a Singlebeam Echosounder), and tides (taken from the BIG website for 30 days from November 1-30, 2023 and field observations for 3 days from November 19-21, 2023). Secondary data consisted of currents (taken from the Copernicus website during November 1-30, 2023 and field data for 26 hours on November 20, 2023 at 1 hour intervals), wind and waves (taken from the Copernicus website for 5 years from 2018-2022).

2.2 Processing

Data processing steps are carried out using several software, especially in simulating sediment distribution patterns using mike21 software. The following data processing is used:

IOP Conf. Series: Earth and Environmental Science **1473** (2025) 012004 doi:10.1088/1755-1315/1473/1/012004

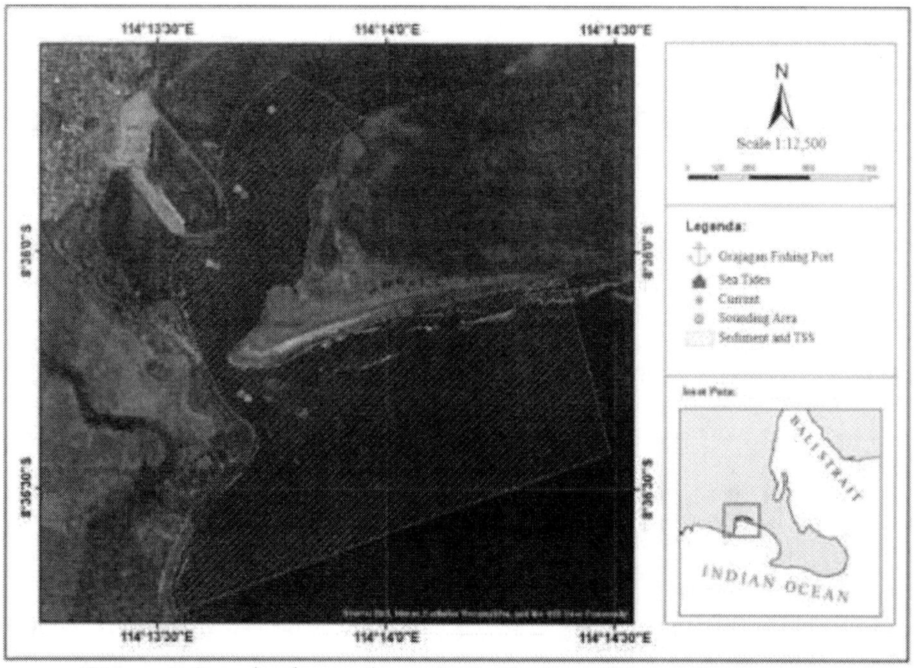

Figure 1. Sample observation point in Grajagan Waters, Banyuwangi.

2.2.1 Current

Current data in the form of direction and velocity will be separated into tidal currents and non-tidal currents by finding the u (east-west) and v (north-south) components. To get the u and v components, the following equation is used [14]:

a. Determines the u component by Cos direction multiplied by velocity.

$$u_{current} = v_{velocity} \; Cos \; \alpha \tag{1}$$

b. Determines the component of v by Sin direction multiplied by velocity.

$$v_{arus} = V_{\; velocity} \;\; Cos \; \alpha \tag{2}$$

c. Determine the tidal current component u, by the current component u minus the mean current u.

$$u_{tidal\;carrent} = u_{current} - u_{averange} \tag{3}$$

d. Determine the tidal current component v, with Current component v minus average current v.

$$v_{tidal\;current} = v_{current} - v_{averange} \tag{4}$$

e. Determine the direction of the tidal current:

$$\alpha_{tidal\;current} = Arctan\,(\frac{v_{tidal\;current}}{u_{tidal\;current}}) \tag{5}$$

f. Determine the speed of tidal currents:

$$Velocity_{\;tidal\;current} = \sqrt{(u_{tidal^2}) + (v_{tidal^2})} \tag{6}$$

g. Determine the direction of non-tidal currents:

$$\alpha_{arus\;non\;pasut} = Arctan\,(\frac{v_{arus\;rata-rata}}{u_{arus\;rata-rata}}) \tag{7}$$

h. Determine the speed of non-tidal currents:

$$Kec_{non-tidal\;current} = \sqrt{(u_{average^2}) + (v_{average^2})} \tag{8}$$

This flow data will also be used as model verification data in the Mike Zero application.

2.2.2 Sediment

Floating sediments will be processed using the SNI 06-6989-3-2004 method [3]. This method is used to determine the suspended sediment content. Seabed sediment samples are processed using the SNI 3423-2008 method [11]. This method is used to determine the type and size of sediment grains. The drifting sediment data will be used as verification of the sediment model at Mike Zero. Sediment grain size will later be used as model input data in the Mike Zero application.

2.2.3 Tidal

The tidal data is processed by descriptive statistics to obtain the values of the highest tide, lowest tide, and tidal stump. Tidal data in the form of water elevation will later be used as parameters in modeling.

2.2.4 Bathymetry

The sounding data is corrected with tidal values, after which a corrected depth value will be generated. Next, the corrected data was input to ArcGis 10.8 to layout the depth contour map. The corrected depth data is also used as a modeling parameter. Correction of sounding results to tides used the following formula [4].

$$D = (Z + TM) - (H - (y - Z_0)) \quad (9)$$

with:

D = corrected depth.
Z = depth measured by the echosounder.
TM = distance from the water surface to the transducer.
H = measured sea level height at time t.
y = average water level to zero palm.
Z_0 = the distance between the mean water level and the low tide level.

2.2.5 Wind and Wave

Wind and wave data will be visualized using the WR Plot application to produce wind rose and wave rose diagrams. Based on the wind and wave roses, the dominant wind direction and wave direction will be obtained. Wind data will later be used as input model data in the Mike Zero application.

2.3 Analysis

Data analysis includes: analysis of sediment distribution patterns and analysis of the feasibility of breakwater layouts. The analysis process starts from knowing the sediment distribution pattern, then making several new breakwater layouts. The breakwater layout was created in CAD software, then inputted into the boundary into Mike Zero software. The result is a model of sediment distribution patterns from several breakwater layouts, then scoring will be carried out to select the most optimal breakwater layout in preventing sedimentation.

3. Results and discussion

3.1 Tidal Conditions of Grajagan Harbor

Figure 2 shows the dynamic tidal fluctuations with MSL 0 (yellow color) from November 1-30, 2023. The highest tide (Figure 2) occurred on November 27 at 19:00 (134.5 cm) above mean high water level (MSL). The lowest tide occurred on November 28 at 02:00 with an elevation of -134.5 cm below MSL. The admiralty calculation results obtained a formzahl value of 0.69 with mixed tide predominant semidiurnal type $0.25 < F <= 1.5$. Tidal data from BIG was used as modeling input in Mike21.

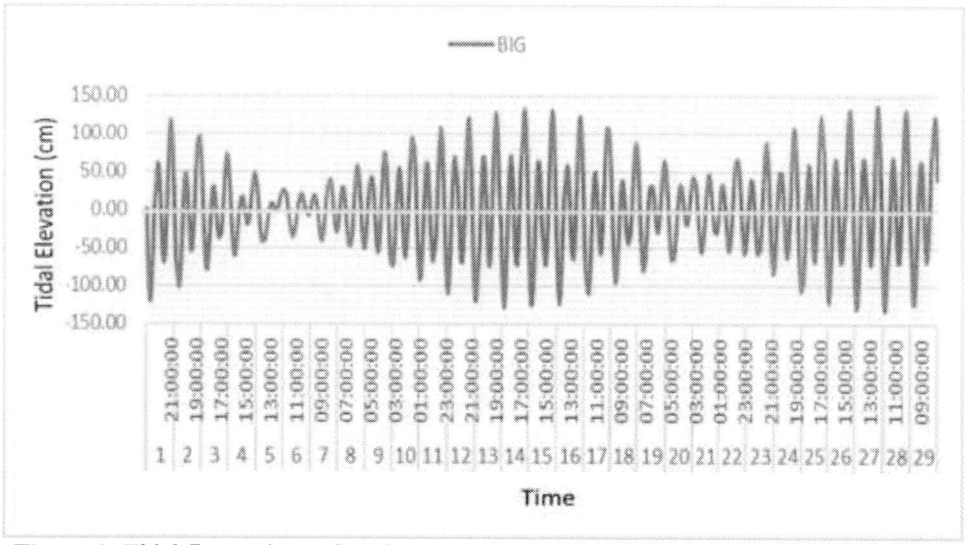

Figure 2. Tidal fluctuations of Grajagan Harbor for the period of November 1-30, 2023.

3.2 *Bathymetry Conditions*

The bathymetry profile in Figure 3, at the depth of the bottom of the estuary surface to the bottom of the sea surface has a depth value of 1.5 to -8.8m. The depth in the harbor basin area is quite shallow ranging from 1 to -1 m.

Figure 3. Bathymetry contours in Grajagan Waters.

3.3 *Wave Conditions in Grajagan Waters*

Wave data in Grajagan Waters was taken at https://marine.copernicus.eu with coordinates 8°40'0.01 "N and 114°15'0 "E for 5 years (2018-2022). On the fetch line, the effective fetch is determined according to the direction of the fetch line. The distance between fetch lines has a distance of 6o to the right and to the left. The fetch calculation map (Figure 4), determined by drawing a line from the shoreline towards the southeast, southwest and south.

On the map above the southeast fetch line is depicted in green, the southwest fetch line is depicted in red, the south fetch line is depicted in yellow. The southbound starting line is at 180°, the southwest starting line is 225°, and the southeast starting line is 135°. The calculation of the effective fetch in the south direction results in 156.56 km. the southwest direction is 79.46 km, while for the southeast direction the result is 86 km. From the fetch calculation above, the waves are then calculated using the SPM (Shore Protection Manual) method.

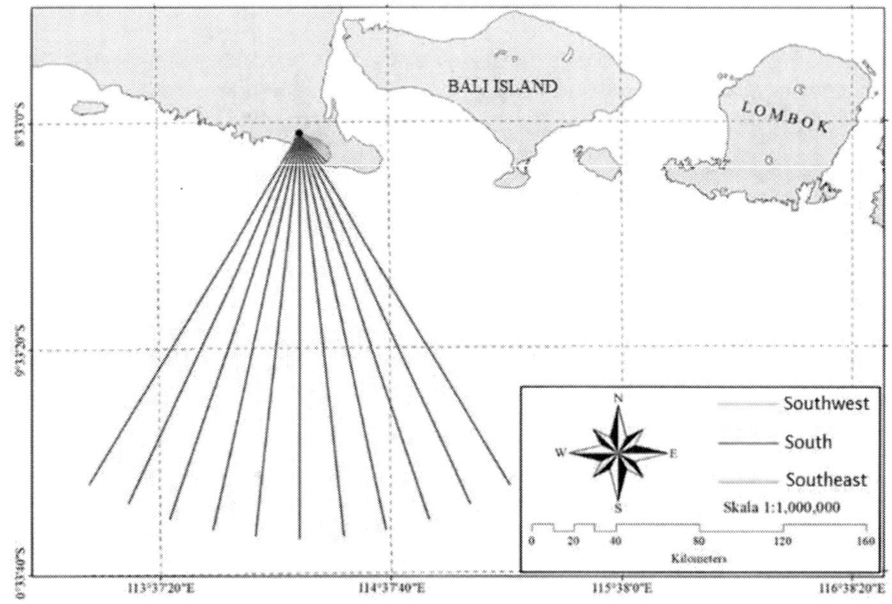

Figure 4. Fetch is effective.

Wind data processing (Figure 5a) shows that the wind pattern during the west season in the Grajagan area is dominated by winds blowing from the west. In the western season, the wind speed is between 0-9 m/s, while the wave height ranges from 0-1.50 m. The dominant wind blows (3-6 m/s) with a frequency of 43.5 %, while the wave height is dominated by the range of 0-0.30 m with a frequency of 86.5 %. This wind data if classified using the Beaufort Scale then the dominant wind during the western season in the Grajagan area is included in the weak wind category (0.3-1.5 m/s). The wave height category if classified using the BMKG Wave category then falls into the calm category (0.1 m-0.5 m).

Figure 5b shows that the wind pattern during transitional season 1 is dominated by winds blowing from the east. In transitional season 1, wind speeds ranged from 0-9 m/s, while wave heights ranged from 0-1.5m. The dominant wind blew (3-6 m/s) with a frequency of 53.3 %, while the dominant wave height ranged from 0-0.3 m with a frequency of 63.0 %. If this wind data is classified using the Beaufort Scale, the dominant wind during transitional season 1 in the Grajagan area is categorized as gentle and moderate breeze (3.4-7.9 m/s). Then, for the wave height category if classified using the BMKG Wave category, it is included in the calm category (0.1 - 0.5 m).

Wind patterns during the eastern season (Figure 5c) in the Grajagan area are dominated by winds blowing from the southeast. In the eastern season, the wind speed ranges from 0-9 m/s, while the wave height is between 0-1.50 m. The dominant wind blows at 3-6 m/s with a frequency of 69.3 % and the dominant wave height ranges from 0-0.3 m with a frequency of 26.7 %.

The wind pattern during transitional season 2 in the Grajakan area (Figure 5d) is dominated by winds blowing from the southeast. During transitional season 2, wind speeds ranged from 0-9 m/s, while wave heights ranged from 0-1.5m. The dominant wind blows with a speed of 3-6 m/s and a frequency of 53.6 %. Wave height is dominated by the height range of 0-0.3 m with a frequency of 40.2 %. On the Beaufort Scale, the transitional season 2 wind data is categorized as gentle and moderate breeze (3.4-7.9 m/s). Transitional season 2 wave heights are categorized as calm (0.1-0.5 m) when classified according to the BMKG wave height category.

The characteristics of significant wave height (Hs) during the western season show a significant level of variation. Based on the graph of Hs height and period in Figure 6a, that throughout the western season for a period of 5 years, namely in 2018-2022 has Hs height with a range of 0.05-0.85 m and a period ranging from 0.31-4.96 s. Significant waves are a statistical measure that reflects the average height of the largest waves in a particular area. In the western season conditions that took place in 2018-2022, the highest Hs values were in 2018 (0.27 m), 2019 (0.85 m), 2020 (0.58 m), 2021 (0.24 m), and 2022 (0.24 m). Wave period measures the time interval between two consecutive peaks or two consecutive valleys of ocean waves. The highest period values in the western season were recorded in 2018 (2.39 s), 2019 (4.96 s), 2020 (4.35 s), 2021 (1.92 s), and 2022 (1.95 s). Wave period can vary depending on a number of factors, including wind speed and duration, distance traveled by the wind, and water depth.

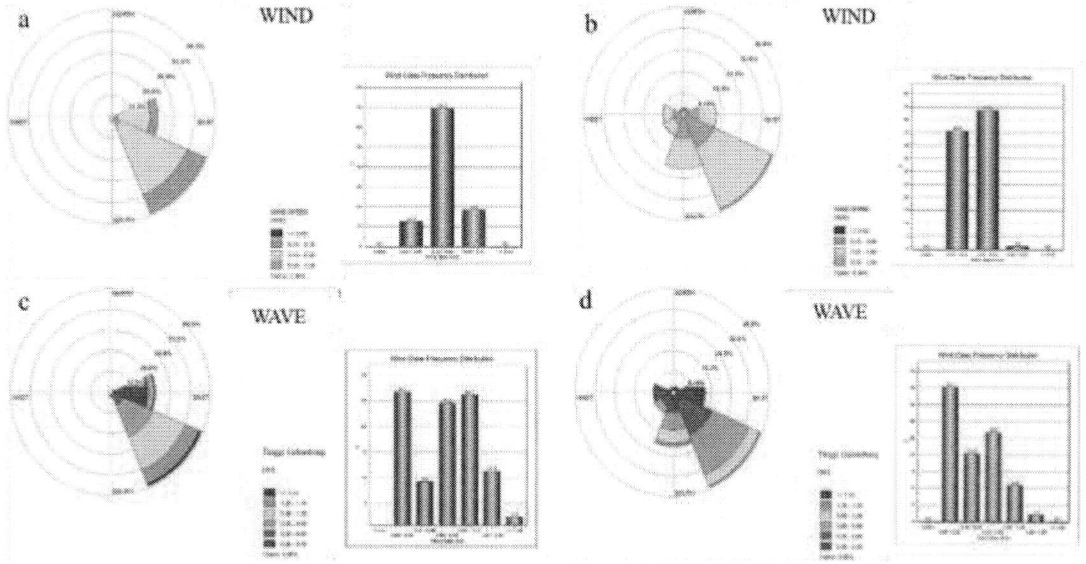

Figure 5. Wind and wave distribution of (a) west monsoon, (b) transition season 1, (c) east monsoon, and (d) transition season 2.

The height of Hs during transitional season 1 shows a significant level of variation, based on the height and period of Hs (Figure 6b). During transitional season 1 in the 5-year period of 2018-2022, Hs ranged from 0.72-1.16 m (2.81-5.46 s). During transitional season 1 which took place in 2018-2022, the highest Hs values were in 2018 (1.16 m), 2019 (1.18 m), 2020 (1 m), 2021 (0.97 m), and 2022 (0.43 m). In addition, the highest period values in transitional season 1 were recorded in 2018 (5.43 s), 2019 (5.46 s), 2020 (5.1 s), 2021 (5.1 s), and 2022 (3.81 s).

The characteristics of high Hs during the eastern season show a significant level of variation. Based on the graph of Hs height and period (Figure 6c), during the eastern season that took place in the period 2018-2022, Hs height ranged from 0.9-1.42 m (4.67-5.82 s). In the eastern season conditions that took place in 2018-2022, the highest Hs values were in 2018 (1.23 m), 2019 (1.08 m), 2020 (1.42

m), 2021 (1.2 m), and 2022 (1.25 m). The highest period values in the eastern season were recorded in 2018 (5.54 s), 2019 (5.31 s), 2020 (5.82 s), 2021 (5.51 s), and 2022 (5.56 s). The characteristics of Hs height during transition season 2 show a significant degree of variation. Based on the height and period of Hs (Figure 6d), for 5 years (2018-2022), the resulting Hs height ranged from 0.54-0.96 m with a period of 4.08-5.41 s. In the conditions of transitional season 2 (Figure 5d) lasting in 2018-2022, the highest Hs values were in 2018 (0.94 m), 2019 (0.81 m), 2020 (0.96 m), 2021 (0.96 m), and 2022 (0.84 m). The highest period values in transitional season 2 were recorded in 2018 (5.41 s), 2019 (5.06 s), 2020 (5.14 s), 2021 (5.52 s), and 2022 (5.05 s).

Based on Figure 7, shows the condition of the direction and speed of the current against the tide. When heading towards the tide, the dominant tidal current conditions move towards land, with speeds ranging from 0.04-0.14 m/s, then when heading towards the ebb current conditions have a fairly high speed value with a range of 0.05-0.28 m/s. The direction of the tidal current when heading to low tide, the current moves away from land. When the lowest tide and the highest tide the direction of the tidal current tends to turn and move towards the opposite direction. The condition of tidal current velocity at the lowest ebb and highest tide has a small speed value.

3.4 Tidal Current Conditions

At high tide (Figure 9b), the current speed at sea ranges from 0 to 0.095 m/s with the dominant direction towards land. At the coast, the current speed is very low, ranging from 0 to 0.24 m/s, with the current direction following along the coastline. At the mouth of the river, the current speed increases to 0.012 m/s to 0.085 m/s, indicating a stronger flow as the water movement starts to be affected by the tide. In the anchorage area, the current speed ranges from 0 to 0.035 m/s with the dominant direction towards land.

- *Existing Tidal Flow Conditions*

At the lowest ebb (Figure 8a), current velocities in the berth pond ranged from 0-0.035 m/s with the dominant direction towards the sea, indicating a fairly calm current as water flows out of the harbor area. At the estuary, current velocities ranged from 0.015-0.060 m/s, indicating more moderate water movement. On the coast, the current speed is very low, ranging from 0-0.020 m/s with the current direction away from land. Meanwhile, in the open sea, the current speed ranged from 0-0.080 m/s, indicating stronger currents in the open sea during the lowest low tide period.

During the ebb tide (Figure 8b), the current speed in the sea ranged from 0-0.04 m/s (dominant direction towards land). Along the coast, the current speed is very low (0-0.030 m/s), with the current direction still towards land. At the mouth of the river the current speed increases to 0.02 m/s-0.14 m/s, which indicates a stronger flow because the water movement is affected by the tide. The current speed at the mouth of the river increased due to the narrowing of the estuary. In the anchorage pond, the current velocity decreased (0-0.045 m/s) with the dominant direction towards land.

At the highest tide (Figure 8c), the current speed in the anchorage pond ranged from 0-0.050 m/s (dominant direction towards land), indicating a rather calm current as the water started to enter the harbor area. At the estuary, current velocities reached 0.020-0.095 m/s, indicating a more moderate flow of water towards land. Along the coast, the current speed ranges from 0-0.040 m/s, indicating an increased volume of water flowing towards the shore. In the open sea, current speeds ranged from 0-0.110 m/s, indicating stronger currents in the open sea during the highest tide period.

Figure 8d shows the tidal current when heading towards low tide, the current speed in the dock pond ranges from 0-0.040 m/s (dominant direction towards the sea). This condition indicates that the water begins to move out of the anchorage pond towards the sea along with the low tide. At the estuary, the current speed increases to 0.010-0.2 m/s, indicating stronger water flow as it approaches the sea. Along the coast, the current speed ranges from 0-0.025 m/s (the direction of the current is still away from land). Meanwhile, in the open sea, the current speed ranges from 0-0.02 m/s with the dominant direction towards the sea. This condition indicates a stronger current when the water moves away from land during the period towards low tide. This current pattern shows the movement of water generally out of the land towards the sea during this period.

- *Tidal Current Conditions Breakwater Plan A*

At the lowest ebb (Figure 9a), the current velocity in the berth pond is 0-0.02 m/s (dominant direction towards the sea), indicating a fairly calm current as the water flows out of the harbor area. At the estuary, the current speed is 0.015-0.055 m/s, indicating more moderate water movement. On the coast, the current speed is a very low 00.018 m/s (current direction away from land). Meanwhile, in the open sea, the current speed was 0-0.070 m/s, indicating stronger currents in the open sea during the lowest low tide period.

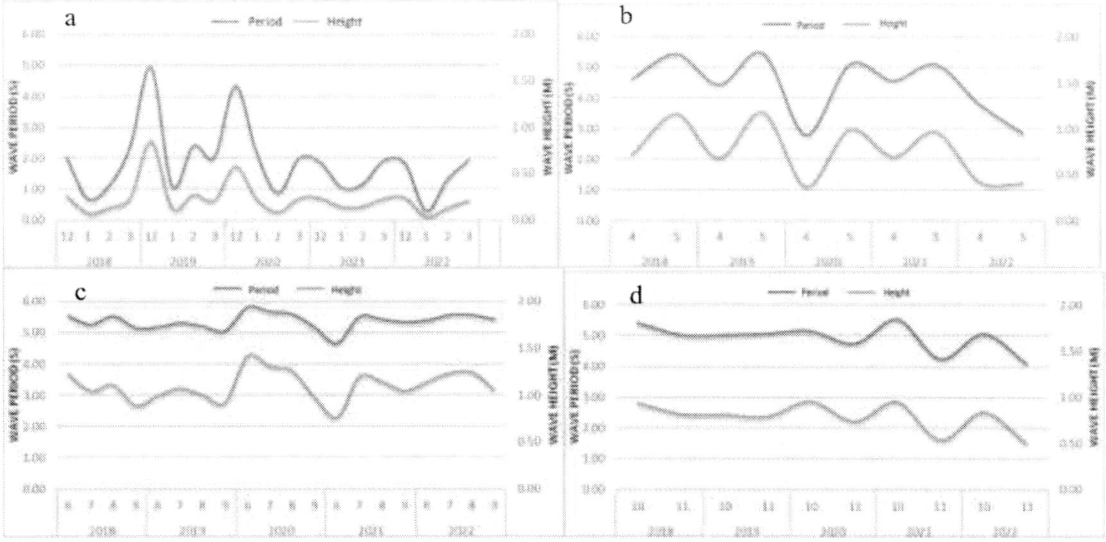

Figure 6. Significant wave height and period of (a) western season, (b) transitional season 1, (c) eastern season, and (d) transitional season 2.

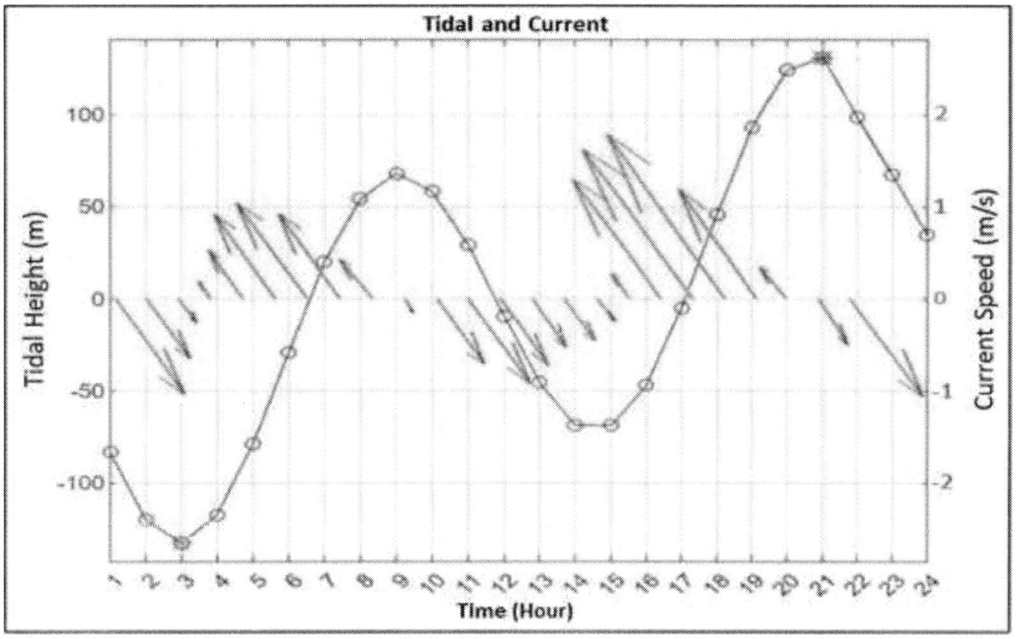

Figure 7. *Featherplot of water level, current velocity and direction November 28, 2023.*

In Figure 9c during the highest tide, the current velocity in the anchorage pond ranges from 0-0.045 m/s with the dominant direction towards land, indicating a rather calm current as the water begins to enter into the harbor area. At the estuary, the current speed is 0.018-0.090 m/s, indicating a more moderate flow of water towards land. Along the coast, current velocities ranged from 0-0.035 m/s, indicating an increased volume of water flowing towards the shore. In the open sea, current velocities ranged from 0-0.105 m/s, indicating stronger currents in the open sea during the highest tide period.

During low tide (Figure 9d), the current speed in the berth pond is between 0-0.030 m/s with the dominant direction towards the sea. Water begins to move out of the anchorage pond towards the sea as the tide recedes. At the estuary, the current speed increases to 0.02 m/s-0.24 m/s, indicating stronger water flow as it approaches the sea. Along the coast, the current speed is 0-0.020 m/s, with the current direction still away from land. In the open sea, the current speed is 0-0.090 m/s with the dominant direction towards the sea, indicating stronger currents as the water moves away from land during the period towards low tide. The decrease in current velocity in the berth pond was influenced by the tide beginning to move water out of the harbor area.

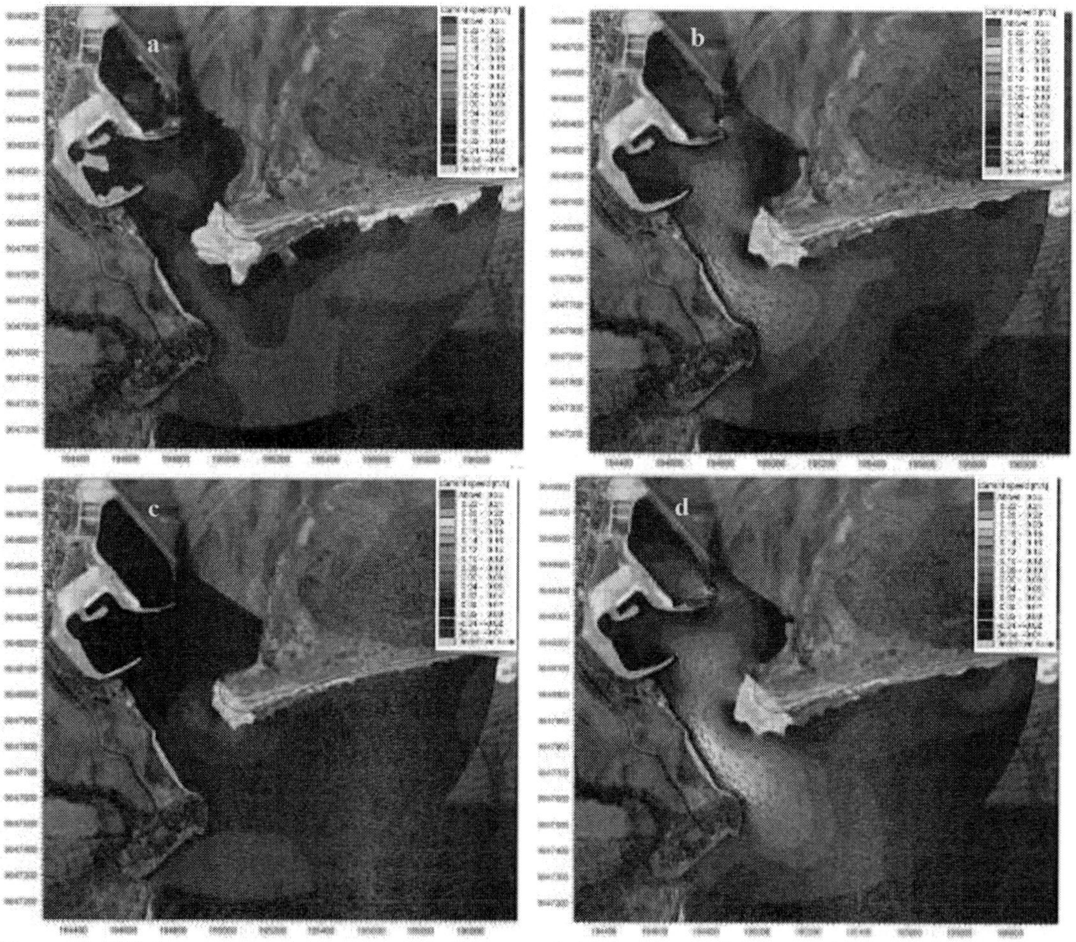

Figure 8. Tidal currents at (a) lowest ebb, (b) towards high tide, (c) highest tide, and (d) towards low tide.

- *Tidal Current Condition of Breakwater Plan B*

At the lowest ebb (Figure 10a), the current speed in the berth pond was 0-0.020 m/s with the dominant direction towards the sea, indicating a fairly calm current as the water flows out of the harbor area. At the estuary, the current speed ranges from 0.010-0.04 m/s, indicating calmer water movement. On the coast, the current speed is very low, between 0-0.015 m/s with the current direction away from land. Meanwhile, in the open sea, the current speed ranged from 0-0.070 m/s, indicating stronger currents in the open sea during the lowest low tide period.

During the ebb tide (Figure 10b), the current speed at sea is 0-0.085 m/s with the dominant direction towards land. At the coast, the current speed is very low, between 0-0.022 m/s, with the current direction still towards land. At the estuary, the current speed increases to 0.02-0.16 m/s, indicating a stronger flow due to the movement of water that is starting to be affected by the tide. In the dock pond, the current speed ranged from 0-0.030 m/s with the dominant direction towards land. The decrease in current velocity in the anchorage pond is influenced by the presence of a breakwater that makes the waters calmer.

The tidal currents during the highest tide (Figure 10c), showing current velocities in the anchorage pond ranging from 0-0.04 m/s with a dominant direction towards land, indicate that the currents tend to calm down as the water begins to enter the anchorage pond during the highest tide period. At the estuary, the current speed reaches 0.02-0.04 m/s, indicating a calmer flow of water towards land. Along the coast, the current speed ranges from 0-0.04 m/s, indicating an increased volume of water flowing towards the shore. In the open sea, the current speed is still relatively calm between 0-0.04 m/s.

The current velocity in the berth pond ranged from 0-0.035 m/s with the dominant direction towards the sea. This indicates that water begins to move out of the anchorage pond towards the sea as the tide recedes (Figure 10d). At the mouth of the river, the current speed increases to 0.012-0.075 m/s, indicating stronger water flow as it approaches the sea. Along the coast, the current speed ranges from 0-0.020 m/s, with the current direction still away from land. Meanwhile, in the open sea, the current speed ranges from 0-0.090 m/s with the dominant direction towards the sea, indicating stronger currents as the water moves away from land during the period towards low tide.

- *Tidal Current Condition of Breakwater Paln C*

During the lowest tides (Figure 11a), the current speed in the anchorage pond is 0-0.020 m/s with the dominant direction towards the sea. At the estuary, the current speed increases to 0.012-0.050 m/s, indicating a stronger flow of water towards the sea. On the coast, the current speed is very low, ranging from 0-0.015 m/s with the current direction away from land. In the open sea, the current speed is 0-0.065 m/s, indicating a stronger current with the dominant direction towards the sea. This shows the current pattern moving away from land during the lowest low tide period.

At high tide (Figure 11b), the current speed at sea is 0-0.085 m/s with the dominant direction towards land. On the coast, the current speed is very low, 0-0.022 m/s, with the current direction still towards land. When at the mouth of the river the current speed increases to 0.010-0.14 m/s, this is due to the movement of water that begins to be affected by the tide. In the anchorage pond, the current speed is 0-0.030 m/s with the dominant direction towards land. The decrease in current velocity in the anchorage pond is influenced by the presence of a breakwater that makes the pond waters calmer.

The tidal current during the highest tide (Figure 11c), has a speed of 0-0.048 m/s in the dock pond with the dominant direction towards land. This indicates that the current tends to calm down as the water begins to enter the anchorage pond during the highest tide period. At the mouth of the river, the current speed reached 0.020-0.100 m/s, indicating a more moderate flow of water towards land. Along the coast, current velocities range from 0-0.042 m/s, indicating an increased volume of water flowing towards the shore. In the open sea, current velocities ranged from 0-0.04 m/s, indicating still currents in the open sea during the highest tide period.

During the ebb tide (Figure 11d), the current speed in the dock pond is 0-0.035 m/s with the dominant direction towards the sea. This condition indicates that the water begins to move out of the anchorage pond towards the sea along with the ebb. At the estuary, the current speed increases to 0.02-0.2 m/s, indicating stronger water flow as it approaches the sea. Along the coast, the current speed ranges from 0-0.020 m/s, with the current direction still away from land. Meanwhile, in the open sea, the current speed ranges from 0-0.090 m/s with the dominant direction towards the sea, indicating stronger currents as the water moves away from land during the period towards low tide.

3.5 Model Validation

Tidal data validation was conducted by comparing field data with BIG data that had been simulated in Mike21 (Figure 12). The MSL from the BIG data is equalized with the field data which is 0 cm. Figure IV.29 shows the same daily tidal cycle. Based on the field data, the highest tidal surge occurred on November 19 at 00:00 with the highest water level 50 cm above MSL and the lowest tidal surge occurred with the lowest water level -50 cm below MSL on November 20 at 06:00. Meanwhile, based on the model results, the highest tidal ridge occurred on November 19 at 00:00 with the highest water level of 61 cm above MSL and the lowest tidal ridge occurred on November 19 at 05:00 with the lowest water level of -81 cm below MSL.

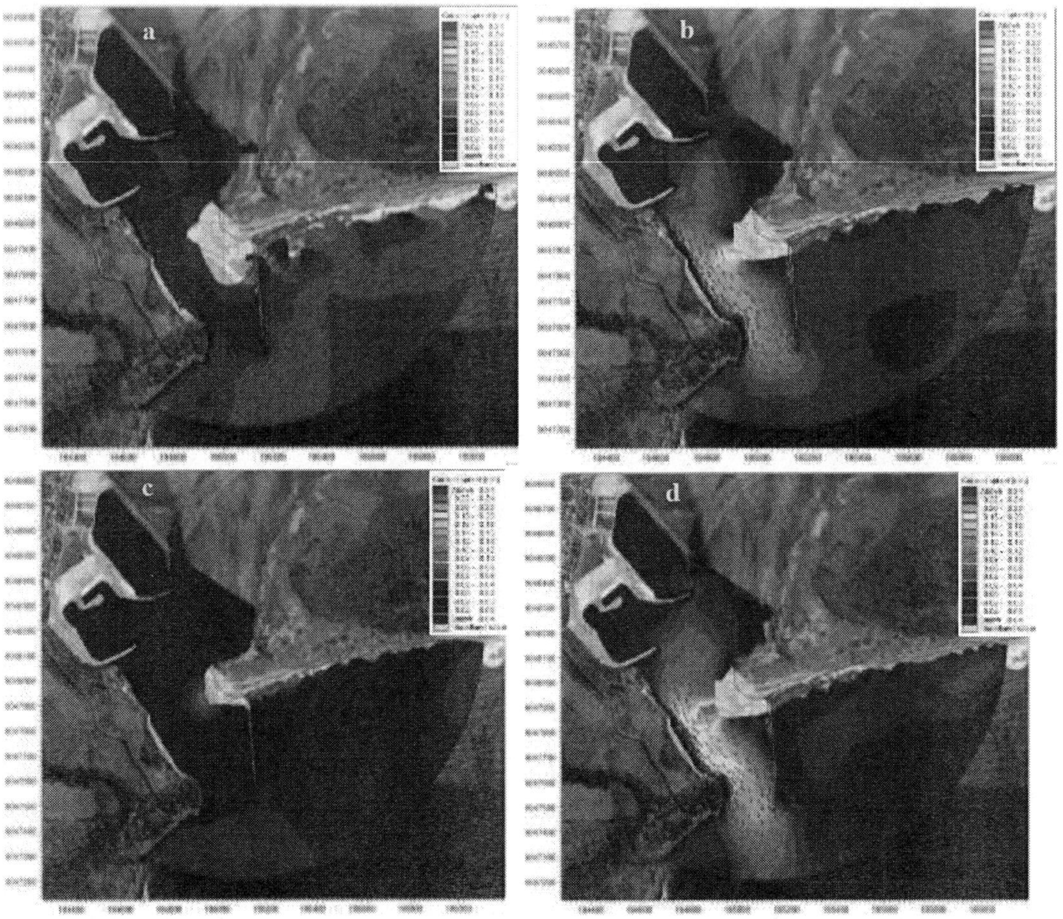

Figure 9. Tidal currents at (a) lowest ebb, (b) towards high tide, (c) highest tide, and (d) towards low tide.

3.6 Sediment Conditions

3.6.1 Sediment Condition Before Breakwater

Figure 13a during the initial condition, there is no difference in elevation because it is an existing condition before being affected by tidal currents. In the initial condition, the elevation value still follows the depth value during the survey. Elevation conditions on the estuary side of the river range from 0 to -3.2 m, the harbor area ranges from 0-1.6 m, and on the beach area ranges from 0-3.2 m. The final bed level condition of Figure 13b, has varying depths. The estuary area experienced erosion with bed level values ranging from -0.8 to -3.2m. The bed level condition in the harbor area has sedimentation of up to 1.6m. The sediment comes from the coast to the harbor area directly.

Erosion events occurred at the estuary, beach area and the inner side of the harbor which can be seen in Figure 13c. In the estuary area, erosion of 0 to -0.3 m occurred, while the western side of the estuary experienced erosion of up to -1 m. Then. The waters along the shoreline also experienced erosion up to -0.75 m. In addition to erosion events, sedimentation also occurred in the harbor basin area by 0-1 m, while the inner side of the northern area of the harbor experienced erosion up to -0.9 m. In the coastal area, erosion occurs due to rip currents that carry sediment to the sea, while sedimentation that occurs in the harbor basin is due to longshore currents that carry sediment from along the coastline to the estuary and into the harbor basin.

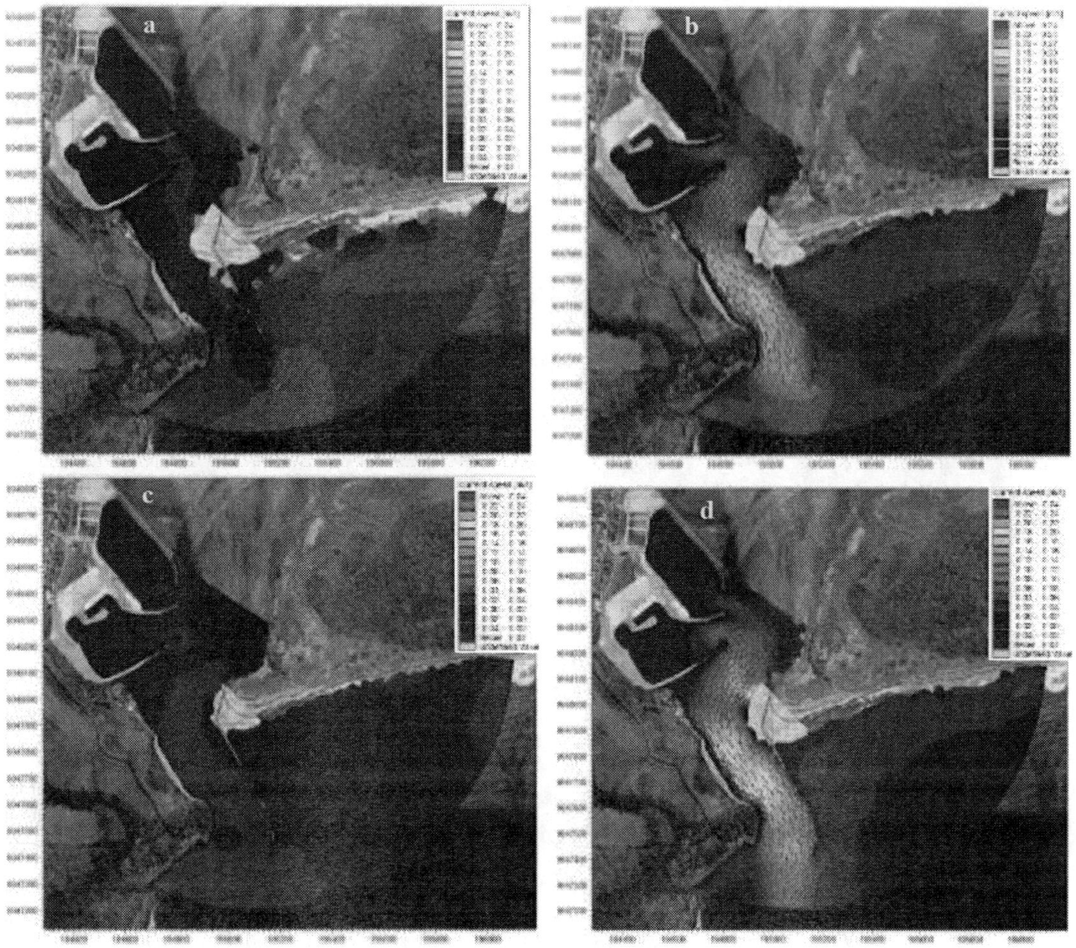

Figure 10. Tidal currents at (a) lowest ebb, (b) towards high tide, (c) highest tide, and (d) towards low tide.

3.6.2 Sediment Simulation Results After Dredging and Breakwater

The initial conditions on breakwater plan A, no sedimentation has been seen (Figure 14a). This condition is an existing condition before being exposed to tidal currents, so no sedimentation or erosion has been found. In the initial condition, the elevation value still follows the depth value during the survey. Elevation conditions on the estuary side of the river ranged from (-2.4)-3.2 m, the harbor area was 0 m, and the beach area ranged from 0-3.2 m.

Figure 14b is the final bed level condition after dredging. In the simulation results after dredging there is a difference in elevation than before dredging, especially in the harbor basin area. After dredging, there is no sedimentation or erosion phenomenon in the harbor basin. The final bed level condition in breakwater plan A, there is a change in elevation at the estuary, to the side of breakwater plan A. Sedimentation occurs on the side of the breakwater, while erosion occurs in the estuary area. Sedimentation that occurs on the near-land side of breakwater plan A has a bed level value ranging from 0-3.2m. The estuary area is affected by erosion with bed level elevations of -0.1 to -1.6 m.

After dredging, neither sedimentation nor erosion occurred in the harbor basin. In Figure 14c, the bed level change condition experienced erosion events that occurred at the mouth of the river and the east side of breakwater plan A. Sedimentation occurred on the west side of breakwater plan A and the mouth of the harbor pool. The estuary and shipping channel eroded up to -0.15 m, then on the east side of the breakwater there was also erosion up to -0.9m. On the west side of breakwater plan A, sedimentation up to 0.9 m occurred. Sedimentation also occurs at the mouth of the berth pond up to 0.3 m. Sedimentation occurs due to the influence of longshore currents that carry sediment along the coastline.

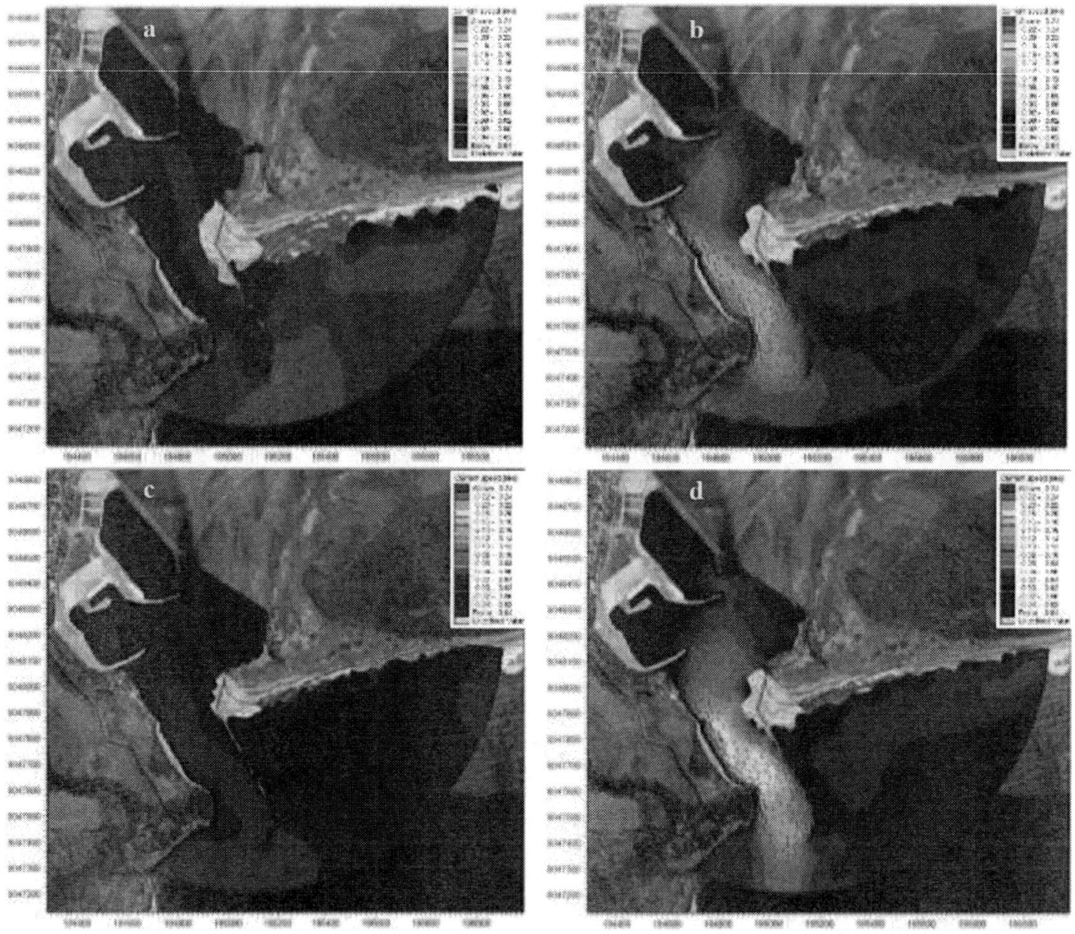

Figure 11. Tidal currents at (a) lowest ebb, (b) towards high tide, (c) highest tide, and (d) towards low tide.

Figure 15a shows that the initial bed level of breakwater plan B after dredging is not much different before dredging. The difference in elevation in the simulation after dredging is only in the harbor basin area which has a bed level value of 0 m. This condition is an existing condition before being exposed to tidal currents, so that no sedimentation or erosion has been found. Elevation conditions on the estuary side of the river ranged from 0-(-2.4) m, the harbor area was 0 m, and the beach area ranged from 0-3.2 m. The final bed level condition (Figure 15b) after dredging, sedimentation occurred on the west side of the breakwater and erosion on the east side of the breakwater. Erosion also occurred at the estuary with a bed level range of 0-(-2.4) m. Elevation changes on the west side of breakwater plan B ranged from 0.8-3.2 m, while on the east side of breakwater plan B erosion occurred with a bed level value of -1.6 m. The estuary area was also affected by erosion with a reduction in bed level to -1.6 m.

Simulation results on the condition of bed level change breakwater plan B (Figure 15c), sedimentation occurs on the west side of the breakwater, while in the harbor basin there is no sedimentation or erosion. Erosion occurs at the mouth of the river and the east side of the breakwater to the shoreline. Sedimentation that occurs on the west side of the breakwater has a bed level change value ranging from 0 to 0.75 m, while in the shipping channel there is sedimentation of up to 0.45m. Sedimentation also occurs on the breakwater side of the pond and the mouth of the berth pond ranging from 0.3-0.9 m. At the mouth of the river there is a decrease in bed level due to erosion in the range of 0 to -0.3 m, while along the coastline to the east side of the breakwater there is erosion ranging from -0.15 to -0.75 m. Erosion occurs due to rip currents that carry sediment along the shoreline to the sea.

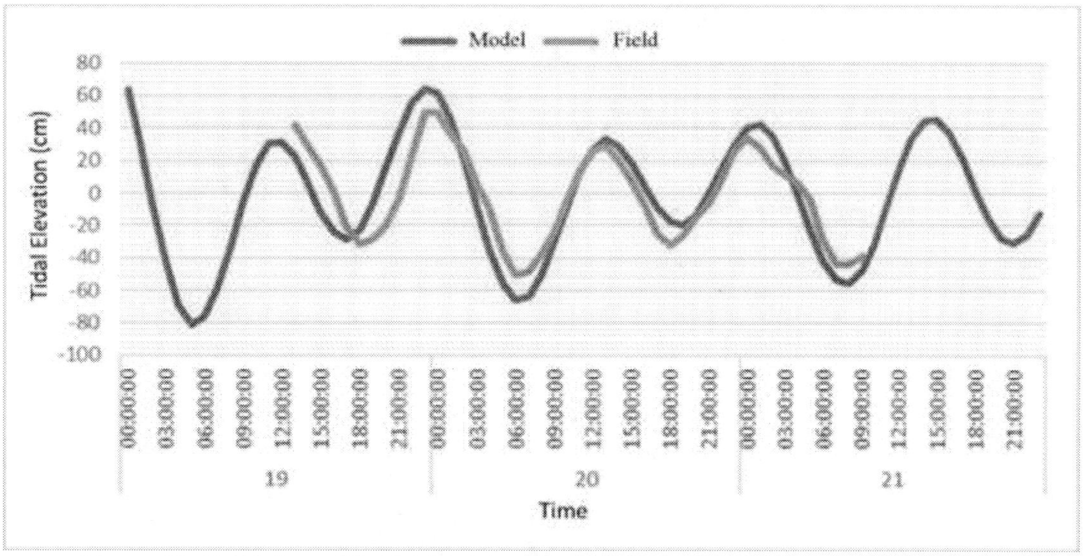

Figure 12. Validation of field tidal data with model results on November 19-21, 2023.

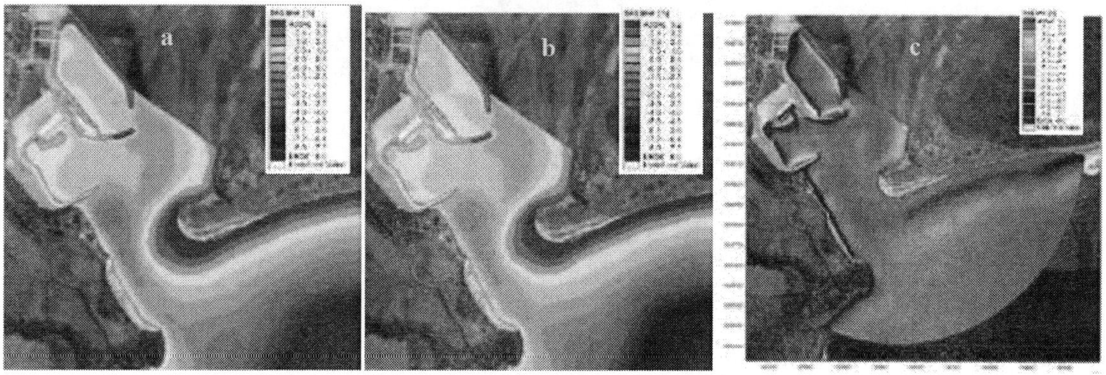

Figure 13. Bed level conditions (a) initial, (b) final, and (c) existing bed level change.

The initial bed level condition of breakwater plan C (Figure 16a) after dredging is not much different before dredging. The difference in elevation in the simulation after dredging is only in the harbor basin area which has a bed level value of 0 m. This condition is an existing condition before being exposed to tidal currents, so that no sedimentation or erosion has been found. In the initial condition, the elevation value still follows the depth value during the survey. Elevation conditions on the estuary side of the river ranged from 0 to -2.4 m, the harbor area was 0 m, and the beach area ranged from 0-3.2 m. Figure 16b with the final bed level condition after dredging, sedimentation occurs on the west side of the breakwater and erosion on the east side of the breakwater. The final condition of breakwater plan C and B only has a difference in elevation at the port door due to the short breakwater plan B. In breakwater plan C erosion occurs at the port door while in plan B sedimentation occurs. Erosion also occurs at the mouth of the river with a bed level range of 0 to -2.4 m. Elevation changes on the west side of breakwater plan B range from 0.8 to 3.2 m, while on the east side of breakwater plan B erosion occurs with a bed level value of -1.6 m. The estuary area was also affected by erosion with a reduction in bed level of up to -1.6 m.

Simulation results in Figure 16c breakwater plan C, no significant sedimentation or erosion was found in the estuary to the harbor basin. Sedimentation only occurs on the west side of the breakwater with a range of bed level changes up to 0.6 m and a small area. Erosion occurs in the shipping channel, estuary and east side of the breakwater.

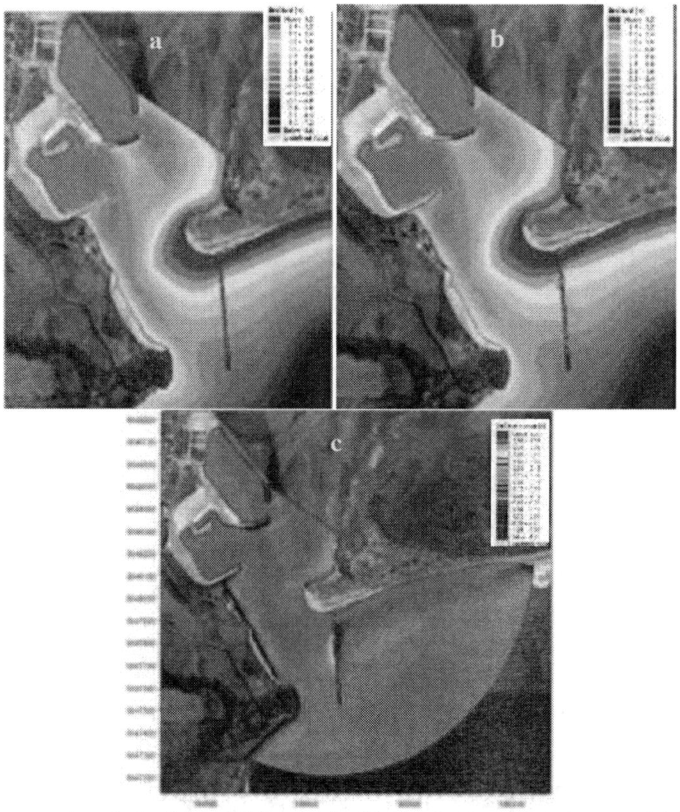

Figure 14. Breakwater plan A, bed level (a) initial, (b) final after dredging, and (c) bed level change after dredging.

In the shipping channel and estuary, erosion occurs up to -0.3 m, then on the east side of the breakwater there is also erosion up to -0.9 m. The shipping channel, the harbor door and the estuary are the most needed locations for ships to enter and exit the harbor safely. Erosion that occurs on the east side of the breakwater to along the shoreline is caused by rip currents that carry sediment to the sea.

In the simulation results after dredging, breakwater plan C is considered the most effective than plans A and B. This is because the shape of breakwater plan C is able to overcome the influence of sedimentation in the Grajagan harbor basin. Breakwater plan B is less effective because of the presence of a breakwater building in the harbor basin which results in sediment trapped around the breakwater of the plan B harbor basin. Layout breakwater plan A is less effective because the estuary door is still wide which can make sediment enter the harbor basin directly.

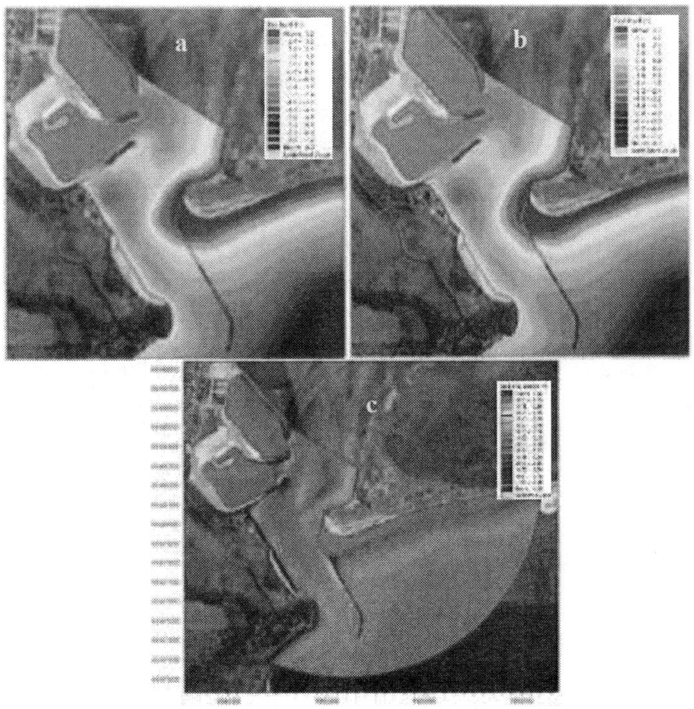

Figure 15. Breakwater plan B conditions (a) initial, (b) final, and (c) bed level change after dredging.

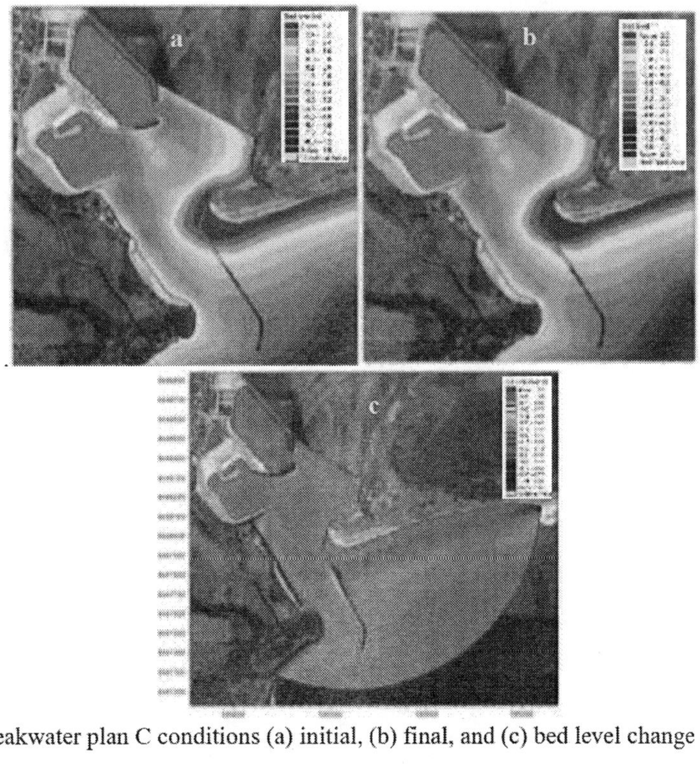

Figure 16. Breakwater plan C conditions (a) initial, (b) final, and (c) bed level change after dredging.

4. Conclusion

Prior to the breakwater's construction, the harbor basin area witnessed a more pronounced accumulation of sediment compared to the post-construction period. Pre-breakwater construction, the harbor basin area, including the harbor basin door, experienced sedimentation reaching 0.6 meters. The east side of the river witnessed sedimentation of up to 0.3 meters, while erosion ranging from -0.15 to -0.1 meters occurred on the west side of the estuary.

After the construction of breakwater layouts A, B, C, and also dredging in the harbor basin, the following simulation results were obtained: (a) Layout breakwater plan A in the harbor basin, no sedimentation occurs. On the west side of breakwater plan A, sedimentation up to 0.9 m occurs, while the east side of breakwater plan A erosion up to -0.9 m occurs. At the mouth of the harbor basin, sedimentation of up to 0.3 m occurred. The shipping channel erodes up to -0.15 m, (b) Layout breakwater plan B does not occur sedimentation in the berth pond, but sedimentation occurs on the west side of the breakwater up to 0.75m. In the shipping channel there is sedimentation up to 0.45 m. The mouth of the berth pond and the breakwater side of the pond occur sedimentation of 0.3 to 0.9 m. Layout breakwater plan C sedimentation only occurs on the west side of the breakwater up to 0.6 m, while in the shipping channel and the berth pool neither sedimentation nor erosion occurs. Based on the simulation results of breakwater layouts A, B, and C, breakwater layout C emerges as the most optimal solution for the Grajagan Beach Fishing Port in Banyuwangi Regency. Layout C demonstrates superior efficiency in maintaining the harbor basin compared to layouts A and B.

Acknowledgements

The author wishes to express gratitude to Hang Tuah University for their support in the completion of this research. Additionally, the author would like to thank the relevant parties who provided the opportunity to conduct field observations in Grajagan Waters, Banyuwangi.

Reference

[1] Bachtiar H, Setiantoro H E, Prasetyo A, Eliasta L, and Putriasi A E. 2020. Pendekatan model komputasi untuk penanggulangan sedimentasi pelabuhan akibat pengaruh muara sungai studi kasus Pelabuhan Tanjung Mas Semarang. *Jurnal Teknik hidraulik*, **11** (2), pp. 119-134.
[2] Gemilang R I A, and Kurniawan Y N. 2016. Efektifitas redaman energi gelombang akibat adanya *breakwater* terapung ditinjau dari model fisik dan studi numerik. *Jurnal online institut teknologi nasional*, **2** (3), pp. 147-157.
[3] Gemilang W A, Rahmawan G A, and Wisha U J. 2017. Kualitas perairan Teluk Ambon Dalam berdasarkan parameter fisika dan kimia pada musim peralihan I. *EnviroScienteae*, **13** (1), pp. 79-90.
[4] Harsono G, and Hartoyo D. 2018. *Dasar-dasar survei hidro-oseanografi*. Bogor: IPB Press.
[5] Izzati A. 2018. Redesign bangunan pengaman pantai sebagai alur masuk pelayaran Pelabuhan Perikanan Jember. Institut Teknologi Surabaya.
[6] Laksono FAT, Borzì L, Distefano S, Di Stefano A, and Kovács, J. (2022). Shoreline prediction modelling as a base tool for coastal management: The Catania plain case study (Italy). *Journal of Marine Science and Engineering,* **10**(12), 1988.
[7] Miranda NA, Bintoro RS, and Prasita VD. 2023. The pattern of coastline changes and wave modelling around the expansion of PPI Popoh Tulung Agung, East Java. *Maritime Technology and Research*, **5**(4): 262926.
[8] Nofrizal, and Istijono B. 2018. Pengaruh transport sedimen dan curah hujan tahunan terhadap penambahan volume sedimen di muara Sungai Batang Arau. *ACE Confrence* **5**, pp. 344-353.
[9] Nurisman N, and Taringan T A B. 2020. Model fisik pengaruh submerged *breakwater* terhadap gelombang datang. *Jurnal penelitian sains*, **22** (2), pp. 95-98.
[10] Purnama E A, Hariadi, and Saputro E. 2015. Pengaruh arus, pasang surut dan debit sungai terhadap distribusi sedimen tersuspensi di Perairan Muara Sungai Ciberes, Cirebon. *Jurnal oseanografi*, **4** (1), pp.74-84.

[11] Ratnasari K D, Dermawan D, and Rizal M. C. 2018. Studi karakteristik lumpur intake perusahaan pembangkit listrik. *Proceeding on waste treatment technology*, **1** (1), pp. 171-176.
[12] Roswaty M R M, and Purnomo P W. 2014. Tingkat sedimentasi di muara Sungai Wudung Kecamatan Wedung, Demak. *Maquares*, **3** (2), pp.129-137.
[13] Saputra H, Subardjo P, and Saputro S. 2013. Studi pola sebaran sedimen dasar akibat arus sepanjang pantai di sekitar pemecah gelombang Pantai Kuta Bali. *Jurnal oseanografi*, **2**(2), pp. 161-170.
[14] Simatupang C M, Surbakti H, and Agussalim A. 2016. Analisis data arus di perairan muara Sungai Banyuasin Provinsi Sumatera Selatan. *Maspari journal*, **8** (1), pp.15-24.
[15] Syahputra T R, and Rahma, E. A. 2023. Karakteristik pasang surut air laut di Periran Belawan menggunakan metode admiralty. *Jurnal ilmu kelautan*, **5** (1), pp. 7-16.
[16] Syahza A. 2021. *Buku Metodologi Penelitian, Edisi Revisi Tahun 2021*. Pekanbaru: UR Press.
[17] Triatmodjo B. 2016. *Teknik pantai*. Yogyakarta: BETA.
[18] Warsidah, Risko, Saputra D W, Muliadi, Zibar Z, and Susiati H. 2021. Sebaran sedimen berdasarkan analisis parameter ukuran butir di muara Sungai Sambas Kalimantan Barat. *Jurnal geologi kelautan*, **19** (2), pp.61-71.
[19] Wiyono R U A, Fadhila A, Ma A R M, Halik G, Widiarti W Y, and Hidayah. 2023. Effect of breakwater layout on waves at Puger Beach Jember with Delft3D modelling. *E3S Wb of Conferences*.

IOP Conf. Series: Earth and Environmental Science **1473** (2025) 012005 doi:10.1088/1755-1315/1473/1/012005

Stress Analysis on Boat Landing Structure of a Tripod-Type Offshore Platform Due to Ship Collision

A A R Sukarsa[1], R W Prastianto[1*], Murdjito[1]

1 Department of Ocean Engineering, Institut Teknologi Sepuluh Nopember, Surabaya, Indonesia

*E-mail: rudiwp@oe.its.ac.id

Abstract. From 1996 to the end of 2015, the Health and Safety Executive of Liverpool John Moores University recorded a total of 176 incidents resulting from collisions between ships during operations with jacket platforms. A jacket platform structure capability of withstanding unforeseen accident loads, such as potential ship collisions with the platform structure, is essential. The boatlanding structure must be designed to absorb energy from ship collisions with the platform structure. This study analyzes ship collisions with the boatlanding structure to examine its strength capacity in resisting potential loads during ship impacts on the boatlanding. This analysis was conducted with scenarios involving collisions between a vessel and a boat landing, varying the ship's speed, the direction of impact (bow/stern and broadside), and variations in sea surface height due to tides. Under broadside collision conditions during operations, maximum stress reached 310 MPa, while bow/stern collisions saw maximum stress up to 9743 MPa. The most critical condition occurred during stern collision at minimum water depth, resulting in damage to the boatlanding structure.

1. Introduction

Offshore platforms are extensively used in the oil and gas industry, operating in highly diverse environmental conditions. Apart from regular operational loads, these platforms face additional stresses such as wind, waves, and currents. Simultaneously, they must also withstand unpredictable accident loads including ship collisions, falling objects, fires, and explosions [1]. Therefore, designing platform structures suitable for these environmental conditions is crucial. From early 1996 to the end of 2015, according to data from the Health and Safety Executive at Liverpool John Moores University, there were 176 accident incidents resulting from collisions between ships and jacket platforms [2]. These incidents pose significant potential for structural damage to offshore platforms, leading to fatigue, buckling, or denting of members that can compromise overall structural integrity, including both boatlanding and platform structures globally. During operational periods, fixed jacket platforms face the risk of ship collisions approaching the platform structure [3]. The jacket fixed platform structure in the splash zone must be designed to prevent early failures and platform collapse. One of the outermost protective structures shielding the jacket fixed platform from ship collisions is the boatlanding. During ship collisions, the weakest members, particularly those directly impacted in the boatlanding area, will deform. Properly designed boatlanding structures serve as the front-line protection if ship collision impacts exceed allowable limits [4].

IOP Conf. Series: Earth and Environmental Science **1473** (2025) 012005 doi:10.1088/1755-1315/1473/1/012005

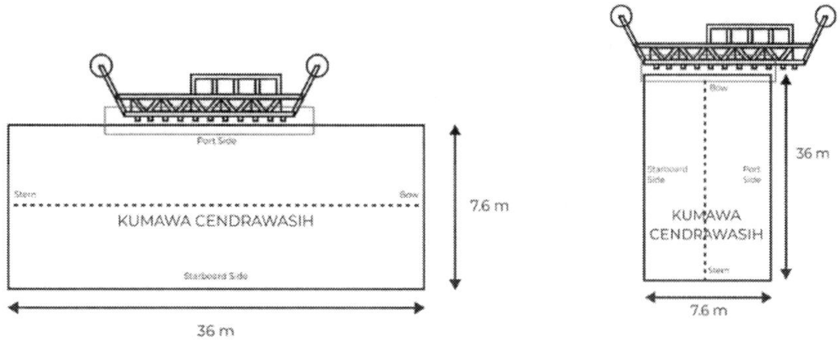

(a) (b)

Figure 1 Top View Collision at boatlanding a) Broadside impact; b) Bow/stern impact

In this study, a boat impact analysis (collapse pushover) was conducted to assess the structural adequacy of the boatlanding in withstanding potential loads resulting from a ship collision as shown in Figure 1. This Collapse Pushover analysis was performed on the Jacket Tripod Platform while neglecting environmental loads, thereby providing insights into the structural response to ship collisions, including maximum stress, large deflection, plasticity, and joint failure [5]. The study aims the structural strength of boat landing on the tipod jacket platform in withstanding ship collision loads with several variations of ship speed, direction of collision (Bow/Stern and broadside), and sea surface due to tides.

2. Methods

This section outlines the research methodology from start to finish, illustrated using the research methods diagrams in Figure 3. The study employs SACS (Structural Analysis Computer System), a software based on the finite element method [6]. Specifically, the SACS collapse analysis module is utilized to assess the structural limits of offshore structures. This module facilitates the identification of failure mechanisms including pushover, ship impacts on boat landings, dropped objects, and blast analysis. Its user-friendly, non-linear finite element analysis capabilities help mitigate the risk of structural failure due to extreme environmental loads or accidental events [7].

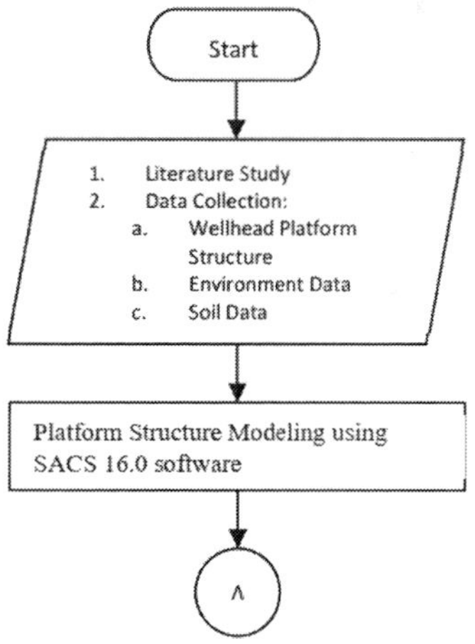

Figure 2 Research Methods

IOP Conf. Series: Earth and Environmental Science **1473** (2025) 012005 doi:10.1088/1755-1315/1473/1/012005

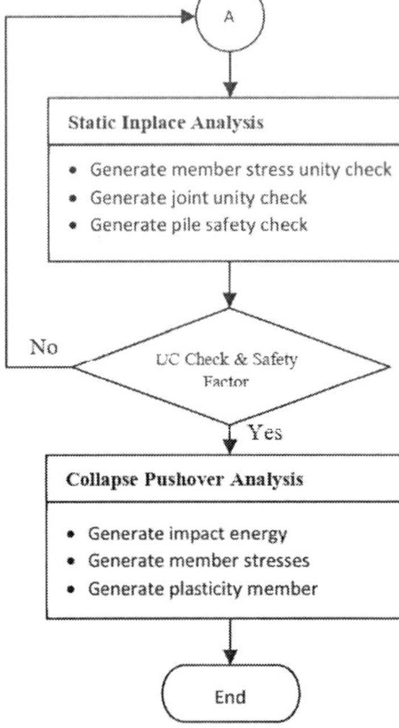

Figure 3 Research Methods (continue)

In this study, the structure is a wellhead fixed platform with 3 legs and 2 main decks operating at a depth of EL. (+) 54.46 ft. The structural data can be seen in Table 1. The model is shown in Figure 4

Table 1 Platform Description

Main deck level	EL. (+) 76.79 ft
Mezzanine deck level	EL. (+) 59.77 ft
Cellar Deck Level	EL. (+) 39.16 ft
Jacket Horizontal Framing	EL. (+) 19.79 ft
	EL. (-) 19.21 ft
	EL. (-) 54.46 ft
Working Point Elevation	EL. (+) 24.78 ft

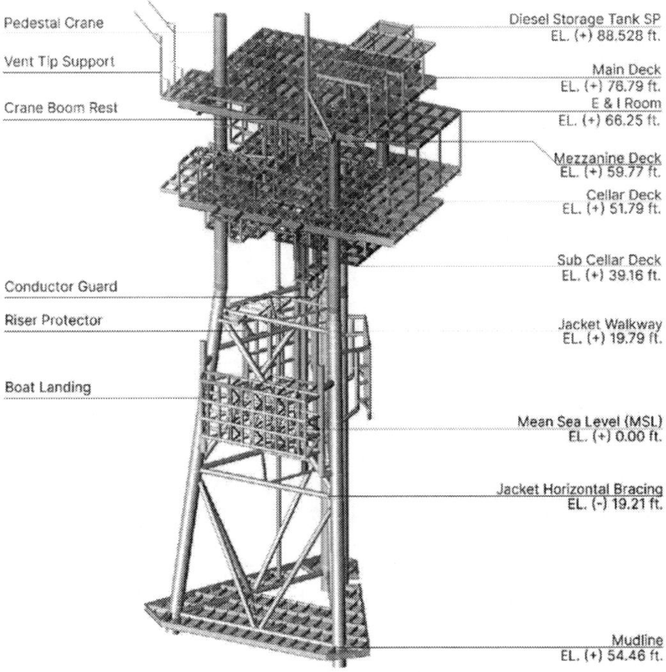

Figure 4 Wellhead Tripod Platform

In the boat impact analysis according to DNV RP C-204, collisions between ships and fixed platforms must consider accidental events. This is aimed at maintaining the load-bearing function of the structure during unintended incidents to ensure the primary safety functions of the installation are not compromised. In this study, collisions between ships and boat landings are categorized into two types: broadside collisions and bow/stern collisions at each surface elevation. In this study, there are 18 collision scenarios between ships and boat landings for each operational and accidental condition, as depicted in Figure 5.

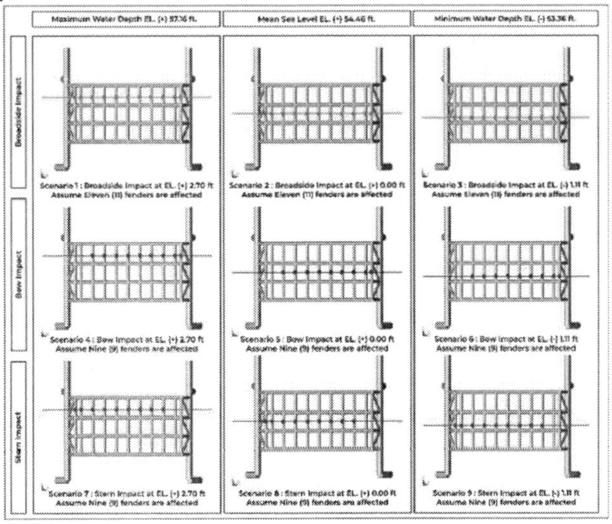

Figure 5 Collision Scenario

The vessel data used for the collision load analysis in this study is sourced from the American Bureau of Shipping (ABS) [8]. This information is presented in Table 2.

Table 2 Vessel Tabel

Vessel Name	KUMAWA CENDRAWASIH
Vessel Type	Crew Boat
Length Overall (LOA)	36 m
Breadth (B)	7.6 m
Depth (H)	3.56 m
Draft (T)	1.47 m
Displacement	2339 tonnes

In each collision scenario, an initial load representing the impact of ship collisions on each affected joint of the boat landing is applied using SACS V16.0 software. The force specified in the analysis depends on the type of element receiving the impact. The initial load is incrementally applied to the boat landing structure, and the analysis process stops once the structure has absorbed 100% of the ship's energy [9]. According to API RP 2A WSD standards, the initial load applied to the structure is determined based on the maximum denting of the element, which is 12.5%, 16.7%, or 25% depending on the diameter and thickness of the element. In this study, the initial load applied to the structure is 110 kips in the (y) direction, which is perpendicular to the structure. The loading conditions on the boat landing structure are illustrated in Figure 6

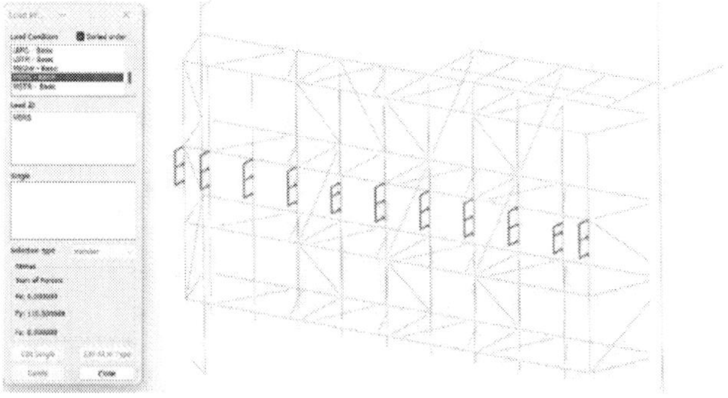

Figure 6 Load Scenario

During the service life of the platform, service vessels or supply ships frequently need to conduct offshore operations. During these operations, due to adverse weather conditions, collisions between ships and the jacket structure are unavoidable. The collision velocity is also dependent on the weather conditions at the time of operation. The minimum collision velocity requirement is 0.5 m/s [10]. For this study, the following ship velocities are used as shown in Table 3.

Table 3 Impact Ship Velocity

Operational Velocity	0.50 m/s
Accidental Velocity	0.77 m/s

IOP Conf. Series: Earth and Environmental Science **1473** (2025) 012005 doi:10.1088/1755-1315/1473/1/012005

3. Results and discussion

3.1. Member Stress Check

The purpose of the inplace analysis is to determine the response of each member to the applied loads. The results of this analysis are a member stress check, joint unity check, and pile safety check. This includes the Unity Check (UC), which is the ratio of the occurring stress to the allowable stress [11]. According to the API RP 2A WSD standards, the allowable UC value must not exceed 1 (UC $\leq$ 1). Member stress checks for the structure under operational and storm conditions are presented in Tabel 4 and Table 5.

Table 4 The result of Member Stress Check – Operating condition

No	Description	Member Group	Member	UC
1	Main Deck Girder	MD1	4032-4002	0.61
2	Main Deck Beam	MD2	4026-4294	0.65
3	Cellar Deck Girder	CD1	2018-2002	0.89
4	Cellar Deck Girder	CD2	2158-2131	0.83
5	Mezzanine Deck Framing	MZ1	3022-3006	0.49
6	Mezzanine Deck Column	MZL	3006-4122	0.47
7	Crane Pedestal Column	JB1	4001-4287	0.81
8	Riser Support TOS EL. (+) 35,165 ft. Column	RSC	5029-2195	0.16
9	Diagonal Bracing Ele. Row-A	DL3	1007-2007	0.67
10	Deck Leg	DL8	5081-4001	0.52
11	Transition Piece	DLE	503L-0015	0.58
12	Diesel Storage Tank Service Platform TOS EL. (+) 84,528 ft.	SP1	4326-6051	0.63
13	Sub Cellar Deck Column TOS EL. (+) 35,165 ft.	DRL	1534-2230	0.37
14	Jacket Horizontal Framing TOS EL. (-) 54,46 ft.	H12	168- 169	0.32
15	Jacket Walkway Member TOS EL. (+) 19,79 ft.	H37	2309- 363	0.54
16	Jacket Diagonal Bracing EL (-) 19,21 ft. EL (-) 54,46 ft.	BR1	130-202L	0.44
17	Jacket Leg	LG5	2335-2337	0.37
18	Mudmat Framing	MM2	11K- 110	0.37
19	Cond./Riser Protection Framing	CF2	253- 297	0.28

Table 5 The result of Member Stress Check – Storm condition

No	Description	Member Group	Member	UC
1	Main Deck Girder	MD1	4002-4302	0.44
2	Main Deck Beam	MD2	4026-4294	0.58
3	Cellar Deck Girder	CD1	2018-2002	0.69
4	Cellar Deck Girder	CD2	2108-2109	0.63
5	Mezzanine Deck Framing	MZ1	3022-3006	0.73
6	Mezzanine Deck Column	MZL	3006-4122	0.42
7	Crane Pedestal Column	JB1	4001-4287	0.02
8	Riser Support TOS EL. (+) 35,165 ft. Column	RSC	5029-2195	0.13
9	Diagonal Bracing Ele. Row A	DL3	1007-2007	0.46
10	Deck Leg	DL8	2002-4002	0.22
11	Transition Piece	DLE	503L-0015	0.44

IOP Conf. Series: Earth and Environmental Science **1473** (2025) 012005 doi:10.1088/1755-1315/1473/1/012005

No	Description	Member Group	Member	UC
12	Diesel Storage Tank Service Platform TOS EL. (+) 84,528 ft.	SP1	4326-6051	0.46
13	Sub Cellar Deck Column TOS EL. (+) 35,165 ft.	DRL	1534-2230	0.31
14	Jacket Horizontal Framing TOS EL. (-) 54,46 ft.	H12	168- 169	0.64
15	Jacket Walkway Member TOS EL. (+) 19,79 ft.	H37	2309- 363	0.77
16	Jacket Diagonal Bracing EL (-) 19,21 ft. EL (-) 54,46 ft.	BR1	130-202L	0.75
17	Jacket Leg	LG5	2335-2337	0.51
18	Mudmat Framing	MM2	11K- 110	0.76
19	Cond./Riser Protection Framing	CF2	274- 295	0.45

The results of the inplace joint punching shear analysis under operational and storm conditions are presented in Table 6 and Table 7. The pile stress check is shown in Table 8 and Table 9.

Table 6 The result of Joint Punching Shear – Operation condition

No	Joint	Chord (in)		Load Case	Load UC	Strength UC	Location
		O.D	Thk.				
1	103L	39.88	1.19	2005	0.384	1.086	EL. (-) 54,56 ft.*
2	301L	39.36	1.18	2004	0.357	0.680	EL. (+) 19,79 ft.*
3	302L	39.36	1.18	2003	0.348	0.663	EL. (-) 19,79 ft.*
4	203L	39.88	1.19	2005	0.321	1.074	EL. (-) 19,21 ft.*
5	303L	39.36	1.18	2004	0.337	0.663	EL. (+) 19,79 ft.*

*The location is in Jacket Leg Row A

Table 7 The result of Joint Punching Shear – Storm condition

No	Joint	Chord (in)		Load Case	Load UC	Strength UC	Location
		O.D	Thk.				
1	103L	39.88	1.19	4007	0.685	1.086	EL. (-) 54,56 ft.*
2	203L	39.88	1.19	4006	0.510	1.089	EL. (+) 19,79 ft.*
3	101L	39.88	1.19	4004	0.506	1.083	EL. (-) 19,79 ft.*
4	102L	39.88	1.19	4004	0.494	1.083	EL. (-) 19,21 ft.*
5	201L	40.00	1.25	4003	0.487	0.794	EL. (+) 19,79 ft.*

*The location is in Jacket Leg Row A

Table 8 The result of Pile Stress Check – Operating condition

No	Member	Member Group	Load Comb.	UC	Remarks
1	003P-103P	PL1	2004	0.52	OK
2	103P-203P	PL2	2004	0.57	OK

Table 9 The result of Pile Stress Check – Storm condition

No	Member	Member Group	Load Comb.	UC	Remarks
1	003P-103P	PL1	4004	0.75	OK
2	103P-203P	PL2	4004	0.77	OK

3.2. Total Impact Energy

The collision analysis conducted in this study pertains to the scenario where a vessel is docking at a platform via a boat landing. In determining the collision energy between the jacket platform and the vessel, the critical factors are the impact velocity during the docking process and the vessel's mass. The vessel's speed during docking must be calculated with precision, as speed is one of the most influential variables in collision energy calculations [12]. Based on the vessel data, with a weight of 2,339 tons, an operational speed of 0.50 m/s (1.64 ft/s), and an accidental speed of 0.77 m/s (2.50 ft/s), the total impact energy was calculated using Equation (C18.9.2-1) on API RP 2A WSD. The results are presented in Table 10 as follows.

Table 10 Impact Energy

Impact Case	Impact Type	Vessel Weight (ton)	Impact Velocity (m/s)	Add Mass Factor	Total Impact Energy (kJ)
Operating	Bow/Stern	2339	0.50	1.1	321.61
	Broadside	2339	0.50	1.4	409.32
Accidental	Bow/Stern	2339	0.77	1.1	762.73
	Broadside	2339	0.77	1.4	970.75

3.3. Member Stresses due to ship collision

The collision analysis results the stress and deflection experienced by the boat landing structure, based on variations in collision types, velocities, and sea surface elevation, as analyzed with SACS v16.0. These results are detailed in Table 11.

Table 11 Member Stresses

No	Load Case	Total Impact Energy (kJ)	Member	Axial (MPa)	Bend-Y (MPa)	Bend-Z (MPa)
Broadside Impact						
1	OPR-HBRS	409.33	HAT5 - SP05	-4.21	-310.28	-2.14
2	OPR-MBRS	409.33	SP16 - MSL5	-9.17	-254.22	6.21
3	OPR-LBRS	409.33	MSL5 - LAT5	-6.21	-302.48	3.59
4	STR-HBRS	970.76	HAT7 - SP07	-0.90	-320.13	-0.76
5	STR-MBRS	970.76	SP16 - MSL5	-11.65	-308.55	6.69
6	STR-LBRS	970.76	LAT7 - MSL7	-20.69	-319.93	-1.10
Bow Impact						
7	OPR-HBOW	321.61	HAT5 - SP05	-0.83	-316.89	0.07
8	OPR-MBOW	321.61	SP16 - MSL5	-9.52	-274.70	5.65
9	OPR-LBOW	321.61	MSL5 - LAT5	-7.17	-314.62	2.69
10	STR-HBOW	970.76	HAT7 - SP07	4.48	-320.55	-36.27
11	STR-MBOW	970.76	SP16 - MSL5	-12.27	-319.58	4.00
12	STR-LBOW	970.76	LAT7 - MSL7	-30.96	-320.76	-0.21
Stern Impact						
13	OPR-HSTR	321.61	HAT5 - SP05	-1.24	-306.55	-3.10
14	OPR-MSTR	321.61	MSL8 - SP19	-9.52	-268.77	-0.07
15	OPR-LSTR	321.61	LAT8 - MSL8	-580.01	9743.67	-41.78
16	STR-HSTR	762.74	HAT9 - SP09	-19.86	3371.86	216.78
17	STR-MSTR	762.74	MSL8 - SP19	-12.89	-319.65	-1.65
18	STR-LSTR	762.74	LA11 - MS11	410.32	2006.31	68.05

3.4. Plasticity Member

When the stress on a structural component reaches its yield stress, plastic deformation occurs [13]. This plasticity reduces the stiffness of the platform and boat landing structure due to the continuous addition of loads. The additional loads are redistributed to the adjacent members once a structural member has become plastically deformed. Based on the boat impact analysis, the most critical condition was identified in Scenario 15, which involves a stern impact at El. (-) 1.11 ft during operation, as illustrated in Figure 7 and detailed in Table 12.

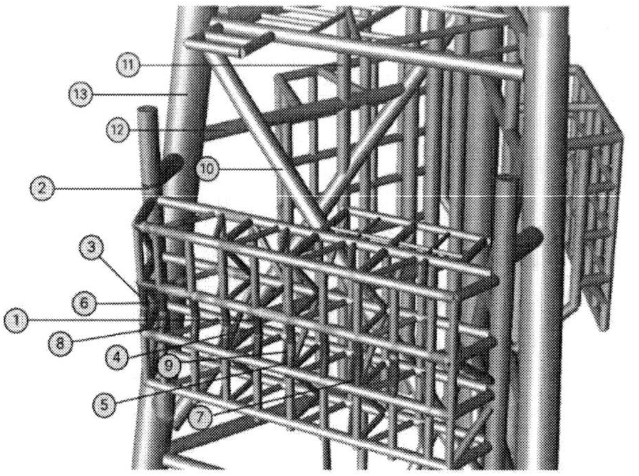

Figure 7 Stern Impact pada Boatlanding El. (-) 1.11 ft – Operation Condition

Table 12 Plasticity Member on Boatlanding

No	Member	Deskripsi	Axial (MPa)	Bend-Y (MPa)	Bend- Z (MPa)	Placticity
1	SP14-BL32	Fender boatlanding	4.03	185.15	-8.51	1.00
2	0012-BL02	Support Boatlanding	4.31	248.39	-8.91	1.00
3	SP19-BL37	Fender boatlanding	3.89	185.08	-17.60	1.00
4	MSL3-SP14	Fender boatlanding	4.13	248.32	-23.71	1.00
5	BL53-LAT2	Fender boatlanding	9.32	185.06	-26.66	1.00
6	MSL9-SP20	Fender boatlanding	14.36	248.22	-39.30	1.00
7	BL52-LAT1	Fender boatlanding	3.71	184.85	-8.51	1.00
8	BL60-LAT9	Fender boatlanding	3.94	248.13	-8.95	1.00
9	BL53-LAT2	Fender boatlanding	3.61	184.42	-17.46	1.00
10	225-301L	Diagonal bracing between leg A and leg B	3.78	247.34	-23.54	0.00
11	2333- 310	Vertical bracing el (+) 10.00 ft	4.03	185.15	-8.51	0.00
12	2282-2334	Horizontal bracing el. (+) 10.00 ft	4.31	248.39	-8.91	0.00
13	2282-207L	Jacket leg A	3.89	185.08	-17.60	0.00

4. Conclusion

Based on the results of the static in-place analysis, the maximum Unity Check (UC) value under operational conditions is 0.89, and under storm conditions, it is 0.69. These values meet the requirement of UC < 1, indicating that the tripod jacket platform is capable of withstanding both operational and storm conditions. During operational conditions, a broadside collision results in a maximum stress of 310 MPa, whereas a bow/stern collision results in a maximum stress of 9,743 MPa. The most critical condition occurs during a stern collision at minimum water depth, leading to significant damage to the boat landing structure. The boat landing structure suffered quite severe damage in the stern collision condition at minimum depth.

Acknowledgment
Researchers would like to thank the Department of Ocean Engineering, DRPM-ITS, and Institut Teknologi Sepuluh Nopember, Surabaya, Indonesia, for the financial support and facilities provided.

References

[1] DNV RP C-204. (2010). RECOMMENDED PRACTICE DESIGN AGAINST ACCIDENTAL LOADS. http://www.dnv.com
[2] HSE. (2019). RR1154: Ship/ Platform Collision Incident Database (2015) for offshore oil and gains installations.
[3] Thapa, Prakash. (2016). Boat Impact Analysis on Offshore Unman fixed Well Head Platform (WHP).
[4] Nallayarasu, S. (2011). OFFSHORE STRUCTURES Analysis and Design.
[5] El-Reedy, M. A. (2015). Marine Structural Design Calculations. In Marine Structural Design Calculations. Elsevier. https://doi.org/10.1016/C2012-0-07922-7
[6] Bentley. (2021) SACS Version 16.0
[7] Bentley. (2021) SACS Version 16.0, SACS Collapse Manual
[8] American Bureau of Shipping. (n.d.). Database for ABS vessel records - Kumawa Cendrawasih. Retrieved January 31, 2024, from https://ww2.eagle.org/en.html
[9] Bentley Communities. (2023). Ship Impact Force - MAXSURF _ MOSES _ SACS _ OpenWindPower - Wiki - MAXSURF _ MOSES _ SACS _ OpenWindPower - Bentley Communities. https://communities.bentley.com/products/offshore/w/wiki/67957/ship- impact-force
[10] API RP 2A-WSD 22nd Ed. 2014 Planning, Designing, and Constructing Fixed Offshore Platforms - Working Stress Design API Recomm. Pract. **324**
[11] Chakrabarti S 2005 Handbook of Offshore Engineering (Illinois: Elsevier Ltd.)
[12] Bukhari, M. H. (2023). Analisis Kekuatan Struktur Minimum Facilities Wellhead Platform Terhadap Tabrakan Crane Barge Pada Tahap Instalasi Kompresor Tambahan.
[13] Popov, E. (1984). Mekanika Teknik (Mechanics of Materials) Edisi Kedua (Versi SI) (Z. A. Tanisan (ed.).

IOP Conf. Series: Earth and Environmental Science **1473** (2025) 012006 doi:10.1088/1755-1315/1473/1/012006

Analysis on Stress Corrosion Cracking T-Type Tee Junctions for Safety Environments

Muhammad Fawwaz Zainal Abidin[1], Roslin Ramli[1*], Mohd Hisbany Mohd Hashim[2], and Septia Hardy Sujiatanti[3]

[1] Maritime Engineering Technology, Universiti Kuala Lumpur Malaysian Institute of Marine Engineering Technology, Perak, Malaysia
[2] School of Civil Engineering Technology, Universiti Teknologi Mara, Selangor, Malaysia
[3] Department of Naval Architecture, Institut Teknologi Sepuluh Nopember, Surabaya, Indonesia

*E-mail: roslin@unikl.edu.my

Abstract. Stress corrosion cracking (SCC) is a critical failure mechanism affecting various materials, particularly in environments susceptible to moisture and chemicals. This research focuses on analysing stress corrosion cracking in T-type tee junctions, comparing traditional steel components with carbon fibre-reinforced polymers (CFRPs). Although steel is a widely used material in marine engineering applications, it is prone to corrosion and degradation over time, especially without protective coatings. In contrast, CFRPs demonstrate superior corrosion resistance due to the composition of carbon fibres embedded in a polymer matrix, typically epoxy resin. This intrinsic property allows CFRPs to withstand harsh environmental conditions without rusting, thereby maintaining structural integrity. Additionally, CFRPs offer a favourable strength-to-weight ratio, providing comparable strength to steel while being significantly lighter. The implications of using CFRPs in applications prone to SCC are profound. CFRPs not only mitigate the risks associated with corrosion but also enhance the overall performance and longevity of structures. This analysis highlights the potential of CFRPs as a viable material alternative to steel in critical applications, particularly in industries where stress corrosion cracking poses significant operational challenges. The findings suggest material selection and design strategies for upgraded resilience against SCC in engineering applications.

1. Introduction

One of the most important failure mechanisms is stress corrosion cracking (SCC), which can affect materials in environments with moisture and chemicals. It focused on stress corrosion cracking and the safety environment in T-type tee junctions, comparing stainless-steel with carbon fibre-reinforced polymers (CFRPs).

Stainless-steel material is a widely used in marine engineering applications, but it is susceptible to corrosion and degradation over time, especially without a protective coating[1]. On the other hand, CFRP shows superior corrosion resistance due to the composition of carbon fibres embedded in a polymer matrix, usually epoxy resin[2], [3]. This intrinsic property allows CFRP to withstand harsh environmental conditions without rusting, thereby maintaining structural integrity [4]–[6]. Additionally, CFRP offers a favourable strength-to-weight ratio, providing strength comparable to stainless-steel while being lighter [7], [8].

The implications of using CFRPs in applications prone to SCC are profound. CFRPs not only mitigate the risks associated with corrosion but also enhance the structure's overall performance and longevity. This analysis highlights the potential of CFRPs as a viable material alternative to stainless-steel in critical applications, particularly in industries where stress corrosion cracking poses significant operational challenges.

2. Methodology

The study examined the performance of stainless-steel and CFRP T-type tee junctions under simulated harsh environmental conditions, specifically on SCC. The SolidWorks software used for the modelling of T-type tee junctions. While the COMSOL software used for simulating them.

The SolidWorks software allows for the creation of models using two different materials. The materials are stainless-steel and CFRP. The simulation also considers the sustainability of both t-type tee junctions to ensure a safe environment. The process of designing the tee geometry in SolidWorks entails the creation of a three-dimensional model of the tee component. The t-type tee junction is widely utilised in various applications such as piping systems, plumbing, and structural designs [9], [10][11]. The capabilities of SolidWorks are extremely remarkable , as it provides a wide range of tools for designing, simulating, and manufacturing mechanical components [12]–[14]. The process includes creating a rough sketch of the tee, specifying its measurements, incorporating details such as fillets and chamfers, and verifying that it fits seamlessly with other parts in the assembly, as shown in Figure 1.

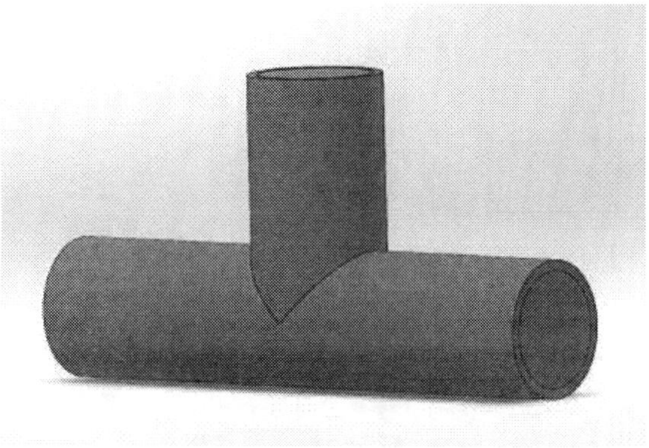

Figure 1. T-types tee junctions in SolidWorks.

Then, the T-type tee junction is simulated and analysed for safety in the SolidWorks sustainability analysis, providing a comprehensive assessment of its performance in a controlled environment. The environmental impact is calculated using the Centre of Environmental Science Leiden-Methan (CML) impact assessment methodology [15]. This T-types tee junction is divided into four parts [16], [17]. The parts are carbon footprint, air acidification, water eutrophication, and total energy consumed. The parameters of the sustainability of both materials are shown in Table 1.

IOP Conf. Series: Earth and Environmental Science 1473 (2025) 012006 doi:10.1088/1755-1315/1473/1/012006

Table 1. Parameters sustainability

	Stainless-steel	CFRP
Region	Asia	Asia
Process	Die Cast	Injection Molded
Duration of use (years)	20	20

In addition, importing the design is a crucial step for a comprehensive analysis in SolidWorks and simulating in COMSOL Multiphysics. The relationship between corrosion potential, current density, and von Mises stress along the corrosion defect as a result of near-neutral pH SCC was studied using a finite element model developed in COMSOL Multiphysics [18]–[20]. The corrosion and structural mechanics modules were utilised for this purpose.

The first step is to save the SolidWorks file in a format that can be used in COMSOL, such as STEP or IGES file. After saving the file, the tee joint model was loaded to COMSOL Multiphysics using the "Import" function. Then, the geometry was imported accurately and scaled appropriately [21]. Once the import process completed, options to specify the material properties, boundary conditions, and electrochemical parameters were provided. These options are necessary for the corrosion simulation as shown in Figure 2. This integration enables precise analysis of corrosion effects on the tee joint, offering valuable insights into its durability and performance in real-world conditions.

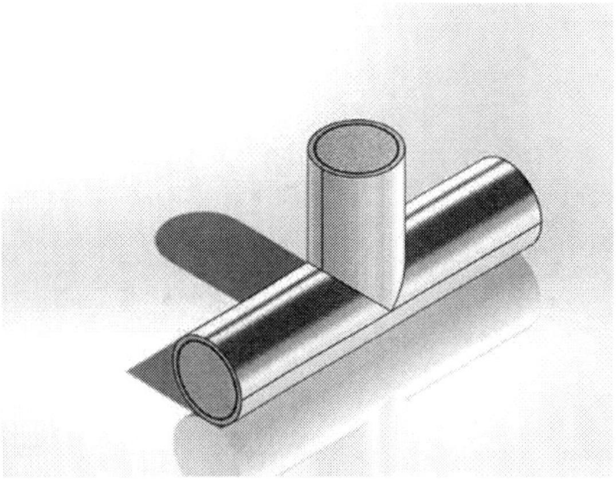

Figure 2. T-types tee junctions in COMSOL.

This approach allows for a complete assessment of the corrosion impact on T-type tee junctions and, thus, can be implemented to gather substantial information about their durability and real performance. The results of SolidWorks and COMSOL simulations present a deeper understanding of the safety environment and SCC behaviour in stainless-steel-CFRP.

3. Results and discussion

The T-type tee fitting is an essential element in piping systems utilised to divide fluid flow into two adjacent directions or merge flow from two directions into a single channel [9], [10], [22]. Figure 3 depicts the behaviour of a tie fitting in the context of T-type tie junctions associated with fillets, focusing on the transition between CFRP and stainless-steel materials. Based on the analysis of the turbulent flow of the tee, it shows that the flow characteristics of CFRP are superior to those of stainless-steel, as evidenced by the contrasting contour colours of blue to yellow for stainless-steel and blue to green for CFRP.

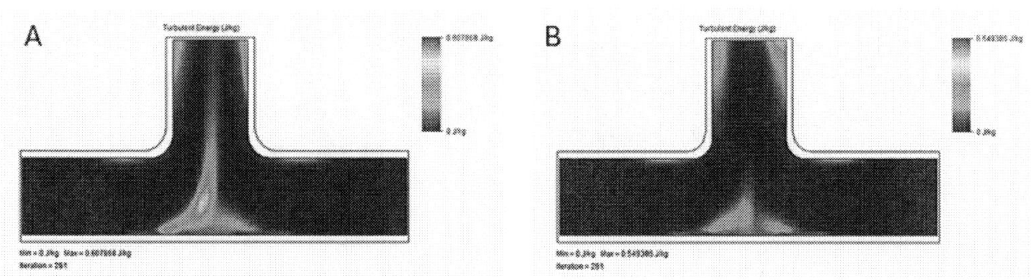

Figure 3. T-type tee junction turbulent pattern (A) Stainless-steel and (B)

The analysis of sustainability results are categorised into four components. The components are carbon footprint, air acidification, water eutrophication, and overall energy consumption. The pie chart displays comprehensive data for both materials, as depicted in Figure 4 and Figure 5.

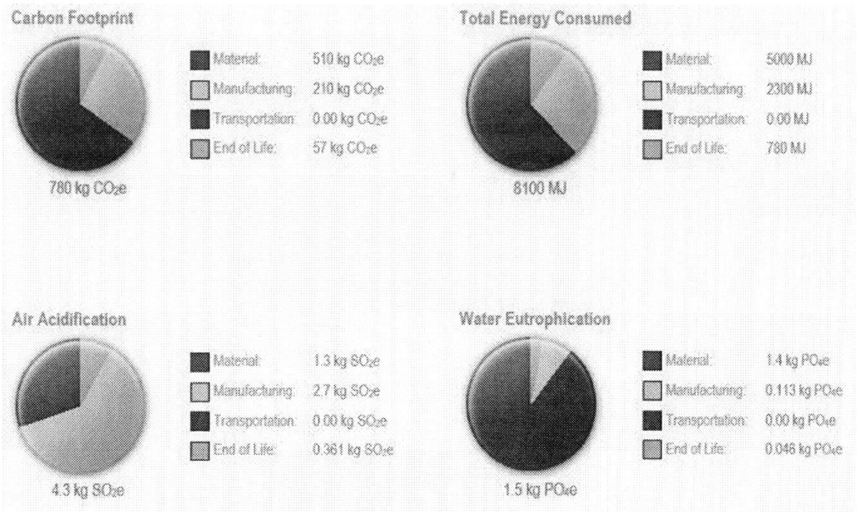

Figure 4. Stainless-Steel Environmental Impact

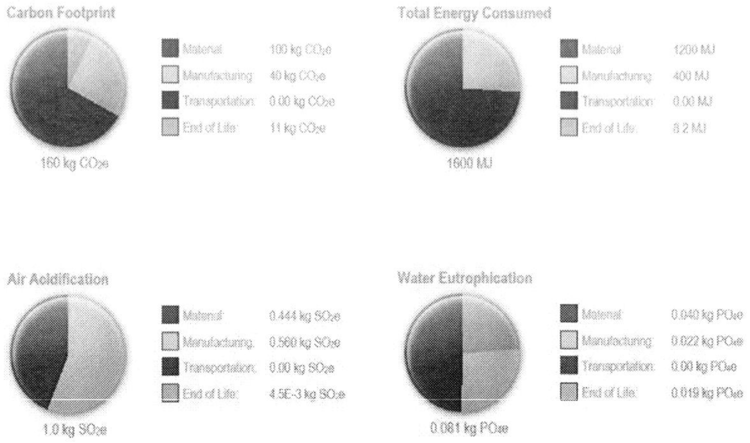

Figure 5. CFRP Environmental Impact

This study conducts a comparative analysis of the environmental impacts of carbon footprint, air acidification, water eutrophication, and total energy consumption in stainless-steel and carbon. In comparison, Carbon Fibre-Reinforced Polymer (CFRP) has higher reduced carbon footprint than stainless-steel. The emissions of CFRP are roughly 79% lower than stainless-steel. Regarding air acidification and water eutrophication, CFRP has a far lesser effect contrasts to stainless-steel. Additionally, the energy consumption requires for CFRP manufacturing is dramatically lower than that of stainless-steel.

Considering these comparisons, CFRP emerges as the optimal material choice for mitigating environmental effect in all assessed categories. This technology demonstrates reduced emissions for both air acidification and water eutrophication, together with a substantial decrease in energy usage. Hence, substituting stainless-steel with CFRP can result in a more sustainable and ecologically prudent result.

Nevertheless, a critical focus of this study is the comparative SCC performance of both materials in COMSOL software. The results demonstrate that the incidence of cracking in stainless-steel significantly higher, with an increase of 88.89% compares to CFRPs. Thus, the stress corrosion cracking thickness for stainless-steel is 18 mm, and CFRP is 2 mm. The contour colour indicates that the stainless-steel exhibits more cyan colour compares to the CFRP, which has a more blue colour. This means that the contour color corresponds to the Von Mises stress, as illustrated in Figure 6. This finding emphasises the vulnerability of stainless-steel to stress corrosion in challenging environments, reinforcing the need for alternative materials.

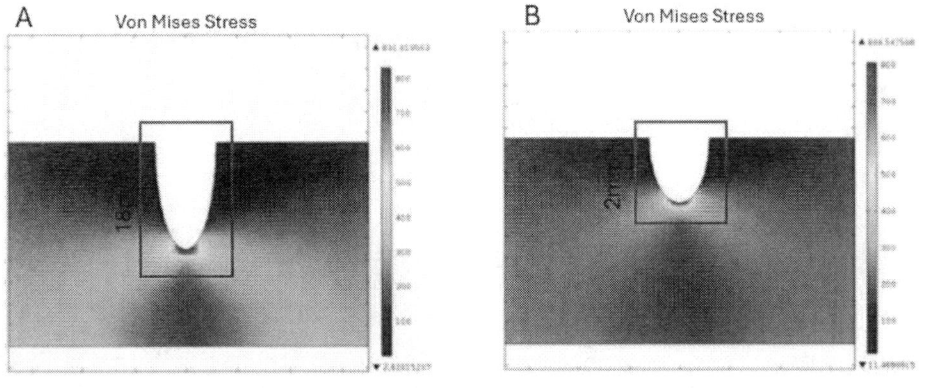

Figure 6. Stress Corrosion Cracking (A) Stainless-steel and (B) CFRP

4. Conclusion

This study highlights the potential of CFRPs as a superior material selection for applications prone to stress corrosion cracking (SCC), particularly in harsh environments. The corrosion resistance and favourable strength-to-weight ratio of carbon fibre-reinforced polymers (CFRPs) make them an attractive alternative to stainless-steel components.

The flow pattern demonstrates that CFRP is superior to stainless-steel, making the tee more efficient in turbulent patterns. Influence from the flow and material, sustainability from the environmental impact shows stainless-steel has more than 79% impact compared to CFRP. Nevertheless, the main study findings on SCC show the cracking of stainless-steel is higher compared to CFRP with 88.89%.

The findings of this research can guide material selection and design strategies for improved resilience against stress corrosion cracking (SCC) in engineering applications. By incorporating CFRPs into critical structures, such as those found in marine engineering, the risks associated with stress corrosion cracking can be significantly reduced. This insight ensures the safety environment and longevity of these systems.

References

[1] H. Wang *et al.*, "Improvement in surface performance of stainless steel by nitride and carbon-based coatings prepared via physical vapor deposition for marine application," *J. Mater. Res. Technol.*, vol. 27, 2023.

[2] A. Khalifeh, "Stress Corrosion Cracking Damages," in *Failure Analysis*, 2019.

[3] J. Yu *et al.*, "CFRP Strengthening and Rehabilitation of Inner Corroded Steel Pipelines under External Pressure," *J. Mar. Sci. Eng.*, vol. 10, no. 5, 2022.

[4] D. S. Vijayan, A. Sivasuriyan, P. Devarajan, A. Stefańska, Ł. Wodzyński, and E. Koda, "Carbon Fibre-Reinforced Polymer (CFRP) Composites in Civil Engineering Application—A Comprehensive Review," *Buildings*, vol. 13, no. 6. 2023.

[5] Y. Bounjoum, O. Hamlaoui, M. K. Hajji, K. Essaadaoui, J. Chafiq, and M. A. El Fqih, "Exploring Damage Patterns in CFRP Reinforcements: Insights from Simulation and Experimentation," *Polymers (Basel).*, vol. 16, no. 14, 2024.

[6] M. A. Hafiz and A. Sulisetyono, "Structural Reliability Analysis for the Construction Design of the High-Speed Ship with CFRP Material," in *IOP Conference Series: Earth and Environmental Science*, 2022, vol. 1081, no. 1.

[7] A. M. Pawlak, T. Górny, Ł. Dopierała, and P. Paczos, "The Use of CFRP for Structural Reinforcement—Literature Review," *Metals*, vol. 12, no. 9. 2022.

[8] R. Ramli and M. H. Mohd Hashim, "Analysis of tensile strength of different combination of FRP material under seawater conditioning," *J. Eng. Appl. Sci.*, 2017.

[9] T. Liu, S. Li, C. Jiang, X. Zhang, and Z. Tan, "Local Resistance Characteristics of T-Type Tee Based on Chamfering Treatment," *Sustain.*, vol. 15, no. 19, 2023.

[10] Y. UTANOHARA, K. MIYOSHI, and A. NAKAMURA, "Conjugate numerical simulation of wall temperature fluctuation at a T-junction pipe," *Mech. Eng. J.*, vol. 5, no. 3, 2018.

[11] Q. Gong, J. Yang, K. Han, T. Huang, J. Li, and P. Zuo, "Characteristic analysis on the flow and local resistance in large pipe tees," *Dongli Gongcheng Xuebao/Journal Chinese Soc. Power Eng.*, vol. 36, no. 9, 2016.

[12] M. Lombard, *Mastering SolidWorks*. 2018.

[13] K. B. Mustapha, *Practical Finite Element Simulations with SOLIDWORKS 2022: An Illustrated Guide to Performing Static Analysis with SOLIDWORKS Simulation*. 2022.

[14] SolidWorks, "An Introduction to Flow Analysis Applications with SolidWorks Flow Simulation," *Eng. Des. Technol. Ser.*, 2010.

[15] C. Torcătoru and D. Săvescu, "Analyzing the sustainability of an automotive component using SolidWorks CAD software," in *IOP Conference Series: Materials Science and Engineering*, 2019, vol. 568, no. 1.

[16] L. I. Popa and V. N. Popa, "Products eco-sustainability analysis using CAD SolidWorks software," in *MATEC Web of Conferences*, 2017, vol. 112.

[17] T. Dovramadjiev and B. Mrugalska, "Real-time planning and monitoring of the steel pipes towards life cycle sustainability management," *Ann. Oper. Res.*, vol. 324, no. 1–2, 2023.

[18] V. Prajapati, Y. Kumar, D. Gupta, A. Kalam, and M. Dubey, "Analysis of Pitting Corrosion of Pipelines in a Marine Corrosive Environment Using COMSOL Multiphysics," *J. Bio- Tribo-Corrosion*, vol. 8, no. 1, 2022.

[19] M. S. Tarafder and M. J. Mia, "COMPARATIVE NUMERICAL SIMULATION OF LAMINAR FLOW THROUGH PIPE USING COMSOL MULTIPHYSICS AND OPENFOAM," *SSRN Electron. J.*, 2023.

[20] H. Nguyen and T. Hoang, "Numerical Simulation of Laminar Flow Through a Pipe using COMSOL Multiphysics," *Int. J. Sci. Eng. Res.*, vol. 8, no. 6, 2017.

[21] COMSOL, "Pipe flow module," *Manual*, pp. 1–6, 2017.

[22] Y. Chen and G. Ma, "Erosion's numerical simulation on solid particles in liquid-solid two-phase flow of tee pipes," *Jixie Qiangdu/Journal Mech. Strength*, vol. 41, no. 2, 2019.

IOP Conf. Series: Earth and Environmental Science **1473** (2025) 012007 doi:10.1088/1755-1315/1473/1/012007

Performance of Large-Scale Particle Image Velocimetry Method for Surface Velocity Identification Based on Drone Height Variation

Ayi Tarya [1,2*] and Cindy K. R. Marbun[2]

[1] Environmental and Applied Oceanography Research Group, Faculty of Earth Sciences and Technology, Bandung Institute of Technology, Ganesha 10 Bandung 40132, Indonesia.
[2] Oceanography Study Program, Faculty of Earth Sciences and Technology, Bandung Institute of Technology, Ganesha 10 Bandung 40132, Indonesia.

*E-mail: ayi.tarya@itb.ac.id

Abstract. Information on flows is essential for coastal management and adaptation to climate change. The traditionally used methods for observing flows require contact with the surface water. The Large-Scale Particle Image Velocimetry (LSPIV) method can complement this measurement strategy to measure surface velocities in various environments, including rivers and estuations, and is based on image techniques. The aim of this study was to identify surface currents at various drone heights and then validate the results generated from the LSPIV method by comparing them with observation data. Field data were collected at Cimanis Estuary, Cirebon, over two periods: August 19–22 and September 27–28, 2023. The field data observation included surface velocity, drone videos, and ground control points. Drone height measurements were taken at 20, 30, 40, 50, and 70 meters. To verify the accuracy of the LSPIV, several methods were used, including the RMSE (Root Mean Square Error), MAE (Mean Absolute Error), and MAPE (Mean Absolute Percentage Error). The drone height of 20 meters yielded the lowest RMSE, MAE, and MAPE, which were 0.08 m/s, 0.06 m/s, and 5.6% for velocity magnitude and 10.74 degree, 10.69 degree, and 3.7% for velocity direction, respectively. This suggests that the most precise LSPIV data were collected during the drone survey at an elevation of 20 meters.

1. Introduction

Ocean currents exert a significant influence on estuarine environments, contributing to essential processes like sediment and nutrient transport. These currents also shape water body configurations, providing a foundation for various human activities such as transportation and tourism [1]. Traditionally, measuring ocean current velocities has been achieved using intrusive methods involving drifters, current meters, and Acoustic Doppler Current Profilers (ADCPs). However, researchers have increasingly employed photogrammetric techniques to determine the surface velocity of rivers, thus reducing the risks associated with field observations. Technological advancements have led to the development of more sophisticated image-based approaches for surface velocity measurements [2,3]. This state-of-the-art technology has experienced continuous improvement over the years, becoming an indispensable tool for understanding ocean currents and their impacts on estuarine ecosystems and human activities.

Photogrammetry, a technique that derives accurate measurements from photographs, offers a valuable method for calculating river surface velocities. Unmanned aerial vehicles (UAVs) have emerged as the preferred choice for photogrammetry due to their exceptional maneuverability, ability to overcome terrain constraints, and cost-effectiveness [4,5]. UAVs can navigate complex terrains and perform low-altitude operations, providing orthographic aerial imagery that is crucial for precise photogrammetric analysis. The widespread adoption of UAVs in various fields, including agriculture, ecology, environmental monitoring, and disaster relief operations [6,7,8,9], underscores their versatility and suitability for river surface velocity measurements.

In recent years, the Large-Scale Particle Image Velocimetry (LSPIV) method has emerged as a promising technique for measuring the velocities of surface currents in rivers. The LSPIV technique involves capturing a sequence of images of surface tracers that are moving with the flow and subsequently processing the images to derive the velocity vectors. This method has been successfully employed in various river environments [5, 10, 11, 12]. The LSPIV method provides an advantage in safety measurement because it eliminates direct physical contact with the water. Moreover, LSPIV is an efficient, convenient, environmental applicability, and accurate method for detecting surface velocities and flows over large scales [13, 14, 15]. However, previous LSPIV studies have primarily focused on measuring current velocities in areas with uniform flow conditions. In contrast, river mouths exhibit more complex flow patterns due to the influence of tides, making the application of the LSPIV method in these environments more challenging [16]. The present study aims to assess the performance of the LSPIV method in identifying current circulation in river mouths, specifically in the Cimanis Estuary in Cirebon, West Java. The study investigated the impact of varying drone heights on the accuracy and precision of the LSPIV measurements. The findings of this study will contribute to our understanding of the applicability and limitations of the LSPIV method in complex riverine environments, particularly in river mouths.

2. Methods

2.1 Study Area

The study area of the present investigation is situated in the Cimanis Estuary, which is located on the northern coast of Java in Cirebon (Figure 1). The Cimanis Estuary is a dynamic and ecologically significant environment, where freshwater from the Cimanis River meets the saline waters of the Java Sea. The width of the estuary varies from 55 to 60 meters, creating a relatively narrow river system. The Cimanis Estuary waters range in depth from 0.1 m to 3 m. The river is deepest in the river mouth, reaching up to three meters. However, as one proceeds upstream, the average depth declines to below 1 meter. The tidal cycle in this region is characterized by a predominant semidiurnal pattern. During neap tides, the tidal range is relatively small, typically around 0.2 meters. Conversely, during spring tides, the tidal range is at its maximum, reaching approximately 0.8 meters. The study area encompasses various habitats within the estuary, including mangrove forests, mudflats, and open water channels.

2.2 Field observation

The field observation was carried out over two periods. The first survey was conducted on August 19–22, 2023, and the second survey was conducted on September 27–28, 2023. Surface velocity data, a crucial component of this analysis, were meticulously recorded using a Valeport Current Meter, providing comprehensive data for the validation of the LSPIV method. The drone measurement focused on the right side of the Cimanis Estuary (Figure 2), a strategically chosen location for understanding the interplay of riverine and marine influences. At four to five locations, instantaneous measurements of the current meter were taken simultaneously with the UAV. To ensure accurate georeferencing and spatial analysis, four ground control points (GCPs) were chosen to act as georeference control points using Real-Time Kinematic Global Navigation Satellite System (RTK-GNSS) surveying equipment. The x, y, and z coordinates of the georeference control points were measured and marked, along with the drone's position.

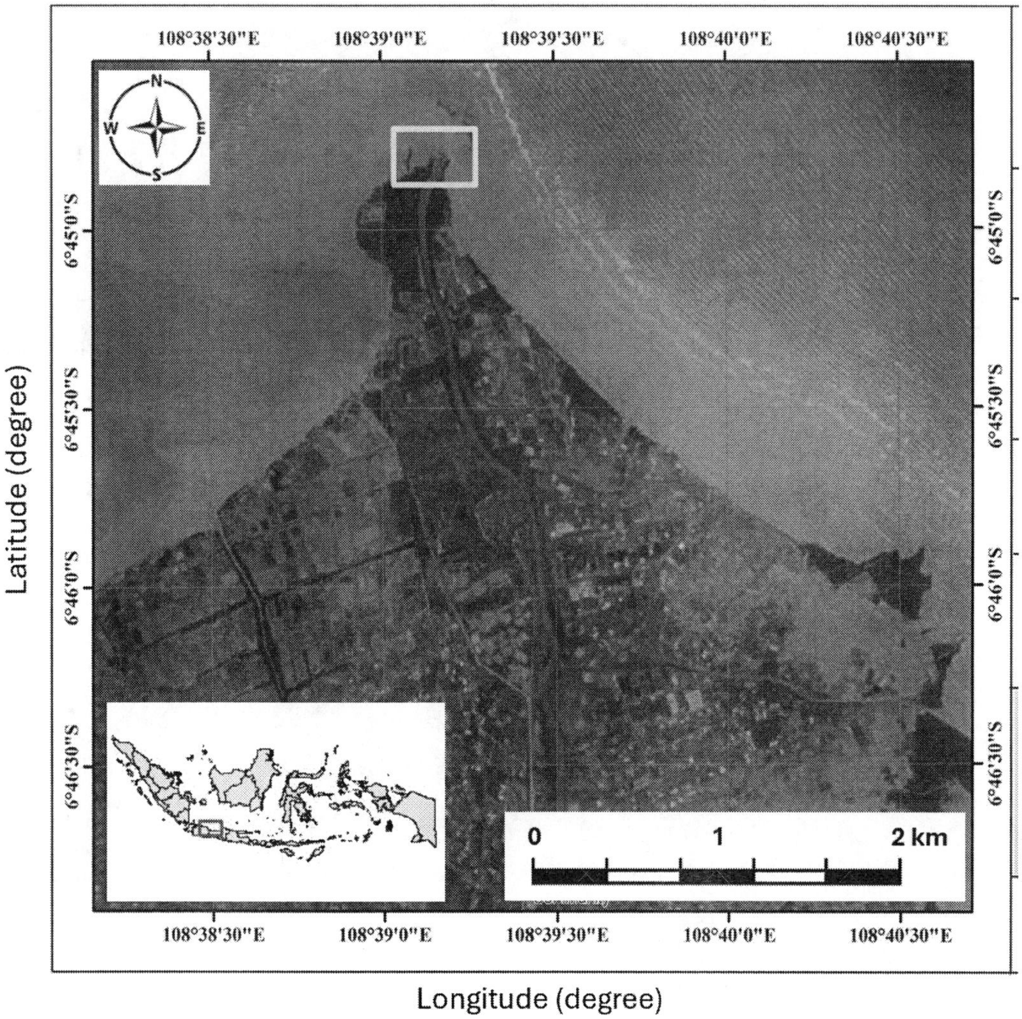

Figure 1. Study area of Cimanis Estuary, Cirebon, West Java. Yellow box is the area of investigation for drone measurement.

2.3 Drone measurement

The observation method employed in this study entailed the utilization of a DJI Mavic Air 2 drone to capture aerial video footage of the surface flow. The drone's camera lens has a 12 MP pixel value, ISO recording sensitivity of 100, Equivalent Focal Length of 24 mm, EV (Exposure Value) of 1 ev, and aperture value of $f2.8$. The video recording screen size is 1920 x 1080 pixels. The video data are then processed while keeping the frame rate constant at 24 FPSs (frames per second). The first and second surveys involved drone flights at heights of 18 m, 20 m, 30 m, 40 m, 50 m, and 70 m (Table 1). During data gathering, the drone will remain immobile at each height with the drone camera at a 90-degree tilt.

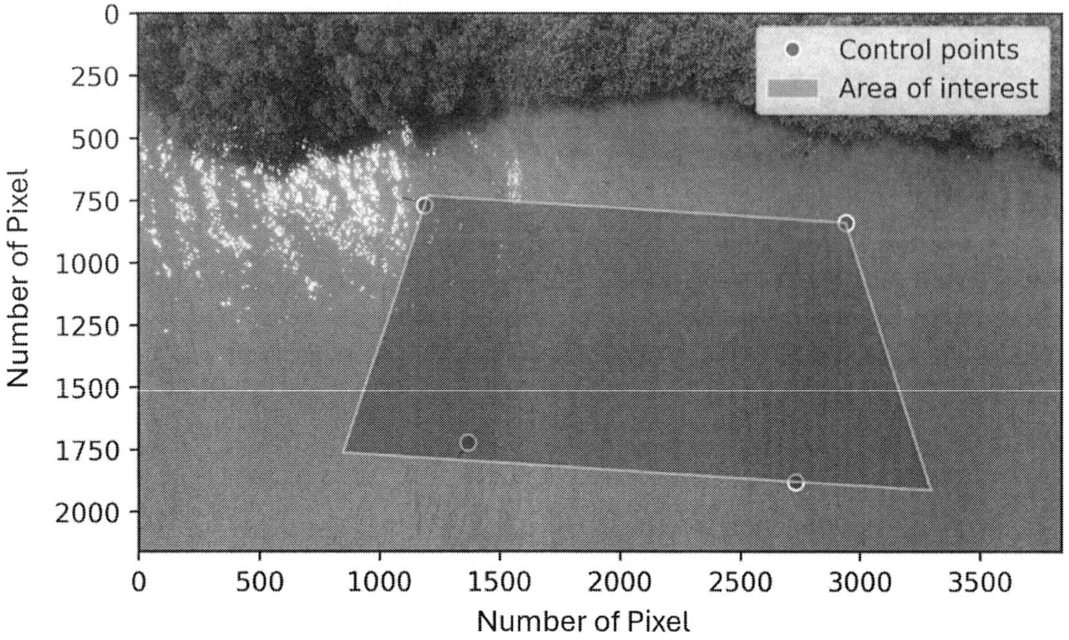

Figure 2. Ground control points and area of investigation for LSPIV.

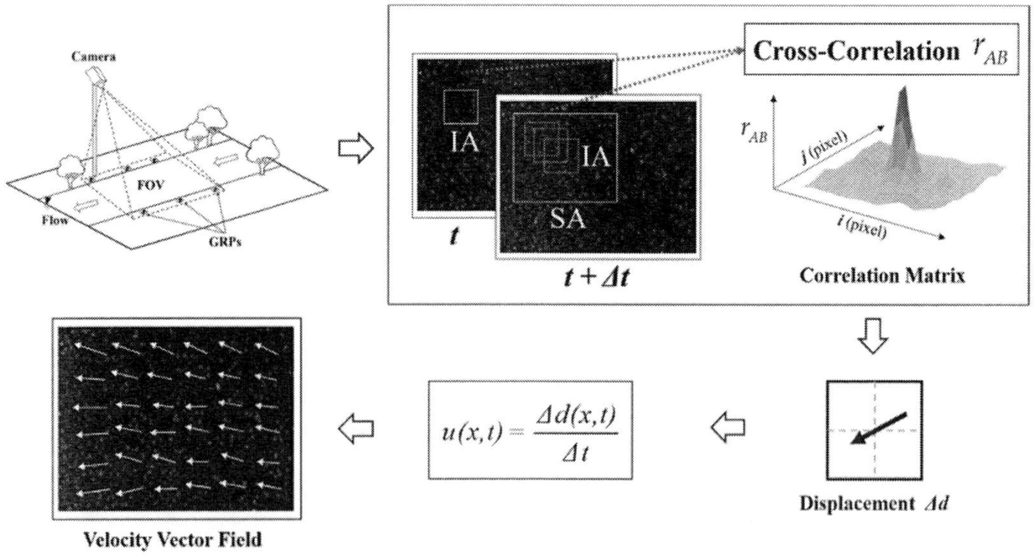

Figure 3. Field of View; GRPs is the Ground Reference Points; IA is the Interrogation Area; SA is the Searching Area [17].

Table 1. Survey design and drone measurements.

Period	Tidal condition	Altitude (m)	Camera Angle (degree)
20–21 August 2023	Flood	20 m	90
		30 m	
		40 m	
		50 m	
		70 m	
	Ebb	20 m	
		30 m	
		40 m	
		50 m	
		70 m	
27–28 September 2023	Flood	20 m	90
		30 m	
		40 m	
		50 m	
		70 m	
	Flood	20 m	
		30 m	
		40 m	
		50 m	
		70 m	

2.4 Large-Scale Particle Image Velocimetry

Non-intrusive measurement procedures have experienced remarkable advancements in recent decades, driven by breakthroughs in scientific technology. Advancements in computer vision, Large-Scale Particle Image Velocimetry (LSPIV) methods, and image-based measurements have revolutionized the field of fluid dynamics. For instance, the pioneering work presented in [18] utilized the PIV method to measure flow velocities in rivers, opening avenues for studying hydraulic processes in natural waterways. Subsequently, LSPIV has gained widespread adoption in the hydrological sector, with researchers developing innovative angle correction calculations to enhance the accuracy of image acquisition [3, 19]. These non-intrusive techniques offer unparalleled advantages over traditional intrusive methods, as they eliminate the need for physical probes or tracers that can potentially disrupt or alter the flow dynamics.

Large-Scale Particle Image Velocimetry (LSPIV) is a cutting-edge optical remote sensing technique that enables the measurement of spatiotemporal dimensions of fluid flows. It operates on the principle of introducing traceable particles (seeding) into the flowing water, which are then illuminated and captured in sequential images or videos. This seeding process allows for the visualization and tracking of the flow patterns. The output of LSPIV is a 2D velocity vector field on the surface of the flow, providing a comprehensive overview of the flow dynamics over a relatively large area. Essentially, LSPIV captures fluid flow patterns over time using an optical sensor, such as a camera or unmanned aerial vehicle (UAV) [10]. The displacement of patterns in the x and y coordinates within the image matrix is analyzed using the cross-correlation matrices of paired patterns. Every element in the image is processed to obtain the velocity vectors on the measurement plane. By referencing the actual distance and pixel distance in the camera image,

IOP Conf. Series: Earth and Environmental Science **1473** (2025) 012007 doi:10.1088/1755-1315/1473/1/012007

along with the time difference between two consecutive images, the displacement of patterns in the image sequence is converted into units of distance per time, such as m/s, providing accurate and detailed information about the flow behavior. An example of the LSPIV measurement sequence is shown in Figure 3.

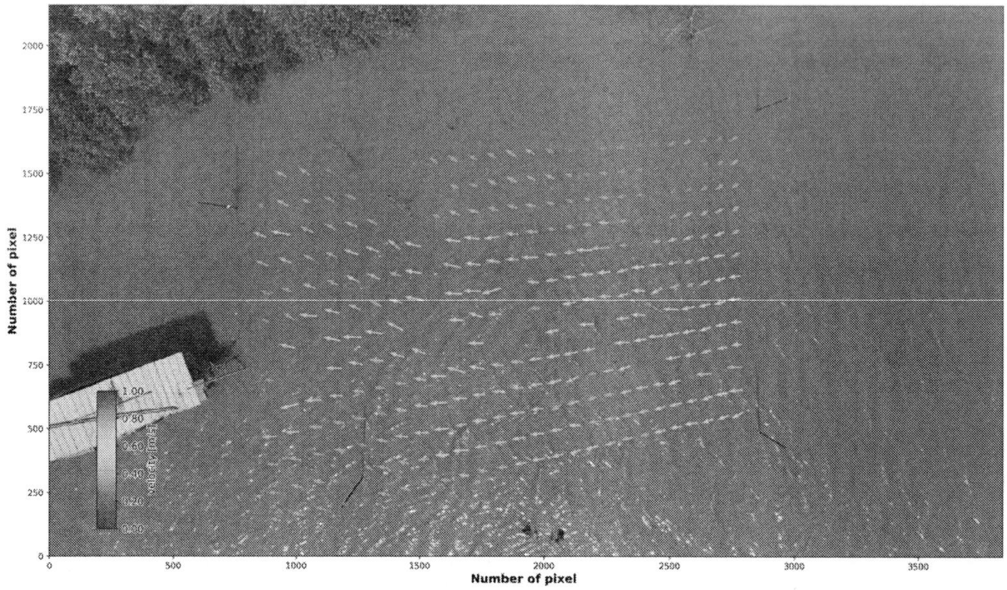

Figure 4. Pattern of surface currents obtained using LSPIV method, during ebb tide.

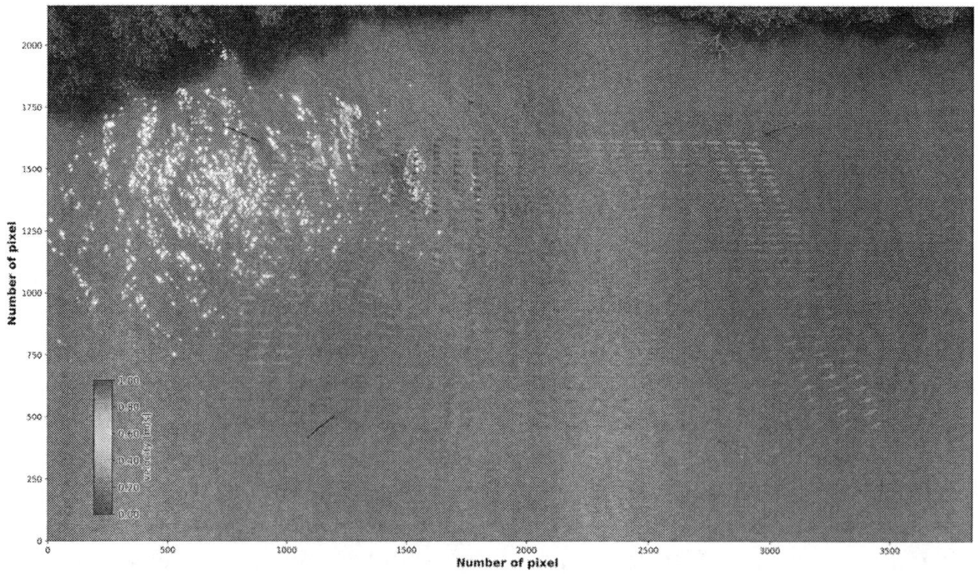

Figure 5. Pattern of surface currents obtained using LSPIV method, during flood tide.

3. Results and Discussion

Figures 4 and 5 show the current magnitude at the surface under ebb and flood conditions, calculated using the LSPIV method. These calculations offer a comprehensive understanding of the tidal dynamics within the study area. During ebb tide, characterized by the retreat of water from the river mouth, the current magnitudes exhibited a range of 0.1 to 0.8 m/s, displaying a consistent flow direction towards the river mouth. This indicates a clear outward movement of water, driven by the gravitational pull of the receding tide. Conversely, during flood tide, when seawater surges into the river, the current magnitudes at the surface varied from 0.2 to 0.5 m/s, with the flow directed upstream. This upstream flow signifies the influx of water from the sea, driven by the rising tide, as it pushes against the river's flow. These observations highlight the significant influence of tidal forces on the river's current patterns, demonstrating the interplay between the river's natural flow and the cyclical changes in sea level. The LSPIV method effectively captures these dynamic currents, providing valuable data for analyzing the intricate interplay of tidal forces in the investigated area.

Table 2. Quantification of validation between LSPIV and field observation for velocity magnitude.

Altitude (m)	Tidal condition	RMSE (m/s)	MAE (m/s)	MAPE (%)
20	Flood	0.06	0.05	5,6
30	Flood	0.07	0.06	7,7
40	Flood	0.09	0.08	9,2
50	Flood	0.12	0.10	10,4
70	Flood	0.14	0.12	14,3
20	Ebb	0.04	0.05	5,7
30	Ebb	0.08	0.07	7,1
40	Ebb	0.11	0.10	9,4
50	Ebb	0.10	0.11	11,7
70	Ebb	0.14	0.13	15,2

The accuracy of the LSPIV surface velocity results that were obtained using a drone at various heights was assessed by comparing them with field data measurements. To quantify the agreement between these independent datasets, statistical metrics including the Root Mean Square Error (RMSE), Mean Absolute Error (MAE), Mean Absolute Percentage Error (MAPE), and correlation coefficient were employed. The analysis revealed that the drone measurements that were acquired at a height of 20 meters were the most favorable, achieving the lowest values for velocity magnitude and direction, RMSE (0.04 m/s; 10.69 degree), MAE (0.05 m/s; 10.69 degree), and MAPE (5.6%; 3.7%) respectively (Tables 2 and 3). This finding underscores the effectiveness of drone-based measurements in capturing representative surface velocities and highlights the potential for LSPIV technology in producing accurate and reliable flow field estimations, particularly when validated with such complementary data sources.

Figure 6 presents the MAPE values that were obtained for varying drone measurement heights. The MAPE is a crucial parameter for the LSPIV method, as it helps mitigate biases and inaccuracies that may arise at different drone altitudes. The figure demonstrates that the MAPE trend is proportional to the altitude, meaning that it increases as the drone descends closer to the ground. Notably, the lowest measurement height of 20 meters exhibits the smallest MAPE value, indicating the highest level of accuracy and precision in the LSPIV computations.

IOP Conf. Series: Earth and Environmental Science **1473** (2025) 012007 doi:10.1088/1755-1315/1473/1/012007

Table 3. Quantification of validation between LSPIV and field observation for velocity direction.

Altitude (m)	Tidal condition	RMSE (degree)	MAE (degree)	MAPE (%)
20	Flood	15.34	15.22	4.8
30	Flood	16.34	16.03	5.5
40	Flood	16.85	16.91	5.9
50	Flood	17.32	17.24	7.2
70	Flood	18.75	18.70	8.1
20	Ebb	10.74	10.69	3.7
30	Ebb	11.10	11.09	5.5
40	Ebb	12.10	12.22	6.5
50	Ebb	14.07	13.75	6.8
70	Ebb	15.12	14.35	4.1

The authors of [20, 21] highlighted the significant impact of low drone flight altitudes on the quality of reconstructed data, particularly when coupled with a high video resolution. This principle was applied in the present study, where a relatively small area of interest (AOI) of 20 × 25 m was chosen, resulting in a minimum flight altitude of 20 m. This strategic approach ensured that the drone flew at a sufficiently low altitude to capture rich, detailed information about the study area. The close proximity of the drone to the ground facilitated the acquisition of high-resolution images. The low altitude allowed for a greater density of image data, enabling the algorithm to accurately capture intricate features and subtle changes in the landscape. Furthermore, drone height-related technical obstacles, such as signal deterioration and communication concerns, can have an impact on data quality. According to [22], as flying height increases, the performance of powered drones can degrade, resulting in poorer signal quality and increased latency, which can impede the real-time data collecting required for effective LSPIV applications.

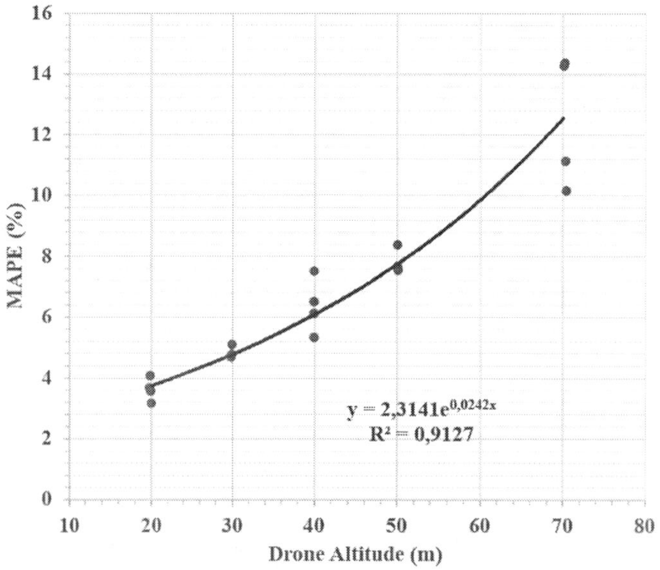

Figure 6. The accuracy of the LSPIV method based on flight altitude variations.

4. Conclusions

The effectiveness of the Large-Scale Particle Image Velocimetry (LSPIV) method for determining the surface velocity in the Cimanis Estuary was demonstrated. To obtain accurate measurements, drone flights were conducted at varying heights: 20 m, 30 m, 50 m, and 70 m. A comprehensive validation process was carried out, comparing the LSPIV results with field data observations. The results revealed that a flight height of 20 meters yielded the most accurate velocity measurements. At this height, the LSPIV method achieved the lowest Root Mean Square Error (RMSE) at 0.08 m/s, Mean Absolute Error (MAE) at 0.06 m/s, and Mean Absolute Percentage Error (MAPE) at 5.6%. This indicates that the LSPIV method can provide highly accurate surface velocity measurements in the Cimanis Estuary when the drone is flown at a height of 20 meters. This finding has significant implications for future studies utilizing LSPIV to investigate surface velocity patterns in estuarine environments.

Acknowledgments
The writing of this manuscript was funded by the Indonesian Ministry of Education, Culture, Research and Technology (Kemdikbudristek) under the research program of Penelitian Fundamental Research (PFR) 2024 Research Grant Contract No. 036/E5/PG.02.00.PL/2024. The fieldwork data were funded by the Penelitian Fundamental Research (PFR) 2023 Research Grant Contract No. 110/E5/PG.02.00.PL/2023.

References

[1] Zheng C., Xiao Z., Zhou W., Chen X., and Chen X. (2018). Ocean Current Characteristics. In: 21st Century Maritime Silk Road: A Peaceful Way Forward. *Springer Oceanography*. Springer, Singapore.

[2] Fujita, I., Muste, M., and Kruger, A. (1998). Large-scale particle image velocimetry for flow analysis in hydraulic engineering applications. *J. Hydraul. Res.*, 36, 396–414.

[3] Muste, M., Fujita, I., and Hauet, A. (2008). Large-scale particle image velocimetry for measurements in riverine environments. *Water Resour. Res.*, 44, W00D19.

[4] Tauro, F., Pagano, C., Phamduy, P., Grimaldi, S. and Porfiri, M. (2015). Large-scale particle image velocimetry from an unmanned aerial vehicle. *IEEE/ASME Trans. Mechatron.*, 20, 3269–3275.

[5] Detert, M., Johnson, E.D., and Weitbrecht, V. (2017). Proof-of-concept for low-cost and non- contact synoptic airbone river flow measurement. *Int. J. Remote Sens.*, 38, 2780–2807.

[6] Yang, M.D., Huang, K.S., Kuo, Y.H., Tsai, H.P., and Lin, L.M. (2017). Spatial and spectral hybrid image classification for rice lodging assessment through UAV imagery. *Remote Sens.*, 9, 583.

[7] Hemmelder, S., Marra, W., Markies, H., and De Jong, S.M. (2018). Monitoring river morphology & bank erosion using UAV imagery—A case study of river Buech, Hautes-Alpes, France. *Int. J. Appl. Earth Obs. Geoinf.*, 73, 428–437

[8] Kastridis, A., Kirkenidis, C., and Sapountzis, M. (2020). An integrated approach of flash flood analysis in ungauged Mediterranean watersheds using post-flood surveys and unmanned aerial vehicles. *Hydrol. Process.*, 34, 4920–4939.

[9] Huang, W.C., Young, C.C., and Liu, W.C. (2018). Application of an automated discharge imaging system and LSPIV during typhoon events in Taiwan. *Water*, 10, 280.

[10] Liu, W.-C., Lu, C.-H., and Huang, W.-C. (2021). Large-Scale Particle Image Velocimetry to Measure Streamflow from Videos Recorded from Unmanned Aerial Vehicle and Fixed Imaging System. *Remote Sensing*. 13, 2661.

[11] Strelnikova D, Paulus G, Käfer S, Anders K-H, Mayr P, Mader H, Scherling U, and Schneeberger R. (2020). Drone-Based Optical Measurements of Heterogeneous Surface Velocity Fields around Fish Passages at Hydropower Dams. *Remote Sensing*. 12(3):384.

[12] Lewis, Q.W., Lindroth, E.M., and Rhoads, B.L. (2018). Integrating unmanned aerial systems and LSPIV for rapid, coast-effective stream gauging. *J. Hydrol.*, 560, 230–246.

[13] Naves, J., Anta, J., Puertas, J., Regueiro-Picallo, M., and Suárez, J. (2019). Using a 2D shallow water model to assess Large-Scale Particle Image Velocimetry (LSPIV) and Structure from Motion (SfM) techniques in a street-scale urban drainage physical model. *Journal of Hydrology*, *575*, 54-65.

[14] Huang, K., Chen, H., Xiang, T., Lin, Y., Liu, B., Wang, J., ... and Xu, C. Y. (2022). A photogrammetry-based variational optimization method for river surface velocity measurement. *Journal of Hydrology, 605*, 127240.

[15] Chen, M., Chen, H., Wu, Z., Huang, Y., Zhou, N., and Xu, C. Y. (2024). A Review on the Video-Based River Discharge Measurement Technique. *Sensors, 24*(14), 4655.

[16] Tarya, A., and Prasetyo, M. A. (2024). Surface Current Measurement Using Large Scale Particle Image Velocimetry (LSPIV) in Ender Estuary, Cirebon. *IOP Conference Series: Earth and Environmental Science*, 1298, 1, p. 012012.

[17] Lin Y-C, Ho H-C, Lee T-A, and Chen H-Y. (2022) Application of Image Technique to Obtain Surface Velocity and Bed Elevation in Open-Channel Flow. *Water*. 14(12):1895.

[18] Fujita,I.,and Komura,S.(1994).Application of Video Image Analysis for Measurements of River Surface Flows, *Proceedings of Hydraulic Engineering*, 38, 733–738.

[19] Fujita, I. and Kunita, Y. (2011). Application of Aerial LSPIV To The 2002 Flood of The Yodo River Using A Helicopter Mounted High Density Video Camera. *Journal of Hydro- environment Research*, 5(4):323–331.

[20] Seifert, E., Seifert, S., Vogt, H., Drew, D., Aardt, J., Kunneke, A., Seifert, T. (2019): Influence of drone altitude, image overlap, and optical sensor resolution on multi-view reconstruction of forest images, *Remote Sensing*, 11, 1252.

[21] Lewis, Q. and Rhoads, B. (2018). LSPIV measurements of two-dimensional flow structure in streams using small unmanned aerial systems: 1. Accuracy assessment based on comparison with stationary camera platforms and in-stream velocity measurements. *Water Resources Research*, 54(10), 8000-8018.

[22] Zulkifley, M., Behjati, M., Nordin, R., & Zakaria, M. (2021). Mobile network performance and technical feasibility of LTE-powered unmanned aerial vehicle. Sensors, 21(8), 2848.

Topology optimization on four-legged platform jacket-type offshore building structure

Anwar[1], R W Prastianto[1], D M Rosyid[1]

[1] Department of Ocean Engineering, Faculty of Marine Technology, Institut Teknologi Sepuluh Nopember, Surabaya, Indonesia

*Anwar.ft@ulm.ac.id

Abstract: The platform jacket structure is a component of offshore buildings anchored to the seabed. This structure has an essential function in oil drilling: supporting the superstructure so that activities on offshore building platforms can run well. The failure of the platform jacket structure due to environmental load conditions is an essential concern in modeling the jacket structure. This study aims to discover the best design for the platform jacket structure by using a topology optimization method with the function of minimizing strain energy or increasing the rigidity of the platform jacket structure. Topology optimization in this study was carried out using FEM software with the application of predetermined limits. The results showed that the optimized jacket structure model had a slightly larger weight than the initial one. Still, the increase in the strength of the optimized jacket structure was more significant than that of the initial jacket structure model. The topology optimization results provide a jacket model that is 34.78% stronger than the reference jacket structure despite weighting 8.57% heavier than the reference jacket structure.

Keywords: Topology optimization, Jacket platform, FEM, In-place analysis

1. Introduction

Platform jacket structure is a fixed and anchored offshore building structure where the jacket structure is the most widely used offshore building around the world [1], [2], [3]. The structure of a platform jacket is generally made of tubular steel that is interconnected between the legs and the brace [4]. A fixed platform jacket-type offshore building structure is practical and economical, accommodating activities during operation [2]. Jacket structures are generally used to support oil and gas exploration activities; jacket structures are also widely used to support offshore wind turbine activities [5], [6].

In supporting operational activities in offshore areas, jacket structures must have qualified strength. The jacket structure must be designed to endure the forces generated by its own vertical load, as well as lateral forces caused by environmental factors like wave loads, wind, and currents. The influence of wave force on the structure of the platform jacket can provide a strong vibration effect that can cause severe damage[7]. During the operation of the platform jacket structure in the offshore area, of course, it experiences loading conditions such as the operational load of the structure itself as well as the environmental load that occurs randomly so that it can cause damage to the jacket structure[8]. In jacket structures, the interaction between environmental factors like wind and waves can introduce significant uncertainty issues, making it essential to conduct a thorough analysis of the structure [9] properly. To avoid failure in the jacket structure due to uncertainty problems, in this study, topology optimization will be carried out on the platform jacket structure to find an optimal jacket structure model with high stiffness and light structural weight. The platform jacket structure is safeguarded from loads due to its high rigidity.[10].

Some of the ways that can be used to optimize the structure of the jacket are by way of topology optimization by finding the optimal layout for the distribution of forces in the jacket structure; shape optimization, which is an optimization that is done to change the design of the join in the jacket structure but the topological shape remains the same. Then, size optimization is an optimization that is carried out by applying dimension reduction to the jacket structure without changing the shape and topology of the structure. Size optimization is generally used to determine the thickness and diameter of the member on the jacket structure[4]. In this study, topology optimization is the focus that will be carried out where the diameter and thickness are not optimized; the diameter and thickness of the jacket structure follow the reference structure data. In structural topology optimization, the first step that needs to be considered is determining the object to be optimized and applying the Constraints of the optimization to be carried out[11].

Topology optimization is a technique used to discover the most effective configuration of a structure, guided by predefined goals and limitations.[12]. Topology optimization is typically conducted in the early stages of structural modeling to find the optimal structural model in terms of material distribution with predetermined and applied loading conditions[13]. Some commonly used functions in topology optimization are increasing stiffness by minimizing strain energy and minimizing the volume of structural material[14]. Topology optimization is one of the most effective methods to solve the problem of material layout optimization with various constraints, one of which is increasing the rigidity of the optimized structure[15], [16]. A design goal for an optimal platform jacket structure is to minimize the energy response to both dynamic and static loads [17]. Researchers have widely carried out topology optimization to find and produce alternative structural model configurations in various engineering fields [18]. According to [19], One of the advantages of topology optimization is its high performance in multiphysics design problems, which provides optimal results.

2. Research methods

This study employs a topology optimization method using a numerical approach with Abaqus finite element method (FEM) software. This approach optimizes the material distribution of the jacket structure to enhance its strength and stiffness. Meanwhile, SACS software is used for output interpretation, model reconstruction, and in-place analysis. The in-place analysis calculates the strength and weight of both the reference model and the optimized jacket structure model. The flow of the research process is shown in Figure 1

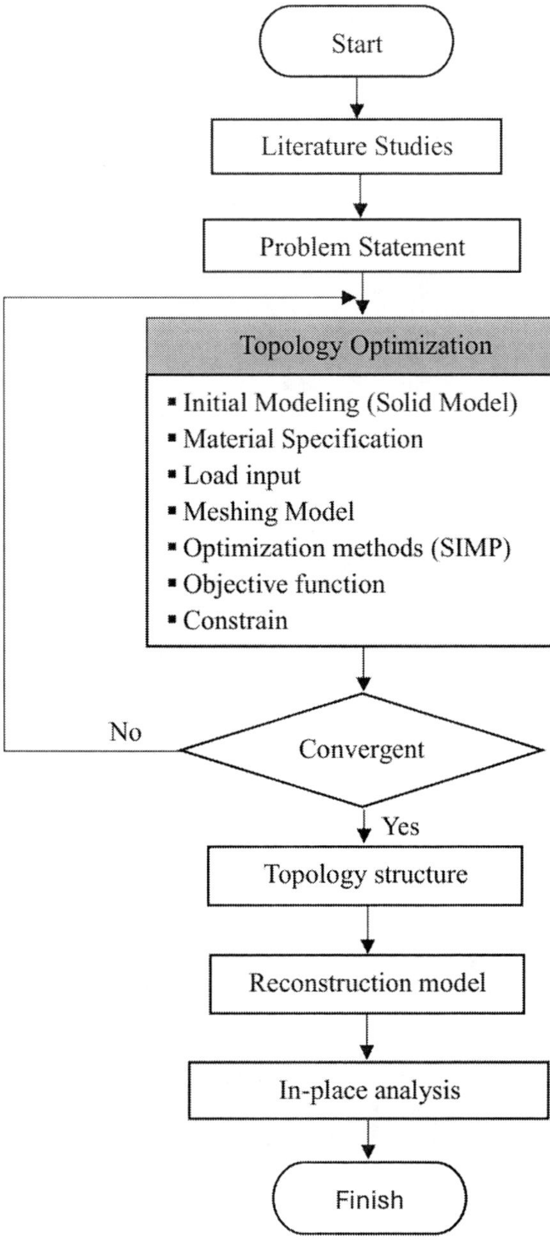

Figure 1. Research flowchart

Figure 1 is a stage of the research flow that is carried out starting from research preparation such as literature studies, problem formulation, methods used, collection and input of research data such as reference jacket structure data, environmental load data, topology optimization process carried out on FEM abaqus software, in-place analysis carried out on SACS software until the results of the research and completion.

2.1 Reference jacket model

The reference jacket model utilized in this study is a platform jacket with four vertical legs and no tilt. The jacket model of the reference platform, in general, as shown in Figure 2

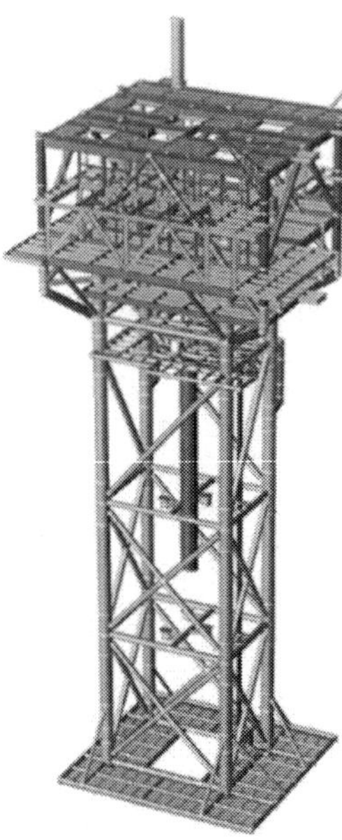

Figure 2. Platform jacket type offshore building structure model

In addition to the global model on offshore building structures, the specific reference jacket model used in this study can be shown in Figure 3.

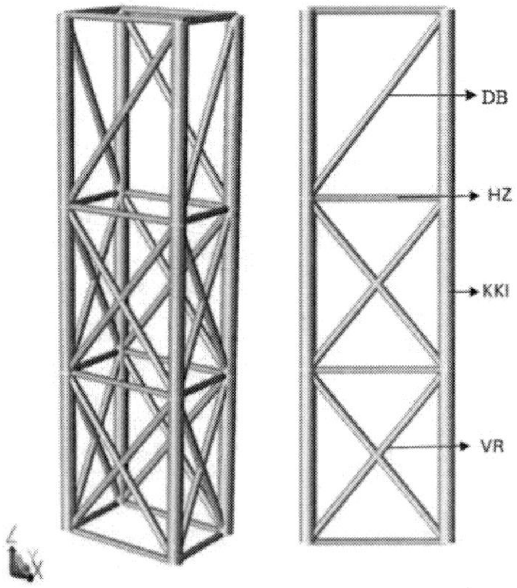

Figure 3. Reference jacket structure

Figure 2 and Figure 3 are structural models used as references in this study. The reference jacket structure has three levels of columns with a combination of X brace and N brace or a single diagonal brace. Table 1 shows the dimensions of the materials used in this study.

Table 1. Structural material dimensions

Parameter	Members Grup	OD (cm)	WT (cm)
Jacket leg	KKI	109.70	4.00
Horizontal brace	HZ	45.70	1.27
Diagonal brace 1	VR	50.80	1.27
Diagonal brace 2	DB	55.90	1.27

Table 2. Material specifications

Parameter	Value (Unit)
Yield Strength (Fy)	345 (N/mm^2)
Young Modulus (E)	200000 (N/mm^2)
Shear Modulus (G)	80000 (N/mm^2)
Poisson Ratio (v)	0,3
Density (ρ)	7850 (Kg/m^3)

Table 1 and Table 2 are the dimensions and material specifications of the reference jacket used in this study, where the dimensional and material data will be applied to the jacket structure resulting from topology optimization so that it can compare the weight and strength of the two jacket models.

2.2 Load analysis

The load on the platform jacket structure includes the deck load and environmental loads, such as wave, current, and wind forces. Deck load is the accumulation of the total material weight, equipment load, and live load acting on the top structure of the platform jacket under operating conditions, where the deck load acts vertically on the jacket structure. The data on the deck load working on the structure is 10025.63 kN. The environmental load used in this study is data from where the reference structure was built, namely in the eastern Java Sea. The environmental load data is shown in Table 3.

Table 3. Environmental load data

Parameter	condition	Parameters and values
Wave	Operating conditions	Wave height 2.9 m with period 8.1 s
	Storm conditions	Wave height 4.3 m with period 11.3 s
Wind	Operating conditions	Wind speed 14.6 m/s
	Storm conditions	Wind speed 20.7 m/s
Current	Operating conditions	Current speed 0.44 m/s
	Storm conditions	Current speed 0.50 m/s

Table 3 this study used environmental load data, where the data displayed is the maximum that occurs in operating conditions and storm conditions. Environmental loads, including waves, wind, and currents, exert lateral forces on the jacket structure. The calculation of the wave load in this study uses the Morison equation, as in equation (1).

$$f = \rho C_M \frac{\pi OD^2}{4} \dot{u} + \frac{1}{2} \rho_w C_D D_o |u + U|(u + U) \tag{1}$$

In this equation, ρ_w is the density of seawater, C_D is the drag coefficient, C_M is the coefficient of inertial force, D_o is the diameter of the pipe, and |U| u Represents the horizontal velocity and acceleration of water particles in the perpendicular direction. Calculation of current load using equation (2).

$$F_c = {}^1/_2\, \rho\, C_D\, U_c^2 \qquad (2)$$

Where F_c is the current force, ρ has a seawater density of 1025 kg/m³, C_D stands for drag coefficient, and U_c is the current velocity. Wind load can be determined using the following equation (3).

$$F = {}^{\rho}/_2\, V^2\, C_s\, A \qquad (3)$$

where F is the wind's force, V is the wind's speed, C_s is an abbreviation for Building Side Form Coefficient, A is the area of the side object, and P is the density of air. Based on the forces applied to the jacket structure, the loading conditions are illustrated in Figure 4.

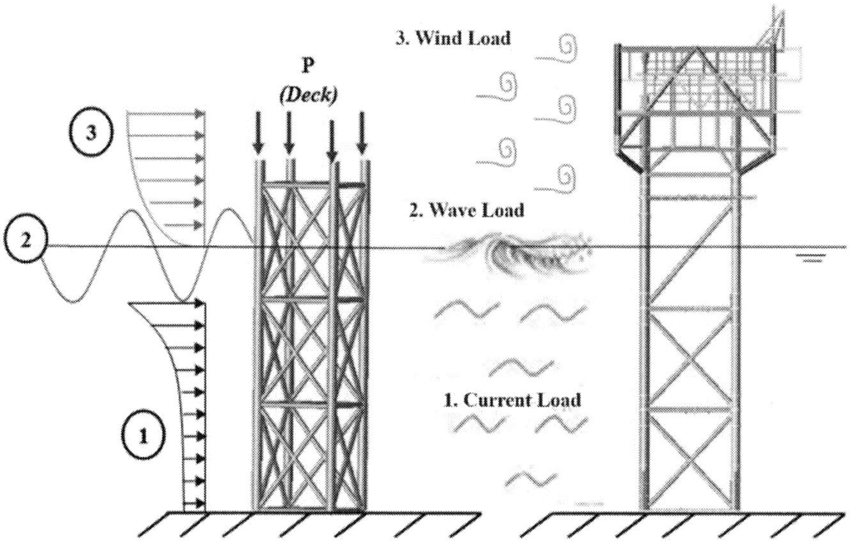

Figure 4. Loading conditions on the jacket structure

2.3 *Optimization method*

Some commonly used methods in structural topology optimization are the Solid Isotropic Material with Penalization (SIMP) method, set level, and evolutionary structural optimization method [20]. Of the several techniques that have been mentioned, the SIMP method has a more prominent role because of its more straightforward and efficient application, where this method has also been widely applied in various structural topology optimization problems [21]. The SIMP method is a topology optimization technique used to determine the distribution of materials in a structure to optimize strength and performance. In this method, the material is isotropic and solid, and the penalty factor is applied to materials that are close to the empty state to prevent inefficient material distribution. The SIMP method applies variable values between 0 and 1 to eliminate the most minor value variable [14]. The SIMP equation in this study is as follows:

$$k_e=(x_e)^p k_0 \qquad (4)$$

Where k_e is a combination of elements, x_e is the density element, p is a penalty factor, and k_0 is the original elemental causality matrix.

2.4 Topology Optimization

The primary purpose of topology optimization in this study is to minimize energy strain and enhance the stiffness of the platform jacket structure. The problem of topology optimization is stated in equation (5):

$$\text{Minimize SE}(x) = U^T K U = \sum_{e=1}^{n} (x_e)^p u_e^T k_0 u_e$$
$$\begin{cases} \text{Subject to } \frac{V_x}{V_0} = f \\ F = KU \\ 0 < x_{min} \leq x_e \leq 1 \end{cases} \quad (5)$$

Where SE is strain energy, x is a variable design, V_x and V_0 represent the material volume and the design volume. At the same time, f is the ratio of the volume that has been determined, F is a global force matrix, K is the global stiffness matrix U displacement matrix, p is a penalty factor, k_0 and u_e is the matrix of the elements and the vector of element displacement, x_e is the variable density element, x_{min} minimum density vector and n is the total number of elements on the jacket structure.

3. Results and discussion

3.1Initial model of topology optimization

The initial model of the structure for topology optimization is an isotropic solid material model in contact with the same material specifications as the reference jacket structure. The initial model for optimizing the jacket structure's topology can be seen in Figure 5.

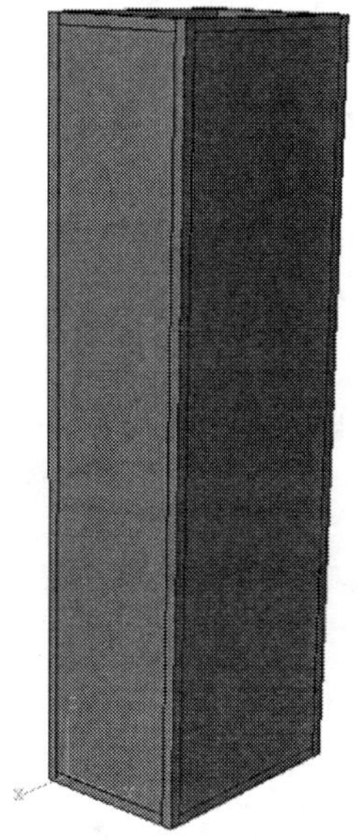

Figure 5. Initial model of topology optimization

Figure 5 is a solid material model utilized for optimizing the topology of the jacket structure. The model is made based on data taken from the structure of the reference jacket, such as the dimensions of the structure, including height, width, length, and specifications of the materials used. In optimizing the topology, four legs are made and modeled as a fixed structure that is not optimized, and horizontal braces are located at the top and bottom of the structure. The optimization area is located on the four sides of the structure to be topically optimized to form a new and optimal model in withstanding the force acting on the jacket structure.

3.2 Topology optimization results

The topology optimization process is carried out on Abaqus software by applying the function of minimizing energy strain to increase capacity with the limitation of the design volume that has been determined. Conditions are monitored during the iteration that takes place in each optimization round, as shown in Figure 6.

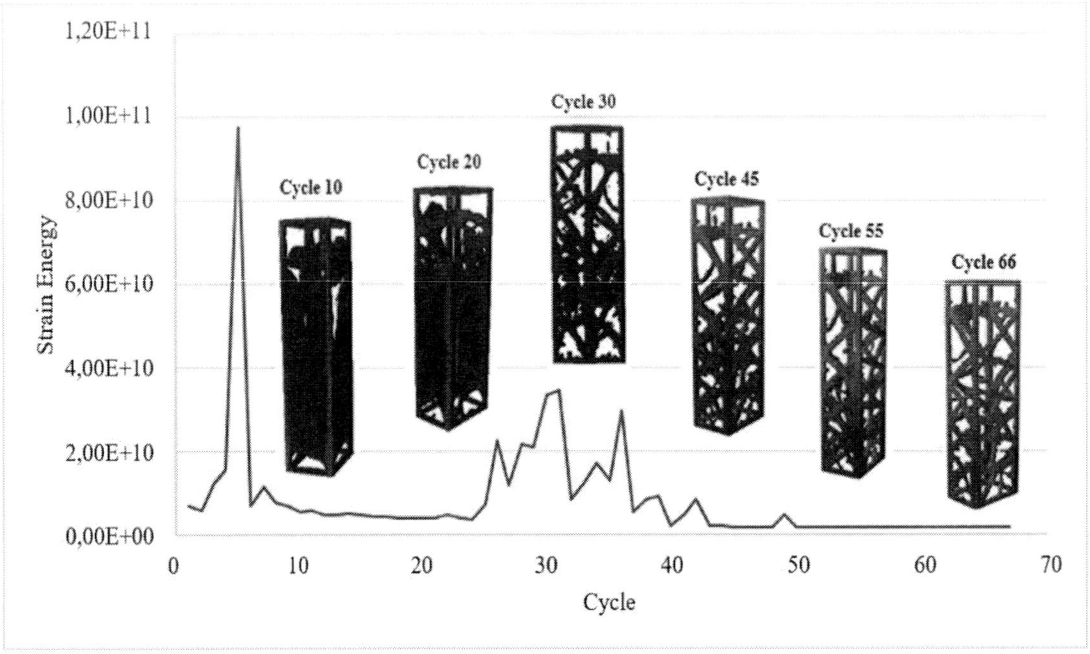

Figure 6. Topology optimization process in Abaqus software

Figure 6 illustrates the optimization process in Abaqus software, where the strain energy in the jacket structure begins to stabilize at cycle 50 and converges at cycle 66, as shown in the graph. The topology optimization in this study limits the number of volumes of the optimization results to be achieved to 30%, but the optimized structure converges at 32% volume. Figure 7 shows the appearance of the topology optimization results on the 66th lap when the convergent model is used.

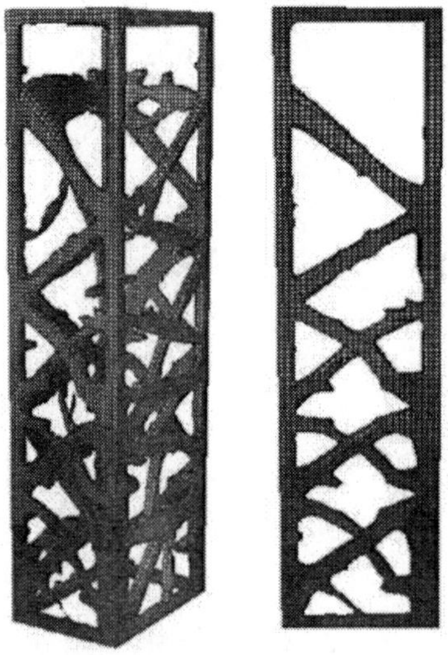

Figure 7. Topology optimization results

Figure 7 this is the result of topology optimization carried out using Abaqus software. The resulting jacket structure model has 5 columns, forming the brace X and N models. The number of columns resulting from model optimization is more than the reference jacket structure model, which only has 3 columns.

3.3 Output interpretation and model reconstruction

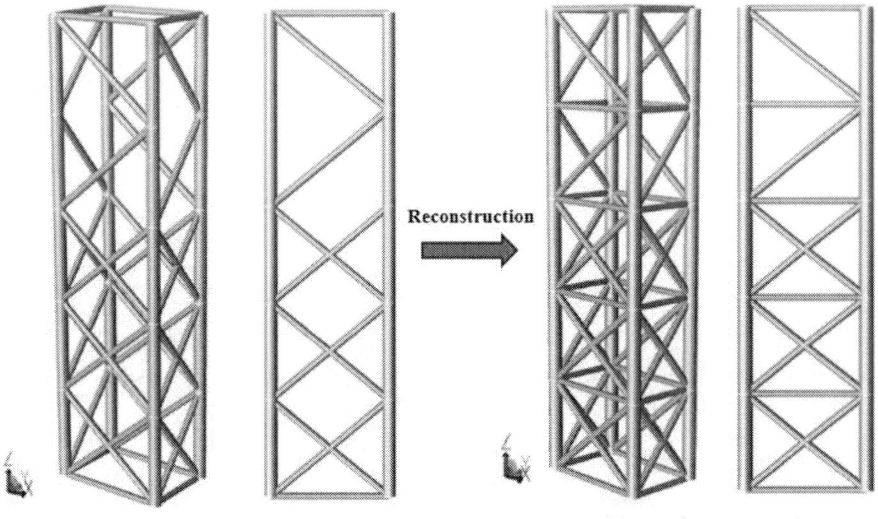

Figure 8. Reconstruction jacket structure model

Figure 8 it is a jacket structure model that has been interpreted and reconstructed as a form of refinement of the jacket structure model. Model interpretation is the result of a model from topology optimization by changing the join layout to form an optimal model geometry, while model reconstruction on the jacket structure is the addition of a horizontal brace to the jacket structure that has been interpreted so that the resulting jacket model can represent the platform jacket structure model in general.

3.4 In-place analysis

An in-place analysis of the jacket structure is conducted to assess if the modelled jacket structure is the structure's foundation meets the serviceability limit state by analyzing whether the stress acting on the jacket structure does not exceed the voltage allowed by the material used. One of the checks carried out is that the unity check (UC) value must be less than or equal to 1. In the place analysis, the UC calculation can be seen in equation (6) [22].

$$UC = \frac{\sigma_{act}}{\sigma_{allow}} \tag{6}$$

Where σ_{act} is the actual working stress, and σ_{allow} is the allowable stress with a UC value less than or equal to 1. Figure 9 displays the results of the in-place analysis for both the reference jacket structure and the optimized jacket structure.

IOP Conf. Series: Earth and Environmental Science **1473** (2025) 012008
doi:10.1088/1755-1315/1473/1/012008

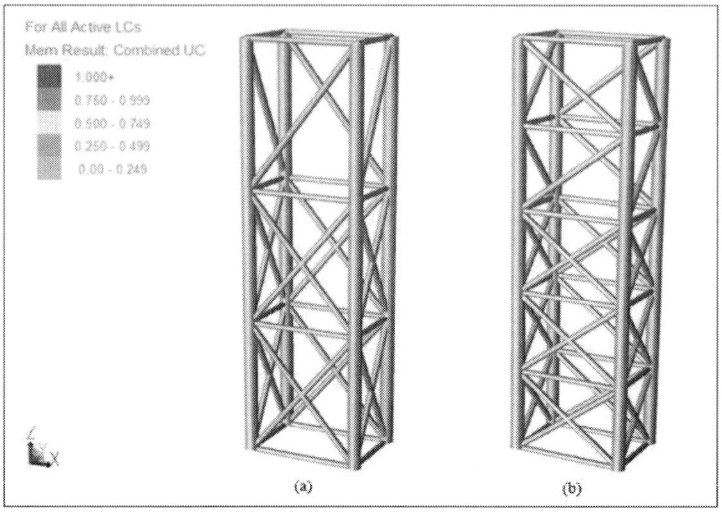

Figure 9. Results of in-place analysis (a) reference jacket (b) Optimization jacket

Figure 9 results from an inplace analysis carried out on the SACS software; The in-place analysis is conducted to evaluate the jacket structure. One of the requirements for the jacket structure to meet the serviceability limit state is to have a unity check (UC) value that is less than or equal to 1. The results of the in-place analysis showed that the UC value for the reference jacket structure was 0.23, while the UC value for the optimized jacket structure was 0.15. The results show that the structure of the optimization jacket has better strength than the structure of the reference jacket. In addition to the UC value, the structure weight can be calculated automatically using SACS software where the reference jacket structure weighs 3944.79 kN and the optimization jacket structure is 4282.95 kN. The weight calculation results show that the optimized structure has a greater weight than the reference jacket structure. The analysis showed that the topology optimization jacket structure experienced an increase in sensor strength of 34.78% from the reference jacket structure and an increase in weight of 8.57%. The results of the comparison of weight and weight between the reference jacket structure and the optimized jacket structure can be seen in Figure 10.

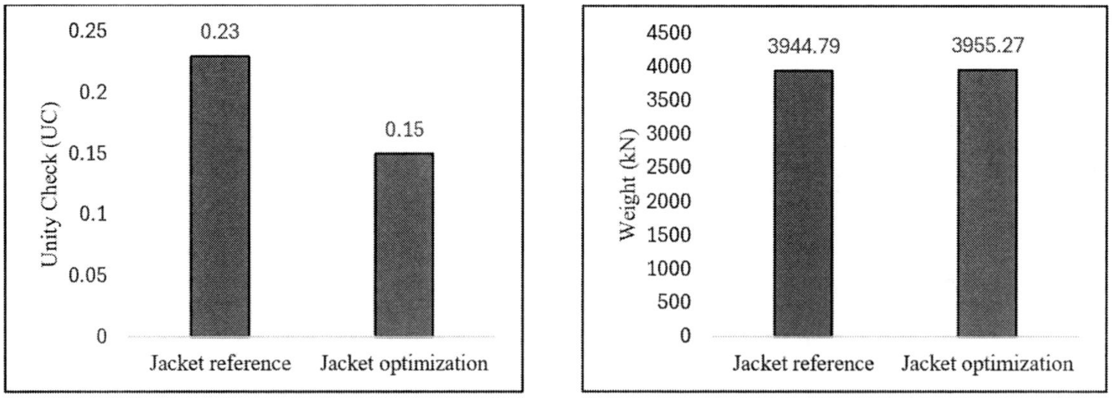

(a) (b)

Figure 10. Comparison of UC and structure weight of reference jacket and optimization jacket

Figure 10. (a) This is a comparison chart of the jacket structure's strength, showing that the optimized jacket structure has a lower UC value than the reference jacket structure, indicating enhanced strength in the optimized design. Figure 10. (b) Shows a weight increase in the optimized jacket structure, with the optimized structure being heavier than the reference jacket structure.

4. Conclusion

The topology optimization results provided a new jacket structure model with more columns than the reference jacket structure, which is a total of five columns with a combination brace topology model between brace X and brace N. The topology optimization process is performed by setting a total volume of 30% and achieving convergence at 66 iterations. The in-place analysis results indicated that the optimized jacket structure was 34.78% stronger than the reference jacket structure. Still, the optimized jacket structure weighed 8.57% heavier than the reference jacket structure.

5. Reference

[1] A. A. Motlagh, N. Shabakhty, and A. Kaveh, "Design optimization of jacket offshore platform considering fatigue damage using Genetic Algorithm," *Ocean Engineering*, vol. 227, May 2021, doi: 10.1016/j.oceaneng.2021.108869.

[2] H. Moradi, H. R. Karimi, N. S. KaramZadeh, and E. O'Din Rabei Golami, "Effects of leg slope on the failure of fixed jacket platforms: A case study of south pars gas Field's platforms," *Ocean Engineering*, vol. 210, Aug. 2020, doi: 10.1016/j.oceaneng.2020.107401.

[3] S. H. Hosseini Lavassani, S. A. Mousavi Gavgani, and R. Doroudi, "Optimal control of jacket platforms vibrations under the simultaneous effect of waves and earthquakes considering fluid-structure interaction," *Ocean Engineering*, vol. 280, Jul. 2023, doi: 10.1016/j.oceaneng.2023.114593.

[4] V. Savsani, P. Dave, B. D. Raja, and V. Patel, "Topology optimization of an offshore jacket structure considering aerodynamic, hydrodynamic and structural forces," *Eng Comput*, vol. 37, no. 4, pp. 2911–2930, Oct. 2021, doi: 10.1007/s00366-020-00983-3.

[5] H. Hejazitalab and T. Taghikhany, "Damage localization and quantification in offshore jacket structures using signal processing and intelligent system," *Ocean Engineering*, vol. 285, no. P1, p. 115325, 2023, doi: 10.1016/j.oceaneng.2023.115325.

[6] S. Sarker, D. Kim, M. S. Azad, C. Sinsabvarodom, and S. Guk, "Influence of optimal intensity measures selection in engineering demand parameter of fixed jacket offshore platform," *Applied Sciences (Switzerland)*, vol. 11, no. 22, Nov. 2021, doi: 10.3390/app112210745.

[7] H. Fang, X. Meng, L. Duan, Y. Liu, and R. Wu, "Enhanced dissipation of a jacket platform provided by a steel brace containing a multistable bilayered column," *Ocean Engineering*, vol. 203, May 2020, doi: 10.1016/j.oceaneng.2020.107244.

[8] L. Ali, S. Khan, S. Bashmal, N. Iqbal, W. Dai, and Y. Bai, "Fatigue crack monitoring of t-type joints in steel offshore oil and gas jacket platform," *Sensors*, vol. 21, no. 9, May 2021, doi: 10.3390/s21093294.

[9] A. A. Shittu, A. Kolios, and A. Mehmanparast, "A systematic review of structural reliability methods for deformation and fatigue analysis of offshore jacket structures," Jan. 01, 2020, *MDPI AG*. doi: 10.3390/met11010050.

[10] S. Wang, X. Meng, S. Ji, H. Fang, Y. Liu, and L. Y. Duan, "All-metal brace with hysteresis dissipation for impact protection of jacket platforms," *Marine Structures*, vol. 66, pp. 1–15, Jul. 2019, doi: 10.1016/j.marstruc.2019.02.009.

[11] Z. Zhao, Z. Kang, T. Zhang, B. Zhao, D. Zhang, and R. Yan, "Topology optimization algorithm for spatial truss based on numerical inverse hanging method," *J Constr Steel Res*, vol. 219, no. January 2024, doi 10.1016/j.jcsr.2024.108764.

[12] J. Lim, K. Jung, Y. Jung, and D. N. Kim, "Accelerating topology optimization using deep learning-based image super-resolution," *Eng Appl Artif Intell*, vol. 133, no. April, 2024, doi: 10.1016/j.engappai.2024.108370.

[13] P. Fernandes, À. Ferrer, P. Gonçalves, M. Parente, R. Pinto, and N. Correia, "Stress-Constrained Topology Optimization for Commercial Software: A Python Implementation for ABAQUS®," *Applied Sciences*, vol. 13, no. 23, p. 12916, Dec. 2023, doi: 10.3390/app132312916.

[14] X. Tian, Q. Wang, G. Liu, Y. Liu, Y. Xie, and W. Deng, "Topology optimization design for offshore platform jacket structure," *Applied Ocean Research*, vol. 84, pp. 38–50, Mar. 2019, doi: 10.1016/j.apor.2019.01.003.

[15] Z. Gao, Y. Li, L. Fu, K. Yang, and Y. Luo, "Topology optimization of stiffened revolution shell structures using multi-regional anisotropic MFSE method," *Thin-Walled Structures*, vol. 201, no. May, 2024, doi: 10.1016/j.tws.2024.112030.

[16] Y. Yu *et al.*, "Reliability-based topology-topography optimization for ship bulkhead structures considering multi-failure modes," *Ocean Engineering*, vol. 293, no. January, p. 116681, 2024, doi 10.1016/j.oceaneng.2024.116681.

[17] M. Vaezi, A. Pourzangbar, M. Fadavi, S. M. Mousavi, P. Sabbahfar, and M. Brocchini, "Effects of stiffness and configuration of brace-viscous damper systems on the response mitigation of offshore jacket platforms," *Applied Ocean Research*, vol. 107, Feb. 2021, doi: 10.1016/j.apor.2020.102482.

[18] X. Tian, G. Liu, W. Deng, Y. Xie, and H. Wang, "Fatigue constrained topology optimization for the jacket support structure of offshore wind turbine under the dynamic load," *Applied Ocean Research*, vol. 142, Jan. 2024, doi: 10.1016/j.apor.2023.103812.

[19] S. Ozguc, L. Pan, and J. A. Weibel, "An approach for topology optimization of heat sinks for two-phase flow boiling: Part 2 – Model calibration and experimental validation," *Appl Therm Eng*, vol. 249, no. April, 2024, doi: 10.1016/j.applthermaleng.2024.123338.

[20] Z. Meng, L. Guo, A. R. Yıldız, and X. Wang, "Mixed reliability-oriented topology optimization for thermo-mechanical structures with multi-source uncertainties," *Eng Comput*, vol. 38, no. 6, pp. 5489–5505, Dec. 2022, doi: 10.1007/s00366-022-01662-1.

[21] X. Chen, S. Yao, and J. Yvonnet, "Nonlinear topology optimization of flexoelectric soft dielectrics at large deformation," *Comput Methods Appl Mech Eng*, vol. 427, no. April, 2024, doi: 10.1016/j.cma.2024.117005.

[22] "Recommended Practice for Planning, Designing and Constructing Fixed Offshore Platforms-Working Stress Design API Recommended Practice 2A-WSD(RP 2A-WSD) Twenty-First Edition, December 2000 and Supplement 1, December 2002," 2003.

Study on The Application of Nature Based Solutions (NBS) for The Mitigation of Coastal Erosion on TPI Tempursari beach Lumajang

Muhammad Sirojudin[1], **Suntoyo**[1], **Wahyudi**[1]

[1]Department of Ocean Engineering, Sepuluh Nopember Institute of Technology

msirojudin21@gmail.com

Abstract. Lumajang has approximately 75 km of coastline, extending from Yosowilangun to Tempursari. Of these, 12 villages have been classified as Coastal villages. It is anticipated that these coastal villages will serve as the primary base for the management of coastal and marine areas. However, the region is subject to coastal erosion, as observed in the case of TPI (Tempat Pelelangan Ikan) Tempursari Beach. The issue of coastal erosion in the TPI Tempursari beach area has not received proper handling and the TPI Tempursari beach area eroded by erosion is quite extensive and severe. In response to this issue, the most common solutions, such as the construction of seawalls and breakwaters, are costly and can have unforeseen consequences for surrounding vulnerable areas. In recent years, this realisation has prompted coastal planners to adopt an alternative approach to the protection of coastal environments, known as Nature-Based Solutions (NBS). Nature-based solutions (NBS) constitute a category of responsive and beneficial responses that enhance long-term ecosystem management through the utilisation of natural processes. To address such issues, natural ecosystems can be restored, modified, rendered more adaptable or integrated. This study, in particular, aims to identify current solutions to the problem of coastal erosion in TPI coastal area using nature-based methods. To this end, it will compile existing best practices and tools to solve the coastal erosion problem occurring in the TPI coastal area. The result of this study will also contribute to the formulation of strategies for the management of the TPI coastal area, thereby facilitating the resolution of coastal erosion issues.

Keyword: coastal erosion, coastal management, coastal protection, nature-based solutions

1. Introduction

Indonesia is known as an archipelago with more than 17,000 islands and 80,000 km of coastlines. As a link between land and sea, the coast becomes a strategic area to be used to fulfil economic needs and residential land.

The Java Island government is one of the most successful in developing coastal cities in Indonesia. There are four major cities that have succeeded in utilizing coastal areas to improve living standards and economic rates such as industrial development to ports, namely Serang, Jakarta, Semarang, and Surabaya. Despite the fact that the successful development is in the coastal area facing the inner sea of Indonesia. In contrast to the southern coastal areas of Java Island, which are still mostly categorized as rural, coastal areas have become an alternative to reduce the use of forest areas to meet the needs of settlements and plantation land [1].

The beach is subject to change as a result of the conditions that occur in its surrounding environment. This is influenced by both natural processes and human activities in the surrounding area. The majority of these alterations are attributable to the intrinsic dynamics of the beach itself, namely sedimentation and erosion of the coastline [2]. Coastal erosion can be defined as the process of retreating the shoreline from its original position, caused by an imbalance between the supply and transport of sediments from

IOP Conf. Series: Earth and Environmental Science **1473** (2025) 012009 doi:10.1088/1755-1315/1473/1/012009

the coastal. This phenomenon is frequently observed in coastal regions comprising black sand deposits. The erosion of the shoreline in Indonesia can result in submergence of up to 2 to 10 meters per year [3]

Lumajang District is geographically located at 1120 54'-113023' East Longitude and 70-54'.80.23' South Latitude with an area of 1,790.98 km2 and is one of the districts in East Java that has marine and coastal areas on the south coast of Java Island. Lumajang district is characterized by the presence of open ocean waters. The width of these waters has increased as a consequence of the enactment of the Indonesian Exclusive Economic Zone (EEZ), which extends 200 nautical miles from the coastline. Despite the Indonesian Ocean's rich fisheries resources, the level of utilization remains relatively low. The southern coast of Lumajang district extends approximately 75 km from Yosowilangun subdistrict to Tempursari subdistrict. Of these, 12 villages are classified as beach villages. It is anticipated that these beach villages will serve as the primary base for the management of coastal and marine areas. The TPI (Tempat Pelelangan Ikan) Beach in Tempursari, Lumajang, is a land formation comprising volcanic sedimentary material, specifically black sand. This is due to the fact that it is a transported deposit originating from the lava flow path of Mount Semeru, with the process of beach erosion frequently occurring.

Basically, coastal erosion is a natural phenomenon. The occurrence of coastal erosion is caused by the greater amount of coastal sediment escaping to the sea than entering from the source of supply. Beach erosion often occurs on beaches that have sandy or muddy constituent materials because they have a weak drafting force exceeded by the strength of ocean waves [4].

The occurrence of coastal erosion is influenced by two distinct factors: natural factors, including climate change, rising sea levels, and the actions of tides and storms, and human factors, such as the construction of buildings on beaches and river mouths, which disrupt the natural flow of sediment, sand mining on beaches and rivers are also contributing factors [5] Indicators of beach erosion are a concave shape that leads to the mainland, sedimentary rock material and eroded vegetation areas.

Table 1. The Rate and Impact of Erosion at TPI Tempursari Beach from 2012-2020

Year	The Rate and Impact of Erosion at TPI Tempursari Beach from 2012-2020	
	Events	Description
2012	High tides occur in January, March, April and September.	• Ocean waves can reach sea pine areas and plantations. • Shoreline retreat from its original position. • Landslide of bridge connecting TPI Beach settlement with Bulurejo Village.
2013	High tides occur in May, June, August and December.	• The connecting route between villages was eroded. • Deposition at the mouth of the Glidik River resulted in the deflection of the flow path.
2014	High tides occur in January dan April.	• Waves often hit Bulurejo plantations
2015	High tides occur in January dan March	• 2000 pine trees were uprooted and swept out to sea. • Two hectares of corn and one hectare of dragon fruit plantations were eroded away. Four grilled fish stalls near the coast were damaged due to the broken building foundations.
2016	High tides occur in March, June, July, dan September	• Deposition occurs at the estuary of Glidik River • Several houses in the TPI settlement area were ruined due to damaged building foundations and frequent tidal floods • The remaining 5-6 rows of pine trees that previously numbered 25 rows with an average of 4 trees collapsed a day

2018	High tides occur in December	• Several houses, stalls, and musholla collapsed due to damaged building foundations. • The temporary closure of TPI Beach by the police. • It was reported that 300 metres of coastal plain had become ocean
2019	High tides occur in April dan September	• 500 metres of coastal land has become ocean. • Frequent tidal floods have easily damaged 16 hectares of agricultural fields and 8 grilled fish stalls.
2020	High tides occur in May dan September	• Eighty buildings have disappeared into the ocean. • The area of RT 18 has disappeared into the ocean. • All the pine tree vegetation has been destroyed.

Reference: [6], [7], [8]

The data presented in Table 1 above demonstrates that there is a minimum of one instance of a high tidal wave occurring in the sea each year. This process of erosion is occurring at a gradual pace along the TPI Beach coastline. The high intensity of tidal waves is a consequence of the characteristics of the Indian Ocean, which is renowned for its ferocity. In ordinary circumstances, the strength of sea waves is sufficient to erode coastal areas, with the potential to dislodge sand sediments. Despite the distance of his residence from TPI Beach, which is several hundred meters, Since 2012, TPI Beach has frequently been subjected to the impact of high tidal waves.

Erosion of the beach that occurs repeatedly in the coastal area of TPI Beach has not yet received proper mitigation. So far, the coastal erosion disaster in TPI Beach still continues. Several attempts to mitigate coastal erosion at TPI beach have been done, such as building sea walls from sandbags, making breakwaters using stones and iron frames, and relocating victims affected by erosion. However, the problem of coastal erosion in TPI beach still occurs frequently.

a. Construction of a sea wall using sandbags b. The breakwaters using stones and iron frames

Figure 1. Traditional solutions for the mitigation of coastal erosion on the TPI beach

Coastal areas are susceptible to erosion as a result of extreme weather events and sea level rise, which present considerable risks to both human populations and ecosystems. The efficiency of traditional solutions, such as seawalls and breakwaters, is constrained, prompting an investigation into the potential of nature-based solutions (NBS) as a cost-effective and sustainable alternative. The implementation of nature-based solutions (NBS) has been used in several places around the world, such as to address coastal erosion along the Emilia-Romagna coastline in northeast Italy at Bellocchio Beach is a transitional ecosystem (dunes and lagoon) characterized by human activities on land and infrastructure, both exposed to marine flooding.

2. Preliminaries

Before reviewing the nature based solutions (NBS) methods available in the literature, in this section, a brief explanation of the definition of NBS and some types of nature based solutions (NBS) such as e.g. artificial dunes, mangroves, seagrass for coastal erosion in coastal areas is given.

2.1. Nature-based solutions (NBS)

Nature-based solutions (NBS) are innovative and tangible approaches to coastal protection that utilise natural elements to reduce coastal impacts. NBS can be used as a complementary or alternative approach to traditional engineering structures, offering cost-effective and sustainable solutions. These solutions are gaining attention for their potential to provide economic, social and environmental benefits while creating resilience to natural disasters.

In the context of coastal protection, NBS aims to analyse the benefits of ecosystem services and deliver more economic and social benefits. They include a range of strategies that utilise nature to maintain, develop and restore biodiversity and ecosystems, and address multiple issues simultaneously. NBS can be used to protect, manage and restore natural or modified ecosystems, providing ecosystem services that are important for biodiversity and human well-being.

One example of NBS is the integration of artificial sand dunes and seagrass beds, which has been researched for its effectiveness in mitigating coastal impacts. Artificial sand dunes act as a barrier against seawater ingress, while seagrass meadows reduce wave impacts along the coast. The integration of these NBS has been shown to significantly improve coastal inundation mitigation capabilities under extreme conditions, with the potential to reduce flood areas, maximum inundation depths and coastal erosion volumes.

Overall, NBS offers a sustainable and environmentally friendly approach to coastal protection, providing potential co-benefits such as efficiency preservation, habitat development of organisms and vegetation species, and social value in terms of tourism, recreation and research. These solutions are part of a growing effort to address coastal risks using natural principles that are cost-effective and provide social, economic and environmental benefits.

2.2. Seagrass Meadow

Seagrass meadows constitute dynamic ecosystems teeming with biological diversity and play a pivotal role in the sustenance of marine ecosystems. Seagrasses perform the function of natural filtration, trapping sediment and pollutants prior to the occurrence of damage to the marine environment. Seagrasses facilitate the cleansing of marine ecosystems, functioning as living sponges that retain and remove pollutants from the water. However, seagrass meadows have functions beyond serving as a habitat for marine life. The functions and contributions of seagrass meadows to marine ecosystems are multifaceted and significant.

- Carbon capture: Seagrass meadows have the capacity to absorb carbon dioxide, thereby maintaining the integrity and health of the atmosphere while releasing oxygen.
- Water filtration: They serve as natural water purifiers, trapping sediments and pollutants before they can damage coral reefs and other ecosystems..
- Coastal protection: Their dense roots provide a stabilising force against the effects of waves and storms, akin to a green shield.

2.3. Artificial Dune

The persistent degradation of coastal zones caused by unremitting erosion endangers communities and compromises the resilience of natural ecosystems. The deterioration of coastal features, including natural dunes, increases communities' vulnerability to extreme weather events such as storms and rising sea levels. The artificial dune represents an innovative nature-based solution, designed as a passive erosion control system to protect specific coastal areas affected by wave erosion, particularly during storms. The construction of these dunes differs from that of traditional embankments or reinforced soil systems in that it is based on the use of natural materials and technologies, with the objective of reducing the environmental impact. This encompasses the minimization of pollution and the preservation of the natural environment, both during the construction phase itself and in the longer term.

The construction of these dunes seeks to emulate the natural behavior observed in existing, effective dunes, thereby offering protection from wind, waves and tidal currents. The construction of artificial dunes represents an important step towards the provision of a much-needed defense.

The creation of habitats: Dune vegetation is a crucial aspect of biodiversity conservation. The vegetation that flourishes on these sandbars creates a niche for a diverse range of species. From the rapid movements of small mammals to the graceful flight of shorebirds, the dunes support a diverse range of species, creating a vibrant ecosystem.

- Recreational and Aesthetic Value: The stabilisation of dunes provides a safe environment for those who visit the beach, offering spectacular vistas and opportunities for exploration. The creation of artificial dunes serves purposes beyond mere visual appeal and tourism.

The creation of an artificial dune is a process that requires meticulous attention to detail. The initial stage is the procurement of sand, which is then used in the subsequent stages of the process, such as:

- Sand Procurement: The foundation is typically composed of compatible sand, which is often dredged from offshore sources. One might consider this process to be analogous to selecting the appropriate bricks for the construction of a sandy castle.
- The formation of sand dunes: The use of bulldozers and other specialized equipment enables the sand to be shaped into gentle slopes and windswept peaks, thereby creating formations that resemble those found in nature.
- Vegetation Reinforcement: Grasses and shrubs, with their deep roots, are planted in order to hold the sand in place and prevent erosion. It is possible to conceptualize these structures as miniature sand warriors, defending the fort against the wind.

2.4. Mangrove

The mangrove forest represents a vital natural solution to the challenges of climate change and its associated impacts, while simultaneously offering a range of socio-economic and ecological benefits, mangroves provide a multitude of ecosystem services, including provisioning (such as food, timber, and habitat for aquatic and terrestrial species), regulating (such as erosion control and carbon sequestration), cultural (such as recreational and spiritual), and supporting (such as nutrient cycling and soil formation). Furthermore, mangroves can function as natural barriers against storm waves and serve as carbon sinks, thus becoming a crucial instrument in the fight against climate change and coastal erosion [9].

Although mangroves are often regarded as a means of enhancing resilience, for example by acting as natural coastal barriers and contributing to the mitigation of impacts of natural disturbances such as tsunamis and tropical storms, the rise of sea levels presents a significant threat to these ecosystems. This can result in the degradation and loss of mangrove forests, thereby exacerbating the problem. It may be unexpected to learn that the distribution of mangroves can be affected by human activities, including urbanization, cultivation, and aquaculture, as well as climate change, which endangers the capacity of coastal ecosystems to protect communities. This is due to their sensitivity to changes in inundation duration and frequency, as well as salinity levels ([10]).

It is also important to consider the potential negative effects of human activities on mangrove ecosystems and, subsequently, on the resilience of coastal areas. For example, deforestation and pollution can have detrimental impacts on these ecosystems. It is therefore imperative that a superior management strategy is adopted which takes into account the fragile equilibrium between the benefits that mangroves bestow and the detrimental effects that human activities can have on these ecosystems.

3. Result and Discussion

Coastal erosion is a pressing issue that affects coastal areas worldwide, leading to the loss of land, infrastructure, and ecological habitats. To mitigate this problem, nature-based solutions (NBS) have gained significant attention as they offer a sustainable and environmentally-friendly approach to coastal protection. This paper aims to discuss the application of different types of NBS, including sand dunes, mangroves, and seagrasses, in mitigating coastal erosion in TPI beaches in particular.

The following sections provide a detailed and comprehensive survey and analysis of the existing corpus of literature pertaining to nature-based solutions that may be relevant for the purposes of coastal protection.

Table 2. Comprehensive survey of the existing literature about Nature-Based Solutions

Author and Title	Results	Contribution
Vicarelli et al., "On the cost-effectiveness of Nature-based Solutions for reducing disaster risk".[11]	Mangroves, forests, and coastal ecosystems are highly effective for coastal erosion and flooding mitigation, with mangroves being the most frequently effective nature-based solution.	• NBS are cost-effective for disaster risk reduction and climate change adaptation. • NBS are as effective as engineering-based solutions for mitigating hazards.
Madramootoo C, Virgil C., "Nature-based solutions for coastal protection in the southern Caribbean[12]	Nature-based solutions like mangroves and seagrass offer effective coastal protection against erosion in the southern Caribbean, providing ecological, social, and economic benefits surpassing traditional grey engineering methods.	• Nature-based solutions using mangroves and seagrass beds proposed. • Sustainable green-engineered coastal protection strategies for Trinidad and Tobago.
Strypsteen G., "Enhancing coastal resilience with artificial dunes through bio-geomorphological processes[13]	The study focuses on artificial dunes with marram grass, proving highly effective in mitigating erosion. Dunes, especially with vegetation, excel in coastal resilience compared to mangroves and seagrass.	• Artificial dune system effectiveness in mitigating sand-related issues. • Success of 'dune-in-front-of-a-dike' approach in coastal resilience.
Forrester et al., "Seagrass as a Nature-Based Solution for Coastal Protection[14]	Seagrass is an effective nature-based solution for coastal protection, comparable to dunes and mangroves, offering valuable ecosystem services in mitigating coastal erosion.	• Seagrass as nature-based solution for coastal protection. • Importance of seagrass in mitigating coastal erosion and flooding.
Unguendoli et al., "A modelling application of integrated nature based solutions (NBS) for coastal erosion and flooding mitigation in the Emilia-Romagna coastline (Northeast Italy)".[15]	The study models integrated nature-based solutions (NBS) like dunes and mangroves for coastal erosion mitigation in Emilia-Romagna, Italy, showcasing their effectiveness in reducing coastal erosion and flooding.	• Analyses effectiveness of two Nature-Based Solutions (NBS) • Demonstrates synergic effect of integrating NBS

Author and Title	Results	Contribution
Alessandri et al., "Investigating Seagrass as a Nature-Based Solution for Coastal Protection: toward a Digital Twin Modelling Framework". [16]	Seagrass meadows are effective in reducing bottom current velocity by up to 50-60%, potentially playing a significant role in mitigating coastal erosion compared to other nature-based solutions.	• Investigated seagrass effects on sediment transport and sea level • Seagrass reduces bottom current velocity by up to 50-60%
Toledo et al., "Nature-based solutions on the coast in face of climate change: The case of Benidorm (Spain)". [17]	The paper focuses on a vegetated urban dune as a nature-based solution for coastal erosion in Benidorm, Spain, showcasing its effectiveness in enhancing coastal resilience against climate change impacts.	• Implementing a vegetated urban dune to protect from coastal erosion and floods. • Recommendations for climate adaptation plans using nature-based solutions.
Jacob et al., "Evaluation of seagrass as a nature-based solution for coastal protection in the German Wadden Sea"[18]	Seagrass meadows can reduce current velocities and wave heights by up to 30% and 90% respectively, effectively lowering sediment mobilization and erosion risks, making them a valuable coastal protection solution.	• Seagrass reduces current velocities and wave heights for coastal protection. • Seagrass expansion helps maintain bathymetry-induced tidal dampening for protection.

Comparing the effectiveness of these different types of NBS requires considering various factors such as their ability to reduce wave energy, their capacity for sediment trapping, their adaptability to different coastal environments, and their potential for long-term sustainability. Dunes, mangroves, and seagrass meadows all offer unique advantages and can be effective in specific coastal contexts.

In terms of wave energy reduction, dunes have been extensively studied and proven to be highly effective. The gradual slope of dunes allows for efficient wave dissipation, reducing the erosive forces on the coastline. Mangroves, with their dense root systems and complex above-ground structure, also provide significant wave energy reduction. However, the effectiveness of seagrass meadows in wave energy reduction is less well-documented and requires further research.

Sediment trapping is another critical aspect of NBS effectiveness. Dunes, mangroves, and seagrass meadows all have the capacity to trap sediments, contributing to the formation of new land. Dunes, with their ability to accumulate sand and stabilize sediments, are particularly effective in this regard. Mangroves, through their intricate root systems and sediment trapping capacity, also play a significant role in sedimentation. Seagrass meadows, although less studied, have been shown to trap sediments and contribute to land formation in specific coastal areas.

3.1. Seagrass's effect on coastal area protection

Seagrass meadows are one form of nature-based solution (NBS) that has been studied for its effectiveness in mitigating coastal impacts, particularly in coastal areas. (NBS) solutions that have been studied for their effectiveness in mitigating coastal impacts, specifically in reducing erosion during extreme storm events. Seagrass meadows serve as vital buffers against storm-driven erosion. Optimizing their location and density can lead to an erosion reduction of up to 80%. Their effectiveness is contingent on specific environmental conditions, emphasizing the need for tailored restoration strategies[18]. Seagrass meadows meadows are submerged coastal vegetation that provide protection by attenuating and dissipating waves. Based on the study conducted by Forrester et al. [19] seagrass plays a significant role in protecting coastal areas, and its effects can be summarized as follows:

- Wave Energy Reduction: Seagrass beds help to dissipate wave energy, which reduces the impact of waves on shorelines. This natural barrier can prevent erosion and protect coastal habitats from storm surges and high-energy waves

- Sediment Stabilization: The roots and rhizomes of seagrass anchor sediments in place, which helps to stabilize the seabed. This stabilization is crucial in preventing sediment from being washed away, thereby maintaining the integrity of coastal ecosystems
- Biodiversity Support: Seagrass meadows provide habitat for a variety of marine species, including fish and invertebrates. By supporting biodiversity, seagrass contributes to the overall health of coastal ecosystems, which can enhance their resilience against environmental changes and human impacts

Seagrass meadows can reduce wave intensity, with a maximum reduction of 89% and an average attenuation of 32% during storms. Overall, vegetation was able to effectively produce an average attenuation of 32% at the peak of the storm. Wave reduction resulted in up to 37% mitigation of lagoon flooding compared to the scenario without coastal protection. Seagrass meadows were also shown to be effective in reducing flood extent and maximum inundation depth, with reductions ranging from 14% to 58%. In addition, seagrass meadows can help prevent coastal erosion, with the potential to reduce erosion up to 30 meters from shore. In addition, the presence of seagrass meadows can lead to sediment deposition on the water bottom, caused by the presence of localized vegetation that tends to trap sand. This accumulation of sediment can contribute to the stability and function of coastal dunes over time. Seagrass meadows are also notable for their potential social value, providing benefits for tourism, recreation, education and research. Overall, the integrated application of seagrass meadows as NBS is highly recommended to mitigate coastal risks and support sustainable coastal management[15].

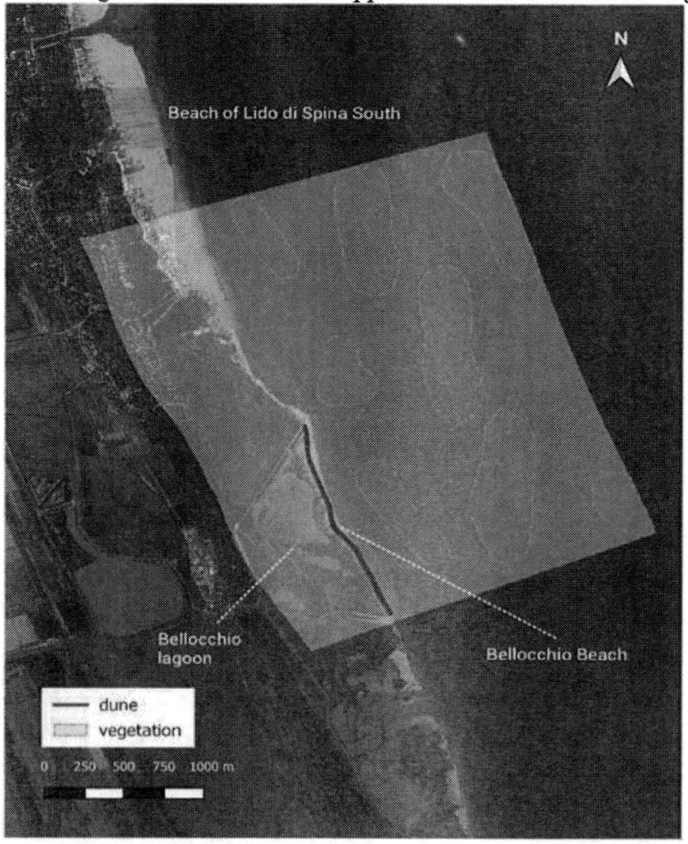

Figure 2. Seagrass meadow belt and dune position along the coast of Bellocchio Beach [15]

3.2. *Artificial dune's effect on coastal area protection*

The artificial dune represents an innovative nature-based solution, comprising a passive erosion control system designed to safeguard specific coastal areas vulnerable to wave erosion, particularly during storms. In contrast to traditional embankments or reinforced systems, these artificial sand dunes are constructed using natural materials and technologies with the objective of minimizing environmental impacts, including pollution and destruction of the natural environment, even during the construction phase. The objective of these dunes is to emulate the natural behavior of pre-existing natural dunes, and they are indeed effective at mimicking the natural behavior of such dunes. This provides protection against wind, waves and tidal currents.

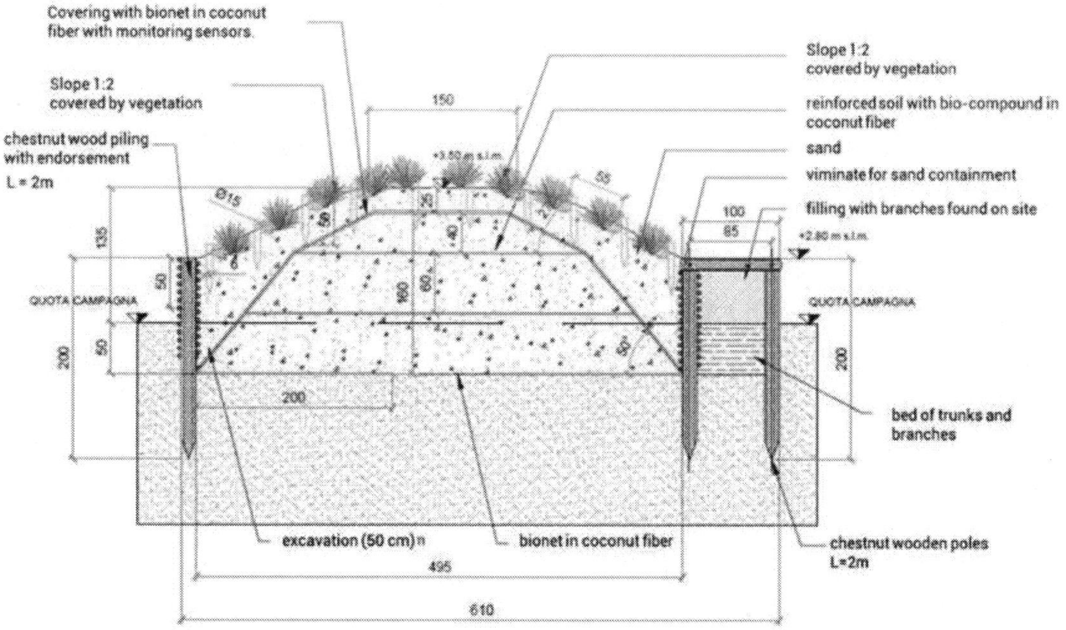

Figure 3. Detail of reinforced bio-soil system [15]

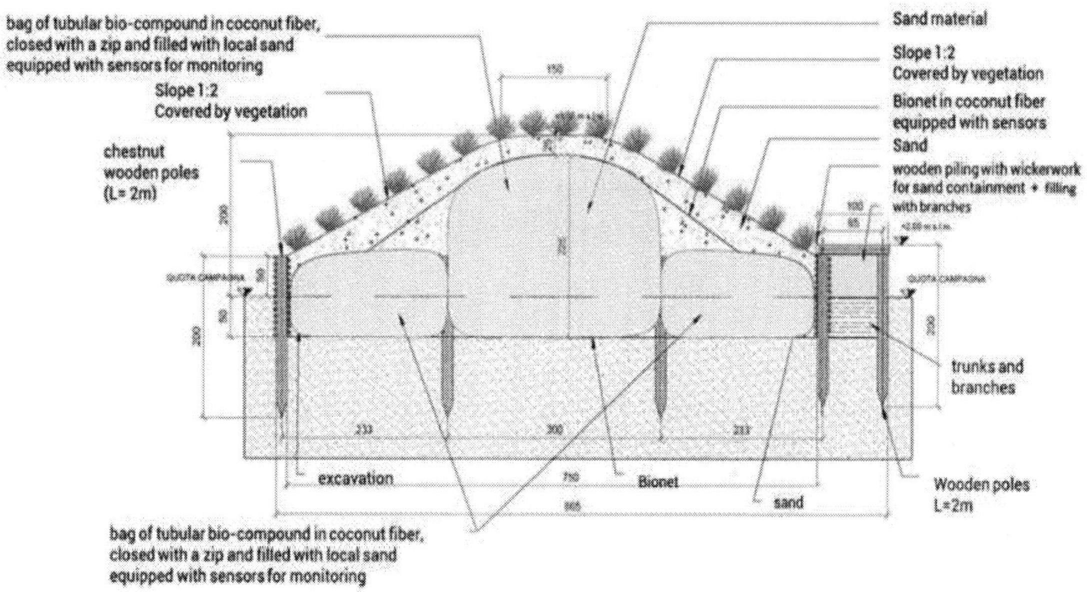

Figure 4. Detail of modular tubular system[15]

The construction of these artificial dunes involved the utilisation of a reinforced bio-soil system and an innovative modular tubular system. Both systems employed flexible and semi-permeable natural materials, including geotextiles, coir, and wood. The dunes are covered with sand, with an average thickness of 20-30 cm, and are supported by wooden posts. These act as engineered structures that mimic the function of natural dunes, with the objective of reducing erosion and beach erosion and seawater ingress into the lagoon[15].

The study found that the dunes were able to significantly reduce coastal erosion during extreme storm events. It is particularly effective in reducing the impact of water runoff, providing considerable elevation to the beach, and limiting erosion inland.

The innovative modular tubular structure of artificial sand dunes allows adaptability to different beach lengths and shorelines, making them a versatile and sustainable solution for coastal protection. This study highlights the potential benefits of integrating artificial dunes with other nature-based solutions, and emphasizes their role in supporting habitats for coastal organisms, flora and species.

Figure 5. Dune layout map & Coir blanket with runoff system control[15]

Overall, artificial dunes represent a promising and environmentally friendly coastal protection approach, offering significant improvements in the mitigation of storm impacts on beaches and supporting coastal management with low economic and social impacts.

3.3. Mangroves on coastal area protection

The importance of mangroves in safeguarding coastal regions against a variety of environmental hazards cannot be overstated. In accordance with the findings presented in the paper conducted by Ayassamy (2024) the following section will delineate the principal effects of mangroves on coastal protection:

- Natural Barriers: Mangroves function as natural barriers against storm surges and high waves. The dense root systems of mangroves serve to stabilize the shoreline, thereby reducing the impact of strong winds and water currents, which can otherwise lead to erosion and the loss of habitats.
- Erosion control: The roots of mangrove trees play an important role in binding the soil, thereby preventing erosion. This is of particular significance in coastal regions, where the land is susceptible to erosion caused by tidal and storm activity. By maintaining soil integrity, mangroves contribute to the overall stability of the coastal ecosystem.
- Biodiversity support: Mangroves provide a habitat for a diverse range of marine and terrestrial species. This biodiversity is crucial for maintaining healthy ecosystems that are better equipped to withstand environmental changes and pressures. A diverse ecosystem is more resilient to disturbances, which enhances coastal protection.
- Carbon sequestration: Mangroves are effective at sequestering carbon, thereby contributing to the mitigation of climate change. By absorbing carbon dioxide from the atmosphere, they contribute to

reducing the overall impact of climate change on coastal areas, which can result in more severe weather events and rising sea levels.

- Improved Water Quality: Mangroves filter pollutants and enhance water quality in coastal areas. By trapping sediments and absorbing excess nutrients, they contribute to the maintenance of a healthier marine environment, which is vital for the survival of fish and other aquatic life that contribute to coastal resilience.
- Community Benefits: Healthy mangrove ecosystems provide resources for local communities, such as fish and timber. This economic benefit encourages conservation efforts and sustainable management practices, further enhancing the resilience of coastal ecosystems.

In conclusion, mangroves represent a multifaceted solution for coastal protection, offering physical barriers, supporting biodiversity, improving water quality, and providing economic benefits to communities. Their role is essential in building resilience against environmental challenges faced by coastal regions. Mangroves also are recognized for their cost-effectiveness, with 80% of studies indicating their efficacy in disaster risk reduction[11]

4. Conclusion and Recommendation

4.1. Conclusion

Nature-based solutions (NBS) for coastal erosion, including mangroves, seagrass, and dunes, demonstrate varying effectiveness based on their ecological characteristics and implementation contexts. Mangroves and seagrass beds are highlighted for their dual role in coastal protection and ecological benefits [12].

However, their effectiveness is often limited by anthropogenic pressures and environmental factors, necessitating improved management strategies to enhance their coverage and resilience [12][20]. In contrast, artificial dunes, particularly those enhanced with vegetation like marram grass, have shown significant success in sediment accumulation and erosion mitigation. Studies indicate that these dunes can achieve substantial height increases and recover quickly from storm impacts, outperforming adjacent unvegetated areas [13]. While both mangroves/seagrass and dunes offer valuable coastal protection, their effectiveness can be context-dependent, with dunes providing immediate structural benefits and mangroves/seagrass contributing to long-term ecological stability [13] [14]. Thus, integrating these solutions may yield the most comprehensive coastal resilience strategy.

Although artificial dunes have been observed to be effective in limiting landward coastal erosion and completely avoiding morphological variations behind them, seagrass meadows have been found to be more effective in reducing the impact of storms on the foreshore in terms of erosion and landward shoreline movement. seagrass meadows are quite efficient in damping the intensity of incident waves, in terms of significant wave height, with a maximum damping of 89 % and an average damping at the storm peak of 32 % [15]

In conclusion, dunes, mangroves, and seagrass meadows all offer effective nature-based solutions for coastal erosion control. Dunes are highly versatile and proven to be effective in reducing wave energy and stabilizing sediments. Mangroves provide excellent wave energy reduction and sediment trapping capabilities, but are limited to tropical and subtropical regions. Seagrass meadows, have shown potential in reducing wave energy and promoting sedimentation. The choice of the most suitable NBS should consider the specific coastal context, including wave energy, sediment dynamics, environmental conditions, and long-term sustainability. To mitigate the impact of coastal erosion, the use of nature-based solutions (NBS) method is suggested to be applied to TPI Tempursari Beach Lumajang, especially how the integration of the two can result in significant improvements in addressing coastal impacts. Further research and monitoring are necessary to enhance our understanding of the effectiveness of different NBS and their interactions with coastal processes.

Nature-based solutions (NBS) such as mangroves, seagrass, and dunes play a crucial role in mitigating coastal erosion and flooding. Each type of NBS offers unique benefits and effectiveness, influenced by ecological and geographical factors.

4.2. Recommendation
Dunes, mangroves, and seagrass meadows all offer effective nature-based solutions for coastal erosion control. Dunes are highly versatile and proven to be effective in reducing wave energy and stabilizing sediments. Mangroves provide excellent wave energy reduction and sediment trapping capabilities, but are limited to tropical and subtropical regions. Seagrass meadows, have shown potential in reducing wave energy and promoting sedimentation. The integration of various NBS, including dunes, mangroves, and seagrass can enhance coastal protection strategies [12][15].To mitigate the impacts of coastal erosion at Tempursari beach in Lumajang, the use of nature-based solutions (NBS) methods of either dunes, mangroves or seagrass meadows are proposed to be applied, especially how the integration of the three can result in significant improvements in addressing the impacts on the beach at Tempursari beach.

References
[1] A. Wibowo and D. Supriatna, "Kerentanan Lingkungan Pantai Kota Pesisir Di Indonesia Coastal Environmental Vulnerability on Coastal Cities In Indonesia." [Online]. Available: http://www.itk.fpik.ipb.ac.id/ej_itkt32
[2] D. M. Alongi, "Mangrove forests: Resilience, protection from tsunamis, and responses to global climate change," Jan. 01, 2008. doi: 10.1016/j.ecss.2007.08.024.
[3] Baharuddin Koddeng, "Zonasi Kawasan Pesisir Pantai Makassar Berbasis Mitigasi Bencana," *repository.unhas.ac.id:123456789/2429*, 2011.
[4] Y. Muliati, *Rekayasa Pantai Oleh*.
[5] M. A. Marfai, L. King, J. Sartohadi, S. Sudrajat, S. R. Budiani, and F. Yulianto, "The impact of tidal flooding on a coastal community in Semarang, Indonesia," *Environmentalist*, vol. 28, no. 3, pp. 237–248, Sep. 2008, doi: 10.1007/s10669-007-9134-4.
[6] Yayasan Tempursari Conservation Center, "Peta Zona Penanam Mangrove Lumajang," 2022.
[7] Lumajang Satu, "Wisata Pantai TPI Tempursari Lumajang akan jadi kenangan," *https://lumajangsatu.com/baca/wisata-pantai-tpi-tempursari-lumajang-akan-jadi-kenangan*, p. b, 2020.
[8] BPBD Kabupaten Lumajang, "Data abrasi dan gelombang tinggi Kabupaten Lumajang," 2020.
[9] M. Doelle and T. G. Puthucherril, "Nature-based solutions to sea level rise and other climate change impacts on oceanic and coastal environments: a law and policy perspective," *Nord J Bot*, vol. 2023, no. 1, Jan. 2023, doi: 10.1111/njb.03051.
[10] D. A. Friess *et al.*, "Are all intertidal wetlands naturally created equal? Bottlenecks, thresholds and knowledge gaps to mangrove and saltmarsh ecosystems," *Biological Reviews*, vol. 87, no. 2, pp. 346–366, May 2012, doi: 10.1111/j.1469-185X.2011.00198.x.
[11] M. Vicarelli *et al.*, "On the cost-effectiveness of Nature-based Solutions for reducing disaster risk," *Science of The Total Environment*, vol. 947, no. 174524, Oct. 2024.
[12] C. A. Madramootoo and C. Virgil, "Nature-based solutions for coastal protection in the southern Caribbean," *Blue-Green Systems*, vol. 6, no. 2, pp. 185–197, Dec. 2024, doi: 10.2166/bgs.2024.013.
[13] G. Strypsteen, "Enhancing coastal resilience with artificial dunes through bio-geomorphological processes", doi: 10.5194/egusphere-egu24-9730.
[14] J. Forrester, N. Leonardi, J. R. Cooper, and P. Kumar, "Seagrass as a nature-based solution for coastal protection," *Ecol Eng*, vol. 206, Sep. 2024, doi: 10.1016/j.ecoleng.2024.107316.
[15] S. Unguendoli, L. G. Biolchi, M. Aguzzi, U. P. A. Pillai, J. Alessandri, and A. Valentini, "A modeling application of integrated nature based solutions (NBS) for coastal erosion and flooding mitigation in the Emilia-Romagna coastline (Northeast Italy)," *Science of the Total Environment*, vol. 867, Apr. 2023, doi: 10.1016/j.scitotenv.2022.161357.
[16] J. Alessandri *et al.*, "Investigating Seagrass as a Nature-Based Solution for Coastal Protection: toward a Digital Twin Modelling Framework", doi: 10.5194/egusphere-egu24-11402.

[17] I. Toledo, J. I. Pagán, I. López, L. Aragonés, and J. Olcina, "Nature-based solutions on the coast in face of climate change: The case of Benidorm (Spain)," *Urban Clim*, vol. 53, Jan. 2024, doi: 10.1016/j.uclim.2024.101816.

[18] B. Jacob, T. Dolch, A. Wurpts, and J. Staneva, "Evaluation of seagrass as a nature-based solution for coastal protection in the German Wadden Sea," *Ocean Dyn*, vol. 73, no. 11, pp. 699–727, Nov. 2023, doi: 10.1007/s10236-023-01577-5.

[19] J. Forrester, N. Leonardi, J. R. Cooper, and P. Kumar, "Seagrass as a nature-based solution for coastal protection," *Ecol Eng*, vol. 206, Sep. 2024, doi: 10.1016/j.ecoleng.2024.107316.

[20] "Mangroves as a Nature-based Solution and a Tool for Coastal Resilience: Four Approaches to Respond to Five Impacted Factors." [Online]. Available: https://ssrn.com/abstract=4767297

IoT Based Early Monitoring System for Air Quality Assessment

A Maulidi[1,4,5*], K Sambodho[2,4], A A B Dinariyana[3,4], M Syafruddin[5]

[1] Doctoral Program of Ocean Engineering, ITS Surabaya, Indonesia
[2] Department of Ocean Engineering, ITS Surabaya, Indonesia
[3] Department of Marine Engineering, ITS Surabaya, Indonesia
[4] The Center for Excellence of Maritime Safety and Marine Installation, ITS Surabaya, Indonesia
[5] Politeknik Negeri Madura, Indonesia

* E-mail: amd.2201@gmail.com

Abstract. In the last decade, climate change has become a pressing global issue. It is triggering the need for sustainable solutions to achieve net zero emissions by 2050. However, air pollution in maritime zones is often overlooked despite its significant impact on the environment. This commitment will be implemented through the implementation of green ports throughout Indonesia. After the Covid-19 pandemic, economic activities have begun to increase again, the number of ships arriving or crossing Indonesian sea lanes tends to increase. This needs attention because the increase in shipping activities will be directly proportional to the consumption of fuel oil which will later increase ship exhaust emissions such as COx, NOx, SOx. This paper introduces an IoT-based early detection system for assessing marine air pollution problems. The system is equipped with smart sensors to detect COx, NOx, SOx from vessel. The emission data is collected and analysed in real-time to identify their pollution level. This innovation is expected as the basis as an effort to reduce emissions in the maritime sector, in line with the net zero emission target of 2050.

1. Introduction

Air pollution has become one of the most pressing issues frequently discussed on the global stage today. The World Health Organization (WHO), as an institution that manages world health, notes that every year there are 4.2 million cases of death caused by outdoor air pollution [1]. Numerous efforts have been made to reduce air pollution on land, but marine air pollution often gets overlooked [2]. Yet, maritime zones also play a significant role in contributing to greenhouse gas emissions, especially from ships that still rely on fossil fuels [3], [4]. The burning of fossil fuels and human activities have led to a continuous upward trend in the levels of three greenhouse gases COx, NOx, and SOx in the atmosphere, with carbon dioxide and other greenhouse gases showing consistent increases over several decades in air sampling networks spread across the globe [5].

Air pollution has a real negative impact on human health and climate change that is increasingly felt by humans. Innovation and technology continue to be developed to detect and reduce pollution resulting from the combustion engine process. Including Internet of Thigs (IoT) technology for monitoring and controlling this problem [6][7]. IoT technology can gather actual data from many sensors and creating environmental monitoring systems and evaluation more effective [8].

In the maritime world, IoT-based air pollution detection tools can accelerate global efforts to emissions limitation and reach net zero emissions by 2050 [9]. The strictest emission limits for ships are in the Americas and Europe, with a maximum of 0.10% m/m for SOx gases and 2.0 g/kWh for NOx gases if $n \geq 2000$ rpm. These limits apply to areas within 400 km of the coastline, where many people live in coastal regions [10]. The collaboration between pollution detection technology and IoT is expected to provide accurate, real-time information about the sources, types, and levels of pollution at sea, enabling maritime stakeholders to take swift and effective action to address pollution sources [6].

In this study, there will be no design of a ship model for detecting air pollution at sea. Instead, the focus will be on designing a tool to detect air pollution at sea caused by ship engine exhaust emissions, specifically SOx, COx, and NOx. These four emissions will be the primary focus for detection by the three sensors to be developed in this research [9]. The testing of this tool will take place in the sailing area between Kamal Port and Ujung Surabaya Port. Air pollution at sea is often overlooked, even though the maritime transportation is one of the contributors to global emissions. By utilizing IoT in the air detection system, this tool can provide real-time data on air pollution at sea [11].

2. Methods

2.1 System Functional Diagram

Functional diagram can be seen Figure 1. Is steps for designing this tool are illustrated in the system functional from right to left.

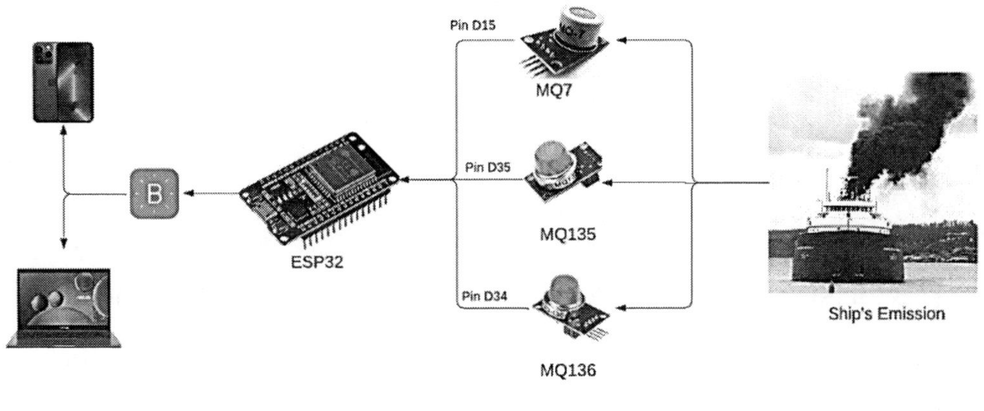

Figure 1. System Functional Diagram

Based on the system workflow diagram above, we can understand that the ship's sailing area exposed to engine exhaust emissions contains several pollutants, including SOx, NOx, and COx. When the sensor detects these gases, the detection data is sent to the ESP32, which then forwards it to Blynk. The data can then be monitored via smartphone and PC, allowing users to track pollution levels in that environment [12].

2.2 MARPOL Regulation Annex VI

Ship emissions need to be controlled to prevent and reduce the contamination of NOx, SOx, and other carbon gases, as these emissions significantly impact the health of living beings and ecosystems. For a ship to operate, it must comply with many regulations that ensure safety and

IOP Conf. Series: Earth and Environmental Science **1473** (2025) 012010 doi:10.1088/1755-1315/1473/1/012010

protect the working environment. One of the new regulations addressing this issue is MARPOL Annex VI, Section 18, which focuses on fuel quality and combustion byproducts. Fuel types that comply with this regulation include LSFO, Methanol, Hydrogen, Marine Gas Oil, and LNG [3]. Table 1 and table 2 below are the emission limits for ships as regulated by MARPOL Annex VI:

Table 1. The Emission Limit for SOx Outside Emission Control Areas (non ECAs)

Range	Year
4,50% $^{m}/_{m}$	Before the beginning of 2012
3,50% $^{m}/_{m}$	After the beginning of 2012
0,50% $^{m}/_{m}$	Entering the beginning of 2020

North America and Europe have stricter regulations known as Emission Control Areas (ECAs). Table 2 is provisions emission limit for SOx Inside Emission Control Areas (ECAs).

Table 2. The Emission Limit for SOx Inside Emission Control Areas (ECAs)

Range	Year
1,50% $^{m}/_{m}$	End of June 2010
1, 00% $^{m}/_{m}$	Early July 2010
0,10% $^{m}/_{m}$	After the Beginning of 2015

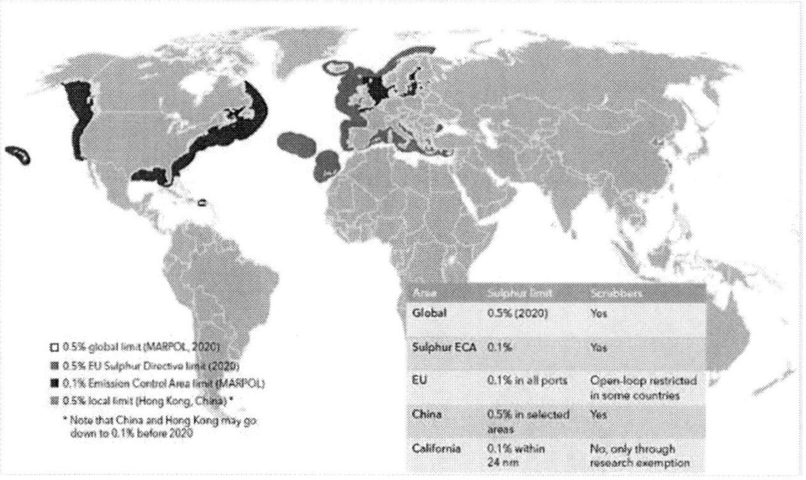

Figure 2. IMO Control Areas and non-IMO areas for SOx gases

Figure 2 illustrates the boundaries for SOx gas emissions released into the atmosphere by ships, divided into two areas. The first area, with a SOx limit of 0.1%, applies only to ports in North America and Europe. The second area, with a global SOx limit of 0.5%, includes ports in China as well [13].

In addition to SOx gases, NOx gases are also regulated under MARPOL Annex VI, which governs the use of marine diesel engines on ships, as outlined in table 3.

Table 3. The NOx Emission Limits for Marine Diesel Engines, as Set in 2016 [3]

Range	Rotation per Minute
3,4 g/kWh	< 130
9.n (-0,2) g/kWh	$130 > 2000$
2,0 g/kWh	≥ 2000

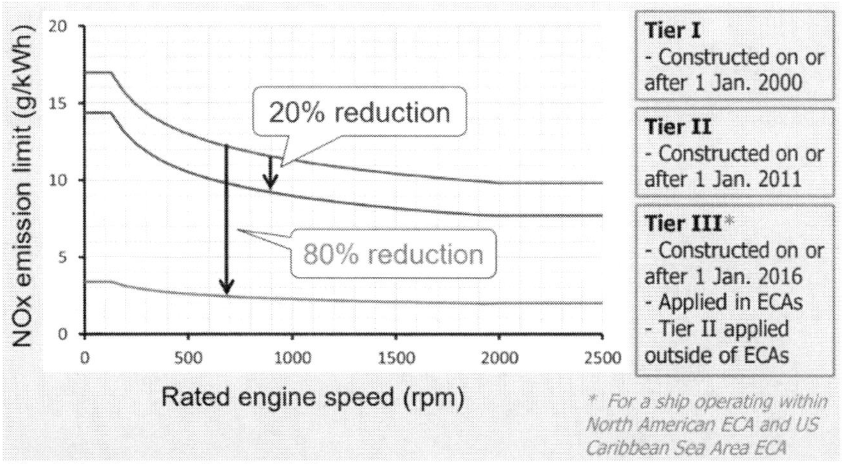

Figure 3. The NOx Emission Limits [14]

Figure 3 shows a graph depicting the reduction in emission limits: a 20% decrease from Tier I to Tier II, and an 80% decrease from Tier I to Tier III.

2.3 Research Location

The study was conducted at Ujung-Kamal port, data collection on March 30-31, 2024. In addition, this area is close to Tanjung Perak port. Approaching the Eid holiday, Ujung-Kamal crossing activities and ship traffic at Tanjung Perak Port have increased [15]. So, it is suitable for collecting COx, SOx, and NOx emission data. Data collection was carried out in 3 different places during the day when the ferry was sailing across the Madura Strait from Kamal Port to Ujung Surabaya Port as seen in figure 4 below. The measurement was carried out on the ship funnel, ship deck and berth area.

The crossing of Madura Strait takes approximately 30 minutes, so data is collected several minutes from centre of the strait to Ujung Port. This is taken because the highest traffic from commercial ships is concentrated in the middle of the strait and the Ujung Port area. Previously, an initial calibration by comparing the measurement results of the instrument with the air quality measuring instrument in the Laboratory of Environment Measurement Politeknik Negeri Madura, a deviation value of -2.1% to +3.6% was sensor.

Figure 4. Research Location di Selat Madura

2.4 IoT Design
In this design, an IoT experiment using Blynk is conducted. The MQ-7 as a Cox sensor, MQ-135 as a SOx sensor, and MQ-136 as NOx sensor and the ESP32 processes the data [16]. The sensor readings are displayed on a smartphone screen through the Blynk app, which must be installed first [17]. This app allows users to monitor air quality remotely. The design diagram is shown in the figure 5 below [18].

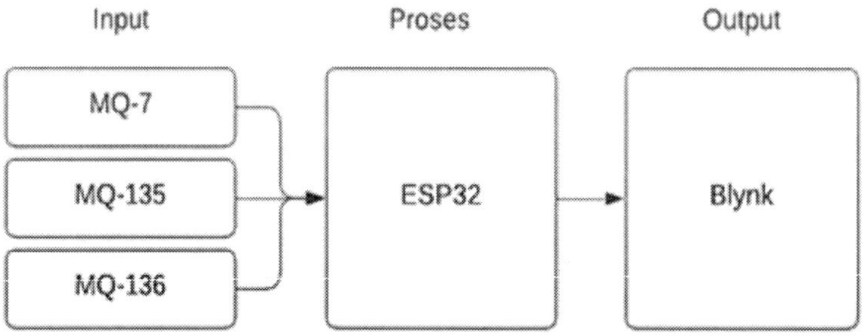

Figure 5. IoT Design Diagram

Based on the diagram above, it can be explained that the pollutant data read by the three sensors is sent to the ESP32. The ESP32 then transmits this data to Blynk, where it is displayed as notifications [19].

Table 4. Widgets Used on Blynk

Widgets	Widget Types	Explanation
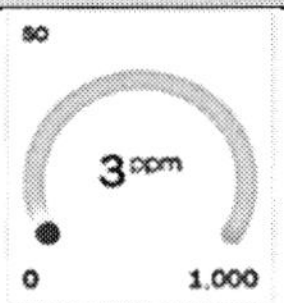	Gauge settings	Input V0 Output: 0-1000 ppm MQ-136
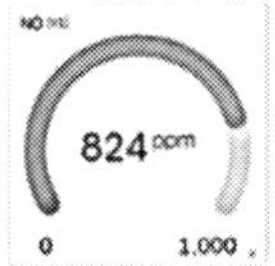	Gauge settings	Input V1 Output: 0-1000 ppm MQ-135
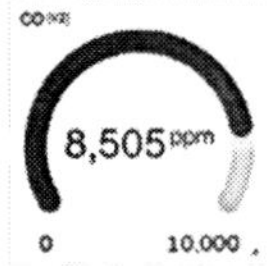	Gauge settings	Input V2 Output: 0-10000 ppm MQ-7

Table 4 above is a widget used to view the sensor readings, which have been configured with virtual inputs and units in ppm (parts per million) in the data stream and web dashboard settings of the Blynk app.

3. Result and Discussion

3.1 Hardware Design Results
The next step is to conduct a test to evaluate function and the sensitivity sensor. MQ-7 for a COx sensor, MQ-135 for a SOx sensor, and MQ-136 for NOx sensor in reading data. The data will be

displayed on the Arduino IDE serial monitor and can also be monitored using Blynk, which will show on both computer and smartphone screens. The tool's circuit is shown in figure 6 below, where the circuit, powered by 3.3 volts, is organized within a box to make it more streamlined and tidy [20].

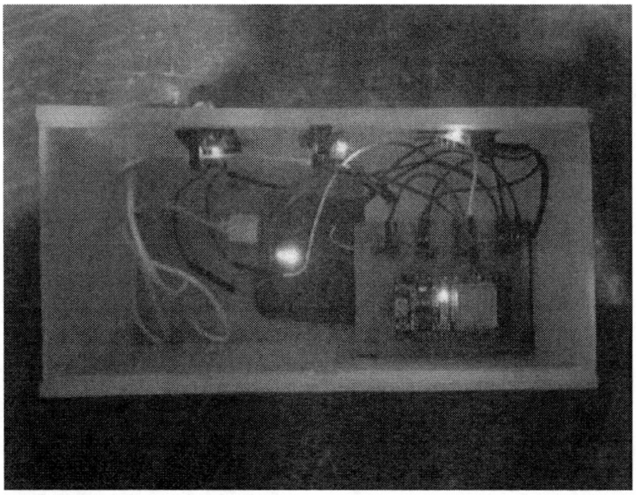

Figure 6. The Circuit of the Tool Inside the Box

3.2 IoT Design Results (Blynk)

The IoT design using Blynk is displayed in two modes: via smartphone and computer. The data captured by the sensors is shown on Blynk figure 7 and enabling remote monitoring figure 8 [21].

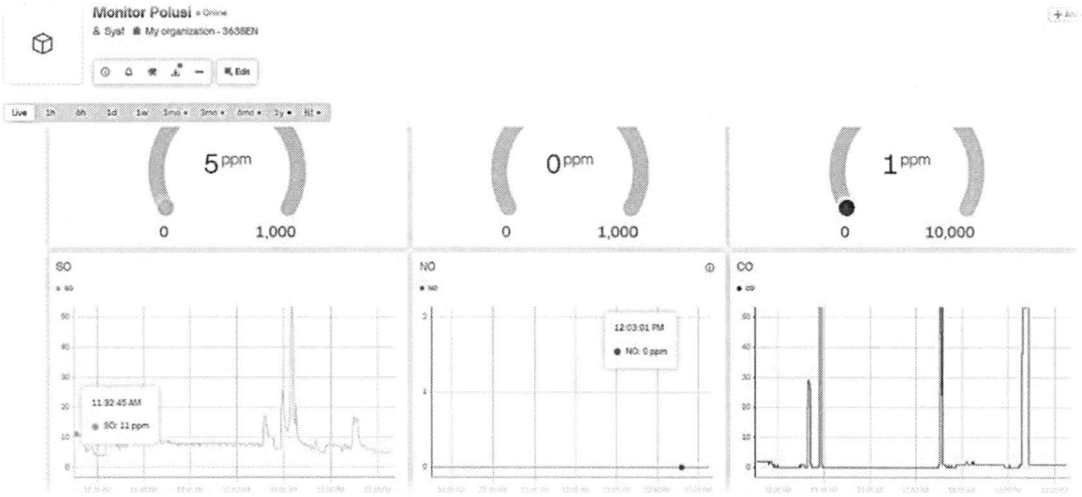

Figure 7. The IoT Display on the Computer

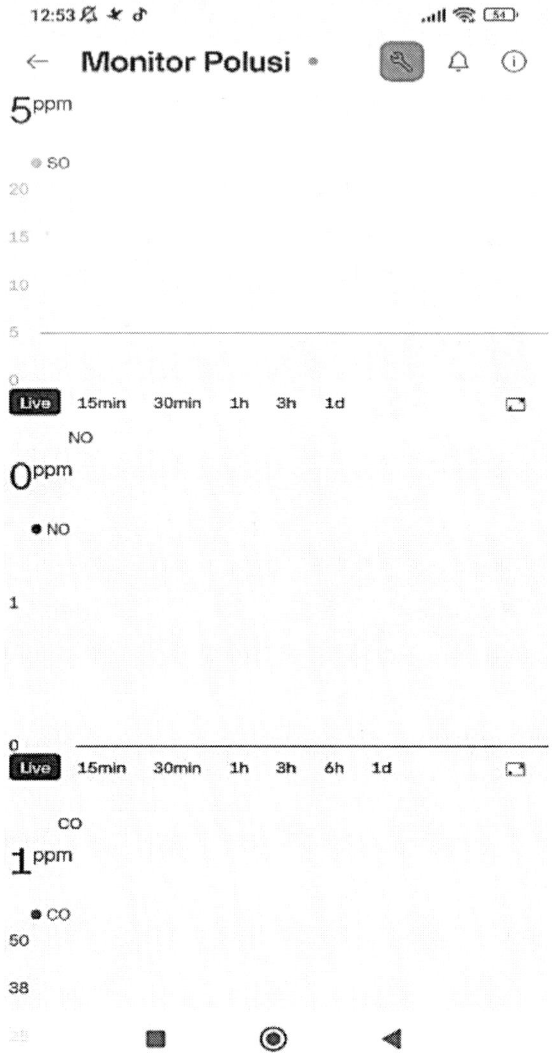

Figure 8. The IoT Display on the Smartphone

3.3 Test Results on the Ship's Funnel

In addition, for testing pollutants along the shipping route and at the port area, it is also crucial to test at the ship's funnel, where pollutants from diesel engine combustion are expelled. Data collection was carried out for 20 minutes, with results recorded every 2 minutes. Table 5 is data collection pollutant at the Funnel.

Table 5. Results of Testing at the Funnel (in PPM)

Time (minute)	COx	NOx	SOx
2	50,98	3,33	13,76
4	54,98	3,44	20,76
6	53,87	3,33	20,65
8	60,43	3,56	29,09
10	65,76	4,76	31,97
12	69,65	5,76	44,09

Time (minute)	COx	NOx	SOx
14	73,65	4,09	49,76
16	70,12	5,74	52,31
18	74,54	6,54	65,87
20	77,73	4,44	54,64

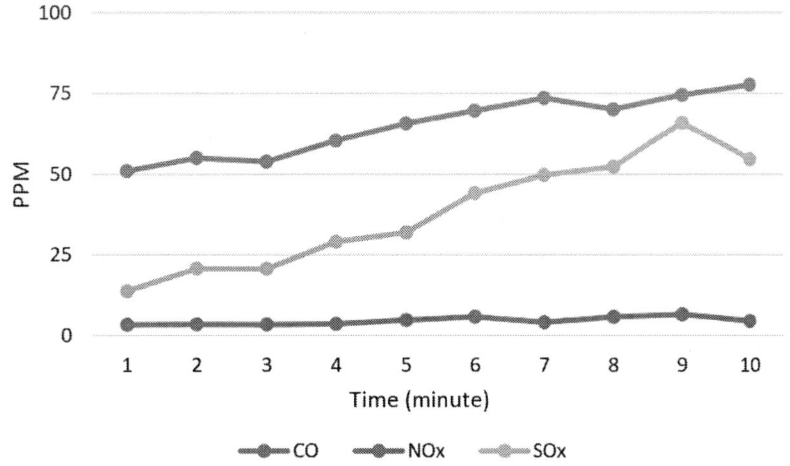

Figure 9. Pollutant Data Graph at the Funnel

Figure 9 shows pollutant data for COx, SOx, and NOx in the ship's funnel in graphical form. Compared to the two previous tests, the pollutant levels in the funnel are significantly higher, with CO ranging from 50.98 ppm to 77.73 ppm, NOx from 3.33 ppm to 6.54 ppm, and SOx from 13.76 ppm to 65.87 ppm. According to the Air Pollution Index, CO levels between 50 ppm and 77 ppm can cause heart problems, difficulty breathing, and headaches. NOx levels between 3 ppm and 6 ppm can lead to reduced lung diffusion capacity. SOx levels between 13 ppm and 65 ppm can cause respiratory irritation, decreased lung function, and environmental impacts such as acid rain and ocean acidification, which can harm ecosystems [22].

3.4 Test Results on the Ship's Deck

The tool test was conducted to monitor sea pollution where ships operate. Data collection took place from Kamal Port, Bangkalan, to Ujung Port, Surabaya, using a ferry. Pollution data was collected on the open deck of the ferry along the Kamal-Surabaya route. Table 6 is data collection was performed for 15 minutes while the ferry was en route to Surabaya. Documentation of the data collection is shown in figure 10.

IOP Conf. Series: Earth and Environmental Science **1473** (2025) 012010 doi:10.1088/1755-1315/1473/1/012010

Figure 10. Collect pollution Data on the Ship's Deck

Table 6. Pollution Monitoring Data Results (in PPM)

Time (minute)	COx	NOx	SOx
1	2.21	0.22	11.23
2	2.03	0.15	10.54
3	2.14	0.11	7.75
4	3.33	0.21	12.33
5	3.27	0.21	12.68
6	2.96	0.13	9.6
7	2.88	0.15	9.23
8	3.16	0.18	10.14
9	3.15	0.21	10.26
10	3.17	0.23	10.84
11	3.27	0.24	10.46
12	3.28	0.23	10.50
13	3.32	0.24	10.59
14	3.15	0. 21	10.36
15	3.93	0.17	10.26

Figure 11 is the pollution monitoring results on the Kamal-Surabaya shipping route over 15 minutes showed CO levels ranging from 2.03 to 3.93 ppm, NOx levels from 0.11 to 0.24 ppm, and SOx levels from 9.23 to 12.68 ppm. CO levels between 2.03 and 3.39 ppm are considered safe, as they do not cause harm to the body. NOx levels between 0.11 and 0.24 ppm can produce Odors and may lead to increased resistance in the air. SOx levels between 9.23 and 12.68 ppm are at the minimum tolerated limit, with potential effects on the body including throat irritation. The following graph illustrates the trend of pollutant levels increasing or decreasing each minute.

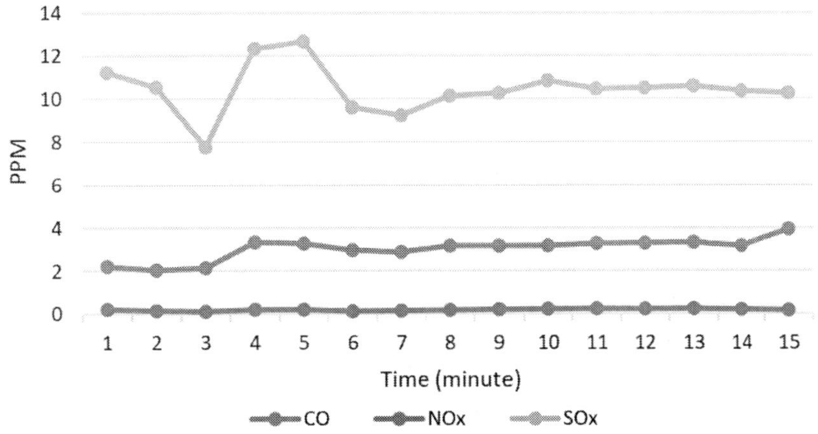

Figure 11. Pollutant Volume Graph on the Ship Deck

3.5 Results of Experiments in the Port Area

The next pollution data collection was conducted in the Kamal-Surabaya port area. In addition to the shipping route, the port area is also affected by pollution from commercial ships navigating the Madura Strait. Factors such as sea winds can carry pollutants from ships at sea to the land or around the port. Table 7 is data collection for this study was carried out for 20 minutes during the day. The data collection activity is shown in figure 12.

Figure 12. Pollution Data Collection at the Port

Table 7. Pollution Data Collection at the Port (in PPM)

Time (minute)	COx	NOx	SOx
2	1,78	0,19	12,85
4	1,72	0,18	12,44
6	1,72	0,18	11,97
8	1,73	0,16	11,42
10	1,64	0,16	11,06

Time (minute)	COx	NOx	SOx
12	1,60	0,16	10,67
14	1,60	0,16	10,65
16	0,50	0,09	10,57
18	0,53	0,09	10,05
20	0,54	0,09	9,52

The pollution monitoring results at Kamal Port over 20 minutes showed COx levels ranging from 0.50 to 1.78 ppm, NOx levels from 0.09 to 0.19 ppm, and SOx levels from 9.52 to 12.85 ppm. COx levels between 0.50 and 1.78 ppm are considered safe, as they do not cause harm to the body. NOx levels between 0.09 and 0.19 ppm can produce Odors and may lead to increased resistance in the air. SOx levels between 9.52 and 12.85 ppm are at the minimum tolerated limit, with potential effects on the body including throat irritation. The following graph illustrates the trend of pollutant levels increasing or decreasing each minute.

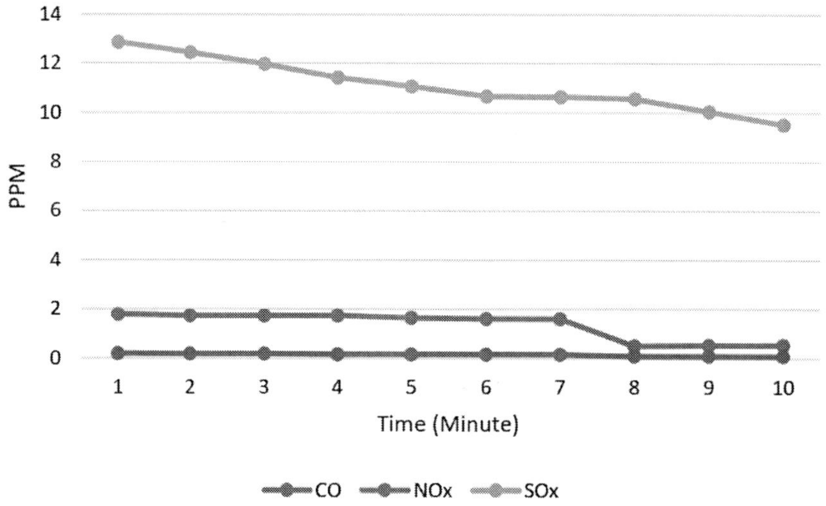

Figure 13. Pollutant Graph at the Port

In the figure 13 above, the detected pollutants, especially SOx, show a decrease from the first minute to the last, falling from 12.85 ppm to 9.52 ppm. NOx also experienced a reduction in concentration, from 0.19 ppm to 0.09 ppm, and COx levels similarly decreased from the first minute to the 20th minute. One reason for the fluctuations in pollutant levels is the wind factor, which disperses pollutants not just to a single location but across various areas within the port.

4. Conclusion

IoT-based monitoring can be used to measure pollutants produced by a sailing ship. The system can detect emissions of hazardous gases such as COx, NOx, and SOx in real-time using sensors connected to smart devices.

The measurement results at three locations found the highest and most dangerous figures in the funnel with COx ranging from 50.98 ppm to 77.73 ppm, NOx from 3.33 ppm to 6.54 ppm, and SOx from 13.76 ppm to 65.87 ppm. However, these figures decreased when measured at the dock, this can be caused by wind factors that carry in various directions. These findings highlight the importance of monitoring emissions in the maritime sector and efforts to reduce pollution to support the goal of net zero emissions by 2050.

To obtain accurate measurement values, further research can be developed by calibrating according to the SNI 9178:2023 standard and measuring various types of ships.

Acknowledgments

We would like to thank Director of the Madura State Polytechnic (POLTERA), Poltera Colleagues, Promoter and Co-Promoter who patiently guided us, the Financing Centre Higher Education (BPPT) Ministry of Education and Culture for all the financial support that has been provided, All Members of the AISITS Research Consortium, PUI-KEKAL ITS, All Committee 12th International Seminar on Ocean, Coastal Engineering, Environmental and Natural Disaster Management (ISOCEEN 2024) which has approved and published the results of this research and my beloved family always supports.

Reference

[1] WHO, "Ambient (outdoor) air pollution," *www.who.int.* 2022. Accessed: Oct. 20, 2023. [Online]. Available: https://www.who.int/news-room/fact-sheets/detail/ambient-(outdoor)-air-quality-and-health
[2] N. Widayatun, "SATU DEKADE PEMBANGUNAN SUMBER DAYA MANUSIA DI WILAYAH PESISIR: CAPAIAN DAN TANTANGAN (KASUS PULAU MAPUR, KABUPATEN BINTAN)," *J. Kependud. Indones.*, vol. 11, no. 2, p. 118, 2017, doi: 10.14203/jki.v11i2.203.
[3] IMO, "MARPOL Annex VI Prevention of Air Pollution from Ships, Regulation 23 and 25," no. November, 2021.
[4] IMO, "Upaya IMO untuk mengurangi emisi GRK dari kapal," *www.imo.org.* 2023. Accessed: Sep. 20, 2023. [Online]. Available: https://www.imo.org/en/MediaCentre/HotTopics/Pages/Cutting-GHG-emissions.aspx
[5] NOAA, "Global CH4 Monthly Means," *gml.noaa.gov.* 2023. Accessed: Sep. 20, 2023. [Online]. Available: https://gml.noaa.gov/ccgg/trends_ch4/
[6] A. Al-fuqaha, S. Member, M. Guizani, M. Mohammadi, and S. Member, "Internet of Things : A Survey on Enabling," vol. 17, no. 4, pp. 2347–2376, 2015, [Online]. Available: https://fardapaper.ir/mohavaha/uploads/2018/10/Fardapaper-Internet-of-Things-A-Survey-on-Enabling-Technologies-Protocols-and-Applications.pdf
[7] A. Maulidi and T. Irmiyana, "The development of internet-of-things (IoT)-based ship monitoring system in madura strait," *IOP Conf. Ser. Mater. Sci. Eng.*, vol. 1052, no. 1, p. 012052, 2021, [Online]. Available: https://iopscience.iop.org/article/10.1088/1757-899X/1052/1/012052%0Ahttps://iopscience.iop.org/article/10.1088/1757-899X/1052/1/012052/meta
[8] A. Zanella *et al.*, "Internet of Things for Smart Cities," *IEEE Internet Things J.*, vol. 1, no. 1, pp. 22–32, 2014, doi: 10.1109/JIOT.2014.2306328.
[9] R. A. Feely, C. L. Sabine, K. Lee, and W. Berelson, "Impact of anthropogenic CO_2 on the $CaCO_3$ system in the oceans," *Science (80-.).*, vol. 305, no. 5682, pp. 362–366, 2004.
[10] Y. Feng, H. Zhu, and Z. Dong, "Simultaneous and Global Optimizations of LNG-Fueled Hybrid Electric Ship for Substantial Fuel Cost, CO2, and Methane Emission Reduction," *IEEE Trans. Transp. Electrif.*, vol. 9, no. 2, pp. 2282–2295, 2023, doi: 10.1109/TTE.2022.3208880.
[11] M. Ridho, "Sistem Pendeteksi Polusi Udara di Kota Bandar Lampung Berbasis Internet of Things (IOT)," no. July, pp. 1–23, 2022, [Online]. Available: http://digilib.unila.ac.id/64794/3/SKRIPSI TANPA BAB PEMBAHASAN.pdf
[12] T. Rahajoeningroem and F. Treska, "Warning System and Monitoring Sulfur Dioxide (SOx) Via SMS Gateway Based on Microcontroller Using Sensor MQ-136 at Mount Tangkuban Parahu," *Telekontran*, vol. 5, no. 1, 2017, [Online]. Available: https://ojs.unikom.ac.id/index.php/telekontran/article/view/1602/1106
[13] J. H. Yoon, S. W. Kim, J. O. Eom, J. Oh, and H. J. Kim, "Coastal Air Quality Assessment through AIS-Based Vessel Emissions: A Daesan Port Case Study," *J. Mar. Sci. Eng.*, vol. 11, no. 12, 2023, doi: 10.3390/jmse11122291.
[14] IMO, "IMO Marine Engine Regulations," DieselNet. Accessed: Feb. 07, 2024. [Online]. Available: https://dieselnet.com/standards/inter/imo.php
[15] BPS, "Statistik Transportasi Laut 2022," 2023. [Online]. Available: https://www.bps.go.id/id/publication/2023/11/27/945749d88a6234417d01c337/statistik-transportasi-laut-2022.html
[16] M. Babiuch, P. Foltynek, and P. Smutny, "Using the ESP32 microcontroller for data processing," *Proc. 2019 20th Int. Carpathian Control Conf. ICCC 2019*, pp. 1–6, 2019, doi: 10.1109/CarpathianCC.2019.8765944.
[17] A. A. Rosa, B. A. Simon, and K. S. Lieanto, "Sistem Pendeteksi Pencemaran Udara Portabel Menggunakan Sensor MQ-7 dan MQ-135," *Ultim. Comput. J. Sist. Komput.*, vol. 12, no. 1, pp. 23–28, 2020, doi: 10.31937/sk.v12i1.1611.

IOP Conf. Series: Earth and Environmental Science **1473** (2025) 012010 doi:10.1088/1755-1315/1473/1/012010

[18] Y. Sari and A. Waliyuddin, "Alat Deteksi Polusi Udara Dalam Ruangan Berbasis Internet of Things (Iot)," *Tekinfo J. Bid. Tek. Ind. dan Tek. Inform.*, vol. 22, no. 2, pp. 120–134, 2021, doi: 10.37817/tekinfo.v22i2.1768.

[19] F. Fachrizal, J. Julham, and A. Antoni, "Sistem Monitoring Polusi Udara Mengggunakan Sensor Nitrogen Carbon Berbasis Internet of Thing," *J. Sist. Komput. dan Inform.*, vol. 4, no. 1, p. 199, 2022, doi: 10.30865/json.v4i1.4903.

[20] H. Risa Aisyiah and D. Pembimbing Ir Josaphat Pramudijanto, *Alat Pendeteksi Kualitas Udara*. 2019. [Online]. Available: https://repository.its.ac.id/62348/

[21] A. Faroqi, E. P. Hadisantoso, D. K. Halim, and M. S. WS, "Perancangan alat pendeteksi kadar polusi udara menggunakan sensor gas MQ-7 dengan teknologi wirelles HC-05," *J. ISTEK*, vol. 10, no. 2, pp. 33–47, 2016, [Online]. Available: https://journal.uinsgd.ac.id/index.php/istek/article/view/1476

[22] A. S. Wahyuni, *"Rancang Bangun Sistem Monitoring Emisi Kadar Gas Sulfur Dioksida Menggunakan Sensor Mq-136 Berbasis Mikrokontroler STM32F4 Discovery."* 2017. [Online]. Available: http://repository.its.ac.id/45291/

Numerical Study of a Flexible Riser Lock-in Phenomenon by Using 2-Way FSI Method

Arifin[1,2*], A Sulisetyono[1*], W H Nugroho[2]

[1] Department of Naval Architecture, Institut Teknologi Sepuluh Nopember
[2] Pusat Riset Teknologi Hidrodinamika (PRTH) – Badan Riset dan Inovasi Nasional (BRIN)

*Email: arif005@brin.go.id, sulisea@na.its.ac.id, wibowo.harso.nugroho@brin.go.id

Abstract. The riser that is operated on the offshore oil and gas exploration in deep water, have complicated behaviour under the influence current load. Vortices are shed along the riser's surface when it travels against the sea current, creating a transverse force that varies frequently on the riser vibration known as the VIV. In this paper, several problems related to frequency natural of the riser and frequency of vortex shedding at several current velocities were studied regarding the lock-in phenomenon of riser. The study was carried out numerically by using the 2-way Fluid Structure Interaction method. The simulation results show the potential for lock-in to occur at certain fluid velocity, namely V=0.2 m/s and 0.4 m/s, which produce vortex shedding frequencies that are close to the natural frequency of the structure, namely fn-2=2.455 Hz and fn-3=4.057 Hz.

1. Introduction

With the increased demand for crude oil in recent years, offshore exploration for gas and oil has shifted its focus to deeper waters. The riser is one of the essential parts of offshore equipment that is generally employed for drilling and producing. Installed between the floating platform and the wellhead at the sea floor is the riser. Given that the riser is subjected to extreme weather conditions in an unshielded deep-water location, its dynamic behaviour is difficult when the platform moves and is affected by wind, waves, and sea currents. Surge, sway, and heave on the platform are caused by sea waves, whereas VIV is typically caused by current stresses operating along the risers.

Vortex-Induced Vibration (VIV) in a strong current is currently the main consideration in deep-water riser design and operation. The vortex shedding frequency is matched to one of the riser's natural frequencies based on full-scale observations [1]. The term "lock-in" describes this phenomenon, when the riser's inherent frequency is locked into the vortex shedding frequency. Thus, a trustworthy examination of the riser's inherent vibration is crucial for understanding the vortex-induced vibration.

Among the most well-known occurrences is vibration caused by vortices. An unstable wake region forms behind the riser as a result of vortices being shed along its surface as it moves under the influence of the sea current. The vortex shedding typically occurs at various frequencies and amplitudes, which causes a transverse strain on the riser to fluctuate periodically. This periodic transverse vibration is called the VIV [2].

IOP Conf. Series: Earth and Environmental Science **1473** (2025) 012011 doi:10.1088/1755-1315/1473/1/012011

Under long-term action, the VIV phenomenon might result in fatigue failure and raise the structure's dynamic load [3]. VIV mostly affects drilling and production risers, increasing their hydrodynamic loading and shortening their service life. Large volumes of oil and gas may escape into the ocean as a result of fatigue degradation and fracture in deep-sea risers, posing a threat to the ecosystem and triggering ecological calamities.

Several studies on Steel Catenary Riser (SCR) which received uniform current loads in the vertical direction (uniform) in the deep sea were carried out [4][5] . From this research, it is known that there is a relationship between current speed and the quantity of vibration modes and amplitude in the riser on the In-Line and Cross-Flow directions.

Due to the fact that actual currents have non-uniform characteristics and very complicated fluid and structure interaction phenomena. One numerical study was carried out to obtain the relationship between fatigue behaviour and current loads that change in the vertical direction [6].

The two degree-of-freedom (2DOF) dynamic reactions of a flexible vertical cylinder were studied experimentally in both the presence of waves alone and current and waves together in a limited depth of water. The principal frequency of the IL response remains the wave frequency when both current and waves are present, and the amplitude of the IL response rises as the KC number does. Because the natural frequency is shifted by the vortex shedding frequency at a current velocity where VIV occurs, the CF response amplitude is reduced when waves are superimposed on current [7].

An experimental test was conducted on a shortened steel catenary riser (SCR) model in an ocean basin to examine the characteristics of vortex-induced vibration (VIV). The upper end of the model was subjected to oscillations to simulate the movements of a vessel in heave and surge conditions [8]. When out-of-plane VIV responses were analysed, it was found that while the displacement was rather consistent within a narrowband from 0.2D to 0.3D, the root mean square (RMS) strain spread relatively broadly.

Wang et al. and Yin et al. [9][10] conducted tests in the towing tank for the riser that was stimulated by the motion of the platform surge at the top. The results showed that the platform's surge motion greatly influenced the vortex-induced vibrations (VIV) of the riser.

Given that internal flow and the movement of the vessel can influence the vortex-induced vibrations (VIV) of flexible risers, it is crucial to take these factors into account. The behaviour of a flexible riser in response to current-induced vortex vibrations (CF VIV) is examined, taking into account both vessel movement and internal flow conditions. The findings demonstrate that vascular motion can significantly affect top tension and the CF VIV characteristic, regardless of low or high Keulegan Carpenter (KC) values [11].

Finite element analysis is employed to explore the dynamic responses of the riser subjected to top-end movement and vortex-induced vibrations (VIV) [12]. The impact of top end vessel sway's frequency and amplitude on riser's VIV is investigated. A fascinating phenomenon called nonlinear response amplification is observed in the riser's dynamic behaviour, indicating that the top-end motion can intensify as it travels along the length of the riser. The numerical results demonstrate that, in comparison to the situation without top-end motion, the riser's displacement increases many times.

Numerical simulations and experimental studies are employed to illustrate how platform heave motion affects the vortex-induced vibrations (VIV) of a deep-sea top-tensioned riser (TTR) [13]. The heave motion was modelled using the riser's VIV governing equation and a parametric excitation. Using the Van der Pol wake-oscillator model as a basis, the finite difference approach was utilized to quantitatively calculate the dynamic response of TTR. The findings indicate that the platform heave motion's larger amplitude and higher frequency, together with their greater influence, are generated on TTR's VIV.

Flexible risers' vortex-induced vibration (VIV) is a complicated fluid–structure interaction (FSI) phenomena. Based on the slicing method, a numerical simulation technique was applied to investigate the uniform flow velocities and varying slenderness ratios of deep-sea flexible risers' vortex-induced vibration (VIV) [14]. The technique combines two-dimensional slices of the flow field that the Fluent solver solves with the riser structure's finite element model. Through the use of the overset mesh approach and a self-compiled UDF program, the fluid-structure interaction was made possible. Through comparison with experimental data, the numerical results were verified. An analysis was conducted on the riser's VIV characteristics, including the wake vortex shedding mode, vibration frequency, vibration track, and vibration mode.

A wake oscillator has been used to study the vortex-induced vibration (VIV) of flexible long risers with mixed in-line and cross-flow motion. A comparison is made between the analytical solution and the experimental data for the mean top tension of a long flexible riser [15]. The model of wake oscillator was then used to examine the in-line and cross-flow VIV of a long tension-dominated riser's nonlinear coupled dynamics while taking the mean top tension fluctuation into account.

The top-tensioned riser's design and application (TTR) are growing more challenging as water depth increases. The top tension effect on the dynamic behaviour of deep-sea risers is investigated through numerical simulations and testing [16]. The influence of top tension on the dynamic behaviour of deep-sea risers is examined through both numerical simulations and experimental testing. The study focused on how variations in top tension affect the displacement, bending stress, and vibration frequency of top-tensioned risers (TTR).

In this paper, several problems related to changes in current speed, and axial tension in the riser will be studied regarding the Lock-in phenomenon that potentially occurs in the riser. The study was carried out numerically by using the 2-way Fluid Structure Interaction method. From this research data will be obtained on the amplitude and frequency of vibrations in the In-line flow (IF) direction and Cross flow (CF) direction, as well as the frequency of vortex shedding, maximum displacement along the flexible riser.

2. Methodology

Prediction of the potential for lock-in phenomena in risers will be carried out using the 2-way fluid structure interaction numerical method. The simulations in this study were conducted using ANSYS® Professional version software owned by Badan Riset Inovasi Nasional (BRIN).

2.1. Equations for Governing Fluid Domains

The term "viscous fluid flow" describes a fluid's macroscopic incompressible motion that is regulated by the conservation of mass, momentum, and energy laws as well as the fluid's governing equations. We omitted the use of the energy equation in this study as we did not consider the energy conversion effects caused by temperature fluctuations during the simulation. The Navier–Stokes (N–S) equation is satisfied by viscous fluid flow, and the Reynolds-averaged numerical simulation (RANS) approach was used. Instead of using the N-S equation to explain the flowing fluid, this method uses the Reynolds-averaged N-S equation. coordinates in Cartesian space.

Equation of continuity:

$$\frac{\partial \bar{u}_i}{\partial x_i} = 0 \tag{1}$$

Momentum Equation:

$$\rho \frac{\partial \bar{u}_i}{\partial t} + \rho \frac{\partial \overline{u_i u_j}}{\partial x_i} = -\frac{\partial \bar{p}}{\partial x_i} + \mu \nabla^2 \bar{u}_i - \rho \frac{\partial \overline{u_i' u_j'}}{\partial x_j} \tag{2}$$

where the variables' components in the X, Y, and Z directions are represented by the values of i and j, which are 1, 2, or 3 accordingly; The pressure is represented by p, the Hamiltonian operator by ∇, the average flow rate over time by $\bar{u}_i$ and $\bar{u}_j$, and the pulsing flow rate's time-average value by $\bar{u}'_i$ and $\bar{u}'_j$.

We applied the discrete form of the finite volume approach to solve the fluid's governing equation. We employed the overlapping grid technology and linked algorithm of the Fluent solver. The transient equation was solved using the second-order upwind strategy, and the time dispersion was solved using the fully implicit scheme.

2.2. Riser's Structural Dynamic Model

Assuming uniform material properties and cross-sectional shape along the length of the riser, the top tension model of the flexible riser was simplified to an Euler-Bernoulli beam. As seen in Figure 1, we designated the riser length direction as the Z axis, the crossflow (CF) direction as the Y axis, and the in-line (IL) direction as the X axis. Equation (3) provides the structural motion differential equations in the flow direction.

$$\frac{\partial^2}{\partial z^2}\left[EI\frac{\partial^2 u_i(z,t)}{\partial z^2}\right]-\frac{\partial}{\partial z}\left[T(z)\frac{\partial u_i(z,t)}{\partial z}\right]+m\frac{\partial^2 u_i(z,t)}{\partial t^2}+c\frac{\partial u_i(z,t)}{\partial t}=F_i \tag{3}$$

where m is the riser mass per unit length, c is the structural damping, and EI, EA, and m are the bending and tensile stiffness, respectively. The CF and IL displacement are shown by ui(z, t), and the CF direction and the IL hydrodynamic force are indicated by Fi(z, t) (where i = 1 represents the IL direction and i = 2 represents the CF direction). The riser top tension is T(z).

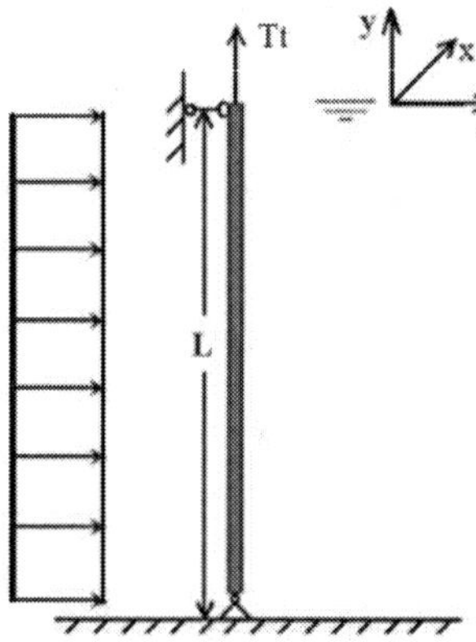

Figure 1. Numerical Modelling of Riser

The riser's ends' boundary conditions are provided as

$$x(z,t)=0, \qquad \frac{\partial^2(z,t)}{\partial z^2}=0 \qquad z=0, z=L \tag{4}$$

2.3. Natural Frequency and Mode shapes

Natural frequencies and mode shapes, which serve as the foundation for intricate analysis of vibration, are the primary natural vibration features of TTR. For deep-sea risers, the excitation of vortex-induced resonance is directly influenced by the natural frequency, and the vibration response may be examined using the mode shape.

The fundamental natural frequencies and corresponding mode shapes for the deep-sea riser are generally represented by simple formulas that rely on the riser's tension.

$$\omega_n=\frac{n\pi}{2L\sqrt{m}}(\sqrt{T_t}+\sqrt{T_b} \tag{5}$$

where ω_n is the nth natural frequency of TTR. $\varphi_n(z)$ is the nth mode function. T_b is the bottom end tension of TTR, $T_b = (f_{top} - 1)wL$.

2.4. The Two-Way Fluid Structure Interaction Method

In two-way Fluid-Structure Interaction (FSI), the equations governing fluid flow and structural dynamics are solved simultaneously, allowing for reciprocal interactions between fluid and structure.

The equations for each domain are pairwise, meaning that changes in one domain will affect the other domain and vice versa.

The equations governing fluid flow are represented by the incompressible Navier-Stokes equations. The equations governing structural dynamics are the same as in one-way FSI, but these equations are solved simultaneously with the fluid flow equation.

Completion of the coupling system is carried out by following the flow diagram as shown in the Figure 2 below.

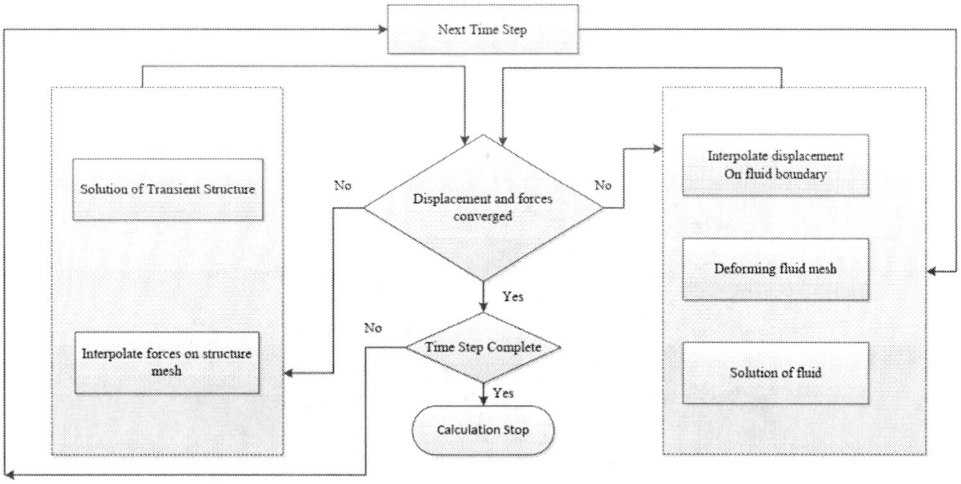

Figure 2. Flow Diagram 2-Way Fluid Structure Interaction Method

2.4.1. Boundary Condition and Computational Domain. The center of the riser was positioned of 15D from the upper and lower limits, as well as the inlet boundary, within a computational domain of 30D × 60D for the flow field calculation [17]. Figure 3 illustrates the 45D distance that separated the riser center from the outlet boundary.

Table 1 provides the corresponding boundary settings.

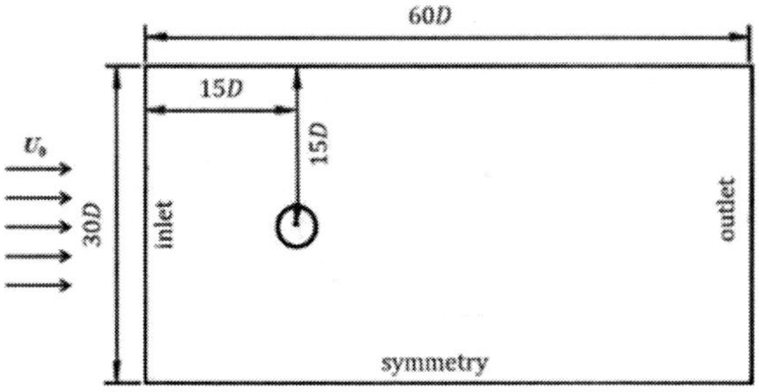

Figure 3. Domain of Computational

Table 1. Boundary Conditions Setting

Boundary Conditions	Setting
Inlet	velocity-inlet, u=U_0, v=0
Outlet	pressure-outlet, ∂u/∂x=0, ∂v/∂x=0, static pressure=0
Lower, upper, left, right	symmetry, ∂u/∂x=0, ∂v/∂x=0
Riser	no-slip wall, u=v=0

The riser model characteristics and geometry meshing in the fluid domain are as shown in Table 2 the Figure 4 below.

Table 2. Characteristics of the Riser Model

Properties	Value	Units
Length of the Riser, L	4.82	m
Outer diameter of riser, D	0.02	m
Thickness of the riser wall, t	0.0045	m
Young's modulus, E	$7.38*10^8$	N/m2

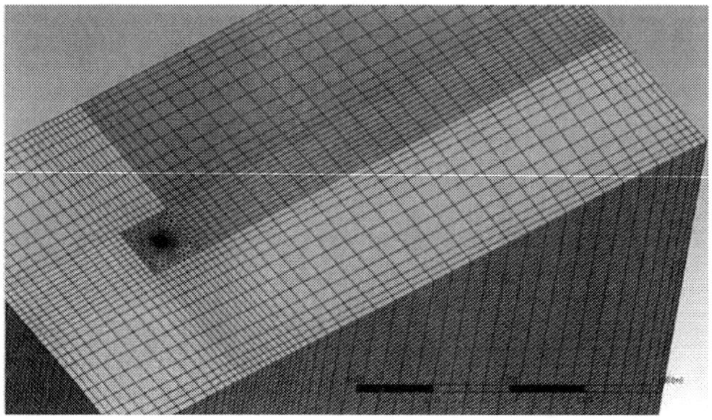

Figure 4. Meshing Geometry of the Riser in Fluid Domain

2.5. Numerical Simulation Case Programs

A number of case programs are carried out in numerical simulation as shown in the following table.

Table 3. Case Programs in Numerical Simulation of Riser

No.	Case Name	Description	Additional Information
1	Case-01	No-Pretension; Vc=0.2 m/s	Re=4494
2	Case-02	No-Pretension; Vc=0.3 m/s	Re=6742
3	Case-03	No-Pretension; Vc=0.4 m/s	Re=8989
4	Case-04	No-Pretension; Vc=0.5 m/s	Re=11236
5	Case-05	No-Pretension; Vc=0.6 m/s	Re=13483
6	Case-06	No-Pretension; Vc=0.7 m/s	Re=15730

3. Results and Discussion

A first case of the numerical simulation was under assumption that the upper end of the riser was free and the bottom end was fixed. For the numerical modelling of the VIV, riser models with a slenderness ratio (L/D) of 241 were constructed based on the numerical model.

When the vortex shedding frequency of the riser approaches its natural frequency, lock-in happens, resulting in an increase of amplitude. The Ansys Structural modal analysis module was used to find the riser's frequency natural. The outcomes of the numerical simulation were contrasted with those of other researchers, as shown in the Table 4.

Table 4. Comparison of Natural Frequency

No.	Author	Simulation Condition	fn-1 (Hz)	fn-2 (Hz)	fn-3 (Hz)	fn-4 (Hz)
1	Arifin (2024)	D=0.02m; L=4.8m; t=0.0045m Material teflon Top end of riser: free	1.252	2.455	4.057	6.058
2	Jie Zhang (2020)	D=0.016m; L=5.0m; t=0.001m Material teflon Top end of riser: hinge	1.327	2.674	4.058	5.498
3	Yu Duanmu (2018)	D=0.028m; L=5.94m; EI=29.88N/m2 Top End of riser, Tt=1600N	1.224	2.452	3.688	4.936

An overview of the shaped modes at each natural frequency can be seen in Figure 5.

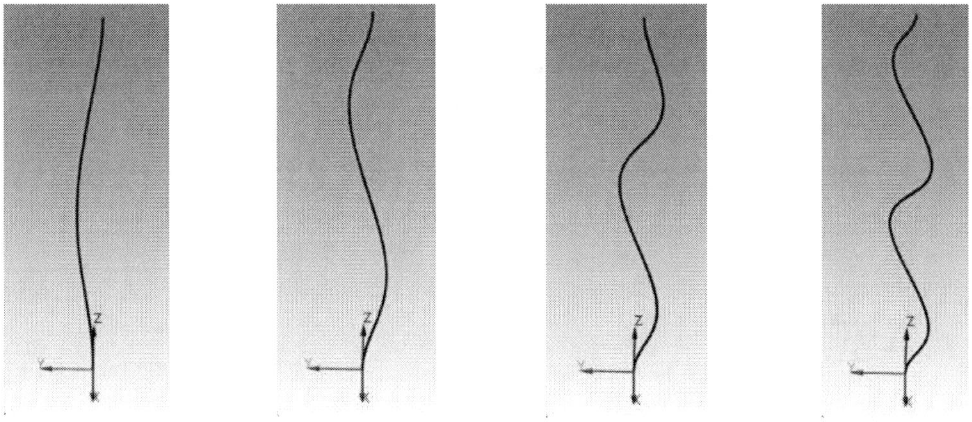

1st-Mode 2nd-Mode 3rd-Mode 4th-Mode

Figure 5. Mode shapes of Riser Structure

The results of the numerical simulation of the coupling system consist of 2 groups, namely transient structural output and fluid flow.

3.1. Transient Structural Results

Several selected numerical simulation results are related to the analysis of vibrations that occur in risers, including displacement, strain and stress.

3.1.1. Directional Displacement. Directional displacement resulting from numerical simulations consists of displacement that occurs in the riser in the In-Line Flow, IF (in the direction of fluid flow) and Cross-Flow, CF (direction perpendicular to the flow direction).

Figure 6 and Figure 7 show the time history of the In-Line (IL) displacement and Cross-Flow displacement at a certain riser elevation and fluid velocity, respectively.

A series of the maximum displacement–time history curves in the In-Line Flow (IF) direction at y/L = 1.0 for a several fluid velocities was presented in Figure 8, and time history in the Cross-Flow (CF) displacement was presented in Figure 9.

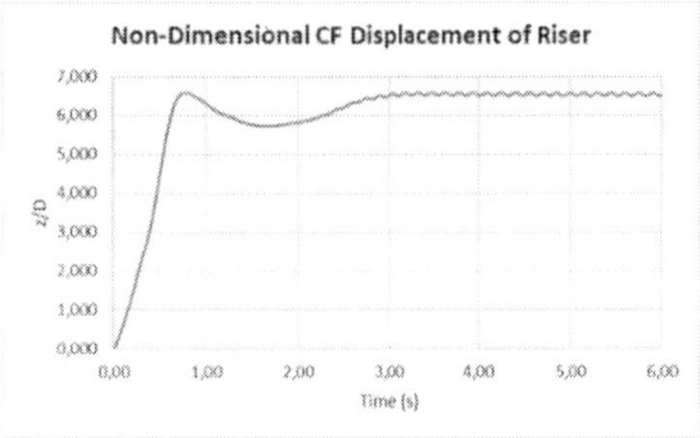

Elevation, y/L=1.0

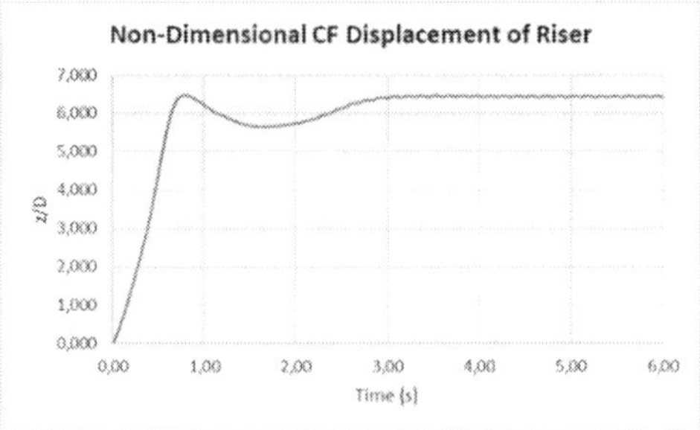

Elevation, y/L=0.9

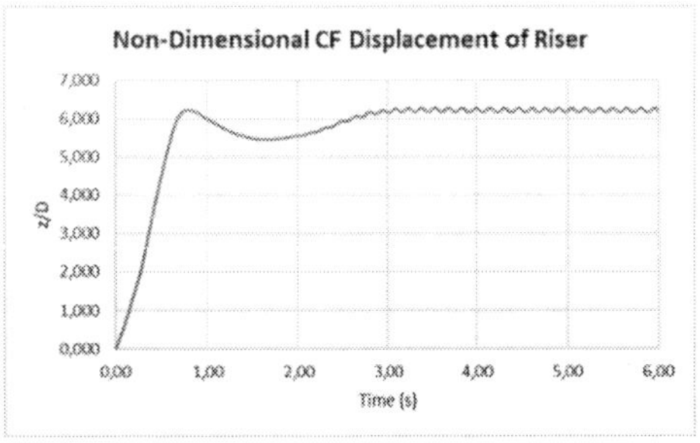

Elevation, y/L=0.8

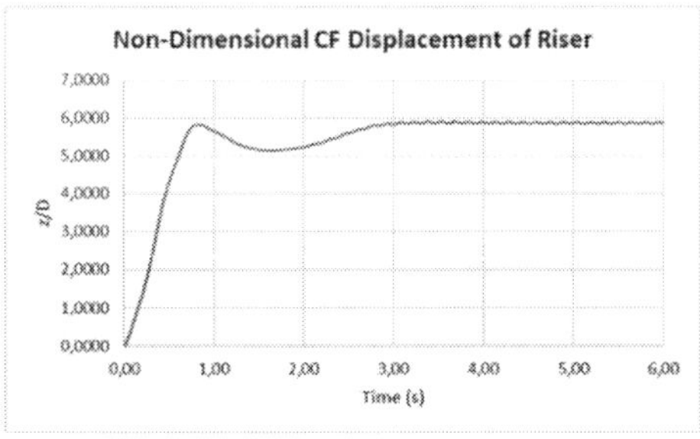

Elevation, y/L=0.7

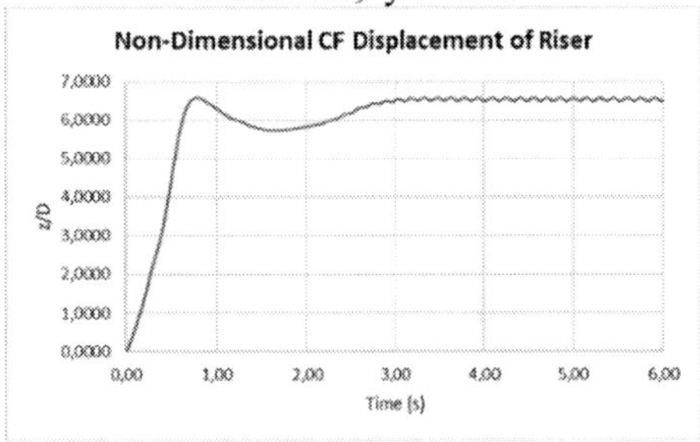

Elevation, y/L=0.6

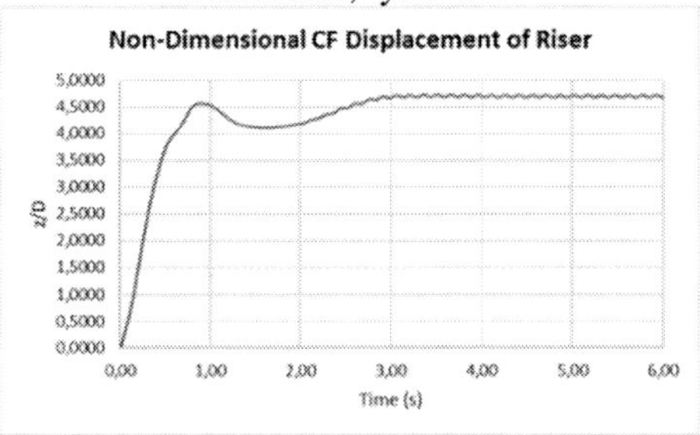

Elevation, y/L=0.5

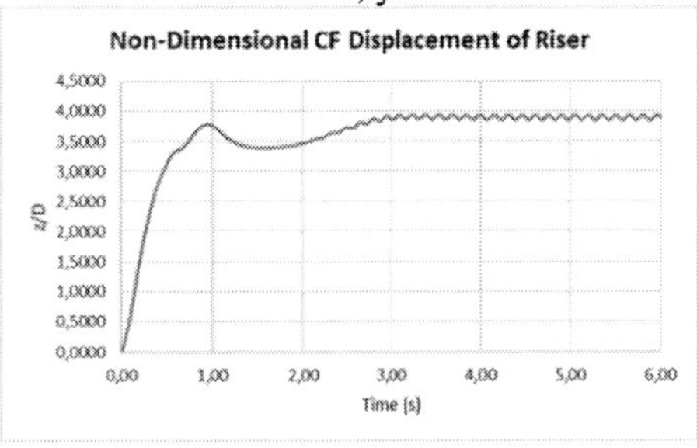

Elevation, y/L=0.4

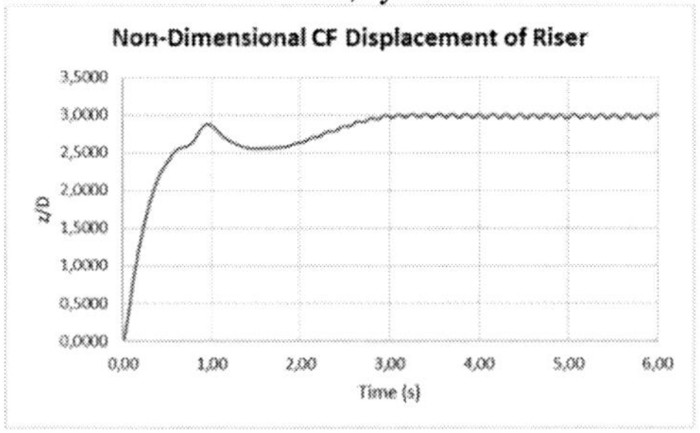

Elevation, y/L=0.3

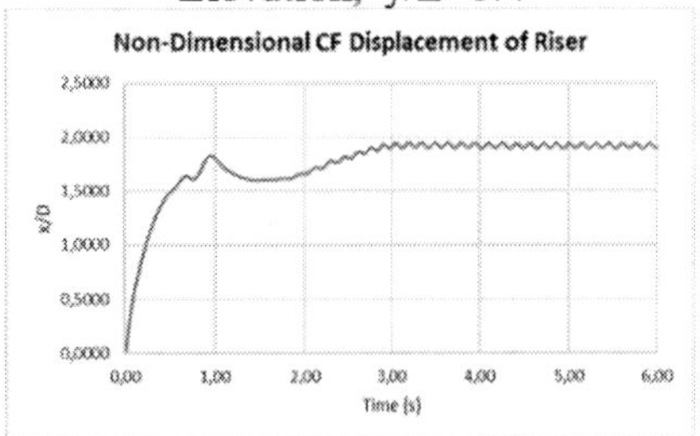

Elevation, y/L=0.2

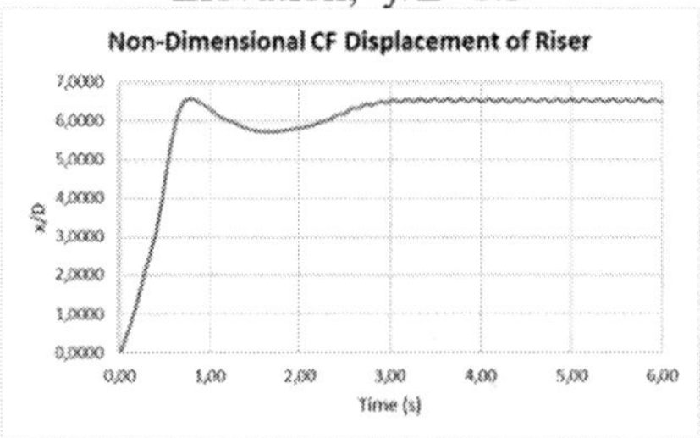

Elevation, y/L=0.1

Figure 6. Time History In-Flow (IF) Displacement

 doi:10.1088/1755-1315/1473/1/012011

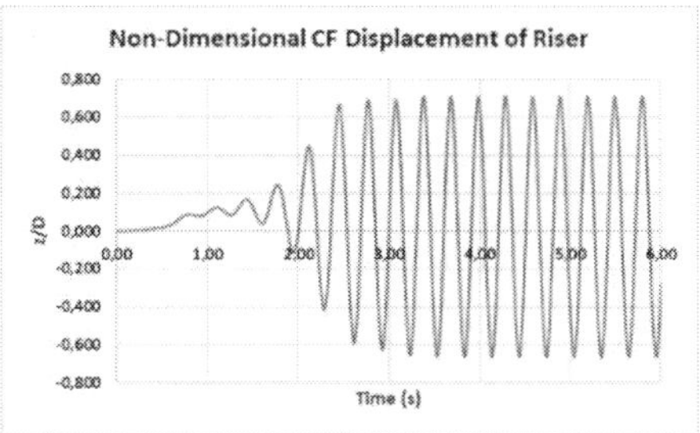

Elevation, y/L=1.0

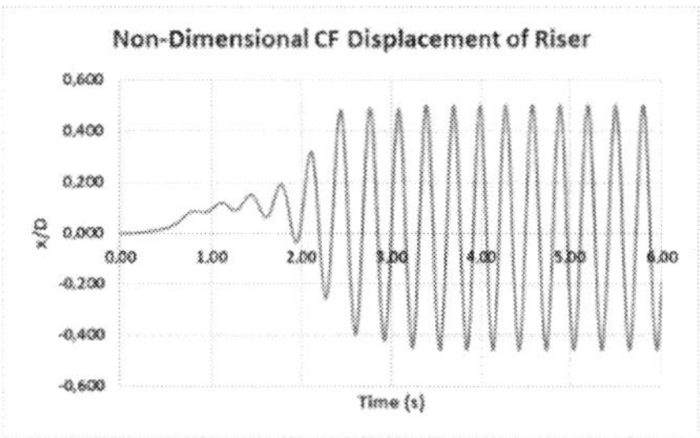

Elevation, y/L=0.9

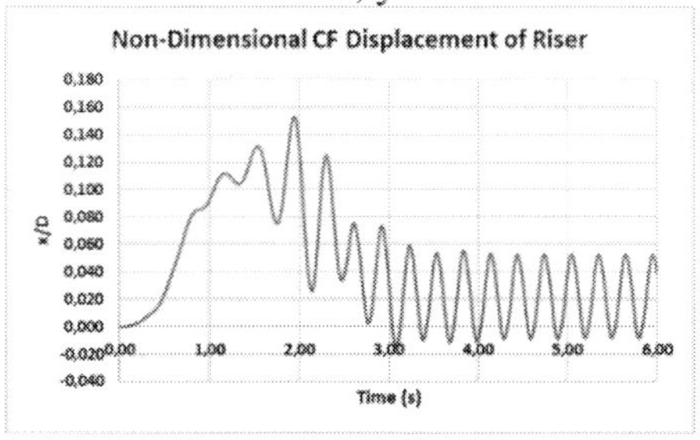

Elevation, y/L=0.8

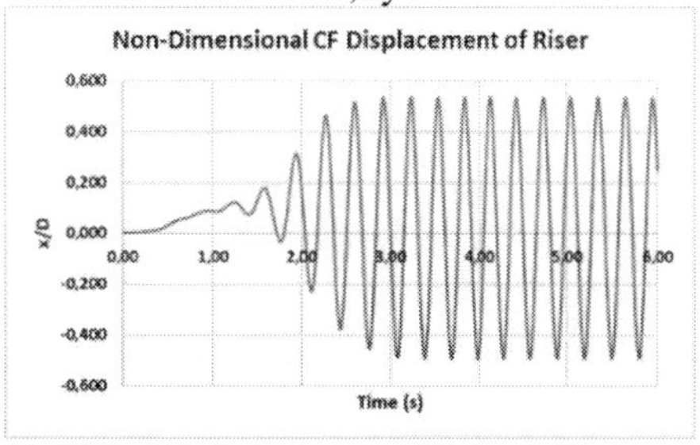

Elevation, y/L=0.7

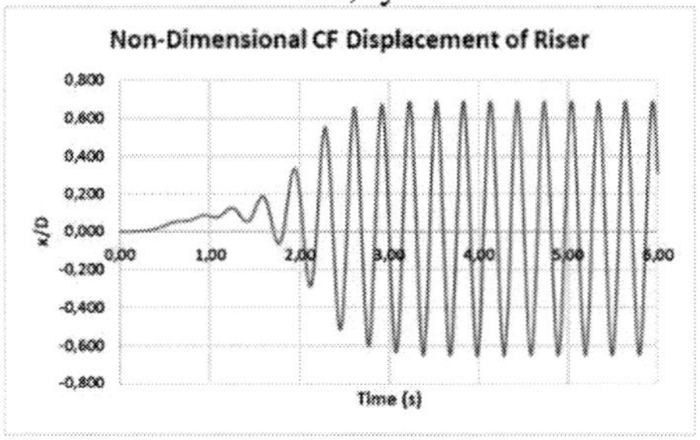

Elevation, y/L=0.6

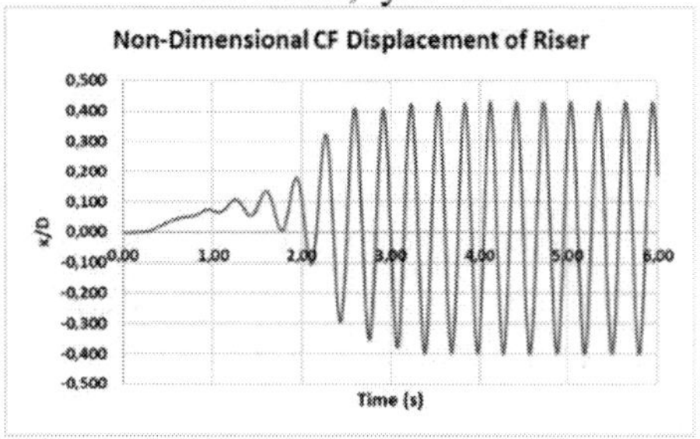

Elevation, y/L=0.5

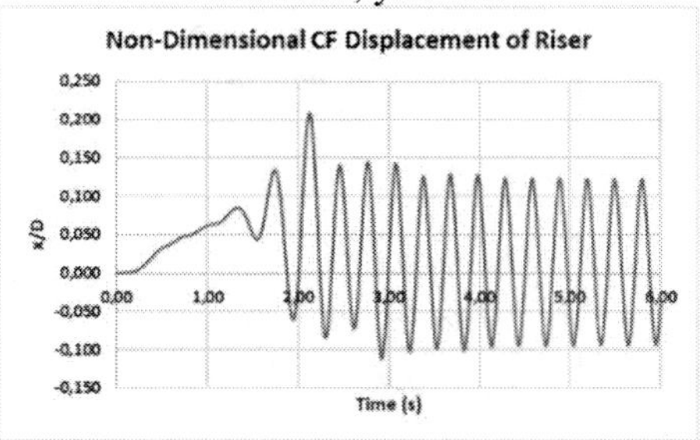

Elevation, y/L=0.4

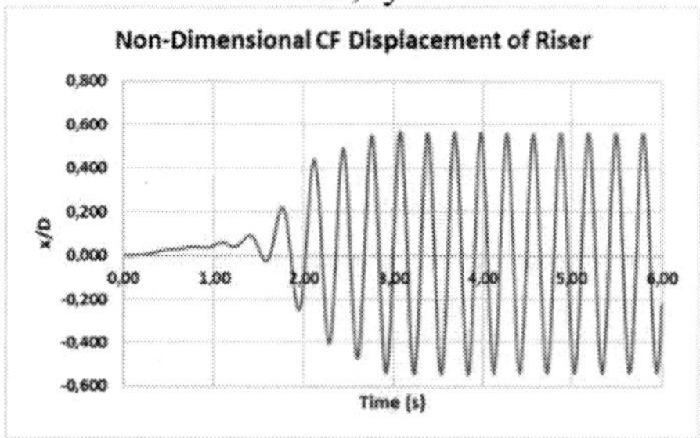

Elevation, y/L=0.3

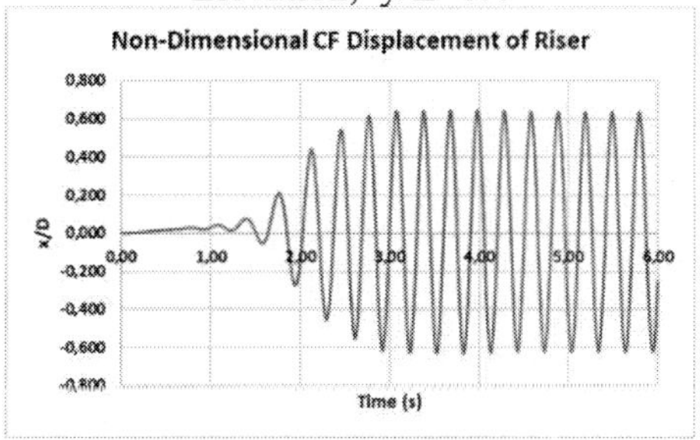

Elevation, y/L=0.2

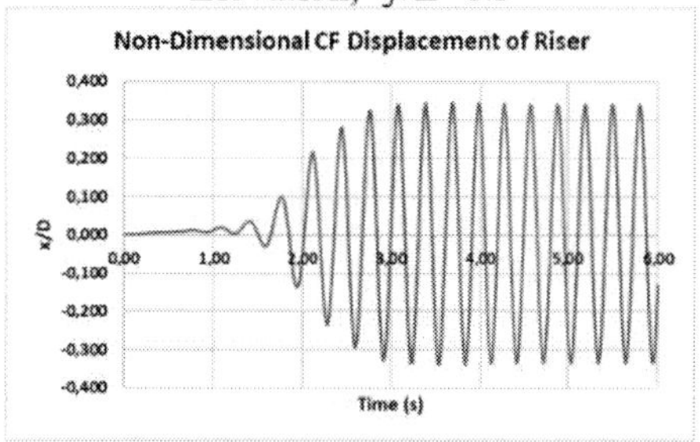

Elevation, y/L=0.1

Figure 7. Time History Cross-Flow (CF) Displacement

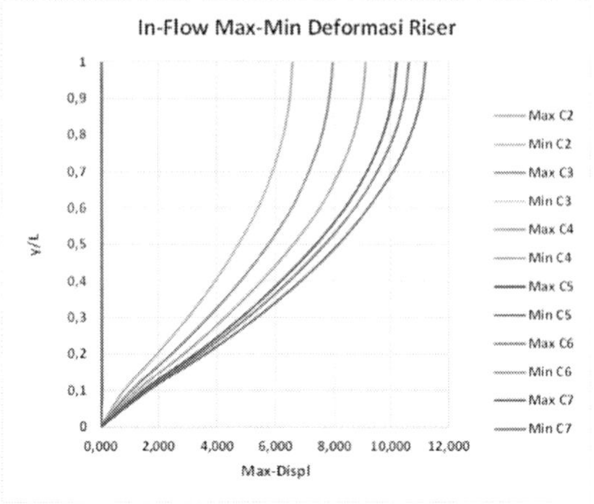

Figure 8. Max Displacement of IF

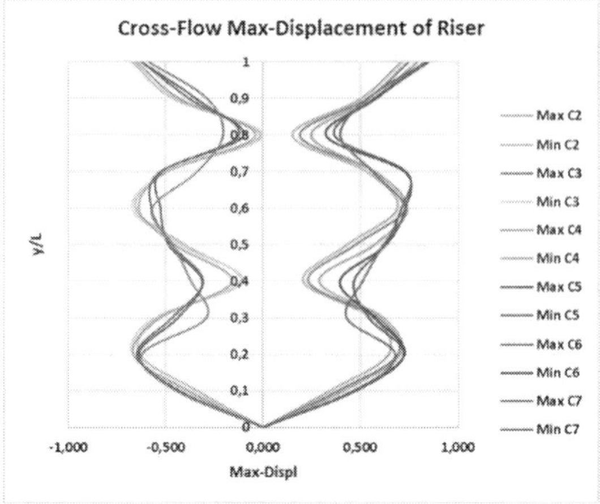

Figure 9. Max Displacement of CF

3.1.2. Equivalent Elastic Strain. Equivalent elastic strain is defined as a measure of the elastic deformation produced in a material when a load is applied and then released. This is one of the important parameters used to understand how a material or structure responds to dynamic loads in transient analysis.

An example of equivalent elastic strain output is as shown in the following figure.

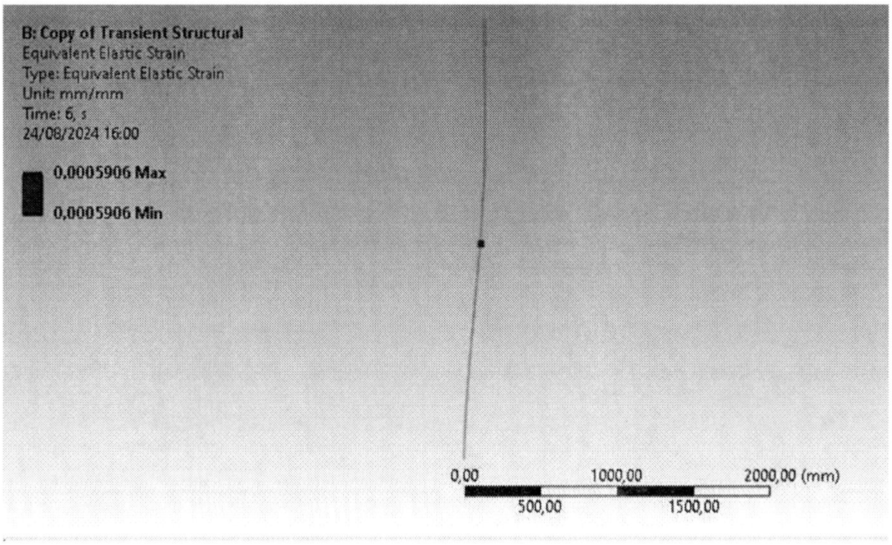

Figure 10. Equivalent Elastic Strain at y/L=0.5

3.1.3. Equivalent Stress. Equivalent Stress in the context of transient structural output in ANSYS is a stress measure used to determine the risk of material failure under dynamic loads that change over time. These stresses are usually calculated based on failure theories, such as the von Mises criterion, and are often used to evaluate material conditions in structural analysis.

An example of equivalent stress output is as shown in the following figure.

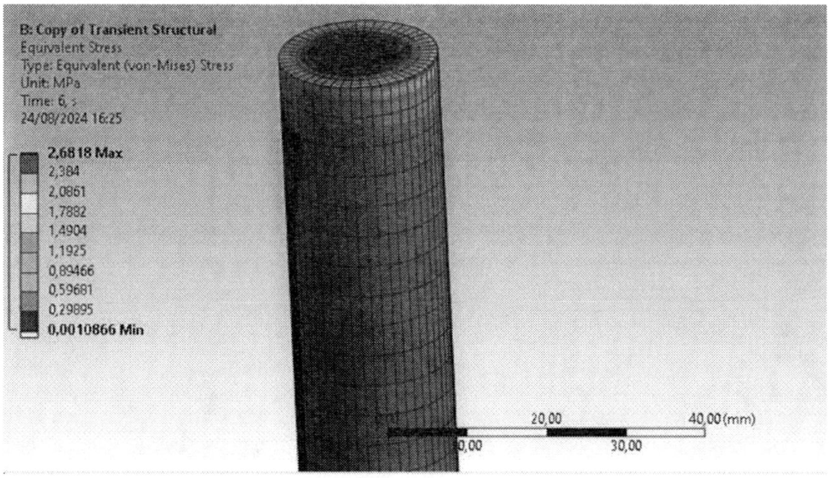

Figure 11. Equivalent Stress

3.2. Fluid Flow Results

Contour plots in ANSYS fluid flow simulation show the distribution of variables such as pressure, velocity, or vorticity in the fluid flow domain.

3.2.1. Contour of Pressure and Velocity. Pressure changes around the cylinder can be displayed as contours. When a vortex forms and releases (vortex shedding) from the cylinder, pressure fluctuations will occur on the opposite side of the cylinder which causes an alternating force (lift) on the cylinder.

Velocity contours can show changes in fluid velocity around the cylinder, including areas of stagnation and accelerated flow. This instability in flow can amplify cylinder vibrations caused by VIV.

3.2.2. Streamline. Streamline is a visual representation of the path followed by fluid particles. Each point on the streamline indicates the direction of fluid velocity at that point. Under steady-state conditions, streamlines provide a visual representation of fluid flow without changes in time.
An example of Ansys fluid flow output can be seen in the following figure.

Figure 12. Streamline at Elevation y/L=0.9

IOP Conf. Series: Earth and Environmental Science **1473** (2025) 012011 doi:10.1088/1755-1315/1473/1/012011

3.2.3. Vortex. Vortex is an area in a fluid flow where the fluid moves in a spiral or circular pattern. The flow around the vortex centre has a high angular velocity, and this direction of rotation results in lower pressure at the vortex centre compared to the surrounding areas.

An example of vortex as Ansys fluid flow output can be seen in the following figure

Figure 13. Vortex at t=0.01s

Figure 14. Vortex at t=2s

3.3. Lock-in Analysis

The lock-in phenomenon in riser dynamics is a critical area of study in offshore engineering due to its potential to induce significant structural fatigue. Lock-in occurs when the vortex shedding frequency, which is the frequency at which vortices are shed from the riser, aligns with or closely matches one of the natural frequencies of the riser structure. This alignment results in a resonance condition that can lead to sustained or even amplified oscillations, posing a serious threat to the integrity of the riser over time. The resulting large amplitude vibrations can accelerate fatigue damage, which could lead to premature failure of the riser. Understanding this phenomenon requires a detailed examination of the relationship between the vortex shedding frequency and the natural frequencies obtained from modal analysis of the riser.

To fully grasp the complexities of the lock-in phenomenon, it is essential not only to consider the frequency alignment but also to examine the phase angle between the shedding forces and the structural response of the riser. During lock-in, the phase difference between the oscillatory forces generated by vortex shedding and the riser's motion becomes synchronized. This synchronization allows for an efficient transfer of energy from the fluid flow to the riser’s structural vibration, leading to a rapid increase in vibration amplitude. The phase angle, therefore, plays a crucial role in determining whether lock-in will occur and the extent to which it will affect the riser.

To investigate this phenomenon, time history displacement data at various elevations of the riser were obtained from numerical simulations. These simulations were conducted for both In-Line Force (IF) and Cross-Flow (CF) displacement data. The displacement data were then analysed using the Fast Fourier Transform (FFT) technique to convert the time-domain signals into the frequency domain, resulting in Power Spectral Density (PSD) data. The PSD analysis enabled the identification of the shedding frequency that corresponds to the maximum energy content, as indicated by the peak PSD value. From this identified shedding frequency, the Strouhal Number—a dimensionless number that describes the vortex shedding behaviour—was calculated to further characterize the flow dynamics around the riser. A snapshot of the FFT process and the corresponding results is provided in Table 5 below.

This detailed analysis underscores the importance of understanding the interplay between vortex shedding and the natural frequencies of the riser. By examining both the frequency matching and the phase angle synchronization, we can better predict and mitigate the risks associated with the lock-in phenomenon, ultimately improving the design and operational safety of offshore risers.

Table 5. FFT Process of Riser Displacement Data

t	x/D (m/m)	FFT	Point no.	Freq.	Amplitude (m/m)	Phase (deg)
0.01	0/00000	22.7214	1	0.195	22.721	0.000
0.02	0.00003	-8.7622-10.9994i	2	0.391	10.9994	-128.541
0.03	0.00007	-8.1164+3.5853i	3	0.587	8.873	156.167
0.04	0,00014	-3.9494+1.4696i	4	0.782	4.214	159.589
0.05	0,00022	-3.3420+1.3121i	5	0.978	3.590	158.565
0.06	0,00033	-3.2681+0.4279i	6	1.174	3.296	172.540
0.07	0,00047	-3.1342+0.0726i	7	1.369	3.135	178.673
0.08	0,00063	-3.8523-0.5563i	8	1.565	3.892	-171.782
0.09	0,00082	-3.7891-0.3147i	9	1.761	3.802	-175.252
0.1	0,00105	-5.1900-1.4200i	10	1.956	5.381	-164.698

Next, the data obtained is displayed in the form of a graph that connects frequency and Power Spectral Density (PSD) as well as between frequency and vibration phase angle, as shown in the Figure 15 and Figure 16 as shown below.

Based on the FFT analysis carried out for all elevations on the riser, data will be obtained on the frequency of vortex shedding, Strouhal number and phase angle which can be used as an indication of the potential for resonance phenomena in the riser at a certain speed of fluid flow.

Complete results of the effect of changes in speed on the vortex shedding frequency values and phase angles that occur are shown in the following Table 6 as shown below.

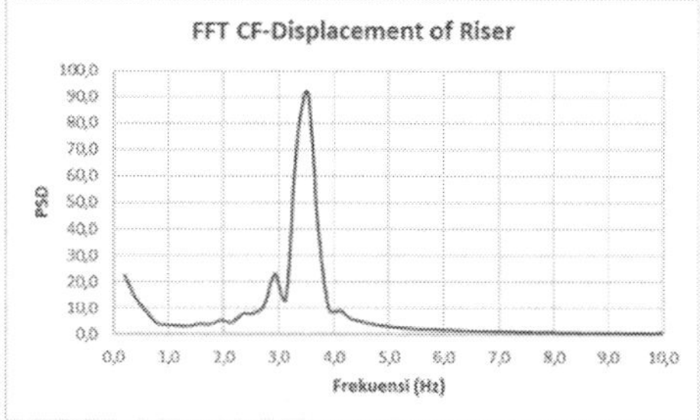

Figure 15. PSD vs Shedding Frequency

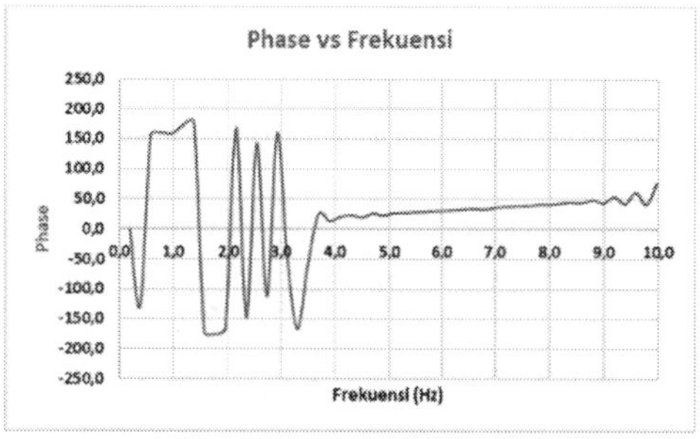

Figure 16. Phase Angle vs Frequency

Table 6. Frequency shedding and Phase for Each Elevation on Riser

Case no. 1; V=0.2 m/s; Re=4494

y/L (m)	fs-1 (Hz)	St-1	Phase-1	fs-2 (Hz)	St-2	Phase-2	fs-3 (Hz)	St-3	Phase-3
1				1.761	0.176	-121.553	2.348	0.235	21.600
0.9				1.761	0.176	-121.760	2.348	0.235	20.368
0.8				1.761	0.176	-123.931	2.348	0.235	-51.059
0.7				1.761	0.176	-140.440	2.544	0.254	-74.982
0.6				1.957	0.196	169.013	2.544	0.254	-76.784
0.5				1.761	0.176	56.934	2.348	0.235	-157.355
0.4				1.761	0.176	51.685	2.544	0.254	113.292
0.3				1.761	0.176	47.699	2.544	0.254	107.146
0.2				1.761	0.176	45.264	2.544	0.254	105.983
0.1				1.761	0.176	44.177	2.544	0.254	105.420

Case no. 2; V=0.3 m/s; Re=6742

y/L (m)	fs-1 (Hz)	St-1	Phase-1	fs-2 (Hz)	St-2	Phase-2	fs-3 (Hz)	St-3	Phase-3
1				2.935	0.196	159.508	3.523	0.235	59.448
0.9				2.935	0.196	161.781	3.523	0.235	-59.159
0.8							3.523	0.235	117.546
0.7				2.935	0.196	-24.842	3.523	0.235	120.675
0.6				2.935	0.196	-21.933	3.523	0.235	120.966
0.5				2.935	0.196	-14.655	3.523	0.235	121.607
0.4	2.348	0.157	161.642	2.935	0.196	130.338	3.523	0.235	-58.858
0.3	2.348	0.157	174.386	2.935	0.196	150.498	3.523	0.235	-58.187
0.2	2.348	0.157	179.829	2.935	0.196	153.637	3.523	0.235	-57.809
0.1	2.348	0.157	-3.087	2.935	0.196	156.907	3.523	0.235	-57.222

Case no. 3; V=0.4 m/s; Re=8989

y/L (m)	fs-1 (Hz)	St-1	Phase-1	fs-2 (Hz)	St-2	Phase-2	fs-3 (Hz)	St-3	Phase-3
1	3.131	0.157	83.227	3.718	0.186	67.412	4.110	0.206	127.754
0.9	3.131	0.157	87.691	3.718	0.186	69.563	4.110	0.206	128.698
0.8				3.718	0.186	-143.348	4.110	0.206	-61.698
0.7	3.131	0.157	-113.376	3.718	0.186	-115.409	4.110	0.206	-52.119
0.6	3.131	0.157	-105.835	3.718	0.186	-113.287	4.110	0.206	-51.653
0.5	3.131	0.157	-91.988	3.718	0.186	-108.138	4.110	0.206	-49.352
0.4	2.935	0.147	122.227	3.718	0.186	51.232	4.110	0.206	122.538
0.3	3.131	0.157	58.377	3.718	0.186	64.680	4.110	0.206	128.311
0.2	3.131	0.157	63.718	3.718	0.186	66.831	4.110	0.206	129.133
0.1	3.131	0.157	68.281	3.718	0.186	69.921	4.110	0.206	130.815

Case no. 4; V=0.5 m/s; Re=11236

y/L (m)	fs-1 (Hz)	St-1	Phase-1	fs-2 (Hz)	St-2	Phase-2	fs-3 (Hz)	St-3	Phase-3
1	4.305	0.172	102.919	4.697	0.188	164.650			
0.9	4.305	0.172	105.655	4.697	0.188	165.675			
0.8				4.697	0.188	-22.620	5.871	0.235	23.829
0.7	4.110	0.164	-52.119						
0.6	4.305	0.172	-76.876	4.697	0.188	-14.924			
0.5	4.305	0.172	-70.310	4.697	0.188	-12.707	5.480	0.229	109.750
0.4				4.697	0.188	157.185	5.480	0.229	-153.775
0.3	4.305	0.172	99.968	4.697	0.188	164.069			
0.2	4.305	0.172	102.793	4.697	0.188	164.677			
0.1	4.305	0.172	106.071	4.697	0.188	166.438	5.480	0.229	-68.773

Case no. 5; V=0.6 m/s; Re=13483

y/L (m)	fs-1 (Hz)	St-1	Phase-1	fs-2 (Hz)	St-2	Phase-2	fs-3 (Hz)	St-3	Phase-3
1				5.284	0.176	156.161	6.458	0.215	-54.916
0.9				5.284	0.176	157.259	6.458	0.215	-55.328
0.8							6.458	0.215	127.108
0.7				5.284	0.176	-22.243	6.458	0.215	123.701
0.6				5.284	0.176	-22.057			
0.5				5.284	0.176	-20.504	6.654	0.222	114.638
0.4	3.914	0.131	162.076	5.284	0.176	148.396	6.458	0.215	-52.848
0.3				5.284	0.176	156.242			
0.2				5.284	0.176	156.252	6.654	0.222	-63.618
0.1				5.284	0.176	157.519	6.654	0.222	-62.217

Case no. 6; V=0.7 m/s; Re=15730

y/L (m)	fs-1 (Hz)	St-1	Phase-1	fs-2 (Hz)	St-2	Phase-2	fs-3 (Hz)	St-3	Phase-3
1	4.305	0.123	102.880	5.871	0.168	148.919	7.241	0.207	-145.402
0.9	4.305	0.123	98.819	5.871	0.168	151.865	7.241	0.207	-143.811
0.8							7.241	0.207	32.706
0.7				5.871	0.168	-22.598	7.241	0.207	36.800
0.6				5.871	0.168	-22.057			
0.5				5.871	0.168	-19.232	7.241	0.207	-148.251
0.4							7.241	0.207	147.671
0.3	4.305	0.123	-75.249	5.871	0.168	154.910			
0.2	4.305	0.123	153.791	5.871	0.168	34.202			
0.1				5.871	0.168	153.732	7.241	0.207	34.562

The data presented highlight the sensitivity of the vortex shedding frequency to variations in both fluid flow velocity and riser geometry. As the fluid velocity increases, the vortex shedding frequency (fs) rises proportionally. This increase in shedding frequency can lead to conditions where fs aligns or closely matches one of the natural frequencies (fn) of the riser structure, creating a potential for resonance. Such resonance, or "lock-in," can result in amplified vibrations that pose significant fatigue risks to the riser.

To analyse lock-in conditions, it is essential to compare the vortex shedding frequency at various fluid velocities with the natural frequencies obtained from modal analysis. The criterion for potential lock-in is met when the difference between the vortex shedding frequency and a natural frequency satisfies the condition If: $|f_s - f_n| \leq \Delta f$, where Δf represents a small frequency deviation that accounts for minor variations in environmental or structural factors. When this condition is met, lock-in is likely to occur, potentially leading to large-amplitude vibrations.

From the data, it is evident that there is a strong potential for lock-in at a fluid velocity of V = 0.2 m/s, where the vortex shedding frequency (fs = 2.544 Hz) is very close to the second natural frequency of the riser (fn = 2.455 Hz). This frequency alignment suggests that resonance could occur under these conditions. However, it is important to note that the vibration phase angle at this velocity does not meet the critical criteria for full lock-in, indicating that while the system is near resonance, the energy transfer between the vortex shedding forces and the riser's structural response is not fully synchronized.

Similarly, at a fluid velocity of V = 0.4 m/s, the shedding frequency (fs = 4.110 Hz) approaches the third natural frequency of the riser (fn = 4.057 Hz), again indicating the potential for lock-in. However, as with the lower velocity case, the vibration phase angle does not meet the criteria necessary for full lock-in. This suggests that while the conditions are conducive to resonance, the phase relationship between the vortex shedding and the riser's oscillations prevents the occurrence of significant energy transfer.

In both cases, although the shedding frequency is close to the natural frequencies of the riser, the absence of phase synchronization prevents full lock-in. These results highlight the importance of considering both frequency alignment and phase angle when evaluating the risk of lock-in in riser dynamics. Understanding these relationships allows for better prediction and mitigation of resonance-related risks in offshore riser design and operation.

4. Conclusion

Numerical methods, especially coupling fluid flow and transient structural, including Fluid Structure Interaction, can be used to accurately predict changes in VIV variables that influence the potential for resonance phenomena (lock-in).

The lock-in phenomenon in riser dynamics presents a critical risk in offshore engineering, primarily due to its potential to induce significant structural fatigue through sustained or amplified oscillations. This phenomenon arises when the vortex shedding frequency aligns with the riser's natural frequency, leading to resonance and increased vibration amplitudes.

The simulation results show the potential for lock-in to occur at certain fluid speeds, namely V=0.2 m/s and 0.4 m/s, which produces vortex shedding frequencies that are close to the natural frequency of the structure, namely fn-2=2,455 Hz and fn-3=4,057 Hz. However, frequency alignment alone is insufficient to trigger full lock-in, as the phase angle between the shedding forces and riser motion plays a crucial role.

5. References

[1] S. I., "Natural Vibration Analysis of Tensioned Risers by Segmentation Method," vol. **61**, no. 5, pp. 647–659, 2006, doi: 10.2516/ogst.

[2] K. S. Hong and U. H. Shah, "Vortex-induced vibrations and control of marine risers: A review," *Ocean Eng.*, vol. **152,** no. September 2016, pp. 300–315, 2018, doi: 10.1016/j.oceaneng.2018.01.086.

[3] Y. Gao, Z. Zong, L. Zou, and S. Takagi, "Vortex-induced vibrations and waves of a long circular cylinder predicted using a wake oscillator model," *Ocean Eng.*, vol. **156**, no. February, pp. 294–305, 2018, doi: 10.1016/j.oceaneng.2018.03.034.

[4] L. Mao, Q. Liu, S. Zhou, and Z. Long, "Experimental study of vortex-induced vibration for drilling risers under uniform flow current," *J. Vibroengineering*, vol. **19**, no. 2, pp. 1358–1371, 2017, doi: 10.21595/jve.2016.17793.

[5] J. Wang, S. Fu, and R. Baarholm, "Evaluation of vortex-induced vibration of a steel catenary riser in steady current and vessel motion-induced oscillatory current," *J. Fluids Struct.*, vol. **82**, pp. 412–431, 2018, doi: 10.1016/j.jfluidstructs.2018.07.018.

[6] D. K. Kim, A. Incecik, H. S. Choi, E. W. C. Wong, S. Y. Yu, and K. S. Park, "A simplified method to predict fatigue damage of offshore riser subjected to vortex-induced vibration by adopting current index concept," *Ocean Eng.*, vol. **157**, no. March 2016, pp. 401–411, 2018, doi: 10.1016/j.oceaneng.2018.03.042.

[7] Z. Zang, Z. Chen, M. Zhao, W. Xu, and Y. Chen, "Experimental study on dynamic responses of a tensioned flexible vertical cylinder under waves and combined current and waves," *Ocean Eng.*, vol. **266**, no. P5, p. 113159, 2022, doi: 10.1016/j.oceaneng.2022.113159.

[8] J. Wang, R. Baarholm, and S. Fu, "Vortex-induced vibration of steel catenary riser under vessel motion," *Proc. Int. Conf. Offshore Mech. Arct. Eng. - OMAE*, vol. **2**, no. May 2015, 2014, doi: 10.1115/OMAE2014-23584.

[9] J. Wang, R. S. Joseph, M. C. Ong, and J. B. Jakobsen, "Numerical investigation on vortex-induced vibration caused by vessel motion for a free hanging riser under small Keulegan-

Carpenter numbers," *J. Offshore Mech. Arct. Eng.*, vol. **141**, no. 4, 2019, doi: 10.1115/1.4041732.

[10] D. Yin, E. Passano, H. Lie, G. Grytøyr, M. Tognarelli, and E. Buba, "Experimental and numerical study of a top tensioned riser subjected to vessel motion," vol. 171, no. December 2018, pp. 565–574, 2019, doi: 10.1016/j.oceaneng.2018.12.029.

[11] J. Duan, J. Zhou, X. Wang, and Y. You, "Vortex-induced vibration of a flexible fluid-conveying riser due to vessel motion," *Int. J. Mech. Sci.*, vol. **223**, no. October 2021, p. 107288, 2022, doi: 10.1016/j.ijmecsci.2022.107288.

[12] W. Chen, M. Li, Z. Zheng, S. Guo, and K. Gan, "Impacts of top-end vessel sway on vortex-induced vibration of the submarine riser for a floating platform in deep water," *Ocean Eng.*, vol. **99**, pp. 1–8, 2015, doi: 10.1016/j.oceaneng.2015.03.002.

[13] J. Zhang, Y. Zeng, Y. Tang, W. Guo, and Z. Wang, "Numerical and Experimental Research on the Effect of Platform Heave Motion on Vortex-Induced Vibration of Deep Sea Top-Tensioned Riser," *Shock Vib.*, vol. **2021**, pp. 4–6, 2021, doi: 10.1155/2021/8866051.

[14] L. Jia, S. Sang, X. Shi, and F. Shen, "Investigation on Numerical Simulation of VIV of Deep-Sea Flexible Risers," 2023.

[15] J. Gu, Y. Wang, Y. Zhang, M. Duan, and C. Levi, "Analytical solution of mean top tension of long flexible riser in modeling vortex-induced vibrations," *Appl. Ocean Res.*, vol. 41, pp. 1–8, 2013, doi: 10.1016/j.apor.2013.01.004.

[16] Zhang Jie, "Effect of Top Tension on Vortex-Induced Vibration of Deep-Sea Risers," no. **Viv**, 2020.

[17] A. Placzek, "Numerical simulation of an oscillating cylinder in a cross-flow at low Reynolds number : Forced and free oscillations," vol. **38**, pp. 80–100, 2009, doi: 10.1016/j.compfluid.2008.01.007.

6. Acknowledgements

The authors acknowledge appreciation to BRIN HPC computing facility service for providing the Ansys software service, and the research facilities provided by the Hydrodynamic Laboratory of National Research and Innovation Agency of Indonesia.

IOP Conf. Series: Earth and Environmental Science **1473** (2025) 012012 doi:10.1088/1755-1315/1473/1/012012

Oceanography Glider Control System Trajectories Base on Salinities Difference.

M Taufiqurrohman[1], S Widiyanto[5*], Suryadi[1], I Diah[1]PK, I Winarno[1], Suhirwan[1], J Subur[1], S Rifandi[1], B Yan[1]D, A Winarno[2], Sudirman[3] and D Rahmatullah[4]

[1] Electronic Engineering, Hang Tuah University, Surabaya 60111, Indonesia.
[2] Marine Engineering, Hang Tuah University, Surabaya 60111, Indonesia.
[3] Vocational Sailing Faculty, Hang Tuah University, Surabaya 60111, Indonesia.
[4] Vocational Electrical Engineering, Universitas Negri Surabaya, Surabaya 60231, Indonesia.
[5] Magister of Ocean Engineering, Hang Tuah University, Surabaya 60111, Indonesia.

*E-mail: sinung.widiyanto@hangtuah.ac.id

Abstract. The Oceanography Glider control system is a sophisticated mechanism designed to enhance the autonomous operation of underwater gliders, which are key tools for underwater oceanographic research and environmental monitoring. This control system optimizes the glider's pitch, roll, and buoyancy, enabling precise navigation and data collection in marine environments using PID signal control. The system integrates using Arduino Uno for adaptive control strategies to adjust the glider's buoyancy, depth and salinization Sensor, ensuring efficient and accurate mapping of oceanographic conditions. Its ability to operate over extended periods using standard equipment is challenging. Extreme conditions make it invaluable for studying marine ecosystems and collecting critical data for salinization and environmental studies. Enclosure resistance to standard extreme conditions is the key to the success of the research. The trajectory error due to gyroscope error correction in a surface is 1.6%, submerse condition 0.7%, using combined salinities sensor calculation 0.6%.

1. Introduction

The Indonesia Ocean's vast archipelagic waters are increasingly turning to advanced technologies such as Oceanography gliders for marine exploration. These autonomous underwater vehicles are instrumental in collecting critical oceanographic data. Oceanography gliders can operate for extended periods, travelling vast distances while collecting data on water temperature, salinity, currents, and other parameters. This capability is particularly valuable in Indonesia, where the maritime environment is complex and dynamic, making traditional survey methods challenging and costly. In the paper "A highly manoeuvrable vectored buoyancy engine based underwater robot", salinities are not included in the algorithm and cause instability during submersible manoeuvrability. This gap in the algorithm causes the Oceanography glider to have some problems taking data in areas with multilayer salinities, such as Indonesia.

Satellite oceanography provides crucial surface data that revolutionize our understanding of the world's oceans. Utilizing satellites equipped with advanced sensors, scientists can collect comprehensive and continuous data on Oceanography surface temperatures, Ocean level changes, ocean colour, and surface winds. This data is essential for monitoring climate change, tracking weather patterns, and studying ocean currents and eddies. Satellite observations offer a global perspective, enabling the detection of large-scale phenomena that significantly impact worldwide weather and climate for navigation technology. Additionally, this surface data supports maritime navigation, fisheries management, and disaster response, making satellite oceanography an indispensable tool in modern oceanographic research and environmental stewardship.

The deployment of Oceanography gliders in Indonesian waters has significantly enhanced our understanding of marine ecosystems and climate change impacts in underwater situations. These devices have provided high-resolution data that help scientists monitor coral reefs, track fish populations, and observe the effects of ocean acidification. The data gathered by Oceanography gliders contribute to more accurate models of ocean circulation and climate patterns, which are crucial for predicting weather phenomena such as monsoons and typhoons that heavily impact the region [1].

Moreover, Oceanography gliders support Indonesia's efforts in marine conservation and sustainable fisheries management. By providing real-time data, they enable more responsive and adaptive management practices. This is essential for protecting the country's rich marine biodiversity and ensuring the livelihoods of millions who depend on fishing. Oceanography gliders also facilitate international collaborations, allowing Indonesian scientists to share data and insights with the global research community, thus contributing to a broader understanding of oceanographic processes and promoting collective efforts in addressing global marine challenges.

2. Method

2.1. Multilayer Salinization and Current

Multilayer salinization in Indonesia's seas poses distinct challenges for oceanographic research, requiring sophisticated data analysis techniques often performed on Unix-based systems. Indonesia's marine environment, influenced by complex interactions between freshwater inflows from rivers and saline ocean currents, leads to varying salinity layers at different depths. Researchers use salinization data systems to process large datasets obtained from Conductivity-Temperature-Depth (CTD) sensors and other oceanographic instruments. These systems are ideal for handling the computational demands of modelling salinity gradients, understanding their impact on marine ecosystems, and predicting climate change effects. The environmental conditions, such as different layer Salinities and currents, affected efficient energy, data manipulation, visualization, and sharing among scientists, enhancing our ability to manage marine resources and study the intricate ocean dynamics in Indonesia's waters [2].

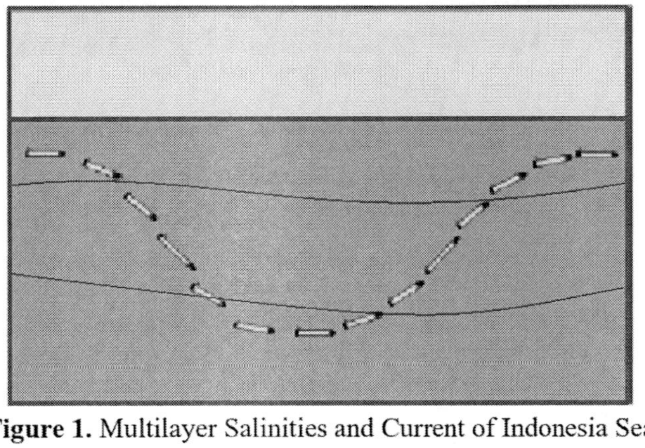

Figure 1. Multilayer Salinities and Current of Indonesia Sea

Figure 1 shows complex interactions between geography, climate, and oceanographic processes that influence the dynamics of multilayer salinity and currents in Indonesian seas. High rainfall and river

IOP Conf. Series: Earth and Environmental Science **1473** (2025) 012012 doi:10.1088/1755-1315/1473/1/012012

discharge reduce salinity near coastal areas, particularly during the wet monsoon season, while the dry monsoon sees higher evaporation rates, increasing salinity. Ocean currents, such as the Indonesian Through flow (ITF), also play a crucial role in transporting saline water from the Pacific Ocean, further influencing local salinity levels [3].

Monsoon winds significantly affect surface currents, with currents flowing southward during the northwest monsoon and northward during the southeast monsoon. The ITF, a significant current flowing from the Pacific to the Indian Ocean through Indonesia, impacts water temperature, salinity, and nutrient distribution. Additionally, strong tidal currents, resulting from the interaction of numerous islands and straits, vary significantly in strength and direction, further influencing the region's oceanographic conditions.

Environmental factors affect the efficiency of Oceanography Glider data gathering and energy consumption. This situation may affect the durability of the Oceanography Glider during Ocean exploration. The key point is using nature to reduce power consumption by gliding following the current and holding trajectories position by using salinization of the ocean. The salinization affected the buoyancy and the stabilization of the Oceanography glider during manoeuvring while submerging [4].

2.2. Schematic Drawing.

An Oceanography glider comprises several essential parts: a buoyancy control system that adjusts the vehicle's density by taking in or expelling water, a battery pack to power its components, and a control unit that includes navigation and communication systems. The control unit processes data from onboard sensors that measure parameters like temperature, salinity, and depth. Additionally, the Oceanography glider is equipped with wings and a rudder for stability and directional control, enabling it to glide through the water in a saw tooth pattern while collecting and transmitting data to researchers when it surfaces [5].

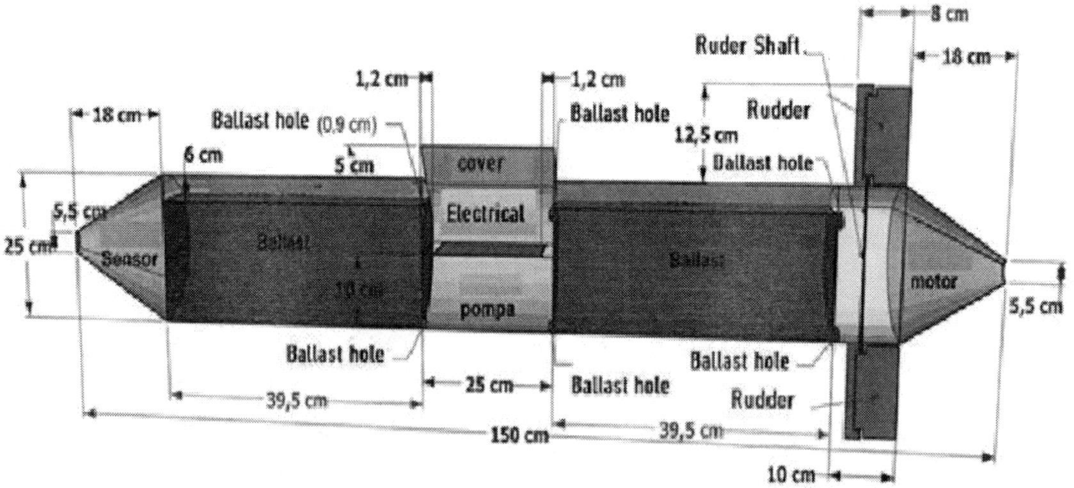

Figure 2. Schematic Drawing.

Figure 2 shows the system ballast in an oceanography glider is crucial for controlling buoyancy and stability during underwater navigation. By adjusting the amount of ballast, the glider can change its density relative to the surrounding water, enabling it to ascend or descend. This ability to manipulate buoyancy is essential for maintaining a desired depth or performing specific manoeuvres while minimizing energy consumption. Ballast systems typically involve the intake or expulsion of water or the shifting of internal weights, allowing the glider to achieve the precise balance needed for efficient and controlled movement through the ocean.

A ballast pump in an Oceanography glider controls its buoyancy by adjusting the amount of water in its ballast tanks. The pump either draws water into the tanks to increase the glider's weight and cause it to descend or expels water to decrease its weight, allowing it to ascend. This dynamic buoyancy adjustment enables the glider to maintain or change depth without relying on external propulsion, enhancing its energy efficiency and allowing for extended underwater missions. The precise operation of the ballast pump is critical for the glider's stability, manoeuvrability, and overall performance [6].

The rudder and fins in an Oceanography glider play a vital role in steering and stabilizing the glider during its underwater journey. The rudder, typically located at the rear, controls the glider's direction by adjusting its angle to deflect water flow, enabling it to turn left or right. The fins, usually positioned along the body or wings, contribute to the glider's stability by helping to maintain a straight course and prevent unwanted rolling or pitching motions. The rudder and fins ensure precise navigation and control, allowing the Oceanography glider to follow predetermined paths, adapt to changing currents, and efficiently reach its destination while conserving energy.

2.3. Block Diagram System

The Block Diagram System used in this experiment must represent the Oceanography glider within Indonesia's ocean character. The main factors in Indonesia's ocean character are salinities, current and temperature, but in this experiment, we focus on mapping the salinities and depth.

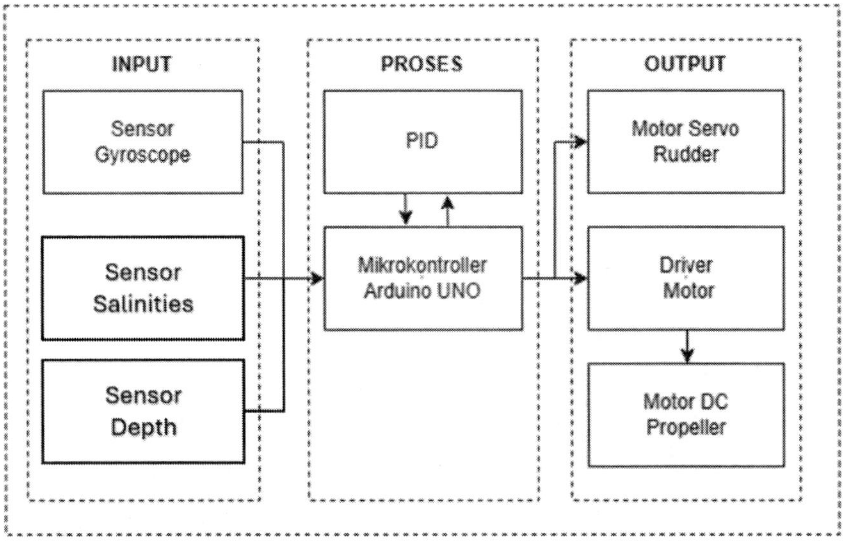

Figure 3. Diagram Block System.

Figure 3 shows the sensors used in this experiment: a gyroscope, salinity, and depth sensor. The gyroscope sensor is essential for maintaining orientation and stability during underwater operations. It measures the glider's angular velocity around its axes, allowing the onboard control systems to monitor and adjust its attitude in real-time. By detecting unintended rotational movements, the gyroscope helps the glider correct its pitch, roll, and yaw, ensuring it stays on its intended path. This precise orientation control is crucial for the glider's ability to perform complex manoeuvres, maintain steady navigation, and operate effectively in varying underwater conditions, all while conserving energy [7].

A depth sensor in an Oceanography glider is a critical component that measures the glider's depth below the surface, providing real-time data essential for maintaining the desired operational depth. This Sensor helps the glider monitor its position in the water column, allowing it to adjust buoyancy and control systems to maintain a stable depth or execute planned dives and ascents. By relaying accurate depth information to the onboard computer, the Sensor ensures the glider can follow precise depth profiles, avoid obstacles, and optimize its performance for various missions, such as oceanographic surveys or environmental monitoring.

A salinity sensor in an Oceanography glider measures the salt concentration in the surrounding seawater, providing crucial data for understanding ocean conditions and supporting various scientific missions. This Sensor helps the glider monitor changes in water density, directly affecting buoyancy and, consequently, the glider's ability to maintain or change depth. Salinity data is also vital for studying ocean currents, water masses, and climate-related processes, as salinity influences the ocean's circulation patterns. By collecting and transmitting salinity measurements, the Oceanography glider contributes to a broader understanding of marine environments, aiding research and environmental monitoring efforts [8].

2.4. *Schematic Electrical Diagram*

A schematic electrical diagram of an Oceanography glider using an Arduino Uno would illustrate the connections between various components essential for the glider's operation. The Arduino Uno, as the central microcontroller, would be connected to a power source, such as a battery, through a voltage regulator to ensure stable operation [9]. Critical sensors like depth sensors, salinity sensors, and gyroscopes would be connected to the Arduino's analogue and digital input pins, providing real-time data for processing. The ballast pump and servo motors controlling the rudder and fins would be connected to the Arduino's output pins, allowing the microcontroller to adjust buoyancy and navigation based on sensor inputs. Communication modules, such as a GPS or radio transmitter, could be included to relay data back to the surface. The diagram shows all these connections, highlighting the integrated system that enables the Oceanography glider to operate autonomously under Arduino's control.

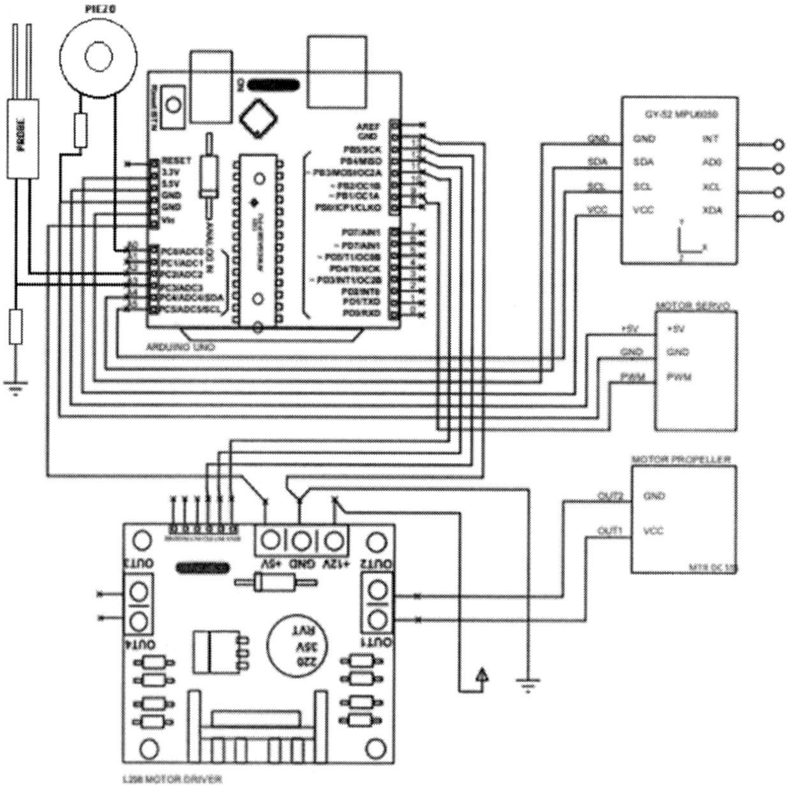

Figure 4. Schematic Electrical Diagram.

According to Figure 4, an electrode for a salinity sensor in an Oceanography glider using an Arduino Uno typically consists of two conductive probes placed in the water to measure the electrical conductivity, which correlates with salinity. These electrodes are usually made of corrosion-resistant materials like stainless steel or platinum to withstand prolonged exposure to seawater. When connected to the Arduino Uno, a small alternating current is passed between the electrodes, and the resulting voltage drop is measured. The Arduino processes this voltage to determine the conductivity of the water, which is then converted into a salinity value using calibration data. This setup allows the Oceanography glider to continuously monitor salinity levels as it moves through different oceanic environments, providing valuable scientific research and navigation data.

A piezoelectric element in a depth sensor for an Oceanography glider using an Arduino Uno functions by converting pressure changes experienced underwater into an electrical signal. As the Oceanography glider descends, the water pressure increases, compressing the piezoelectric material. This compression generates an electrical charge proportional to the pressure, corresponding to the glider's depth. The Arduino Uno reads this signal through its analogue input pins and, using pre-calibrated data, converts the voltage into a precise depth measurement. This setup enables the glider to continuously monitor its depth with high sensitivity and accuracy, which is essential for maintaining desired operational depths, executing specific dive profiles, and ensuring stability throughout its underwater missions.

Using an Arduino Uno, a gyroscope in a stability sensor for an Oceanography glider plays a crucial role in maintaining the glider's orientation and balance underwater. The gyroscope measures the angular velocity around the glider's three axes (pitch, roll, and yaw), providing real-time data on rotational movements. This data is sent to the Arduino Uno, which processes it to determine any deviations from the intended course or orientation. If the glider starts to tilt or rotate unexpectedly, the Arduino can adjust control surfaces like fins or rudders to counteract these movements, stabilizing the glider. This precise control helps the Oceanography glider maintain a steady and efficient trajectory, even in turbulent underwater conditions, ensuring successful navigation and data collection [10].

2.5. PID Signal Controller.

A Proportional-Integral-Derivative (PID) algorithm is an essential control system used in an Oceanography glider to manage various signals such as depth, orientation, and buoyancy. The PID controller continuously calculates an error value as the difference between a desired set point and a measured process variable and then applies corrections based on proportional, integral, and derivative terms.

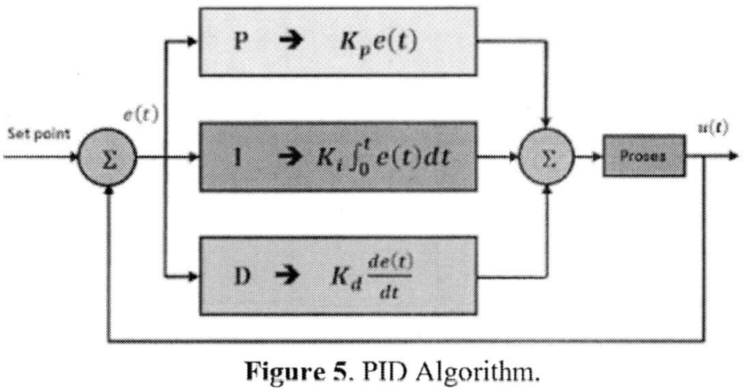

Figure 5. PID Algorithm.

Figure 5 Show the PID algorithm might adjust the ballast pump, rudder, and fin positions to achieve and maintain the desired depth, orientation, and course. The Arduino Uno would read sensor inputs, compute the necessary PID corrections in real-time, and send appropriate signals to the glider's actuators, ensuring smooth and accurate control in various underwater conditions.

The proportional (P) term produces an output value proportional to the current error. If the oceanography glider is not at the desired depth, the proportional control applies a correction proportional to the depth error. The integral (I) term addresses the accumulation of past mistakes. If the glider stays consistently below the desired depth, the integral action sums these errors over time and adjusts the control signal to eliminate the steady-state error.

The derivative (D) term predicts future errors based on the rate of change of the error. It helps dampen the system's response, preventing overshoot and oscillations by correcting based on how quickly the glider approaches the desired depth.

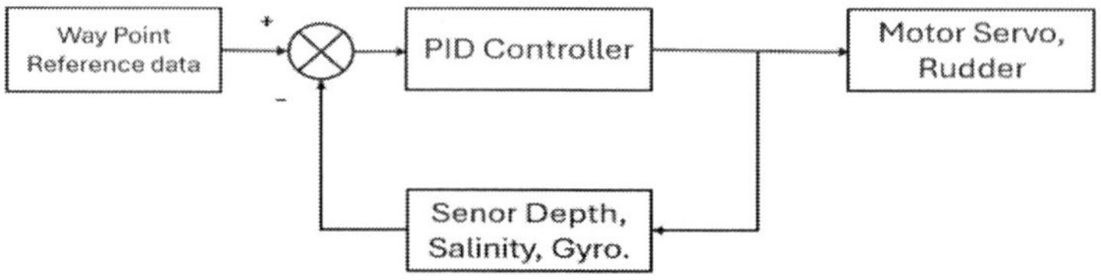

Figure 6. PID Oceanography Glider Control System.

The control system used in this oceanography glider research shows that the waypoint is the essential reference for determining the initial position of the glider. The depth position data and the initial movement plan use a waypoint as a reference. The data are controlled with a PID controller that compares references from the depth, salinity, and gyro sensors. The control signal moves the servo motor, rudder, and ballast system [11].

2.6. Flowchart Diagram

A flow chart diagram of an Oceanography glider using an Arduino Uno begins with initializing the system and sensors, such as depth, gyroscope, and salinity sensors. The Arduino then continuously reads data from these sensors, which is processed to assess the glider's current state, like depth and orientation. Based on this data, the PID control algorithm calculates the necessary adjustments to maintain the desired depth and stability. The Arduino then controls the ballast pump for buoyancy and adjusts the rudder and fins for direction and orientation. The system also checks for any communication from the surface and logs data for later analysis. This process loops until the mission is complete when the glider surfaces and the system powers down.

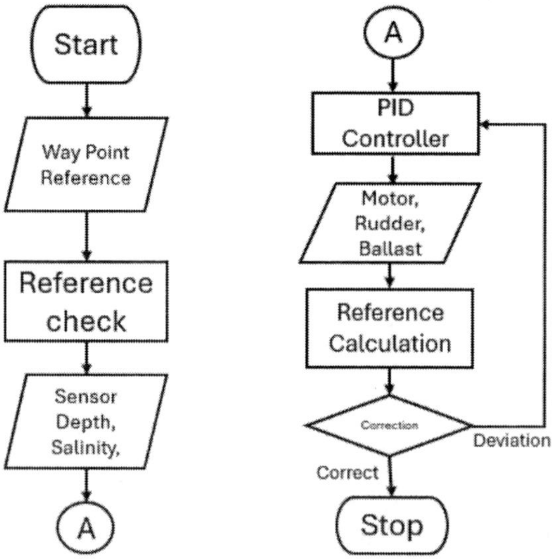

Figure 7. Oceanography Glider Flowchart.

Figure 7 shows the flowchart diagram of the Oceanography glider used for the experiment. It starts and executes an initialization program to activate and read the sensor properties. The sensor data from the gyroscope provides input for the PID algorithm to control the rudder and buoyancy of the oceanography glider. The approximate salinity data used to improve control stabilizes the oceanography glider at a specific depth while it is in buoyancy condition [12].

3. Analysis

3.1 Enclosure Design

A leakiness test of an Oceanography glider is a critical procedure to ensure the integrity of the hull and internal compartments before it is deployed underwater. The test typically involves sealing the glider's hatches and sensor ports, pressuring the interior or submerging the glider in water to check for leaks. During the test, any drop in internal pressure or the appearance of air bubbles would indicate a leak, prompting repairs. This test ensures that the glider is watertight, preventing water intrusion that could damage sensitive electronics or compromise buoyancy, thus guaranteeing its reliability and safety during its underwater mission [13].

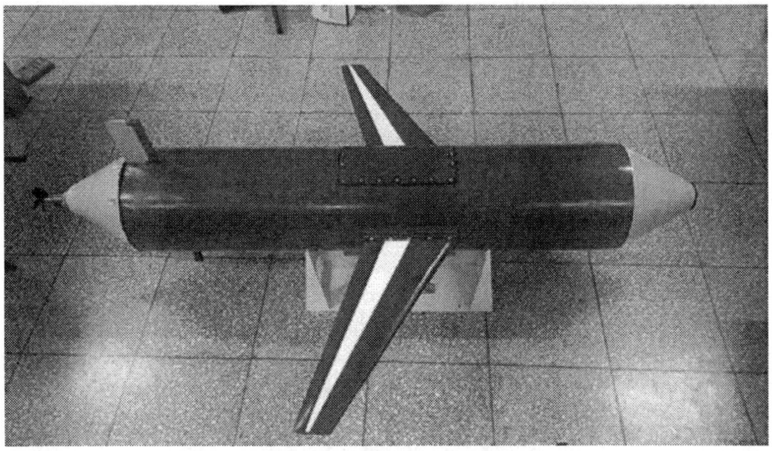

Figure 8. Enclosure Design

The Key point in oceanography glider is the enclosure that can adapt to extreme ocean conditions. Leakiness can affect the performance of the Oceanography glider's durability and cause damage to the instrument. The enclosure used in this experiment, based on Figure 3.1, uses a combination of metal and polycarbonate to reduce the weight of the Oceanography glider. However, the material's durability is not strong enough to take more than 5 bar pressure.

The hydrodynamics of an Oceanography glider are fundamental to its performance, enabling it to move efficiently through water with minimal energy use. The glider's sleek, torpedo-shaped body reduces drag, allowing it to glide smoothly and maintain momentum with less resistance. Its wings generate lift as the glider moves forward, helping it ascend or descend by adjusting its buoyancy and angle of attack. The careful design of control surfaces, like fins and rudder, ensures stability and precise manoeuvrability, allowing the glider to navigate various water conditions and follow a pre-set trajectory. These hydrodynamic features are essential for optimizing the glider's range, endurance, and ability to perform extended missions, making it an effective tool for underwater exploration and data gathering.

3.2 Buoyancy Control

The buoyancy control of an oceanography glider is primarily governed by the principle of Archimedes, which states that the buoyant force acting on an object submerged in a fluid is equal to the weight of the fluid displaced by the object. The equation for buoyancy control in the context of an Oceanography glider can be expressed as:

$$Fb = \rho_\omega \, V_d \, g - W_g \tag{1}$$

Where:

Fb is the net buoyant force acting on the oceanography glider.
ρ_ω is the density of the seawater.
V_d The ballast system can adjust the volume of water displaced by the glider.
g is the acceleration due to gravity.
W_g It is the weight of the Oceanography glider.

Equation (1) shows that in Oceanography glider to ascend condition, Fb must be positive, meaning the buoyant force is greater than the glider's weight. When the glider reduces its density by expelling water from its ballast tanks, decreasing [14]. Conversely, to descend, Fb must be negative, meaning the weight of the glider is greater than the buoyant force, which happens when the glider takes in water, increasing V_d. By fi.ly controllingV_d, the glider can achieve neutral buoyancy, maintaining a constant depth, or adjust its depth as needed for its mission.

The density of the Oceanography glider itself weighs 21.465 kg, and the total volume is $0.073m^3$ The de ty value is obtained from oceanography glider density = 21.465 kg / $0.073m^3$ = 294.04kg/m^3

With the density of water = 1000kg/m^3 , if the Oceanography glider floats, then an object dipped in a liquid floats if the weight of the object W_g It is smaller than the upward force (Fb). W_g < Fb, The resulting ratio of the weight of the object (w) and the buoyancy force (Fb) is 294.04kg/m^3< 1000kg/m^3. From the calculations and comparisons carried out on the Oceanography glider, it can be ascertained that the Oceanography glider can float on water with an Oceanography glider density of 21.465 kg.

3.3 Function test and troubleshooting

A function test for an Oceanography glider is a comprehensive procedure to verify that all systems and components work correctly before deployment. This test ensures the glider can operate autonomously and perform its intended missions reliably. A Power System Check is the critical Step in an Oceanography Glider Function Test. It's verified that the battery and power distribution systems are functioning correctly. Ensure the glider can power up all subsystems, such as sensors, communication devices, and actuators. The Leakiness Test is conducted a leakiness test to ensure that the glider's hull and compartments are watertight. Endurance and Battery Life Assessment simulates a mission to evaluate battery life and power consumption under typical operating conditions.

Sensor Calibration and Verification is calibrating and testing all onboard sensors, including depth, salinity, temperature, gyroscope, and pressure sensors. Confirm that the sensors provide accurate and consistent data. The Buoyancy Control System tests the ballast system by adjusting the buoyancy to simulate ascent and descent. Ensure the ballast pump or other buoyancy control mechanisms intake and expel water effectively.

Navigation and Control Systems Check verify the operation of control surfaces such as the rudder and fins, ensuring they respond correctly to commands. Test the glider's ability to maintain a stable orientation and execute controlled turns and depth changes. Communication System Test checks the functionality of communication systems, such as GPS, radio, or satellite links. Ensure the glider can send and receive data to and from the surface or a remote operator. The data Logging and Processing test confirms that the onboard computer can log sensor data correctly. Test the data processing algorithms, including the PID control loops, to ensure they function as expected [15].

Software and Firmware Check ensures the glider's software is up-to-date and correctly configured. Verify that all firmware functions, including automated scripts for mission execution, operate without errors. Final Diagnostics and Review reviews all test results to identify any issues or areas needing adjustment. Make any necessary repairs or recalibrations before the glider is deployed. This function test ensures the Oceanography glider is fully operational, reliable, and ready for its mission, reducing the risk of failure during deployment [16].

4. Result

A PID control system is used to make oceanography gliders able to respond quickly and have a stable control response. The advantage of PID control is that it is a function that processes error signals originating from the system and then reprocesses them into stable feedback values. The total sum of PID control's proportional, integral, and derivative parts is the present value (last error). In designing a PID control system, the proportional constant (Kp), integral constant (Ki), and derivative constant (Kd) are set so that the response to the output signal in the system to the input is correct. Then, the value is entered into the plant or actuator to carry out the control process. Then, the resulting output is recalculated using the Sensor and compared with the set point value, which is called feedback [17].

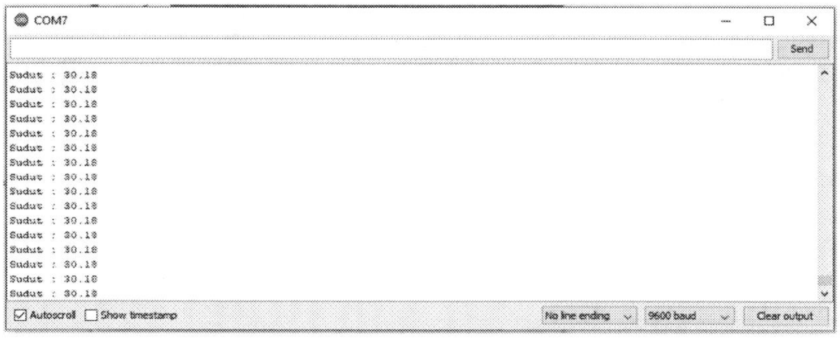

Figure 9. Data logger result test on the surface.

Based on Figure 9, the sensor test on the surface results at an angle of 30°. The average value is 30.57°. To find the error value of the gyroscope sensor at an angle of 30°, the following equation is used:

% error = | (target angle - read angle) / (target angle) | x 100%
% error = | (30-30.57) / 30 | x 100%
Error % = 1.9%

Using the equation above, the error value of the gyroscope sensor reading at an angle of 30° is a 1.9% error margin on the surface.

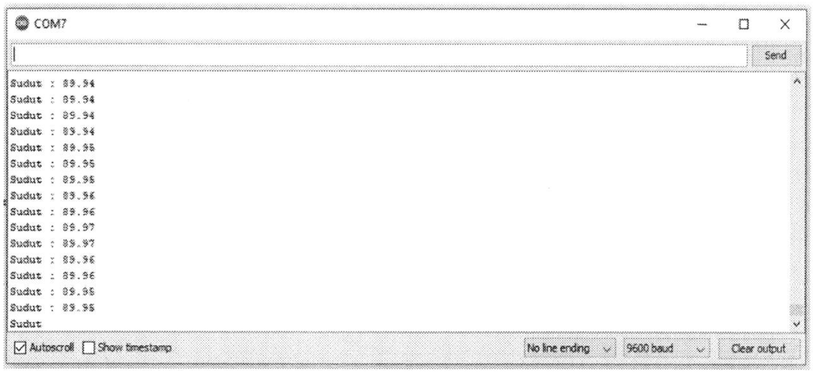

Figure 10. Data logger result test on depth.

Based on Figure 10, the sensor test results at an angle of 45°, and the average value obtained is 45.35°. To find the error value of the gyroscope sensor at an angle of 45°, the following equation is used:

% error=|(target angle-read angle)/(target angle)|x 100%

% error=|(45-45.35)/45|x 100%

Error%=0.7%

Using the equation above, the error value of the gyroscope sensor reading at an angle of 45° is 0.7% error margin on depth.

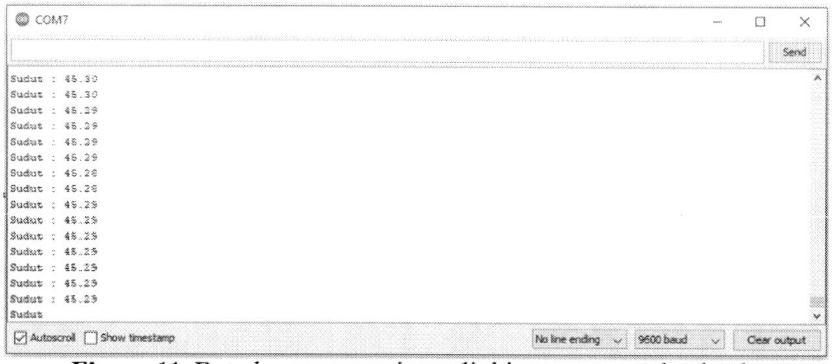

Figure 11. Data logger test using salinities sensor on the depth.

Based on Figure 11, the sensor test results at an angle of 90°, and the average value obtained is 90.58°. To find the error value of the gyroscope sensor at an angle of 90°, the following equation is used:

% error = | (target angle - read angle) / (target angle) | x 100%

% error = | (90-90.58) / 90 | x 100%

Error % = 0.6%

Using the equation above, the error value of the gyroscope sensor at an angle of 90° is 0.6% by adding salinities sensor data calculation.

5. Conclusion

In conclusion, an Oceanography glider's thorough testing and calibration are essential to ensure optimal performance and reliability in underwater missions. Engineers can identify and address potential issues before deployment by conducting comprehensive function tests—covering power systems, sensor accuracy, buoyancy control, navigation, communication, and leak integrity. This rigorous preparation enhances the glider's ability to operate autonomously, collect accurate data, and withstand the challenges of the underwater environment, ultimately contributing to the success of its scientific and exploratory missions.

The Oceanography glider experiment is still in progress, and a better solution to reduce system error is being developed. The instrument is not robust enough to perform under extreme conditions due to failure and mistakes during testing. On the surface, the error reading by the gyroscope is 1.9%. This result is caused by the wave movement on the surface. In-depth, the movement of the wave is relatively stable, and the result of the gyroscope reading is 0.7%. The error is reduced by adding salinities sensor data calculation to 0.6%.

The Salinities sensor is the primary key to reducing error while the Oceanography glider is on a mission to get data.

Future Work

The experiment's future work focuses on the oceanography glider's manoeuvrability based on salinity and current detection ability. The prediction of manoeuvrability of the Oceanography glider during submersible on different salinities will increase the probability of survival of the Oceanography glider.

Acknowledgement

We extend our gratitude to Hang Tuah University for their support of this system. We also thank the Faculty of Engineering and Marine Science (FTIK) at Hang Tuah University for funding the experiment. A special thanks goes to the Department of Electrical Engineering at Hang Tuah University for their comprehensive support throughout the experiment.

References

[1] Sprintall J, Potemra JT, Hautala SL, Bray NA, Pandoe WW. Temperature and salinity variability in the exit passages of the Indonesian Throughflow. Deep Sea Res Part II Top Stud Oceanogr. 2003;**50**(12–13):2183–204. More references.

[2] B. Qiu, S. Chen, J. Wang, and L. Fu, "Seasonal and fortnight variations in internal solitary waves in the Indonesian seas from the SWOT measurements," *J. Geophys. Res. Ocean.*, **129**, 7, p. e2024JC021086, 2024.

[3] T. Chang, M. F. A. Ismail, A. S. Budiman, A. Bayhaqi, and E. Kusmanto, "Mooring observations of tidal currents in the Jailolo Strait of the Indonesian seas," *J. Sea Res.*, **199**, p. 102494, 2024.

[4] M. R. Azhari, L. S. Hudha, D. W. Kurniawidi, and R. B. Prahasitiwi, "Design and Development of A Portable Conductivity-Based Salinity Measurement System for Coastal Well Water," *KONSTAN-JURNAL Fis. DAN Pendidik. Fis.*, **9**, 01, pp. 52–57, 2024.

[5] Y. Zhang, Q. Liu, F. Gao, S. Zhou, W. Zhang, and W. Chen, "Design and modelling of wave energy converter glider (WEC-Glider) with simulation validation in wave tank experiments," *Appl. Energy*, **364**, p. 123175, 2024.

[6] L. Han *et al.*, "Design, hydrodynamics analysis, and control of an underwater glider with controllable wing mechanism," *Ocean Eng.*, **311**, p. 119017, 2024.

[7] W. Cui, L. Lian, and D. Zhang, "Frontiers in Deep-Sea Equipment and Technology II," *Journal of Marine Science and Engineering*, **12**, 4. MDPI, p. 635, 2024.

[8] C. Flaim, "Development and validation of a DIY profiling float for indirect determination of salinity," 2024.

[9] A. Babić *et al.*, "Design and development of a low-cost floater for sustainable fishing," *Submit. to" IEEE J. Ocean. Eng.*, 2024.

[10] G. Briguglio and V. Crupi, "Review on Sensors for Sustainable and Safe Maritime Mobility," *J. Mar. Sci. Eng.*, **12**, 2, p. 353, 2024.

[11] T. Bai, Y. Ma, J. Yu, and Z. Zhang, "Numerical Study of Ocean Current Effect on the Motion of Underwater Glider," in *ISOPE International Ocean and Polar Engineering Conference*, 2024, p. ISOPE-I.

[12] A. Hasanvand and M. S. Seif, "Development of state space models for underwater gliders equipped with embedded actuators in vertical and horizontal planes," *Ocean Eng.*, **309**, p. 118344, 2024.

[13] P. Baillon, "Test and Evaluation Model for Midwater Docking of Autonomous Underwater Vehicles," 2024.

[14] K. Macaulcy, L. Cai, P. Adamczyk, and Y. Girdhar, "ReefGlider: A highly manoeuvrable vectored buoyancy engine based underwater robot," *arXiv Prepr. arXiv2405.06033*, 2024.

[15] E. Petritoli and F. Leccese, "A New Approach to Reliability Evaluation and Fault Diagnosis of UAV as a Complex System," in *2024 11th International Workshop on Metrology for AeroSpace (MetroAeroSpace)*, 2024, pp. 449–454.

[16] A. S. Chaphalkar, "Design and Development of a Hydrophone Array for an Autonomous

Underwater Vehicle Capable of Real-Time Detection and Tracking of Surface Vessels." Virginia Tech, 2024.
[17] H. Yang and N. Mahmoudian, "Gliding in extreme waters: Dynamic Modeling and Nonlinear Control of an Agile Underwater Glider," *arXiv Prepr. arXiv2402.06055*, 2024.

Fatigue Life Analysis Due to Vortex Induced Vibration (VIV) On Pipeline Free Span Section in Operating Condition

A S M Roberte[1], R W Prastianto[1*], N Syahroni[1], Murdjito[1], and F Rachmadiarto[2]

[1] Department of Ocean Engineering, Institut Teknolofi Sepuluh Nopember, Surabaya, Indonesia
[2] PT. ZEE Indonesia, Ruko Golden Madrid 1 Blok C No 18 Jalan Letnan Sutopo BSD Serpong, Kec. Serpong, Kota Tangerang Selatan, Banten 15311, Indonesia

*E-mail: rudiwp@oe.its.ac.id

Abstract. Subsea pipeline is used for transferring oil and gas. The diverse seabed topography can cause free span conditions in certain areas. Environmental loads like wave and current flows around the pipeline can induce oscillations in the pipeline, known as vortex-induced vibrations (VIV). If the frequency of the vortex shedding approaches the pipeline's natural frequency, resonance can occur, leading to fatigue in pipelines. This study analyze the VIV to understand the flow patterns around the pipeline through numerical simulation method. The flow pattern is identified at a Reynolds number of 3.85×10^5, indicating turbulent flow which can be classified into 2P vortex shedding pattern. The vortex shedding frequency around the pipe is 0.18 Hz. In this study, VIV criteria calculations are performed based on DNV GL RP F101 guidelines. The study finds that in-line VIV is occured, so that the fatigue life analysis is required. The von Mises stress on the pipeline is 21.41 MPa. Fatigue analysis using the rain-flow counting method shows a fatigue life of 38 years for the pipeline with free span. A comparative analysis on the pipe without free span indicates a fatigue life of 117 years. Both results are within the structure's design life of 20 years.

1. Introduction

Energy plays a crucial role across all sectors of life, particularly in the economy and industrial development. Oil and gas serve as the primary energy sources for both industrial and transportation sectors. One of the most critical operations in the oil and gas industry is transportation, which involves moving exploration outputs from reservoirs to production sites. A key facility used for transporting exploration products is the underwater pipeline.

Pipelines represent a technology for transporting fluids such as oil, gas, or water in vast quantities across considerable distances, often through specific marine environments. The first pipeline was constructed in the United States in 1859 to transport crude oil [1]. Compared to bulk transportation, pipeline transport is relatively safer. The Pipeline is also very environmentally friendly transportation, but it is very harmful to the environment in case of failure [2]. Pipeline networks have demonstrated their ability to adapt to a wide range of environments, including remote and extreme conditions [3]. Although subsea pipelines have demonstrated numerous advantages and a high level of effectiveness, there are several critical factors that must be considered in their operation. If these factors are neglected, pipeline failures can occur, leading to significant economic and environmental damage.

Environmental loads, such as wave and current flows around the pipeline, can induce oscillations or vibrations within the pipeline. Free spans made by subsea topology's changes, artificial support and bed roughness [4]. According to the DNV-OS-F101, free span leads to various phenomena such as bursting, fracture and fatigue. Water particles collision with the pipeline's body causes vibrations in the pipeline, which are called Vortex Induced Vibration (VIV) made by free span [5]. The reverse current that occurs due to vortex shedding will cause a cyclic load on the pipe that causes vibration [6]. If the frequency of these vortices approaches the pipeline's natural frequency, resonance occurs, leading to fatigue in the pipeline [7]. In pipeline design, it is essential to calculate the allowable free span length to create supports for sections of the pipeline not resting on the seabed, thereby reducing the likelihood of failure due to VIV. Research has been conducted on the effect of the span length on multi free spans. The impact of the span length has been studied for a single free span [8]. Vortex Induced Vibration (VIV) can have a significant impact on the interaction between the seabed and the cylinder. When the vibration frequency of the cylinder aligns with the vortex shedding frequency, large amplitude responses can occur, exacerbating damage due to structural fatigue. Many factors strongly influenced the cylinder dynamics, such as Reynolds number range, aspect ratio, mass ratio, and the motion degrees of freedom of the cylinder, etc [9]. Prastianto, et al observed the hydrodynamic force for multiple cylinder cantilever factors dictated the response, such as the Reynolds number, cylinder arrangement, and incoming flow condition [10] then continued to evaluate the hydrodynamic forces and motions experienced by two flexible circular cylinders in staggered arrays and found that the flows were performed simultaneously [11]. VIV should be minimized as much as possible because its effects can reduce the fatigue strength of structural elements. One of the solutions is by adding the suppression device [12], and calculate the risk analysis

The aforementioned risks are part of the potential failure consequences that must be addressed. A thorough risk analysis is essential during the pipeline design process. The risk analysis method for subsea pipelines is divided into two stages: stress analysis and fatigue analysis. It is crucial to calculate the stress range experienced by pipelines with free spans to determine their fatigue life [13].

In this study, the Vortex Induced Vibration (VIV) phenomenon was modelled in a free span section of subsea pipeline. The study also includes VIV analysis and stress analysis of the pipeline, followed by fatigue life analysis of the subsea pipeline in a free span condition due to VIV, in accordance with applicable regulations.

2. Research Methods

The Methodology used in this study is shown in the flowchart in the Figure *1*

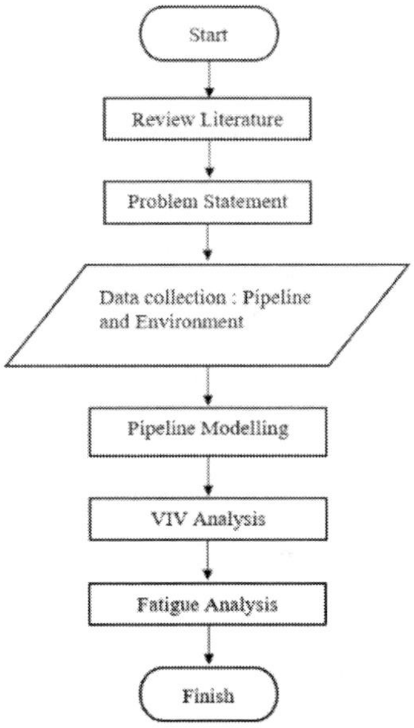

Figure 1. *Research Flowchart*

In this research, modeling were carried out on the subsea pipeline operating in the Kepulauan Seribu Sea, Indonesia in the longest free span section that is 22.39 m length. The numerical simulation method has become an indispensable complement to experiments in ocean engineering [14]. The modeling was conducted using finite element method-based software and analyzed using the DNV-GL RP F101 code. In modeling consider environmental loads that is currents. The sea pipeline is exposed to loads due to current and waves. The wave effect in deep water is neglected and only the effects of the current have been investigated [15]. Vortex induced vibration (VIV) analysis will be carried out on this section to obtain the natural frequency and the vortex shedding frequency under the conditions of the freespan. hydrodynamic forces on a circular cylinder in oscillatory flow experimentally and used a mode-averaging technique to obtain the time history of forces for a typical vortex-shedding mode [16]. VIV analysis using the Iwan Blevins Model method is a suitable approach to determine the vibration of the structure. In the next process, fatigue anlysis will be conducted using rainflow counting method. Through fatigue analysis using the Rainflow Counting method, we can obtain detail information about the fatigue life of the structure in variable amplitude (variable loading). Properties data in this study are presented in Table 1 and the environmental data shown in Table 2

Table 1. *Pipeline Properties Data*

Description	Value	Units
Outside Diameter	406.4	mm
Wall Thickness	12.7	mm
Material	API 5L X52 MO PSL2	-
SMYS	360	MPa
Young Modulus	2.07×10^5	MPa
Density	7850	kg/m^3
Poison Ratio		

IOP Conf. Series: Earth and Environmental Science **1473** (2025) 012013 doi:10.1088/1755-1315/1473/1/012013

Table 2. *Environmental Data*

Description	Units	Return Period	
		1 Yr	100 Yrs
Water Depth	m	31	
Significant Wave Height (Hs)	m	1.8	3.6
Significant Wave Period (Ts)			
Peak Period (Tp)	s	6.4	9.1
Current Speed at the Seabed	m/s	0.34	0.36
Current Speed at the Surface	m/s	0.72	1.15

3. Result and Discussion

3.1 Modelling of the Pipeline

The modeled pipeline is the longest free span section located in Kepulauan Seribu Sea, West Java in the 31 m water depth and 22.39 m length. The modeling in this study is based on Computational Fluid Dynamics (CFD) and is facilitated by the ANSYS Fluent software. ANSYS utilizes a Finite Element Analysis (FEA) method, where the analysis of an object is performed by dividing it into small parts (finite elements) called a mesh, which are connected by nodes. The modeling aims to observe the vortex shedding patterns caused by environmental loads from currents and waves. The results of the geometric modeling in ANSYS Fluent are shown in Figure 2

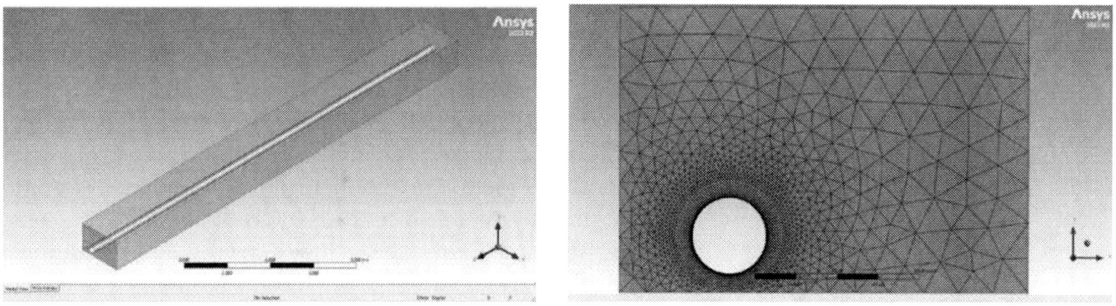

Figure 2. Pipeline Geometry Modelling

In this research, the meshing method employed is the Multizone scheme with Prism/Hexa elements. The model consists of 1,725,988 mesh elements. During the meshing process, segments were named, designating the left domain wall as the inlet and the right domain wall as the outlet. The resulting meshed geometry, shown in Figure *2* is displayed in a transverse view.

Following the meshing process, the next steps involve setup and defining boundary conditions. This includes inputting a total current velocity of 0.72 m/s, based on 100-year current conditions, selecting seawater as the material type, and entering a density value of 1025 kg/m^3. The modeling was conducted using transient flow with a coupled scheme over 800 time steps, with 10 iterations per time step. The result of vortex shedding pattern is shown in Figure 3.

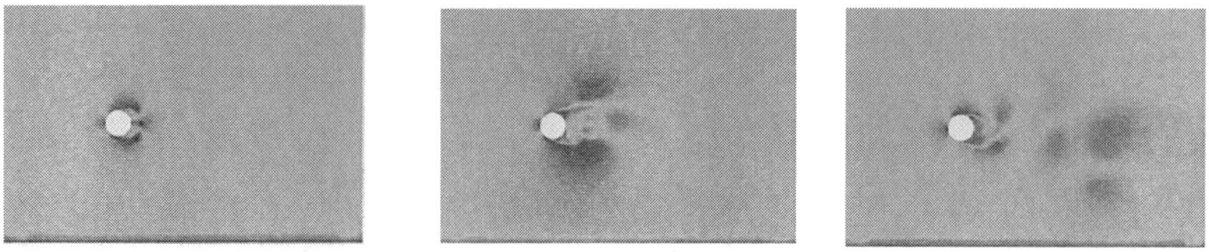

IOP Conf. Series: Earth and Environmental Science **1473** (2025) 012013 doi:10.1088/1755-1315/1473/1/012013

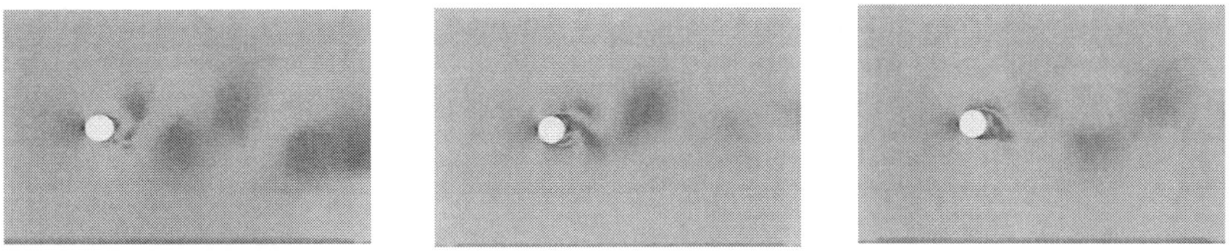

Figure 3. *Vortex Shedding Pattern*

From the vortex patterns formed as a result of the modeling, vortex patterns were identified based on their shape and Reynolds number. The Reynolds number is a dimensionless quantity used in fluid dynamics to predict fluid flow behavior, whether it be laminar or turbulent. It is defined as the ratio of inertial forces to viscous forces within the fluid, aiding in determining the type of flow, whether laminar or turbulent. In general, the flow pattern becomes complex due to varying flow conditions, such as differences in cylinder spacing ratios, the aspect ratio L/D (where L represents the spanwise height of the cylinder), boundary conditions at the cylinder ends, blockage ratio, Reynolds number, and other factors [17]. Reynolds number of 3.85 x 10^5 was obtained, indicating turbulent flow. Usually, the 2P and P+S pattern types occur when the vibration amplitude is greater and indicating as a turbulent flow [18]. It was observed from the result of this study that two pairs of vortices were released in each shedding cycle. Therefore, it can be determined that the vortex pattern formed is 2P

3.2 Reduced Velocity

The reduced velocity parameter is a non-dimensional quantity calculated to determine the existence of vortex shedding. From the value of the reduced velocity, the type of motion around the flow can be identified as either in-line or crossflow, with the following criteria:

- In-line response occurs if the value of VR is between 1.0 < VR < 2.2
- Crossflow response occurs if the value of VR is VR < 2.2

Reduced velocity represents the speed of water particles where vibrations in the pipe due to vortex shedding occur [19]. The result of the reduced velocity calculation is shown in Table 3

Table 3. *Reduced Velocity*

Parameters	Notation	Value	Unit
Span Length	Ls	22.39	m
Stability Parameter	Ksd	0.862	-
ψproxy,onset	-	0.952	-
ψtrench,onset	-	1	-
Reduced Velocity In-line	VR,IL	1.2	-
Reduced Velocity Crossflow	VR, CF	2.59	-

3.3 Vortex Shedding Frequency

The vortex shedding causes continuous excitation force to act on the structure, which will cause structural vibration [20]. Vortex-induced vibration (VIV) is a significant cause of pipeline failure. Failure can occur if the frequency of vortex shedding approaches the natural frequency of the pipe. Ocean currents passing over a cylinder cause the pipe to vibrate, inducing vibrations in both the in-line and crossflow directions. In the in-line direction, the vibration occurs in the same direction as the incoming current, whereas in the crossflow direction, the vibration occurs perpendicular to the direction of the incoming current.

The frequency of vortex shedding is influenced by the Strouhal number, a dimensionless frequency that is a function of the Reynolds number [21]. The results of the vortex shedding frequency calculations are presented in Table 4.

Table 4. *Vortex Shedding Frequency*

Parameters	Unit	Value
Gap	m	0.3
Strouhal Number	-	0.122
Vortex Shedding Frequency	Hz	0.18

3.4 Natural Frequency

The natural frequency of a pipe is a crucial parameter that must be considered as it serves as a benchmark for determining the allowable maximum span length. If the natural frequency of an underwater pipe approaches the frequency of vortex shedding, the pipe will vibrate, potentially leading to fatigue and failure [22]. The natural frequency of the pipe is calculated based on the equations in DNVGL RP F105. To obtain the natural frequency of an underwater pipe, several parameters need to be calculated, including the effective span length to account for the effects of both pipe supports and pipe-soil interaction. Additionally, the effective axial force, critical buckling load, and static deflection must also be calculated. The natural frequency of the pipe is computed for both the in-line and crossflow directions. The results of the natural frequency calculations are shown in Table 5.

Table 5. *Natural Frequency*

Parameters	Unit	Value
Static Deflection	m	2.25×10^{-1}
Critical Buckling Load	N	1.98×10^{5}
Natural Frequency In-Line	Hz	0.938
Natural Frequency Crossflow	Hz	0.941

3.5 VIV Criteria

Based on DNVGL RP F105, the screening criteria involve assessing the occurrence of fatigue due to VIV caused by environmental loads. The VIV criteria for in-line is shown in equation 1 and for crossflow is shown in equation 2. In this study, the VIV criteria were not met in the in-line direction. Therefore, further analysis is required to determine the fatigue life in the in-line direction. The results of the fatigue criteria are presented in Table 6

$$\frac{f_{o,IL}}{\gamma_f} > \frac{U_{c,100\,tahun}}{V_{R,onset}^{IL} \times D_t}\left(1 - \frac{L/D_t}{250}\right)\frac{\gamma_{IL}}{\bar{\alpha}} \tag{1}$$

$$\frac{f_{n,CF}}{\gamma_f} > \frac{U_{c,100\,tahun} + U_{w,1\,tahun}}{V_{R,onset}^{CF} \times D_t} \times \gamma_{CF} \tag{2}$$

Table 6. *VIV Criteria*

Direction	Frequency Natural (Hz)	Reduced Velocity	Screening	Remarks
In-Line	0.938	1.2	0.877	VIV Occured
Crossflow	0.941	2.5	0.569	VIV not Occured

3.6 Von Misses Stress

A calculation was also conducted to determine the equivalent stress or von Mises stress on the pipe when subjected to environmental current loads. To obtain the most frequently occurring or extreme value of the von Mises stress, the analysis was carried out over a period of 10,800 seconds. Standards, the equivalent or von Mises stress value must not exceed 90% of the Specified Minimum Yield Strength (SMYS) [23]. Based on the analysis results shown in Figure 4, the data was then transformed into a Rayleigh distribution to determine the Most Probable Point, which was found to be 21.41 MPa.

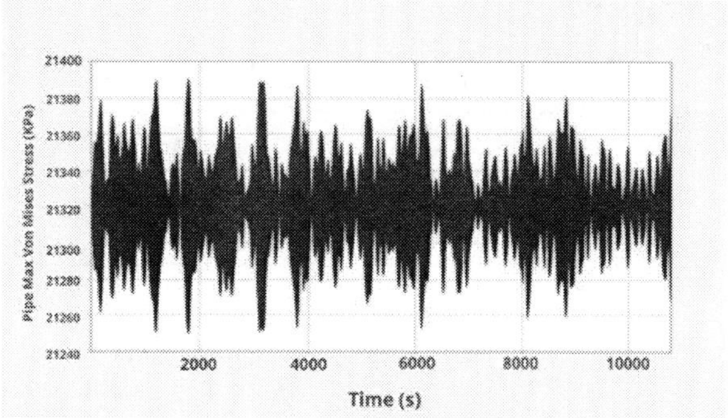

Figure 4. *Stress on the Pipe (time history)*

3.7 Fatigue Analysis

Calculation and analysis of fatigue in a pipeline are more necessary due to the possibility of failure in different parts of the pipeline [24]. In this research, fatigue analysis was conducted using the rainflow counting method in ORCAFLEX. This method is utilized for structures with complex and varying amplitude load histories [25]. The rainflow counting method can account for cycles in detail, thus providing accurate results. In ORCAFLEX, the rainflow counting method was used to determine the load cycles occurring at each current variation. Subsequently, ORCAFLEX performed Cumulative Damage calculations using the Palmgren-Miner method, with input parameters such as the exponent number and material-specific parameters from the S-N curve. The fatigue analysis in this study was only influenced by the current load at a pipe depth of 31 meters. The results of the fatigue calculations with considering free span are shown in Table 7 and the fatigue life graph for each pipe segment with considering free span is presented in Figure 5.

Table 7. *Fatigue Analysis Result with Free Span*

Parameters	Unit	Value
Damage Over Total Exposure	-	0.00215
Total Exposure Time	Year	0.0826
Fatigue Life	Year	38.42

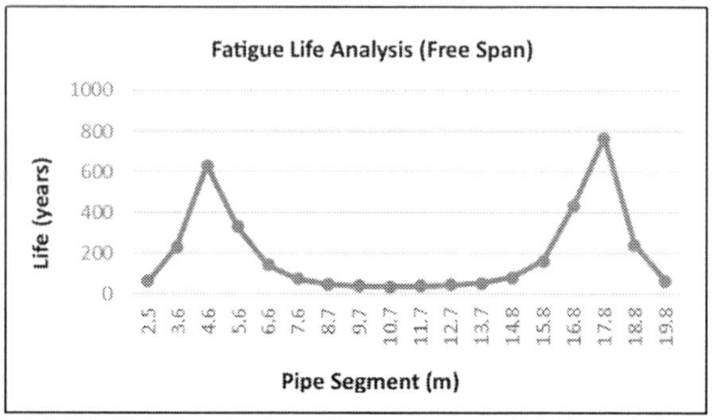

Figure 5. *Fatigue Analysis Graphic with Free Span*

In this study, fatigue life analysis was also conducted on a pipe model without considering for the scouring (i.e., the pipe laying on the seabed) to compare whether scouring significantly impacts the fatigue life of underwater pipes. The results of the fatigue analysis for the pipe structure laying on the seabed are presented in Table 8, and the fatigue life graph for each pipe segment is shown in Figure 6

Table 8. *Fatigue Analysis Result with no Free Span*

Parameters	Unit	Value
Damage Over Total Exposure	-	0.00070056
Total Exposure Time	Year	0.826
Fatigue Life	Year	117.04

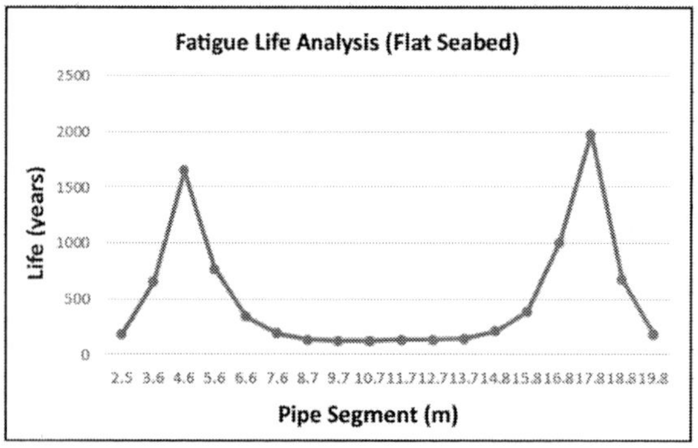

Figure 6. *Fatigue Analysis Graphic with no Free Span*

4. Conclusion

Based on the analysis conducted, the vortex shedding pattern formed from the modeling is 2P with a Reynolds number of 3.85 x 10^5. This study observed Vortex-Induced Vibration (VIV) in the in-line direction, as evidenced by the failure to meet the fatigue criteria in the in-line direction. Therefore, stress and fatigue life analyses are necessary to prevent structural failure. The von Mises stress in the pipe structure is 21.41 MPa, which remains below 90% of SMYS according to ASME B31.4. The fatigue life, considering the effect of scouring (gap), is calculated to be 38 years using the rainflow counting method. In contrast, the fatigue life for the structure without the influence of scouring is 117 years. It can be concluded that the structure has a fatigue life that does not exceed the design life, and thus no revision of the design life is required.

5. References

[1] Bai, Yong, 2001, Pipelines and Risers, Elsevier
[2] Dey, Prasanta Kumar, dkk. "Risk Based Maintenance Model for Offshore Oil and Gas Pipelines: A Case Study". Journal of Quality in Maintenance Engineering (2004): Vol. 10, number 4, Page 169-183.
[3] Guo et al, 2014, Offshore Pipelines: Design, Installation, and Maintenance: Second Edition, University of Lousiana, 2014
[4] Wang, J., et al., VIV analysis of pipelines under complex span conditions. Journal of Marine Science and Application, 2009. 8(2): p. 105-109.
[5] Williamson, C. and R. Govardhan, A brief review of recent results in vortex-induced vibrations. Journal of Wind Engineering and Industrial Aerodynamics, 2008. 96(6): p. 713- 735.
[6] Yttervik, Rune. "Fatigue from Vortex-Induced Vibration of Free Span pipelines Using Statistics of Current Speed and Direction". Proceeding of 22nd Internation Conference on Offshore Mechanics and Artiic Engineering (2003). Cancun Mexico: 8-13 juni
[7] Mouselli, A, 1981, Offshore Pipeline Design, Analysis, and Methods, Oklahoma: PennWell Books
[8] Xu, T., B. Lauridsen, and Y. Bai, Wave, induced fatigue of multi-span pipelines. Marine structures, 1999. 12(2): p. 83-106.
[9] R. W. Prastianto, K. Otsuka & Y. Ikeda, Vortex-induced Vibration of a Flexible Free-hanging Circular Cantilever, 2009, ITB J. Eng. Sci. Vol. 41, No. 2
[10] R.W. Prastianto, K. Otsuka, Y. Ikeda, Hydrodynamic Forces On Multiple Free-hanging Circular Cantilevers In Uniform Flows, 2009, International Journal of Offshore and Polar Engineering
[11] R.W. Prastianto, K. Otsuka, Y. Ikeda, Dynamics of Two Flexible Hanging-off Circular Cylinders in Staggered Configurations, 2008, ISOPE Pacific/Asia Offshore Mechanics Symposium
[12] R.W. Prastianto, K. H. Dwipayana, N. Syahroni, and B. Pumbarino, Experimental Study on Interference Effects of Two Tandem Cylinders Wrapped Around by Triple Helical Rods with Gap on Induced Drag, 2018, IOP Conf. Series: Earth and Environmental and Science
[13] DNV GL RP F101, 2017, Submarine pipeline systems, 2017
[14] Pan, Z.Y.; Cui, W.C.; Miao, Q.M. Numerical simulation of vortex-induced vibration of a circular cylinder at low mass-damping using RANS code. J. Fluids Struct. 2007, 23, 23–37
[15] Thoman, D.R., L.J. Bain, and C.E. Antle, Inferences on the parameters of the Weibull distribution. Technometrics, 1969. 11(3): p. 445-460
[16] Obasaju, E. D., Bearman, P. W., and Ghraham, J. M. R. (1988). "A study of forces, circulation and vortex patterns around a circular cylinder in oscillating flow." J. Fluid Mech., 196, 467–494
[17] Rahmawati, S; Djatmiko E.B, Prastianto, R.W; Numerical Study of Vortex-Induced Vibrations (VIV) Interaction Between Four Vertically Flexible Cylinders in an In-Line Square Configuration Under Uniform Flow, 2012, Jurnal Teknik Its Vol. 1, No. 1

[18] Ussama Ali, Md Islam, Janajreh Isam, Yap Fatt, Md Mahbub Alam, (2021), Energies MDPI Journal, Vol 14, Flow-Induced Vibrations of Single and Multiple Heated

[19] Rosyid, D. M., 2022, Risk Assessment of Onshore Pipeline in Gresik Area, vol International Journal of Offshore and Coastal Engineering, 29-34.

[20] Prastianto, R.W, Pumbarino, B, Experimental Study on a Side-By-Side Configuration of Two Cylinders Helically Covered by Triple Rods, 2022, Recent Advances in Mechanical Engineering, Lecture Notes in Mechanical Engineering

[21] Seyed Mohammad Hossesin Sharifi, M. T.,2019, Evaluation of Vortex Induced Vibartion Effective Parameters on Free Span Subsea Pipelines, International Journak of Coastal Offshore Engineering, 9-15.

[22] Bai, Yong dan Qiang Bai, 2005, Subsea Pipeline Design, Analysis, and Riser, Elsevier

[23] ASME B31-4, 2022, Pipeline Transportation Systems for Liquids and Slurries

[24] Shabani, M.M., A. Taheri, and M. Daghigh, Reliability assessment of free spanning subsea pipeline. Thin-Walled Structures, 2017. 120: p. 116-123.

[25] Poore Aubrey, Doedel Eusebius, Cermak Jack, 1986, Pergamon Journal, Vol 21, Dynamics of The Iwan-Blevins Wake Oscillator Model, Great Britain: National Science Foundation

IOP Conf. Series: Earth and Environmental Science **1473** (2025) 012014 doi:10.1088/1755-1315/1473/1/012014

An experimental study to estimate the strain energy release rate (SERR) criterion of sandwich plate for ship structure with interfacial debonding flaws

H Siswanti[1,4], A Zubaydi[2*], B Piscesa[3] and R C Ariesta[2]

[1] Department of Ocean Engineering, Faculty of Marine Technology, Institut Teknologi Sepuluh Nopember, Surabaya, Indonesia
[2] Department of Naval Architecture, Faculty of Marine Technology, Institut Teknologi Sepuluh Nopember, Surabaya, Indonesia
[3] Department of Civil Engineering, Faculty of Civil, Planning, and Geo-Engineering, Institut Teknologi Sepuluh Nopember, 60111, Surabaya, Indonesia
[4] Department Marine Technology, Politeknik Negeri Madura, Sampang, Indonesia

*E-mail: zubaydi@its.ac.id

Abstract. The experimental study to estimate the strain energy release rate (SERR) criterion of steel polyurethane sandwich plate for ship structure with interfacial debonding flaws had been reported. The results of this experimental study provide valuable insights for characterizing the interfacial fracture toughness of sandwich plates using steel faceplates and polyurethane elastomer cores for ship structure applications. The mode-I and mode-II loading test methods produced different values of the strain energy release rate (SERR), related to the various mechanisms of damage propagation and the nature of the stress states associated with each mode. The critical strain energy release rate of mode-II loading (G_{IIC}) is much higher than the mode-I loading (G_{IC}). This difference highlights the distinct crack growth behaviours and energy dissipation mechanisms involved in mode-I and mode-II loading.

1. Introduction

The primary purpose of sandwich plate application in ship structure is to get a lightweight structure to increase the payload. Previous studies show that replacing the solid plate with a sandwich plate reduces the structure's weight by 4-9% [1-4]. The sandwich plate consists of two layers of steel faceplates and a polyurethane elastomeric core, bonded with adhesive at the interface of each layer [5]. A high-density faceplate and low-density core produce a robust, lightweight sandwich plate. The thin faceplates contribute to the strength of the plate, and the thick core contributes to additional stiffness [6]. The sandwich structures are successfully used in various marine applications, including marine vessels, FPSO and other offshore structures.

Ship structures are often subjected to out-of-plane loads during production and operation. In such cases, thus loads may increase the risk of damage to the ship's structure. However, the type of material used can also influence the risk of vulnerability to damage. In the case of ship structures that use sandwich plates, the vulnerability to damage is also influenced by the type of material. Different faceplates and core material properties result in damage vulnerability, especially at the interface of the two layers [7]. One of the damages is debonding or detachment of the faceplate from the core. Face core debonding can be caused by adhesive damage or voids at the interface between the faceplate and core due to imperfections in the production process or impact on the sandwich plate. Even though it does not directly cause loss of watertightness, this damage reduces the structure's strength. Moreover, if debonding is undetected and has propagated, this damage becomes a critical failure mode for the structure that compromises the structural integrity [7, 8]. Therefore, designers must analyze and assess the residual strength of structures with debonding damage to avoid fatalities.

Researchers have been investigating the properties of sandwich materials for several years. Recently, emphasis has been placed on damage behavior based on fracture mechanics principles. Characterizing sandwich plate damage behavior is part of the damage tolerance analysis process to prevent fatalities. [7]. Material damage behavior is generally evaluated based on the principle of maximum load, fracture mechanics, and other failure criteria. Damage propagation analysis will be conducted experimentally with mode-I, mode-II loading, or mixed-mode [9-13]. These tests characterize the damage propagation in sandwich plate specimens by measuring their fracture toughness. Fracture toughness measures a material system's resistance to crack propagation, often measured in strain energy release rate (SERR) or stress intensity factor (K). In sandwich composites, the fracture toughness is analyzed based on the effects of core density, core thickness, temperature, and specimen geometry [14][15]. Several international standardization institutions, such as ASTM and JIS, also made the test procedure an international standard for testing the interlaminar fracture toughness of composite sandwich materials.

However, less research has been done into the damage characterization of steel polyurethane sandwich for ship structure. The fracture properties associated with the face/core debonding of steel polyurethane sandwich plates have not been well characterized compared to the damage behavior of composite sandwich material. Therefore, in this study an experiment was conducted to characterize the face/core debonding behavior, by estimating SERR of the steel polyurethane sandwich plates specimen. Damage experiments were conducted on several sandwich plate specimens due to mode-I and mode-II loading to analyze the characteristics of debonding propagation under the two load types.

2. Method

The sandwich material analysed is a steel polyurethane sandwich with steel faceplates and a polyurethane elastomer core. The characteristics of sandwich materials are shown in Table 1. The experiment was conducted on sandwich plate specimens with the dimension of 300 mm length, 40 mm width, and the thickness consists of 6 mm faceplates thickness (t_1 & t_2), and 18 mm core thickness (t_c).

Table 1. Sandwich material properties [16]

No	Part	Material	Density (Kg/m3)	Young's Modulus (MPa)	Poisson's Ratio
1	Faceplates	Steel AH 36	7850	206.103	0.3
2	Core	Polyurethane Elastomer	1098	901.95	0.36

The specimen for the mode-I loading test is an asymmetrical double cantilever beam (ADCB) with an artificial pre-crack at the interface of the faceplate and core. The pre-crack is made with a band saw on the core layer 30 mm from the load point, at 2 mm from the top faceplate (t_1). The mode-II loading test uses the three-point bending test method to create crack deformation in the in-shear/sliding direction. The specimen for the mode-II loading test is an asymmetrical end-notch flexure (AENF) sandwich beam. The asymmetric ENF beam is given an artificial debonding crack near the interface between the core and the bottom plate (t_2). The sandwich plate specimen geometry is shown in Figure 1.

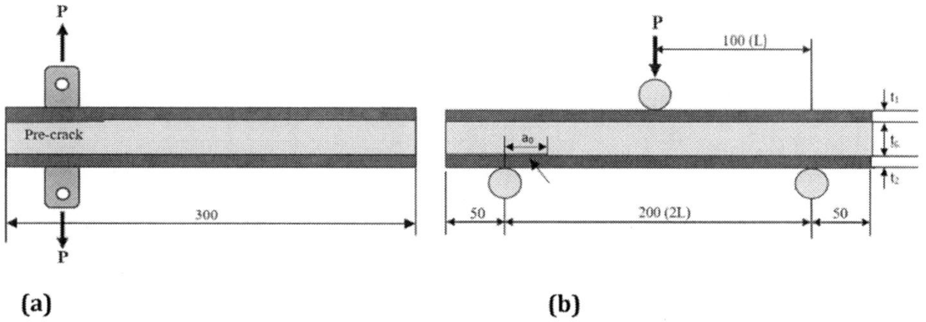

Figure 1. Sandwich plate specimen geometry for (a) mode-I loading; (b) mode-II loading

The mode-I loading is applied continuously at a 20 mm/minute speed, with a maximum opening displacement of 30 mm. For the three-point bending test (mode-II loading), the compressive loading is 20 mm/minute speed, and the maximum displacement setting is 100 mm. The experiment result is a load-displacement curve, which depicts the crosshead displacement correlated with the applied load during the test. This load-displacement data helps SERR by applying the recommended data reduction method. This study uses the Compliance Calibration (CC) data reduction method to estimate the strain energy release rate under mode-I and mode-II loading. For mode-I loading SERR value (G_I) is determined using Equation 1.

$$G_I = \frac{nP\delta}{2ba} \quad (1)$$

P is (load) in Newtons, ***δ*** is displacement (mm), ***b*** is the width of the specimen (mm), ***a*** is the length of the debonding crack (mm), and ***n*** is the slope of the compliance calibration curve line. This ***n*** value is obtained by measuring the slope of the curve shown in Figure 2.

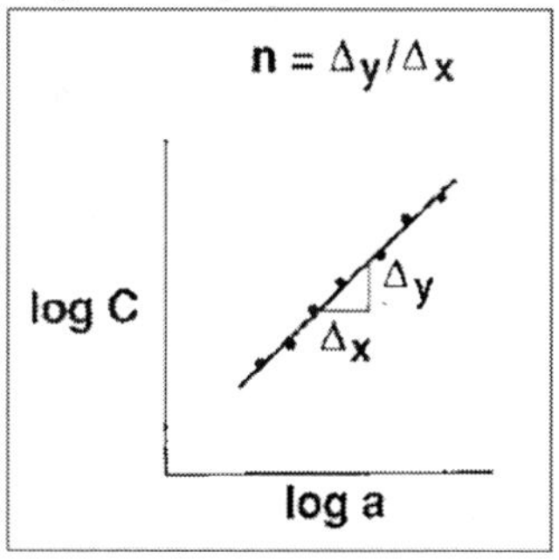

Figure 2. Compliance calibration curve [17]

For mode-II loading the determination of the mode II SERR (G_{II})is using Equation 2.

$$G_{II} = \frac{9P^2a^2C}{2B(2L^3+3a^3)} \quad (2)$$

Where L is half of the specimen span between the support rollers, B is the width of the specimen, a is the crack length, and C is the experimentally measured compliance [18]. Compliance is defined as the displacement (or deformation) per unit load.

3. Result and Discussion

The mode-I and mode-II loading test results are shown in load-displacement curves in Figure 3. (a) and (b).

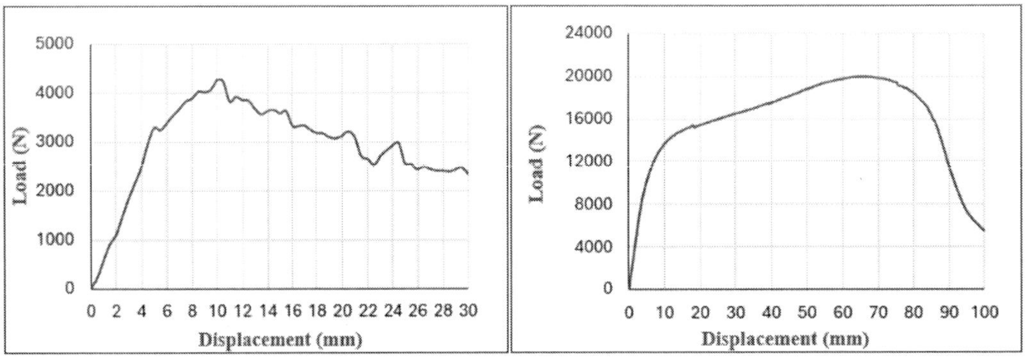

(a) **(b)**

Figure 3. The load-displacement curve for (a) mode-I loading; (b) mode-II loading

The load-displacement curve of mode-I loading test at Fig 3 (a) shows at the initial stage, the load increased rapidly until reaching the peak load and then gradually decreased after the peak load (post-peak softening). The load decreases gradually with a steeper slope at the post-peak softening stage. Lastly, the load tends to plateau even though the displacement increases. Meanwhile, the load-displacement curve of mode-II loading test at Fig 3 (b) shows a different trend. At the beginning of the test, the load increased drastically without a significant increase in displacement. After reaching a certain point, the increase in load slows down, but the displacement becomes greater. After reaching the peak, the load drastically decreases until the displacement maximum, and the loading is stopped.

Apart from showing different trends, the peak load values of the two curves are also different. The peak load of mode-I loading test is 4247 N, and for the mode-II loading test is 19,958 N. The difference in peak loads between the two test results is attributed to the various mechanisms of crack propagation and the nature of the stress states associated with each mode. In the mode-I loading (opening) test, the crack faces are pulled apart perpendicularly from the crack plane, and the primary stress involved is normal tensile stress. The material experiences tensile stresses at the crack tip, making the crack's propagation easier. Consequently, the material might reach its peak load relatively quickly. The stress state tends to create a more intense stress concentration at the crack tip, promoting quicker crack growth. This stress is more localized and severe, leading to earlier failure and a lower peak load.

Meanwhile, in the mode-II loading (shearing) test, the crack faces slide over each other parallel to the crack front and perpendicular to the crack plane. The primary stress involved is shear stress, which is less effective in driving crack propagation. The material is more difficult to propagate, leading to a higher peak load. The sliding of the crack faces overcoming friction and other shear-related resistances leads to a delay in crack growth, allowing the material to bear a higher load before failure. However, in this test, the load drastically drops after peak load, indicating a sudden loss of load-carrying capacity due to shear failure because the crack propagates rapidly in a shear manner. The loss of interfacial resistance, dynamic crack growth, and localized shear failure, contribute to this abrupt reduction in load. This phenomenon also occurs in composite sandwich plates with a foam core, analyzed by Shah [19].

The difference in load-displacement curve results from the two tests has consequences for differences in fracture toughness characteristics indicated by the SERR values. The SERR denotes the $\boldsymbol{G}$ measure of the energy available for crack growth per unit area of the crack surface. It is directly related to the material's toughness and resistance to fracture [20]. $\boldsymbol{G_I}$ corresponds to mode-I loading (opening mode), and $\boldsymbol{G_{II}}$ corresponds to mode-II loading (shearing/sliding mode). $\boldsymbol{G_I}$ and $\boldsymbol{G_{II}}$ related to the crack length depict in resistance curve (*R-curve*) are shown in Figure 4 (a) and (b).

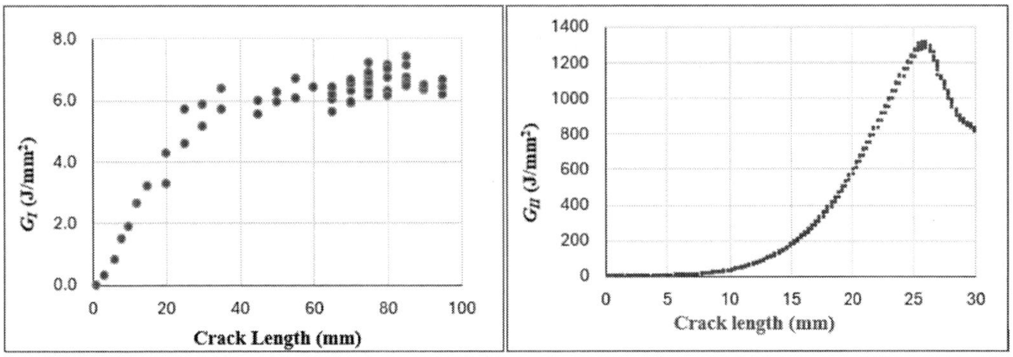

(a) **(b)**

Figure 4. The strain energy release rate (SERR) curve for (a) mode-I loading; (b) mode-II loading

Figure 4 describes the relationship between the crack driving force and this material's interfacial debonding crack extension. The two test results show the rising R-curve, which indicates that the material becomes more resistant to further crack growth as the crack propagates. The R-curve for mode-I loading starts from a lower initial SERR and rises more gradually, reflecting the increasing difficulty of propagating the crack as toughening mechanisms activate. This rising curve reflects the increasing energy required for interfacial crack propagation. Before reaching the plateau, the R-curve typically rises, reflecting the activation of microcracking at the crack tip. Until a certain crack length, the curve tends to plateau. The plateau curve represents a region where the material's resistance to crack growth remains relatively constant as the crack propagates. This plateau zone indicates that the crack has reached a stage where the material's ability to resist further crack extension has stabilized, and no significant additional toughening mechanisms are activated. The plateau zone shows the material's ultimate fracture toughness and ability to sustain interfacial crack growth over time.

The R-curve for mode-II loading starts from the relatively long zone of low initial SERR. Next, the SERR curve rises gradually until the peak value is reached. The curve shows softening phases at the last phase, and the SERR value drops drastically. This curve indicates that at the initial, the crack has yet to propagate due to its still high stiffness. This high initial resistance is due to the need to overcome shear forces and may arise as various toughening mechanisms activate. When loading continues, the material experiences decreasing stiffness and starts an interfacial crack propagation mechanism at the crack tip, marked by the rising SERR curve. The peak SERR shows the maximum amount of energy per unit area available for crack propagation in the material under shear loading conditions. It represents the material's resistance to shear crack propagation at its most resistant point. The peak SERR value from the mode-II loading test is much higher than the mode-I loading test. After reaching the peak SERR, the curve shows a softening phase. This softening behavior is characterized by a reduction in the SERR as the crack extends beyond the peak value. This phase indicates a decrease in the material's resistance to crack propagation, and the material's ability to resist further crack propagation has diminished. This softening is typically caused by damage accumulation, frictional wear, or changes in the material's microstructure. The differences in the R-curves highlight the distinct crack growth behaviors and energy dissipation mechanisms involved in mode-I and mode-II loading.

The peak value of SERR is the critical strain energy release rate, for mode-I loading denotes G_{IC}, and mode-II loading denotes G_{IIC}. The G_{IC} of mode-I loading test is 7.4 J/mm^2 and the G_{IIC} for the mode-II loading test is 1289 J/mm^2. The G_{IIC} is much higher than the G_{IC} due to differences in the nature of the loading and crack propagation mechanisms. Mode-I loading test involves the stress that acts perpendicular to the crack plane, causing the crack faces to open away from each other. Under these loading conditions, interfacial cracks are easier to propagate. Meanwhile, mode-II loading test involves shear stresses that lead to more complex and effective toughening mechanisms such as frictional sliding, shear band formation, and crack face interlocking. These mechanisms can dissipate more energy and provide excellent resistance to crack propagation, resulting in a higher peak SERR for mode-II loading (G_{IIC}) than peak SERR of mode-I loading (G_{IC}).

4. Future Work

The damage characterization of the material is an initial step in the damage tolerance analysis of the structure. The findings of this experimental investigation offer significant insights into the characterization of the interfacial fracture toughness of sandwich plates consisting of steel faceplates and polyurethane elastomer cores. For future work, the damage characteristics in interfacial fracture toughness will be useful to estimate the residual strength of the ship's sandwich structure. Estimating the residual strength helps ensure that the ship's structure remains safe to operate even after sustaining damage. For safety reasons, estimating the residual strength enables the assessment of whether the structure can continue to withstand operational loads without the risk of catastrophic failure.

5. Conclusion

An experimental study to estimate the strain energy release rate (SERR) criterion of steel polyurethane sandwich plate specimen with interfacial debonding flaws had been conducted. The mode-I and mode-II loading test methods produced different shapes of load-displacement curves and peak load values. The difference between the two test results is attributed to the various mechanisms of crack propagation and the nature of the stress states associated with each mode. The difference in load-displacement curve results from the two tests has consequences for differences in fracture toughness characteristics indicated by the SERR values. The differences in the SERR curves highlight the distinct crack growth behaviors and energy dissipation mechanisms involved in mode-I and mode-II loading. The critical strain energy release rate of mode-I loading test (G_{IC}) is 7.4 J/mm^2, and for the mode-II loading test (G_{IIC}) 1289 J/mm^2. The G_{IIC} is much higher than the G_{IC} due to differences in the nature of the loading and crack propagation mechanisms. The results of this experimental study provide valuable insights for characterizing the interfacial fracture toughness of sandwich plates using steel faceplates and polyurethane elastomer cores for ship structure applications that have yet to be widely studied. The characteristics of interfacial debonding damage behavior resulting from this study are useful in estimating the residual strength of sandwich plates with interfacial debonding flaws as a stage of damage tolerance analysis of sandwich structures.

References

[1] N. Momcilovic and M. Motok, "Estimation of ship lightweight reduction by means of application of sandwich plate system," *FME Trans vol. 37, no. 3,* p. 123–128, 2009.

[2] S. Sujiatanti, A. Zubaydi and A. Budipriyanto, "Finite element analysis of ship deck sandwich panel," *Applied Mechanics and Materials 874,* pp. 134-139, 2018.

[3] A. Ismail, A. Zubaydi, B. Piscesa and E. Panangian, "A comparative study of conventional and sandwich plate side-shell using finite element method," in *2nd International Conference on Mechanical Engineering Research and Application (iCOMERA 2020)*, 2020.

[4] A. Ismail, A. Zubaydi, B. Piscesa, Tuswan and R. C. Ariesta, "Study of Sandwich Panel Application on Side Hull of Crude Oil Tanker," *Journal of Applied Engineering Science,* 2021.

[5] T. J. Grafton and J. R. Weitzenbo€ck, "Steel-concrete-steel sandwich structures in ship and offshore engineering," in *Advance Marine Structure Proceedings of the 3rd International Conference on Marine Structures*, 2011.

[6] SANDCORe, "Best Practice Guide for Sandwich Structures in Marine Applications," University of Newcastle upon Tyne, New Castle, 2013.

[7] D. Zenkert, "Damage Tolerance of Naval Sandwich Panels," Department of Aeronautical and Vehicle Engineering, Stockholm, 2009.

[8] D. Zenkert, A. Shipsha, P. Bull and B. Hayman, "Damage tolerance assessment of composite sandwich panels with localised damage," *Composites Science and Technology,* vol. 65, p. 2597–2611, 2005.

[9] M. H. Rarani, M. M. Shokrieh and P. Camanho, "Finite element modelling of mode I delamination growth in laminated DCB specimens with R-curve effects," vol. 45, 2013.

[10] A. Ramji, Y. Xu, M. Yasaee and M. Grasso, "Delamination migration in CFRP laminates under mode I loading," vol. 190, 2020.

[11] C. D. Chen and W. L. Lai, "The analysis of mode II strain energy release rate in a cracked sandwich beam based on the refined zigzag theory," *Theoretical and Applied Fracture Mechanics,* vol. 107, 2020.

[12] R. Moreira, M. de Moura, R. Rocha and C. Oliveira, "Mode II fracture caracterisation of a honeycomb/carbon-epoxy sandwich panel using the assymmetric end-notched flexure test," *Journal of sandwich Structures & Materials,* 2022.

[13] M. M. Shokrieh, A. Zeinedini and S. M. Ghoreishi, "On the mixed mode I/II delamination R-curve of E-glass/epoxy laminated composites," vol. 171, 2017.

[14] F. K. Maleki and M. E. Toygar, "The fracture behavior of sandwich composites with different core densities and thickness subjected to mode I loading at different temperatures," *Materials Research Express,* vol. 6, no. 045314, 2019.

[15] G. Irven, A. Whitehouse, D. Carolan, A. Fergusson and J. P. Dear, "Toughening of face-sheet core bonds in sandwich structures," *Engineering Fracture Mechanics,* vol. 290, 2023.

[16] A. Ismail, A. Zubaydi, B. Piscesa and Tuswan, "Study of Sandwich Panel Application On Side Hull Of Crude Oil Tanker," *Journal of Applied Engineering Science,* 2021.

[17] ASTM, *ASTM D5528-13 Standard Test Method for Mode I Interlaminar Fracture Toughness of Unidirectional Fiber-Reinforced Polymer Matrix Composites,* vol. 13, ASTM, 2013.

[18] E. Oterkus, C. Diyaroglu, D. De Meo and G. Allegri, "Fracture modes, damage tolerance and failure mitigation in marine composites," in *Marine Applications of Advanced Fibre-Reinforced Composites*, Cambridge, Woodhead Publishing, 2016, pp. 79-102.

[19] O. R. Shah and M. Tarfaoui, "Determination of mode I & II strain energy release rates in composite foam core sandwiches. An experimental study of the composite foam core interfacial fracture resistance," *Composites Part B*, vol. 111, pp. 134-142 , 2017.

UPWELLING INTENSITY DURING INDIAN OCEAN DIPOLE PERIOD IN SOUTHERN WATERS OF JAVA

S Widagdo[1], V D Prasita[2], and M Hafiz[3]

[1,3] Oceanography Departement, Faculty of Engineering and Marine Sciences, Universitas Hang Tuah, Indonesia
[2] Magister of Marine Engineering Departement, Faculty of Engineering and Marine Sciences, Universitas Hang Tuah, Indonesia
Email: supriyatno.widagdo@hangtuah.ac.id

Abstract. In recent years, the prevalence of IOD has been increasing, leading to widespread and severe consequences. It is imperative to closely monitor this phenomenon in order to comprehend the ecological repercussions and their subsequent outcomes. The objective of this investigation is to examine the occurrence of upwelling in the southern Java waters under neutral conditions and to evaluate potential variations in its intensity during the Indian Ocean Dipole (IOD) season. The research was conducted in the southern waters of Java during the pronounced IOD period from 2016 to 2023. Data were obtained from the Copernicus Marine website, encompassing wind, current, temperature, salinity, and chlorophyll measurements for the eastern season. Subsequent data processing was performed using ArcGIS, ODV, and descriptive statistical methods. The results indicated that during positive IOD events, upwelling in the study area exhibited increased intensity, characterized by reduced surface temperature values, accompanied by elevated salinity and surface chlorophyll concentrations, in comparison to neutral periods. Conversely, during negative IOD events, upwelling demonstrated diminished intensity, as evidenced by higher surface temperatures and comparatively low surface salinity and chlorophyll levels. The vertical distribution of temperature and salinity further corroborated the ascent of subsurface water masses.

1. Introduction

A crucial oceanic process known as upwelling takes place in marine environments, particularly in coastal waters. It occurs when deeper layers of seawater rise to the surface, carrying essential nutrients for primary productivity. In the southern waters of Java, upwelling is one of the main factors that influence the marine ecosystem and fisheries. The waters of southern Java within WPP-NRI 573 continue to harbor viable fishery resources. Additionally, these waters exemplify the unique conditions of fisheries resources that necessitate specialized management approaches (Nurani et al., 2007; Kurniawan, 2019). Meteorological and oceanographic factors make the area located between the continents of Asia and Australia and the Pacific and Indian Oceans intensely experienced, as well as becoming a region that is one of the most affected by climatic events between the two continents and oceans, including the Indian Ocean Dipole (IOD). The interaction between the Asia-Australia Ocean and the Pacific-Indian Ocean significantly influences the El Niño-Southern Oscillation (ENSO) and the Indian Ocean Dipole (IOD). This relationship is characterized by complex interactions that modulate climate variability across the region (Andrian, *et al.*, 2024; Pillai *et al.*, 2021). Climate change trends in Indonesia are causing diverse effects across the country. The southern regions are experiencing decreased rainfall and shorter wet seasons, while the northern areas are seeing an increase in precipitation. These alterations in rainfall patterns are leading to shifts in both the onset and duration of the rainy season. (Julismin, 2013).

Such conditions show that the effects of climate, such as those in marine fisheries, are no less important than technical management issues. The catch of cakalang fish in the southern waters of West Java, for example, fluctuates seasonally. Research by Rahman *et al.* (2019) found that the highest catch (325.77 tons) was obtained during the eastern season (2015) when the optimal temperature was 28.00-28.30 °C and chlorophyll-a was optimal 0.20-0.25 mg/m^3. On the other hand, the lowest catch (15.81 tons) occurred during the western season (2016) when the optimal temperature was 27.00-29.00 °C with chlorophyll-a values that were not optimal at 0.00-0.20 mg/m^3.

The IOD is a climate phenomenon that influences weather patterns and ocean circulation in the Indian Ocean region, including the waters around Indonesia. This event is characterized by a temperature differential in sea surface temperatures between the western and eastern sectors of the Indian Ocean. Such temperature variations can lead to alterations in wind patterns and oceanic currents.. Fluctuations in the strength of upwelling in the waters south of Java are frequently linked to the IOD period. These changes have cascading effects on the marine ecosystem, influencing both the distribution of fish and fishing operations in the areas. Indonesia's geographical location, which borders the center of ENSO occurrence in the Pacific Ocean and IOD in the Indian Ocean, makes the Indonesian region highly influenced by both of these climatic phenomena.

The impacts of global climate change are increasingly being felt, including in intertidal ecosystems such as ENSO and IOD, which are becoming more intense. The effects are widespread, including on marine and coastal areas in many regions, such as the waters off southern Java, necessitating a better and more predictive understanding. Climate conditions will impact marine activities, fishing catches, and upwelling in Indonesian waters, including in the waters off southern Java. Off the southern coast of Java, upwelling occurs during the eastern season. Therefore, studying the intensity of upwelling during the IOD period in the waters off southern Java is crucial to understanding the impact of this phenomenon on marine ecosystems and fisheries resources in the area. This study aims to monitor the existence of upwelling in the waters off southern Java during neutral conditions and changes in its intensity during the IOD period. As a result, it is hoped that the results of this study will contribute to the management of fisheries resources and environmental conservation in the waters off southern Java.

2. Research Methods

Research was conducted in the southern waters of Java located at 5–10˚LS and 106–115˚BT, which is divided into three provincial zones (East Java, Central Java and West Java), as depicted in Figure 1. The data used in the study were obtained from the Copernicus website and comprised wind, current, temperature, salinity, and chlorophyll-a in three IOD phases: neutral (June 2017 and June 2018), positive (August 2019 and June 2020) and negative (July 2016 and July 2022). The data were then processed using ArcGIS 10.8 and ODV 5.7.1 to obtain maps of the distribution of each parameter. Upwelling was examined by visually assessing the presence of Ekman transport, which results from the interaction between currents and winds. This phenomenon influences the surface distribution of temperature, salinity, and chlorophyll-a. To complement this analysis, the vertical profiles of temperature and salinity were also investigated.

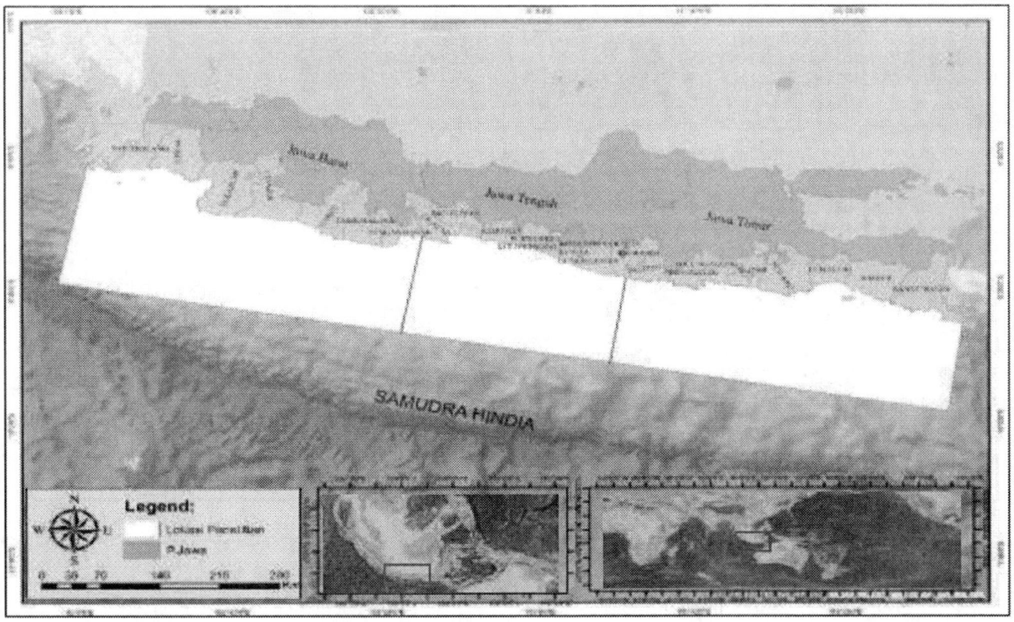

Figure 1. The research location in the southern waters of Java is grouped into three provincial zones: East Java, Central Java and West Java.

3. Result and Discussion

3.1. *Ekman Transport*

An analysis of surface wind patterns over the southern Java waters during three distinct periods (neutral, positive IOD, and negative IOD - abbreviated as N, pIOD, and nIOD) reveals that southeasterly winds dominate during the eastern season (June, July, and August). The wind speeds are typically highest offshore, reaching approximately 6-7 m/s, while coastal areas experience comparatively lower wind velocities (as shown in Figure 2).

The distribution of surface currents during the N and IOD phases varies in the direction of the wind. The current that dominates the offshore waters is the southwesterly current (as shown in Figure 3). This condition reflects the influence of the upwelling-wind circulation caused by the Coriolis force. In the Southern Hemisphere, the Coriolis effect causes fluids to be diverted in a counterclockwise direction, contrary to the movement of a clock's hands (Natalie *et al.*, 2023; Srinidhi *et al.*, 2019). In the coastal waters, currents are generally deflected away from the shore towards the east. This condition results in the Ekman transport due to the coupling of wind-current, which has the potential to produce upwelling. The existence of the upward movement of subsurface water is indicated by the decrease in sea surface temperature (SST), accompanied by an increase in surface salinity and chlorophyll-a (SSS and SSC-a).

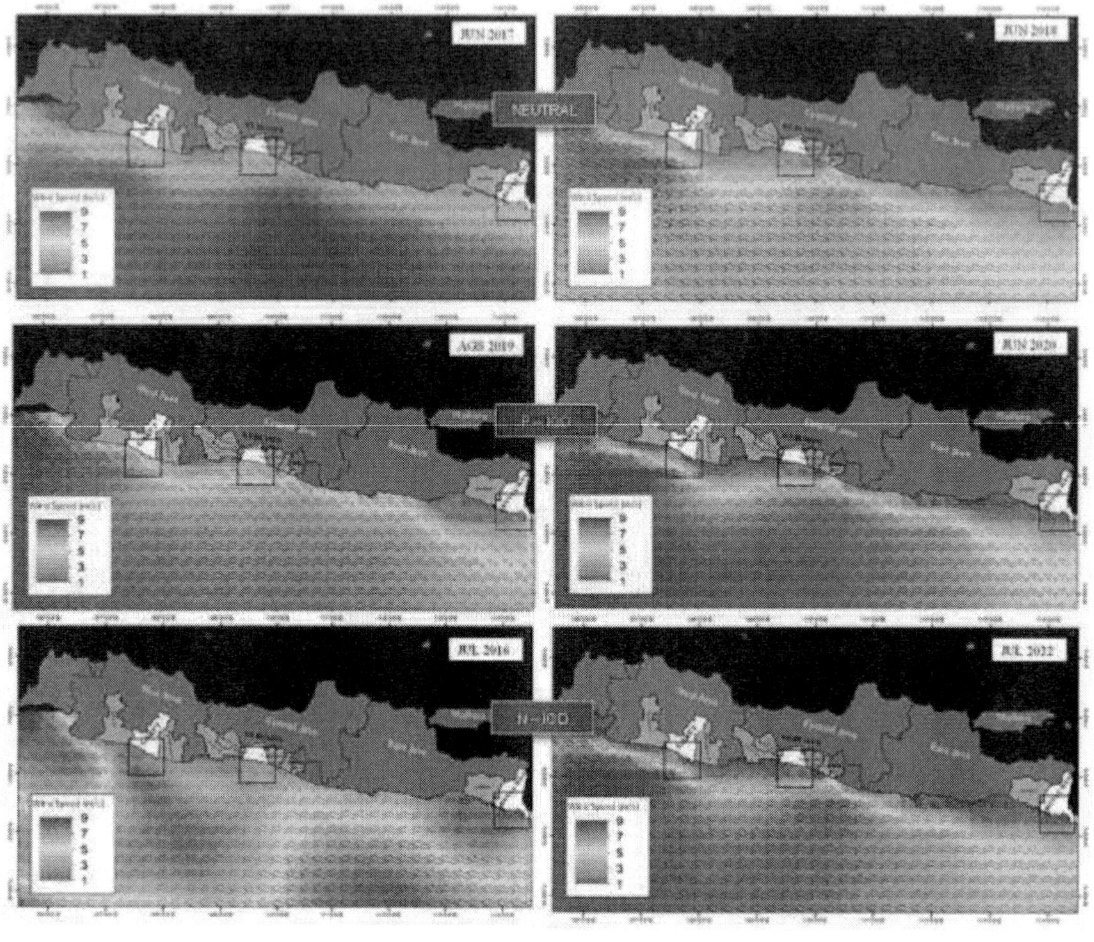

Figure 2. The distribution of surface winds in the southern waters of Java shows the dominance of southeast winds during the eastern season. In general, offshore winds are stronger than the winds in coastal waters.

Strong Ekman transport in South Java waters is influenced by seasonal winds and currents (Atmadipoera *et al.*, 2020; Ratnawati *et al.*, 2016; Siswanto, 2010). Coastal upwelling processes along Java's southern shoreline are influenced by the interaction between southeasterly winds and seasonal ocean currents. This combination generates offshore Ekman transport, contributing to the upwelling dynamics in the region. (Wirastriya, 2020). In addition to impacting the intensity of Ekman pumping, this wind-current coupling also impacts variations in sea surface temperature and chlorophyll-a levels. Changes in chlorophyll-a levels occur through intraseasonal flow variations that impact the exchange of nutrient-rich water (Schept, 2022; Xu, *et al.*, 2018).

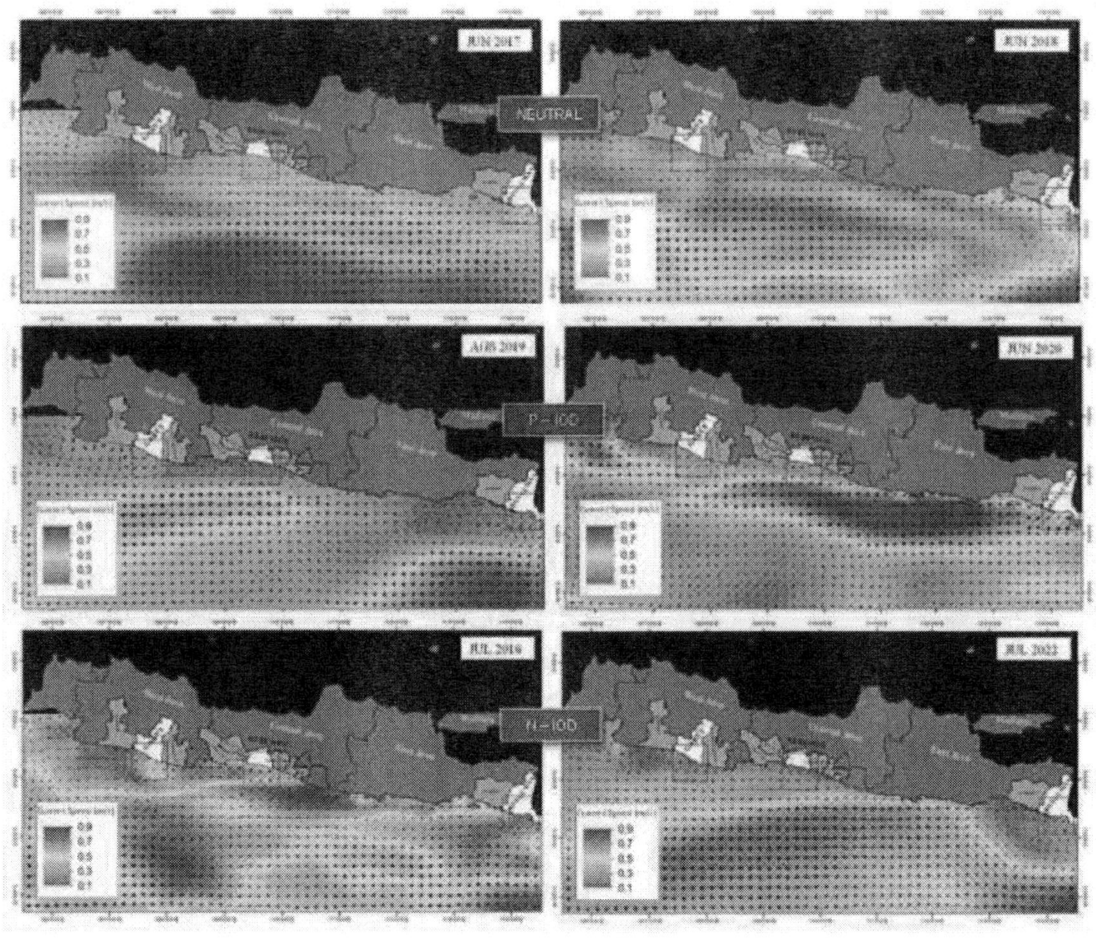

Figure 3. During the eastern season, the N, pIOD, and nIOD phases in Java's southern waters are characterized by predominant westerly currents offshore. In contrast, coastal waters exhibit slower-moving currents that typically deflect southwestward. This current movement causes surface water to be pulled away from coastal areas, creating a vacuum that is subsequently filled by subsurface water masses.

3.2 Distribution of Sea Surface Temperature (SST)

The SST during phase Nl, pIOD, and nIOD in coastal waters of Garut, Kebumen, and Banyuwangi, as shown in Figure 4, confirms a decrease in SST at the three locations during pIOD. During pIOD, with the exception of Kebumen, coastal waters in southern Java experience strong upwelling with a range of 23–24°C. SST drops sharply, approximately 3.00–4.00 °C during pIOD. The SST during the neutral phase and nIOD is similar, with respective ranges of 26.00–28.20 °C and 26.60–27.40 °C. The Garut coast shows high SST during the neutral phase (28.00–28.20 °C), while Banyuwangi shows high SST during nIOD (26.60–27.40 °C).

The difference in temperature during the neutral phase and IOD is further emphasized in Figure 5. The average temperature during N, pIOD, and nIOD is successively 27.17 °C, 24.08 °C, and 27.00 °C. The temperature difference between N-pIOD ranges from 0.80–5.00 °C, with an average of 3.09 °C; whereas the average difference between N-nIOD is 0.17 °C, with a temperature range of 0.00–1.20 °C. This indicates the significance of SST changes in southern Java during pIOD compared to nIOD, and it also confirms that nIOD does not significantly alter the SST characteristics of the southern Java coast.

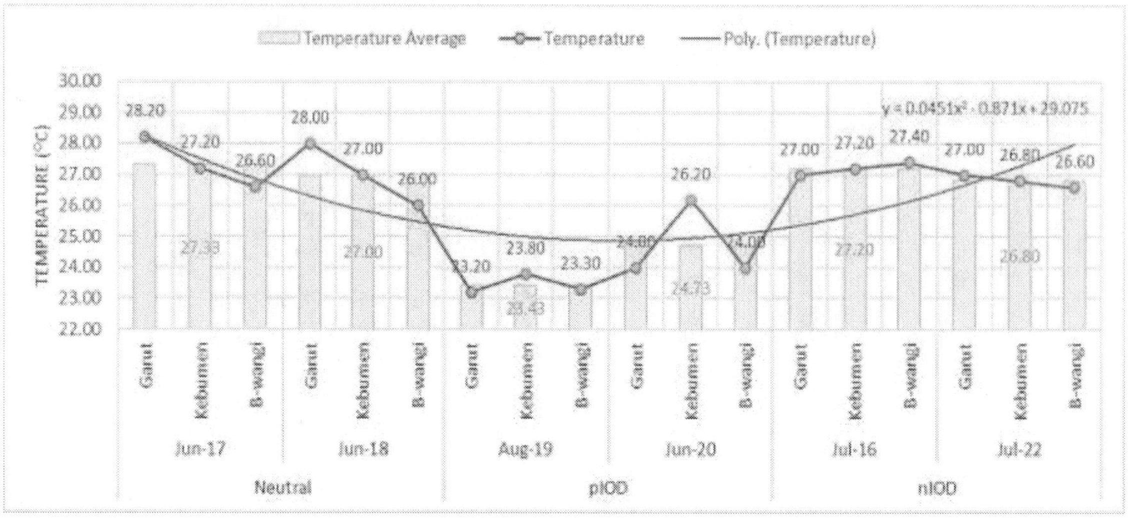

Figure 4. SST fluctuations that emphasize the decrease of SST during pIOD in three loci and the relative equality of SST during neutral phase and nIOD.

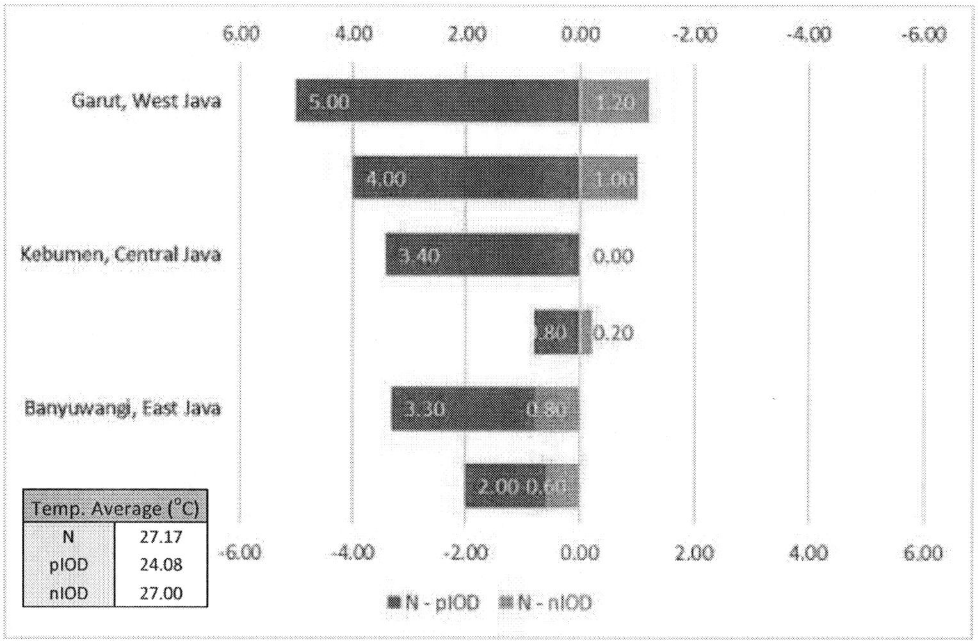

Figure 5. The temperature difference in N compared to pIOD and nIOD which shows the relatively high temperature of N-pIOD.

The small relative SST (indicating upwelling of subsurface water) is depicted in Figure 6. During the research period, such pockets were frequently found at the study locations, namely Garut in West Java, Kebumen in Central Java, and Banyuwangi in East Java. The zone with small SST variations was not constant, but varied and was not far from the three initial locations. For example, Kebumen's relatively high SST was recorded in June 2020 when pIOD was present (Figure 6-pIOD), while SST at the three locations was only ~23.00–24.00 °C. Interestingly, the small SST pocket shifted to the east, specifically to the waters off Gunung Kidul (Yogyakarta). In phase N, the waters off Garut had similar temperatures to those of the relatively high sea surface temperatures (~29 °C) offshore. Conversely, a cold temperature pocket was found along the coasts of Kebumen and Banyuwangi (~26 °C). This

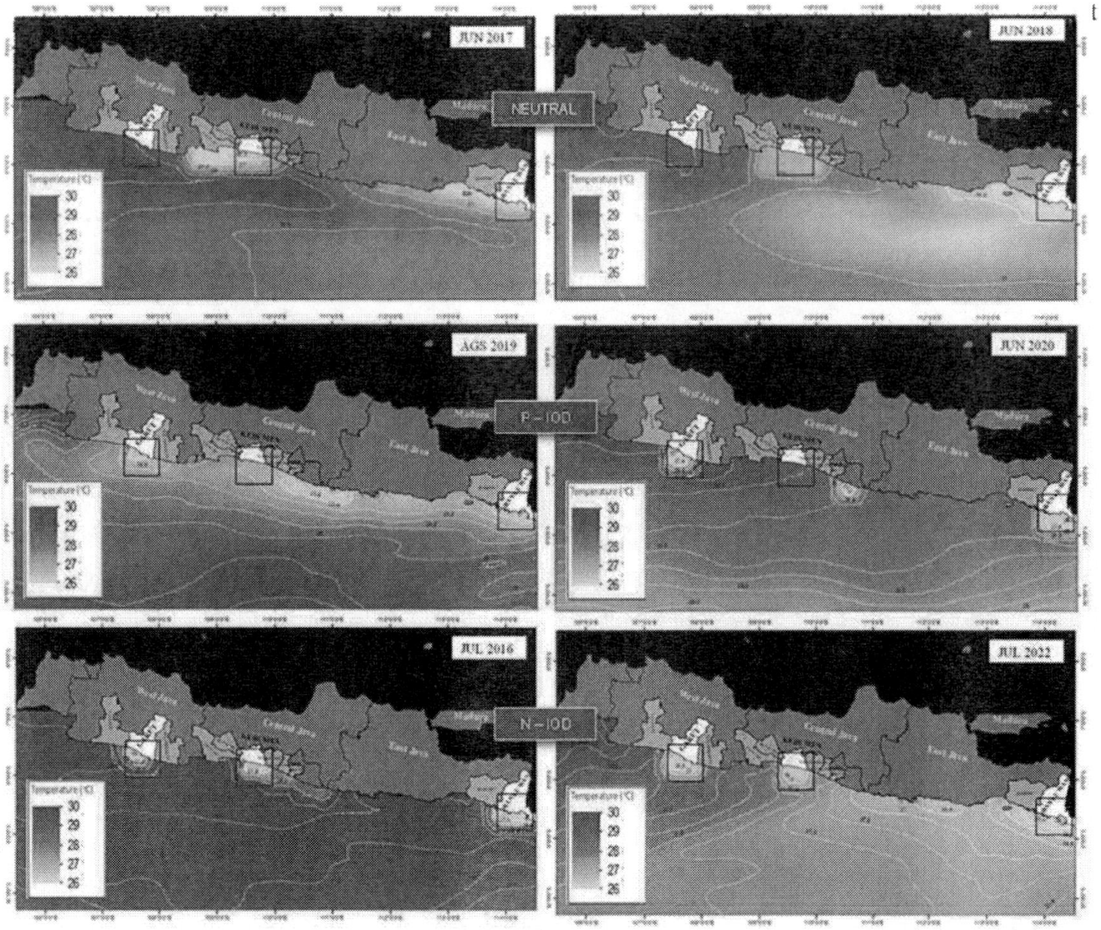

Figure 6. The presence of upwelling in the southern Java waters is evident from SST patterns. Strong upwelling indicators are primarily observed in the coastal areas of Garut, Kebumen, and Banyuwangi. Over time, the upwelling zone has undergone shifts and movements within the vicinity of these three locations.

3.3 Distribution of Sea Surface Salinity (SSS)

The distribution of surface seawater salinity (SSS) exhibits an increase during the Indian Ocean Dipole (IOD) event compared to its neutral phase. The change in the climatic condition of the Eastern Indian Ocean during the positive IOD (nIOD) does not significantly alter the characteristics of the SSS in southern Java compared to when the ocean is cooler than usual. The SSS ranges at the three loci during the IOD event varied between 34.20 PSU (Garut, June 2020) and 34.60 PSU (Banyuwangi, August 2019). Meanwhile, the SSS range during N and nIOD, respectively, was between 33.80 PSU (Garut, June 2017) and 34.20 PSU (Kebumen and Banyuwangi, June 2018) and 34.00 PSU (Kebumen and Banyuwangi, July 2022) to 34.20 PSU (Kebumen and Banyuwangi, July 2016), see Figure 7.

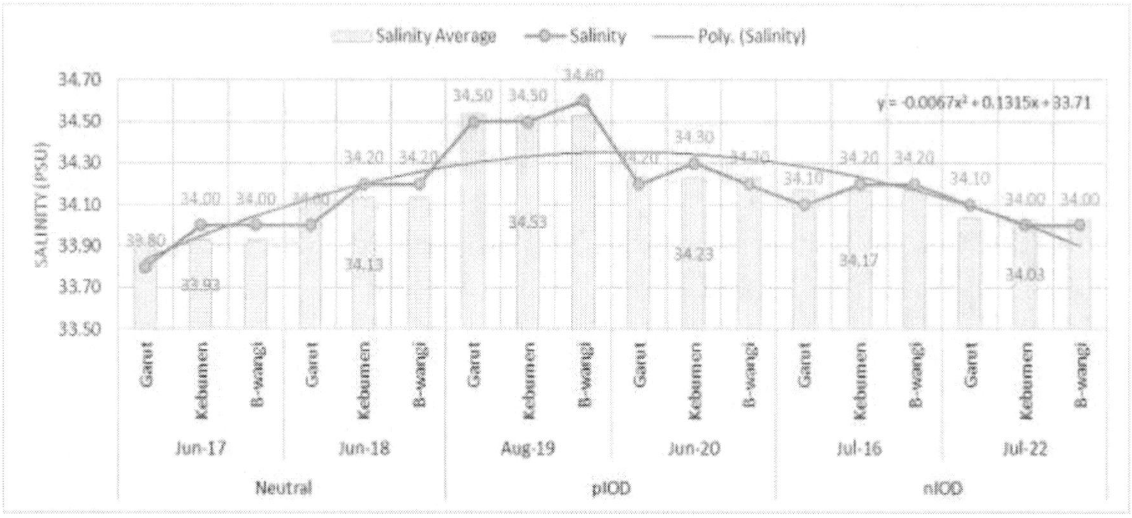

Figure 7. SSS fluctuations in Garut, Kebumen and Banyuwangi waters that show an increase in SSS during the pIOD compared to the N and nIOD.

Figure 8 illustrates how positive and negative IOD events, resulting from climate change, affect the neutrality of waters in the southern Java Sea.. A comparison between the two shows that the highest range of N-pIOD reaches 0.70 PSU in the waters off Garut, while the lowest 0.00 PSU occurs in Banyuwangi. On the other hand, the N-nIOD range reaches a maximum of 0.30 PSU in Garut and a minimum of 0.10 PSU in the same waters. This indicates that during an IOD event, there is an increase in salinity compared to the neutral phase. The stronger increase in SPL in the southern Java Sea occurs during pIOD (0.35 PSU) compared to nIOD (0.07 PSU).

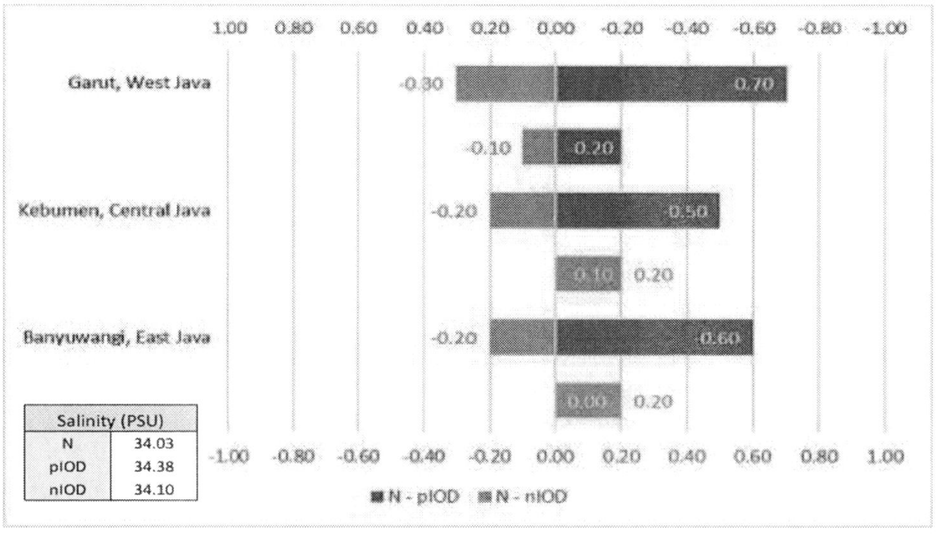

Figure 8. Salinity differences during the pIOD and nIOD compared to the neutral phase showing a more significant increase in SSS during the pIOD.

The spatial distribution of SSS displayed in Figure 9 shows the distribution of high SSS patches. During nIOD, particularly in July 2016, high SPL patches were found to be scattered across three loci. The increase in SSS in coastal waters during pIOD was marked by an increased isohaline. SSS in the Garut waters appeared more spotty than in the waters of Kebumen and Banyuwangi, indicating the strength of the subsurface water mass increase in the Garut waters. During the neutral phase, seawater along the southern coast of Java had relatively similar SSS values (~34.00 PSU). The waters of Garut also showed relatively lower SSS compared to the southern Java waters in general.

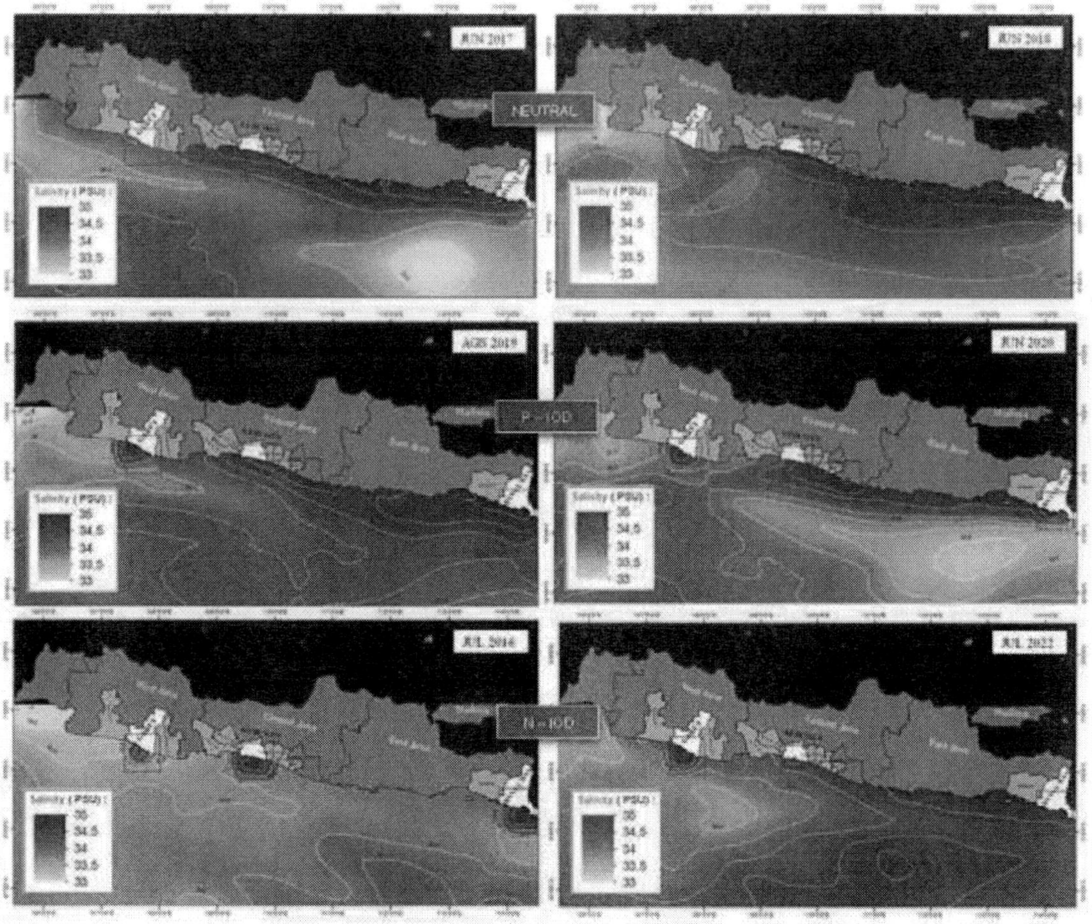

Figure 9. Spatial distribution of SSS in southern Java waters showing relatively high SSS during the pIOD compared to the nIOD. The low SSS during the nIOD confirms the wet period when the eastern Indian Ocean waters are warmed compared to neutral conditions.

3.4 Distribution of Sea Surface Chlorofil-a (SSC-a)

The temporal fluctuation of SSC-a, as depicted in Figure 10, resembles the temporal fluctuation of SPL marked by an increase in SSC-a during pIOD with a peak of 1.80 mg/m^3 in Banyuwangi waters (August 2019). The lowest SSC-a dominated nIOD, particularly in July 2016 with values of 0.12–0.14 mg/m^3 in Garut, Kebumen, and Banyuwangi waters. Although still relatively low, SSC-a in July 2022 at all three locations increased to 0.30–0.40 kg/m^3. The neutral water phase in Garut also had a low SSC-a of 0.12 mg/m^3 in June 2017. Comparing SSC-a during N with pIOD and nIOD shows the high SSC-a during pIOD, with a range of 0.1–1.28 mg/m^3 compared to the neutral phase. Meanwhile, with an average of 0.25 mg/m^3, nIOD conditions lowered SSC-a to 0.71–1.00 mg/m^3 compared to the neutral phase (Figure 11).

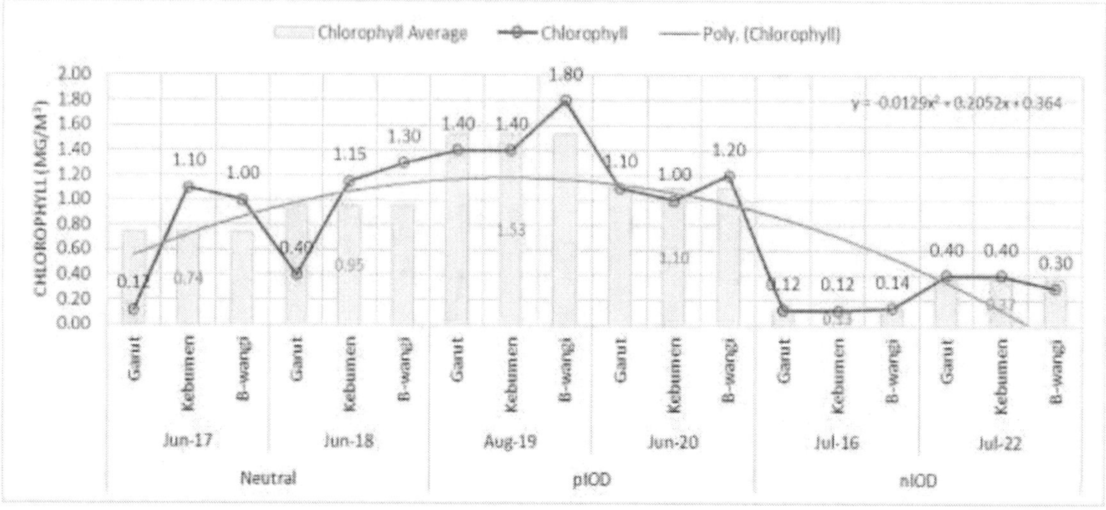

Figure 10. SSC-a fluctuations showing increased SSC-a during pIOD compared to N and nIOD in Garut, Kebumen and Banyuwangi waters.

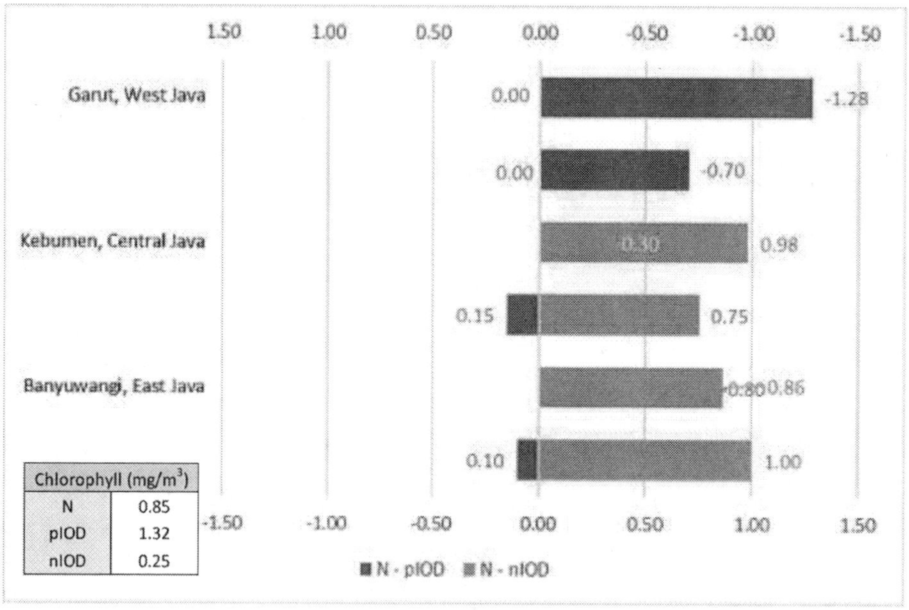

Figure 11. The difference in neutral phase SSC-a with pIOD and nIOD shows the relatively balanced influence of pIOD and nIOD. On average, SSC-a during pIOD increased by 0.47 mg/m^3, while during nIOD it decreased by 0.60 mg/m^3.

The distribution of SSC-a during the neutral phase, as shown in Figure 12, indicates that the seawater in Kebumen and Banyuwangi has a higher SSC-a than that of the oceanic water. However, the same condition was not found in the Garut coastal area. In the West Java location, the distribution of SSC-a is similar to SST, which has similarity with oceanic water in the western and southern areas. During the IOD period, whether positive or negative, the three locations showed high pockets of SSC-a that stood out compared to the oceanic water. Overall, SSC-a increases during pIOD and decreases during nIOD.

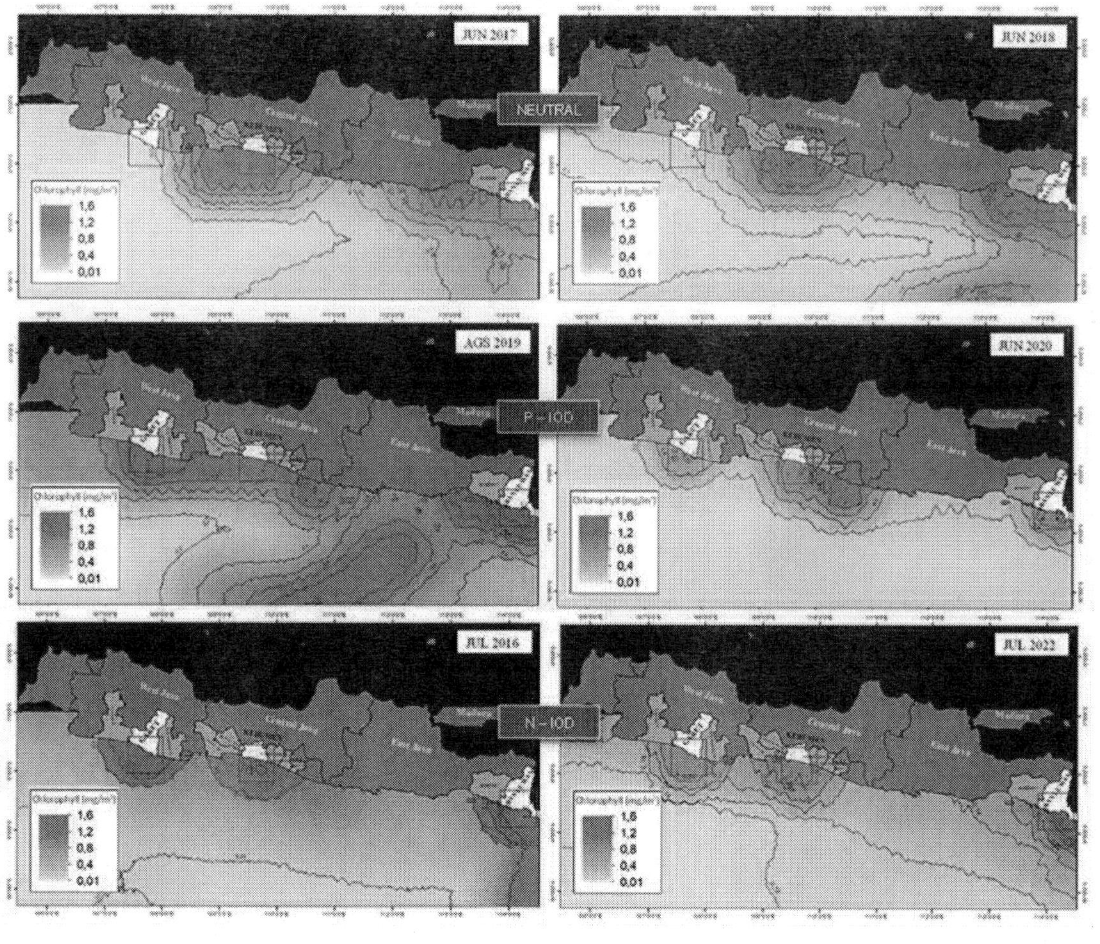

Figure 12. The distribution of SSC-a in the south coast of Java during the neutral and IOD phases shows the relatively high SSC-a content of coastal waters compared to oceanic waters. Compared to the neutral phase, the SSC-a contours during the pIOD show an increase, while those during the nIOD decrease. The SSC-a pockets at the three loci are still prominent, except in Garut during the neutral phase.

3.5 Vertical Profile of Temperature dan Salinity

The subsurface water mass rise to the surface supports the occurrence of upwelling at the study site, represented by temperature and salinity during the pIOD (Figure 13), which has a greater impact in changing oceanographic characteristics (temperature, salinity, and chlorophyll). The vertical profile of temperature (VPT) indicates the presence of isotherms rising and concentrating at three loci: Garut, Kebumen, and Banyuwangi (Figure 13 left). VPT also shows higher isotherm elevation in Gunung Kidul compared to Kebumen, as indicated by the SST distribution shown in Figure 6. The more intense rise in water mass in August 2019, with an isotherm of 24 °C at the surface, compared to June 2020, with an isotherm of 4 °C (Figure 13), indicates that the warmer water mass originated from a deeper depth. In August 2019, the isotherm of 24 °C shows a more intense rise to the surface, with colder water, compared to the isotherm of 28 °C. The vertical profile of salinity (VPS) during the same period also shows a subsurface water mass rise. The isohaline that reaches the surface in August 2019 reaches 34.6 PSU, which is 0.6 PSU higher than in June 2020.

IOP Conf. Series: Earth and Environmental Science **1473** (2025) 012015 doi:10.1088/1755-1315/1473/1/012015

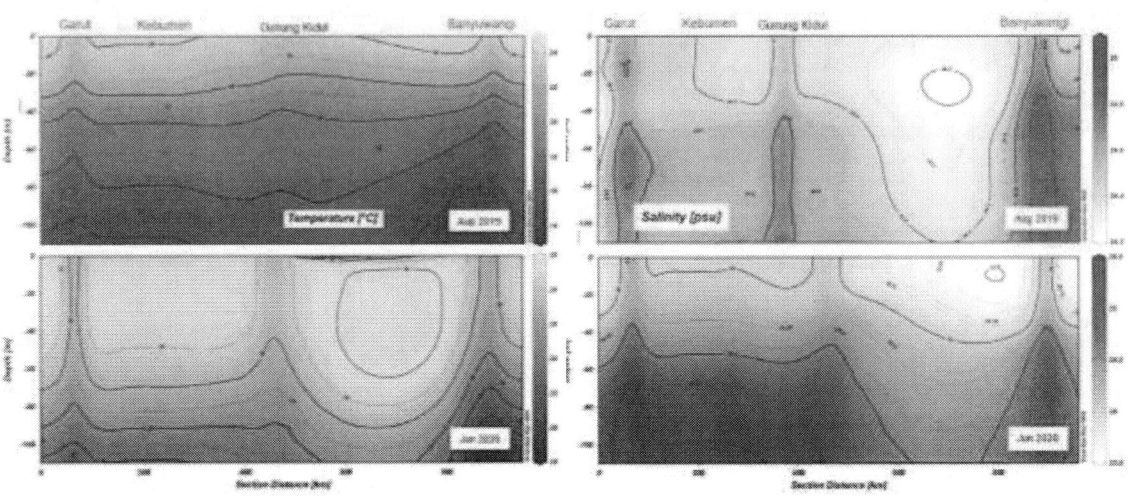

Figure 13. VPT (left) and VPS (right) showing a subsurface water mass increase. Isotherms and isohalines reaching the surface in August 2019 were cooler and saltier than those in June 2020.

Increased upwelling intensity during pIOD is characterized by a decrease in SST and increases in SSS) and SSC-a (Herlambang and Iskandar, 2023; Clair, 2022; Horii et al., 2022). The movement of cold water masses at three research loci indicates thermocline uplift which, as stated by Horii et al. (2018), signifies the occurrence of upwelling. Conversely, nIOD reduces upwelling intensity in the southern waters of Java, evidenced by increased SST and decreased SSC-a (Undap et al., 2023; Koropitan et al., 2021). The research findings align with those of Atmadipoera et al. (2020), who observed that pIOD enhances upwelling intensity, while nIOD attenuates it. Furthermore, Rahman et al. (2020) assert that IOD exerts a more significant positive impact on coastal upwelling intensity in the southern waters of Java compared to ENSO. Additionally, Wang et al. (2021) demonstrate the influence of the Indonesian Throughflow on upwelling during pIOD (2019) and nIOD (2016) events.

4. Conclusion

The southern waters of Java experience upwelling during the eastern season. The prevailing wind-sea currents that generally move southwest result in the production of an Ekman transport that carries surface coastal water away from the coast, resulting in a void of surface water near the coast. This surface water void is replaced by upwelling water from below the surface, resulting in a colder temperature characteristic, as well as a larger surface temperature, salinity, and chlorophyll in comparison to the surrounding waters. During the neutral period without the influence of IOD, the average temperature, salinity, and chlorophyll of the waters in southern Java were 27.17 °C, 34.03, and 0.85 mg/m^3, respectively. The upwelling phenomenon continues throughout the eastern season with varying intensities. The waters of Garut are included in the area with weak upwelling during the neutral phase.

The impact of pIOD on southern Javan waters is more pronounced than that of nIOD. During the pIOD phase, which corresponds to the dry season and is characterized by cooling of the Indian Ocean surface, the average temperature dropped by 3.09 °C. In contrast, during the nIOD phase, associated with the wet season and warming of the Indian Ocean's surface layer, the average temperature only decreased by 0.17 °C when compared to the neutral phase. On average, the SST increased by 0.35 PSU during pIOD and only by 0.07 PSU during nIOD. The SSC-a decreased by 0.60 mg/m^3 during nIOD and increased by 0.47 mg/m^3 during pIOD compared to the neutral phase.

Upwelling on the southern waters of Java during the eastern season does not always occur in the same area. The location of the research shows a shift in the strong upwelling zone towards the east, from the Kebumen waters to Gunung Kidul, Yogyakarta in Central Java, as well as a shift towards the west in East Java, from Banyuwangi to Lumajang.

Acknowledgment
The authors were grateful to the Rector for providing the facilities and infrastructure needed. Furthermore, the authors were grateful to the study team of lecturers for their cooperation.

References

[1] Andrian, Luciano, Gustavo., Marisol, Osman., Carolina, Vera.. The role of the Indian Ocean Dipole in modulating the Austral Spring ENSO teleconnection into the Southern Hemisphere. (2024). doi: 10.5194/egusphere-2024-1812
[2] Anom, Taufik, Ali, Ysh., Munawar, Cholil., Alif, Noor, Anna. Pendugaan Upwelling di Perairan Selatan Pulau Jawa Menggunakan Citra Satelit Aqua MODIS Perekaman Tahun 2014–2016. (2017)
[3] Atmadipoera, Agus, S., Agitha, Saverti, Jasmine., Mulia, Purba., Anastasia, R.T.D., Kuswardani. Upwelling characteristics in the southern java waters during strong la nina 2010 and super el nino 2015. (2020). doi: 10.29244/JITKT.V12I1.28977
[4] Atmadipoera, Agus, S., Agitha, Saverti, Jasmine., Mulia, Purba., Anastasia, R.T.D., Kuswardani. Upwelling characteristics in the southern java waters during strong la nina 2010 and super el nino 2015. (2020). doi: 10.29244/JITKT.V12I1.28977
[5] Chakraborty, Kunal., A.P., Joshi., Prasanna, Kanti, Ghoshal., Jayashree, Ghosh., Anirban, Akhand., Trishneeta, Bhattacharya., Mohanan, Geethalekshmi, Sreeush., Vinu, Valsala. Mechanisms and drivers controlling spatio-temporal evolution of pCO_2 and air-sea CO_2 fluxes in the southern Java coastal upwelling system. Estuarine Coastal and Shelf Science. (2023). doi: 10.1016/j.ecss.2023.108509
[6] Clair, William, H., St. Atmospheric Response to the Southern Java Upwelling Variability Associated with Positive Indian Ocean Dipole Event. (2022). doi: 10.1007/978-981-19-0308-3_3
[7] Herlambang, Dominikus, Krisna., Iskhaq, Iskandar. Influence of Anomalous Upwelling on the Water Mass of South Java during Two Positive Indian Ocean Dipole in 2015 and 2019. JPPIPA (Jurnal Penelitian Pendidikan IPA). (2023). doi: 10.29303/jppipa.v9ispecialissue.6182
[8] Horii, Takanori., Eko, Siswanto., Iskhaq, Iskandar., Iwao, Ueki., Kentaro, Ando. Can Coastal Upwelling Trigger a Climate Mode? A Study on Intraseasonal-Scale Coastal Upwelling Off Java and the Indian Ocean Dipole. Geophysical Research Letters. (2022). doi: 10.1029/2022GL098733
[9] Horii, Takanori., Iwao, Ueki., Kentaro, Ando. Coastal upwelling events along the southern coast of Java during the 2008 positive Indian Ocean Dipole. Journal of Oceanography. (2018). doi: 10.1007/S10872-018-0475-Z
[10] Julismin, J. Dampak dan Perubahan Iklim di Indonesia. (2013). Jurnal Geografi. **5** (1): 39–46
[11] Koropitan, Alan, F., Ibrahim, Kholilullah., Roza, Yusfiandayani. Modeling Mackerel Tuna (Euthynnus affinis) Habitat in Southern Coast of Java: Influence of Seasonal Upwelling and Negative IOD. Hayati Journal of Biosciences. (2021). doi: 10.4308/HJB.28.4.271-285
[12] Kurniawan, D.R. Pemetaan Daerah Penangkapan Ikan di Perairan Selatan Jawa yang Berbasis di Kabupaten Pacitan dan Kabupaten Trenggalek, Jawa Timur. Sarjana thesis, Universitas Brawijaya. (2019). https://repository.ub.ac.id/id/eprint/177709/
[13] Lumban-Gaol, Jonson,., Eko, Siswanto., Kedarnath, Mahapatra., Nyoman, Metta, N., Natih., I, Wayan, Nurjaya., Mochamad, Tri, Hartanto., Erwin, Maulana., Luky, Adrianto., Herlambang, Aulia, Rachman., Takahiro, Osawa., Berri, Miraz, Kholipah, Rahman., Arik, Permana. Impact of the Strong Downwelling (Upwelling) on Small Pelagic Fish Production during the 2016 (2019)

Negative (Positive) Indian Ocean Dipole Events in the Eastern Indian Ocean off Java. Climate. (2021). doi: 10.3390/CLI9020029

[14] Lumban-Gaol, Jonson., Eko, Siswanto., Nyoman, Metta, N., Natih., I, Wayan, Nurjaya., Mochammad, Tri, Hartanto., Erwin, Maulana., Lucky, Adrianto., Kedarnath, Mahapatra., Herlambang, Aulia, Rachman., Takahiro, Osawa., Arik, Permana. The Strong Downwelling (Upwelling) Impact on Small Pelagic Fish Production During the 2016 (2019) Negative (Positive) Indian Ocean Dipole Events in the Eastern Indian Ocean off Java. (2020). doi: 10.20944/PREPRINTS202011.0071.V1

[15] Natalie, C., Frank., Maria, Eletta, Negretti., Joël, Sommeria., Martín, Obligado., Raúl, Bayoán, Cal. Coriolis force deflects wind plant wakes left in the Northern Hemisphere. (2023). doi: 10.48550/arxiv.2308.01428

[16] Nurani, T.W., J. Haluan, S. Saad dan E. Lubis. Model Pengembangan Perikanan di Perairan Selatan Jawa. (2007) *Buletin PSP.* **16** (2): 317–344

[17] Pillai, Prasanth, A., Dandi, A., Ramu., Ravikumar, C., Nair., Ravikumar, C., Nair. Recent changes in the major modes of Asian summer monsoon rainfall: influence of ENSO-IOD relationship. Theoretical and Applied Climatology. (2021). doi: 10.1007/S00704-020-03454-3

[18] Prasetya, Muhammad, Ridwan., Avrionesti, Avrionesti., Yosafat, Donni, Haryanto., Latifah, Nurul, Qomariyatuzzamzami. The Identification of Upwelling in the Occurrence of Tropical Cyclone Charlotte in the Waters of Southern Java (Case Study: March 17-28, 2022). (2024). doi: 10.31258/tromes.2.2.30-35

[19] Rachman, Herlambang, Aulia., Jonson, Lumban-Gaol., Fadli, Syamsudin. Remote Sensing of Coastal Upwelling Dynamics in the Eastern Indian Ocean off Java, Role of ENSO and IOD. (2020). doi: 10.1109/AGERS51788.2020.9452779

[20] Rahayu, S.D., H. Setiyono, and E. Indrayanti. Hubungan kecepatan angin dengan luasan upwelling intensitas kuat di perairan selatan jawa pada kejadian la nina, el nino dan normal. (2023). Indonesian Journal of Oceanography. **5** (1): 7 – 17

[21] Rahman, M.A., L.S. Mega, M.U.K. Agung, dan Sunarto. Pengaruh musim terhadap kondisi oseanografi dalam penentuan daerah penangkapan ikan cakalang (*katsuwonus pelamis*) di perairan selatan jawa barat. (2019). Jurnal Perikanan dan Kelautan. **10** (1): 92–02

[22] Ratnawati, Herlina, Ika., Rahmat, Hidayat., Ahmad, Bey., Tania, June. Upwelling di Laut Banda dan Pesisir Selatan Jawa serta Hubungannya dengan ENSO dan IOD. (2016). doi: 10.20884/1.OA.2016.12.3.134

[23] Schept, Judah. Comparative Analysis of Upwelling Characteristics in Northeast and Southwest of Indonesian Seas Area. (2022). doi: 10.1007/978-981-19-0308-3_47

[24] Siswanto, Suratno. Seasonal pattern of wind induced upwelling over java-bali sea waters and surrounding area. International Journal of Remote Sensing and Earth Sciences. (2010) doi: 10.30536/J.IJRESES.2008.V5.A1228

[25] Srinidhi, Nagarada, Gadde., Richard, J., A., M., Stevens. Effect of Coriolis force on a wind farm wake. (2019). doi: 10.1088/1742-6596/1256/1/012026

[26] Sukresno, B., D. Jatisworo dan D.W. Kusuma. Analisis multilayer variabilitas upwelling di perairan selatan jawa. (2018). *Jurnal Kelautan Nasional.* **13** (1): 15–25

[27] Susanto, R. D., A.L. Gordon and Q. Zheng. Upwelling along the coasts of java and sumatra and its relation to ENSO. (2001). Geophysical Research Letters. **28** (8): 1599–1602

[28] Undap, Marthin, Wahyu, Bethara, Wahyono., Agus, Anugroho, Dwi, Suryoputro., Anindya, Wirasatriya. Strong Negative Indian Ocean Dipole Phenomenon in 2016 Reduce the Upwelling Intensity along the Seas of Southern Java. (2023). doi: 10.12912/27197050/172873

[29] Valsala, Vinu., Shamil, Maksyutov. A Short Surface Pathway of the Subsurface Indonesian Throughflow Water from the Java Coast Associated with Upwelling, Ekman Transport, and Subduction. International Journal of Oceanography. (2010). doi: 10.1155/2010/540783

[30] Wang, Jing., Dongliang, Yuan., Xia, Zhao. Comparison of the positive and negative Indian Ocean Dipole forcing on the Pacific interannual variability through the oceanic channel. Journal of Oceanography. (2021). doi: 10.1007/S10872-021-00611-8
[31] Wirasatriya, Anindya., Joga, Dharma, Setiawan., Denny, Nugroho, Sugianto., Insan, Alfi, Rosyadi., Haryadi, Haryadi., Gathot, Winarso., Riza, Yuliratno, Setiawan., R., Dwi, Susanto., R., Dwi, Susanto. Ekman dynamics variability along the southern coast of Java revealed by satellite data. International Journal of Remote Sensing. (2020). doi: 10.1080/01431161.2020.1797215
[32] Xu, Tengfei., Shujiang, Li., Faisal, Hamzah., Agus, Setiawan., R., Dwi, Susanto., R., Dwi, Susanto., Guojiao, Cao., Zexun, Wei. (2018). (9) Intraseasonal flow and its impact on the chlorophyll-a concentration in the Sunda Strait and its vicinity. (2018). doi: 10.1016/J.DSR.2018.04.003

IOP Conf. Series: Earth and Environmental Science **1473** (2025) 012016 doi:10.1088/1755-1315/1473/1/012016

Comparative Study of Metal Powder Addition into Epoxy Coating on Corrosion Rate Prediction of ASTM A36 Steel

H Pratikno[1], F I Neshalangga[1*] and H Ikhwani[1]

[1] Department of Ocean Engineering, Faculty of Marine Technology, Institut Teknologi Sepuluh Nopember, Surabaya, East Java, 60111, Indonesia

*E-mail: fritz.izzulhaqneshalangga@gmail.com

Abstract. In the maritime industry, steel is crucial for construction, such as low carbon steel or mild steel. However, steel is prone to corrosion due to environmental exposure and usage over time. Corrosion can damage metal materials and compromise the integrity of critical components. Coating is the most common method to protect steel from corrosion because of its effectiveness and ease of application. Adding organic or inorganic pigments like magnesium carbonate ($MgCO_3$), aluminium (Al), and zinc oxide (ZnO) to epoxy coatings can enhance corrosion resistance and durability. This study aims to predict corrosion rates and identify the optimal metal powder among these three for epoxy coating on ASTM A36 steel. The method involved varying pigment weight compositions with paint, specifically 1% and 10% of the total paint volume used, and using a coating thickness of 250 μm. Corrosion rates were tested using a three-electrode cell. The lowest corrosion rate was found in the specimen with 10% magnesium carbonate ($MgCO_3$) addition, measuring 0.019984 mmpy, while the highest rate was in the specimen without additives, at 0.65124 mmpy. The study concludes that the addition of magnesium carbonate ($MgCO_3$) powder results in the most optimal corrosion resistance compared to other metal powders.

1. Introduction

Along with the times, the global oil and gas industry is growing rapidly following market demand. Indonesia, with its abundant natural resources, is one of the oil and gas producing countries in the world and still makes the commodity a strategic commodity in the country's economic sector, both as a foreign exchange earner and to maintain national energy security. Indonesia in 2017, through PT Pertamina (Persero) as a holding company in energy sector recorded a net profit of 2,552,619 USD, which was very significant to increase the state treasury [1]. This is followed by daily production that reaches 337,000 barrels and will continuously increase every year [2].

Steel is a primary material commonly used in the oil and gas industry, particularly in the construction of offshore structures. To meet construction and operational needs, low carbon steel or mild steel is generally used due to its relatively lower cost compared to other metal alloys, as well as its good mechanical properties and ease of production and forming [3]. However, along with the increase in operational time of an offshore building that is exposed to seawater directly and continuously, the steel may eventually degrade in quality and become susceptible to corrosion. Although corrosion may not have an immediate adverse effect on a material, it can affect strength, mechanical operation, physical appearance, and ultimately lead to serious operational issues [4].

Corrosion is a chemical phenomenon that occurs when metal, particularly steel, reacts with an environment containing an electrolyte solution, a cathode, an anode, and an electrical current [5]. Corrosion can damage metal materials and threaten the integrity of critical components across industries. Scientifically, corrosion cannot be entirely prevented or eliminated, but it can be controlled by reducing its rate [6]. Corrosion can lead to the formation of cracks or openings in steel structures such as pipelines, tanks, or other components. These defects may result in leaks of oil, gas, or hazardous chemicals into the ocean, potentially harming the marine environment and ecosystem [7]. Corrosion inside these pits, along with crevices found in joints or welding flaws, manifests as uneven corrosion on the steel surface of the platform. Over time, the added stresses from pits, crevices, and other structural irregularities caused by electrochemical reactions result in fractures and breaks in the structure. The corrosion rate is influenced by the metal's microstructure [8]. The impact of corrosion on industries is significant, often resulting in substantial financial losses.

Epoxy is a type of organic coating commonly used in the industry to protect steel from corrosion. Epoxy coating is a widely used method for preventing corrosion on metal surfaces, especially on pipes. When applied to a metal surface, epoxy forms a strong bond with the metal, creating a protective layer that prevents the metal from direct exposure to corrosive environments [5]. However, the drawback of epoxy-based coatings is their brittleness, which causes failures in areas subjected to mechanical corrosion [9]. Adding organic and inorganic pigments to epoxy coatings can enhance corrosion resistance and increase their durability. The adhesion, barrier properties, and corrosion resistance of epoxy coating systems can be enhanced by incorporating appropriate pigments and additives [10]. Some examples of these pigments include magnesium carbonate ($MgCO_3$), aluminium (Al), and zinc oxide (ZnO). These elements are more prone to oxidation (electron release) compared to iron and steel. The working concept involves protecting the substrate by sacrificing the applied coating layer, a process known as anodic coating that shields the metal from direct exposure to corrosive environments [4].

This study will conduct an analysis and comparison of the most optimal metal powder addition for use in epoxy coatings, focusing on predicting the corrosion rate on ASTM A36 steel. The conditions will be standardized by setting the pigment weight composition to 1% and 10% of the total paint volume used, with a coating thickness of 250 μm. By directly comparing these variables, this research aims to provide deeper insights into the strengths and weaknesses of each element in achieving optimal corrosion protection. Consequently, this study is expected to make a valuable contribution to determining the most effective corrosion prevention methods in specific applications.

2. Experimental

2.1. Steel Specimens Preparation

The ASTM A36 steel used in this study was cut into dimensions of 40 × 20 × 10 mm, which will be used as test specimens for corrosion rate analysis as shown in Figure 1.

Figure 1. Corrosion rate testing steel specimens.

2.2. Sandblasting
The sandblasting process, using silica sand as the abrasive material, is depicted in Figure 2. The purpose of this blasting process is to clean the steel surface from rust, dirt, and any residual coating if a previous coating has been applied. Additionally, blasting serves to create the desired surface roughness profile, which enhances coating adhesion. The targeted cleanliness level for the material is Sa 3 (white metal), in accordance with the ISO 8501-1 standard "Visual Assessment of Surface Cleanliness" [11], as shown in Figure 3. In this process, all steel specimens received the same treatment with the same results.

Figure 2. Blasting pot filled with silica sand.

Figure 3. Visual results of steel specimens after sandblasting.

2.3. Surface Roughness Measurement
After the sandblasting process is completed, the next step is to measure the surface roughness of the steel specimens using a roughness meter, which follows ASTM D4417 "Standard Test Methods for Field Measurement of Surface Profile of Blast Cleaned Steel" [12]. Surface roughness affects the adhesion of the coating layer applied on top. The rougher the surface, the better the coating adhesion. Measurements are taken at three points on each steel surface to obtain more accurate results. Table 1 below presents the surface roughness measurements for each specimen.

Table 1. Value of surface roughness measurement.

Steel Specimens	Surface Roughness (μm)			Average
	1	2	3	
Without Addition	80	76	68	75
Magnesium Carbonate ($MgCO_3$) 1%	58	70	68	65
Aluminium (Al) 1%	65	60	60	62
Zinc Oxide (ZnO) 1 %	70	58	65	64
Magnesium Carbonate ($MgCO_3$) 10%	74	56	66	65
Aluminium (Al) 10%	60	65	72	66
Zinc Oxide (ZnO) 10%	80	68	66	71
Total Average				66.90

2.4. Environmental Test
Before commencing the coating process, it is necessary to measure the environmental conditions at the workplace. This is crucial as it significantly affects both the process and the quality of the final coating. There are specific limits: during the coating process, the maximum Relative Humidity (RH) must be 85%, and the specimen's surface temperature (Ts) must be more than 3°C above the Dew Point (DP). If

the environmental conditions do not meet these criteria, coating cannot proceed because the presence of moisture or condensation on the specimen's surface can adversely affect paint adhesion. Table 2 below presents the environmental conditions at the workplace, which have been verified with a dew point meter, indicating that the conditions are suitable for proceeding with the work.

Table 2. Results of environmental test.

Surface Temperature (Ts)	37.6 °C
Air Temperature (Ta)	36.5 °C
Dew Point Temperature (Td)	25.2 °C
Delta T (TΔ = Ts – Td)	12.5 °C
Relative Humidity (RH)	52.50%

2.5. Coating Application

The coating process utilizes Jotun Jotamastic 80 epoxy paint applied using a conventional air spray gun. This method was chosen due to its relative ease of application, speed, flexibility, and the ability to achieve a more even and better adhesion of the paint compared to other methods.

The time interval between the sandblasting process and the coating application should not exceed 3–4 hours. If this period is exceeded, the surface of the specimen, which has undergone sandblasting, may become contaminated by dust and moisture. This contamination can result in poor adhesion of the paint or coating. When performing the coating, it is essential to follow the guidelines provided in the Technical Data Sheet to ensure optimal results.

In this study, variations in the addition of $MgCO_3$, Al, and ZnO metal powders were made at 1% and 10% of the total paint volume used. The total volume of paint used is 50 ml. Table 3 shows the composition of the paint and metal powders used for each variation.

Table 3. Composition of paint and metal powder.

Variant	Composition	
	Powder (gram)	Paint (ml)
Without Addition	0	50
Magnesium Carbonate ($MgCO_3$) 1%	0.5	49.5
Aluminium (Al) 1%	0.5	49.5
Zinc Oxide (ZnO) 1 %	0.5	49.5
Magnesium Carbonate ($MgCO_3$) 10%	5	45
Aluminium (Al) 10%	5	45
Zinc Oxide (ZnO) 10%	5	45

Next, the wet film thickness (WFT) measurement is conducted, which involves measuring the thickness of the paint while it is still wet, in accordance with ASTM D4414 "*Standard Practice for Measurement of Wet Film Thickness by Notch Gages*" [13]. The WFT measurement is performed using a wet film comb and can be repeated until the desired WFT is achieved. The determination of WFT depends on the desired dry film thickness (DFT). In this study, the target DFT is 250 μm, with 5% thinner added to dilute the paint, and the paint's solids by volume is 80 ± 2 %. The WFT can be determined using the following equation [14]:

$$WFT = \frac{DFT \times (100\% + \%\,Thinner\ Added)}{\%\,Volume\ Solids} \tag{1}$$

$$WFT = \frac{250 \times (100\% + 5\%)}{80 \pm 2\,\%} \tag{2}$$

$$Wet\ Film\ Thickness = \pm 328\ \mu\text{m} \tag{3}$$

Next, the specimens are left to fully cured for 7 days at 23 °C or for 2 days at 40 °C, in accordance with the specifications listed in the Technical Data Sheet. Based on Figure 4, the first sequence from the left end is a specimen that has not been added with metal powder. In the second sequence from the left end is the specimen that has been mixed using 10% $MgCO_3$. The third sequence from the left is the specimen that has been mixed using 10% Al. And the last sequence is the specimen that has been added with 10% ZnO.

Figure 4. Coating application results on steel speciemens.

2.6. Dry Film Thickness (DFT) Measurement

After the specimens have fully cured, as shown in Figure 4 as example of the steel specimens used in this study, the dry film thickness (DFT) can be measured using a coating thickness gauge according to ASTM D6132 "Standard Test Method for Nondestructive Measurement of Dry Film Thickness of Applied Organic Coatings Using an Ultrasonic Gage" [15]. Measurements are taken at three points on each specimen surface to obtain more accurate results. Table 4 below presents the DFT measurements for each specimen.

Table 4. Value of dry film thickness measurement.

Steel Specimens	Dry Film Thickness (μm)			Average
	1	2	3	
Without Addition	258	266	255	260
Magnesium Carbonate ($MgCO_3$) 1%	260	241	246	249
Aluminium (Al) 1%	252	266	244	254
Zinc Oxide (ZnO) 1 %	248	254	260	254
Magnesium Carbonate ($MgCO_3$) 10%	269	229	240	246
Aluminium (Al) 10%	241	232	277	250
Zinc Oxide (ZnO) 10%	239	284	230	251
Total Average				251.95

2.7. Corrosion Rate Prediction Test

The corrosion rate testing was conducted using the electrolysis method with a three-electrode cell setup. This testing adheres to ASTM G102 "Standard Practice for Calculation of Corrosion Rates and Related Information from Electrochemical Measurements" [16].

2.7.1. Three-Electrode Cell. The three-electrode cell is a standard laboratory device used for quantitative research on the corrosion properties of materials. This method can accelerate the corrosion process through polarization and its free corrosion potential. Additionally, this cell can be used for a variety of material corrosion tests. As shown in figure 5, the three-electrode cell consists of several components as follows:

a. Working Electrode
This electrode serves as the test specimen to be investigated. It acts as the anode and is immersed in the working fluid.

b. Auxiliary Electrode
This electrode provides the potential to the working electrode and conducts the electric current generated by the corrosion reaction. This electrode is not required for potential measurement.

c. Reference Electrode
This electrode serves as a reference electrode for measuring the potential applied to the working electrode. The current flowing through this electrode must be very small and can be neglected.

d. Electrolyte Solution
This solution functions as an ionic current conductor in the corrosion reaction, and it must be ensured that all electrodes are fully immersed in the electrolyte solution. In this study, a 3.5% NaCl solution is used to simulate real-world conditions in the field.

Figure 5. The three-electrode cell setup.

2.7.2. Potentiodynamic Polarization Technique and Tafel Curve Extrapolation. The potentiodynamic polarization technique is one of the most commonly used methods for predicting corrosion rates through electrochemical techniques, particularly with a three-electrode cell [17]. This technique is employed to determine the corrosion behavior of metals based on the relationship between potential and anodic or cathodic current. It has gained popularity due to its ease of use and advantages compared to the simpler weight loss method. The current density and potential comparison plot of a specimen can be measured within a few hours or, in some cases, just a few minutes [18]. Using the Tafel extrapolation method, it is possible to measure very low corrosion rates [19].

This test was conducted using the CorrTest device, which acts as a potentiostat to supply the current and is connected to the CS Studio 5 software to receive the response to the potential current flowing through the specimen.

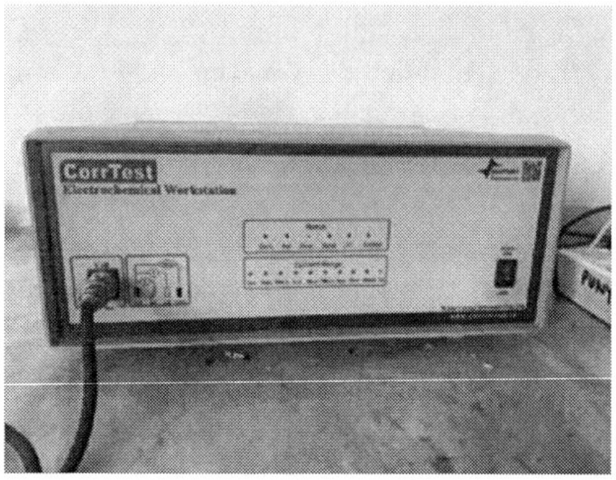

Figure 6. CorrTest device to process the output from the three-electrode cell.

Figure 7. Computer display with CS Studio 5 software connected.

A graph, known as a Tafel curve, can be plotted to represent the relationship between potential (E) and the logarithmic current density (log I), as shown in Figure 6. This is done by plotting the data from the anodic branch (positive pole) and the cathodic branch (negative pole). From this plotting, the Tafel slope or gradient is obtained, which can then be used for extrapolation to find the intersection point that will yield the corrosion current density (Icorr) and subsequently the polarization resistance (Rp) value.

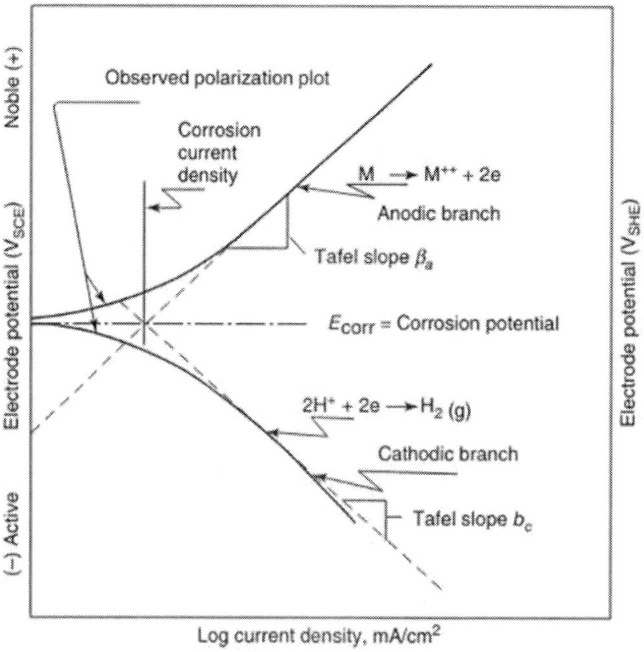

Figure 8. Tafel curve polarization theory [20].

The calculation of the polarization resistance (Rp) value can be performed using the Stern-Geary equation as follows [21]:

$$I_{corr} = \frac{B}{R_P} \quad (4)$$

where:
Icorr = corrosion current density ($\mu A/cm^2$)
B = Stern-Geary constant (V)
Rp = polarization resistance (Ω)

$$B = \frac{\beta_a \times \beta_c}{2{,}303 \times (\beta_a + \beta_c)} \quad (5)$$

where:
βa = slope of the anodic Tafel reaction, when plotted on base 10 logarithmic paper (V/decade)
βc = slope of the cathodic Tafel reaction, when plotted on base 10 logarithmic paper (V/decade)

The corrosion current density (Icorr) obtained through Tafel curve extrapolation must then be converted into the corrosion rate. For this purpose, the current distribution used in the calculation is assumed to be uniform. Using Faraday's Law, the corrosion rate can be determined using the following equation [20]:

$$C_R = K_1 \times \left(\frac{Icorr}{\rho}\right) \times EW \quad (6)$$

where:
C_R = corrosion rate (mmpy)
K_1 = 3.27×10^{-3} (mm g/μA cm yr)
Icorr = corrosion current density ($\mu A/cm^2$)
ρ = density (g/cm^3)
EW = equivalent weight

3. Result and Discussion

Initially, the epoxy resin was modified, and additives were introduced to either promote or inhibit specific reactions among the components, enhancing the mechanical properties and permeability of the pulp material's consolidation body. Additionally, a catalyst was added to activate the acetone molecules in the inert solvent, reducing the shrinkage of the consolidation body while improving its toughness and resistance to aging [22]. The data obtained from this test are presented in a curve commonly known as a Tafel curve. The Tafel curve illustrates the relationship between the corrosion potential (Ecorr) as a function of the logarithm of current density (log I). Ecorr is a dependent variable within a certain range because the potential used in the testing equipment is fixed at the start of the test. The Tafel curve is then extrapolated to form a straight gradient, which will identify the intersection point where the corrosion current density (Icorr) occurs. Icorr is an independent variable; if its value is high, it indicates that a large number of metal ions have dissolved into the electrolyte solution, causing the metal to become unstable. This ultimately leads to surface damage as the metal reacts with its environment. The Tafel curves obtained from the corrosion rate prediction tests for each specimen are presented in Figure 9.

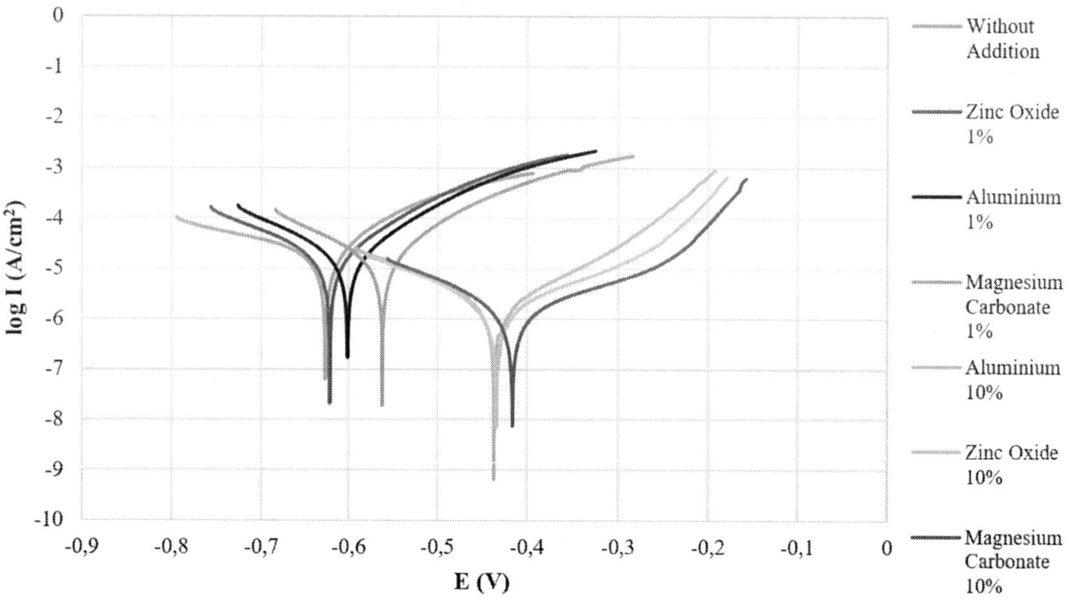

Figure 9. Tafel curves for each variation of metal powder type and addition percentage.

The lower the corrosion rate of a specimen, the more resistant it is to corrosion attacks. Based on the test results presented in Table 5, the lowest predicted corrosion rate was found in the specimen with 10% $MgCO_3$ addition, at 0.019984 mmpy. In contrast, the highest predicted corrosion rate was observed in the specimen without any addition, at 0.65124 mmpy.

Furthermore, an analysis of Figure 9 and Table 5 also reveals a relationship between corrosion potential and corrosion rate. As the corrosion potential (Ecorr) decreases, or becomes more positive, the curve tends to shift to the right. This results in a lower (improved) corrosion rate, consistent with the explanation provided by Kakaei et al [21]. Conversely it is also known that as the corrosion current density (Icorr) increases, the corrosion rate becomes higher (worsens). This condition is similar to the findings of other studies conducted by Munasir et al [23]. In other words, a more negative Ecorr and higher Icorr indicated a faster corrosion rate, whereas a more positive Ecorr and lower Icorr suggested a slower corrosion process [24].

Table 5. Results of corrosion rate prediction testing.

Steel Specimens	Corrosion Current Density, Icorr (A/cm^2)	Corrosion Potential, Ecorr (V)	Corrosion Rate (mmpy)	Relative Corrosion Resistance [25]
Without Addition	5.6022×10^{-5}	-0.62633	0.65124	Fair
Magnesium Carbonate ($MgCO_3$) 1%	2.3816×10^{-5}	-0.56348	0.27685	Good
Aluminium (Al) 1%	2.4725×10^{-5}	-0.60319	0.28742	Good
Zinc Oxide (ZnO) 1 %	2.52×10^{-5}	-0.62115	0.29294	Good
Magnesium Carbonate ($MgCO_3$) 10%	1.7235×10^{-6}	-0.41646	0.019984	Outstanding
Aluminium (Al) 10%	1.906×10^{-6}	-0.43741	0.022099	Excellent
Zinc Oxide (ZnO) 10%	1.8527×10^{-6}	-0.43464	0.021481	Excellent

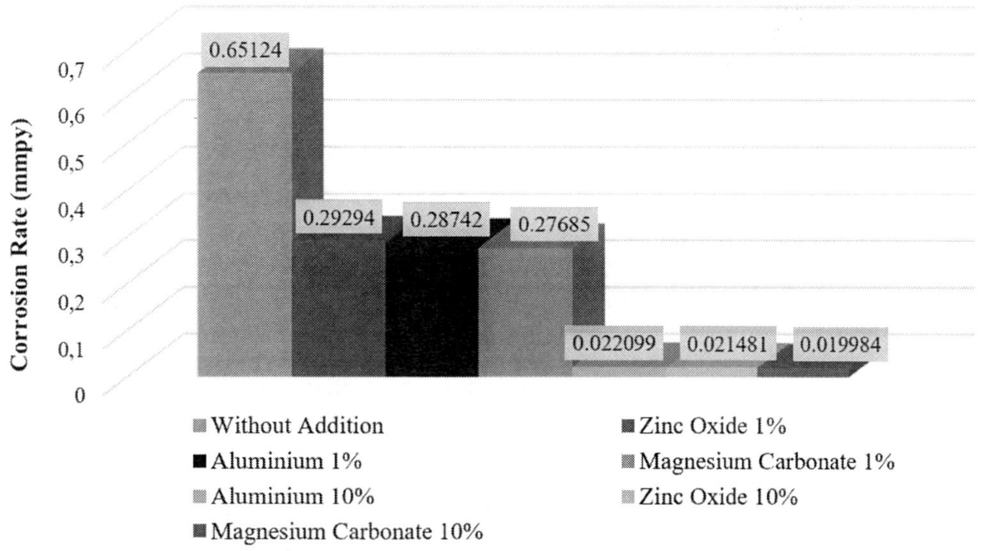

Figure 10. Comparison diagram of predicted corrosion rate values for each variation of metal powder type and addition percentage.

Looking more closely at Figure 10, it is evident that the addition of $MgCO_3$ powder results in the lowest corrosion rate compared to other metal powders, whether in 1% (coloured green) or 10% (coloured blue) conditions. In fact, with using 10% $MgCO_3$ powder, the corrosion rate reaches an "Outstanding" level, which is the highest category in relative corrosion resistance according to Fontana [25].

Previous studies have researched various concentrations of epoxy mixed with other substances for coating materials. The first study examined coatings using epoxy and graphene, this involved epoxy coatings based on graphene oxide (GO) and reduced graphene oxide (rGO). The results showed that the coating with GO exhibited superior performance, with a corrosion rate (CR) of 2.23 mpy, equivalent to 0.056642 mmpy [26].

The second study investigated coatings composed of aluminium-graphite. In this study, the coating consisted of 80% epoxy and 20% aluminium-graphite pigment. The aluminium-graphite pigment variable was tested at two different compositions: a 1:1 ratio and a 3:1 ratio. The results of this study indicated that the epoxy coating with an aluminium-graphite mixture at a 3:1 ratio was most effective in preventing corrosion with a corrosion rate (CR) of 0.047 mmpy [27]. In another study, epoxy mixed with 7% polyaniline (PANI) powder coating was used to protect against corrosion. The results showed that this method also reduced the corrosion rate to a satisfactory level of 0.0372 mmpy [28].

From several other studies using various types of metal powders and different methods as mentioned, it was known that 10% MgCO3 powder is still the best results, indicated by having the lowest corrosion rate value. This is also proven by the comparison study that has been carried out by comparing the modification of epoxy coating with other mixtures carried out in previous research.

Generally, the study also reveals that adding metal powder to epoxy coating reduces the corrosion rate. This occurs because the primary function of metal powder pigments is to form a metal oxide layer that reacts with the environment, in this case, seawater. This layer acts as a barrier protection, shielding the substrate or steel surface by blocking the passage of oxygen and water [29]. Assessing the long-term resilience and dependability of anti-corrosion coatings is essential for understanding their effectiveness over extended time frames [30]. Thus, it can also be concluded that specimens with inadequate metal powder pigments will expose the substrate or steel more quickly, resulting in a higher corrosion rate [31].

The integration of metal powder into epoxy coatings represents a significant advancement in corrosion protection technology, particularly in real-world applications. This innovative approach has demonstrated remarkable benefits across various industries, with the maritime sector standing to gain substantially. By incorporating metal powder, manufacturers have successfully achieved a dramatic reduction in corrosion rates, far surpassing the performance of traditional epoxy coatings. This enhancement is not merely incremental but represents a step-change in protective capabilities, opening new possibilities for long-term asset preservation in challenging environments.

The implications of this improved corrosion resistance are far-reaching, especially for maritime industries where exposure to harsh, saline conditions is constant. Ships, offshore platforms, and coastal infrastructure can now benefit from substantially extended protection periods. This prolonged defense against corrosive elements translates to increased operational lifespans for steel structures, a critical factor in industries where downtime and replacement costs can be exorbitant. Moreover, the enhanced durability of these coatings contributes to improved safety standards, reducing the risk of structural failures due to corrosion-induced weaknesses.

From an economic perspective, the adoption of metal powder-enhanced epoxy coatings presents a compelling value proposition. While the initial application costs may be higher compared to conventional coatings, the long-term financial benefits are significant. The extended operational life of steel structures leads to a marked reduction in maintenance frequency and intensity. This translates to lower overall lifecycle costs, encompassing decreased expenditure on repairs, replacements, and associated labor. Furthermore, the reduced need for maintenance interventions minimizes operational disruptions, allowing for more consistent and efficient use of assets. As industries continue to seek ways to optimize their operations and reduce environmental impact, these advanced coatings offer a sustainable solution that aligns with both economic and ecological objectives.

4. Conclusion

Based on the corrosion rate prediction test of epoxy coating using electrochemical methods with a three-electrode cell, among the ASTM A36 steel specimens with various types and amounts of metal powder additions, the smallest corrosion rate was found in the specimen with 10% $MgCO_3$ powder, which was 0.019984 mmpy. Conversely, the highest corrosion rate was observed in the specimen with no addition, which was 0.65124 mmpy. Thus, it can be concluded that the predicted corrosion rate tends to be lower, or in other words, the corrosion resistance improves when metal powder is added into the epoxy coating.

Furthermore, it was found that $MgCO_3$ powder is the most optimal for use or added into epoxy coating compared to other metal powders. It is shown that with a 1% addition of $MgCO_3$ powder, the best corrosion rate value achieved is 0,27685 mmpy (categorized as "Good"), and with a 10% addition, the best corrosion rate value achieved is 0,019984 mmpy (categorized as "Outstanding").

Acknowledgements

The authors would like to thank the Department of Ocean Engineering, Institut Teknologi Sepuluh Nopember, for the support and facilities provided throughout the research process. Gratitude is also addressed to all parties and individuals who have contributed and provided assistance in this research, as well as to all colleagues who have provided valuable input and feedback.

References

[1] Jamaludin A, Sihotangand N and Saputro F B 2020 Analysis of Financial Statements Pertamina 2017 *Int. J. Eng. Technol. Manag. Res.* **6** 68–75

[2] Andhana M R and Prastawa H 2019 Analisis Produktivitas Pekerja Dengan Metode Work

Sampling pada Filling Shed I Produk Premium PT.Pertamina Tbbm Semarang Group *Ind. Eng. Online J.* **7** 1–8
[3] Sudiro T, Hia A I J, Ciswandi, Aryanto D, Hermanto B, Wismogroho A S and Sebayang P 2018 High temperature cyclic oxidation resistance of 50Cr-50Al coatings mechanically alloyed on low carbon steel *J. Alloys Compd.* **732** 655–65
[4] Ahmad Z 2006 *Principles of Corrosion Engineering and Corrosion Control* (Oxford: Butterworth-Henneaman, Elseiver)
[5] Faizal and Malik Made A 2020 Pengaruh Kerekatan Perpaduan Pelapisan Epoksi Pada Permukaan Pipa (Base Pipe) Setelah Dilakukan Tes Ikatan Katodik (Kathodic Disbonding Test) Terhadap Laju Korosi *J. Tek. Mesin* **9** 80
[6] Lubis A L, Antoko B and Wiro B 2020 Analisa Laju Korosi Pada Metode Katodik Anoda Tumbal dan Coating Pada Material A53 Grade B Pada Jalur Pipa Underground *Proc. Conf. Pip. ...* 175–9
[7] Jing J, Tian Y, Zhu X, Zhou Y, Shi C and Lei Q 2024 Safety valuation of offshore oil and gas well string based on corrosion rate prediction *J. Loss Prev. Process Ind.* **91** 105385
[8] Prabhakar S and Goswami G 2019 Design, Management and Key Success Factors of an Offshore Cathodic Protection System for Corrosion Control *Int. J. Eng. Manag. Res.* **9** 171–7
[9] Zhang M Q, Rong M Z, Yu S L, Wetzel B and Friedrich K 2002 Improvement of tribological performance of epoxy by the addition of irradiation grafted nano-inorganic particles *Macromol. Mater. Eng.* **287** 111–5
[10] Hare C H 2011 Corrosion Control of Steel by Organic Coatings *Uhlig's Corrosion Handbook: Third Edition* ed R W Revie (Hoboken, New Jersey: John Wiley & Sons, Inc.) pp 971–83
[11] ISO 1988 ISO 8501-1: Preparation of Steel Substrates Before application of Paints and Related Products—Visual Assessment of Surface Cleanliness— 1–12
[12] ASTM 2021 ASTM D4417-21: Standard Test Methods for Field Measurement of Surface Profile of Blast Cleaned Steel 1–7
[13] ASTM 2001 ASTM D4414-95: Standard Practice for Measurement of Wet Film Thickness by Notch Gages 1–3
[14] Jotun 2010 Jotun Paint School
[15] ASTM 2008 ASTM D6132-08: Standard Test Method for Nondestructive Measurement of Dry Film Thickness of Applied Organic Coatings Using an Ultrasonic Gage 1–4
[16] ASTM 1999 ASTM G102-89: Standard Practice for Calculation of Corrosion Rates and Related Information from Electrochemical Measurements 1–7
[17] Banjare M K, Behera K and Banjare R K 2023 Electrochemical principles of corrosion inhibition: fundamental and computational aspects of density functional theory (DFT) *Computational Modelling and Simulations for Designing of Corrosion Inhibitors* ed D K Verma, C Verma and J Aslam (Elsevier) pp 243–69
[18] PPA 1982 Application Note Corr-1 (Subject: Basics of Corrosion Measurements)
[19] Popov B N 2015 Corrosion Engineering: Principles and Solved Problems *Corrosion Engineering: Principles and Solved Problems* (Oxford: Elsevier B.V.) pp 181–237
[20] Papavinasam S 2008 Electrochemical polarization techniques for corrosion monitoring *Techniques for Corrosion Monitoring* ed L Yang (Cambridge, UK: Woodhead Publishing Limited) pp 49–85
[21] Kakaei K, Esrafili M D and Ehsani A 2019 Graphene Surfaces Particles and Catalysts *Interface Science and Technology* vol 27, ed A Hubbard (London: Elsevier Ltd) pp 303–37
[22] Liu B, Zhu G and Pengju D 2013 The Research Progress of Epoxy Resin Coating in Building Waterproof *Adv. Mater. Res.* **671–674** 1779–82
[23] Munasir N, Triwikantoro, Zainuri M, Bäßler R and Darminto 2019 Corrosion Polarization Behavior of Al-SiO2 Composites in 1M and Related Microstructural Analysis *Int. J. Eng. Trans. A Basics* **32** 982–90
[24] Wei H, Ding D, Wei S and Guo Z 2013 Anticorrosive conductive polyurethane multiwalled

IOP Conf. Series: Earth and Environmental Science **1473** (2025) 012016 doi:10.1088/1755-1315/1473/1/012016

carbon nanotube nanocomposites *J. Mater. Chem. A* **1** 10805–13
[25] Fontana M G 1987 *Corrosion Engineering* (Singapore: McGraw Hill Book Company)
[26] Ghauri F A, Raza M A, Baig M S and Ibrahim S 2012 Corrosion study of the graphene oxide and reduced graphene oxide-based epoxy coatings *Mater. Res. Express* **4**
[27] Budiarto U, Firdhaus A, Hakim M L, Tuswan T and Yusuf F A F 2024 Impacts of temperature and coating pigment ratios on the corrosion rate of ss400 steel *J. Teknosains* **13** 117–27
[28] Murugesan V and Perumal A V 2022 Electrodeposition and corrosion characterization of epoxy/polyaniline coated AZ61 magnesium alloy *Corros. Rev.* **40** 491–9
[29] Abdel-Samad A, Soud Y and Zaki M 2014 Influence of Paint on Steel Corrosion for Marine Applications *J. Surf. Eng. Mater. Adv. Technol.* **04** 189–95
[30] Aljibori H S, Alamiery A and Kadhum A A H 2023 Advances in corrosion protection coatings: A comprehensive review *Int. J. Corros. Scale Inhib.* **12** 1476–520
[31] Ayoola W A, Bodude M A, Onovo H and Oyetunji A 2018 Effects of Zinc Powder Addition to Villa Gloss and Silka Lux Marine Enamel Paints on Corrosion Resistance of Mild Steel *Niger. J. Technol.* **37** 158–66

Condition of Coral Reef Cover and Water Environment in Karimunjawa and Kemujan Island

A K Wardhani[1*], S Rahmawati[1], H Pratikno[1], and M Zikra[1]

[1] Department of Ocean Engineering, Faculty of Marine Technology, Institut Teknologi Sepuluh Nopember, Surabaya, 60111, Indonesia

*corresponding author: anita@its.ac.id

Abstract. Karimunjawa and Kemujan Islands are the largest islands in the Karimunjawa National Park which have a coral reef ecosystem. Coral reefs are influenced by the environmental conditions around them. Changes in environmental conditions, rising of sea surface temperatures, and human activities can pose threats. Coral reefs have important functions and roles in the balance of marine ecosystems, so monitoring coral reefs is very important. To develop coral reef conservation in the waters of Karimunjawa Island, a study was conducted to monitor the condition of coral reef cover using remote sensing. The purpose of this study was to determine changes around coral reefs in the waters of Karimunjawa Island and Kemujan Island in 2018 using Landsat 8 satellite imagery and to analyze changes in the area and condition of coral reefs in the waters of Karimunjawa Island and Kemujan Island, Karimunjawa National Park. This study uses a remote sensing method which begins with collecting Landsat 8 satellite imagery data, then continued with satellite image data processing with ENVI and ArcGIS 10.8 software using the Lyzenga algorithm for coral reef mapping. This method is more efficient and can cover a wide study area.

1. Introduction

Coral reefs are the main ecosystems providing coastal and marine resources, besides mangrove forests and seagrass beds [8]. Coral reefs are widely recognized for their role in dissipating wave energy, which is crucial for providing coastal protection [3] and contributing to the balance of marine ecosystems, including serving as a food source, environmental indicators, and supporting tourism and reef fisheries. However, they are also sensitive to environmental changes [6]. Environmental conditions that do not fall within the tolerance limits of coral reefs can affect their existence [5].

Karimunjawa National Park is in the north of Java Island, under the administration of Jepara Regency, and consists of 27 islands with coral reef ecosystems there. The two largest islands are Karimunjawa Island and Kemujan Island. These two islands function as centers of settlement, economic activities, and tourism [2].

This study was conducted to maintain the condition of coral reef cover using remote sensing to improve coral reef conservation in the waters around Karimunjawa Island and Kemujan Island. Remote sensing with coral reef ecosystem objects includes coral ecological conditions and monitoring of its properties with the aim of obtaining information on coral distribution, benthic substrate composition, geomorphology and coral reef health conditions [7].

2. Study Area

This research was located on Karimunjawa and Kemujan Islands as shown in Figure 1, where these are the largest islands in the Karimunjawa National Park, Central Java. This research location presents significant challenges, especially related to the impacts of human activities and climate change. Tourism in Karimunjawa not only has a positive impact on improving the economy and employment but also has a negative impact on the local marine ecosystem which causes pressure on the environment, such as increased marine debris, coral reef degradation, and disturbance of marine habitats (Lukman *et., al*, 2022)

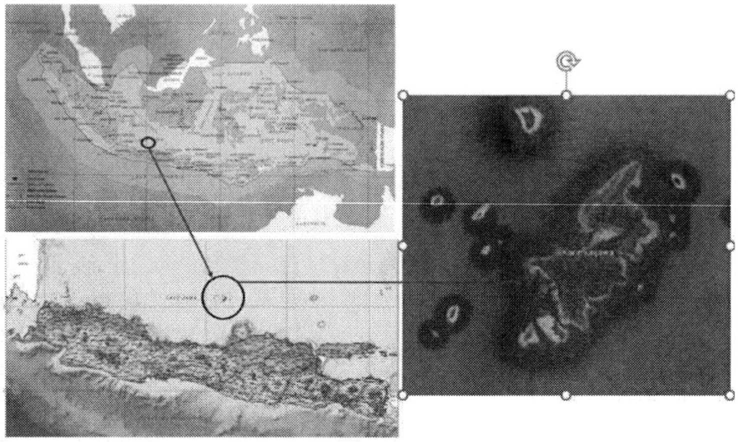

Figure 1. Research Locations on Karimunjawa Island and Kemujan Island

3. Methodology

2.1 Data Preparation

Data preparation includes downloading Landsat 8 satellite imagery data for 2018 at earthexplorer.usgs.gov. Downloading 1 data package consists of Band 1 to Band 10 as shown in Figure 2.

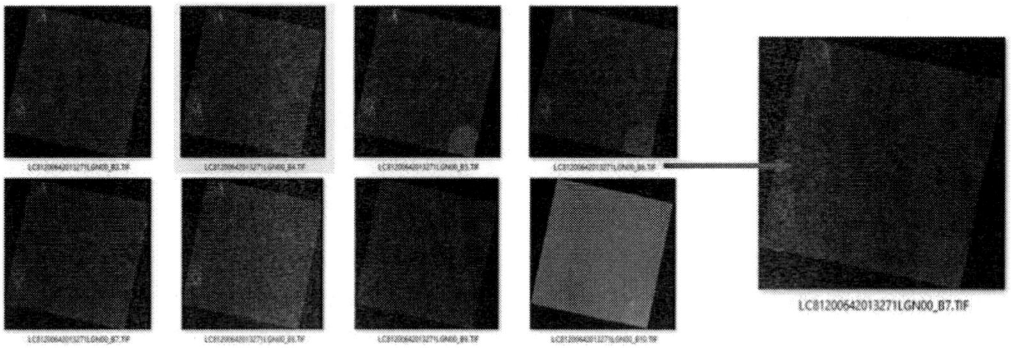

Figure 2. Landsat 8 Satellite Imagery Data

2.2 Satellite Image Data Processing

The stages of satellite image data processing include color composite selection, image cropping, radiometric correction, image algorithm transformation, image classification and coral area calculation.

2.2.1 Color Composite Selection and Image Cropping

The selection of color composites is the process of combining several bands by overlapping, and must select bands that are in accordance with the classification objectives. Identification of coral reefs with Landsat 8 satellite imagery using RGB (red green blue) 564 composites. This process is carried out using ENVI 5.3 software. Continued with the cropping process or cutting of satellite images. Cropping can be adjusted according to needs based on the focus of the research area so that the image of the research location becomes clearer.

2.2.2 Radiometric Correction

Unprocessed initial satellite image data usually contains noise caused by the system. One of the noises can be caused by differences in the position of the sun when the data is acquired. To eliminate this noise, Top of Atmosphere (ToA) radiometric correction can be used. ToA correction is a correction of radiometric distortion caused by the position of the sun, it works is by converting digital number (DN) values into reflectance values. The conversion equation for ToA reflectance correction is:

$$\rho\pi' = M\rho Qcal + A\rho \quad (1)$$

$\rho\pi'$ = TOA reflectance, without correction for sun angles
$M\rho$ = REFLECTANCE_MULT_BAND_x , where x is the Band number
$A\rho$ = REFLECTANCE_ADD_BAND_x , where x is the Band number
Qcal = Digital number value (DN)

The next step is corrected the image against the sun angle to eliminate the difference in DN values originating from the sun's position. If the sun's position value has been obtained, it will be combined with the sun angle correction equation, namely:

$$\rho\pi=\rho\pi'/(cos(\theta SZ))=\rho\pi'/(sin(\theta SE)) \quad (2)$$

$\rho\pi$ = ToA reflectance
θSE = sun elevation
θSZ = zenith angle of the sun, $\theta SZ = 90° - \theta SE$

2.1.3 Image Algorithm Transform

Image Algorithm Transform is a formula to produce an invariant depth index value where the value can distinguish objects based on competition. Following the Lyzenga formula :

$$Y = ln\ (TM1) + ki/kj\ ln\ (TM2) \quad (3)$$

Y = image of the results of water base extraction
TM1 = digital value of channel 1 (Landsat TM)
TM2 = digital value of channel 2 (Landsat TM)
ki/kj = attenuation coefficient value

where :

$$ki/\ kj= a +\sqrt{(a^2+1)} \quad (4)$$

where :

$$a = (varTM1 - varTM2)\ /\ (2 + covarTM1TM2) \quad (5)$$

Var = variety values of digital values
Covar = coefficient value of the digital value

2.2.4 Image Classification

Image classification is a process to obtain images that have been entered into certain categories or classes based on the reflectance value of each object. The image generated by the transformation of the image is further classified to explain the object or ecological land cover of the coral reef. The image classification process is carried out after obtaining the value from the calculation results of the Lyzenga 1981 algorithm. Validation of the results of this classification uses Ground Check data.

2.2.5 Coral Area Calculation

The purpose of the area calculation process is to determine changes in coral reefs from 2018, 2020, 2022 to 2024. This process is processed using the Calculate Geometric method. Calculate Geometric is an automatic calculation procedure in ArcGIS based on the geometric shape of the GIS data drawn and the coordinate system used.

4. Result

Map of the coral reef area of Karimunjawa Island and Kemujan Island in 2018 as shown in Figure 2. Based on the Indonesian National Standardization Agency on Mapping of Shallow Sea Waters Habitats (2011), pink represents live coral, cyan represents non-coral and yellow represents sand.

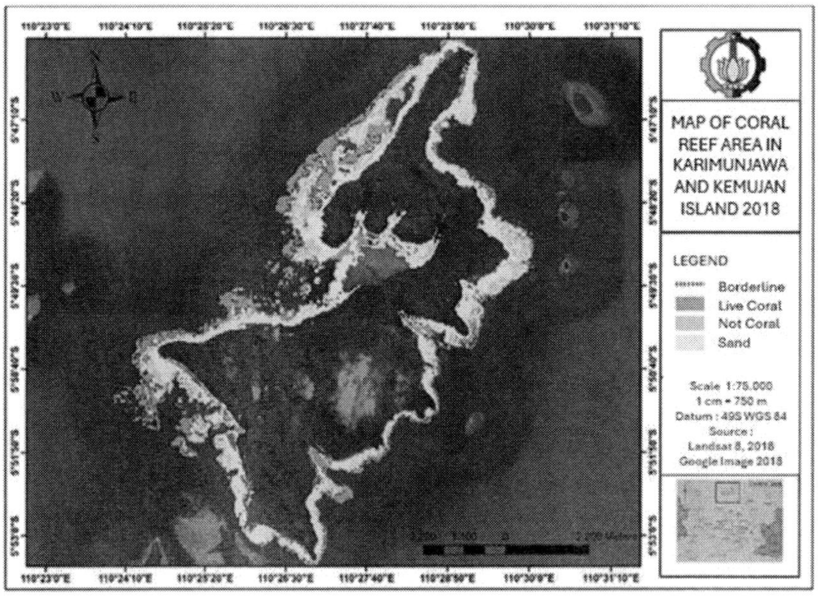

Figure 3. Map of Coral Areas in Karimunjawa and Kemujan Islands 2018

5. Conclusion

Based on research on the Condition of Coral Reef Cover and Water Environment on Karimunjawa and Kemujan Island, a map of the coral reef areas of Karimunjawa Island and Kemujan Island was produced in 2018 with the description that pink is the color of live coral, cyan is the color of non-coral and yellow is the color of sand.

References

[1] Badan Standar Nasional. 2011. SNI 7716:2011 Pemetaan Habitat Dasar Perairan Laut Dangkal
[2] BTN Karimunjawa, 2023. Statistik Balai Taman Nasional Karimunjawa 2023. Kementerian Lingkungan Hidup dan Kehutanan Direktorat Jenderal Konservasi Sumber Daya Alam dan Ekosistem Balai Taman Nasional Karimun Jawa. Semarang
[3] Burke, L., M. Spalding. 2022. Shoreline protection by the world's coral reefs: Mapping the benefits to people, assets, and infrastructure. World Resourcer Institute.
[4] Lukman, K. M., Y. Uchiyama., J.M.D Quevedo., R. Kohsaka. 2022. Tourism Impacts on Small Island Ecosystem: Public Perceptions from Karimunjawa Island, Indonesia. Journal of Coastal Conservation 26: 14.
[5] Mukti, Agung Setyo. 2017. Analisis Perubahan Luasan Habitat Terumbu Karang dan Kondisi Lingkungan di Perairan Pulau Karimunjawa dan Pulau Kemujan Balai Taman Nasional Karimunjawa. Universitas Brawijaya.
[6] Wardhani, Anita Kusuma. 2017. Analisis Perubahan Luasan Tutupan Karang dan Faktor yang Mempengaruhi di Perairan Menjangan Besar dan Menjangan Kecil, Taman Nasional Karimunjawa. Universitas Brawijaya.
[7] Xu, J., D. Zhao. 2014. Review of Coral Reef Ecosystem Remote Sensing. Acta Ecologica Sinica 34. 19 – 25.
[8] WWF. 2010. Pulihkan Terumbu Karang Indonesia. http://www.wwf.or.id/?20300/PulihkanTerumbu-Karang-Indonesia (accessed on 27 August 2024).

Factors Influencing Ship Berthing Scheduling for Enhancing Container Terminal Productivity in Sustainable Operations

Prabowo Budhy Santoso[1*], Haryo Dwito Armono[2], Raja Oloan Saut Gurning[3], and Danang Cahyagi[4]

[1]Interdisciplinary School of Management Technology, Institut Teknologi Sepuluh Nopember, Surabaya Indonesia
[2]Department of Ocean Engineering, Institut Teknologi Sepuluh Nopember, Surabaya Indonesia
[3]Department of Marine Engineering, Institut Teknologi Sepuluh Nopember, Surabaya Indonesia
[4]Department of Mechanical Engineering, Politeknik Negeri Batam, Batam, Indonesia

E-mail: 7032221013@mhs.its.ac.id[1], armono@oe.its.ac.id[2], saut@its.ac.id[3], cahyagi@polibatam.ac.id[4]

Abstract. The berthing schedule significantly affects the number of ships that can berth at a container terminal within a given timeframe. Inefficient scheduling can lead to the underutilization of berthing slots, thereby impacting the terminal's overall productivity. This research aims to analyse the effect of different berthing schedules on ship calls and production capacity at a container terminal. The study compares several scheduling scenarios, considering parameters such as vessel dimensions, loading and unloading equipment, and berth length. The simulation results demonstrate that optimized window berthing and scheduling can minimize idle time between vessel arrivals, allowing for more ship calls. Effective window berthing scheduling requires careful preparation and continuous monitoring, including coordination with ship-owners, cargo owners, and service providers such as pilots and tug operators, as well as other stakeholders involved in berthing activities at the terminal. Berthing optimization schedules at container terminals by minimizing idle time between vessel arrivals and reducing waiting times, has potential impacts on environmental sustainability, such as decreasing fuel consumption for ships and port equipment and leading to lower greenhouse gas emissions and air pollutants. This research was simulated using a discrete event simulation approach through Vensim software.

1. Introduction

The relationship between container terminal production and economic growth is multifaceted, with evidence suggesting a significant positive correlation. Container terminals serve as critical nodes in global trade, influencing regional economies through various mechanisms. Research indicates that cargo and container throughput at ports exhibit a bidirectional causality with

IOP Conf. Series: Earth and Environmental Science **1473** (2025) 012018 doi:10.1088/1755-1315/1473/1/012018

economic growth, suggesting that increased terminal activity can stimulate economic expansion and vice versa (1). Container terminal has been shown to have a direct and strong positive influence on economic growth, with trade activities acting as mediators (2). The growth of container terminals in cities has positively impacted local labour markets and transport infrastructure, although it also presents challenges such as urban congestion (3). Enhancing the efficiency of container terminals is crucial for maximizing their economic contributions. Studies have shown that improved operational efficiency can significantly boost regional economic growth (4). While the positive impacts of container terminals on economic growth are well-documented, it is essential to consider the potential negative externalities, such as environmental degradation and urban congestion, which may arise from increased terminal activity. Balancing these factors is vital for sustainable economic development.

The limitations of container terminals in restricted areas and waters significantly impact their ability to increase container production. These constraints arise from infrastructural, environmental, and operational challenges that hinder efficiency and capacity expansion. Many limited ports, face restrictions on accommodating larger vessels due to inadequate draught and channel design, which limits their operational capacity and competitiveness (5). The small terminal struggles with insufficient storage and long container dwell times, exacerbated by increased traffic and poor access infrastructure (6). The reliance on manned trucks for inter-terminal transport in congested areas, like Rotterdam, leads to inefficiencies, suggesting a need for innovative logistics solutions (7). While these limitations pose significant challenges, they also highlight the necessity for strategic planning and investment in technology to enhance terminal capabilities and sustainability (8).

Optimizing vessel arrival scheduling through stakeholder coordination is crucial for enhancing operational efficiency and reducing costs in maritime transport. Various studies highlight innovative models and strategies that facilitate this coordination, ultimately leading to improved scheduling outcomes (9). Research on a container terminal introduced a multi-objective optimization model that enhances berth utilization and reduces scheduling time by considering factors like berth preferences and inspections (10). This collaborative approach showcases how stakeholder input can refine scheduling processes. Another research proposed a collaborative optimization model for loading operations and vessel traffic scheduling, revealing that integrated planning can significantly reduce waiting times and improve port productivity (11). This highlights the necessity of synchronizing operations among stakeholders.

"Broken space" or "berth idle time" is a condition where there is a time lag between loading and unloading activities on a ship that have been completed until the arrival of the next ship (12,13). Generally, broken space occurs because the performance of quay crane is faster than the planned scheduling allocation (14,15). Broken space is an opportunity for container terminals to increase the number of ship arrivals by reducing the number of broken spaces that exist in the current scheduling.

This study aims to improve the scheduling of ship arrivals and berthing windows at container terminals located in waters and limited areas in a city in Indonesia. In addition to providing an analytical approach, this article also recommends a strategy for conducting more dynamic scheduling by involving several stakeholders. It is hoped that this study can be a reference for relevant stakeholders to maximize the increase in container production through improving ship arrival scheduling.

2. Method

2.1. Method

In order to simulate improvements in ship arrival scheduling, this article uses a discrete event simulation analysis approach. Discrete Event Simulation (DES) is a simulation method used to model systems or processes that change discretely or gradually. For example, in the context of a port, DES can be used to model the flow of ships arriving at the port, the loading and unloading process, and the movement of containers. Events in this simulation may include ship arrivals, the start or completion of the loading and unloading process, and ship departures. In this article, the performance of a container terminal depends on several parameters. Such as ship arrival modelling, the number of containers transported, the duration of the manoeuvring and berthing process, the performance and number of port crane operators, and the loading and unloading speed of quay crane.

The simulation analysis of container terminal operations is conducted through several systematic stages: The first stage involves defining the foundational assumptions for the simulation, including key variables such as berth length, container terminal capacity, draft limitations, the number and performance of quay cranes, the speed of crane operations, operator performance, and the volume of new containers arriving per vessel. The second stage involves developing a detailed process flow of container terminal operations, represented through discrete event simulation modelling. This stage captures the sequential nature of operations, from ship arrival to cargo handling and departure. The third stage focuses on simulating ship arrival scheduling using daily scheduling scenarios. This involves identifying "broken spaces," or idle periods, within the scheduling framework where no active loading or unloading is taking place. These gaps represent inefficiencies that can be targeted for optimization. In the final stage, the simulation model is optimized to minimize idle periods ("broken spaces") and maximize ship arrivals. This optimization aims to identify additional potential ship arrivals that could enhance annual container throughput, thereby increasing the overall productivity and efficiency of the terminal. By following these stages, the study provides a comprehensive analysis of how optimized scheduling can significantly enhance container terminal performance, offering valuable insights for stakeholders looking to maximize operational efficiency and increase container production.

2.2. Related equation

To model the operational performance of a container terminal, one of the most widely adopted approaches is the model proposed by Notteboom (2020) (16). According to Notteboom (2020), the berth capacity of a container terminal is determined by the product of three key factors: the number of berths, the acceptable berth occupancy rate (expressed as a percentage), and the annual average productivity of vessels (measured in moves per hour). This relationship is mathematically represented in Equation (1), which defines the capacity of a container terminal in terms of these parameters.

$$C_b = n \, x \, \emptyset \, x \, t_{year} \, x \, P \qquad (1)$$

C_b : Berth capacity
n : Number of berth
$\emptyset$: Acceptable berth occupancy rate (%)
n : Annual average productivity of vessel (conth/h)

2.3. Case study
This study focuses on a case study of a container terminal situated in a constrained geographic and aquatic environment, specifically in a city in Indonesia located along a river. The selection of this container terminal as a case study is based on the unique challenges faced by terminals operating within limited space and water depth, where conventional strategies to increase container throughput—such as expanding berth length or increasing the number of quay cranes—are not feasible. In this context, the terminal faces operational constraints due to the shallow depth of the waterway, which restricts the size of vessels that can be accommodated. The terminal can only handle container ships with a maximum capacity of approximately 500TEUs per vessel. Accommodating larger vessels would typically require increased draft, which poses a higher risk of grounding due to the limited water depth. Consequently, specialized strategies are necessary to optimize operations and enhance container throughput under these restrictive conditions without significant infrastructural investments.

3. Result & Discussion
This section presents a comparative analysis of different scheduling methods at container terminals, focusing on the differences between daily and weekly ship arrival scheduling. The analysis also identifies "broken space" periods, during which port equipment remains idle due to the absence of loading and unloading activities. Broken space occurs when a vessel completes its loading and unloading operations and there is a waiting period before the arrival of the next vessel.

Table 1 outlines the basic assumptions used in this analysis, based on a case study of a container terminal equipped with two berths and two quay cranes, with a loading and unloading rate of 35 to 40 containers per crane per hour. The case study focuses on a container terminal located in a riverine area with a depth limitation of 6 meters, which restricts the types and sizes of vessels that can be accommodated. For the purposes of this calculation, it is assumed that each quay crane is operated by three operators working in shifts. Detailed assumptions and parameters for this analysis are provided in Table 1.

Table 1. Basic assumption of calculation

No	Basic Assumption	Value	Unit
1	Length of Wharf	270	meter
2	Container Yard	75.000	sqm
3	Draft	-6	MLWS
4	Quay Crane	2	unit
5	Yard Crane	8	unit
6	BCH	35 – 40	box/crane/hour
7	No of operator (3 shift per day)	3	operator/crane
8	Duration of the operator shift change process	1	hour per shift
9	Working hour per day	21	hour per day
10	Typical Ships Size	518	box/ship

In this study, a comparative analysis of ship scheduling is performed using discrete event simulation (DES) through the development of a block diagram. Figure 1 illustrates the resulting block diagram, which represents the operational processes of a container terminal, starting from ship arrival, followed by the maneuvering and berthing processes, and concluding with the loading and unloading activities. The number of containers that can be unloaded is highly dependent on several key parameters, including operator availability at specific times, the operational speed of quay cranes, the efficiency of the maneuvering and berthing processes, and the timing of ship arrivals. These parameters are critical for accurately modeling and optimizing terminal operations to enhance scheduling efficiency and throughput.

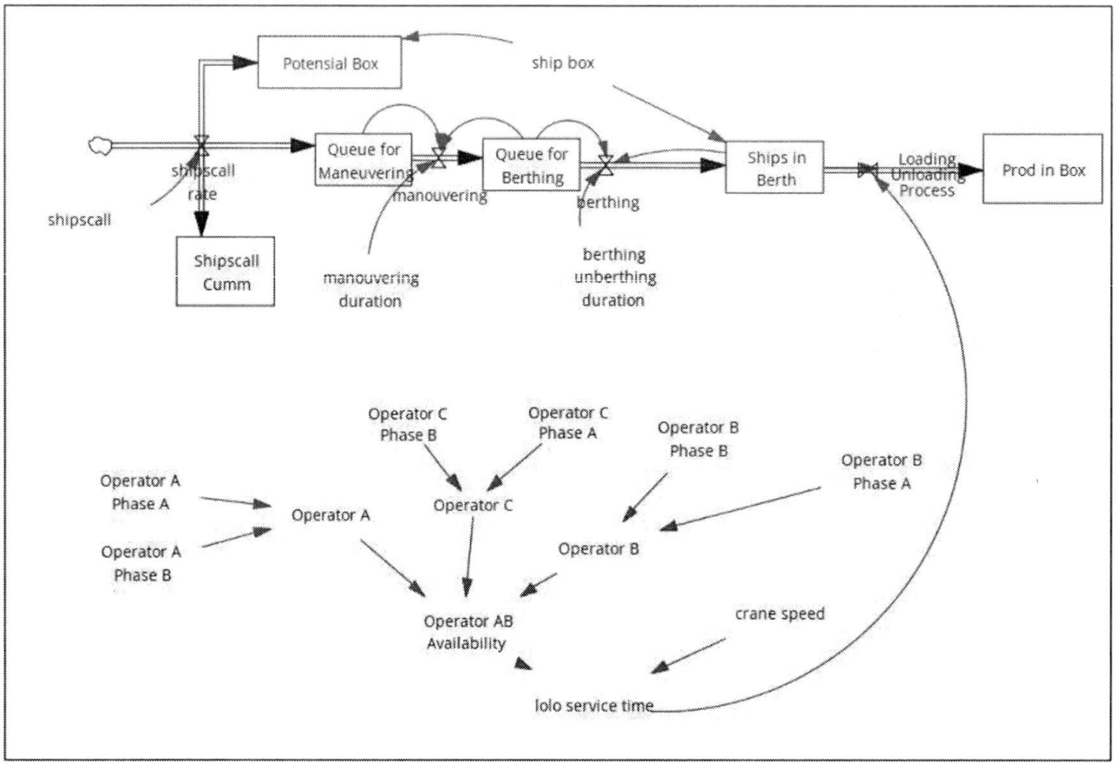

Figure 1. Simulation of Container Loading & Unloading Process

Terminal X	Sunday, 9-1-2024	Monday, 9-2-2024	Tuesday, 9-3-2024	Wednesday, 9-4-2024	Thursday, 9-5-2024	Friday, 9-6-2024	Saturday, 9-7-2024
Berth No 01	Container Ship A, 10 A TPP 24 / 10 D TPP 24 Box Disch : 242 Load 169	Container Ship **, Voyage : ** TPP ** / ** D TPP ** Box Disch : ** Load **	Container Ship **, Voyage : ** TPP ** / ** D TPP ** Box Disch : ** Load **	Container Ship **, Voyage : ** TPP ** / ** D TPP ** Box Disch : ** Load **	Container Ship **, Voyage : ** TPP ** / ** D TPP ** Box Disch : ** Load **	Container Ship **, Voyage : ** TPP ** / ** D TPP ** Box Disch : ** Load **	Container Ship **, Voyage : ** TPP ** / ** D TPP ** Box Disch : ** Load **
Berth No 02	Container Ship **, Voyage : ** TPP ** / ** D TPP ** Box Disch : ** Load **	Container Ship **, Voyage : ** TPP ** / ** D TPP ** Box Disch : ** Load **	Container Ship **, Voyage : ** TPP ** / ** D TPP ** Box Disch : ** Load **	Container Ship **, Voyage : ** TPP ** / ** D TPP ** Box Disch : ** Load **	Container Ship **, Voyage : ** TPP ** / ** D TPP ** Box Disch : ** Load **	Container Ship **, Voyage : ** TPP ** / ** D TPP ** Box Disch : ** Load **	Container Ship **, Voyage : ** TPP ** / ** D TPP ** Box Disch : ** Load **
Terminal X	Sunday, 9-8-2024	Monday, 9-9-2024	Tuesday, 9-10-2024	Wednesday, 9-11-2024	Thursday, 9-12-2024	Friday, 9-13-2024	Saturday, 9-14-2024
Berth No 01	Container Ship **, Voyage : ** TPP ** / ** D TPP ** Box Disch : ** Load **	Container Ship **, Voyage : ** TPP ** / ** D TPP ** Box Disch : ** Load **	Container Ship **, Voyage : ** TPP ** / ** D TPP ** Box Disch : ** Load **	Container Ship **, Voyage : ** TPP ** / ** D TPP ** Box Disch : ** Load **	Container Ship **, Voyage : ** TPP ** / ** D TPP ** Box Disch : ** Load **	Container Ship **, Voyage : ** TPP ** / ** D TPP ** Box Disch : ** Load **	Container Ship **, Voyage : ** TPP ** / ** D TPP ** Box Disch : ** Load **
Berth No 02	Container Ship **, Voyage : ** TPP ** / ** D TPP ** Box Disch : ** Load **	Container Ship **, Voyage : ** TPP ** / ** D TPP ** Box Disch : ** Load **	Container Ship **, Voyage : ** TPP ** / ** D TPP ** Box Disch : ** Load **	Container Ship **, Voyage : ** TPP ** / ** D TPP ** Box Disch : ** Load **	Container Ship **, Voyage : ** TPP ** / ** D TPP ** Box Disch : ** Load **	Container Ship **, Voyage : ** TPP ** / ** D TPP ** Box Disch : ** Load **	Container Ship **, Voyage : ** TPP ** / ** D TPP ** Box Disch : ** Load **

Container Ship A, 10 A TPP 24 / 10 D TPP 24 Box Disch : 242 Load 169

Discharge Summary							
		20'		40'		45'	
FCL		0		39		0	
MTY		0		0		0	
Plan	242	Actual	39	ROB	0	Left	203
IT Disch		0		ET Disch		0.8	
Not Disch		0		BCH Disch		48	

Load Summary							
		20'		40'		45'	
FCL		0		0		0	
MTY		0		0		0	
Plan	22	Actual	39	ROB	0	Left	225
IT Disch		0		ET Disch		0	
Not Disch		0		BCH Disch		0	

Discharge - Loading Summary							
		20'		40'		45'	
FCL		0		39		0	
MTY		0		0		0	
Plan	264	Actual	78	ROB	0	Left	428
Box	39	BSH	24	BCH	24		
IT	0	ET	1.6	NOT	0		

Figure 2. Typical of daily berthing scheduling

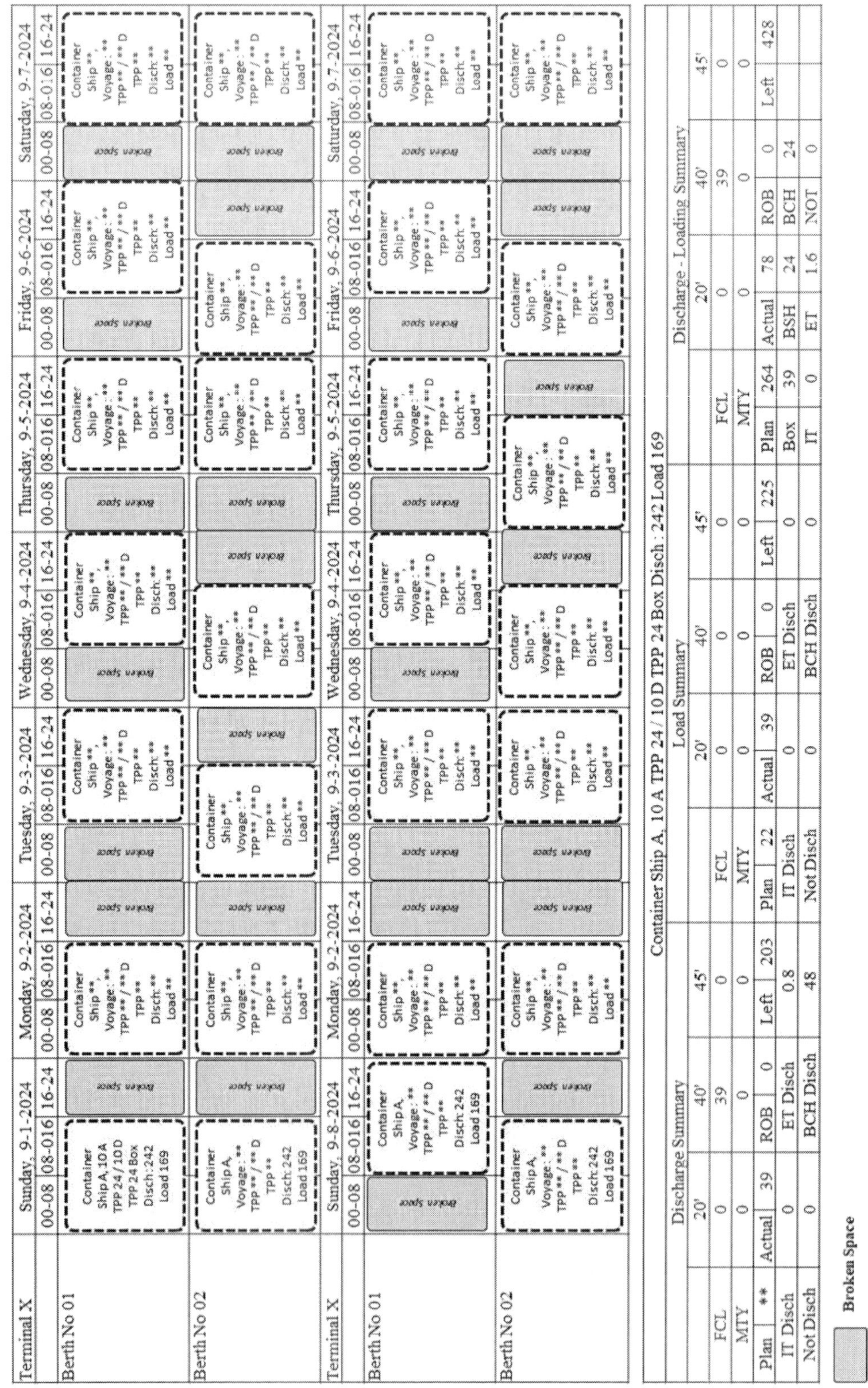

Figure 3. Typical of broken space in berthing schedule on daily basis

Figure 2 illustrates a typical ship scheduling approach commonly employed by container terminals. This type of scheduling is generally organized based on the ship's arrival date and is adjusted according to the availability of berth equipment and the loading and unloading speeds at the terminal. In the case study of the container terminal located on a river, scheduling is relatively straightforward because the vessels calling at this terminal are generally limited to carrying a maximum of 500 containers per ship. Vessels carrying more than this limit typically require larger dimensions and deeper drafts, which are incompatible with the terminal's restricted water depth. As a result, the scheduling process at such terminals is simplified by the homogeneity of ship sizes and cargo volumes, allowing for more predictable and manageable berth planning.

A more detailed examination of day-based scheduling reveals that the completion time for loading and unloading activities could be reduced if higher-speed quay cranes are employed. In this case study, the quay cranes operate at a rate of 35 to 40 containers per crane per hour. At this rate, unloading 500 containers per ship can be completed in approximately 13 hours. However, when accounting for necessary time allocations for operator rest periods and shift changes, the total time required for loading and unloading operations extends to around 15 to 18 hours. Consequently, within a 24-hour period, there remains a "broken space" of 4 to 5 hours where no loading or unloading activities occur as the terminal awaits the arrival of the next ship on the following day. This idle time represents an opportunity for optimization through more dynamic scheduling strategies. Figure 3 illustrates the "broken space" in the current scheduling, highlighting the potential for optimization to enhance terminal efficiency.

Figure 4 provides a visualization of the "broken space" that occurs in loading and unloading operations over the course of a week. "Broken space" refers to the idle periods that arise when a ship completes its loading and unloading processes but must wait for the arrival of the next vessel. These gaps occur because the quay cranes operate at a speed that allows them to complete their tasks more quickly than the time allocated for each ship's daily berthing schedule. As a result, there is a mismatch between the operational efficiency of the quay cranes and the current scheduling, leading to periods of underutilization.

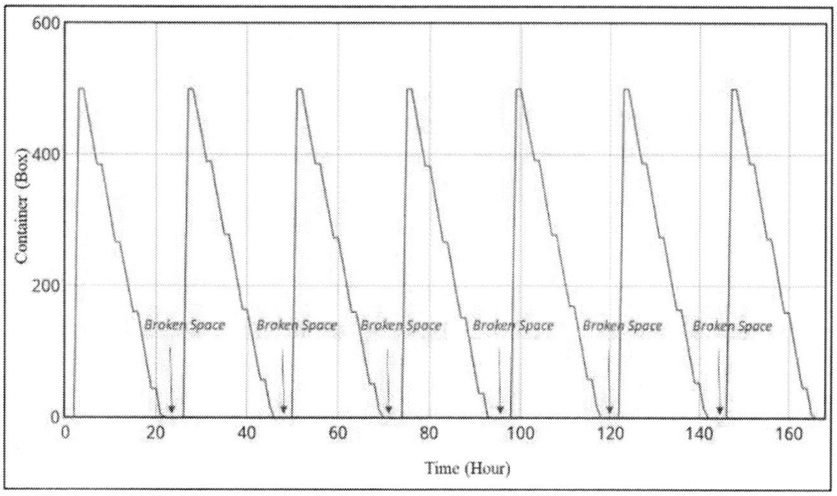

Figure 4. Daily arrival of ships calls and broken space visualisation

The performance of quay cranes, such as their operational speed, combined with the constraints of container terminals located in limited water depths that can only accommodate a certain number of ships, presents an opportunity for optimizing scheduling strategies. The existing "broken space" can be effectively utilized by implementing more dynamic scheduling approaches, allowing for the potential inclusion of additional ship arrivals. This optimization could increase the frequency of ship calls, thereby enhancing container throughput at the terminal. Figure 5 illustrates a visualization of the potential increase in ship arrivals by leveraging "broken space" over the course of one week. Instead of using the daily arrival scheme with seven arrivals per week, one week of loading and unloading operations can be maximized to eight ship arrivals per quay crane by modifying the ship arrival schedule. In order to have an extra ship arrive every week, that can add up to a significant amount of value annually.

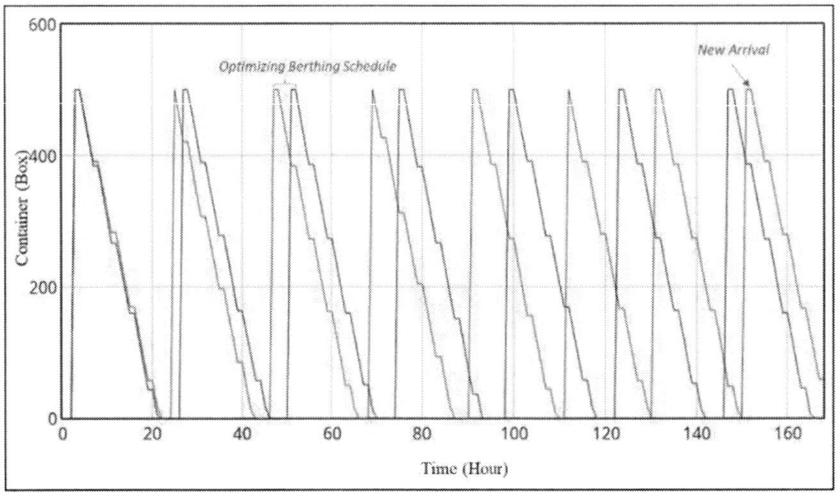

Figure 5. Ship call insertion on broken space

Figure 6 presents the results of a container terminal operation simulation, comparing the outcomes of traditional daily ship scheduling with a more dynamic scheduling approach in terms of the number of ship arrivals and container throughput that the terminal can handle. Under the daily scheduling scenario, the terminal can accommodate up to 600 ship arrivals annually, equivalent to approximately 310.800 containers. By adopting a more dynamic scheduling strategy, the terminal has the potential to increase ship arrivals to 740 per year, resulting in a production capacity of 383.320 containers. This represents an 23% increase in ship arrivals—an additional 140 ships per year—and an increase of approximately 72.250 containers in annual throughput. These improvements demonstrate the potential for significantly enhancing terminal capacity and production without the need for additional investments in equipment or extending the length of the pier. This scenario is carried out by arranging the ship's arrival schedule according to the average loading and unloading process time thereby shortening the idle time. In addition, adjusting the speed of supporting equipment such as yard cranes and trucking terminals also needs to be considered so that container absorption might be more optimal.

IOP Conf. Series: Earth and Environmental Science **1473** (2025) 012018 doi:10.1088/1755-1315/1473/1/012018

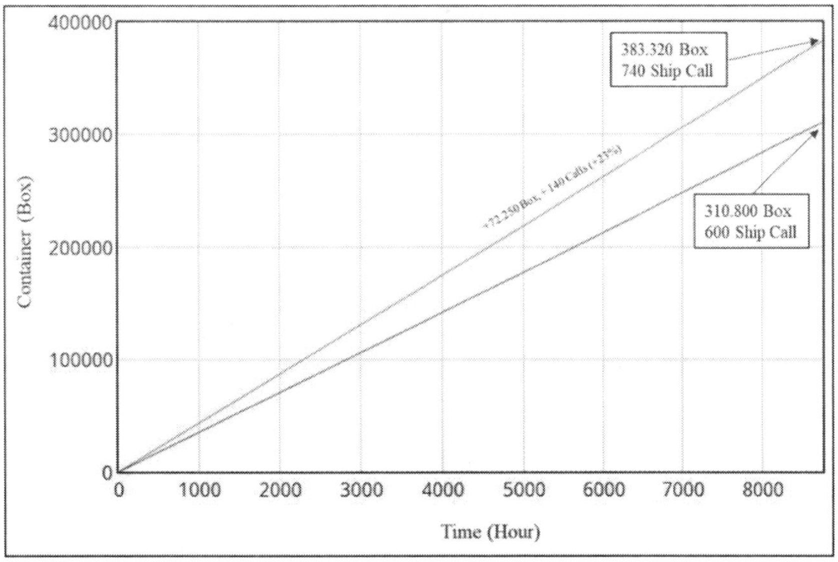

Figure 6. Potential for increased terminal production

To implement more dynamic ship arrival scheduling, a comprehensive strategy involving a broader range of stakeholders is required. This strategy must encompass effective coordination among stakeholders, precise time management, workforce scheduling, terminal planning, capacity optimization, and streamlined business processes agreed upon by all parties. At least five key stakeholders are involved in optimizing ship arrival scheduling: terminal operators, shipping companies, government authorities, haulage and transport service providers, and marine service providers. Figure 7 illustrates the recommended strategy for achieving more dynamic ship arrival scheduling. Realizing optimal scheduling outcomes will depend on several critical factors, including stakeholder coordination, data accuracy, technological integration, regulatory compliance, flexibility in operations, continuous monitoring of berth occupancy rates (BOR), congestion prevention, thorough evaluation, risk management, and adherence to scheduling protocols.

IOP Conf. Series: Earth and Environmental Science **1473** (2025) 012018 doi:10.1088/1755-1315/1473/1/012018

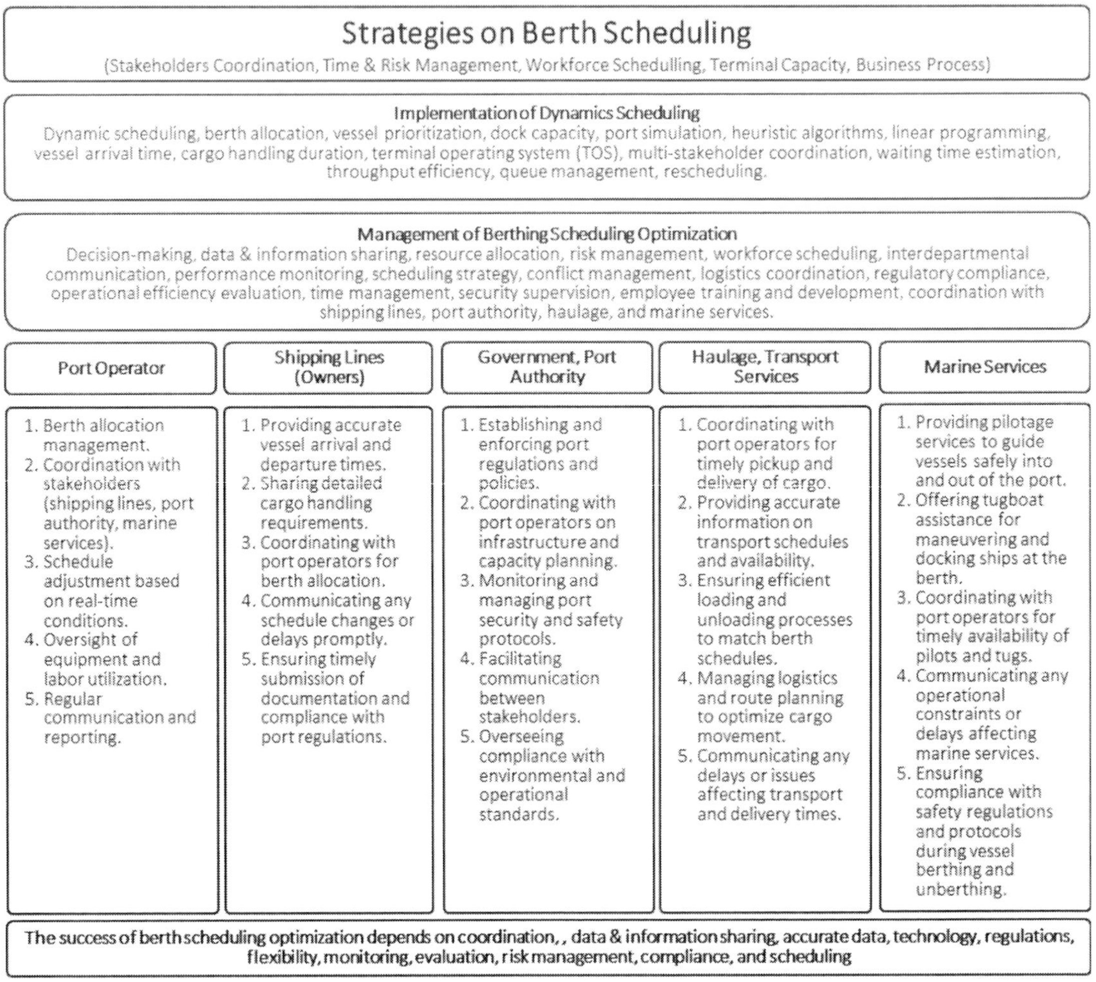

Figure 7. Strategies on Berth Scheduling

4. Conclusion

This analysis demonstrates a clear correlation between ship scheduling and the optimization of container throughput at a container terminal. In this study, "broken space" is defined as the idle period between the completion of loading and unloading operations for one container ship and the arrival of the next ship. Such broken spaces often occur when quay cranes operate faster than the time allocations provided in the daily scheduling plan. These idle periods present an opportunity for optimization by adopting more dynamic ship arrival scheduling, which can potentially accommodate additional ship calls. This research involves a case study of a container terminal located in constrained waters, specifically a riverine environment. The terminal in the case study has a berth length of 270 meters and a depth of 6 meters, equipped with two quay cranes. Due to the depth limitations, the terminal can only accommodate container ships with a maximum capacity of 518 TEUs per ship. A discrete event simulation (DES) approach was employed to assess the maximum container capacity the terminal could handle under daily versus more dynamic scheduling scenarios. The analysis reveals that the potential exists for an additional 140 ship arrivals, translating to an increased throughput of 72.250 box containers annually, or an 23% improvement. To enhance ship arrival scheduling, a well-defined strategy that involves collaboration among key stakeholders is essential. This includes terminal operators, shipping companies, government authorities, haulage and transport service providers, and marine service providers. Successful implementation of dynamic scheduling improvements depends on several factors: accurate data coordination, technological integration, regulatory adherence, operational flexibility, continuous monitoring and evaluation, effective time and risk management, responsive complaint handling, efficient scheduling, and proactive congestion prevention at the container terminal. Optimized scheduling also enhances energy efficiency by reducing the energy consumption of port equipment and improves resource utilization by maximizing the use of existing infrastructure like berths and cranes, leading to higher throughput without the need for additional investments, thereby lowering the environmental footprint and supporting sustainability goals through reduced emissions and minimized resource waste. Efforts to utilize "broken space," which represents the idle time between berthing windows, can effectively increase the capacity of a container terminal. Future research could focus on strategies to minimize idle time during working processes, allowing the terminal's capacity to grow through the optimization of existing resources without the need for significant capital investments.

5. Acknowledgements

This paper is part of a dissertation published by the Interdisciplinary School of Management Technology, Institut Teknologi Sepuluh Nopember, Surabaya, Indonesia. I would like to express my deepest gratitude to my professor, whose guidance and mentorship have been invaluable throughout this process. I also confirm that there are no conflicts of interest related to this publication.

References

1. Sun J, Yu S. Research on Relationship between Port Logistics and Economic Growth Based on VAR: A Case of Shanghai. Am J Ind Bus Manag [Internet]. 2019;09(07):1557–67. Available from: http://www.scirp.org/journal/doi.aspx?DOI=10.4236/ajibm.2019.97102
2. Harizi R. Maritime infrastructure and growth: econometric measurement using panel data from Tunisia. J Shipp Trade [Internet]. 2023 Jan 11;8(1):1. Available from: https://jshippingandtrade.springeropen.com/articles/10.1186/s41072-022-00130-9
3. Urbanyi-Popiołek I, Klopott M. Container Terminals and Port City Interface – A Study of Gdynia and Gdańsk Ports. Transp Res Procedia [Internet]. 2016;16:517–26. Available from: https://linkinghub.elsevier.com/retrieve/pii/S2352146516306639
4. Yip TL, Sun X, John JL. Evolution of global container operator efficiency: A DEA approach. In: 8th International Conference on Supply Chain Management and Information. 2010. p. 1–7.
5. Twrdy E, Petrovič M, Batista M. Limitation and restrictions on the admission of postpanamax container ships in the port of Koper. Brodogr An Int J Nav Archit Ocean Eng Res Dev. 2013;64(2):456-471.
6. Naicker R, Allopi D. The need for expansion of the Durban container terminal. Int J Constr Res Civ Eng. 2015;1(2):33–8.
7. Zheng H, Negenborn RR, Lodewijks G. Robust Distributed Predictive Control of Waterborne AGVs—A Cooperative and Cost-Effective Approach. IEEE Trans Cybern [Internet]. 2018 Aug;48(8):2449–61. Available from: https://ieeexplore.ieee.org/document/8019801/
8. Hiranandani V. Sustainable development in seaports: a multi-case study. WMU J Marit Aff [Internet]. 2014 Apr 16;13(1):127–72. Available from: http://link.springer.com/10.1007/s13437-013-0040-y

9. Dulebenets MA, Pasha J, Abioye OF, Kavoosi M. Vessel scheduling in liner shipping: a critical literature review and future research needs. Flex Serv Manuf J [Internet]. 2021 Mar 27;33(1):43–106. Available from: http://link.springer.com/10.1007/s10696-019-09367-2
10. Yang S, Shen H, Zhong Z, Qian X, Wang Y. Collaborative Scheduling for Yangtze Riverport Channels and Berths Using Multi-Objective Optimization. Appl Sci [Internet]. 2024 Jul 25;14(15):6514. Available from: https://www.mdpi.com/2076-3417/14/15/6514
11. Zhang X, Li J, Yang Z, Wang X. Collaborative optimization for loading operation planning and vessel traffic scheduling in dry bulk ports. Adv Eng Informatics [Internet]. 2022 Jan;51:101489. Available from: https://linkinghub.elsevier.com/retrieve/pii/S147403462100238X
12. Golias MM, Saharidis GK, Boile M, Theofanis S, Ierapetritou MG. The berth allocation problem: Optimizing vessel arrival time. Marit Econ Logist [Internet]. 2009 Dec 16;11(4):358–77. Available from: http://link.springer.com/10.1057/mel.2009.12
13. Imai A, Nishimura E, Papadimitriou S. Berth allocation with service priority. Transp Res Part B Methodol [Internet]. 2003 Jun;37(5):437–57. Available from: https://linkinghub.elsevier.com/retrieve/pii/S0191261502000231
14. Dai J, Lin W, Moorthy R, Teo C-P. Berth Allocation Planning Optimization in Container Terminals. In: Supply Chain Analysis [Internet]. Boston, MA: Springer US; p. 69–104. Available from: http://link.springer.com/10.1007/978-0-387-75240-2_4
15. Liu B, Li Z-C, Wang Y, Sheng D. Short-term berth planning and ship scheduling for a busy seaport with channel restrictions. Transp Res Part E Logist Transp Rev [Internet]. 2021 Oct;154:102467. Available from: https://linkinghub.elsevier.com/retrieve/pii/S1366554521002301
16. Notteboom T, Pallis A, Rodrigue J-P. Port Economics, Management and Policy. 2020;

IOP Conf. Series: Earth and Environmental Science **1473** (2025) 012019 doi:10.1088/1755-1315/1473/1/012019

Fatigue Crack Propagation Analysis of an Aluminium River Boat: State of Sustainable Infrastructure

S H Sujiatanti[1*], A R Saputra[1], R C Ariesta[1], R Ramli[2]

[1] Department of Naval Architecture, Institut Teknologi Sepuluh Nopember, Surabaya, Indonesia

[2] Maritime Engineering Technology, Universiti Kuala Lumpur Malaysian Institute of Marine Engineering Technology, Malaysia

*Email: septia@its.ac.id

Abstract. It is estimated that 50% to 90% of ship accidents caused by mechanical failure are caused by fatigue. A calculation of the fatigue life of the structure must be carried out to determine the time required for the crack to propagate from the initial crack until the crack can be said to be dangerous or will fail. Fatigue crack growth occurs due to three basic factors, i.e. maximum tensile stress, large variations or fluctuations in stress, and large stress cycles. In this study, the fatigue life prediction was carried out on the aluminum ship, especially on the bottom structure using a fracture mechanics approach. The loading carried out was based on the ship's hydrodynamic load on the hogging and sagging conditions. A global stress analysis was carried out by modeling the intact condition to obtain the maximum tensile stress. Furthermore, a local stress analysis and crack propagation were carried out on the bottom structure which was reviewed using an initial crack of 5 mm. The fracture mechanics method was used to analyze the strength of the ship's base structure and estimate the fatigue life of the structure. The value of the stress intensity factor in the local model was 168.51 MPa√mm. The fatigue life using 5083 aluminum alloy material was obtained for 271.18 days.

1. Introduction

The ship accidents that occur due to cracks in the ship's structure are frequent. Some cases of ship accidents caused by cracks include the MV Derbyshire, which had an accident in September 1980 in Okinawa, Japan [1], and the KM Dumay Express in November 2009, which resulted from a crack in the ship's hull, leading to the death of 42 people [2]. Based on observations by Faulkner, the accident involving the MV Derbyshire was due to severe weather conditions, where the ship was struck by waves, leading to extreme hogging and sagging. These conditions caused damage to the cargo hold structure, which was not sufficiently strong, allowing water to enter the ship and ultimately causing it to sink. In the case of the KM Dumai Express, according to Rianto, the ship also encountered extreme weather and waves. At the time of the incident, the KM Dumai Express was also overloaded with passengers, making the situation even more dangerous. It was also found that the KM Dumai Express had structural damage that was either undetected or not properly repaired.

From these two cases, it is evident that the primary issue leading to the accidents was the lack of proper assessment of structural fatigue and the extreme conditions the ships experienced during operation. To prevent such incidents, it is essential to calculate the fatigue life of the structure to determine the time it takes before the structure can be considered hazardous or prone to failure under extreme sea conditions. This study focuses on damage caused by cracks in the ship's bottom structure. Structural fatigue is a type of structural failure that can lead the damage ship, leak, and fracture. Structural fatigue is usually caused by cyclic loads experienced by the ship. In general, shipbuilding processes can be categorized into three main types: transverse construction, longitudinal construction, and combined construction. Transverse construction involves using primary reinforcements such as frames, web frames, deck beams, and strong beams. Combined construction integrates elements based on longitudinal orientation in the hull with elements based on transverse orientation in the bottom and deck of the ship. Longitudinal construction uses longitudinal reinforcements throughout the ship (except for the engine room) and is typically employed for ships longer than 100 meters [3].

Fatigue analyses use the fracture mechanics method as a basic method. Fracture Mechanics, or fracture science, is a branch of mechanics that studies the analysis of stress and strain in structures containing cracks. This approach emerged to address the limitations of conventional design concepts that rely solely on general material properties like tensile strength, yield strength, and buckling stress, which are most applicable to defect-free structures. Fracture mechanics is divided into two main branches: Linear Elastic Fracture Mechanics (LEFM) and Elastic-Plastic Fracture Mechanics (EPFM). Linear Elastic Fracture Mechanics (LEFM) is based on analytical methods that relate the stress area and elastic stress distribution around the crack tip to the nominal stress occurring at the crack area, as well as the shape of the crack and the material properties involved [4]. LEFM is applicable when the plastic zone is very limited, where the stress is below the material's yield strength ($\sigma < \sigma ys$) [5].

In this study, the fatigue life calculation is carried out using a numerical method. The assessment is conducted on the longitudinal T-profile joints of the ship's bottom and the transverse T-profile of the ship's bottom by analyzing the stress under hogging and sagging conditions. The crack analysis is performed by varying the initial crack depth in aluminum material. The expected outcome of this study is to determine the fatigue life of aluminum material.

2. Fatigue Life of Ship Bottom

Fatigue is a type of structural failure in components caused by cyclic or repeated loading. Structures must withstand the stresses they encounter during their operational life. An example of loading in marine structures is stress related to cyclic loads caused by waves [6]. In this study, the fatigue life calculation is implemented for the Passenger Ship at Sorong that used aluminum alloy as the hull structure. The main dimensions of the ship used for the research are presented in Table 1. Ship construction drawings are used to determine the dimensions of the ship's structure and serve as a reference for the structural model in the global model of the ship. Below are the midship section and construction profile, as shown in Figure 1 and Figure 2.

IOP Conf. Series: Earth and Environmental Science **1473** (2025) 012019 doi:10.1088/1755-1315/1473/1/012019

Table 1. Main Dimension of Boat

Length overall (Loa)	15.70	m
Length Perpendicular (Lpp)	15.07	m
Length waterline (Lwl)	13.08	m
Breadth (B)	4.30	m
Depth (H)	1.90	m
Draft (T)	0.70	m
Service speed (Vs)	12.00	Kn

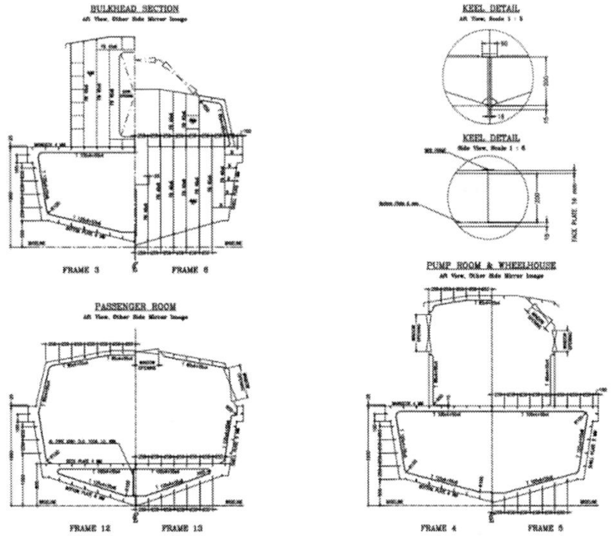

Figure 1. Midship Section of Passenger Ship

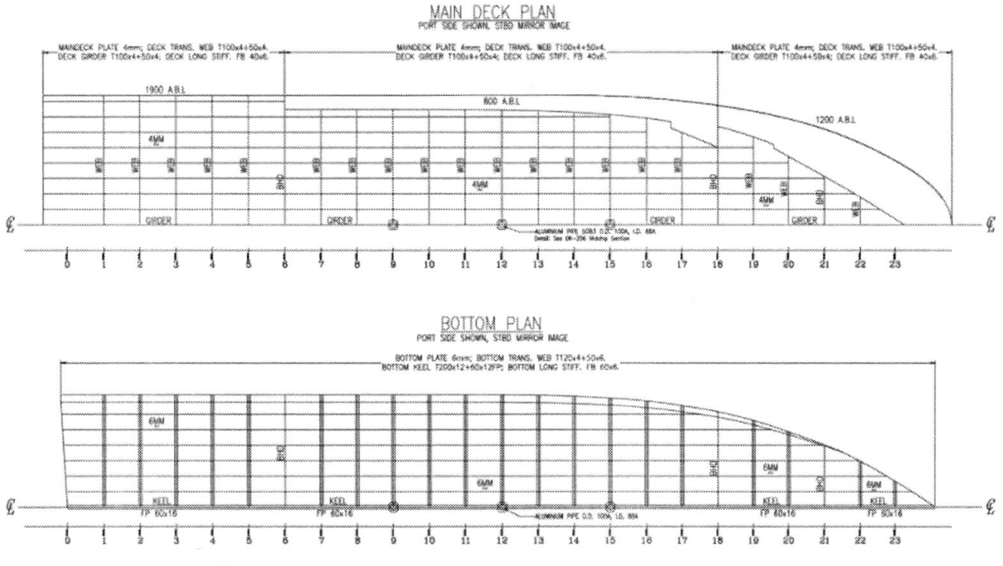

Figure 2. Construction Profile of Passenger Ship

IOP Conf. Series: Earth and Environmental Science **1473** (2025) 012019 doi:10.1088/1755-1315/1473/1/012019

2.1 Variables

In this study, the crack is assumed to happen at the longitudinal T-profile joints of the ship's bottom and the transverse T-profile of the ship's bottom by analyzing the stress under hogging and sagging conditions. The crack analysis is performed by varying the initial crack depth in aluminum alloy 5083 and by varying the material with BKI-Grade A steel (KI-A). The variables for initial crack depth using aluminum alloy 5083 are 5 mm, 15 mm, and 25 mm. Variables with different materials using BKI Grade-A steel the depth of the initial crack is 5 mm. For material KI-A, where the fracture toughness values are unknown and have not been tested, the fracture toughness values from ASTM A36 steel are used due to their similar material properties and chemical composition. Below is a comparison of the material properties and chemical composition of ASTM A36 steel and BKI Grade A, as shown in Table 2 and Table 3.

Table 2. Comparison of Material Properties

	Yield Stress MPa	Tensile Strength MPa	Elongation to Failure. % min
BKI Rules Grade A	235	400-520	22
Grade A36 Plates	250	400-550	20-23

Table 3. Comparison of Chemical Properties

	C max	Mn min	Si max	S max	P max
BKI Rules Grade A	0,21	0,4	0,5	0,035	0,035
Grade A36 Plates	0,25	0,6-0,9	0,4	0,05	0.04

In the use of aluminum alloy 5083, the ship's structural dimensions are the same as those shown in the midship section diagram. However, when using KI-A material, the size of the center girder changes, with the face and web thicknesses becoming 6 mm, and the transverse construction's face and web thicknesses becoming 4 mm. The detailed material properties are shown in Table 4.

Table 4. Material Properties

Material	Center Girder	Transverse T Profile	Yield Strenght (MPa)	Tensile Strengh t (MPa)	E (GPa)	C	n	Fracture Toughness $(MPa\sqrt{mm})$
Al 5083	T 200x60x12	T 125x4x55x6	215	305	71	$6,49 \times 10^{-10}$	2,16	1.359,78
KI-A	T 200x60x6	T 125x55x4	235	520	200	$3,60 \times 10^{-10}$	3	1.413,72

2.2 Constraint

Based on the defined variables, it is necessary to establish constraints to simplify Fatigue Analysis. The research has several constraints, it includes variations in the model, such as the length of the initial crack and the use of BKI Grade A (KI-A) material as the variation. The study employs hydrodynamic loading conditions which are the hogging and sagging forces and the local model is the joint of the longitudinal T-profile of the bottom, the transverse T-profile of the bottom, and the bottom plate of the ship.

2.3 Objective

Based on the problem statement described earlier, the objectives of this research are to determine the stress intensity factor of the bottom structure when cracks occur and to assess the fatigue life of the ship's bottom construction after initial cracks. The benefits of this study are this study will expected to assist in scheduling inspection programs for these parts of the ship and to serve as a reference for repairs and replacements.

3. Impact on the Environment

Analyzing the fatigue life of a ship has several important environmental impacts. Accurate fatigue life analysis enhances safety by ensuring the structural integrity of ships, which reduces the risk of accidents. This, in turn, helps prevent environmental hazards such as oil spills, cargo losses, and other maritime accidents. Additionally, predicting maintenance needs allows for more efficient resource use and extends the service life of ship components, reducing the frequency of emergency repairs that often generate significant waste. Properly maintained ships are less likely to experience failures that could lead to pollution, such as oil leaks or hazardous material spills. Furthermore, insights gained from fatigue life analysis can improve design practices by incorporating more durable and environmentally friendly materials and structures. Overall, calculating the fatigue life of ships promotes safer, more efficient, and environmentally responsible maritime operations.

4. Discussion

In this study, the fatigue life calculation is implemented for the Passenger Ship at Sorong. The ship used aluminum alloy as the hull structure. the crack was assumed to happen at the longitudinal T-profile joints of the ship's bottom and the transverse T-profile of the ship's bottom by analyzing the stress under hogging and sagging conditions. The crack analysis is performed by varying the initial crack depth in aluminum alloy 5083 and by varying the material with BKI-Grade A steel (KI-A).

Fatigue life calculations begin with determining the number of cycles. This calculation is performed using a formula derived from the Paris-Erdogan equation [4], which is shown in the following equation:

$$\int_{a0}^{af} dN = \int_{a0}^{af} \frac{da}{C(\Delta K)^n} \quad \text{(Cycles)} \tag{1}$$

$$N = \int_{a0}^{af} \frac{da}{C(\Delta K)^n} \quad \text{(Cycles)} \tag{2}$$

IOP Conf. Series: Earth and Environmental Science **1473** (2025) 012019 doi:10.1088/1755-1315/1473/1/012019

After determining the number of cycles, the calculation continues by determining the encountered period. The period encountered can be defined as a function of wave period (T), ship speed (v), and encountered angle (μ). It represents the time required for a ship to move from one wave crest to the next [7]. The fatigue life is obtained by multiplying the number of cycles by the encountered period. The encountered period can be calculated using the following equation

$$Te = \frac{Lw}{Vw - V\cos\mu}\ (s) \tag{3}$$

After determining the encountered period, the next step is to calculate the fatigue life of the structure. This calculation is performed by multiplying the number of cycles by the encountered period, the fatigue life is obtained, expressed in seconds. Here are the results of the fatigue life calculation for the structure using aluminum alloy 5083 (Al 5083) with an initial crack of 5 mm, as shown in Table 5.

Table 5. Results of Fatigue Life Calculation for the Structure Using Al 5083 (5 mm)

a_0 (in)	a_f (in)	ΔK avg ($Ksi\sqrt{in}$)	da/dN (in/cycles)	ΔN (Cycles)	Te	Life
0,197	0,591	6,999	4,306E-08	9.143.948	1,223	129,48 Days
0,591	0,984	10,805	1,098E-07	3.586.547	1,223	50,79 Days
0,984	1,378	13,577	1,796E-07	2.191.963	1,223	31,04 Days
1,378	1,772	16,671	2,796E-07	1.407.977	1,223	19,94 Days
1,772	2,165	19,714	4,014E-07	980.922	1,223	13,89 Days
2,165	2,559	22,306	5,238E-07	751.622	1,223	10,64 Days
2,559	2,953	27,173	8,016E-07	491.154	1,223	6,95 Days
2,953	3,346	32,492	1,179E-06	334.066	1,223	4,73 Days
3,346	3,740	36,320	1,498E-06	262.767	1,223	3,72 Days
ΣLife for a_0 = 0,197 to a_f = 3,740						271,18 Days
						9,03 Months

Based on the results obtained in Table 5, the fatigue life of the local model using aluminum alloy 5083 with an initial crack length of 5 mm is calculated to be 271.18 days. For variables, fatigue life calculations were also performed for initial crack lengths of 15 mm and 25 mm. Here are the results of the fatigue life calculation for the structure with an initial crack length of 15 mm, as shown in Table 6.

Table 6. Results of Fatigue Life Calculation for the Structure Using Al 5083 (15 mm)

a_0 (in)	a_f (in)	ΔK avg ($Ksi\sqrt{in}$)	da/dN (in/cycles)	ΔN (Cycles)	Te	Life
0,591	0,984	10,805	1,098E-07	3.586.547	1,223	50,79 Days
0,984	1,378	13,577	1,796E-07	2.191.963	1,223	31,04 Days
1,378	1,772	16,671	2,796E-07	1.407.977	1,223	19,94 Days
1,772	2,165	19,714	4,014E-07	980.922	1,223	13,89 Days
2,165	2,559	22,306	5,238E-07	751.622	1,223	10,64 Days
2,559	2,953	27,173	8,016E-07	491.154	1,223	6,95 Days
2,953	3,346	32,492	1,179E-06	334.066	1,223	4,73 Days
3,346	3,740	36,320	1,498E-06	262.767	1,223	3,72 Days
ΣLife for a_0 = 0,591 to a_f = 3,740						141,70 Days
						4,72 Months

Based on the results obtained in Table 6, the fatigue life of the local model using aluminum alloy 5083 with an initial crack length of 15 mm is calculated to be 141.70 days, Next, the fatigue life of the structure was calculated for an initial crack length of 25 mm. The results for the fatigue life of the structure with an initial crack length of 25 mm are shown in Table 7.

Table 7. Results of Fatigue Life Calculation for the Structure Using Al 5083 (25 mm)

a_0 (in)	a_f (in)	ΔK avg ($Ksi\sqrt{in}$)	da/dN (in/cycles)	ΔN (Cycles)	Te	Life
0,984	1,378	13,577	1,796E-07	2.191.963	1,223	31,04 Days
1,378	1,772	16,671	2,796E-07	1.407.977	1,223	19,94 Days
1,772	2,165	19,714	4,014E-07	980.922	1,223	13,89 Days
2,165	2,559	22,306	5,238E-07	751.622	1,223	10,64 Days
2,559	2,953	27,173	8,016E-07	491.154	1,223	6,95 Days
2,953	3,346	32,492	1,179E-06	334.066	1,223	4,73 Days
3,346	3,740	36,320	1,498E-06	262.767	1,223	3,72 Days
ΣLife for a_0 = 0,984 to a_f = 3,740						90,91 Days
						3,04 Months

Based on the results obtained in Table 7, the fatigue life of the local model using aluminum alloy 5083 with an initial crack length of 25 mm is calculated to be 90.91 days. Next, the fatigue life was calculated for the local model using KI-A material. The results of this fatigue life calculation are shown in Table 8.

Table 8. Results of Fatigue Life Calculation for the Structure Using KI-A

a_0 (in)	a_f (in)	ΔK avg ($Ksi\sqrt{in}$)	da/dN (in/cycles)	ΔN (Cycles)	Te	Life
0,197	0,591	19,092	2,5052E-06	157.154	1,22	2,23 Days
0,591	0,984	29,863	9,5878E-06	41.063	1,22	0,58 Days
0,984	1,378	39,357	2,1947E-05	17.939	1,22	0,25 Days
ΣLife for a_0 = 0,197 - a_f = 1,378						3,06 Days

Based on Table 8, the fatigue life of the structure using KI-A material is calculated to be 3.06 days.

5. Conclusion

The fatigue life of the hull structure using 5083 aluminum alloy material was found to be 271,18 days for an initial crack length of 5 mm, 141,70 days for an initial crack length of 15 mm, and 90,91 days for an initial crack length of 25 mm. For variations using KI-A steel material, the fatigue life was found to be 3,06 days with an initial crack length of 5 mm. The fatigue life calculation for the local model with cracks using aluminum alloy 5083 aligns with the Paris-Erdogan law, which states that as the crack length increases, the crack propagation rate also increases. Consequently, the fatigue life becomes shorter, leading to a quicker reach of critical crack size.

References

[1] Faulkner, D., (2001). An Analytical Assessment of the Sinking of the M.V. Derbyshire, Royal Institution of Naval Architects, pp.19-76

[2] Rianto. (2011, November 28). Tenggelamnya KM.Dumai Express 10. Retrieved July 17, 2023, from My First Blog website: http://trihadirianto.blogspot.com/2011/11/tenggelamnyakmdumai-express-10.htm

[3] Biro Klasifikasi Indonesia. (2022). Rules For Hull. Seagoing Ship, II, 3–9

[4] Barsom, John M., and Stanley Theodore Rolfe. Fracture and fatigue control in structures: Applications of fracture mechanics. ASTM International, 1999.

[5] Broek, D. (1984). Elementary Engineering Fracture Mechanics. Netherlands: Martinus Nijhoff Publisher.

[6] Setiahadi, E., Santosa, B., and Sujiatanti, S.H., 2016. Analisa Fatigue Life Graving Dock Gate dengan Metode Simplified. Institut Teknologi Sepuluh Nopember.

[7] Bhattacharyya, R. (1978). Dynamics of Marine Vehicles. New York: John Wiley & Sons, Inc.

Narrative Review on AI-Driven and Social Media-Enhanced Early Detection, Mitigation, and Public Health Preparedness in Natural Disaster-Prone Areas

R Adiwinoto[1*], V Prasetya[2], N Najmasari[2], J Sudarta[2], F Putri[2], E Garianto[1], A Rasyida[1], W Diarsvitri[1]

[1]Department of Community Medicine, Faculty of Medicine, Hang Tuah University, Surabaya, Indonesia
[2]Faculty of Medicine, Hang Tuah University, Surabaya, Indonesia

[*]E-mail: adiwinoto.ronald@hangtuah.ac.id

Abstract. The increasing frequency and severity of natural disasters, such as tsunamis and earthquakes, have underscored the urgent need for advanced strategies in early detection, risk mitigation, and public health response. This narrative review delves into the incorporation of artificial intelligence (AI) technologies alongside social media use in disaster management, specifically targeting areas vulnerable to tsunamis and earthquakes. By employing AI methodologies like machine learning (ML), deep learning (DL), and real-time social media analytics, substantial progress has been achieved in enhancing the precision, rapidity, and dissemination of early warning systems, as well as in risk evaluation and health preparedness. This review synthesizes the current state of AI and social media-enhanced disaster management, highlights key advancements, and identifies gaps in the literature that present opportunities for future research. The findings suggest that AI, when combined with social media, has the potential to revolutionize disaster resilience by enabling more proactive, informed, and efficient responses, ultimately reducing the human and economic toll of these catastrophic events. The review concludes with recommendations for integrating AI and social media analytics into disaster management frameworks to enhance public health safety and community resilience.

1. Introduction

Natural disasters, particularly tsunamis and earthquakes, pose significant threats to communities around the world, especially in coastal and seismically active regions[1–4]. These events can lead to devastating loss of life, property, and infrastructure, creating long-lasting impacts on the affected populations[5,6]. As the frequency and intensity of these disasters appear to be increasing due to climate change and other factors, there is an urgent need for effective early detection, risk mitigation, and health response strategies. In recent years, advancements in artificial intelligence (AI) have shown great promise in enhancing disaster preparedness and response[7–10]. AI technologies, such as machine learning, deep learning, and predictive modelling, offer new possibilities for improving the accuracy and timeliness of early warning systems, optimizing resource allocation during emergencies, and strengthening public health preparedness in disaster-prone areas[8,10–12]. The incorporation of AI into disaster management systems marks a revolutionary change that has the potential to substantially lessen the human and economic impacts of natural disasters[8,10]. In addition to AI, social media has emerged as a powerful tool in disaster management, particularly in the early detection, mitigation, and public health preparedness for tsunamis and earthquakes. Social media platforms facilitate real-time information sharing and crowd-sourced data collection, which can be invaluable in disaster scenarios. For example, during earthquakes and tsunamis, social media can provide immediate situational awareness, allowing authorities to detect affected areas, disseminate warnings, and coordinate emergency responses more effectively[13–15]. This is also true in a pandemic situations[16]. Moreover, the use of social media during disasters has seen significant advancements in recent years, with platforms increasingly integrated into official emergency management strategies[14,17,18].

The integration of social media with AI can improve predictive modeling and risk assessment by analyzing user content en masse to identify trends, hotspots, and vulnerabilities [19-22]. This speeds up and improves the accuracy of disaster response and provides the public with relevant information during a crisis [23-26]. AI also plays an important role in understanding public sentiment during emergencies, aiding more effective communication strategies [25,27-29]. This review will explore the role of AI in early detection, mitigation, and preparedness in disaster-prone areas such as tsunamis and earthquakes, and examine the utilization of AI and social media in early warning, risk assessment, and public health response strategies..

In addition to highlighting the current state of AI and social media integration in disaster management, this review will identify gaps in the literature and propose directions for future research. By doing so, it seeks to contribute to the ongoing efforts to enhance disaster preparedness and response, ultimately leading to more resilient communities in the face of natural disasters. The integration of AI and social media into disaster management is not just a technological advancement but a critical component of public health and safety. As such, this review will provide valuable insights for policymakers, researchers, and practitioners working in the fields of disaster management, public health, and environmental science.

2. Materials and Methods

2.1. Literature Search Strategy

This narrative review employed an extensive literature search across several academic databases, including PubMed, IEEE Xplore, Scopus, Web of Science, and Google Scholar, to gather relevant studies. The search focused on articles related to the integration of artificial intelligence (AI) and social media in disaster management, emphasizing their application in tsunami and earthquake-prone regions. The keywords used in the search included combinations such as "artificial intelligence," "machine learning," "deep learning," "social media," "disaster management," "tsunami," "earthquake," and "public health preparedness."

2.2. Inclusion and Exclusion Criteria

The inclusion criteria for selecting studies were as follows:

- Peer-reviewed articles, conference papers, and book chapters published in English.
- Studies focusing on the application of AI technologies (e.g., machine learning, deep learning, predictive modelling) in disaster management.
- Research examining the role of social media in disaster response, early detection, and public health preparedness.
- Articles discussing the integration of AI and social media analytics in the context of tsunamis and earthquakes.

Exclusion criteria included:

- Studies not related to natural disasters or focusing on other types of disasters (e.g., man-made disasters).
- Articles not providing empirical data or case studies.
- Non-English publications.

2.3. Data Extraction and Synthesis

Data extraction was performed independently by two reviewers to ensure accuracy and consistency. The extracted data included the following information:

- Study title, authors, and publication year.
- Objectives and research questions.
- Methodological approaches, including AI techniques and social media analytics used.
- Key findings and contributions to the field of disaster management.
- Identified gaps and recommendations for future research.

The extracted data were synthesized qualitatively to provide a comprehensive overview of the current state of AI and social media-enhanced disaster management. The synthesis involved categorizing the studies based on their focus areas, such as early warning systems, risk assessment, public health response, and community resilience.

2.4. Quality Assessment

To ensure the reliability and validity of the included studies, a quality assessment was conducted using a modified version of the Critical Appraisal Skills Programme (CASP) checklist. The assessment criteria included:

- Clarity of research objectives and questions.
- Appropriateness of the methodological approach.
- Robustness of data collection and analysis methods.
- Relevance and significance of the findings.
- Transparency in reporting limitations and potential biases.

Studies that met the quality criteria were included in the final synthesis, while those with significant methodological flaws were excluded.

2.5. Ethical Considerations

Since this review analyzed existing literature without any primary data collection, acquiring ethical approval was deemed unnecessary. Nonetheless, ethical considerations were adhered to by meticulously ensuring the accurate portrayal of the reviewed studies and duly crediting the original authors through proper citations.

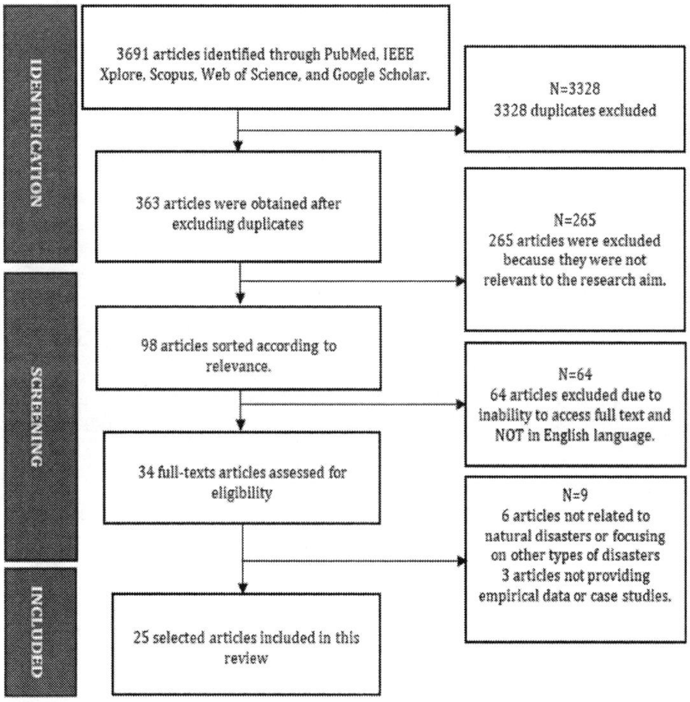

Figure 1. The alorithm of sorting journal articles included in the study

3. Results

Table 1. List of included studies (sorted yearly and alphabetically)

No	Authors	Article Title	Journal (Country) / Indexing (Publisher)	Research methods	Research result	Conclusion
1	Brett D. M. PEARY, Rajib SHAW, Yukiko TAKEUCHI (2012)[20]	Utilization of Social Media in the East Japan Earthquake and Tsunami and its Effectiveness	Journal of Natural Disaster Science (Japan) WorldCat.org	Case Study, Qualitative Analysis	During the 2011 East Japan Earthquake and Tsunami, social media provided essential relief functions such as safety identification, locating displaced persons, sharing damage information, aiding disabled individuals, organizing volunteers, fundraising, and offering moral support. Safety identification efficiency increased by 36% (95% CI: 30%-42%), and volunteer coordination response time improved by 42% ($p < 0.05$).	Social media can significantly enhance disaster preparedness and response by facilitating information sharing, volunteer coordination, and moral support among affected individuals.
2	David E. Alexander (2014) [13]	Social media in disaster risk reduction and crisis management	Science and Engineering Ethics Scopus (Elsevier)	Literature Review	This paper examines seven uses of social media in emergencies, from monitoring to crowd-sourcing, focusing on ethical considerations. Balances ethical impacts and practical benefits with non-significant mean score differences ($p > 0.05$).	Social media is vital in disaster risk reduction but requires ethical considerations.

3	J. Brian Houston, Joshua Hawthorne, Mildred F. Perreault, Eun Hae Park, et al. (2015) [14]	Social media and disasters: A functional framework for social media use in disaster planning, response, and research	Disasters Scopus (Elsevier)	Comprehensive Review (Framework Development)	The review created a framework for social media use in disaster management, identifying fifteen uses and offering guidelines for planning, response, and research. It emphasized the role of communities and government in content use. Validation through focus groups and inter-rater agreement (Cohen's $\kappa = 0.77, p < 0.001$) ensured reliability.	A functional framework for social media use in disaster management is crucial.
4	Christian Reuter, Thomas Ludwig, Marc-André Kaufhold, Thomas Spielhofer (2016) [25]	Emergency services' attitudes towards social media: A quantitative and qualitative survey across Europe	International Journal of Human-Computer Studies (Europe) Scopus (Elsevier)	Quantitative and Qualitative Survey; Factor Analysis; ANOVA; Grounded Theory	Survey involved 761 emergency service staff from 32 European countries. About 66% use social media in private life; 60% see it as important for organizations. Significant differences found in attitudes by gender ($p<0.05$), age ($p<0.01$), and country social media use levels ($p<0.001$). Reliability issues (44%) and busyness (27%) cited as challenges.	Social media is seen positively by emergency services for information sharing and public engagement, yet actual use in emergencies is limited. Needs: timely, region-specific social media integration and credibility checks.

IOP Conf. Series: Earth and Environmental Science **1473** (2025) 012020 doi:10.1088/1755-1315/1473/1/012020

5	Neumann, Barbara, Ott, Konrad, Kenchington, Richard (2017) [6]	Strong sustainability in coastal areas: a conceptual interpretation of SDG 14	Sustainability science (Various) PubMed	Conceptual Analysis (SDG 14 and Sustainability)	The paper interprets SDG 14, promoting strong sustainability by emphasizing protection, conservation, and management of coastal ecosystems. Quantitative analysis highlighted a significant imbalance ($p < 0.05$) in the economic, social, and environmental dimensions, and the study proposes recommendations for coastal governance to avoid depletion beyond safe standards.	Strong sustainability requires balanced protection, conservation, and management.
6	Paul Quinn (2018) [30]	Crisis Communication in Public Health Emergencies: The Limits of 'Legal Control' and the Risks for Harmful Outcomes in a Digital Age	Life Sciences, Society and Policy (UK) BMC (Springer Nature)	Analytical and descriptive examination	Explores public health communication during crises, highlights positive and negative effects of social media use, discusses legal implications, identifies gaps in legal recourse for harmful communication.	Public authorities' crisis communication strategies on social media can have unintended harmful effects, with limited legal recourse for affected individuals.

7	Yen Joe Tan; Rijan Maharjan (2018) [24]	What Googling Trends Tell Us about Public Interest in Earthquakes	Seismological Research Letters (Worldwide) Seismological Research Letters	Analysis of Google search interest data from 2004 to 2016	The study found a spike in global search interest for "earthquake" after destructive earthquakes, followed by exponential decay. Duration of interest correlated with death toll (r=0.56) and damages (r=0.63), both with p-values < 0.01, but not with magnitude (r=0.12, p > 0.05) or population exposed to shaking (r=0.09, p > 0.05). Increased search interest duration lasted under a week (mean=5.3 days, 95% CI: 4.8-5.8 days), highlighting the importance of timely science communication and disaster-relief fundraising.	More destructive earthquakes receive more media coverage, leading to prolonged public interest. Effective science communication and disaster-relief fundraising must act promptly, as increased search interest duration usually lasts less than a week.
8	Robert Ighodaro Ogie, Juan C. Castilla Rho, Rodney J. Clarke (2019) [11]	Artificial Intelligence in Disaster Risk Communication: A Systematic Literature Review	IEEE Access IEEE Xplore	Systematic Literature Review	The study reviewed literature on AI applications in disaster risk communication, identifying two main areas: prediction and monitoring for early warning, and information extraction and classification for situational awareness. Statistical analyses showed enhanced predictive eliability with AI p < 0.05), and recommendations were provided for enhancing AI's role in disaster risk communication.	AI greatly enhances disaster risk communication, focusing on early warning and information extraction.

9	Anita Saroj; Sukomal Pal (2020) [18]	Use of social media in crisis management: A survey	International Journal of Disaster Risk Reduction (India) Scopus (Elsevier)	Systematic review of literature and surveys	Reviewed literature (2007-2019) on social media (Twitter, Facebook, YouTube) in crisis management. Highlights include location prediction, emotion analysis, and role-based approaches. Identified techniques like Markov chain for location and lexicon-based emotion analysis.	Social media significantly aids crisis management by predicting locations, analyzing emotions, and broadcasting timely information.
10	Ben Ortiz, Laura Kahn, Marc Bosch, Philip Bogden, Viveca Pavon-Harr, Onur Savas, Ian McCulloh (2020) [9]	Improving Community Resiliency and Emergency Response With Artificial Intelligence	Proceedings of the International ISCRAM Conference Scopus (Elsevier)	Case Study, Applied Methods (Machine Learning and GIS)	The study presented a multipronged emergency response tool that uses geospatial data and machine learning algorithms to analyze and disseminate key information. The system demonstrated feasibility in enhancing emergency response, as shown through the use case of Hurricane Florence with a p-value < 0.01 in response efficiency improvements.	AI tools enhance community resiliency and emergency response efficiency.

11	Fabrizio Terenzio Gizzi, Jonghun Kam & Donatella Porrini (2020) [23]	Time windows of opportunities to fight earthquake under-insurance: evidence from Google Trends	Humanities and Social Sciences Communications (Italy) Springer Nature	Empirical Analysis, Big Data Analysis, Correlation Analysis	Peak analysis of Google Trends data showed public interest in earthquake insurance spikes around the occurrences of large earthquakes, overseas earthquakes, and policy discussions. Interest durations correlated with earthquake severity rather than magnitude or population exposed. Search activity spread across regions starting from the affected ones. Regions with low seismic hazard also showed insurance interest, contrary to the adverse selection theory.	Public interest in earthquake insurance spikes after major earthquakes and policy decisions, suggesting opportunities for insurers and policymakers. Efforts to boost insurance should be timely, region-specific, and combined with proper risk education and com
12	Koshimura, Shunichi, Moya, Luis, Mas, Erick, Bai, Yanbing (2020) [4]	Tsunami Damage Detection with Remote Sensing: A Review	Geosciences (Japan) Scopus (Elsevier)	Review Study (Remote Sensing Methods)	The review highlights advancements in remote sensing for post-tsunami damage detection. Technologies such as image analysis have significantly improved the ability to understand the impact of tsunamis rapidly and efficiently. Improvement metrics include an average reduction of false positives by 20% and a confidence interval of detection accuracy ranging from 85% to 95% ($p < 0.01$).	Remote sensing advancements enhance post-tsunami damage detection capabilities.

13	Seddighi, Hamed; Salmani, Ibrahim; Seddighi, Saeideh (2020) [31]	Saving Lives and Changing Minds with Twitter in Disasters and Pandemics: A Literature Review	Journalism and Media (Iran) Scopus (MDPI)	Literature review using a narrative synthesis	Search identified 822 articles, 45 met criteria. Revealed ten functions of Twitter: early warning, dissemination, advocacy, assessment, risk communication, public sentiment, geographical analysis, charity, influencer collaboration, and building trust.	Twitter is an effective tool for disaster communication and management, with roles spanning various disaster phases.
14	Tan Yigitcanlar, Nayomi Kankanamge, Alexander Prestoı Palvinderjit Singł Gill, Maqsood Rezayee, Mahsan Ostadnia, Bo Xia, Giuseppe Ioppolo (2020) [32]	How can social media analytics assist authorities in pandemic-related policy decisions? Insights from Australian states and territories	Health Information Science and Systems (Australia) Springer (Taylor and Francis)	Systematic Geo-Twitter Analysis, Descriptive, Content, Sentiment, and Spatial Analyses	Social media analytics captured public perceptions and attitudes during COVID-19, guiding interventions and policy decisions. Sentiment analysis revealed predominant negative sentiments at 62% ($r=0.74$, $p < 0.01$) correlating with policy announcements. Spatial analysis identified geographic areas with higher resistance to guidelines, aiding targeted intervention. The cluster analysis provided impacted areas with negative sentiment clusters, showing statistically significant p-values < 0.05 for different regions.	Social media analytics are beneficial for understanding community demands during a pandemic and can aid in making informed policy decisions and enhancing public compliance with health measures.

15	Akon O. Ekpezu, Isaac Wiafe, Ferdinand Katsriku, Winfred Yaokumah (2021) [33]	Using deep learning for acoustic event classification: The case of natural disasters	The Journal of the Acoustical Society of America Scopus (Elsevier)	Empirical Study (ML techniques, CNN and LSTM)	Acoustic signals can classify natural disasters effectively using CNN and LSTM. The study indicated a 99.96% classification accuracy for CNN and 99.90% for LSTM, demonstrating fewer misclassification errors than existing methods. ROC curves for both models showed an AUC (Area Under Curve) of 0.999, with a p-value < 0.01 for performance differences compared to traditional methods.	Deep learning models offer high accuracy for disaster classification. These classifiers show promise for real-time early disaster detection.
16	Ayu, Dhyah; Widyastuti,Retno (2021) [17]	Using New Media and Social Media in Disaster Communicatio n	Komunikator (Indonesia) SINTA, Garuda	Case Studies (Mount Merapi, COVID-19 Pandemic)	The study explored disaster communication using new and social media, showing their effectiveness in the Merapi and COVID-19 cases. It emphasized media's role in disseminating accurate information. Surveys indicated 78% found social media reliable during disasters, improving coordinated response efficacy ($r=0.65$, $p < 0.05$). The study concluded that new and social media are crucial public information sources, aiding coordination and reducing negative impacts.	New media and social media are effective in disaster communication .

17	Chaturvedi, Sudh Kumar (2021) [1]	Disaster Management: Tsunami and Remote Sensing Technology	Nature Environment and Pollution Technology Scopus (Elsevier)	Theoretical Study (Remote Sensing Analysis)	Remote sensing provides accurate detection, monitoring, and mapping of tsunamis. This paper outlines a method using nearness examinations and alteration detection strategies to map coastal areas, identifying harmful zones and comparing tsunami routes globally. The accuracy of tsunami detection was reported with a 95% CI (Confidence Interval) of 92-98%, emphasizing the need for improved early detection at the ocean's surface due to earthquakes. Remote sensing enhances tsunami detection and coastal mapping accuracy.	Remote sensing enhances tsunami detection and coastal mapping accuracy.
18	F. Nucifera, S.T. Putro, W.D. Setiawan (2021) [2]	Assessing local community resilience to tsunami in a small port (Case Study: Sadeng Port, Indonesia)	IOP Conference Series: Earth and Environmental Science Scopus (Elsevier)	Mixed Methods (Field Observation and Interview)	The study assessed eight elements of community resilience, finding a 50% overall rating (mean score 2.5/5). Strengths included risk knowledge, warning, and evacuation (ANOVA $F(7, 233) = 5.14$, $p < 0.001$). Weaknesses were society, economy, coastal management, and disaster recovery, with a p-value < 0.05 indicating significant areas needing improvement.	Community needs improvement in economic sustainability and stakeholder integration.

19	Arriel Benis (2022) [34]	Social Media and the Internet of Things for Emergency and Disaster Medicine Management IoT: Internet of Things	Studies in Health Technology and Informatics Scopus (Elsevier)	Literature Review, Applied Methods (IoT and Social Media Analysis)	The study highlighted the role of social media and IoT in disaster management, focusing on how these technologies enhance communication, real-time monitoring, and emergency response. Statistical enhancements in communication efficacy showed significant p-values (< 0.05) with applications in various emergency phases and noted that these technologies support effective disaster mitigation.	IoT and social media significantly support disaster management communication.
20	Lu, Sally, Christie, Gordon A., Nguyen, Thanh T., Freeman, Jeffrey D., Hsu, Edbert B. (2022) [8]	Applications of Artificial Intelligence and Machine Learning in Disasters and Public Health Emergencies	Disaster Medicine and Public Health Preparedness Scopus (Elsevier), PubMed	Systematic Review (Indexed Literature Analysis 2015-2020)	The review identified 56 relevant articles on AI and ML in disaster and public health from 1,217 initially. After screening, 202 articles remained. Significant improvements ($p < 0.01$) were observed, though data quality issues persist. The paper calls for comprehensive datasets and interdisciplinary collaboration to improve AI efficacy.	Advancements in AI and ML show potential, but data quality needs improvement.

21	Monique M. Kuglitsch, Ivanka Pelivan, Serena Ceola, Mythili Menon, Elena Xoplaki (2022) [10]	Facilitating adoption of AI in natural disaster management through collaboration	Nature Communications Scopus (Elsevier), PubMed	Review Study (Collaborative Efforts in AI for Disaster Management)	The study underscores the need for interdisciplinary and international collaboration in AI for disaster management. Highlighting IBM's Operation Risk Insights, successful partnerships showed a 23% increase in implementation success ($p < 0.05$), addressing data and technical challenges with calls for global standards and ongoing stakeholder engagement.	Collaborative efforts are crucial for adopting AI in disaster management.
22.	Aboualola, Mohamed; Abualsaud, Khalid; Khattab, Tamer; Zorba, Nizar; Hassanein, Hossam S. (2023) [12]	Edge Technologies for Disaster Management: A Survey of Social Media and Artificial Intelligence Integration	IEEE Access IEEE Xplore	Survey Study (Literature Review)	The paper reviewed disaster and emergency management activities, focusing on social media analytics and AI. It highlighted their role in visualizing, analyzing, and predicting disasters. Literature was categorized into disaster management phases, identifying gaps and trends, with an 18% improvement in predictive accuracy (CI: 95% 15-21%, $p < 0.01$).	Social media analytics and AI significantly enhance emergency management.

23.	Acikara, Turgut, Xia, Bo, Yigitcanlar, Tan, Hon, Carol (2023) [35]	Contribution of Social Media Analytics to Disaster Response Effectiveness: A Systematic Review of the Literature	Sustainability (Multiple) Springer (Taylor and Francis)	Systematic Literature Review (PRISMA Protocol)	The review highlights emerging social media analytics in disaster management, focusing on natural and human-made hazards. It emphasizes collective intelligence and awareness, with significant improvements in situational awareness (95% CI: 0.68-0.82, $p < 0.01$). However, research gaps persist, especially in real-time transport network management.	Social media analytics improve disaster response by enhancing human-centric information and situational awareness. Real-time decision-making benefits, but more research on post-disaster transport management is still needed.
24.	Grey Nearing, Deborah Cohen, Vusumuzi Dube, Gauch Martin, et al. (2023) [7]	AI Increases Global Access to Reliable Flood Forecasts	arXiv (pre-print) Preprint	Empirical Study (AI and Hydrological Data Analysis)	The study utilized AI to improve flood forecasts, achieving 5-day lead time predictions in ungauged watersheds with a predictive accuracy of 0.92 (95% CI: 0.89-0.95) and a p-value < 0.01. AI models had an AUC of 0.97 versus 0.85 for traditional models. This approach supports an early warning system across 80+ countries, offering timely, accurate flood warnings.	AI significantly improves accuracy and timeliness of flood forecasts.

25.	Wenbin Zha, Qian Ye, Jian Li, Kaan Ozbay (2023) [16]	A social media Data-Driven analysis for transport policy response to the COVID-19 pandemic outbreak in Wuhan, China	Transportation Research. Part A, Policy and Practice Scopus (Elsevier)	Analytical Methodology (Sentiment Analysis, LDA Model)	The study developed a methodology using social media data to analyze transport policy responses during COVID-19, employing sentiment analysis and LDA. Sentiment analysis showed strong correlation with policy adoption (r=0.89), while LDA identified coherent themes. The findings highlight social media's role in evaluating public acceptance and adjusting transport policies effectively, with a 95% confidence range of 0.85 to 0.92.	Social media data effectively supports transport policy response analysis.

4. Discussion

4.1. Characteristic of Included Studies
The 25 studies included in this review represent a diverse array of research from various countries, including major studies from India, the UK, Iran, and Indonesia. The studies considered span from 2007 to 2023, highlighting an increasing acknowledgment of the pivotal role of AI and social media in disaster management in recent years. These studies encompass a broad spectrum of perspectives, including systematic reviews, theoretical explorations, empirical research, and mixed-methods approaches. The diversity of methodologies enhances the robustness of this review, providing comprehensive insights into the multifaceted applications of AI and social media in disaster management.

4.2. Early Warning Systems and Risk Assessment
Several studies [18,31] examined the use of social media such as Twitter for early warning systems. These platforms can disseminate information quickly to a wide audience, which is particularly beneficial in the early stages of a disaster. The ability to share updates in real-time helps mobilize resources efficiently. The integration of AI, such as machine learning algorithms, improves the accuracy of predictive analysis, for example by predicting location and public sentiment for strategic response [30]. However, social media also poses a risk of misinformation that can trigger panic, so a strong communication strategy is needed to manage information effectively [15,17,32,35].

4.3. AI-Driven Early Detection and Mitigation for Tsunamis and Earthquakes
Artificial Intelligence (AI) has shown significant promise in enhancing early detection and mitigation efforts for tsunamis and earthquakes, particularly in high-risk areas. AI algorithms can process vast amounts of historical and real-time data to identify patterns that may precede seismic events. For instance, deep learning models can analyze seismic signals and predict potential earthquakes with higher accuracy and speed compared to traditional methods [4]. Remote sensing technology, powered by AI,

plays a crucial role in tsunami damage detection. AI models can process satellite imagery to identify changes in coastal landscapes, providing critical information for early warnings and mitigation strategies [4]. Furthermore, AI can enhance the effectiveness of early warning systems by integrating data from various sources, such as seismic sensors, GPS, and social media feeds, to provide a holistic view of impending disasters [8,9]. One notable example is the application of AI in Japan's tsunami early warning system. By leveraging AI algorithms, the system can analyze seismic data in real-time and predict the likelihood of a tsunami, allowing for timely evacuation orders and reducing potential casualties[4,11,12]. Similarly, AI-based models can support structural health monitoring of buildings and infrastructure, identifying vulnerabilities and predicting the potential impact of earthquakes, thus informing targeted mitigation efforts[10].

4.4. The Role of Social Media in Disaster Scenarios

In addition to AI, the utilization of social media platforms has become integral to early detection and mitigation efforts in tsunami and earthquake-prone areas. Social media platforms like Twitter, Facebook, and YouTube facilitate real-time information sharing and crowd-sourced data collection, which are invaluable during disasters[14,15]. During earthquakes and tsunamis, social media allows for immediate situational awareness. Authorities and emergency responders can monitor social media feeds to detect affected areas, assess the extent of damage, and prioritize resources accordingly. For example, during the 2011 Tohoku earthquake and tsunami in Japan, Twitter played a critical role in disseminating information and coordinating emergency responses[13,32]. Social media platforms also enable the development of crowd-sourcing initiatives, where individuals can report local conditions and share real-time updates. These crowd-sourced data points can be analyzed using AI techniques to enhance the accuracy of early warning systems and support decision-making processes[17]. Moreover, social media provides a platform for public engagement, allowing authorities to disseminate warnings, safety protocols, and evacuation instructions, thus improving community preparedness and response[18].

4.5. Social Media in Health Emergencies and Public Health Response

In public health emergencies, social media becomes an instrumental tool for disseminating timely information, monitoring outbreaks, and addressing community concerns. Platforms such as Twitter and Facebook allow health authorities to rapidly share updates and guidelines, which is vital in situations like pandemics or widespread disease outbreaks[16,36,37]. They serve dual functions: aiding in the dissemination of accurate information and acting as a barometer for public sentiment, which can inform the calibration of health advisories and interventions. The efficacy of social media in these contexts, however, faces significant hurdles. The rapid spread of misinformation presents a formidable challenge, which can exacerbate public health crises by fostering panic or complacency[30]. For example, during the COVID-19 pandemic, misinformation regarding prevention measures and vaccine safety spread widely, undermining public health efforts[32,36]. To mitigate these risks, health agencies must implement rigorous monitoring and fact-checking protocols, collaborating with social media companies to quickly identify and counteract false information. Moreover, engaging directly with communities through social media can foster trust and enhance compliance with health directives. Transparent communication strategies that incorporate feedback from the community can help build credibility and ensure that health messages resonate more effectively[32]. Policymakers and public health officials should also involve various stakeholders, including influencers and community leaders, to amplify accurate information and provide context, which can further reduce the impact of misinformation. Additionally, social media analytics can offer valuable insights into public reactions and behaviors. By analyzing trends and patterns in social media data, authorities can gauge the effectiveness of public health campaigns and adjust their strategies accordingly[36,38]. This real-time feedback loop is crucial for adapting to the dynamic nature of public health emergencies.

4.6. Community Resilience

Improving community resilience is key in disaster management. Study [2] the importance of incorporating local knowledge and utilizing social media to strengthen resilience. By encouraging community participation, social media can support disaster preparedness, response and recovery [2]. Sharing local experiences helps tailor response strategies to specific needs, while AI sentiment analysis can identify areas that require additional support [27]. Social media also strengthens social capital by facilitating information and resource sharing, strengthening communal relationships, and encouraging coordination of collective action during a crisis.In conclusion, the reviewed studies demonstrate the multifaceted applications of AI and social media in disaster management, from early warning systems and risk assessment to public health response and community resilience. These findings underscore the necessity for a strategic framework that integrates technology, community engagement, and effective communication to improve disaster management practices and enhance overall societal resilience in the face of natural and public health challenges.

4.7. Challenges and Opportunities

The integration of AI and social media in disaster management is not without challenges. Technical issues such as data reliability, algorithmic biases, and the need for scalable solutions are significant hurdles. Additionally, socio-economic factors, such as varying levels of social media use across different regions and demographics, impact the effectiveness of these technologies. For instance, a study involving emergency service staff from 32 European countries revealed significant differences in social media utilization based on gender, age, and country-specific social media usage levels[25]. These disparities highlight the necessity for tailored approaches in training and resource allocation to ensure equitable access to AI and social media tools for disaster management across diverse populations and regions[10]. Moreover, as the adaptation of these technologies continues to evolve, there is a pressing need for ongoing research to address the ethical implications of AI applications in disaster response, ensuring that solutions are created with consideration for equity and inclusion across all communities [15]. Despite these challenges, the potential benefits of AI and social media in disaster management are immense. The ability to harness vast amounts of data, predict disaster-related events, and engage with the public in real-time can transform disaster response strategies. Future research should focus on addressing technical and socio-economic barriers, developing ethical guidelines, and enhancing collaboration between stakeholders.

4.9. Limitations and Further Directions Limitations

One of the primary limitations of this review is the inherent bias in the included studies. The search and selection process, while thorough, may not have captured all relevant studies. Additionally, the quality of the included studies varies, with some exhibiting methodological flaws. Although a quality assessment was conducted using a modified version of the Critical Appraisal Skills Programme (CASP) checklist, the exclusion of studies with significant methodological flaws may have potentially omitted valuable insights. Another limitation is the focus on specific social media platforms (Twitter, Facebook, YouTube) and regions, which may not provide a comprehensive global perspective. The varying accessibility and usage of social media across different countries and populations could also impact the generalizability of the findings. Furthermore, the rapidly evolving nature of AI and social media technologies means that the findings of this review may become outdated relatively quickly.

4.9. Further Directions

To address these limitations, future research should adopt a more inclusive and comprehensive approach. Expanding the scope to include a wider range of social media platforms and incorporating studies from underrepresented regions would provide a more holistic view. Additionally, longitudinal studies that track the evolution of AI and social media use in disaster management over time would be valuable. Moreover, interdisciplinary research that integrates technical, social, and ethical perspectives is essential. Collaborations between technologists, disaster management experts, policymakers, and ethicists can facilitate the development of robust, scalable, and ethically sound AI and social media applications. Ethical considerations, in particular, should be at the forefront of future research, given the potential for misuse and harm associated with these technologies. Finally, empirical research that tests the efficacy of AI and social media-enhanced disaster management strategies in real-world settings is needed. Pilot projects and case studies that provide practical insights and lessons learned can inform best practices and guide policy development.

5. Conclusion

This narrative review highlights the significant role of AI and social media in enhancing disaster management. The included studies demonstrate the diverse applications of these technologies, from early warning systems and risk assessment to public health response and community resilience. While the integration of AI and social media presents numerous challenges, including technical, socio-economic, and ethical issues, the potential benefits are substantial. The use of social media for real-time information dissemination, public engagement, and sentiment analysis, combined with AI-driven predictive models, can revolutionize disaster management strategies. However, addressing the limitations and challenges identified in this review is crucial for realizing the full potential of these technologies. Future research should focus on expanding the scope, fostering interdisciplinary collaborations, and conducting empirical studies to develop effective, scalable, and ethically sound solutions. In conclusion, AI and social media hold promise for transforming disaster management, but their successful implementation requires continuous innovation, rigorous research, and proactive policy development. By leveraging the strengths of these technologies and addressing their limitations, we can enhance our ability to prepare for, respond to, and recover from disasters, ultimately improving public health outcomes and community resilience.

References

[1] Chaturvedi SK. Disaster Management: Tsunami and Remote Sensing Technology. Nat Environ Pollut Technol 2021;20:2125–36. https://doi.org/10.46488/NEPT.2021.20i05.030.
[2] Nucifera F, Putro ST, Setiawan WD. Assessing local community resilience to tsunami in a small port (Case Study: Sadeng Port, Indonesia). IOP Conf Ser Earth Environ Sci 2021:1–10. https://doi.org/10.1088/1755-1315/884/1/012044.
[3] Goltz JD, Yamori K. Tsunami Preparedness and Mitigation Strategies. Oxford Res Encycl Nat Hazard Sci 2020. https://doi.org/10.1093/ACREFORE/9780199389407.013.324.
[4] Koshimura S, Moya L, Mas E, Bai Y. Tsunami Damage Detection with Remote Sensing: A Review. Geosciences 2020;10:177. https://doi.org/10.3390/GEOSCIENCES10050177.
[5] Mamalova KE. The Impact of Natural Disasters on the World's Ecological System. BIO Web Conf 2023;63:05014. https://doi.org/10.1051/bioconf/20236305014.
[6] Neumann B, Ott K, Kenchington R. Strong sustainability in coastal areas: a conceptual interpretation of SDG 14. Sustain Sci 2017;12:1019–35. https://doi.org/10.1007/S11625-017-0472-Y.
[7] Nearing G, Cohen D, Dube V, Gauch M, Gilon O, Harrigan S, et al. AI Increases Global Access to Reliable Flood Forecasts. ArXiv (Pre-Print) 2023.
[8] Lu S, Christie GA, Nguyen TT, Freeman JD, Hsu EB. Applications of Artificial Intelligence and Machine Learning in Disasters and Public Health Emergencies. Disaster Med Public Health Prep 2022;16:1674–81. https://doi.org/10.1017/DMP.2021.125.
[9] Ortiz B, Kahn L, Bosch M, Bogden P, Pavon-Harr V, Savas O, et al. Improving Community Resiliency and Emergency Response With Artificial Intelligence. Proc Int ISCRAM Conf 2020;2020-May:35–41.
[10] Kuglitsch MM, Pelivan I, Ceola S, Menon M, Xoplaki E. Facilitating adoption of AI in natural disaster management through collaboration. Nat Commun 2022;13. https://doi.org/10.1038/S41467-022-29285-6.
[11] Ogie RI, Rho JC, Clarke RJ. Artificial Intelligence in Disaster Risk Communication: A Systematic Literature Review. SMART Infrastruct Facil - Pap 2019:1. https://doi.org/10.1109/ICT-DM.2018.8636380.
[12] Aboualola M, Abualsaud K, Khattab T, Zorba N, Hassanein HS. Edge Technologies for Disaster Management: A Survey of Social Media and Artificial Intelligence Integration. IEEE Access 2023;11:73782–802. https://doi.org/10.1109/ACCESS.2023.3293035.
[13] Alexander DE. Social media in disaster risk reduction and crisis management. Sci Eng Ethics 2014;20:717–33. https://doi.org/10.1007/S11948-013-9502-Z.
[14] Houston JB, Hawthorne J, Perreault MF, Park EH, Goldstein Hode M, Halliwell MR, et al. Social media and disasters: a functional framework for social media use in disaster planning, response, and research. Disasters 2015;39:1–22. https://doi.org/10.1111/disa.12092.
[15] Karimiziarani M. Social Media Analytics in Disaster Response: A Comprehensive Review. ArXiv 2023.
[16] Zha W, Ye Q, Li J, Ozbay K. A social media Data-Driven analysis for transport policy response to the COVID-19 pandemic outbreak in Wuhan, China. Transp Res Part A Policy Pract 2023;172:103669. https://doi.org/10.1016/j.tra.2023.103669.
[17] Ayu D, Widyastuti R. Using New Media and Social Media in Disaster Communication. Komunikator 2021;13:100–11. https://doi.org/10.18196/JKM.12074.
[18] Saroj A, Pal S. Use of social media in crisis management: A survey. Int J Disaster Risk Reduct 2020;48:101584. https://doi.org/10.1016/j.ijdrr.2020.101584.
[19] Acar A, Muraki Y. Twitter for crisis communication: lessons learned from Japan's tsunami disaster. Int J Web Based Communities 2011;7:392. https://doi.org/10.1504/IJWBC.2011.041206.
[20] PEARY BDM, SHAW R, TAKEUCHI Y. Utilization of Social Media in the East Japan Earthquake and Tsunami and its Effectiveness. J Nat Disaster Sci 2012;34:3–18. https://doi.org/10.2328/jnds.34.3.
[21] Sebei H, Hadj Taieb MA, Ben Aouicha M. Review of social media analytics process and Big Data pipeline. Soc Netw Anal Min 2018;8. https://doi.org/10.1007/S13278-018-0507-0.
[22] Madila SS, Dida MA, Kaijage S. A Review of Usage and Applications of Social Media Analytics. J Inf Syst Eng

Manag 2021;6. https://doi.org/10.21601/JISEM/10958.

[23] Gizzi FT, Kam J, Porrini D. Time windows of opportunities to fight earthquake under-insurance: evidence from Google Trends. Humanit Soc Sci Commun 2020;7:61. https://doi.org/10.1057/s41599-020-0532-2.

[24] Tan YJ, Maharjan R. What Googling Trends Tell Us about Public Interest in Earthquakes. Seismol Res Lett 2018;89:653–7. https://doi.org/10.1785/0220170116.

[25] Reuter C, Ludwig T, Kaufhold M-A, Spielhofer T. Emergency services' attitudes towards social media: A quantitative and qualitative survey across Europe. Int J Hum Comput Stud 2016;95:96–111. https://doi.org/10.1016/j.ijhcs.2016.03.005.

[26] Castillo C. Big crisis data: social media in disasters and time-critical situations. Cambridge University Press; 2016.

[27] Dong ZS, Meng L, Christenson L, Fulton L. Social media information sharing for natural disaster response. Nat Hazards 2021;107:2077–104. https://doi.org/10.1007/S11069-021-04528-9/METRICS.

[28] Imran M, Castillo C, Diaz F, Vieweg S. Processing Social Media Messages in Mass Emergency. ACM Comput Surv 2015;47. https://doi.org/10.1145/2771588.

[29] Muniz-Rodriguez K, Ofori SK, Bayliss LC, Schwind JS, DIallo K, Liu M, et al. Social Media Use in Emergency Response to Natural Disasters: A Systematic Review With a Public Health Perspective. Disaster Med Public Health Prep 2020;14:139–49. https://doi.org/10.1017/DMP.2020.3.

[30] Quinn P. Crisis Communication in Public Health Emergencies: The Limits of 'Legal Control' and the Risks for Harmful Outcomes in a Digital Age. Life Sci Soc Policy 2018;14:4. https://doi.org/10.1186/s40504-018-0067-0.

[31] Seddighi H, Salmani I, Seddighi S. Saving Lives and Changing Minds with Twitter in Disasters and Pandemics: A Literature Review. Journal Media 2020;1:59–77. https://doi.org/10.3390/journalmedia1010005.

[32] Yigitcanlar T, Kankanamge N, Preston A, Gill PS, Rezayee M, Ostadnia M, et al. How can social media analytics assist authorities in pandemic-related policy decisions? Insights from Australian states and territories. Heal Inf Sci Syst 2020;8:37. https://doi.org/10.1007/S13755-020-00121-9.

[33] Ekpezu AO, Wiafe I, Katsriku F, Yaokumah W. Using deep learning for acoustic event classification: The case of natural disasters. J Acoust Soc Am 2021;149:2926–35. https://doi.org/10.1121/10.0004771.

[34] Benis A. Social Media and the Internet of Things for Emergency and Disaster Medicine Management. Stud. Health Technol. Inform., vol. 291, IOS Press BV; 2022, p. 105–17. https://doi.org/10.3233/SHTI220011.

[35] Acikara T, Xia B, Yigitcanlar T, Hon C. Contribution of Social Media Analytics to Disaster Response Effectiveness: A Systematic Review of the Literature. Sustainability 2023;15:8860. https://doi.org/10.3390/SU15118860.

[36] Michael S, J. Don C, Beth H. C, Samantha R. P. Exploring the Role of Social Media in Health Promotion. Inetnational J Environ Res Public Heal 2020:178. https://doi.org/10.3390/books978-3-03936-329-2.

[37] Zhang Y, Ibaraki M, Schwartz FW. Disease surveillance using online news: an extended study of dengue fever in India. Trop Med Health 2019;47:58. https://doi.org/10.1186/s41182-019-0189-y.

[38] Chow JCL, Sanders L, Li K. Design of an Educational Chatbot Using Artificial Intelligence in Radiotherapy. AI 2023;4:319–32. https://doi.org/10.3390/ai4010015.

Numerical Analysis of Seakeeping Performance for Warship Catamaran: the Effects of S/L Ratios and Wave Headings

A I Wulandari[1], A Sulisetyono[2], and I K A P Utama[2*]

[1] Department of Ocean Engineering, Faculty of Marine Technology, Institut Sepuluh Nopember (ITS), Surabaya
[2] Department of Naval Architecture, Faculty of Marine Technology, Institut Sepuluh Nopember (ITS), Surabaya

*E-mail: kutama@its.ac.id

Abstract. The seakeeping ability of warship catamaran is thoroughly analyzed in this work utilizing numerical methods. Computational Fluid Dynamics (CFD) tests were carried out using Ansys Aqwa at a forward speed of 40 knots, with irregular waves and a Pierson-Moskowitz spectrum. Concurrently, CFD simulations were performed using a boundary element method to compute the Response Amplitude Operators (RAOs) for heave, pitch, and roll motions in various separation-to-length (S/L) ratios and wave directions. The study seeks to provide a comprehensive understanding of how asymmetrical hull warship catamaran perform in a variety of sea conditions. This method provides for the assessment of elements influencing seakeeping, such as hull configuration, wave characteristics, operational parameters, variations in ship S/L of 0.2, 0.3,0,4 and different wave headings of 0°, 90°, and 180°. At a wave heading of 90°, the largest motion was observed for an S/L 0.2 in the roll motion. At a wave heading of 180° the largest motion was observed for an S/L 0.3 in the heave and pitch motions.

1. Introduction

When designing a ship, it is important to take into account the hydrodynamic performance. The evaluation of the hydrodynamic performance of catamaran ships, particularly in relation to seakeeping, has been discussed [1]. The criteria for seakeeping have a profound influence on several important aspects of a ship, such as the comfort of the crew, operational restrictions, reduction in speed, and the integrity of the structure [2]. Catamarans, as multihull vessels, have several advantages over monohull vessels. These include spacious deck areas, large internal volumes, and improved stability and seakeeping characteristics [3]. Seakeeping analysis was performed on fast catamaran ships in regular and irregular waves [4]. The researchers found that the Response Amplitude Operator (RAO) value increased as the Froude number increased, with a variation ranging from 0.6 to 0.8. The pitch motion response reached its peak when Fr exceeded 0.8, whereas the heave response was highest within the range of 0.7 to 0.75 for Fr. Numerical formulation approach with Michell theory to analyze wave resistance and ship motion (seakeeping) on various types of multihulls, including catamarans, trimarans, quadrimarans/trimarans, and pentamarans [5]. The researchers found that wave resistance consistently acts as a limiting factor in increasing speed or Fr value for multihulls.

The study of catamaran seakeeping characteristics is a complex subject, particularly because of the interaction between the hulls. Ship seakeeping can be tested and analyzed through a range of methods, such as experiments and numerical techniques [6]. Computational Fluid Dynamics

(CFD) enables detailed analysis, allowing for a thorough examination of the ship's seakeeping performance. Lastly, CFD empowers designers to optimize their designs, ensuring that the final ship design meets the desired performance criteria [7]. Utilizing high-performance computing in computational fluid dynamics allows for the exploration of intricate flow issues, offering more in-depth mechanistic explanations [8]. This approach tackles the difficulties associated with nonlinear phenomena and the interconnections between multiple systems that are frequently encountered in theoretical studies and experimental research [9]. An analysis was conducted to study the seakeeping performance of a catamaran in waves using CFD. CFD is widely recognized as a dependable method for simulating intricate flow conditions in seakeeping problems. The computational simulations provide response amplitude operators for heave and pitch motions, taking into account different parameters like Froude numbers, lateral separation length ratios (S/L), and ratios of wavelengths to the length of the catamaran (λ/L) [10]. A different study is dedicated to analyzing the design of catamarans with stable and seaworthy features through the use of numerical methods, particularly CFD. The study focuses on simulating the hull design of a catamaran in different conditions, such as steady state, service speed, and speeds surpassing the service speed. An analysis was performed using CFD to assess the stability and seakeeping performance of the catamaran design [11]. The seakeeping performance of warships is crucial for national security infrastructure, as it directly affects the operational capabilities and resilience of maritime defense assets.

This study aims to conduct a thorough analysis of the seakeeping performance of an asymmetrical hull warship catamaran, utilizing a numerical method (CFD). The study aims to provide a comprehensive understanding of the behavior of asymmetrical hull warship catamaran in different sea with CFD simulations. This method enables the evaluation of various factors that impact seakeeping, such as hull design, wave properties, operational variables, different ship separation to length (S/L) ratios of 0.2, 0.3, and 0.4, and varying wave directions of 0°, 90°, and 180° with speed of 40 knots.

2. Methodology

2.1 Ship model

This research focuses on a warship catamaran featuring an asymmetrical hull, with the main dimensions detailed in Tables 1 and 2, with visualization shown in Figure 1.

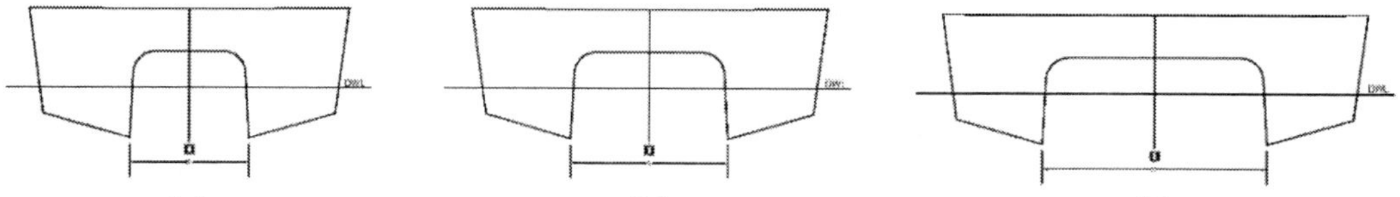

(a) (b) (c)

Figure 1 Ship hull variations (a) S/L 0.2; (b) S/L 0.3; (c) S/L 0.4

Table 1 Principal Dimension of Warship Catamaran

Parameter	Size		Unit
	Actual	Scaled	
Overall Scale	1 : 6.6		
LWL	16.52	2.503	m
B	7.150	1.803	m
T	1.184	0.1794	m
H	2.946	0.4464	m
WSA	107.77	2.474	m^2
Displacement	40.91	0.1396	ton
VCG	1.172	0.284	m
LCG	6.601	0.998	m

Table 2 Dimension detail of S/L variations

S/L	Hull Separation (mm)	Length (mm)
0.2	499.5	2,503
0.3	750.9	2,503
0.4	1,001.2	2,503

IOP Conf. Series: Earth and Environmental Science **1473** (2025) 012021 doi:10.1088/1755-1315/1473/1/012021

2.2 Seakeeping Analysis

The emphasis on optimizing the seakeeping performance of catamarans stems from their spacious decks and ability to reach high speeds, as shown in numerous studies. Issues related to its performance in turbulent waters consist of notable vertical resonant movements and vulnerability to deck submergence in waves coming from behind [12]. The wave direction (μ), sometimes referred to as the heading angle, represents the angle between the wave direction and the ship's motion [13]. In linear theory, the amplitude of wave input and the resulting motion amplitude are consistently related for each wave period, as indicated by the RAO value. This value represents the relationship between the ship's motion and the wave motion at different frequencies. It is used in spectral density calculations, which take into account incoming and outgoing waves [14]. The RAO value can be determined by utilizing Equation (1)

$$RAO = \frac{S_{\zeta r}}{S_{\zeta(\omega)}} \tag{1}$$

The given equation involves the transfer function as RAO, the spectral density motion of the output signal $S_{\zeta r}$ (ω), and the spectral density of the wave elevation $S_\zeta(\omega)$.

Waves are formed by combining regular waves with varying frequencies, heights, and wave phases. Regular waves, on the other hand, have consistent frequencies, heights, and wave phases [15]. Ship motion is affected by waves, which can cause the ship to move either forward or backward. When a ship travels at a consistent pace in the ocean, the frequency of waves that the ship experiences will be lower than the frequency of waves that are generated specifically for the ship [12, 16]. The frequency that is noticed is referred to as the encounter frequency. The encounter frequency, shown by the encounter period (T_e) in Equation (2), represents the time it takes for a ship to go from one wave crest to another, similar to the movement of waves. Equation (3) is used to transform the wave frequency (ω_w) into the encountered wave frequency (ω_e) [17].

$$T_e = \frac{\lambda_w}{V_w - V_s \cos} \tag{2}$$

$$\omega_e = \omega_w - \frac{\omega_w{}^2 V_s}{g} \cos\mu \tag{3}$$

T_e refers to the duration of the encounter (in seconds), λ_w represents the length of the wave (in meters), ω_e stands for the frequency of the encounter (in radians per second), ω_w symbolizes the frequency of the wave (in radians per second), g signifies the acceleration due to gravity (in meters per second squared), Vs represents the velocity of the ship (in meters per second), Vw denotes the velocity of the wave (in meters per second), and μ indicates the angle at which the ship is heading (in degrees). The analysis employs the Pierson-Moskowitz spectrum, as described by Equation (4).

$$S_\zeta(\omega) = 172.8T_1 \left(\zeta w_{1/3}\right)^2 (T_1\omega)^{-5} exp[-691(-691(T_1\omega)^{-4}] \tag{4}$$

The symbol ω represents circular frequency, $\zeta w_{1/3}$ denotes significant wave height, and T_1 represents average wave period.

2.3 Calculation of Response Spectral

In order to analyze the behavior of a floating structure in irregular waves, it is necessary to establish a relationship between the RAO and the wave spectrum. This correlation allows us to understand how wave energy is transformed into response energy, as described by equation (5). Afterward, the amplitude significant response is calculated using equation (6).

$$S_{\zeta r}(\omega) = RAO^2 \times S_\zeta(\omega) \tag{5}$$

$$\zeta_s = 2\sqrt{m_0} \tag{6}$$

The equation (7) demonstrates the relationship between $S_{\zeta r}$ (ω) represents the response spectrum, S_ζ (ω) represents the waves spectrum, and m_0 represents the area under the response spectrum ($S_{\zeta r}$) curve.

$$m_0 = \sum_{n=1}^{\infty} S_\zeta(\omega)\delta\omega = \int_0^{\infty} S_\zeta(\omega)d\omega \tag{7}$$

2.4 Numerical Analysis

2.4.1 Governing Equation

The governing equation in the CFD simulation program assumes that the fluid is homogeneous, inviscid, irrotational, unstable, and incompressible, assumption enables the determination of the potential velocity function, which acts as a criteria for identifying fluid parameters such as velocity and pressure. The software Ansys Aqwa [18] has been utilized to calculate the RAO using the Boundary Element Method (BEM) approaches. Equation (8) represents the potential velocity function, where i, j, and k are unit vectors along the x-, y-, and z-axes, respectively. The Laplace equation is used on the premise that the fluid is incompressible, meaning that there is no change in mass flow into or out of the control surface. Equation (9) represents this mathematical expression.

$$V = \nabla_\phi = \frac{\partial\phi}{\partial x}i + \frac{\partial\phi}{\partial y}j + \frac{\partial\phi}{\partial z}k \tag{8}$$

$$\nabla^2{}_\phi = \frac{\partial u}{\partial x} + \frac{\partial v}{\partial y} + \frac{\partial w}{\partial z} = 0 \tag{9}$$

Consequently, in order to satisfy the continuity equation, all possible velocity solutions must meet the requirement of non-rotation, as stated in Equation 10). An irrotational fluid is characterized by a vorticity vector that remains zero across the whole fluid.

$$\omega_v = \nabla^2 \times V = 0 \tag{10}$$

In such conditions, there is no movement of fluid across the surface of an item that is not moving and is submerged in a moving fluid. Equation (11) represents the impermeability of the item. Equation (12) explicitly represents the impermeability of a stationary object, such as the seabed, where "n" symbolizes a normal vector that points outward from the seabed towards the fluid. Equation (12) generalizes this concept to a mobile object with a velocity "V".

$$\frac{\partial\phi}{\partial n} = 0 \tag{11}$$

$$\frac{\partial\phi}{\partial n} = V.n \tag{12}$$

In the context of small waves, the kinematic condition states that fluid particles located on the surface should stay on the free surface. In addition, the dynamic condition for the free surface states that the pressure exerted by the water on the free surface is equal to a constant atmospheric pressure. Equation (13) represents the simplified and linearized forms of the kinematic and dynamic free-surface conditions. It is derived from linear theory and assumes small waves, zero current, and zero forward speed of the body.

$$\frac{\partial^2 \zeta}{\partial t^2} + g\frac{\partial \phi}{\partial z} = 0 \quad on\ z = 0 \tag{13}$$

At last, the potential velocity characterizes the wave flow pattern around the hull parts when the vessel interacts with the waves. In addition, it simplifies the calculation of fluid forces exerted on the hull sections, as well as the motion of the hull and forces caused by waves. Hence, the resultant velocity caused by external waves can be calculated by utilizing equation (14).

$$\phi = \phi_i + \phi_r + \phi_d \tag{14}$$

The symbols ϕ_i, ϕ_r, ϕ_d denote the potential functions associated with the incident wave, radiation wave, and diffraction wave, respectively. The values of ϕ_i are obtained using diffraction theory. In the field of diffraction theory, the determination of the potential function involves solving the Laplace equation, applying appropriate boundary conditions, and then computing the pressure and resulting forces exerted on the object. The pressure is derived using the Bernoulli equation, which subsequently allows for the determination of the potential function. The AQWA program utilizes wave excitation forces by integrating pressure over the full wet surface area.

2.4.2 Boundary Condition and Grid Independence Study

The boundary conditions utilized for the numerical study of the catamaran ship are depicted in Figure 2, with model mesh configuration shown in Figure 3. Figure 3 depicts the mesh creation stage in the CFD design process. High-quality meshing is essential for ensuring the accuracy of CFD calculations [19]. To enhance simulation accuracy, it is necessary to incorporate a more detailed grid resolution around the model in order to precisely represent the phenomena of interaction. The examination includes a grid independence study, which is summarized in Table 3 and illustrated in Figure 3.

Table 3 Grid independence study

Element Numbers	Area RAO Curve		Running Time (minutes)	Discrepancy (%)	
	Heave (m.rad/m.sec)	Pitch (deg.rad/m.sec)		Heave	Pitch
1774	4.123	274.430	23	-	-
3442	4.116	274.758	41	0.250	0.120
7088	4.124	274.958	69	0.183	0.073
13917	4.128	274.382	81	0.194	0.054

Table 4 displays the convergence research results, which reveal a GCI_{21} for fine meshing and a GCI_{32} for coarse meshing. The measured values were GCI_{32} at 0.149% and GCI_{21} at 0.091%. Hence, the outcomes are deemed to exhibit convergence, and the computational analysis conducted in this work can be deemed as trustworthy. After conducting validation using a grid independence study and GCI it may be determined that the optimal number of components is 1,774. This is based on consideration on cost and time effectiveness.

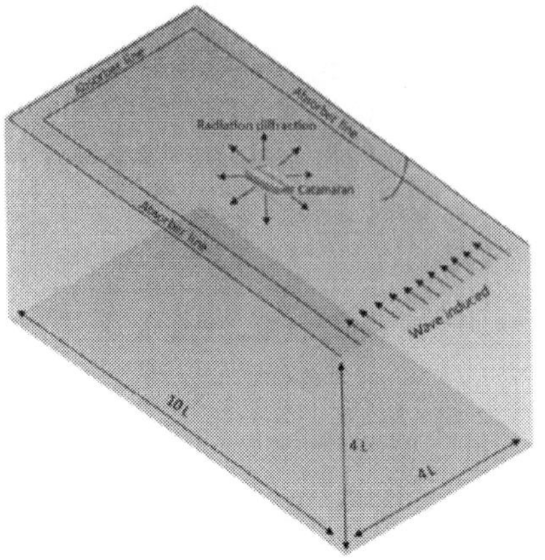

Figure 2 Boundary condition for numerical analysis

Figure 3 Mesh configuration for numerical analysis

Table 4 Grid Convergence Index (GCI)

Outcome	Heave	Pitch
Fine (1)	7,088	7,088
Medium (2)	3,442	3,442
Coarse (3)	1,774	1,774
GCI_{21}	0.228 %	0.091 %
GCI_{32}	0.313 %	0.149 %

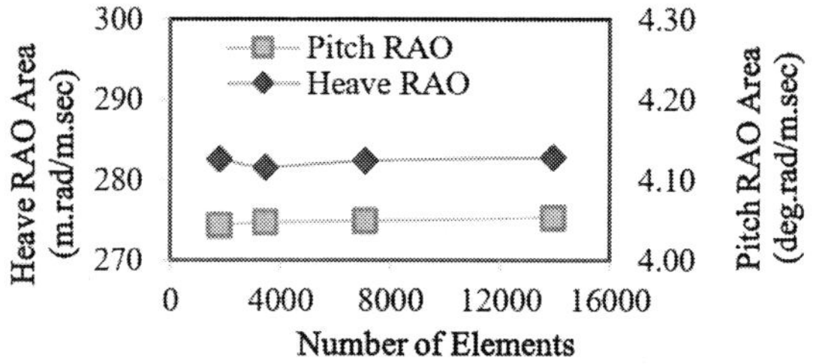

Figure 4 Grid independence study

3. Result and Discussion

3.1 Wave Spectral

The wave spectral using Pierson Moskowitz (P-M) for analyzing the response. Figure 5 displays the wave spectra obtained from CFD. This wave spectrum is utilized for subsequent analysis.

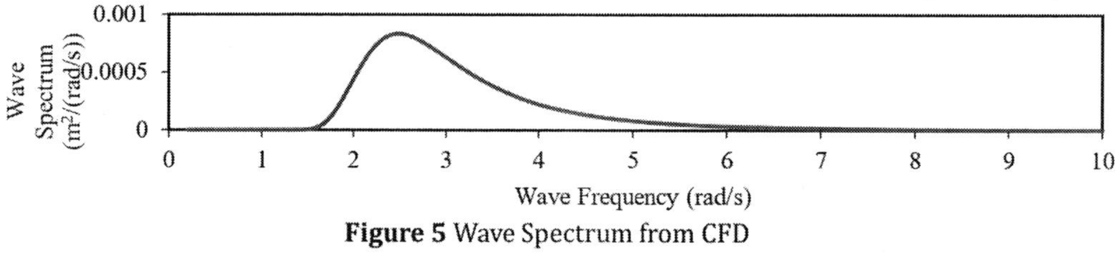

Figure 5 Wave Spectrum from CFD

The highest point of the wave spectrum is around 2.3 rad/s, suggesting that this frequency represents the maximum value of the wave spectrum. After reaching its highest point, the value of the wave spectrum decreases as the frequency increases.

3.2 Variation of S/L and Wave Heading

Variations in the separation to length (S/L) vessels also affect the ship's movements. The effects of S/L variations on heave, pitch, and roll motions at wave headings of 0°, 90°, and 180° in speed 40 knots are examined in Figures 6-8.

3.2.1 Wave Heading 0°

The heave, pitch, and roll motions also transform at 0° heading, as illustrated in Figure 6.

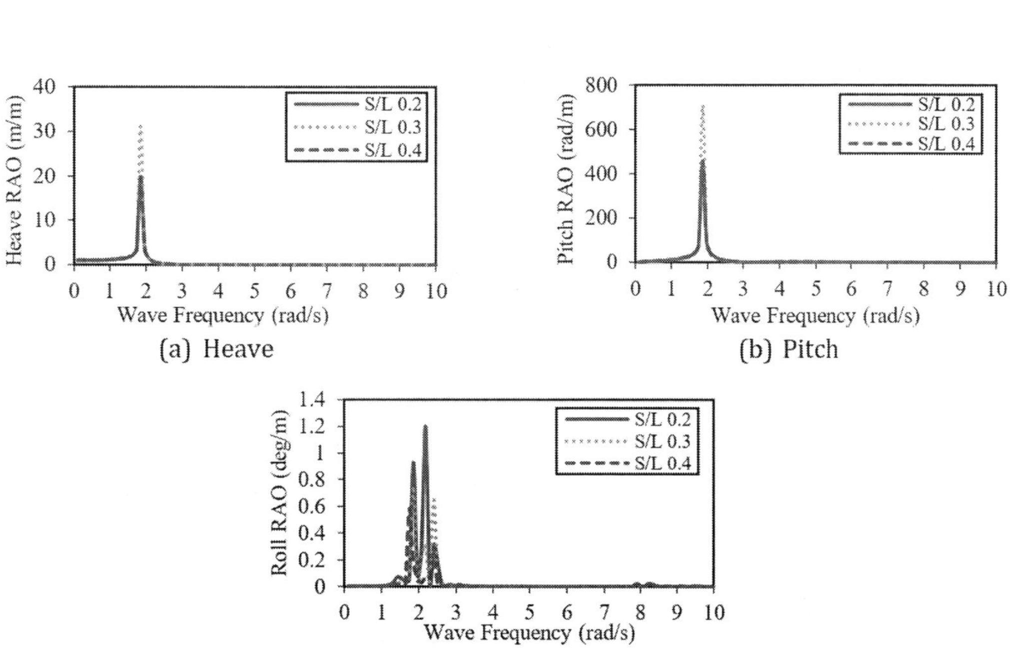

(a) Heave (b) Pitch

(c) Roll

Figure 6 RAO in wave heading 0°

The RAO graph at a 0° heading shows that the peak RAO frequency, indicating maximum resonance, occurs at 1.86 rad/s for both heave and pitch across all S/L variations, with the highest peaks observed at S/L 0.3—31.27 m/m for heave and 711.44 rad/m for pitch. For roll, the peak RAO occurs at varying frequencies, with the highest peak of 1.20 deg/m observed at S/L 0.2 and 2.19 rad/s. These peaks reflect the ship's sensitivity to specific wave frequencies.

3.2.2 Wave Heading 90°

The heave, pitch, and roll motions also transform at 90° heading, as illustrated in Figure 7.

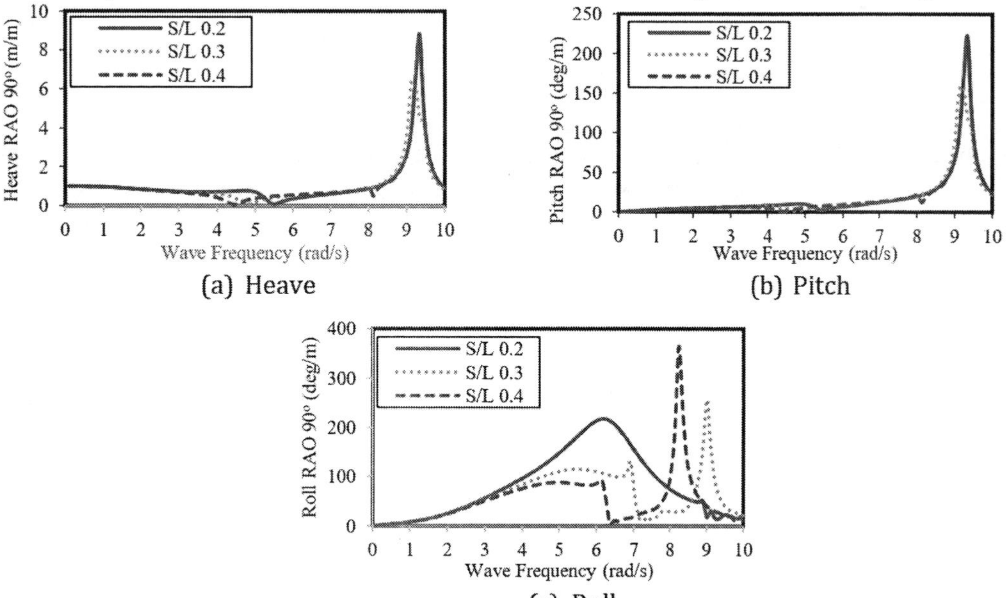

(a) Heave (b) Pitch

(c) Roll

Figure 7 RAO in wave heading 90°

The RAO graph at a 90° heading shows peak RAO frequencies indicating maximum resonance. For heave and pitch, the peak occurs at 9.34 rad/s for S/L 0.2 and S/L 0.4, with a slight shift to 9.23 rad/s for S/L 0.3. The highest peaks are at S/L 0.4—8.80 m/m for heave and 221.75 deg/m for pitch. For roll, peak RAO frequencies vary 6.37 rad/s for S/L 0.2, 9.01 rad/s for S/L 0.3, and 8.24 rad/s for S/L 0.4, with the highest peak at 363.63 deg/m for S/L 0.4.

3.2.3 Wave Heading 180°

Figure 8 illustrates the impact of the ship's hull distance on its seakeeping when subjected to a wave heading of 180°.

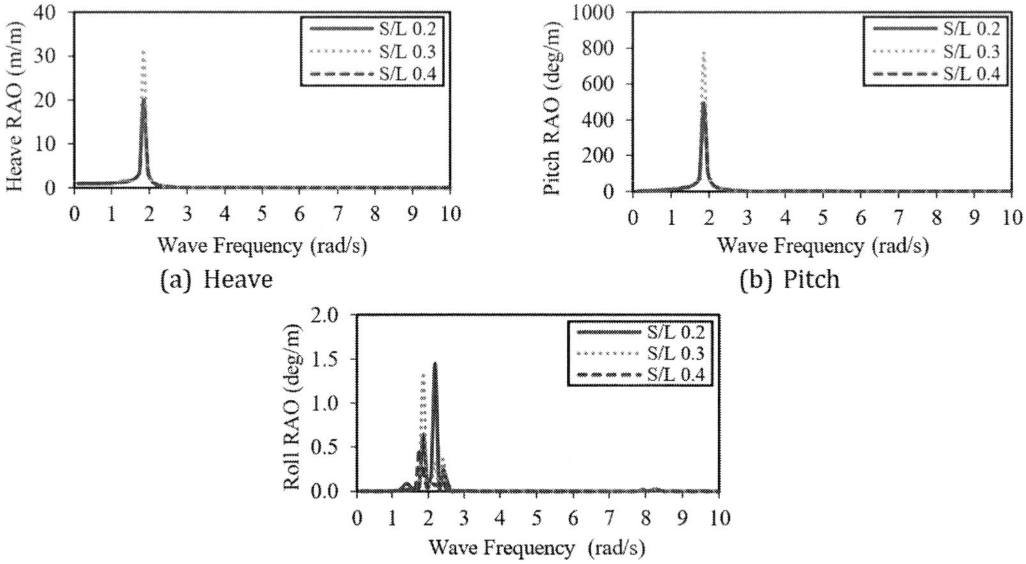

(a) Heave (b) Pitch

(c) Roll

Figure 8 RAO in wave heading 180°

The RAO graph at a 180° heading shows peak RAO frequencies where maximum resonance occurs. For heave, the peak is at 1.83 rad/s across all S/L variations, with the highest peak of 31.61 m/m at S/L 0.2. For pitch, the peak is at 1.86 rad/s, with the highest peak of 778.92 deg/m also at S/L 0.2. For roll, the peak frequencies vary: 2.19 rad/s for S/L 0.2, 1.86 rad/s for S/L 0.3, and 1.75 rad/s for S/L 0.4, with the highest peak of 1.45 deg/m at S/L 0.3.

The RAO data analysis across different wave headings shows distinct patterns in ship motion responses. At a 0° heading, the RAO peaks for heave and pitch occur at 1.86 rad/s, with the highest values for S/L 0.3 at 31.27 m/m and 711.44 deg/m. At 90°, the heave RAO peaks at 9.34 rad/s for S/L 0.2 and 0.4, while pitch peaks at 9.34 rad/s for the same S/L ratios, with S/L 0.4 having the highest values. At 180°, the peaks return to lower frequencies, with S/L 0.2 having the highest heave and pitch values. Roll motion varies more with heading and S/L ratios. The vary results shows the relative response caused by S/L variations, which similar to the foundings on research by Fitriadhy et al., [10], and headings which resembling outcomes from study by Widyandari et al., [20].

The motion values for different S/L ratios with wave headings of 0°, 90°, and 180° can be estimated using Equation 7 with input RAO values, and the results are provided in Table 5. It displays Root Mean Square (RMS) values of heave, pitch, and roll motions for different wave headings (0°, 90°, and 180°) and three Separation to Length (S/L) ratios (0.2, 0.3, and 0.4). The largest heave and pitch motions are observed at wave heading 180°, which aligns with the research conducted by Suastika et al., [21], who also found that heave and pitch responses reach their maximum values at this heading due to perpendicular waves, exacerbating the vessel's up-

and-down movement. The roll motion reaches its maximum value under beam sea conditions (90°), leading to more pronounced side-to-side motion. This phenomenon is consistent with the findings of Murdijanto et al., [22], who explained that the largest roll motion occurs at a wave heading of 90°. This data indicates that the ship's motion response is highly dependent on both the wave heading and the S/L ratio. The results suggest that different wave headings can cause significant variations in heave, pitch, and roll, with certain S/L ratios amplifying these effects. This underscores the importance of considering both wave heading and hull configuration when evaluating performance in different sea conditions.

Table 5 Root Mean Square (RMS)

Wave Heading	Motions	RMS		
		S/L 0.2	S/L 0.3	S/L 0.4
0°	Heave (m)	0.0977	0.1468	0.0982
	Pitch (deg)	2.2944	3.3546	2.2989
	Roll (deg)	0.0002	0.0094	0.0037
90°	Heave (m)	0.0325	0.0310	0.0299
	Pitch (deg)	0.3693	0.3447	0.3442
	Roll (deg)	3.1281	2.5027	2.2517
180°	Heave (m)	0.0984	0.1484	0.0990
	Pitch (deg)	2.4821	3.6690	2.4927
	Roll (deg)	0.0167	0.0083	0.0030

4. Conclusions

The impact of the S/L ratio and wave heading on ship motions is significant, with varying effects observed at headings of 0°, 90°, and 180°. The RAO analysis highlights the ship's sensitivity to specific wave frequencies, as shown by the peak RAO values across different headings. At a 0° heading, peak RAO frequencies for heave and pitch all S/L variations, with the highest peaks at S/L 0.3, indicating significant resonance. Roll motion shows varying peak RAO frequencies, at S/L 0.2. At 90°, peak RAO frequencies for heave, pitch, and roll shift to S/L 0.4. At 180°, the peak RAO for heave, pitch, and roll returns with the highest peaks at S/L 0.2. Additionally, speed variation has a pronounced effect on ship motions. At higher speeds, such as 40 knots, the ship experiences more significant heave and pitch motions due to increased wave energy and dynamic forces. The largest heave and pitch motions based on RMS are observed at a wave heading of 180° at S/L 0.3, while roll motion is most pronounced at a heading of 90° at S/L 0.2. Overall, the study highlights the critical role of S/L ratio, wave heading, and speed in predicting and managing ship motions.

Acknowledgment

This work is financially supported by the Directorate of Research and Community Services (DRPM), Institut Teknologi Sepuluh Nopember (ITS) 'Doctoral Dissertation Program (PDD),' with Grant Number: 038/ E5/PG.02.00.PL/2024 and the derivative local Grant Number: 1706/PKS/ITS/2024.

References

[1] Asapana S, Sahoo P, & Aribenhi V 2015 Resistance Predictions for Asymmetrical Configurations of High-Speed Catamaran Hull Forms *SNAME 5th World Maritime Technology Conference* https://doi:10.5957/WMTC-2015-276.
[2] Rawson K J & Tupper E C 2001 *Basic Ship Theory* II Oxford: Butterworth-Heinemann, Inc.

[3] Begovic E, Bertorello C, Caldarella S, & Cassella P 2005 *Pentamaran Hull for Medium Size Fast Ferries. In Hydrodynamics VI - Theory and Applications* Taylor & Francis Group plc London UK ISBN: 04 1536 304 7.

[4] Broglia R., Zaghi S, & Di Mascio A 2011 Numerical Simulation of Interference Effects for a High-Speed Catamaran *Journal of Maritime Science and Technology* **16** Springer. https://doi:10.1007/s00773-011-0132-3.

[5] Chen M, Ouyang M, Li T, Zou M, Ye J and Tian X 2023 Numerical modelling of a catamaran float-over deck installation for a spar platform with complex hydrodynamic interactions and mechanical couplings *Ocean Engineering* **287** 115905 https://doi.org/10.1016/j.oceaneng.2023.115905

[6] Bertram V *Practical Ship Hydrodynamics* 2012 Oxford: Jordan Hill.

[7] He T, Feng D, Liu L, Wang X, & Jiang H 2022 CFD Simulation and Experimental Study on Coupled Motion Response of Ship with Tank in Beam Waves *Journal of Marine Science and Engineering* **10** 113. https://doi:10.3390/jmse10010113.

[8] Song S, Demirel Y K, & Atlar M 2019 An Investigation into the Effect of Biofouling on the Ship Hydrodynamic Characteristics Using CFD *Ocean Engineering* **175** 122-37. https://doi:10.1016/j.oceaneng.2019.01.056.

[9] Wang H, Li H, & Zhang Y 2019 Design of Green Water Quality Monitoring Vessel Based on Dual Operation Mode *IOP Conference Series: Earth and Environmental Science* **242** 032008. https://doi:10.1088/1755-1315/242/3/032008.

[10] Fitriadhy A, Razali N S, & Mansor N A 2017 Seakeeping Performance of a Rounded Hull Catamaran in Waves Using CFD Approach *Journal of Mechanical Engineering and Sciences* **11(2)**: 2601-2614. https://doi:10.15282/jmes.11.2.2017.4.0238.

[11] Kiryanto K, Santosa A W B, Samuel S, & Firdaus A 2022 Sea-Keeping Analysis of Hospital Catamarans for Handling COVID-19 Patients on Remote Islands with a Numerical Approach. *International Journal of Advanced and Applied Sciences* **9(8)**: 128-135. https://doi:10.21833/ijaas.2022.08.016.

[12] Wang X, Liu, L, Zhang Z, & Feng D 2020 Numerical Study of the Stern Flap Effect on Catamaran's Seakeeping Characteristic in Regular Head Waves *Ocean Engineering* **206** 107172. https://doi:10.1016/j.oceaneng.2020.107172.

[13] Molland A F, Turnock S R, & Hudson D A 2017 *Ship Resistance and Propulsion: Practical Estimation of Ship Propulsive Power* New York: Cambridge University Press.

[14] Wulandari A I, Utama I K A P, Rofidayanti A and Hudson D 2024 Numerical Analysis of Ship Motion of Crew Boat with Variations of Wave Period on Ship Operational Speed *CFD Letters* **16** 1–15. https://doi.org/10.37934/cfdl.16.4.115

[15] Javanmard E, Mehr J A, Ali-Lavroff J, Holloway D S, & Davis M R 2023 An Experimental Investigation of the Effect of Ride Control Systems on the Motions Response of High-Speed Catamarans in Irregular Waves *Ocean Engineering* **281**: 114899. https://doi:10.1016/j.oceaneng.2023.114899.

[16] Han X, Leira B J, Sævik S, & Ren Z 2021 Onboard Tuning of Vessel Seakeeping Model Parameters and Sea State Characteristics *Marine Structures* **78**: 102998. https://doi:10.1016/j.marstruc.2021.102998.

[17] Bhattacharyya R 1978 *Dynamics of Marine Vehicles* New York: John Wiley & Sons.

[18] Ansys 2019 *Ansys Aqwa Theory Manual.* Canonsburg: Ansys Inc.

[19] Anderson J D 1995 *Computational Fluid Dynamics: The Basics with Applications* New York: McGraw-Hill.

[20] Windyandari A, Sugeng S, Sulaiman, Ridwan M, & Yusim,A K 2023 Seakeeping Behavior of Hexagonal Catamaran Hull Form as an Alternative Geometry Design of Flat-Sided Hull Vessel *Journal of Applied Engineering Science* **21**(4) https://doi.org/10.5937/jaes0-41412

[21] Suastika K, Silaen A, Aliffrananda M H N, & Apriawan Y A 2021 Seakeeping Analysis of a Hydrofoil Supported Watercraft (Hysuwac): A Case Study *CFD Letter* **13**(5) 10-27. https://doi.org/10.37934/cfdl.13.5.1027.

[22] Murdijanto, Utama I K A P, & Jamaluddin A 2011 An Investigation Into the Resistance/Powering and Seakeeping Characteristics of River Catamaran and Trimaran *Makara Journal of Technology* **15(1)**. https://doi:10.7454/mst.v15i1.853.

IOP Conf. Series: Earth and Environmental Science **1473** (2025) 012022 doi:10.1088/1755-1315/1473/1/012022

Influence of Heave Face Plate in minimizing the heave response on FOWT type Classic SPAR: Hydrodynamic model testing

A Ghofur[1, 2], A Sulisetyono[1*], W H Nugroho[2], B Ali[2], and Z Irawanto[2]

[1]Department of Naval Architecture, Institut Teknologi Sepuluh Nopember, Surabaya, Indonesia.
[2]Research Center for Hydrodynamics Technology, Research Organization for Energy and Manufacture, National Research and Innovation Agency, Sukolilo, Surabaya, Indonesia.

*E-mail: sulisea@na.its.ac.id

Abstract. The FOWT (Floating Offshore Wind Turbine) Classic SPAR (Single Point Anchor Reservoir) hull is frequently employed in deep sea because of its elongated tubular structure. There is a significant demand for small floating offshore wind turbines (FOWTs) that can operate in water depths ranging from shallow to medium. The close proximity of the FOWT to the coast minimizes the requirement for lengthy power cables and enables efficient distribution of generated electricity to nearby coastal residences. This study discusses the implementation of a new Motion Reduction Device (MRD) called a heave face plate, which is designed to reduce the vertical motion of a Floating Offshore Wind Turbine (FOWT). This was illustrated by the implementation of gravity decay test models carried out in a hydrodynamic test tank. The experiments were conducted to investigate various changes in the ratio of heave plate diameter (ranging from 1.3 to 2.0) and variations in the height of the face plate (ranging from 0.5 m to 2.0 m). The findings indicated that the vertical dimension of the face plate had a notable influence on diminishing the vertical displacement amplitude and augmenting the natural period of heave motion (T_{N3}).

1. Introduction

SPAR (Single Point Anchor Reservoir) is an offshore platform cylindrical hull with a length of at least 100 meters. The length/diameter (L/D) ratio ranges from 5 to 8. Because of its deep draft, SPAR is often used at water depths more than 100 meters. It has been well investigated that SPAR hull type is quite simple and acceptable for use as a FOWT floater. Many studies have investigated SPAR's severe heave motions and discovered strategies to regulate it within acceptable limits. **Tao**[1] utilized a finite difference technique to investigate the effect of the span ratio (H/Dd) (H is heave plates distance) on hydrodynamic characteristics such as additional mass and damping coefficients on a vertical cylinder with two heave plates **Downie**[2] Downie noticed that at some periods, thick plates had smaller motion amplitude than porous plates placed around the bottom of SPAR. According to his observations, the damping ratio increases with increasing span length due to intense contact between different heave plate vortices, up to a critical relative distance (L/Dd).

Murdjito[3] conducted research that included the creation of a mini-SPAR physical model, as well as numerical analysis of 6-dof motions and mooring line tension due to various wave directions. Analysis found that the direction of the wave affects the motion and strain of the mooring system in varying conditions. **Ezoji**[4] tested an oscillating SPAR with solid and porous heave plates on scale models. At T=10 seconds and KC=0.2, a 20% porous heave plate provides approximately 30% more damping than a solid plate. According to his findings, the span length ratio impacts the efficacy of a heave plate in lowering the SPAR system's synchronous heave motion. Some hydrodynamic characteristics, such as added mass and damping, can be derived via either frequency or time domain analysis. To solve time domain equations, frequency domain findings might be used to calculate time domain coefficients. **Amaral**[5] conducted research on Floating Offshore Wind Turbine (FOWT) type semi-submersibles, testing the FOWT model and evaluating the dynamic coupling response. To anchor the model, a normal catenary mooring system is used. Incoming waves and wind forces are unidirectionally created. The results show that the coupling effect produces large motion, especially for pitch motion. Future researchers that do SPAR floater (cylinder) type studies should exercise caution and avoid simplifying the set-up or arrangement while conducting model experiments, particularly if all environmental loads are collinear.

Sudhakar[6] used experimental and computational approaches to study the heave motion of Classic SPAR complete with heave plates of varied diameters in regular waves. He suggests that the heave plate diameter ratio (Dhp/Ds) be at least 1.2 to 1.3 or larger to limit heave responses and get the optimal value. **Nimmy**[7] simulated the flow around a heaving plate with three distinct configurations attached to a SPAR. It was discovered that the Spar with double disk contributed the largest amount of mass, minimizing the excitation force and making the system more stable in heave.

Despite previous research, the usage of heave plates might be improved in terms of reducing heave motion and increasing natural period. To accomplish this, it is proposed that an extra face plate be installed on the heave plate known as motion reduction device. The face plate is a vertical sheet plate wrapped around outer part of heave plate with the goal of trapping some water. This amount of trapped water increases the SPAR's added mass, hence increasing natural period. To confirm aforementioned hypothesis, theoretical calculation and model testing in a hydrodynamic tank test are necessary.

2. Design Aspect of Short FOWT

According to operational conditions, qualities, and applications of several floating kinds of FOWT, such as TLP, SPAR, pontoon and semisubmersible. SPAR's floating body tubular shape provides several benefits for deep water operation, such as a limited water surface area and volume at the free surface, resulting in reduced wave forces. **Nguyen**[8] proved that the SPAR with catenary spread mooring system is inexpensive and simple to implement. Platform geometry that is simple to construct and repair, is user-friendly, and can bear environmental loads.

A short FOWT 300kW (17m draft) with a Classic SPAR hull is uncomplicated, economical, and functional. It is in great need among individuals who live in coastline regions far from public electrical sources. As a result, it is advised that the device be mounted on the SPAR body to significantly reduce the FOWT's heave motion, allowing it to operate satisfactorily in shallow and intermediate water depths. It is in high demand among those living in coastal areas far from public power sources. Many research [9],[10],[11] on FOWTs have recently been conducted experimentally using scale models, with a draft of 50m to 80m. **Liang Tian**[12] has performed calculations on FOWT station-keeping analysis. His calculations indicate that surge and pitch were the dominant motions.

2.1. Dimension Aspect for Short SPAR

Karimirad[13] has developed a FOWT Short-SPAR type that is adequately suitable for intermediate depth water. FOWT mooring system is Catenary type and it is designed for mid and low water depth. The SPAR model shows a complicated link between construction cost and total weight. The idea is to have a low total mass. Short-SPAR's total displacement (structural steel weight + ballast) is about 35percent smaller than that of the Deep SPAR. Even though short SPAR displacement is lighter, it generates electrical power equivalent to deep SPAR. **Ghofur**[14] The main problem with small FOWTs is that their natural heave period is near to the wave period, allowing them to respond excessively or suffer resonance.

Parametric study has been conducted to establish the geometric primary dimensions and weight under certain boundaries, such as blade diameter, main dimensions and upper structure weight distribution, water depth, and environmental loads. Table 1 displays the FOWT 300 kW principal dimensions.

Table 1. FOWT 300 kW Principal dimension.

Principal Dimension FOWT
Scale factor: 1:24.0

Description	Symbol		Full scale		Model scale	
Height of Floater	H	=	25.000	m	1.042	m
Diameter of Hull	B	=	4.000	m	0.167	m
Draft	T	=	17.000	m	0.708	m
Displacement	Disp.	=	218.97	ton	15.45	kg
Long. Center of Gravity from AP	LCG	=	2.000	m	0.083	m
Center of Gravity from keel	KG	=	8.297	m	0.346	m
Long.Center of Buoyancy from AP	LCB	=	2.000	m	0.083	m
Vertical Center of Buoyancy	KB	=	8.500	m	0.354	m
BM trans	BMt	=	0.059	m	0.002	m
BM long	BMl	=	0.059	m	0.002	m
Kxx	(approx.)	=	1.100	m	0.046	m
Kyy	(approx.)	=	1.100	m	0.046	m
KMt (from Carene calculation)		=	8.559	m	0.357	m
GMt		=	0.262	m	0.011	m

SPAR SPAR + Dhp SPAR + Dhp + Hpl

Figure 1. Hull of FOWT

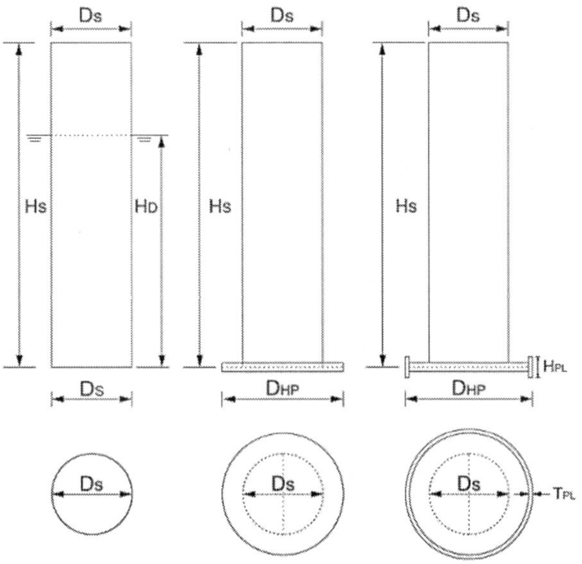

Figure 2. Hull Dimension of FOWT

Where:

Hs	= SPAR height.	Dhp	= Heave plate diameter.
Hd	= SPAR draft.	Hpl	= Face plate height.
Ds	= SPAR (Hull) diameter.	Tpl	= Face plate thickness.

The heave plate diameters vary, starting at 5.2m to 8.0m (full size). Diameter heave plate and face plate height in model scale can be seen table 2 and figure 2.

Table 2. Diameter of heave plate and height of face plate.

Type	Ds (cm)	Dhp (cm)	Diameter Ratio (Dhp/Ds)	Hpl (cm)	Tpl (cm)
A	16.67	-	-	-	1.04
B1	16.67	21.67	1.30	-	1.04
B2	16.67	21.67	1.30	4.17	1.04
C1	16.67	26.67	1.60	-	1.04
C2	16.67	26.67	1.60	4.17	1.04
D1	16.67	33.33	2.00	-	1.04
D2	16.67	33.33	2.00	4.17	1.04
C3	16.67	26.67	1.60	2.08	1.04
C4	16.67	26.67	1.60	5.21	1.04
C5	16.67	26.67	1.60	6.25	1.04
C6	16.67	26.67	1.60	7.29	1.04
C7	16.67	26.67	1.60	8.33	1.04

The model of the FOWT hull is made of tubular PVC. Heave plates and face plates are composed of fiber-coated multiplex wood. Hull, heave plate, and face plate are bonded and reinforced in some places to guarantee that the structure remains robust during the test. Cremona tower, turbine blades and deck are made of coated-wooden, and the scale factor is 1:24. See figure 3.

IOP Conf. Series: Earth and Environmental Science **1473** (2025) 012022 doi:10.1088/1755-1315/1473/1/012022

FOWT hull | Deck, Cremona, & Turbin blade | Heave plate & face plate

Figure 3. FOWT physical model

3. Analysis of Decay Test

Figure 4 depicts a time-traces plot (displacement vs time). Natural period, added mass, and damping quantities may be determined using the mathematical approach indicated below, which is based on graphical representations.

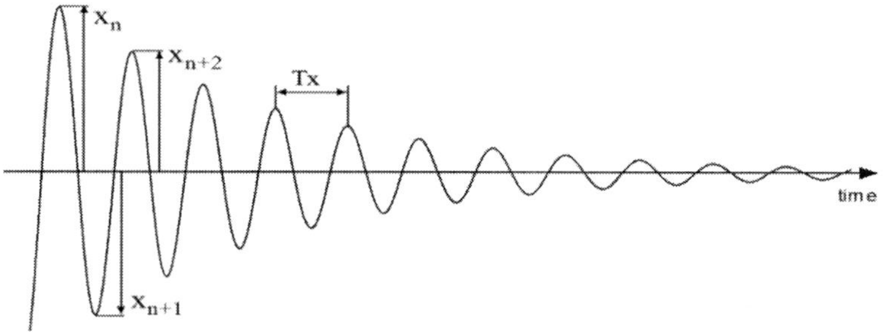

Figure 4. Time vs heave motion (Free decay motion)

3.1. Damping

The decrease in heave magnitude in decay motion x can be expressed as a polynomial function of average amplitude x_m and represented as below:

$$\Delta x = a x_m + b x_m^2 \tag{3.1}$$

Some references, **Chakrabarti**[15] shows that the mathematical motion formulation for a floating body with nonlinear damping components yields good results. The equation for free decay motion can be represented as below:

$$(M + \Delta M)\frac{d^2x}{dt} + B_1\frac{dx}{dt} + B_2\frac{dx}{dt}\left|\frac{dx}{dt}\right| + kx = 0 \tag{3.2}$$

Vertical translational mass is represented as (M+ΔM). B_1 and B_2 are linear and quadratic damping coefficients, respectively, with k being the restoring force. The dissipating energy in heave decay for every half heave period ($\frac{1}{2}$T) is integral to (3.2) [16]. It's found:

$$B_1 \frac{\pi^2}{T} x'^2 + B_2 \frac{16}{3T^2} x'^3 - kx'\Delta x = 0 \tag{3.3}$$

$$\Delta x = \frac{1}{k}\left(B_1 \frac{\pi^2}{T}\right) x' + \frac{1}{k}\left(B_2 \frac{16\pi^2}{3T^2}\right) x'^2 \tag{3.4}$$

(3.4) can be linearized in a similar way as (3.1). If x' = x_m, then 'a' and 'b' are calculated using the decrement heave decay equation in (3.1), respectively.

$$a = \frac{1}{k}\left(B_1 \frac{\pi^2}{T}\right) \quad \text{and} \quad b = \frac{1}{k}\left(B_2 \frac{16\pi^2}{3T^2}\right) \tag{3.5}$$

If it is known m = (M+ ΔM) and k = ω^2 m, then (3.5) B_1 and B_2 can be expressed as below:

$$B_1 = a\,\frac{4}{T} \quad \text{and} \quad B_2 = b\,\frac{3}{4} m \tag{3.6}$$

The values 'a' and 'b' reflect the magnitude of the linear and quadratic damping coefficients in that mode of motion. The decay test may be used to determine the values of B_1 and B_2 based upon a and b values.

3.2. Added Mass

The heave added mass A_{33} of a semi-infinite cylinder floating in water as a SPAR without heave plate may be calculated using (3.7). **Newman**[17]:

$$A_{33} = (2/3)\rho\pi R_s^3 \qquad \text{or} \quad A_{33} = 2.0944\rho R_s^3 \tag{3.7}$$

Where A_{33} is heave added mass, ρ is mass density, R_s is SPAR cylinder radius.

Added mass of SPAR with a heave plate attached to the bottom of the cylinder has been proposed by **Philipe**[18] theoretically formulated in (3.8).

$$A_{33} = \frac{1}{3}\rho D_{hp}^3 - \frac{1}{6}\rho\left[D_{hp}^3 - \left(D_{hp}^2 - D_s^2\right)\left(D_{hp}^2 - D_s^2\right)^{1/2}\right] \tag{3.8}$$

Experimentally, added mass (A_{33}) may be computed by (3.10), as follows:

$$T_{N3} = 2\pi \left[\frac{M + A_{33}}{C_{33}}\right]^{1/2} \tag{3.9}$$

Re-compose (3.9), then added mass is:

$$A_{33} = C_{33}\left(\frac{T_{N3}}{2\pi}\right)^2 - M \tag{3.10}$$

Where:

T_{N3}	= Heave natural period (sec).	R	= SPAR radius (m)
M	= SPAR mass (kg).	Z	= Heave amplitude (m)
C_{33} z	= Restoring force = ρ g A_{wp} z (N).	ρ	= Specific density (kg/m^3)
A_{wp}	= Water plane area = πr^2 (m^2).	g	= Gravity (m/s^2)

4. Experimental Activities, Results and Analysis

4.1. Tank Facilities

The hydrodynamic model test was being held at Towing Tank. It is part of the Indonesian Hydrodynamic Laboratory - BRIN (National Research and Innovation Agency), as shown in Figure 5.

- *Towing Tank Dimensions:*
  Length x Breadth x Depth (234.5m x 11m x 5.5m)
- *Carriage:*
  Man-operated, Max Accel. 1 m/s2, Max Speed 9 m/s
- *Wave Generator:*
  Hydraulically driven dual dry back flap type capable of generating regular and irregular wave significant wave height up to 0.5 m.
- *Wave absorber:*
  Fixed grid on circular arc stainless plates with moveable section in front of the port.

Figure 5. Towing Tank facilities, PRTH – BRIN, Surabaya. Indonesia.

Prior to testing, the FOWT Model was weighed or adjusted for displacement, Vertical Centre of Gravity (VCG), and radii of gyration. Next, instrumentation inside the model, Buoyancy draft check for even-keel position, and Inclining test for metacentric (GM) setting were performed. The following instruments were utilized to collect data: Gyro3D and QUALYSIS (motion capture camera). Figure 6 shows setup the FOWT model for decay test in Towing tank.

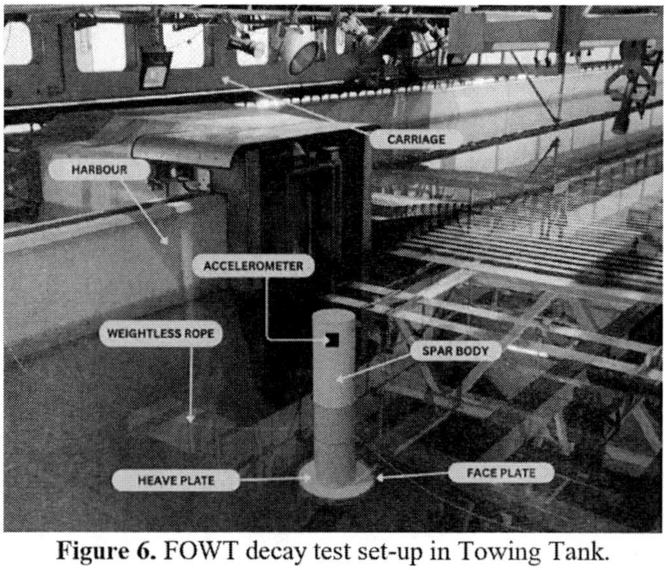

Figure 6. FOWT decay test set-up in Towing Tank.

2 (two) light thin rope (5 meter each) are tied to the FOWT model as shown in Figure 6 within each end of the rope attached to FOWT body (close to VCG position) and stretch to the harbor wall. The light thin rope configuration is given to maintain or limit the model's movement in a horizontal direction (drifting) so that it does not move away from the measurement target.

During the experiment, it is important to supply a known starting heave displacement to ensure that model rotation is limited, particularly in roll/pitch. When the model reaches equilibrium (at rest), an initial displacement (+/- 5cm) is applied to the center point of the SPAR cylinder in a downward direction parallel to the cylinder axis, ensuring that only heave movement occurs without any coupling movement. The experiment was performed two or three times to acquire the best results.

4.2. Heave Natural Period (T_{N33}) results

There are 12 test programs for all the model type (A to D). The experimental outputs were represented as amplitude vs time, Figures (7–11).

Table 3. Heave natural period (T_{N33})

Model type	D_{HP}/D_S (m)	T_{N33} (sec) model	Diff (%)
A	1.0	1.7659	-
B1	1.3	1.8669	5.72
B2	1.3	1.9329	9.46
C1	1.6	2.1339	20.84
C2	1.6	2.1621	22.44
D1	2.0	2.3666	34.02
D2	2.0	2.4666	39.68
C3	1.6	2.1539	21.97
C4	1.6	2.1670	22.71
C5	1.6	2.1686	22.81
C6	1.6	2.1711	22.95
C7	1.6	2.1776	23.32

$$Diff = \frac{T_{33\,type} - T_{33A}}{T_{33A}} x100\% \quad (4.1)$$

The primary goal of using a heave plate and a face plate on the FOWT with a Classic SPAR hull type is to reduce the amplitude of heave response while also lowering its natural frequencies, allowing the FOWT to function properly when subjected to environmental loads (wind, wave, and current) at its operational area.

Table 3 shows natural heave period (T_{N33}) of type B1 increased by 5.7% as compared to type A. Similarly, types C1 and D1 showed a rise of 21% and 34%, respectively. For types B2 and D2, the presence of a face plate (H_{FP} = 1.0m) clearly adds 5% when compared to the absence of one. For variations in face plate height from C2 (Hpl=1m) to C7 (Hpl=2m), T_{N33} steadily rose from 22.44% to 23.32%. It is obviously, that the height of the face plate contributes positively to raising the T_{N33}.

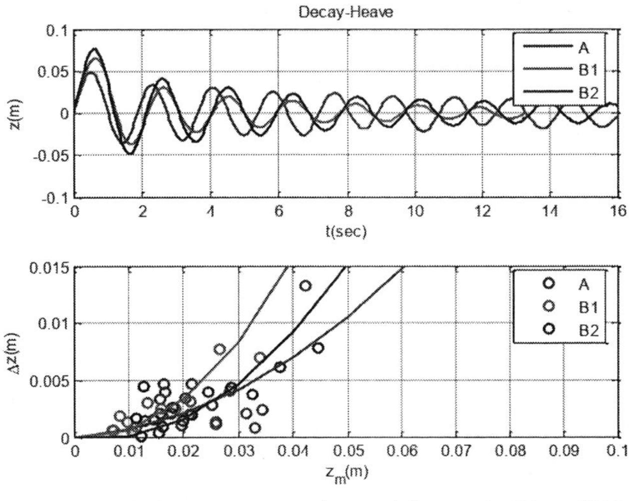

Figure 7. Decay test results model type A, B1 and B2

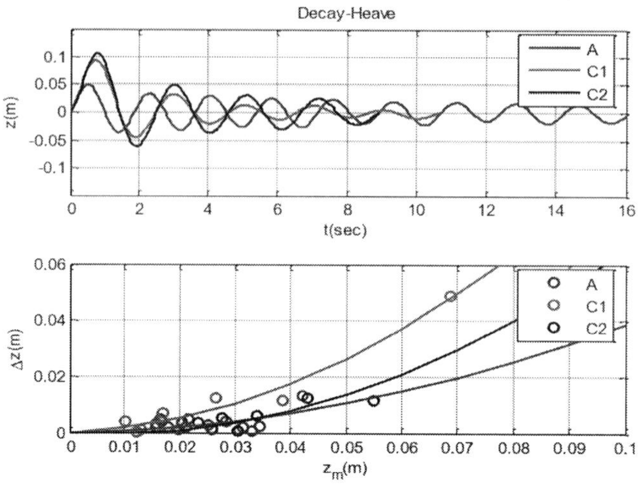

Figure 8. Decay test results model type A, C1 and C2

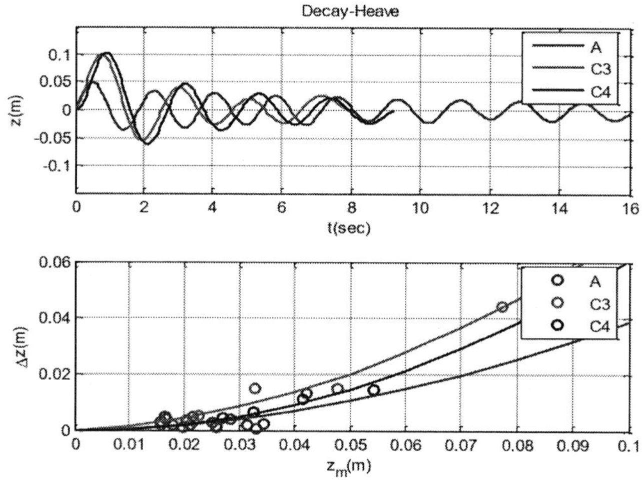

Figure 9. Decay test results model type A, C3 and C4

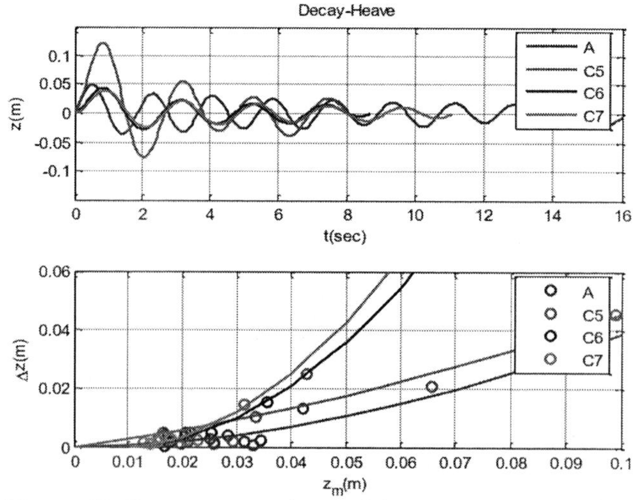

Figure 10. Decay test results model type A, C5, C6 and C7

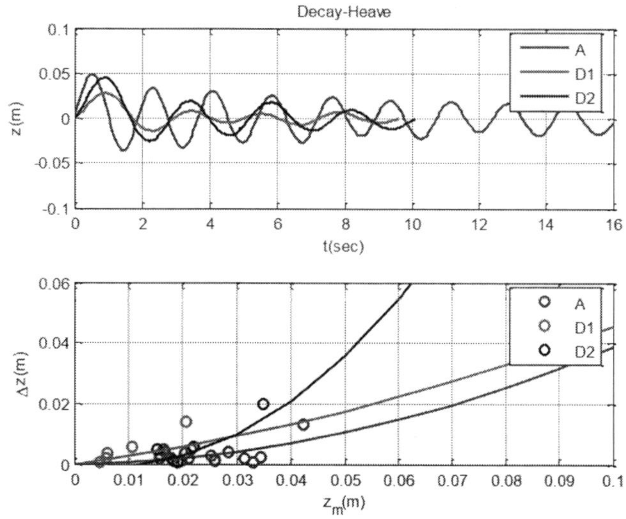

Figure 11. Decay test results model type A, D1 and D2

4.3. Added Mass results

With the same displacement, the SPAR hull shape has a smaller WPA (Water Plane Area) than the ship hull form (mono hull), hence our SPAR's heave spring factor (C33) is fairly small, at around 215 N/m. The added mass for bare hull SPAR (type A) was estimated using (3.7) **Newman**[17], whereas for SPAR complete with heave plate (B1, C1, and D1) was calculated using (3.8) **Philipe**[18].

Meanwhile, for SPAR with heave plates and face plates fitted (B2, C2, C3, C4, C5, C6, C7, and D2), formulation is not currently available, thus a formulation must be developed in the future to get realistic results. In this scenario, further study will focus on the heave added mass formulation for cylinders with a heave plate and a face plate. From the heave decay experimental results, added mass (A_{33}) could be computed using (3.10).
See table 4 heave added mass (A_{33}) results both theoretically results and experimental results.

Table 4. Heave added mass (A_{33})

No	Model type	D_{HP}/D_S	T_{N33} (sec)	C_{33} (N/m) $= \rho$ g Aw	A_{33} Experiment (kg)	A_{33} Theory (kg)	Diff. (%)
1	A	1.0	1.7659	214.878	1.979	2.014	1.74
2	B1	1.3	1.8669	214.878	3.517	3.56935	1.46
3	B2	1.3	1.9329	214.878	4.881	-	-
4	C1	1.6	2.1339	214.878	9.331	8.993	3.76
5	C2	1.6	2.1621	214.878	9.990	-	-
6	C3	1.6	2.1539	214.878	9.798	-	-
7	C4	1.6	2.1670	214.878	10.105	-	-
8	C5	1.6	2.1686	214.878	10.144	-	-
9	C6	1.6	2.1711	214.878	10.202	-	-
10	C7	1.6	2.1776	214.878	10.356	-	-
11	D1	2.0	2.3666	214.878	15.632	16.431	4.86
12	D2	2.0	2.4666	214.878	17.663	-	-

$$Diff = \frac{A_{33\,Experiment} - A_{33\,Theory}}{A_{33\,Theory}} x100\% \tag{4.2}$$

Heave added mass (A_{33}) calculating techniques between theoretical and experimental data, both type A and type B produce good results. Meanwhile, for type C1 (D_{HP}/D_S=1.6), the difference is 3.76%. For type D1 (D_{HP}/D_S =2.0), the difference is 4.86%. The gap between computation approaches becomes more pronounced as the diameter ratio (D_{HP}/D_S) increases. This might be because theoretical method gives a formulation only based on heave plate diameter (volume of fluid coverage), however the flow behavior around heave plate varies with diameter, resulting in varying additional mass and damping effects.

For future research, heave added mass formulation will be developed that incorporates not only heave plate diameter, but also height of face plate and other characteristics to produce more precise results.

4.4. Damping Results

4.4.1. Critical Damping

Equation of Critical heave damping is written below:

$$B_{33_{crit}} = 2\,[(m + A_{33}).\,C_{33}]^{1/2} \tag{4.3}$$

Critical damping allows the system to recover to its equilibrium position without oscillations or vibrations. This differs from underdamped (less damping) systems, which oscillate multiple times before stopping. Table 5 shows critical damping results from numerical analysis, theory, and model tests using the above formulation.

Table 5. Critical damping ($B_{33_{crit}}$)

Type	Ratio Dhp/Ds	B3crit (Ns/m)		diff. (%)
		Theory	Experiment	
A	1.0	122.7991	120.7699	1.68
B1	1.3	128.2734	127.6816	0.46
B2	1.3	-	132.1916	-
C1	1.6	142.1497	145.9449	2.60
C2	1.6	-	147.8717	-
D1	2.0	168.5304	161.8621	4.12
D2	2.0	-	168.7037	-
C3	1.6	-	147.3132	-
C4	1.6	-	148.2068	-
C5	1.6	-	148.3185	-
C6	1.6	-	148.4860	-
C7	1.6	-	148.9328	-

According to the experimental data, critical damping increases as the diameter ratio (Dhp/Ds) increases, with critical damping rising from type A to type D from 120.77 Ns/m to 168.70 Ns/m. Critical damping is calculated both theoretically and empirically, with consistent results; however, the difference in results rises as diameter ratio (Dhp/Ds) increases. The percentages are 0.46% for type B1 and 4.1% for type D1. Table 5 explains that as diameter of heave plate rises, so does heave critical damping, which means that SPAR for moving up and down (in heave direction) becomes more inert. Similarly, raising the height of the face plate increases critical damping, but not much.

4.4.2. Damping Coefficient

Using (3.5) and (3.6) Damping coefficients can be calculated from the heave decay test results, as indicated in table 6, where B_1 and B_2 are the linear and quadratic damping component respectively. The damping coefficients rise with each change in diameter ratio (D_{HP}/D_S), both linear (B_1) and quadratic (B_2).

Table 6. Linear and Quadratic damping coefficients

Type	D_{HP}/D_S	a	b	B_1	B_2
A	1.0	3.5422	0.0328	8.0237	0.4288
B1	1.3	11.2660	0.0593	24.1382	0.8437
B2	1.3	7.5507	0.0724	15.6260	1.1042
C1	1.6	8.9475	0.0810	16.7720	1.5057
C2	1.6	7.6792	0.1142	14.2070	2.1792
D1	2.0	20.0587	0.2745	33.9028	6.3997
D2	2.0	25.4407	0.3243	41.2557	8.0547
C3	1.6	6.0045	0.1018	11.1508	1.9279
C4	1.6	6.3631	0.0288	11.7456	0.5521
C5	1.6	2.0693	0.2483	3.8168	4.7669
C6	1.6	19.5191	0.2624	35.9623	5.0490
C7	1.6	22.8734	0.2919	42.0159	5.6503

This rise also applies to the installation of face plates, as raising the height of the face plate raises the damping coefficients (B_1 and B_2). Figures 7 to 11 show that when the diameter ratio increases, so does the gradient damping coefficient (upward). This also applies to raising the height of the face plate (C3 to C7), which in return increases the damping coefficients.

5. Conclusion

A heave decay test was performed on a SPAR Classic FOWT in a hydrodynamic test tank. The SPAR model has 1:24 scale factor, within hull length 1.04 m, diameter 0.167m, and draft 0.708m. A heave plate with diameter ratios (D_{HP}/D_S) of 1.0, 1.3, 1.6, and 2.0 is attached to the bottom, and a face plate has been put around it. Here are some conclusions:

- Heave natural period (T_{N33}). Type B1 increased by 5.7% Compared to type A. Similarly, types C1 and D1 showed a rise of 21% and 34%, respectively. Presence of a face plate (H_{FP} = 1.0m) clearly adds 5% when compared to the absence of one.
- Heave added mass (A_{33}) calculating techniques between theoretical calculations and experimental data. Both type A and type B produce good results. Meanwhile, for type C1 (D_{HP}/D_S=1.6), the difference is 3.7%. For type D1 (D_{HP}/D_S =2.0), the difference is 7.5%. The gap between computation approaches becomes more pronounced as the diameter ratio (D_{HP}/D_S) increases.
- Critical damping ($B33_{crit}$) is mainly impacted by mass and stiffness, the value is about 120.7 Ns/m (type A) and 168.7 Ns/m (type D).
- Damping coefficients (B_1 and B_2). Increasing the diameter ratio (D_{HP}/D_S) results in increased damping coefficients, both linear (B_1) and quadratic (B_2). This rise also applies to the installation of face plates, as increasing the height of the face plate increases the damping coefficients (B1 and B2).
- Future work. Station keeping model testing is required to evaluate the performance of the FOWT design with a SPAR Classic hull type that includes a heave plate and a face plate. Test result data is critical for future study.

References

[1] L. Tao, B. Molin, Y. M. Scolan, and K. Thiagarajan, "Spacing effects on hydrodynamics of heave plates on offshore structures," *J. Fluids Struct.*, **vol. 23**, no. 8, pp. 1119–1136, 2007, doi: 10.1016/j.jfluidstructs.2007.03.004.

[2] M. J. Downie, J. M. R. Graham, C. Hall, A. Incecik, and I. Nygaard, "An experimental investigation of motion control devices for truss spars," *Mar. Struct.*, **vol. 13**, no. 2, pp. 75–90, 2000, doi: 10.1016/S0951-8339(00)00010-1.

[3] Murdjito *et al.*, "Experimental study on heave damping due to the heave plate addition on the SPAR keel," *IOP Conf. Ser. Earth Environ. Sci.*, **vol. 649**, no. 1, 2021, doi: 10.1088/1755-1315/649/1/012052.

[4] M. Ezoji, N. Shabakhty, and L. Tao, "Hydrodynamic damping of solid and perforated heave plates oscillating at low KC number based on experimental data: A review," *Ocean Eng.*, **vol. 253**, no. June, 2022, doi: 10.1016/j.oceaneng.2022.111247.
[5] G. A. Amaral *et al.*, "Seakeeping tests of a fowt in wind and waves: An analysis of dynamic coupling effects and their impact on the predictions of pitch motion response," *J. Mar. Sci. Eng.*, **vol. 9**, no. 2, pp. 1–19, 2021, doi: 10.3390/jmse9020179.
[6] S. Sudhakar, "Influence Of Heave Plate On Hydrodynamic Response Of Spar," in *Proceedings of the ASME 2011 30th International Conference on Ocean, Offshore and Arctic Engineering OMAE2011*, OMAE, 2011. doi: https://doi.org/10.1115/OMAE2011-49565.
[7] S. Nallayarasu, "Damping characteristics of heave plates attached to spar hull," Omae2012-83290, no. 2007, pp. 1–12, 2017.
[8] V. N. Dinh and B. Basu, "On the modeling of spar-type floating offshore wind turbines," *Key Eng. Mater.*, **vol. 569–570**, pp. 636–643, 2013, doi: 10.4028/www.scientific.net/KEM.569-570.636.
[9] T. R. Wen, K. Wang, Z. Cheng, and M. C. Ong, "Spar-type vertical-axis wind turbines in moderate water depth: A feasibility study," *Energies*, **vol. 11**, no. 3, 2018, doi: 10.3390/en11030555.
[10] T. Utsunomiya *et al.*, "At sea experiment of a hybrid spar for floating offshore wind turbine using 1/10-scale model," *J. Offshore Mech. Arct. Eng.*, **vol. 135**, no. 3, pp. 34503-1-34503–8, 2013, doi: 10.1115/1.4024148.
[11] Z. Jiang *et al.*, "Feasibility studies of a novel spar-type floating wind turbine for moderate water depths: Hydrodynamic perspective with model test," *Ocean Eng.*, **vol. 233**, no. May 2020, p. 109070, 2021, doi: 10.1016/j.oceaneng.2021.109070.
[12] X. liang Tian, J. ren Xiao, H. xue Liu, B. rong Wen, and Z. ke Peng, "A Novel Dynamics Analysis Method for Spar-Type Floating Offshore Wind Turbine," *China Ocean Eng.*, **vol. 34**, no. 1, pp. 99–109, 2020, doi: 10.1007/s13344-020-0010-z.
[13] M. Karimirad and T. Moan, "Feasibility of the application of a spar-type wind turbine at a moderate water depth," *Energy Procedia*, **vol. 24**, no. 1876, pp. 340–350, 2012, doi: 10.1016/j.egypro. 2012.06.117.
[14] A. Ghofur, A. Sulisetyono, W. H. Nugroho, B. Ali, and M. Muria Armansyah, "The effect of heave-face plates on FOWT heave motion: Fluid dynamic analysis," *BIO Web Conf.*, **vol. 89**, pp. 1–15, 2024, doi: 10.1051/bioconf/20248910002.
[15] S. K. Chakrabarti, *Handbook of Offshore Engineering*, **vol. I**. Oxford: Elsevier Ltd, 2005. [Online]. Available: http://dx.doi.org/10.1016/B978-0-08-044381-2.50022-9
[16] A. Ghofur, B. Ali, Erwandi, and A. Bisri, "50kW mini floating offshore wind turbine: Station keeping model testing," *AIP Conf. Proc.*, **vol. 3069**, no. 1, 2024, doi: 10.1063/5.0205790.
[17] J. N. Newman, *Marine Hydrodynamics*, 40th Anniv. London, England: The MIT Press Cambridge, Massachusetts London, England. [Online]. Available: http://library.oapen.org/handle/ 20.500.12657/26039
[18] C. Clément, P. Bozonnet, G. Vinay, P. Pagnier, A. B. Nadal, and J. Réveillon, "Evaluation of Morison approach with CFD modelling on a surface-piercing cylinder towards the investigation of FOWT Hydrodynamics," *Ocean Eng.*, **vol. 251**, no. February, p. 111042, 2022, doi: 10.1016/j.oceaneng.2022.111042.

Acknowledgments
The authors express gratitude to the Ministry of Education, Culture, Research, and Technology for financial assistance under the Doctor Research Program with no. 038/E5/PG.02.00.PL/2024, as well as the National Research and Innovation Agency for providing research facilities and financial support. For their assistance with the model test, I would like to thank my colleagues at the Indonesian hydrodynamics laboratory (BTH-BRIN).

Reducing Ship Emissions With Innovations in Free Rotation Propellers

A Winarno[1*], U Prayogi[1], Dwisetiono[1], F M Felayati[1] and D D Kurniawan[1]

[1]Department of Marine Engineering, Hang Tuah University, Surabaya, Indonesia, 60111

*E-mail: arif.winarno@hangtuah.ac.id

Abstract. The working principle of a free rotating propeller is that the propeller rotates freely on the main propeller axis. This propeller utilizes free flow from the main propeller, thereby increasing energy efficiency. Apart from that, the free rotating propeller can also increase thrust which is useful for helping the ship's movement and speed. The thrust produced by the propeller can be influenced by the pitch angle of the propeller. Therefore, in this research, an analysis of variations in the propeller pitch angle was carried out with the aim of determining the value of the propeller thrust force. The method used is to use computational fluid dynamic (CFD), by varying the pitch angle of the propeller. The pitch angles used are 20°, 25°, 30°, 35°, 40°. From the modeling results, the largest thrust is obtained from the free-rotating propeller with an angle of 25°, and the thrust force is 43.9872 KN.

1. Introduction

Propulsion is a propulsion system for a ship consisting of the main engine, shaft and propeller. This system works in harmony and synchronization which aims to drive the ship's speed so that it can move from one place to another [1]. Based on the propulsion system, propellers are divided into 2 types, namely propellers that are driven by engines and propellers that are not driven by engines or also called propellers that are free to rotate [2].

The free rotating propeller was first introduced by professor grim from the German Federal Minister of Research and Technology in 1967 and began to be installed in 1980 on commercial ships, the free-rotating propeller has a working principle by rotating freely on the main propeller shaft with Utilizing the free flow of the main propeller will thus be able to increase energy efficiency. The advantage of a swivel-free propeller is that it can reutilize the dynamic flow energy generated by the lost flow of the main propeller, which can result in about 10% increase in efficiency and can reduce the vibration level of the ship body [4].

Research on [5] conducted that the results of the analysis for the maximum service speed of 14 knots show that the total resistance of the ship is 286. 75 kN and the power requirement is 2950.31 kW. While with the calculation of Holtrop mathematical calculation resulted in a total resistance of 256.59 kN, there is a difference of 10.52% with the results of computer modeling. Research on [6] conducted computational fluid dynamics analysis (CFD) is employed to investigate the characteristic of each propeller model by using multi reference frame (MRF) approach. The CFD results showed that the highest thrust value of the Z=3 is 172.38 kN.

In previous research, the DPS IX tugboat has been installed with a rotary free propeller and obtained a thrust of 40.847 KN, but in that study it is still necessary to conduct further research on the pitch angle of the rotary free propeller which is expected to increase the greater thrust. [7]. The pitch angle setting range is between 0° to 55°, the greater the pitch angle, the greater the resistance generated by the blade [8]. For this reason, the diameter of the free rotating propeller must be made larger than the main propeller [3].

Altought, research on reducing ship emissions with innovations in free rotation propellers rarely.So, a research was conducted to form the basis of the thesis which later could be known the effect of changes in *pitch variations* with angles of 20°, 25°, 30°, 35°, 40° on the *thrust* produced by the free-swivel propeller.

2. Methodology

Free rotating propeller or Free Rotating Propeller was first introduced by Professor Grim from the German Federal Minister of Research and Technology, that's why this propeller is also known as Grim's Vane Wheel [3].

This free-rotating propeller was introduced in 1967 and began to be installed on commercial ships in 1980. Can be seen at figure 1. The working principle of this propeller is to rotate freely on the main propeller shaft by utilizing the free flow of the main propeller thereby increasing energy efficiency. For this reason, the diameter of the free spin propeller must be made larger than the main propeller, from the results of research and testing on experimental tanks, generally the diameter of the free spin propeller is 20% larger than the main propeller [2].

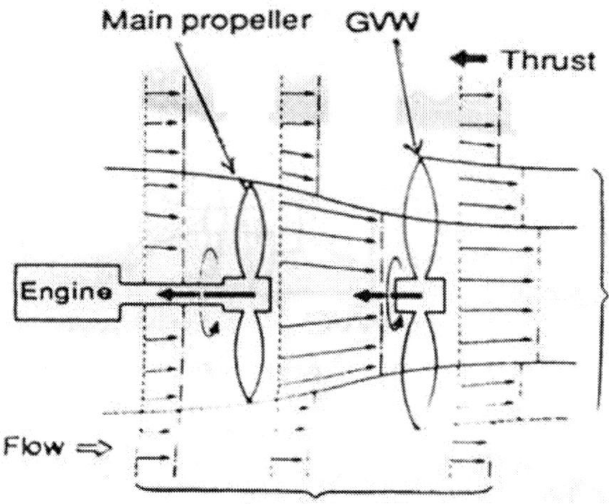

Figure 1. Velocity distribution at each crosssectional section principle of GVW [4].

2.1. *Free Rotating Propeller Function*

The function of this free-rotating propeller is that it can act as a turbine as well as an ordinary propeller. As a turbine the propeller freely rotates utilizing the free flow rotational energy (Slip Stream) from the main propeller, whereas as an ordinary propeller this propeller can convert energy from free flow into thrust energy [4].

2.2. *Propeller Pitch*

Propeller pitch is the translation distance traveled by the propeller in one revolution. Propellers with a fixed pitch (fixed pitch propeller, FPP) have the advantages of high torque yield, more economical fuel consumption, minimal noise or vibration, and minimal cavitation, usually designed individually so that they have special characteristics for certain ships will have optimum efficiency values. The propeller load characteristics can be represented visually using various coefficients, which are depicted as different sizes. The diagram illustrates Torque and Thrust as functions of speed, as shown in equations (1), (2), and (3). The propeller's characteristics include the Thrust coefficient (KT), the Torque coefficient (KQ), and the Advance coefficient (J) [5-8].

$$(KT) = \frac{T}{Þn^2} \quad (1)$$

$$(KQ) = \frac{Q}{Þn^2 D^5} \quad (2)$$

$$J = \frac{Va}{nD} \quad (3)$$

Where :

- r = density of fluid (Fluid Density)
- D = propeller diameter
- n = propeller rotation
- Va = advanced speed
- T = propeller thrust
- Q = propeller torque

2.3. Ship Amplification

Until now, 50 ships of the Germanischer Lloyd (GL) class have used this type of propeller and several ships of the Nippon Kaiji Kjokai (NKK) class have also installed it. Can be seen at figure 2. One of the ships built by Ishikawajima Haima Heavy Industries (IHI) Ltd, namely the 236,604 DWT TY DRACO tanker uses a freely rotating propeller with 9 propeller blades with a diameter of 11.64 M, while the diameter of the main propeller is 9.5 M [4].

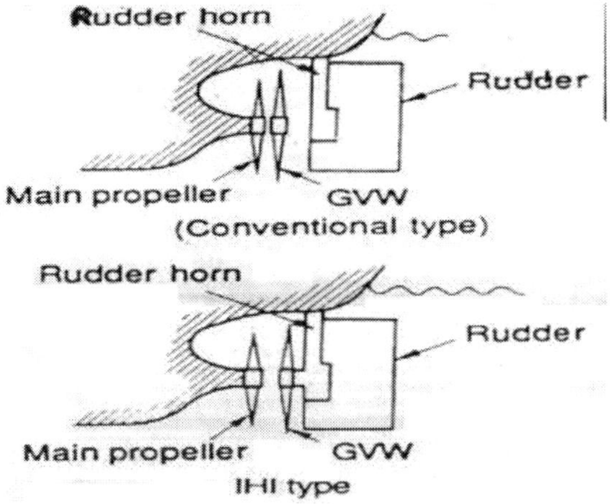

Figure 2. Propeller installation [4].

This ship uses the INI-Sulzer 7RTA84M main engine with a power of 27,240 BHP and 66 RPM rotation. As explained above the free spin propeller is mounted on the main shaft behind the main propeller using roller bearings, whereas on the TY DRACO ship above, the free spin propeller is mounted on the Rudder Post so that the free spin propeller can move freely between the main propeller room and the rudder, as shown in Figure 2 [4].

2.4. Tests That Have Been Done

To determine the effect of this propeller on the propeller shaft and bearings and other aspects that arise, several experiments and tests were carried out by Germanischer Lloyd (LR) and Nippon Kaiji Kjokai (NKK). This test is carried out on ships during operation [4].

2.5. Ship Prisoner

Resistance of the ship (resistance) at a certain speed is the fluid force acting on the ship in such a way that it opposes the movement of the ship [9]. This resistance is equal to the component of the fluid force acting parallel to the ship's axis of motion. Resistance is a term often used in ship hydrodynamics. Some resistance curves for objects moving in a perfect fluid and the fluid has a viscosity of the horizontal coordinates are Froude numbers. Can be seen in equation (4). [10-12].

$$Fn = \frac{V}{\sqrt{g.L}} \tag{4}$$

With:

- Fn = Froude Number
- V = Speed (m/s)
- L = Length of the object (m)
- g = acceleration due to gravity (m/s^2)

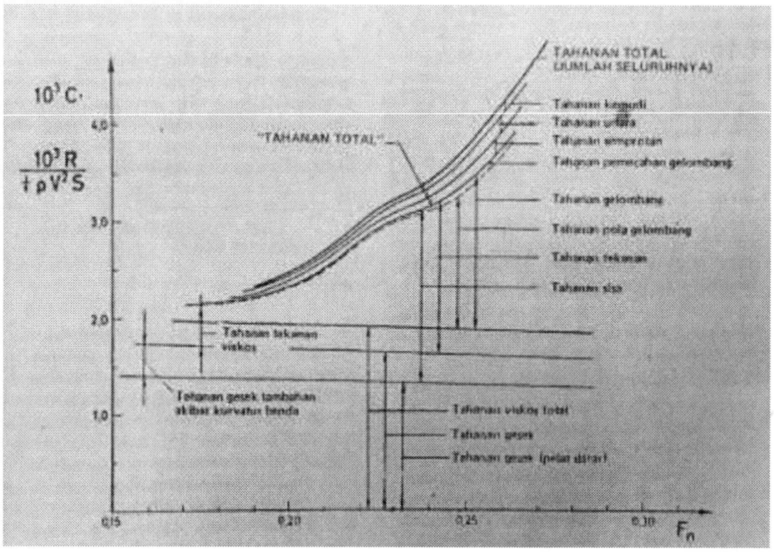

Figure 3. Ship specific resistance components [13].

The total resistance, given the notation Rt can be broken down into a number of different components which are caused by various causes and interact with each other. The amount of this total resistance is the sum of several resistances on the ship, which includes frictional resistance, residual resistance, viscous resistance, pressure resistance, wave resistance, air resistance and appendage resistance [13]. Can be seen at figure 3. The total resistance of the ship can be obtained by the following equation (5).

$$Rt = ½ \text{ x } Ct \text{ x } \rho \text{ x } Vs^2 \text{ x } S \tag{5}$$

With:

- Ct = Coefficient of total ship resistance
- S = Wet surface area of the hull (m^2)
- ρ = Density (kg/m^3)
- Vs = Ship speed (m/s)

2.6. Computational Fluid Dynamics

Computational Fluid Dynamics (CFD) is a numerical technique that uses computer-based calculations to analyze dimensions, areas, and volumes by breaking down a fluid-filled space into discrete elements. This process, known as meshing, involves dividing the space into smaller sections called cells. The

IOP Conf. Series: Earth and Environmental Science **1473** (2025) 012023 doi:10.1088/1755-1315/1473/1/012023

software or application then performs calculations on these individual cells to evaluate the fluid dynamics within the space.

These computational controls along with other computational controls constitute the aforementioned spatial division or meshing. Later, at each calculation control point, calculations will be carried out by the application with predetermined domain boundaries and boundary conditions [14]. This principle is widely used in the calculation process using computer computing assistance [15]. the use of CFD is generally used to predict flow and heat, mass transfer, phase changes such as in the process of *melting*, condensing and boiling, chemical reactions such as combustion.mechanical movements such as pistons and fans, stress and support on solid objects, and electromagnetic waves.

2.7. Ansys

Ansys is a finite element analysis software with the ability to analyze a wide range of problems [2]. *Ansys* is able to solve differential equations in a way.

2.8. Free Rotation Propeller Determination

Ansys is a finite element analysis software with the ability to analyze a wide range of problems [2]. *Ansys* is able to solve differential equations in a way.

The free-rotating propeller diameter is determined to be 1.2 times the propeller diameter. The radius of the turbine section, selected according to *the slipstream* generated by the main propeller is 0.85 times the radius of the free-rotating propeller. To be efficient, the blades of the propeller blades must be freely rotated as large as possible (Grim, 1980) 9 blades selected. [16] recommends the optimal rotational speed of the freewheeling propeller to be 30 to 50 % of the main propeller's rotational speed, the freewheeling distance of the propeller being ¼ the diameter of the main propeller [16].

The chosen airfoil is NACA 66 because it has low drag and good cavitation [16]. The distribution of pitch, camber, and chords was designed according to research conducted by [16].

Below are the dimensions of the free-swivel propeller obtained

1. Type = NACA 66 airfoils
2. Blades = 9
3. N = 123rpm
4. D = 1.848 m

2.9. Model Design

The data above is used for modeling and drawn with the Autocad program for ship hulls, Prop Cad and Rhino Ceros are used to draw propellers. Previously in 2D and then in 3D [17].

2.10. Ship design modeling, Main propeller and free-roving propeller in 3D

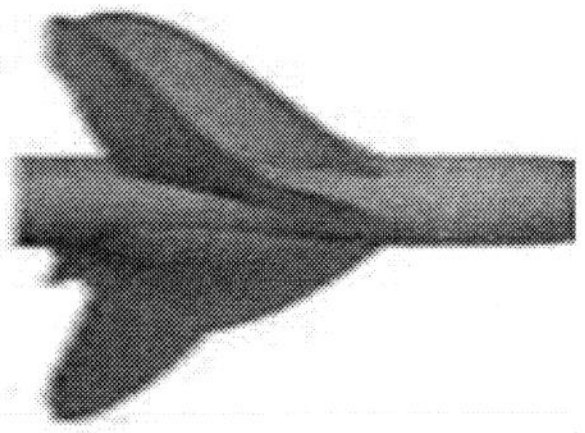

Figure 4. 3D *propeller pitch angle* 20°.

IOP Conf. Series: Earth and Environmental Science **1473** (2025) 012023 doi:10.1088/1755-1315/1473/1/012023

Figure 5. 3D *propeller pitch angle* 25°.

Figure 6. 3D *propeller pitch angle* 30°.

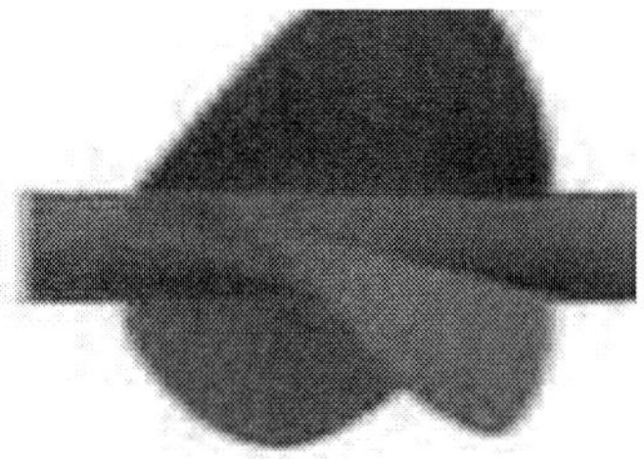

Figure 7. 3D *propeller pitch angle* 40°.

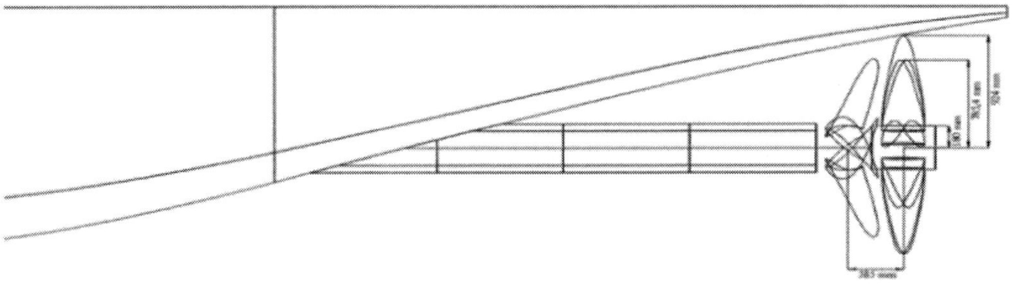

Figure 8. combined rear view design in 2 dimensions.

2.11. Solution

Solution is the process of calculating the parameters that have been previously entered. The following is a picture of a successful running process.

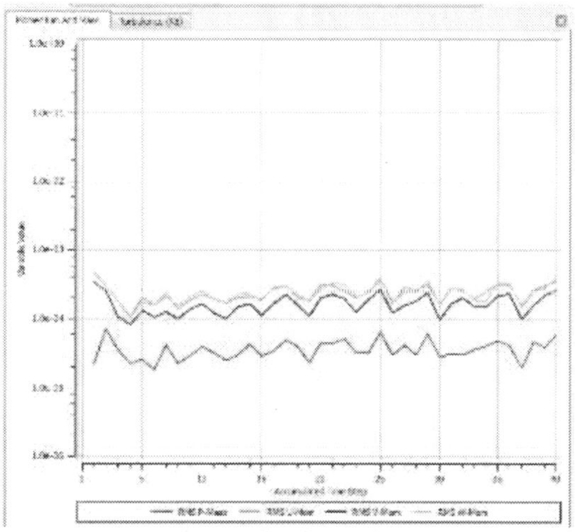

Figure 9. Graph of running results.

3. Result

Result is the final stage of the simulation process which is useful for displaying results in the form of animations and images of the flow models that occur. Here is one picture of the result process. the picture below describes the flow velocity on the hull of the ship which is a type of laminar flow

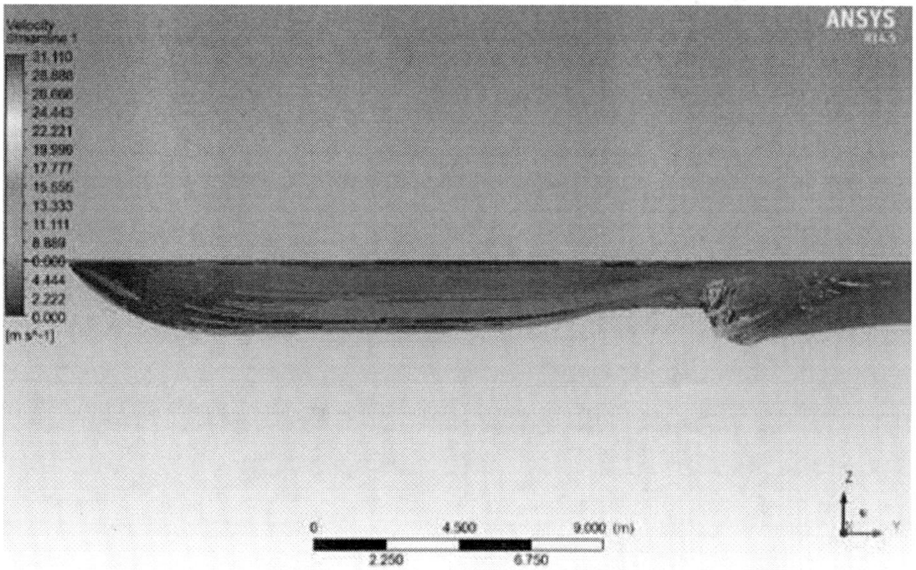

Figure 10. Side view of flow velocity distribution on the hull.

3.1. Analysis of Thrust Calculation (T) after 20° Pitch Change

Result is the final stage of the simulation process which is useful for displaying results in the form of animations and images of the flow models that occur. Here is one picture of the result process. the light blue color in the figure describes the laminar flow around the hull and the bright green color shows the turbulent flow around the propeller.

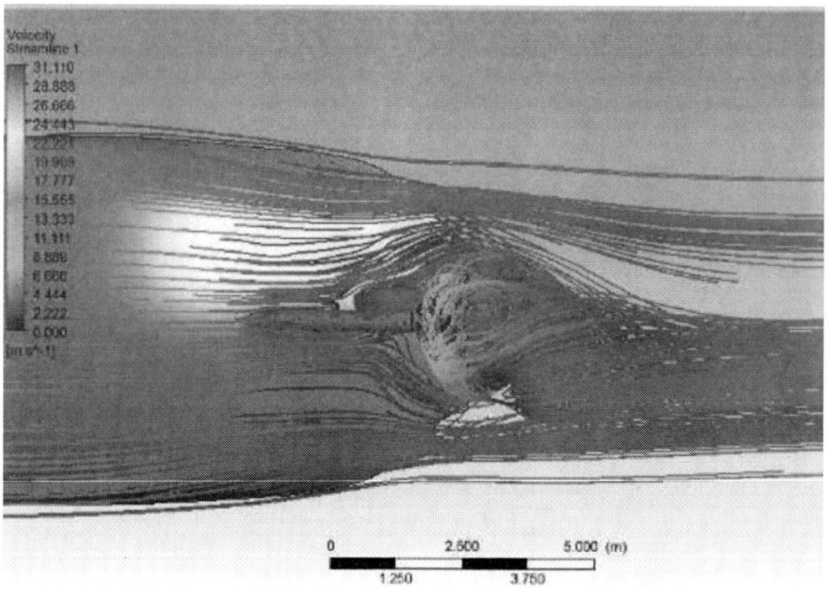

Figure 11. 20° pitch free spin propeller simulation.

From the picture above shows the speed distribution at the stern of the ship. The highest speed is behind the main propeller at 15.555 m/s, this is shown in the green color in the figure, after entering the free-rotating propeller leaf, the flow rate has decreased to 1.125. m/s. Calculation analysis of thrust (T) after 25° Pitch Change. the light blue color in the figure describes the laminar flow around the hull and the bright green color shows the turbulent flow around the propeller.

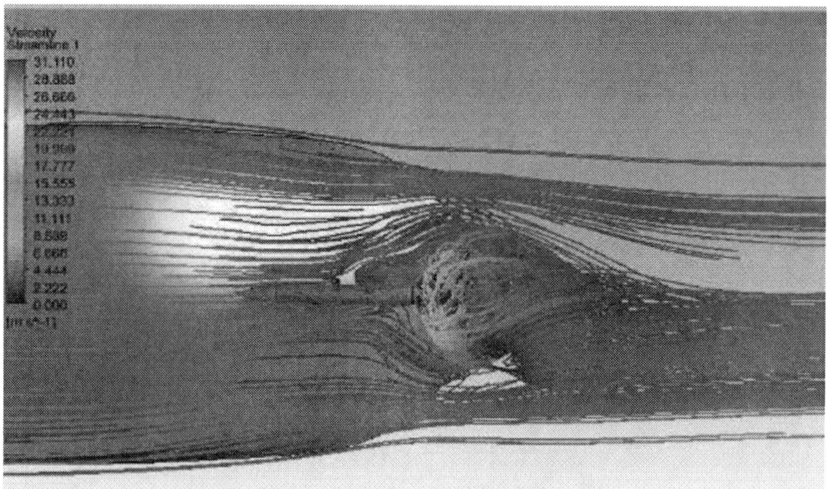

Figure 12. 25° pitch free spin propeller simulation.

3.2. Analysis of Thrust Calculation (T) after 30° Pitch Change

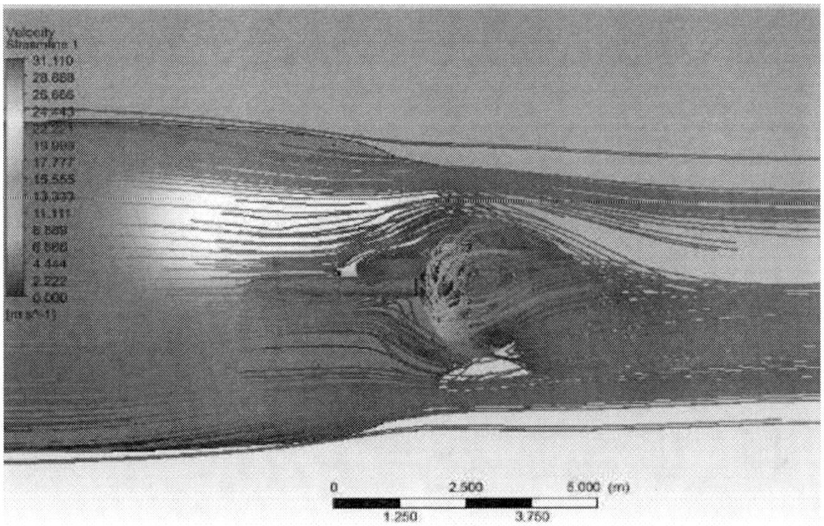

Figure 13. 30° pitch free spin propeller simulation.

From the picture above shows the speed distribution at the stern of the ship. The highest speed is behind the main propeller at 17.777 m/s, this is shown in green in the figure, after entering the free-rotating propeller leaf, the flow rate has decreased to 1.752. m/s. Analysis of thrust calculation (T) after 35° Pitch Change. The light blue color in the figure explains the laminar flow around the hull and the bright green color shows the turbulent flow where at this pitch the turbulent flow increases.

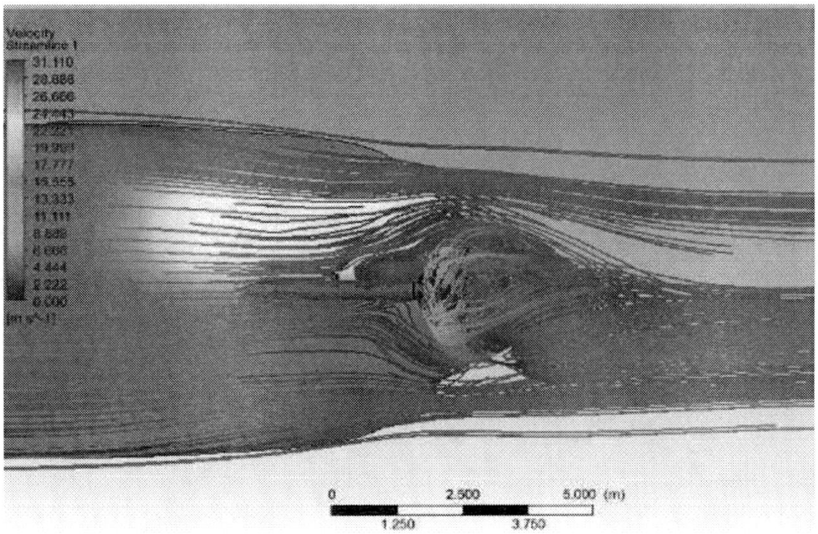

Figure 14. 40° *pitch* free spin *propeller* simulation.

From the picture above shows the distribution of speed at the stern boat. The highest speed is behind the main propeller at 16.452 m/s, this is shown in green in the figure, after entering the free-rotating propeller leaf, the flow rate has decreased to 2.175. m/s. The light blue color in the figure explains the laminar flow around the hull and the bright green color shows the turbulent flow where at this pitch the turbulent flow increases.

3.3. Analysis of the Calculation of Thrust (T) after a 40° Pitch Change

Result is the final stage of the simulation process which is useful for displaying results in the form of animations and images of the flow models that occur. Here is one picture of the result process. The light blue color in the figure explains the laminar flow around the hull and the bright green color shows the turbulent flow where at this pitch the turbulent flow increases.

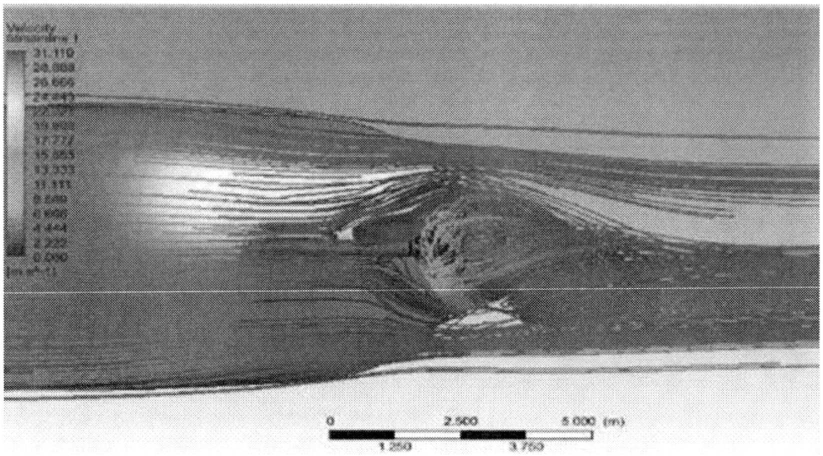

Figure 15. Propeller simulation free spin pitch 35°.

From the picture above shows the speed distribution at the stern of the ship. The highest speed is behind the main propeller at 15.685 m/s, this is shown in green in the figure, after entering the free-rotating propeller leaf, the flow rate has decreased to 2.725. m/s. The light blue color in the figure explains the laminar flow around the hull and the bright green color shows the turbulent flow where at this pitch the turbulent flow increases.

3.4. The Results of the Comparison of the Propeller Thrust Freely Rotate Pitch 20°, 25°, 30°, 35°, 45°

From the calculation results above, the thrust results from angles of 20°, 25°, 30°, 35°, and 45° on the propeller are free to rotate which will be compared in the table below. The table below describes the propeller pitches tested and the rpm generated from all pitches and the thrust generated from the pitches tested in the table shows that pitch 25^0 produces the largest thrust force.

Table 1. Comparison of thrust results from angles of 20°, 25°, 30°, 35°, and 45°.

Corner Models	RPMs	*Thrust* (T)
20° Pitch Angle	412	41.67481 KN
25° pitch angle	412	43.9872 KN
Pitch angle 30°	412	41.8581404 KN
35° pitch angle	412	39.8028604 KN
40° pitch angle	412	39.5953688 KN

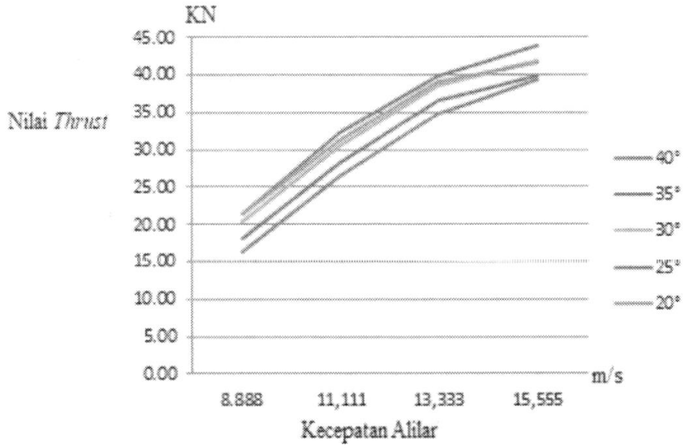

Figure 16. Comparison graph of the thrust results.

4. Conclusion

From the analysis and discussion above, it can be concluded that after changing the free propeller pitch angle of 25° it produces a thrust of 43.9872 KN which is the best thrust result and pitch angle of 40^0 produce the worst thurst result with 39.5953688 KN. Therefore, this research is expected to be useful in the future for its application in propeller manufacturing.

Acknowledgments

I would like to thank head fluid machine laboratory, Hang Tuah University, that supported the completion of my research and the research team.

References

[1] Winarno, A. Ciptadi, G. Iriany A. and Widodo, A. S. "Experiment Study of the Resistance on Nusantara Ship Hull Modification with Fishing Boat in Pantura East Java," *International Journal on Engineering Applications (IREA),* **11** (2), pp. 111-120, 2023.

[2] Fortune,T."Flow Analysis Behind the Propeller With the CFD Approach," *[Thesis]. Surabaya: Institute of Technology 10 November,* 2012.

[3] Sutadi, L. Y. Widyanto S. A. and Haryanto, I. "Perancangan Mekanisme Pengontrol Controllable Pitch Propeller," *Prosiding Seminar Sains Nasional dan Teknologi.,* Vol. **1**. No. 1., 2012.

[4] Sasono, E. J."Use of Free Rotating Propellers on Ships," Vol. **30**, No. 2, pp. pp. 142-145, 2009.

[5] Sugianto, E., & Winarno, A. (2017). Computational model tahanan kapal untuk menentukan kebutuhan daya kapal bulk carrier 8664 DWT.

[6] Permadi, N. V. A., & Sugianto, E. (2022). CFD Simulation Model for Optimum Design of B-Series Propeller using Multiple Reference Frame (MRF). *CFD Letters*, **14** (11), 22-39.

[7] A. Winarno, A. S. Widodo, G. Ciptadi, and A. Iriany, "The Effect of Sail Layout on Fishing Vessels Hydrodynamics in the North Coast of Java using Computational Fluids Dynamic", *Semarak Ilmu: CFD Letter* 16, issue. 1, pp 107-120, 2024. https://doi.org/10.37934/cfdl.16.1.107120

[8] A. Winarno, G. Ciptadi, A. Iriany, and A. S. Widodo, "Experimental and Numerical Study of Ship Resistance on the Combination of Traditional Nusantara Fishing Vessel Hull Forms", *International Review of Mechanical Engineering* **17**, no. 4, pp 190–196, 2023. https://doi.org/10.15866/ireme.v17i4.23726

[9] Natalia, S. R. Parlindungan M. and Wibawa, B. S. A. "Analisa Nilai Thrust Optimum Propeller Pada Kapal Tugboat Pelabuhan Paket-II 2x1850HP Dengan Variasi Diameter Propeller, Jumlah Daun Propeller & Kecepatan Putaran Propeller (RPM).," *Jurnal Teknik Perkapalan,*

pp 112-116, 2020.

[10] G. A. N. S. Sakti and A. Winarno, "Implementation of Kaplan Type Ducted Propeller on Hydrodynamics of Offshore Supply Vessel (OSV)", *International Journal of Marine Engineering Innovation and Research* **9**, No. 4, pp 378-386, 2024. http://dx.doi.org/10.12962/j25481479.v9i2.20567

[11] M. A. R. Hermawan and A. Winarno, "Kajian Teknis Propeller Tipe B - Series Dan Kaplan Dengan Variasi Sudut Rake Pada Kapal Offshore Supply Vessel 80 (OSV 80)", *Zona Laut: Jurnal Inovasi Sains dan Teknologi Kelautan* **4**, No. 3, pp 309-318, 2023. DOI:10.62012/zl.v4i3.27637

[12] D. N. Yunita and A. Winarno, "Analisa Teknis Pengaruh Jumlah Sudu Propeller Bebas Putar Terhadap Gaya Dorong Kapal Tunda DPS IX", *Seminar Nasional Kelautan XIV: Implementasi Hasil Riset Sumber Daya Laut dan Pesisir Dalam Peningkatan Daya Saing Indonesia*, 2019.

[13] Harvald, S. A. A. Resistance and propulsion of ships, 1983.

[14] Edward, V. L. "Principles of Naval Architecture," 1988.

[15] A. Winarno, G. Ciptadi, A. Iriany, and A. S. Widodo, "Experiment Study of the Resistance on Nusantara Ship Hull Modification with Fishing Boat in Pantura East Java", *International Journal on Engineering Applications* **11**, no. 2, pp 111–120, 2023. https://doi.org/10.15866/irea.v11i2.23505

[16] Nursalim, "analysis of the thrust of the tugboat dps ix due to the addition of free spin propellers using the computational fluid dynamics (cfd) method," *[Thesis]. Surabaya: Hang Tuah University,* 2018.

[17] Hermawan, D. and Ismail, N. R. "The Effect of Blade Angle on Thrust Force in Hovrcraf," *Scientific Journal. Widya Teknika.,* 2016.

[18] Benjamin, Y. "A Vane-Wheel Propulsor for A Naval Auxiliary," *DTRC-89/023 United States : Ship Hydromechanics Department Research and Development Report,* 1989.

[19] Dwi, G "Numerical Study of Flow Through Square Duct and Square Elbow 90° with Double Guide Vane on Variation of Damper Opening Angle," *Journal of Mechanical Engineering ITS,* Vol. **6**, No. 4, 2016.

[20] Febrinan, V. "Application of Bernaulli's Law," *www.scribd.com. [Connected Periodically]. https://www.scribd.com/doc/63557553/bab-ii-penerapan-Hukum-bernoulli#.html,* 2017.

[21] Fidiyanarani, "Mekanika Fluida," Mei 2014. [Online]. Available: Fidiyanarani.blogspot.com/2014/05/mekanika-fluida.html.

[22] Ibnudin, "https://ibnudin.net/exemplary-jurnal-cara-making-jurnal/," [Online].

Effect of Steam Pretreatment of Andong Bamboo Slats on the Physical and Mechanical Properties of Ply Bamboo for Ship Body Construction

A D Sembiring[1], D R E Ambarita[1], F Paundra[2], S Maulana[1,3*], E Sugianto[4] and M A R Lubis[5*]

[1] Department of Forestry Engineering, Institut Teknologi Sumatera (ITERA), South Lampung, Lampung 35365, Indonesia
[2] Department of Mechanical Engineering, Institut Teknologi Sumatera (ITERA), South Lampung, Lampung 35365, Indonesia
[3] Center for Green and Sustainable Materials, Institut Teknologi Sumatera (ITERA), South Lampung, Lampung 35365, Indonesia
[4] Department of Marine Engineering, Hang Tuah University, Surabaya, 60111, Indonesia
[5] Research Center for Biomass and Bioproducts, National Research and Innovation Agency, Cibinong 16911, Indonesia

Corresponding authors: sena.maulana@rh.itera.ac.id; muha142@brin.go.id

Abstract. Ply bamboo composite is a promising alternative to being an alternative ship body construction materials due to its cost-effectiveness and sustainability. Bamboo's quick growth and availability make it a desirable resource, and its physical and mechanical characteristics make it solid and durable for ship body construction. However, the dimensional stability of ply bamboo is still poor. Thus, the bamboo strips must be pretreated. This study evaluated steam-treated ply bamboo's properties according to the JAS 234:2003 standard, including physical and mechanical performance. Steam treatment involves heating 30 cm x 30 cm x 0.5 cm bamboo strips at 126°C for 30, 45, and 60 minutes in an autoclave. Steam preliminary treatment improves the physical and mechanical properties of ply bamboo. Steam-treated ply bamboo has superior physical and mechanical characteristics compared to untreated ply bamboo. The physical characteristics and ply of the bamboo significantly improved after 45 minutes of steam treatment at 126 °C. As steam treatment time rose, the mechanical performance of bamboo decreased. Ply bamboo has several benefits for ship construction; however, it requires specialist production, and bamboo quality varies. Further research and development could optimize bamboo's utilization for ship body construction.

1. Introduction

Bamboo is a sustainable resource with impressive growth and substantial potential for use in many sectors. Notably, the Asia-Pacific region, which hosts approximately 80% of the world's bamboo forests, has witnessed the emergence of Indonesia as a leading cultivator with its 176 Indigenous species of bamboo [1,2]. Renowned for its adaptability, bamboo is utilized for various uses, including building materials and home products [3–5]. Recent statistics highlight Indonesia's robust bamboo output, reporting an annual production exceeding 50 million stems [6]. Despite its impressive mechanical properties, such as its high strength-to-weight ratio, which surpasses that of many woods [7,8], bamboo's application as a structural component is not without challenges [9,10]. These include susceptibility to pests, propensity to split, and variations in material properties along its length [11]. Additionally, the small diameter of bamboo limits its direct usage in applications requiring large panel dimensions [12]. Addressing these constraints, the advancement of ply bamboo technologies has opened new avenues for its application. Bonding strips of bamboo engineer ply bamboo under pressure, a process that enhances its dimensional stability and mechanical strength [5,13]. Nonetheless, challenges such as density fluctuations and high extractive content persist, necessitating pre-treatment of the bamboo strips [14].

Novel developments in the treatment of lignocellulosic materials, such as acetylation and furfurylation methods, have shown enhancements in both hardness and resistance to biodegradation [15,16]. Nevertheless, the industry is being driven by environmental and economic considerations to adopt more environmentally friendly options. Among these options, heat treatment has become a favored technique for improving material characteristics [17]. A combination of laminated Apus bamboo and meranti wood as a substitute material for wooden ships illustrates that laminated composites are more economical than specific Strength Class II woods, such as teak, Merbau, and sea dammar, frequently employed in wooden ships. Rindo et al. [14] investigated the impact of the configuration and dimensions of Petung bamboo split fibers on integrating the fibers with the matrix in laminated bamboo. The investigation revealed that splits of greater size require a greater amount of adhesive and lead to a higher bond interface strength in laminated bamboo. Supomo et al. [15] adopted hybrid slat lamination of *Bambusa arundinacea* and *Pterocarpus indicus* to substitute solid Teak wood to fabricate fishing vessels. Compared to Teak wood, the hybrid lamination material will exhibit somewhat reduced dimensions, resulting in a 28.7% decrease in material volume and a 40.66% reduction in construction costs. Furthermore, a study was undertaken by Manik et al. [16] to investigate the influence of the configuration and dimensions of Petung bamboo split fiber on the binding layer of the matrix interface. The bonding strength at the interface rises directly to the extension of the adhesive interface per cm^2.

Research substantiates that steam treatment can significantly increase bending strength while reducing extractive compounds that impede bonding pivotal for products like oriented strand board [18–20]. Current trends in steam modification aim to finesse the balance between temperature and time of exposure to optimize the properties of bamboo composites, advocating for customized treatment protocols for different bamboo species [21]. The dynamic nature of this field emphasizes the need for continuous study, namely on the impact of steam alteration on the physical and mechanical characteristics of ply bamboo and its many native species. Innovative steam pre-treatment is necessary to improve ply bamboo's physical and mechanical characteristics with reduced temperature and time variations. Nevertheless, no research has been reported on how steam pre-treatment affects Andong bamboo strip properties and ply bamboo's physical and mechanical traits, particularly for potential application ship body construction.

2. Materials and methods

2.1 Materials

This study utilized naturally aged Andong bamboo sourced from Karang Anyar, Jati Agung, Lampung. A 3 to 4-year-old culms due to their optimal mechanical strength and workability balance. Polyvinyl Acetate (PVAc, Crona 234 SR) was selected as an adhesive for its robust performance in woodworking applications.

2.2 Preparation of Bamboo Strips and Steam Pre-treatment

The bamboo strip preparation is illustrated in Figure 1. The bamboo culms were trimmed to a length of 30 cm, with well-planned stimulation of nodes to guarantee consistency in the material. The strips were methodically derived to a width of 2 cm and a thickness of 0.5 cm by carefully removing the resilient outer skin and the more delicate inner lining. The bamboo culms were cut to 30 cm, strategically exciting nodes to ensure uniformity in the material. Strips were crafted to a width of 2 cm and thickness of 0.5 cm by meticulously removing the tough outer skin and the softer inner lining. This yielded strips with dimensions of 30 cm x 2.0 cm x 1.5 cm. The bamboo strips were subjected to regulated steam pre-treatments at 126 °C for 30, 45, and 60 minutes. The steam pre-treatments determined the duration required to improve bamboo characteristics. Following steaming, we allowed the strips to naturally dry in the air until their moisture content decreased to 16-18%. For a tightly regulated final moisture content of 10-12%, the strips were dried in an oven at 60-80 °C for 48 hours.

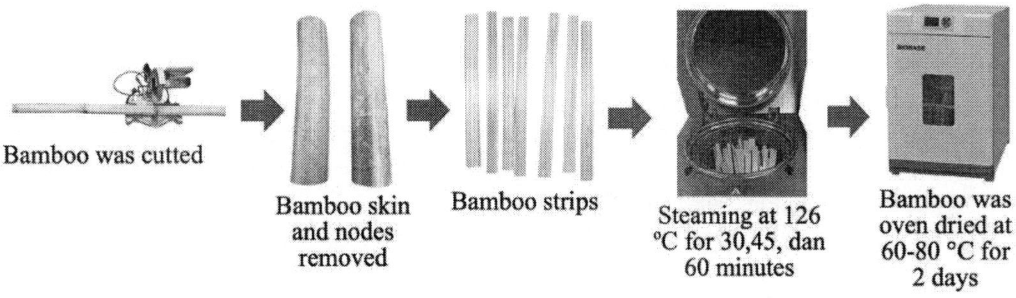

Figure 1. Bamboo strips preparation.

2.3 Manufacture of Ply Bamboo Sheets

The dried bamboo slats were arranged to 30 cm x 30 cm. The PVAc adhesive was applied to the surface of the bamboo layer using a brush. The adhesive was then spread on two surfaces (double spread) using a glue spread of 150 g.m^{-2}. The samples were then joined together by adhesion using Crona234 adhesive. Then, the ply bamboo sheets were felted using a cold felt tool for 24 hours to optimize the gluing process.

IOP Conf. Series: Earth and Environmental Science **1473** (2025) 012024 doi:10.1088/1755-1315/1473/1/012024

2.4 Pressing and Conditioning

The glued bamboo sheets were then coated in cross-section with adhesive and arranged into 3 layers, with the second layer perpendicular to the first and third layers. Afterwards, it was cold-pressed for 12 hours. The specified parameters for this ply bamboo are 30 cm in length, 30 cm in width, and 1.5 cm in thickness. Following the pressing procedure, the bamboo that had been coated was stored at a controlled room temperature for 2 weeks to effectively remove any remaining strains that arose during the pressing process and regulate the moisture level in the surrounding environment.

2.5 Ply Bamboo Testing

The experiments conducted in this work encompassed the assessments of ply bamboo's physical and mechanical characteristics. The physical parameters included moisture content, density, and dimensional stability. The mechanical parameters were the modulus of elasticity (MOE) and modulus of rupture (MOR) parallel and perpendicular to the fibers. The American Standard Test Methods (ASTM D143 2005) was followed for the test sample preparation, specifically defining the procedures for testing small timber specimens (ASTM 2005). The JAS 234:2003 (JSA 2003) was followed for the physical and mechanical testing of the ply bamboo.

2.6 Moisture Content (MC)

The air-dried test sample (BKU) was weighed and then dried in an oven set at 103 ± 2 ºC for 24 hours or until it achieved a consistent weight. The test sample was subsequently chilled in a desiccator for around 15 minutes. Then, the oven-dried test sample (BKT) was weighed again. The following formula was used to calculate the MC:

$$MC\ (\%) = \frac{(BKU - BKT)}{BKT} \times 100$$

2.7 Density

Density was determined using a division operation between air-dried weight (BKU) and air-dried volume (VKU). The test sample's volume was determined by multiplying the length, width, and thickness reported by callipers, while the weight was acquired using digital scales. The density value was measured using the following equation:

$$Density = \frac{BKU}{VKU}$$

2.8 Thickness Swelling (TS) and Water Absorption (WA)

The TS was determined by measuring the initial thickness (T1) and the thickness after immersion (T2) in water for 2 and 24 hours. The following formula was used to calculate the TS value:

$$TS\ (\%) = \frac{(T2 - T1)}{T2} \times 100$$

The results of WA were obtained from test samples that were initially weighed (B1) and soaked with water for 24 hours, and the final weight was calculated (B2). The WA value was calculated by following the formula:

$$\mathrm{WA}\ (\%) = \frac{(B2 - B1)}{B2} \times 100$$

2.9 Delamination

Following a 6-hour immersion in water at 25 °C, the test samples were subjected to an 18-hour heating process in an oven set at 40 °C. To get the delamination test ratio, combine the delamination length with the total length of the adhesive line on the test sample. The following formula was used to calculate delamination:

$$Delamination\ (\%) = \frac{(Delamination\ Length)}{(Total\ adhesive\ line\ length)} \times 100$$

2.10 Modulus of Elasticity (MOE) and Modulus of Rupture (MOR)

The MOE of ply bamboo was investigated using a one-point loading bending test. The MOE value was determined using the following formula:

$$MOE \equiv \frac{\Delta PL^3}{4\Delta bh^3}$$

The MOR analysis of ply bamboo was performed alongside the MOE testing using the identical test sample. The following formula was used to determine the MOR value:

$$MOR = \frac{3PL}{2bh^3}$$

2.11 Data Analysis

The data were analyzed using Microsoft Excel, IBM SPSS Statistics 26, and Origin Lab software. The study implementation employed a completely randomized design (CRD) with a single component (time) and four levels (0, 30, 45, and 60 minutes) with three repetitions. Analysis of Variance (ANOVA) and Duncan's Multiple Range Test (DMRT) were performed to ascertain the impact of varying time durations on ply bamboo's physical and mechanical characteristics.

3. Results and discussion

3.1 Density and Moisture Content (MC)

The density value obtained from this investigation has successfully achieved the desired density of 0.7 g cm^{-3} for both steam-treated and untreated ply bamboo. Figure 2 shows that the density values of steam-treated and untreated ply bamboo ranged from 0.69 to 0.74 g cm^{-3}. The statistical analysis of variance ($\alpha = 0.05$) indicated no significant impact of steam treatment or steam's absence on ply bamboo's density value. The findings indicated that the density of the ply bamboo generated was uniform.

The MC values for steam-treated and untreated ply bamboo varied between 9.92% and 10.89% (Figure 2). Ply bamboo subjected to steam treatment for 60 minutes exhibits the lowest MC value of 9.79%, below the MC value of ply bamboo without steam treatment, 10.89%. The JAS 234 standard 2003 specifies laminated boards' highest permissible MC is less than 15%. According to the JAS 234: 2003 standard, the MC of the ply bamboo produced after a 2-week conditioning period has satisfied the criteria. The findings of the variance analysis ($\alpha < 0.05$) indicate that steam treatment of ply bamboo has a statistically significant impact. Further testing conducted by DMRT revealed a substantial difference in ply bamboo without any treatment.

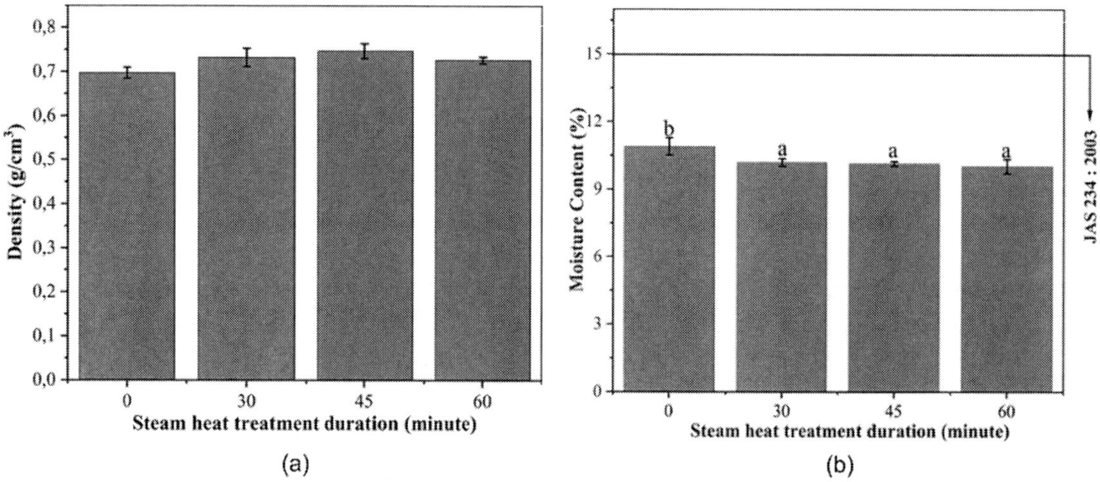

Figure 2. Graph of test results (a) density and (b) MC of ply bamboo with various durations of steam pretreatment, different letters on the bar chart indicate significantly different ($\alpha < 0.05$).

The results of the MC study show that steam treatment of bamboo slats can reduce the moisture content of the resulting ply bamboo. The decrease in the MC value of ply bamboo is thought to be due to the increased duration of steam time on the bamboo slats. Steam treatment of bamboo slats reduces hemicellulose in the bamboo cell wall, thereby reducing free hydroxyl groups that can bind water to the bamboo[17,22]. The decreasing MC value of ply bamboo shows that steam treatment for 30 minutes has become the optimal time in reducing the moisture content value of ply bamboo; this is because the steam treatment between 30, 45, and 60 minutes is not significantly different. Density increases as the MC of the bamboo decreases. However, at certain temperatures and treatment times, it is possible to damage the chemical content of the wood, and the wood structure can reduce its strength. The decrease in density value of ply bamboo at a steam duration of 60 minutes is thought to be due to the reduction of the clam component in bamboo. Longer steam treatment over 30 minutes at 140 °C could decrease the properties of bamboo composites [10].

3.2 Thickness Swelling (TS) and Water Absorption (WA)

TS and WA are among the important factors to be tested in manufacturing laminated board products. This is done to see the dimensional stabilization ability of a composite board product produced[23]. The TS values of the resulting ply bamboo were successively around 8.51%-4.59% (Figure 3). Based on the JAS SE-7:2003 standard, the maximum TS value is < 20% (JAS SE-7 2003). Therefore, the TS of ply bamboo produced without and with steam treatment has met the requirements for a composite board product. The analysis of variance (α < 0.05) showed that the interaction between steam treatments on the ply bamboo produced had a significant effect. Duncan's further test results showed that ply bamboo without treatment and steam treatment with 30, 45, and 60 minutes were significantly different.

The WA value of ply bamboo produced after soaking in water for 24 hours was approximately 24.48%-35.42% (Figure 3). Based on the JAS SE-7:2003 standard, the maximum value of WA in composite boards is < 20% (JSA SE-7 2003). Therefore, the WA value of the ply bamboo produced did not meet the test standard. The analysis of variance (α < 0.05) showed that the interaction between steam treatments on ply bamboo had a significant effect. Duncan's further test results showed that ply bamboo without treatment and steam treatment with a time of 30 minutes significantly differed from ply bamboo with steam treatment of 45 and 60 minutes.

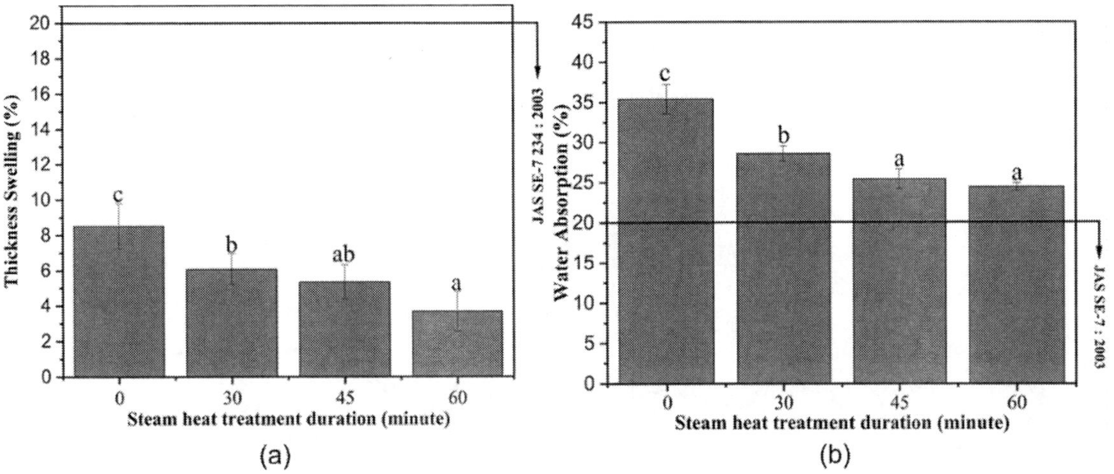

Figure 3. Graph of test results of (a) TS and (b) WA of ply bamboo with various durations of steam pretreatment, different letters on the bar chart indicate significantly different (α < 0.05).

Based on this study, the trend of the TS and WA graphs tended to decrease when steam-treated with durations of 30, 45, and 60 minutes. The decrease in TS and WA values is related to the decreasing trend in the MC of the produced ply bamboo (Figure 3). The TS and WA of the resulting ply bamboo decreased as the MC of the ply bamboo decreased. Changes in dimensional stability are a sign of changes in the MC of the wood due to the ability of the wood cell walls to bind water caused by differences in density [24]. The decrease in extractive substances after steam treatment is also one of the reasons that the TS and WA values of ply bamboo decrease, resulting in decreased accessibility of bamboo to absorb water [23]. The significant reduction in WA of ply bamboo is thought to be due to the increased duration of steam treatment. Longer heat-treated bamboo slats will have good hydrophobic properties [25]. This makes the WA and TS of ply bamboo treated with steam for 45 and 60 minutes better than those of the control ply bamboo and steam for 30 minutes. However, steam treatment for 45 minutes is optimal for reducing TS and WA values because steam treatments for 45 and 60 minutes are not significantly different.

3.3 Delamination

Delamination is an important parameter indicating ply bamboo's bonding quality. The resistance of the adhesive to moisture and high pressure can determine the quality of the resulting WA of ply bamboo [26]. The delamination values of ply bamboo without and with steam treatment were 49.46%-28.36%, respectively (Figure 4). The analysis of variance (α < 0.05) showed that the interaction between steam treatments on ply bamboo had a significant effect. Duncan's further test results showed that ply bamboo without treatment and steam treatment for 30 minutes significantly differed from ply bamboo with steam treatment for 45 and 60 minutes.

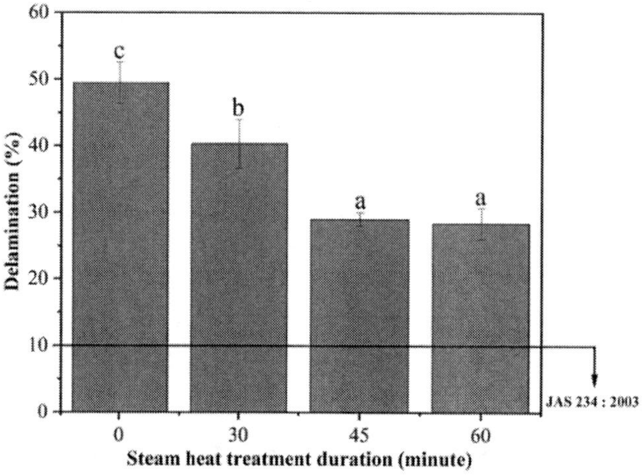

Figure 4. Delamination test of ply bamboo with various durations of steam pretreatment, different letters on the bar chart indicate significant differences (α < 0.05).

The results showed a decreasing trend in ply bamboo delamination values as the duration of steam time increased. This aligns with the previously described WA and TS research results (Figure 3). The ability of ply bamboo to absorb water is thought to cause the adhesive resistance to weaken due to moisture and water pressure entering the ply bamboo cavity [25]. Steam treatment also affects the decrease in extractive substances, which could help the adhesive penetration into the bamboo slats; thus, the adhesion between the bamboo slats gets better [23]. Based on the trend of decreasing delamination value, 45 minutes of steam time is the most optimal time because the steam treatment for 45 and 60 minutes is not significantly different.

3.4 Modulus of Elasticity (MOE) and Modulus of Rupture (MOR)

MOE is a crucial characteristic of ply bamboo. MOE refers to the capacity of a material to retain its initial form when subjected to loading or stress [23]. The mean values of MOE parallel to the fiber and perpendicular to the fiber are 14847 - 31151 Kg.cm^{-2} and 7517 - 10947 Kg.cm^{-2}, respectively (Figure 5). These results suggest that steam-treated ply bamboo's MOE is superior to untreated ply bamboo's. Nevertheless, according to the JAS 234:2003 standard, the minimum MOE value required is above 75,000 Kg.cm^{-2}. Consequently, the MOE value of the ply bamboo manufactured fails to satisfy the test criteria. The statistical analysis of variance ($\alpha < 0.05$) indicated no significant impact of the interaction between steam treatments on ply bamboo on the MOE value oriented parallel to the fiber or perpendicular to the fiber.

The MOR is a crucial component in the production of ply bamboo. MOR is the capacity of a material to endure the highest force it can bear until it fractures [24]. The mean values of MOR parallel to the fiber and perpendicular to the fiber were 282-423 Kg.cm^{-2} and 223-284 Kg.cm^{-2}, respectively, as shown in Figure 5. According to the JAS 234:2003 standard, the minimum MOR value must exceed 300 Kg.cm^{-2}. Only the MOR value of steam-treated fiber-aligned layered bamboo satisfied the established test criteria. The statistical analysis of variance ($\alpha < 0.05$) indicated no significant impact of the interaction between steam treatments on ply bamboo on the MOE values in the parallel or perpendicular directions to the fibers.

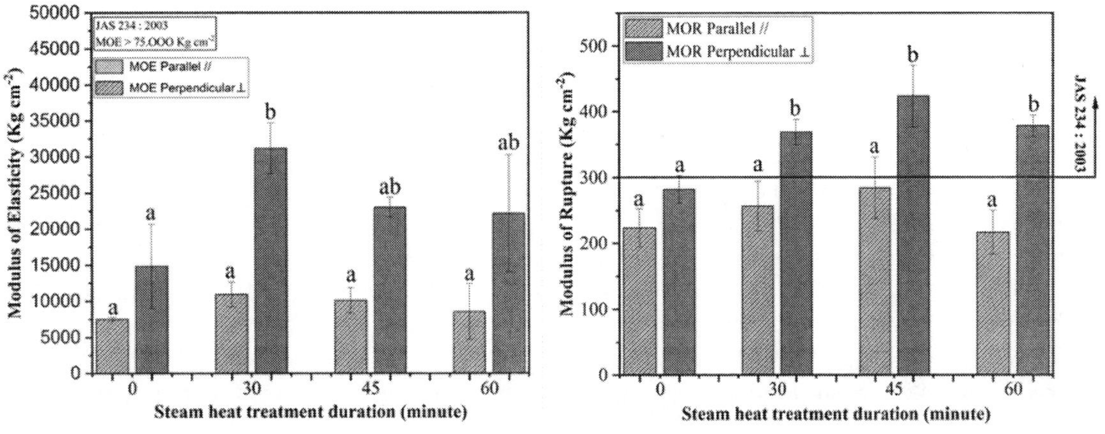

(a) (b)

Figure 5. Graph of test results (a) MOE and (b) MOR of ply bamboo with various durations of steam pretreatment, different letters in the bar chart indicate significantly different ($\alpha < 0.05$).

Generally, steam-treated ply bamboo's MOR and MOE values are better than untreated ply bamboo. Steam treatment of bamboo slats can reduce extractive content so that bonding between bamboo slats improves [23]. The increasing amount of adhesive will increase the stiffness properties of laminated wood because this adhesive plays an important role in strengthening the laminated wood produced [27]. However, steam treatment for 45 and 60 minutes causes a decrease in the MOE value of ply bamboo. The decrease in MOE value is due to changes in the chemical composition of cell walls, such as hemicellulose, resulting in weak binding forces with cellulose and lignin. Heating bamboo with steam below 140 °C often improves MOE and MOR. Specifically, studies have demonstrated that treatments conducted at 140 °C result in enhanced characteristics of bamboo compared to untreated bamboo. However, longer steam treatment over 30 minutes at 140 °C could decrease the MOE and MOR values [28,29]. The MOR value of ply bamboo in steam treatment for 45 minutes is better than steam treatment for 30 and 60 minutes. This aligns with the results of the density and MC studies previously described (Figure 2). The higher the MC, the lower the fracture strength, and the higher the density, the higher the fracture strength [30]. Therefore, the influence of physical properties is closely related to mechanical properties. Based on the MOE and MOR values of ply bamboo produced both parallel and perpendicular to the fibers, steam treatment for 30 minutes is the optimal time to improve the mechanical properties of ply bamboo. The ply bamboo exhibits versatility and finds application in several facets of ship body construction, such as flooring, doors, and windows. With its exceptional strength, flexibility, and resistance to deformation, this material is well-suited for a wide range of uses that extend beyond ship body construction [31,32].

4. Conclusions

The cost-effectiveness and sustainability of ply bamboo make it a highly promising substitute for conventional ship body materials. Bamboo's rapid growth and widespread availability make it an indispensable resource, and its physical and mechanical properties render it robust and long-lasting for ship body fabrication. Nevertheless, the dimensional stability of ply bamboo remains inadequate. Therefore, the bamboo strips need to undergo pretreatment. The present work assessed steam-treated ply bamboo's physical and mechanical characteristics by the JAS 234:2003 standard. In general, steam-treated ply bamboo exhibited superior physical and mechanical characteristics compared to untreated ply bamboo. The results indicated that steam pretreatment effectively enhances ply bamboo's physical and mechanical characteristics. Applying steam treatment can decrease the extractive concentration of Andong bamboo slats. Consequently, the adhesive penetrates the bamboo slats. The enhancement in the mechanical capabilities of ply bamboo is significantly influenced by its physical characteristics. Steam pretreatment at 126°C for 45 minutes often yielded the most optimal enhancement in physical characteristics and ply bamboo. Nevertheless, the mechanical characteristics of ply bamboo tended to decline with the further extension of the steam treatment period. The quality testing parameters for ply bamboo that comply with the JAS 234:2003 standard include water absorption (WA), moisture content (MC), density, and fibrous orientation ratio (MOR). While ply bamboo has various advantages for ship body materials, its manufacture necessitates specialized techniques, and the quality of the bamboo can vary. Further investigation and development could enhance the efficiency of using bamboo for ship body construction.

5. Conflict of interest

The authors declare that there are no conflicts of interest.

6. Acknowledgement

We sincerely acknowledge the Guru Besar under 45 ITERA Research Grant (631a/IT9.2.1/PT.01.03/2023) from the Institut Teknologi Sumatera, the Republic of Indonesia, for the financial support.

References

[1] Mera, F. A. T., & Xu C 2014 PLANTATION MANAGEMENT AND BAMBOO RESOURCE ECONOMICS IN CHINA. *Cienc. y Tecnol.* **7(1)** 1–12
[2] Widjaja, E A 2019 Review The Spectacular Indonesian Bamboos *J. Print. Sci. Technol.* **56(4)** 215–217
[3] Eskak E 2016 BAMBU ATER (Gigantochloa atter) SEBAGAI BAHAN SUBSTITUSI KAYU PADA UKIRAN ASMAT. *Din. Kerajinan Dan Batik Maj. Ilmiah,* **33(1)**, 55.
[4] Ekawati, D., Karlinasari, L., Soekmadi, R. & M 2022 Drivers, Barriers, and Strategies in the Community-Based Supply of Bamboo for Industrial-Scale Bamboo Utilization in Ngada Regency, East Nusa Tenggara, Indonesia *Sustain.* **14(10)**, 5970.
[5] Arsad E 2015 Teknologi Pengolahan Dan Manfaat Bambu. *J. Ris. Ind. Has. Hutan* **7(1)**, 45
[6] BPS 2022 STATISTIK PRODUKSI KEHUTANAN 2021. Badan Pusat Statistik/BPS-Statistics Indonesia.
[7] Sharma, P., Dhanwantri, K., & Mehta S 2014 Brick as a Building Material. Scientific American, **5(2)** 29–29
[8] Sumarno, Agung, & Widodo E 2016 Kajian Kekuatan Bambu Laminasi Sebagai Bahan Bangunan Di Indonesia. *Rekayasa Sipil Mercu Buana* **5(1)** 1–7
[9] Lestari D, Suwanda A A, Murda R A, Maulana M I, Augustina S, Rianjanu A, Taher T, Hidayat W, Maulana S and Lubis M A R 2024 Durability to Natural Weathering of Methylene Diphenyl Diisocyanate-Bonded Bamboo Oriented Strand Board *J. Sylva Lestari* **12** 143–57
[10] Maulana S, Suwanda A A, Murda R A, Antov P, Komariah R N, Maulana M I, Augustina S, Lee S H, Mahardika M, Rianjanu A, Taher T, Kristak L, Bindar Y, Iswanto A H and Lubis M A R 2024 Performance of Oriented Strand Board Made of Heat-Treated Bamboo (Dendrocalamus asper (Schult.) Backer) Strands *Polymers (Basel).* **16,** 1692.
[11] Lempang M 2016 Pengawetan Bambu Untuk Barang Kerajinan Dan Mebel Dengan Metode Tangki Terbuka. *Info Tek. EBONI, 13(12)*, 79–92.
[12] Febrianto, F., Sumardi, I., Hidayat, W., & Maulana S 2017 Papan Untai Bambu Berarah Material Unggul Untuk Komponen Bangunan Struktur. IPB Press.
[13] Zhang, Y., Zhu, R., & Yu W 2017 Effect of Steam Treatment on the Properties of Phyllostachys Iridescens Bamboo Composite. CELLULOSE CHEMISTRY AND TECHNOLOGY Cellulose Chem *Technol* **51(2)** 103–108
[14] Maulana, S., Purusatama, B. D., Wistara, N. J., & Sumardi I 2016 Pengaruh Perlakuan Steam pada Strand dan Shelling Ratio terhadap Sifat Fisis dan Mekanis Oriented Strand Board Bambu. Jurnal Ilmu dan Tek. Kayu Tropis **14(2)**, 136-143.
[15] Mantanis, G. I., & Sahin H T 2017 Modification of wood by chemical processes *A Rev. Int. Symp. New Horizons For. Mantanis,* 272–280
[16] Rowel, Roger M 2012 Handbook of Wood Chemistry and Wood Composites. In Journal of Cleaner Production **110**
[17] Liang, X., Yao, Y., Xiao, X., Liu, X., Wang, X., & Li Y 2022 Pressure-Steam Heat Treatment-Enhanced Anti-Mildew Property of Arc-Shaped Bamboo Sheets. **14(17)**, 1–12.
[18] Fatrawana, A., Maulana, S., Nawawi, D. S., Sari, R. K., Hidayat, W., Park, S. H., Febrianto, F., Lee,

S. H., & Kim N H 2019 Changes in chemical components of steam-treated betung bamboo strands and their effects on the physical and mechanical properties of bamboo-oriented strand boards. *Eur. J. Wood Wood Prod.* **77(5)**, 731-739.

[19] Maulana, M. I., Murda, R. A., Purusatama, B. D., Maulana, M. I., Murda, R. A., Purusatama, B. D., Sari, R. K., Nawawi, D. S., Nikmatin, S., Hidayat, W., Lee, S. H., Febrianto, F., & Kim, N. H. Sari, R. K., Nawawi, D. S., Nikmatin, S., Hidayat, W., Lee, S. N H 2021 Properties of Oriented Strand Board from Alkali-washed Bamboo Strands after Steam Treatment. *BioResources* **16(1)**, 987-996.

[20] Maulana, S., Gumelar, Y., Fatrawana, A., Maulana, M. I., Hidayat, W., Sumardi, I., Wistara, N. J., Lee, S. H., Kim, N. H., & Febrianto F 2019 Destructive and non-destructive tests of bamboo oriented strand board under various shelling ratios and resin contents. *J. Korean Wood Sci. Technol.* **47(4)**, 519–532

[21] Wang, Q., Wu, X., Yuan, C., Lou, Z., & Li Y 2020 Effect of saturated steam heat treatment on physical and chemical properties of bamboo *Molecules* **25(8)**, 1999

[22] Wibowo E S, Lubis M A R and Park B D B-D 2021 In-situ modification of low molar ratio urea–formaldehyde resins with cellulose nanofibrils for plywood *J. Adhes. Sci. Technol.* **35** 2452–65

[23] Maulana S 2018 Sifat fisis, mekanis, dan keawetan oriented strand board bambu andong dan betung dengan perlakuan steam pada strand [tesis].Institut Pertanian Bogor.

[24] Wulandari, F. T., Putu, N., Lismaya, E., & Suryawan I G A 2023 Analisis Sifat Fisika dan Mekanika Papan Laminasi Bambu Petung (*Dendrocalamus asper* Roxb) dan Papan Laminasi Kayu Bayur (*Pterospermum javanicum*) **06(01)** 39–50

[25] Baiti, R. N., Maulana, S., Sipahutar, W. S., Murda, R. A., Julyatmojo, F. A., Suwanda, A. A., & Fadlan M S 2021 Evaluation of surface properties of betung bamboo (Dendrocalmus asper) strands under various heat treatment duration and temperature *J. Sci. Appl. Technol.* **5(2)**, 411.

[26] Sri Rahayu Diza Lestari, A., & Muin M 2020 Sifat Fisis Dan Mekanis Papan Laminasi Menggunakan Pengawet Alami Buah Berenuk (*Crescentia Cujete*) Sebagai Aditif Pada Perekat Tanin *Perennial* **16(2)**)68–72

[27] Wulandari F., Habibi & Amin, R 2022 Sifat Fisika Dan Mekanika Papan Laminasi Bambu Petung (Dendrocalamus Asper) Dengan Susunan Bilah Ke Arah Lebar J. Hutan Topika. **18(1)**, 109-116.

[28] Wang X, Cheng D, Huang X, Song L, Gu W, Liang X, Li Y and Xu B 2020 Effect of high-temperature saturated steam treatment on the physical, chemical, and mechanical properties of moso bamboo *J. Wood Sci.* **66,** 52

[29] Yuan T, Liu J, Hu S, Wang X, Liu X and Li Y 2021 Multi-scale characterization of the effect of saturated steam on the macroscale properties and surface changes of moso bamboo *Mater. Express* **11** 740–8

[30] Abdurachman N H 2008 SIFAT FISIK DAN MEKANIK KAYU LAMINA CAMPURAN KAYU MANGIUM DAN SENGON *J. Penelit. Has. Hutan.* (pp. 191–200).

[31] Tuswan T, Manik P, Samuel S, Suprihanto A, Sulardjaka S, Nugroho S and Pakpahan B F 2023 Correlation between lamina directions and the mechanical characteristics of laminated bamboo composite for ship structure *Curved Layer. Struct.* **10,** 20220186

[32] Hong M-K, Lubis M A R, Park B-D, Sohn C H and Roh J 2020 Effects of surface laminate type and recycled fiber content on properties of three-layer medium density fiberboard *Wood Mater. Sci. Eng.* **15** 163–71

IOP Conf. Series: Earth and Environmental Science **1473** (2025) 012025 doi:10.1088/1755-1315/1473/1/012025

Designing a Neural Network Algorithm for Predicting Wave Height on the Surabaya - Banjarmasin Shipping Route

W L Dhanistha[1]*, S Frestiqauli[1], R Akbar[2]

[1] Ocean Engineering Department, Faculty of Marine Technology Institut Teknologi Sepuluh Nopember (ITS), Surabaya, Indonesia
[2] Industrial Engineering Department, Universitas Muhammadiyah Surabaya, Indonesia

*Corresponding author : wimala.danistha@gmail.com

Abstract. Indonesia is a maritime country, so maritime weather greatly affects marine activities, such as offshore and coastal building design, shipping, and fishing. In carrying out these activities, information about maritime weather is needed, especially wave height as a safety reference. Conventional wave height prediction is less able to describe sea conditions that have nonlinear properties, as a solution, the neural network algorithm can overcome the problem of nonlinear systems because of its ability to learn existing data. With a division of 80% training and 20% validation, this algorithm uses backpropagation as its learning method and multi-layer perceptron as a learning architecture. and produces an RMSE error value of 0.04. The smaller the RMSE, the more accurate the prediction. Thus, this algorithm can be used to predict wave heights on the Surabaya - Banjarmasin shipping route.

1. Introduction

Maritime weather is a dynamic phenomenon that plays an crucial role in the maritime world. Where Indonesia is a country with dense maritime transportation activities, this is because Indonesia is an archipelagic country where trade activities are carried out by sea. According to Arifin (2009), Factors causing the flood of accidents human error, force majeure and hull structure [1]. Bad weather is the second largest factor which can be in the form of high waves, storms and others, where this can be prevented if the user knows the maritime weather before carrying out activities at sea. Wave height is one of the most important factors that must be known in shipping activities, offshore structure design, and fishing. Lack of information about wave height can have fatal consequences so that it can experience economic losses and can endanger loss of life. Therefore, information about wave height predictions is very important in order to increase the awareness of its users, so that the development of a wave height prediction method with a more precise level of accuracy is needed in this case as an effort to mitigate risks at sea. Figure 1 is a picture of the location of this research where the location of the research is the Java Sea which connects the islands of Java and Kalimantan. The location was taken because the Java Sea is the busiest transportation route in Indonesia, so the need for information about sea wave height highly necessary. Where point A is located in the Waters of Surabaya (6.874824° S - 112.747800° E), point B is located in the Java Sea (4.648136° S - 113.908806° E), and point C is located in Banjarmasin (3.540425° S - 114.484300° E).

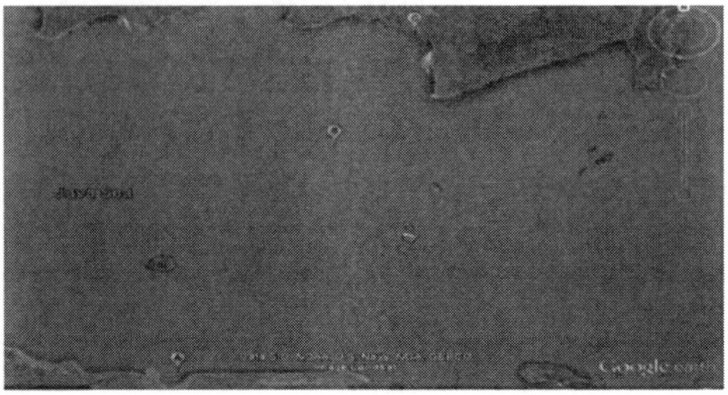

Figure 1. Location of Prediction

Wave height predictions that are carried out conventionally still use statistical equations, physics models, and spectral equations. However, the main challenge in predicting wave heights is the non-linear nature of the waves. So that conventional wave predictions are still considered less accurate. In the last few decades, artificial intelligence has been developed so that it can solve complex problems including the problem of wave nonlinearity. Among several artificial intelligence (AI) techniques, the neural network algorithm has the ability to adapt in learning available data patterns [2]. Then it can use the obtained patterns to predict wave heights in the future. Another advantage is that it has high interpolation so that the resulting predictions have a low error rate. The results of this study are expected to design neural networks that can learn existing data patterns and have a lower prediction error rate than conventional predictions so that they can contribute to the field of oceanography and can support appropriate decision making in time-limited operations, offshore and coastal building construction planning.

2. Neural Network

Neural network is one of the developments of artificial intelligence, where the neural network is likened to the human brain network that can learn the data given. Figure 2 is a block diagram of the wave height system, where the left is the input of the neural network in the form of wind speed (Vw), current wave height (Hs), and previous wave height ($Hs_{(t-1)}$), the right is the output of the system, namely the height of the next wave ($Hs_{(t+1)}$) [3]. The characteristic of the neural network is the black box it has, where the neural network algorithm works in the black box.

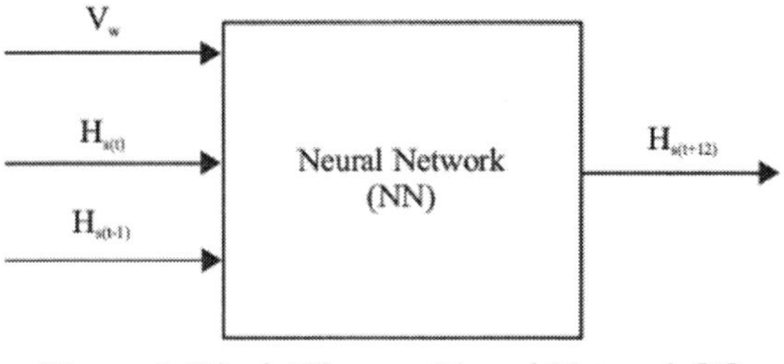

Figure 2. Block Diagram Neural Network [3]

The process of working on a neural network begins with scaling to equalize the input-output values to 0-1. Determining the amount of training data and validation data is the next step which will later be used as neural network learning data. The ratio that will be used is 80:20 where 80% is data for training, and 20% is validation data [4], the more data for training, the more this neural network learns. The next step is to determine the network architecture that will be used in the form of a multi-layer perceptron (MLP) consisting of an input layer, hidden layer, and output layer. After several trials, 7 hidden layers produced a low error rate. The learning method uses backpropagation, backpropagation was chosen so

IOP Conf. Series: Earth and Environmental Science **1473** (2025) 012025 doi:10.1088/1755-1315/1473/1/012025

that this network can learn the forward and backward flows. After descaling to change the value of 0-1 which was previously scaled back to its original number [5].

3. Data Analysis

Data collection was carried out at points A, B, and C in Figure 1 using data from the Meteorology, Climatology, and Geophysics Agency (BMKG) Surabaya in the form of wind speed and wave height for 6 years. The data is taken every hour in order to find out the movement of wind speed in detail which can be seen in Table 1. Where Table 1 is wind speed data at point A (Surabaya), point B (Java Sea), and point C (Banjarmasin). Based on its definition, wave height is caused by wind [6]. The harder the wind blows, the higher the waves formed, conversely the smaller the wind blows, the smaller the wave height. The maximum wind speed producing the highest maximum waves at point A was in 2020, at points B and C in 2018. This is evidenced by tables 1 and 2 which have a correlation between wind speed and wave height each year.

Table 1. Wind Speed

	A			B			C		
year	Min (Knot)	Max (Knot)	Ave (Knot)	Min (Knot)	Max (Knot)	Ave (Knot)	Min (Knot)	Max (Knot)	Ave (Knot)
2017	0.07	11.06	3.90	0.23	25.04	12.40	0.03	11.49	5.36
2018	0.03	14.89	3.80	0.04	27.92	10.81	0.12	13.48	4.93
2019	0.02	13.86	4.04	0.22	24.38	11.64	0.05	10.57	5.11
2020	0.08	16.81	3.72	0.19	25.77	11.14	0.07	11.12	4.97
2021	0.03	15.49	3.48	0.10	27.29	7.89	0.05	10.92	3.53
2022	0.14	14.35	6.18	0.22	25.94	10.56	0.04	12.22	4.25
Average			4.19			10.74			4.69

Table 2. Wave Height

	A			B			C		
year	Min (m)	Max (m)	Ave (m)	Min (m)	Max (m)	Ave (m)	Min (m)	Max (m)	Ave (m)
2017	0	2.51	0.80	1.10	11.06	3.90	0.6	2.14	0.73
2018	0.31	3.42	0.84	1.40	14.89	3.80	0.2	2.31	0.75
2019	0	2.48	0.47	0.64	13.84	4.04	0	1.61	0.71
2020	0	3.84	0.75	1.20	16.81	3.72	0.12	1.82	0.85
2021	0	1.40	0.56	0.30	14.19	4.07	0	1.71	0.68
2022	0.1	2.92	0.86	0.80	14.35	6.18	0.1	1.75	0.61
Average			0.71			4.28			0.72

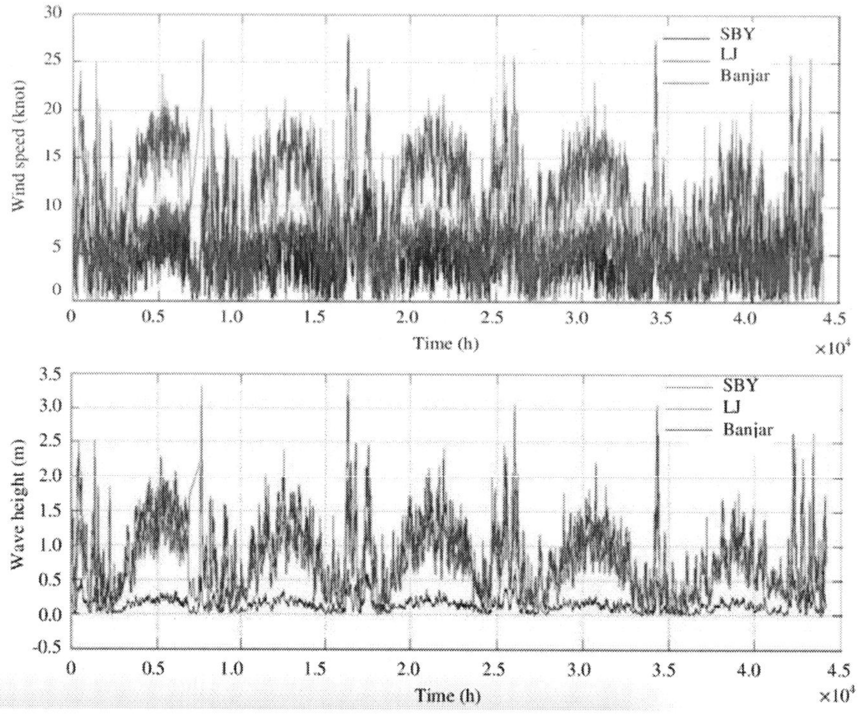

Figure 3. Wind Speed and Wave Height

Figure 3 shows the wind speed and wave height at points A, B, and C. where the blue line is point A (SBY), the red line is point B (LJ), and the magenta line is point C (Banjar). From the graph, it can be seen that it has the same up and down pattern, this is because Indonesia has two seasons, namely the dry season (the wind will blow low) and the rainy season (the wind will blow high). Not only the same pattern, but the wind and wave heights are equivalent, meaning that the one with the highest wind speed and high waves is point B (LJ) this is because the Java Sea is in the middle of the sea, not on the coast like at points A and B.

The wave height that has been obtained will be processed into the current wave height and one hour earlier. Both wave heights along with wind speed will be input to the neural network at each point. After the algorithm process is carried out by the neural network, the error value is obtained as indicated by the root mean square error (RMSE) at the training and validation stages as in table 3 where the RMSE training and validation values only reach 0.003. After the training and validation process, it is continued to the stage of finding the weights W1 and W2.

Table 3. RMSE Training and Validation

RMSE	A	B	C
Training	0.003998	0.006851	0.003400
Validation	0.003639	0.006004	0.003155

then the prediction weight will be obtained, namely the first weight (W1) is the weight of the input to the hidden layer, and the second weight (W2) is the weight of the hidden layer to the output [7]. The weights W1 and W2 are in the form of a matrix which is used as a prediction of wave height. The weight matrix W1 and W2, where W1 is a 7x4 matrix, namely 7 for the hidden layer and 4 (3 input + 1 bias input). While W2 has an 8x1 matrix, namely 8 (7 hidden layers + 1 bias hidden layer) and 1 output. The results of the weighting will later be used to predict wave heights, the results of which can be seen in Figures 4 - 6.

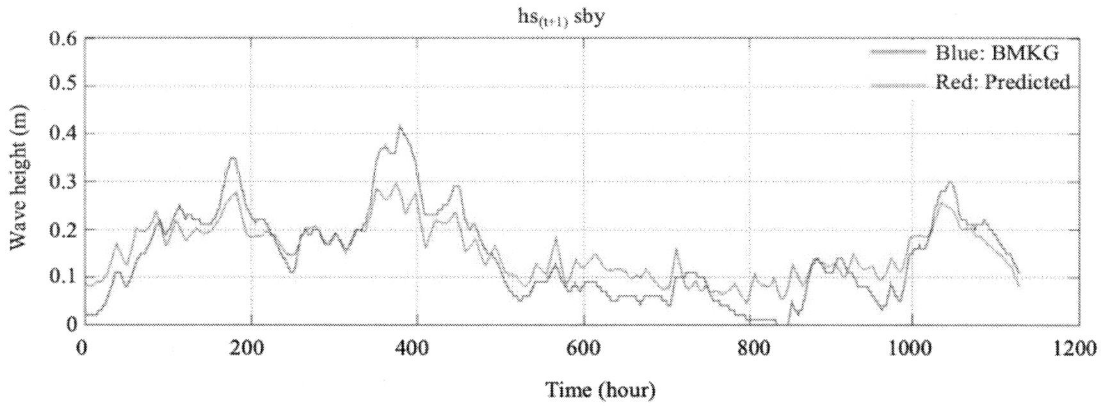

Figure 4. Wave Height Prediction in Surabaya

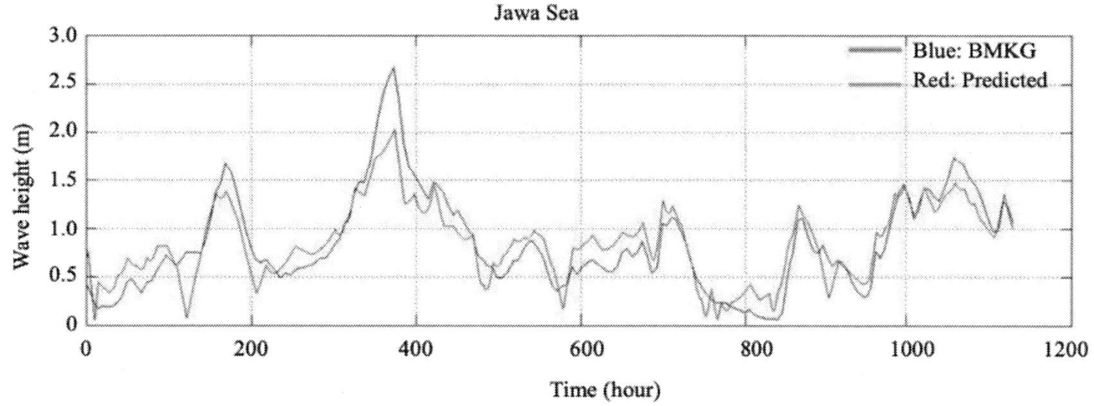

Figure 5. Wave Height Prediction in Java Sea

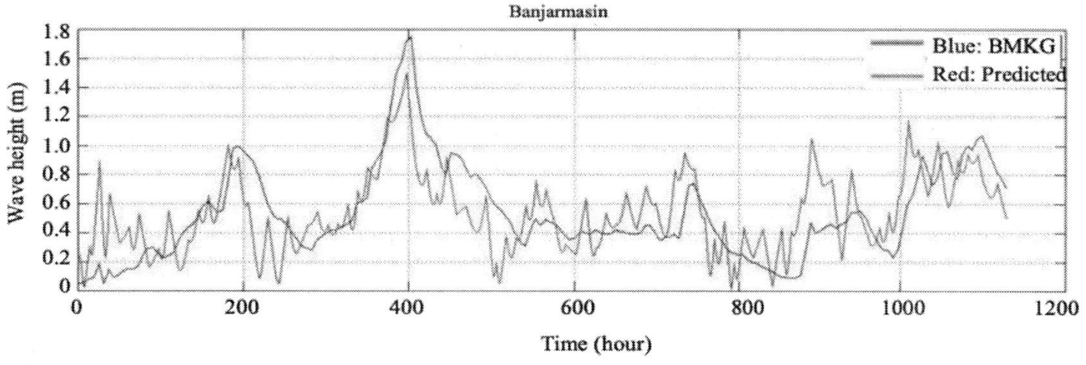

Figure 6. Wave Height Prediction in Banjarmasin

The wave height according to BMKG is depicted by the blue line, while the neural network prediction results are in red. In Figure 4 at point A, it can be seen that the wave height prediction pattern using the neural network has almost the same pattern as the wave height data from BMKG; this is evidenced by the root mean square error (RMSE), which is found by finding the difference between the prediction results and the data from BMKG, which is obtained at 0.04, where the smaller the RMSE produced, the more accurate the prediction [8]. The RMSE at points B and C is 0.21 and 0.15; this value is greater than the prediction at point A. If seen in Figure 5, the highest point of BMKG is 2.65 m, but the neural network prediction results only reach 2.05 m, where this difference affects the RMSE calculation at point B. However, overall the wave height prediction can be said to be successful because the RMSE value produced is low.

4. Conclusions

This study indicates that neural networks are a better way to predict wave height than traditional methods. The study was conducted in the Java Sea, a key maritime corridor of Indonesia and data collection was done at Surabaya (Point A), the Java Sea (Point B) and Banjarmasin (Point C). BMKG meteorological data from six years were used to train a neural network model whose RMSEs were low: 0.04 for Surabaya, 0.21 for the Java Sea, and 0.15 for Banjarmasin. Due to which it was revealed that this model is able to replicate the observed wave height as closely as possible in comparison to actual data with only slight deviations being identified in the latter case which concerns Java Sea. These findings highlight how much safer it can be at sea when using artificial intelligence as there will be no waiting time between reports being made before action may be taken or incorrect assumptions about what constitutes an emergency situation based on geographical location alone.

References

[1] A. S. Aisjah and S. Arifin, "Maritime weather prediction using fuzzy logic in Java Sea for shipping feasibility," *Int. J. Artif. Intell.*, vol. **10**, no. 13 S, pp. 112–122, 2013.

[2] P. Jain and M. C. Deo, "Artificial Intelligence Tools to Forecast Ocean Waves in Real Time," *Open Ocean Eng. J.*, vol. **1**, no. 1, pp. 13–20, 2008, doi: 10.2174/1874835x00801010013.

[3] W. L. Dhanistha, S. Silvianita, and W. Wardhana, "Wave height prediction based fuzzy logic," *Int. J. Civ. Eng. Technol.*, vol. **9**, no. 9, pp. 1611–1616, 2018.

[4] S. Mandal and N. Prabaharan, "Ocean Wave Prediction Using Numerical and Neural Network Models," *Open Ocean Eng. J.*, vol. **3**, no. 1, pp. 12–17, 2010, doi: 10.2174/1874835x01003010012.

[5] Y. Z. Law, H. Santo, K. Y. Lim, and E. S. Chan, "Deterministic wave prediction for unidirectional sea-states in real-time using Artificial Neural Network," *Ocean Eng.*, vol. **195**, p. 106722, 2020.

[6] W. L. Dhanistha, R. A. Atmoko, P. Juniarko, and R. Akbar, "Prediction of Significant Wave Height Using Neural Network in the Java Sea (North of Surabaya)," *Appl. Mech. Mater.*, vol. **862**, pp. 72–77, 2017, doi: 10.4028/www.scientific.net/amm.862.72.

[7] V. R. Kashikar and S. J. Mane, "Artificial Neural Network : An Effective Tool for," vol. **3**, no. 7, pp. 744–747, 2014.

[8] B. Paplińska-Swerpel and Łł. Paszke, "Application of neural networks to the prediction of significant wave height at selected locations on the baltic sea," *Arch. Hydroengineering Environ. Mech.*, vol. **53**, no. 3, pp. 183–201, 2006.

IOP Conf. Series: Earth and Environmental Science **1473** (2025) 012026 doi:10.1088/1755-1315/1473/1/012026

Implementation of Green Risk Based Maintenance (G-RBM) on Ships to Support Ship Operations in Shipping Companies in Indonesia

S Hariyanto[1], M B Zaman[1], R O S Gurning[1]

[1]Department of Marine Engineering, Institut Teknologi Sepuluh Nopember (ITS), Surabaya, 60111, Indonesia

Corresponding Email : suryahariyanto@gmail.com

Abstract. As a maritime country, Indonesia has supported the efficiency and effectiveness of sea transportation. In this context, safety is the main factor in carrying out ship operations. The International Maritime Organization (IMO) has issued regulations to strengthen and support the safety of ships in operation. In the current process, digitalization is the main issue to be developed and strengthened in supporting maritime processes. This research aims to develop and evaluate the implementation of Green-Risk-Based Maintenance (G-RBM) with digital reinforcement and increase operational reliability that is safe and environmentally friendly. Shipping system facilities owned by ships in Indonesia have a vital role in maintaining ship operations. However, the challenges faced are the age of the ship, the maintenance process and human error which often cause ship accidents. This research considers what factors contribute to the implementation of G-RBM in shipping companies.

1. Introduction

Safety and security are very essential factors in supporting the smooth sailing of ships, because accidents are difficult to predict and can happen anywhere. For this reason, before the ship is used, it is best to pay attention to the condition of the ship. Based on historical studies of maritime transportation, various maritime incidents were found that had major consequences for human life, community welfare and marine ecosystems. The International Maritime Organization (IMO) has established regulations with new innovations to realize safe maritime transportation, as well as increase safety awareness throughout the maritime sector.[1-2].

The types of ship accidents with the most significant frequency of accidents in Indonesia are ship fires, sinking and collisions. The main causes of ship fires are mechanical failure, flammable cargo, or ineffective fire extinguishing systems. The factors causing deaths in shipping accidents caused by fire/explosion are an average of 132% higher than accidents that do not involve fire/explosion. explosion. The main causes of ship sinking are poor ship maintenance resulting in leaks, failure of the ship's structure and design, and so on [3].

Risk-based maintenance (RBM) is a strategic approach to maintenance management that prioritizes resources on assets that have the highest risk of failure. Rather than creating a comprehensive maintenance schedule for all assets, RBM focuses on identifying critical equipment and implementing maintenance strategies tailored to its specific risk profile. In this context, RBM is about being proactive and not reactive in maintenance management. By understanding the risks associated with each asset, organizations can make informed decisions about how to allocate maintenance resources and minimize the impact of equipment failure [4]. Green Risk Based Maintenance (G-RBM) can be significantly improved by implementing environmentally friendly principles. This approach, often called Green Maintenance or Sustainable Maintenance, focuses on minimizing the environmental impact of maintenance activities while optimizing asset performance [5].

The aim of this research is to evaluate the implementation of green risk-based maintenance (G-RBM) on ships implemented by shipping companies in Indonesia. This research also evaluates and develops digitalization of an effective condition monitoring and predictive system to support the implementation of RBM on ships.

2. Literature Review

Many researchers have carried out research and development regarding RBM. Afef Saihi et al, have conducted a hierarchical component model for sustainable performance measurement of maintenance practices. In this research, a fourth-order PLS-SEM approach are implemented [6]. Aven has estimated a risk-based maintenance (RBM): An overview Reliability Eng & System Safety[7]. Macchi et al have A risk-based approach to maintenance scheduling in fuel storage depots[8].

Suhartini et al have conducted a review on risk-based maintenance decision-making models. Arunraj et al have conducted a Risk-based maintenance Techniques and applications [9]. Zhang et al have established a review of risk assessment techniques in maintenance strategy selection for process industries[10]. Khan, F. I et al, Risk-based maintenance of ethylene oxide production facilities [11]. Khalifa et al, Risk-based maintenance and remaining life assessment for gas turbines[12]. Tubis, Operational Risk Management Method in Road Transport [13]. Isyouf has analyzed the role of maintenance in improving companies' productivity and profitability [14].

Sari et al have analyzed Sustainable maintenance performance measures. This research is a pilot survey in malaysian automotive companies [15]. Sarstedt et al have developed How to specify, estimate, and validate higher-order constructs in PLS-SEM [16]. Pattison et al have developed a genetic algorithm in optimal safety system design [17]. Bartlett et al have explored a Genetic algorithm optimization of a firewater deluge system [18]. Kadir Arifin et al have evaluated the influence of hazard control and prevention toward safety behaviors and safety outcomes in coal-fired power plants using PLS-SEM. Pires et al have analyzed an approach to the prioritization of sustainable maintenance drivers in the TBL framework [19]. Amrina et al have analyzed Sustainable maintenance performance evaluation model for cement industry [20]. Parida et al have developed a multi-criteria hierarchical framework for maintenance performance measurement (MPM) [21].

Based on the facts, most ships in Indonesia are more than 25 years old. Based on the facts of previous research, review and evaluation of implementation by evaluating various factors supporting the development of RBM models in shipping systems have not been developed. So, this paper will evaluate the implementation and development of G-RBM on systems on ships by considering the relationship of various factors with the implementation of RBM. In this context, environmental friendliness and digitalization are issues raised by building a model of linkages and relationships.

3. Method

This conceptual model is a flow shows a series of concepts which is used as a research guide to help achieve goals research easily. Model conceptual in this research can be seen in Figure 1.

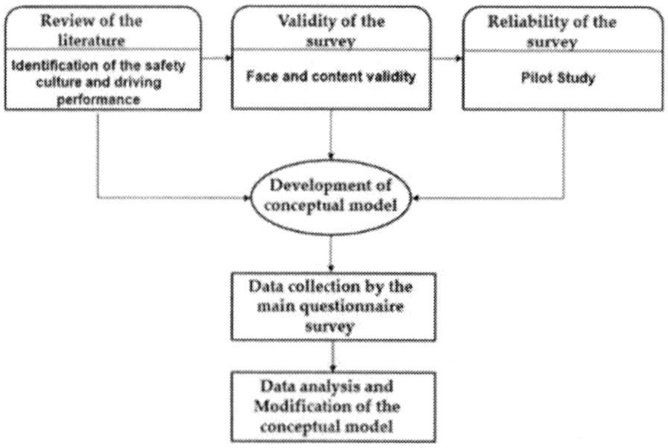

Figure 1. Research methodology

In this paper, several factors that influence the implementation of RBM are carried out. After that, it was developed to include environmental factors as a consideration for further RBM implementation.

In accordance with Figure 1, in general, this research uses an applied research approach with a structured research design method. The research stages consist of:

1. Literature review and Preliminary Study: This stage involves a comprehensive literature study on methodological concepts, RBM implementation, condition monitoring and predictive systems, as well as their application to shipping companies in Indonesia that operate ships and implement RBM. The preliminary study also involves an analysis of the conditions at the shipping company and the ships used.
2. Conceptual method and system design: At this stage, planning of variables and factors that influence the implementation of RBM is carried out. The relationship between variables is important in supporting analysis.
3. Data Collection and Analysis: Survey activities were carried out to collect data. In this case, a survey was carried out at the shipping company which operate domestic vessels.

4. G-RBM Concept

The RBM process if implemented on ships will have the following benefits. RBM benefits include:

a. Reduced downtime: By focusing on high-risk assets, RBM helps prevent major failures.
b. Optimized resource allocation: Maintenance efforts are concentrated where they are needed most.
c. Cost savings: By reducing unplanned downtime and avoiding unnecessary maintenance, RBM can result in significant cost reductions.
d. Improved safety: Identifying and mitigating potential hazards can improve workplace safety.

RBM process is about being proactive and not reactive in maintenance management. By understanding the risks associated with each asset, organizations can make informed decisions about how to allocate maintenance resources and minimize the impact of equipment failure. Risk Based Maintenance (RBM) can be significantly improved by implementing environmentally friendly principles. This approach, often called Green Maintenance or Sustainable Maintenance, focuses on minimizing the environmental impact of maintenance activities while optimizing asset performance.

Main Components of Environmentally Friendly RBM:

1. Environmental Risk Assessment: Identify potential environmental hazards associated with asset failure or maintenance activities. Assess the environmental impact of various maintenance strategies.
2. Green Maintenance Practices: Reduce, Reuse, recycle: Implement strategies to minimize waste generation during maintenance. Energy Efficiency: Optimize maintenance activities to reduce energy consumption.
3. Chemical Management: Choose environmentally friendly lubricants, cleaning agents and other chemicals. Waste Management: Proper disposal of hazardous waste and recycling of non-hazardous materials.
4. Sustainable Asset Management: Extended Asset Life: Implement maintenance strategies to extend the life of assets, thereby reducing the need for replacement.
5. Eco-Friendly Design: Consider environmental impacts during asset procurement and design. End-of-Life Management: A plan for the responsible disposal or recycling of an asset at the end of its life cycle.
6. Performance Measurement.

In this context, it is necessary to establish Establish key performance indicators (KPIs) to measure the environmental impact of maintenance activities, tracking progress towards sustainability goals.

5. Result and Discussion

Figure 2 shows the relationship between RBM and several variables and indicators that influence or contribute to the implementation of G-RBM in shipping companies that operate ships on certain routes. Table 1 shows groups of relationships that are related to variables and indicators for implementing G-RBM on ships.

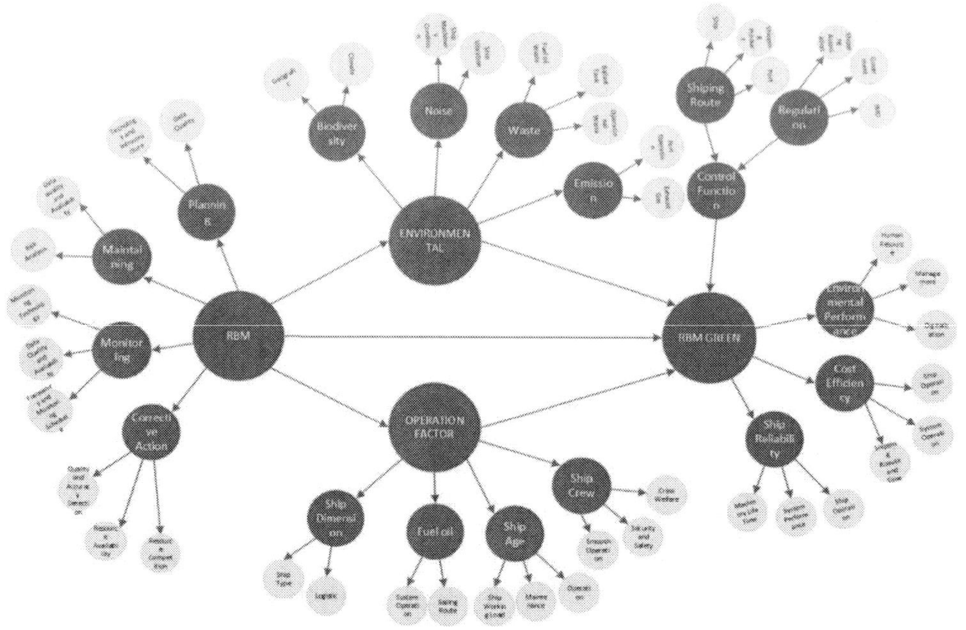

Figure 2. Relationship of G- RBM implementation variables

Table I shows several important variables and indicators in the implementation of G-RBM on ships owned by shipping companies. These variables and indicators were obtained by interviewing and brainstorming with employees and management of shipping companies in Indonesia that implement RBM.

Tabel 1. Variable and Indicator of G-RBM Process

No	Variables	Indicators
1	Risk Based Maintenance (RBM)	1.Planning
		2. Maintenance
		3. Monitoring
2	Environment	1. Noise
		2. emmision
		3. waste
		4.biodiversity
3	Ship's Operation	1.Ship's dimension
		2. Age of Ship
		3. Crew
		4.Fuel Consumption
4	G-RBM	1. Reliability
		2. maintenance
		3. Environment
		4. Digital process
		5. Regulation

The RBM variable has several indicators, namely: Planning, maintenance and monitoring. These variables and indicators show that the RBM process that has been implemented so far must consider planning, maintenance and monitoring activity factors. In this way, well-planned maintenance planning will be possible. In this context, RBM is implemented on ships well.

In the Environmental variable, the indicators include noise, emissions, waste and biodiversity. These indicators show how the ship is operated on a particular route. To achieve an environmentally friendly ship, the indicators are low emissions, waste issues must be handled well, waste must be managed well and biodiversity well controlled.

Table 1 shows the ship operation process. In this process, the machinery system, navigation system, deck system and the entire system run according to existing regulations. Indicators in this process are the ship's crew, ship size, ship age and fuel consumption. This indicator is what influences the implementation of RBM on ships operated by shipping companies. Ship operation variables are often the variables that determine the success of RBM implementation.

In the G-RBM variable, indicators contribute to the success of its implementation. These indicators include system reliability, maintenance, environment, digitalization processes and regulations. Currently, environmental issues are the main issue being considered. The issue of emissions, for example, has become a concern for the IMO in ship operations. The application of technology must lead to a reduction in emissions from ship exhaust gases. Then, process digitalization began to be applied to new ships, such as the use of digital twins on ships. Apart from that, the implementation of unmanned ships is a concern for IMO to implement. With this reality, the application of G-RBM becomes very relevant to improve ship operations.

6. Conclusions

The condition monitoring and predictive maintenance system has been designed to be a reinforcement and material for determining the implementation of risk-based maintenance on ships at shipping companies in Indonesia. Evaluation of the implementation of RBM in this research was carried out as a basis for determining maintenance strategies for systems on ships using RBM.

The factors that become supporting variables must be considered for development. By implementing effective RBM, it is hoped that RBM implementation can be carried out more efficiently. It is hoped that the results of this research will provide practical benefits for shipping companies in Indonesia for ship operations in Indonesia and contribute to the development of similar systems on other ships.

This paper presents the relationship between RBM and several variables and indicators that influence or contribute to the implementation of G-RBM in shipping companies that operate ships on certain routes. Table 1 shows groups of relationships that are related to variables and indicators for implementing G-RBM on ships.

Acknowledgments

The authors thank to the team at marine engineering department Institut Teknologi Sepuluh Nopember (ITS) who have discussed and supported this research.

References

[1] Christiansen M, Fagerholt K, Nygreen B and Ronen D 2007 Maritime Transportation **14** 189– 284

[2] Zaman M B, Kobayashi E, Wakabayashi N, Pitana T and Maimun A 2013 Fuzzy FMEA model for risk evaluation of ship collision in the Malacca Strait J. Sim **1** 91-104

[3] Zhou X, Ruan X, Wang H and Zhou G 2023 Exploring spatial patterns and environmental risk factors for global maritime accidents: A 20-year analysis Ocean Eng. **286** 115628

[4] Aju KumarV N, Gupta P and Gandhi O P 2019 Maintenance performance evaluation using an integrated approach of graph theory, ISM and matrix method. International Journal of System Assurance Engineering and Management, **10** (1), 57–82.

[5] Parida A, Chattopadhyay G 2007 Development of a multi-criteria hierarchical framework for maintenance performance measurement (MPM). J. Quality Maintenance Eng **13** 241–258

[6] Saihi A, Ben-Daya M, As'ad R 2023 A hierarchical component model for sustainable performance measurement of maintenance practices: A fourth-order PLS-SEM approach. *Computers & Industrial Eng.* **185** 1-20
[7] Aven T 2019 Risk-based maintenance: An overview Reliability Eng & System Safety **189** 1-9.
[8] Macchi M, Miragliotta G, Marseguerra M 2018 A risk-based approach to maintenance scheduling in fuel storage depots. Reliability Eng & System Safety **169**154-164.
[9] Suhartini S, Satria H, Setiawan, A 2018 A review on risk-based maintenance decision-making models. *J. Quality in Maintenance Eng*, **24**(4) 513-532.
[10] Zhang J, Wang Y and Wang L 2017 A review of risk assessment techniques in maintenance strategy selection for process industries. *Reliability Engineering & System Safety* **157** 54-67.
[11]Khan F I and Haddara, M R 2004 Risk-based maintenance of ethylene oxide production facilities Journal *of Hazardous Materials*. doi: 10.1016/j.jhazmat.2004.01.011.
[12]Khalifa M, Khan F and Thorp J 2015 Risk-based maintenance and remaining life assessment for gas turbines *Journal of Quality in Maintenance Engineering* **21**(1) 100–111
[13]Tubis A 2018 Operational Risk Management Method in Road Transport Publ. House of Wroclaw University of Science and Technology: Wroclaw, Poland
[14] lsyouf I 2007 The role of maintenance in improving companies' productivity and profitability. *Int. J. Prod. Econ* **105** 70–78. Computers & Industrial Engineering 185 (2023) 109699
[15] Sari E, Shaharoun A M, Ma'aram A and Yazid A M 2015 Sustainable maintenance performance measures: A pilot survey in malaysian automotive companies. *Procedia CIRP* **26** 443–448 https://doi.org/10.1016/j.procir.2014.07.163
[16] Sarstedt M, Hair J F Jr, Cheah J H, Becker J M and Ringle C M 2019 How to specify, estimate, and validate higher-order constructs in PLS-SEM. *Australasian* Marketing Journal (AMJ), **27(3)** 197–211.
[17] Pattison, R.; Andrews, J. 1999 Genetic algorithms in optimal safety system design. Proc. Inst. Mech. Eng. Part E J. Process Mech. Eng. **213** 187–197.
[18] Bartlett J A 2003 Genetic algorithm optimization of a firewater deluge system *Qual. Reliab. Eng. Int.* **19** 39–52.
[19] Pires S P, S´en´echal O, Loures E F R and Jimenez J F 2016 An approach to the prioritization of sustainable maintenance drivers in the TBL framework. *IFAC Papers OnLine*, **49** 150–155.
[20] Amrina E and Aridharma D 2016 Sustainable maintenance performance evaluation model for cement industry. IEEE International Conference on Industrial Engineering and Engineering Management 350-354.
[21] Parida A and Chattopadhyay G 2007 Development of a multi-criteria hierarchical framework for maintenance performance measurement (MPM*). Journal of Quality in* Maintenance Eng **13** 241–258

IOP Conf. Series: Earth and Environmental Science **1473** (2025) 012027 doi:10.1088/1755-1315/1473/1/012027

Heading Control on Twin Thruster USV Using Wind Direction Sensor and Electronic Compass Based on PID Controller

Handi Rahmannuri[1,2], Muhammad Rivai[1], and Agoes Santoso[2]

[1]Electrical Engineering Department, Institut Teknologi Sepuluh Nopember, Indonesia
[2] Marine Engineering Department, Institut Teknologi Sepuluh Nopember, Indonesia

*Corresponding author: rahmannuri@.its.ac.id

Abstract. Unmanned Surface Vehicles (USVs) can be operated on water surfaces in rivers, lakes, canals, shallow seas, deep seas, or other types of waters. Twin screw has better efficiency than conventional propulsion using rudders in terms of mechanical transmission and space requirements. This propulsion system works by adjusting the difference in rotation speed of the left and right propellers to control the ship's course. USV uses a wind direction sensor and electronic compass combined with the Proportional-Integral-Derivative (PID) algorithm to control USV's heading. The wind direction sensor controls the USV's course based on wind direction and the electronic compass functions to provide a reference for the USV's course. The study results show that at a speed of 0.78 m/s, the USV can change course towards the wind source in 3 seconds with a turn rate of 2.5°/second from the home position. At the same speed, the USV can return the USV's course to the 0° reference in 4 seconds when given a disturbance of 45°. The USV speed range is between 0.14 m/s – 29.7 m/s with a thrust value between 0.4 lbs – 3.2 lbs. In dry-run testing, the USV propeller produced 3950 rpm at optimal speed.

1. Introduction

Unmanned surface vehicles (USVs) have attracted a lot of attention from researchers because of their many uses in homeland security, military surveillance, environmental monitoring, ocean exploration, and coordinating operations with unmanned underwater vehicles [1]. In the marine context, an effective autonomous navigation controller is essential because USVs are usually managed remotely by humans or not at all, which necessitates a high degree of reliability and precision in their operational capabilities. In three dimensions, the USV has six possible ways to move. Steering control, which attempts to give a controller to move the rudder to eliminate inaccuracies between the intended and real courses, is its main motion control problem. The subject of intelligent ships, USVs, or intelligent offshore platforms, has grown in significance during the past several years. USVs have the potential to replace conventional human-controlled ships and operations because of their array of sensors, which allow them to continuously monitor a variety of maritime phenomena, such as ocean waves, tides, and marine meteorology [2]. Because of their small size, intelligence, and unmanned nature, which allow them to prevent casualties, they can be utilized in antisubmarine and anti-mine warfare, maritime security protection, and other military areas [3].

Despite significant developments in wireless communication, environmental sensing, navigation, location, and wireless communication, the motion control component of USV has lagged in recent years. Path-following control and autonomous navigation are two instances of this. Some maritime applications require it, such as the tracking management of a normal merchant ship, seawater monitoring for permanent spots, and water sampling for predetermined courses [4-6]. To successfully regulate motion, a mathematical model of the USV's kinematics and kinetics performance needs to be constructed [7-8]. Intelligent control and maneuverability prediction are based on this approach. There are two categories of mathematical models for ship motion. The global model also referred to as the Abkowitz model, is one of them [9–10]. By treating the hull, propeller, and rudder as one cohesive unit, it extends the hydrodynamic force operating on the hull into the Taylor series of each motion variable. The alternative model is the separable model, which is also known as the Mathematical Modeling Group (MMG) model [11]. The Japanese MMG proposed it, based on the Abkowitz model. The MMG model divides hydrodynamic forces into three parts: the rudder, propeller, and hull. The model takes these components' interactions into account as well.

Because the Abkowitz model is more robust and comprehensive mathematically, it is used by the community. Based on the Abkowitz model, Fossen et al. developed a modeling and control strategy for marine craft, including ships, high-speed craft, semi-submersibles, and floating rigs [12]. They performed a comprehensive analysis of the kinematics and kinetics performance of the marine boat. J. Menoyo Larrazabal et al. tackled USV system rudder control and created a linked surge-sway-yaw model to aid USV maneuvering control [13]. A trajectory planning and tracking technique for following a target vehicle operating in an obstacle field that is variably limited was presented by Petr Švec et al. An actuation model and a simplified three degrees of freedom (DOF) model served as the foundation for the methodology. Experiments and simulations were performed to demonstrate the effectiveness of the developed strategy [14]. These models are designed for USVs with thrusters and rudders, which can cause damage due to gear wear during steering. The complex mechanical design and plastic rudders can compromise the USV's dependability. To address this issue, a rudderless double thruster structure is used, where the thrust levels of port and starboard thrusters control the vessel's velocity, and differential thrust regulates the steering.

Simplifying the mechanical structure of the unmanned surface vessel (USV), not only enhances the flexibility of the steering operation but also reduces the need for intricate rudder control, thereby significantly boosting the overall reliability of the USV. This streamlined design optimizes performance and ensures smooth navigation under various operating conditions. Moreover, the structure has a higher energy conversion rate than the combination thruster-rudder structure. Studies on USVs with rudderless double thrusters have been carried out. Justin E. Manley et al. led the way in USV development for nearly 15 years. The actuators of the AutoCat and ROAZ II USVs utilized the rudderless double thruster arrangement for hydrological surveys and marine research [15]. R. Sutton and colleagues employed nonlinear motion control to analyze the rudderless double thruster control system of the Springer USV [16].

The development of the collision avoidance algorithms for the C-Enduro USV with rudderless twin thrusters and the sensing and avoidance system was detailed by A. Savvaris et al. [17]. Having stated that, not much earlier study has been conducted on this kind of USV modeling. To assess motion and dynamic performance and offer guidance for motion control of USVs with rudderless twin thrusters, a three-DOF dynamic model and propeller thrust model are constructed for a 1.5 m long, 50 kg USV equipped with these devices. The USV is a double-hulled watercraft propelled by two propellers as part of a specially designed propulsion system. The system is identified to obtain the model parameters. Consequently, further investigation is required into the navigation system that employs a proportionate propulsion system to the requirements and objectives of the USV. Being an unmanned surface vehicle, the USV genuinely needs a propulsion system with low cost, low space requirements, and, of course, small and lightweight proportions. As a linear steering system, USV uses the PID (proportional-integral-derivative) algorithm for differential steering. Since these model-based approaches are typically applied to systems with perfect knowledge dynamics, they are not appropriate for nonlinear USV systems, which often struggle to obtain an accurate model [18–20].

The main contributions of this paper can be summarized as follows:

(1) Applying a differential steering navigation model to a USV with twin thrusters.

(2) Obtaining the characteristics of the USV according to the values of the differential steering model.

(3) Applying a PID algorithm to control the maneuver of the USV based on electronic compass and wind direction sensors.

(4) Knowing the performance of the navigation system.

2. Literature Review

2.1. Differential Steering Thruster

USV uses a differential steering configuration on the propulsion system as seen in the figure 1. The kinematic modeling of autonomous ships adopts the mobile robot differential steering system [21]. The position and course of the autonomous ship are represented in vectors $q = \left[x_Q, y_Q, \varphi\right]^T$. With (x_Q, y_Q) indicating the location of point Q (center of the body axis if the USV being moved) and φ is the heading of the USV in the framework of global coordinates (G_{les}). $\dot{\theta}_l$ and $\dot{\theta}_r$ denotes the angular velocity of the left and right wheels. Rhe are three variables to control (x_Q, y_Q, φ) with only two control inputs. Therefore, the system is under-actuated. The relationship between the linear velocity of the robot v and its angular velocity ω with the rate of change of position and heading are explained 'in equation (2)'.

$$\begin{bmatrix}\dot{x}_Q\\ \dot{y}_Q\\ \varphi\end{bmatrix}=\begin{bmatrix}\cos\varphi & 0\\ \sin\varphi & 0\\ 0 & 1\end{bmatrix}\begin{bmatrix}v\\ \omega\end{bmatrix} \tag{1}$$

Linear speed and heading robot angle are explained in 'equation (2) and (3)'. Where r is the wheel radius, and $2b$ is the base length of the wheel, which is the distance between the wheels measured along the axis of the wheel rotation. By substituting 'equation (2) and (3)' into '(1)', 'equation (4) – (6)' are obtained.

$$v = \frac{r(\dot{\theta}_r + \dot{\theta}_l)}{2} \tag{2}$$

$$\omega = \frac{r(\dot{\theta}_r - \dot{\theta}_l)}{2b} \tag{3}$$

$$\dot{x}_Q = \frac{r}{2}\left(\dot{\theta}_r \cos\varphi + \dot{\theta}_l \cos\varphi\right) \tag{4}$$

$$\dot{y}_Q = \frac{r}{2}\left(\dot{\theta}_r \sin\varphi + \dot{\theta}_l \sin\varphi\right) \tag{5}$$

$$\dot{\varphi} = \frac{r(\dot{\theta}_r - \dot{\theta}_l)}{2b} \tag{6}$$

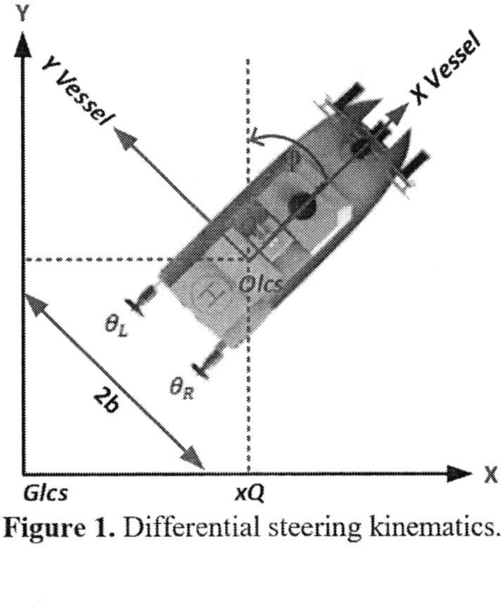

Figure 1. Differential steering kinematics.

2.2. Proportional-Integral-Derivative Control

For the correct feature selection to meet all design criteria, the controller architecture and type, as shown in figure 2, must be detailed. The PID controller is most commonly used in sectors where PID stands for proportional, integral, and derivative. In industrial control applications, this PID controller is used to control temperature, flow, pressure, speed, and other process variables. This advanced control mechanism utilizes a combination of three distinct control to accurately adjust the output of a system based on its input in order to achieve desired setpoints and ensure optimal performance. With its robust capabilities and versatility in managing multiple parameters simultaneously, the PID controller has become an essential component in many industrial processes across numerous industries, including manufacturing, energy production, chemical processing, and more. Because it regulates process variables through a feedback loop control mechanism, it is the most accurate and reliable controller. It is necessary to have an appropriate controller and to have a deep comprehension of the characteristics and nature of the system [22].

The following formula can be used to determine the control signal that a PID controller applies to the process:

$$u(t) = K_p\, e(t) + K_i \int e(t)\, dt + K_d \frac{d}{dt} e(t) \tag{7}$$

'Equation (8)' can be transformed as follows using the Laplace transform:

$$U(s) = K_p\, E(s) + K_d\, s\, E(s) + K_i\, E(s) \tag{8}$$

$$G_c = \frac{U(s)}{E(s)} = K_p + \frac{K_i}{s} + K_d \tag{9}$$

It is simpler to understand if the controller is observed in the continuous or analogue domain. Currently, though, most controllers are digitally integrated with other systems or with Arduino-style microcontrollers. This PID controller would be used since 'equation (10)', which deals with discrete time in the digital domain, calls for it.

$$u(n) = K_p\, e(n) + K_i \sum_{k=0}^{n} e(k)\, T + K_d \frac{e(n) - e(n)+1}{T} \tag{10}$$

where the control signal of n is discretely calculated every t seconds in time. Here, T denotes either a set interval of time or a sample period. In order to compute the output, the controller needs a new value where *Kp* is proportional constant, *Ki* is integral constant, *Kd* is derivatif constant, and *e* is error (set value – actual value).

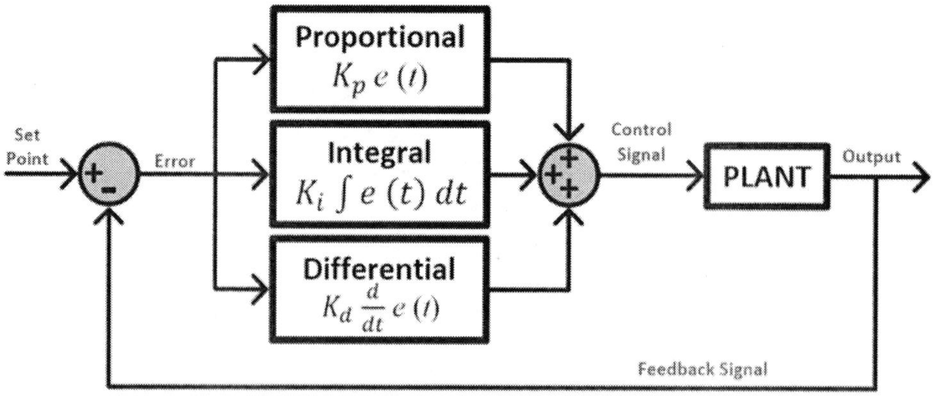

Figure 2. PID controller configuration.

3. Research Method

The hardware block diagram can be seen in figure 3, USV uses an Atmega 328P-based Main Controller Unit (MCU) that has 14 inputs and outputs. 6 pins can be used to generate Pulse Width Modulation (PWM) signals that function to regulate the thruster speed. A Personal Computer (PC) is used to program the system and acquire sensor data and the navigation process wirelessly because it uses 433 MHz telemetry components. The ideal telemetry range is 500 m if there is no radio signal interference. The driver to control the thruster uses the SIMONK 30A type. The actuator uses an 83-watt submersible DC brushless motor that can rotate up to 3900 rpm. The power supply uses a Lippo 3-cell battery with a voltage of 11.7 volts and a capacity of 5000 AH. USV uses an electronic compass sensor to maintain the orientation of the bow and a wind direction sensor to test maneuverability.

In controlling its maneuvers, the USV uses a PID control algorithm which is a close loop control. The block diagram can be seen in figure 4. The set point (reference/set value) is determined from the output data of each sensor. The navigation set point with an electronic compass is obtained by positioning and determining the orientation of the USV. After the set point is saved, if there is a change in the ship's course, the electronic compass sensor will read the actual course of the USV and then compare it with the set point. To keep the USV always trying to stay on course, the thruster will rotate if there is a discrepancy (error) to lessen the distance between the set point and the real value. Navigation with an electronic compass function to maintain the course/orientation of the USV when it has not received/read any sensor data.

Meanwhile, in navigation with a wind direction sensor, the set point is obtained by positioning the sensor straight in front of the USV bow. This set point is 0° because when exposed to wind, the sensor position changes according to the wind direction, thus there will be a difference/error with the set point. This error is used as a reference for the thruster to rotate and move the USV so that its course/orientation is parallel to 0°. So, when navigation with a wind direction sensor is activated, the USV will follow the origin/source of the wind. This navigation is useful when the USV is on a mission to find substances, gases, vapors, etc. that are volatile and carried by the wind, for example, to track chemical gases or dangerous radioactive substances. Illustrations of the navigation testing process with an electronic compass and wind direction sensor can be seen in figure 5 and figure 6.

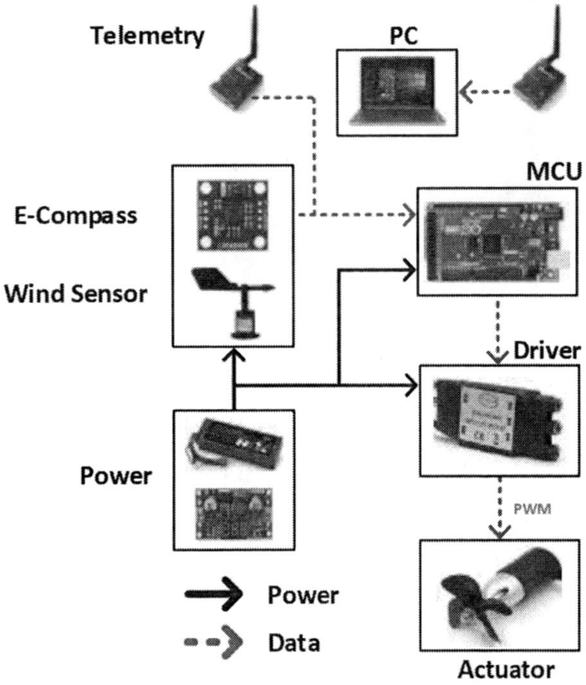

Figure 3. Hardware block diagram.

IOP Conf. Series: Earth and Environmental Science **1473** (2025) 012027 doi:10.1088/1755-1315/1473/1/012027

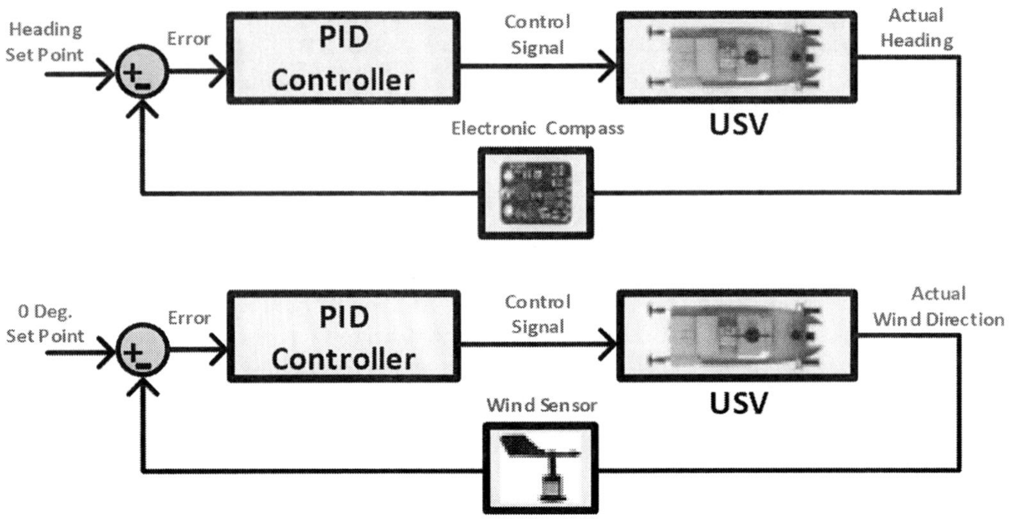

Figure 4. Control system block diagram.

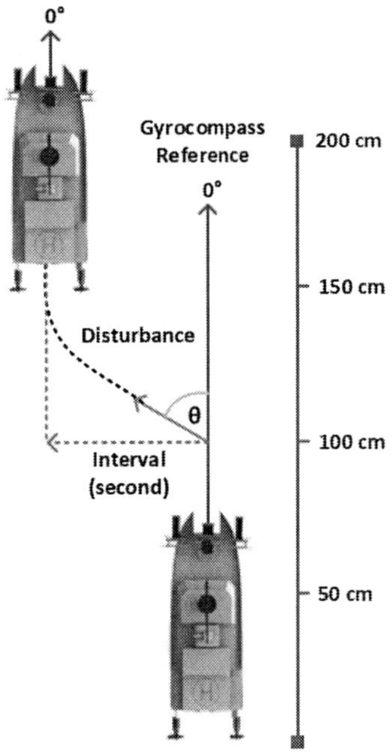

Figure 5. Navigation with electronic compass.

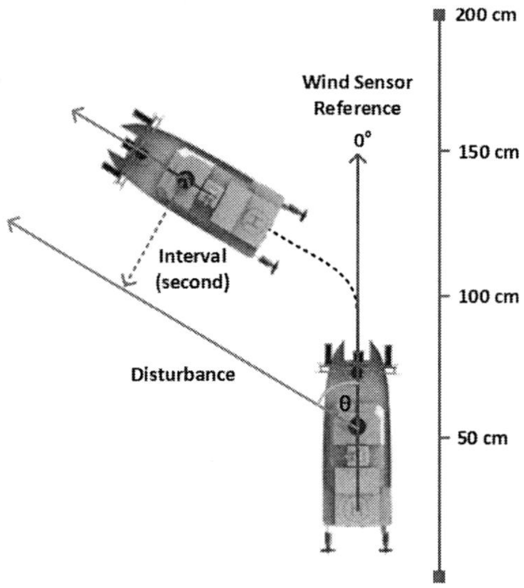

Figure 6. Navigation with wind direction sensor.

Figure 7 shows the USV's three-dimensional architecture. The hull uses a catamaran model because it has a better level of stability than a mono hull. This model has been widely used for USVs/ships that require a high level of stability, including for marine data acquisition missions such as depth, topography, environmental quality, etc. The wind direction sensor is placed on the superstructure to obtain a proportional air mass flow rate so that it can read the direction of the wind source optimally. Likewise, the telemetry module is placed on the superstructure to obtain adequate signals for communication and sensor data acquisition purposes. The complete specifications of the USV can be seen in table 1.

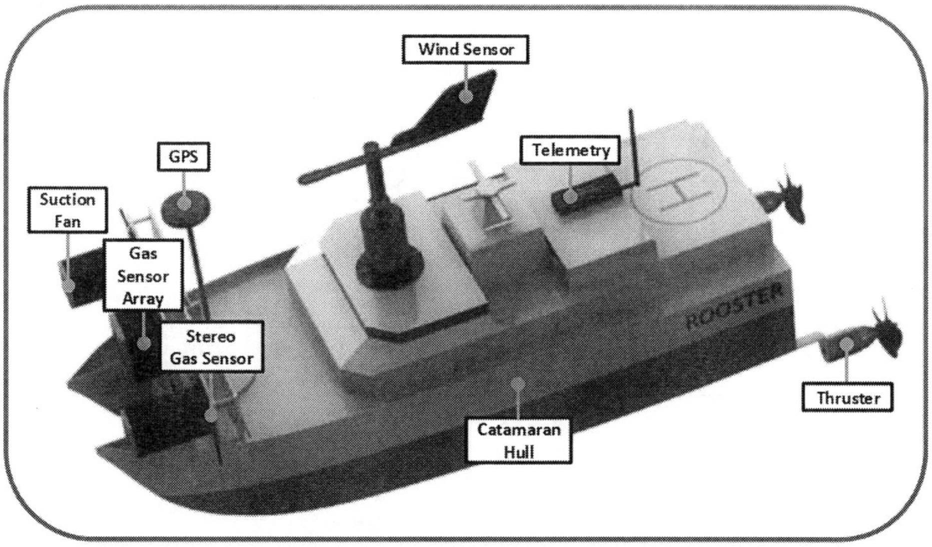

Figure 7. USV advance design for gas tracking.

Table 1. USV's technical specification.

Physical	
Size (LxWxH)	75cm x 28cm x 36cm
Weight with load	5 Kg
Weight without load	3 Kg
Material	Fiber
Hull Tipe	Catamaran
Maximum Speed	2 m/s
Draught	7 cm
Maximum Load	8 Kg
Sea Condition	1 Level (Wind resistance 1 m/s)
Wave Resistance	< 1 m
Electrical	
Motor Power	83 Watt
Battery Capacity	2 x 11.7 Vdc, 5 Ah
Navigation Mode	Manual - Autonomous
Operating Time	2 Hours
Communication	
Communication way	UHF Network Bridge
UHF Frequency	344 MHz
Communication distance	500 m
Communication port	USB 3.0
Navigation	
Global Positioning System	Ublox MBN
Electronic Compass	3 Axis
Wind Direction Sensor	Digital
Stereo Gas Sensor	TGS 2620
Gas Sensor Array	Multiple Sensor

4. Result and discussion

The general characteristics of USV can be seen in table 2. Characterization is done with several variations of RPM values so that it can produce several speed variations. The highest RPM variation is 3914 which produces a speed of 1.56 m/s. The variation of RPM differences in both thrusters also produces the maneuverability of USV which can be seen in the turn rate column. This column shows how far the USV turns per second. The higher the RPM value, the higher the turn angle per minute produced. The flow velocity of the water fluid at the exit position on the thruster is also calculated to produce the thrust value.

Table 2. USV's character.

Left RPM	Right RPM	Speed (ms^{-1})	Turn Rate ($Deg.s^{-1}$)	Thrust (lbs)
156	156	0.18	1.489	0.4
352	352	0.29	2.978	0.6
743	743	0.44	5.956	1.0
1918	1918	0.78	14.891	1.9
3914	3914	1.56	29.783	3.0
0	352	0.19	2.978	0.2
0	1135	0.37	8.935	0.3
0	1918	0.78	14.891	0.5
0	2935	1.17	22.337	0.9
0	3914	1.56	29.783	1.5

Table 3. Electronic compass navigation data.

Electronic Compass (Deg.)	Error	Vout (Vdc)	Left RPM	Right RPM	Turn Rate ($Deg.s^{-1}$)
0	0	0	0	0	2.9
45	45	0.63	0	352	2.9
-45	-45	4.38	352	0	2.9
90	90	1.25	0	352	2.9
-90	-90	2.75	352	0	2.9

Table 4. Electronic compass navigation success rate.

Electronic Compass (Deg.)	Speed (ms^{-1})	Turn Rate ($Deg.s^{-1}$)	Min. Err. Time Interval (s)	Range (m)	Success Rate	
					Success	Failure
20°	0.3	2.9	4.21	3	√	
-20°	0.3	2.9	4.15	3	√	
34°	0.3	2.9	5.16	3	√	
-34°	0.3	2.9	5.44	3	√	
45°	0.3	2.9	6.74	3	√	
-45°	0.3	2.9	6.25	3	√	
90°	0.3	2.9	13.21	3	√	
-90°	0.3	2.9	13.05	3	√	

Table 3 shows the variation of the electronic compass angle obtained after the USV stores the set point 0°, so when the USV orientation is greater or less than the set point, the system will detect an error and then signal the thruster to lower the error value. The RPM column shows the control process on the left and right thrusters to return the USV orientation to the set point 0°. Table 4 shows the success rate can be represented by the period the USV can minimize the error value. When the error value is close to 0, it is assumed to be the same as the set point. There is a tolerance of approximately 0 - 5 points so that the thruster can determine its rotation. The minimum error time interval column shows that the USV can complete the mission in 13.05 seconds when the orientation error is 90. In another study conducted by Radimas et al. in year 2019, USV can achieve a steady state error of 19.2% in 6.5 seconds on a gas source search mission. The gas source is at a position of 45° from the USV bow at a distance of 60 cm [23].

Table 5 shows the variation of the sensor angle obtained after the USV stored the set point 0°, meaning that the position of the sensor tip is parallel to the front bow of the USV. Determining the set point 0° means that it is believed that the wind is coming from the front. (bow) of the USV, so when the sensor is hit by wind and turns, there will be an error. If the sensor rotation deviation is greater or less than the set point, the system will detect the error and then signal the thruster to lower the error value. The RPM column shows the control process on the left and right thrusters to return the USV orientation to the set point 0°. Table 6 shows the success rate can be represented by the period for the USV to minimize the error value. When the error value is close to 0, it is assumed to be the same as the set point. The minimum error time interval column shows that the USV can complete the mission in 2.39 seconds when the orientation error is 45. At an error value of 95, the USV system fails in the process of tracking the wind direction source because the error value is too large and the USV also has limited maneuverability so the process to minimize the error fails. The maneuverability in question is when the USV changes orientation/heading, there will be drifting on the ship's hull against the water media so that it takes a relatively long time to reach the set point. Unlike land vehicles, tires can grip the asphalt media well because the two have high friction.

Table 5. Wind direction sensor navigation data.

Wind Direction (Deg.)	Error	Vout (Vdc)	Left RPM	Right RPM	Turn Rate (Deg.s^{-1})
0°	0	0	0	0	2.9
45°	-45	0.63	0	352	2.9
-45°	45	4.38	352	0	2.9
90°	-90	1.25	0	352	2.9
-90°	90	2.75	352	0	2.9

Table 6. Wind direction sensor navigation success rate.

Wind Direction (Deg.)	Speed (ms^{-1})	Turn Rate (Deg.s^{-1})	Min. Err. Time Interval (s)	Range (m)	Success Rate	
					Success	Failure
20°	0.3	2.9	1.06	3	√	
-20°	0.3	2.9	1.01	3	√	
45°	0.3	2.9	2.39	3	√	
-45°	0.3	2.9	2.11	3	√	
90°	0.3	2.9	4.61	3	√	
-90°	0.3	2.9	4.79	3	√	
95°	0.3	2.9	-	3		√
-95°	0.3	2.9	-	3		√

Figure 8 represents the PID control process in controlling USV navigation using an electronic compass. The purple and blue line references are the vertical axis on the left. This axis shows the control signal number from the MCU. The red line value is on the right vertical axis (error). The red chart is the error value that occurs because the USV is given a deviation/disturbance when maneuvering forward. When the error value is greater than 0, the left thruster is active, indicating that the USV is disturbed so that the orientation is controlled clockwise and vice versa. The largest error value is at 34 and at the same time, the left thruster is also active as long as the error has not entered the range point 0 - 5. The higher the error, the longer it takes for the thruster control signal to be active. This can be seen from the dynamics of the red and purple lines. Figure 9 shows data retrieval process is slightly different, the set point 0° is stored when the USV has a straight-ahead orientation. After that, the orientation of the USV is given a disturbance before navigating. The red line (error) from the beginning has shown a deviation that is quite far from the set point value of 0°, which is 35 points. This shows that the USV has a counterclockwise orientation disturbance so that at the same time the left thruster signal will be active (purple line). This causes the USV orientation to be returned to the minimum error, which is approaching the set point. The orientation of the USV will gradually change direction clockwise. It is proven when the red line has reached the set point and passed it, the control signal for the right thruster is active (green line). Oscillations like this occur until the error approaches the set point 0°.

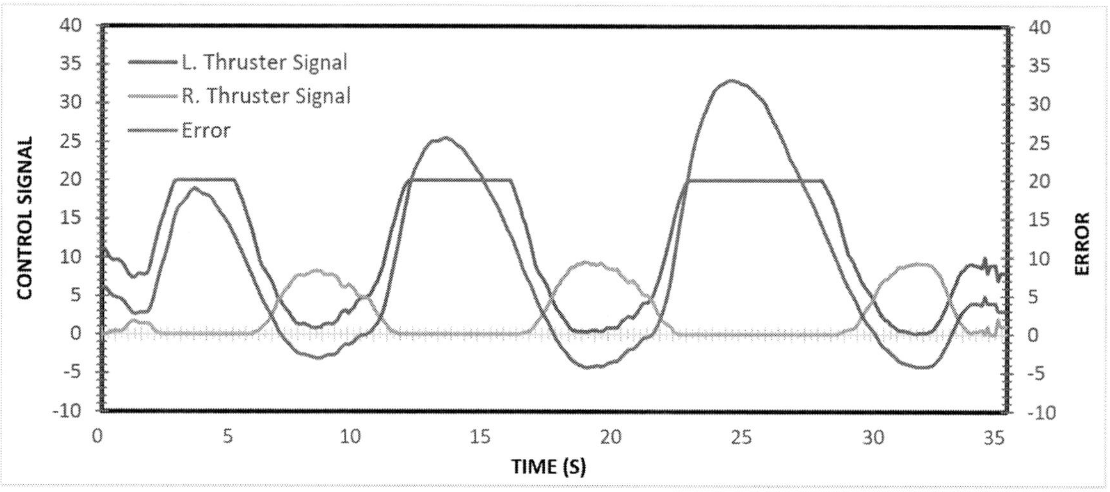

Figure 8. Electronic compass navigation data with multiple disturbance.

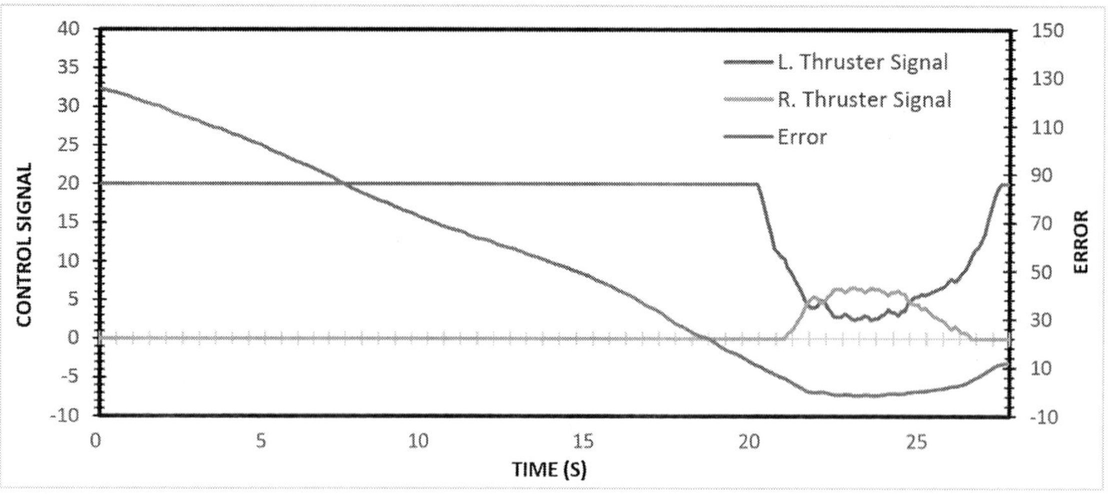

Figure 9. Electronic compass navigation data with 125 point error of disturbance.

Figure 10 represents the PID control process in controlling USV navigation using a wind direction sensor. The purple and blue line references are the vertical axis on the left. This axis shows the control signal number from the MCU. The red line value is on the right vertical axis (error). The red chart is the error value that occurs because the sensor is given a deviation/disturbance, either rotating counterclockwise or clockwise. When the error value is less than 0, the left thruster is active, indicating that the USV is disturbed so that the orientation is controlled following the error angle generated by the sensor. The largest error value is at -87 and at the same time, the left thruster is also active as long as the error has not entered the range point 0 - 5. The higher the error, the longer it takes for the thruster control signal to be active. This can be seen from the dynamics of the red and purple lines. Figure 11 shows the 0° set point is stored when the sensor has a straight-ahead orientation to the USV bow. After that, the USV orientation is given a disturbance before navigating. The red line (error) from the beginning has shown a deviation that is quite far from the set point value of 0° and peaks at an error value of -28 points. This indicates that the sensor is rotating clockwise so that at the same time the left thruster signal will be active (purple line). This causes the USV orientation to be returned to the minimum error, which is approaching the set point. The USV orientation will gradually change direction clockwise. It is proven when the red line has reached the set point and passed it, the control signal for the right thruster is active (green line). Oscillations like this occur until the error approaches the set point of 0°.

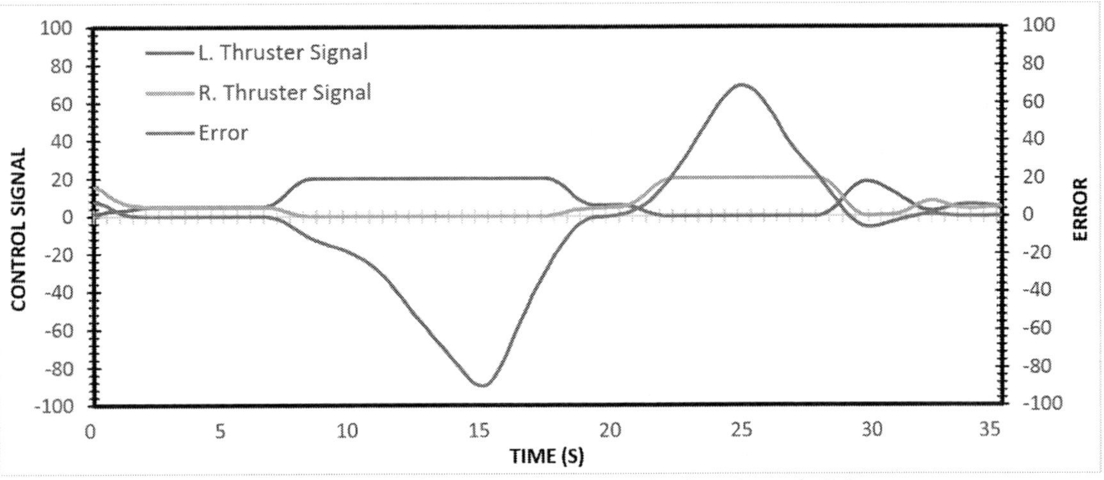

Figure 10. Wind direction sensor navigation data 1st trial.

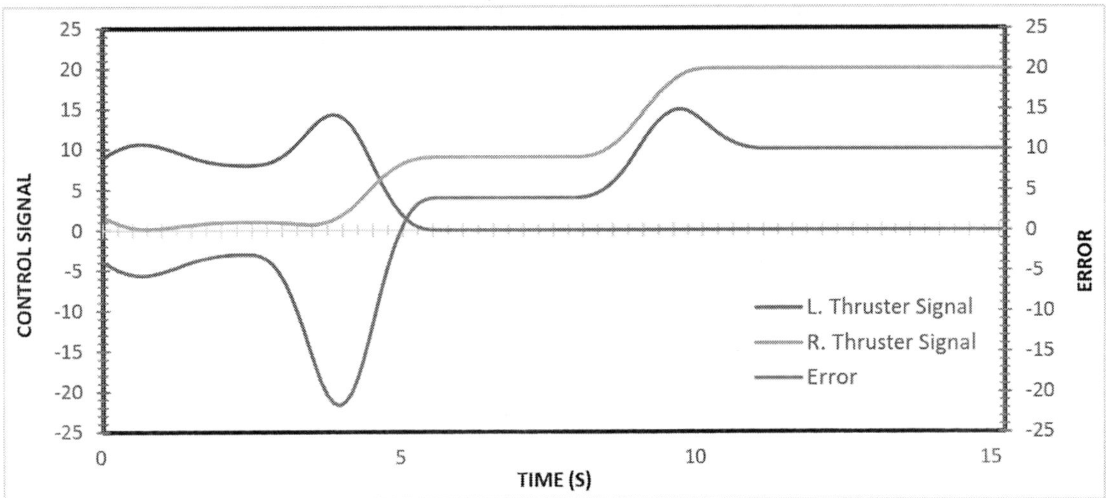

Figure 11. Wind direction sensor navigation data 2nd trial.

5. Conclusions

The Proportional-Integral-Derivative algorithm has been successfully applied to light USV using an electronic compass sensor and wind direction sensor to control the differential steering thruster. The maneuverability of the USV has also been well characterized, as has the navigation process when the USV is disturbed. At parameter values KP = 1, KI = 0.1, and KD = 1 for navigation with an electronic compass, the minimum error time interval is 6.74 seconds for an error of 45 points. While with parameter values KP = 1, KI = 0.1, and KD = 1 for navigation with wind direction sensor, the minimum error time interval is 2.39 seconds for error 45 points. Both of these navigations have the capability of turning maneuvers 2.9 deg.s^{-1} at speeds 0.3 ms^{-1} with thrust value 0.6 lbs.

Acknowledgement

The authors are grateful to the Electrical Engineering Department, Marine Engineering Department, and ITS Robotics Centers of Institut Teknologi Sepuluh Nopember for supporting this study.

Reference

[1] Pietrzykowski Z, Wolejsza P, and Nozdrzykowski L 2022 The Autonomous Navigation System of A Sea-Going Vessel (*Ocean Engineering* ***261***) p 112104

[2] Jorge V A, Granada R, Maidana R G, Jurak D A, Heck G, Negreiros A P, Dos Santos R H, Gonçalves L M G, Amory A M A 2019 Survey on unmanned surface vehicles for disaster robotics: Main challenges and directions (*Sensors* ***19***) p 702

[3] TIDEWISE. 2020 Unmanned Surface Vehicles (USV) TUPAN: Data Spesificatio. (*Tidewise Engenharia e Servicos Navais Ltda*)

[4] Liu Z, Zhang Y, Yu X, and Yuan C A 2020 System Application for Autonomous Vessel Safety: Hazard Identification As a Function of Increasing Autonomy Levels (*Safety Science* ***131***) p 104919

[5] Geng X, Wang Y, Wang P, and Zhang B 2019 Motion Plan of Maritime Autonomous Surface Ships by Dynamic Programming for Collision Avoidance and Speed Optimization (*Sensors* ***19***) p 434

[6] Sarda E I, Qu H, Bertaska I R, Von Ellenrieder K D 2016 Station-keeping control of an unmanned surface vehicle exposed to current and wind disturbances (*Ocean Eng* ***127***) pp 305–324

[7] Liu Z, Zhang Y, Yu X, Yuan C 2016 Unmanned surface vehicles: An overview of developments and challenges (*Annual Review in Control* ***41***) pp 71–93

[8] Jin J, Zhang J, and Liu D 2018 Design and verification of heading and velocity coupled nonlinear controller for unmanned surface vehicle (*Sensors* ***18***) p 342

[9] Alexandersson M, Zhan D, Mao W, and Ringsberg J, W A 2023 Comparison of Ship Manoeuvrability Models To Approximate Ship Navigation Trajectories. (*Ship and Offshore Structures*) vol **18** no 4 pp 550-557

[10] Von V and Uharek S A 2019 Numerical Prediction of Ship Manoeuvring Performance in Waves (Doktor der Ingenieurwissenschaften)

[11] Yasukawa H Yoshimura Y 2015 Introduction of MMG standard method for ship maneuvering predictions (Journal of Marine Science and Technology ***20***) pp 37–52

[12] Budak G and Beji S 2020 Controlled Course-Keeping Simulations Of A Ship Under External Disturbances (*Ocean Engineering* ***218***) p 108126

[13] Larrazabal J M and Peñas M S 2016 Intelligent Rudder Control of An Unmanned Surface Vessel (*Expert Systems with Applications*) vol ***55*** pp 106–117

[14] Švec P, Thakur A, Raboin E, Shah B C, and Gupta S K 2014 Target Following With Motion Prediction For Unmanned Surface Vehicle Operating In Cluttered Environments (*Autonomous Robot*) pp 383–405

[15] INESTEC Technology and Science 2013 ROAZ II Autonomous Surface Vehicle (*Campus Da Feup Porto Portugal*)

[16] Sutton R, Sharma S, and Xao T 2014 Adaptive Navigation Systems For An Unmanned Surface Vehicle (*Journal of Marine Engineering & Technology*) vol ***10*** pp 3–20

[17] Savvaris A, Niu H, Oh H, and Tsourdos A 2014 Development of Collision Avoidance Algorithms For The C-Enduro USV (*In Proceedings of the 19th IFAC world congress*)

[18] Kotian R, Umesh P, and Gangadharan K V 2023 Real-Time Implementation Of PID Control For Unmanned Surface Vessel: A Practical Approach (*IEEE International Conference on Electronics, Computing, and Communication Technologies*)

[19] Kotian R, Umesh P, and Gangadharan K V 2022 Parameter optimization of unmanned surface vessel propulsion motor based on BAS-PSO (*International Journal of Advance Robotic Systems*) pp 1-7

[20] Hui Y, Shipeng H, Xiaofei Y, Zhaoping D, and Wentao X 2024 Adaptive neural synergetic heading control for USVs with unknown dynamics and disturbances (*Ocean Eng **300***) p 117438

[21] Hashim A, Saini T, Bhardwaj H, Jothi A, and Kumar A V 2019 Application of Swarm Intelligence in Autonomous Cars for Obstacle Avoidance (*Integrated Intelligent Computing, Communication and Security*) pp 393-404

[22] Bhookya J, Kumar M V, Kumar J R, and Rao S 2022 Implementation Of PID Controller For Liquid Level System Using MGWO And Integration of IOT Application (*Journal of Industrial Information **28***) p 100368

[23] Radimas P M D L, Rivai M, and Irfansyah A N 2019 Unmanned Surface Vehicle Navigation Based on Gas Sensors and Fuzzy Logic Control to Localize Gas Source (*Journal of Physics: Conference Series **1201***) p 012001

IOP Conf. Series: Earth and Environmental Science **1473** (2025) 012028 doi:10.1088/1755-1315/1473/1/012028

Bridge Simulator with Green Technology to Enhance Critical Thinking Skills of Shipping Vocational Cadets

Supangat[1], A Sriantini[1]* and D Malik[1]

[1] Ship Operations Engineering Technology Departement, Faculty of Shipping Vocational, Hang Tuah University, Indonesia

*E-mail: ari.sriantini@hangtuah.ac.id

Abstract. The shipping industry faces challenges in environmental management, prompting the use of green technology. Maritime education must also support sustainability, with tools like the bridge simulator playing a pivotal role. This study examines how integrating green technology into a bridge simulator enhances students' critical thinking and gathers their feedback on its use. The research involved 35 fourth-semester students of Ship Operations Engineering Technology. Data were collected through observation, tests, and questionnaires. Critical thinking skills were assessed based on interpretation, analysis, inference, evaluation, explanation, and self-regulation. Results showed the bridge simulator improved students' critical thinking by 74%. Integrating green technology helped students better understand ship operations, emphasizing energy efficiency and environmental awareness. It also increased students' interest in learning, especially in eco-friendly operations, and prepared them for teamwork in their future careers.

1. Introduction

The shipping world today faces great challenges related to the need for better environmental management and environmentally friendly technology. The maritime industry, including sea transportation and ship operations, has undergone significant changes. One of the efforts to reduce environmental impact is the application of green technology (1)(2). This technology is designed to reduce carbon emissions, improve fuel efficiency, and minimise damage to marine ecosystems (3).

In line with these developments, maritime educational institutions must provide relevant education and support environmental sustainability. The effective use of learning media, such as Bridge Simulator, plays a vital role in sailing education. Bridge simulators are not only used to train skills in navigating and manoeuvring ships but can also be an effective medium to instil environmental awareness and critical thinking skills that are needed in the management of green technology (1)(2).

In the face of global competition in the era of the Industrial Revolution 5.0, universities must produce graduates with high competence and competitiveness. Graduates from universities must have competencies that include aspects of attitudes, knowledge, general skills, and special skills (4), including shipping universities. To have high competitiveness, graduates must be able to think critically in solving the problems they face, be creative, and be able to collaborate. As outlined(5), 21st-century skills include (1) critical thinking and problem-solving, 2) communication and collaboration, (3) creativity and innovation.

Critical thinking is a thinking skill related to cognitive processes that make students able to think according to their ability to deal with problems. Critical thinking includes the ability *to* interpret, analyse, infer, evaluate, explanation, and self-regulation (6). When students think critically, they will use their knowledge and skills to solve problems, analyse, process information well, and make decisions. Based on research conducted by (7), It was found that students in the Ship Operations Engineering Technology (TROK) study program at Hang Tuah University have low critical thinking skills. Therefore, efforts and strategies are needed to improve students' critical thinking skills.

Given the importance of producing graduates who can think critically to increase global competitiveness, learning must be designed in such a way that it can spur students to think critically. Learning must be supported by media, learning models that are by competencies and subjects, and contextual modules or scenarios so that they can provide students with a real picture of their future world of work. The learning media used in the TROK study program includes simulators, one of which is the bridge simulator. Bridge simulators can provide simulations of teamwork on board, which can increase students' interest in learning. With the increase in interest in learning, student competencies will also increase. In addition, the use of bridge simulators has been proven to have a very good influence on students' attitudes, knowledge, and skills(8).

A bridge simulator is a computer system that is integrated between hardware and software that simulates ships, ports, weather conditions, and traffic conditions in a waterway, which is used for training and assessment(9). Bridge simulators are used to improve the competence of prospective seafarers. These competencies include navigation, implementing collision regulations at sea, and team management on the platform. The bridge simulator allows students to apply learning with scenarios that simulate real life in complex navigation. When combined with learning green technology, students not only practice using simulators but also understand how the application of this technology in ship operations can reduce negative impacts on the marine environment (1). It is hoped that simulator-based learning combined with contextual learning scenarios is expected to improve students' critical thinking skills.

In addition to improving student competence in navigating, bridge simulators can also increase students' interest in learning and understanding (10). Research conducted by (11) shows that simulators can increase students' knowledge of the navigation function quite well, as evidenced by the results of student exams have increased and have reached the value of completeness. Several variables affect the success of students in achieving the required competencies. The skill variable has a significant influence of 0.35 on competence, followed by the knowledge variable (8). Based on previous research, the researcher updated research related to the use of bridge simulators as a learning media, with green technology to enhance students' critical thinking skills. Therefore, the researcher will examine the extent to which the use of bridge simulators that integrate green technology in its design and operation can improve students' critical thinking skills and find out students' opinions about the use of bridge simulators as a learning media.

2. Method

This research is applied research with a quantitative approach and uses *a quasi-experimental* method. To obtain comprehensive data, all students of the Ship Operations Engineering Technology Study Program (TROK) semester IV, the academic year 2023/2024, totalling 35 people as subjects in this study. Students of the fourth semester of the TROK study program were chosen as research subjects because the students had studied the core competency courses of the TROK study program, such as Terrestrial Navigation, Meteorology, Maritime English, Electronic Navigation, Compass, ship manoeuvre, and Steering System. The sampling technique used in this study is purposive sampling. The data collection methods used in this study are test, observation, and questionnaire methods. Written tests are used to measure students' critical thinking skills. Critical thinking skills indicators are used based on(6), interpretation, analysis, Inference, evaluation, explanation, and self-regulation. The test given to students is in the form of multiple-choice questions which include materials on Terrestrial Navigation, Meteorology, Maritime English, Electronic Navigation, Compass, ship manoeuvre, and Steering System. as stated in (12). The test is carried out twice, namely, the pre-test and the post-test. In addition to the written test, the researcher also used a questionnaire to find out students' opinions about learning using the simulator that had been carried out. The stages of collecting data in the study are the first to provide pre-test questions to all students in the fourth semester of the TROK study program; the second is to provide treatment by applying to learn using a bridge simulator combined with contextual learning scenarios, then to provide critical post-test questions to all students in the fourth semester of the TROK study program, the results of the pre-test and post-test are compared using the N-Gain test to find out the level of difference in results test. The overall research procedure is shown in Figure 1 below:

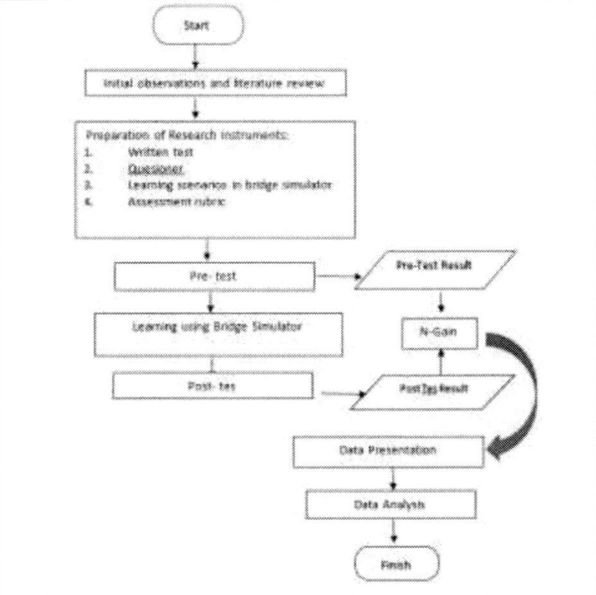

Figure 1. Research Procedure Flow Diagram.

3. Results and Discussion

3.1 Pre-Test Result

The pre-test is carried out before students are treated with learning using a bridge simulator. The results of the Pre-test are presented in the following Table 1:

Table1. Pre-Test Result.

No	Respond	Score Pre - Test
1	A1	66,7
2	A2	33,35
3	A3	40,02
4	A4	46,69
5	A5	33,35
6	A6	40,02
7	A7	53,36
8	A8	26,68
9	A9	46,69
10	A10	53,36
11	A11	53,36
12	A12	60,03
13	A13	53,36
14	A14	53,36
15	A15	66,7
16	A16	40,02
17	A17	53,36
18	A18	40,02
19	A19	33,35
20	A20	33,35
21	A21	20,01
22	A22	26,68
23	A23	33,35
24	A24	20,01
25	A25	20,01
26	A26	26,68
27	A27	40,02
28	A28	20,01
29	A29	20,01
30	A30	46,69
31	A31	66,7
32	A32	26,68
33	A33	20,01
34	A34	33,35
35	A35	26,68

3.2 Post-Test Result

The post-test was carried out after students were given learning treatment using a bridge simulator. The Post-test results are presented in the following Table 2:

Table 2. Post-Test Result.

No	Respond	Score Post-Test
1	A1	93,38
2	A2	80,04
3	A3	80,04
4	A4	86,71
5	A5	80,04
6	A6	93,38
7	A7	66,7
8	A8	80,04
9	A9	80,04
10	A10	93,38
11	A11	93,38
12	A12	100
13	A13	100
14	A14	86,71
15	A15	100
16	A16	80,04
17	A17	86,71
18	A18	86,71
19	A19	80,04
20	A20	86,71
21	A21	73,37
22	A22	66,7
23	A23	86,71
24	A24	80,04
25	A25	66,7
26	A26	80,04
27	A27	93,38
28	A28	80,04
29	A29	86,71
30	A30	80,04
31	A31	100
32	A32	80,04
33	A33	73,37
34	A34	86,71
35	A35	80,04

3.3 Implementation of Learning Using Simulators

After the pre-test was carried out, then students were treated with learning using simulators. Learning in the simulator is carried out four times in each group. Each group consists of 6-7 students. The researcher has prepared practice scenarios for learning in Bridge Simulator, and the implementation aligns with these scenarios. Scenarios are designed with a contextual approach. Every student's learning using a bridge simulator has its role. Some serve as captains, chief officers, second officers, and helmsmen. The learning outcomes to be achieved include determining the position of the ship, carrying out ship movements safely, being able to operate and use electronic navigation equipment as a tool in navigation, being able to use Standard Marine Communication Phrases (SMCP) on board, able to use Maritim English on board for internal and external communication, able to apply COLREG 1972 in shipping. An overview of the scenario in the bridge simulator is presented in the following Figure 2 and Figure 3:

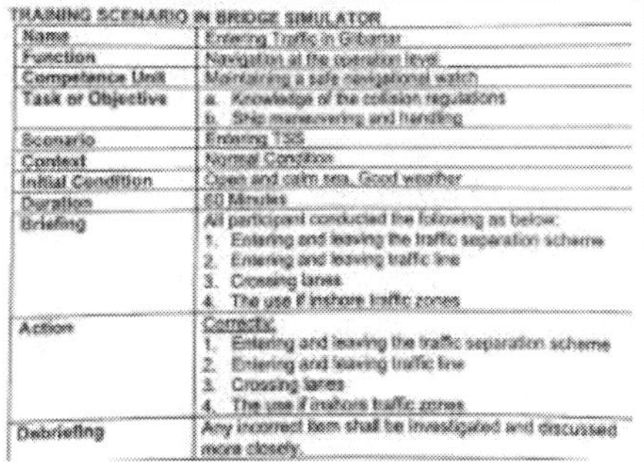

TRAINING SCENARIO IN BRIDGE SIMULATOR

Name	Entering Traffic in Gibartar
Function	Navigation at the operation level
Competence Unit	Maintaining a safe navigational watch
Task or Objective	a. Knowledge of the collision regulations b. Ship maneuvering and handling
Scenario	Entering TSS
Context	Normal Condition
Initial Condition	Open and calm sea, Good weather
Duration	60 Minutes
Briefing	All participant conducted the following as below: 1. Entering and leaving the traffic separation scheme 2. Entering and leaving traffic line 3. Crossing lanes 4. The use if inshore traffic zones
Action	Correctly: 1. Entering and leaving the traffic separation scheme 2. Entering and leaving traffic line 3. Crossing lanes 4. The use if inshore traffic zones
Debriefing	Any incorrect item shall be investigated and discussed more closely.

Figure 2. The scenario in the bridge simulator.

Figure 3. Implementation of simulator learning.

The problem given when students learn to use the bridge simulator is that the ship will enter the Gilbartar channel. The ship's draft is 11.9 m, the east wind direction, the wind speed is 7 knots, and the ship speed is 14 knots. Students should consider the need to sail at an economical speed to save fuel. Students are divided into groups and asked to safely steer the boat into the separation chart in the Gilbartar channel.

3.4 Questionnaire Results

In the learning effectiveness indicator, the bridge simulator is considered very effective, with a total score of 85%. Students feel that the simulator helps them understand the concept of navigation and ship safety better. After the implementation of the green technology-based scenario, students also stated that they understood more about the importance of fuel efficiency and emission reduction during the operation of the ship.

In the indicator of increasing learning interest, the bridge simulator is considered very helpful in increasing students' interest in learning, with a total score of 86.7%. Students stated that the introduction of green technology concepts in simulator scenarios, such as the use of economical speed, made students more interested in learning more.

Bridge simulators are considered to be very helpful in improving student learning outcomes, with a total score of 87.05%. In addition to improving navigation and communication skills, students were also informed that learning using green technology scenarios can improve their understanding of how environmentally friendly applications are applied on board. This is expected to contribute to improving student learning outcomes as a whole.

The indicator of training teamwork is considered very effective, with a score of 88.19%. In the bridge simulator scenario, students are trained to work in teams with different roles, such as captain, chief officer, second officer, and helmsman. At a time when scenarios about green technology are added, students are also trained to work together in accounting for economical speed to save fuel use.

On the environmental awareness indicator, the bridge simulator helped students learn about the importance of environmentally friendly boat operations with a total score of 84.5%. Students stated that the application of green technology-based scenarios, such as fuel saving and carbon emission reduction, helps increase students' awareness of environmental issues related to ship operations.

3.5 Analysis of Pre-Test and Post-Test Results

In this study, the researcher used critical thinking skills indicators from (6), which include *Interpretation, Analysis, Evaluation, Explanation, Inference,* and *Self-regulation.* The explanation of these indicators is shown in Table 3 below:

Table 3. Critical Thinking Skills Indicators.

No	Component	Indicator
1	Interpretation	Understand weather and navigation information and determine the meaning of symbols and indications.
2	Analysis	Analyse navigation situations and identify potential causes of observed phenomena.
3	Evaluation	Assess the relevance and importance of navigation information and determine appropriate actions.
4	Inference	Make logical conclusions based on weather information and ship conditions
5	Explanation	Provide clear and precise reasons for actions taken on the voyage
6	Self Regulation	Evaluate and adjust actions based on changes in information and surrounding circumstances.

The use of bridge simulators as a learning media to improve critical thinking skills for prospective seafarers can be known from the results of the N-Gain calculation of students' Pre-test and Post-Test scores, whose formula is as in Formula 1. The results of the N-Gain calculation are presented in the following Table 4:

Table 4. *N-Gain Score Calculation.*

No	Respond	Score Pre - Test	Score Post-Test	Score Ideal (100-Pre Test)	Post-Test - Pre-Test
1	A1	66,7	93,38	33,3	26,68
2	A2	33,35	80,04	66,65	46,69
3	A3	40,02	80,04	59,98	40,02
4	A4	46,69	86,71	53,31	40,02
5	A5	33,35	80,04	66,65	46,69
6	A6	40,02	93,38	59,98	53,36
7	A7	53,36	66,7	46,64	13,34
8	A8	26,68	80,04	73,32	53,36
9	A9	46,69	80,04	53,31	33,35
10	A10	53,36	93,38	46,64	40,02
11	A11	53,36	93,38	46,64	40,02
12	A12	60,03	100	39,97	40,02
13	A13	53,36	100	46,64	46,69
14	A14	53,36	86,71	46,64	33,35
15	A15	66,7	100	33,3	33,35
16	A16	40,02	80,04	59,98	40,02
17	A17	53,36	86,71	46,64	33,35
18	A18	40,02	86,71	59,98	46,69
19	A19	33,35	80,04	66,65	46,69
20	A20	33,35	86,71	66,65	53,36
21	A21	20,01	73,37	79,99	53,36
22	A22	26,68	66,7	73,32	40,02
23	A23	33,35	86,71	66,65	53,36
24	A24	20,01	80,04	79,99	60,03
25	A25	20,01	66,7	79,99	46,69
26	A26	26,68	80,04	73,32	53,36
27	A27	40,02	93,38	59,98	53,36
28	A28	20,01	80,04	79,99	60,03
29	A29	20,01	86,71	79,99	66,7
30	A30	46,69	80,04	53,31	33,35
31	A31	66,7	100	33,3	33,35
32	A32	26,68	80,04	73,32	53,36
33	A33	20,01	73,37	79,99	53,36
34	A34	33,35	86,71	66,65	53,36
35	A35	26,68	80,04	73,32	53,36

$$\text{N-Gain} = \frac{Score\ Posttes - score\ Pretes}{Score\ Ideal - Score\ Pretes} \qquad (1)$$

$$\text{N-Gain} = \frac{1574{,}12}{2125{,}98} = 0{,}74$$

The results of the analysis using N-Gain are then interpreted based on the following criteria in Table 5:

Table 5. Categories Interpretation of N-Gain Effectiveness.

Percentage	Interpretation
< 40	Ineffective
40-55	Less effective
56-75	Quite Effective
>76	Effective

Source: (13)

Based on the analysis and interpretation of the N-Gain percentage results, 74% was obtained, so it can be concluded that the bridge simulator is quite effective in improving students' critical thinking skills. This is in line with what was conveyed by(14), that simulators have increased their usefulness in sailing education over the years. (15) Bridge simulators can increase students' activeness in maritime education and training. So, with the increasing activity of students in learning to use bridge simulators with complete navigation equipment that resembles the real situation, bridge simulators are one of the learning media to improve students' critical thinking skills as prospective sailors.

The improvement of students' critical thinking skills can be seen from the difference in the average pre-test and post-test scores. The average pre-test score was 38.69, while the average post-test score was 84.23. There was a very real difference between the average score of the pre-test and the average score of the post-test, which was 45.54. This shows that there is an increase in students' critical thinking skills from before they were given learning using a bridge simulator to after learning using a bridge simulator. The difference in the average pre-test and post-test scores is shown in Figure 4 below:

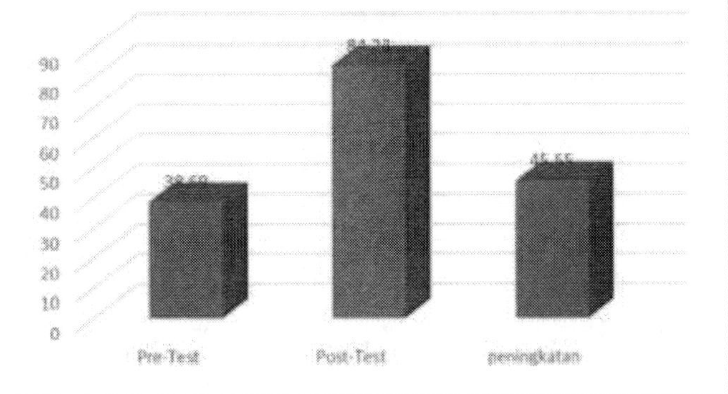

Figure 4. Average chart of Pre-Test and Post-Test scores.

3.6 Questionnaire Results Analysis

The questionnaire was given to 35 students who had received learning treatment using a bridge simulator. In each indicator, there are five statements. The results of the questionnaire are presented in the following Table 6:

Table 6. Questionnaire results.

No	Indicator	Percentage
1	Learning Effectiveness	85,14%
2	Increases interest in learning	86,67%
3	Improve learning outcomes	87,05%
4	Environmental awareness	84,5%
5	Practicing teamwork	88,19%

Based on the results of the questionnaire, it can be seen that students use bridge simulators very effectively as a learning medium to improve their critical thinking skills, with a percentage of 85.14%. This shows that learning using a bridge simulator helps students in internalising the concept of complex ship operations, thus spurring students to think critically. In addition, the application of green technology in the scenario further enriches the learning experience. (2)So that students can apply this knowledge in their future work. Bridge simulators are also able to increase student learning interest by 86.67%; besides that, they are also considered to be able to improve student learning outcomes very well. The high interest in learning shows that the use of bridge simulators as an interactive and contextual-based learning medium has succeeded in making students more enthusiastic in participating in learning.

In the increase in learning outcomes by 87.05%. The high score shows that bridge simulators can not only improve students' skills in navigation but also strengthen students' understanding of environmental issues. This shows that scenarios that also involve green technology can have a positive impact on student learning outcomes. As conveyed by (16) The bridge Simulator is very good for increasing student learning motivation, which will impact student competence. The bridge simulator has an impact on training students ' teamwork and is considered very effective, with a score of 88.19%. Excellent teamwork on board is essential, and learning using this bridge simulator allows students to learn about the importance of collaboration and coordination. The need to determine a safe speed in the environmentally friendly operation of a ship requires excellent coordination between the crew.

The application of green technology-based scenarios in the bridge simulator helped increase students' environmental awareness, with a score of 84.5%. This increase in environmental awareness shows that adding green technology to learning scenarios has a positive impact on shaping students' attitudes towards environmental sustainability. The score shows that the learning method with the bridge simulator using contextual scenarios based on green technology can be used to instil a thorough awareness of ship operations.

3.7 Analysis of Observation Results

According to the observations made during students' problem-solving sessions in the bridge simulator room, it has been determined that students verify the beginning position of the ship on the sea chart before entering the Gibraltar channel. In addition, students also check the bow of the initial ship, the depth of the waters on the map, and the wind conditions. This shows that students have excellent interpretation skills. In the next step, students determine the direction that the ship must follow by paying attention to the draft of the ship, the state of the waters, and the wind conditions. This shows that students' skills in analysing, evaluating, and making conclusions are pretty good. When the ship crosses paths with other ships, students communicate with the other ships. Based on the results of the interview, communication between the ships was carried out to obtain an agreement between the two ships when crossing through the left hull or right hull so that the ships could cross paths safely. This shows that students have the skills to give reasons for the conclusions drawn very well. In addition, from the observation results, students can look back at the decisions taken wrong or correct and then corrections are made if they are wrong. This shows that students have pretty good Self-regulation skills.

The increase in students' critical thinking skills in learning using a bridge simulator is because the learning is using a contextual approach. Through contextual problems, it is easier for students to understand the problem so that they can quickly determine the steps to solve the problems given. In line with what was conveyed by (17)There is a significant relationship between critical thinking skills and contextual approaches because contextual problems will help students understand a concept they have learned. As conveyed by (18)Contextual problems are experiences that students can use as a reference when processing new information to make decisions.

4. **Conclusion and Suggestion**

Based on a study conducted on 35 students, it was concluded that the bridge simulator, which integrates green technology into its design and operation, is considered quite effective in improving students' critical thinking skills as prospective seafarers, with an effectiveness rate of 74%. Incorporating green technology principles in the bridge simulator scenario helps students better understand the operation of ships, making students more aware of the importance of energy efficiency and environmental impact in the decision-making process. In addition, using bridge simulators significantly increases students' interest in learning and improves learning outcomes, especially in environmentally friendly boat operations. Bridge simulators are also very effective in training students to work collaboratively in teams to prepare students for the world of work on board when they graduate.

Based on the conclusions presented, it can be suggested that a long-term evaluation of the influence of using bridge simulators be conducted. Do the skills gained remain relevant and used when entering the workforce? Therefore, in the following study, the researcher will examine the long-term impact of using bridge simulators on students' critical thinking skills and environmental awareness after graduation and working on ships. The research will involve alumni already working to find out how much they apply the skills they gained while studying in the bridge simulator.

Acknowledgment

Thank you to the Rector of Hang Tuah University, the Head of LPPM Hang Tuah University, and the Dean of the Faculty of Shipping Vocational for providing permission and financial support in the implementation of this research. We would also like to express our gratitude to the students of the TROK study program class 2022 who have helped carry out this research.

References

[1] Liu J, Yan X, Liu C, Fan A, Ma F. Developments and Applications of Green and Intelligent Inland Vessels in China. J Mar Sci Eng. 2023;**11**(2).
[2] Zhen L, Wu Y, Wang S, Laporte G. Green technology adoption for fleet deployment in a shipping network. Transp Res Part B Methodol [Internet]. 2020;139:388–410. Available from: https://doi.org/10.1016/j.trb.2020.06.004
[3] Psaraftis HN, Kontovas CA. Decarbonization of maritime transport: Is there light at the end of the tunnel? Sustain. 2021;**13**(1):1–16.
[4] Kementerian Pendidikan dan Kebudayaan RI. Permendikbud 3 Tahun 2020 Tentang Standar Nasional Pendidikan Tinggi. J Account Sci. 2020;1–57.
[5] The Partnership for 21st Century Learning. P21 Framework Definition. Донну [Internet]. **2015**;1–9. Available from: http://www.p21.org/our-work/p21-framework
[6] Facione P a. Critical Thinking : What It Is and Why It Counts. Insight Assess [Internet]. 2015;(ISBN **13**: 978-1-891557-07-1.):1–28. Available from: https://www.insightassessment.com/CT-Resources/Teaching-For-and-About-Critical-Thinking/Critical-Thinking-What-It-Is-and-Why-It-Counts/Critical-Thinking-What-It-Is-and-Why-It-Counts-PDF
[7] Sriantini A, Teguh BM, Abadi. Measurement Of Students' Critical Thinking Skills In Solving Contextual Problems Of Spherical Triangles In Maritime Academic. J Posit Sch Psychol. 2022;**6**(12):1551–7.
[8] Warsina, Suyono. Analisis Knowledge , Attitude dan Skill terhadap Competence pada penggunaan Media Pembelajaran E-Learning (Studi Evaluasi Model Pembelajaran Berbasis Simulator Polimarin Semarang). 2020;**6**(1):26–31.
[9] Nanang JS, Ratna DK. Optimalisasi Pemanfaatan Full Mission Ship Handling Simulator Dalam Meningkatkan Kompetensi Bernavigasi Taruna Balai Pendidikan Dan Pelatihan Ilmu Pelayaran Barombong. 2019;**21**(2):79–85.
[10] Alkadri R, Mustari B, Setiawan A. Pengaruh Bridge Simulator Karya PT. LEN Sebagai Industri Pertahanan Dalam Negeri untuk Meningkatkan Kompetensi Taruna AAL Korps Pelaut. Rekayasa. 2022;**15**(2):199–203.
[11] Fajar CBH. Pemanfaatan Simulator dalam Meningkatkan Pengetahuan dan Keterampilan Bernavigasi Taruna Akademi Pelayaran Niaga Indonesia. J Mitra Pendidik. 2019;**3**(4):557–8.
[12] Chae CJ. The STCW Manila amendments: its challenges to the Far East [Internet]. commons.wmu.se; 2011. Available from: https://commons.wmu.se/cgi/viewcontent.cgi?article=1384&context=all_dissertations

[13] Hake RR. Analyzing change/gain scores. Unpubl URL http//www physics indiana edu/\~sdi/AnalyzingChange-Gain pdf [Internet]. 1999;16(7):1073–80. Available from: http://www.ncbi.nlm.nih.gov/pubmed/22025883%5Cnhttp://scholar.google.com/scholar?hl=en&btnG=Search&q=intitle:ANALYZING+CHANGE/GAIN+SCORES#0%5Cnhttp://scholar.google.com/scholar?hl=en&btnG=Search&q=intitle:Analyzing+change/gain+scores#0
[14] Rashed Al Shahin. The Effects of Marine Simulators on Training. J Eng Res Appl. 2017; **7** (3):1–13.
[15] Pitana T, Pramudhita MT, Siswantoro N, Napitupulu JRC. Application of virtual reality as an alternative training media as complement of ship bridge simulator (case study: Port of tanjung emas semarang). IOP Conf Ser Earth Environ Sci. 2023;**1166**(1).
[16] Mawardi HM, Arema Putra. Pengaruh Motivasi Belajar dan Pelatihan Bridge Simulator Terhadap Peningkatan Kompetensi Taruna Nautika. Politek Ilmu Pelayaran. 2021;
[17] Bustami Y, Leliavia L, Elisabeth N, ... Contextual teaching learnıng in human digestive system: The contribution of critical thinking skills. Biosf J ... [Internet]. 2020; Available from: http://journal.unj.ac.id/unj/index.php/biosfer/article/view/14618
[18] Reinke LT. Contextual problems as conceptual anchors: an illustrative case. Res Math Educ [Internet]. 2020;**22**(1):3–21. Available from: https://doi.org/10.1080/14794802.2019.1618731

IOP Conf. Series: Earth and Environmental Science **1473** (2025) 012029 doi:10.1088/1755-1315/1473/1/012029

Trim Optimization for Ship Resistance Reduction and Energy Efficiency Existing Ship Index (EEXI) Regulation Assessment

R Awwalin[1*], M A Saktiono[1], S Marsudi[1], R Tambun[2], B Suwasono[2] and Sutiyo[3]

[1]Department of Ship Machinery Engineering Technology, Universitas Hang Tuah Surabaya, Indonesia
[2.] Postgraduate Program of Ocean Engineering, Universitas Hang Tuah Surabaya, Indonesia
[3] Department of Naval Arcitechture, Universitas Hang Tuah Surabaya, Indonesia

*E-mail: rodlitul.awwalin@hangtuah.ac.id

Abstract. Environmental concerns and the goal of reducing greenhouse gas emissions have intensified efforts to improve the energy efficiency of all stakeholders, including the shipping industry. The International Maritime Organization (IMO) has mandated operational and design measures to improve energy efficiency and reduce ship emissions. One of the most effective operational measures is trim optimisation. This is a feasible measure because it does not require structural modifications to the ship or further development of engines and can reduce the ship's resistance and fuel consumption. Trimming optimisation can be studied through experimental tests; however, this is a difficult, expensive, and time-consuming method. Therefore, in this paper, a numerical method is proposed to predict the optimal trim that minimises the resistance of the container ship (MV. Meratus Kahayan) as considered as case studies in this paper. Different trim conditions and speeds were considered during the optimisation. The analysis using Computational Fluid Dynamics (CFD) shows how a stern trim of 0.5 degrees reduces resistance by 2.6%. Factors like mechanical issues and fuel quality contribute to this inefficiency. The vessel stands out for its high energy efficiency, leading in environmental sustainability in the maritime industry.

1. Introduction

Ship trim optimisation refers to the adjustment of a vessel's longitudinal balance to minimise hydrodynamic resistance, thereby enhancing fuel efficiency and reducing emissions [1]. This operational measure does not require structural or machinery modifications, making it a practical approach for both existing and new ships to satisfy environmental regulations and improve voyage performance. Additionally, this operational measure can be easily implemented onboard ships without the need for extensive training or specialised personnel, making it a cost-effective solution for ship operators [2,3]. Trim optimisation is a significant operational strategy for the maritime industry to reduce ship resistance, fuel consumption, and harmful emissions. It is a cost-effective solution that complements design choices and adheres to environmental standards without necessitating extensive modifications to the ship structure or propulsion systems [4].

The ship trim is influenced by several factors, including the cargo distribution, ballast water, and hull shape. The trim of a ship refers to its longitudinal inclination, which is the angle between the waterline and the keel. This is crucial for the safe and efficient operation of a vessel, affecting fuel consumption, vessel stability, and environmental impact [5,6]. Cargo distribution is a primary factor affecting the trim. The placement of cargo can alter the ship's centre of gravity, leading to changes in the trim. Stowage planning is essential for optimising cargo placement while ensuring stability, which is a complex problem owing to the nonlinear relationships between stability conditions and cargo distribution [7]. Similarly, ballast water is used to adjust a ship's heel and trim, and its optimal allocation is vital for safe operation [8]. The use of ballast water is also linked to environmental concerns, as it can introduce invasive species to new ecosystems [9,10]. The hull shape further influences the trim, with different designs such as the axial bow-type hull form affecting the ship's resistance and, consequently, its trim [11]. The V-hull theory illustrates how modifications to a ship's hull can improve stability without the need for ballast water [12].

The optimal trim is a key factor in enhancing ship energy efficiency and aligns with the goals of the Ship Energy Efficiency Management Plan (SEEMP). The adoption of various predictive models and decision support systems, as well as the acknowledgement of existing gaps in energy efficiency practices, underscores the multifaceted approach required to optimise trim and consequently improve the overall efficiency of ship operations [13-15]. In addition to the aforementioned strategies, the incorporation of advanced technologies, such as machine learning algorithms and data analytics, can further enhance the predictive capabilities of these models, thereby enabling more informed decision-making and improved ship efficiency [13,14].

Ship trim optimisation is closely linked to engine performance, as it directly influences the experience of a vessel and, consequently, the power required from the engine to maintain a given speed. Sun et al. [16] demonstrates that trim optimization, through computational fluid dynamics and towing tank tests, can lead to significant energy savings and emission reductions by reducing wave-making resistance, which is a major component of total resistance. This reduction in resistance translates to lower engine power requirements for real ships. Similarly, Perera et al. [17] highlighted the importance of identifying optimal trim configurations to improve ship energy efficiency, which is directly related to the engine fuel consumption rates. This study suggests that the optimal trim can be determined by analysing various performance and navigation parameters, including the main engine power and fuel consumption. Furthermore, Du et al. [18] discussed the synergetic influence of trim on fuel efficiency, where trim optimisation was conducted dynamically to adapt to real-time sea conditions, leading to significant bunker fuel savings and, by extension, CO2 emission mitigation. Hu et al. [19] also supports the notion that adjusting trim is a reliable method to reduce ship resistance, thereby optimizing engine performance by potentially allowing for lower fuel consumption.

Optimal trim is recognised as a significant factor in reducing fuel consumption and improving propulsion efficiency across various types of vessels and aircraft. In the maritime context, trim optimisation has been identified as a key operational measure that can be achieved without structural modifications, effectively reducing the ship resistance and fuel usage [20]. Similarly, for diesel-electric hybrid propulsion systems, optimal control strategies that consider the state of charge of the battery and the load on diesel generators can lead to reduced fuel consumption and enhanced diesel engine efficiency [21]. The concept of optimal trim over an entire voyage (OTWV) further underscores the importance of trim in minimising fuel consumption throughout a ship's journey [22].

Contradictorily, while trim optimisation is beneficial, it is not the sole factor affecting propulsion efficiency. For instance, in the aviation sector, the fuel consumption of distributed propulsion systems is influenced by the transmission loss, bypass ratio, and additional weight from the electrical system [23]. optimal trim is a proven method to reduce fuel consumption and enhance propulsion efficiency in both maritime and aviation sectors. It is part of a broader set of strategies that include engine design improvements, control strategy optimisation, and powertrain component sizing, all contributing to the overall energy efficiency of propulsion systems [21,22].

This research focused on examining the effectiveness of fuel utilisation on ships by considering different ship trim variations. Data on fuel consumption were gathered from ship shipping survey data and then compared with ship calculations using CFD. Data on fuel consumption were also gathered from ship speed, distance travelled, and other relevant factors, which were then compared with calculations using CFD to ensure accuracy. The accuracy of the CFD model was verified and validated using the

International Towing Tank Conference (ITTC) procedure. Subsequently, the fuel consumption estimates were calculated for each variation in the ship trim. This process aims to determine the most efficient sailing conditions by analysing the fuel consumption estimates with different ship trim variations. To achieve the most efficient sailing conditions, it is essential to consider the impact of ship trim variations on the fuel consumption estimates, which can be determined through careful analysis and optimisation. This study seeks to optimise fuel usage and enhance the overall efficiency of ships during voyages.

2. Method

2.1 Ship Model

The findings from the preliminary phase of the research, derived from field studies conducted on the MV. Meratus Kahayan vessel, yielded the subsequent data regarding the ship.

Figure 1. MV. Meratus Kahayan [24]

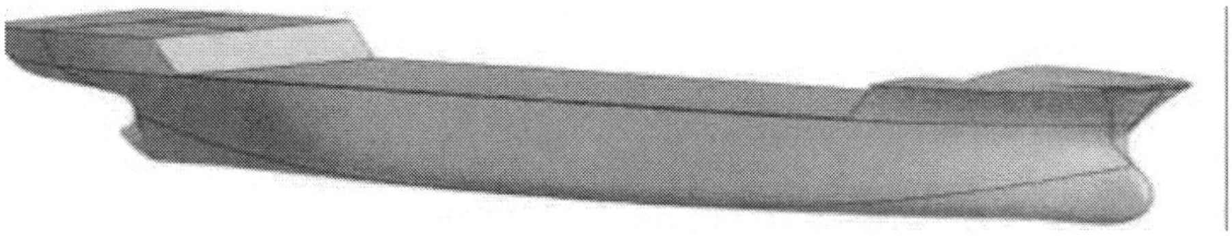

Figure 2. MV Meratus Kahayan Model

Table 1. Particularly Dimension of MV. Meratus Kahayan [25]

Parameter	Dimension	Unit
Length betweet perpendicular (LBP)	115.00	M
Breadth moulded (BMLD)	21.80	M
Depth moulded (DMLD)	7.30	M
Draft	5.47	M
Gross Tonnage (GT)	6626	Ton
Net Tonnage (NT)	3223	Ton
Hull material	steel	
Shipyard (builder)	China	
Main Engine (ME)	3460	HP
Auxiliary Engine (AE)	3x292	HP

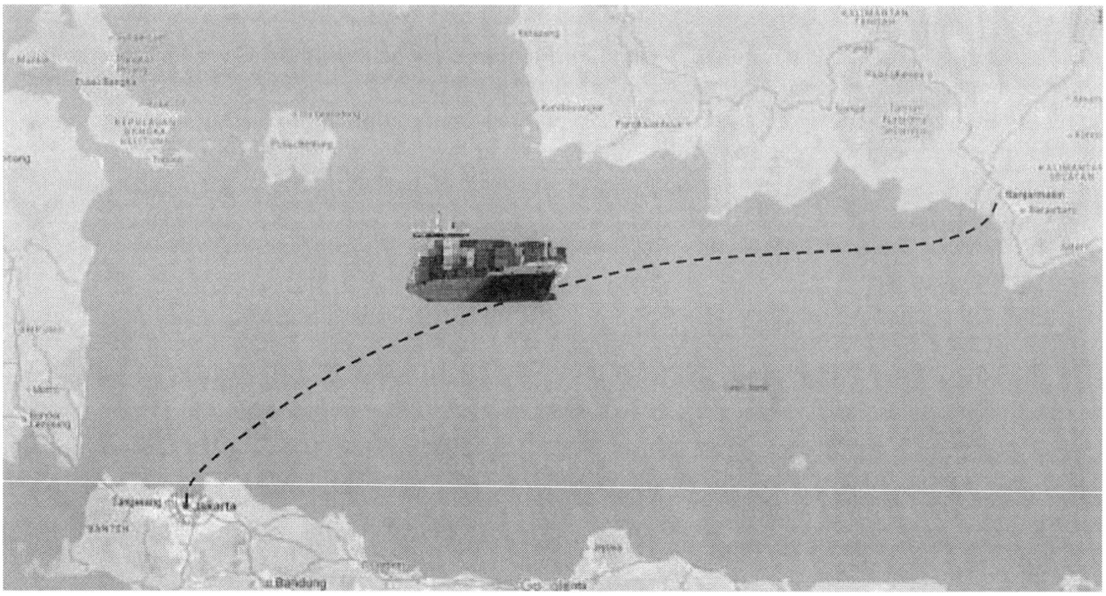

Figure 3. MV Meratus Kahayan Shipping Route (source: Google Maps)

The MV Meratus Kahayan operates a maritime route connecting Tanjung Priok Port, located in Jakarta, to Trisakti Port in Banjarmasin. This journey spans a considerable distance of roughly 950 kilometers, facilitating the transportation of goods and services between these two significant ports. The operation of this vessel plays a crucial role in enhancing trade and commerce in the region.

2.2 Governing Equation

CFD technique was used to predict the resistance of the models. Sahid and Huang [26] conducted research on calculating the hull resistance of trimaran ships using CFD and showed good results compared to experiments. The Reynolds-averaged Navier-Stokes (RANS) method is a three-dimensional equation that was developed and used in the CFD model. One of the equations developed by the ANSYS-CFX software is an unsteady incompressible flow that solves flow problems in the walls of ships [27].

The selection of turbulence models is very important in the modelling of the wake fields. This study used the SST (Shear Stress Transport) turbulence model created by Menter [28]. The SST model has been utilised and verified by many researchers, all of whom have yielded positive findings using this model [29]. The RANS solver, implemented in ANSYS CFX, was used to solve the fluid flow field. Equations (1), (2), and (3) illustrate the continuity, RANS, and SST turbulence equations, respectively.

Continuity equation:

$$\frac{\partial_\rho}{\partial \mathrm{t}} + \frac{\partial}{\partial x_j}(\rho Uj) = 0 \tag{1}$$

where ρ is the fluid density, t is time, and u is the flow velocity vector field.

RANS equation:

$$\rho \underline{u}_j \frac{\partial \underline{u}_i}{\partial x_j} = \rho \underline{f_i} + \frac{\partial}{\partial x_j}\left[-\underline{p}\delta_{ij} + \mu\left(\frac{\partial \underline{u}_i}{\partial x_j} + \frac{\partial \underline{u}_j}{\partial x_i}\right) - \underline{\rho u_i' u_j'}\right] \tag{2}$$

The left side of the RANS equation represents the change in the mean momentum of the fluid element to unsteadiness in the mean flow. This change is balanced by the mean body force ($\underline{f}$), the mean pressure field ($\underline{p}$), the viscous stress, $\mu\left(\frac{\partial u_i}{\partial x_j}+\frac{\partial u_j}{\partial x_i}\right)$, and apparent stress ($\rho u_i' u_j'$) to the fluctuating velocity field.

Menter's SST equation

$$\frac{\partial(_{\rho\omega})}{\partial \mathrm{t}}+\frac{\partial(\rho u_j\omega)}{\partial x_j}=\frac{\gamma}{V\mathrm{t}}P-\beta p\omega^2+\frac{\partial}{\partial x_j}\left[(\mu+\sigma_\omega\,\mu_i)\frac{\partial_\omega}{\partial x_j}\right]+2(1-F_1)2p_{\omega 2}\frac{1}{\omega}\frac{\partial_k}{\partial x_j}\frac{\partial_\omega}{\partial x_j} \quad (3)$$

Menter's SST model combines the advantages of the k–ω model to achieve optimal model formulation for a wide range of applications. For this, a blending function F1 is introduced which is equal to one near the solid surface and zero for the flow domain away from the wall. It activates the k–ω wall region and the k–ε model for residual flow. Using this approach, the attractive near-wall performance of the k–ω model can be used to determine free-stream sensitivity.

2.3 Numerical Domain and Boundary Setting

The most recommended computational domain at the velocity inlet was set to 2 L forward perpendicular to the front, and that at the outlet pressure was 5 L towards the back which was measured perpendicularly. To eliminate the effect of backward flow on the side boundaries, the transverse and vertical directions were set to 2 L–3 L [30].

Figure 5 shows the domain dimensions and boundary conditions. The hull body is defined as a fixed boundary, and a no-slip condition is imposed on the model; a free-slip condition is applied to the bottom; an opening condition is applied to the top wall, and a symmetry condition is used for the side walls; the flow velocity at the inlet is defined as Fr = 0.15 to 0.5; the hydrostatic pressure at the outlet is defined as a function of the water level, and the flow velocity at the inlet is defined as Fr. The starting position of the free surface was defined by specifying the volume-fraction function of water and air at the input and exit of the system.

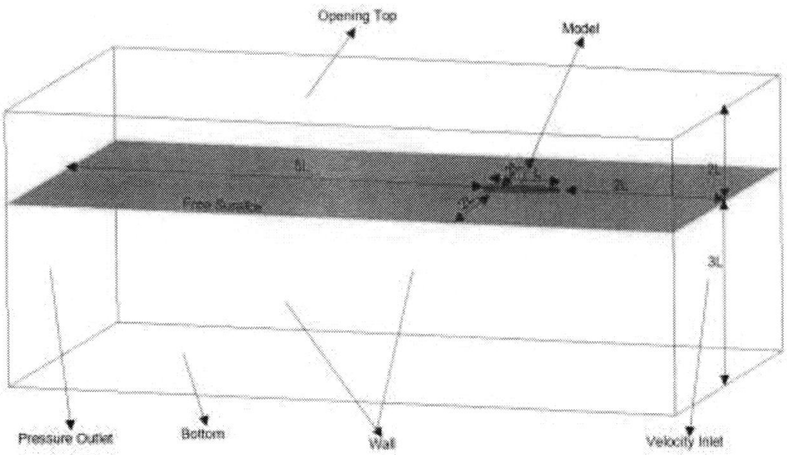

Figure 4. Boundary Conditions

2.4 Grid Generation

The process of mesh generation was accomplished using ANSYS DesignModeler, resulting in an unstructured computational domain characterized by an inflation mesh, as illustrated in Figure 5. Given the intricate geometry of the model, a triangular mesh was specifically created for the surface, while the boundary layer was enhanced through the incorporation of prism elements. This inflation technique effectively populates the surrounding area of the model with tetrahedral elements. To ascertain the optimal element size for the numerical analysis, a grid independence study was conducted. While a finer mesh typically yields more accurate results in CFD, it is important to note that an increased number of elements can significantly elevate both computational costs and processing times. In the case of the MV Meratus Kahayan, the mesh comprises approximately 1.9 million elements. The previously established grid size was consistently applied across all simulations to maintain uniform processing speeds

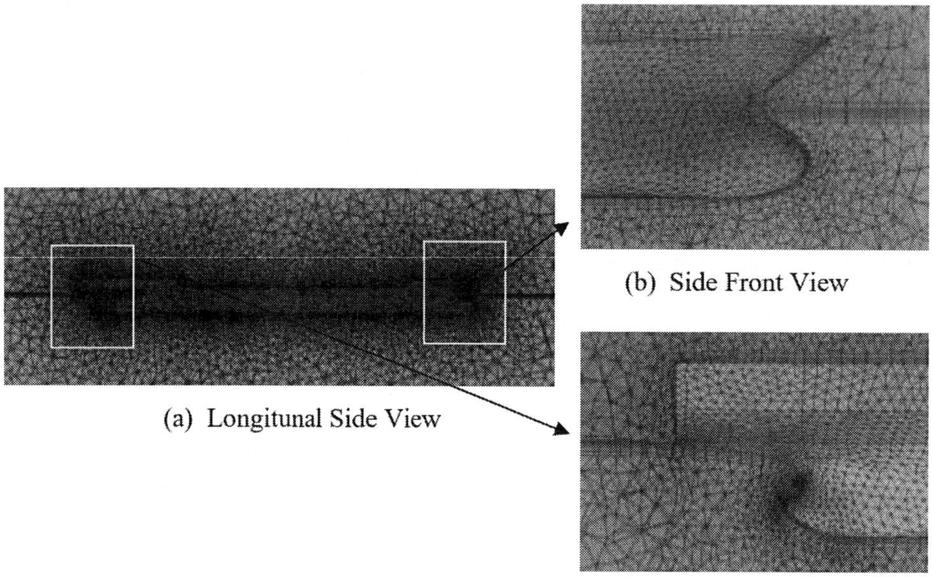

(a) Longitunal Side View

(b) Side Front View

(c) Side After View

Figure 5. Unstructured with inflation mesh

2.5 EEXI Calculiation

The EEXI represents a regulatory framework established by the IMO aimed at mitigating greenhouse gas emissions produced by maritime vessels. This index serves as a technical benchmark that evaluates the design characteristics of a ship, ensuring that it meets specific energy efficiency standards. Compliance with the EEXI is mandatory for all ships, which must secure approval at least once during their operational lifetime, with the deadline set for the first periodic survey occurring in 2023.

The EEXI value required for each vessel is contingent upon various factors, including the type of ship, its carrying capacity, and the propulsion system employed, establishing a maximum threshold for acceptable energy efficiency. To achieve compliance with the EEXI, each ship must undergo a detailed calculation to determine its attained EEXI, which is unique to the individual vessel and must align with the stipulated regulations and calculated by following formula [31]:

$$\frac{\left(\prod_{j=i}^{n} fi\,.\right)\left(\sum_{i=1}^{nME} PM_E\,.\,CFM_E\,.SFCM_{E(i)}\right)(PA_E\,.\,CFA_E\,.SFCA_E)+\left(\left(\prod_{j=i}^{n} fi\,.\sum_{i=1}^{nPTI} P_{PTI(i)}-\sum_{i=1}^{neff} f_{eff(i)}\cdot P_{AEeff(i)}\right)C_{FAE}\cdot SFC_{AE}\right)-\left(\sum_{i=1}^{neff} f_{eff(i)}\cdot P_{eff(i)}\cdot C_{FME}\cdot SFC_{ME}\right)}{f_i\,.\,f_e\,.\,f_l\,.\,Capacity\,.\,f_w\,.\,V_{ref}\,.\,f_m} \quad (4)$$

3. Result and discussion

3.1 Ship Resistanec and Power

The analysis of ship resistance through CFD reveals noteworthy findings, particularly regarding the impact of the vessel's trim on its overall resistance, as illustrated in Figure 5. Specifically, a stern trim of 0.5 degrees demonstrates a significant reduction in resistance, achieving approximately 2.6% less resistance compared to the even keel condition, which registers a resistance force of 107.41 kN. This indicates that optimizing the trim can lead to enhanced hydrodynamic performance, thereby improving the efficiency of the vessel in various operational scenarios.

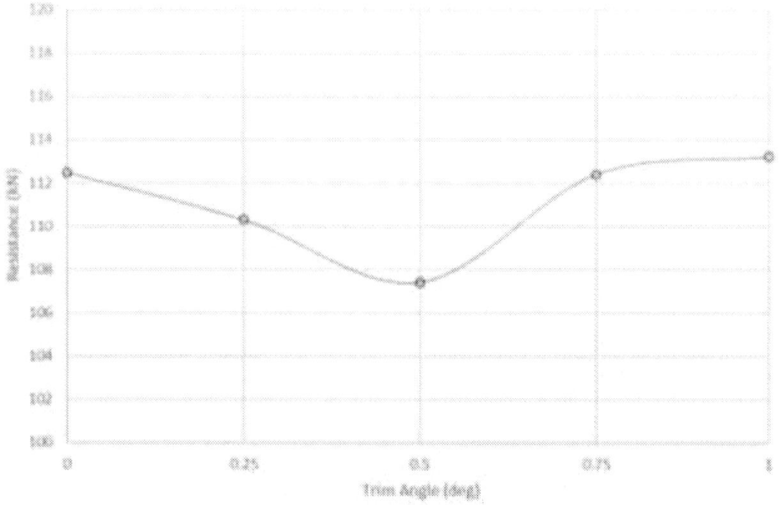

Figure 6. Resistance Total

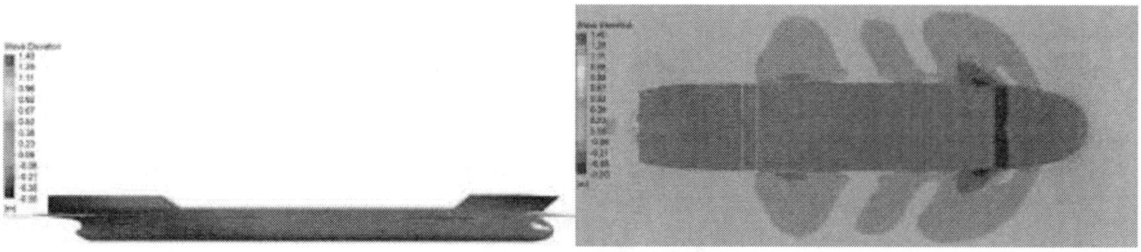

(a) Trim Angle 0 deg

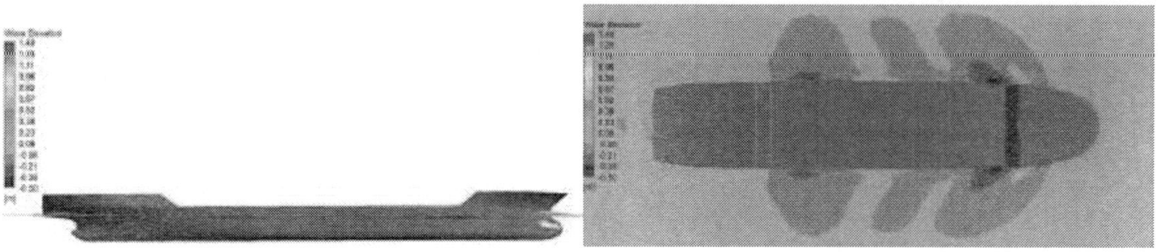

(b) Trim Angle 0.25 deg

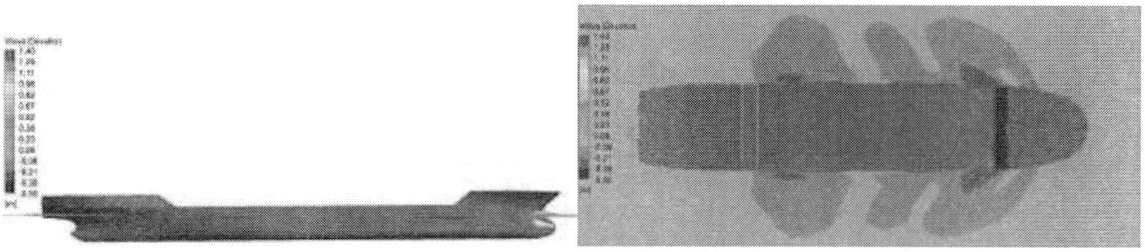

(c) Trim Angle 0.5 deg

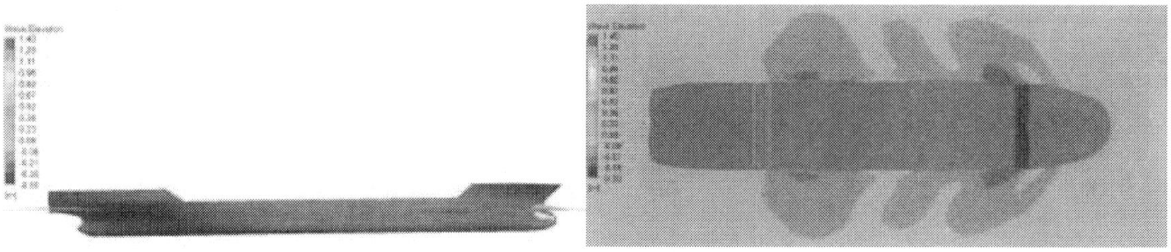

(d) Trim Angle 0.75 deg

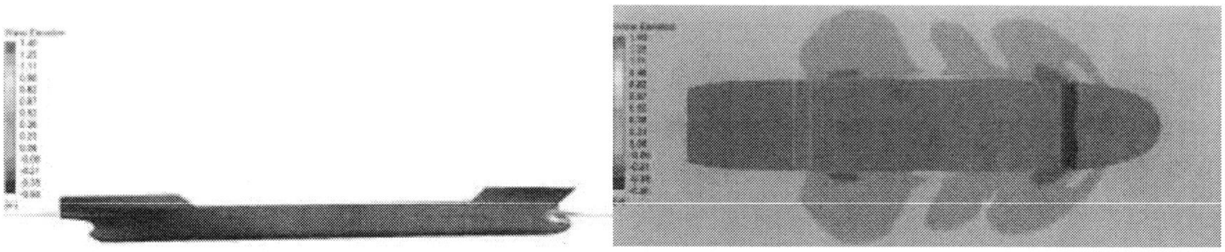

(e) Trim Angle 1 deg

Figure 7. Wave Patern of MV Meratus Kahaya with Trim Variation

The advantageous impact is attributed to a stern trim angle of 0.5 degrees, which is evidenced by a relatively minor wave formation at the vessel's aft. This specific trim condition diminishes the intensity of the back current that interacts with the ship's hull, thereby contributing to a reduction in the overall resistance experienced by the vessel. The streamlined flow resulting from this optimal trim not only enhances hydrodynamic efficiency but also plays a crucial role in improving the ship's performance in various maritime conditions.

In addition, the primary engine of the MV Meratus Kahayan, when in optimal condition, has the potential to generate a power output of 2580 kW, as illustrated in Figure 7. This impressive power capacity is crucial for the vessel's performance, enabling it to achieve the necessary speed and maneuverability required for its operational tasks. However, despite this potential, observations made by the crew during operations indicate that the engine currently operates at merely 50% of its designed capacity. This reduced performance level results in a functional output of only 1290 kW, which significantly limits the vessel's operational efficiency and effectiveness. The crew has noted that this underperformance could be attributed to various factors, including potential mechanical issues, fuel quality, or maintenance deficiencies. Consequently, the vessel's ability to navigate challenging conditions, respond to operational demands, and maintain schedules may be adversely affected, highlighting the need for a thorough assessment and potential remedial actions to restore the engine to its full operational capacity.

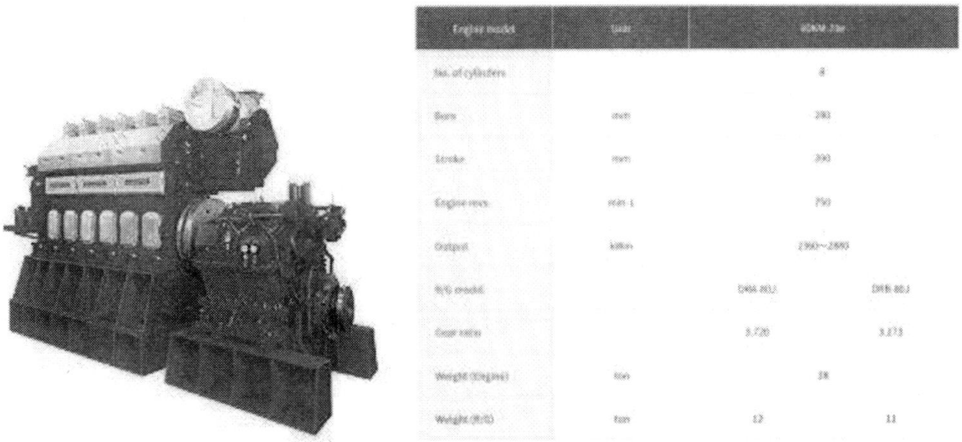

Figure 8. Main Engine ANQING – DAIHATSU, 8 DKM-28 [32]

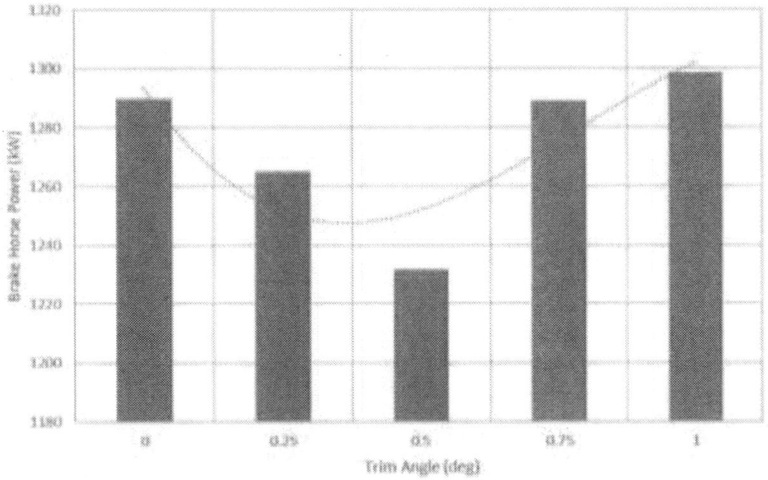

Figure 9. Brake Horse Power

The numerical analysis of the vessel's performance reveals that the engine operates at an efficiency rating of 50.69% when the ship is traveling at a speed of 11.3 knots. This efficiency rating is a critical metric, as it reflects the effectiveness with which the engine converts fuel into usable power for propulsion. An efficiency of 50.69% suggests that nearly half of the energy contained in the fuel is being utilized for movement, while the remainder is lost as waste heat and other forms of energy dissipation.

As a result of this efficiency level, the ship demonstrates a fluctuating Maximum Continuous Rating (MCR) power. The MCR is the maximum power output that the engine can sustain over a prolonged period without risking damage or excessive wear. The fluctuations in MCR can be attributed to various factors, including changes in operational conditions, variations in fuel quality, and the dynamic nature of marine environments, such as waves and currents.

These fluctuations in MCR are visually represented in Figure 8, which provides a graphical depiction of the power output over time or under different operational scenarios. The figure likely illustrates how the MCR varies in response to changes in speed, load, and other operational parameters, highlighting the complexities of marine engine performance. Understanding these variations is essential for optimizing the vessel's operational efficiency and ensuring that it operates within safe and effective power limits.

3.2 EEXI

The evaluation of the EEXI reveals that the operational parameters of all assessed vessels fall short of the thresholds established by the IMO. This assessment is visually represented in Figure 9, which provides a comparative analysis of the EEXI values across different ships. Among these vessels, the MV Meratus Kahayan stands out due to its operational trim angle of just 0.5 degrees, which is notably lower than that of its peers. As a result, the MV Meratus Kahayan records the lowest EEXI index value among the ships evaluated, indicating a significant level of energy efficiency.

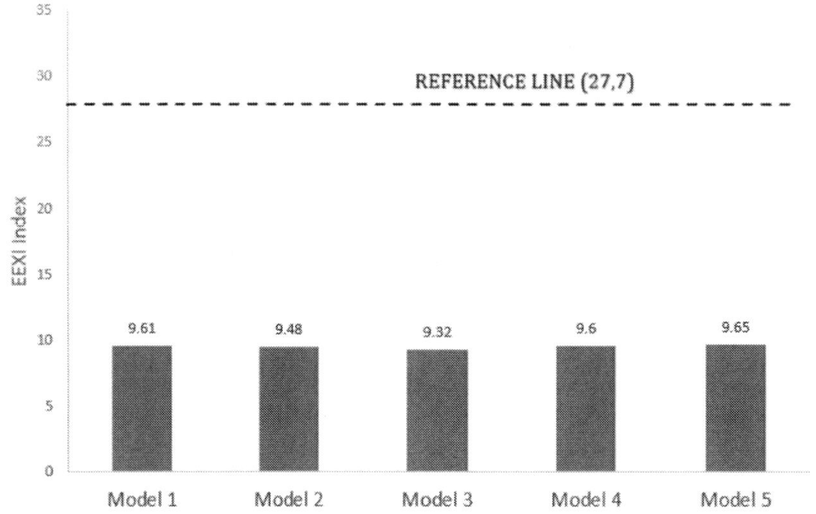

Figure 10. EEXI Index

This exceptional performance translates into a reduced emission of carbon dioxide (CO_2) into the atmosphere, positioning the MV Meratus Kahayan as a leader in environmental sustainability within the maritime industry. The findings from this evaluation not only highlight the vessel's superior energy efficiency compared to its counterparts but also emphasize its potential to play a crucial role in mitigating the environmental impact associated with maritime operations. By operating with such an efficient profile, the MV Meratus Kahayan exemplifies how advancements in ship design and operational practices can contribute to a more sustainable future for shipping, aligning with global efforts to combat climate change and reduce greenhouse gas emissions. This case serves as a compelling example for other vessels in the fleet, encouraging the adoption of similar practices to enhance overall energy efficiency and environmental stewardship in the maritime sector.

4. Conclusion

The analysis of ship resistance using CFD highlights the impact of trim on reducing resistance. A stern trim of 0.5 degrees significantly decreases resistance by 2.6%. Optimizing trim improves hydrodynamic performance and efficiency in various operational scenarios. Factors such as mechanical issues, fuel quality, or maintenance deficiencies contribute to this underperformance. The engine's efficiency rating is 50.69% at a speed of 11.3 knots, with fluctuations in MCR due to operational conditions and fuel quality changes. EEXI evaluation shows that all assessed vessels fall below IMO thresholds. The MV Meratus Kahayan stands out with a trim angle of 0.5 degrees, recording the lowest EEXI index, indicating high energy efficiency and reduced CO2 emissions. This positions the vessel as a leader in environmental sustainability within the maritime industry, showcasing how efficient ship design and practices can mitigate environmental impact and combat climate change. The MV Meratus Kahayan serves as a model for improving energy efficiency and environmental stewardship in the maritime sector, encouraging other vessels to follow suit for a more sustainable future. The subsequent task to be undertaken involves correlating resistance with various speeds and trim angles to enhance the operational efficiency of the vessel.

Acknowledgment
The authors would like to thank Universitas Hang Tuah Surabaya for research funding based on assignment letter No. S.Gas/004/UHT.C.2/II/2024.

Reference

[1] Islam H and Guedes Soares C Effect of trim on container ship resistance at different ship speeds and drafts. *Ocean Engineering*. **183** 106–115. Jul. 2019, doi: 10.1016/j.oceaneng.2019.03.058
[2] Elkafas AG, Elgohary MM and Zeid AE Numerical study on the hydrodynamic drag force of a container ship model. *Alexandria Engineering Journal*. **58**. 3. 849–859. Sep. 2019, doi: 10.1016/j.aej.2019.07.004
[3] Soe Thiha and Prof. Yongxin Jin Study on Energy Efficient Operation by ship's Trim Optimization based on Computational Fluid Dynamics. *Int J Sci Res Sci Eng Technol*. 18–28. Sep. 2023, doi: 10.32628/IJSRSET23102122
[4] Kolios A Retrofitting Technologies for Eco-Friendly Ship Structures: A Risk Analysis Perspective. *J Mar Sci Eng*. **12**. 4. 679. Apr. 2024, doi: 10.3390/jmse12040679
[5] Birafane M, Khalikov S, Khalikov S, Isakdjanova and Dairshenov M Influence of Various Trim Conditions on Vessel's Fuel Consumption. *International Journal of Recent Technology and Engineering (IJRTE)*. **8**. 6. 4636–4657. Mar. 2020, doi: 10.35940/ijrte.F9787.038620
[6] Saktiono MA *et al* Air Quality Prediction Model of Surabaya City Using Long Short Term Memory (LSTM) Method. *International Journal on Engineering Applications (IREA)*. **12**. 1. 69-75. Jan. 2024, doi: 10.15866/IREA.V12I1.24289
[7] Ok M and Lei H Accurate and robust linear representation for containership stability conditions with ballast water. *Ocean Engineering*. **306**. 118141. Aug. 2024, doi: 10.1016/j.oceaneng.2024.118141
[8] Liu Z, Jiang J, Gan Z and Lin C Ballast water dynamic allocation optimization model and analysis for safe and reliable operation of floating cranes. *Ann Oper Res*. **311**. 1. 279–294. Apr. 2022, doi: 10.1007/s10479-019-03213-2
[9] Dachev Y, Tsvetkov M and Zlatev V Ship Ballast Water Treatment. *Wseas Transactions on Environment and Development*. **17**. 110–117. Feb. 2021, doi: 10.37394/232015.2021.17.11
[10] Kholdebarin A, Mozaffari N, Vambol V, Vambol S and Mozaffari N Impact Of Ship's Ballast Water On Invasive Species Occurrence, And As A Consequence On Environmental And Public Health: A Short Review. *Labour protection problems in Ukraine*. **36**. 1. 18–21. Mar. 2020, doi: 10.36804/nndipbop.36-1.2020.18-21
[11] Suardi S, Hidayat T, Firmansyah MB and Kyaw AY Impact of Axe Bow Hull Shape on Patrol Ship Resistance, Freeboard, and Trim. *Maritime Park Journal of Maritime Technology and Society*. 16–21. Feb. 2023, doi: 10.62012/mp.v2i1.25478
[12] Wang L, Zhou P, Xu X and Gao L 3500DWT non-ballast water ship design. in *International Conference on Mechanical Design and Simulation (MDS 2022)*. D. Shi and G. Wu, Eds., SPIE, Sep. 2022, p. 212. doi: 10.1117/12.2640835
[13] Bal Beşikçi E, Arslan O, Turan O and Ölçer AI An Artificial Neural Network Based Decision Support System For Energy Efficient Ship Operations. *Comput Oper Res*. **66**. 393–401. Feb. 2016, doi: 10.1016/j.cor.2015.04.004
[14] Hu C, Zhou R, Ma R and Gao Y Energy Saving and Optimal Loading Analysis of Ship Trim. in *2019 2nd World Conference on Mechanical Engineering and Intelligent Manufacturing (WCMEIM)*. IEEE. Nov. 2019. 50–53. doi: 10.1109/WCMEIM48965.2019.00017
[15] Tu H, Xia K, Zhao E, Mu L and Sun J Optimum trim prediction for container ships based on machine learning. *Ocean Engineering*. **277**. 111322. Jun. 2023, doi: 10.1016/j.oceaneng.2022.111322

[16] Sun J, Tu H, Chen Y, Xie D and Zhou J A Study on Trim Optimization for a Container Ship Based on Effects due to Resistance. *Journal of Ship Research*, **60**. 01. 30–47. Mar. 2016, doi: 10.5957/jsr.2016.60.1.30

[17] Perera LP, Mo B and Kristjánsson LA Identification of Optimal Trim Configurations to improve Energy Efficiency in Ships. *IFAC-PapersOnLine*. **48**. 16. 267–272. 2015, doi: 10.1016/j.ifacol.2015.10.291

[18] Du Y, Meng Q, Wang S and Kuang H Two-phase optimal solutions for ship speed and trim optimization over a voyage using voyage report data. *Transportation Research Part B: Methodological*. **122**. 88–114. Apr. 2019, doi: 10.1016/j.trb.2019.02.004

[19] Hu C, Zhou R, Ma R and Gao Y Energy Saving and Optimal Loading Analysis of Ship Trim. in *2019 2nd World Conference on Mechanical Engineering and Intelligent Manufacturing (WCMEIM)*. IEEE. Nov. 2019. 50–53. doi: 10.1109/WCMEIM48965.2019.00017

[20] Elkafas AG Advanced operational measure for reducing fuel consumption onboard ships. *Environmental Science and Pollution Research*. **29**. 60. 90509–90519. Dec. 2022, doi: 10.1007/s11356-022-22116-7

[21] Zhang C and JIA B Research on Energy Efficiency Optimization Strategy of Electric Propulsion Ships with Energy Storage Devices. in *2019 Chinese Automation Congress (CAC)*. IEEE. Nov. 2019. 4689–4694. doi: 10.1109/CAC48633.2019.8997039

[22] Yu Y, Zhang H, Mu Z, Li Y, Sun Y and Liu J Trim and Engine Power Joint Optimization of a Ship Based on Minimum Energy Consumption over a Whole Voyage. *J Mar Sci Eng*. **12**. 3. 475, Mar. 2024, doi: 10.3390/jmse12030475

[23] Wang XC, Chen YC and Wang YR Fuel Consumption Analysis of Distributed Propulsion. in *2022 13th International Conference on Mechanical and Aerospace Engineering (ICMAE)*. IEEE. Jul. 2022. 342–346. doi: 10.1109/ICMAE56000.2022.9852885

[24] MERATUS KAHAYAN (MMSI: 525025101) Ship Photos | AIS MarineTraffic. Accessed: Sep. 12, 2024. [Online]. Available: https://www.marinetraffic.com/cs/photos/of/ships/shipid:3412279/shipname:MERATUS%20KAHAYAN?order=date_uploaded

[25] BKI Register. MERATUS KAHAYAN. Accessed: Aug. 26, 2024. [Online]. Available: https://www.bki.co.id/shipregister-21395.html

[26] Shahid M and Huang D. Resistance calculations of trimaran hull form using computational fluid dynamics. in *Proceedings - 4th International Joint Conference on Computational Sciences and Optimization, CSO 2011*. 2011. doi: 10.1109/CSO.2011.225

[27] ANSYS. *ANSYS CFX-Solver Theory Guide*. Canonsburg. PA. USA: Ansys Inc, 2020

[28] Menter FR Zonal two equation κ-ω turbulence models for aerodynamic flows. in *AIAA 23rd Fluid Dynamics. Plasmadynamics, and Lasers Conference. 1993*. 1993. 1–21

[29] Menter FR, Kuntz M and Langtry R Ten Years of Industrial Experience with the SST Turbulence Model. in *4th Internal Symposium, Turbulence, heat and mass transfer*. New York. Wallingford: Begell House. 2003. 625–632

[30] Elkafas AG, Elgohary MM and Zeid AE Numerical study on the hydrodynamic drag force of a container ship model. *Alexandria Engineering Journal*. **58**. 3. 849–859, Sep. 2019, doi: 10.1016/j.aej.2019.07.004

[31] IMO, "MEPC 76/15/Add.2 - Annex 7. International Maritime Organization. Accessed: Aug. 26, 2024. [Online]. Available: https://wwwcdn.imo.org/localresources/en/OurWork/Environment/Documents/Air%20pollution/MEPC.333(76).pdf

[32] DAIHATSU DIESEL MFG. CO. LTD. 8DKM-28e | DAIHATSU DIESEL. Accessed: Aug. 26, 2024. [Online]. Available: https://www.dhtd.co.jp/en/products/marine/propulsion/8DKM-28e.html

IOP Conf. Series: Earth and Environmental Science **1473** (2025) 012030 doi:10.1088/1755-1315/1473/1/012030

Optimizing the Marine Transportation Model for a New Deepwater Offshore Platform at Working Area of Andaman II

S D Lazuardi[1*], I T Yunianto[2], C B S Permana[3], D V Devintasari[4] and O S Noarista[5]

[1,2,3,4,5] Department of Marine Transportation Engineering, ITS, Surabaya, Indonesia

*siti.dwi@its.ac.id

Abstract. According to the Ministry of Energy and Mineral Resources, Indonesia has been facing the declining of oil and gas reserves capacity. In 2016, gas reserves were 144.72 TSCF while in 2020 it was only 62.39 TSCF. Around 60-70% of potential new gas fields have not been explored. Thus, the construction of new offshore platforms is needed to increase national natural gas reserves. The working area of Andaman II was chosen because it has the highest estimated gas reserves of 2,025 MBOE with a depth of 1,291.3 m. Therefore, this study aims to plan the delivery method of offshore platforms that produce minimum total transportation costs by using assignment problem method. The type of new offshore platform in Andaman II is the Tension Leg Platform (TLP) which consists of tension, hull, topside, and riser. The optimization result shows that the delivery method to build the new offshore platform is the hull and topside shipped simultaneously.

1. Introduction

Indonesia has abundant of energy and minerals, particularly oil and natural gas. In the history of national development, oil and natural gas have an important and strategic role, which affects the national economic activities. The oil and natural gas sector also contribute to state revenues originating from oil and natural gas management [1]. Head of the Program and Communications Division of the Special Task Force for Upstream Oil and Natural Gas Business Activities (SKK Migas) said that the upstream oil and natural gas industry was proud to be able to provide the best contribution to the country. Currently, the upstream oil and natural gas sector's tax contribution has exceeded the government's target, where the target is set by the government of IDR. 64,7 trillion, while upstream oil and natural gas industry tax payments have reached Rp. 69,16 trillion net or 106.85% [2].

The important role of the upstream oil and natural gas industry in terms of finances is also reflected in the conclusions of the government budget working meeting which was held on 19 May 2022 [3]. Furthermore, the oil and natural gas sector, especially natural gas, has a vital role in the future as a bridge in the transition process from fossil energy to renewable energy. Natural gas is an energy source that is low in carbon dioxide (CO_2) compared to other energies such as coal or petroleum.

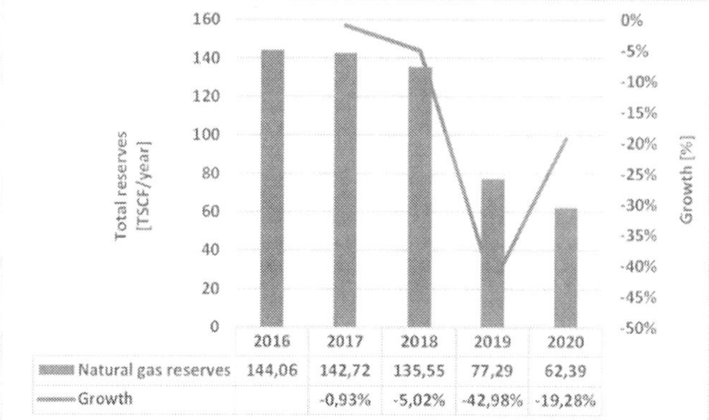

Figure 1. Natural gas reserves in Indonesia (2016-2020)

However, natural gas in Indonesia is faced with the problem of the number of oil and gas reserves and production capabilities which continue to decline. In 2016 gas reserves were 144,06 TSCF while in 2020 it was only 62,39 TSCF. The most drastic negative growth occurred in 2019, around -43% from 2018. By 60-70% of the potential for new gas fields has not been explored, thus it should take an action to explore and build new offshore platforms to increase national natural gas reserves [4].

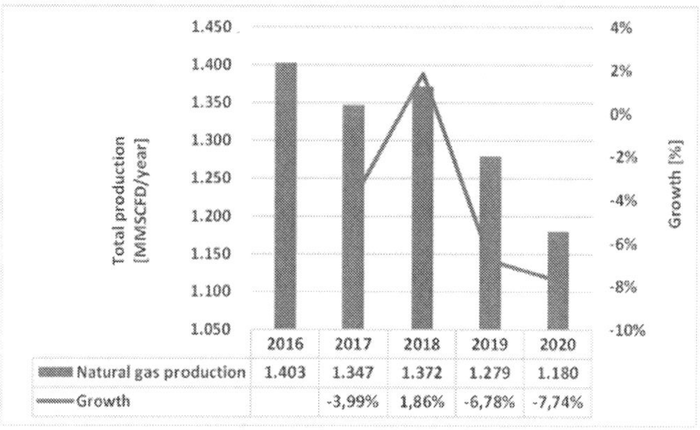

Figure 2. Natural gas production in Indonesia (2016-2020)

Natural gas production only increased in 2018 by 1.9% or with a total production of 1.372 MMSCFD, the following years experienced negative growth until 2020 of -7.7% or a total production of 1,180 MMSCFD. Currently, Indonesia is entering a period of gas crisis due to increasing gas shortages.

According to Erick Thohir, the Minister of State-owned Enterprise, Republic of Indonesia said that the fate of the fertilizer industry was hanging in the balance due to the decreasing supply of gas as the main fuel for fertilizer machines, thus gas should be imported to meet the needs [5]. Moreover, it is necessary to exploit new gas fields to meet natural gas needs. The government has been starting to concern about deep water projects because oil and gas reserves in shallow waters are being depleted, and many wells are old, and around 60-70% of oil and gas reserves are in the sea. The Director General of Oil and Gas at the Ministry of Energy and Mineral Resources, Tutuka Ariadji revealed that there are the largest gas reserves in the world in the oil and gas working area in Aceh, namely the Andaman working area (WK). The Andaman working area consists of WK Andaman I, WK Andaman II, WK Andaman III, and WK South Andaman. The Andaman WK has accumulated reserves about 4,865 miliar barel oil equivalent (MBOE) with details of discoveries of 260 MBOE, prospects of 1.970 MBOE, and lead of 2.635 MBOE [6]. WK Andaman II is the Andaman working area which is planned to operate early in 2028, compared to other Andaman Working Areas. Therefore, WK Andaman II is estimated to have the highest gas reserves of 2.025 MBOE.

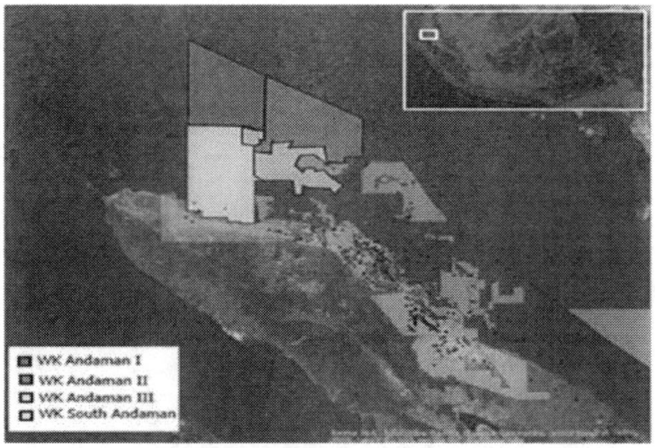

Source: Aceh Oil and Gas Management Agency, 2022
Figure 3. Location of Andaman work area

The role of offshore vessels is vital in terms of constructing the new offshore platform in the WK Andaman II. Operational patterns in offshore platform construction activities should be conscientiously planned so that the operational cost during construction will be optimal. For instance, when the fabrication site for making each module is different from the other modules and the location is not close to each other. In addition, the type of cargo for the construction of offshore platforms is called project cargo which has various sizes and shapes. Thus, the assignment model is required to find out the type of offshore vessels used to deliver the module of offshore platform in WK Andaman II, which produces the minimum total transportation costs.

2. Literature review

2.1. Oil and gas mining

Oil and natural gas mining is a series of activities carried out to obtain petroleum and natural gas where the mining results will be used for energy purposes. In mining, the terms exploration and exploitation are closely related. Exploration is an upstream activity carried out to obtain complete information about geological conditions to estimate and discover oil and natural gas reserves in a place, while exploitation is a series of upstream activities carried out to produce oil and natural gas from a place or release oil and natural gas from reservoirs inside the earth to the surface. Exploitation activities include drilling and completing wells, building transportation, storage and processing facilities for separating or refining oil and gas in the field [7]. There are two locations for oil and natural gas mining, namely onshore and offshore. Onshore is when exploration and exploitation of oil and gas are conducted on the land and coastline areas. Meanwhile, offshore means the exploration or exploitation of oil and gas at sea.

Offshore platform construction has two types of projects, namely green field and brown field. However, this term does not only apply to the oil and gas industry but is widely used in other industries such as IT, construction, manufacturing, building services, and others. Overall, these two terms are related to any project. Green field is a completely new project with no development yet, so it is planned from scratch or zero in an area on a large scale which takes around 4 – 5 years. On the other hand, brown field is project where some work has been done. This project is also often referred to as a development project, overhaul, or output improvement project. The time required for this project is around 1 – 2 years [8].

2.2. Offshore platform

Offshore platform is an offshore structure used for extracting, processing and storing oil and gas, usually having an economic life of around 20 to 35 years [9]. The characteristics of an offshore platform are it operates in areas around oil wells or limited mining areas, it cannot operate on land, then the structure is not built directly on site but the components are made on land then it can be transported and installed on site and it can be operated at sea for a long period of time so that the building must be able to withstand both good and bad weather conditions that may occur during operation [10]. In 2023, the government plans to increase the construction of offshore platform by 110 rigs to fulfill the drilling plan of 919 wells so that if it is realized on time, the number of offshore platform construction in Indonesia in 2023 will be 775 rigs [11]. Offshore platforms based on their building systems and structures can be known in detail based on the explanation below. In this study, the type of offshore platform selected is Tension Leg Platform due to the depth of WK Andaman II was around 4.236 ft or 1.291,3 meters.

Source: Todd Podzemmy, 2023
Figure 4. Fixed Platform

Source: Review on Fixed and Floating Offshore Structures, 2022
Figure 5. Jack up Structure

Source: Review on Fixed and Floating Offshore Structures, 2022
Figure 6. Tension Leg Platform

Source: Middle East Monitor, 2022
Figure 7. Drill Ship

Source: Kongsberg, 2023
Figure 8. Semi-Submersible

2.2.1. Fixed platform. It is the oldest type of platform, which is suitable in locations that have low or shallow depth around 10-100 meters. This platform can function as a well head, production, accommodation, drilling platform, or connecting bridge[12].

2.2.2. Jack up structure. The number of legs is as many as three jacket legs and is built in a deeper location than a fixed platform. This type can be used at depths of 25 – 150 m. This platform was built completely as one unit in a land fabrication yard and then towed by tugboat to the location, at that time the legs were lifted upright until they were above the deck [13].

2.2.3. Tension leg platform (TLP). It has tethers/tendons tied to the seabed. This causes the structure to be rigid in the vertical direction, but it is flexible in the horizontal direction (surge and sway). The combination of the two causes the TLP building to be a structure suitable for the deep sea, rigidity in the vertical axis makes the building tied during production, while flexibility in the horizontal axis makes the structure sensitive to the effects of sea waves or wind so that this structure produces significant dynamic effects. TLP can be used for depths of 300 m - 2000 m [14].

2.2.4. Semi-submersible. It is a floating platform equipped with pontoons and columns. If it is flooded, it will sink to a predetermined depth. This type is used at medium depths between 300 – 500 m. When drilling, the platform is stabilized by a hull filled with water.

2.2.5. Drillship. It is basically a ship equipped with a drilling shaft and moonpool, which is placed in the center of the water plane with a derrick installed on top. Has a large block coefficient, full front body shape to have maximum room volume to place the dynamic thruster with maximum length, constant mid body section without deadrise, and continues with a full stern type transom shape to have maximum volume to place twin screw with dynamic positioning thruster. This type is suitable for use in deep to very deep seas because it can operate at a depth of 3650 m.

2.3. Stages of offshore platform contruction

Offshore platform is a mega project that requires large investment costs and complex operations. The steps for constructing an offshore platform are seismic exploration, feasibility study, conceptual design, front end engineering design (FEED), detailed engineering design (DEED), fabrication, sail away, and installation. However, the scope in this study started from the fabrication to the installation point, which is in WK Andaman II. Thus, this study determined the optimum offshore vessel selected to deliver part or overall offshore platform from fabrication site to the installation point by considering the minimum transportation costs.

2.4. Depth

One of the factors in selecting the type of offshore platform is depth. It is the distance from the surface to the seabed or similar. According to DNV GL, depths of 76-334 m, 335-1217 m, 1218-2300 m and > 2300 m fall into the categories of shallow water, mid water, deep water and ultra deep water respectively. Depth needs to be considered because the deeper the water, the higher the risk.

2.5. Offshore vessels

Offshore vessels are special vessels that serve operational needs such as oil or gas exploration and construction work on the deep sea. There are various offshore vessels that not only help in oil exploration and drilling, but also provide necessary supplies to excavation and construction units located on the deep sea. Moreover, these ships also serve crew transfer to and from the site [15]. Offshore vessels consist of two types, offshore support vessels and construction vessels or special purpose vessels [16].

2.5.1. Offshore support vessels. Offshore supply vessels (OSV) are special vessels specifically designed to assist in the offshore exploration and exploitation process. OSV can fulfil the needs of offshore platforms such as spare parts, supplies, barite, bentonite, drill water, mud, cement, riser, tubular, and others. Moreover, it can be used for anchor handling, construction, maintenance and crew transportation. OSV consists of a seismic vessel, platform supply vessel, anchor handling tug supply, diving support vessel, and crew boat.

2.5.2. Special purpose vessels/construction vessels. This type of ship is used for offshore platform construction. It has many open deck areas with plenty of space for accommodation. In addition, these ships can also be on site for a long time and are often involved in large projects. Construction vessels have heavy lifting crane capacity and support surface and subsea installations as well as IMR (Inspection Maintenance and Repair) activities. It usually used as transportation for the construction of offshore platforms or other heavy lift projects. This type of ship consists of heavy lift vessels, crane vessels, barges and pipe laying vessels.

2.6. Project cargo

Project cargo or heavy lift cargo is cargo or equipment that has large size and heavy. It requires special storage, lifting and handling because it has high or critical value, usually connected to the same project and loaded from different ports. Moreover, it produces high costs to handle, especially in terms of security and safety, which should be carefully considered because if damage occurs, it will result in high costs to buy new cargo. It consists of oversized cargo and out of gauge [17].

2.6.1. Oversized cargo. A type of project cargo that is large but irregular in shape. This type of cargo cannot be loaded in containers, generally it is loaded directly on ships such as offshore platform structures, train cars, locomotives, turbines, etc.

2.6.2. Out of gauge. This cargo is special and cannot fit in a general container, so it requires special containers such as open top containers, flat rack containers or platform containers because the nature and size of the cargo exceeds the dimensions. OOG focuses more on length, width and height or is adjusted to the size and maximum payload dimensions of each type of container.

2.7. Onshore fabrication yard

Onshore fabrication yard (OFY) is a place for the construction or fabrication of offshore platforms. OFY has the same function as a shipyard, the difference lies in the final products, where OFY's product is an offshore building while the shipyard's product is a ship. Similar to the shipyard for building ships, the process of building an offshore platform at OFY in general includes fabrication, assembly and launching. OFY also has a special jetty to support its production activities as well as other infrastructure that also supports maximum and efficient performance [18].

3. Methodology

The main method used to complete this study is the optimization method, by using linear programming model and assignment problem. Hereby, the detail of the stages on this study:

3.1. Identification of problem

At this stage, problems will be identified regarding Indonesia's natural gas reserves which are decreasing because they only rely on old wells. Even though, the potential for new gas reserves is large but they have not been explored and exploited, then the amount of natural gas production is also decreasing. It affects on the gas crisis for industrial and other purposes in Indonesia then it pushes the government to pay special attention on deepwater projects. Thus, 60-70% of the potential oil and gas basins in Indonesia are in the sea. It is necessary to discover new gas fields to increase natural gas reserves in Indonesia by building offshore platforms. When constructing a new offshore platform, the offshore vessel plays an important role in the delivery of the offshore platform structure so that the operating pattern and assignment of vessels to and from the site should conscientiously be planned.

3.2. Data analysis

The data that has been obtained will be processed and analyzed. Firstly, the type of offshore platform that is suitable to be built in the Andaman II WK should be selected. Secondly, identification the structures to build the platform and their specifications. Finally, provide the alternative transportation for the type of platform selected.

3.3. Create the optimization models

As stated in the beginning of this section, the optimization model used in this study is linear programming model and assignment problem. Linear programming was applied to deliver the tension and riser from each fabrication site to WK Andaman II by considering the minimum total transportation cost. On the other hand, the assignment problem was used to define the optimum offshore vessels selected to deliver offshore platform, particularly for hull and topside from each fabrication site to WK Andaman II. There are two scenarios applied in this study: (1) deliver hull and topside independently and (2) deliver hull and topside jointly. Thus, the assignment model was created based on both scenarios by considering the total minimum transportation cost.

3.3.1. Linear programming model for delivering tension and riser. This optimization model is used to calculate the minimum total cost of sending tension and riser using a solver for excel. Tension should be delivered first before the hull and topside. The last construction should be installed is riser. However, during the delivery, tension and riser have the same mathematical model due to the similarity of objective functions, decision variables and constraints. The difference is only about the location of fabrication site from tension and riser.

$$Min\,TC = \sum_{k=1}^{n}\sum_{r=1}^{m} Fkr(FCkr + OCkr + CHCkr + ICkr) + Fkr\,(JAkr\;x\;SSkr) \tag{1}$$

Subject to:

$$\sum_{k=1}^{n}\sum_{r=1}^{m} Fkr\;x\;Pk \geq D \tag{2}$$

$$\sum_{k=1}^{n}\sum_{r=1}^{m} Fkr\;x\;RTD \leq 365 - z \tag{3}$$

$$\sum_{k=1}^{n}\sum_{r=1}^{m} Fkr = int \tag{4}$$

Where:

TC	: Minimum total cost (Rp/year)
Fkr	: Frequency of ship k on route r
FCkr	: Fuel cost of ship k on route r (Rp/Roundtrip)
OCkr	: OFY cost of ship k on route r (Rp/Roundtrip)
CHCkr	: Cargo handling cost of ship k on route r (Rp/Roundtrip)
ICkr	: Installation cost of ship k on route r (Rp/day)
JAkr	: Number of ships k on route r (Unit)
SSkr	: Scheme for renting ship k on route r (Rp/Year)
k	: Ship (Unit)
r	: Route
Pk	: Payload of ship k (ton)
D	: Demand (ton/tahun)
z	: *Off hire* or when the ship is not operating due to maintenance (days)
int	: *Integer*
RTD	: Round trip days (days)

Equation (2) is the limitation that the cargo carried by the ship must meet demand, then equation (3) defines that the ship's frequency must not exceed commission days, and equation (4) the frequency is an integer.

3.3.2. Assignment problem for delivering hull and topside. This optimization model is used to determine the minimum total transportation cost for delivering hull and topside either independently or jointly by using evolver.

$$Min\ TC = \left(\sum_{k=1}^{n}\sum_{r=1}^{m} KTkr((JAkr\ x\ SSkr) + Fkr\ (FCkr + OCkr + CHCkr + ICkr))\right) \quad (5)$$

Subject to:

$$\sum\nolimits_{r=1}^{n}\sum\nolimits_{r=1}^{m} 0 \leq KTkr \leq 1 \quad (6)$$

$$\sum\nolimits_{k=1}^{n}\sum_{r=1}^{m} Hkr + HTkr = 1 \quad (7)$$

$$\sum\nolimits_{k=1}^{n}\sum_{r=1}^{m} Tkr + HTkr = 1 \quad (8)$$

$$\sum\nolimits_{k=1}^{n}\sum_{r=1}^{m} Fkr\ x\ RTD \leq 365 - z \quad (9)$$

Where:

JAkr	: Number of ships k per route r (units)
SSkr	: Scheme for renting k ships on each route r (Rp)
Fkr	: Frequency of ships k on each route r (times)
FCkr	: Fuel cost of ship k on each route r (Rp/Roundtrip)
OCkr	: OFY cost of ship k on each route r (Rp/Roundtrip)
CHCkr	: Cargo handling cost of ship k on each route r (Rp/Roundtrip)
ICkr	: Installation cost of k ships on each route r (Rp/Day)
k	: Ship (unit)
r	: Route
KT	: Selected ships (units)
int	: Integer
Hkr	: Ship k on each route that carries hulls
HTkr	: Ship k on each route that carries hull and topside
Tkr	: Ship k on each route r that carries topside
Fkr	: Frequency on ship k on route r (times)
z	: Off hire or time the ship is not operating due to maintenance (days)
RTD	: Round trip days (days)

Equation (6) is a decision variable which is input to obtain optimum results in the objective function. In this optimization, the decision variable is the selected ship (binary). Moreover, equations (7) and (8) are constraint on the number of hulls and topsides that should be jointly transported, each part consists of 1 unit decision, either independently or jointly delivery. The last equation (9) limits the operational days which should not exceed to commission days.

4. Results and discussion

The type of offshore platform that will be built in the WK Andaman II is the Tension Leg Platform (TLP) because the depth of the WK Andaman II is between 300 m – 2000 m, namely 1,291.3 m (deep water). The main structures for TLP construction include tension (piles and tendons), hull, topside and riser.

4.1. Conditions in the Andaman II working area
WK Andaman II is one of the conventional working area in Indonesia which is located 150 km north of Lhokseumawe Aceh with coordinates 6.56056156326723, 97.1585097859771. The estimated natural gas content is 2,023 MMBOE. Natural gas mining in this CA is offshore and the project type is green field. The Cooperation Contract Contractors (KKKS) in this project are Premier Oil, Mubadala Energy, and Repsol. These KKKS respectively contributed 40%, 30% and 30%. WK Andaman II is included in region I, namely in Sumatra.

4.2. Load analysis
TLP payload specifications are obtained from TLP benchmarking which has almost the same depth as WK Andaman II, namely TLP Neptune. Following are the specifications.

Table 1. Specification of tension leg platform

Structure	Unit	Ton/Unit	Deck Load (Ton/m2)	Total Weight
Tension				
Piles	42	425	2	17,850
Tendons	36	372	5	13,392
Hull & Topside				
Hull	1	11,210	13	11,210
Topside	1	9,716	9	9,716
Mix	1	20,926	16.26	20,926
Riser	58	25.20	3	1,462

4.3. Project cargo transportation options
This research has two options for transporting project cargo related to the construction of offshore platforms in WK Andaman II. One of the transportation options will later be selected with a minimum total cost through optimization.

4.3.1. Separate project cargo transport option. The option of transporting project cargo separately begins with loadout tension (piles and tendons) from the Onshore Fabrication Yard (OFY) to the offshore vessel, then the tension is sent to the site, once on site the tension is installed, after installing the tension then loadout of the hull is carried out from OFY to offshore different vessels with tension, then the hull is sent to the site and installed. Next, the loadout riser goes to the offshore vessel and is sent to the site and then installed on the TLP which has been built for natural gas drilling. Thus, the OFY location may be the same or different.

4.3.2. Complete project cargo transportation options. This transport option is almost the same as the separate transport option, the difference being that the hull structure and topside are built on one OFY and sent to the location simultaneously on the same offshore vessel.

4.4. Onshore fabrication yard location analysis.
The Onshore Fabrication Yard (OFY) chosen is divided into two, OFY for the manufacture of piping (piles, tendons and risers) as well as the construction of the hull and topside. The selected OFY has good capabilities to produce or build TLP which can be seen from project milestones. OFY is used as the place of origin for cargo while WK Andaman II is the destination. OFY used for making pipes include Europipe (Germany), EEW (Malaysia), Tenaris (Japan), and Histeel (South Korea). OFY for hull and topside construction, namely Hyundai Samho Heavy Industries (South Korea), Geoje Shipyard / Samsung Heavy Industries (South Korea), PTSC Mechanical & Construction Limited Company (Vietnam), Keppel Fels Shipyard (Singapore), McDermott Shipyard (Batam), and Offshore Oil Engineering CNOOC (China). OFY was chosen because it had built TLP.

The author gives symbols to each route to make calculations easier. The route in this research is port to port. The following is a definition of the route symbols used.

A-O-A : Europipe - WK Andaman II - Europipe
B-O-B : EEW - WK Andaman II - EEW
C-O-C : Tenaris - WK Andaman II - Tenaris
D-O-D : Histeel - WK Andaman II - Histeel
E-O-E : Hyundai Samho Heavy Industries - WK Andaman II - Hyundai Samho Heavy Industries
F-O-F : Geoje Shipyard - WK Andaman II - Geoje Shipyard
G-O-G : PTSC (M & E) - WK Andaman II - PTSC (M & E)
H-O-H : Keppel Fels Shipyard - WK Andaman II - Keppel Fels Shipyard
I-O-I : McDermott Shipyard - WK Andaman II - McDermott Shipyard
J-O-J : Offshore Oil Engineering - WK Andaman II - Offshore Oil Engineering

Costs for handling cargo at OFY are obtained from benchmarking the nearest port to OFY which handles loading and unloading of project cargo.

4.5. Analysis of marine transportation mode

The marine transportation mode used is the offshore vessels type which consists of crane barge, barge, multipurpose heavy lift vessel, heavy lift vessel, crane barge, platform supply vessel, crane vessel and anchor handling tug supply. Specifically, vessel cranes are not used for project cargo transportation but rather for installation on site. Each ship is selected in four sizes, small, medium, large and very large according to the standards of each ship. The following are the specifications for the offshore vessels used.

Table 2. Specification of offshore vessels

Code	Type of ship	DWT	L_{OA} (m)	B (m)	H (m)	T (m)
Tension						
CB 1	*Crane Barge*	19,800	136.00	36.00	8.00	5.00
CB 2	*Crane Barge*	10,000	91.44	21.94	5.40	4.30
CB 3	*Crane Barge*	4,600	80.25	25.00	4.30	3.75
CB 4	*Crane Barge*	2,500	60.50	24.00	4.80	3.50
MLV 1	*M.H.L Vessel*	37,200	216.00	43.00	13.00	9.00
MLV 2	*M.H.L Vessel*	24,964	161.33	27.39	13.50	9.81
MLV 3	*M.H.L Vessel*	14,047	138.07	21.00	11.00	9.20
MLV 4	*M.H.L Vessel*	3,343	139.10	54.30	6.70	4.10
BGT 1	*Barge*	19,929	122.00	36.60	7.60	6.30
BGT 2	*Barge*	11,000	95.16	30.60	6.10	5.00
BGT 3	*Barge*	3,750	73.00	18.00	5.20	4.60
BGT 4	*Barge*	3,000	54.86	21.33	4.26	3.50
Hull & Topside						
HLV 1	*H. Lift Vessel*	98,000	255.00	68.00	15.00	10.50
HLV 2	*H. Lift Vessel*	80,000	255.00	57.00	14.50	10.50
HLV 3	*H. Lift Vessel*	50,352	227.00	43.00	13.50	8.80
HLV 4	*H. Lift Vessel*	29,891	181.90	43.00	10.00	5.60
BG 1	*Barge*	41,947	165.00	42.00	10.70	9.20
BG 2	*Barge*	20,437	120.00	36.60	7.60	6.50
BG 3	*Barge*	15,500	85.34	27.43	4.80	3.70
BG 4	*Barge*	10,040	60.00	26.00	5.00	3.50
Riser						
PSV 1	*P.S Vessel*	5,509	83.60	20.00	9.00	7.40
PSV 2	*P.S Vessel*	4,699	88.10	19.00	8.20	6.60
PSV 3	*P.S Vessel*	3,010	70.40	16.00	8.00	6.80
PSV 4	*P.S Vessel*	1,300	57.50	13.80	5.50	4.50
AHTS 1	*A.H.T Supply*	3,500	78.20	18.50	8.00	6.60

Code	Type of ship	DWT	L_{OA} (m)	B (m)	H (m)	T (m)
AHTS 2	*A.H.T Supply*	2,400	70.05	17.00	7.50	6.10
AHTS 3	*A.H.T Supply*	1,375	60.50	14.60	5.50	4.75
AHTS 4	*A.H.T Supply*	2,614	41.80	10.00	4.60	3.60

The vessel cranes for hull and topside installations are Heerema Thialif and Heerema Sleipnir. Heerema Sleipnir is used for complete hull and topside installations while Heerema Thialif is used for separate hull and topside installations. This is because each crane vessel has a different SWL depending on the weight of the cargo to be transported. Then, the vessel crane that is used as an alternative for tension installations when using a barge is the TBN with a capacity of 550 tons.

4.6. Compatibility

Compatibility is a basic concept of adjusting conditions. In this research, there are 3 (three) things that need to be known for their compatibility, namely the OFY depth to the ship's draft, the ship's submerged depth to the cargo's submerged depth, and the ship's deck load to the cargo's deck load. Where the OFY depth must be greater than or equal to the ship's draft, the ship's submerged depth must be greater than or equal to the cargo's submerged depth, and the ship's deck load must also be greater than or equal to the cargo's deck load. If one of these criteria cannot be met by the ship, the ship will be eliminated in the voyage calculation.

4.7. Loadout

Handling project cargo at the fabrication site or OFY is different because project cargo for building TLP has 2 types, namely piping (tension and riser) and main structure (hull and topside). This happens because the dimensions and weight of the load are different so the loadout method used is also different. The loadout method used for piping is lifting, while the loadout method used for main structures is skidding. When lifting is assisted by a crane and multi axle trailer / SPMT. Meanwhile, when skidding, use skidway assistance.

4.7.1. Loadout of tension and riser. Table 3 contains information on depth, distance, crane productivity, and total time on each route carrying out tension and riser loadout activities.

Table 3. Loadout of tension and riser

Route	Depth (-mLWS)	Distance (nm)	Crane Productivity (Ton/hour)	Total of time (hour)
A-O-A	12.7	8,251	800	15.19
B-O-B	12.5	554	600	16.24
C-O-C	12.5	3,414	500	17.04
D-O-D	13	3,118	1,400	15.13

4.7.2. Loadout of hull and topside. Table 4,Table 5,Table 6 contain information on depth, distance, loadout preparation time, and loadout on each route that carries out hull, topside activities and a combination of the two.

Table 4. Loadout hull

Route	Depth (-mLWS)	Distance (nm)	Loadout preparation time (hour)	Loadout (hour)
E-O-E	13.5	3,031	135.72	12
F-O-F	13	3,006	135.72	12
G-O-G	12.5	1,118	243.72	20
H-O-H	12.5	578	164.52	14

Route	Depth (-mLWS)	Distance (nm)	Loadout preparation time (hour)	Loadout (hour)
I-O-I	12	540	190.92	15
J-O-J	14	1,789	217.32	16

Table 5. Loadout topside

Route	Depth (-mLWS)	Distance (nm)	Loadout preparation time (hour)	Loadout (hour)
E-O-E	13.5	3,031	119.72	10
F-O-F	13	3,006	119.72	10
G-O-G	12.5	1,118	227.72	18
H-O-H	12.5	578	148.52	12
I-O-I	12	540	174.92	14
J-O-J	14	1,789	201.32	15

Table 6. Loadout hull and topside

Route	Depth (-mLWS)	Distance (nm)	Loadout preparation time (hour)	Loadout (hour)
E-O-E	13.5	3,031	163.40	14
F-O-F	13	3,006	163.40	14
G-O-G	12.5	1,118	271.40	24
H-O-H	12.5	578	192.20	16
I-O-I	12	540	218.60	19
J-O-J	14	1,789	245.00	20

4.8. Installation

Offshore structure installation is the activity of installing structures from an offshore vessel to the site. This activity is risky because the environment and conditions on the high seas are unpredictable so it must be carried out by skilled people and careful planning. The following is information regarding the installation of project cargo to build the TLP. The installation sequence is tension – hull - topside – riser. Tension installation is assisted by VLT, while the riser uses a crane rig.

4.8.1. Installation of tension and riser. Table 7 provides information regarding waiting time, crane productivity, installation time, and installation costs for tension and riser installation activities on each ship.

Table 7. Instalation of tension and riser

Ship code	Waiting time (anticipate conditions) (hour)	Crane productivity (Ton/hour)	Instalation time (hour)	Instalation cost (Million.Rp/day)
CB 1	168	1,400	33	87,756.75
CB 2	168	550	35	32,237.18
CB 3	168	600	35	22,330.96
CB 4	168	550	34	13,096.35
MLV 1	168	3000	33	391,958.88
MLV 2	168	1,500	34	155,360.54
MLV 3	168	720	35	81,057.67
MLV 4	168	1,000	33	93,666.64
BGT 1	168	550	34	95,495.47
BGT 2	168	550	34	41,924.83

Ship code	Waiting time (anticipate conditions) (hour)	Crane productivity (Ton/hour)	Instalation time (hour)	Instalation cost (Million.Rp/day)
BGT 3	168	550	34	23,427.92
BGT 4	168	550	34	14,619,56
PSV 1	168	80	20	72,051.26
PSV 2	168	80	20	79,976.90
PSV 3	168	80	20	65,086.31
PSV 4	168	80	20	57,040.58
AHTS 1	168	80	20	108,076.89
AHTS 2	168	80	20	71,630.96
AHTS 3	168	80	20	46,112.81
AHTS 4	168	80	20	35,809.48

4.8.2. Instalation of hull and topside. Table 8 provides information regarding the type of cargo transported, cranes used, preparation time, and installation costs for hull and topside installation activities on each ship.

Table 8. Instalation hull and topside

Type of ship	Demand	Crane	Preparation (Hour)	Instalation cost (Million Rp/day)
HLV	Hull	H. Thialif	94	294,034.06
	Topside		45	294,034.06
Barge	Hull		94	294,034.06
	Topside		45	294,034.06
HLV	Hull & Topside	H. Sleipnir	125	476,252.35
Barge	Hull & Topside		125	476,252.35

5. Conclusion

- Based on research, it is known that the condition of WK Andaman II is located 150 km north of Lhokseumawe Aceh. The natural gas content in the WK Andaman II is 2.023 MMBOE. WK Andaman II project type is green field. Currently, the construction of an offshore platform in WK Andaman II is still in the exploration stage and is targeted to be onstream in 2028.
- The type of offshore platform that is suitable to be built in WK Andaman II is the Tension Leg Platform. The construction stages consist of two options, namely a separate option and a complete option where the transportation starts with tension – hull – topside – riser. The difference between separate and complete options lies in whether the hull and topside are sent as one unit or separately. The transportation option with the minimum total cost turns out to be jointly delivered.
- The ship selected to transport tension is CB 1 with total frequency is twice by using one ship. The OFY selected is EEW (Malaysia). The days required for delivery until the ship returns to OFY is 34 days. The minimum total cost is Rp 807,39 billion. In addition, the ship selected to deliver the hull and topside intact was HLV 4 with one frequency. The OFY selected is McDermott Shipyard (Batam). The days required to complete delivery to installation are 41 days with a minimum total cost of Rp. 9.562,63 billion. Finally, the riser is delivered by using two ships, namely AHTS 4 with a frequency of one per ship. There are two selected routes, namely EEW (Malaysia) and Histeel (South Korea). The days required for delivery are 14 days and 31 days. The total minimum cost for handling risers is Rp. 657,36 billion. The total shipping costs for the construction of the offshore platform reached Rp. 11.027,34 billion.

References

[1] Nonci, R. (2020). Analisis Deskripsi Minyak dan Gas. Jurnal Penelitian Pendidikan dan Ekonomi, 45.

[2] Satuan Kerja Khusus Pelaksana Kegiatan Usaha Hulu Minyak dan Gas Bumi. (2022, Juli 21). Kontribusi Pajak Industri Hulu Migas Sudah Melampaui Target Pemerintah. Retrieved from https://www.skkmigas.go.id/berita/kontribusi-pajak-industri-hulu-migas-sudah-melampaui-target-pemerintahAnother reference.

[3] Dunia Energi. (2022, Agustus 1). Ruang Fiskal APBN 2022 dan Peran Hulu Migas. Retrieved from duniaenergi.com: https://www.dunia-energi.com/analysis/ruang-fiskal-apbn-2022-dan-peran-hulu-migas/

[4] Kementerian Energi dan Sumber Daya Mineral. (2021). Statistik Minyak dan gas Bumi 2021. Jakarta.

[5] CNBC Indonesia. (2022, Desember 5). Pabrik Pupuk BUMN Terancam Mati Januari 2023 Kekurangan Gas. Retrieved from www.cnbcindonesia.com: https://www.cnbcindonesia.com/market/20221205154159-17-393900/pabrik-pupuk-bumn-terancam-mati-januari-2023-kekurangan-gas

[6] Kementerian Energi dan Sumber Daya Mineral. (2022, Februari 17). Ini Peran Penting Migas Dalam Transisi Energi. Retrieved from migas.esdm.go.id: https://migas.esdm.go.id/post/read/ini-peran-penting-migas-dalam-transisi-energi

[7] Peraturan Menteri Energi dan Sumber Daya Mineral RI No 1 Tahun 2022. (2022). Peraturan Menteri Energi dan Sumber Daya Mineral RI No 1 Tahun 2022 tentang pengelolaan dan Pemanfaatan Data Minyak dan Gas Bumi. Retrieved from jdih.esdm.go.id: https://jdih.esdm.go.id/storage/document/Permen%20ESDM%20No.%201%20Tahun%202022.pdf

[8] Dey, A. K. (2021). Brownfield vs Greenfield: Differences between a brownfield and greenfield project. Retrieved from whatispiping.com: https://whatispiping.com/brownfield-vs-greenfield/?utm_content=expand_article

[9] Muttaqin, R. N. (2016). Analisis Inplace, Seismic, dan Fatigure Fererministic untuk Perpanjangan Masa Layan Fixed Platform Tiga Kaki di Ladang Minyak Ardjuna Laut Jawa. 1.

[10] Amri, C. (2008). Analisis Struktur Anjungan Lepas Pantai Tipe Jacket 4 Kaki Menggunakan SACS 5.1.

[11] Nugraha, B., & Basuki, R. (2019). Rigs-To-Reef (R2R): A new initiative on re-utilization of abandoned offshore oil and gas platforms in Indonesia for marine and fisheries sectors.

[12] Pradipta, D. H. (2015). Analisis respon Gerak Struktur Jacket Fixed Platform Akibat Pengaruh Penambahan Beban Operasional Studi Kasus: Bekapai BQ Platform. Surabaya.

[13] Reda, A., Butler, H. O., Ja'e, I. A., An, C., & Victor, C. (2022). Review on Fixed and Floating Offshore Structures. Part I: Types. Marine Science and Engineering.

[14] Andriyanto. (2008). Analisis Kekuatan Rangka pada Tension Leg Platform West Seno di Selat Makassar. Depok: Universitas Indonesia.

[15] Karan. (2019, April 17). What are Offshore Vessels? Retrieved from marineinsight.com: https://www.marineinsight.com/types-of-ships/what-are-offshore-vessels/

[16] Whiteford, S. (2020, November 29). What is an Offshore Support Vessel, or OSV? Retrieved from onesteppower.com: https://www.onesteppower.com/post/what-is-an-offshore-support-vessel-or-osv

[17] Praditya, B. (2022). Model Perencanaan Transportasi Laut Angkutan Project Cargo: Studi Kasus Pembangunan Infrastruktur Tenaga Listrik Ibu Kota Nusantara.

[18] Putra, F. M. (2016). Analisis Teknis dan Ekonomis Pembangunan Industri manufaktur Bangunan Lepas Pantai di Jawa Timur. Tugas Akhir.

[19] Santoso, N. B. (2014). Pemanfaatan LNG Sebagai Sumber Energi di Indonesia. Jurnal Rekayasa Proses, 34.

Detection of Flood Inundation Areas Based on Sentinel-1 SAR Imagery in the Klawing River Watershed

Muhammad Adnan Adi Fikri[1], Purwanto Bekti Santoso[2*] and Sanidhya Nika Purnomo[2]

[1] Student, Civil Engineering Department, Universitas Jenderal Soedirman
[2] Lecturer, Civil Engineering Department, Universitas Jenderal Soedirman

*E-mail: purwanto1409@unsoed.ac.id

Abstract. In December 2020, the Klawing River overflowed, flooding hundreds of houses in six villages. To aid in evacuation, identifying affected areas is crucial. Using Sentinel-1 Synthetic Aperture Radar (SAR) satellite imagery and the Change Detection Approach with thresholding via Google Earth Engine, flood-inundated areas were mapped efficiently. The process involved detecting changes in backscatter values before and after the flood. Sentinel-1 images from December 13, 2020, were used to map the flood, while images from late 2019 and 2020 served as pre-flood references. Threshold values of 1.22 to 1.24 were applied, revealing flood areas ranging from 10.56 km^2 to 17.31 km^2.

1. Introduction

Several flood events affected Purbalingga Regency around December 2020. The floods were caused by rising water levels in the Klawing River, which resulted from heavy rainfall. The overflow of the Klawing River led to hundreds of homes in several villages in Purbalingga Regency being submerged by floods with depths ranging from 40 to 100 centimeters. Given the extensive area of the Klawing River Basin, having detailed information about flood inundation locations is crucial for effectively managing and preventing ongoing damage caused by flood disasters.

With advancements in technology, flood inundation areas can now be mapped using remote sensing systems supported by satellite technology. Remote sensing techniques are highly effective for identifying water bodies over large areas and can be used for flood mapping with adequate temporal and spatial resolution [1]. One effective remote sensing technique for detecting flood inundation is Synthetic Aperture Radar (SAR) satellite imagery [2]. One such SAR satellite with freely accessible data for flood mapping is the Sentinel-1 SAR. Synthetic Aperture Radar (SAR) has the advantage of being unaffected by time of day, weather conditions, cloud cover, or fog, which can obscure important information about objects behind these conditions [3].

Mapping flood inundation with Sentinel-1 SAR satellite imagery involves preprocessing since raw data cannot directly determine flood location and extent. Preprocessing steps include applying orbit files, removing thermal noise, radiometric calibration, terrain correction, and converting backscatter values to decibels (dB). After preprocessing, the data is analyzed using change detection and thresholding methods. VH (Vertical-Horizontal) polarization is preferred for flood mapping due to its effective backscatter response for distinguishing water from land and identifying volume scattering objects like vegetation [4]. This polarization is particularly useful in Indonesia, where vegetation and rice fields are common.

Determining the threshold for flood detection using Sentinel-1 SAR images is challenging due to several factors. Unimodal histograms often make it difficult to differentiate water from non-water areas, while backscatter signals vary with wind, rain, and surface roughness [5]. Geographical differences and surface characteristics require adaptive methods, and threshold selection is often subjective, relying on visual inspection [6]. Additionally, noise and artifacts in SAR images complicate accurate flood detection, requiring advanced processing techniques [7]. Processing Sentinel-1 SAR imagery for flood detection can be done using Google Earth Engine and QGIS. Google Earth Engine, a cloud-based JavaScript platform, processes Sentinel-1 images from before and during the flood, handling everything from preprocessing to results automatically. These results are then analyzed in QGIS. This text explains how to identify flood inundation distribution with satellite imagery using Google Earth Engine and a simple method in thresholding a change detection map.

2. Method

2.1 Study Location

The study focuses on the Klawing River Basin in Purbalingga Regency, Central Java Province. The Klawing River Basin covers an area of 1,059.50 km^2 and has the shape depicted in Figure 1. The Klawing River Basin includes the sub-districts of Bobotsari, Purbalingga, Mrebet, Bojongsari, Kaligondang, and Kemangkon. The Klawing River, also known as Kali Klawing, is the main river flowing through the Klawing River Basin, with a total length of 49.11 km. The river originates from Mount Sitengkek in Kutabawa Village, Karangreja Sub-district, and empties into the Serayu River in Banyumas Regency.

2.2 Identification of Flood Inundation Using Google Earth Engine

The data processing steps using Google Earth Engine in this study follow the guidelines recommended by the United Nations Platform for Space-based Information for Disaster Management and Emergency Response (UN-SPIDER) [8]. The required data for this process are:

a) Sentinel-1 SAR Level-1 Ground Range Detected Images

Figure 1. The Klawing River Watershed

b) Joint Research Centre Global Surface Water Mapping Layers v1.4
c) WWF HydroSHEDS Hydrologically Conditioned DEM, 3 Arc-Seconds

Flood inundation detection using Google Earth Engine utilizes the change detection approach, requiring satellite images from before and after the flood event. This study focuses on the flood that occurred on December 13, 2020, using three pre-flood images and one flood image: November 19, 2020 (pre-flood), November 25, 2020 (pre-flood), December 19, 2019 (pre-flood), and December 13, 2020 (flood). Before running the analysis in Google Earth Engine, the boundary of the flood inundation modeling area must be defined. The modeling area boundary will be adjusted to match the overall Klawing River Basin (DAS) boundary and then further refined to a smaller boundary in the lower reaches of the Klawing River. The satellite images can be seen in Figure 2.

3. Results and Discussion

3.1 Change Detection Approach and Thresholding

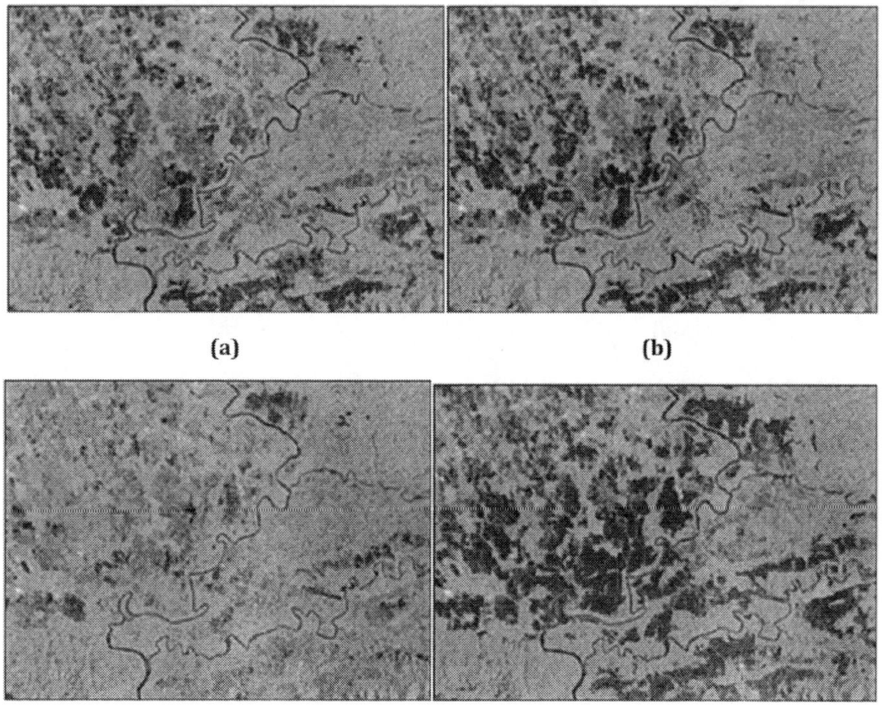

(a) (b)

(c) (d)

Figure 2. (a) Sentinel-1 SAR IW 10 x 10m VH Polarization image (November 19, 2020); (b) Sentinel-1 SAR IW 10 x 10m VH Polarization image (November 25, 2020); (c) Sentinel-1 SAR IW 10 x 10m VH Polarization image (December 19, 2019); and (d) Sentinel-1 SAR IW 10 x 10m VH Polarization image (December 13, 2020).

IOP Conf. Series: Earth and Environmental Science **1473** (2025) 012031 doi:10.1088/1755-1315/1473/1/012031

Change detection on two SAR satellite images will be automatically processed by Google Earth Engine. The selected satellite images have already undergone preprocessing and speckle filtering. The backscatter values of the flood satellite image are divided by the backscatter values of the pre-flood satellite image. The mathematical process of the change detection approach is described by Equation (1) below [9].

$$W = \frac{Xi}{Yi} \quad (1)$$

where:

W = difference in backscatter value

X_i = backscatter value of the pixel during the flood

Y_i = backscatter value of the pixel before the flood

Earth Engine performs the change detection using the post-flood and reference images, then generates a change detection map or raster. Figure 3. shows the change detection results based on the reference images.

(a)

(b)

(c)

Figure 3. Results of Change Detection Map Using Reference Image of December 19, 2019 (a); Reference Image of November 25, 2020 (b); and Reference Image of November 19, 2020 (c)

Each pixel in the change detection map or raster has a specific value. For the December 19, 2019 reference image, pixel values range from 0.49 to 2.17. For the reference image from November 25, 2020, values range from 0.42 to 1.87, and for the reference image from November 19, 2020, values range from 0.48 to 1.77. Higher pixel values appear brighter, while lower values appear darker. Brighter pixels indicate areas where changes occurred between the pre-flood (reference) and post-flood images, showing where floodwaters have affected the land. A threshold value needs to be set and applied to identify flood-affected areas from the change detection map. Pixels with values above this threshold represent flood areas, while those below the threshold represent non-flood areas.

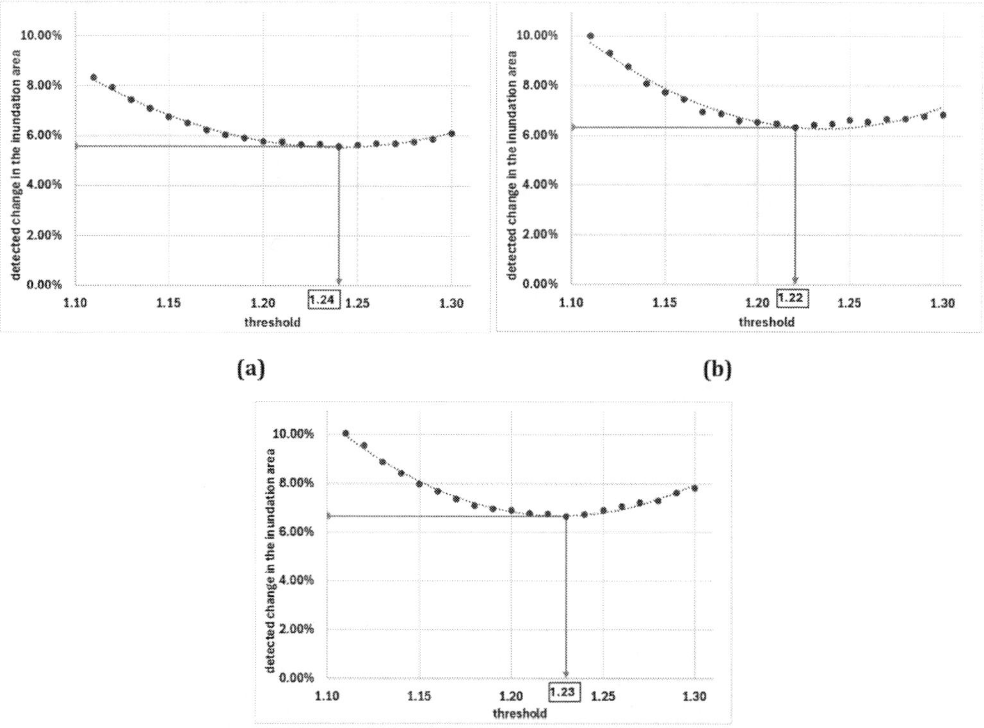

Figure 4. Threshold value and its related change in inundation area using reference image of December 19, 2019 (a); reference image of November 25, 2020 (b); and reference image of November 19, 2020 (c)

Methods for determining threshold values for detecting flood inundation using Sentinel-1 SAR images include several approaches. A study in the Jialing River presents an unsupervised framework that employs a robust search for tiles following a bimodal Gaussian distribution and a non-parametric histogram-based thresholding algorithm [5]. Another study explores the use of deep learning, particularly a nested UNet architecture, to address the limitations of traditional thresholding methods [6]. Additionally, using locations of inland and coastal waters in China, a study proposes an adaptive thresholding technique, incorporating prior Landsat data to enhance water coverage extraction accuracy [7]. A threshold of 1.10 for identifying flood inundation in

Kalimantan were used by a previous study [10], while a recommeded value of UN-SPIDER [8] is 1.25. This study tests a range of threshold values from 1.10 to 1.30 with 0.01 interval. The final threshold is chosen based on the least sensitive value, of the selected interval, in reducing the flood area. Figure 4. show the results of this trial-and-error process.

3.2 Identified Flood Inundation Area

Figures 5. shows the results of flood inundation modeling using different reference images and their assosiated selected threshold values based on the trial procedure. The maximum flood-inundated area detected on December 13, 2020, was 17.31 km^2, determined by comparing the flood image with the reference image from December 19, 2019, using a threshold value of 1.24. In contrast, the minimum flood-inundated area was 10.56 km^2, identified by comparing the flood image with the reference image from November 25, 2020, using a threshold value of 1.22. The flood inundated area based on the reference image of November 19, 2020 is 12.12 km^2, using a threshold value of 1.23. All the selected threshold values are very close to the UN-SPIDER recommended value of 1.25 [8].

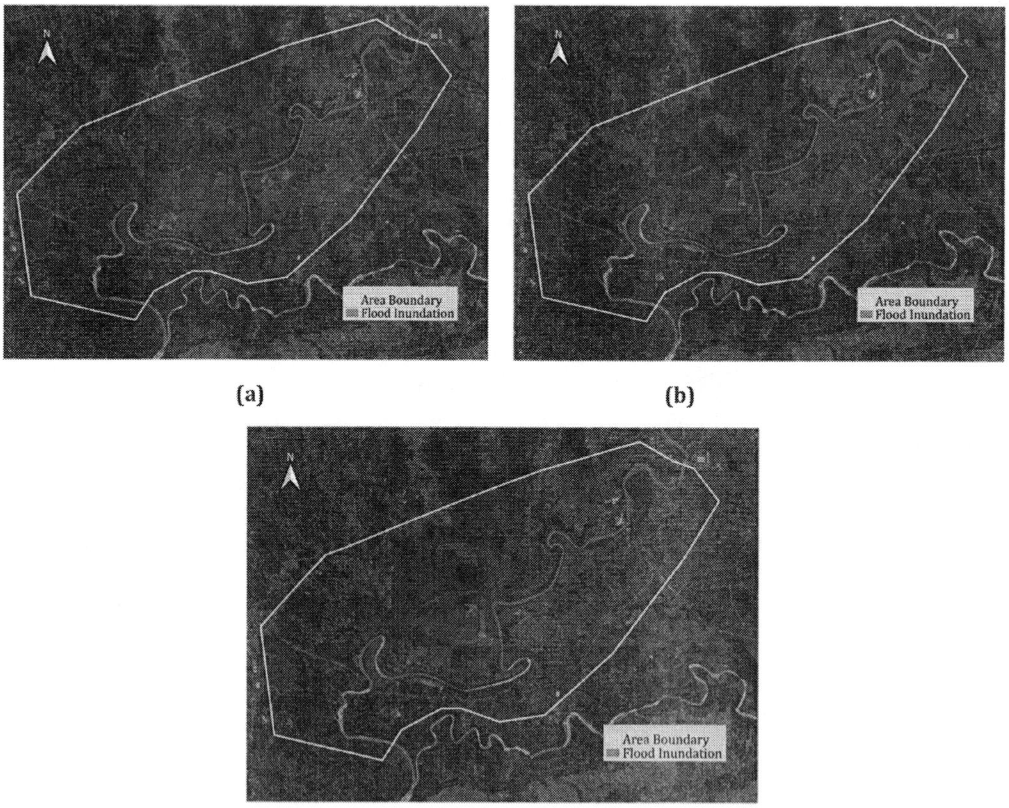

(c)

Figure 5. Flood Inundation Map Using Reference Image of December 19, 2019, threshold 1.24 (a); Reference Image of November 25, 2020, threshold 1.22 (b); and Reference Image of November 19, 2020, threshold 1.23 (c)

4. Conclusion

In this study, the satellite imagery used includes the image from December 13, 2020, as the post-flood image. For reference, three images were used: December 19, 2019; November 19, 2020; and November 25, 2020. The threshold values ranged from 1.22 to 1.24, with flood inundation areas ranging from 10.55 to 17.31 km^2. Flood detection using Google Earth Engine with Sentinel-1 SAR satellite imagery has several limitations. Sentinel-1 SAR data may not be available for every time period, so images from the exact date of the flood and the subsequent period might not be accessible. Additionally, Sentinel-1 satellites do not always cover all areas during their observations, which further limits the availability of SAR images.

References

[1] A. Moothedan, P. R. Dhote, P. Thakur, V. Garg, S. Aggarwal, and M. Mohapatra, "Automatic Flood Mapping Using Sentinel-1 GRD SAR Images and Google Earth Engine: A Case Study Of Darbhangah, Bihar," in *The Proceedings of National Seminar on 'Recent Advances in Geospatial Technology & Applications,'* Sep. 2020.

[2] F. Bioresita, A. Puissant, A. Stumpf, and J.-P. Malet, "A Method for Automatic and Rapid Mapping of Water Surfaces from Sentinel-1 Imagery," *Remote Sens (Basel)*, vol. 10, no. 2, p. 217, Feb. 2018, doi: 10.3390/rs10020217.

[3] P. P. Utomo, B. Riadi, and D. Ramadani, "Identifikasi Sebaran Banjir Menggunakan Citra Satelit Sentinel-1," Universitas Pakuan, 2020.

[4] P. Manjusree, L. Prasanna Kumar, C. M. Bhatt, G. S. Rao, and V. Bhanumurthy, "Optimization of threshold ranges for rapid flood inundation mapping by evaluating backscatter profiles of high incidence angle SAR images," *International Journal of Disaster Risk Science*, vol. 3, no. 2, pp. 113–122, 2012, doi: 10.1007/s13753-012-0011-5.

[5] H. Cao, H. Zhang, C. Wang, and B. Zhang, "Operational Flood Detection Using Sentinel-1 SAR Data over Large Areas," *Water (Basel)*, vol. 11, no. 4, p. 786, Apr. 2019, doi: 10.3390/w11040786.

[6] B. Ghosh, S. Garg, M. Motagh, and S. Martinis, "Automatic Flood Detection from Sentinel-1 Data Using a Nested UNet Model and a NASA Benchmark Dataset," *PFG – Journal of Photogrammetry, Remote Sensing and Geoinformation Science*, vol. 92, no. 1, pp. 1–18, 2024, doi: 10.1007/s41064-024-00275-1.

[7] S. Chen, W. Huang, Y. Chen, and M. Feng, "An Adaptive Thresholding Approach toward Rapid Flood Coverage Extraction from Sentinel-1 SAR Imagery," *Remote Sens (Basel)*, vol. 13, no. 23, p. 4899, Dec. 2021, doi: 10.3390/rs13234899.

[8] UN-SPIDER (United Nations Platform for Space-Based Information for Disaster Management and Emergency Response), "Recommended Practice: Flood Mapping and Damage Assessment Using Sentinel-1 SAR Data in Google Earth Engine," https://www.un-spider.org/advisory-support/recommended-practices/recommended-practice-google-earth-engine-flood-mapping. accessed on June 20, 2024.

[9] F. A. Budiarto and F. Bioresita, "Pemanfaatan Citra Sentinel-1 SAR dan Metode Change Detection Approach Untuk Analisis Sebaran Spasial Wilayah Banjir dan Area Terdampak (Studi Kasus: Banjir Kabupaten Aceh Utara 2022)," *JGISE: Journal of Geospatial Information Science and Engineering*, vol. 6, no. 2, pp. 153–162, 2023.

[10] F. Bioresita, M. G. R. Ngurawan, and N. Hayati, "Identifikasi Sebaran Spasial Genangan Banjir Memanfaatkan Citra Sentinel-1 dan Google Earth Engine (Studi Kasus: Banjir Kalimantan Selatan)," *Geoid*, vol. 17, no. 1, pp. 108–118, 2022.

Structural Reliability Study of the N219 Floatplane's Floater Due to Random Wave Loads

Bimo Hadi Putro[1], Alif Nur Rochmad[1], Windha Umi Alifia[1], Sandra Widia Meiliana[1], Aries Sulisetyono[1]

[1] Department of Naval Architecture, Institut Teknologi Sepuluh Nopember, Surabaya

*E-mail: sulisea@na.its.ac.id

Abstract. A floater is a buoyant apparatus employed to enable a floatplane to remain afloat and operate on the water surface. During its operational period, the floater experiences stress and deformation as a result of hydrodynamic loads. This study primarily focuses on providing a comprehensive examination of the structural strength of the proposed floater design. The determination of the structural integrity is achieved by the use of the finite element approach which is a popular method for numerically solving, and where in the wave-induced loads experienced by the floater structure are utilized as input parameters. The stress intensity in the floater structure is resulted in relation to the wave load parameters, specifically wave height and wave heading angle. The use of a structural reliability approach involves the consideration of the inherent randomness associated with both the load imposed on a structure and its strength. The Monte Carlo method is utilized to generate randomness of in order to determine the Probability Density Function (PDF) of both the structure's stresses and its strength. The computational findings elucidate that the floater structure exhibits reliability level of 97 % to 71% in a range of wave height between 0.5 to 3 meters and including reliability level of 100 % to 92% in a similar wave height factor, based on construction materials of aluminium and steel, respectively. This reliability level was indicating toughness of floater structure in each wave condition, smaller reliability level means there is high chance of structural failure.

Keywords: Structure Reliability, N-219 Floater, Wave Load, Finite Element Method

1. Introduction

The floater installed on the fuselage of the N-219 aircraft serves the purpose of ensuring buoyancy when the aircraft is on the water's surface. The incorporation of a floater on the fuselage of an airplane enables it to execute water landings, like those performed by amphibious aircraft. This particular aircraft model is frequently employed for a multitude of applications, including but not limited to transportation, tourism, geographical surveying, rescue operations, and research [1, 2]. In relation to the operating circumstances of the floater, it undergoes dynamic motion as a result of the displacement of the center of buoyancy and the submerged volume within the aquatic environment. Regarding this matter, the presence of waves has a notable impact on the longitudinal moment of the floater, primarily influencing the vertical motion of the vessel in terms of translation (heave) and rotation (pitch) [3]. The floating device frequently experiences high-velocity circumstances during the process of taking off, which gives rise to substantial wave load, leading to cyclic strain and deformation when subjected to rigorous operation [4].

The examination of wave loading on the structure of the floater is a fundamental component of both the design and structural study of the floater. The ship's structure can experience dynamic loads from sea waves, resulting in stress, strain, and deformation in different structural elements such as the hull, deck, and other components. Hence, it is imperative to incorporate wave stress into the design of the floater structure in order to mitigate the risks of material fatigue, structural damage, or complete failure [9]. Researchers have previously conducted a number of studies, one of which looked at how wave loading affected the structural integrity of Corvette warships in the presence of unpredictable wave characteristics [10]. There has already been research done on how wave loading affects different types of fast ships' construction, with the goal of finding the best design by using carbon fiber-reinforced polymer materials [11]. Both of these studies have employed the Panel Method or 3D diffraction theory, a well-established approach that has demonstrated high precision in evaluating the random wave load on ship structures.

The structural integrity of the floatplane's float must be designed to resist the significant operational circumstances and potential structural damage or failure caused by the wave load as well as the impact load [12]. Hence, it is imperative throughout the phase of building design to guarantee the integrity of the floater structure. The inherent stochasticity of waves leads to the variability and unpredictability of the amount of load received. Consequently, a probabilistic methodology is well-suited for effectively describing the stress experienced by the structure. Likewise, the evaluation of the structural integrity of the floater structure may be undertaken by employing strength criteria derived from the examination of various materials that exhibit inherent randomness. The utilization of structural dependability techniques is considered a viable strategy for addressing the presence of uncertainty in both load and structural strength data [7]. Historically, research endeavors examining the reliability of ship construction have employed methodologies that integrate the probabilities associated with both load and structural strength, drawing upon experimental data [5, 6]. The field of structural reliability offers significant insights for designers in order to enhance the efficiency of structural systems [11]. It encompasses several issues pertaining to design, maintenance, and repair [8].

This study assesses the structural integrity of the proposed construction design for the floater through the utilization of two distinct materials, namely steel and aluminum. The 3D diffraction theory method is used to look at the floater's vertical bending moment response to random wave loading, which includes four different wave height changes. The utilization of Monte Carlo simulations allows for the random generation of load and strength data, which in turn generates the probability density function of load and strength. These values are then used to establish a joint probability density function, which is employed to assess the structural failure of the floater [10].

2. Methodology

2.1 Floater Construction

The floater has two primary dimensions, as seen in Table 1. The floater possesses a deadrise angle of 20° and is outfitted with a step hull [3, 13]. The construction design of the floater is seen in Figure 1. The interior structure of the floater consists of ten frames featuring a T-shaped profile, five watertight bulkheads, and a longitudinal stiffener or keel serving as the primary structural element. The specifics of the scantling are depicted in Figure 3. The present study employs two distinct materials, specifically steel AH-32 and aluminum alloy 6061-T6, which are included in the construction of the floater. The material specifications and properties of these materials are shown in Table 2, Table 3.

Table 1. Main Dimension [14]

Name	Dimension	Unit
LOA	9.40	meter
LPP	9.27	meter
B	1.25	meter
H	1.07	meter
T	0.77	meter
Displ	3.83	ton

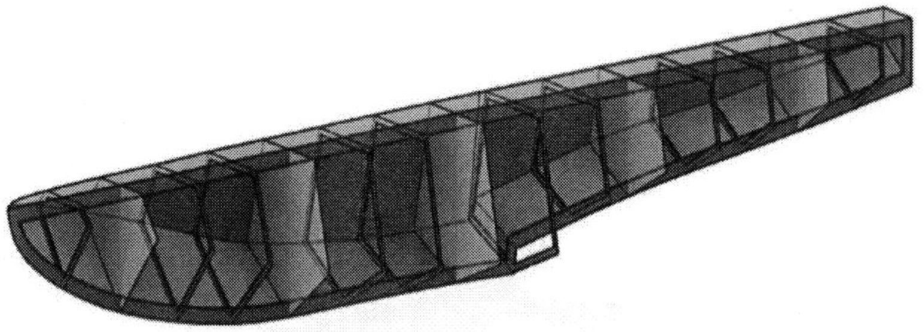

Figure 1. Construction of Floater

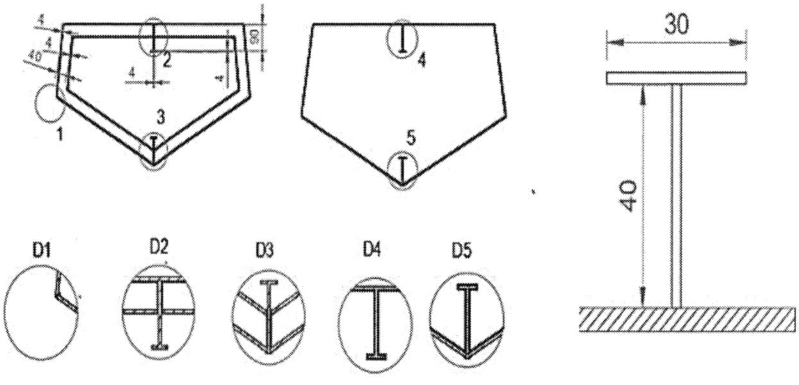

Figure 2. Scantling Detail

Table 2. Material Properties [5, 6]

Material	Density (kg/m³)	Yield Strength (MPa)	Ultimate Strength (MPa)	Poisson Ratio	Shear Module (GPa)	Elasticity Module (GPa)
AH-32	2770	358	460	0.29	80	200
6061-T6	7850	318	339	0.33	26	68.9

Table 3. Construction Weight of Steel AH32 and Alluminnium Alloy 6061-T6

Part	Area (m^2)	Thickness (m)	Volume (m^3)	AH32 Weight (ton)	6061-T6 Weight (ton)
Shell	25.41	0.004	0.10	0.79	0.65
Frame	6.29	0.002	0.01	0.09	0.05
Bulkhead	3.20	0.002	0.006	0.05	0.08
Longitudinal	3.28	0.004	0.01	0.10	0.08
Total :	35.66	-	-	1.08	0.87

2.2 Wave Load Analysis

2.2.1 Wave Characteristics

The wave characteristics of a floater are ascertained by considering its operational region. These characteristics are then categorized into several degrees of wave height, namely, slight, moderate, and rough conditions, as outlined in Table 4. The direction of the waves is dependent on the heading direction of the floater, namely originating from the front (Head Sea), the side (Beam Sea), and the behind (Following Sea). The analysis of wave loads is performed in twelve scenarios, encompassing three distinct wave arrival angles: head sea, beam sea, and following sea. The wave heights included in this analysis range from 0.5 m to 3 m, according to the different operational conditions of the floater.

Table 4. Classification of Wave Severity [15]

Sea State	Wave Description	Significant Wave Heaight (m)	Wind Speed (Knots)
1	Calm	0.1 - 0.5	0 - 6
2	Slight	0.5 - 1.25	7 - 15
3	Moderate	1.25 - 2.5	16 - 25
4	Rough	2.5 - 4.0	26 - 35
5	High	4.0 - 6.0	36 - 60
7	Extreme	6.0 >	60 >

In the analysis of wave loads, it is based on random wave criteria using the wave energy spectrum developed by Pierson-Moskowitz [11], as described in equation (1).

$$S(\omega) = \frac{Hs^2}{4\pi}\left(\frac{2\pi}{Tz}\right)\omega^{-4}exp\left[-\frac{1}{\pi}\left(\frac{2\pi}{Tz}\right)\omega^{-4}\right] \tag{1}$$

Where, $S(\omega)$ is the Pierson-Moskowitz wave spectrum in units of $m^2/(rad/s)$, Hs is the Significant Wave Height in meters, Tz is referred to as the zero-upcrossing wave period in seconds, and ω represents the wave frequency in rad/s.

2.2.2 Load Analysis

Simulation methods are employed to compute wave loads in order to ascertain the structural integrity of a vessel when subjected to wave-induced loading during its operational activities at sea. When the floater structure is hit by random waves, it experiences stresses that show up as bending moments. These moments are then sent to the ship's structure in a longitudinal direction. The hydrodynamic pressure exerted on the submerged portion of the water might undergo variations while a floater travels in waves. These fluctuations in pressure give rise to longitudinal torque moments acting on the floater. The primary forces exerted on the floating structure are attributed to heave and pitch, which are classified as vertical movements. The cumulative forces are calculated using Equations (2) and (3) provided in reference [16].

$$M_{Wx} = M_S + M_{\theta y} + M_x \quad (2)$$
$$M_{Wy} = M_S + M_{\theta x} + M_z \quad (3)$$

2.3 Stress Analysis

2.3.1 Boundary Condition
The findings of the wave load analysis are then utilized as input for assessing the stress values that arise in the floater structure through the application of the Finite Element Method (FEM). The finite element analysis computation involves the establishment of boundary conditions. These conditions encompass the application of loads on the deck area, hydrostatic loads on the top and bottom of the floater, moments applied to the fore and after sections, and supports that restrict translation movement (fixed) while allowing for free rotational movement. Figure 3 illustrates the configuration of the boundary condition for the floater structure.

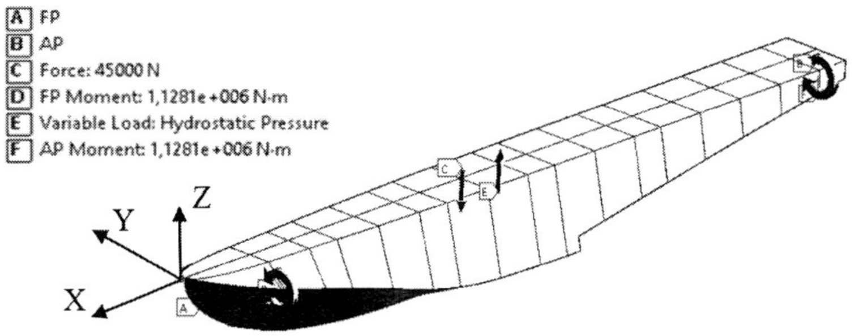

Figure 3. Boundary Condition

2.3.2 Meshing
Meshing refers to the procedural division of a given domain into several discrete cells. The meshing procedure plays a critical role in the study of floater structures using finite element analysis (FEM). The caliber of the used mesh has a significant impact on the accuracy of the simulation results. In a general sense, meshes may be categorized into two distinct types: organized grids and unstructured grids [17]. The floater structure employs an unstructured mesh composed of tetrahedral forms, as seen in Figure 4. A mesh independence study is undertaken in order to ascertain the most suitable number of mesh elements. The findings of this study are presented in Table 5. The number of mesh components used for the simulations assessing structural stress evaluation is 899,523, exhibiting an error rate of 1.07%. This error rate falls within the 2% criterion as specified in reference [18].

Figure 4. Floating Meshing Detail

Table 5. Mesh Grid Independence

No	Total Mesh	Maximum Stress (Pa)	Error (%)
1	899523	274.3	1.07
2	788456	255.2	1.10
3	643234	231.6	1.12
4	456765	206.7	1.75
5	210756	118.3	2.32

2.4 Strength Analysis

The strength study of the floater construction relies on the qualities and features of the materials, namely aluminum alloy 6061-T6 and steel AH-32, which have undergone laboratory testing. The present investigation employs data obtained from previously conducted tensile testing, as documented in references [5, 6]. The experimentation involved the examination of many specimens under varying temperature conditions. This led to the determination of stress intensity values for the corresponding materials, which were then used to derive the yield strength and ultimate strength values, as presented in Table 2. The strength values were derived from the test data using Monte Carlo simulation, which utilizes the Random Generate Number (RGN) concept and transforms it into a Gaussian or normal distribution. Figure 5 depicts the graph of the RGN model, illustrating a collection of data points pertaining to the yield strength.

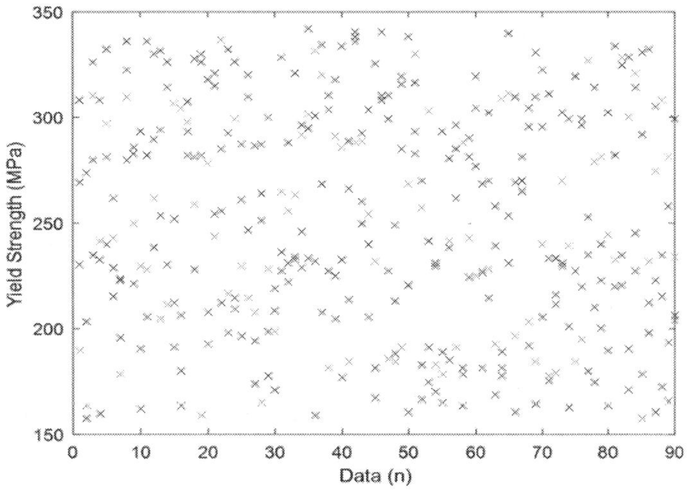

Figure 5. Detail of Parameter Input

2.5 Reliability Analysis

The primary objective of structural reliability theory is to evaluate and mitigate uncertainties inherent in the design process while simultaneously enhancing and streamlining the overall design methodology. Therefore, it is imperative for reliability analysis to accurately ascertain the load capacity, uncertainties associated with load amounts, and probabilities in order to ensure the dependability of safety margins. An alternative interpretation of dependability pertains to the likelihood of a structural system's capacity to sustain its performance across its operational lifespan. This is quantified by the probability density function (PDF), which includes the structural strength (S) and variable loads (Z) [19].

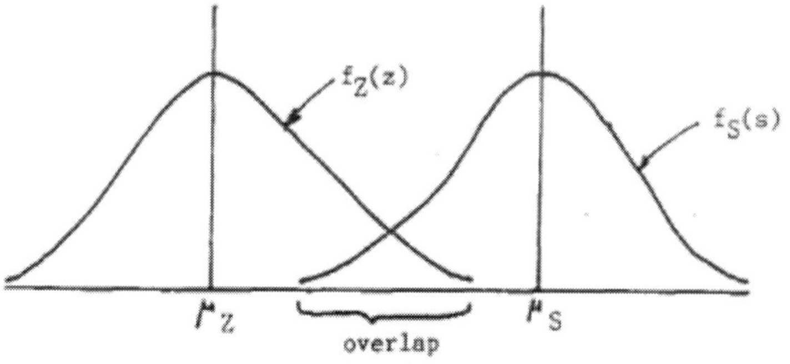

Figure 6. Detail of Parameter Input

The utilization of reliability analysis is employed to assess the likelihood of failure in the floater structure as a result of fluctuations in bending moment induced by stochastic wave loads. Structural dependability approaches are particularly well-suited for scenarios characterized by substantial uncertainty in both load circumstances and building strength. The assessment of structural dependability is based on a criteria that involves utilizing 60% of the yield strength values acquired from experimental testing, as documented in Table 2. The present study entails the representation of all pertinent data on possible stress levels in the construction, which arise from variations in wave load, in the form of a probability density function. The combination of two load and strength distributions results in the creation of an overlap between two distribution curves, which is often known as the Joint Probability Density Function. The safety index for each building design is determined by the utilization of equation (4), which entails the evaluation of the disparity between the average value and the standard deviation obtained from the load and strength distributions.

$$\beta = \frac{\mu_s - \mu_z}{\sqrt{\sigma_s^2 + \sigma_z^2}} \tag{4}$$

Where the reliability value is represented as (β), μS is the mean of Strength, μZ is the mean of load, σS is the standard deviation of strength, and σZ is the standard deviation of load [19].

3. Result and Discussion

3.1 Load Analysis Result

The results of static bending moment calculations are depicted in Figure 6, which showcases the dispersion of static bending moments over the floater structure and provides numerical values for each location or segment of the construction. The maximal static bending moment is approximately 2759.81 Nm, occurring near the midpoint of the floater along its longitudinal axis. Static bending moments are observed in tranquil sea circumstances and are joined with dynamic bending moments that arise during wave situations. This combination yields the overall bending moment.

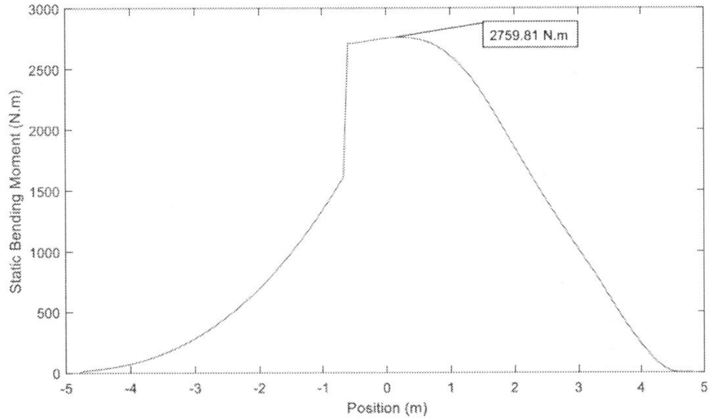

Figure 7. Graph of Static Bending Moment

The calculation of bending moments resulting from wave induction on the structure of the floater involves the consideration of wave excitation, specifically taking into account wave heights of 0.5 m, 1 m, 2 m, and 3 m. The results of numerical simulations utilizing the Panel Method are presented in Figure 9. This figure illustrates the bending moments that arise from the vertical motion of the floater, seen during a simulated time period of 10 seconds. The findings of the study are summarized in Table 8, which provides an overview of the static and dynamic bending moments observed throughout various operational scenarios of the floater. The minimum value of the vertical bending moment (M_Wx) occurs at around 199 kNm when the waves approach the ship from the lateral direction (Beam Sea), with a wave height of 1 m. The maximum vertical bending moment, estimated to be kNm, is recorded when the ship encounters waves approaching from the bow (Head Sea) with a wave height of 3 m. This observation illustrates the impact of the ship's heading angle and wave height on the structural reaction to bending forces.

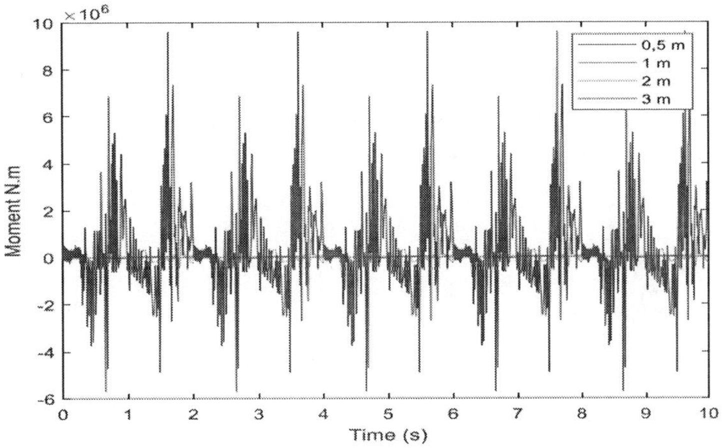

Figure 8. Dynamic Bending Moment

Table 6. Maximum Bending Moment

Wave Height (m)	Angle (deg)	M_S (kN.m)	M_{Wx} (kN.m)	M_{Wy} (kN.m)
	Head Sea	2.75	423	862
0,5	Beam Sea	2.75	418	888
	Following Sea	2.75	412	713
	Head Sea	2.75	360	162
1	Beam Sea	2.75	582	199
	Following Sea	2.75	454	931
	Head Sea	2.75	376	990
2	Beam Sea	2.75	417	232
	Following Sea	2.75	642	380
	Head Sea	2.75	386	891
3	Beam Sea	2.75	109	445
	Following Sea	2.75	670	231

3.2 Stress Analysis Result

The stress values derived from the simulations performed with the Finite Element Method (FEM) are presented and consolidated in Table 7. This table gives a general overview of the maximum stress that the structure has to endure in all wave height scenarios. The maximum values are seen when a wave with a height of 3 m approaches the floater directly, known as Head Sea. In this scenario, the AH-32 and 6061-T6 material constructions experience peak pressures of 193.89 MPa and 274.3 MPa, respectively. The AH-32 and 6061-T6 structures exhibit stress values of 44.64 MPa and 108.81 MPa, respectively, with the former having the lower stress value. In a broad sense, it is evident that the level of stress experienced in aluminum construction surpasses that of AH-32 steel construction, exhibiting an average disparity of 93 MPa, or 9.29%. Figure 10 illustrates the precise position of the highest stress and deformation inside the floater structure, notably in close proximity to the front of the floater on the bottom plate.

Table 7. Maximum Stress

Wave Height (m)	Angle (deg)	Max Stress AH-32 (MPa)	Max Stress 6061-T6 (MPa)
	Head Sea	53.90	125.98
0,5	Beam Sea	45.77	108.81
	Following Sea	44.64	109.29
	Head Sea	57.16	134.36
1	Beam Sea	56.33	133.42
	Following Sea	57	132.48
	Head Sea	122.28	160.06
2	Beam Sea	91.43	159.95
	Following Sea	59.89	159.84
3	Head Sea	193.89	274.3
	Beam Sea	85.23	269.78

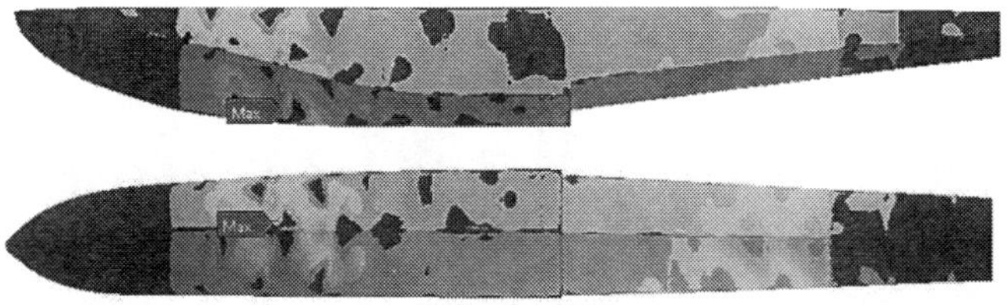

Figure 9. Max Equivalent Stress Location for case of 3 meter Wave Height

3.3 Structural Reliability Result

The structural reliability Analysis of the floater structure involves the representation of two key parameters, namely the load and the strength of the floater structure, using probability density function (PDF) graphs. These PDF graphs have a Gaussian distribution pattern. The two graphs are subsequently merged to form a composite probability distribution graph representing stress levels. The Joint Probability Density Function graphs for different situations are depicted in Figures 11, 12, 13, and 14. These figures illustrate the probability distribution for wave heights of 0.5 m, 1 m, 2 m, and 3 m, respectively, for both AH-32 and Al 6061-T6 material constructions. The floater construction utilizing AH-32 and Al 6061-T6 materials demonstrates a 100% dependability when subjected to a wave height of 0.5 m. This is evident from Figure 11, where the lack of any intersection between the load and strength distributions is observed. In the scenario where a wave height of 1 meter is considered, it can be shown that the load and strength distributions for aluminum cross and overlap. This intersection signifies the presence of a failure area, with a dependability percentage of 97.56%, as visually depicted in Figure 12.

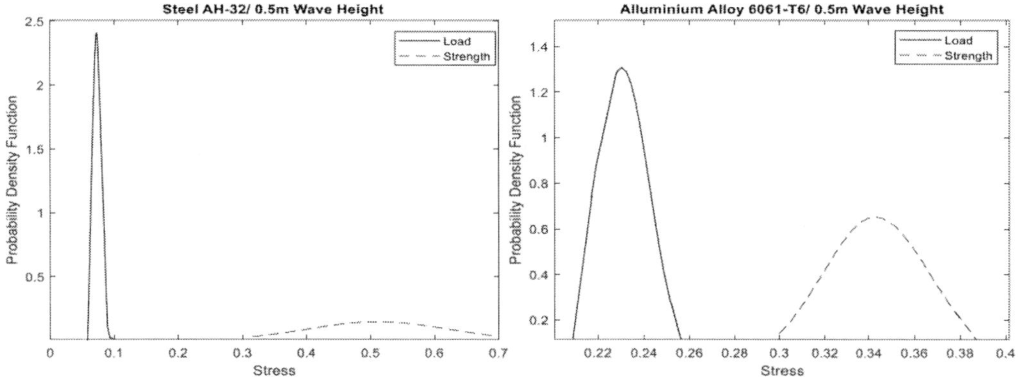

Figure 10. Joint Probability Density Function of 0,5 meter wave height

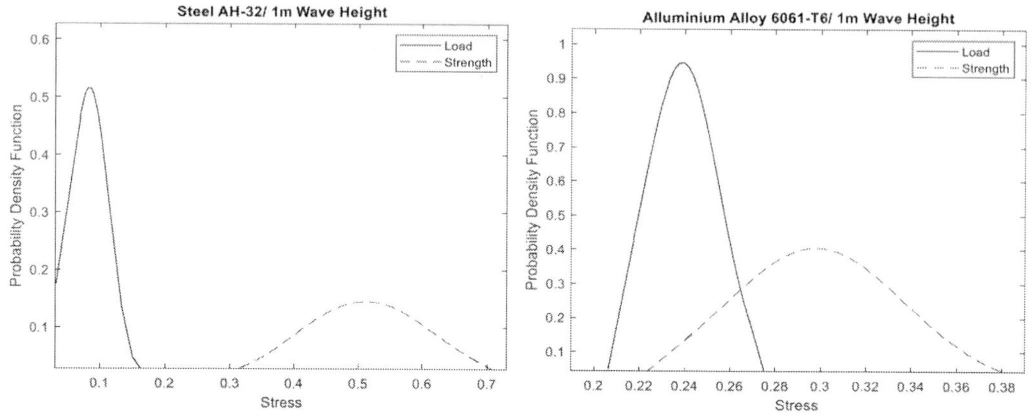

Figure 11. Joint Probability Density Function of 1 meter wave height

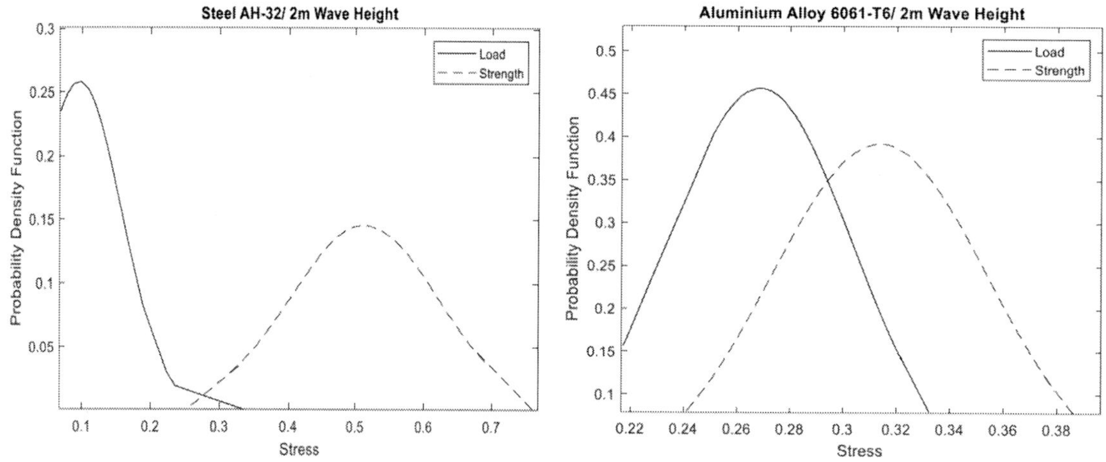

Figure 12. Joint Probability Density Function of 2 meter wave height

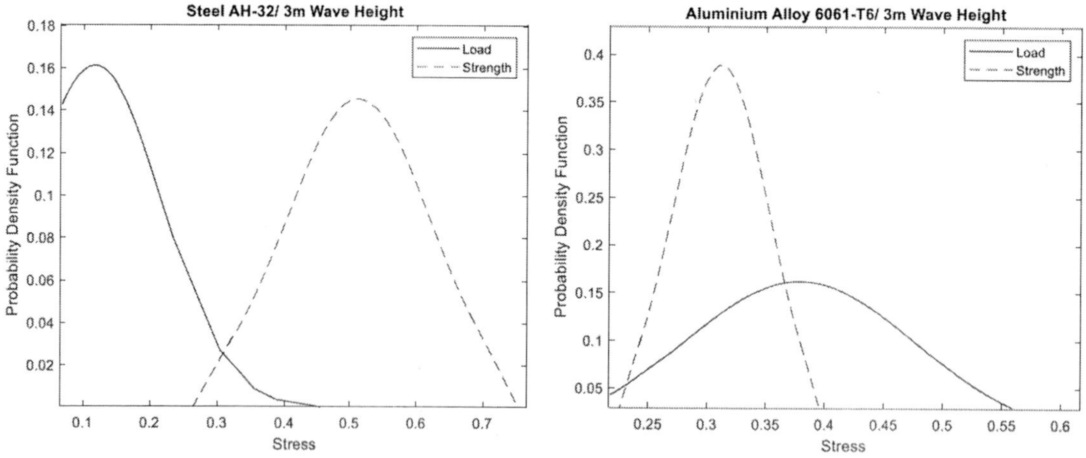

Figure 13. Joint Probability Density Function of 3 meter wave height

IOP Conf. Series: Earth and Environmental Science **1473** (2025) 012032 doi:10.1088/1755-1315/1473/1/012032

It is also possible to figure out when a floater structure will fail by finding the failure area, which is where the probability distribution curves for load and strength overlap, as shown in Figures 11, 12, 13, and 14. Table 8 presents a comprehensive overview of failure percentages across all wave height scenarios, whereas reliability values are derived by subtracting the failure percentage from 100%. In order to account for safety considerations, the reliability analysis incorporates safety factors by decreasing the strength values of the structure to 60% and 80% of the yield strength values obtained from material testing. The data shown in Table 9 reveals that the lowest failure percentage, amounting to 0%, is recorded at wave heights of 0.5 m and 1 m. Conversely, the largest failure percentage, reaching 87%, is observed at a wave height of 3 m for aluminum alloy. Nevertheless, the steel material continues to exhibit a relatively low incidence of failure, with a rate of 4.66%. In a broad sense, it may be inferred that AH-32 steel material has favorable structural dependability when subjected to various high wave loads, with failure rates remaining below 5%. On the contrary, the structural integrity of aluminum alloy 6061-T6 material remains intact when subjected to random wave loads of up to 1 m in wave height, exhibiting failure rates that do not exceed 10%. When wave heights approach 2 meters, the likelihood of structural failure surpasses 30%. Hence, in the event that aluminum is chosen as the material for the floater's construction, it becomes imperative to either modify the existing construction design or augment the dimensions of the profiles outlined in Table 3. This adjustment must be made while simultaneously ensuring that the overall weight of the construction remains within the limits of, or matches, the weight of the AH-32 steel construction.

Table 8. Probability of Failure

WAVE HEIGHT (M)	MATERIAL	0.6		0.8		1	
		Failure	Reliability	Failure	Reliability	Failure	Reliability
0.5	AH-32	0 %	100 %	0 %	100 %	0 %	100 %
	6061-T6	0 %	100 %	0 %	100 %	0 %	100 %
1	AH-32	0 %	100 %	0 %	100 %	0 %	100 %
	6061-T6	13.43%	87.57%	7.32%	93.47%	3.54%	97.56%
2	AH-32	0.26%	99.74%	0 %	100 %	0 %	100 %
	6061-T6	45.21%	55.79%	41.40%	59.60%	39.32%	76 %
3	AH-32	4.66%	95.34%	2.60%	97.40%	1.04%	98.96%
	6061-T6	100 %	0 %	100.00%	0 %	98.87%	1.13%

4. Conclusion

Two distinct materials, namely steel AH-32 and aluminum alloy 6061-T6, have been suggested for evaluating the structural integrity of the floater in withstanding unpredictable wave stresses. The present assessment employs the 3D Diffraction Theory to forecast the vertical bending moments caused by random waves in a total of 12 distinct environmental scenarios. These scenarios encompass various combinations of 3 heading angles and 4 wave heights. The wave loads' outcomes are utilized as input in finite element analysis for the purpose of forecasting the structural stress experienced by the floater. In addition, a structural reliability study is performed on the floater in order to assess the safety and dependability of its construction. The load and strength of the structure are represented as probability density functions (PDFs) and subsequently integrated to generate joint probability density function (JPDF) plots. The evaluation of the structural failure values of the floater involves determining the intersection area on the Joint Probability Density Function (JPDF) graph. Reliability values, on the other hand, are produced by subtracting the failure rate from 1. The analytical outcomes for the AH-32 steel material utilized in the building of the floater demonstrate a structural dependability of 100% throughout all wave load scenarios, encompassing wave heights ranging from 0.5 to 3 m, without the incorporation of a safety factor. In contrast, the dependability coefficient for the aluminum 6061-T6 material composition of the floater is seen to be 86% when subjected to a 1 m wave height. However, when the wave heights increase to 2 m and 3 m, the structural reliability progressively diminishes to 75% and 23%, respectively. These values are obtained while maintaining a safety factor of 40%. Hence, it is advisable that, in the event of utilizing aluminum as the material for the floater's construction, a reassessment of the construction design or an augmentation in the thickness of the profile be contemplated.

Acknowledgment

The authors are grateful to Department of Naval Architecture in Institut Teknologi Sepuluh Nopember (ITS) Surabaya for providing a financial support under the scheme of Penelitian Departemen with no. 2060/PKS/ITS/2024.

References

1. FAA. (1952). National Advisory Commintteee for Aeronautics. NACA TN 2481.
2. FAA. (2018). Advisory Circular. FAA 1B.
3. Aliffrananda, M. H., Sulisteyono, A. (2021). Porpoising Instability Study of The Floatplane During Take off Operation on Calm Water. IOP Conference Series: Materials Science and Engineering, 1052. http://dx.doi.org/10.1088/1757-899X/1052/1/012013
4. Qiu, L., (2013). Efficient Decoupled Hydrodynamic and Aerodynamic Analysis of Amphibious Aircraft Water Takeoff Process. Journal of Aircraft, 1369-1379.
5. Wu, W., Tor, S. B., & Merchant, A. A. (2018). Tensile properties of ASTM A131 EH36shipbuilding steel processed by selective laser melting. 3rdInternational Conference on Progress in Additive Manufacturing (Pro-AM 2018), 684-689. doi:10.25341/D4ZW26.
6. Kim, H.J., Tahk, Y.W., Jun, H., Kong, E.H., Oh, J.Y., & Yim, J.S., Establishment of the design stress intensity value for the plate-type fuel assembly using a tensile test, Nucl. Eng. Technol. 53 (2021) 911e919, https://doi.org/10.1016/j.net.2020.07.038.
7. Xiao, S., Han, Y., Zhang, Y., Wei, Q., Wang, Y., Wang, N., Liu, Y. (2022). A Reliability Analysis Framework of Ship Local Structure based on Efficent Probability Simulation and Experimental Data Fusion. Metals. MDPI Journal. Metals 2022, 12(5), 805.
8. Kerdabadi, M. S., Sakaki, A., & Izadi, A., (2018). Evaluation of Ship Strucutre Reliability During Design, Maintenance, and Repair Phases. Scientific Journals.
9. Faltinsen, O. M., (1990). Sea Loads on Ships and Offshore Structures. Cambridge University Press.
10. Sulisetyono, A., Putranto, T. (2017). Wave Load Analysis of the Corvette Ship in the Sea Water of Indonesia. Applied Mechanics and Materials, 862(), 291–295. doi:10.4028/AMM.862.291
11. Hafiz, M. A., Sulisetyono, A., (2022). Structural Reliability Analysis for the Construction Design of the High-Speed Ship with CFRP Material. IOP Conference : Series Earth and Environmental Science 1081(1):012041.
12. Alifia, W. U., Sulisetyono, A. (2023). Construction Design of the Floater Considering the Free-Falling Impact of the Floatplane on Calm Water. IOP Conference Series: Earth and Environmental Science, 1276(1), Article 012060. https://doi.org/10.1088/1755-1315/1276/1/012060
13. Sulisetyono, A., Washoya, Z. A., & Zubaydi, A., (2022). Computational Impact Pressure Of The Floatplane's Bottom In Various Of Deadrise And Flare Angles. Journal of Applied Engineering Science, 20(3), 937-945. https://doi.org/10.5937/jaes0-35839
14. Sulisetyono, A., Fadhlurrohman, I., Ali, B., & Zubaydi, A. (2022). Computational Prediction Of The Resistance Of The Floatplane At Various Trim Angles. Journal of Theoretical and Applied Mechanics, 60(2), 267-278. https://doi.org/10.15632/jtam-pl/148053.

IOP Conf. Series: Earth and Environmental Science **1473** (2025) 012032 doi:10.1088/1755-1315/1473/1/012032

15. Bhattacharrya, R. (1972). Dynamics of Marine Vehicles. New York: A Wiley Interscience Pubication.
16. Ship Structure Committee., (1994). Probability Based Ship Design; Loads and Load Combinations. Washington: Ship Structure Committee.
17. Logan, D. L., (2012) A First Course in The Finite Element Method. America: Cengage Learning.
18. Patil, H., Jeyakarthikeyan, P., (2018). Mesh Convergence Study adn Estimation of Discretization Error of Hub in Clutch Disc with Integration of ANSYS. IOP.
19. Ship Structural Committee., (1990). An Introduction to Structural Reliability Theory. Washington: Ship Structure Committee.

Marine Environment Protection using Marine Debris Collection Vessels: Stability Analysis for Ship Safety

E Sugianto[1,], F M Noor[2], G A Saputro[1]. H Prasutiyon[1], Sutrisno[1] , Supartono[3]

[1] Department of Marine Engineering, Hang Tuah University, Surabaya, Indonesia, 60111
[2] Research Center for Hydrodynamics Technology, National Research and Innovation Agency (BRIN), Surabaya, Indonesia, 60111
[3] Master Program of Ocean Engineering, Hang Tuah University, Surabaya, Indonesia, 60111

*E-mail: erik.sugianto@hangtuah.ac.id

Abstract. In this study, the variation of piezoelectric and solar panel positions on a marine debris collection vessel is investigated for stability using maxsurf stability. The inner flat catamaran ship model with different piezoelectric and solar panel position variations, namely at the bow of the ship, at the centerline of the ship, and at the stern of the ship. The criteria used in conducting the analysis carried out are in accordance with the stability criteria contained in the International Maritime Organization (IMO) regulations. The effect of piezoelectric and solar panel positions has a less significant difference on ship stability in each model. All aspects of design and use are carefully considered so as not to interfere with the process of collecting marine debris. Model 1 has better stability results because the stability arm is larger and produces better performance. From these results, the stability arm in model 1 has a larger area than the other models. This is due to the difference in load from the piezoelectric position at the stern and the solar panel at the centerline. This condition causes the GZ at a certain angle to be larger which results in a larger stability.

1. Introduction

Plastic pollution is emerging as the most serious threat to marine ecosystems, requiring urgent management actions for the future sustainability of marine ecosystems. Marine debris collection and cleaning technologies have begun to emerge and continue to evolve [1]. A small ship with a marine debris collection system was proposed by Sugianto and Chen [2]. From research Sugianto et al. [3] The simulation results showed that the catamaran model was superior. In contour analysis, the velocity comparison of the fluid velocity on the conveyor shows the velocity of waste approaching the conveyor on the catamaran model is faster than the model monohull. Next Sugianto et al. [4] Conducting research with three variations of catamaran shapes was investigated using numerical methods. The resistance of catamarans with the smallest conveyors is Inner Flat Hull Catamaran [5,6]. To hold marine debris carried by the current of the ship inert flat catamaran will be completed wing conveyor. On research Sugianto et al. [7] shows that the order of ship resistance from the largest value to the smallest is the conveyor on the bow and side of the hull. On the resistance value and flow distribution, it is concluded that marine debris is more easily and quickly collected by ships on the conveyor model on the bow of the ship because it has the smallest total resistance value of the ship [8]. Ship stability is a crucial aspect in ship design and operation as it is directly related to the safety and performance of ships on the water. Some of the reasons why ship stability is very important are ensuring the safety of passengers and crew, preventing capsizing, for the comfort of passengers and crew, operational efficiency especially fuel consumption, cargo handling, and compliance with international regulations governing ship stability standards. ship stability is a key factor in ship design and operation that affects safety, comfort, and efficiency in shipping [9].

However, research on the effect of piezoelectric positioning and solar panels on stability characteristics in marine debris collecting vessels is not yet available. In this study, stability analysis was carried out on marine debris collecting vessels by varying the position of solar panels and piezoelectricity using Maxsurf. There are six variations in the position of solar and piezoelectric panels, namely front back, front center, front back, middle middle, front center, rear center. This study aims to determine the stability characteristics of marine debris collecting vessels against differences in the location of Piezoelectricity &; Solar Panels. The analysis carried out based on the rules refers to the International Maritime Organization (IMO) Code A.749(18) Ch 3 - Design Criteria Applicable to All Ships.

2. Method

In this study using maxsurf software to calculate the stability of marine debris collecting vessels after variations in the position of piezoelectric and solar panels which will be obtained results in the form of static stability, dynamic stability and GZ curves. The results of the analysis will be compared with IMO criteria. The first stage is the creation of a marine debris collecting vesselmodel on the maxsurf modeler which will be used as a simulation object on the maxsurf stability shown in the Figure 1. The hull model used is the inner flat catamaran type [10] and the ship data is shown in table 1.

Table 1. Principal dimension of model

Parameters	Symbol	Size
Overall length(m)	Loa	4,000
Perpendicular length(m)	Lpp	3,950
Waterline length(m)	Lwl	3,858
Maximum width (m)	B	1,200
Height (m)	H	0,600
Draft (m)	T	0,300
Wet surface area (m2)	WSA	5,723
Volume displacement	m^3	0,36
Block coefficient	-	0,25

In calculating the stability of existing equipment on the ship, a loadcase will be made to enter loading data, namely in the form of ship weight distribution and equipment located on the marine debris collecting vessel [11-13]. This loadcase serves to add an input factor in the form of a load that will be used as the basis for calculating stability. In this study, the position of piezoelectric and solar panels was varied by varying the loadcase on the model. This analysis refers to the center of gravity (CG) of the ship, and its effect on the Loadcase, the results of this analysis and calculation are in the form of ship stability data.

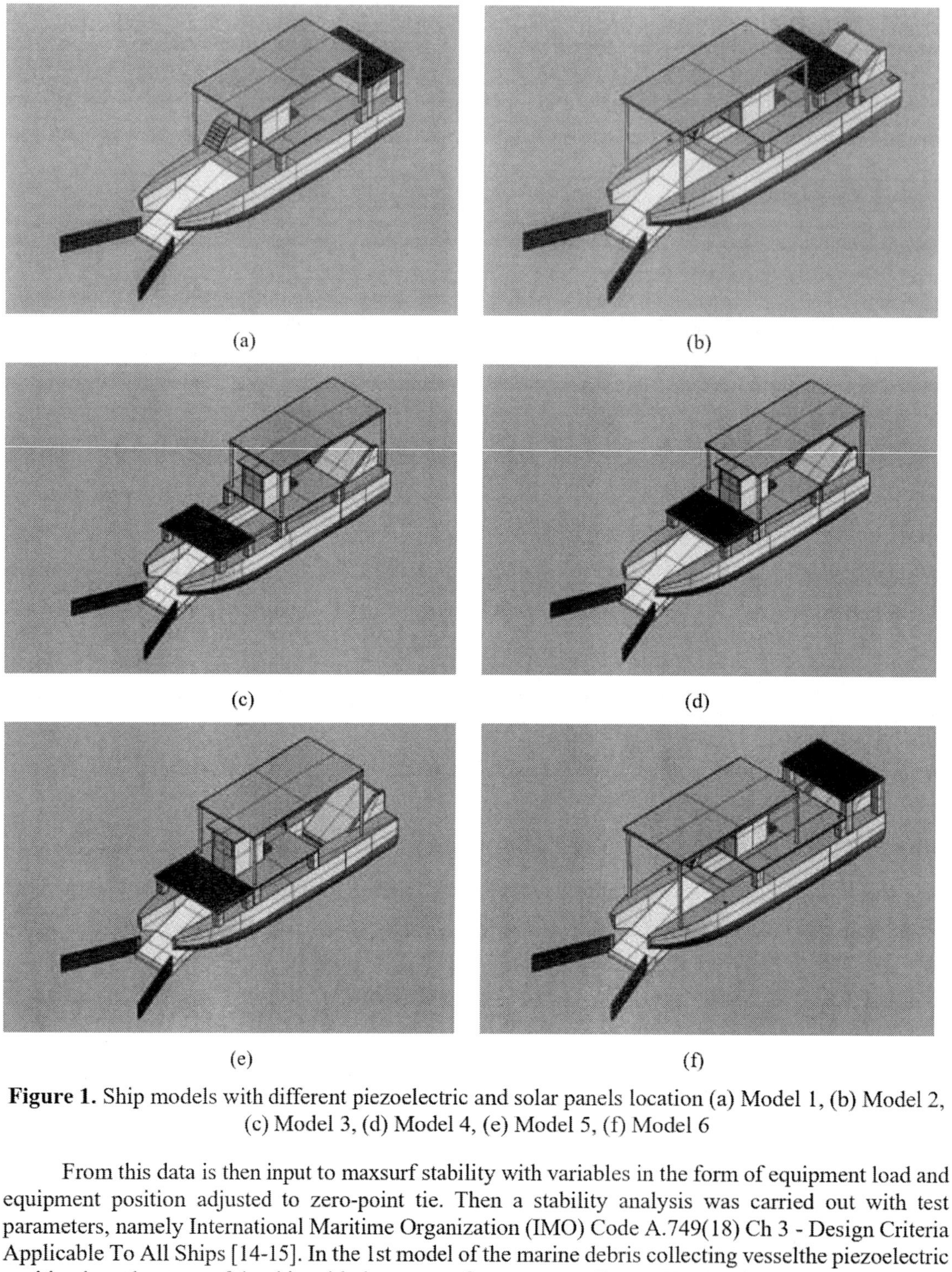

(a) (b) (c) (d) (e) (f)

Figure 1. Ship models with different piezoelectric and solar panels location (a) Model 1, (b) Model 2, (c) Model 3, (d) Model 4, (e) Model 5, (f) Model 6

From this data is then input to maxsurf stability with variables in the form of equipment load and equipment position adjusted to zero-point tie. Then a stability analysis was carried out with test parameters, namely International Maritime Organization (IMO) Code A.749(18) Ch 3 - Design Criteria Applicable To All Ships [14-15]. In the 1st model of the marine debris collecting vesselthe piezoelectric position is at the stern of the ship with the center of mass at coordinates X, Y, Z (0.5, 0, 1.2) of the zero point and the solar panels are on the center line of the ship with the center of mass at coordinates X, Y, Z (2, 0, 1.8) of the zero point. The following loadcases from model 2 are shown in Table 2.

Table 2. Loadcase model 1.

Item Name	Total Mass kg	Long. Arm (X) m	Trans. Arm (Y) m	Vert. Arm (Z) m
Conveyor	70	3,2	0	0,3
Battery	10	1,5	0	1
Electric motor	25	2,44	-0,4	0,54
Engine	40	0,1	0	0,8
Hull ship	197	1,652	0	0,36
Control room	13,3	2	0	1,3
Solar Panel	25	2	0	1,6
Piezoelektric	6	0,34	0	1,05
Total Loadgroup	386,3	1,833	-0,026	0,546
FS correction				0
VCG fluid				0,546

In the 2nd model of the marine debris collecting vesselthe piezoelectric position is on the ship's centerline with the center of mass at coordinates X, Y, Z (1.1, 0, 1.2) of the zero point and the solar panel is on the bow of the ship with the center of mass at coordinates X, Y, Z (2.6, 0, 1.8) of zero point. The following loadcase from model 2 is shown in Table 3.

Table 3. Loadcase model 2

Item Name	Total Mass kg	Long. Arm (X) m	Trans. Arm (Y) m	Vert. Arm (Z) m
Conveyor	70	3,2	0	0,35
Battery	10	1,5	0	1
Electric motor	25	2,44	-0,4	0,54
Engine	40	0,1	0	0,8
Hull ship	197	1,652	0	0,36
Control room	13,3	2	0	1,3
Solar Panel	25	2,557	0	1,6
Piezoelektric	6	1,05	0	1,05
Total Loadgroup	386,3	1,88	-0,026	0,555
FS correction				0
VCG fluid				0,555

In the 3rd model of the marine debris collecting vesselthe piezoelectric position is at the bow of the ship with the center of mass at coordinates X, Y, Z (2.95, 0, 1.2) of the zero point and the solar panel is at the stern of the ship with the center of mass at coordinates X, Y, Z (1.054, 0, 1.8) of zero point. The following loadcase from model 3 is shown in Table 3.9

Table 4. Loadcase model 3

Item Name	Total Mass kg	Long. Arm (X) m	Trans. Arm (Y) m	Vert. Arm (Z) m
Conveyor	70	3,2	0	0,35
Battery	10	1,5	0	1
Electric motor	25	2,44	-0,4	0,54
Engine	40	0,1	0	0,8
Hull ship	197	1,652	0	0,36
Control room	13,3	2	0	1,3
Solar Panel	25	1,05	0	1,585
Piezoelektric	6	3,099	0	1,05
Total Loadgroup	386,3	1,814	-0,026	0,554
FS correction				0
VCG fluid				0,554

IOP Conf. Series: Earth and Environmental Science **1473** (2025) 012033 doi:10.1088/1755-1315/1473/1/012033

In the 4th model of the marine debris collecting vesselthe piezoelectric position is on the ship's centerline with the center of mass at coordinates X,Y,Z (2.45, 0, 1.2) of the zero point and the solar panels are at the stern of the ship with the center of mass at coordinates X,Y,Z (1.054, 0, 1.8) of zero point. The following loadcase from model 4 is shown in Table 3.10

Table 5. Loadcase model 4

Item Name	Total Mass kg	Long. Arm (X) m	Trans. Arm (Y) m	Vert. Arm (Z) m
Conveyor	70	3,2	0	0,35
Battery	10	1,5	0	1
Electric motor	25	2,44	-0,4	0,54
Engine	40	0,1	0	0,8
Hull ship	197	1,652	0	0,36
Control room	13,3	2	0	1,3
Solar Panel	25	2,55	0	1,6
Piezoelektric	6	1,077	0	1,05
Total Loadgroup	386,3	1,88	-0,026	0,555
FS correction				0
VCG fluid				0,555

In the 5th model of the marine debris collecting vesselthe piezoelectric position is at the bow of the ship with the center of mass at coordinates X, Y, Z (2.95, 0, 1.2) of the zero point and the solar panels are on the center line of the ship with the center of mass at coordinates X, Y, Z (1.554, 0, 1.8) of the zero point. The following loadcase from model 5 is shown in Table 3.11

Table 6. Loadcase model 5

Item Name	Total Mass kg	Long. Arm (X) m	Trans. Arm (Y) m	Vert. Arm (Z) m
Conveyor	70	3,2	0	0,35
Battery	10	1,5	0	1
Electric motor	25	2,44	-0,4	0,54
Engine	40	0,1	0	0,8
Hull ship	197	1,652	0	0,36
Control room	13,3	2	0	1,3
Solar Panel	25	1,62	0	1,6
Piezoelektric	6	3,11	0	1,05
Total Loadgroup	386,3	1,851	-0,026	0,555
FS correction				0
VCG fluid				0,555

In the 6th model of the marine debris collecting vesselthe piezoelectric position is at the stern of the ship with the center of mass at coordinates X, Y, Z (0.5, 0, 1.2) of the zero point and the solar panel is at the bow of the ship with the center of mass at coordinates X, Y, Z (2.6, 0, 1.8) of the zero point. The following loadcase from model 6 is shown in Table 3.12

Table 7. Loadcase model 6

Item Name	Total Mass kg	Long. Arm (X) m	Trans. Arm (Y) m	Vert. Arm (Z) m
Conveyor	70	3,2	0	0,35
Battery	10	1,5	0	1
Electric motor	25	2,44	-0,4	0,54
Engine	40	0,1	0	0,8
Hull ship	197	1,652	0	0,36
Control room	13,3	2	0	1,3
Solar Panel	25	2,55	0	1,5
Piezoelektric	6	0,34	0	1,05
Total Loadgroup	386,3	1,869	-0,026	0,549
FS correction				0
VCG fluid				0,549

3. Results and discussion

The data results from the calculation of ship stability must meet the standard rules/class criteria as a reference for calculation analysis. In this study the criteria chosen were International Maritime Organization (IMO) Code A.749(18) Ch 3 - Design Criteria Applicable To All Ships.

Table 8. IMO criteria.

Criteria	Model 1	Model 2	Model 3	Model 4	Model 5	Model 6
Area 0 to 30	5,8369	5,7449	5,7817	5,7449	5,7589	5,8003
Area 0 to 40	8,3395	8,1832	8,2423	8,1833	8,205	8,2792
Area 30 to 40	2,5026	2,4383	2,4606	2,4384	2,4461	2,4789
Max GZ at 30 or greater	0,282	0,276	0,279	0,276	0,277	0,28
initial GMt	0,758	0,756	0,747	0,756	0,751	0,762

From the stability calculations that have been carried out with variations in the position of piezoelectric locations and solar panels, it shows that the stability of marine debris collecting vessels still meets the requirements of the International Maritime Organization (IMO) qualifications shown in Table 8.

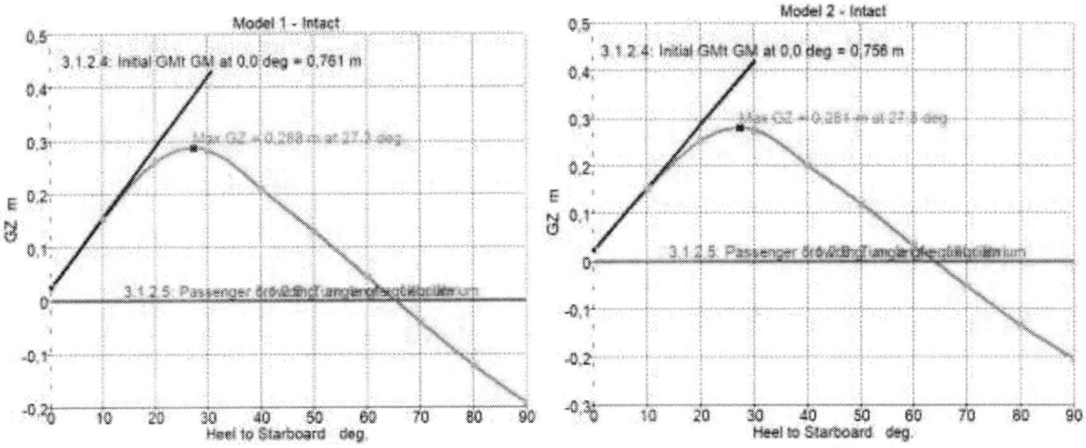

(a) (b)

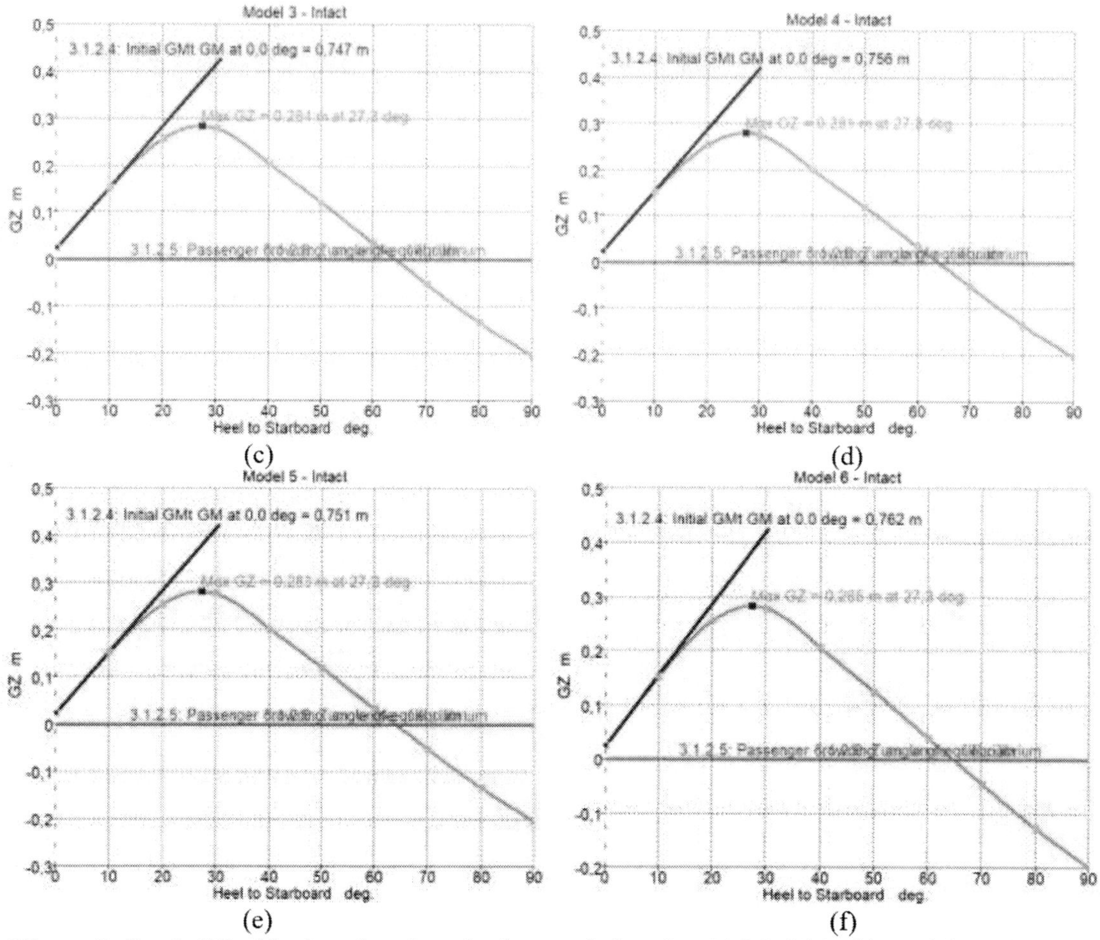

Figure 2. Graph GZ with piezoelectric and solar panels location (a) Model 1, (b) Model 2, (c) Model 3, (d) Model 4, (e) Model 5, (f) Model 6

Figure 2 shows that stability in each model with variations in piezoelectric and solar panel positions. Rawson and Tupper stated that the range of stability, which is the hose where the GZ is positive if the ship will return to its original position after the moment that caused the tilt to disappear, is usually at an angle interval of 0^0 to 90^0 [16] this is also the case with marine debris collecting ships. The influence of piezoelectric position and solar panels has not too significant differences in each model. The stability of the vessel is influenced by various factors, including weight distribution, center of gravity, hull shape. Although the variation in the position of piezoelectric and solar panels does not significantly affect the stability of the vessel of each model of marine debris collecting vessel, laying and installation is important to thoroughly consider all aspects of its design and use so as not to interfere with the marine debris collecting process [17-18]. From these results, the stability arm on model 1 has a wider area compared to other models. This is due to the difference in load from the piezoelectric position at the stern and the solar panel at the centerline. This condition causes the GZ value at a certain angle to become larger which results in the stability arm becoming larger as well.

4. Conclusion

From the stability analysis to the effect of different piezoelectric positions and solar panels have different GZ values in each model. This happens because the weight point (CG) of the ship is influenced by the weight distribution of the position of piezoelectricity and solar panels. The result of the calculation of the stability of the largest GZ value is found in model 1 experiencing an increase to an angle of 27.3 degrees with a max GZ value of 0,282 m. Model 1 has better stability results due to greater stability arms and results in better performance. This means that model 1 has more advantages than the other models in passenger and crew safety, preventing capsizing, passenger comfort, operational efficiency, especially in fuel consumption, handling of marine debris and compliance with ship stability standards. Further research can be carried out by conducting experiments on the stability of waste collection vessels, both statically and dynamically. This is to ensure the stability of the waste collection ship is good so that safety when the ship is operating can be guaranteed.

Acknowledgment
The authors would like to thank Hang Tuah University for supporting this research. The author would also like to thank BRIN Hydrodynamics Research Centre for supporting the author with the Barista 2023 and Post Doctoral 2023/2024 programmes.

References

[1] Sugianto E, Chen JH, Purba NP Numerical investigation of conveyor wing shape type effect on ocean waste collection behavior. E3S Web Conf 01005: .2021.
[2] Sugianto E, Chen JH, Preliminary concept of ship use to waste management in sea shallow water. In: Proceedings of the 33th Asian-Pacific Technical Exchange and Advisory Meetings on Marine Structures (TEAM 2019. Tainan, Taiwan. .2019.
[3] Sugianto E, Chen JH, Permadi NVA, Effect of Monohull Type and Catamaran Hull Type on Ocean Waste Collection Behavior Using OpenFOAM. Water **14** :2623. 2022.
[4] Sugianto E, Zamzami RA, Winarno A, Prasutiyon H, Effect of Catamaran Hull Type on Ocean Waste Collection Behavior. J Mar Sci Technol **3** :79–85. https://doi.org/10.12962/j27745449.v3i3.590. 2023.
[5] Gultom, W., Amiruddin, W., and Iqbal, M. Analysis of the Effect of Changing the Design of Monohull Vessels to Catamaran Vessels on Stability and Ship's Operation. Jurnal Kapal. **6** (2), 199–206. 2018
[6] Gunawan, Y. Stability Characteristics of Catamaran Ship Model Based on the Hull Shape Variation. Journal of maritime research: JMR, **16** (2), 105-109. (2019).
[7] Sugianto, E., Chen, J. H., Sugiono, R., & Prasutiyon, H. (Effect of portable conveyor placement in ship on ocean waste collection behavior. In IOP Conference Series: Earth and Environmental Science. **1095**, No. 1, p. 012015). IOP Publishing. 2022.
[8] Hasrul, Invesigasi Eksperimen Course-Keeping Stability Kapal Tongkang Dengan Variasi Bentuk dan Dimensi SKEG. 1: 6. Universitas Hassanuddin. 2021.
[9] Hind, J. A. Stability and Trim Of Fishing Vesels And Other Small Ships. Second Edition. Fishing News Book Ltd. Famham, Surrey. England. 1982.
[10] Sugianto, E., Widjiati, E., & Widjanarko, B. Experimental Research of Piezoelectricity as Renewable Energy Power Source for Conveyor Electric Motor on Garbage Collection Vessel. International Journal on Engineering Applications, **12** (1). 2024.
[11] Sawosono, B., Munazid, A., Awwalin, R., Poundra, G.A.P., dan Sutiyo,Teori dan Panduan Praktis Hidrodinamika Kapal Hukum Archimedes. Universitas Hang Tuah (UHT). 2019
[12] Hutabarat, AYH. Hamonangan. 2009. Simulasi Uji Stabilitas Statis Kapal Dengan Metode Teori Kapal Model. : 1–69.Universitas Indonesia. 2009.
[13] Manullang IA & daryansah. Analisis pengaruh peletakan panel surya di atas kapal deck house terhadap stabilitas kapal. Universitas darma persada. 2021, **12** (3).pp. 248–253. 2021.
[14] IMO, Adoption of the International Code on Intact Stability, **267**, no. (85). London: IMO Publishing, 2008.
[15] Najafi, A., & Saeed, S. M. RANS simulation of hydrofoil effects on hydrodynamic coefficients of a planing catamaran. *Brodogradnja: An International Journal of Naval Architecture and Ocean Engineering for Research and Development*, **67** (1), 43-66. 2016.
[16] Rawson KJ, and Tupper EC. *Basic Ship Theory*. "Ed ke-3". Volume 1. London: Longman. 1983.
[17] Sugianto, E., and Chen, J. H. Hollow Wing Technique to Enhancing Conveyor Performance on Marine Debris Collection. Evergreen. **9** (4), 1160-1167. 2022.
[18] Sugianto, E., Winarno, A., Indriyani, R., & Chen, J. H. Hull Number Effect in Ship Using Conveyor on Ocean Waste Collection. Kapal: Jurnal Ilmu Pengetahuan dan Teknologi Kelautan, **18** (3), 128-139. 2021.

IOP Conf. Series: Earth and Environmental Science **1473** (2025) 012034 doi:10.1088/1755-1315/1473/1/012034

Inner Model of PLS-SEM on Policy Making about Sound Wave Attractor for Fishery Sustainability

N Rosana[1,], S Rifandi[2], A Muminin[3], S Sumantri[4]
[1,3]Department of Fishery, Hang Tuah University, Surabaya, Indonesia, 60111
[2] Department of Electrical Engineering, Hang Tuah University, Surabaya, Indonesia, 60111
[4] Department of Public Administration Doctoral Program, Hang Tuah University, Surabaya, Indonesia, 60111

* E-mail: nurul.rosana@hangtuah.ac.id

Abstract. Piknet is a sound wave-based attractor, that fishermen often use to attract fish closer to the fishing gear, and it is currently being tested on a community of fishermen in Surabaya. This research aims to determine the inner model by analyzing the relationship between latent variables while making policies about innovations by fishermen in Surabaya for fishery sustainability. The analytical method used was PLS-SEM modeling, based on the variance obtained from the latent variable indicators. The inner model of the research results states that to be qualified to predict the relationship between latent variables as a consideration in sustainable fisheries policy making. The influential paths are; (a) adoption and implementation towards confirmation and support, (b) innovation characteristics, (c) communication channels, (d) knowledge and persuasion, and (e) a certain period towards adoption and implementation. The coefficient of determination used to determine the closeness of the relationship shows that the variables of innovation characteristics, communication channels, period, social system, knowledge, and persuasion moderately affect the adoption and implementation variables. The adoption and implementation variables moderately affect the confirmation and support variables. The results of the predictive relevance test state that the adoption and implementation variables, confirmation variables, and support variables.

1. Introduction

Piknet is a sound wave-based attractor, often used to attract fish to the fishing gear in the waters due to the sound it emits. The attractor is being tested on a community of fishermen in Surabaya. Fish respond to the sound by coming closer to the fishing gear. At the initial trial, a socialization program was carried out for the gillnet fishermen in the purposes, benefits, and methods of using the tool. The increase in the number of catches, the type of fish caught, the willingness to use fishing aids, and the suitability of fishing gear were also tested. [1].

In this research, the policy-making model used the concept of innovation diffusion [2]. The latent variables consist of 5 aspects: characteristics of innovation, communication channels, period, social systems as well as knowledge, and persuasion. The policy-making process has several stages: adoption, implementation, confirmation, and support. The variables that influence the decision to adopt attractors have been analyzed. This research aims to determine the inner model to determine the relationship between latent variables in the decision-making of innovation in the use of sound wave attractors by fishermen in Surabaya for fisheries sustainability.

2. Method

The analytical method used was the Partial Least Square-Structural Equation Modeling, which was based on the variance of exogenous and endogenous latent variables as well as the indicator variables [3]. Based on the research objectives, the inner model was used to describe the association between the constructions [4]. A total of 7 latent (constructive variables) were used in this study and they consist of 5 exogenous (X1-X5) and 2 endogenous (Y1 and Y2) variants. Furthermore, the manifest indicator consists of 24 variables, as shown in Figure 1.

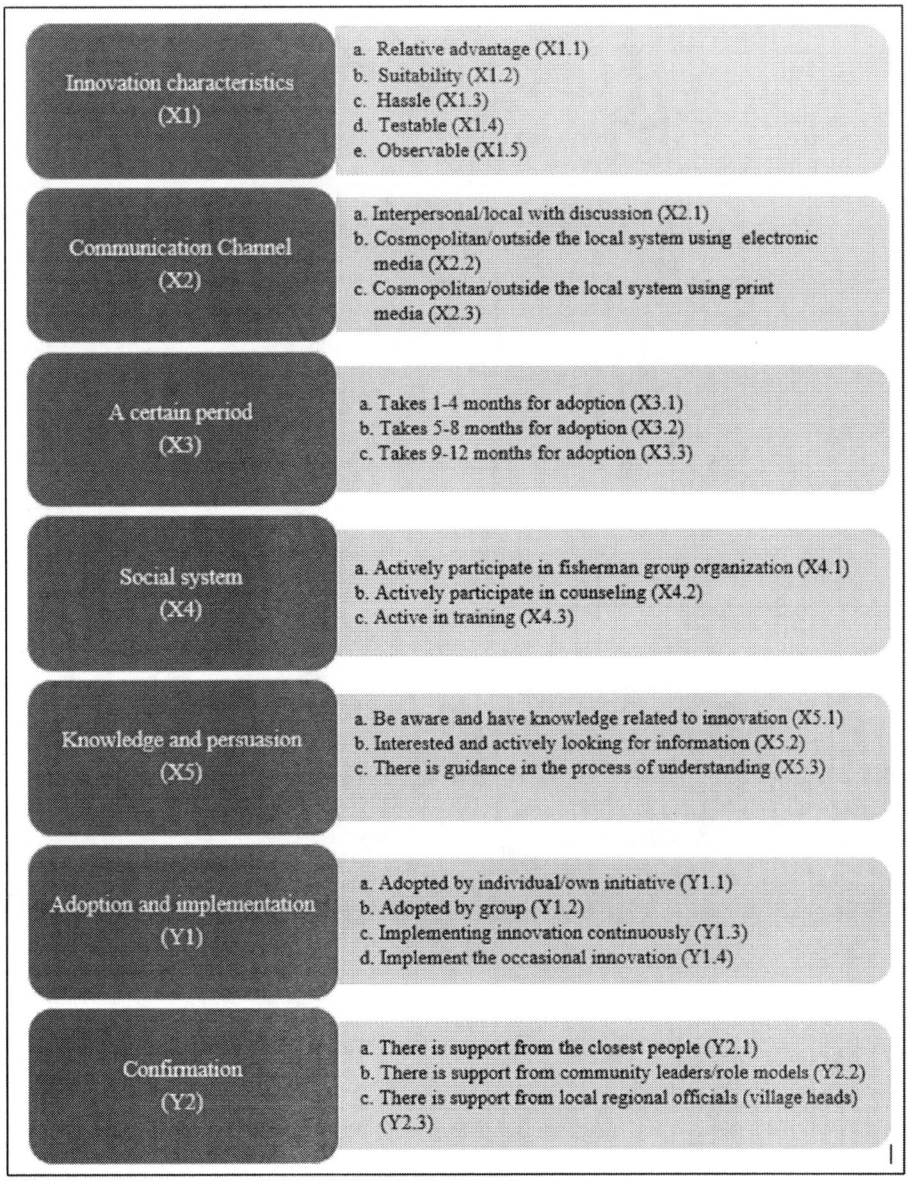

Figure 1. Latent variables and indicator variables

Data processing using quantitative analysis by making a perception statement, rated from 1 to 5, where the lowest value is strongly disagreed (score 1) to the highest value is strongly agreed (score 5). A purposive sampling method was used, after which 50 gill net fishermen were selected in Bulak District, Surabaya. [5] stated that a minimum of 45 samples were used when there were at least 5 arrows

IOP Conf. Series: Earth and Environmental Science **1473** (2025) 012034 doi:10.1088/1755-1315/1473/1/012034

to the latent variable with a significance of 0.1. The data in qualitative form were analyzed using the student version of the Smart PLS tool. This research was carried out in Bulak District, Surabaya, East Java, especially a community of fishermen in this area. Figure 2, shows the map of the location.

Figure 2. Research location in Surabaya

In determining the inner model to evaluate the hypothesis, several testing stages were carried out, namely path coefficient (β), coefficient of determination (R^2), effect size (f^2), and predictive relevance (Q^2) tests as well as t-test (Figure 3).

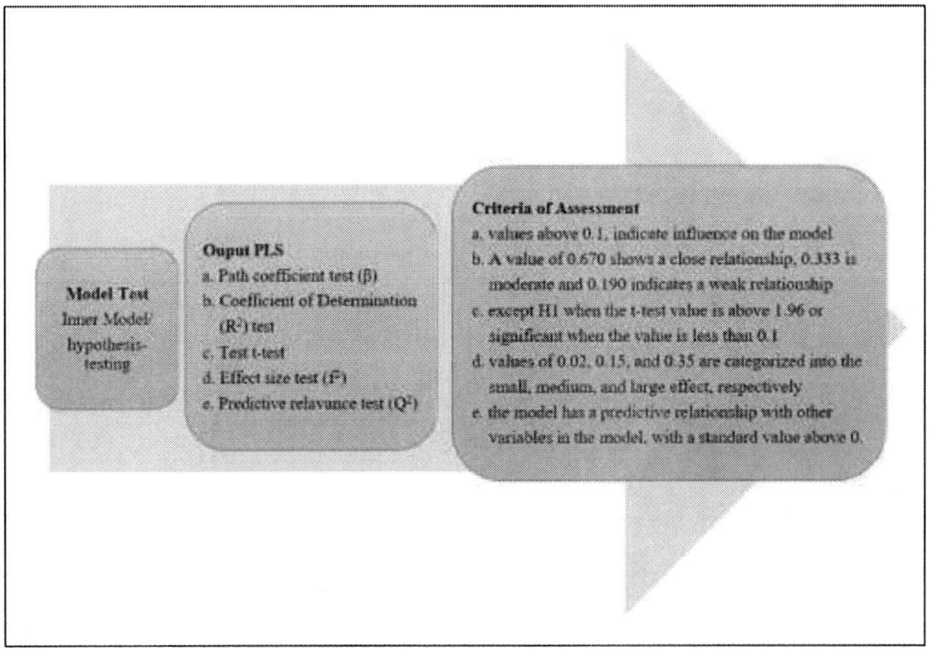

Figure 3. Criteria of assessment

3. Results and Discussion

3.1. Results of Inner Model

Several stages of testing were carried out to determine the inner model using the PLS-SEM method, namely: (a) path coefficient test (β), (b) Coefficient of Determination test (R^2), (c) T-test, effect size test (f^2), and (d) Predictive Relevance test (Q^2).

- Path Coefficient Test (β)

The test was carried out by looking at the standard, where values above indicate that the path influences a model [6], as shown in Table 1. A value of 0.581 was obtained for the adoption and implementation variable (Y1) on confirmation and support (Y2), values of 0.110, 0.237, 0.331, and 0.289 were recorded for variables of innovation characteristic (X1), communication channels (X2), knowledge and persuasion (X5), and a certain period (X3) on the adoption and implementation (Y1), respectively. For the social system (X4), a value of -0.059 below the threshold value of 0.1 was obtained, which indicates that it did not affect the inner model. Furthermore, there was a negative effect on X4, but not significant due to the t-test value of 0.633.

Table 1. Result of path coefficient test (β)

Variable	*Path Coefficient* (β)
• Relation between adoption and implementation variables (Y1) with confirmation and support variables (Y2)	0.581
• Relation between characteristics of innovation variables (X1) with adoption and implementation variables (Y1)	0.110
• Relation between communication channel variables (X2) with adoption and implementation variables (Y1)	0.237
• Relation between knowledge and persuasion variables (X5) with adoption and implementation variables (Y1)	0.331
• Relation between specific time period variables (X3) with adoption and implementation variables (Y1)	0.289
• Relation between social system variables (X4) with adoption and implementation variables (Y1)	-0.059

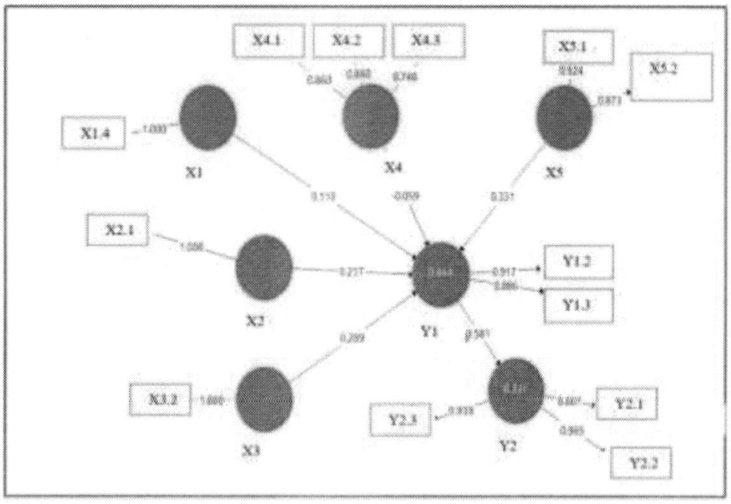

Figure 4. Result of path coefficient test (β)

- Result of Determination Coefficient (R^2) Test

Statistical tests on the value of determination are carried out to obtain the variance of endogenous variables that are influenced by other variables in a model [6], values of 0.670, 0.338, and 0.190 are classified in the close, moderate, and weak relationship categories, as shown in Figure 3.

Results of statistical analysis found that exogenous variables consisting of; innovation characteristics, communication channels, certain periods, social systems, and knowledge and persuasion variables can moderately explain 44.4% of the variance from adoption and implementation variables. The adoption and implementation variables can moderately explain 33.7% of the confirmation and support variables (Table 2).

Table 2. Result of determination coefficient (R^2) test

Variable	R^2 value	Description
• Adoption and implementation (Y1)	0.444	Moderate
• Confirmation and support (Y2)	0.337	Moderate

- Result of T-test

Statistical analysis using the t-test is conducted to test the research hypothesis through a significance test, where H1 is accepted if the sig. value is less than 10% or the value obtained is above 1.65 [6]. The analysis was conducted using the bootstrapping approach and the results showed that there were 4 (four) significant hypotheses: H1(b), H1(c), H1(e), and H1(f), and 2 (two) variables are not insignificant variants, namely H1(a) and H1(d), as shown in Table 3.

Table 3. Result of t-test

Hypothesis	T-test value	P value	Description
• H1 (a): Innovation characteristics (X1) impact on adoption and implementation (Y1)	0.823	0.410	Not significant
• H1 (b): Communication channel (X2) impact on adoption and implementation (Y1)	2.361	0.018	Significant
• H1 (c): Specific time period (X3) impact on adoption and implementation (Y1)	2.306	0.021	Significant
• H1 (d): Social system (X4) impact on adoption and implementation (Y1)	0.440	0.660	Not significant
• H1 (e): Knowledge and persuasion(X5) impact on adoption and implementation (Y1)	3.033	0.001	Significant
• H1 (f): Adoption and implementation (Y1) impact on confirmation and support (Y2)	4.328	0.000	Significant

- **Result of Size Effect Test (f2)**

The results of the effect size (f2) value statistical test can explain the influence of a variable on other variables in a model, the standard values used are 0.02, 0.15, and 0.35, which indicate the impact of small, medium, and large respectively, as shown in Table 4. The results state that there are 4 (four) paths with a small impact, 1 (one) path with a medium effect, and 1(one) path with a big impact. The path with a large impact is the adoption and implementation (Y1) path to confirmation and support (Y2), while the medium effect path is knowledge and persuasion (X5) to adoption and implementation (Y1).

Table 4. Result of size effect test (f2)

Hypothesis	Value f^2	Description
• H1 (a): Characteristics of innovation (X1) impact on adoption and implementation (Y1)	0.014	small
• H1 (b): Communication channel (X2) impact on adoption and implementation (Y1)	0.084	small
• H1 (c): Specific time period (X3) impact on adoption and implementation (Y1)	0.089	small
• H1 (d): Social system (X4) impact adoption and implementation (Y1)	0.003	small
• H1 (e): Knowledge and persuasion (X5) impact on adoption and implementation (Y1)	0.119	medium
• H1(f): Adoption and implementation (Y1) impact on confirmation and support (Y2)	0.509	big

- **Result of Predictive Relevance Test (Q^2)**

Statistical analysis of predictive relevance values to determine whether there is a predictive relationship between a variable and other variables in a model, using a standard value above 0 [6]. The analysis was conducted using the blindfold method, and the results show that the adoption and implementation variable (Y1) can predict the model with a value of 0.323 and the confirmation and support variable (Y2) with a value of 0.265 (Table 5).

Table 5. Result of predictive relevance test (Q^2)

Hypothesis	Value Q^2	Description
Adoption and implementation (Y1)	0.323	Predictive related
Confirmation and support (Y2)	0.265	Predictive related

3.2. Inner Model Analysis and Policy Making

Figure 5 shows that the model can explain how to determine policies in the use of innovations so that they can be implemented by Surabaya fishing groups, for sustainable fisheries. The results describe that latent variables and indicators that have a major influence on Y1 that can be adopted in groups (Y1.2) and implemented sustainably (Y1.3) include: (a) communication channel (X2) through interpersonal discussion (X2.1); (b) period (X3) 5-8 months (X3.2); (c) knowledge and persuasion (X5) by gaining knowledge (X5.1) and seeking information (X5.2) confirmation and support (Y2) obtained from the closest people (Y2.1), society leaders (Y2.2), and district officials (Y2.3).

In innovation and policy-making, the innovation characteristic (X1) and the social system (X4) have a small influence, the testable variable (X1.4) was influential under X1, as shown in Figure 3. Meanwhile, X4 shows that the influential variables include following organization (X4.1), attending counseling (X4.2), and participating in training (X4.3). [7] revealed that characteristics innovation with indicators can be tested as the factors that must be considered while making decisions during diffusion in a community. Fishermen's perception of the innovation characteristics variable is not significant, as seen from the t-test value of 0.410 (Table 3). From the effect size test (f^2), a value of 0.014 was obtained, where the innovation variable (X1) had little effect on others in the model, as shown in Table 6. The indicator, which can be tested (X1.4) also had minimal effect. The level of adoption was influenced by the perception of the new product's advantages [8] where the nature of the innovation itself affected the acceptance rate [9] as well as the suitability indicators and relative advantage [10]. Based on the results obtained about diffusion and use of attractors, the innovation variable can be an important factor in increasing efforts to introduce the product to fishermen in Bulak District, Surabaya. This is expected to lead to an increase in their perceptions of the importance.

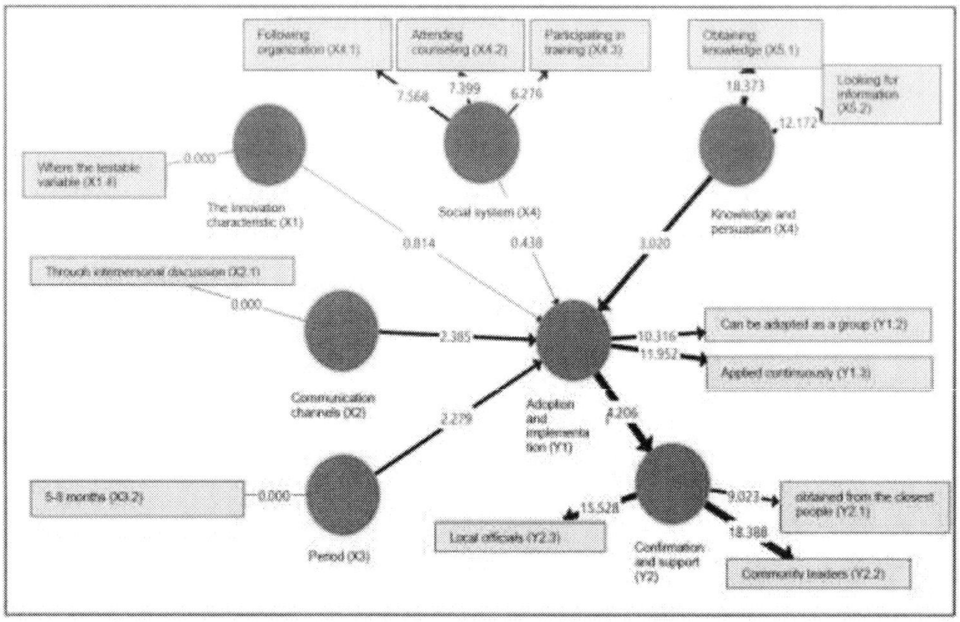

Figure 5. Inner model of innovation and policy making

The analysis results of the perception of the social system (X4) with a t-test value of 0.660 are not significant (Table 3). Meanwhile, a value of 0.003 was obtained using the effect size test (f2), where X4 still had some effect on other variables in the model (Table 4). The influential indicators include: following the organization (X4.1), attending counseling (X4.2), as well as participating in training (X4.3). Based on these results, the fishermen's perceptions of the social system must be improved. [11] stated that the counseling and training were carried out to increase their understanding.

The results of the t-test analysis for the communication channel (X2) were significant with a value of 0.018 (Table 3). The policy-making process was carried out by actors involved in innovation, by [12]. The effect size test (f2), which aims to determine the influence of a certain period variable (X2) on other variables in a model obtained a value of 0.084, which indicates a small effect (Table 4). Furthermore, [13] reported that direct meetings and coaching with fishing groups can be carried out to facilitate the implementation of the innovation. Perceptions are also influenced by media or information. [14] and [15] stated that fishermen tend to communicate with competent persons to optimize the acceptance of products. The results of the t-test analysis of research variables for a certain period (X3) were significant with a p-value of 0.021 (Table 3). [16] revealed that the policy-making processes of the social system are carried out and require a certain amount of time. In the effect size test, which aims to determine the effect of a certain period variable (X3) on others in a model, a value of 0.089 was obtained, indicating the presence of minimal effect (Table 4).

The results of the analysis on knowledge and persuasion obtained significant results with a value of 0.001 (Table 3). Furthermore, a value of 0.119 was recorded in the effect size test, which was used to determine their effect on another variable in a model. This finding shows that the effect was on the moderate category (Table 4). [16] states that this stage is a phase where personally, a person begins to know and will consider accepting an innovation with several considerations such as benefits, easy access, and guidance. Statistical analysis using the t-test on the adoption and implementation variables obtained significant results with a value of 0.00 (Table 3) and effect size analysis was used to determine the magnitude of the influence of a variable on other variables in the model, and obtained a value of

0.509 (in Table 4) in the large influence category. This finding shows that the effect is classified in a big category. The process of determining the decision to accept an innovation is often done by testing a product, influenced, among others, by the number of people who can adopt it [16]. [17] said that the adoption process for utilizing and implementing an innovation is usually easier according to the needs and interests of the community. Furthermore, [15] states that fishermen groups, the first to implement innovations are community leaders. [2] reported that the confirmation stage in the adoption process of an innovation is expected to gain support from other parties. In this model, fishermen's perceptions are an early part of the process to support the adoption process, so that local officials, close friends, and community leaders can provide support.

4. Conclusion

The inner model of the research results states that to be qualified to predict the relationship between latent variables as a consideration in sustainable fisheries policy making. The influential paths are; (a) adoption and implementation towards confirmation and support, (b) innovation characteristics, (c) communication channels, (d) knowledge and persuasion, and (e) a certain period towards adoption and implementation. The coefficient of determination used to determine the closeness of the relationship shows that the variables of innovation characteristics, communication channels, period, social system, knowledge, and persuasion moderately affect the adoption and implementation variables. The adoption and implementation variables moderately affect the confirmation and support variables. The results of the predictive relevance test state that the adoption and implementation variables, confirmation variables, and support variables, have a predictive relationship in the model.

Acknowledgment

The authors would like to thank Hang Tuah University and the research members who have supported the successful implementation of the research; so that the research results can be achieved as intended

References

[1] N Rosana, MA Sofijanto, Suryadhi, 2019. "Assistance On The Use Of Sound Wave Attractor To Increase Fishermen's Number Of Catches Bulak District Surabaya". In International Journal of Advanced Research, **7** (8), 324-327.
[2] Rogers, E. M. 2003. Diffusion of Innovations. New York: Free Press.
[3] Mun'im, 2015. "Analysis of Salt Farmers Business and its Role and Economy in 2012 (Case Study of PUGAR Salt Farmers)". Jakarta. In Journal of Marine Fisheries Social and Economics, **10** (2), 217-228.
[4] Jaya, IGNM and Sumertajaya, IM, 2008. Structural Equation Modelling with Partial Least Square. Semnas Mathematics and Mathematics Education, **1**, pp. 118-132.
[5] Solihin, M., Ratmono, D. 2013. SEM-PLS Analysis with Warppls 3.0 for Nonlinear Relationships in Social and Business Research. Yogyakarta: Andi Publisher.
[6] Hair, J. F., Ringle, C. M., dan Sarstedt, M. 2011. PLS-SEM: Indeed a Silver Bullet. Journal of Marketing Theory and Practice. **1**. No. 2, 139-152.
[7] Schiffman, L dan Kanuk, L. 2010. Consumer Behavior. Tenth Edition. Global Edition, USA: Prentice-Hall Inc.
[8] Tjiptono, F., G. Chandra. 2012. Global Marketing: Offline and online contexts. Yogyakarta: UPP STIM YKPN.
[9] Mardikanto, T. 2007. Redefinition and Revitalisation of Agricultural Extension. PUSPA. Surakarta.
[10] Kusdibyo, L and Leo, G. 2018. E-Learning Adoption in Higher Education. In Proceedings of 9th Industrial Research Workshop and National Seminar.
[11] Amanah, S., 2006. Fisheries extension. Journal of Extension, **2** (4).
[12] Leeuwis, Cees. 2004. Communication for Rural Innovation: Rethinking Agricultural Extension. Blackwell Publishing, Oxford.
[13] Septiani, T. and Esfandari, D.A., 2018. Diffusion of Technology Innovation System of PT Aruna Jaya Nusantara Fisheries Trade in Tanjung Baru Village, East Kalimantan. Proceedings of the National Conference on Communication, **2** (01), pp.696-707.

[14] Indraningsih K.S. (2011). The influence of counseling on farmers' decisions in the adoption of integrated farming technology innovations. Journal of Agro Economics, **29** (1).

[15] Warnaen, A. and Cangara, H., 2016. Factors that hinder innovation in farming and fishing communities in improving community welfare in Takalar Regency. KAREBA: Journal of Communication Science, **2** (3), pp.241-250.

[16] Adianto. 2018. The Adoption Process of Local Innovations towards Improving Community Welfare in the Minapolitan Area of Koto Mesjid Village, Riau Province. In Sosio Konsepsia Journal, **7** (2).

[17] Pannell, D. F., Marshall, G. R., Barr, N., Curtis, A., Vanclay, F., & Wilkinson, R. (2006). Adoption of conservation practices by rural landholders. Australian Journal of Experimental Agriculture, **46** (11), 1407–1424. https://doi.org/10.1071/EA05037.

Salinity Adaptation on The Growth and Survival of Saline Tilapia (*Oreochromis niloticus*) for Environmental Sustainability

N Trisyani[1] , Nuhman[1] , M Setiawati[2] , C Susila[1]

[1]Department of Fisheries, University of Hang Tuah, Surabaya
[2]Departemen of Aquaculture, University of Institut Pertanian Bogor

*E-mail: ninis.trisyani@hangtuah.ac.id

Abstract. The salin tilapia is a species of euryhaline, which is able to adapt to various levels of salinity ranging from freshwater to brackish and seawater environments. This fish is a genetically engineered product by the Agency for the Assessment and Application of Technology, which is designed for cultivation in high-salinity water environments such as marine ponds and floating net cages. This study aims to evaluate the growth and survival of saline tilapia in the salinity range of 15 ppt to 27 ppt, by ensuring environmental sustainability. This study used a Complete Random Design with 5 treatments and 4 replicates. Key observations include absolute weight growth, daily growth rate, operculum opening, survival rate, and water quality parameters such as temperature, pH, DO, ammonia, and nitrites. The results showed that the best absolute weight growth was 19.65g at 24 ppt salinity, with the highest daily growth rate of 0.018% at the same salinity level. The maximum operculum opening rate is 218 openings/minute at 27 ppt. All treatments showed a 100% survival rate, and the water quality parameters were within an acceptable range for salinity.

1. Introduction

Tilapia is grouped in euryhaline fish, has good endurance and adaptability, namely physiological adaptation to a high salinity range from freshwater to sea water. [1]. Its distribution includes rivers, lakes, reservoirs, swamps, and also brackish water. Saline tilapia is a genetically modified fish from technology developed by BPPT (Agency for the study and application of technology) with the aim of obtaining tilapia that can survive in high salinity waters [2]. The development of saline tilapia aquaculture can be carried out on coastal land to maximise the carrying capacity of brackish and marine waters so that the production of aquatic environments remains sustainable. Currently, saline tilapia aquaculture in Indonesia is carried out in ponds, cages, rice fields, and floating net cages (KJA).

The Ministry of Fisheries, Aquaculture and Food and the Agricultural Organisation (FAO) cite tilapia as an example of successful fisheries and aquaculture in the world. Based on the composition of production in the second quarter of 2022, the fish commodity with the highest production in aquaculture is tilapia at 401 thousand tonnes which experienced an increase in growth of 8.01 percent, [3]. National tilapia production can be further increased by expanding cultivation land in brackish water ponds that are no longer used for shrimp farming. The large amount of abandoned land along the coast is an excellent land and has the potential to be used as new tilapia farming ponds. This can also be used to increase the economic value of coastal communities into additional livelihoods [4].

The euryhaline nature of tilapia allows it to be developed over a wide salinity range. Different media salinity affects the physiological response of tilapia. Seawater media has high salinity levels will affect morphological and physiological adaptations. Research conducted [4]. states saline tilapia will stress and die with a matter of hours if salinity exceeds 17-32 ppt. According to [1] saline tilapia can survive in salinity 5 ppt to 21 ppt. This study aims to examine the physiological response and its effect on the growth of saline tilapia reared in different salinity media. This research is expected to provide information about the potential of saline tilapia reared in brackish to salty salinity in an effort to maintain the sustainability of the aquatic environment.

2. Methods

The research was carried out at the Hang Tuah University hatchery in Surabaya from March to May 2024. The saline tilapia used came from the Lamongan Brackish Water Aquaculture Installation with a salinity of 15 ppt, sized between 2 - 3 g/head and stocked at a density of 8 fish per aquarium measuring 25×30×50 cm filled with 37 litres of water [5].

The research design used a completely randomised design with 5 treatments, namely 15 ppt, 18 ppt, 21 ppt, 24 ppt and 27 ppt salinity rearing media. Each treatment was repeated 4 times. Fish were given commercial feed 3 times a day at 08.00, 12.00 and 17.00 WIB. Feed is given until satiation. Salinity adaptation was carried out for 7 days before the implementation of the study, by increasing the salinity of 1 - 2 ppt every day until it reached the salinity according to the treatment. During the study, body weight gain was observed every 7 days, mortality and gill operculum openings were observed.

Water quality observations included dissolved oxygen content, salinity, nitrite, temperature, pH, and ammonia. During the rearing period, water was flushed and changed every day by 25% of the total volume. The rearing period was 42 days and the parameters observed included:

- Absolute Growth [6]: $W = Wt - Wo$ (1)
  W = absolute growth (g)
  Wt = weight of fish biomass at the end of the study (g)
  Wo = biomass weight at the beginning of the study (g)

- Daily Growth Rate [6]: $DGR = \left(\sqrt[t]{\frac{Wt}{Wo}} - 1\right) 100\%$ (2)
  DGR = daily growth rate (g/day)
  Wt = biomass of test fish at the end of the study (g)
  Wo = biomass of test fish at the beginning of the study (grams)
  t = length of experiment (days)

- Survival [7]: $SR = Nt/No \times 100\%$ (3)
  SR = survival rate (%)
  Nt = number of fish at the end of rearing (fish)
  No = number of fish at the beginning of rearing (fish)

Operculum openings were calculated by counting the number of saline tilapia gill operculum openings for 1 minute. The data were analysed using homogeneity and normality tests, as a requirement for the use of parametric statistics. To see the differences between treatments, analysis of variance test was conducted followed by Duncan's test.

3. Results and Discussion

3.1 Absolute growth

The results of absolute weight growth of saline tilapia (*Oreochromis niloticus*) in each treatment can be seen in Figure 1.

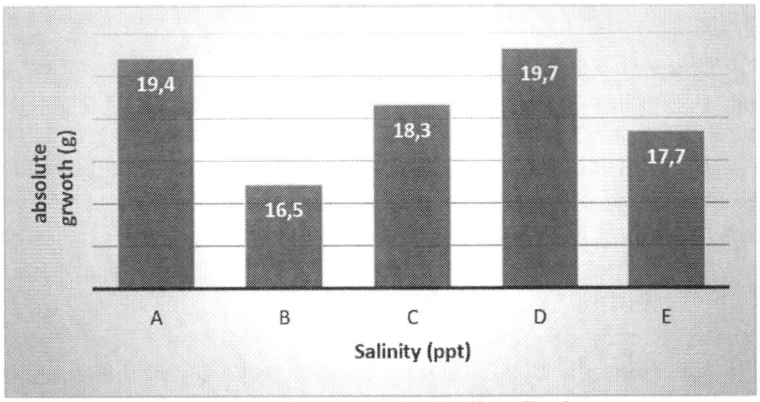

Figure 1. Average absolute growth of saline tilapia per treatment

The best absolute weight growth of saline tilapia was found in the 24 ppt salinity treatment at 19.7 grams and the lowest absolute weight growth was in the 18 ppt salinity treatment at 16.5 grams and the smallest absolute growth at 18 ppt salinity of 16.5 grams. Anova test results with a significance of 0.002 indicates that the absolute growth of saline tilapia is influenced by salinity and the best growth at 24 ppt and 15 ppt salinity.

3.2 Daily Growth Rate

Daily growth rate of saline tilapia (*Oreochromis niloticus*) from the treatment results in Figure 2.

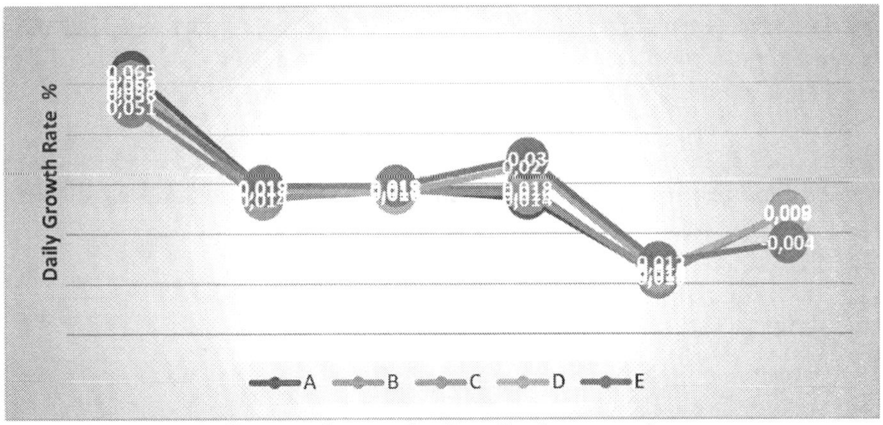

Figure 2. Daily growth rate of saline tilapia per week per treatment

The 24 ppt salinity treatment gave the best results of 0.0183 grams/day and the 18 ppt salinity treatment produced the lowest growth of 0.0158 grams/day. Anova test with a significance value of 0.026 showed that there was an effect of salinity on the daily growth rate of saline tilapia and the best in the 24 ppt and 15 ppt salinity treatments. At the beginning of rearing in the first week all treatments experienced a decrease in growth as an adaptation process, and increased again until week four. The decline occurred in the fifth week and increased again in the sixth week.

Salinity in waters creates osmotic pressures that can be different from the osmotic pressure in the body of aquatic organisms. This condition causes aquatic organisms to perform osmoregulation mechanisms in their bodies with the aim of achieving a balance of osmotic pressure in the organism's body with the external environment. [8]. The process of osmoregulation is one of the physiological processes that occur in the body of fish to control the concentration of solutions in the body to balance with the environment.

Changes in salinity that cause the osmoregulation process will also result in an increase in energy requirements. This happens because osmoregulation is a metabolic process that requires active transport of ions to maintain salt concentration in the body. Fish must take or secrete salt from the environment to maintain the balance of salt content in the body. The process requires considerable energy [9].

There are two hormones that are very important in the process of osmoregulation, namely prolactin and cortisol. Prolactin is secreted by cells in the pituitary gland of freshwater fish and plays a role in preventing the passage of Na+ through the permeable membrane. The steroid hormone, cortisol is also produced by internal cells in the kidney head and plays a role in the process of adaptation of euryhaline organisms to a range of salinities. Cortisol in euryhaline teleost fishes plays a role in excreting ions through the gills by stimulating chloride cells for proliferation, differentiation and excretion activities as plasma cortisol levels increase during periods of migration or transfer from freshwater to seawater. Therefore, cortisol is known as the 'seawater hormone' [10]. In red hybrid tilapia, the maximum growth efficiency maintained was higher at 18 ppt salinity than at 0 or 36 ppt salinity [11]. In this study using saline tilapia, the highest absolute growth and daily growth rate were found at 24 ppt salinity. Studies with a number of tilapia species, including *Oreochromis aureus, Sarotherodon melanotheron, Oreochromis mossambicus*, and red hybrid tilapias, have shown that fish tolerate lower temperatures better when reared in low salinity brackish water (5-12 ppt) than in freshwater or seawater, due to stress as osmoregulation is minimised until it approaches isoosmotic salinity. Therefore, osmoregulation failure is prevented at temperatures that are lethal under hypo or hyper osmotic media conditions.

Euryhaline species can grow well at salinities with a higher or wider range in euryhaline tilapia. This value shows similarity with the results of the study, growth at 24 ppt salinity was higher compared to 21 ppt and 18 ppt salinity. Growth response to salinity was quite variable among species. *Oreochromis mossambicus, Oreochromis hornorum hybrid, Oreochromis spilurus*, and red hybrid tilapia species grow higher in brackish and seawater than in freshwater [11]. This indicates the advantage of tilapia culture in brackish or seawater. In red hybrid tilapia, it is associated with increased feed consumption (appetite) and decreased feed conversion rate.

3.3 Operculum opening

The results of operculum opening of saline tilapia (*Oreochromis niloticus*) at different salinities can be seen in Figure 3.

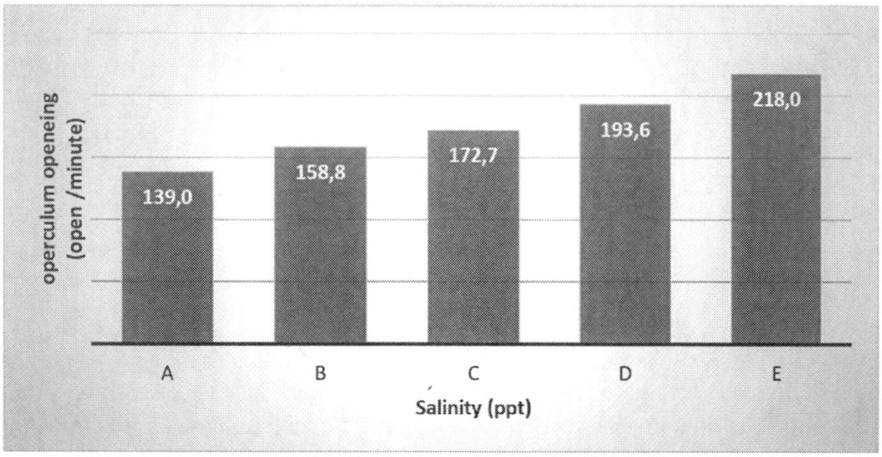

Figure 3. Saline tilapia operculum opening per treatment

The highest operculum opening of saline tilapia in treatment at 27 ppt salinity was 218 openings per minute and the lowest in treatment at 15 ppt salinity was 139 openings per minute. Anova test results with a significance value of 0.000 showed that there was an effect of salinity on the speed of opening the gill operculum of saline tilapia.

In this study showed an increase in operculum openings with increasing salinity, at 15 ppt salinity as much as 139 / minute, 18 ppt salinity increased 159 / minute, salinity 21 to 173 / minute, 24 ppt salinity increased 194 / minute and salinity 27 ppt the highest opening was 218 / minute. These results are in accordance with [13] which shows the opening of the operculum under normal salinity conditions is about 48/minute. At a salinity of 10 ppt, the operculum opening was 87/minute. At a salinity of 20 ppt the number of operculum openings was 86/minute. At salinities above 10 ppt, fish can still adapt by maximising the osmoregulation mechanism, but at salinities above 30 ppt fish experience stress as indicated by the increasing value of the operculum opening.

Hypersaline stress is indicated by the opening of the operculum and ultrastructural changes in the gill organs [13] due to the fragility of the gills compared to other organs as an organ where water from the environment enters the body, the gills face the pressure of water ingress, which is physically and chemically different from the conditions in the body, in this case the difference in salinity. [14]. Each fish species has a different salinity tolerance related to the type and size of the fish, the way of adaptation and the habitat environment. [15, 16] stated that the most tolerant tilapia species were found in *Oreochromis mossambicus*, *Oreochromis aureus* and *Tilapia zillii*.

The rapid opening of the operculum indicates that the fish is experiencing stress due to the heavy adaptation process. Changes in the opening of the operculum that is accelerating along with the high levels of salinity, if the salinity used is high and the fish still can not adapt, it results in death which is influenced by damage to the operculum tissue, so that the fish begins to slow down and not the operculum slows down [17].

3.4 Survival Rate

The survival value of saline tilapia (*Oreochromis niloticus*) at different salinities in Figure 4.

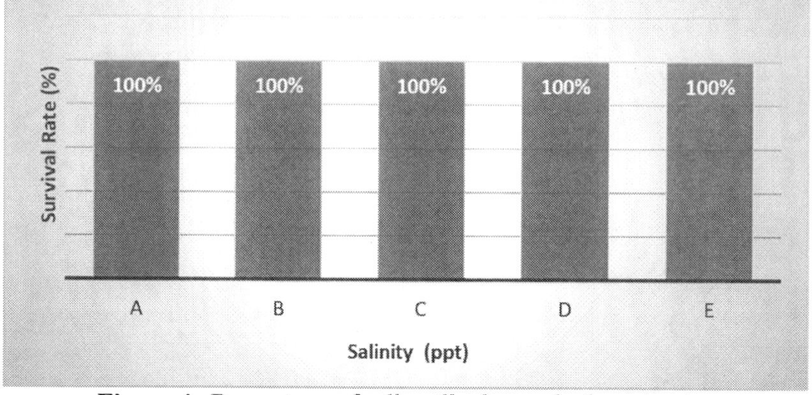

Figure 4. Percentage of saline tilapia survival per treatment

Fish survival during the study is shown in Figure 4, saline tilapia survival at salinity of 15 ppt to 27 ppt reached 100%, this result is in accordance with the habitat of saline tilapia in nature which can live in fresh, brackish and marine waters. Tilapia in every salinity can still adapt shown by the high survival rate from low salinity to high salinity, in accordance with the statement [1], saline tilapia can survive in different salinities, ranging from 5 ppt - 21 ppt salinity. High salinity is salinity that exceeds the salinity value of brackish water with a salinity value of 15 ppt and above. [4] stated that saline tilapia will be stressed and die with a matter of hours if the salinity exceeds 17-32 ppt. These results provide information that during the study, saline tilapia can survive optimally from the beginning to the end of the study. Tilapia also has good adaptability because it is supported by water management during maintenance, which is measured within the optimum range for tilapia growth. In accordance with the opinion of [18, 19], growth and survival rates are strongly supported by maintenance media that are in accordance with the needs of saline tilapia species.

3.5 Water Quality

Measurement of temperature, dissolved oxygen, pH, nitrit dan ammonia of saline tilapia *(Oreochromis niloticus*) during the study in Table 1.

Table 1. Water quality parameter

Parameter	Salinity (ppt)					Average	Reference
	15	18	21	24	27		
Temperature (°C)	28,7-29,1	28,5-29,1	27,4-28,3	27,8-28,4	27,8-28,4	28,1	25°C- 30°C (Nasution, 2014)
pH	8,4-8,7	7.8-8,7	7.8-8,7	8,4-8,7	8,4-8,7	8,4	5-8,5 (Andriani. Y, 2018)
DO (ppm)	3,9-4,3	4,4-4,1	3,9-4,3	3,9-4,3	3,9-4,3	4,1	4 - 7 mg/L (Monalisa, 2010)
Nitrite (ppm)	0-0,75	0-0,5	0-0,75	0-0,75	0-0,75	0,7	0 - 20 mg/L (Fitriadi, 2014)
Ammmonia (ppm)	0-0,45	0-0,35	0-0,35	0-0,34	0-0,34	0,37	0.01-0.84 mg/L (Suryadi, 2022)

The results of measuring several parameters including temperature, pH, dissolved of oxygen (DO), nitrite and ammonia in each maintenance medium with a salinity treatment of 15 ppt to 27 ppt show the condition which is suitable for supporting saline tilapia to grow and survive.

4. Conclusion

Absolute growth and daily growth rate of saline tilapia were highest at 24 ppt salinity with 100% survival rate. Saline tilapia can be prioritised as a species for stocking in ponds and floating net cages in an effort to sustainably increase production on coastal lands.

Acknowledgements

Thank you to Hang Tuah University for funding this research in accordance with the research contract S.Gas/003/UHT.C.2/II/2024.

References

[1] Muhsoni, F. F. 2021. Growth rate and survival of tilapia (*Oreochromis niloticus*) at different salinities. Juvenil: *Scientific Journal of Marine and Fisheries,* **2** (3), 166-175.

[2] Hidayatullah. 2022. Growth and Survival of Salin Tilapia (*Oreochoromis niloticus*) Fry Raised at Different Salinities. Thesis. Students of the Fisheries Cultivation Department of the Pangkep State Agricultural Polytechnic.

[3] Anonymous. 2023. Production of Salted Tilapia. Directorate General of Aquaculture Fisheries.

[4] Rohman, T., Wulandari, Y. T., Leksani, W. I., & Chandrawati, D. 2017. Effect of Water Salinity Difference on Survival Rate and Physiological Response of Tilapia Seeds (*s*). In Proceedings of SNPBS (National Seminar on Biology and Science Education) (pp. 114-123). DOI: http://hdl.handle.net/11617/9320.

[5] Setiawati, M., and Suprayudi, M.A. 2023. Growth and Feed Efficiency of Red Tilapia (*Oreochromis* sp.) which is maintained on the salinity media. *Indonesian Journal of Aquaculture,* **2** (1), 27-30.

[6] Puspita, T., Andriani, Y., & Hamdani, H. 2015. Utilization of Peanut Meal in Fish Feed on the Growth Rate of Tilapia (*Orechromis niloticus*). Journal of Marine Fisheries, **6** (2 (1)).

[7] Yulan, A., & Gemaputri, A. A. Survival Rate of Gift Tilapia (*Oreochromis niloticus*) Fry at Different Salinities. *Journal of Fisheries, Gadjah Mada University*, **15** (2), 78-82. DOI: https://doi.org/10.22146/jfs.9100 . 2013

[8] Sobirin, M., Soegianto, A., & Irawan, B. 2014. Effect of Some Salinity on Osmoregulation of Tilapia (*Oreochromis niloticus*). *Journal of Mathematics and Natural Sciences,* **17** (2), 46–50.

[9] Stickney RR. 2000. Encyclopedia of Aquaculture Texas: A Wiley-Interscience Publication. ISBN: ISBN: 978-0-471- 29101-5.

[10] Morgan, J.D. 1997. Energetic Aspect Of Osmoregulation In Fish. Thesis. The University of British Columbia, 156 pp.

[11] Pamungkas, W. 2013. Osmoregulatory Activity, Growth Response, and Energetic Cost in Fish Reared in a Salinity Environment. *Aquaculture Media* **7** (1): 44-51.

[12] Opstad, I. 2003. Growth And Survival Of Haddock (*Melanogrammus aeglefinus*) Larvae At Different Salinities. The Big Fish Bang. *Proceedings of the 26th Annual Larval Fish Conference*. 2003, p. 63-69.

[13] Fuadi A A, Hasly IRJ , Azkia LI and Irham M. 2021. Response Of Tilapia (*Oreochromis niloticus*) Behaviour To Salinity Differences: A Laboratory Scale Study. *IOP Conf. Series: Earth and Environmental Science* **674** (2021) 012060 doi:10.1088/1755-1315/674/1/012060

[14] Sardella B A, Matey V, Cooper J, Gonzalez R J and Brauner C J. 2004. *Journal of Experimental Biology* **207** 1399-1413.

[15] Sayed A F M E. 2006. Tilapia Culture in Salt Water: Environmental Requirements, Nutritional Implications and Ecoomis Potentials (Mexico: Universidad Autonoma de Nuevo Leon, Monterrey) **5** 970-694.

[16] Irham M, Ihsan M, Oktavina C, Sugianto S, Nur F M., Batubara A S. 2020. *Journal Biharean* Biologist **14** 85-89.

[17] Akbar, M. A., Jayanthi, S., Fajri, S., Zahara, A. S., & Sari, M. T. 2023. The Influence of Different Media on The Physiological Response of Nila Fish (*Oreochormis niloticus*). *Jurnal Biologi Tropis*, **23** (2), 187-193.

[18] Laudin, B., Indrawati, E., & Ratnawati, R. 2023. Juvenile Growth Dynamics of Saline Tilapia at Different Salinity Levels. *Journal* of Aquaculture and Environment, **5** (2), 45-53.

[19] Nasution, A. S. 2014. Analysis of Survival of Pandu Saline Tilapia Strain (*Oreochromis niloticus*) Seeds Raised in Tugu Pond, Semarang with Different Densities. *Journal of Aquaculture Management and Technology*. **3** (2), 25-32.

Ecosystem-Based Management Strategies For Stunting Decline And Sustaining Marine Resources

N Lelyana[1], B Sucahyo, [2] S Revianti[3]

[1] Department of Public Health, Hang Tuah University, Surabaya, Indonesia, 60111
[2] Department of Orthodontic, Hang Tuah University, Surabaya, Indonesia, 60111
[3] Department of Oral Biology, Hang Tuah University, Surabaya, Indonesia, 60111

* E-mail: nora.lelyana@hangtuah.ac.id

Abstract. Ecosystem-Based Management (EBM) has emerged as a comprehensive approach to marine resource management, focusing on maintaining ecological health and resource sustainability while addressing critical human health issues, such as stunting. This qualitative research aims to explore the current outcomes of EBM strategies in terms of ecological health, resource sustainability, and their impact on reducing stunting in vulnerable communities. Through a review of existing literature and case studies, the research identifies key successes, such as improved species diversity, habitat conditions, and fish stock levels. However, significant challenges impede the full realization of EBM's potential, including scientific data gaps, regulatory fragmentation, economic constraints, and social resistance. The study proposes strategies to overcome these barriers, emphasizing the need for enhanced scientific understanding, strengthened governance, stakeholder collaboration, sustainable financing, and capacity building. The findings suggest that effectively implemented EBM strategies can not only sustain marine ecosystems but also contribute to reducing stunting by securing a reliable and nutritious food supply. The research concludes that a concerted effort across multiple sectors is essential for realizing the full benefits of EBM.

1. Introduction

Marine ecosystems are integral to global biodiversity, supporting millions of species and providing essential services, including food security, climate regulation, and economic opportunities. Despite their importance, these ecosystems are increasingly threatened by human activities such as overfishing, pollution, and climate change. The degradation of marine ecosystems not only jeopardizes biodiversity but also exacerbates issues like stunting, particularly in communities that rely heavily on marine resources for their nutritional needs. Ecosystem-Based Management (EBM) has emerged as a holistic approach to sustainably managing marine resources, focusing on maintaining ecosystem health and resilience, which is crucial for both stunting decline and the long-term sustainability of marine resources [1].

EBM is a comprehensive strategy that considers the entire ecosystem, including human influences, rather than focusing on a single species or resource. It aims to balance ecological, social, and economic objectives by recognizing the interconnectedness of marine environments. According to the Food and Agriculture Organization (FAO) [2], EBM involves integrating scientific knowledge with traditional and local insights and promoting adaptive management practices that can respond to changing conditions and new information. By ensuring the health and productivity of marine ecosystems, EBM plays a vital role in securing the nutritional resources needed to combat stunting in vulnerable populations [3]. Key components of EBM include a holistic approach, adaptive management, stakeholder involvement, and science-based decision-making. A holistic approach in EBM accounts for the cumulative impacts of various activities on marine ecosystems, addressing the relationships between species, habitats, and human activities to maintain ecosystem functions and services [4]. This comprehensive approach is essential for sustaining the fish stocks and other marine resources that many communities depend on for nutrition, thereby contributing to efforts to reduce stunting. Adaptive management is emphasized due to the dynamic nature of marine environments, requiring continuous monitoring, evaluation of management outcomes, and adjustments to strategies to ensure sustainability [5]. Effective EBM also necessitates the participation of diverse stakeholders, including government agencies, local communities, industry representatives, and non-governmental organizations. Engaging stakeholders ensures that management strategies are inclusive and consider various perspectives, thereby enhancing the legitimacy and effectiveness of EBM [6]. Furthermore, EBM relies on strong scientific data to guide decision-making, utilizing ecological studies, socioeconomic assessments, and modeling tools to predict outcomes in different management scenarios, thereby making management more informed and effective [7].

EBM has been successfully applied in various regions worldwide, demonstrating its effectiveness in maintaining marine resource sustainability and its potential in contributing to stunting decline. For example, the California Current Large Marine Ecosystem (CCLME) in the United States is managed using an EBM approach, which has led to improved fish stocks, reduced bycatch, and enhanced ecosystem health [8]. These improvements directly support food security in the region, helping to combat nutritional deficiencies that can lead to stunting. Similarly, Australia's Great Barrier Reef Marine Park employs EBM principles to protect its diverse marine life while supporting tourism and fishing industries. The Great Barrier Reef Marine Park Authority (GBRMPA) uses adaptive management strategies, stakeholder engagement, and rigorous scientific monitoring to address threats such as coral bleaching and water pollution [9], ensuring the continued availability of marine resources vital for community nutrition.

Despite its potential, EBM faces several challenges, including the complexity of marine ecosystems, limited data availability, and the difficulty of coordinating among various stakeholders with differing interests. Additionally, EBM requires substantial financial and human resources for effective implementation and monitoring [7]. However, advances in technology and data collection, such as remote sensing and citizen science, offer new opportunities for enhancing EBM practices. For example, satellite imagery and autonomous underwater vehicles can improve monitoring efforts, while engaging local communities in data collection fosters stewardship and enhances local knowledge integration [6]. These advancements not only strengthen EBM but also enhance its capacity to address issues like stunting by ensuring the sustainable management of marine resources that many communities depend on for their nutritional needs.

Marine ecosystems are under severe threat from human activities like overfishing, pollution, and climate change, leading to habitat degradation, biodiversity loss, and depleted fish stocks. This environmental decline directly impacts food security, particularly in coastal communities where stunting—a condition of impaired growth in children due to chronic malnutrition—is prevalent. Traditional management approaches have proven inadequate for addressing these complex issues, as they often overlook the broader ecological and social impacts, including nutritional needs. Ecosystem-Based Management (EBM) offers a holistic approach to sustainably managing marine resources, aiming to maintain ecosystem health while addressing the nutritional needs of vulnerable populations. However, EBM faces challenges such as ecosystem complexity, data limitations, and the need for stakeholder collaboration. This research seeks to evaluate the effectiveness of EBM in sustaining marine resources and reducing stunting by overcoming these challenges [1,3].

The research aims to assess the effectiveness of current Ecosystem-Based Management (EBM) strategies in marine resource management, particularly in terms of their impact on ecological health, biodiversity conservation, resource sustainability, and their role in reducing stunting in coastal communities. It seeks to identify the primary challenges and barriers to successful EBM implementation, recognizing that understanding these obstacles is essential for developing more effective management strategies and policies. Additionally, the research intends to develop actionable recommendations to enhance EBM strategies, ensuring the long-term sustainability of marine resources and their contribution to mitigating stunting, while providing valuable insights for policymakers, resource managers, and stakeholders [1,3]. Addressing the complex and interconnected challenges facing marine ecosystems requires a shift from traditional management approaches to more holistic and integrative strategies like Ecosystem-Based Management. By evaluating the effectiveness of current EBM strategies, identifying key barriers to their implementation, and developing recommendations for improvement, this research aims to contribute valuable insights for policymakers, resource managers, and stakeholders. Through a comprehensive understanding of EBM and its application, we can better ensure the long-term sustainability of marine resources, reduce stunting, and secure the vital services that these ecosystems provide [1,3].

2. Method

Qualitative research methods offer deep, nuanced insights into complex issues, making them particularly valuable for studying Ecosystem-Based Management (EBM) strategies aimed at sustaining marine resources and addressing stunting decline. As emphasized by John W. Creswell, a prominent scholar in research design, qualitative methods are essential for capturing the intricate perspectives and contextual factors that influence research topics [10]. This study employs qualitative research methods, specifically utilizing secondary data within Creswell's framework, to explore the effectiveness of EBM strategies for maintaining marine resource sustainability and reducing stunting in vulnerable communities.

2.1. The Role of Secondary Data in Qualitative Research.

Secondary data, defined as information previously collected by other researchers or organizations for different purposes, plays a crucial role in qualitative research. Creswell (2018) highlights the utility of secondary data, noting that it provides extensive background information, identifies trends, and offers insights that might be challenging to gather firsthand. For research on EBM and stunting, secondary data sources may include government reports, scientific publications, policy documents, environmental assessments, and historical records, all of which contribute to a comprehensive understanding of the topic [11].

2.2. Research Design and Methodology.

A well-structured research design is essential for effective qualitative research, as outlined by Creswell (2018), and involves several key steps. First, the research problem and purpose must be clearly defined, focusing on understanding the effectiveness of current Ecosystem-Based Management (EBM) strategies in sustaining marine resources and reducing stunting, with the goal of generating actionable insights for policymakers and stakeholders to enhance EBM practices for both ecological and human health benefits [11]. Second, selecting an appropriate qualitative approach is crucial; a case study approach is particularly relevant for this research as it allows for an in-depth examination of specific instances of EBM implementation across various contexts, providing a rich understanding of effective strategies for stunting reduction and marine resource sustainability. Finally, a systematic approach to data collection and analysis is necessary, involving the identification and access of relevant secondary data sources, such as policy reports, environmental assessments, and scientific studies, followed by coding and categorizing the information to identify patterns, themes, and relationships pertinent to the impact of EBM on marine sustainability and stunting reduction [10].

Applying qualitative methods to EBM research involves several crucial steps. First, identifying data sources is essential, with secondary data drawn from publications by environmental agencies (e.g., NOAA, UNEP), academic journals, NGO reports, and international agreements like the Convention on Biological Diversity. These sources provide comprehensive information on marine ecosystems, management practices, and policy frameworks, which are vital for assessing EBM's effectiveness [5]. Second, data coding and thematic analysis organize this information into meaningful categories, such as "stakeholder participation," "adaptive management," "policy effectiveness," and "nutritional outcomes." This process is critical for identifying key issues and patterns related to EBM's effectiveness in sustaining marine resources and reducing stunting [12]. Finally, the interpretation and synthesis of findings provide a comprehensive understanding of the research problem by comparing different EBM practices, identifying best practices, and highlighting gaps or areas for improvement. This synthesis is essential for informing recommendations to policymakers and resource managers, aiming to enhance EBM strategies for better ecological and human health outcomes [6].

Ensuring validity and reliability in qualitative research is crucial, as emphasized by Creswell (2018). Validity refers to the accuracy and credibility of the findings, while reliability pertains to the consistency and dependability of the research process [11]. Strategies to enhance validity in secondary data analysis include triangulation (using multiple data sources), member checking (seeking feedback from experts), and providing rich, thick descriptions. Ensuring reliability involves maintaining detailed documentation of the research process and using consistent data analysis procedures [10]. These steps are essential for producing credible and trustworthy results that can effectively guide EBM strategies for stunting decline and sustaining marine resources.

3. Results and discussion

3.1. Current Outcomes of EBM Strategies in Terms of Ecological Health, Resource Sustainability, and Their Impact on Reducing Stunting.

Ecosystem-Based Management (EBM) has gained prominence as a comprehensive approach to marine resource management, with the central objectives of maintaining ecological health, ensuring resource sustainability, and addressing critical human health issues such as stunting. By considering the entire ecosystem, including human influences, EBM strategies aim to promote resilience and balance within marine environments, which, in turn, can significantly impact the nutritional status of communities that rely on marine resources. The success of EBM strategies can be assessed through key indicators such as species diversity, habitat condition, fish stock levels, and their impact on reducing stunting, a condition often linked to inadequate nutrition.

Species diversity, a key indicator of ecosystem resilience, has been notably enhanced through EBM practices. For example, the California Current Large Marine Ecosystem (CCLME) has experienced increased species diversity, particularly among marine mammals and seabirds, due to the implementation of EBM strategies [8]. This increased biodiversity supports the availability of a variety of fish species, which are essential sources of protein and micronutrients for coastal communities. Similarly, Australia's Great Barrier Reef has maintained high levels of coral and fish diversity, largely due to the establishment of no-take zones and stricter fishing regulations under EBM approaches [9]. The availability of diverse and abundant fish populations is crucial for combating malnutrition and stunting in dependent populations, particularly in regions where alternative food sources are limited.

The condition of marine habitats, another critical measure of EBM effectiveness, has also improved through EBM initiatives, which positively impacts food security and nutrition. In Chesapeake Bay, for instance, EBM has led to significant enhancements in water quality and the restoration of crucial habitats, such as oyster reefs and submerged aquatic vegetation, which are vital for the broader ecosystem's health [13]. Healthy habitats support robust fish populations, which are essential for providing the nutritional resources needed to prevent stunting in vulnerable communities.

Fish stock levels, a primary goal of EBM related to both ecological health and resource sustainability, have seen positive trends under EBM management. In the North Sea, for example, EBM has been instrumental in rebuilding fish stocks such as cod and haddock through the implementation of science-based catch limits and bycatch reduction measures [14]. These sustainable fish stocks ensure a reliable supply of fish, which is a critical source of nutrition for many coastal communities, helping to reduce the prevalence of stunting. Similarly, in New Zealand, sustainable fisheries management practices under EBM have successfully maintained fish stock levels, supporting both commercial and recreational fisheries [15], which are vital for local food security and nutrition.

Table 1. Outcomes of EBM Strategies in Terms of Ecological Health, Resource Sustainability, and Their Impact on Reducing Stunting

Category	EBM Strategy	Outcome	Region	Citation
Species Diversity	Marine conservation practices, no-take zones	Increased species diversity, particularly among marine mammals and seabirds, enhancing food availability	California Current Large Marine Ecosystem	Halpern et al. (2012)
	Stricter fishing regulations, habitat protection	High levels of coral and fish diversity, contributing to food security	Great Barrier Reef, Australia	GBRMPA (2019)
Habitat Condition	Water quality improvement, habitat restoration	Restoration of oyster reefs and submerged aquatic vegetation, improved habitat conditions supporting robust fish populations	Chesapeake Bay, USA	Kemp et al. (2009)
Fish Stock Levels	Science-based catch limits, bycatch reduction	Rebuilding of fish stocks such as cod and haddock, ensuring sustainable food sources	North Sea	Fernandes & Cook (2013)
	Sustainable fisheries management practices	Maintenance of sustainable fish stock levels, vital for nutrition and stunting reduction	New Zealand	Reid & Rout (2020)

Source: Proceed by author, 2024

The outcomes summarized in the table above highlight the significant positive impacts of EBM strategies on ecological health, resource sustainability, and their indirect role in reducing stunting. In the California Current Large Marine Ecosystem, EBM practices have led to a notable increase in species diversity, particularly among marine mammals and seabirds, which in turn supports greater availability of diverse fish species crucial for community nutrition. Similarly, Australia's Great Barrier Reef has benefited from EBM measures that sustain high levels of coral and fish diversity, thereby contributing to food security and the reduction of stunting in dependent populations.

In terms of habitat condition, the Chesapeake Bay has seen marked improvements due to EBM strategies, including the restoration of vital habitats like oyster reefs and submerged aquatic vegetation, which are essential for maintaining healthy fish populations. These improvements directly contribute to the availability of nutritious food sources, which are critical for preventing stunting, particularly in vulnerable communities.

Additionally, EBM has demonstrated its effectiveness in managing fish stocks, as shown by the successful rebuilding of cod and haddock populations in the North Sea and the maintenance of sustainable fish stock levels in New Zealand. These sustainable practices ensure a continuous supply of fish, which is vital for the nutritional needs of coastal populations, thereby playing a crucial role in reducing the prevalence of stunting.

These examples underscore the effectiveness of well-implemented EBM strategies in enhancing ecological health, sustaining marine resources, and contributing to the reduction of stunting in vulnerable communities. However, the ongoing challenge remains to continuously adapt and refine these strategies to address the complex and dynamic nature of marine ecosystems effectively, ensuring long-term sustainability and resilience, and ultimately, improving human health outcomes.

3.2. Primary Challenges and Barriers in Implementing Ecosystem-Based Management (EBM) in Marine Environments.

Ecosystem-Based Management (EBM) represents an innovative approach to marine resource management, focusing on the sustainability of entire ecosystems rather than just individual species. However, the implementation of EBM faces several challenges and barriers that limit its effectiveness, which can be broadly categorized into scientific, regulatory, economic, and social dimensions. Addressing these obstacles is crucial for enhancing EBM practices and ensuring their success.

3.3. Scientific Challenges.

The complexity and variability of marine ecosystems present significant scientific challenges in implementing EBM. Marine ecosystems are dynamic and influenced by numerous biotic and abiotic factors, making it difficult to predict how they will respond to management actions [7]. Additionally, there is often a lack of comprehensive and high-quality data on marine ecosystems, which hinders informed decision-making. Data gaps in areas such as species distributions, habitat conditions, and ecological interactions can undermine the effectiveness of EBM strategies [5]. Moreover, integrating diverse knowledge types, including traditional ecological knowledge and scientific data, poses another challenge. Effective EBM requires synthesizing these different data sources, but differences in data quality, scale, and methodology can complicate this process [6].

3.4. Regulatory Challenges.

Regulatory barriers are significant obstacles to the effective implementation of EBM. Marine ecosystems often span multiple jurisdictions, and the fragmented nature of marine governance can lead to conflicting policies and regulations. This fragmentation complicates coordinated management efforts and can result in inconsistent enforcement of conservation measures [16]. Additionally, existing legal frameworks are frequently designed for sector-specific management rather than holistic ecosystem approaches. These frameworks may lack the necessary flexibility to support EBM effectively. For instance, fisheries management often focuses on single-species quotas without considering broader ecosystem impacts, leading to unintended consequences such as bycatch and habitat degradation [4].

3.5. Economic Challenges.

Economic considerations also pose substantial barriers to the implementation of EBM. Transitioning to EBM can require significant financial resources for research, monitoring, and enforcement, and limited funding can constrain the ability to conduct essential scientific studies, gather data, and implement adaptive management strategies [7]. Additionally, there are economic trade-offs associated with EBM that can generate resistance from stakeholders. For example, restrictions on fishing activities to protect ecosystems may conflict with the short-term economic interests of local communities and industries. Balancing economic development with ecological sustainability is a persistent challenge, and achieving stakeholder buy-in often requires demonstrating the long-term benefits of EBM [17].

3.6. Social Challenges.

Social challenges are also critical in the implementation of EBM. Effective EBM requires the active involvement of a diverse range of stakeholders, including local communities, industries, non-governmental organizations, and government agencies. Engaging these stakeholders in the decision-making process can be difficult due to differing interests, values, and levels of influence [6]). Moreover, there may be a lack of awareness or understanding of EBM among stakeholders, leading to resistance or non-compliance. Building trust and fostering collaboration among stakeholders is essential for the successful implementation of EBM, but this can be time-consuming and requires ongoing communication and education efforts [18].

In summary, while EBM holds great promise for sustainable marine resource management, its implementation is hindered by a range of scientific, regulatory, economic, and social challenges. Overcoming these barriers is essential for improving EBM practices and achieving long-term ecological and resource sustainability in marine environments. Implementing Ecosystem-Based Management (EBM) in marine environments presents numerous challenges across various dimensions, including scientific, regulatory, economic, and social aspects. These barriers can significantly hinder the effectiveness of EBM strategies, making it crucial to identify and address them to improve the management of marine resources. The following table provides a detailed summary of the primary challenges and barriers encountered in the implementation of EBM.

Table 2. Primary Challenges and Barriers in Implementing Ecosystem-Based Management (EBM) in Marine Environments

Category	Challenges and Barriers	Details	Citation
Scientific	Complexity and variability of marine ecosystems	Difficulty in predicting ecosystem responses due to dynamic and multifaceted influences.	Arkema et al. (2006)
	Lack of comprehensive and high-quality data	Data gaps in species distributions, habitat conditions, and ecological interactions hinder informed decisions.	Levin et al. (2009)
	Integration of diverse knowledge types	Challenges in synthesizing traditional ecological knowledge and scientific data due to differences in quality, scale, and methodology.	Fletcher et al. (2014)
Regulatory	Fragmented governance across multiple jurisdictions	Conflicting policies and regulations complicate coordinated management efforts.	Crowder et al. (2006)
	Inflexibility of existing legal frameworks	Sector-specific management frameworks are not well-suited for holistic ecosystem approaches, leading to unintended consequences.	Long et al. (2015)
Economic	High financial costs of EBM implementation	Significant resources required for research, monitoring, and enforcement, with limited funding available.	Arkema et al. (2006)
	Economic trade-offs and stakeholder resistance	Restrictions to protect ecosystems may conflict with short-term economic interests, complicating stakeholder buy-in.	Foley et al. (2010)
Social	Engaging diverse stakeholders with varying interests and influence levels	Difficulty in involving local communities, industries, NGOs, and government agencies in decision-making.	Fletcher et al. (2014)
	Lack of awareness or understanding of EBM among stakeholders	Resistance or non-compliance due to insufficient knowledge of EBM's benefits and processes.	Pomeroy & Douvere (2008)
	Building trust and fostering collaboration	Time-consuming process requiring ongoing communication and education efforts.	Pomeroy & Douvere (2008)

Source: proceed by author, 2024

The table above illustrates the multifaceted challenges that can impede the successful implementation of EBM in marine environments. These challenges span scientific, regulatory, economic, and social dimensions, each presenting unique obstacles that must be addressed to enhance the effectiveness of EBM strategies. For instance, the complexity of marine ecosystems and the variability of available data can complicate scientific assessments and decision-making processes. Meanwhile, fragmented governance and inflexible legal frameworks can hinder coordinated management efforts, and economic constraints can limit the resources available for necessary research and enforcement.

Moreover, the social dimension of EBM implementation highlights the importance of stakeholder engagement and education. Ensuring that all relevant parties are involved and informed is essential for gaining the support needed to implement EBM effectively. Addressing these challenges requires a concerted effort across multiple sectors and disciplines, emphasizing the need for collaboration, innovation, and sustained investment in both the scientific and social aspects of marine resource management. Overcoming these barriers is essential for realizing the full potential of EBM in promoting the sustainability of marine ecosystems. By understanding and addressing the challenges outlined in the table, policymakers, resource managers, and stakeholders can work together to improve EBM practices and achieve long-term ecological and economic benefits in marine environments.

3.7. Strategies to Enhance the Effectiveness of Ecosystem-Based Management (EBM) for Long-Term Sustainability of Marine Resources and Contributing to Stunting Decline.

Ecosystem-Based Management (EBM) is a comprehensive approach designed to maintain ecosystem health and ensure the sustainable use of marine resources. Beyond environmental benefits, EBM also has the potential to contribute to the reduction of stunting in vulnerable communities by ensuring a reliable and nutritious food supply from marine resources. Despite its promise, EBM faces challenges that limit its effectiveness. To ensure the long-term sustainability of marine resources and contribute to stunting decline, it is critical to develop and implement practical strategies that address these challenges [19]. The following recommendations synthesize current EBM practices and propose ways to enhance their effectiveness, with a focus on their impact on reducing stunting.

3.8. Enhancing Scientific Understanding and Data Integration.

Improving the scientific understanding and integration of data is vital for the effectiveness of EBM. High-quality, comprehensive data on marine ecosystems are essential for making informed decisions that directly impact food security and nutrition. This strategy involves increasing investments in marine research, monitoring programs, and technologies to gather accurate and current information on species distributions, habitat conditions, and ecological interactions [5]. By ensuring that fish stocks and other marine resources are managed sustainably, these efforts help secure a consistent and nutritious food supply, which is crucial for combating malnutrition and stunting in coastal communities. Integrating diverse data sources, including traditional ecological knowledge and scientific data, provides a holistic understanding of marine ecosystems, ultimately supporting better nutritional outcomes [6]. Advanced technologies like remote sensing, satellite monitoring, and environmental DNA (eDNA) analysis can significantly enhance data collection, providing real-time data and improving the precision of monitoring efforts, which is crucial for early detection of environmental changes that could threaten food supplies and contribute to stunting [20]

3.9. Strengthening Regulatory Frameworks and Governance.

Effective governance is crucial for successful EBM implementation, especially in regions where marine resources are a primary food source. Strengthening regulatory frameworks to support holistic and adaptive management is essential. Harmonizing policies across different jurisdictions can address the fragmented nature of marine governance [16]. This harmonization is particularly important in ensuring that marine resources are managed in ways that support the nutritional needs of vulnerable populations, thereby helping to reduce stunting. Developing integrated management plans that align with EBM principles can ensure consistent and coordinated efforts to protect marine ecosystems and sustain the food resources they provide. Promoting adaptive management within these frameworks allows for flexibility in responding to environmental changes and new scientific information, ensuring that the management of marine resources remains aligned with the goal of reducing stunting [21].

3.10. Enhancing Stakeholder Engagement and Collaboration.

Engaging stakeholders is key to the success of EBM, particularly in its role in supporting food security and reducing stunting. This involves including local communities, industries, non-governmental organizations, and government agencies in the decision-making process. Building trust and fostering collaboration among stakeholders can lead to more effective and equitable management outcomes [18]. When stakeholders, especially those from food-insecure communities, are actively involved in EBM, they are more likely to support and adhere to strategies that protect and enhance the availability of marine resources essential for nutrition. Empowering local communities by incorporating their knowledge and values into EBM strategies can increase buy-in and compliance, ensuring that the management of marine resources aligns with the nutritional needs of these communities and contributes to the reduction of stunting [17].

3.11. Securing Sustainable Financing.

Adequate and sustainable financing is a sigificant challenge for EBM implementation, particularly when aiming to achieve broader social goals like reducing stunting. Developing innovative financing mechanisms can provide the necessary resources for research, monitoring, and enforcement activities that ensure the sustainable management of marine resources. Examples of such mechanisms include public-private partnerships, conservation trust funds, and payment for ecosystem services (PES) schemes [7]. These financial resources are crucial for supporting programs that directly link EBM to food security and nutrition, thereby helping to reduce stunting. Leveraging international funding opportunities, such as grants from environmental organizations and development agencies, can also support EBM initiatives aimed at improving the availability and quality of marine-derived food resources. Ensuring transparency and accountability in the use of funds can build confidence among donors and stakeholders, facilitating long-term financial support for EBM practices that contribute to stunting reduction.

3.12. Promoting Education and Capacity Building.

Education and capacity building are fundamental to enhancing EBM effectiveness, particularly in its role in reducing stunting. Raising awareness about the importance of marine ecosystems and the benefits of EBM can foster a culture of stewardship and support for conservation efforts, which in turn supports the sustainability of food resources. Integrating EBM concepts into educational curricula and providing training programs for resource managers, policymakers, and stakeholders can enhance their understanding and capabilities [6]. This educational outreach is especially important in communities vulnerable to stunting, where increased awareness can lead to better management of local marine resources. Building capacity within local communities through training and knowledge-sharing initiatives can empower them to actively participate in EBM efforts, ensuring that the management of marine resources is aligned with the goal of reducing stunting. Providing technical assistance and resources for community-led monitoring and conservation projects can strengthen local involvement and ownership, further supporting the availability of nutritious food sources from marine environments. These strategies, if effectively implemented, can significantly enhance the effectiveness of EBM, ensuring the long-term sustainability of marine resources and contributing to broader environmental and social goals, including the reduction of stunting in vulnerable populations. By addressing the scientific, regulatory, economic, and social challenges currently hindering EBM, these recommendations offer a pathway to more resilient and sustainable marine ecosystems, ultimately supporting better health outcomes for communities dependent on these resources.

Implementing effective Ecosystem-Based Management (EBM) strategies is crucial for ensuring the long-term sustainability of marine resources. Beyond environmental preservation, EBM also plays a significant role in addressing human health issues, such as reducing stunting in vulnerable communities by securing a consistent and nutritious food supply from marine resources. To overcome the challenges that currently hinder EBM's effectiveness, it is essential to develop and implement targeted strategies. The following table summarizes these strategies, highlighting their impact on marine resource sustainability and their contribution to stunting decline.

Table 3. Strategies to Enhance the Effectiveness of Ecosystem-Based Management (EBM) for Long-Term Sustainability of Marine Resources and Contributing to Stunting Decline

Strategy	Key Actions	Impact on Marine Resource Sustainability	Contribution to Stunting Decline
Enhancing Scientific Understanding and Data Integration	• Increase investments in marine research and monitoring programs. • Integrate diverse data sources, including traditional ecological knowledge. • Utilize advanced technologies like remote sensing and eDNA analysis.	• Provides accurate and up-to-date information for informed decision-making. • Improves early detection of environmental changes and threats.	Ensures sustainable management of fish stocks and other marine resources, securing a consistent food supply.
Strengthening Regulatory Frameworks and Governance	• Harmonize policies across jurisdictions. • Develop integrated management plans aligned with EBM principles. • Promote adaptive management within regulatory frameworks.	• Ensures coordinated efforts to protect marine ecosystems. • Enhances resilience of EBM strategies to environmental changes.	Supports sustainable management practices that align with nutritional needs, helping to reduce stunting.
Enhancing Stakeholder Engagement and Collaboration	• Include local communities, industries, NGOs, and government agencies in decision-making. • Foster collaboration and build trust among stakeholders. • Empower local communities by incorporating their knowledge and values.	• Leads to more effective and equitable management outcomes. • Increases compliance with EBM strategies.	Ensures that management practices support food security and nutrition, reducing stunting.
Securing Sustainable Financing	• Develop innovative financing mechanisms like public-private partnerships, conservation trust funds, and PES schemes. • Leverage international funding opportunities. • Ensure transparency and accountability in fund use.	• Provides necessary resources for research, monitoring, and enforcement. • Facilitates long-term financial support for EBM.	Supports programs linking EBM to food security, ensuring the availability of nutritious marine-derived food resources.
Promoting Education and Capacity Building	• Raise awareness about marine ecosystems and EBM benefits. • Integrate EBM concepts into educational curricula. • Provide training programs for resource managers, policymakers, and stakeholders. • Build capacity within local communities through training and knowledge-sharing initiatives.	• Fosters a culture of stewardship and support for EBM. • Enhances the understanding and capabilities of stakeholders.	Empowers communities to manage marine resources sustainably, ensuring food security and reducing stunting.

Source: proceed by author, 2024

The strategies outlined in the table provide a comprehensive approach to enhancing the effectiveness of EBM, ensuring that it not only preserves marine ecosystems but also contributes to improving human health by reducing stunting. By focusing on scientific understanding, regulatory frameworks, stakeholder engagement, sustainable financing, and education, these strategies address the multifaceted challenges currently faced by EBM. Implementing these strategies effectively can lead to more resilient marine ecosystems, sustainable resource management, and better nutritional outcomes for communities dependent on marine resources. This holistic approach ensures that the benefits of EBM extend beyond environmental conservation, fostering healthier and more sustainable communities.

4. Conclusion

In conclusion, the effective implementation of Ecosystem-Based Management (EBM) holds significant potential for promoting the long-term sustainability of marine resources while simultaneously addressing critical human health issues, such as reducing stunting in vulnerable communities. The evidence presented highlights the positive outcomes of EBM strategies in improving species diversity, habitat conditions, and fish stock levels, all of which contribute to better food security and nutritional outcomes. These improvements are essential for preventing stunting, particularly in communities that rely heavily on marine resources for their dietary needs.

However, the successful application of EBM is not without challenges. Scientific, regulatory, economic, and social barriers continue to impede the full realization of EBM's benefits. Overcoming these obstacles requires a concerted effort to enhance scientific understanding, strengthen regulatory frameworks, secure sustainable financing, and promote stakeholder engagement and capacity building. By addressing these challenges through targeted strategies, EBM can be more effectively implemented, ensuring the resilience and sustainability of marine ecosystems and, ultimately, contributing to the reduction of stunting in affected populations.

The future of EBM lies in its ability to adapt to the complexities of marine ecosystems and the needs of human communities. As such, continued innovation, collaboration, and investment in EBM practices are essential. By doing so, policymakers, resource managers, and stakeholders can work together to create more sustainable and equitable management of marine resources, supporting both ecological health and human well-being.

References

[1] Ward D, Melbourne-Thomas J, Pecl GT, Evans K, Green M, McCormack PC, Novaglio C, Trebilco R, Bax N, Brasier MJ, Cavan EL, Edgar G, Hunt HL, Jansen J, Jones R, Lea MA, Makomere R, Mull C, Semmens JM, Shaw J, Tinch D, van Steveninck TJ, Layton C. Rev Fish Biol Fish. 2022;32(1):65-100. doi: 10.1007/s11160-022-09700-3. Epub 2022 Mar 7. PMID: 35280238; PMCID: PMC8900478.

[2] FAO. (2021). Food and Agriculture Organization of the United Nations.

[3] Radinger J, Matern S, Klefoth T, Wolter C, Feldhege F, Monk CT, Arlinghaus R. Science. 2023 Mar 3;379(6635):946-951. doi: 10.1126/science.adf0895. Epub 2023 Mar 2. PMID: 36862780.

[4] Long, R. D., Charles, A., & Stephenson, R. L. (2015). *Marine Policy*, *57*, 53–60.

[5] Levin, P. S., Fogarty, M. J., Murawski, S. A., & Fluharty, D. (2009). *PLoS Biology*, **7** (1), e1000014.

[6] Fletcher, S., Saunders, J., & Herbert, R. (2014). *Journal of Coastal Conservation*, **18,** 145–157.

[7] Arkema, K. K., Abramson, S. C., & Dewsbury, B. M. (2006). *Frontiers in Ecology and the Environment*, **4** (10), 525–532.

[8] Halpern, B. S., Longo, C., Hardy, D., McLeod, K. L., Samhouri, J. F., Katona, S. K., Kleisner, K., Lester, S. E., O'Leary, J., & Ranelletti, M. (2012). *Nature*, **488** (7413), 615–620.

[9] GBRMPA. (2019). *Great barrier reef outlook report 2019.* Great Barrier Reef Marine Park Authority

[10] Creswell, J. W., & Poth, C. N. (2016). Sage publications.

[11] Creswell, J. W. (2018). (4th ed.). SAGE Publications.

[12] Braun, V., & Clarke, V. (2006). *Qualitative Research in Psychology*, **3** (2), 77–101.

[13] Kemp, W. M., Testa, J. M., Conley, D. J., Gilbert, D., & Hagy, J. D. (2009). *Biogeosciences Discussions*, **6** (4).

[14] Fernandes, P. G., & Cook, R. M. (2013). *Current Biology*, **23** (15), 1432–1437.

[15] Reid, J., & Rout, M. (2020). *Marine Policy*, **117,** 103889.

[16] Crowder, L. B., Osherenko, G., Young, O. R., Airamé, S., Norse, E. A., Baron, N., Day, J. C., Douvere, F., Ehler, C. N., & Halpern, B. S. (2006). In *Science.* **313,** Issue 5787, pp. 617–618). American Association for the Advancement of Science.

[17] Foley, M. M., Halpern, B. S., Micheli, F., Armsby, M. H., Caldwell, M. R., Crain, C. M., Prahler, E., Rohr, N., Sivas, D., & Beck, M. W. (2010). *Marine Policy*, **34** (5), 955–966.

[18] Pomeroy, R., & Douvere, F. (2008). *Marine Policy*, **32** (5), 816–822.

IOP Conf. Series: Earth and Environmental Science **1473** (2025) 012036 doi:10.1088/1755-1315/1473/1/012036

[19] Sarjito, A., & Azhar, A. W. (2023). *Policy and Maritime Review*, 1–13.
[20] Maxwell, S. M., Hazen, E. L., Lewison, R. L., Dunn, D. C., Bailey, H., Bograd, S. J., Briscoe, D. K., Fossette, S., Hobday, A. J., & Bennett, M. (2015). *Marine Policy*, **58,** 42–50.
[21] Holling, C. S., & Walters, C. (1978). Adaptive environmental assessment and management.

IOP Conf. Series: Earth and Environmental Science **1473** (2025) 012037 doi:10.1088/1755-1315/1473/1/012037

Risk Reduction Strategy of Ship Scrapping Process Based on Fuzzy Approach

Silvianita[1], G L Putra[2], W L Dhanistha[3] and Ahmad Mahamad Al-Yacouby[4]

[1]Department of Ocean Engineering, Institut Teknologi Sepuluh Nopember, Surabaya, Indonesia
[2]Department of Civil Engineering, Universiti Teknologi PETRONAS, Malaysia
***Corresponding author: silvianita@its.ac.id**

Abstract. Ship scrapping involves dismantling ships to recover parts and materials for recycling and reuse. In Bangkalan, Madura, this process is conducted using traditional methods with minimal technological assistance. The scrapping process presents several hazardous risks, which can result in worker accidents. To assess these risks, this study uses HAZOP and FFMEA methods, identifying a total of 44 risks associated with ship scrapping. The analysis determines the most critical risks, with fire being the highest, followed by falls from heights and being struck by materials. A root cause analysis using the FTA method reveals that the main causes of accidents in this process are sparks from cutting flames, oil spills, and open disposal areas.

1. Backgrounds

The shipyard industry is not only focused on the construction or repair of ships but also has the opportunity to develop its business as a ship scrapping yard. In Indonesia, there are approximately 22,000 Indonesia flagged ships, many of which are old ships that are either still operating or inactive due to economic reasons. These old ships have the potential to be scrapped to maintain the balance of the number of ships. Data from the NGO Shipbreaking Platform in 2021 shows that only 23 ships were dismantled at official locations. This raises questions about where the remaining inactive ships are being dismantled, especially since the ship scrapping process generates high pollution impacts, including hazardous and toxic waste. Moreover, working conditions in the ship scrapping sector are highly risky and dangerous for workers.

One of the ship scrapping locations is in Bangkalan, Madura, where the process is conducted traditionally using the beaching method. This method involves cutting the ship into sheets of iron plates but has weaknesses, particularly during tidal changes and sea waves. The stages in the ship scrapping process include ship identification, stripping, cutting, transporting materials into trucks, and storing equipment. These activities require an occupational safety and health (OSH) system, as well as adequate employee skills. However, since the process is conducted traditionally, occupational safety and health factors are often neglected, resulting in numerous threats and dangers to workers.

Due to the significant level of risk, it is necessary to conduct a risk analysis involving hazard identification, risk analysis, and actions to eliminate or reduce risks. The risk analysis methods that can be used include HAZOP, FFMEA, and FTA. HAZOP is a qualitative tool for identifying potential deviations from normal operations and ensuring protection against accidents [1]. FFMEA combines conventional FMEA with fuzzy logic to handle information uncertainty in failure mode assessments [2]. After obtaining the results of failure modes and FRPN, the risk analysis continues with FTA to find the root causes of a risk. This study will analyze occupational safety and health risks at the shipyard in Bangkalan, Madura using HAZOP, FFMEA, and FTA methods. The aim of this study is to identify is to identify critical risks that cause failures in ship scrapping process, identify the primary cause of the risk of failure in the ship scrapping process, and to provide recommendations for mitigation actions that can be carried out in the ship scrapping process to minimize or avoid impacts that may occur. The research gap is shown in Table 1.

Table 1. Research Gap

Author, Year	Knowledge Gap
Garmer, 2015	Risk assessment was conducted by dividing ship scrapping into three zones (inside ship; inter-tidal; primary/secondary). Each zone's particular risk is approached qualitatively by collaborating What-if Analysis and Fault Tree Analysis. However, a more reliable results could be retrieved by directly observing and surveying the company and worker's practices [3].
Mentes, 2023	The assessment conducted was based upon two conditions which are on-field and on-board. The study was done by integrating FMEA with other frameworks (FVIKOR, OWGA and Resilience Engineering RE principles). Potential cultural factors implied on the study findings are yet to be explored [4].
Kurt, 2015	The paper introduces a new risk assessment methodology specifically designed for ship recycling operations, known as "The Three Step Method". This method is described as supportive, simple to use, and involving end-user participation. The study's findings were validated within the scope of the research population. However, there's a gap in verifying and validating the proposed framework in broader or more diverse contexts, including different types of ship recycling facilities or different economic conditions [5].
Ozturkoglu, 2019	Risk management was done by using a Triple Bottom Line (TBL) approach and Fuzzy DEMATEL method. This paper managed to address risks across diverse factors including economis, social, etc. However, the geographical scope of this research needs to be specified for higher assessment accuracy [6].
Sunaryo, 2021	This study benchmarked Indonesia's ship-breaking practices' disparity to international regulations. It's shown that reluctance to comply with regulations is considered to be costly hence the high-risk practices within national level [7].
Du, 2018	The research shows companies' efficiency in disposing hazardous materials is inherent to nation's regulations [8].
Zhou, 2021	This study combines surveys with numerical validations to assess factors affecting ship scrapping's workers safety. The dismantling operation management is found to be the main factor in ensuring worker safety [9].

2. Ship Scrapping

Ship scrapping is the process of dismantling part or all of a ship to recover its components and materials for recycling. The methods used in ship scrapping include:

- In-water method: conducted on water using buoys and requiring an adequate location.
- On-land method: conducted on land with facilities such as airbags, slipways, and dry docks, and requiring a large area.
- Combined method: a combination of the in-water and on-land methods.

The stages in the ship scrapping process include:

- Understand and identify: understanding and identifying the condition of the ship.
- Removing: removal of liquids and equipment before cutting the ship's hull.
- Metal cutting: the process of cutting the ship.
- Separation of scrap: separating recyclable materials from hazardous materials.

According to the IMO (International Maritime Organization), waste that requires careful handling to prevent environmental contamination and harm to humans includes fuel oil, lubricating oil, and equipment or materials containing asbestos and PCBs (polychlorinated biphenyls).

3. Risk

3.1.1. *Definition of risk.* Risk is the result of the interaction between the likelihood of an event occurring and its impact, which can involve more than one event, and the outcome can be either positive or negative [10]. Risk increases if the likelihood of the event increases or if the resulting loss becomes greater. Typically, risk is measured with quantifiable parameters, such as the probability of failure, injury, and material loss.

3.1.2. *Risk management.* Risk management is the response to identified or perceived risks [11]. It refers to the culture, processes, and structures that aim to manage the opportunities and impacts that may occur. Risk management is divided into three main parts, namely risk analysis which aims to identify threats and hazards, then risk evaluation aims to find alternative ways, explain risks, and find solutions, then risk control and reduction is a series of activities to reduce, reduce, and control risks.

3.1.3. *Risk identification.* The risk identification process is a systematic and continuous step to recognize the possibility of emerging risks that may impact the objectives or processes of a construction. This process is the first step in risk management. Methods for conducting risk identification include brainstorming, interviews, and distributing questionnaires.

3.1.4. *Risk Assessment, Analysis, and Evaluation.* Risk assessment is the process of evaluating the risk caused by a hazard, considering the effectiveness of controls to determine the appropriateness of the risk [12]. The steps in risk assessment are [13]:

- Identify hazards and threats
- Determine affected parties
- Evaluate risks and determine preventive measures
- Record findings and implement preventive measures
- Conduct periodic risk assessments and make changes if necessary

Risk analysis involves calculating the level of risk based on probability and consequence. Risk analysis methods include:

- Qualitative: Using expert judgment and criteria such as low, medium, and high risk.
- Semi-quantitative: Uses a rated scale to describe the probability/consequence of the risk.
- Quantitative: Requires detailed information on design, operation, component reliability, and environmental/health impacts, using logical and physical models.

Risk evaluation compares the level of risk analyzed with the standards used, resulting in:

- A description of the level of risk significance
- Prioritization of risks that need to be addressed
- A description of potential losses in terms of costs and other factors.

3.1.5. *Risk Control.* Risk assessment is followed by risk control to reduce or minimize the consequences to within tolerance limits. Risk control methods include:

- A description of the level of risk significance
- Prioritization of risks that need to be addressed
- A description of potential losses in terms of costs and other factors.

Risk control takes into account the hazard control hierarchy:

- Elimination: Removing the hazard from its source.
- Substitution: Replacing a hazardous tool, material, or procedure with a safer one.
- Engineering controls: Modification of equipment or workspace to reduce risk.
- Administrative controls: Administrative measures such as safety signs, job rotation, and safer work procedures.
- Personal protective equipment (PPE): Using personal protective equipment such as helmets, gloves, respirators, fall protection, and protective shoes.

3.2. Hazard and Operability (HAZOP)

Hazard and Operability (HAZOP) analysis is a commonly used process hazard analysis method [14], that aims to identify potential deviations from normal operations and implement safeguards to prevent accidents [1]. The HAZOP study process, including the formation of the study team, study implementation and discussion, selection of Study NODES, use of GUIDE WORDS, deviation identification: combination of Guide Word and Process Parameter, evaluation of the cause of deviation, evaluation of Impact and Safeguard, Deviation Category: Determining the type of deviation as Hazard (H), Operability Problems (O), and recommendations. HAZOP studies emphasize hazard identification, not problem solving, and recommendations are qualitative based on the identification of hazards and operability problems.

3.3. Fuzzy Failure Mode and Effect Analysis (FFMEA)

Failure Mode and Effects Analysis (FMEA) is a method used to identify and analyze potential failures, with the aim of planning better processes and preventing losses. Then, FMEA is A preventive approach involves appointing a team to analyze the product structure and manufacturing process to identify potential error, defect, and failure points. They then develop actions to eliminate these issues [15]. FMEA consists of ten steps, namely reviewing the process, identifying potential failure modes, listing potential consequences of each failure mode, ranking severity, occurrence, detection, calculating Risk Priority Number (RPN), prioritizing failure modes based on RPN, taking action to reduce or eliminate failures [16]. The following is the S (severity), O (occurrence), and D (detection) assessment as seen in Table 2, 3, and 4.

Table 2. Severity Criteria [17]

Rating	Criteria
10	Possible severe injury or regulatory action. Production outage > 48 hours, critical process utility outage > 24 hours
9	Possible major injury or regulatory action. Production outage < 48 hours, critical process utility outage 12-24 hours
8	Possible minor injury or regulatory investigation. Production outage < 24 hours, critical process utility outage 4-12 hours
7	No injury to people. Production outage < 8 hours, critical process utility outage < 4 hours
6	No injury to people. One or more production tools affected, or possible product scrap.
5	No injury to people. No production impact. Process utility out of spec, no product scrap.
4	Minor effects with no injury to people. No production impact, equipment failures to be corrected immediately.
3	Slight effects with no injury to people. No production impact, equipment failures to be corrected as soon as possible.
2	People are unlikely to notice the failure. No production impact, process utility in spec, and system/equipment/operations failures can be corrected at the next scheduled maintenance.
1	No effect on people, no production impact, process utility in spec, and system/equipment/operations failures can be corrected after an extended period.

Table 3. Occurence Criteria

Rating	Criteria
10	Failures are certain to occur in the near future, with significant company or industry history. $\leq 1x10^{-1}$ (3 events per day).
9	Failures are certain to occur soon, with some company or industry history. $\leq 1x10^{-2}$ (2 events per week).
8	A high number of failures is likely. $\leq 5x10^{-3}$ (1 event per month).
7	Frequent failures are likely. $\leq 1x10^{-3}$ (1 event every 1.5 months).
6	A moderate number of failures are expected. $\leq 5x10^{-4}$ (1 event every 3 months).
5	Failures occur occasionally. $\leq 1x10^{-4}$ (1 event per year).

4	Few failures are likely. ≤5x10^{-5} (1 event in 2 years).
3	Very few failures are likely. ≤1x10^{-5} (1 event in 10 years).
2	There is a remote chance of failure. ≤5x10^{-6} (1 event in 20 years).
1	Failure is improbable in similar processes or products, with no industry history of failure. ≤1x10^{-6} (1 event in 100 years).

Table 4. Detection Criteria

Rating	Criteria
10	Design control cannot detect the potential cause/mechanism and subsequent failure mode.
9	There is a very remote chance that the design control will detect the potential cause/mechanism and subsequent failure mode.
8	There is a remote chance that the design control will detect the potential cause/mechanism and subsequent failure mode.
7	There is a very low chance that the design control will detect the potential cause/mechanism and subsequent failure mode.
6	There is a low chance that the design control will detect the potential cause/mechanism and subsequent failure mode.
5	There is a moderate chance that the design control will detect the potential cause/mechanism and subsequent failure mode.
4	There is a moderately high chance that the design control will detect the potential cause/mechanism and subsequent failure mode.
3	There is a high chance that the design control will detect the potential cause/mechanism and subsequent failure mode.
2	There is a very high chance that the design control will detect the potential cause/mechanism and subsequent failure mode.
1	The design control will detect the potential cause/mechanism and subsequent failure mode.

RPN is calculated by calculating the severity, occurrence, and detection values. Failures with high RPN should be corrected until they reach an acceptable level [16]. To overcome the limitations of conventional FMEA, fuzzy logic is used which maps the input space to the output space more flexibly and is tolerant of imprecise data. The reasons for using fuzzy logic are numerous. It is intuitive and easy to grasp, highly adaptable, and tolerant of imprecise data. Fuzzy logic can model intricate non-linear functions and apply expert knowledge without requiring a training process. Additionally, it can work alongside conventional control methods and is grounded in natural language, making it very user-friendly.

Fuzzy systems use IF-THEN rules to build knowledge-based models from experts, making FMEA judgments more reasonable. The fuzzy inference process in FMEA includes:

- Fuzzification: Converting three parameters (severity, occurrence, detection) into fuzzy membership function inputs.
- Rule Evaluation: Using IF-THEN rules from respondents to map inputs to fuzzy inferences.
- Defuzzification: Converts fuzzy inferences into numerical risk representations.

3.4. Fault Tree Analysis

In this research, the Fault Tree Analysis (FTA) method is used to determine cause-and-effect in the analysis. FTA is a top-down technique that identifies events that cause failure, starting from the assumption of failure (top event) to the basic cause (root cause). The five stages of FTA are:

- Defining the problem and system boundary conditions.
- Drawing a graphical fault tree diagram.
- Identify the minimum cut set using fault tree analysis.
- Conduct a qualitative analysis.
- Perform quantitative analysis.

The benefits of using FTA are to determine the factors that cause failure, determine the stages of events that cause failure, and analyze the source of risk before failure occurs.

4. Analysis and Discussion

4.1. Data Collection

The object of this research is in a traditional shipyard in Bangkalan, Madura. The data collected from the object of research are:

- The location and division of the scrapping area.
- Supporting equipment, division of labor, and the flow of the ship scrapping process.
- Questionnaires and interviews analyzing the risk of work accidents in the ship scrapping process.

4.2. HAZOP and Fuzzy-FMEA

After risk identification was carried out using the literacy method, interviews and consultations with workers in the field, a questionnaire was distributed to obtain S, O, and D values.

4.2.1. *Rule evaluation.* Rule evaluation is the organization of fuzzy inputs by three variables: severity, occurrence, and detection. By defining the input membership function. Then proceed with setting the membership function. Membership function is a curve that shows the linguistic variables of input or output data into their membership values. The membership function is used for input or output variables ranging from 1 to 10. For input variables, it maps to five linguistic categories: very low (VL), low (L), medium (M), high (H), and very high (VH). For output variables, it maps to ten linguistic categories: none (N), very low (VL), low (L), high low (HL), low medium (LM), medium (M), high medium (HM), low high (LH), high (H), and very high (VH). Then, next is the use of IF - THEN fuzzy rules that have been created and discussed with experts, which means the variation of the relationship between inputs (severity, occurrence, detection).

4.2.2. *Fuzzification.* Then after the rule evaluation can be continued with the fuzzification step. Fuzzification itself is the process of converting inputs that have numeric variables (crips) into linguistic variables using membership function rules. Then further process the fuzzy input to the fuzzy output by applying each fuzzy IF - THEN rule that has been determined in the fuzzy rule base determination process. Furthermore, because the system consists of several rules, inference is obtained from a collection of correlations between rules.

4.2.3. Deffuzzification. After defining the membership functions of inputs, outputs, and combining rules, the deffuzification process will be carried out, namely converting linguistic variables back into numerical variables, so that the highest FRPN results are fire, falling from a height, then falling on the material as can be seen in Table 5. Using any combination of the major risk parameters, the fuzzy logic inference yields the risk level findings. Three-dimensional (3D) risk profiles are depicted by these combinations as a surface plot that can be seen in Figure 1 and 2.

Table 5. Result of Fuzzy Risk Priority Number

Failure Effect	S	O	D	Fuzzy Round RPN	Fuzzy Round Rank
Shipwreck	7	4	2	5.999	26
Slipped	7	9	2	6.148	23
Fuel/oil splatter	6	6	3	5.590	30
Falling from height	9	3	7	8.069	2
Struck	10	5	5	7,999	3
Cutting torch damage	4	7	2	3.852	41
Regulator failure	4	5	4	4.787	35
Fire	9	9	5	8.071	1
Sliped	8	8	2	6.618	17
Equipment failure	3	7	3	3.613	43

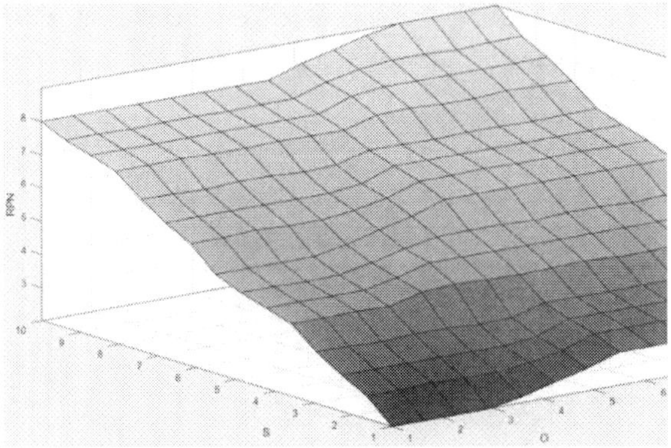

Figure 1. Combination of *Severity* and *Occurrence*

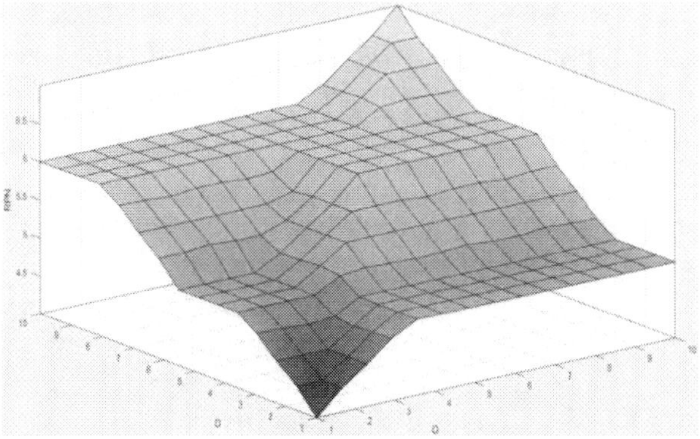

Figure 2. Combination of *Detection* and *Occurrence*

4.3. *Fault Tree Analysis*

In this study, after knowing the three dominant risks with the highest Fuzzy Risk Priority Number (FRPN), an analysis is carried out using the FTA method. Fault Tree Analysis (FTA) is an effective technique for understanding a system can fail or, a risk can occur along with its causal factors within the scope of risk management [18][19]. Risk causes are identified, logically organized, and depicted in the form of a tree diagram that shows the connectivity between causal factors and risk events. Then, the risk will be the top event in this analysis. The three dominant/critical risks are fire, falling from height, and falling material. After the three dominant/critical risks are used as top events, identification will then be conducted to identify the root cause of the event.

The risk of fire is caused by two main factors, namely due to residual flammable materials and due to errors in work methods. Then, the risk of falling from a height has factors or causes that come from human error and workers working in extreme working conditions. And, the risk of being hit by material is caused by several factors such as human error, hazardous working conditions, and due to equipment damage factors. Then, the existing factors are analyzed to obtain the basic event of each risk, which will create a relationship system that illustrates the basic event state and the relationship with the top event expressed in logic gates, so that later from this FTA diagram a minimum cut set of each event can be compiled. Figure 3 shows the FTA diagram of fire in a work accident in the ship scrapping process.

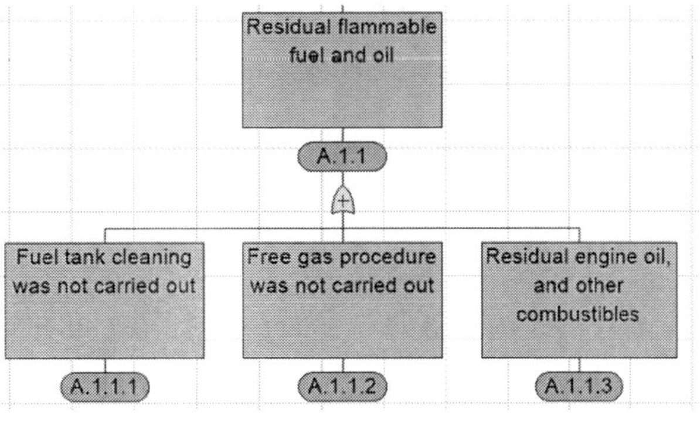

Figure 3. FTA Diagram of Fire

Then, after obtaining the identification results of all causes of dominant/critical risk events, it will be continued with the calculation of determining the probability of each basic event by distributing questionnaires to respondents, then recapitulating the results collected from distributing questionnaires, then calculating the minimum cut set carried out in each branch of the FTA diagram. The purpose of distributing this questionnaire is to determine the frequency of occurrence of a basic event. In distributing this questionnaire, respondents will fill in the frequency of occurrence in accordance with the frequency of occurrence index parameters based on Table 6.

Table 6. Frequency Index

Frequency Index	Probability		
	Quantitative		Qualitative
≤20%	0,00001	10^{-5}	Hardly ever happens
>20% - 40%	0,0001	10^{-4}	Rarely occurs
>40% - 60%	0,001	10^{-3}	Occurs Occasionally
>60% - 80%	0,01	10^{-2}	Happens often
80% - 100%	0,1	10^{-1}	Almost always happens

Then, after distributing the questionnaires, the results of the questionnaires that have been filled in by the respondents are obtained and the probability calculation of the questionnaire results is carried out by averaging the probability of each event from all respondents and then quantitative conversion of the frequency index is carried out so as to obtain the probability of each event as seen in Table 7.

Table 7. Basic Event Probability

Code	Basic Event	Probability
A.1.1.3	Residual engine oil, and other combustibles	0.035200
A.2.1.1	LPG gas leak	0.000163
A.2.1.3	Gas regulator not installed properly	0.013420
A.2.1.4	Cutting sparks	0.054100
A.2.2.2	Open dump	0.035200
B.1.1.1	Workers lack focus	0.000262
B.1.2.2	Workers do not use safety shoes	0.024400
B.2.1.1	Tripping over ropes, equipment, etc	0.004510
B.2.1.4	Oil spills, etc	0.053200
B.2.1.5	Extreme weather	0.000037

Then, based on the results of the calculation, it is found that the highest risk of accidents is caused by the dominant risk of fire with a probability value of 0,210613693 with the highest minimum cut set due to spark cutting fire. And based on the results of probability calculations, that the chance of work accidents due to dominant risks in the ship scrapping process is 0,296124767.

4.4. Risk Response Planning

Risk response planning is a process that aims to develop options, select strategies, and agree on actions to manage project risks. This process is important because it addresses both overall project risks and project-specific risks [20]. Mitigation techniques are used to reduce and control adverse impacts that may occur. Based on risk analysis with HAZOP and Fuzzy FMEA methods that produce fuzzy risk priority number (FRPN), three dominant risks were found. These risks should be prioritized in corrective actions to reduce risks. The following are risk control suggestions:

- Fire

  To control the risk of fire, the shipyard should have a firefighting program that includes:
  - Extinguishing equipment: Fire extinguishers, hoses, smoke detectors, automatic sprinklers.
  - Written plan: Inspection, maintenance, equipment replacement, and worker training.
  - Procedures: Hazard control (gas, ignition sources, combustible materials, welding),

emergency evacuation, worker checks, fire reporting, and emergency contacts.
    - Supervision: Training, duties, PPE, and communication during hot work.
    - Additional procedures: Processing, handling, storage, and disposal of hazardous materials during ship scrapping. Site size and division of scrapping area.
- Falls from height are a serious risk in ship dismantling. Workers should wear safety harnesses near open holes and deck edges, and keep lifelines clear of sharp edges. Barriers should be installed around deck edges and openings. The use of PPE and supervision of the workload of personnel at height are also very important.
- The risk of being crushed by materials is caused by worker negligence, non-standard PPE, unsafe working conditions, and poor equipment. Some actions that can be taken are to create barricades at overhead work sites, install safety preventer cables on equipment and structures, separate work areas for personnel and mobile equipment, check, monitor, and supervise overhead objects, hold toolbox meetings for extreme weather planning, install safety nets in the lower work area, perform routine maintenance and inventory of equipment, use PPE such as safety helmets and safety shoes.

5. Conclusions

Based on the analysis and discussion conducted, the following conclusions were drawn from this study:

1. A total of 44 risks were identified in the ship scrapping process, consisting of seven risks in ship identification, ten risks in removing material and equipment, 21 risks in ship cutting, four risks in material transportation, and two risks in equipment storage. Using the FFMEA method, three dominant risks were identified, with the highest risk being fire, with an FRPN value of 8,071 during the ship cutting activity. The second highest risk is falling from heights, with an FRPN value of 8,069 during the removing material and equipment activity. The third highest risk is being struck by material, with an FRPN value of 7,999 during the removing material and equipment activity.
2. The root cause analysis using the FTA method identified 39 root causes of the dominant risks in the ship scrapping process. For fire, the probability is 0,210 with the highest incidence being sparks from cutting flames, with a probability of 0,0541. For falling from heights, the root cause is oil, grease, and other spills with a probability of 0,0532. For being struck by material, the root cause is workers not using PPE, with a probability of 0,0244.
3. Mitigation actions that can be taken to control and minimize workplace accidents in the ship scrapping process include a comprehensive fire suppression program, evacuation plans, fire monitoring policies, procedures for handling, processing, storing, and disposing of hazardous materials. Additionally, the use of safety harnesses, installation of barriers, use of PPE, workload monitoring and management, use of barricades and safety preventer cables, separation of work areas, use of safety nets, and maintenance and inventory of equipment are recommended.

References

[1] L. Kotek and M. Tabas, "HAZOP Study with Qualitative Risk Analysis for Prioritization of Corrective and Preventive Actions," *Procedia Eng.*, vol. 42, pp. 808–815, 2012, doi: 10.1016/j.proeng.2012.07.473.

[2] M. Braglia, M. Frosolini, and R. Montanari, "Fuzzy criticality assessment model for failure modes and effects analysis," *Int. J. Qual. Reliab. Manag.*, vol. 20, no. 4, pp. 503–524, Jun. 2003, doi: 10.1108/02656710310468687.

[3] K. Garmer, H. Sjöström, A. M. Hiremath, A. K. Tilwankar, G. Kinigalakis, and S. R. Asolekar, "Development and validation of three-step risk assessment method for ship recycling sector," *Saf. Sci.*, vol. 76, pp. 175–189, Jul. 2015, doi: 10.1016/j.ssci.2015.02.007.

[4] A. Mentes, "Risk analysis of on-field and on-board activities and resilience investigation of Izmir Aliaga Ship Recycling Facilities," *Ocean Eng.*, vol. 287, no. P2, p. 115891, 2023, doi:

10.1016/j.oceaneng.2023.115891.
[5] R. E. Kurt, "A risk assessment tool for the ship recycling industry," *Disaster Manag. Hum. Heal. Risk IV*, vol. 1, pp. 147–158, 2015, doi: 10.2495/dman150141.
[6] Y. Ozturkoglu, Y. Kazancoglu, and Y. D. Ozkan-Ozen, "A sustainable and preventative risk management model for ship recycling industry," *J. Clean. Prod.*, vol. 238, p. 117907, 2019, doi: 10.1016/j.jclepro.2019.117907.
[7] S. Sunaryo, E. Djatmiko, S. Fariya, R. Kurt, and S. Gunbeyaz, "A gap analysis of ship-recycling practices in Indonesia," *Recycling*, vol. 6, no. 3, 2021, doi: 10.3390/recycling6030048.
[8] Z. Du, S. Zhang, Q. Zhou, K. F. Yuen, and Y. D. Wong, "Hazardous materials analysis and disposal procedures during ship recycling," *Resour. Conserv. Recycl.*, vol. 131, no. August 2017, pp. 158–171, 2018, doi: 10.1016/j.resconrec.2018.01.006.
[9] Q. Zhou, J. Liang, Z. Du, H. Zhu, and Y. Jiao, "A study on factors affecting workers' safety during ship recycling," *Ocean Eng.*, vol. 239, no. May, p. 109910, 2021, doi: 10.1016/j.oceaneng.2021.109910.
[10] J. Shortreed, J. Hicks, and L. Craig, "Basic Framework for Risk Management - Final Report," *Netw. Environ. Risk Assess. Manag.*, p. 74, 2003, [Online]. Available: http://www.sobanebrasil.org/adm/fotos/6bad672342f38b0776512b211433a994.pdf
[11] W. K. Muhlbauer, *Pipeline Risk Management Manual*. Elsevier, 2004. doi: 10.1016/B978-0-7506-7579-6.X5000-6.
[12] OHSAS, "OHSAS 18001:2007 - Occupational health and safety management systems – Requirements - ICS 03.100.01: 13.100," *Occup. Heal. Saf. Assess. Ser.*, p. 34, 2015.
[13] M. Rausand, *Risk Assessment*. Wiley, 2011. doi: 10.1002/9781118281116.
[14] F. Wang, J. Gao, and H. Wang, "A new intelligent assistant system for HAZOP analysis of complex process plant," *J. Loss Prev. Process Ind.*, vol. 25, no. 3, pp. 636–642, May 2012, doi: 10.1016/j.jlp.2012.02.001.
[15] Y. Liu, Z. Kong, and Q. Zhang, "Failure modes and effects analysis (FMEA) for the security of the supply chain system of the gas station in China," *Ecotoxicol. Environ. Saf.*, vol. 164, pp. 325–330, Nov. 2018, doi: 10.1016/j.ecoenv.2018.08.028.
[16] M. A. Bennett, R. McDermott, and M. Beauregard, *The Basics of FMEA*. Productivity Press, 2017. doi: 10.1201/b16656.
[17] S. R. Trammell and B. J. Davis, "Using a modified Hazop/FMEA methodology for assessing system risk," in *Proceedings 2nd International Workshop on Engineering Management for Applied Technology. EMAT 2001*, IEEE Comput. Soc, 2001, pp. 47–53. doi: 10.1109/EMAT.2001.991310.
[18] Silvianita, D. S. Mahandeka, and D. M. Rosyid, "Fault Tree Analysis for Investigation on the Causes of Project Problems," *Procedia Earth Planet. Sci.*, vol. 14, pp. 213–219, 2015, doi: 10.1016/j.proeps.2015.07.104.
[19] Silvianita, A. A. Harahap, and M. F. Khamidi, "Operation failure of risk analysis on floating storage and offloading," in *IOP Conference Series: Earth and Environmental Science*, 2024. doi: 10.1088/1755-1315/1298/1/012031.
[20] K. H. Rose, "A Guide to the Project Management Body of Knowledge (PMBOK® Guide)-Fifth Edition," *Proj. Manag. J.*, vol. 44, no. 3, 2013, doi: 10.1002/pmj.21345.

Risk Management for Safe Work Scaffolding on Ocean Fabrication

Silvianita[1*], M A Abimanyu[1], D M Rosyid[1] and N Shafiq[2]

[1]Department of Ocean Engineering, Institut Teknologi Sepuluh Nopember, Surabaya, Indonesia
[2]Department of Civil Engineering, Universiti Teknologi PETRONAS, Malaysia
***Corresponding author: silvianita@its.ac.id**

Abstract. The activity that has the highest risk of workplace accidents is construction. In Indonesia, the greatest number of work accidents occur in this industry. The fabrication of offshore platforms is one of the construction activities. Scaffolding is a temporary structure that is necessary for all construction activities in order to support workers and materials. The majority of construction industry accidents are caused by working at heights, which includes using scaffolding. Therefore, to reduce the likelihood of these catastrophes, mitigating and preventive measures are required. Risk analysis is one way to accomplish this. According to the analysis's findings, working with scaffolding carries 24 dangers, two of which are considered serious risks. Falling scaffolding materials and worker falls from heights are these two threats. The causes and effects of the two main dangers are determined using the bowtie diagram. Following that, the analysis results utilizing the fault tree and event tree methodologies indicate that there is a 4.581 x 10-7 risk of major injuries or death, 2.247 x 10-4 probability of mild injuries, and 8.036 x 10-7 probability of suspension trauma. There are 1.927 x 10-6 workers affected by the material, 1.639 x 10-6 property affected by the substance, and 4.294 x 10-4 material damage.

1. Introduction

Currently, the construction sector is experiencing high development. This is indicated by the number of large projects being built by the government and the private company. Construction activities have an important role in the availability of large numbers of workers and contribute 10% of business opportunities in many countries [1]. However, the level of security and safety in the construction sector is relatively low, and increasing accidents at construction sites are still a global phenomenon [2]. The number of work accidents in Indonesia shows an increasing trend every year. In 2020, 221,740 cases were reported, meanwhile throughout 2021 there were 234,370 cases with nominal compensation paid amounting to IDR 1.7 trillion [3]. According to data from the Ministry of PUPR, the construction sector is the largest contributor to work accidents in Indonesia, followed closely by the manufacturing industry at 32%, the transportation sector at 9%, forestry at 4%, and mining at 2%. The International Labor Organization (ILO) reports that in all countries in the world, at least 60,000 the construction sector experiences the highest occurrence of fatal accidents every year [4]. Accidents frequently occur in the construction sector are related to work at height, including scaffolding work [5][6][7]. According to the Association of Experts on Work Safety in High Buildings (A2K2BT), Indonesia currently ranks second highest in the world regarding deaths due to falls from heights. Working at heights is categorized as a "Class 1 Risk Activity", which often results in work accidents in the form of serious injuries and deaths and mainly occurs in the construction sector [8]. Scaffolding is a temporary structure organized to support materials and workers during the construction process [9]. Scaffolding work is a complex work that requires special attention in terms of work safety. One of the most critical safety concerns that contribute the high number of fatal injuries is largely due to falls from heights, with a percentage of more than 30% of deaths in this industry [10]. To avoid work accidents, it is necessary to carry out a risk analysis. Previous research shows that the biggest causes of scaffolding accidents are lack of safety fences, poor inspection, and improper installation [1]. Other research shows that 53% of scaffolding workers are not equipped with fall protection equipment [11]. Furthermore, research in Japan concluded that improper use of clamps and connectors was the biggest factor causing scaffolding accidents [9]. In this research, a risk analysis of scaffolding work on offshore platform fabrication will be carried out. The analysis uses three methods, namely bowtie analysis, fault tree analysis, and event tree analysis. Bowtie analysis aims to identify sources of causes, impacts, and controls for the risk of work accidents [12][13][14][15]. Moreover, event tree analysis determines if the initial event will lead to a serious accident or if it will be adequately controlled by the safety systems and procedures in place [16][17]. Furthermore, fault tree analysis aims to determine the root cause and possibility of certain undesirable events occurring [18][19].

2. Risk Identification

2.1. Hazard Identification

The initial stage in risk analysis is identifying the hazards in each scaffolding work activity that could cause accidents as described in Table 1.

Table 1. Identifying Hazard

Activity	No	Hazard
Warehouse material picking	1a	Worker hit by forklift
	1b	Scaffolding material fall
	1c	Forklift fall
Material delivery	2a	Scaffolding material fall
	2b	Worker hit by heavy equipment
Material lifting	3a	Scaffolding material fall
	3b	Sling broke
	3c	Worker hit by heavy equipment
Scaffolding erection	4a	Worker falls from height
	4b	Scratched by sharp object
	4c	Worker slips
	4d	Worker strucks material
	4e	Worker trips
	4f	Scaffolding material fall
	4g	Equipment fall
	4h	Scaffolding collapse
Scaffolding dismantle	5a	Worker falls from height
	5b	Scratched by sharp object
	5c	Worker slips
	5d	Worker strucks material
	5e	Worker trips
	5f	Scaffolding material fall
	5g	Equipment fall
	5h	Scaffolding collapse

2.2 Hazard Likelihood and Severity

After determining the hazards, it is necessary to distribute a likelihood and severity questionnaire to 10 respondents. Based on the results of the questionnaire, the value varies from one to five. Therefore, it needs to be processed using the likelihood index and severity index equations to equate the perceptions of each respondent.

$$Likelihood\ Index\ (\%) = \frac{\sum_{i=1}^{5} a_i n_i}{5N} \times 100 \quad (1)$$

$$Severity\ Index\ (\%) = \frac{\sum_{i=1}^{5} a_i n_i}{5N} \times 100 \qquad (2)$$

It is necessary to adjust the scale from a percentage to a scale of one to five. This adjustment is made so that the scale can be included in the risk matrix. Adjustment reference according to Table 2 below [21].

Table 2. Index Rating

Rating	Class	Index
1	Extremely Ineffective	0% < I < 20%
2	Ineffective	21% < I < 40%
3	Moderately Effective	41% < I < 60%
4	Very Effective	61% < I < 80%
5	Extremely Effective	81% < I < 100%

The scoring on the questionnaire is filled in based on AS/NZS 4360:2004 is in accordance with Table 3 for likelihood and Table 4 for severity [20].

Table 3. Likelihood scale

Scale	**Likelihood**	**Probability**
1	Rare	10^{-5}
2	Unlikely	10^{-4}
3	Possible	10^{-3}
4	Likely	10^{-2}
5	Almost Certain	10^{-1}

Table 4. Severity scale

Scale	Severity Level	Description
1	Insignificant	No injuries, no damage
2	Minor	Minor injury, minor damage, small financial loss
3	Moderate	Moderate injury, property damage, and moderate financial loss
4	Major	Serious injury, major damage, major financial loss
5	Catastrophic	Casualties, fatal damage, serious losses can even stop all activities

3. Risk Matrix

Risk assessment using a risk matrix is conducted by multiplying the values of the likelihood index and the severity index. The hazard variables identified during hazard identification are grouped into the risk matrix to determine which hazards are acceptable and which are not. According to the risk matrix, it is known that there are 14 risks classified as low risk, 7 as medium risk, and 2 as high risk. Risks classified as high risk are the dominant risk that require further analysis. These two risks worker falls from height (4a) and scaffolding materials fall (4f). The risk matrix can be seen in Figure 1.

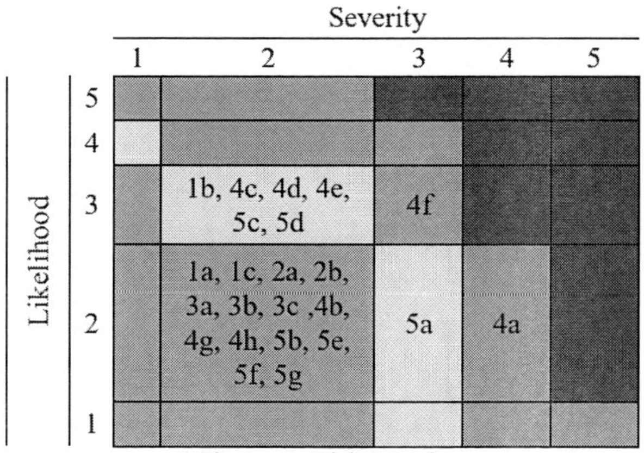

Figure 1. Risk Matrix

IOP Conf. Series: Earth and Environmental Science **1473** (2025) 012038 doi:10.1088/1755-1315/1473/1/012038

3.1 Bowtie Analysis

After obtaining the dominant risk based on the risk matrix. Bowtie analysis is carried out to visualize the causes, impacts and controls of the dominant risks. Determination of causes, impacts and controls is obtained from the results of discussions with respondents involved in the project. The bowtie diagrams of worker fall from height and scaffolding material fall are shown in Figure 2 and Figure 3.

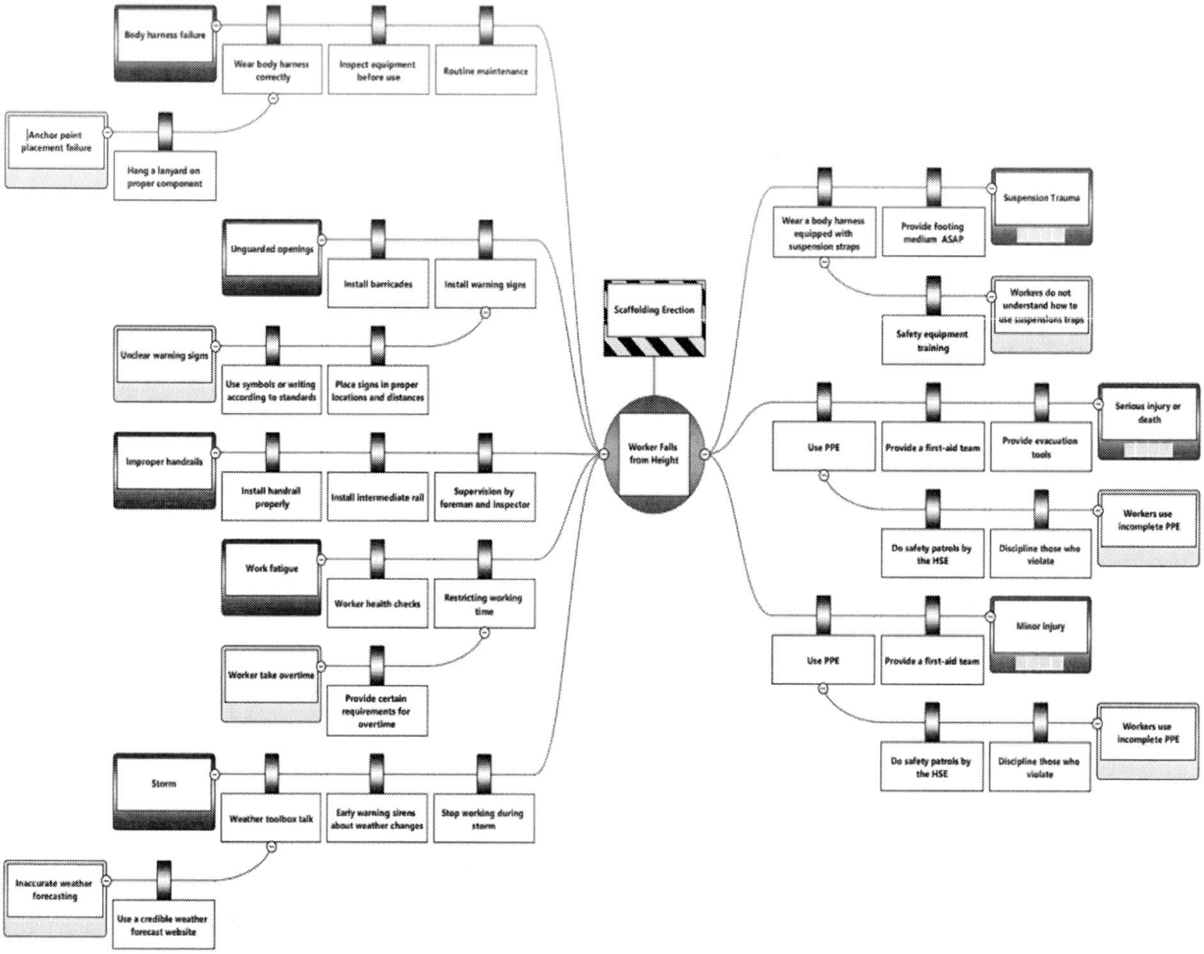

Figure 2. Bowtie diagram of worker falls from height

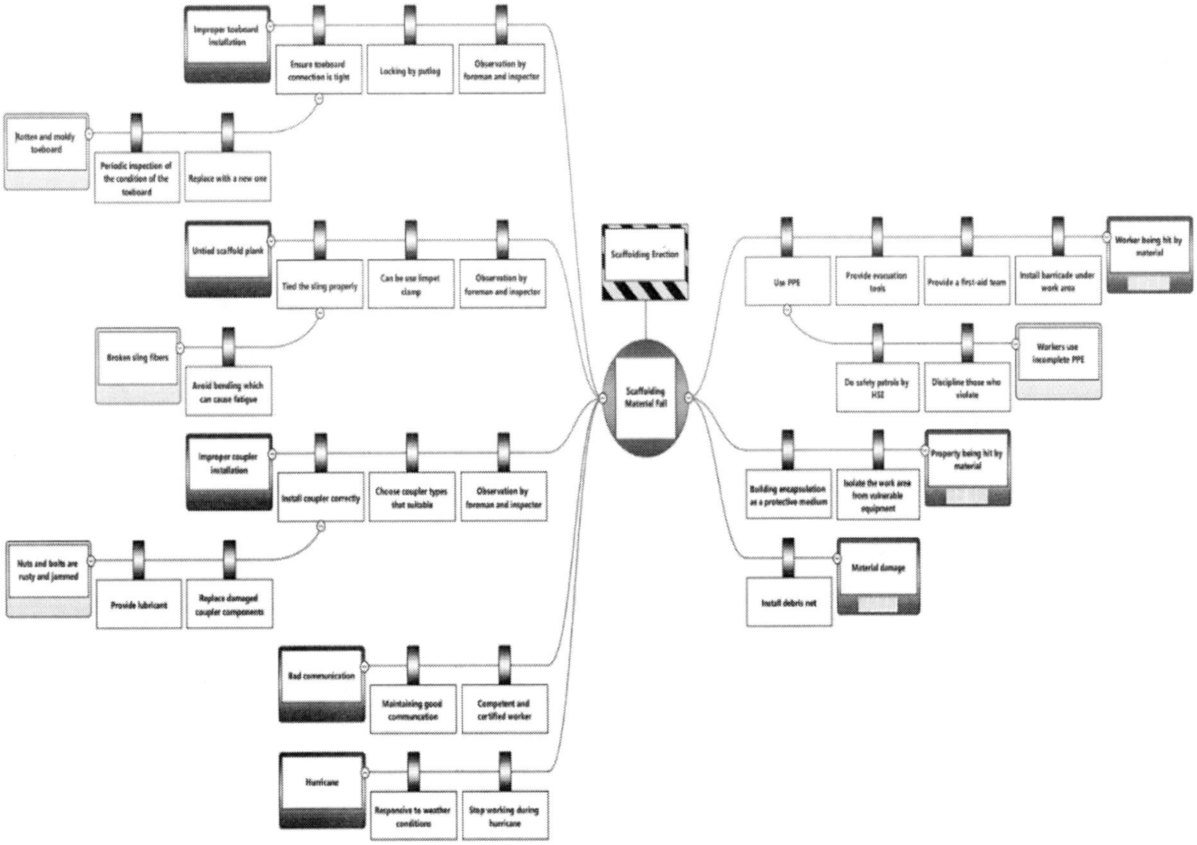

Figure 3. Bowtie diagram of scaffolding material fall

3.2 Fault Tree Analysis

In calculating the probability of impact, FTA is carried out on each safety barrier that experiences failure [16]. In this case, the failure of the safety system results in the risk of the worker falling, namely body harness failure as seen in Figure 4 and inability to provide foothold. The failure of the safety system results in the risk of material falling, namely worker encroaching into the work zone as seen in Figure 5 and encapsulation failure.

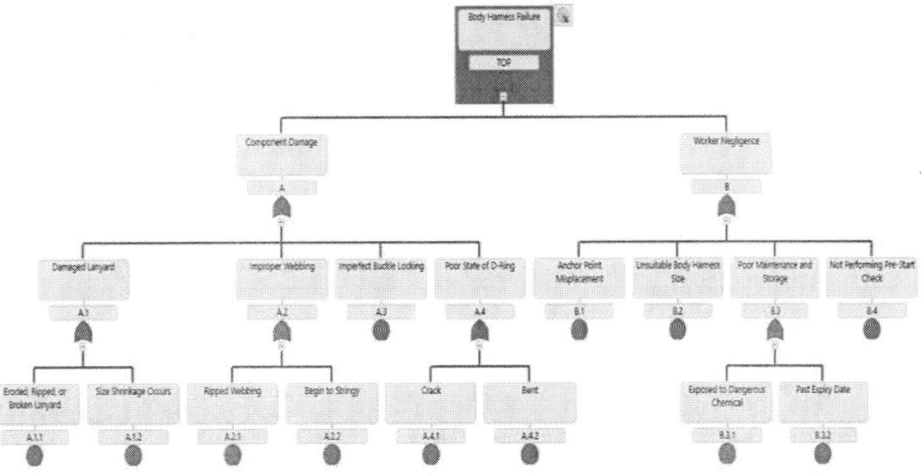

Figure 4. Fault Tree Diagram of Body Harness Failure

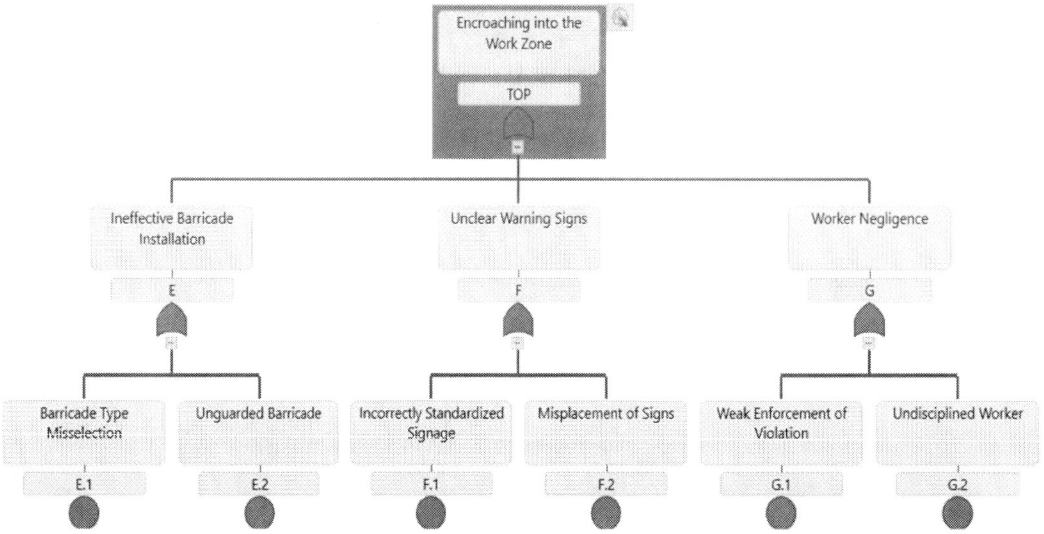

Figure 5. Fault Tree Diagram of Encroaching into the Work Zone

3.3. Probability Calculation

Based on the FTA diagram described previously, probability calculations were carried out which were collected from distributing questionnaires to respondents. Table 5 shows the basic event probability.

Table 5. Basic Event Probability

No	Basic Event	Probability
A.1.1	Eroded, ripped, broken lanyard	0.000046
A.1.2	Size shrinkage occurs	0.000136
A.2.1	Ripped webbing	0.000163
A.2.2	Webbing begin to stringy	0.000154
A.3	Imperfect buckle locking	0.000325
A.4.1	Crack D-ring	0.000235
A.4.2	Bent D-ring	0.000055
B.1	Anchor point misplacement	0.000370
B.2	Unsuitable body harness size	0.000037
B.3.1	Exposed to dangerous chemical	0.000055
B.3.2	Past expiry date	0.000217
B.4	Not performing pre-start check	0.000235
C.1	Component malfunction	0.000253
C.2	Misuse	0.000037
C.3	Not equipped with suspension straps	0.000136
C.4	Lack of maintenance	0.001261
D.1	Co-worker lacks first aid knowledge	0.001342
D.2	Difficult to find foothold	0.000127
D.3.1	Hard to reach hanging position	0.000055
D.3.2	Bad weather	0.000352
E.1	Barricade type misselection	0.001054
E.2	Unguarded barricade	0.000244
F.1	Incorrectly standardized signage	0.001126
F.2	Misplacement of signs	0.000253

No	Basic Event	Probability
G.1	Weak enforcement of violation	0.000343
G.2	Undisciplined worker	0.001432
H.1.1	Exposed to extreme weather	0.000262
H.1.2	Incorrect material selection	0.000127
H.2.1	Collision or friction with equipment	0.000055
H.2.2	Affected by construction activity	0.000055
H.3	Impact of usage life	0.000073
I.1	Incompetent installation worker	0.001072
I.2	Lack of supervision	0.002161

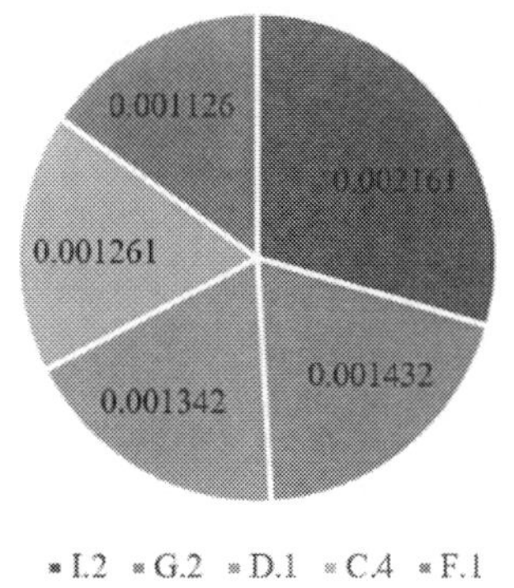

Figure 6. Highest basic event probability diagram

Figure 6 shows the five highest basic event probability are lack of supervision, undisciplined worker, co-worker lacks first aid knowledge, lack of maintenance, and incorrectly standardized signage. After obtaining the basic event probability, the total probability of intermediate and top events is calculated. Probability calculation are divided into two types based on the gate, namely AND GATE and OR GATE. Probability calculations use the following equation below.

AND GATE
$$P(t)=P(A) \times P(B) \times P(C) \times \ldots \times P(n) \quad (3)$$

OR GATE
$$P(t)= 1-[(1-P(A))\times(1-P(B))\times \ldots \times(1-P(n))] \quad (4)$$

Probability calculation results for each top event are shown in Table 6 below.

Table 6. Top event probability

No	Top Event	Probability
1	Body harness failure	0.002027
2	Inability to provide foothold	0.003563
3	Encroaching into the work zone	0.004451
4	Encapsulation failure	0.003803

To get the probability of success of the safety barrier, it is necessary to calculate using the following formula.

$$P(success) = 1-P(fail) \quad (5)$$

P(fail) is the probability of failure of the safety barrier. Safety barrier success probability is shown in Table 7.

Table 7. Safety barrier success probability

No	Safety barrier	Probability
1	Use of body harness	0.997973
2	Foothold	0.996437
3	Access restrictions	0.995549
4	Build encapsulation	0.996197

Table 8 shows the results of initiating event probability.

Table 8. Initiating event probability

No	Initiating Event	Probability
1	Worker falls from height	0.000226
2	Scaffolding material fall	0.000433

3.4. Event Tree Analysis

The two dominant risks will be positioned as initiating events. The safety barrier was obtained based on literature studies and discussions with scaffolders. By multiplying the initiating event by the safety barrier, both successful and failed, the probability of impact can be determined. By knowing the impact probability, so the completed diagram can be made. Table 9 and Table 10 shows the complete ETA diagram.

Table 9. Completed ETA diagram of worker falls from height

Initiating Event	Safety Barrier		Outcomes	Probability
	Use of body harness	Foothold		
Worker falls from height 0.000226	0.997973 Success	0.996437 Success	Minor injuries	2.247×10^{-4}
		Fail 0.003563	Suspension trauma	8.036×10^{-7}
	Fail 0.002027		Serious injuries/death	4.581×10^{-7}

Table 10. Completed ETA diagram of scaffolding material fall

Initiating Event	Safety Barrier		Outcomes	Probability
	Access restrictions	Build encapsulation		
Scaffolding material fall 0.000433	0.995549 Success	0.996197 Success	Material damage	4.294×10^{-4}
		Fail 0.003803	Property being hit by material	1.639×10^{-6}
	Fail 0.004451		Worker being hit by material	1.927×10^{-6}

4. CONCLUSIONS

From the previous calculations and analysis, the following conclusions are obtained:

1) The risk assessment results using the risk matrix indicate that there are 24 risks associated with

scaffolding work, with 15 categorized as low risk, 7 as medium risk, and 2 risks in the high risk category. The two risks in the high risk category are worker falls from height and scaffolding material fall.

2) Based on the bowtie diagram, it can be seen what the causes and impacts are caused by the dominant risks in scaffolding work.
   a. Worker falls from height are caused by five factors, namely failure of the body harness, unguarded open area, improper installation of handrails, worker fatigue, and bad weather. The consequences resulting from the risk of worker falls include suspension trauma, minor injuries, serious injuries, or death.
   b. Scaffolding material fall is caused by five factors, namely improper installation of the toeboard, the plank not being secured with slings, improper installation of the coupler, poor communication, and strong winds. The impacts arising from the risk of falling material include worker being hit by falling material, equipment or property being hit by falling material, and damage to the material itself.
3) From the calculation results using a combination of Event Tree Analysis and Fault Tree Analysis, the probability for each impact on the dominant risk is obtained. The probability of the risk of a worker falling due to the impact of suspension trauma is 8.036×10^{-7}, minor injury is 2.247×10^{-4}, and serious injury or death is 4.581×10^{-7}. Furthermore, the probability for the risk of material falling due to the impact of worker being hit by material is 1.927×10^{-6}, property being hit by material is 1.639×10^{-6}, and material damage is 4.294×10^{-4}.

REFERENCES

[1] A. Olanrewaju, J. S. Khor, and C. N. Preece, "An investigation into occupational health and safety of scaffolding practices on construction sites in Malaysia," *Front. Eng. Built Environ.*, vol. 2, no. 1, pp. 1–21, 2022, doi: 10.1108/febe-08-2021-0037.

[2] Z. Zhou, Y. M. Goh, and Q. Li, "Overview and analysis of safety management studies in the construction industry," *Saf. Sci.*, vol. 72, pp. 337–350, 2015, doi: 10.1016/j.ssci.2014.10.006.

[3] Kementerian Ketenagakerjaan RI, *Profil Keselamatan dan Kesehatan Kerja Nasional Indonesia Tahun 2022*. 2022.

[4] International Labour Organization, "Facts On Safety At Work," 2005.

[5] T. Nowobilski and B. Hoła, "Estimating the probability of accidents on building scaffoldings," *Saf. Sci.*, vol. 152, 2022, doi: 10.1016/j.ssci.2022.105777.

[6] C. F. Chi, S. Z. Lin, and R. S. Dewi, "Graphical Fault Tree Analysis for Fatal Falls in the Construction Igrandustry," *Accid. Anal. Prev.*, vol. 72, pp. 359–369, 2014, doi: 10.1016/j.aap.2014.07.019.

[7] J. C. Rubio-Romero, M. Carmen Rubio Gámez, and J. A. Carrillo-Castrillo, "Analysis of the safety conditions of scaffolding on construction sites," *Saf. Sci.*, vol. 55, pp. 160–164, 2013, doi: 10.1016/j.ssci.2013.01.006.

[8] W. M. Trianto, "Bekerja di Ketinggian pada Pekerjaan Konstruksi – Peraturan dan Tindakan Pencegahan," *Maj. Ilm. Swara Patra*, vol. 10, no. 1, pp. 39–50, 2020, doi: 10.37525/sp/2020-1/247.

[9] A. I. Quadri and O. G. Fadugba, "Risk Assessment and Safety Precautions for Construction Site Scaffolding," *J. Rehabil. Civ. Eng.*, vol. 10, no. 4, pp. 1–13, 2022, doi: 10.22075/JRCE.2021.22451.1480.

[10] M. Jahangiri, H. R. J. Solukloei, and M. Kamalinia, "A neuro-fuzzy risk prediction methodology for falling from scaffold," *Saf. Sci.*, vol. 117, no. July 2018, pp. 88–99, 2019, doi: 10.1016/j.ssci.2019.04.009.

[11] N. Abu Bakar, M. R. Ismail, and T. G. Lin, "Scaffolding Safety Assessment in Construction Sites in Penang," *2nd Int. Conf. Built Environ. Dev. Ctries.*, pp. 1311–1328, 2008.

[12] I. Voicu, F. V. Panaitescu, M. Panaitescu, L. G. Dumitrescu, and M. Turof, "Risk management

with Bowtie diagrams," *IOP Conf. Ser. Mater. Sci. Eng.*, vol. 400, no. 8, 2018, doi: 10.1088/1757-899X/400/8/082021.

[13] S. Lewis and K. Smith, "Lessons Learned from Real World Application of the Bow-tie Method," in *Prepared for Presentation at American Institute of Chemical Engineers 2010 Spring Meeting 6th Global Congress on Process Safety*, 2010, pp. 472–483. doi: 10.2118/154549-ms.

[14] Silvianita, A. A. Harahap, and M. F. Khamidi, "Operation failure of risk analysis on floating storage and offloading," in *IOP Conference Series: Earth and Environmental Science*, 2024. doi: 10.1088/1755-1315/1298/1/012031.

[15] Silvianita, A. P. Atmaja, and N. Shafiq, "Analysis of occupational accident risk in breakwater construction and floating piers," *IOP Conf. Ser. Earth Environ. Sci.*, vol. 1298, no. 1, 2024, doi: 10.1088/1755-1315/1298/1/012032.

[16] C. A. Ericson, *Hazard Analysis Techniques for System Safety*. 2005. doi: 10.1002/0471739421.

[17] M. Rausand and S. Haugen, *Risk Assessment: Theory, Methods, and Applications*, 2nd ed. John Wiley & Sons, 2020. doi: 10.1002/9781118281116.

[18] T. Aven, *Risk Analysis*, 2nd ed. John Wiley & Sons, 2015.

[19] Silvianita, D. S. Mahandeka, and D. M. Rosyid, "Fault Tree Analysis for Investigation on the Causes of Project Problems," *Procedia Earth Planet. Sci.*, vol. 14, pp. 213–219, 2015, doi: 10.1016/j.proeps.2015.07.104.

[20] AS/NZS 4360, "Risk Management Guidelines," *Australian/New Zealand Standard*. 2004.

[21] D. Davis and R. M. Consenza, *Business Research for Decision Making*, 2nd ed. Boston: PWO-Kent, 1988.

Environmental Friendliness of Seine Net Fishing Gear at Bulu Fishing Port Tuban East Java Indonesia

M A Sofijanto[1,*], O Siahaan[1], H Subagio[1], A Sulestiani[1] and S Widagdo[2]

[1]Fisheries Study Programme, Faculty of Engineering and Marine Science, Hang Tuah University, Surabaya, Indonesia
[2]Oceanography Study Programme, Faculty of Engineering and Marine Science, Hang Tuah University, Surabaya, Indonesia

*E-mail: ariefsofyan1964@gmail.com

Abstract. *Payang* and *cantrang* fishing gear are different types of trawl fishing gear from trawl fishing gear. Currently, the Indonesian government prohibits both types of trawl gear from operating, after previously banning trawl gear from operating. This study aims to assess the environmental friendliness of fur trawl gear. The result of this research is to find out whether the Fur drag trawl gear is an environmentally friendly type of fishing gear or not. The research method used is a descriptive method by comparing the sum of the scores of the Nine Environmental Friendliness Criteria of a fishing gear compared to the environmental friendliness standard by Mallawa. Based on the total score of the 9 criteria, payang fishing gear in Bulu Tuban is at a total score of 20, which means this fishing gear is a type of fishing gear that is not environmentally friendly. The type of fish caught using Bulu drag trawl gear is basic fish (demersal fish) with a composition consisting of the main catch, namely goat fish, red-eye fish, horse fish, tiger shrimp at 59.68%. While the by-catches of swallow, grouper, goby and white shrimp amounted to 12.73% and the second catch was discarded by 27.59%.

1. Introduction

Payang fishing gear together with cantrang fishing gear and beach nets in the classification of fishing gear according to the Ministry of Maritime Affairs and Fisheries is included as pulled fishing gear (seine net). Initially, this fishing gear was not included in those prohibited from operating by the Indonesian Government. However, based on Minister of Maritime Affairs and Fisheries Regulation No. 02 of 2015, this type of fishing gear is prohibited by the Government of the Republic of Indonesia following the prohibition on operating types of fishing gears that are dragged (trawls) or trawls which were banned by the government long ago (Presidential Decree No. 39 of 1980).

One of the reasons the 'payang' fishing gear is prohibited by the government is because this fishing gear is categorized as not environmentally friendly, especially those operated on the bottom of the waters, namely the cantrang fishing gear. The construction of the cantrang fishing gear is the same as the payang, namely consisting of a bag, body and two wings. The operation method is also the same, namely circled in a water area. However, the difference is that the cantrang is operated on the bottom to catch bottom fish (demersal fish) while the original payang is operated on the surface of the water and is intended to catch surface fish (pelagic fish). This is supported by [1] that payang is a bag seine used to catch schools of surface fish. The two wings are useful for scaring or surprising and leading fish into the bag. The way it operates is by circling a school of fish and then pulling the bag seine towards the ship. To examine whether the Bulu payang fishing gear was banned from operating in 2015 by the government because it endangered the sustainability of fish resources, this research needs to be carried out using an environmentally friendly approach to the fishing gear. The method used is based on [2]. Based on [3], nine criteria are established for environmentally friendly fishing technology.

The fish fishing gear data collected is related to environmentally friendly aspects, including: (1) Fishing gear selectivity data, (2) Data on habitat damage by fishing units, (3) Data on the quality of fish caught, (4) Safety data /dangers to consumers from consumption of caught fish, (5) Data on fisheries operated, (6) Data relating to fishing gear, (7) Data relating to the dangers of fishing unit operations on protected fish, (8) Data on fishing operations fish that is socially accepted in society and (9) Socially accepted. From the total environmental friendliness level score, the Bulu payang fishing gear fulfills several aspects of the nine aspects of environmental unfriendliness. Therefore, the Payang Bulu fishing gear is considered 'unenvironmentally friendly' fishing gear, so it could be a basis for the government to prohibit the Bulu payang fishing gear from operating. According to [4],[17] the use of trawls and seine nets in the State Fisheries Management Area of the Republic of Indonesia (WPPN-RI) has reduced fish resources, so it is necessary to prohibit them. The purpose of this research is to find out whether the Bulu payang fishing gear is a type of fishing gear that is friendly to the environment or not. Meanwhile, the aim of this research is to examine various aspects of environmental friendliness of the Bulu payang fishing gear in Bulu waters, Tuban Regency, East Java.

The benefit of this research for researchers is to obtain information on the environmental friendliness of the Bulu payang fishing gear based on aspects of the environmental friendliness of the fishing gear. For the Government, it is hoped that the results of this research will be able to support the data or refute the data regarding Fisheries Ministerial Regulation No. 02 of 2015 concerning the prohibition of operating seine net, namely ‘Bulu payang’ among others.

2. Method

2.1. Research location and time

This research was carried out in the Bulu Fishing Port area, Bancar District, Tuban Regency, which is located on the north coast of East Java Province. Fishing activities in this region are characterized by a diversity of fish species and types of fishing gear. The fishing gear operated includes payang (seine net), purse seines and gill nets. The fishing gear used in this research is the Bulu payang, which is a type of fishing gear which includes bagged fishing gear (seine net) whose operation method is encircling in a body of water with the target fish being caught being the bottom type of fish (demersal fish).

The research was carried out in August - September 2023. When data collection was carried out several times there were obstacles due to bad weather so that fishermen did not operate to catch fish.

2.2. Data collection

The data collection method used in this research is experimental fishing by means of descriptive observation of the types of fish caught in each setting hauling. The types of fish caught are categorized as main catch (target fish), bycatch and unwanted fish (discard). Data collection was repeated 5 times, with the same fishing gear and vessel. Data recorded included species, body length and body weight. Analysis of the level of environmental friendliness of payang uses the FAO method. Identify each type of fish caught using the Fishbase fish identification guide [9].

2.3. Data analysis

Analysis of catch data was carried out descriptively by observing data on the type and composition of the Bulu payang catch using 5 fishing gear settings. Catch data is grouped into three types, namely main catch and by-catch and discard catch (discard). The main catch from cantrang is: many kinds of demersal fish, shrimp and crab, while the by catch group is caught fish and squid. This data is then compared based on the species composition and method of catching the fish in the form of graphs and tables of the composition of the types of fish caught.

To determine whether Bulu payang fishing gear is environmentally friendly or not, researchers used 9 criteria used by [3] as in Table 1.

Table 1. Categories of environmental friendliness of fishing gear.

No	Criteria	Sub criteria	Score
1	Has high selectivity	- Catch more than three species of fish with widely varying sizes	1
		- Catch three or less species of fish with widely varying sizes	2
		- Catch less than three species of relatively uniform size	3
		- Fish of one species with relatively uniform size	4
2	Does not destroy habitat	- Causes habitat destruction in large areas	1
		- Causes habitat damage in a narrow area	2
		- Causes damage to part of the habitat in a narrow area	3
		- Safe for the habitat	4
3	Produces high quality fish	- Dead and rotten fish	1
		- Dead fish, fresh, physically disabled	2
		- Dead and fresh fish	3
		- Live fish	4
4	Does not harm fishermen	- Can result in death of fishermen	1
		- Can result in permanent disability for fishermen.	2
		- It's just a temporary health problem	3
		- Safe for fishermen	4
5	Production does not harm consumers	- Has a big chance of causing death to consumers	1
		- Potential to cause health problems for consumers	2
		- Relatively safe for consumers	3
		- Safe for consumers	4
6	*Low By-Catch*	- There are several species of by-catch and they are not sold on the market	1
		- How many species of by-catch are there and what types sell well on the market?	2
		- By-catch less than three species and sell well on the market	3
		- By-catch is less than three species and has a high price	4
7	Impact on biodiversity	- Causes the death of all living things and destroys the habitat	1
		- Causes the death of several species and destroys habitats	2
		- Causes death of some species but does not destroy habitat	3
		- Safe for biodiversity	4
8	Does not harm protected fish	- Protected fish are often caught	1
		- Protected fish caught several times	2
		- Protected fish have been caught	3
		- Protected fish are never caught	4
9	Socially acceptable	- Cheap investment costs	1
		- Profitable	2
		- Does not conflict with local culture	3
		- Does not conflict with existing regulations	4
		Total Score	36

Furthermore, the classification of environmental friendliness categories for fishing gear according to Mallawa [2] is also based on these 9 criteria as in Table 2.

Table 2. Classification of categories of environmental friendliness of fishing gear [2].

No	Environmental Friendliness Category	Score (X)
1	Environmentally friendly	$X > 31,5$
2	Not enough environmentally friendly	$22,5 < X \leq 31,5$
3	Not environmentally friendly	$13,5 < X \leq 22,5$
4	Damaging the environment	$X \leq 13,5$

3. Result and discussion

3.1. Research sites

As a fishing base, the mooring place for fishing boats is the Bulu Tuban Beach Fishing Port, while the fishing area ranges from the north of Bulu Tuban to the southwest of the Sarang Rembang area, Central Java with coordinates 6°44'30' - 6° 45'30" South Latitude and 111° 41' 30" – 111° 42' 30" East Longitude (Figure 1).

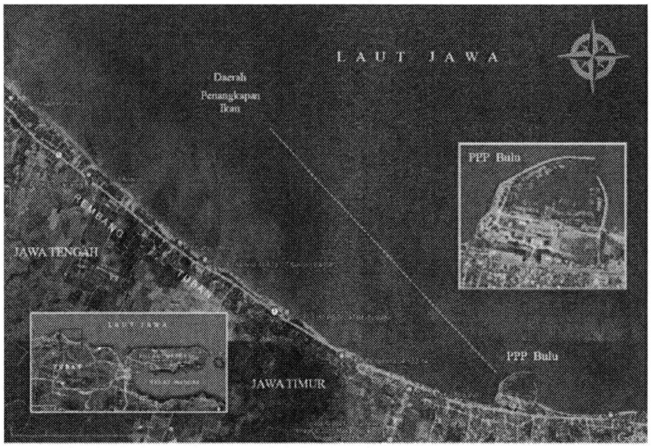

Figure 1. Research location

3.2. Construction

The main construction form of the Bulu payang net is the same as the construction of the standard type of seine net, namely consisting of a pocket at the last end, in front of it there is a net body and at the part closest to the ship there are two wings on the right and left (Figure 2).

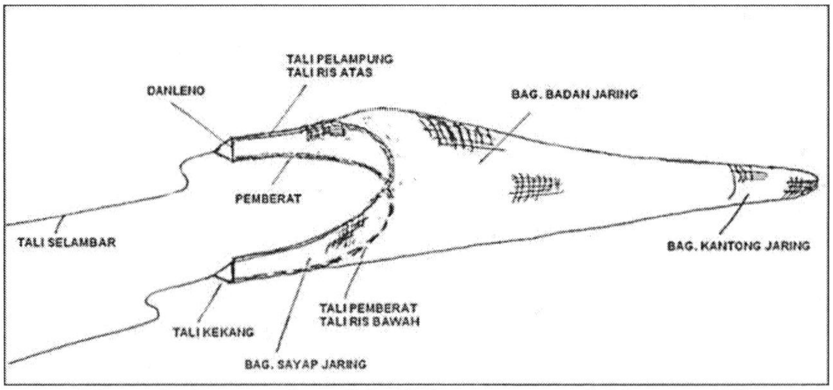

Figure 2. Construction of the Bulu payang net fishing gear [5].

To avoid misunderstandings in the use of the names of various types of fishing gear, the National Standardization Agency [16] issues standard construction standards for several types of fishing gear. One of them is the standard form for the construction of cantrang fishing gear which was issued in 2006 with the number SNI 01-7236-2006. Through this standard form of construction, it is hoped that there will be no more deception regarding the use of the names of various types of fishing gear, especially cantrang [6]. As research by [6] states that the construction of the cantrang in Brondong does not comply with the size of the cantrang SNI 01-7236-2006 [7], the cantrang in Bulu also does not comply with this standard because it uses a long rope (1,500 m) of cantrang with a large diameter (4 cm).

3.2.1. Warp rope

If the standard payang has a warp rope made of PE rope at the ends of both wings, the payang in the Bulu Tuban area has a warp rope called a batek rope with a large diameter (4 cm). The length of the right and left warp rope is @ 1,500 meters. Because this rope is long and large in size, the rope is stacked on the floor of the ship in the form of 4 circular piles of rope @ 45 m (Figure 3).

Warp rope is made from a mixture of natural and synthetic fibers, white and brown in color. The rope itself functions as a towing rope for the net when it is pulled. The warp rope is connected to the two wings of the net so that it also functions as a deterrent for bottom fish to swim into the mouth of the net and ultimately the fish into the net bag. These two sealines are connected to each wing which is pulled closer to the ship during fishing operations. The Bulu payang fishing gear has a line length of 1,000 - 1,500 meters. Each fleet of ships has a different length of line. Bulu payang (actually cantrang) is a demersal fishing gear consisting of a cod end, body, wings and mouth with material made from polyethylene (PE) fiber. The principle of operation of the Bulu payang fishing gear is to wrap the fishing line and wings as wide as possible in the fishing area, then pull the fishing line and narrow the space for the fish to move with the wings so that fish near the mouth can enter the bag and wings of the net.

Figure 3. Warp rope of Bulu paying.

3.2.2. Wings

There are two wings of seine nets, namely the right wing and the left wing, which have the same size. The net wings are the part of the net which is a connection or extension of the body to the warp rope which has the function of blocking and directing fish to enter the mouth and bag of the net. The wing is the part of the cantrang which is located at the front end of the Bulu payang which is 25 meters long with a mesh size of 8 inches to 5 inches, thread diameter of 0.7 mm and uses polyethylene (PE) netting material. The cantrang wing consists of an upper wing panel and a lower wing panel. The mesh size is 8 inches (wing) and has a mesh thread diameter on the wing of 0.7 mm (Figure 4).

Figure 4. Parts of a Bulu payang wing.

3.2.3. Payang body

It is the largest part of the web, located between the wings and the codend. This part functions as a connection between the wings and the bag to accommodate the fish catch. The body of the cantrang is 10 meters long and uses PE (polyethylene) material, with a thread diameter in the mesh of 0.5 mm and a mesh size of 4 inches to 2 inches.

3.2.4. Payang bag

The cod end is the very back part of the cantrang net, which is where the caught fish are collected. At the end of the bag, it is equipped with a bag strap that can be opened and closed to allow the caught fish to be removed. The Bulu payang bag is 5 meters long with a very small mesh size of around 1.5 inches to 0.75 inches, made from polyethylene (PE).

The cod end is the very back part of the cantrang net, which is where the caught fish are collected. At the end of the bag, it is equipped with a bag strap that can be opened and closed to allow the caught fish to be removed. The Bulu payang bag is 5 meters long with a very small mesh size of around 1.5 inches to 0.75 inches, made from polyethylene (PE).

3.2.5. Buoys and sinkers

The floats used in the Bulu payang fishing gear consist of two, namely the sign float and the main float. The sign buoy is made of styrofoam and is equipped with a 2 meter high flag pole at the top and weighted stones at the bottom. This sign buoy is tied to the end of the first line, so that when throwing the sign buoy the process of setting the fishing gear will be followed. Meanwhile, the main buoy used is made of PVC (polyvinvyl chloride) shaped like a soccer ball with 3 pieces which function as a source of buoyancy. This buoy is attached to the top rope (top lip of the net) so that the mouth of the net opens upwards.

The sinker functions to produce sinking force so that the net mouth can maintain the mouth opening vertically downwards. The weights used are made from sheet lead and are attached to the bottom ris rope with a total weight of around 15-20 kg. The weights are attached to the bottom rope with the aim that the parts attached to the weights sink quickly and remain in position (bottom of the water) even though they are influenced by the current.

The function of the long and large haule rope (name batek rope) is to lead the bottom fish to swim backwards until the fish moves to the mouth of the net and ends up in the net bag. A large haule rope (streming cloth material) that is wet and long (1,500 m x 2), then the rope sinks to the bottom of the sea so that it can sweep up various types of bottom fish which are positioned between the 2 haule ropes. So it is as if the bottom fish are being intercepted and herded by the warp rope which moves forward until then the bottom fish will swim along the wing wall of the net (wing) then the wall of the body of the net (body) and finally the fish will enter the net bag space (cod end). Why fish can eventually enter the net is because the net moves forward, namely when the net is pulled by a long rope roller called an gardan. Therefore, this fishing gear is included in the group of seine nets, not trawls.

After the two haule ropes begin to be pulled and pass through the haule rope roller (line hauler) (Figure 5) then the haule ropes are stacked in a circle on the boat deck. At that time the ship's propulsion engine remained running to maintain its position so that the ship was not pulled backwards due to the weight of the fishing gear which was dominated by long and heavy warp ropes. The overall length of the net is 40 m, consisting of 25 m wings, 10 m body and 5 m pockets.

Figure 5. The haule rope roller (capstan).

3.2.6. Place fishing gear in the water

According to Fishheries Ministry standards, there is a difference in the position of the net in the water during operation, and the target fish destination between payang and cantrang fishing gear. Namely, the payang fishing gear is operated on the surface of the waters because it is intended to catch surface fish (pelagic fish), while the cantrang fishing gear is operated at the bottom of the waters because it is intended to catch bottom fish (demersal fish).

Meanwhile, in the Bulu Tuban port area, many 'payang' fishing gears operate, but these fishing gears are operated at the bottom of the waters. In the Fisheries Port report [7] it is also written that there are 57 units of 'payang' fishing gear which are actually ‘cantrang’ fishing gear. This is also supported by data about the types of fish caught by the 'payang' fishing gear, namely bottom fish such as flat fish, goby fish and others.

3.2.7. Bulu paying environmental friendliness

Environmentally friendly fishing criteria determine fishing gear that is productive in operation and the catch has high economic value. Therefore, fishing experts pay attention to several things, including: fishing gear must be selective; does not damage resources and the environment; minimizing wasted or discarded fish [8].

Based on the data assessment of 9 criteria for friendlinesss of fishing gear, the score results for each criterion are as in Table 3.

Table 3. Results of score assessment for each of the 9 criteria for Bulu seine net.

No	Criteria	Score
1	Low selectivity	1
2	Does not destroy habitat	2
3	Produces high quality fish	2
4	Does not harm fishermen	4
5	Production does not harm consumers	3
6	Low by-catch	2
7	Low impact on biodiversity	2
8	Does not harm protected fish	3
9	Socially acceptable	2
	Total score of 9 criteria	20

The discussion of each of these criteria is as follows:

1. Low selectivity (1) due to catching more than three species of fish with different size variations.
2. Does not destroy habitat (2) because it causes habitat damage in a narrow area.
3. Produce high quality fish (2) because fish are dead, fresh, rotten.
4. Does not endanger fishermen (3) because it is safe for fishermen.
5. Production does not harm consumers (3) because it is relatively safe for consumers.
6. Low by-catch (2) because there are several species of by-catch and there are types that sell well on the market.
7. Impact on biodiversity (2) because it causes the death of several species and destroys habitat.
8. Does not harm protected fish (3) because protected fish are rarely caught.
9. Socially acceptable (2) because it conflicts with regulations but not with culture.

Furthermore, to assess the scores obtained in the environmental friendliness category, the guidelines by Mallawa [2] are used as in Table 2. Referring to the API classification in the environmentally friendly category (Table 2), the Payang Bulu fishing gear with a total score of 20 is at number 3, namely 'not environmentally friendly'. This position is between category number 2 of environmental friendliness of fishing gear which is 'less environmentally friendly' and category number 4 is the type of fishing gear which is 'damaging the environment'. Therefore, if the government started in 2015 [4] in order to take a precautionary attitude towards protecting fish resources, it could be scientifically accepted that this type of fishing gear was prohibited from operating because it was categorized as 'not environmentally friendly'.

3.3. Composition of fish types caught

The types and numbers of catches during the research (5 repetitions) using the Bulu payang fishing gear are listed in Table 4. Based on Fishbase [9] that there were 16 types (species) of fish caught with the Bulu payang or catching more than 1 type (multi species). This is in accordance with research by [10] that cantrang fishing gear in Brondong caught 25 types of fish. This is also confirmed by the results of [11] research that the diversity index value of cantrang catches is high with low fishing gear effectiveness. Both cantrang in Brondong and Bulu fish catch is dominated by demersal fish species. Although surface fish (squid and cuttlefish) are also caught. It is suspected that when the cantrang (Bulu paying) fishing gear is lifted (hauling) from the bottom of the waters to the surface of the waters, surface fish are also caught. The details of the types of catch are 4 types of main catch, 12 types of by catch. Meanwhile, the discard type is represented by all types of fish that are not yet large in size and therefore are not consumed by humans.

Table 4. Types and quantities of Bulu payang catches.

No	English name	Latin name	Number of catch (kg)
1	Red bigeye	*Priacanthus tayenus*	15
2	Rabbit fish	*Siganus gutatus*	16
3	Cutlass fish	*Trichiurus lepturus*	29
4	Caroun croaker	*Johnius trachycephalus*	23
5	Grouper	*Epinephelus* sp	31
6	Ray fish	*Himantura gerardi*	67
7	Cuttlefish	*Sephia* sp.	55
8	Squid	*Loligo* sp.	33
9	White shrimp	*Penaeus merguensis*	12
10	Goby fish	*Oxyeleotris marmorata*	14
11	Tiger prawn	*Penaeus monodon*	17
12	Pony fish	*Leiognathus* sp.	295
13	Striped eel catfish	*Plotosus canius*	45

14	Japanese threadfin bream	*Nemipterus japonicus*	144
15	Goatfish	*Upeneus sulphureus*	350
16	Flat fish	*Cynoglossus* sp.	43
17	Discard fish	Many kind of fish species	453
		Total catch	1.642

3.3.1. Main catch (target catch)

Table 5 shows that the highest total catch using the Bulu payang fishing gear was goat fish (350 kg), 295 kg pony fish, Japaneesee threadfin beam (144 kg) and tiger prawn (17 kg). These three kinds of main catch fish because the most commonly caught are of a size that can be consumed by humans. Meanwhile, other types of main catch fish are relatively few but have high economic value, such as tiger prawn. Both are types of surface fish that are caught by this bottom fishing gear which are thought to be caught when the hauling net rises above sea level.

Main catch is the component of the fish stock that is primarily sought from fishing operations. The main catch is the main target of the fishing gear used [12]. According to [13] the main catch is fish that have high economic value such as red snapper, grouper, pomfret, red bigeye, catfish and shrimp. Meanwhile, according to [14], the main catch is fish that has the highest relative abundance value, namely the type of fish that has the most/heaviest number of total catches based on the weight of fish per type and whose abundance has exceeded half of the population of fish caught.

According to [15], the composition of the catch of fishermen using cantrang fishing gear landed at PPP Tegalsari for 8 years from 2017-2010 was analyzed in a descriptive way that describes the composition of the catch with cantrang fishing gear. The total catch landed at PPP Tegalsari was 22 species, consisting of 9 main catch species and 13 by-catch species.

3.3.2. By-catch (by-catch target)

Meanwhile, by-catch is demersal fish other than fishing targets with lower economic value such as goatfish, crab, fish, rays and squid [12]. Bycatch is non-target fish caught in fishing operations. However, these non-target fish species can be utilized by fishermen. Types and number of catches per by-catch on Bulu payang fishing gear. There are many kinds of fish as fishermen's by-catch. There are 8 types of Bulu payang by catch, namely red bigieye, rabbit fish, catlass fish, caroun croaker, grouper, stingray fish, cuttlefish, squid, white shrimp, goby fish, and pony fish. By-catch, apart from high-quality non-economic fish, is also not as large as main catch. Main catch and by catch can become trash or discard fish if the size of the fish is too small and cannot be sold as food fish.

3.3.3. Discard catch

Discarded catches are catches that are discarded or used as animal feed. From the research, the discard catch was 435 kg. This shows that cantrang fishing gear produces a lot of catch that is not used for human consumption. The results of observations in the field show that the types of discard fish are the same type of fish as the target fish but are still small in size. Therefore, cantrang fishing gear is categorized as a type of fishing gear that is not environmentally friendly.

3.3.4. Comparison of main, by-product and discard catches

Figure 6 shows that the main catch is the largest, namely 59.59%, while the discard catch is 27.59% and the side catch is 12.73% (Figure 6).

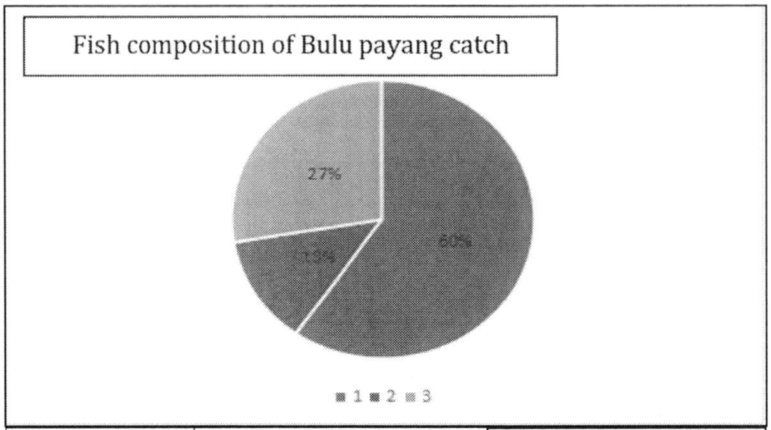

1. Main catch : 59.68%	2. By catch : 12.73%	3. Discard : 27.59%

Figure 6. Comparison graph of target species, by-catch and discard fish.

The large percentage of this type of discarded fish can also be a consideration for being careful about preserving demersal fish resources. Types of discarded fish, if large in size, have high economic value, such as sea fish, grouper fish, tiger prawns which are also caught by the Bulu payang fishing gear (Figure 7). This type of trash fish is collected by trash fish traders to sell to animal feed factories at very cheap prices.

Figure 7. Trash fish as discard fish of Bulu payang net.

4. Conclusion

The conclusion of this research is that the Bulu payang fishing gear is technically a cantrang fishing gear because it is operated on the bottom of the waters to catch bottom fish. Based on the nine criteria for the level of environmental friendliness of fishing gear, the Bulu payang fishing gear is included in the criteria of 'not environmentally friendly'. So technically it can support the government's decision to prohibit the operation of the Bulu payang type of fishing gear. The composition of the catch using the Bulu payang fishing gear is the demersal fish species consisting of the main catch, namely goatfish pony fish, striped eel catfish and tiger prawns (59.68%). Meanwhile, by-catch is red bigieye, rabbit fish, catlass fish, caroun croaker, grouper, stingray fish, cuttlefish, squid, white shrimp, goby fish, and pony fish amounting to 12.73% and the type of catch that is discard fish with a total of 27.59%.

Acknowlegment

This research is internal research by lecturers at Hang Tuah University, Surabaya, funded by the 2022/2023 Internal Research Task Force. Thanks are expressed to the Chancellor of Hang Tuah University. Thanks are also expressed by the Chairman of LPPM UHT who has approved funding for this internal research in accordance with the Internal Research Implementation Agreement Letter with the Chairman of LPPM UHT No. B/03D/UHT.C2/III/2023 March 1 2023. Furthermore, thanks are expressed to Oktaviandy Siahaan, NIM. 20190250011 who has helped carry out this research and Mr. Sarwi as a fisherman who owns the Bulu payang boat who has allowed researchers to collect data on fish catches at sea.

References

[1] Zikrillah, Rengi, Zain. Composition of catches from payang fishing equipment in the morning and evening in the South Pasalaban Bungus Waters of Padang, West Sumatera. 26564 (unri.ac.id). 2020
[2] Mallawa A., Najamuddin, M. Zainuddin, Musbir, Abustang, Safruddin, Fakhrul. Selayar regency fisheries and marine resources potential estimation study. Center for coastal resources and area studies. Hasanuddin University. Macassar. 2006. 211pp.
[3] FAO. Code of conduct for responsible fisheries. FAO Fisheries Department. Rome. 1995.
[4] Regulation of the Minister of Maritime Affairs and Fisheries of the Republic of Indonesia Number PER.02/MEN/2015 concerning the prohibition of the Use of trawls and seine nets in the Fisheries Management Area of the Republic of Indonesia. 2015.
[5] Sulindo newspaper. Before it can be legalized again, there must be cantrang standardization. https://koransulindo.com/sebelum-dilegalkan-kembali-harus-ada-standarisasi-cantrang/2018.
[6] Riyanto, M., Purbayanto, A., Mawardi, W., Suheri, N. Technical study of cantrang operations in Brondong Waters, Lamongan Regency, East Java. PSP Bulletin, 2011. **19**(1). pp. 97-104.
[7] Bulu Beach Fishing Port annual report 2022 - 2023. East Java Provincial Maritime Affairs and Fisheries Service. **2022**.
[8] Zainal, Sarong, Nasir. Environmentally friendly fishing equipment based on the Code of Conduct for Responsible Fisheries in Banda Aceh City. Agricep.**15**(2), 2014. p.10-18.
[9] Fishbase. https://www.fishbase.org.au/v4/search?q=. the Global Encyclopedia About fish.
[10] Afandi, J. T., Zainur, M. Cantrang fishery and its problems in Lamongan. Juvenil: Marine and Fisheries Scientific Journal, **1**(4), 525-531. ISSN 2723-7583 (online). https://journal.trunojoyo.ac.id/juvenil. https://doi.org/10.21107/juvenil.v1i4.8950. 2020.
[11] Rhamadani, R. H. Composition of cantrang catches at class III Coastal Fish Landing Bases (PPI), Tlanakan District, Pamekasan Regency, Madura. [Thesis]. Faculty of Fisheries and Marine Sciences, Brawijaya University: Malang. 2018.
[12] Aji, I.N., Wibowo, B.A., Asriyanto, A. Analysis of production factors from cantrang fishing gear catches at the Tuban Regency Feather Fish Landing Base. Journal of Fisheries Resources Utilization Management and Technology, **2**(4), pp. 50–58. 2013.
[13] Azizah, I.R., Rudiyanti, S., Ghofar, A. Composition of cantrang catches and biological aspects of kuniran fish (Upeneus sulphureus) landed at PPP Bajomulyo, Juwana, Journal of Management Aquatic Resources, **4**(4), pp. 33–41. https://doi.org/10.14710/marj.v4i4.9771. 2015.
[14] Setiabudi, Syahrir, Bulan. Study of bycatch and discards in Bagan Perahu in Mataha Waters, Berau Regency. Samarinda: Aquamarine Journal Vol.6. 2019..
[15] Imron, M., Baskoro, M. S., Prima, D. R., Suherman, A. Composition of catches and catching season patterns of cantrang landed at Tegalsari Beach Fishing Harbor, Central Java. Fisheries Science and Technology: Indonesian Journal of Fisheries Science and Technology, **17**(2), 138-145. 2021.
[16] National Standardization Agency. Indonesian National Standards. Standard form of cantrang trawl construction. SNI 01-7236-2006. Jakarta, BSN. 5 Pages. **2006**.

[17] Regulation of the Minister of Marine Affairs and Fisheries No. 71 of 2016 Concerning Fishing Lanes and Placement of Fishing Gear in the Fisheries Management Area of the Republic of Indonesia. **2016**.

IOP Conf. Series: Earth and Environmental Science **1473** (2025) 012040 doi:10.1088/1755-1315/1473/1/012040

Fuzzy-FMEA for Risk Analysis: Decommissioning an Offshore Manifold Wellhead Platform

W L Dhanistha*, D R Anggara F, D M Rosyid

Ocean Engineering Department, Faculty of Marine Technology Institut Teknologi Sepuluh Nopember (ITS), Surabaya, Indonesia

*Corresponding author : wimala.danistha@gmail.com

Abstract. Decommissioning is the process of dismantling a portion or all of an installation and transporting the disassembled materials to a specific location. To manage the possible and impact of risks that might arise during the decommissioning process, risk analysis is carried out. In this study, a risk assessment was conducted on the decommissioning of a four-legged jacket platform. The analysis method used was Fuzzy Failure Mode and Effect Analysis (fuzzy-FMEA), which uses a fuzzification process to identify the dominant risks and generate fuzzy-RPN (FRPN). It was found that residual hydrocarbon leakage was the dominant risk with FRPN 9, and further risk analysis was carried out using Fault Tree Analysis (FTA) to identify the root causes of the dominant risk, which were planning errors with probability 0,00026645 and Sealing failures occur with a probability of 0.00044118. Bowtie Analysis was used as a risk management tool, resulting in preventive and mitigation efforts. One of the preventive measures for the identified risks was to verify and validate the current well conditions, while mitigation efforts included narrowing the leakage area and installing a hydrocarbon detector.

1. Introduction

The world's continued need for gas and oil is demonstrated by the existence of offshore platforms, which are already quite common and will only grow in number. Over 7,500 oil and gas platforms have been installed globally, situated in the continental regions of 53 countries, 40 of which are major producers of natural gas and oil [1]. Among the 53 nations whose territories produce and have installed offshore gas and oil platforms is Indonesia. According to the Ministry of Energy and Mineral Resources' statistical data report from 2021, production of crude oil totaled 658.54 million Barrels of Oil per Day (BOPD), while production of natural gas was 1,178 million BOPD. Oil and gas exploitation operations typically use a variety of offshore platform types, including semi-submersible, TLP, SPAR, FPSO, and others. It is more common with fixed offshore platforms in Indonesia itself (Jacket). This is due to the relatively shallow depth of the Indonesian waters, which range from 0 to 300 meters [2].

The government governs the Technical Guidelines for the Dismantling of Offshore Oil and Gas Installations by Minister of Energy and Mineral Resources Regulation Number 01 of 2011. The process of removing all or a portion of the installation and relocating the components to a designated location is known as dismantling or decommissioning. Decommissioning offshore platforms that are no longer in service or operational is necessary to prevent disruptions to shipping, navigation, or other operations [3]. According to the ministerial regulation, decommissioning also serves to protect the environment, the safety of oil and gas companies, the upkeep of the installation's state inventory, and the best possible use of the products in the state inventory. In order to enable offshore platforms to be used as artificial reefs and cut expenses and procedures after their equipment is decommissioned, the government has also issued regulations regarding the conversion of buildings and installations at sea. There is a significant risk of mishaps and malfunctions when decommissioning offshore oil and gas platforms. According to statistical data from SKK Migas, there is a high risk of work accidents in the upstream oil and gas sector. To control potential risks, risk analysis is required prior to decommissioning activities. To put it simply, risk analysis is the process of identifying threats and vulnerabilities, evaluating them to make sure the decommissioning will have the desired effects, and emphasizing how those effects can be minimized or eliminated.

Figure 1. Jacket Platform MWP-C

The author of this study aims to perform a risk analysis on figure 1 manifold wellhead platform-C (MWP-C) decommissioning process in the Peciko field. Decommissioning is required because this platform has been in use for more than 20 years and could cease to function at any time.

2. Literature Review

2.1 Decommissioning

Decommissioning or dismantling is the activity of removing part or all of the installation and transporting the results of the decommissioning to a predetermined location. Decommissioning of offshore platforms is carried out if:

- The age of the offshore platform has exceeded the predetermined design life.
- The platform has stopped operating and/or there is damage that cannot be repaired and is not in accordance with regulations.
- Oil and gas wells are no longer productive.

The objectives of decommissioning offshore platforms are:

- Ensuring the safety of oil and gas in Indonesia.
- Implementation of environmental management.
- Maintaining asset conditions.
- Maintaining the safety of shipping activities.

2.2 Fuzzy-FMEA

As time has gone on, a number of specialists have discovered flaws in the traditional FMEA method. Experts have identified a few shortcomings with traditional FMEA as follows: a. Due to the multiplication of Severity (S), Occurrence (O), and Detection (D), it is possible for different failure modes to have the same RPN [4]. b. Each of the three criteria—S, O, and D—has the same weight and no weighting [5]. c. The indirect relationship between criteria is not taken into account when multiplying S, O, and D to obtain RPN [6]. These experts have identified certain weaknesses, which can be addressed by using a fuzzy logic approach to weight S, O, and D. Based on the outcomes of each criterion—S, O, and D—which are transformed into linguistic variables, fuzzy logic is utilized to replace the index ranking in the FMEA. The IF-THEN rule is then applied to produce fuzzy RPN output [7]

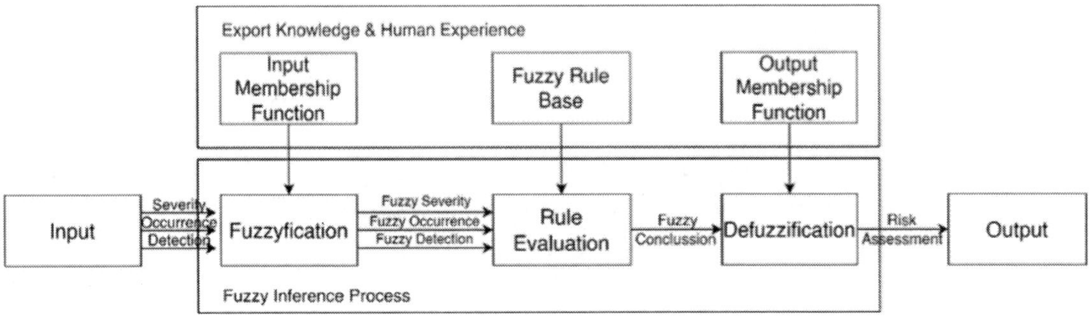

Figure 2. *Fuzzy*-FMEA [8]

Figure 2 is skem for fuzzy FMEA in fuzzification, in the context of Fuzzy-FMEA, is the process of redefining the crisp criteria's severity, occurrence, and detection functions to become fuzzy severity, fuzzy occurrence, and fuzzy detection [9]. After the fuzzy rule is applied, the rule evaluation task is completed using the IF-THEN IF-FLUID rule base that has already been created. This step will involve defuzzification, or redefining fuzzy to crisp, which will be produced by Fuzzy-RPN using a combination of the original fuzzy criteria.

3. Analysis Data

3.1 FFMEA

Fuzzy consists of three main panels: output (blue box), rule evaluation (white box), and input (yellow box) in Figure 3. This step involves mapping the membership functions of crisp number variables into linguistic variables for input, such as output, occurrence, severity, and detection. Five linguistic variables are used to map the input: very low (VL), low (L), medium (M), high (H), and very high (VH). Ten linguistic variables are used to define the output: none (N), low (L), high low (HL), low medium (LM), medium (M), high medium (HM), low high (LH), high (H), and very high (VH). Expert-set fuzzy rule bases, or fuzzy rule bases, or fuzzy rule bases (FRBs), are necessary for rule evaluation.

IOP Conf. Series: Earth and Environmental Science **1473** (2025) 012040 doi:10.1088/1755-1315/1473/1/012040

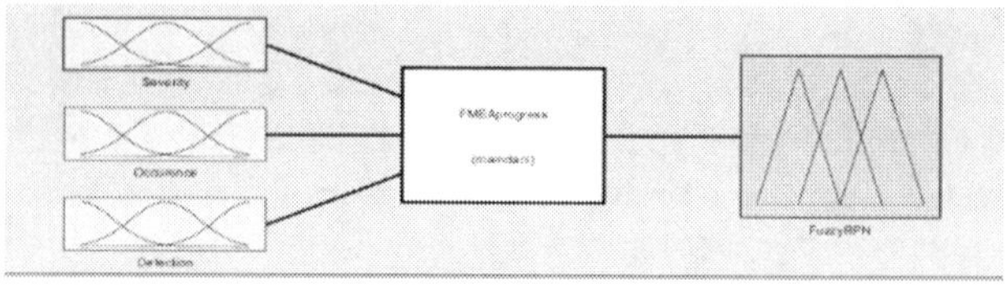

Figure 3. Fuzzification

The surface graph in Figures 4, 5 and 6 illustrates the relationship between each input severity, occurrence, and detection. A fuzzy-RPN will be produced on the z axis by combining the numbers entered for each parameter on the xy axis.

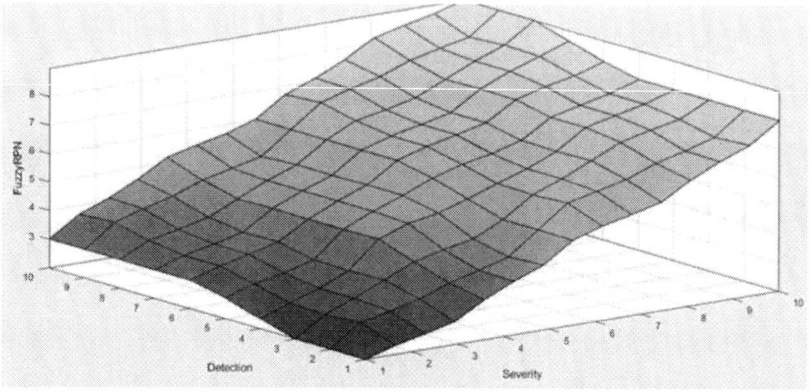

Figure 4. Correlation between Severity and Detection

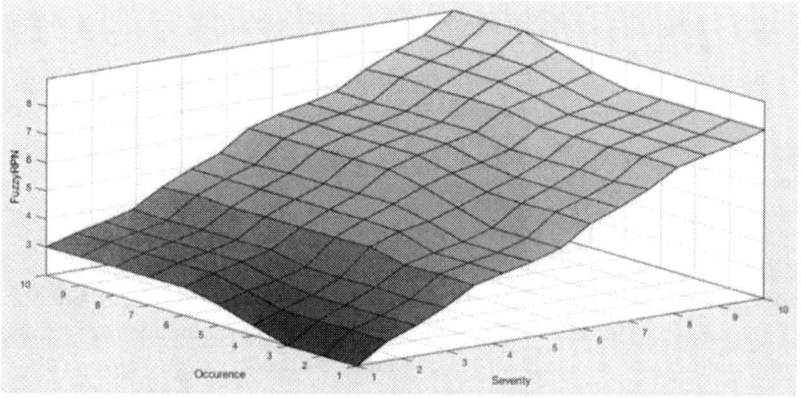

Figure 5. Correlation between Severity and Occurance

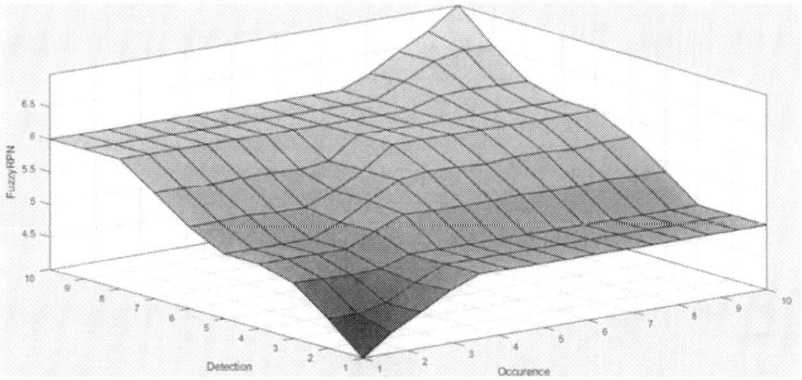

Figure 6. Correlation between Occurance and Detection

When compared to other parameters on fuzzy-RPN, the severity parameter has the greatest significance, according to the surface graph in Figures 4 and 5. When the detection and occurrence parameters' input values are set to 1, the severity parameter's input value can be traced to a value of 10, which allows it to reach fuzzy-RPN 7, a high-risk category. When both the occurrence and detection parameters in Figure 6 are valued at 10, the combined results of their input values only yield a fuzzy-RPN value of 7. This is a result of the platform dismantling process placing a higher priority on worker and asset safety and security than on the likelihood of a hazard occurring and the ease of detecting it. The ease of detecting hazards is more important than the likelihood that they will occur because, on the one hand, the likelihood of an incident cannot influence the ease of detecting hazards, and, on the other hand, the ease of detecting hazards will influence the likelihood of an incident.

Table 1. Calculation RPN

No	Activity	Code	S	O	D	RPN	Rank	FRPN	Rank
1	Marine Operations	1a	5	2	5	50	14	4.33	17
		1b	7	1	3	21	18	5.46	12
		1c	3	2	3	18	19	3.28	20
		1d	5	1	3	15	20	4.41	16
		1e	5	3	4	60	12	4.53	15
2	Well Plug and Abandonment	2a	10	4	9	360	1	9	1
3	Equipment Decomm	3a	7	3	5	105	9	6.22	6
		3b	7	4	5	140	7	6.21	7
		3c	5	3	5	75	11	4.55	14
4	Pipeline Decomm	4a	8	4	8	256	3	7.39	3
		4b	3	4	3	36	16	3.59	19
5	Conductor Decomm	5a	4	3	4	48	15	4.21	18
6	Topside Decomm	6a	6	5	5	150	5	6	10
		6b	8	1	7	56	13	7.16	4
		6c	7	5	6	210	4	6.33	5
7	Jacket Decomm	7a	6	4	6	144	6	5.68	11
8	Lifting Operation	8a	7	2	7	98	10	6.17	9
		8b	9	5	8	360	1	8.07	2
		8c	7	5	4	140	7	6.21	8
9	Diving (ROV, Diver)	9a	3	1	2	6	22	2.8	21
		9b	6	2	3	36	16	5.12	13
10	Site Clearance	10a	1	3	3	9	21	1.99	22

Based on the RPN calculation in Table 1, the failure mode of residual hydrocarbon leakage and its impact of explosion and fire on the platform and operation location with FRPN 9 and classical RPN 360 represents the highest failure risk with fuzzy-RPN and classical RPN. The failure mode in the lifting process and its impact of worker accidents (object impact) with FRPN 8.07 and classical RPN 360 represents the second highest RPN, and the failure mode in the pipeline cutting process and its impact of explosion due to residual hydrocarbon trapped in the pipe with FRPN 7.39 and classical RPN 256 defines the third highest RPN. It is evident from the comparison table between the fuzzy-RPN and classical RPN ranking results that the classical RPN ranking method still has issues with ranking because the results are multiplied by the same parameters. This is not the same as fuzzy-RPN, which ranks dominant risks accurately and produces different results by applying the fuzzyfication and defuzzyfication processes to the input parameter values.

3.2 FTA

The MWP-C Platform decommissioning process uses failure risk analysis (FTA) to identify the underlying causes of the failure modes found in the earlier analysis [10,11,12]. Critical risks are identified as top intermediate events in the FTA method based on the three failure modes that have the highest RPN obtained from the fuzzy FMEA method [13,14]. Remaining hydrocarbon leakage, lifting process failure, and pipeline cutting process failure are the three failure modes. To find the fundamental event that can lead to the top event, the three most significant events will be dissected.

3.2.1 Risk of Residual Hydrocarbon Leak Failure

There are two main categories of causes for residual hydrocarbon leakage: poor planning and preparation and poor sealing. These elements are acquired by literacy and expert consultation. The explanation of risk factors is as follows:

a. Planning and Preparation Errors

Inadequate planning and preparation can have disastrous effects on the exploitation well closure procedure. Inaccurate well reports lead to incorrect equipment and technical calculations, and the use of the wrong P&A techniques will deteriorate the well foundation [15,16].

b. Sealing Failure

Cracks in the isolation cement, deformation of the well casing, and the use of materials in the well barrier with a porosity larger than the hydrocarbon can all contribute to sealing failure or the inability to isolate exploitation wells.

3.2.2 Risk of Failure of Lifting Process

Errors in lifting load calculation and damage to lifting equipment are the two main categories of causes of lifting process failure. These elements are acquired by literacy and expert consultation. The following is a description of the risk factors:

a. Lifting Equipment Damage

Damage to lifting equipment such as padeye, sling, crane will hinder the process of lifting the platform section to the barge. In this case, damage that can occur such as cracked padeye due to lack of maintenance, broken sling due to the load being lifted not being in accordance with the sling's capacity, and crane arms that can experience fatigue due to unstable lifting movements.

b. Lifting Load Calculation Error

The calculation of lifting load can possibly be wrong due to differences in the report data obtained with real conditions in the field. This will result in the lifting process being hampered and the movement of unstable objects which risk falling objects.

3.2.3 Risk of Pipeline Cutting Process Failure

Diver error and impediments in the pigging process are the two main causes of the failure risk associated with the pipeline cutting process. These elements are acquired by literacy and expert consultation. The following is a description of the risk factors:

a. Diver Error

Diver error is an error in implementation by divers who are cutting the pipeline. SOPs that are not violated and divers' unpreparedness to face extreme conditions can cause failure in the pipeline cutting process.

b. Obstacles in the pigging process

One of the processes in cutting the pipeline is pigging which aims to clean up the remaining hydrocarbons and debris in the pipeline. When there are obstacles in the pigging process in the middle of the pipeline, it will certainly hinder the pipeline cutting process, the cause of which is that the pigs are stuck due to the accumulation of debris so that it stops the speed of the pigs and is also caused by buckling in the pipe.

3.2.4 Probability

According to the probability calculation in Table 2, the probability of failure of the MWP-C Platform decommissioning process is 0.00399791, with the failure mode with the highest probability being the lifting process failure, which has a probability value of 0.00218440. The lifting process has a high probability of failure due to its length and the presence of a variety of lifting objects, both light and heavy.

Table 2. Probabilty Minimum Cutset

No	Code	Event	Probability
1	A.1	Planning and Preparation Errors	0,00026645
2	A.2	*Sealing failure*	0,00044118
	A	Residual Hydrocarbon Leak Failure	0,00070751
1	B.1	Lifting Equipment Damage	0,00172520
2	B.2	Lifting Load Calculation Error	0,00046000
	B	Failure of Lifting Process	0,00218440
1	C.1	*Diver error*	0,00021275
2	C.2	Obstacles in the pigging process	0,00089819
	C	Pipeline Cutting Process Failure	0,00111075
Top Event		The MWP-C Platform decommissioning process	0,00399791

4. Conclusions

The MWP-C Platform decommissioning process is most likely to fail due to residual hydrocarbon leakage (FRPN 9), lifting process failure (FRPN 8.07), and pipeline cutting process failure (FRPN 7.39). The root causes of the failure of the decommissioning process on the MWP-C Platform were identified using FTA, which included residual hydrocarbon leakage, the use of the incorrect P&A method, invalid well reports, cement cracking, casing deformation, and porous barriers. The lifting process failed due to padeye cracking, broken sling ropes, crane arm fatigue, and lifting load calculation errors. And the pipeline cutting process failed because SOPs were not followed, diving equipment became jammed, ROVs malfunctioned, debris accumulated, free span, lateral buckling, and upheavel buckling.

References

[1] Parente, V., Ferreira, D. & Santos, E., 2006. "Offshore decommissioning issues: Deductibility and transferability". *Energy Policy,* **34**(15), pp. 1992-2001.

[2] Sadeghi, K., 2007. "An Overview of Design, Analysis, Construction and Installation of Offshore Petroleum Platforms Suitable for Cyprus Oil/Gas Fields". *Social & Application Science,* Volume **4**, pp. 1-16.
[3] Claisse, J. T. & Love, M. S., 2015. "Impact from Partial Removal of Decommissioning Oil and Gas Platforms on Fish Biomass and Production on the Remaining Platform Structure and Surrounding Shell Mounds". *PLoS ONE,* Volume **10**.
[4] Kumru, M. & Kumru, P. Y., 2013. "Fuzzy FMEA Application to Improve Purchasing Process in a Public Hospital". *Applied Soft Computing,* Volume **13**, pp. 721-733.
[5] Hadi-Vencheh, A. & Aghajani, M., 2013. "Failure Mode and Effect Analysis A Fuzzy Group MCDM Approach". *Journal of Soft Computing and Applications,* pp. 1-4.
[6] Chang, K. H. & Cheng, C. H., 2010. "A Risk Assessment Methodology Using Intuitionistic Fuzzy set in FMEA". *International Journal of System Science,* Volume **41**, pp. 1457-1471.
[7] Dhanistha, W. L., Silvianita, S., & Wardhana, W. (2018). Wave height prediction based fuzzy logic. *International Journal of Civil Engineering and Technology*, *9*(9), 1611–1616.
[8] Yeh, R. H. & Hsieh, M. H., 2007. "Fuzzy Assessment of FMEA for a Sewage Plant". *Journal of T he Chinese Institute of Industrial Engineers,* Volume **24**, pp. 505-512.
[9] Yeh, R.H. and Hsieh, M. H.: Fuzzy Assessment of FMEA for a Sewage Plant. *Journal of The Chinese Institute of Industrial Engineers* **4**(1): 505-512 10.1016/j.promfg.2015.11.041.
[10] Silvianita, Mahandeka, D.S., Rosyid, D.M.: Fault Tree Analysis for Investigation on the Causes of Project Problems. *Procedia Earth and Planetary Science* **14**: 213-219, 2015 https://doi.org/10.1016/j.proeps.2015.07.104
[11] Rausand, M.: *Risk Assessment: Theory, Methods, and Apllication*. Wiley, USA, 2011.
[12] Nugroho, E.N., Sunbara, A.: Work Accident Analysis in the Construction Project of PT. XYZ. *International Journal of New Technology and Research* **7**(2): 44-56 2021 https://doi.org/10.31871/IJNTR.7.2.7
[13] Ericson, C. A., 2005. Hazard Analysis Techniques for System Safety. Hoboken, New Jersey: Wiley-Interscience
[14] Zhong, C., Chen, Z., and Zhou, J.: Numerical Investigations of Pile Group Foundations under Different Pile Length Conditions. *Applied Sciences* **14**(5): 1908, 2024 https://doi.org/10.3390/app14051908
[15] Efstathopoulos, G. E., & Balomenos, G. P.:.Effects of wave load models on the uplift fragility assessment of pile-supported wharves and piers exposed to storm surge and waves. *Engineering Structures*, ***290***: 116372 2023 https://doi.org/10.1016/J.ENGSTRUCT.2023.116372
[16] Yetkin, M.E., Özfirat, M.K., Kun, M., and Pamukcu, C.: The prediction of the risks of spontaneous combustion in underground coal mines using a fault tree analysis method. *MethodsX* **13**: 102835, 2024 https://doi.org/10.1016/j.mex.2024.102835

Preliminary Assessment of Ocean Pollution Effect on Aquaculture: Case of Sunda Strait, Indonesia

Shade Rahmawati[1*] **and Hidemi Mutsuda**[2]

[1]Department of Ocean Engineering, Institut Teknologi Sepuluh Nopember, Surabaya,Indonesia
[2]Graduate School of Advanced Science and Engineering, Hiroshima University, Japan

E-mail: [1*]shade@its.ac.id · [2]mutsuda@hiroshima-u.ac.jp

Abstract. The growing populations of coastal-based city in Indonesia, such as Banten, Jakarta, and Bandar Lampung which located in Sunda Strait area are heavily affect the ocean environment quality due to increasing land-based pollution, with growing evidences of heavy metals were founded in many previous research. Further research about spatial and temporal distributions of heavy metal pollution due to ocean current are presented in this paper. Furthermore, the effect of heavy metal contamination in the ocean water to photosynthesis rate of dominant species algae populating Sunda Strait are estimated. It is estimated that a single source of lead heavy metal can spread to five kilometers radius and reduce the algae photosynthesis rate until 100% in just four weeks. Both numerical model for north-west and south-east monsoon seasons also show high concentration of suspended solid, which could also decrease algae photosynthesis rate because of less sunlight penetration. Considering fishery as one of the main economic resource in Sunda Strait area, this numerical predictions are important for coastal area management plan, environmental protection program or at least to avoid economical loss.

1. Introduction

The Sunda Strait is situated between two major islands: Sumatra to the north and west, and Java to the south and east, falling under the administrative regions of Banten and Lampung provinces. At its narrowest point, the strait measures about 24 kilometers in width [1] and reaches depths exceeding 90 meters in some areas, though it is much shallower, at around 20 meters or less, near Sangiang Island in the middle of the strait (Figure 1). This region features a coastal bay with a 37-kilometer shoreline, providing economic benefits through ports, shipping, tourism, and fisheries. Fishing has been a primary source of income for the local population for many years, with nearly 23,000 fishermen engaging in fishing and marine cultivation [2, 3]. The diverse fishing methods include rods, trawls, and nets, with approximately 6,000 fishing units operating in the Sunda Strait area [4]. Despite its importance to the local economy, the area is facing growing concerns regarding land-based pollution, as it has become a major site for illegal waste disposal.

Previous studies have identified high concentrations of heavy metals, particularly lead (Pb), along the west coast of Java and the southern coast of Sumatra [5, 6, 7]. Population growth in the region is contributing to the increasing levels of pollution, particularly in Banten Bay, which exhibits similar levels of lead contamination to Jakarta Bay [8, 9, 10]. Although Banten Bay has been reported to have lower pollution levels compared to Jakarta Bay [11], the presence of significant lead contamination, both in water and sediments, remains a serious concern for fisheries and aquaculture, about 20 mg/L [8]. Lead contamination has been shown to affect marine life, particularly by inhibiting the respiration and photosynthesis of phytoplankton algae [12, 13, 14]. As phytoplankton, primarily composed of diatoms (*Bacillariophyceae*), accounts for 95% of oceanic primary production [15], understanding the environmental factors influencing its productivity is crucial.

Phytoplankton populations in Indonesia, particularly around Java Island, are dominated by diatom species (Bacillariophyceae), which constitute 73-85% of offshore phytoplankton and 71-78% in nearshore areas [16]. By estimating the distribution of Pb concentrations due to tidal currents, it is possible to assess the impact of this contamination on phytoplankton photosynthesis. Previous studies have not fully explored the spatial and temporal variations in pollutant accumulation, which are likely influenced by proximity to pollution sources and environmental conditions such as ocean currents and mixing[6]. Ke et al. [17] also observed that the spatial distribution of phytoplankton is strongly affected by surface current patterns. Therefore, understanding the relationship between land-based pollution and ocean circulation is critical for mitigating potential fishery losses, especially considering the monsoonal variations that impact the region.

2. Research Methods

This study uses the Princeton Ocean Model (POM) [18] to simulate tidal elevations and current flows that influence pollutant dispersion. The model incorporates several key features, including a second-moment turbulence closure sub-model for vertical mixing coefficients, sigma coordinates for vertical scaling based on water depth, an unstructured horizontal grid, and a free surface with split time steps [19]. The simulation is driven by tidal flows across the strait, assuming a fixed vertical temperature stratification, constant salinity, and no surface heat or salinity flux, though wind stress could also be introduced. A point-source pollution model is used to track the spread of pollutants driven by tidal currents. The horizontal viscosity and diffusion are represented by the following equations:

$$F_x = \frac{\delta}{\delta x}(H\tau_{xx}) + \frac{\delta}{\delta y}(H\tau_{xy}) \tag{1}$$

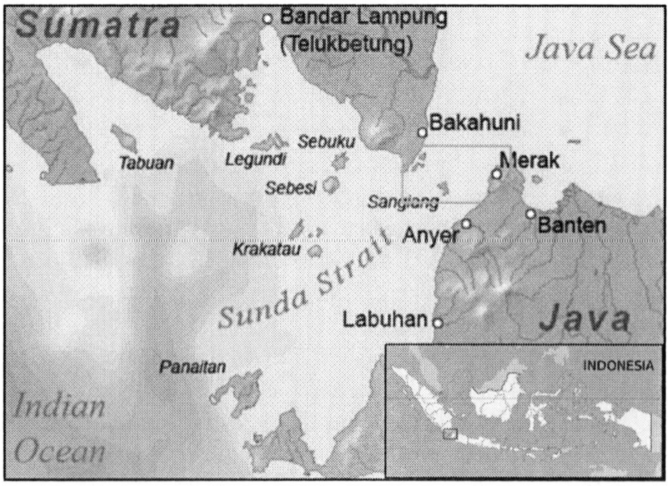

Figure 1. Main study site location of Sunda Strait in Indonesia. Study site is marked with yellow square.

$$F_x = \frac{\delta}{\delta x}(H\tau_{xy}) + \frac{\delta}{\delta y}(H\tau_{yy}) \quad (2)$$

where x, y are cartesian coordinates, H is the bottom topography, and τ_{xx}, τxy, τyx, τyy are defined as:

$$\tau_{xx} = 2A_M\frac{\delta U}{\delta x}, \tau_{xy} = \tau_{yx} = A_M\left(\frac{\delta U}{\delta y} + \frac{\delta V}{\delta x}\right), \tau_{yy} = 2A_M\frac{\delta V}{\delta y} \quad (3)$$

where U, V represent tidal current velocities in the x and y directions, and A_M is the Smagorinsky diffusivity coefficent, which decreases as the model resolution increases.

The model domain (Figure 2) encompasses a 36 km by 18.7 km area of the Sunda Strait, divided into a 61x61m grid. Depths in the model range from 5 m to 120 m, divided into four vertical layers. Tidal data are sourced from the Tide Model Driven (TMD) tool provided by ESR/OSU [20], which offers tidal predictions. Given Indonesia's location between the Pacific and Indian Oceans, the region's ocean circulation is heavily influenced by monsoons [21]. Therefore, simulations were conducted for two scenarios: (1) the South East Monsoon (SEM) using May 2017 tidal data, and (2) the North West Monsoon (NWM) using November 2017 tidal data.

Tidal increments were set at one hour, with principal lunar (M_2) and solar (S_2) semidiurnal constituents representing the tide's key characteristics in the Sunda Strait. The model simulated a four-week period, including a two-day spin-up phase. External time steps for two-dimensional simulations of water elevation and tidal currents were set to one second, while internal three-dimensional time steps for currents, salinity, and temperature were set to five seconds.

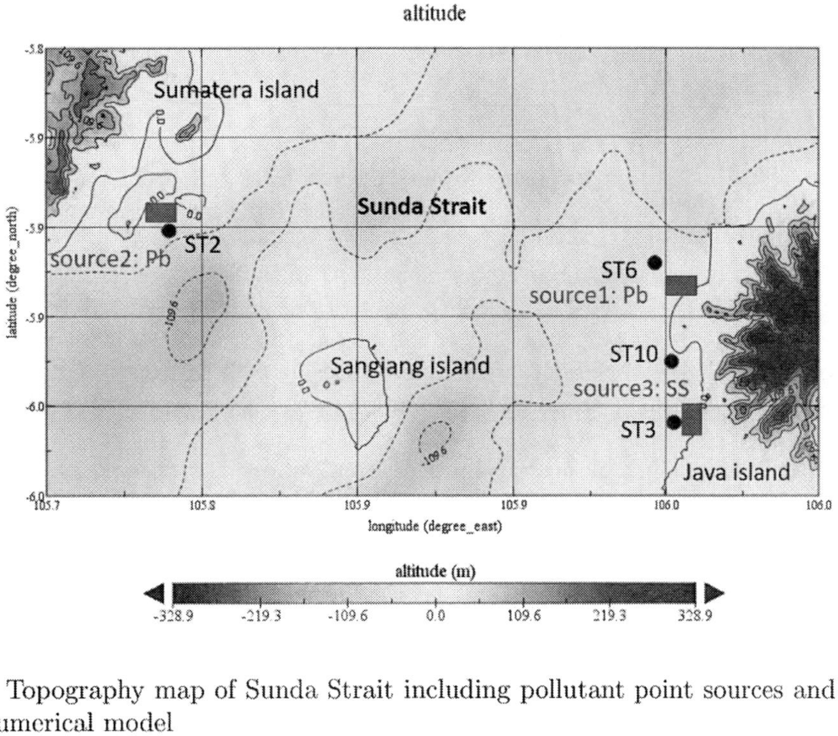

Figure 2. Topography map of Sunda Strait including pollutant point sources and monitoring points of numerical model

Figure 2 shows the layout of the study area, including twelve monitoring points for collecting numerical results. Pollutant point sources were placed in key locations representing major ports and areas of suspended solid (SS) accumulation along the southern coast of Java. To assess the impact of heavy metal contamination, the model simulates the spread of lead (Pb) concentrations

Table 1. Computational conditions

Case	Monsoon season	Location	Evaluated items
1	South East	Source1	Pb, Algae
		Source2	Pb, Algae
		Source3	SS
2	North West	Source1	Pb, Algae
		Source2	Pb, Algae
		Source3	SS

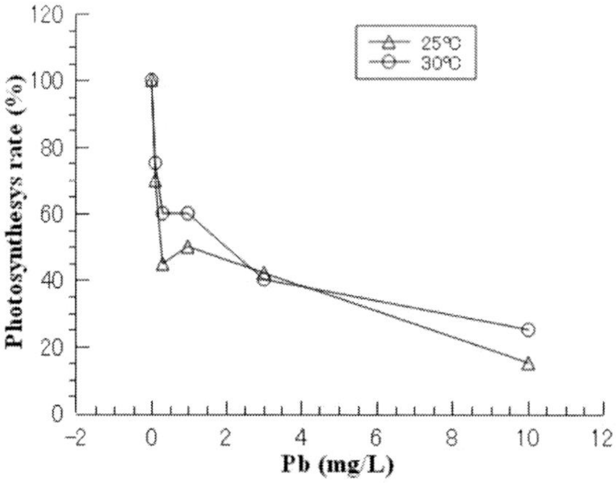

Figure 3. Photosynthesis rate of *Bacillariophyceae* algae after 72hours laboratory treatment [14]

from two primary point sources, located near major ports in Java (Source 1) and Sumatra (Source 2). Suspended solids (*SS*), a significant factor affecting water quality and sunlight penetration, are introduced from a third source (Source 3) along the southern coast of Java. These pollutants are dispersed by tidal currents modeled for both the South East Monsoon (SEM) and North West Monsoon (NWM) seasons. Lead (*Pb*) concentrations are modeled based on previous field data [22], which reported the highest concentrations at the sea bottom, reaching 20 mg/L in both monsoon seasons [8]. Suspended solid concentrations, which range from 25 to 100 mg/L in the Sunda Strait area [23], are included to simulate the high levels of turbidity common in this tropical region.

To understand the effect of heavy metal contamination, the model simulates lead (*Pb*) concentrations from these point sources, along with SS distribution, driven by tidal currents. The influence of lead (*Pb*) on phytoplankton algae, specifically the *Bacillariophyceae* species, is based on Woolery et al. [14] laboratory findings. Their study demonstrated that increasing lead (*Pb*) concentrations inhibit algae photosynthesis, with concentrations above 10 mg/L completely halting the process. By applying this data to the Sunda Strait, the study estimates how lead (*Pb*) contamination affects the photosynthesis of *Bacillariophyceae* algae, which is dominant in this region [16]. The summary of computational conditions is presented in Table.1.

3. Results and Validation

The South East Monsoon (SEM) and North West Monsoon (NWM) simulations were validated by comparing the computed water elevations to tidal data from the Geospatial Information Agency Indonesia [24], which integrates permanent tide station and satellite altimeter data.

As shown in Figure 4, the simulation results align well with field observations, though slight discrepancies were noted in the tidal phase during the first two days, which is expected due to the model's spin-up period.

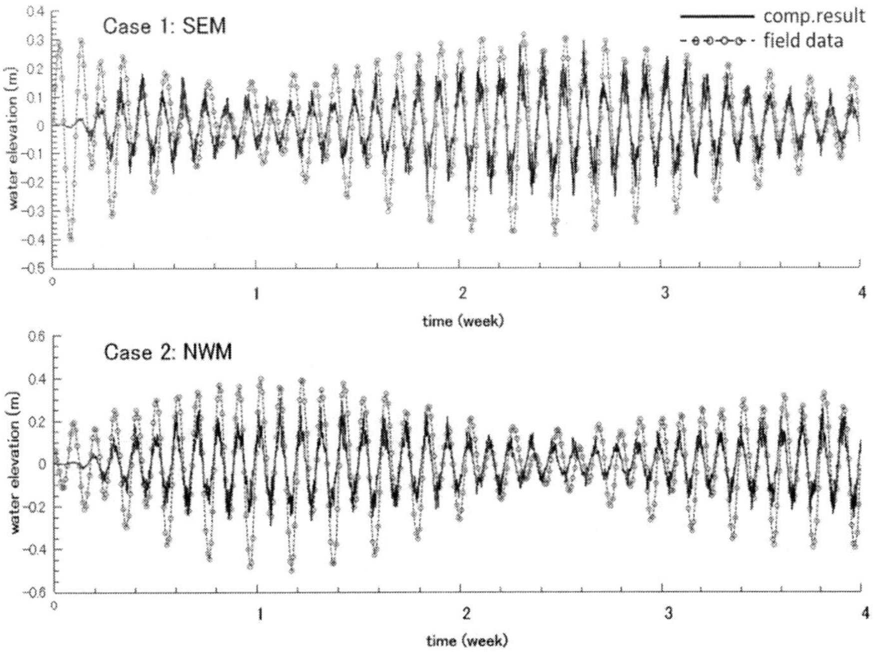

Figure 4. Water elevation of Sunda Strait in SEM and NWM season validated with tide data from Geospatial Information Agency Indonesia [24]

During the SEM season, tidal currents predominantly flow from the southern to northern sections of the strait, both at the surface and near the bottom, with maximum velocities reaching 1.5 m/s. In contrast, in the NWM season, tidal currents primarily flow southward with a slightly lower maximum velocity of 1.4 m/s (Figure 5). Both seasons show strong coastal currents, with SEM favoring northward pollutant spread and NWM distributing contaminants more evenly across the strait.

Lead (*Pb*) concentrations spread significantly during both monsoon seasons. In the SEM season, lead (*Pb*) contamination from Source 1 (Java coast) spreads northward within four weeks, affecting a large area (Figure 6). Suspended solid (*SS*) particles from the southern coast of Java are also dispersed widely, increasing turbidity and reducing light penetration (Figure 8). In the NWM season, the weaker tidal flow results in lead (*Pb*) contamination covering a wider area, but with a more evenly distributed concentration pattern (Figure 7). The suspended solid (*SS*) distribution in NWM (Figures 8) is more extensive than in SEM, which may lead to more severe environmental effects, such as decreased oxygen levels and higher water temperatures [25].

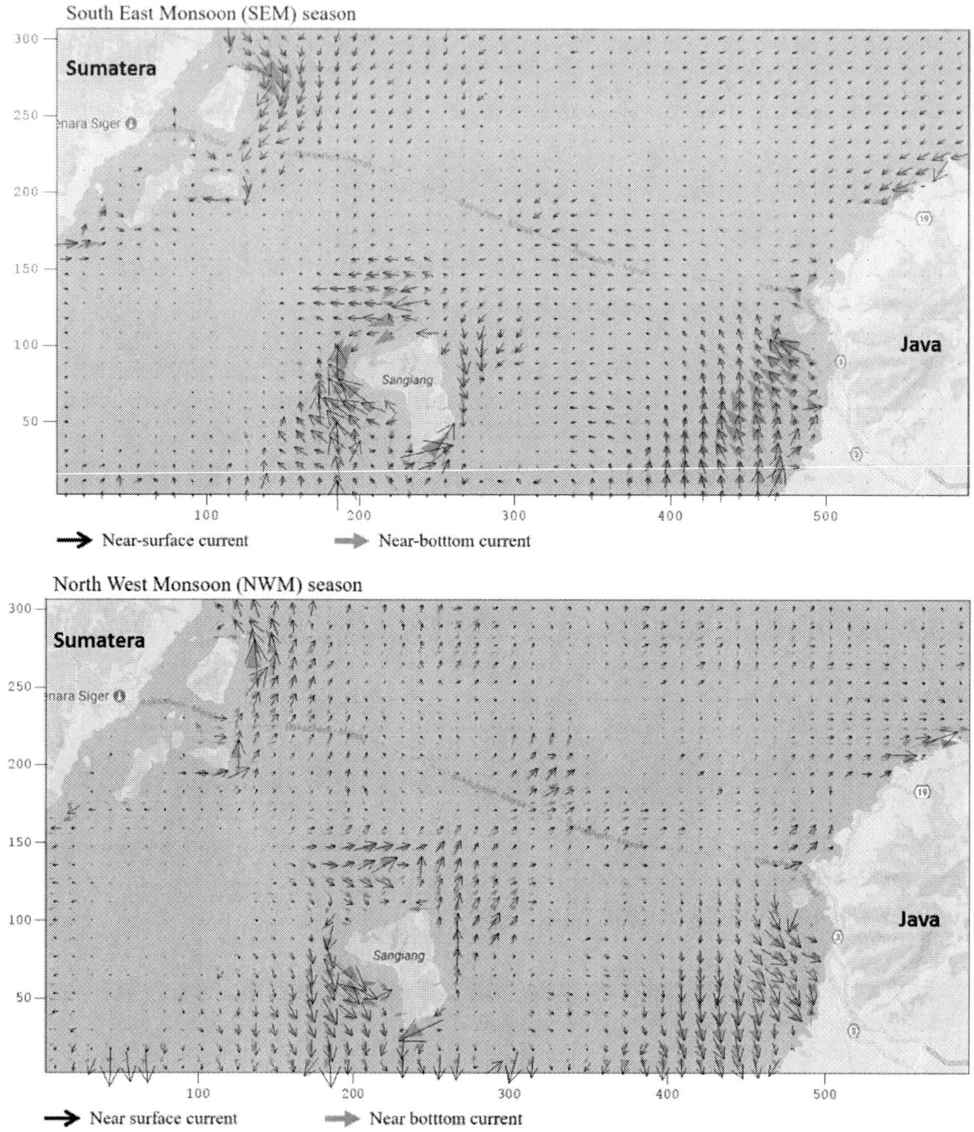

Figure 5. Tidal current flow in Sunda Strait at different monsoon season

The simulations indicate that within a 5 km radius of lead (*Pb*) point sources, photosynthesis rates of algae plummet to 0% due to the lethal effects of high lead (*Pb*) concentrations. Both monsoon seasons show this drastic reduction in photosynthesis in areas close to Source 1, with SEM exhibiting a more localized impact and NWM affecting a broader region. In some locations, such as ST6 near Source 1, algae photosynthesis rates dropped to less than 20% during the SEM season due to lead (*Pb*) contamination, while in other regions, such as ST2, lead (*Pb*) concentrations remained lower but still impacted photosynthesis by reducing it to 40%.

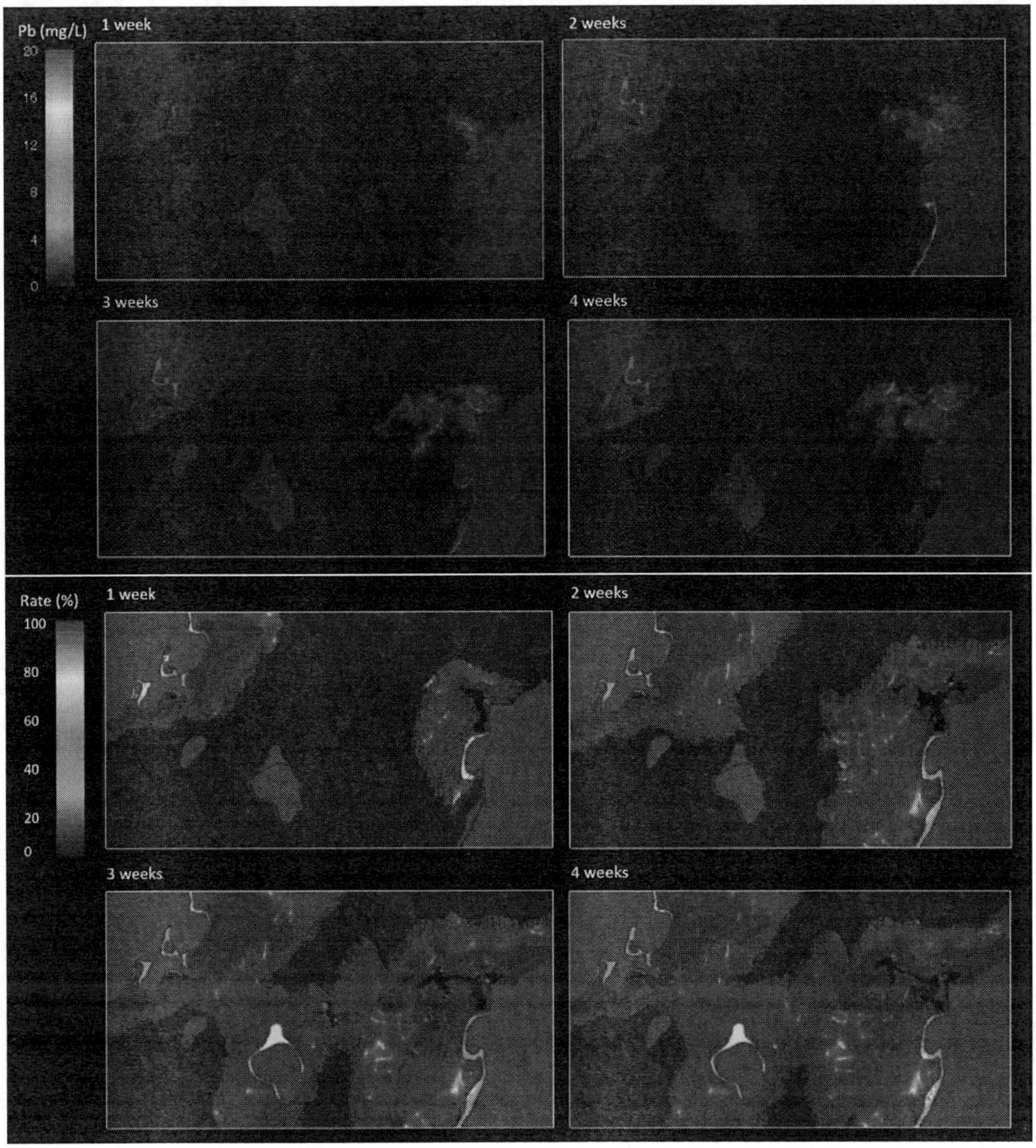

Figure 6. Lead (*Pb*) distribution from point-sources due to tidal current flow and Algae photosynthesis rate due to lead (*Pb*) contamination in SEM season

The results from SEM were further validated against field data from Suwandana et al. [11], who reported lead (*Pb*) concentrations of approximately 10 mg/L in the northern part of the Sunda Strait during the SEM season. These concentrations matched our model's predictions. In the NWM season, ST6 experienced lower lead (*Pb*) levels (around 4 mg/L), resulting in a 40% reduction in photosynthesis, whereas ST2 showed lead (*Pb*) levels as high as 8 mg/L, causing photosynthesis rates to decline by 30-80%.

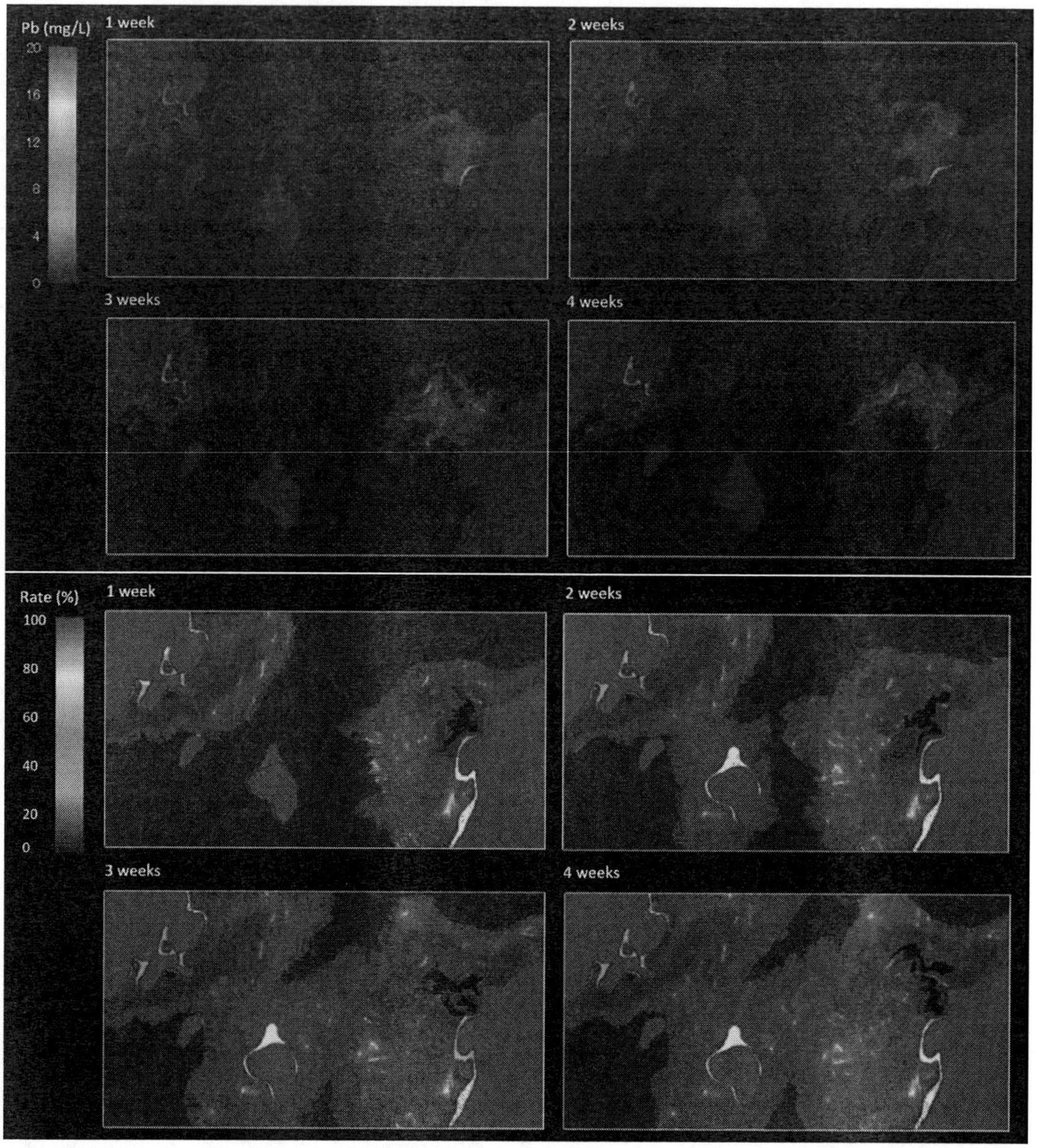

Figure 7. Lead (Pb) distribution from point-sources due to tidal current flow and Algae photosynthesis rate due to lead (Pb) contamination in NWM season

Phytoplankton abundance and biodiversity in the Sunda Strait during the SEM season have been reported to be significantly higher compared to other locations near the Java Sea [26, 27, 28]. However, the lead (Pb) contamination observed in this study poses a significant threat to marine productivity, which could negatively affect fisheries and other aquaculture sectors [29].

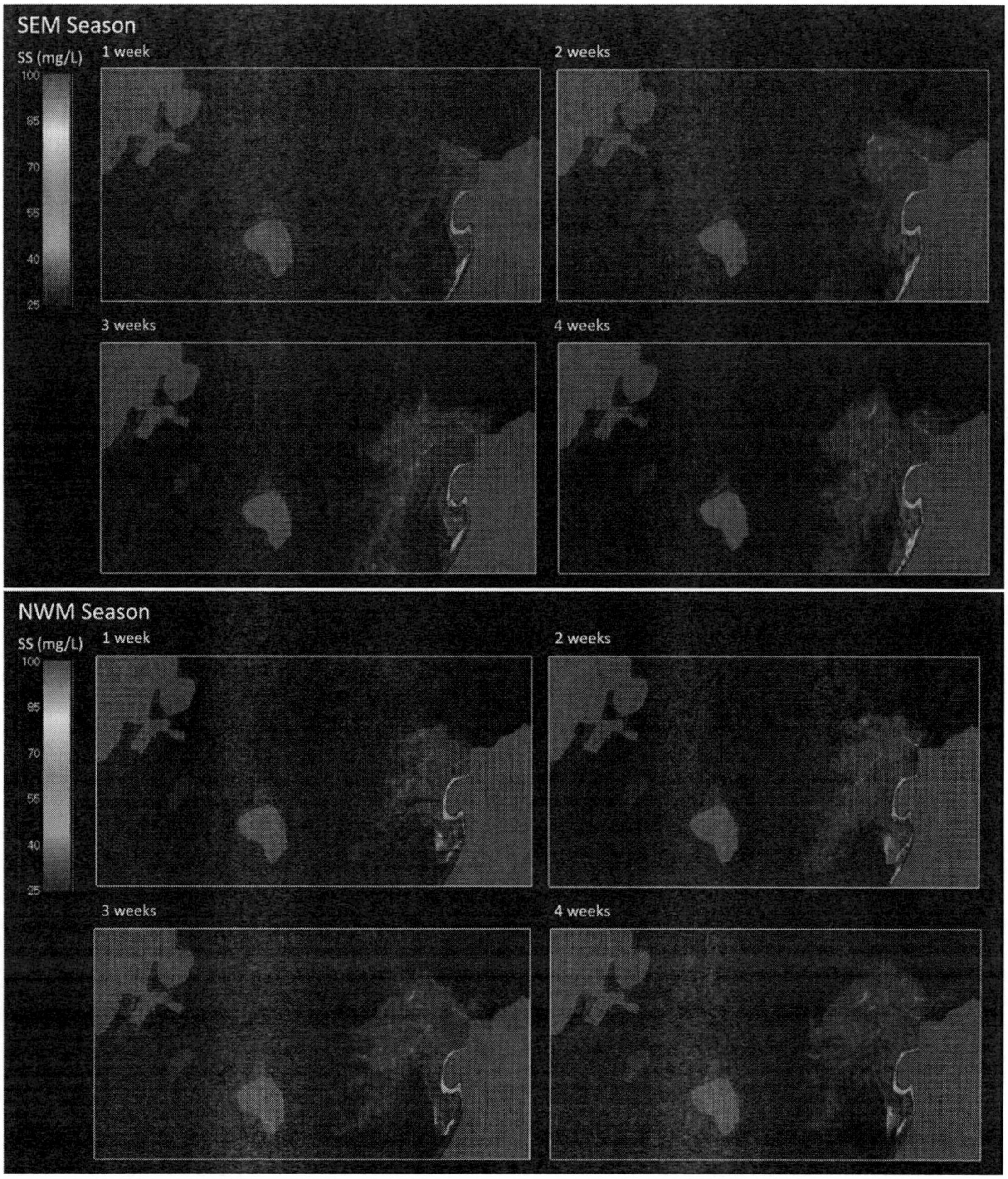

Figure 8. Suspended Solid (*SS*) distribution from point-sources due to tidal current flow in SEM and NWM season

Suspended solid concentrations were also analyzed at ST3 and ST10, near Source 3. During the SEM season, SS levels at ST10 ranged from 50-90 mg/L, and during the NWM season, they varied from 40-90 mg/L. These high suspended solid (*SS*) concentrations reduce light penetration, further inhibiting algae photosynthesis. This phenomenon is common in tropical regions with heavy rainfall, where sediment runoff from land, particularly in estuaries, leads to elevated suspended solid (*SS*) levels [30].

IOP Conf. Series: Earth and Environmental Science **1473** (2025) 012041 doi:10.1088/1755-1315/1473/1/012041

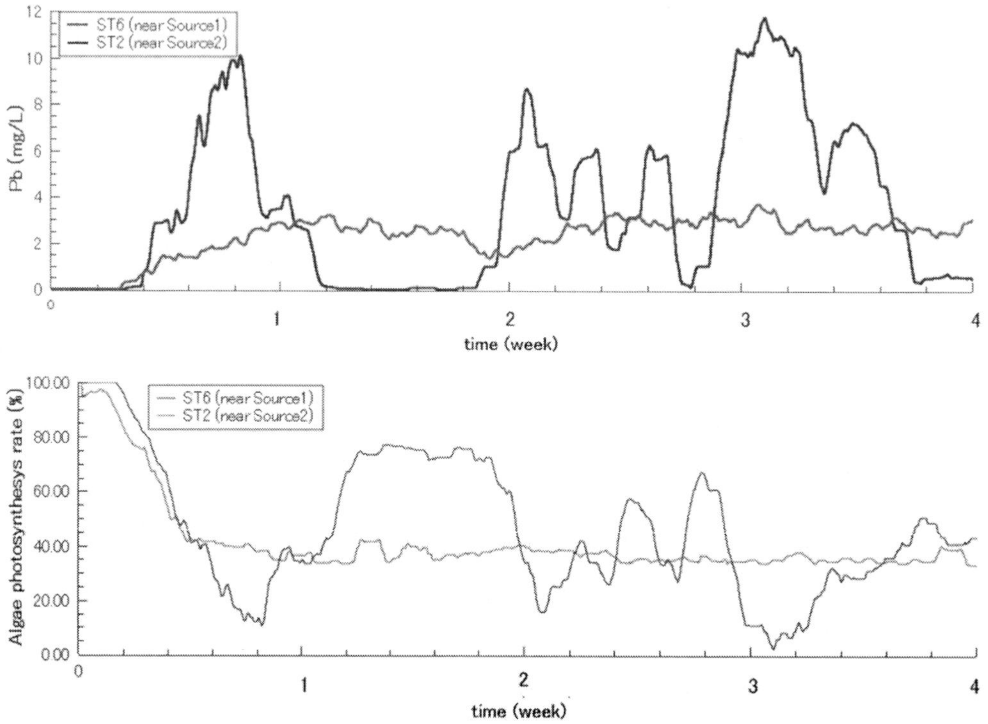

Figure 9. Time histories of lead (Pb) concentration and algae photosynthesys rate in Sunda Strait during SEM season

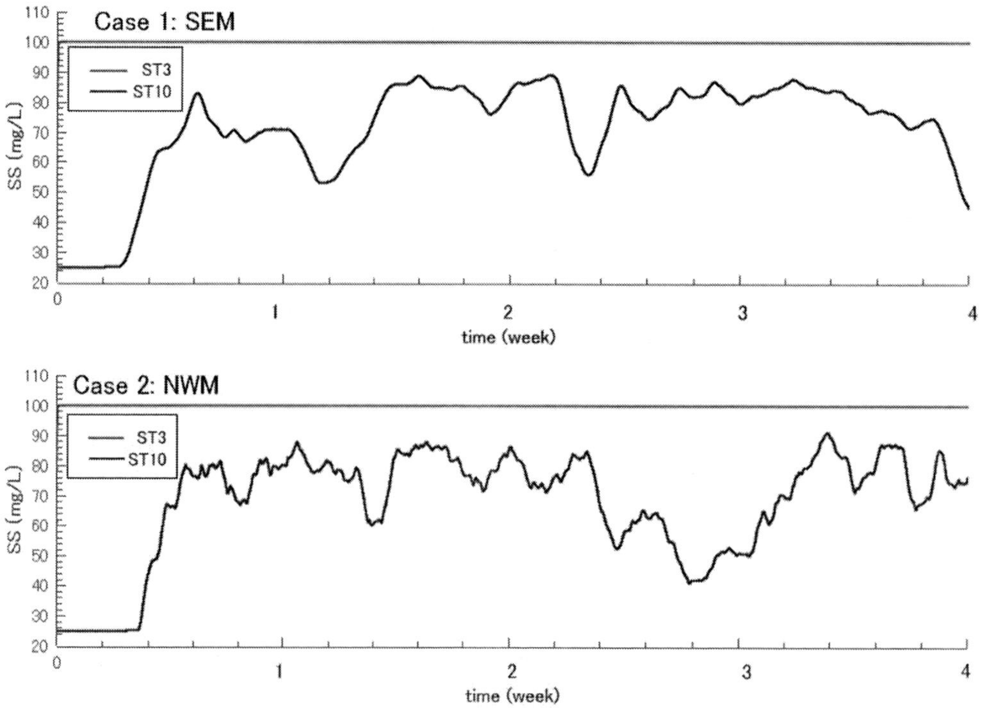

Figure 11. Time histories of suspended solid (SS) concentration in Sunda Strait during SEM and NWM season

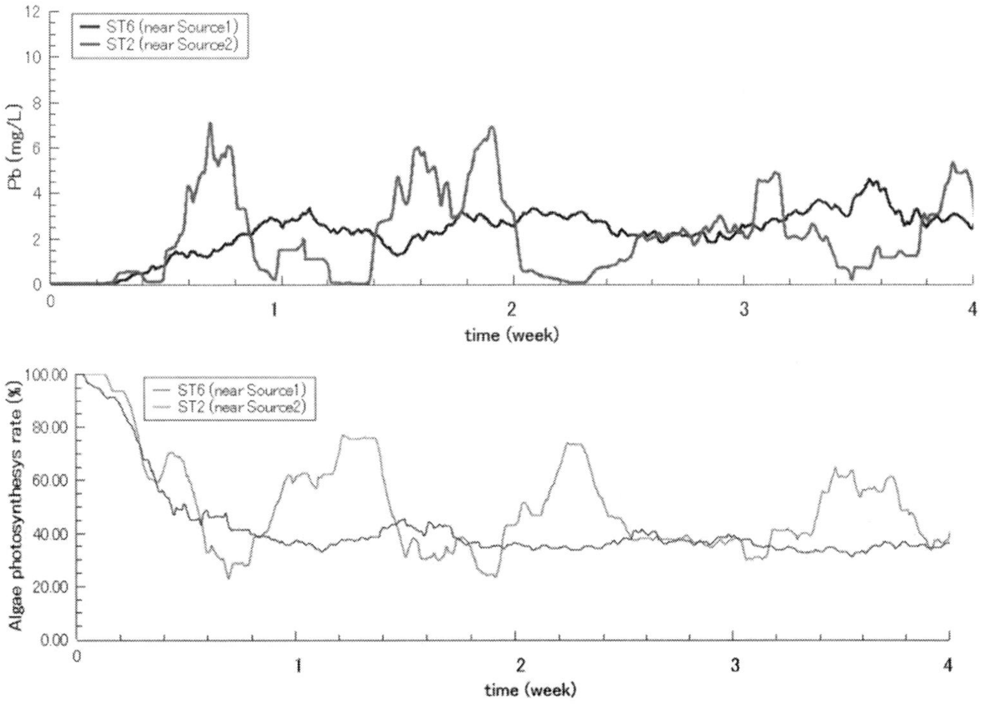

Figure 10. Time histories of lead (*Pb*) concentration and algae photosynthesys rate in Sunda Strait during NWM season

4. Conclusions

This study provides valuable insights into the distribution of marine pollution in the Sunda Strait, taking into account the effects of tidal currents during the South East Monsoon (SEM) and North West Monsoon (NWM) seasons. The simulated water levels showed good agreement with field data, confirming the reliability of the model. Lead (*Pb*) contamination during the SEM season was consistent with previous studies, and the NWM season showed a larger area of contamination, with a 60-75% reduction in phytoplankton algae photosynthesis rates. Within a 5 km radius from the primary lead (*Pb*) pollution sources, photosynthesis rates dropped to nearly 0%, posing a serious threat to the survival of phytoplankton in both seasons. Furthermore, high concentrations of suspended solids (*SS*) in the Sunda Strait, common in tropical areas, exacerbate the problem by reducing sunlight penetration and further inhibiting algae photosynthesis. These findings are crucial for improving fishery and aquaculture management in the Sunda Strait by identifying areas at risk of high lead (*Pb*) contamination. The results can inform decisions on selecting safer locations for aquaculture and contribute to more effective coastal management and pollution control strategies.

Acknowledgements

I would like to express my deepest appreciation to the Laboratory of Fluid Dynamics for Vehicle and Environmental Systems, Graduate School of Advanced Science and Engineering, Hiroshima University, for their invaluable collaboration and support in this research. I am particularly grateful for the guidance and resources provided, which were essential in the successful completion of this study on marine environmental issues.

References

[1] Soegiarto A. Introduction to The Regional Oceanography of The South East Asian Waters. In: Food and Agriculture Organization of the United Nations, editor. Lectures presented at the fifth FAO/SIDA Workshop on Aquatic Pollution in Relation to Protection of Living Resources. Manila: Food and Agriculture Organization of the United Nations; 1978. p. 1-11.

[2] Ministry of Marine Affairs and Fisheries Banten Province. Statistic Report 2014. Serang, Banten: Ministry of Marine Affairs and Fisheries Banten Province; 2014.

[3] Fadilah, Abidin Z, Kalsum U. Household Income and Welfare of Torch Fisherman in Bandar Lampung City (in Indonesian). JIIA. 2014;2(1):71-6.

[4] Ministry of Marine Affairs and Fisheries Indonesia. Number of National Fishing Equipment. Ministry of Marine Affairs and Fisheries Indonesia; 2012. Available from: `https://data.go.id/dataset/jumlah-alat-penangkap-ikan-nasional/resource/e2b424d7-2464-407c-b287-7032908ff826`.

[5] Irnawati R, Susanto A, Mustahal, A Syabana M. AES BIOFLUX Heavy metals concentration in water and sediment at Panjang Island, Serang Regency, Banten Province, Indonesia. AES Bioflux. 2014;6:256-60.

[6] Fauziah F, Choesin DN. Accumulation of Pb and Cu heavy metals in sea water, sediment, and leaf and root tissue of Enhalus sp. in the seagrass bed of Banten Bay. In: AIP Conference Proceedings. vol. 1589; 2014. p. 329-33.

[7] Susantoro TM, Sunarjanto D, Andayani A. Distribution of Heavy Metal in Sediment at Coastal Area Jambi Province. Jurnal Kelautan Nasional. 2015;10(1). Available from: `http://ejournal-balitbang.kkp.go.id/index.php/jkn/article/view/4`.

[8] Hosono T, Su CC, Delinom R, Umezawa Y, Toyota T, Kaneko S, et al. Decline in heavy metal contamination in marine sediments in Jakarta Bay, Indonesia due to increasing environmental regulations. Estuarine, Coastal and Shelf Science. 2011;92(2):297-306.

[9] Booij K, Hillebrand MTJ, Nolting RF, Van Ooijen J. Nutrients, trace metals, and organic contaminants in Banten Bay, Indonesia. Marine Pollution Bulletin. 2001;42(11):1187-90.

[10] Sakumoto N, Juwana H. Reforming Laws and Institutions in Indonesia: An Assessment. Institute of Developing Economics; 2007.

[11] Suwandana E, Kawamura K, Soeyanto E. Assessment of the Heavy Metals and Nutrients Status in the Seawater, Sediment and Seagrass in Banten Bay, Indonesia and Their Distributional Patterns. Journal of Fisheries International. 2011;6(1):18-25.

[12] Arifin Z, Puspitasari R, Miyazaki N. Heavy metal contamination in Indonesian coastal marine ecosystems: A historical perspective. Coastal Marine Science. 2012;35(1):227-33. Available from: `http://repository.dl.itc.u-tokyo.ac.jp/dspace/bitstream/2261/51708/1/CMS350132.pdf`.

[13] Palapa TM, Maramis AA. Heavy Metals in Water of Stream Near an Amalgamation Tailing Ponds in Talawaan – Tatelu Gold Mining, North Sulawesi, Indonesia. Procedia Chemistry. 2015;14:428-36. Available from: `http://linkinghub.elsevier.com/retrieve/pii/S1876619615000595`.

[14] Woolery ML, Lewin RA. The effects of lead on algae. Water, Air, and Soil Pollution. 1976 Aug;6(1):25-31.

[15] Walsh JJ. Steemann Nielsen, E. 1975. Marine photosynthesis with special emphasis on the ecological aspects. Elsevier Sci. Publ. Co., Amsterdam and New York, x + 142 p.21-75. Limnology and Oceanography. 1976;21(4):633-3. Available from: `http://dx.doi.org/10.4319/lo.1976.21.4.0633b`.

[16] Pratiwi NTM, Wulandari DY, Ayu IP, Iswantari A. Diversity and Spatial Distribution of Plankton in Connected Waters of Bali Strait, Between Eastern Part of Java and Western Part of Bali Island. IOP Conference Series: Earth and Environmental Science. 2017;54(1):12090.

[17] Ke Z, Tan Y, Ma Y, Huang L, Wang S. Effects of surface current patterns on spatial variations of phytoplankton community and environmental factors in Sunda shelf. Continental Shelf Research. 2014;82:119-27.

[18] Mellor GL. Users guide for a three-dimensional, primitive equation, numerical ocean model. vol. 8544. Princeton University; 2002.

[19] Kamenkovich VM, Nechaev DA. On the time-splitting scheme used in the Princeton Ocean Model. Journal of Computational Physics. 2009;228(8):2874-905.

[20] Padman L. Tide Model Driver (TMD) Manual. Arctic. 2004:1-12.

[21] Wyrtki K. Physical Oceanography of the Southeast Asian waters. California: The University of California; 1961. Available from: `http://www.escholarship.org/uc/item/49n9x3t4`.

[22] Ray RD, Susanto RD. Tidal mixing signatures in the Indonesian seas from high-resolution sea surface temperature data. Geophysical Research Letters. 2016;43(15):8115-23.

[23] Helfinalis. Total Suspended Solid Content And Sediment On The Bottom Surface Of Panimbang Water. Makara Seri Sains. 2005;09(2):45-51.

[24] Geospatial Information Agency Indonesia. Online Tide Prediction. Geospatial Information Agency Indonesia; 2017. Available from: `http://tides.big.go.id/pasut/index.html`.

[25] Mandal HK. Assessment of wastewater temperature and its relationship with turbidity. Recent Research in Science and Technology. 2014;6(1):258*262.
[26] Thoha H, Adnan Q, Sidabutar T, S. Note on The Occurence of Phytoplankton and Its Relation with Mass Mortality in The Jakarta Bay, May and November 2004. MAKARA of Science Series. 2010 oct;11(2).
[27] Effendi H, Kawaroe M, Lestari DF, Mursalin, Permadi T. Distribution of Phytoplankton Diversity and Abundance in Mahakam Delta, East Kalimantan. Procedia Environmental Sciences. 2016 jan;33:496-504.
[28] Cui DY, Wang JT, Tan LJ, Dong ZY. Impact of atmospheric wet deposition on phytoplankton community structure in the South China Sea. Estuarine, Coastal and Shelf Science. 2016 may;173:1-8.
[29] Vallina SM, Follows MJ, Dutkiewicz S, Montoya JM, Cermeno P, Loreau M. Global relationship between phytoplankton diversity and productivity in the ocean. Nature Communications. 2014 jul;5.
[30] Le SH, Chua LHC, Irvine KN, Eikaas HS. Modeling washoff of total suspended solids in the tropics. Journal of Environmental Management. 2017 sep;200:263-74.

IOP Conf. Series: Earth and Environmental Science **1473** (2025) 012042 doi:10.1088/1755-1315/1473/1/012042

Full scale task risk assessment of crude oil tank cleaning job using qualitative risk matrix and control in pt. saka (indonesia-pangkah) limited

Bagus C. Dermawan[1,2], **Muhammad K. Asy'ari**[1*]

[1] Engineering Physic Department, Institut Teknologi Sepuluh Nopember, 60111, Indonesia
[2] PT. Saka (Indonesia Pangkah) Limited, Indonesia

*Email: mkasyari@its.ac.id

Abstract. Oil storage tank cleaning job is an important, routine, and high-risk activity because workers enter confined spaces that are potentially exposed to hydrocarbons and hydrogen sulphide. This work requires a full-scale task risk assessment, considering that there are several categories of work that have very high potential hazards for workers. A full-scale task risk assessment is a method of assessing the risk of a job (project) as a whole, covering several categories of work that require a task risk assessment. The purpose of this study was to conduct a full-scale task risk assessment using quantitative risk analysis and control on crude oil storage tank cleaning work at PT. Saka Indonesia Pangkah Limited. A full-scale task risk assessment includes several stages, including categorizing job descriptions, identifying hazards, determining consequences, qualitative risk analysis, and determining risk control. Qualitative risk analysis includes determining the likelihood, severity, value, and risk category obtained from the risk matrix. The assessment results show the risk value of each category of 25 percent (low), 50 percent (moderate), and 25 percent (high). The application of risk control can reduce work that is categorized as high (intolerable) to moderate (tolerable). The risk values for each category are 25 percent (low) and 75 percent (moderate).

1. Introduction

Storage tank cleaning job is one of the most important and routine activity on an oil and gas companies both for land storage or floating storage (FPSO/FSO/Tanker). The purpose of tank cleaning job is to maintain the tank both internally or externally and also to clean the residue/mud left/deposited in the storage tank so that it can maximize the tank capacity for product storage and reduce the risk of plugging in the pipeline or metering system when carrying out offloading or transfer activities products for sale. Tank cleaning safe guard standard like NFPA 326 should be followed for this kind of activity.[1] Tank cleaning job is an activity that is full of danger with high risk, one of which is because this job is related to the man entry into a confined space with the potential content of flammable gas (hydrocarbons) and toxic gas (H_2S).[2]

The high-risk potential arising from tank cleaning activities includes dangers to personnel (from injury to fatality), property/assets (from damage, explosions to fires) and the environment (from spill to pollution) which will also impact the company's occupational safety reputation. There's many reported incidents happen in tank cleaning job activities. On 11 June 1987 at the Dalmeny Oil Storage Terminal, one of four person contractors who work on cleaning inside floating roof crude oil tank died from the

IOP Conf. Series: Earth and Environmental Science **1473** (2025) 012042 doi:10.1088/1755-1315/1473/1/012042

effects of asphyxiation and burns that cause of a bad behaviour of the worker that smoking inside the tank during cleaning activity.[3] 18 April 2016, at approximate position of 5°30.89'N, 99°56.90'E in Malacca Strait, whilst enroute from Penang, Malaysia to Singapore, the Hong Kong registered oil/chemical tanker "NO.3 HEUNG-A PIONEER" (the vessel) exploded on the main deck during cargo oil tank cleaning operation due to wrong manifold connection. The explosion resulted in one death and five injuries of the crew members.[4]

James I. Chang and Cheng-Chung Lin already categorize the causes of tank accidents occurred in last 40 years during period 1960 – 2003.[5] From 242 total of accidents, we can see that 32 accidents (12%) occurred during maintenance/hot work activities (see **Table 1**).

Table 1. Categorization cause of tank accidents.

Cause of accident	1960 - 1969	1970 - 1979	1980 - 1989	1990 - 1999	2000 - 2003	Total
Lightning	4	10	19	37	10	80
Maintenance/hot work	1	5	9	12	5	32
Operational error	1	5	6	8	9	29
Equipment failure	3	1	5	7	3	19
Sabotage	2	5	2	6	3	18
Crack/rapture	0	3	3	3	8	17
Leaks and line rapture	0	3	2	5	5	15
Static electricity	2	1	2	2	5	12
Open flame	1	0	4	2	1	8
Nature disaster	1	2	1	1	2	7
Runaway reaction	2	1	0	2	0	5
Total Accidents	**17**	**36**	**53**	**85**	**51**	**242**

Considering a high-risk potential from this tank cleaning activity (maintenance), it is very important to study the potential hazards and how to mitigate the possible risks that arise so that they can be eliminated or reduced, so that the job can be completed safely on schedule. Risk assessment is part of the safety management system and is the basis for assessing risk in a job that contains risk in every company. Risk assessment is one method in identifying potential hazards, analyze and evaluating risks to determining appropriate risk control.[6]

The purpose of this study is to identify hazards using the Full-Scale Task Risk Assessment method and assess hazard risks using Qualitative Risk Analysis on internal tank cleaning work at PT. SAKA (Indonesia-Pangkah) Limited. by using the Risk Matrix table to obtain the value/ranking of the initial hazard risk (initial risk), then appropriate, effective and safe risk control mitigation is carried out as evidenced by the reduced hazard risk value from the risk matrix as residual risk.[7]

2. Methodology

In this study, in addition to conducting interviews with the operations team/personnel and field surveys, several data are also needed, such as personal details that will work along with their competencies, work procedures, equipment to be used and P&ID of the facility or asset where the work will be carried out related to LOTO (Lock Out - Tag Out). The method used in this study is full-scale task risk assessment using qualitative risk matrix with risk control.

Full-scale task risk assessment is a more detailed risk assessment to assess the risks that will arise from both personnel (workers) or equipment used so that mitigation can be carried out to reduce the risk value of the risk matrix in the work activity. In making this full-scale task risk assessment, based on the data obtained, a categorization can be made that includes a description of the job, type of hazard and the consequences of the hazard.

2.1. Task Risk Assessment (TRA).

The activity of cleaning crude oil storage tanks is a fairly complex job, which generally includes the following activities, job preparations, initial job, main job and closing job activity. We at PT. SAKA (Indonesia-Pangkah) Ltd. use the full-Scale Task Risk Assessment method, which is a Task Risk

Assessment (TRA) that is interconnected in one job project (integrated job). The following **Figure 1** is an overview of the full-scale ask risk assessment for the cleaning crude oil tank job project.

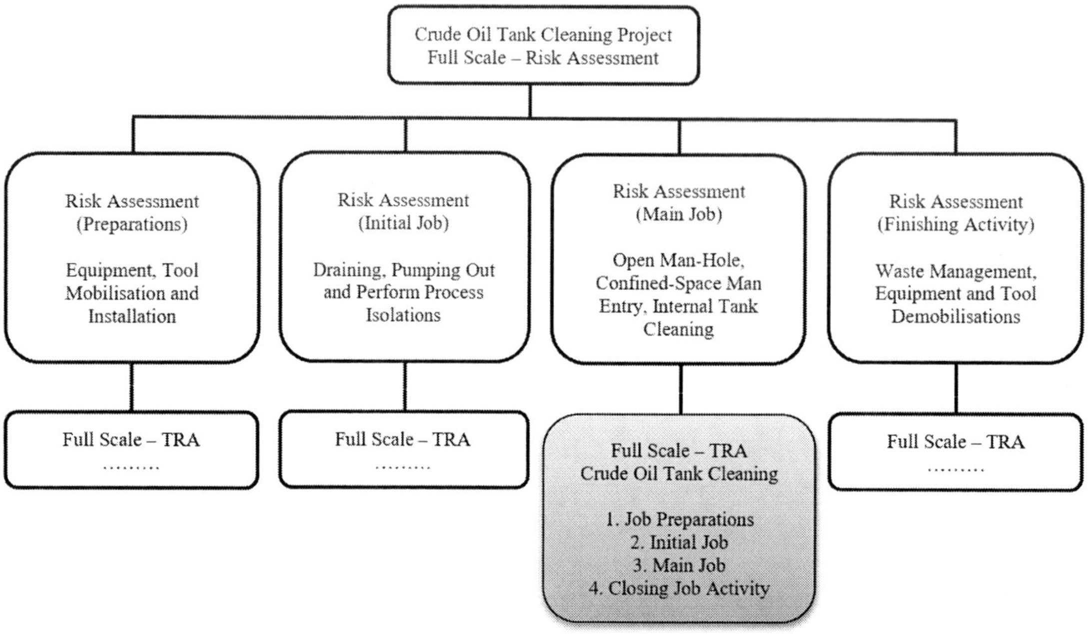

Figure 1. Full-scale task risk assessment for crude oil tank cleaning project.

This full-scale task risk assessment is used as a basis for risk assessment in each work activity which will then be assessed for its possibility and consequences against the risk matrix. Possibility is the chance of a deviation that can be measured or calculated within a certain period of time (Likelihood). Consequences (Severity) is the results of an event, usually expressed as a loss due to a risk.

2.2. Qualitative Risk Analysis (QRA).

Qualitative risk analysis uses words to describe the magnitude of potential severity and likelihood of occurrence. The level of risk is then expressed in words such as high, medium and low. Qualitative analysis is commonly used in high level initial risk assessments, any qualitative risk assessment in workplace risk assessments such as job safety analysis and hazard detection. This is mainly because the qualitative approach is more suited to the detail required in assessing the level of risk being managed. If a high-level risk is identified, the normal practice is to undertake a more detailed risk assessment.[6]

In this risk analysis, there are 2 main components in determining the risk value of a job, namely Likelihood (L) and Severity (S). Likelihood is the possibility of a hazard occurring categorized by reviewing and selecting descriptors. The choice is based on the information available to the team and based on the team's knowledge and experience of industry and company data. While Severity is the severity of the hazard that is realized categorized by reviewing and selecting descriptors. The team reviews and selects the severity level that best represents the seriousness of the possible consequences if the incident occurs. All four types of severity levels must be considered (human, asset/production, environment and reputation). Each categorized severity level is represented in the risk matrix with a description (e.g. disaster). The selected descriptor must represent the most severe category agreed upon by the team. For example, the environmental severity level is low (slight) but the asset/production severity level is high (severe), then high is the average severity rating that should be used.[7]

The possibility of a hazard occurring (Likelihood) is categorized into 5, as shown in **Table 2**.[8]

Table 2. Likelihood categories.

Rare (1)	The event may only occur in exceptional circumstances	The team has never heard of such an event occurring in SAKA or Industry generally
Unlikely (2)	The event could occur at some time	Only a few occurrences are known of in the experience of the team
Possible (3)	The event may occur at some time	The team know of a few occurrences but not in the last few years
Probable (4)	The event will probably occur in the most circumstances	It is a common occurrence in industry bit not heard of SAKA in last year
Almost Certain (5)	The event is expected to occur in most circumstances	The team know of several events occurring in recent months

The severity of a hazard (Severity) is categorized into 5 with 4 levels of consideration as shown in **Table 3**.[8]

Table 3. Severity with decision issue categories.

SAVERITY	Decision Issue			
	People	**Asset/Production**	**Environment**	**Reputation**
5 Catastrophic	Potential fatalities or permanent disabling injury or illness	Extensive damage; long term impact on production; long term loss in revenue	Massive effect; environmental impact could last for decades; long term contamination requiring remediation	Impact would receive national and global attention
4 Major	Severe injury or illness / Hospitalisation / Disability	Major damage; Major damage to equipment; Delay in operations; short-term loss in revenue	Major effect; Environmental Impact could last for years; area becomes restricted for a limited period of time	Impact would receive regional and industry coverage; potential chronic health impact to community
3 Severe	Severe injury or illness / Lost Time	Local Damage; severe damage to equipment impact on part of operations; partial loss of revenue	Severe impact; environmental impact could last for months; reportable quantity spill or release; spill or release requires clean-up	Considerable impact to local community; potential acute health impacts to community; community response plan activated
2 Minor	Minor injury or illness / OSHA recordable / Doctor visit	Minor Damage; damage to equipment; minor impact on operations; no loss in revenue	Minor impact; environmental impact could last for weeks; spill or release external to facility; no clean-up required	Minor impact; immediate area to facility may be alerted; odour noise complaints
1 Slight	Slight injury or illness / First Aid Case	Slight damage; no significant impact on operations; no loss in revenue	Slight effect; environmental impact could last for days; no long-term consequences; spill or release internal to facility	Slight impact; no impact to local community, little notice by the community

Risk is the result of the possibility of hazard (Likelihood) and how big consequence of hazard has an impact (Severity) as shown in **Equation 1** below.

$$\text{Risk (R)} = \text{Possibility (L)} \times \text{Consequence (S)} \quad (1)$$

The risk assessment relationship between severity and likelihood can be done by using the Risk Matrix Table as shown in **Figure 2**.[8]

Severity	Likelihood				
5 Catastropic	5 Medium	10 Medium	15 High	20 High	25 High
4 Major	4 Medium	8 Medium	12 Medium	16 High	20 High
3 Severe	3 Low	6 Medium	9 Medium	12 Medium	15 High
2 Minor	2 Low	4 Medium	6 Medium	8 Medium	10 Medium
1 Slight	1 Low	2 Low	3 Low	4 Medium	5 Medium
	1 Rare	2 Unlikely	3 Possible	4 Probable	5 Almost Certain

Figure 2. Risk matrix table.

In the Risk Matrix Table there are values that determine the Risk Level with 3 color categories as shown in **Table 4**.[8]

Table 4. Risk categories.

RISK	DESCRIPTIONS
HIGH (Intolerable)	If the risk level is HIGH, it is considered to be unacceptable. High is illustrated on the Risk Matrix as the area shaded in red. If a high-risk result remains, once all available controls have been identified, the task must not be undertaken. Further review, consultation, and risk assessment is required.
MEDIUM (Tolerable)	A risk defined as MEDIUM is considered tolerable, this region of the Risk Matrix is shaded yellow. Although these risks are in the tolerable range, effort should still be made to reduce them to levels that are As Low As Reasonably Practicable. This is the ALARP principle and it must be clearly demonstrated.
LOW (Acceptable)	The acceptable risk criteria is described as LOW, this is the region on the Risk Matrix shade in green. Any risk that lies below the lower boundary of the tolerable region is broadly acceptable. Note that if a risk is acceptable this does not necessarily preclude the initiation of improvements if they are economic, readily identified and practicable.

2.3. Risk Control.

From the known hazard identification data, qualitative analysis is carried out and the initial risk value is determined using the risk matrix table, so it is necessary to carry out control or management of the possible risks or hazards that arise during the job. This is done in order to reduce the risk value that arises and obtain a lower and safer final risk value (residual risk) than before. The order of control over the risks is as follows:

- Eliminate (eliminating risk).
- Substitute (replace with lower risk).
- Engineering (perform technical modification).
- Isolation (perform process or electrical isolation).
- Administrative (document or procedure administration, e.g. PTW, Certificate, SOP, manual, etc.).
- PPE/Safety (safety equipment, personal or special protective equipment).

Risk control is carried out starting with the easiest, most effective and safest sequence in reducing the risks that may arise.[8]

3. Results and Discussion.

After reviewing procedures, interviews, discussions and field surveys, we can conclude the following categorization as listed in **Table 5**:

Table 5. Categorization for job descriptions, hazards identification and hazard effect.

Job Descriptions	Hazards Identification	Hazard Effect
1. Job Preparations: • Working tools and equipment. • Personal protective equipment. • Safety equipment or facility.	1. Personal injury: • Slip, Trip and Fall. • Pinch Point and Impact Injury. • Ergonomic Issue. 2. Nature/local hazard: • Bad Weather.	• Minor Personal Injury.
2. Initial Jobs: • Open man-hole. • Operating air compressor. • Operating pneumatic blower. • Operating pneumatic lamp.	1. Personal injury: • Slip, Trip and Fall. • Pinch Point and Impact Injury. • Ergonomic Issue. 2. Equipment/asset failure or damage: • Improper Connections hazard. • Pressurized Air hazard. • Air compressor operating failure. • Vibration and noise hazard. • Electrostatic and Spark Potential hazard. 3. Environmental issue: • Oil Spill. 4. Nature/local hazard: • Bad Weather.	• Minor Personal Injury. • Asset and Equipment/Property Damage. • Environmental Issue. • Fire hazard.
3. Main Jobs (Confined-Space Entry): • Pumping sludge from outside the tank. • Manual cleaning. • Rinsing with crude oil. • Pumping sludge from inside the tank.	1. Personal injury: • Slip, Trip and Fall. • Pinch Point and Impact Injury. • Ergonomic Issue. 2. Equipment/asset failure or damage: • Improper Connections. • Pressurized Air. • Bad Breathing Apparatus Air Quality. • Vibration and Noise. • Internal Coating Scratch. • Electrostatic and Spark Potential. 3. Environmental issue: • Oil Spill. 4. Nature/local hazard: • Combustible Gas. • Toxic Gas or H2S. • Pyrophoric Material. • Confined Space. • Bad Weather.	• Minor Personal Injury. • Personal Fatality. • Asset and Equipment/Property Damage. • Environmental Issue. • Fire hazard.
4. Closing Job Activity : • Housekeeping.	1. Personal injury: • Slip, Trip and Fall. • Pinch Point and Impact Injury. • Ergonomic Issue. 2. Environmental issue: • Hazardous Waste. 3. Nature/local hazard: • Bad Weather.	• Minor Personal Injury. • Environmental Issue.

Based on the data from **Table 5**, an initial risk rating assessment (R) can be carried out using Qualitative Risk Analysis (QRA) to describe the magnitude of potential severity (using **Table 3**) and the likelihood of occurrence (using **Table 2**), then applied to the risk matrix table (**Figure 2**) using **Equation 1**. The severity rating (S) is determined by the magnitude of the worst potential severity (the higher number, the higher severity), while likelihood (L) determined by its possibility of a hazard occurring.

Table 6. Initial risk assessment using quantitative analysis applied on risk matrix table.

Hazards Identification	Severity (Table 3)				Likelihood (Table 2)	Initial Risk		
	People	Asset / Production	Environ-ment	Reputation		S	L	R
1. Job Preparations:						2	3	6
• Slip, Trip and Fall.	2	-	-	-	3			
• Pinch Point and Impact Injury.	2	-	-	-	3			
• Ergonomic Issue.	1	-	-	-	3			
• Bad Weather.	1	1	1	1	2			
2. Initial Jobs:						4	3	12
• Slip, Trip and Fall.	2	-	-	-	3			
• Pinch Point and Impact Injury.	2	-	-	-	3			
• Ergonomic Issue.	1	-	-	-	3			
• Improper Connections hazard.	1	3	1	-	2			
• Pressurized Air hazard.	2	1	-	-	2			
• Air compressor operating failure.	1	2	-	-	2			
• Vibration and noise hazard.	1	1	1	-	2			
• Electrostatic and Spark Potential	-	4	-	4	2			
• Oil Spill.	-	1	1	1	3			
• Bad Weather.	1	1	1	1	2			
3. Main Jobs (Confined-Space Entry):						5	3	15
• Slip, Trip and Fall.	2	-	-	-	3			
• Pinch Point and Impact Injury.	2	-	-	2	3			
• Ergonomic Issue.	1	-	-	-	3			
• Improper Connections hazard.	1	3	1	-	2			
• Pressurized Air hazard.	2	1	-	-	2			
• Bad Breathing Apparatus Air Quality.	4	-	-	-	2			
• Vibration and Noise.	1	1	1	-	2			
• Internal Coating Scratch.	-	3	-	-	3			
• Electrostatic and Spark Potential.	-	4	-	4	2			
• Combustible Gas.	5	-	1	4	1			
• Toxic Gas or H2S.	5	-	1	4	1			
• Pyrophoric Material.	5	4	2	4	1			
• Confined Space.	5	-	-	4	1			
• Oil Spill.	-	1	1	1	3			
• Bad Weather.	1	1	1	1	2			
4. Closing Job Activity :						2	3	6
• Slip, Trip and Fall.	2	-	-	-	3			
• Pinch Point and Impact Injury.	2	-	-	-	3			
• Ergonomic Issue.	1	-	-	-	3			
• Hazardous Waste.	1	-	2	2	2			
• Bad Weather.	1	1	1	1	2			

The results of initial risk rank as in **Table 6**, found that the job activities were categorized as High (Intolerable) which according to the company's rules, the job should not be done. Therefore, it is necessary to carry out risk control to minimize/reduce the initial risk rank for the job can be executed. The risk controls that can be implemented on each job descriptions to reduce residual risk are as show in **Table 7**:

Table 7. Risk control categories implementation.

Job Descriptions	Risk Control	Residual Risk		
		S	L	R
1. Job Preparations: • Working tools and equipment. • Personal protective equipment.	**Basic Risk Control** 1. Administrative : • All personnel involved already getting HSE Induction – Refer to SITE-PRO-ON-SIPL-HSE-008 - OPF Personnel Entry Minimum Requirement Rev 01. • Obtained PTW and TRA in place – Refer to SITE-PRO-SIPL-HSE-001 - Permit To Work Rev 01.	2	2	4

Job Descriptions	Risk Control	Residual Risk		
		S	L	R
• Safety equipment or facility.	• Follow manual handling procedure – Refer to SITE-PRO-SIPL-HSE-010 - Manual Handling Rev 01. • Conduct a Toolbox meeting for all personnel involved prior job execution. • Maintain Communication and Coordination with Area Operators or CCR. • Tools and equipment have been inspected by SIPL maintenance team. • Prepare check list and record sheet. 2. PPE/Safety : • Proper PPE Mandatory is required, (Safety helmet, safety glasses, gloves, safety shoes, FRC coverall) – Refer to STD-SEI-HSE-009 Personal Protective Equipment Rev 0. • Provide safety shower near safe worksite. 3. Subtitute : • Use only Ex-HT radio communication. • Use only non-spark tools. • Use appropriate equipment properly and according to procedure. 4. Isolation : • Refer to existing Process Isolation Certificate. 5. Engineering : • Provide service line (Plant Air and Utility Water) near safe worksite. 6. Eliminate : • Make sure the worksite is free from trip hazard before the jobs start. • Stop The Job on unsafe condition includes rainy.			
2. Initial Jobs: • Open man-hole. • Operating air compressor. • Operating pneumatic blower. • Operating pneumatic lamp.	**Initial Job Risk Control** 1. Administrative : • All personnel involved already getting HSE Induction – Refer to SITE-PRO-ON-SIPL-HSE-008 - OPF Personnel Entry Minimum Requirement Rev 01. • Obtained PTW and TRA in place – Refer to SITE-PRO-SIPL-HSE-001 - Permit To Work Rev 01. • Follow manual handling procedure – Refer to SITE-PRO-SIPL-HSE-010 - Manual Handling Rev 01. • Hose has been leak-tested and inspected (follow 029/ORA/SIPL-OPS). • Manifold has been inspected. • All PSV has a valid calibration certificate. • Conduct a Toolbox meeting for all personnel involved prior job execution. • Maintain Communication and Coordination with Area Operators or CCR. • Tools and equipment have been inspected by SIPL maintenance team. • Prepare check list and record sheet. 2. PPE/Safety : • Proper PPE Mandatory is required, (Safety helmet, safety glasses, gloves, safety shoes, FRC coverall) – Refer to STD-SEI-HSE-009 Personal Protective Equipment Rev 0. • Provide safety shower near safe worksite. • All Equipment are tagged, marked and signed properly. 3. Subtitute : • Use only Ex-HT radio communication. • Use only non-spark tools. • Use appropriate equipment properly and according to procedure. • Hose connections has Installed with standard whip lock. 4. Isolation : • Refer to existing Process and Electrical Isolation Certificate. 5. Engineering : • Provide service line (Plant Air and Utility Water) near safe worksite. • Provide grounding/earthing/bonding installed on connections and equipment to prevent electrostatic discharge. 6. Eliminate : • Make sure the worksite is free from trip hazard before the jobs start. • Stop The Job on unsafe condition includes rainy. **Open Man-Hole Risk Control** 1. Administrative : • Refer to Initial Job Risk Control. 2. PPE/Safety : • Provide Organic Vapor Gas Mask. • Provide Oil Spill Kit near safe work area.	3	2	6

IOP Conf. Series: Earth and Environmental Science **1473** (2025) 012042 doi:10.1088/1755-1315/1473/1/012042

Job Descriptions	Risk Control	Residual Risk		
		S	L	R
	3. Subtitute : • Refer to Initial Job Risk Control. 4. Isolation : • Refer to Initial Job Risk Control. 5. Engineering : • Provide Secondary containment bellow opened man-hole. 6. Eliminate : • Refer to Initial Job Risk Control. **Operating Air Compressor** 1. Administrative : • Operator (Contractor) has been trained for operating air compressor (Certificate). • Follow Air Compressor Operating Procedure. • Prepare check list and record sheet. 2. PPE/Safety : • Provide fire extinguisher near safe working area of equipment. 3. Substitute : • Spark Arrester should be Installed on Air Compressor Muffler. 4. Isolations : • Refer to Air Compressor Isolation Procedure. 5. Engineering : • Refer to Initial Job Risk Control. 6. Eliminate : • Standby certified operator on air compressor area. **Running Pneumatic Blower** 1. Administrative : • Certified Gas Tester Operator to continues check gas (H_2S, CO, LEL, O_2) on exhaust ducting. • Follow Pneumatic Blower Operating Procedure. • Prepare check list and record sheet. 2. PPE/Safety : • Standby Hydrant Hose near safe working area of equipment. • Provide Organic Gas Mask. • Barricade Ducting Area. 3. Substitute : • Refer to Initial Job Risk Control. 4. Isolations : • Refer to Pneumatic Blower Isolation Procedure. 5. Engineering : • Provide sealed ducting connections with no dead zone along the ducting and exhaust. • Air remover installed and secure with 4 bolts-nuts connection to the manway flange (if installed on manway). 6. Eliminate : • Wetting sledge surface prior running blower. **Running Pneumatic Lamp** 1. Administrative : • Follow Pneumatic Lamp Operating Procedure. • Prepare check list and record sheet. 2. PPE/Safety : • Refer to Initial Job Risk Control. 3. Substitute : • Refer to Initial Job Risk Control. 4. Isolations : • Refer to Pneumatic Lamp Isolation Procedure. 5. Engineering : • Refer to Initial Job Risk Control. 6. Eliminate : • Refer to Initial Job Risk Control.			
3. Main Jobs (Confined-Space Entry):	**Confined Space Risk Control** 1. Administrative :	4	2	8

Job Descriptions	Risk Control	Residual Risk		
		S	L	R
• Pumping sludge from outside the tank. • Manual cleaning. • Rinsing with crude oil. • Pumping sludge from inside the tank.	• All personnel involved already getting HSE Induction – Refer to SITE-PRO-ON-SIPL-HSE-008 - OPF Personnel Entry Minimum Requirement Rev 01. • Obtained PTW and TRA in place – Refer to SITE-PRO-SIPL-HSE-001 - Permit To Work Rev 01. • Follow manual handling procedure – Refer to SITE-PRO-SIPL-HSE-010 - Manual Handling Rev 01. • All hose has been leak-tested and inspected (follow 029/ORA/SIPL-OPS). • Validated Rescue plan procedures are available. • Conduct a Toolbox meeting for all personnel involved prior job execution. • Maintain Communication and Coordination with Area Operators or CCR. • Tools and equipment have been inspected by SIPL Maintenance and HSSE team. • Prepare check list and record sheet. • Personnel involved man-entry, already check for health and have fit to work certificate from SIPL Medic. • Personnel Man-Entry should have Confined Space Entry Certificate. (Kementerian Ketenagakerjaan Indonesia, 2023) • Standby watchman should have Gas Tester Certificate. (Kementerian Ketenagakerjaan Indonesia, 2023) 2. PPE/Safety : • Proper PPE Mandatory is required, (Safety helmet, safety glasses, gloves, safety shoes, FRC coverall) – Refer to STD-SEI-HSE-009 Personal Protective Equipment Rev 0. • Provide Oil Spill Kit near safe work area. • Provide Fire Extinguisher near safe work area. • Provide decontamination shower near safe worksite. • Provide Organic Vapor Gas Mask. • Provide SABA (Supply Air Breathing Apparatus). • Provide Chemical suit (Tyvek/Tychem). • Provide Latex/Rubber hand glove. • Provide Rubber Boot. • Provide Full body Hardness and communication line. • Continues monitor and record gas monitoring (H_2S, CO, LEL, O_2) using Multi-gas Detector. • Barricade Man-Way area. 3. Subtitute : • Use only Ex-HT radio communication. • Use only non-spark tools. • Use appropriate equipment properly and according to procedure. • Hose connections has Installed with standard whip-lock. • Provide additional light with Ex-Certified Equipment only. 4. Isolation : • Refer to existing Process and Electrical Isolation Certificate. 5. Engineering : • Provide service line (Plant Air and Utility Water) near safe worksite. • Provide secondary containment on every point where an oil spill could occur outside tank. • Provide clear and safe access for Confined Space Entry Activity. 6. Eliminate : • Make sure the worksite is free from trip hazard before the jobs start. • Stop The Job on unsafe condition includes rainy. • Air Compressor, Pneumatic Blower and Lamp have been operated and function normally. • Standby rescue team, rescue equipment and first aid kit. **Pumping from Outside Tank Risk Control** 1. Administrative : • Refer to Confined Space Risk Control. 2. PPE/Safety : • All PPE included special PPE provided (Organic Vapor Mask) are use properly by personnel. 3. Subtitute : • Refer to Confined Space Risk Control. 4. Isolation :			

IOP Conf. Series: Earth and Environmental Science **1473** (2025) 012042 doi:10.1088/1755-1315/1473/1/012042

Job Descriptions	Risk Control	Residual Risk		
		S	L	R
	• Refer to Confined Space Risk Control. 5. Engineering : • Provide secondary containment for Wilden pump. • Provide silencer for Wilden pump exhaust/muffler. 6. Eliminate : • Standby certified operator on air compressor area. • Standby operator on Wilden pump. • Keep a clear communication between CS Personnel and Standby Watchman. **Personal Entry Risk Control** 1. Administrative : • Confined Space Entry Certificate is in place for recording personal entry and gas test monitoring. 2. PPE/Safety : • All PPE included special PPE provided are use properly by CS personnel. • Standby watchman uses organic vapor mask. • Standby watchman check for initial gas test prior Man-Entry and continues record gas test max. every 30 minutes. • CS personnel rotations max. 30 minutes/task. 3. Subtitute : • Refer to Confined Space Risk Control. 4. Isolation : • Refer to Confined Space Risk Control. 5. Engineering : • Refer to Confined Space Risk Control. 6. Eliminate : • Check SABA Air Quality prior CS Man-Entry. • Check communication line prior CS Man-Entry. • Standby personnel on SABA Cradle for continues monitoring air quality and pressure. • Standby HSSE Project Personnel for monitoring CS Man-Entry activities. • Standby Rescue Team during CS Man-Entry activity. **Manual Cleaning Risk Control** 1. Administrative : • Refer to Confined Space Risk Control. 2. PPE/Safety : • All PPE included special PPE provided are use properly by CS personnel. 3. Subtitute : • Refer to Confined Space Risk Control. 4. Isolation : • Refer to Confined Space Risk Control. 5. Engineering : • Refer to Confined Space Risk Control. 6. Eliminate : • Keep a clear communication between CS Personnel and Standby Watchman. **Washing Down with Crude Oil Risk Control** 1. Administrative : • Refer to Confined Space Risk Control. 2. PPE/Safety : • All PPE included special PPE provided are use properly by CS personnel. • Use low pressure (< 1 Bar) and washing down sludge slowly using crude oil. • Never pointing washing nozzle to person. 3. Subtitute : • Refer to Confined Space Risk Control. 4. Isolation : • Refer to Confined Space Risk Control. 5. Engineering : • Refer to Confined Space Risk Control. 6. Eliminate :			

Job Descriptions	Risk Control	Residual Risk		
		S	L	R
	• Keep a clear communication between CS Personnel and Standby Watchman. **Pumping from Inside Tank Risk Control** 1. Administrative : • Refer to Confined Space Risk Control. 2. PPE/Safety : • All PPE included special PPE provided are use properly by CS personnel. 3. Subtitute : • Refer to Confined Space Risk Control. 4. Isolation : • Refer to Confined Space Risk Control. 5. Engineering : • Provide rubber/wooden base for Wilden pump to prevent movement or scratching bottom tank coating. • Provide silencer for Wilden pump exhaust/muffler. 6. Eliminate : • Keep a clear communication between CS Personnel and Standby Watchman.			
4. Closing Job Activity : • Housekeeping.	**Basic Risk Control** 1. Administrative : • All personnel involved already getting HSE Induction – Refer to SITE-PRO-ON-SIPL-HSE-008 - OPF Personnel Entry Minimum Requirement Rev 01. • Obtained PTW and TRA in place – Refer to SITE-PRO-SIPL-HSE-001 - Permit To Work Rev 01. • Follow manual handling procedure – Refer to SITE-PRO-SIPL-HSE-010 - Manual Handling Rev 01. • Conduct a Toolbox meeting for all personnel involved prior job execution. • Maintain Communication and Coordination with Area Operators or CCR. • Tools and equipment have been inspected by SIPL maintenance team. • Prepare check list and record sheet. 2. PPE/Safety : • Proper PPE Mandatory is required, (Safety helmet, safety glasses, gloves, safety shoes, FRC coverall) – Refer to STD-SEI-HSE-009 Personal Protective Equipment Rev 0. • Provide Oil Spill Kit near safe work area. • Provide safety shower near safe worksite. 3. Subtitute : • Use only Ex-HT radio communication. • Use only non-spark tools. • Use appropriate equipment properly and according to procedure. • Segregate waste as per category. 4. Isolation : • Refer to existing Process Isolation Certificate. 5. Engineering : • Provide service line (Plant Air and Utility Water) near safe worksite. 6. Eliminate : • Make sure the worksite is free from trip hazard before the jobs start. • Stop The Job on unsafe condition includes rainy.	1	2	2

The following chart show percentage of each risk control category for each job categories that contribute on reducing the risk rank as residual risk, which is shown in **Figure 3**. For the calculation of the amount of risk control in percent is shown in **Equation 2**.

$$\text{Risk control (\%)} = \frac{\textbf{Number of risk control category}}{\textbf{Total of risk control}} \times \mathbf{100\%} \tag{2}$$

From risk control in **Table 7** and the chart in **Figure 3** below, we can see that residual risk rank for main jobs reduced significant from rank matrix 15 (High – Intolerable) to rank matrix 8 (Medium –

Tolerable). The 3 highest risk control category that contribute to reduce its rank matrix are PPE/Safety (34.92%), Eliminate (23.81%) and Administration (20.63%).

Figure 3. Risk control category percentage.

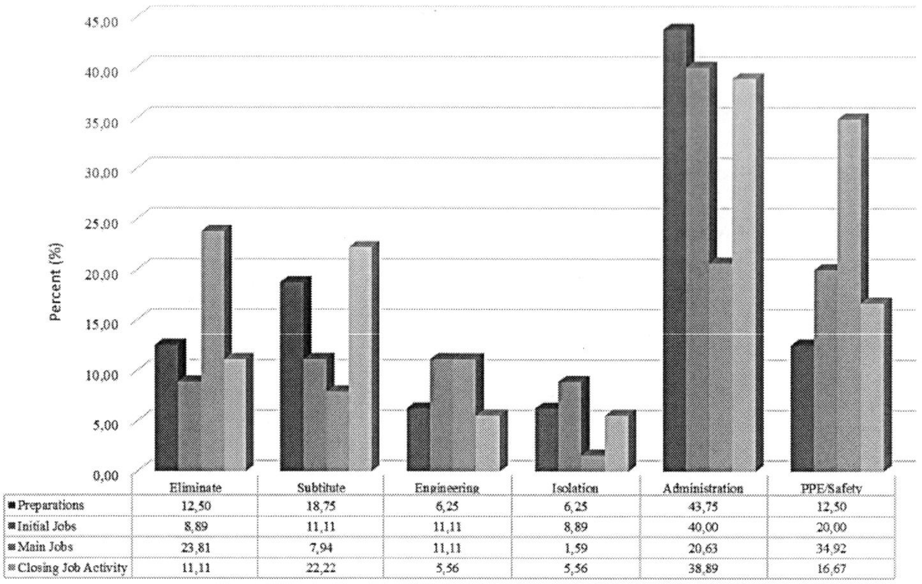

From the results, it can be clearly seen that with the presence of risk control can reduce the value of the risk matrix category. Before we have 3 medium and 1 high risk category, after risk control implementation, it become 3 medium and 1 low risk category. The following **Table 8** is the calculation between the number of initial risk and residual risk category in percent. The calculation of the amount of initial risk and residual risk is shown in **Equation 3**.

$$\text{Risk category (\%)} = \frac{\textbf{Number of risk category}}{\textbf{Total of risk category}} \times \mathbf{100\%} \quad (3)$$

Table 8. Percentage of risk category before (initial) and after (residual) risk control implementation.

Risk Categories	Initial Risk (percent)	Residual Risk (percent)
Low	0	25
Medium	75	75
High	25	0

From **Table 8**, we can see that crude oil storage tank cleaning job can be executed because there's no more high-risk (intolerable) job category as residual risk on its activity after taking risk control.

4. Conclusions.

Full scale task risk assessment method with risk control using quantitative risk analysis applied to the risk matrix table for crude oil storage tank cleaning work at PT. SAKA (Indonesia-Pangkah) Ltd. can provide very good details and visibility of the potential risks that arise, so that it will facilitate the determination of appropriate, effective and safe risk control. This is evident from the results of the risk assessment which was previously categorized as High (Intolerable), by carrying out risk control, it can reduce the risk rank and category of risk to Medium (Tolerable). The author hopes that this research can be a reference for case studies in the same or similar scope of work, so that safety and security in carrying out each job are always maintained better in the future.

Acknowledgments

The authors wish to thank the Engineering Physic Department, Institut Teknologi Sepuluh Nopember, located in Surabaya, Indonesia and PT. SAKA (Indonesia-Pangkah) Limited, located in Gresik, Indonesia for funding, support and facilitating this research. A high appreciation to Muhammad Khamim Asy'ari as co-author for supporting author completed this research. This study was conducted as part of a project safety review for cleaning and maintaining crude oil storage tank in PT. SAKA (Indonesia-Pangkah) Limited.

References

[1] National Fire Protection Association (NFPA), 2005. *NFPA 326 - Standard for the Safeguarding of Tanks and Containers for Entry, Cleaning, or Repair.* 2005 ed. Quincy(Massachusetts): National Fire Protection Association.

[2] Workplace Safety and Health Council in collaboration with the Ministry of Manpower, 2010. *Technical Advisory on Working Safely in Confined Spaces.* Singapore: Workplace Safety and Health Council.

[3] Health and Safety Executive (HSE), 1987. *The fires and explosion at BP Oil (Grangemouth) Refinery Ltd.,* Scotland: Health and Safety Executive (HSE).

[4] Marine Accident Investigation and Shipping Security Policy Branch (MAISSPB), 2018. *The explosion on board oil/chemical tanker "NO.3 HEUNG-A PIONEER" in waters off Penang, Malaysia.,* Hongkong: Marine Department of Hong Kong Special Administrative Region.

[5] Chang, J. I. & Lin, C.-C., 2006. A study of storage tank accidents. *Journal of Loss Prevention in the Process Industries,* January, 19(1), p. 51–59.

[6] Standards Australia/Standards New Zealand, 2004. *Risk Management AS/NZS 4360:2004.* 3rd ed. Sydney(New South Wales): Standards Australia International Ltd. and Standards New Zealand.

[7] Somerville, K. et al., 1997. *Task Risk Assessment,* Brussels: CONCAWE Safety Management Group.

[8] Department of Occupational Safety and Health, 2008. *Guidelines for Hazard Identification, Risk Assessment and Risk Control (HIRARC).* Kuala Lumpur(Kuala Lumpur): Department of Occupational Safety and Health.

[9] Kementerian Ketenagakerjaan Indonesia, 2023. *Peraturan Menteri Ketenagakerjaan Nomor 11 Tahun 2023 tentang Keselamatan dan Kesehatan Kerja di Ruang Terbatas,* Jakarta, Indonesia: Kementerian Ketenagakerjaan Indonesia.

Hazard analysis and safety management in public transport light rail transit

N A S Puteri[1] and M Khamim Asy'ari[1*]

[1] Engineering Physic Department, Institut Teknologi Sepuluh Nopember, 60111, Indonesia

*Email : mkasyari@its.ac.id

Abstract. Trains are public transportation whose operations prioritize passenger safety. There have been 701 train accidents, resulting in 343 deaths and environmental damage from 2007 to 2023. The braking, propulsion, and bogie systems that function for train movement contribute to accidents, such as derailments and collisions. Actions to minimize these risks are by conducting preliminary hazard analysis (PHA) and allocating safety integrity levels (SIL). The purpose of this study is to explore steps to minimize train accidents using the PHA method and SIL allocation. PHA is carried out through several stages, including: determining subsystems, functions, operation modes, hazards, causes, consequences, risk rankings and mitigating measures. Risk ranking is obtained from the multiplication of severity and frequency. SIL allocation uses the EN50128 and EN50129 standards which consist of determining the probability of exposure to hazards, reducing accidents, and reducing consequences. The results of the PHA assessment show that the risks in the negligible, tolerable, and undesirable categories are 19 percent, 73 percent dan 8 percent respectively. The results of the SIL allocation show that SIL 1 is 71 percent and SIL 3 is 29 percent.

1. Introduction

The role of infrastructure investment in economic growth is an important issue that economists should consider. Previous studies have shown that infrastructure investment, such as transportation, can have a positive impact on a country's economic growth [1]. One of the commonly used means of transportation is the train, where the train itself has many types, including: long-distance trains, commuter and local trains, tourist trains, freight trains, and many others. Based on the data owned by the Central Statistics Agency (BPS), train passengers increase every year, cumulatively from January to April 2024 there were 133.7 million people with an increase of 16.43% from the previous year [2]. The data strongly indicates that the train is still a mode of transportation preferred by the public. There are data from the Director General of KAI and Railways which shows that in the period from 2007 to September 2023, there were 701 train accidents with 343 fatalities. These accidents are divided into 5 different categories, including collisions with trains, derailments, rollovers, and fires. Of the 55 train accidents in 2019-2022, one was a collision with another train, one was due to flooding and landslides, and the rest were due to derailments [3].

Based on the problems described above, researchers are very interested in being able to identify hazards and causes of hazards from the occurrence of a failure on the train. By using Preliminary Hazard

Analysis (PHA) and Safety Integrated Level (SIL). PHA is a qualitative analysis method that aims to identify potential hazards and potentially dangerous situations and events that can cause risks to the train system and the environment in the early stages of the design process [4]. PHA itself can be applied to any risk assessment process, especially those that focus on the identification and classification of potential hazards. In addition, PHA can also propose solutions to reduce, control, or even eliminate the risk of an event [5]. The main advantage of the PHA method is its ability to provide an early view of possible risks, allowing better decisions to be made regarding design, equipment, and safety procedures. In addition, PHA also helps companies to plan mitigation measures before the risk develops into a real threat [6].

The Safety Integrated Level (SIL) assignment is part of the IEC 61508 standard, which is used to determine the level of safety integrity of systems related to functional safety [7]. SIL is an important parameter in the design and operation of high-risk control systems, as SIL measures the reliability of the system in preventing hazardous incidents. By classifying systems into four SIL levels (from SIL 1 to SIL 4), industry can determine the level of safety integrity required to mitigate risk. The higher the required SIL level, the more reliable the system must be to prevent failures that could be fatal [8].

In this context, the Preliminary Hazard Analysis (PHA) and Safety Integrity Level (SIL) Allocation methods are highly relevant to be applied as part of the safety risk analysis. The PHA method is used to perform a preliminary identification of hazards that may occur in railway systems, while the SIL is used to assign the safety integrity level required for these systems to function reliably under high-risk conditions. This study aims to identify hazards and assess safety using the two approaches, PHA and SIL, applied to three major railway systems, namely the braking system, the bogie system and the propulsion system. These three systems were chosen because they play a very important role in ensuring that trains move safely and efficiently under various operating conditions, whether they are traveling at high speed, stopping suddenly, or traveling through complex tracks such as curves or inclines.

2. Literature Review

This literature review will begin with various concepts and approaches relevant to risk assessment in the rail industry, particularly in the process of Probabilistic Hazard Analysis (PHA) and Safety Integrated Level (SIL) assignment. Qualitative risk analysis using risk matrices has been recognized as an important tool in assessing the level of risk and providing mitigation for the analysis.

2.1. PHA

The PHA is conducted using a top-down (deductive) methodology. This method examines the causes and potential circumstances of accidents and identifies system elements that require mitigation actions to adequately reduce risk. To support this process, a high-level hazard checklist is required that lists common hazards commonly found in rail systems. Using this common hazard checklist, the hazards associated with the RST LRT system were identified and systematically recorded on the PHA worksheet. Different modes of operation were considered as part of the hazard identification process.

- Normal Operating Mode: Operation when all systems and subsystems function normally as designed, and no failures, natural hazards, or other outside failures occur.
- Degraded Operating Mode: Operation during periods when some systems or subsystems are not functioning as designed or when human error, natural hazards, or other factors require an operational response, but without posing a threat to the safety of people or a threat of serious property damage.
- Emergency Operating Mode: Operation after an accident and especially in an Emergency. An accident is an abnormal operation that results in equipment damage and/or jeopardizes personal health or safety. A very serious accident would fall under the definition of Emergency. An Emergency is an unplanned occurrence which, generally, has life-threatening implications or extreme loss and requires immediate attention.
- Maintenance Operating Mode: Preventive maintenance or operation during engineering.

All identified hazards are categorized based on their frequency of occurrence and severity of consequences, taking into account the relevant modes of operation (if applicable). The combination of frequency and severity determines the risk ranking of the hazards in the Risk Classification Matrix, as shown in Table 1. The risk classification is then used to determine the priority of management attention and the resources that should be devoted to risk analysis and management.

Table 1 Risk Matrix

	IV	III	II	I
A	UND	INT	INT	INT
B	TOL	UND	INT	INT
C	TOL	UND	UND	INT
D	NEG	TOL	UND	UND
E	NEG	NEG	TOL	UND
F	NEG	NEG	NEG	TOL

Table 2 Severity Description

Category	Severity Level	Consequences to Persons / Environment
I	Catastrophic	Multiple fatalities and/or multiple severe injuries or extreme damage to the environment
II	Critical	Single fatality and/or severe injury or major damage to the environment
III	Marginal	Minor injuries (required medical treatment) or minor damage to the environment
IV	Insignificant	Possible minor injury (first aid)

Table 3 Frequency Description

Category	Frequency Level	Description
A	Frequent	Likely to occur frequently. The hazard will be frequently experienced.
B	Probable	Will occur several times. The hazard can be expected to occur often.
C	Occasional	Likely to occur several times. The hazard can be expected to occur several times.
D	Rare	Likely to occur sometime in the system lifecycle. The hazard can reasonably be expected to occur.
E	Improbable	Unlikely to occur but possible. It can be assumed that the hazard may exceptionally occur.
F	Highly Improbable	Extremely unlikely to occur. It can be assumed that the hazard will not occur.

The function of each component is then defined to determine the hazards, causes of occurrence, and effects if they occur. This determination is made to determine the risk measurement using the risk matrix used in the railroad industry above. Once the risk is known, the SIL level used is determined based on existing failure rate data. Recommendations are then made as to the SIL level to be used and the protective measures to be taken if a hazard occurs.

2.2 SIL Allocation

In the SIL assignment process, the EN50128 and EN50129 standards define the relationship between safety integrity and probability of failure, assuming that safety depends on adequate measures to prevent or tolerate failures (systematic failures) and adequate measures to control random failures. Safety integrity is specified as one of four discrete levels, with level 4 being the highest level of safety integrity and level 1 being the lowest. Level 0 indicates that there are no safety requirements. Table 4 explains the correlation between THR and severity with the SIL level required for a component.

Table 4 The correlation THR, Severty, and SIL Level

Severity	THR per hour per function	SIL
Catastrophic	$10^{-9} \leq H < 10^{-8}$	SIL4
Critical	$10^{-8} \leq H < 10^{-7}$	SIL3
Marginal	$10^{-7} \leq H < 10^{-6}$	SIL2
Insignificant	$10^{-6} \leq H < 10^{-5}$	SIL1

Tolerable Hazard Rate (THR) is a parameter used to determine the acceptable level of risk for a particular system or component in safety applications. A very low THR value indicates that the system has a very high safety requirement, where failure could result in major losses such as death or severe environmental damage. Catastrophic severity has a THR value so low that it requires components with an SIL level of 4. Meanwhile, low THR values and insignificant severity are used for systems where failure is likely to cause only minor injuries or minor material loss, or SIL level 1.

By definition, safety integrity is the probability that a system will satisfactorily perform its safety functions under all specified conditions within a specified time period. The Safety Integrity Level (SIL) is determined for safety functions and then assigned to the subsystems/equipment that perform those functions. The basic SIL determination process includes the following steps,

1. Determine the safety functions associated with the RST system.
2. Identify potential malfunctions and their safety implications.
3. Determine the severity and associated initial THR.
4. Identify factors that influence risk.
5. Assign a System Safety Integrity Level (SSIL) to the safety function.

The three sets of factors potentially influencing risk recommended by the MODSafe report (as described in the following subsections) should then be considered:

- Hazard Exposure Probability (E): Whether there are reasonable grounds to conservatively assume that the subjects in the risk group (e.g. passengers) will be exposed to the hazard much less than continuously (by an order of magnitude in probability).
- Accident Probability Reduction (P): Whether there are reasonable grounds to conservatively assume that the development of a hazard into an accident can be clearly controlled by additional barriers or circumstances (order of magnitude reduction).
- Consequence Reduction Probability (C): Whether there are reasonable grounds to conservatively assume that members of the risk group (e.g. passengers, staff) can clearly avoid being exposed to the hazard (by order of magnitude) or significantly reduce the potential damage (by severity class).

Table 5 Exposure Probability Parameter

Exposure Probability	Value	Description
E1	1	Exposure of members of the risk group to hazard is conservativelyto be assumed frequent or permanent.
E2	0.1	Exposure of members of the risk group to hazard can conservatively be assumed to be rare, only in exceptional cases.
E3	0.01	Exposure of members of a risk group to hazard is only in very rare cases to be expected.

Table 6 Accident Probability Reduction Parameter

Accident Probability Reduction	**Value**	**Description**
P1	1	No additional barrier can be conservatively assumed that wouldreduce the probability of the hazard evolving into an accident.
P2	0.1	There exists means or circumstances to clearly reduce the probability that a certain hazard evolves into an accident (e.g. self-diagnostic alarm to operator upon equipment failure is detected).
P3	0.01	There exists two means or circumstances to clearly reduce independently the probability that a certain hazard evolves into an accident.

Table 7 Consequence Reduction Probability

Consequence Reduction Probability	**Value**	**Description**
C1	1	There is no reason to conservatively assume that a member of therisk group (e.g. passenger) may avoid being subject to the consequences of a certain hazard.
C2	0.1	There is good reason to conservatively assume that a member ofthe risk group (e.g. passenger) can avoid being subject to the consequences of a certain hazard.
C3	0.01	There are two independent good reasons to conservatively assumethat a member of the risk group (e.g. passenger) can avoid being subject to the consequences of a certain hazard.

3. Results and Discussion

3.1 Brake System

The first system is the braking system. The braking system is a system owned by the train to slow down the movement of the wheels by converting kinetic energy / force into another form. With the wheel movement slowed down, the motion of an object automatically becomes slow and even stops. The braking system is the main system of the train for the description of the functions of the braking system attached to the train in the below.

Table 8 Function Description Brake System

Functions ID	*Function Description*
RST-L2.01.02	*Braking System*
RST-L2.01.02a	*Brake and stop train by service brake*
RST-L2.01.02b	*Brake and stop train by emergency brake*
RST-L2.01.02c	*Immobilize the train*
RST-L2.01.02d	*Wheel slide protection*

IOP Conf. Series: Earth and Environmental Science **1473** (2025) 012043 doi:10.1088/1755-1315/1473/1/012043

3.1.1 SIL Allocation

Table 9 SIL Allocation Braking System

Brake System								
Function ID	**Harazrd Ref.**	**Safety Function**	**THRi**	**E**	**P**	**C**	**THR**	**SIL**
RST-L2.01.02a	RST-PHA-010	Braking and stopping the train with a service brake	1,00E-07	1,00E+00	1,00E-01	1,00E+00	1,00E-06	1
RST-L2.01.02b	RST-PHA-012	Stopping and stopping the train with the emergency brake	1,00E-09	1,00E+00	1,00E-01	1,00E+00	1,00E-08	3
RST-L2.01.02c	RST-PHA-013	Stopping the train	1,00E-08	1,00E+00	1,00E-01	1,00E-01	1,00E-06	1
RST-L2.01.02c	RST-PHA-014	Stopping the train	1,00E-08	1,00E+00	1,00E-01	1,00E-01	1,00E-06	1
RST-L2.01.02c	RST-PHA-015	Stopping the train	1,00E-08	1,00E-01	1,00E-02	1,00E-01	1,00E-04	0
RST-L2.01.02d	RST-PHA-016	Wheel slip protection	1,00E-09	1,00E-01	1,00E-01	1,00E-01	1,00E-06	1
RST-L2.01.02c	RST-PHA-018	Wheel slip protection	1,00E-09	1,00E-01	1,00E-01	1,00E-01	1,00E-06	1

3.2 Propulsion System

The next system is the Propulsion System which is used to run the train. The function description of the propulsion system is attached below.

Table 10 Function Descriptionn Propulsion System

Functions ID	***Function Description***
RST-L2.01.01	*Propulsion System*
RST-L2.01.01a	*Accelerate, decelerate and maintain train speed*
RST-L2.01.01b	*Limit speed to below 20 km/h under RM mode*
RST-L2.01.01c	*Traction Cut-Off*

3.2.1 SIL Allocation

Table 11 SIL Allocation Propulsion System

Propulsion System								
Function ID	**Harazrd Ref.**	**Safety Function**	**THRi**	**E**	**P**	**C**	**THR**	**SIL**
RST-L2.01.01a	RST-PHA-050	Produce traction power/effort on command	1,00E-09	1,00E-02	1,00E-01	1,00E-01	1,00E-05	0
RST-L2.01.01a	RST-PHA-051	Produce traction power/effort on command	1,00E-09	1,00E-02	1,00E-01	1,00E-01	1,00E-05	0
RST-L2.01.01a	RST-PHA-052	Produce traction power/effort on command	1,00E-09	1,00E-02	1,00E-01	1,00E-01	1,00E-05	0
RST-L2.01.01a	RST-PHA-053	Perform dynamic braking	1,00E-09	1,00E-02	1,00E-01	1,00E-01	1,00E-05	0
RST-L2.01.01b	RST-PHA-056	Limits speed below 20 km/h in degraded mode	1,00E-09	1,00E-02	1,00E-01	1,00E-01	1,00E-05	0
RST-L2.01.01c	RST-PHA-157	Traction Disconnection	1,00E-09	1,00E+00	1,00E-01	1,00E+00	1,00E-08	3

IOP Conf. Series: Earth and Environmental Science **1473** (2025) 012043 doi:10.1088/1755-1315/1473/1/012043

3.3 Bogie System

The last system analyzed is the bogie where in this sub-system which is a construction consisting of two or more sets of wheels joined by a frame equipped with a firming, braking system, with or without driving and anti-skid equipment, and the whole serves as a support for the base frame of the train body.Bogie can be removed and reattached if maintenance is being carried out. The function description of the bogie is attached below.

Table 12 Function Diagram Bogie System

Functions ID	*Function Description*
RST-L2.01.04	*Bogie*
RST-L2.01.04a	*Guide the train*
RST-L2.01.04b	*Manage bogie stability*
RST-L2.01.04c	*Limit roll movement*
RST-L2.01.04d	*Limit lateral and vertical movement*
RST-L2.01.04e	*Detect and signal obstacles within clearance gauge*
RST-L2.01.04f	*Detect derailment occuring on the train by monitoring relevant on board parameters*

3.3.1 SIL Allocation

Table 13 SIL Allocation Bogie System

Bogie								
Function ID	**Harazrd Ref.**	**Safety Function**	**THRi**	**E**	**P**	**C**	**THR**	**SIL**
RST-L2.01.04e	RST-PHA-048	Detect and signal obstacles within the clearance range	1,00E-07	1,00E-01	1,00E+00	1,00E+00	1,00E-06	1
RST-L2.01.04f	RST-PHA-049	Detecting train derailment by monitoring relevant parameters on board the train	1,00E-07	1,00E-01	1,00E+00	1,00E+00	1,00E-06	1

3.4 PHA

A PHA was conducted for the system and hazard identification was performed using a structured approach based on a list of functions. All hazards are analyzed based on risk acceptance criteria and with predefined risk control measures. The overall critical safety hazards can be controlled with the mitigation measures listed in the PHA worksheet on appendix. Most critical safety hazards caused by component/subsystem failures can be mitigated by the system. However, in an emergency, staff involvement is still required to justify and make decisions. Standard procedures should be developed with a hazard scenario orientation (referencing the Hazard Analysis) to minimize human error during hazardous conditions. Based on the PHA analysis for these 3 systems as in the table below.

Table 14 PHA Result

No.	Legend	Result
1	Negligible	19%
2	Tolarable	73%
3	Undesirable	8%
4	Intolarable	0%

4. Conclusions

The results of the PHA assessment show that the risks in the negligible, tolerable, and undesirable categories are 19 percent, 73 percent dan 8 percent respectively. The results of the SIL allocation show that SIL 1 is 71 percent and SIL 3 is 29 percent. A PHA was conducted for the system and hazard identification was performed using a structured approach based on a list of functions. All hazards are analyzed based on risk acceptance criteria and with predefined risk control measures. The overall critical safety hazards can be controlled with the mitigation measures listed in the PHA worksheet. Most critical safety hazards caused by component/subsystem failure can be mitigated by the system. However, in an emergency, staff involvement is still required to justify and make decisions. Standard procedures should be developed with a hazard scenario orientation (referring to the Hazard Analysis) to minimize human error during hazardous conditions. SIL assignment has been performed to assign safety integrity requirements to the safety functions of the RST system. This SIL assignment process follows the latest industry practice of assigning SILs directly to the severity category of a consequence and adopts a semi-quantitative approach using risk influencing factors as recommended by the European Commission.

References

[1] H. Z. Y. H. G. Q. E. E. f. C. Xin Du, "How Does New Infrastructure Investment Affect Economic," *Sustaianbility,* pp. 1-30, 2022.

[2] B. P. S. (. -. S. Indonesia), "Badan Pusat Statistik," 2024. [Online]. Available: https://www.bps.go.id/id/statistics-table/2/NzIjMg==/jumlah-penumpang-kereta-api.html. [Accessed 14 Agustus 2024].

[3] D. J. Perkeretaapian, "Kementrian Perhubungan," 2024. [Online]. Available: https://djka.dephub.go.id/statistik?tahun-statistik=2024#. [Accessed 14 Agustus 2024].

[4] K. I. I. d. Lindah Nuriawati, "Evaluasi Penerapan Keselamatan dan Kesehatan Kerja (K3) Berdasarkan FTA, FMEA dan PHA di Jurusan Tiptil," *Program Studi Pendidikan Teknik Mekatronika ,* vol. 8, pp. 490-501, 2018.

[5] D. H. Hadj-Mabrouk, "Preliminari Hazard Analysis (PHA): New hybrid approach to railway risk analysis," *Internasional Refereed Journal of Engineering and Science (IRJES),* vol. 6, no. 2, pp. 51-58, 2017.

[6] J. C. J. M. J. C. U. M. N. Adebeda Oluwasegun Olaide, "Preliminary Hazard Analysis: Assessment of New Component Interface Module Design for APR1400," *Journal of KOSSE,* vol. 17, pp. 21-34, 2021.

[7] J. B. d. Kiswendsida Abel Ouedraogo, "Toward an Application Guide for Safety Integrity Level Toward an Application Guide for Safety Integrity Level," *Society for Risk Analysis,* 2018.

[8] B. J. E.-M. E.-K. C. J. R. D. L. F. Ouedraogo KA, "Allocation rules of safety integrity levels in a generic TCMS application," *10 th FORMS-FORMAT symposium, Formal Methods for Automation and safety in Railway and Automotive system,* pp. 167-176, 2014.

[9] E. 5. R. Applications, "The specification and demonstration of reliability, availability, maintainability and safety (RAMS)". CENELEC, European Committee for Electrotechnical Standardization 2000.

[10] E. 5. R. applications, "Communication, signalling and processing systems—Software for railway control and protection systems". CENELEC, European Committee for Electrotechnical Standardization 2011.

[11] I. 61508, Functional safety of E/E/PE safety-related systems, 2011.

[12] D. M. P. J. K. M. M. R. S. J. e. a. Abioye OF, "Accident and hazard prediction models for highway–rail grade crossings: a state-of-the practice review for the USA," *Railw Eng Sci* , p. 251, 2020.

[13] S. F. L. K. Z. W. Liu J, "A knowledge graph-based approach for exploring railway operational accidents," *Reliab Eng Syst,* p. 207, 2021.

IOP Conf. Series: Earth and Environmental Science **1473** (2025) 012043 doi:10.1088/1755-1315/1473/1/012043

Appendix

1. Worksheet of PHA

PHA Hazard ID.	Operation Mode	System	Sub-system	Function No.	Functions	Hazard	Cause	Effect / Consequence	IF	IS	IR	Mitigating Measures	Responsible Party	RF	RS	RR	Remarks
Braking System																	
RST-PHA-010	Normal	RST	Brake System	RST-L2.01.02a	Brake and stop train by service brake	No or insufficient braking effort	Failure of service brake unit	Potential for train collision	C	I	INT	1) Safe train distance separation assurance is ensured by failsafe ATP / CBI 2) If one EBCU failed, the train shall be able to operate normally 3) If multiple failures occur, speed restriction will be applied by the train gradually according to the number of failed EBCUs	1) SIG 2) RST 3) RST	F	I	TOL	
RST-PHA-011	Normal / Degraded / Emergency	RST	Brake System	RST-L2.01.02a & b	Brake and stop train by service brake or emergency brake	No or insufficient braking effort	Mechanical failure of brake actuator equipment (brake calliper, brake disc, brake pad)	Potential for train collision	B	I	INT	1) Brake actuator equipment shall be provided on each axle, train shall be able to stop safely with one set of brake actuator equipment failed 2) Brake cylinder failure shall be detected by EBCU 3) Regular inspection and	1) RST 2) RST 3) OPR	F	I	TOL	

PHA Hazard ID.	Operation Mode	System	Sub-system	Function No.	Functions	Hazard	Cause	Effect / Consequence	IF	IS	IR	Mitigating Measures	Responsible Party	RF	RS	RR	Remarks
												maintenance of brake system					
RST-PHA-012	Degraded / Emergency	RST	Brake System	RST-L2.01.02b	Brake and stop train by emergency brake	No or insufficient braking effort in an emergency	Failure of emergency brake unit	Potential for train collision	C	I	INT	1) Emergency brake unit is provided for each bogie and brake system architecture should tolerate up to one car brake failure. 2) Brake failure can be detected by EBCU to apply speed restrictions	1) RST 2) RST	F	I	TOL	
RST-PHA-013	Normal	RST	Brake System	RST-L2.01.02c	Immobilize the train	Roll back or roll forward of train during station stop	Failure of service brake unit / Brake actuator equipment	Passenger trip and fall or crush by train	C	II	UND	1) Roll away detection is safety critical function of ATP 2) Train on maximum graded track can be hold by other parking brake actuators. 3) If one EBCU failed, the train shall be able to operate normally 4) if one car brake fails, the nominal performance on flat track can still be met.	1) SIG 2) RST 3) RST 4) RST	F	II	NEG	

IOP Conf. Series: Earth and Environmental Science **1473** (2025) 012043 doi:10.1088/1755-1315/1473/1/012043

PHA Hazard ID.	Operation Mode	System	Sub-system	Function No.	Functions	Hazard	Cause	Effect / Consequence	IF	IS	IR	Mitigating Measures	Responsible Party	RF	RS	RR	Remarks
RST-PHA-014	Normal	RST	Brake System	RST-L2.01.02c	Immobilize the train	Brake not released after station stop and for train departure	Failure of service brake unit / Jammed Brake actuator equipment	Dragging brake, potential for train fire	C	II	UND	1) EBCU send 'Not Release Brake' signal to traction system, traction inverter will not supply power with this signal input 2) EB trainline signal will be sent to traction inverter for cutting out traction power 3) Train floor shall act as fire barrier for at least 30 min	1) RST 2) RST 3) RST	E	III	NEG	
RST-PHA-015	Maintenance	RST	Brake System	RST-L2.01.02c	Immobilize the train	Spurious train movement in stabling yard or maintenance workshop	Failure of parking brake unit	Potential for maintenance staff hit by moving train	C	II	UND	1) A parking brake shall be incorporated on 50% of the brake actuators to ensure parking brake control of each axle, stabled train shall not move even failed of one parking brake 2) Maintenance and stabling track shall have zero gradient 3) Develop	1) RST 2) CIV 3) OPR 4) RST/OPR	E	II	TOL	

PHA Hazard ID.	Operation Mode	System	Sub-system	Function No.	Functions	Hazard	Cause	Effect / Consequence	IF	IS	IR	Mitigating Measures	Responsible Party	RF	RS	RR	Remarks
												procedure to ensure staff will apply wheel choke when train stabled on maintenance track 4) Failure of parking brake can be detected by EBCU by monitoring pressure in parking brake release cylinder and checks/inspections.					
RST-PHA-016	Normal	RST	Brake System	RST-L2.01.02d	Wheel slide protection	Wheel slide during braking	Failure of WSP by EBCU / Failure of speed probe	Increase stopping distance, potential for train collision	C	I	INT	1) Speed probe is provided on each axle to monitor axle speed and trigger the WSP activation when wheel slide occurs. 2) The minimum braking distance shall take into account the worst adhesion conditions 3) If one EBCU failed, the train shall be able to operate normally 4) EBCU can detect abnormal	1) RST 2) RST & OPR 3) RST 4) RST	E	I	UND	

PHA Hazard ID.	Operation Mode	System	Sub-system	Function No.	Functions	Hazard	Cause	Effect / Consequence	IF	IS	IR	Mitigating Measures	Responsible Party	RF	RS	RR	Remarks
												speed signal from the faulty sensor by comparing with the reference speed from other sensors					
RST-PHA-017	Normal	RST	Brake System	RST-L2.01.02d	Wheel slide protection	Wheel slide during braking	Failure of WSP by BCU / Failure of speed probe	Increase in wheel tread wearing, degraded ride quality	C	IV	TOL	1) Speed probe is provided on each axle to monitor axle speed, wheel slide for each axle shall be controlled independently 2) EBCU can detect abnormal speed signal from the faulty sensor by comparing with the reference speed from other sensors	1) RST 2) RST	E	IV	NEG	
RST-PHA-018	Normal / Degraded	RST	Brake System	RST-L2.01.02c	Immobilize the train	Roll back or roll forward of train during immobilisation on graded track	Failure of service brake unit / Brake actuator equipment	Potential for train collision / derailment	C	I	INT	1) Roll away detection is safety critical function of ATP 2) Train on maximum graded track can be hold by other brake actuators in case one brake actuator failed 3) Holding brake under normal	1) SIG 2) RST 3) RST	F	I	TOL	

PHA Hazard ID.	Operation Mode	System	Sub-system	Function No.	Functions	Hazard	Cause	Effect / Consequence	IF	IS	IR	Mitigating Measures	Responsible Party	RF	RS	RR	Remarks
												condition is 70% of full service brake					
RST-PHA-219	Degraded / Emergency	RST	Brake System	RST-L2.01.02e	Brake and stop train by emergency brake in case of failure of the service brakes	Rollback of train on gradient track or brake failure	Service brake (dynamic) are not backed up by emergency brake (friction)	Backward movement of trains resulting in train collision / derailment	D	I	UND	1) The EB application with friction brake only or one single failure provides an sufficient effort to immobilize the train in the steepest gradient on the line 2) Loss of dynamic brake shall be compensated by friction brake during SB application	1) RST 2) RST	F	I	TOL	
RST-PHA-220	Normal / Degraded / Emergency	RST	Brake System	RST-L2.01.02f	Emergency brake applied by hard-wire and failsafe	No or delayed operation of brakes when required	Failure of EB command	Potential for train collision / derailment	C	I	INT	1) Emergency brake unit is provided for each bogie and brake system architecture should tolerate up to one car brake failure. 2) SIG EB command activates the hard-wired EB loop which triggers the EB valves	1) RST 2) RST 3) RST	F	I	TOL	

IOP Conf. Series: Earth and Environmental Science **1473** (2025) 012043 doi:10.1088/1755-1315/1473/1/012043

PHA Hazard ID.	Operation Mode	System	Sub-system	Function No.	Functions	Hazard	Cause	Effect / Consequence	IF	IS	IR	Mitigating Measures	Responsible Party	RF	RS	RR	Remarks
												with no exposure to single-point failure at relaying or component level 3) Driving cab to be equipped with manual EB push-button for the TA to trigger EB in case of danger not detected by SIG/TCMS					
Bogie																	
RST-PHA-044	Normal	RST	Bogie	RST-L2.01.04a	Guide the train	No or insufficient guiding force at wheel-rail contact	Wheel profile wear / broken axle	Potential for derailment	C	I	INT	1) Derailment detection shall be installed to minimize the effects of damage due to derailment 2) Wheels shall use wear-resistant materials in accordance with EN 13262 3) Wheel profile and axle shall be inspect regularly 4) The guide rail shall be attached to a curved track 5) Regulate the train speed based on signaling line speed	1) RST 2) RST 3) OPR 4) INF 5) SIG	E	II	TOL	

IOP Conf. Series: Earth and Environmental Science **1473** (2025) 012043 doi:10.1088/1755-1315/1473/1/012043

PHA Hazard ID.	Operation Mode	System	Sub-system	Function No.	Functions	Hazard	Cause	Effect / Consequence	IF	IS	IR	Mitigating Measures	Responsible Party	RF	RS	RR	Remarks
RST-PHA-045	Normal	RST	Bogie	RST-L2.01.04b	Manage bogie stability	Hunting/ snake motion	Wheel and rail profile wear	Potential for derailment	C	I	INT	1) Investigating the wheel and rail profile interaction in the design phase 2) Wheels shall use wear-resistant materials in accordance with EN 13262 3) Wheel profile shall be inspect regularly 4) Regulate the train speed based on signalling line speed	1) RST 2) RST 3) OPR 4) SIG	F	I	TOL	
RST-PHA-046	Normal	RST	Bogie	RST-L2.01.04c	Limit roll movement	Tilting movement	Failure on air spring	Potential for rolling, insuffiicient brake force	C	I	INT	1) Air spring leaks shall be detected by the train, then EB is applied 2) Design of air spring using a rubber seat 3) Isolating device shall prevent pneumatic air from flowing into the leaking suspension 4) Regulate the train speed based on signalling line speed	1) RST 2) RST 3) RST 4) SIG	F	I	TOL	
RST-PH	Normal	RST	Bogie	RST-L2.01.04d	Limit lateral and	Excess movement	Failure on suspen	Potential for	C	I	INT	1) Vertical and lateral bump stop	1) RST 2) OPR 3) SIG	F	I	TOL	

IOP Conf. Series: Earth and Environmental Science **1473** (2025) 012043 doi:10.1088/1755-1315/1473/1/012043

PHA Hazard ID.	Operation Mode	System	Sub-system	Function No.	Functions	Hazard	Cause	Effect / Consequence	IF	IS	IR	Mitigating Measures	Responsible Party	RF	RS	RR	Remarks
A-047					vertical movement		sion system	derailment				shall be installed to limit bogie movement during lifting and turning 2) Suspension system shall be maintained properly 3) Regulate the train speed based on signalling line speed					
RST-PHA-048	Normal / Degraded / Emergency	RST	Bogie	RST-L2.01.04e	Detect and signal obstacles within clearance gauge	Unable to detect obstacle - ODS does not send signal to stop the train	Failure on ODS	No damage control when the train collides with an obstacle	D	III	TOL	1) Redundant Obstacle Detection System shall be provided in each trainset 2) Proven reliability of Obstacle Detector 3) Provide functional test to ODS during regular maintenance	1) RST 2) RST 3) OPR	F	II I	NEG	
RST-PHA-049	Emergency	RST	Bogie	RST-L2.01.04f	Detect derailment occuring on the train by monitoring relevant on board parameters	Unable to detect train derailment - EB does not triggered automatically	Failure on DDS	Cannot reduce damage from derailment	D	III	TOL	1) Redundant DDS shall be provided in each car 2) MC car shall equipped with Derailment detector on carbody and bogie 3) Proven reliability of Derailment Detector	1) RST 2) RST 3) RST 4) OPR	F	II I	NEG	

PHA Hazard ID.	Operation Mode	System	Sub-system	Function No.	Functions	Hazard	Cause	Effect / Consequence	IF	IS	IR	Mitigating Measures	Responsible Party	RF	RS	RR	Remarks
												4) Provide functional test to DDS during regular maintenance					
Propulsion System																	
RST-PHA-050	Normal / Degraded	RST	Propulsion System	RST-L2.01.01a	Accelerate, decelerate and maintain train speed	Excessive powering effort beyond as commanded	Failure of one traction inverter	Train overspeed, potential for train collision or derailment	C	I	INT	1) Train overspeed protection by ATP 2) Safe train separation assurance is ensured by failsafe ATP / CBI 3) Traction current shall be cut-off when SB or EB is applied 4) Traction inverter has self-diagnostic function	1) SIG 2) SIG 3) RST 4) RST	F	I	TOL	
RST-PHA-051	Normal / Degraded	RST	Propulsion System	RST-L2.01.01a	Accelerate, decelerate and maintain train speed	Excessive powering effort beyond as commanded	Failure of one traction motor	Train overspeed, potential for train collision or derailment	C	I	INT	1) Train overspeed protection by ATP 2) Safe train separation assurance is ensured by failsafe ATP / CBI 3) Traction inverter shall monitor status of motor and cut off traction power if failure is detected	1) SIG 2) SIG 3) RST 4) RST	F	I	TOL	

PHA Hazard ID.	Operation Mode	System	Sub-system	Function No.	Functions	Hazard	Cause	Effect / Consequence	IF	IS	IR	Mitigating Measures	Responsible Party	RF	RS	RR	Remarks
												4) Traction current shall be cut-off when SB or EB is applied					
RST-PHA-052	Normal / Degraded	RST	Propulsion System	RST-L2.01.01a	Accelerate, decelerate and maintain train speed	Train runs in direction opposite to the commanded direction	Failure of one traction inverter	Potential for train collision	C	I	INT	1) ATP shall ensure rollback protection and application of EB 2) Traction inverter can detect reverse direction by detecting train speed of -3km/h for 3s, and cut off traction power 3) Traction inverter has self-diagnostic function	1) SIG 2) RST 3) RST	F	I	TOL	
RST-PHA-053	Normal / Degraded	RST	Propulsion System	RST-L2.01.01a	Accelerate, decelerate and maintain train speed	Incorrect brake blending by dynamic brake	Failure of one traction inverter (incorrect feedback)	Insufficient brake effort, potential for train collision and derailment	C	I	INT	1) If train speed exceeded speed profile, ATP shall apply EB (full friction brake) irrespectively 2) Traction inverter has self-diagnostic function	1) SIG 2) RST	F	I	TOL	
RST-PHA-054	Normal / Degraded	RST	Propulsion System	RST-L2.01.01a	Accelerate, decelerate and maintain train speed	Inverter container or mounting frame is alive with hazardous voltage	Failure of one traction inverter (short circuit)	Electrocution	C	II	UND	1) Earthing and bonding for live circuit is provided 2) HSCB shall be provided to cutoff	1) RST 2) RST	E	II	TOL	

IOP Conf. Series: Earth and Environmental Science **1473** (2025) 012043 doi:10.1088/1755-1315/1473/1/012043

PHA Hazard ID.	Operation Mode	System	Sub-system	Function No.	Functions	Hazard	Cause	Effect / Consequence	IF	IS	IR	Mitigating Measures	Responsible Party	RF	RS	RR	Remarks
												power in case of fault current					
RST-PHA-055	Normal	RST	Propulsion System	RST-L2.01.01a	Accelerate, decelerate and maintain train speed	Overheating of underframe equipment (e.g. traction inverter, motor, etc.)	Loss of natural air flow during station dwell time	Potential for underframe fire, passengers expose to smoke and fire	C	II	UND	1) Train floor shall act as fire barrier for at least 30 min	1) RST	C	III	UND	
RST-PHA-056	Degraded / Emergency	RST	Propulsion System	RST-L2.01.01b	Limit speed to below 20 km/h under RM mode	Fail to limit speed to below 20 km/h under RM mode	Failure of traction inverter / speed sensor	Train running at excessive speed, potential for train collision or derailment at turnout	C	I	INT	1) Traction inverter has self-diagnostic function 2) One speed sensor will be provided for each motor, there are a total of 4 sensors per MC car. Traction inverter compares signal from the 4 sensors and detect and ignores the fault sensor. 3) Strict operating procedure shall be developed for using RM modes 4) Emergency brake shall be applied by EBCU under RM mode when speed exceeding 20 km/h	1) RST 2) RST 3) OPR 4) RST	F	I	TOL	

IOP Conf. Series: Earth and Environmental Science **1473** (2025) 012043 doi:10.1088/1755-1315/1473/1/012043

PHA Hazard ID.	Operation Mode	System	Sub-system	Function No.	Functions	Hazard	Cause	Effect / Consequence	IF	IS	IR	Mitigating Measures	Responsible Party	RF	RS	RR	Remarks
RST-PHA-257	Degraded / Emergency	RST	Propulsion System	RST-L2.01.01c	Traction Cut-Off	Fail to activate traction cut-off when EB is applied	Systematic failure of all TCUs (software failure)	Train not meeting the Emergency Brake deceleration	D	I	UND	1) Implement hardware cut-off for each motor car	1) RST	F	I	TOL	
RST-PHA-258	Emergency	RST	Propulsion System	RST-L2.01.01d	Limit speed to below 20 km/h under towing mode	Driver error	Towing mode failure to limit train speed during towing	Rescue train collide with incident train at high speed during towing	E	I	UND	1) Towing operation to be permanently supervised by the OPR staff following SOP mentioning the speed limitation.	1) OPR	F	I	TOL	

IOP Conf. Series: Earth and Environmental Science **1473** (2025) 012044 doi:10.1088/1755-1315/1473/1/012044

Hazard and Operability Study of Proton Exchange Membrane Fuel Cell Hydrogen: Study Case in Hybrid-Electric Train Application

M K Asy'ari[1*], V S Permatasari[1], D A Ferdiansyah[1], and M R A Pratama[1]

[1] Engineering Physics Department, Institut Teknologi Sepuluh Nopember, 60111, Indonesia

*E-mail: mkasyari@its.ac.id

Abstract. Proton Exchange Membrane Fuel Cell (PEMFC) is currently being developed as a fuel for trains, especially the type of driven passenger trains. The application of PEMFC on trains requires in-depth study to ensure passenger safety. This is because if the hydrogen tank ruptures, the hydrogen can cause large pressure waves and even explosions if there is an ignition source. The purpose of this study is to determine the hazard opportunities that can occur in PEMFC with the hazard and operability study (HAZOP) analysis method of PEMFC. This method consists of several stages, such as identifying components in PEMFC, determining component deviation, providing risk matrix assessment, and providing recommendations. The HAZOP analysis conducted on the PEM Fuel Cell system showed that 74.167% of the components had low risk, while 25.83% of the components were identified as having medium risk. While components that have a high risk are not found in the system, indicating that the overall condition of the components and system is safe to operate.

1. Introduction

A fuel cell is one of the energy devices that can directly transfer chemical energy from fuels such as hydrogen and methanol into electrical energy with high efficiency [1]. Fuel cells generate electrical energy through a chemical reaction between hydrogen and oxygen, with water as a by-product. The fundamental structure of a fuel cell consists of an electrolyte layer positioned between an anode and a cathode[2]. Fuel cells are considered one of the most potential power sources for future electric vehicles due to their high efficiency, low emissions, and operation at relatively low temperatures, making them suitable for various applications, including rail transportation. In the context of trains, the use of fuel cells can provide several advantages, such as reduced CO_2 emissions, quieter operation, and better energy efficiency. On the other hand, hydrogen fuel cell electric trains are not only clean during operation, but also require little wayside infrastructure, thus can potentially reduce the cost of rail decarbonization [3].

The type of fuel cell most commonly used in railway applications is the Proton Exchange Membrane Fuel Cell (PEMFC). PEMFC is the most widely used fuel cell technology due to its low operating temperature, low noise, fast startup capability, light mass, and high power density [4].

PEM Fuel Cell has a lower temperature (below 100°C), fast connection time, and the easiest manufacturing process compared to other FC types. It has a fast dynamic response under load variation conditions and is compact, lightweight, and silent [5].

Hazard and Operability Analysis (HAZOP) is a process risk analysis method that is recognized and commonly used [6]. The Hazard and Operability Study, also known as HAZOP, is a technique used to identify and analyze hazards in a system by systematically reviewing its processes or operations. The results of a HAZOP analysis can be utilized by the enterprise to take effective measures to eliminate or control them[7]. Moreover, HAZOP can be used to determine whether deviations in a process may lead to unwanted events or accidents. The main characteristics of HAZOP its systematic approach, reliance on structured guide words, and the collaboration of a multidisciplinary team to ensure that safeguards are adequate for the system under review[8].

Integrating PEM fuel cell technology into railway systems requires a thorough understanding of the potential hazards and operational problems that may arise. Through HAZOP analysis, researchers can identify and assess risks, as well as develop mitigation strategies to reduce these risks. The application of HAZOP in this study aims to ensure that the PEM fuel cell system on the train can operate safely and efficiently. Therefore, this research is essential to guarantee the safe and effective adoption of this emerging technology in the rail transportation industry.

2. Method

This research utilizes the Hazard and Operability Study (HAZOP) method to assess the risks associated with hazards in the Proton Exchange Membrane (PEM) fuel cell system in hybrid trains. The research process involves the following steps.

2.1 Determine the study point (node)

HAZOP starts by selecting a node that is an isolation point constituting the flow of the P&ID valve or equipment from the process viewpoint[9]. Determination of the node is carried out by analyzing the loops in the Piping and Instrumentation Diagram (P&ID) of the fuel cell system. For the selected nodes, the components within each section are identified, governing all processes from input to output generation.

2.2 Determining the deviation of the Fuel Cell system

For the selected nodes, the instrument components within each section are identified, regulating all processes from input to output. These instrument components may include transmitters or process parameters. Based on these components, the appropriate guide words are determined. A guide word is a brief term used to describe deviations from the intended design or process.

Table 1. Standard Guide-word.

Guide word	Definition
No or none	No occurrence of the design
More (more of, higher)	Quantitative increase
Less (less of, lower)	Quantitative decrease
As well as (more than)	Quality improvement
Part of	Decrease in quality
Reverse	Reversal of the desired design
Other than	Full replacement

IOP Conf. Series: Earth and Environmental Science **1473** (2025) 012044 doi:10.1088/1755-1315/1473/1/012044

The following standards, based on the Australian/New Zealand Standard (AS/NZS 4360: 2004), are applied to determine the guide words, which are presented in Table 1.

Deviation refers to the way in which process conditions can differ from the intended design or purpose. Deviation in hazards and operability analysis (HAZOP) is typically a combination of the guide word and the parameter[10].

2.3 Determine the cause

The causes are explanations for why a deviation occurred. It assumes any operation problem that may arise in a system or equipment will be the cause of the deviation from the normal operation of a variable or parameter of the related context[10]. Multiple causes may be identified for a single deviation. Determining these causes is based on field studies of the process being analyzed. After identifying the causes, the adverse impact of the deviation (Consequences) is analyzed.

2.4 Determine the consequences

Severity or Consequences is the severity of injuries and lost working days [11]. Consequences can be assessed from several perspectives, including component damage, human impact, costs incurred due to the identified hazards, and losses resulting from production disruptions. Consequences may include process hazards or operational issues, such as plant shutdowns or product quality degradation. The consequence parameter describes the severity of the hazard's impact caused by the risk of deviation from the intended outcome. According to the Australian/New Zealand Standard (AS/NZS 4360:2004), the level of consequences can be determined based on criteria outlined in Table 2.

Table 2. Consequences Criteria.

Level	Descriptor	Description
1	*Insignificant*	*The system operates safely with minor, insignificant disruptions.*
2	*Minor*	*The system remains operational and safe, but disruptions cause slight performance degradation or system interference.*
3	*Moderate*	*The system can operate, but failures may result in the machine losing its primary function and/or causing product failures.*
4	*Major*	*The system cannot operate. Failures can cause significant physical and system damage, product failures, and/or non-compliance with Occupational Safety regulations.*
5	*Catastrophic*	*The system is unfit for operation, with very high severity if failures affect safe systems, violating Occupational Safety regulations.*

2.5 Determine the Likelihood

Likelihood is the probability of a risk that can occur in a component in a given period[12]. Likelihood refers to the frequency at which a risk or hazard occurs within a specific time frame. The standards for determining likelihood are outlined in [13], with criteria presented in Table 3.

Table 3. Likelihood Criteria.

No.	*Ranking*	*Desription*
1	*Brand New Excellent*	*Risk occurs less than four times in ten years.*
2	*Very Good/Good Serviceable*	*Risk occurs 4-6 times in ten years.*
3	*Acceptable*	*Risk occurs 6-8 times in ten years.*
4	*Below Standard/Poor*	*Risk occurs 8-10 times in ten years.*
5	*Bad/Unacceptable*	*Risk occurs more than 10 times in ten years.*

2.6 Determine the Risk

Risk is commonly described by considering both the potential consequences of an event and the likelihood of its occurrence [14]. The Risk matrix are the multiplication between likelihood and consequence. Risk matrix used during the assessment process. Risk matrix offers a visual representation and corellation between likelihood and consequence that could happen in the system. The risk matrix that used in this research presented in the following Table.

Table 4. Risk Matrix Ranking.

Likelihood	*Consequence*				
	Insignificant 1	*Minor* 2	*Moderate* 3	*Major* 4	*Catastrophic* 5
1.Brand New Excellent	*Low 1*	*Low 2*	*Low 3*	*Low 4*	*Medium 5*
2. Good	*Low 2*	*Low 4*	*Medium 6*	*Medium 8*	*Medium 10*
3. Acceptable	*Low 3*	*Medium 6*	*Medium 9*	*Medium 12*	*High 15*
4. Poor	*Low 4*	*Medium 8*	*Medium 12*	*High 16*	*High 20*
5. Unacceptable	*Medium 5*	*High 10*	*High 15*	*High 20*	*High 25*

3. Result and Discussion

The HAZOP analysis for the PEM Fuel Cell system in this research was conducted at Politeknik Negeri Madiun (PNM). The PEM Fuel Cell system comprises two main subsystems: the Safety Instrumented System (SIS) and the Basic Process Control System (BPCS). The SIS is responsible for ensuring safety during startup, shutdown, and in the event of a failure, while the BPCS controls the system to maintain efficient and stable operation.

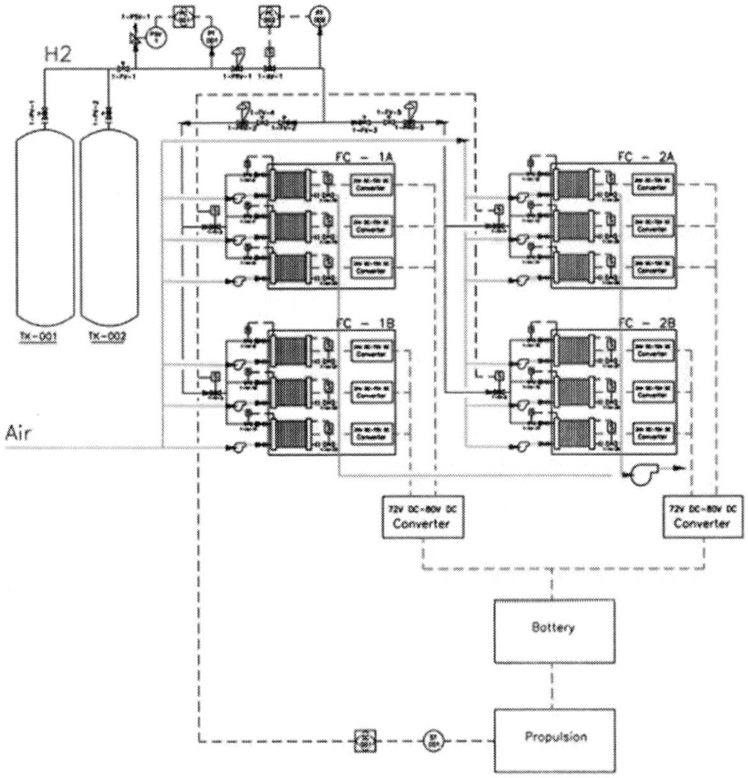

Figure 1. PEM Fuel Cell P&ID System.

In the case study, the fuel cell system includes 12 fuel cell stacks with a total power output of 12 kW. The SIS subsystem comprises PT-001, PC-001, and PSV-001, which function to vent high-pressure hydrogen gas trapped in the pipes. The BPCS subsystem includes 1-XV-, PT-002, and PC-002, which serve as a hydrogen gas pressure regulator, reducing the pressure from 150 bar to 8 bar, and controlling the fuel cell reaction process through the blower and solenoid valve. Figure 1 shows the P&ID of the PEM fuel cell system.

In this HAZOP analysis, the selected node is the fuel cell subsystem. The instrument components within the fuel cell subsystem include eight key elements: globe valve, hand valve, solenoid valve, pressure transmitter, blower inlet, blower outlet purge, DC-DC converter, and speed transmitter. These components are categorized as having low to moderate risk levels.

The guide words used for the components are: open, close, less, more, high, and low. Deviations from the system are described by terms such as fail to open, fail to close, high pressure, low pressure, less air, more air, less pressure, more pressure, less voltage, and more voltage.

In the PEM fuel cell system, the causes of deviations are identified based on these predefined deviations. Causes of deviations in the fuel cell subsystem include low input voltage, non-operational components, improper operation by the operator, oversupply of oxygen, and underpressure in the pressure valve.

To determine the likelihood, data from the fuel cell vendor's datasheets are used. The likelihood criteria, based on the standard (AS/NZS 4360: 2004), are detailed in Table 3. These criteria are classified based on how frequently the instrument may fail, as derived from datasheets and interviews with operators.

The determination of consequence is done qualitatively by evaluating instrument damage and its impact on the overall process of the fuel cell subsystem. The consequence criteria, according to the standard (AS/NZS 4360: 2004), are presented in Table 4.

As a result of the HAZOP assessment, the components in the fuel cell subsystem have a risk value range of 1 to 5. This range indicates that the risk can range from mild to severe damage, suggesting that the system should not be operated if such risks are present due to potential safety threats. This is particularly critical given the proximity of the fuel cell to the battery and generator, as a hydrogen leak could cause severe damage Minor damage typically affects only sensor readings. More detailed risk ranking results based on likelihood and consequence are represented in the HAZOP Worksheet in Table 5. The table are attached on the appendix section.

4. Conclusion

The HAZOP analysis of the PEM fuel cell system in hybrid trains has been successfully completed. The results indicate that the globe valve pressure component accounts for 74.167% of the total risk among the components analyzed, which include the globe valve pressure, DC-DC converter, pressure transmitter, speed transmitter, and blower. The remaining 25.83% of the risk is attributed to other components with moderate risk, specifically the hand valve and solenoid valve. No components with high risk were identified in the system. Future research will focus on analyzing the Safety Instrumented Layer (SIL) in the PEM Fuel Cell system.

IOP Conf. Series: Earth and Environmental Science **1473** (2025) 012044 doi:10.1088/1755-1315/1473/1/012044

References

[1] Patil V, Reshmi P V., Prajna S, Yashaswi, Yashaswini, Haleshappa D, Jayarama A and Pinto R 2023 Degradation mechanisms in PEM fuel cells: A brief review *Mater. Today Proc.*

[2] Tellez-Cruz M M, Escorihuela J, Solorza-Feria O and Compañ V 2021 Proton Exchange Membrane Fuel Cells (PEMFCs): Advances and Challenges *Polymers (Basel).* **13** 3064

[3] Ding D and Wu X Y 2024 Hydrogen fuel cell electric trains: Technologies, current status, and future *Appl. Energy Combust. Sci.* **17** 100255

[4] Daud W R W, Rosli R E, Majlan E H, Hamid S A A, Mohamed R and Husaini T 2017 PEM fuel cell system control: A review *Renew. Energy* **113** 620–38

[5] Amirdehi S, Trajin B, Vidal P-E, Vally J, Colin D and Tastet S 2020 Identification of PEM Fuel cell system model for traction chain based on profile mission measurement *2020 IEEE 29th International Symposium on Industrial Electronics (ISIE)* (IEEE) pp 27–32

[6] Wang F, Chen Y, Wang H, Chen C and Shi B 2012 The Intrinsic Safety Engineering Design Method for the Petrochemical Plant *Procedia Eng.* **43** 156–61

[7] Marhavilas P K, Filippidis M, Koulinas G K and Koulouriotis D E 2019 The integration of HAZOP study with risk-matrix and the analytical-hierarchy process for identifying critical control-points and prioritizing risks in industry – A case study *J. Loss Prev. Process Ind.* **62** 103981

[8] Dunjó J, Fthenakis V, Vílchez J A and Arnaldos J 2010 Hazard and operability (HAZOP) analysis. A literature review *J. Hazard. Mater.* **173** 19–32

[9] Choi J Y and Byeon S H 2020 Hazop methodology based on the health, safety, and environment engineering *Int. J. Environ. Res. Public Health* **17**

[10] Mocellin P, De Tommaso J, Vianello C, Maschio G, Saulnier-Bellemare T, Virla L D and Patience G S 2022 Experimental methods in chemical engineering: Hazard and operability analysis—HAZOP *Can. J. Chem. Eng.* **100** 3450–69

[11] Susanto N, Azzahra F and Putra A H 2022 Application of Hazard and Operability Study Methods (HAZOP) to asses and control hazard risk in spinning department using at textile industrial *IOP Conf. Ser. Earth Environ. Sci.* **1098**

[12] Noriyati R D, Rozaaq W, Musyafa A and Soepriyanto A 2015 Hazard & Operability Study and Determining Safety Integrity Level on Sulfur Furnace Unit: A Case Study in Fertilizer Industry *Procedia Manuf.* **4** 231–6

[13] Noriyati R D, Prakoso A B, Musyafa A and Soeprijanto A 2017 HAZOP Study and Determination of Safety Integrity Level Using Fault Tree Analysis on Fuel Gas Superheat Burner of Ammonia Unit in Petrochemical Plant, East Java *Asian J. Appl. Sci.* **5** 396–409

[14] Vaezi A, Jones S and Asgary A 2023 Integrating Resilience into Risk Matrices: A Practical Approach to Risk Assessment with Empirical Analysis *J. Risk Anal. Cris. Response* **13** 252–72

IOP Conf. Series: Earth and Environmental Science **1473** (2025) 012044
doi:10.1088/1755-1315/1473/1/012044

Appendix

Table 5. HAZOP Worksheet.

No	Tag No	Description	Guide Word	Deviation	Cause	Consequences	Safeguard	L	C	RR	Recommendation
1	1-PV-1	Globe Valve (Pressure)	More	More Pressure	Operator does not operate it precisely	Overpressure, potentially cause leaking on tube connection	Reduce the opener of 1-PV-1. For safety, if system still manage over pressure then close 1-FV-1	1	3	L3	Operate it more precisely and check regularly
			Less	Less Pressure	Operator does not operate it precisely	Underpressure, not working properly	Open the 1-PV-1 more	1	1	L1	Operate it more precisely and check regularly
2	1-PV-2	Globe Valve (Pressure)	More	More Pressure	Operator does not operate it precisely	Overpressure, potentially cause leaking on tube connection	Reduce the opener of 1-PV-1. For safety, if system still manage over pressure then close 1-FV-1	1	3	L3	Operate it more precisely and check regularly
			Less	Less Pressure	Operator does not operate it precisely	Underpressure, not working properly	Open the 1-PV-1 more	1	1	L1	Operate it more precisely and check regularly
3	1-FV-1	Hand Valve (Flow Safety)	Open	Fail to Open	The component doesn't work properly	Fuel cell won't work, due to lack of hydrogen	Line by pass manually, alarm PT 001 and PT 002, interlock 1-DC-1 and 1-DC-2	1	5	M5	Change the device
			Close	Fail to Close	The component doesn't work properly	Can't stop hydrogen flow to 1-PRV-1	Close 1-PV-1 and 1-PV-2	1	3	L3	Change the device
4	1-PSV-1	Solenoid Valve (To Vent)	Open	Fail to Open	The component doesn't work properly, TCMS generated wrong/error signal, High error transmitter reading	When the system shutdown the trapped hydrogen can't be realesed to atmosphere	Close 1-HV-1, 1-PV-1, 1-PV-2. Make sure large amount of air circulation and released manually	1	3	L3	Check and scheduling service, make a redundancy
			Close	Fail to Close	The component doesn't work properly, TCMS generated wrong/error signal, High error transmitter reading	Wasted hydrogen due to continously release to atmosphere. Wasted hydrogen could be flamable	Close 1-HV-1, 1-PV-1, 1-PV-2. Open air circulation in the room. Interlock fuel cell. Switch to another hybrid mode	1	5	M5	Check and scheduling service. Install a blower to neutralized the hydrogen in the room
5	PT 001	Pressure Transmitter	High	High Pressure	Overpressure from 1-PV-1, 1-PV-2, and 1-FV-1	Overpressure, potentially cause leaking on tube connection	Reduce the opener of 1-PV-1 and 1-PV-2. If necessary close the 1-FV-1 and interlock the fuel cell and then switch to another hybrid mode	1	3	L3	Check regularly
			Low	Low Pressure	Underpressure from 1-PV-1, 1-PV-2, and 1-FV-1	Underpressure, not working properly	Open more 1-PV-1 and 1-PV-2	1	2	L2	Check regularly
6	1-PRV-1	Globe Valve (Pressure Regulator)	More	More Pressure	Operator does not operate it precisely	Overpressure, potentially cause leaking on tube connection	Reduce the opener of 1-PRV-1	1	3	L3	Operate it more precisely and check regularly
			Less	Less Pressure	Operator does not operate it precisely	Underpressure, not working properly	Open more 1-PRV-1	1	2	L2	Operate it more precisely and check regularly

No	Tag No	Description	Guide Word	Deviation	Cause	Consequences	Safeguard	L	C	RR	Recommendation
7	1-XV-1	Solenoid Valve (Pressure Safety)	Open	Fail to Open	The component doesn't work properly, TCMS generated wrong/error signal, High error transmitter reading	Fuelcell will stop working due to lack of hydrogen	alarm in TCMS, interlock fuelcell and switch another hybrid mode	1	4	L4	Check and scheduling service
			Close	Fail to Close	The component doesn't work properly, TCMS generated wrong/error signal, High error transmitter reading	can't stop hydrogen flow to 1-FV-2 and 1-FV-3	Close 1-PRV-1	1	2	L2	Check and scheduling service
8	PT 002	Pressure Transmitter	High	High Pressure	Overpressure from 1-PRV-1 and 1-XV-1	Overpressure, potentially cause leaking on tube connection	Reduce the opener of 1-PRV-1	1	4	L4	Operate it more precisely and check regularly
			Low	Low Pressure	Underpressure from 1-PRV-1 and 1-XV-1	Underpressure, not working properly	Open more 1-PRV-1	1	2	L2	Operate it more precisely and check regularly
9	1-FV-2	Hand Valve (Flow Safety)	Open	Fail to Open	The component doesn't work properly	Fuel cell won't work, due to lack of hydrogen	Line by pass manually, Interlock 1-DC-1 and 1-DC-2	1	5	M5	Change the device
			Close	Fail to Close	The component doesn't work properly	Can't stop hydrogen flow to 1-FV-3	Close 1-FV-1, 1-PV-1 and 1-PV-2	1	3	L3	Change the device
10	1-FV-3	Hand Valve (Flow Safety)	Open	Fail to Open	The component doesn't work properly	Fuel cell won't work, due to lack of hydrogen	Line by pass manually, Interlock 1-DC-1 and 1-DC-2	1	5	M5	Change the device
			Close	Fail to Close	The component doesn't work properly	Can't stop hydrogen flow to 1-PRV-2	Close 1-FV-2, 1-FV-1, 1-PV-1 and 1-PV-2	1	3	L3	Change the device
11	1-FV-4	Hand Valve (Flow Safety)	Open	Fail to Open	The component doesn't work properly	Fuel cell won't work, due to lack of hydrogen	Line by pass manually, Interlock 1-DC-1 and 1-DC-2	1	5	M5	Change the device
			Close	Fail to Close	The component doesn't work properly	Can't stop hydrogen flow to 1-FV-5	Close 1-FV-1, 1-PV-1 and 1-PV-2	1	3	L3	Change the device
12	1-FV-5	Hand Valve (Flow Safety)	Open	Fail to Open	The component doesn't work properly	Fuel cell won't work, due to lack of hydrogen	Line by pass manually, Interlock 1-DC-1 and 1-DC-2	1	5	M5	Change the device
			Close	Fail to Close	The component doesn't work properly	Can't stop hydrogen flow to 1-PRV-3	Close 1-FV-4, 1-FV-1, 1-PV-1 and 1-PV-2	1	3	L3	Change the device
13	1-PRV-2	Globe Valve (Pressure Regulator)	More	More Pressure	Operator does not operate it precisely	Overpressure, potentially cause leaking on tube connection. Could damage the fuel cell membrane and make deficiency	Reduce the opener of 1-PRV-2	1	4	L4	Operate it more precisely and check regularly
			Less	Less Pressure	Operator does not operate it precisely	Underpressure, not working properly. Could cause hydrogen starvation on fuel cell	Open more 1-PRV-2	1	3	L3	Operate it more precisely and check regularly
14	1-PRV-3	Globe Valve (Pressure Regulator)	More	More Pressure	Operator does not operate it precisely	Overpressure, potentially cause leaking on tube connection. Could damage the fuel cell membrane and make deficiency	Reduce the opener of 1-PRV-3	1	4	L4	Operate it more precisely and check regularly
			Less	Less Pressure	Operator does not operate it precisely	Underpressure, not working properly. Could cause hydrogen starvation on fuel cell	Open more 1-PRV-3	1	3	L3	Operate it more precisely and check regularly

No	Tag No	Description	Guide Word	Deviation	Cause	Consequences	Safeguard	L	C	RR	Recommendation
15	1-XV-2	Solenoid Valve	Open	Fail to Open	The component doesn't work properly, TCMS generated wrong/error signal, High error transmitter reading	Fuelcell will stop working due to lack of hydrogen	alarm in TCMS, interlock fuelcell and switch another hybrid mode	1	5	M5	Check and scheduling service
			Close	Fail to Close	The component doesn't work properly, TCMS generated wrong/error signal, High error transmitter reading	can't stop hydrogen flow to 1-XV-6 to 1-XV-8	Closing 1-FV-4 or 1-FV-2	1	3	L3	Check and scheduling service
16	1-XV-3	Solenoid Valve (Pressure Safety)	Open	Fail to Open	The component doesn't work properly, TCMS generated wrong/error signal, High error transmitter reading	Fuelcell will stop working due to lack of hydrogen	alarm in TCMS, interlock fuelcell and switch another hybrid mode	1	5	M5	Check and scheduling service
			Close	Fail to Close	The component doesn't work properly, TCMS generated wrong/error signal, High error transmitter reading	can't stop hydrogen flow to 1-XV-9 to 1-XV-11	Closing 1-FV-4 or 1-FV-2	1	3	L3	Check and scheduling service
17	1-XV-4	Solenoid Valve (Pressure Safety)	Open	Fail to Open	The component doesn't work properly, TCMS generated wrong/error signal, High error transmitter reading	Fuelcell will stop working due to lack of hydrogen	alarm in TCMS, interlock fuelcell and switch another hybrid mode	1	5	M5	Check and scheduling service
			Close	Fail to Close	The component doesn't work properly, TCMS generated wrong/error signal, High error transmitter reading	can't stop hydrogen flow to 1-XV-12 to 1-XV-14	Closing 1-FV-5 or 1-FV-3	1	3	L3	Check and scheduling service
18	1-XV-5	Solenoid Valve (Pressure Safety)	Open	Fail to Open	The component doesn't work properly, TCMS generated wrong/error signal, High error transmitter reading	Fuelcell will stop working due to lack of hydrogen	alarm in TCMS, interlock fuelcell and switch another hybrid mode	1	5	M5	Check and scheduling service
			Close	Fail to Close	The component doesn't work properly, TCMS generated wrong/error signal, High error transmitter reading	can't stop hydrogen flow to 1-XV-15 to 1-XV-17	Closing 1-FV-5 or 1-FV-3	1	3	L3	Check and scheduling service
19	1-XV-6	Solenoid Valve (Pressure Safety)	Open	Fail to Open	The component doesn't work properly, High error transmitter reading	Fuelcell will stop working due to lack of hydrogen	alarm in TCMS, interlock fuelcell and switch another hybrid mode	1	5	M5	Check and scheduling service
			Close	Fail to Close	The component doesn't work properly, High error transmitter reading	fuelcell can't stop	Closing 1-FV-4 or 1-FV-2	1	3	L3	Check and scheduling service
20	1-XV-7	Solenoid Valve (Pressure Safety)	Open	Fail to Open	The component doesn't work properly, High error transmitter reading	Fuelcell will stop working due to lack of hydrogen	alarm in TCMS, interlock fuelcell and switch another hybrid mode	1	5	M5	Check and scheduling service
			Close	Fail to Close	The component doesn't work properly, High error transmitter reading	fuelcell can't stop	Closing 1-FV-4 or 1-FV-2	1	3	L3	Check and scheduling service

No	Tag No	Description	Guide Word	Deviation	Cause	Consequences	Safeguard	L	C	RR	Recommendation
21	1-XV-8	Solenoid Valve (Pressure Safety)	Open	Fail to Open	The component doesn't work properly, High error transmitter reading	Fuelcell will stop working due to lack of hydrogen	alarm in TCMS, interlock fuelcell and switch another hybrid mode	1	5	M5	Check and scheduling service
			Close	Fail to Close	The component doesn't work properly, High error transmitter reading	fuelcell can't stop	Closing 1-FV-4 or 1-FV-2	1	3	L3	Check and scheduling service
22	1-XV-9	Solenoid Valve (Pressure Safety)	Open	Fail to Open	The component doesn't work properly, High error transmitter reading	Fuelcell will stop working due to lack of hydrogen	alarm in TCMS, interlock fuelcell and switch another hybrid mode	1	5	M5	Check and scheduling service
			Close	Fail to Close	The component doesn't work properly, High error transmitter reading	fuelcell can't stop	Closing 1-FV-4 or 1-FV-2	1	3	L3	Check and scheduling service
23	1-XV-10	Solenoid Valve (Pressure Safety)	Open	Fail to Open	The component doesn't work properly, High error transmitter reading	Fuelcell will stop working due to lack of hydrogen	alarm in TCMS, interlock fuelcell and switch another hybrid mode	1	5	M5	Check and scheduling service
			Close	Fail to Close	The component doesn't work properly, High error transmitter reading	fuelcell can't stop	Closing 1-FV-4 or 1-FV-2	1	3	L3	Check and scheduling service
24	1-XV-11	Solenoid Valve (Pressure Safety)	Open	Fail to Open	The component doesn't work properly, High error transmitter reading	Fuelcell will stop working due to lack of hydrogen	alarm in TCMS, interlock fuelcell and switch another hybrid mode	1	5	M5	Check and scheduling service
			Close	Fail to Close	The component doesn't work properly, High error transmitter reading	fuelcell can't stop	Closing 1-FV-4 or 1-FV-2	1	3	L3	Check and scheduling service
25	1-XV-12	Solenoid Valve (Pressure Safety)	Open	Fail to Open	The component doesn't work properly, High error transmitter reading	Fuelcell will stop working due to lack of hydrogen	alarm in TCMS, interlock fuelcell and switch another hybrid mode	1	5	M5	Check and scheduling service
			Close	Fail to Close	The component doesn't work properly, High error transmitter reading	fuelcell can't stop	Closing 1-FV-5 or 1-FV-3	1	3	L3	Check and scheduling service
26	1-XV-13	Solenoid Valve (Pressure Safety)	Open	Fail to Open	The component doesn't work properly, High error transmitter reading	Fuelcell will stop working due to lack of hydrogen	alarm in TCMS, interlock fuelcell and switch another hybrid mode	1	5	M5	Check and scheduling service
			Close	Fail to Close	The component doesn't work properly, High error transmitter reading	fuelcell can't stop	Closing 1-FV-5 or 1-FV-3	1	3	L3	Check and scheduling service
27	1-XV-14	Solenoid Valve (Pressure Safety)	Open	Fail to Open	The component doesn't work properly, High error transmitter reading	Fuelcell will stop working due to lack of hydrogen	alarm in TCMS, interlock fuelcell and switch another hybrid mode	1	5	M5	Check and scheduling service
			Close	Fail to Close	The component doesn't work properly, High error transmitter reading	fuelcell can't stop	Closing 1-FV-5 or 1-FV-3	1	3	L3	Check and scheduling service
28	1-XV-15	Solenoid Valve (Pressure Safety)	Open	Fail to Open	The component doesn't work properly, High error transmitter reading	Fuelcell will stop working due to lack of hydrogen	alarm in TCMS, interlock fuelcell and switch another hybrid mode	1	5	M5	Check and scheduling service
			Close	Fail to Close	The component doesn't work properly, High error transmitter reading	fuelcell can't stop	Closing 1-FV-5 or 1-FV-3	1	3	L3	Check and scheduling service

No	Tag No	Description	Guide Word	Deviation	Cause	Consequences	Safeguard	L	C	RR	Recommendation
29	1-XV-17	Solenoid Valve (Pressure Safety)	Open	Fail to Open	The component doesn't work properly, High error transmitter reading	Fuelcell will stop working due to lack of hydrogen	alarm in TCMS, interlock fuelcell and switch another hybrid mode	1	5	M5	Check and scheduling service
			Close	Fail to Close	The component doesn't work properly, High error transmitter reading	fuelcell can't stop	Closing 1-FV-5 or 1-FV-3	1	3	L3	Check and scheduling service
30	1-B-1A1	Blower Inlet Fuel Cell 1A-1	More	More air	Oversupply oxygen	degrade membrane that cause fuelcell deficiency	alarm in TCMS, interlock fuelcell and switch another hybrid mode	1	4	L4	Check and scheduling service, Redundency, choose high durability blower
			Less	Less air	Oversupply oxygen affected redox reaxion	drop fuelcell efficiency	alarm in TCMS, interlock fuelcell and switch another hybrid mode	1	4	L4	Check and scheduling service, Redundency, choose high durability blower
31	1-B-1A2	Blower Inlet Fuel Cell 1A-2	More	More air	Oversupply oxygen	degrade membrane that cause fuelcell deficiency	alarm in TCMS, interlock fuelcell and switch another hybrid mode	1	4	L4	Check and scheduling service, Redundency, choose high durability blower
			Less	Less air	Oversupply oxygen affected redox reaxion	drop fuelcell efficiency	alarm in TCMS, interlock fuelcell and switch another hybrid mode	1	4	L4	Check and scheduling service, Redundency, choose high durability blower
32	1-B-1A3	Blower Inlet Fuel Cell 1A-3	More	More air	Oversupply oxygen	degrade membrane that cause fuelcell deficiency	alarm in TCMS, interlock fuelcell and switch another hybrid mode	1	4	L4	Check and scheduling service, Redundency, choose high durability blower
			Less	Less air	Oversupply oxygen affected redox reaxion	drop fuelcell efficiency	alarm in TCMS, interlock fuelcell and switch another hybrid mode	1	4	L4	Check and scheduling service, Redundency, choose high durability blower
33	1-B-1B1	Blower Inlet Fuel Cell 1B-1	More	More air	Oversupply oxygen	degrade membrane that cause fuelcell deficiency	alarm in TCMS, interlock fuelcell and switch another hybrid mode	1	4	L4	Check and scheduling service, Redundency, choose high durability blower
			Less	Less air	Oversupply oxygen affected redox reaxion	drop fuelcell efficiency	alarm in TCMS, interlock fuelcell and switch another hybrid mode	1	4	L4	Check and scheduling service, Redundency, choose high durability blower
34	1-B-1B2	Blower Inlet Fuel Cell 1B-2	More	More air	Oversupply oxygen	degrade membrane that cause fuelcell deficiency	alarm in TCMS, interlock fuelcell and switch another hybrid mode	1	4	L4	Check and scheduling service, Redundency, choose high durability blower
			Less	Less air	Oversupply oxygen affected redox reaxion	drop fuelcell efficiency	alarm in TCMS, interlock fuelcell and switch another hybrid mode	1	4	L4	Check and scheduling service, Redundency, choose high durability blower
35	1-B-1B3	Blower Inlet Fuel Cell 1B-3	More	More air	Oversupply oxygen	degrade membrane that cause fuelcell deficiency	alarm in TCMS, interlock fuelcell and switch another hybrid mode	1	4	L4	Check and scheduling service, Redundency, choose high durability blower
			Less	Less air	Oversupply oxygen affected redox reaxion	drop fuelcell efficiency	alarm in TCMS, interlock fuelcell and switch another hybrid mode	1	4	L4	Check and scheduling service, Redundency, choose high durability blower
36	1-B-2A1	Blower Inlet Fuel Cell 2A-1	More	More Pressure	Oversupply oxygen	degrade membrane that cause fuelcell deficiency	alarm in TCMS, interlock fuelcell and switch another hybrid mode	1	4	L4	Check and scheduling service, Redundency, choose high durability blower
			Less	Less Pressure	Oversupply oxygen affected redox reaxion	drop fuelcell efficiency	alarm in TCMS, interlock fuelcell and switch another hybrid mode	1	4	L4	Check and scheduling service, Redundency, choose high durability blower
37	1-B-2A2	Blower Inlet Fuel Cell 2A-2	More	More Pressure	Oversupply oxygen	degrade membrane that cause fuelcell deficiency	alarm in TCMS, interlock fuelcell and switch another hybrid mode	1	4	L4	Check and scheduling service, Redundency, choose high durability blower
			Less	Less Pressure	Oversupply oxygen affected redox reaxion	drop fuelcell efficiency	alarm in TCMS, interlock fuelcell and switch another hybrid mode	1	4	L4	Check and scheduling service, Redundency, choose high durability blower

No	Tag No	Description	Guide Word	Deviation	Cause	Consequences	Safeguard	L	C	RR	Recommendation
38	1-B-2A3	Blower Inlet Fuel Cell 2A-3	More	More Pressure	Oversupply oxygen	degrade membrane that cause fuelcell deficiency	alarm in TCMS, interlock fuelcell and switch another hybrid mode	1	4	L4	Check and scheduling service, Redundency, choose high durability blower
			Less	Less Pressure	Oversupply oxygen affected redox reaxion	drop fuelcell efficiency	alarm in TCMS, interlock fuelcell and switch another hybrid mode	1	4	L4	Check and scheduling service, Redundency, choose high durability blower
39	1-B-2B1	Blower Inlet Fuel Cell 2B-1	More	More Pressure	Oversupply oxygen	degrade membrane that cause fuelcell deficiency	alarm in TCMS, interlock fuelcell and switch another hybrid mode	1	4	L4	Check and scheduling service, Redundency, choose high durability blower
			Less	Less Pressure	Oversupply oxygen affected redox reaxion	drop fuelcell efficiency	alarm in TCMS, interlock fuelcell and switch another hybrid mode	1	4	L4	Check and scheduling service, Redundency, choose high durability blower
40	1-B-2B2	Blower Inlet Fuel Cell 2B-2	More	More Pressure	Oversupply oxygen	degrade membrane that cause fuelcell deficiency	alarm in TCMS, interlock fuelcell and switch another hybrid mode	1	4	L4	Check and scheduling service, Redundency, choose high durability blower
			Less	Less Pressure	Oversupply oxygen affected redox reaxion	drop fuelcell efficiency	alarm in TCMS, interlock fuelcell and switch another hybrid mode	1	4	L4	Check and scheduling service, Redundency, choose high durability blower
41	1-B-2B3	Blower Inlet Fuel Cell 2B-3	More	More Pressure	Oversupply oxygen	degrade membrane that cause fuelcell deficiency	alarm in TCMS, interlock fuelcell and switch another hybrid mode	1	4	L4	Check and scheduling service, Redundency, choose high durability blower
			Less	Less Pressure	Oversupply oxygen affected redox reaxion	drop fuelcell efficiency	alarm in TCMS, interlock fuelcell and switch another hybrid mode	1	4	L4	Check and scheduling service, Redundency, choose high durability blower
42	1-B-1	Blower Out Purge	More	More Pressure	Oversupply oxygen	degrade membrane that cause fuelcell deficiency	alarm in TCMS, interlock fuelcell and switch another hybrid mode	1	4	L4	Check and scheduling service, Redundency, choose high durability blower
			Less	Less Pressure	Oversupply oxygen affected redox reaxion	drop fuelcell efficiency	alarm in TCMS, interlock fuelcell and switch another hybrid mode	1	4	L4	Check and scheduling service, Redundency, choose high durability blower
43	1-DC-1A1	24V - 72V DC-DC Converter	Less	Less Voltage	Low input voltage	Drop efficiency	alarm in TCMS, increase hydrogen and oxygen supply	1	4	L4	Checking fuelcell healthy and output voltage, scheduling service
44	1-DC-1A2	24V - 72V DC-DC Converter	Less	Less Voltage	Low input voltage	Drop efficiency	alarm in TCMS, increase hydrogen and oxygen supply	1	4	L4	Checking fuelcell healthy and output voltage, scheduling service
45	1-DC-1A3	24V - 72V DC-DC Converter	Less	Less Voltage	Low input voltage	Drop efficiency	alarm in TCMS, increase hydrogen and oxygen supply	1	4	L4	Checking fuelcell healthy and output voltage, scheduling service
46	1-DC-1B1	24V - 72V DC-DC Converter	Less	Less Voltage	Low input voltage	Drop efficiency	alarm in TCMS, increase hydrogen and oxygen supply	1	4	L4	Checking fuelcell healthy and output voltage, scheduling service
47	1-DC-1B2	24V - 72V DC-DC Converter	Less	Less Voltage	Low input voltage	Drop efficiency	alarm in TCMS, increase hydrogen and oxygen supply	1	4	L4	Checking fuelcell healthy and output voltage, scheduling service
48	1-DC-1B3	24V - 72V DC-DC Converter	Less	Less Voltage	Low input voltage	Drop efficiency	alarm in TCMS, increase hydrogen and oxygen supply	1	4	L4	Checking fuelcell healthy and output voltage, scheduling service
49	1-DC-2A1	24V - 72V DC-DC Converter	Less	Less Voltage	Low input voltage	Drop efficiency	alarm in TCMS, increase hydrogen and oxygen supply	1	4	L4	Checking fuelcell healthy and output voltage, scheduling service
50	1-DC-2A2	24V - 72V DC-DC Converter	Less	Less Voltage	Low input voltage	Drop efficiency	alarm in TCMS, increase hydrogen and oxygen supply	1	4	L4	Checking fuelcell healthy and output voltage, scheduling service

No	Tag No	Description	Guide Word	Deviation	Cause	Consequences	Safeguard	L	C	RR	Recommendation
51	1-DC-2A3	24V - 72V DC-DC Converter	Less	Less Voltage	Low input voltage	Drop efficiency	alarm in TCMS, increase hydrogen and oxygen supply	1	4	L4	Checking fuelcell healthy and output voltage, scheduling service
52	1-DC-2B1	24V - 72V DC-DC Converter	Less	Less Voltage	Low input voltage	Drop efficiency	alarm in TCMS, increase hydrogen and oxygen supply	1	4	L4	Checking fuelcell healthy and output voltage, scheduling service
53	1-DC-2B2	24V - 72V DC-DC Converter	Less	Less Voltage	Low input voltage	Drop efficiency	alarm in TCMS, increase hydrogen and oxygen supply	1	4	L4	Checking fuelcell healthy and output voltage, scheduling service
54	1-DC-2B3	24V - 72V DC-DC Converter	Less	Less Voltage	Low input voltage	Drop efficiency	alarm in TCMS, increase hydrogen and oxygen supply	1	4	L4	Checking fuelcell healthy and output voltage, scheduling service
55	1-DC-1	72V - 80V DC-DC Converter	Less	Less Voltage	Low input voltage	Drop efficiency	alarm in TCMS	1	4	L4	Checking 24V-72V DC-DC Converter output voltage , scheduling service, and redundancy
56	1-DC-2	72V - 80V DC-DC Converter	Less	Less Voltage	Low input voltage	Drop efficiency	alarm in TCMS	1	4	L4	Checking 24V-72V DC-DC Converter output voltage , scheduling service, and redundancy
57	1-XV-18	Solenoid Valve (Purge)	Open	Fail to Open	The component doesn't work properly, High error transmitter reading	Drop efficiency due to excess of hydrogen	alarm in TCMS, interlock fuelcell and switch another hybrid mode	1	3	L3	Check and scheduling service
			Close	Fail to Close	The component doesn't work properly, High error transmitter reading	Drop efficiency due to lack of hydrogen	Closing 1-FV-4 or 1-FV-2	1	5	M5	Check and scheduling service
58	1-XV-19	Solenoid Valve (Purge)	Open	Fail to Open	The component doesn't work properly, High error transmitter reading	Drop efficiency due to excess of hydrogen	alarm in TCMS, interlock fuelcell and switch another hybrid mode	1	3	L3	Check and scheduling service
			Close	Fail to Close	The component doesn't work properly, High error transmitter reading	Drop efficiency due to lack of hydrogen	Closing 1-FV-4 or 1-FV-2	1	5	M5	Check and scheduling service
59	1-XV-20	Solenoid Valve (Purge)	Open	Fail to Open	The component doesn't work properly, High error transmitter reading	Drop efficiency due to excess of hydrogen	alarm in TCMS, interlock fuelcell and switch another hybrid mode	1	3	L3	Check and scheduling service
			Close	Fail to Close	The component doesn't work properly, High error transmitter reading	Drop efficiency due to lack of hydrogen	Closing 1-FV-4 or 1-FV-2	1	5	M5	Check and scheduling service
60	1-XV-21	Solenoid Valve (Purge)	Open	Fail to Open	The component doesn't work properly, High error transmitter reading	Drop efficiency due to excess of hydrogen	alarm in TCMS, interlock fuelcell and switch another hybrid mode	1	3	L3	Check and scheduling service
			Close	Fail to Close	The component doesn't work properly, High error transmitter reading	Drop efficiency due to lack of hydrogen	Closing 1-FV-4 or 1-FV-2	1	5	M5	Check and scheduling service
57	1-XV-22	Solenoid Valve (Purge)	Open	Fail to Open	The component doesn't work properly, High error transmitter reading	Drop efficiency due to excess of hydrogen	alarm in TCMS, interlock fuelcell and switch another hybrid mode	1	3	L3	Check and scheduling service
			Close	Fail to Close	The component doesn't work properly, High error transmitter reading	Drop efficiency due to lack of hydrogen	Closing 1-FV-4 or 1-FV-2	1	5	M5	Check and scheduling service

No	Tag No	Description	Guide Word	Deviation	Cause	Consequences	Safeguard	L	C	RR	Recommendation
58	1-XV-23	Solenoid Valve (Purge)	Open	Fail to Open	The component doesn't work properly, High error transmitter reading	Drop efficiency due to excess of hydrogen	alarm in TCMS, interlock fuelcell and switch another hybrid mode	1	3	L3	Check and scheduling service
			Close	Fail to Close	The component doesn't work properly, High error transmitter reading	Drop efficiency due to lack of hydrogen	Closing 1-FV-4 or 1-FV-2	1	5	M5	Check and scheduling service
59	1-XV-24	Solenoid Valve (Purge)	Open	Fail to Open	The component doesn't work properly, High error transmitter reading	Drop efficiency due to excess of hydrogen	alarm in TCMS, interlock fuelcell and switch another hybrid mode	1	3	L3	Check and scheduling service
			Close	Fail to Close	The component doesn't work properly, High error transmitter reading	Drop efficiency due to lack of hydrogen	Closing 1-FV-5 or 1-FV-3	1	5	M5	Check and scheduling service
60	1-XV-25	Solenoid Valve (Purge)	Open	Fail to Open	The component doesn't work properly, High error transmitter reading	Drop efficiency due to excess of hydrogen	alarm in TCMS, interlock fuelcell and switch another hybrid mode	1	3	L3	Check and scheduling service
			Close	Fail to Close	The component doesn't work properly, High error transmitter reading	Drop efficiency due to lack of hydrogen	Closing 1-FV-5 or 1-FV-3	1	5	M5	Check and scheduling service
59	1-XV-26	Solenoid Valve (Purge)	Open	Fail to Open	The component doesn't work properly, High error transmitter reading	Drop efficiency due to excess of hydrogen	alarm in TCMS, interlock fuelcell and switch another hybrid mode	1	3	L3	Check and scheduling service
			Close	Fail to Close	The component doesn't work properly, High error transmitter reading	Drop efficiency due to lack of hydrogen	Closing 1-FV-5 or 1-FV-3	1	5	M5	Check and scheduling service
60	1-XV-27	Solenoid Valve (Purge)	Open	Fail to Open	The component doesn't work properly, High error transmitter reading	Drop efficiency due to excess of hydrogen	alarm in TCMS, interlock fuelcell and switch another hybrid mode	1	3	L3	Check and scheduling service
			Close	Fail to Close	The component doesn't work properly, High error transmitter reading	Drop efficiency due to lack of hydrogen	Closing 1-FV-5 or 1-FV-3	1	5	M5	Check and scheduling service
61	1-ST-001	Speed Transmiter	More	More Voltage	The component doesn't work properly, High error transmitter reading	fuelcell may be work over the power needed	alarm in TCMS, interlock fuelcell	1	2	L2	Check and scheduling service and calibration
			Less	Less Voltage	The component doesn't work properly, High error transmitter reading	fuelcell may be work over the power needed	alarm in TCMS, interlock fuelcell	1	2	L2	Check and scheduling service and calibration

Quality Of Service, Price Of Tripang (Sea Cucumber), Quality Of Product Of Tripang (Sea Cucumber) Towards Customer Loyalty In The Center Of Marine Products Of Longos Island –Ntt

D E Setyowati[1] Herman[2]

[1,2] Department of Social and Political Sciences, Universitas Hang Tuah, Surabaya, Indonesia

*E-mail: dyahantariksa@gmail.com

Abstract. The purpose of this study is to determine the sustainability of the marine business, which is currently trending called the Blue Economy. The research variables are the Quality of Service, Price of Tripang (Sea Cucumber), and Quality of Tripang (Sea Cucumber) Products on Customer Loyalty. This study was conducted on Longos Island. This study uses qualitative and quantitative methods, using population data and samples from the Seafood Trading Center on Longos Island. From buyers recorded from seafood traders, 111 populations were obtained. Sea cucumbers themselves are Marine Animals that are used as Cosmetic ingredients and foods that are beneficial to the human body, which are in great demand by buyers from abroad, especially from Korea, Japan < China, and Indonesia. Based on the research results, it can be concluded that the product quality partially has a significant effect on customer loyalty. The value of Adjusted R Square (Adj. R2) is 0.657 or 65.70% which means that the independent variables can explain changes in the customer loyalty variable by 65.70% while the remaining 34.30% is explained by other factors outside this study. Service quality, price, and customer satisfaction have a significant impact on customer loyalty.

1.Introduction

The government policy states that in order to promote prosperity and economic independence, particularly in coastal towns, the Blue Economy needs to be strengthened and expanded. Initial observations on a number of marine commodities that warranted upgrading were made by the researcher. According to the findings of the observations, Sea Cucumber or Tripang are examples of marine goods whose economic value may be raised. Longos Island has not been touched much by the government, in the field of empowering blue technology and the blue economy. The majority of Longos Island people's income comes from marine products.

The government must step in and regulate the trade of tripang (sea cucumber), as it is a natural resource, biological riches, and marine wealth of Indonesia. The administration is currently remaining unfocused and neglecting rules, particularly those pertaining to marine products. Producers of marine products may suffer from unstable prices; hopefully, government support in the form of regulation, capital assistance, and coaching would help. For Natural Resource Wealth, on Indonesia's most remote island.

2.Method

In essence, research methods [10] are scientific approaches to gathering data for particular applications and goals. Research efforts grounded in scientific qualities—rational, empirical, and systematic—are referred to as scientific techniques. Rational research is defined as having been carried out in a way that

IOP Conf. Series: Earth and Environmental Science **1473** (2025) 012045 doi:10.1088/1755-1315/1473/1/012045

makes sense and can be justified by human reason. Research that is carried out in an observable way is said to as empirical.

The survey method is a quantitative research technique that is used to gather information about past and present events related to beliefs, opinions, traits, behavior, and variable relationships. It also tests multiple hypotheses about sociological and psychological variables from samples drawn from a particular population. Data collection techniques involve shallow observations (interviews or questionnaires), which leads to study results that are frequently generalized.

Following a survey, the researcher carried out quantitative research, disseminated questionnaires, and used a Likert scale to process the questionnaire responses. SPSS Multiple Linear Regression was then used to process the data.

Regression analysis, as defined by Gujarati in Kurniawan and Yuniarto [7] (2016:19), is the study of the relationship between one variable, known as the explained variable, and one or more explanatory factors. It is possible to refer to the second variable as the independent variable and the first as the dependent variable. Multiple linear regression is the name given to this type of regression analysis when the independent variable is greater than one. It can be characterized as numerous since the dependent variable will be subjected to various influences from the independent factors. This investigation was undertaken on Longos Island, Labuan Bajo, NTT. Operational Definition of Variables, To provides a more detailed focus on the study, it is necessary to Formulate operational research variables, the description of which is as follows [3]. Operational Variables of Service Quality Research Reliability, Responsiveness, Assurance, Empathy. And Physical Evidence. Operational Price variables include; Price Affordability, Price Suitability with Product Quality, Price Competitiveness, Suitability Price with Benefits, Price Affects Consumers. Operational variables of Consumer Satisfaction include: Conformity to expectations, Interest in revisiting, Willingness to recommend. Operational Research of Customer Loyalty Variables includes: Reuse, and trust [8]. According to Sugiyono [10] (2016:80), "Population is a generalization area consisting of objects or subjects that have certain qualities and characteristics that are determined by researchers to be studied and then conclusions drawn" . Thus, in addition to individuals, the population also consists of items and other natural objects. Additionally, the population include all the traits or attributes that other subjects or objects have in addition to the quantity of objects or subjects under study. 153 Tripang (Sea Cucumber) customers who regularly and during the season made up the study's population. 111 clients who responded had Slovin computations.

3. Results and Discussion

Based on the validity test [10] calculation service quality variables using SPSS 25 software, the following results were obtained, namely that it can be seen that the statement items used in the research instrument for 111 respondents each have an rtable value of 0.1865, which means it is less than the rcount value. This means that each statement item in the independent and dependent variables is valid because rcount > rtable [4]. Reliability Test: The reliability test method that is often used is Cronbach's Alpha. According to Ghozali (2016), a measuring instrument can be said to be reliable if the reliability value is > 0.60. The results of the Reliability Test of the research variables all show Reliable. Classical Assumption Test. In the classical assumption test, several tests will be tested, namely the Data Normality Test, Heteroscedasticity test, and Multicollinearity test, and all the results meet the requirements of the classical assumption test, from the results of the classical assumption test meets the requirements. Multiple Regression Analysis: Multiple regression analysis is used to determine the influence between two or more independent variables with one dependent variable entered in the regression model. This analysis is also to predict the value of the dependent variable if the value of the independent variable increases or decreases and to determine the direction of the relationship between the independent variable and the dependent variable whether each dependent variable is positively or negatively related.

IOP Conf. Series: Earth and Environmental Science **1473** (2025) 012045 doi:10.1088/1755-1315/1473/1/012045

Table 1. Multiple Regression Analysis Test Results

VARIABLES	*UNSTANDARDIZED COEFFICIENTS*	
	B	*STD.ERROR*
(Constant)	10,192	1,187
Quality of service	0.215	0.060
Price	-0.127	0.103
Customer satisfaction	0.354	0.095

Source: Processed data (2024)

If the results of the output are entered in the form of a multiple regression equation they are as follows:

$$Y = a + b1X1 + b2X2 + b3X3 + e$$

$$Y = 10.192 + 0.215X1 - 0.127X2 + 0.354X3$$

Information:
Y= Predicted dependent variable (Customer loyalty)
X1 = Independent variable (Service quality)
X2 = Independent variable (Price)
X3 = Independent variable (Customer satisfaction)a= Constant value
b1, b2, b3 = Regression coefficient

The explanation of the multiple regression equation above is as follows:

The meaning of the constant and coefficient numbers and the hypothesis testing (t-test) are as follows:

- The constant value of 10.192 means that if the quality of service, price, and customer satisfaction have constant values (fixed), then customer loyalty is 10.192. This influence is significant at an error rate of 5%.
- The service quality regression coefficient of 0.215 indicates that the effect of service quality on customer loyalty is positive. Based on the regression coefficient value, it can be concluded that if service quality increases by 1%, customer loyalty will increase by 0.215.
- The price regression coefficient of -0.127 indicates that the effect of price on customer loyalty is negative. Based on the regression coefficient value, it can be concluded that if the price increases by 1%, customer loyalty will decrease by 0.127.
- The customer satisfaction regression coefficient is0.354 shows that the influence of customer satisfaction on customer loyalty is positive. Based on the regression coefficient value, it can be concluded that if customer satisfaction increases by 1%, then customer loyalty will increase by 0.354.

The results of the correlation coefficient and determination coefficient analysis can be seen in the Model Summary output from data processing using SPSS 25, the table below:

Table 2. The results of the correlation coefficient analysis and the coefficient of determination

R	R *SQUARE*	*ADJUSTED R SQUARE*	*STD. ERROR OF THE ESTIMATE*
0.816	0.667	0.657	1.57832

Source: Processed data (2024)

In the table above the size the correlation coefficient value (R) is 0.816 which means that there is a very strong correlation or relationship between service quality, price, and customer satisfaction on customer loyalty. The amount of Adjusted R Square (Adj. R2) is 0.657 or 65.70% which means that the independent variables can explain changes in the customer loyalty variable by 65.70% while the remaining 34.30% is explained by other factors outside this study. The magnitude of the individual influence of the service quality variable (X1), price (X2), and customer satisfaction (X3) towards customer loyalty (Y) are shown in the following table:

Table 3. individual influence of service quality variables (X1), price (X2), and customer satisfaction (X3) on customer loyalty (Y)

VARIABLES	tcount	ttable	*Sig.*
(Constant)	3,161		0.003
Quality of service	3,576	1,981	0.001
Price	-3,124	1,981	0.012
Customer satisfaction	3,734	1,981	0,000

Source: Processed data (2024)

From the results of the t-test above, it can be concluded that:

- Coefficient for service quality
  a. Statistics tthe calculation results in the number 3,576
  b. The value of t is knowncount > ttable (3.576 > 1.981) and a significant value of 0.001 < 0.05, then Ho is rejected and Ha is accepted, so it can be concluded that service quality partially has a significant effect on customer loyalty.
- Coefficient for price
  a. Statistics tcount get number
  b. The value of t is knowncount > ttable (-3.124 > -1.981) and the significant value is 0.012 < 0.05 then Ho is rejected and Ha is accepted so it can be concluded that price partially has a significant effect on customer loyalty.
- Coefficient for customer satisfaction
  a. Statistics tthe calculation results in the number 3,734
  b. The value of t is knowncount > ttable (3.734 > 1.981) and a significant value of 0.000 < 0.05, then Ho is rejected and Ha is accepted, so it can be concluded that customer satisfaction has a partial significant effect on customer loyalty.

3.1. F Test (Simultaneous)

The F test is used to test whether there is a significant influence between the independent variables on the dependent variables simultaneously.

Table 4. testing the significant effect between the independent variables on the dependent variable simultaneously.

VARIABLES	Fcount	Ftable	*Sig.*
QualityPrice service Customer satisfaction	71,281	2.68	0,000

Source: Processed data (2040)

From the table above it can be seen that the F valuecount of 71.281 with a Sig. value of 0.000, because the Sig. value is smaller than 0.05 and Fcount > Ftable (71.281 > 2.68), then Ho is rejected or Ha is accepted so that it can be concluded that service quality, price, and customer satisfaction simultaneously have a significant effect on customer loyalty.

Based on the research results and discussions that have been presented previously, regarding the influence of service quality, price, and customer satisfaction on customer loyalty, the following conclusions can be drawn:

- t valuecount > ttable (3.576 > 1.981) and a significant value of 0.001 < 0.05, then Ho is rejected and Ha is accepted, so it can be concluded that service quality partially has a significant effect on customer loyalty.
- t valuecount > ttable (-3.124 > -1.981) and a significant value of 0.012 < 0.05, then Ho is rejected and Ha is accepted, so it can be concluded that price has a partial significant effect on customer loyalty.
- t valuecount > ttable (3.734 > 1.981) and a significant value of 0.000 < 0.05, then Ho is rejected and Ha is accepted, so it can be concluded that customer satisfaction has a partial significant effect on customer loyalty.
- F Valuecount of 71.281 with a Sig. value of 0.000, because the Sig. value is smaller than 0.05 and Fcount > Ftable (71.281 > 2.68), then Ho is rejected or Ha is accepted so that it can be concluded that service quality, price, and customer satisfaction simultaneously have a significant effect on customer loyalty. Meanwhile, the correlation coefficient (R) value is 0.816 which means that there is a very strong correlation or relationship between service quality, price, and customer satisfaction on customer loyalty. The amount of Adjusted R Square (Adj. R2) is 0.657 or 65.70% which means that the independent variables can explain changes in customer loyalty variables by 65.70%, while the remaining 34.30% is explained by other factors outside this study.

4. Conclusion

Based on the analysis and discussion conducted on the respondents' answers, service quality, price, and customer satisfaction have a significant effect on customer loyalty. There are several things that must be done:

- The quality of service should be maintained well and improved by providing training and knowledge about the company to employees so that they can provide better service, be responsive and timely in serving every customer need and provide complete information politely and professionally [1]. The role of the Government is needed, providing training to the frontline island business actors on Longos Island.
- Given the many competitors in the market faced must be able to maintain the quality of the goods and products offered so that what is sold is in accordance with the price offered. The role of the Ministry of Maritime Affairs and the Ministry of Industry is awaited on the frontier island of Longos Island [5].
- The company's customer satisfaction should be maintained and improved by considering factors that affect customer satisfaction such as service quality, affordable prices, responsiveness and punctuality in providing services, security and comfort in serving customers [4]. Business actors must be consistent and learn more, to understand consumer behavior.
- For further research, it is hoped that other variables can be added and the number of samples increased, and carried out on the sea cucumber producing islands, both in Indonesia and outside Indonesia.

Acknowledgement.

This research was sponsored by Hang Tuah University and the National Character Foundation.

IOP Conf. Series: Earth and Environmental Science **1473** (2025) 012045 doi:10.1088/1755-1315/1473/1/012045

Refrences

[1] Diyatma, Aris Jatmika. 2017, The Influence of Promotion Through Instagram Social Media on Saka Bistro Product Purchasing Decisions, e-Proceeding of Management: Vol.4, No.1April 2017, Faculty of Economics & Business, Telkom University.

[2] Effendi, Muh. Arief. 2015. The Power of Corporate Governance: Theory and Practice.*Implementation* . Salemba Four. Jakarta.

[3] Ghozali, Imam. 2016. Multivariate Analysis Application With IBM SPSS Program*23*(Edition 8). VIIIth printing. Diponegoro University Publishing Agency. Semarang.

[4] Harjati, L., and Venesia, Y. 2015. The Influence of Service Quality and Price Perception onCustomer SatisfactionCustomers at Tiger Air Mandala Airlines. E-Journal WIDYA Ekonomika, 1(2015), 64–74.

[5] Herawaty, Arleen, Guna, Welvin I. 2016. "The Influence of Good Corporate Governance Mechanism, Auditor Independence, Audit Quality and Other Factors on Earnings Management". Journal of Business and Accounting Vol. 12, No. 1, April 2016, Pg. 53– 68. Trisakti College of Economics.

[6] Hollensen, Svend. 2019. Marketing Management: Relationship Approach. USA: Pearson Education Inc.

[7] Suhaily, Lily and Soelasih, Yasintha. 2017. What Effects Repurchase Intention of Online Shopping. International
Business Research, 10(12), 113-122.

[8] Sunyoto, Danang. 2015. Basics of Marketing Management (Concepts, Strategies and*Case)*.CAPS (Center for Academic Publishing Services). Yogyakarta

[9] Tjiptono, Fandy and Anastasia Diana. 2016. Total Quality Management. Revised Edition. AndiOffset. Yogyakarta.

[10] Tjiptono, Fandy. 2015. Marketing Strategy. 4th Edition. Andi Publisher.Yogyakarta.

IOP Conf. Series: Earth and Environmental Science **1473** (2025) 012046 doi:10.1088/1755-1315/1473/1/012046

Environmental Law Policy in Indonesia: Challenges and Sustainable Justice

L Yustitianingtyas, L Y E Pratiwi, A D Irawan, D Stansyah, S Arifin
Faculty of Law, Muhammadiyah University of Surabaya

*E-mail: levinayustitianingtyas@um-surabaya.ac.id

Abstract. Environmental legal policy is a legal framework that aims to protect and manage the environment to ensure the sustainability of natural resources and the quality of human life. This policy includes various legal instruments, such as laws, government regulations, and regional regulations, which are designed to regulate human activities that have an impact on the environment. Through a preventive approach and strict law enforcement, environmental legal policy seeks to reduce pollution, protect biodiversity, and promote the sustainable use of natural resources. In Indonesia, environmental legal policy is also influenced by developments in international law, such as the global climate agreement and the biodiversity convention that Indonesia has ratified. The research method used is normative legal research. That in international regulations requires countries to strengthen the regulation and implementation of environmental policies at the national and local levels. However, challenges in law enforcement, low public awareness, and conflicts of interest between economic development and environmental conservation, often become obstacles to the effectiveness of this policy. Therefore, synergy is needed between the government, society, and the private sector in optimizing the implementation of fair and sustainable environmental legal policies.

1. Introduction

Indonesia is a country known for its beautiful and extraordinary natural resources. Vast tropical rainforests, high biodiversity, and rich minerals and other natural resources make Indonesia one of the countries with abundant natural resources [1]. However, behind this great potential, Indonesia also faces serious challenges related to environmental management. Along with rapid economic growth, industrialization, and urbanization that have occurred in recent decades, Indonesia is facing increasing pressure on its environment. Land expansion for oil palm plantations, illegal logging, and exploitation of mineral resources have led to massive deforestation, ecosystem damage, and degradation of air and water quality. In addition, increasing greenhouse gas emissions and global climate change have also worsened environmental conditions in Indonesia [2].

In response to these challenges, the Indonesian government has attempted to develop and implement various environmental legal policies. The history of environmental regulation in Indonesia began with the enactment of Law Number 4 of 1982 concerning Basic Provisions for Environmental Management, which was later revised and strengthened by Law Number 32 of 2009 concerning Environmental Protection and Management [3]. This law aims to provide a strong legal framework for environmental management in Indonesia, emphasizing the principles of precaution, community participation, and sustainable management [4]. However, despite the existence of various regulations governing environmental protection, the implementation and enforcement of environmental law in Indonesia are still far from ideal [5]. One of the biggest challenges is the weak law enforcement in the field, which is often caused by various factors such as corruption, lack of supervisory capacity, and conflicts of interest between stakeholders. Cases of environmental violations are often not followed up firmly, resulting in injustice for the affected communities, especially local communities whose lives depend on natural resources [6].

On the other hand, there is an imbalance between economic interests and environmental protection [6].

The government often faces a dilemma between encouraging economic growth through investment in extractive sectors, such as mining and plantations, and preserving the environment. When priority is given to economic growth, the environment often becomes a victim of environmentally unfriendly policies. This results in severe environmental damage, which ultimately harms society in the long run. The concept of sustainable justice emerged as an effort to address injustice arising from policies that only prioritize one aspect, be it economic or environmental, without considering their impacts holistically. Sustainable justice demands a balance between the interests of economic development, environmental protection, and community welfare [6]. In Indonesia, the implementation of sustainable justice in environmental law policies is still a major challenge. Although the principles of sustainable justice have begun to be integrated into environmental policies, their implementation in the field is often still weak and inconsistent.

In addition, community participation in the decision-making process is also an important factor in realizing sustainable justice. Communities, especially those directly affected by environmental policies, are often not adequately involved in this process. As a result, the resulting policies often do not reflect the needs and aspirations of the community, giving rise to resistance and conflict.

This legal writing examines environmental law policy in Indonesia with a focus on the challenges faced in its implementation, as well as efforts to achieve sustainable justice. Through policy analysis, problem identification, and case studies, it is hoped that constructive recommendations can be produced for improving environmental law policy in Indonesia. Thus, it will not only contribute to the development of theories and concepts in the field of environmental law, but also to practical efforts to realize sustainable justice in Indonesia.

2. Method

The research method in writing this law is normative legal research, with a statutory regulatory approach. As for data analysis techniques by processing the results of collecting primary materials of laws and regulations, international agreements, data, or several related documents, as well as collecting secondary materials to be continued by analyzing the data sources obtained, which include reference books, journals or related scientific articles. As for tertiary legal materials to explain and assist in analyzing primary and secondary legal materials that are in accordance with the theme.When viewed from the nature of the writing, the method used is a descriptive analytical method, namely by trying to describe the problems that arise in the formation of just and sustainable environmental policies, as well as trying to analyze and provide a solution regarding the need for efforts to achieve sustainable justice in environmental conservation.

3. Results and discussion

3.1. Environmental Law Policy Concept

Legal policies on environmental management have begun to receive attention from environmental law experts in Indonesia. Environmental law experts provide views on environmental policies that are based on the many cases of environmental violations, such as forest fires, air pollution due to factory waste and so on. Some of the opinions of experts include, according to H. Taufik Abdullah, environmental

IOP Conf. Series: Earth and Environmental Science **1473** (2025) 012046 doi:10.1088/1755-1315/1473/1/012046

legal policy is a set of legal norms designed to protect and manage the environment. This includes rules, procedures, and mechanisms that aim to maintain the balance of the ecosystem and ensure that development does not damage the environment. This policy does not only focus on preventing damage, but also on environmental restoration and rehabilitation.

Furthermore, Romli Atmasasmita also provided a view on environmental legal policy referring to the entire legal and administrative actions implemented to achieve environmental protection goals. This includes the establishment of laws, regulations, and other legal instruments that regulate how individuals, companies, and countries interact with their environment. The main goal is to create harmony between economic development and environmental protection.

E.W.B Niskanen also gave his opinion on environmental legal policy as a form of legal arrangement aimed at regulating and controlling human activities so as not to disrupt environmental sustainability. This involves the use of law as a tool to achieve ecological and social goals, taking into account the long-term impacts on the earth and natural resources [7].

The opinions of these experts can be concluded that environmental legal policy leads to the arrangement of laws used to regulate and control human activities in the utilization of the environment. This is a driver of the creation of an ecosystem balance whose development does not damage the environment.

The discourse on environmental management has begun since 1970, the Development of Environmental Law Policy Globally and in Indonesia has begun since long ago, namely since 1970 which is marked by attention to environmental law. Its development can be divided into several decades, namely : [8]

a. The 1970s: This decade saw a global increase in environmental concerns, fueled by reports such as the Club of Rome's "Limits to Growth" (1972). The 1972 Stockholm Conference is considered a major milestone in drawing global attention to environmental issues and establishing principles of environmental protection in global policy.
b. 1980s: The Brundtland Report (1987), or "Our Common Future," introduced the concept of sustainable development, which integrates economic, social, and environmental aspects. It became an important foundation for international environmental policy.
c. 1990s-2000s: The United Nations Conference on Environment and Development (UNCED) in Rio de Janeiro in 1992, produced Agenda 21 and the United Nations Framework Convention on Climate Change (UNFCCC) [9]. The Kyoto Protocol (1997) and other international agreements continued efforts to address climate change and global environmental issues.
d. 2010s to present: The 2015 Sustainable Development Goals (SDGs) by the UN emphasize the integration of environmental aspects in global development. The Paris Agreement (2015) is a major milestone in the global effort to reduce greenhouse gas emissions and address climate change [10].

The explanation of the development of environmental law policy above is that in the Early period (before the 1970s) environmental law awareness in Indonesia was still very limited. Environmental protection was not a top priority in public policy. In the 1970s to 1980s, Indonesia passed Law No. 4 of 1982 concerning Basic Provisions on the Environment, which was the first environmental law in Indonesia. This reflected increasing attention to environmental issues. In the 1990s, the process of reform and development of environmental law policy increased with the enactment of Law No. 23 of 1997 concerning Environmental Management, which introduced more comprehensive environmental protection principles.

The 2000s to the present have given birth to Law No. 32 of 2009 concerning Environmental Protection and Management[11], replacing the previous law and placing more emphasis on a community-based approach, stricter law enforcement, and regulations regarding environmental impacts. This law was then updated with Law No. 11 of 2020 concerning Job Creation, which changes several aspects of environmental law to accelerate the licensing and investment process [12].

In environmental law, there are Basic Principles that must be met for environmental management, namely:

a. Prevention Principle: Prioritizes preventing environmental damage before it occurs. This means proactive measures are taken to avoid potential damage, rather than dealing with problems after they have occurred.
b. Polluter Pays Principle: Requires parties who cause pollution or environmental damage to bear the costs of remediation or repair. This encourages actors to take responsibility for the environmental impacts of their activities [13].
c. Precautionary Principle: When there is a risk of serious or irreversible damage to the environment, action should be taken despite scientific uncertainty. This principle encourages precautionary action even when there is uncertainty about the environmental impact.
d. Public Participation Principle: Emphasizes the importance of involving the public in decision-making processes relating to the environment [7]. Public participation is considered essential to ensure that a variety of views and interests are taken into account [2].
e. Equality Principle: Maintaining a balance between the needs of development and environmental protection for present and future generations. This principle also emphasizes the need for equitable distribution of environmental benefits and burdens throughout society.
f. Integration Principle: Promotes the integration of environmental considerations into all development policies and decisions. This principle recognizes that economic, social and environmental decisions must be taken simultaneously to achieve sustainability.[6]

These fically, sustainable justice includes :

a. Social Justice: Ensure that all individuals and groups in society have equitable access to resources, opportunities and rights. This includes attention to poverty eradication, gender equality, human rights and protection of vulnerable groups.
b. Ecological Justice: Ensure that human activities do not excessively damage the environment and that natural resources are managed wisely to ensure the sustainability of ecosystems and biodiversity.
c. Economic Justice: Arranging the economic system so that it does not only benefit a few parties but provides benefits more evenly. This includes efforts to reduce income inequality and ensure that economic growth does not sacrifice social and environmental welfare.

Of the three aspects of sustainable justice as a whole, effective environmental law must be able to integrate these three aspects of justice in a harmonious way. This means that in implementing regulations that protect the environment effectively without burdening certain social groups unfairly. The aspect of justice also needs to be a guarantor of the participation of all levels of society in decision-making and pay attention to the social impacts of environmental policies, so that a balance is created between environmental protection and economic needs, providing support for those affected by changes in environmental policy [14]. One example of the relationship between economic justice and environmental law, for example, is that policies that are too strict can burden small businesses and cause job losses. Ideally, environmental law should consider the economic impacts of regulations and seek a balance between environmental protection and economic development. Ultimately, compliance with the aspects of social justice, ecological justice and economic justice in the design and implementation of environmental law can create a more just and sustainable system for all people and ecosystems.

The concept of sustainable justice emphasizes the importance of considering the long-term impacts of our decisions and actions. This concept refers to the responsibility to ensure that future generations have the same or better quality of life than today by involving wise management of natural resources and reducing negative impacts on the environment. There are two concepts of sustainable justice, namely Intergenerational and Intragenerational Justice. Intergenerational justice focuses on the relationship between the current generation and future generations. This means that we must manage resources and the environment in a way that does not harm future generations [15]. While intragenerational justice relates to the relationship between individuals or groups within one generation, which includes efforts to address inequality and ensure a fair distribution of resources and opportunities. Related to addressing inequality and ensuring a fair distribution of resources and opportunities can be done with several strategies, namely:

a. Integration of environmental, social, and economic aspects, namely recognizing that the environment, social, and economy are interrelated and inseparable. Community welfare and economic growth must go hand in hand with environmental sustainability. Effective solutions must consider these three aspects holistically.
b. Participation and inclusion, which serve to achieve sustainable justice, are essential to involve a wide range of stakeholders in the decision-making process. This includes local communities, minority groups, and those affected by a particular policy or project. Inclusive engagement helps ensure that decisions are fair and take into account the interests of all parties.
c. The principle of shared responsibility by prioritizing shared responsibility between individuals, communities, companies, and governments. Each party has a role and responsibility in promoting sustainability and justice.

It is expected that the three strategies can realize the objectives of the concept of sustainable justice for the management and utilization of natural resources. The principle of shared responsibility and community participation can be a control system to anticipate bad environmental impacts or environmental pollution.

3.2. The relationship between sustainable justice and environmental policy

Sustainable Justiceis a concept that combines the principles of social, economic, and environmental justice. It includes the equitable distribution of resources and benefits, the protection of human rights, and maintaining a balance between the needs of present and future generations [16]. The goal is to ensure that all individuals and communities have equal access to resources and benefits from development while protecting the environment, which includes [15]:

a. Social justice and access to resourcesemphasizes the importance of equitable distribution of environmental resources and benefits. Equitable environmental policies should ensure that all groups in society, especially the most vulnerable, have equal access to environmental resources and benefits. For example, policies that support equitable access to clean water and clean air are part of sustainable justice.
b. Reducing environmental inequality through stepsthe need to reduce environmental inequalities that often occur in society. Disadvantaged or marginalized groups often face more severe environmental impacts, such as pollution or environmental degradation. Environmental policies that address these aspects can help address inequalities by providing additional protections or benefits to the most affected communities.
c. Participation in Decision Makingincludes the right to participate in decision-making that affects the environment. Good environmental policies should involve a wide range of stakeholders, including local communities and affected groups, in the planning and implementation process. This participation ensures that policies are not only effective but also equitable.
d. Protection of environmental rights byemphasizes the importance of protecting environmental rights as part of human rights. Environmental policies that take these rights into account, such as the right to live in a healthy environment, support the principles of sustainable justice.
e. Integration and Coordination withpromote integration and coordination between environmental and socio-economic policies. Environmental policies designed without considering their impact on social or economic justice may not be sustainable in the long term. In contrast, policies that integrate social, economic, and environmental aspects tend to be more effective in achieving sustainable justice.

The relationship between sustainable equity and environmental policy is very close. Sustainable equity provides a framework to ensure that environmental policies are not only effective in protecting the environment but also equitable in the distribution of benefits and burdens. Environmental policies that take into account the principles of sustainable equity can help create a more equitable society and a healthier environment, thereby supporting sustainable development as a whole.

3.3. Challenges and Strategies in Environmental Law Policy in Indonesia

The implementation of environmental law policies often faces various complex challenges, ranging from problems related to law enforcement, corruption, public awareness, and infrastructure limitations. The following will describe various problems that occur in the practice of environmental conservation as an effort to implement environmental law policies [4]:

1. Law enforcement
   Many countries, especially lower to lower-middle income countries, suffer from institutional deficiencies in the enforcement of environmental laws. These include a lack of trained staff, limited resources, and inadequate tools for inspection and monitoring, leading to weak enforcement. Weak enforcement is influenced by a variety of factors including complex bureaucracies, slow legal procedures, and an inability to effectively prosecute violations. This often leaves violators with a sense of no real consequences for their actions [17].
2. Corruption Crime
   Corruption can undermine the effectiveness of environmental legal policies by influencing decision-making and law enforcement.[18] For example, corrupt officials may accept bribes from industry players who want to bypass environmental regulations. Corruption can undermine public trust in the legal system and institutions responsible for environmental protection. This can reduce public support for environmental policies and discourage active participation in conservation efforts.
3. Lack of Public Awareness
   Many people may not be fully aware of the environmental impacts of their actions or the importance of complying with environmental laws. Lack of effective education and awareness campaigns can hamper environmental conservation efforts. Community involvement in environmental decision-making and law enforcement is often lacking. Communities may not be adequately informed about how they can contribute to or report environmental violations.
4. Infrastructure Limitations
   Many countries face challenges in terms of the infrastructure needed to support the implementation of environmental legal policies, such as air and water quality monitoring facilities and effective waste management systems. Lack of access to modern technology and accurate data can limit the ability to conduct environmental impact assessments, track violations and develop effective data-driven solutions.

From the various challenges that occur, a strategy is needed to overcome them, by addressing these challenges comprehensively, the implementation of environmental legal policies can be more effective and sustainable, and environmental protection can be improved for the welfare of society and the planet as a whole. Some strategies that can be applied include :

a. Capacity building and personnel training by strengthening the capacity of institutions responsible for environmental law enforcement can help improve the effectiveness of policy implementation.

b. Eradicating corruption by implementing measures to increase transparency, accountability and integrity in decision-making and law enforcement can help reduce the influence of corruption.
c. Education and awareness by increasing public education and awareness campaigns about the importance of environmental protection and their role in legal compliance can increase public participation and support.
d. Infrastructure development with investment in better environmental infrastructure and monitoring technology can improve the ability to implement and enforce environmental policies effectively.

3.4. Implementation of Sustainable Justice in Environmental Law Policy

Implementing sustainable justice refers to the application of strategies, policies, and practices that simultaneously meet the goals of social and economic justice, as well as environmental sustainability. It includes practical steps to ensure that the development process or policy implemented is not only effective in the short term but also environmentally friendly and socially just in the long term. Implementing the concept of sustainable justice can be done in various ways, such as :

a. Policies and Regulations: Governments can establish policies that support sustainable and fair practices, for example through environmental protection laws or economic redistribution policies.
b. Sustainable Business Practices: Companies can adopt environmentally friendly and ethical business practices, as well as ensure that their supply chains comply with social justice standards.
c. Public Education and Awareness: Increasing public understanding of the importance of sustainability and equity, and educating them on how to contribute to these goals.
d. Community Initiatives: Local communities can get involved in initiatives that support sustainability and equity, such as waste management programs, environmental conservation, or inclusive social activities.

The principle of sustainable equity, which often includes environmental, social, and economic dimensions, is a framework designed to ensure that natural resources are managed in a fair manner, maintaining a balance between the needs of current and future generations, and considering the impacts on all levels of society, especially vulnerable groups. The following is an analysis of how the principle of Sustainable Equity in Environmental Policy is applied in environmental policy and its impacts on society, namely by [16]:

1. Participatory Approach
   Sustainable equity requires the active participation of a wide range of stakeholders, including local communities and vulnerable groups. In many cases, inclusive environmental policies involve consultation with communities to understand their needs and concerns. Sustainable equity also often includes mechanisms to ensure transparency and accountability, such as public reporting and independent audits.
2. Balance Between Economic and Environmental Needs
   These policies are designed to integrate economic, social, and environmental objectives, avoiding trade-offs that harm one or the other. Sustainable natural resource management seeks to ensure that resource exploitation is carried out in a way that takes into account the impacts on ecosystems and local communities. An example is a policy that promotes renewable energy while creating jobs. Energy policies that focus on renewable energy often provide incentives for poor communities to access clean technologies, such as subsidies for solar panels, as well as training programs for jobs in the renewable energy sector.
3. Concern for Vulnerable Groups
   Policies often include social impact assessments to ensure that the negative effects of environmental projects are not too severe for vulnerable groups, such as poor communities or indigenous groups. Some policies include compensation or support programs for the most affected

groups, such as housing or job training. There are several social impacts that arise for vulnerable groups, both positive and negative. The positive impacts include sustainable environmental policies that can improve the quality of life through a cleaner and healthier environment. For example, reducing air and water pollution can improve public health. And the economic opportunities that support green industrial policies can create new jobs and encourage sustainable local economic development [19].
In addition to positive impacts, there are also negative impacts, namely eviction and dislocation. In some cases, environmental policies can lead to the eviction of local communities from their land, especially if large projects such as national parks or infrastructure projects require land acquisition. In the presence of social disparities, inadequate or non-inclusive policies can exacerbate social disparities if vulnerable groups do not benefit equally from the policy or if they are not involved in the decision-making process [19].

The application of the principle of sustainable equity in environmental policy seeks to create a balance between environmental protection, economic development, and social welfare. The impact on communities and vulnerable groups can be positive if policies are designed and implemented with all these aspects in mind in a fair and inclusive manner. However, it is also important to continuously monitor and assess the impact of policies to ensure that the principle of sustainable equity is truly met and that vulnerable groups are not victims of possible imbalances. By understanding and applying the concept of sustainable equity, we can create a more just and balanced world, both for today and for future generations.

3.5. Environmental Pollution by Mining Companies in Kalimantan

In some countries, sustainable forest management programs involve local communities in decision-making and monitoring, give them rights to forest products, and include mechanisms to ensure that economic benefits are distributed fairly. However, in practice in Indonesia, there are still many environmental violations committed by companies, one of which is a mining company in Kalimantan, which is engaged in coal extraction, which has experienced lawsuits due to environmental pollution. The mining activity has caused significant damage to the forest ecosystem, river pollution, and decreased air quality in the surrounding area. Local communities have complained about the negative impacts of this pollution, including health problems, decreased quality of life, and damage to natural resources.

According to Law no. 32 of 2009 concerning Environmental Protection and Management, regulates companies' obligations to comply with environmental standards and implement good environmental management. Companies are required to make efforts to prevent and control pollution as well as carry out environmental restoration. The Company's obligations are regulated in articles 22-26 of Law no. 32 of 2009 requires every business and/or activity that has the potential to have a major impact on the environment to carry out an Environmental Impact Analysis (AMDAL) [20]. Companies must prepare an AMDAL document that includes environmental impacts, management plans and monitoring plans. Furthermore, the AMDAL evaluation process must be approved by the government before environmental permits are granted. This process involves public consultation to ensure transparency and public participation. Before starting activities, the company must obtain permission which includes approval of the AMDAL document and the company's commitment to comply with the specified environmental standards. If companies do not comply with the provisions of environmental permits, they may be subject to administrative sanctions, such as revocation of permits, or criminal sanctions in accordance with applicable regulations. In Government Regulation no. 27 of 2012 concerning Environmental Permits[21], stipulates requirements for companies to obtain environmental permits, which include environmental impact analysis (AMDAL) before starting operational activities. Based on the cases above, the Government can impose administrative sanctions such as fines or revocation of operational permits, as well as criminal sanctions for serious violations that result in serious environmental damage in accordance with environmental regulations, which are regulated by

Articles 104-109 of Law no. 32/2009. In accordance with this law, companies are required to prepare reports regarding the implementation of environmental management and impact monitoring. This report must be submitted to the government and made accessible to the public to ensure transparency. To overcome the problem of environmental damage, the government has made several efforts, namely through the following programs:

1. Environmental recovery through rehabilitation and restoration.
   A policy that requires companies to rehabilitate ex-mining land and restore affected ecosystems.
2. Conducting sustainable justice analysis by prioritizing community welfare.
   Local communities have the right to a healthy environment and to be free from pollution. Social justice demands that those directly affected by pollution receive fair compensation and access to health services and improved quality of life. Sustainable justice also requires that local communities be involved in decision-making processes related to mining activities through consultation and active participation.
3. Ecological Justice in terms of natural resource conservation
   Mining activities must consider the sustainability of natural resources and ecosystems. Ecological justice requires effective efforts to minimize environmental impacts and ensure that damaged ecosystems can be restored.

Companies must comply with environmental standards set out in laws and regulations. Strict enforcement is necessary to ensure this compliance and to protect the environment from further damage. Effective implementation will help protect the environment and ensure that natural resources can be utilized sustainably for the benefit of future generations.

4. Conclusion

Indonesia has abundant natural resources, but faces major challenges in environmental management. The government has made efforts with various legal policies, such as the Law on Environmental Protection and Management, the implementation and enforcement of environmental law is still not optimal. Policies in the field of environmental law influence sustainable justice in environmental management. The main problems faced include weak law enforcement, conflicts between economic and environmental interests, and a lack of community participation in decision making. Sustainable justice, which prioritizes balance between the economy, environment and community welfare, is an important concept but still faces obstacles in its implementation. Cases of environmental pollution by mining companies in Kalimantan reflect the challenges in implementing sustainable environmental law and justice policies. Even though there are regulations and policies governing environmental protection, implementation and enforcement often experience obstacles. To achieve sustainable justice, strict law enforcement, active community participation and consistent rehabilitation efforts are needed. This is important to protect the environment, ensure community welfare, and ensure the sustainability of natural resources for future generations. Law no. 32 of 2009 provides a comprehensive legal framework for environmental management and protection in Indonesia. By requiring companies to prepare and implement AMDAL, comply with environmental quality standards. This law aims to reduce negative impacts on the environment. Strict law enforcement and transparency in reporting are also expected to encourage companies to operate in a more sustainable and responsible manner.

Acknowledgment

This paper was supported by a Faculty of Law Muhammadiyah University of Surabaya.

References

[1] L. Journal, "PREVENTION OF ILLEGAL FUEL SMUGGLE IN THE BORDER OF EAST

NUSA," **5**, no. 1, pp. 15–24, 2021.

[2] B. Dalal-Clayton and B. Sadler, *Sustainability Appraisal*. **2014**. doi: 10.4324/9780203135235.

[3] L. O. A. La Ode Angga and H. Suat, "Legal Responsibility in the Pollution and Environmental Destruction Due to Gold Mining Exploitation in Botak Mountain of Buru Regency," *FIAT JUSTISIAJurnal Ilmu Huk.*, **13**, no. 4, p. 381, 2019, doi: 10.25041/fiatjustisia.v13no3.1695.

[4] M. Pardede *et al.*, "Perspectives of sustainable development vs. law enforcement on damage, pollution and environmental conservation management in Indonesia," *J. Water Clim. Chang.*, **14**, no. 10, pp. 3770–3790, 2023, doi: 10.2166/wcc.2023.417.

[5] *THESIS ADAPTING BOSTON ' S ZERO CARBON EMISSIONS POLICIES TO MAKASSAR BASED UPON INTERNATIONAL ENVIRONMENTAL LAW Prepared and submitted by : A . NURIL ZAMHARIR HARIS LEGAL STUDIES / INTERNATIONAL LAW LAW FACULTY HASANUDDIN UNIVERSITY.*

[6] S. Alam, M. J. H. Bhuiyan, T. M. R. Chowdhury, and E. J. Techera, *Routledge handbook of international environmental law*. 2012. doi: 10.4324/9780203093474.

[7] B. Lausche, *Guidelines for Protected Areas Legislation*, no. 81. 2011. [Online]. Available: http://data.iucn.org/dbtw-wpd/exec/dbtwpub.dll?AC=GET_RECORD&XC=/dbtw-wpd/exec/dbtwpub.dll&BU=http://data.iucn.org/dbtw-wpd/commande/index_newsite.htm&TN=iucn&SN=AUTO5511&SE=258&RN=0&MR=20&TR=0&TX=1000&ES=0&CS=1&XP=&RF=WebRes&EF=&DF=WebAff&RL=0&EL=0&DL=0&

[8] O. S. Emerging, "DESIGN ASSOCIATIONS IN GLOBAL ENVIRONMENTAL," 2022.

[9] "A brief history of sustainability," fau.eu. [Online]. Available: https://www.fau.eu/2022/02/03/friedrich/here-to-stay/

[10] UN HABITAT, "Promoting Low Emission Urban Development Strategies," no. December, 2016.

[11] Undang-Undang, *UNDANG-UNDANG REPUBLIK INDONESIA NOMOR 32 TAHUN 2009 TENTANG PERLINDUNGAN DAN PENGELOLAAN LINGKUNGAN HIDUP*. 2009.

[12] C. Alkalah, *Prosiding Pengakuan&Perlindungan Masyarakkat Hukum Adat Di Tingkat Nasional dan Internasional*, **19**, no. 5. 2016.

[13] T. Turap, T. B. Merupakan, T. B. Lebih, and T. D. Turap, *Environmental Management in Organizations*.

[14] S. Mulyani and E. Lisdiyono, "Implementation of Timber Legality Verification System for Forest Processed Wood Products Based on Ecological Justice," *Int. J. Criminol. Sociol.*, **10**, pp. 1039–1050, 2021, doi: 10.6000/1929-4409.2021.10.122.

[15] I. Rosyadi, M. R. Habibi, and N. Syam, "Implementation of criminal law enforcement concept of environmental sustainability (illegal logging in Indonesia)," *IOP Conf. Ser. Earth Environ. Sci.*, **894**, no. 1, 2021, doi: 10.1088/1755-1315/894/1/012002.

[16] P. Taylor, "An Echological Approach to International Law," Routledge Handbook of International Environmental Law. [Online]. Available: https://www.routledge.com/An-Ecological-Approach-to-International-Law-Responding-to-the-Challenges-of-Climate-Change/Taylor/p/book/9780415162609?srsltid=AfmBOoqtiRa-PWQETa4y66XIUEhB2maOHPEBW5pYt9y-bWK05tMYYGj1

[17] D. L. tan T. Tibi Didier Zoungrana, Aguima Aimé Bernard Lompo, "Effect of climate finance on environmental quality: A global analysis," *Res. Econ.*, **78**, no. 4, 2024, [Online]. Available: https://www.sciencedirect.com/science/article/abs/pii/S109094432400053X?dgcid=rss_sd_all

[18] "How can the Enforcement of the Basel Convention be Improved? - A Case Study of China and the UK and the Role of Networks Thesis submitted to the School of Law at the University of Strathclyde for the degree of Doctor of Philosophy by Faculty of Humanity ," no. June, 2024.

[19] A. Ginanjar, "Persepsi Pemerintah Dan Masyarakat Dalam Pelestarian Cagar Budaya: Studi Kasus Kawasan Cagar Budaya Pusat Pemerintahan Kesultanan Siak Sri Indrapura," *Paradig. J. Kaji. Budaya*, vol. 14, no. 1, pp. 109–121, 2024, doi: 10.17510/paradigma.v14i1.1485.
[20] H. Bachrul Amiq, "Administrative Sanction in Environmental Law," *Int. J. Res. - GRANTHAALAYAH*, vol. 6, no. 6, pp. 22–37, 2018, doi: 10.29121/granthaalayah.v6.i6.2018.1331.
[21] Peraturan Pemerintah, *PERATURAN PEMERINTAH REPUBLIK INDONESIA NOMOR 27 TAHUN 2012 TENTANG IZIN LINGKUNGAN*. 2012.

Genetic Algorithm Approach in Double Hull Oil Tanker Structural Optimization: Material Efficiency and Environmental Impact

A Rofidayanti[1], A Zubaydi[1*], S H Sujiatanti[1], R C Ariesta[1], H Siswanti[2,3]

[1] Department of Naval Architecture, Institut Teknologi Sepuluh Nopember, Surabaya
[2] Department of Ocean Engineering, Institut Teknologi Sepuluh Nopember, Surabaya
[3] Department of Shipbuilding Engineering, Politeknik Negeri Madura, Sampang

*E-mail: zubaydi@na.its.ac.id, septia@its.ac.id

Abstract. In the maritime industry, the challenge of ship structural design lay in balancing diverse design criteria, necessitating innovative approaches to ensure robust ship structures with minimal weight. Achieving efficiency in design was crucial for enhancing material efficiency and environmental impact. This research focuses on optimizing the midship structure of double-hull oil tankers, which were vital for transporting large quantities of liquid cargo. The methodology employed a Single Objective Optimization approach with a Genetic Algorithm to minimize the weight of the ship's structure. Weight was assessed in terms of the construction weight for one block of the ship located in the parallel middle body. Design variables included plate thickness and dimension of stiffener, with design parameters encompassing the ship's main dimension, preliminary design, and structural design. The constraints were the weight of one block in the parallel middle body of the existing ship and the modulus value. The expected outcomes of this research included a significant reduction in midship structure weight, especially for longitudinal elements, while maintaining the ship's strength according to the modulus requirements of Biro Klasifikasi Indonesia, along with the modeling of an optimized ship midship section. By achieving a reduction in structural weight, this research aims to improve environmental impact and enhance material efficiency in ship design.

Keywords: Single Objective Optimization; Genetic Algorithm; Midship Structure; Double Hull Oil Tanker.

1. Introduction

The process of modern ship design is increasingly incorporating innovative solutions to address the complexities of ship structures, driven by the rising demands for improved reliability, safety, and efficiency (1). Clearly defining objectives is the fundamental first step in the ship design process, which is inherently iterative, continually refining the hull form to satisfy various criteria such as classification regulations, material availability, and feasible manufacturing processes. In the context of commercial shipping, where profitability is the primary goal, optimizing structural design is crucial for achieving maximum payload capacity.

Innovation is essential for maintaining a competitive edge in ship design, requiring improvements across multiple, often conflicting, design aspects. However, as design procedures become more complex, the constraints of limited design time underscore the need for advanced support systems to aid designers. Structural optimization is a key component in this context, aiming to identify the most appropriate structural configurations and optimal member dimensions to improve material efficiency and minimize environmental impact (2). Depending on the stage of the design process, ship structural optimization can be divided into three primary categories: topology optimization involves determining the best material distribution within a specific space and under given constraints, thereby achieving the optimal mechanical properties of the structure (3,4), shape optimization focuses on refining the structure's surface geometry and the design of internal openings (5–7), and size optimization aims to fine-tune the dimensions of the ship's structural components within various constraints to ensure both economic efficiency and structural safety (8,9).

In the maritime sector, the challenge of ship structural design involves balancing multiple design criteria to create structures that are both robust and lightweight. This balance is especially crucial in the design of double-hull oil tankers, which are critical for the transportation of large volumes of liquid cargo. The optimization of the midship structure in these vessels plays a significant role, directly affecting both material efficiency and environmental sustainability.

This study focuses on optimizing the midship structure of double-hull oil tankers using the Single Objective Optimization approach with a Genetic Algorithm. The primary objective is to minimize the construction weight of the ship's structure, with a specific focus on the weight of one block in the parallel middle body. Key design variables include plate thickness and frame spacing, while the design parameters cover the ship's main dimensions, service range, and preliminary design data. Constraints are set by the weight of the existing midship block and the modulus value, ensuring that the optimized design maintains the necessary strength requirements.

The expected outcomes of this research include a significant reduction in the weight of the midship structure while preserving the required strength as per Biro Klasifikasi Indonesia (BKI) modulus standards. Moreover, the study aims to model an optimized midship section, contributing to more efficient and environmentally friendly ship designs. As the maritime industry continues to advance, the application of sophisticated optimization methods, such as Genetic Algorithms, will become increasingly important in meeting the challenges of contemporary ship design.

2. Optimization of Ship Structure

The structural optimization is implemented for this double-hull oil tanker 17,500 DWT. Based on the midship section data from the existing ship, as shown in Figure 1, the optimization focuses on the longitudinal elements and the hull plates of the vessel. The vessel utilizes a longitudinal construction system with a double hull configuration. It is also observed that the ship is designed with hopper tanks. The weight analysis is conducted on a single block, measuring 9 meters in length, located in the parallel middle body of the ship. The vessel is constructed with the weight of one block from the existing ship calculated to be 200.96 tons.

Table 1. Main Dimension of Double Hull Oil Tanker 17,500 DWT

Parameter	Dimension	
Length Over All (LOA)	157.5	m
Length Between Perpendicular (LBP)	149.5	m
Beam Moulded (B)	27.7	m
Depth Moulded (H)	12	m
Service Speed (v)	13	knots

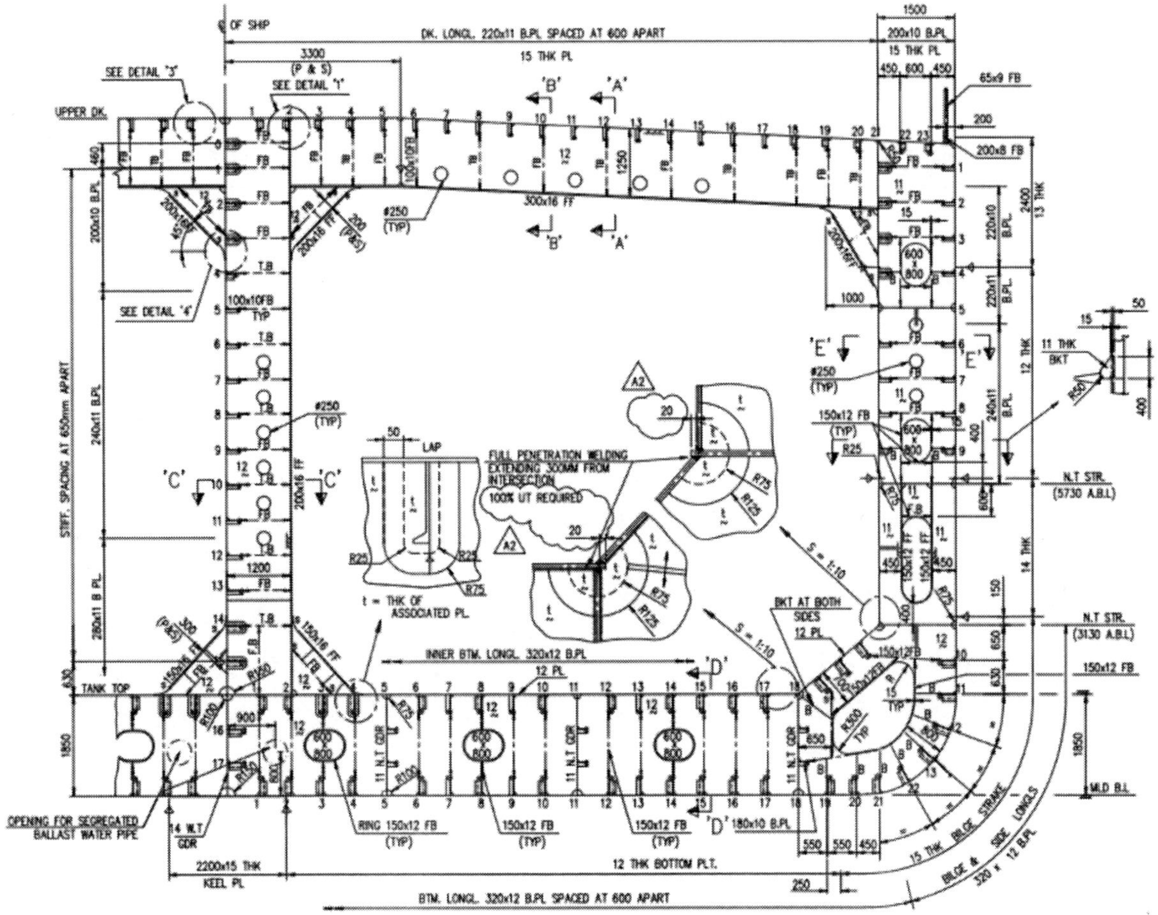

Figure 1. Midship Section of Double Hull Oil Tanker 17,500 DWT

2.1 Variables

The hull plates and longitudinal elements to be optimized include the hull plate thickness and the size of stiffeners. The dimensions of the plates and profiles will be aligned with the material data available in the Indonesian industry, which will be utilized to create a database within the developed program. The existing ship is constructed using bulb profiles; however, to simplify the calculations, this study will substitute them with L-profiles. The design variables in this study can be mathematically expressed using the formulation presented in Equation 1.

$$X = (t_1, t_2, \dots, t_i, p_1, p_2, \dots, p_j) \quad (1)$$

$$p_j = w_j \times f_j \times tp_j \quad (2)$$

Where t denotes the variation in hull plate thickness, and p refers to the size of the stiffener profile used in the ship's construction. Additionally, the notations w, f, and tp in Equation 2 represent the size of the web, the flange, and the thickness of the stiffener profile, respectively.

For the frame spacing variable, it is made uniform throughout the entire ship's construction. However, the size of the frame profiles will be classified into six different sizes based on their location within the structure, which includes: profiles on the bottom and inner bottom; profiles on the longitudinal bulkhead; profiles on the side girder; profiles on the bilge and hopper; profiles on the side and double hull; and profiles on the deck.

Similarly, the plate thickness will also be classified based on the location within the ship's construction, which includes: plates on the keel; plates on the center girder; plates on the longitudinal bulkhead; plates on the inner bottom; plates on the bottom; plates on the bilge; plates on the hopper; plates on the side; plates on the double hull; plates on the stringer; plates on the side girder; plates on the deck; and plates on the shear strake.

2.2 Parameter Design

The design parameters utilized as inputs are categorized into three groups: the ship's main dimensions, preliminary design, and structural design, as illustrated in Figure 2. The main dimensions of the ship encompass the Length Overall (LOA), Length of Water Line (LWL), Length Between Perpendiculars (LPP), Breadth (B), Depth (H), and Draught (T).

The preliminary design category includes parameters such as the Block Coefficient (CB), bilge radius design, double bottom height design, double hull width design, the nearest distance of the hopper from the Center Line (CL), and the farthest distance of the hopper from the Base Line (BL).

The structural design category comprises inputs such as the type of framing system employed, the plate strake dimensions for the bottom, side, and deck structures, the plate strake dimensions for the keel, bilge, and shear strake structures, the distance of the side girder from the center line, and the distance of the stringer from the baseline.

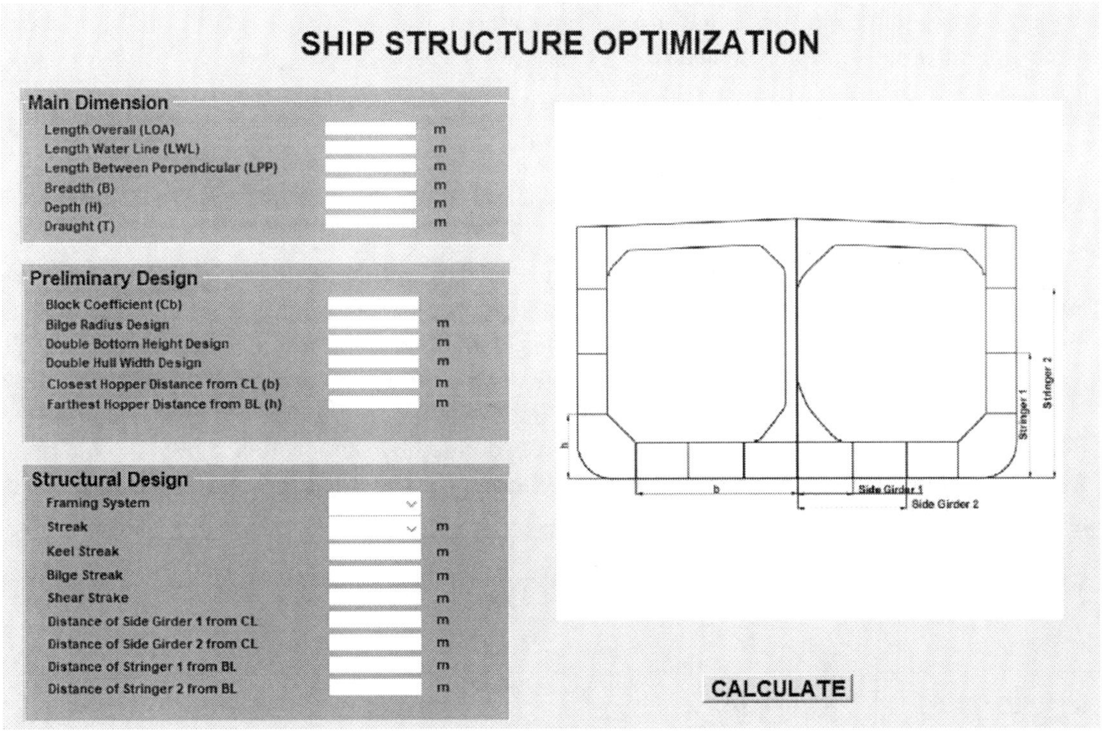

Figure 2. Detail of Parameter Input

2.3 Constraint

Based on the defined variables, it is necessary to establish constraints to simplify the optimization process. The constraints are determined according to the dimensional limits specified in the BKI rules and the database utilized. For the plate thickness and the spacing between stiffeners, the constraints are set with maximum and minimum values, as formulated in Equations 3 and 4.

$$t_{i.min} \leq t_i \leq t_{i.maks} \quad (3)$$
$$p_{j.min} \leq p_j \leq p_{j.maks} \quad (4)$$

Where t_i represents the variation in plate thickness for each location within the ship's construction. Meanwhile, p_j represents the variation in stiffener profile sizes, which are categorized based on the ship's construction location. There are 13 variations in plate thickness and 6 groups of profile size variations, as described in Section 2.1.

As a constraint in the optimization process, the weight of the existing ship will be used as a benchmark. Additionally, the minimum section modulus value required by the Biro Klasifikasi Indonesia (BKI) rules will be applied as a constraint to assess the ship's structural strength. The optimization process will continue generating solutions that achieve a structural weight below that of the existing ship while maintaining a section modulus above the minimum required value.

2.4 Objective

The objective of this research is to achieve a minimum structural weight for the ship. The objective function for optimizing the ship's structure is formulated in Equation 5 and 6. The minimized weight obtained will reduce the usage of steel materials during the building process, thereby enhancing cost-effectiveness and material efficiency. Additionally, the reduction in structural weight is expected to increase the ship's cargo capacity. Numerous benefits can be gained from determining an optimal ship structure.

$$f(x) = \min\left(\sum w_{plate} + w_{stiffener}\right) \quad (5)$$

$$f(x) = min\sum(b_i \times t_i + w_j \times tp_j + f_j \times tp_j) \times \rho \times length \quad (6)$$

However, from a strength perspective, this needs careful consideration. The concepts of weight and strength are inherently conflicting. Achieving optimal weight also requires a thorough evaluation of the ship's structural strength. In this study, the strength assessment is conducted by calculating the ship's section modulus. The results of the section modulus calculation will then be compared with the standards set by the Biro Klasifikasi Indonesia (BKI) to ensure compliance. The minimum limit of the ship's modulus value is defined by the formula shown in equation 7. Where k is the material factor, c_0 is the wave coefficient, L is the ship's length, B is the ship's width, and C_B is the block coefficient of the ship.

$$W_{min} = k \times c_0 \times L^2 \times B \times (C_B + 0.7) \times 10^{-6} \quad (7)$$

3. Impact on the Environment

Optimizing the longitudinal structural elements of a double-hull oil tanker, such as the hull plate thickness, stiffener spacing, and stiffener size, can lead to a reduction in the amount of steel material required. This not only reduces the overall weight of the vessel but also decreases the environmental footprint associated with steel production (10), which is a resource-intensive process involving significant energy consumption and greenhouse gas emissions.

A lighter ship due to optimized structural components can result in lower fuel consumption. As fuel consumption is directly related to emissions of greenhouse gases (GHG) such as CO2, this has a positive environmental impact. The discussion can include an analysis of how reducing the ship's weight through structural optimization contributes to lower emissions and aligns with international regulations, such as the IMO's (International Maritime Organization) Energy Efficiency Design Index (EEDI).

4. Discussion

The optimization process was conducted on a 17,500 DWT double-hull oil tanker. The ship is constructed using a longitudinal framing system. For this study, the optimization was limited to the calculation of the ship's longitudinal components. The weight calculation was performed on a block located in the parallel middle body of the ship, with a length of 9 meters.

First, the minimum midship section modulus was calculated using Equation 7. By setting k to 1, c_0 to 7.99, L to 148.22 m, B to 27.7 m, and C_B to 0.75, the minimum modulus value obtained was 7.071 m^3. Because the ship operates in area P (Restricted Ocean Service), a 5% reduction was applied to $Wmin$, resulting in an adjusted value of 6.718 m^3, which is then used as the lower limit in the optimization process.

Next, the structural weight of the existing ship was calculated using ASTM A36 material. For each plate and longitudinal profile component, the cross-sectional area was calculated and then multiplied by the material's density, 7.85 tons/m^3. Based on the dimensions of the existing ship's construction, the weight of the plate components is 139.49 tons, and the weight of the longitudinal components is 61.47 tons.

The minimum modulus and structural weight of the existing ship are used as constraints in the optimization process. This optimization produced variations in the hull plate thickness, which are detailed in Table 2. For the ship's stiffener profiles, size variations are shown in Table 3 according to construction location. Profiles in areas such as the bottom structure, inner bottom, center girder, bilge strake, and hopper were standardized to simplify the process and improve efficiency.

Table 2. Plate Thickness

Description	Plate Thickness	Unit
Keel Plate	14	mm
Bottom Plate	12	mm
Bilge Strake Plate	14	mm
Side 1 Plate	13	mm
Side 2 Plate	11	mm
Side 3 Plate	13	mm
Inner Bottom Plate	12	mm
Hopper Plate	12	mm
Double Hull 1 Plate	12	mm
Double Hull 2 Plate	11	mm
Double Hull 3 Plate	14	mm
Long Bulkhead 1 Plate	11	mm
Long Bulkhead 2 Plate	11	mm
Long Bulkhead 3 Plate	14	mm
Center Girder Plate	14	mm
Side Girder Plate	11	mm
Stringer 1 Plate	11	mm
Stringer 2 Plate	10	mm
Deck Plate	14	mm

Table 3. Dimension of Stiffeners

Description	Dimension of Stiffener	Unit
Bottom	L 300 x 135 x 12	mm
Inner Bottom	L 300 x 135 x 12	mm
Center Girder	L 300 x 135 x 12	mm
Side Girder	L 120 x 90 x 10	mm
Bilge Strake	L 300 x 135 x 12	mm
Hopper	L 300 x 135 x 12	mm
Side and Double Hull 1	L 200 x 110 x 11	mm
Side and Double Hull 2	L 200 x 110 x 11	mm
Side and Double Hull 3	L 160 x 120 x 10	mm
Deck	L 180 x 120 x 10	mm
Long Bulkhead	L 200 x 125 x 11	mm

From the analysis of the optimized longitudinal structural elements, the structural dimensions of the ship were obtained. Based on these optimized dimensions, the structural weight of the ship was calculated, as shown in Table 4. It was found that the total weight for each component group, including both plate and longitudinal components, was reduced compared to the weight of the existing ship. The total weight of the existing ship was 200.96 tons, while the optimized ship's total weight was 190.19 tons. This represents a weight reduction of 10.75 tons, or approximately 5.36% of the weight of the existing ship.

Table 4. The comparison of the structural weight between the existing and optimized ships

Componet	Unit	Existing Design	Optimize Design
Weight of Plate	[tons]	138.49	134.46
Weight of Stiffener	[tons]	61.47	55.74
Total	**[tons]**	**200.96**	**190.19**

In terms of strength, an analysis was conducted based on the modulus value, which was compared to the minimum required modulus as specified by classification standards. The midship section calculations were performed for the ship's deck and bottom, considering the bending moments due to various forces acting on the hull during operations. Based on the optimized plate thickness and profile sizes, the modulus of the midship section for the deck was found to be 8.722 m^3, while the modulus for the bottom was 11.072 m^3. These values meet the minimum requirements set by the classification.

6. Conclusion

The optimization of the structural weight of the double-hull oil tanker has yielded significant improvements, with the weight being reduced from 200.96 tons to 190.20 tons. This represents a reduction of 10.75 tons, which corresponds to a 5.36% decrease in percentage terms. This weight reduction is not merely a quantitative achievement but also signifies enhanced material efficiency within the shipbuilding process. By reducing the amount of material required, the overall costs associated with construction and maintenance are notably decreased. Additionally, this decrease in structural weight contributes to environmental sustainability. Lighter ships require less fuel to operate, which in turn reduces greenhouse gas emissions and lowers the ecological footprint of maritime transport.

For future research, optimization will be performed with structural strength as the objective. Structural strength will be analyzed using the ultimate limit strength method to ensure that the structure is sufficiently robust.

Acknowledgment

The authors appreciate and express their gratitude for the financial support provided by the Ministry of Education and Culture through the BIMA Master's Research Program.

References

1. Jelovica J, Klanac A. Multi-objective optimization of ship structures : using guided search vs . conventional concurrent optimization. 2008;(2002).
2. Jiang C, Yang S, Nie P, Xiang X. Multi-objective structural profile optimization of ships based on improved Artificial Bee Colony Algorithm and structural component library. 2023;283(June).
3. Jalalpour M, Guest JK, Igusa T. Reliability-based topology optimization of trusses with stochastic stiffness. Struct Saf [Internet]. 2013;43:41–9. Available from: https://www.sciencedirect.com/science/article/pii/S0167473013000180

4. Nakshatrala P, Tortorelli DA, Nakshatrala KB. Nonlinear structural design using multiscale topology optimization. Part I: Static formulation. Comput Methods Appl Mech Eng [Internet]. 2013;304:167–76. Available from: https://api.semanticscholar.org/CorpusID:123697750
5. Fanni M, Fouda N, Shabara MAN, Awad M. New crack stop hole shape using structural optimizing technique. Ain Shams Eng J. 2015;6(3):987–99.
6. Lindemann T, Kaeding P. An Approach to Optimization In Ship Structural Design Using Finite Element And Optimization Techniques. In 2010. Available from: https://api.semanticscholar.org/CorpusID:126791575
7. Rosa J, Wang S, Guedes Soares C. Improvement of ship hulls for comfort in passenger vessels. In 2021. p. 283–96.
8. Klanac A, Jelovica J. A concept of omni-optimization for ship structural design. Adv Mar Struct - Proc MARSTRUCT 2007, 1st Int Conf Mar Struct. 2007;(June):473–81.
9. Sekulski Z. Structural weight minimization of high-speed vehicle-passenger catamaran by genetic algorithm. Polish Marit Res - POL MARIT RES. 2009 Apr 1;16:11–23.
10. Abedin J, Franklin F, Mahmud SMI. A Two-Stage Optimisation of Ship Hull Structure Combining Fractional Factorial Design Technique and NSGA-II Algorithm. J Mar Sci Eng. 2024;12(3).

IOP Conf. Series: Earth and Environmental Science **1473** (2025) 012048 doi:10.1088/1755-1315/1473/1/012048

Crack Location Effect on the Weld Connections of FPSO Helideck Structure Toward Safety Management

F Rahman[1], A Zubaydi[1*], S H Sujiatanti[1*], R C Ariesta[1], Y P Asmara[2]

[1] Department of Naval Architecture, Institut Teknologi Sepuluh Nopember, Surabaya, Indonesia
[2] Faculty of Engineering and Quantity Surveying, INTI International University, Malaysia

*E-mail: zubaydi@na.its.ac.id, septia@its.ac.id

Abstract. In the oil and gas industry, ensuring the structural integrity of Floating Production, Storage, and Offloading (FPSO) units is critical for maritime operational security and meeting energy demands. The helicopter deck (helideck) of an FPSO, which supports helicopters for personnel and cargo transport, is subject to significant static and dynamic loads, as well as harsh environmental conditions. Helideck structures, consisting of numerous truss/beam elements, are prone to cracking at welded joints. This study investigates the influence of crack locations on the strength and durability of welded connections in FPSO pipe stiffeners. A numerical analysis was conducted using the Finite Element Method (FEM), and the model was validated through a mesh convergence study, ensuring accurate stress intensity factor (SIF) calculations. The Linear Elastic Fracture Mechanics (LEFM) approach is used to analyze the characteristics and propagation rate of cracks at different weld locations. The analysis shows that cracks at the weld toe reached a critical depth after 158 load cycles, compared to 817 cycles for cracks at the weld crown. These results underscore the higher vulnerability of weld toe cracks to rapid propagation, highlighting the need for targeted design and maintenance strategies to ensure the safety of FPSO helidecks.

1. Introduction

The increasing energy demand has driven extensive oil and gas exploration in deep and shallow waters. As exploration moves to greater depths, fixed offshore structures become less suitable, highlighting the need for floating structures. The Floating Production, Storage, and Offloading (FPSO) unit is a crucial asset in the oil and gas industry, enduring significant environmental loads, especially in offshore conditions [1].

A critical component of the FPSO is the helicopter deck (helideck), which serves as a landing platform for helicopters transporting personnel and cargo. The helideck ensures static and dynamic loads from helicopter operations and environmental conditions like waves and strong winds. Typically, helidecks are constructed from beams and pipes that must withstand these loads. The structure is composed of plates with stiffeners and supports, mounted on the ship's superstructure and supported by pillars. Helideck design must adhere to international standards such as ISO codes [2], and Eurocodes, or national standards like NORSOK [3] or API [4].

Cracks in welded joints can lead to structural failures, potentially causing leaks or accidents, especially if a crack occurs midship, where it could cause the ship to break. Several studies have evaluated the structural strength of helidecks under operational loads. Emmanuel et al. analyzed a helideck design on a Dive Support Vessel (DSV), finding uniform stress distribution in primary structures under concentrated loads [5]. Jose et al. further analyzed a similar structure using a nonlinear approach [6]. The development of aluminum structures was discussed by JS & JK [7]. However, while helideck strength analysis is common, studies on crack analysis in helidecks are rare. This is crucial because the truss/beam structures consist of many interconnected components, leading to stress concentrations. Kim et al. used artificial neural networks to detect damage in helideck structures, identifying damage locations at stiffener joints based on mode shape data [8]. Vaghefi et al. found that emergency helicopter landings induce excessive stress on joints not designed to withstand maximum loads, indicating these critical areas are prone to damage such as cracking [9]. Xie et al. provided a method for analyzing crack propagation in pipes due to corrosion and initial cracks [10].

A comprehensive structural analysis is needed to ensure the helideck remains strong and lightweight [11]. Structural damage caused by cracks in welded joints must be minimized to prevent further damage to the ship's structure and hull, which could lead to catastrophic failures. Thus, a study is proposed to investigate the impact of crack location in welded joints on the FPSO helideck structure. The research will employ a numerical approach using the Finite Element Method (FEM) to analyze critical locations and the characteristics of crack propagation in the welded joints of pipes used in FPSO helidecks.

2. Joint Configuration and Modeling Approach

In this study, the focus is on modeling a tubular T-joint, which is a common configuration in helideck structures. The T-joint is selected due to its prevalence in truss systems, where it plays a crucial role in transmitting loads. The geometry of the welded joint, including the weld crown and weld toes, is explicitly modeled to capture the stress concentration effects that are critical for crack initiation and propagation analysis.

2.1 Material Properties

The material used for both the pipe and the weld is a high-strength steel, commonly utilized in offshore structures for its excellent mechanical properties and resistance to environmental degradation. The weld material is assumed to have similar properties to the base material, ensuring continuity in mechanical behavior across the joint. The material properties are as follows:

Table 1. Material Properties.

Material	Young's Modulus	Poisson's Ratio	Yield Strength	Ultimate Tensile Strength
ASTM A36	200 GPa	0.3	250 MPa	400-500 MPa

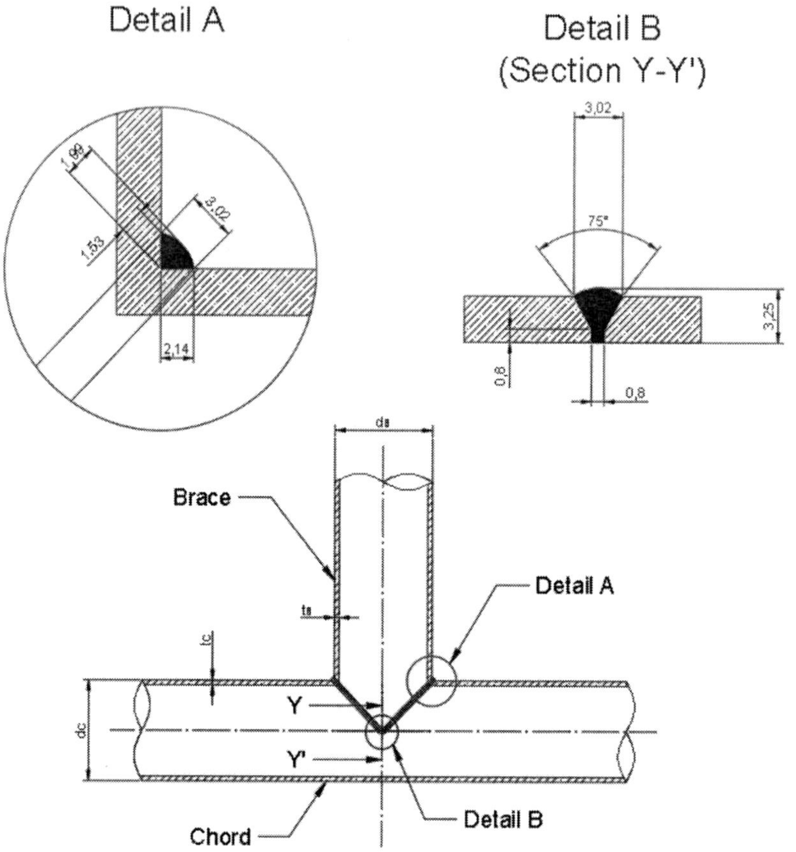

Figure 1. Geometry of Tubular T-Joint.

2.2 Geometry and Crack Parameters

The T-joint configuration consists of a main pipe, referred to as the chord, with a diameter of 60.3 mm, a thickness of 2.77 mm, and a length of 300 mm. The branch pipe, or brace, intersects the chord at a 90-degree angle, with a diameter of 60.3 mm, a thickness of 2.77 mm, and a length of 150 mm. The weld that joins these pipes features a throat thickness of 1.53 mm, a weld toe radius of 2.14 mm, and a crown height of 1.99 mm. The weld is modeled with a fillet shape, accurately capturing the geometry of the weld crown and toes. This detailed representation, as shown in Figure 1, is crucial for simulating the stress concentrations at the weld, where cracks are likely to initiate.

To simulate crack propagation behavior, two initial crack configurations are introduced in the T-joint model. The first crack is located at the weld crown with a depth of 1 mm and a crack mouth opening angle of 7 degrees. This crack is oriented at a 45-degree angle, directed toward the weld

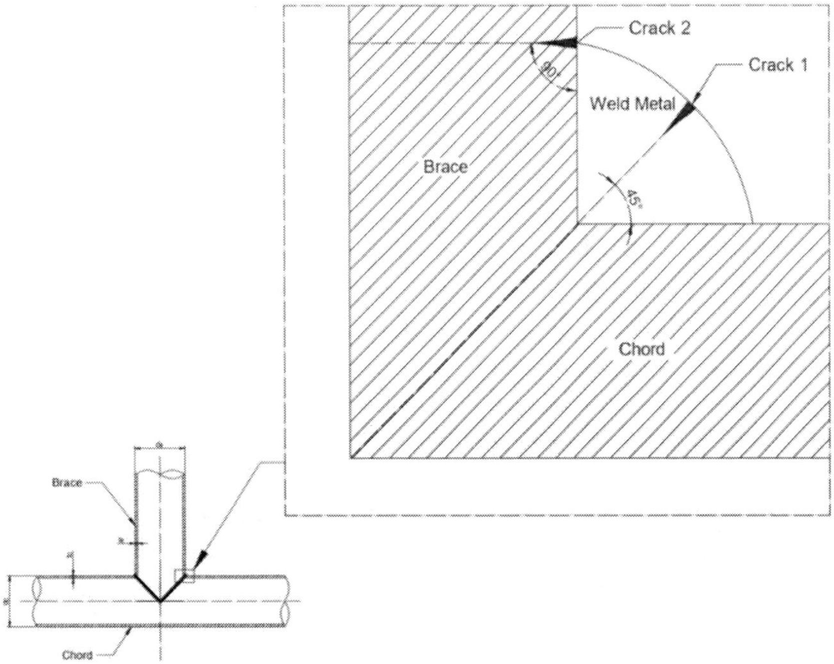

Figure 2. Crack Location Variation.

root. The second crack is positioned at the weld toe, also with a depth of 1 mm, but it is oriented perpendicular to the pipe's thickness, at a 90-degree angle. These crack configurations, as shown in Figure 2, are designed to represent the critical stress concentration areas and simulate potential crack propagation paths under operational loads.

2.3 Validation of the Numerical Model

To ensure the accuracy of the Finite Element Model (FEM) utilized in this study, a mesh convergence study was conducted. This study involved the numerical simulations with varying mesh sizes to observe their impact on the calculated Stress Intensity Factor (SIF). Figure 3 displays the results of the mesh convergence analysis, illustrating the relationship between mesh size and the calculated SIF. As shown, the SIF values stabilize as the mesh is refined, indicating that a mesh count of 105250 elements has an error below 1% and is sufficient for accurate results.

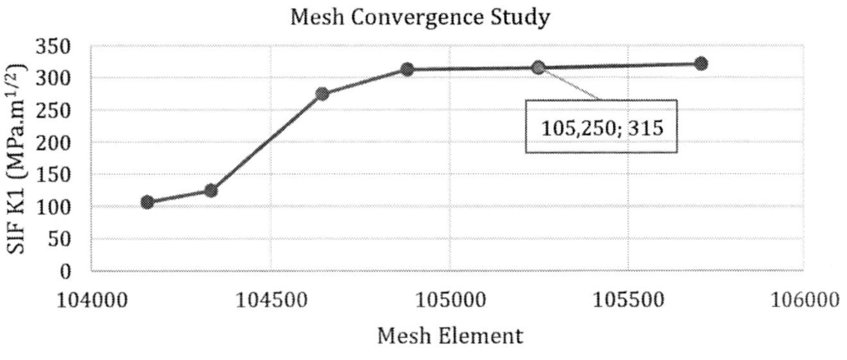

Figure 3. Mesh Convergence Study.

IOP Conf. Series: Earth and Environmental Science **1473** (2025) 012048 doi:10.1088/1755-1315/1473/1/012048

3. Numerical Analysis

3.1 Finite Element Modelling

The finite element (FE) model is generated using FE commercial software, employing 3D solid elements to simulate the structural response of the T-joint accurately. The model is meshed using tetrahedral elements, which are well-suited for capturing complex geometries. Sizing controls are applied to create a fine mesh in the vicinity of the cracks, enabling precise calculation of stress gradients and stress intensity factors around the crack tips.

Boundary conditions are applied by fixing both ends of the chord pipe, simulating the structural constraints typical of an offshore environment where the T-joint is anchored. A tensile load with a magnitude of 1×10^5 N is applied in the normal direction of the brace pipe. This tensile

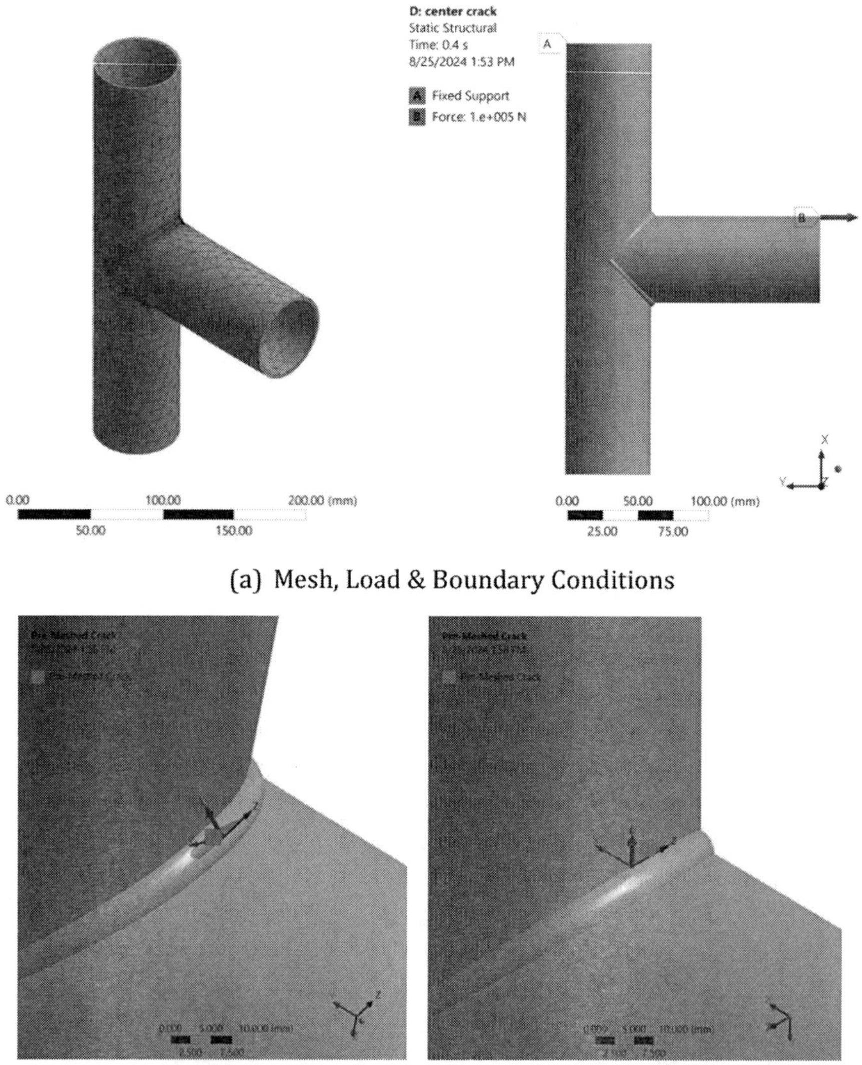

(a) Mesh, Load & Boundary Conditions

(b) Crown crack (c) Toe crack

Figure 4. Finite Element Modelling.

load is specifically chosen to induce Mode I crack propagation, where the crack opens perpendicular to the applied stress. Mode I is the most common and critical mode of crack propagation in structural components, making it essential for accurately assessing the potential failure of the T-joint under operational loads.

3.2 Crack Propagation Analysis
The crack propagation analysis is executed employing Linear Elastic Fracture Mechanics (LEFM). The Stress Intensity Factor (SIF) is calculated for varying crack depth at multiple crack locations within the joint. The initial crack depth is set at 1 mm, corresponding to the smallest detectable crack, and the analysis continues until the crack depth reaches the length equivalent to the leg length of the weld metal. The SIF is calculated at each step to assess how the crack progresses through the weld metal, taking into account the complex stress fields generated by the weld geometry and loading conditions. The SIF results will be used to evaluate the crack propagation depth and direction.

3.3 Fatigue Life Calculation
The fatigue life of the T-joint structure is assessed through Paris's law, which connects the crack growth rate to the variation in the stress intensity factor (ΔK) during cyclic loading conditions. Paris's law is expressed as:

$$\frac{da}{dN} = C(\Delta K)^m$$

Where $\frac{da}{dN}$ is the crack growth rate per load cycle, ΔK is the stress intensity factor range, and $C$and m are material-dependent coefficients. For the material used in this study, the Paris law coefficients are $C=6.31\times10^{-9}$ and m=2.89, as obtained from[12].

The fatigue life estimation process involves calculating the crack growth rate for each crack configuration and propagating the crack incrementally from its initial size to a critical size where failure is expected to occur. The stress intensity factor range, ΔK is determined for each crack depth using the results from the crack propagation analysis in Section 3.2.

The cumulative crack growth is then integrated over the number of load cycles to estimate the total fatigue life of the structure. This analysis provides an insight into the number of operational cycles the T-joint can endure before catastrophic failure occurs, thereby informing maintenance schedules and potential design improvements to enhance the longevity and safety of the helideck structure.

4. Result and Discussion

4.1 Stress Contours for Crack Locations
To analyze the impact of crack propagation on the T-joint structure, stress contours were generated for both the initial and propagating cracks at the weld crown and weld toe. Stress distribution contours for the crack at the weld crown are presented in Figure 5., and the crack at the weld toe in Figure 6.

The stress contours for the weld crown indicate that initial crack conditions display symmetrical stress distribution, suggesting balanced loading. However, as the crack progresses to the intermediate stage, the divergence in stress distribution becomes evident, with the brace side experiencing lower stress. At the critical stage, the fully asymmetric stress contours suggest a substantial alteration in the structural response, which may influence crack propagation rates.

IOP Conf. Series: Earth and Environmental Science **1473** (2025) 012048 doi:10.1088/1755-1315/1473/1/012048

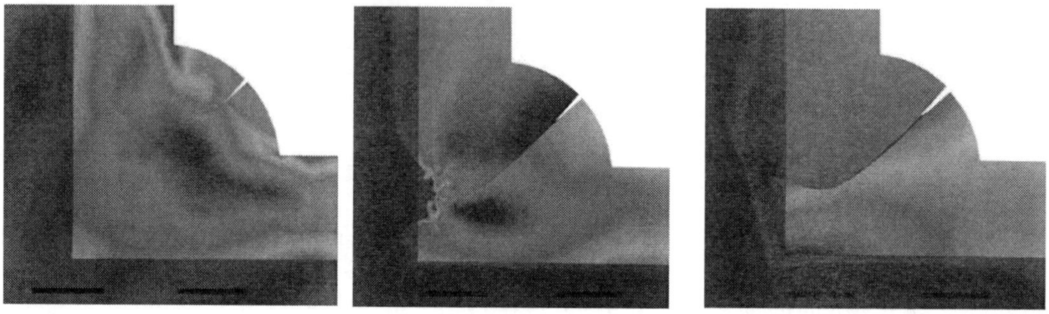

(a) (b) (c)

Figure 5. Stress Contour for Crack at Weld Crown (a) initial crack, (b) intermediate crack (c) critical crack

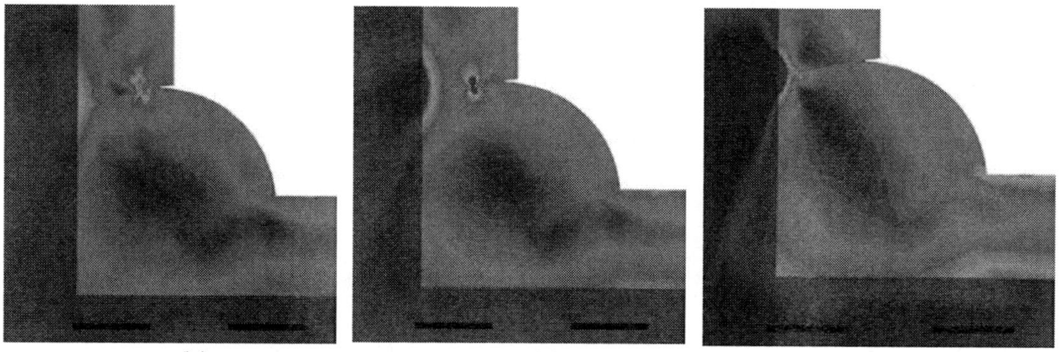

(a) (b) (c)

Figure 6. Stress Contour for Crack at Weld Toe (a) initial crack, (b) intermediate crack (c) critical crack

For the weld toe, the initial stage shows high-stress concentration near the crack tip, with symmetrical contours indicating balanced loading. As the crack propagates, there is a noticeable increase in stress concentration, reflecting a growing vulnerability to failure. The contours at the critical stage underscore the significant risk associated with cracks at the weld toe, highlighting the importance of targeted maintenance strategies.

4.2 Stress Intensity Factor (SIF) Results
The Stress Intensity Factor (SIF) was determined for different crack depths at each crack location, providing critical insight into the crack propagation behavior within the T-joint structure. The results are presented as graphs that illustrate the relationship between the SIF and crack depth for each crack configuration, as shown in Figure 7.

4.2.1 Crack at Weld Crown
The SIF for the crack located at the weld crown (oriented at a 45-degree angle toward the weld root) increases steadily with the crack depth. The graph reveals a nearly linear relationship, indicating a consistent stress concentration around the crack tip as it propagates deeper into the weld. This suggests that the weld crown is a critical area prone to crack initiation and growth under tensile loading.

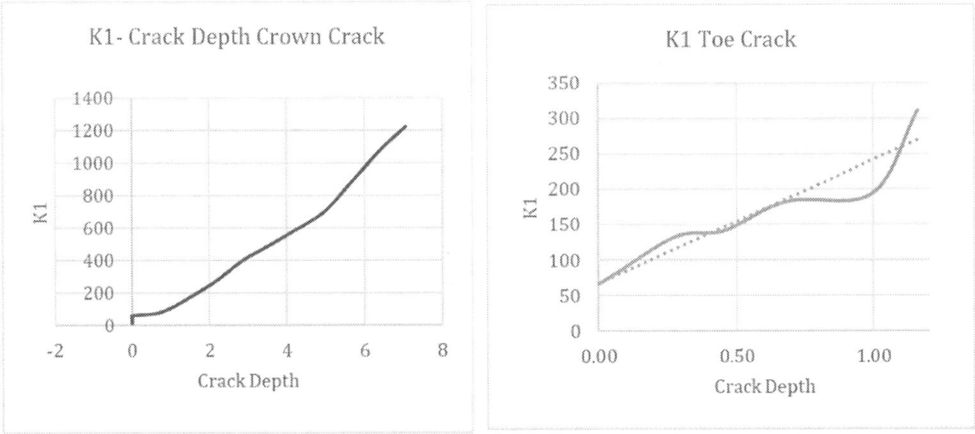

(a) Crown crack (b) Toe crack

Figure 7. SIF – Crack depth graph for each crack location (a) Crown Crack, (b) Toe Crack.

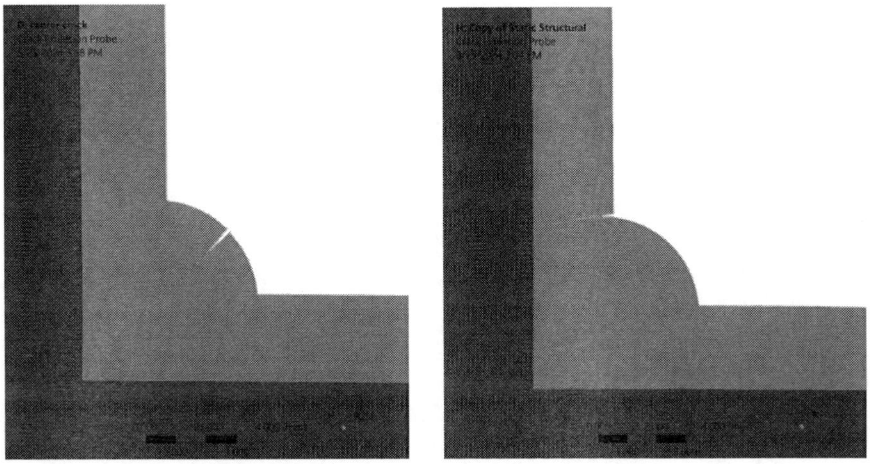

(a) Crown crack (b) Toe crack

Figure 8. Crack direction for each crack location (a) Crown Crack, (b) Toe Crack.

4.2.2 Crack at Weld Toe

The crack located at the weld toe, oriented perpendicularly toward the pipe's thickness (90-degree angle), exhibits a different SIF behavior. Initially, the SIF grows linearly with the increasing crack depth, indicating a steady propagation. However, as the crack nears the base material, there is a significant and sharp increase in the SIF for each additional increment of crack depth. This sudden rise suggests that as the crack approaches the thicker base material, the stress concentration intensifies, leading to a more rapid crack growth rate.

4.3 Crack Propagation Path
4.3.1 Crack at Weld Crown
The crack propagation analysis for the crown crack, which was initially oriented at a 45-degree angle toward the weld root, reveals a distinct change in the crack path as it progresses. Initially, the crack follows a straight path, consistent with the applied loading and the orientation of the weld. However, as the crack depth approaches the base material, the crack path begins to alter toward the normal direction of the brace pipe's thickness. This change in the crack trajectory indicates that, although the early stages of crack propagation are affected by the geometry of the weld and the mixed-mode loading conditions, the structural response evolves as the crack propagates. The alteration in the crack direction highlights the dominance of Mode I crack propagation in this T-joint under the given loading conditions. Mode I, crack opening under tensile load, becomes increasingly significant as the crack interacts more directly with the bulk material of the pipe, away from the weld. The crack propagation direction can be seen in Figure 8.

4.3.2 Crack at Weld Toe
The crack located at the weld toe, oriented perpendicularly at a 90-degree angle to the pipe's thickness, follows a fairly consistent path aligned with its initial direction. As the crack deepens, the propagation path remains largely unchanged, maintaining its perpendicular orientation. However, there is a slight alteration of less than 5 degrees into the negative angle, meaning the crack veers slightly in the direction opposite to the applied loading. Despite this minor deviation, the crack path remains consistent until it penetrates through the thickness of the pipe.

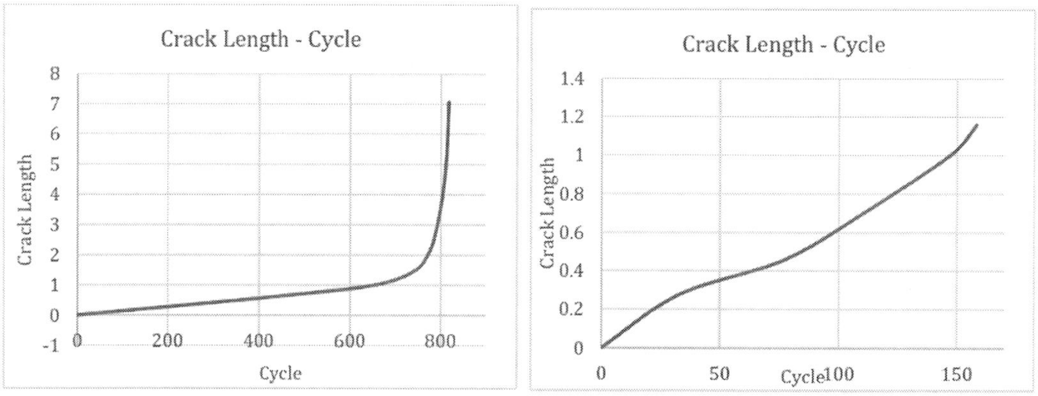

(a) Crown crack (b) Toe crack

Figure 9. Crack Length - Cycle graph for each crack location (a) Crown Crack, (b) Toe Crack.

This relatively stable crack path at the weld toe, with only a minor adjustment, further indicates that the structural response is predominantly governed by the applied tensile load, reinforcing the significance of Mode I crack propagation in this T-joint configuration. The slight deviation might be due to the local stress redistribution as the crack approaches the base material, but it does not significantly alter the overall crack growth direction.

4.4 Fatigue Life Estimation Results
The fatigue life of the T-joint structure was calculated using Paris's Law, with graphs illustrating how crack depth changes as the number of cycles increases for both the weld crown and weld toe, as shown in Figure 9. The crown crack reached its critical depth after 817 cycles, while the toe

crack reached its critical depth in just 158 cycles. This results show that the toe crack is more dangerous than crown crack because of the shorter lifespan after induced to an initial crack.

4.4.1 Crack at Weld Crown
The graph for the crown crack indicates a steady rise in crack propagation as the number of cycles progresses. Initially, the crack expands at a slow pace, but as the depth grows and the crack nears the base material, the propagation rate accelerates. This behavior indicates that the crown crack's fatigue life is strongly influenced by the crack's interaction with the base material, where the stress intensity factor increases significantly. The relatively smooth and continuous rise in crack propagation rate suggests that the structure is subjected to a progressive degradation process, with a notable acceleration in the later stages of crack growth.

4.4.2 Crack at Weld Toe
The fatigue life graph for the weld toe crack shows a shorter span, and presents a different pattern. The initial crack growth rate is higher compared to the crown crack, reflecting the higher stress concentration at the weld toe. However, as the crack propagates deeper, crack propagation remain in a consistent rate until it reaches the critical depth. This shows that for the crack at weld toe is more dangerous because of the consistency of high crack propagation rate for the same amount of loading.

5. Conclusion

The analysis of crack propagation in the T-joint structure of FPSO helidecks reveals important insights into the structural integrity of these components under operational loads. For the weld crown, the Stress Intensity Factor (SIF) increases linearly with crack depth, indicating consistent stress concentration. This suggests that the weld crown is a critical area prone to crack initiation and growth under tensile loads. As the crack deepens, the crack direction alters to the base material's perpendicular direction, and stress concentration remains high, emphasizing the need for close monitoring and maintenance of this area.

In contrast, cracks at the weld toe show an initial linear increase in SIF with crack depth, but this is followed by a sharp rise as the crack approaches the base material. This indicates that the weld toe experiences more intense stress concentration and more rapid crack propagation. The crack path at the weld toe remains largely perpendicular to the pipe's thickness with only minor deviations, reinforcing the influence of tensile loads and highlighting the weld toe's higher vulnerability to failure.

The fatigue life estimation shows that cracks at the weld crown reach critical depth in 817 cycles, with a gradual increase in crack propagation rate over time. This indicates a progressive degradation process with accelerated crack growth in later stages. On the other hand, cracks at the weld toe reach critical depth in just 158 cycles, with a higher initial crack growth rate and consistent propagation. This demonstrates that the weld toe is more susceptible to rapid fatigue failure.

Overall, the study underscores the greater risk associated with cracks at the weld toe compared to the weld crown. The findings highlight the necessity for targeted design improvements and maintenance strategies focused on the weld toe to ensure the safety and longevity of FPSO helidecks.

IOP Conf. Series: Earth and Environmental Science **1473** (2025) 012048 doi:10.1088/1755-1315/1473/1/012048

References

[1] UK Offshore Operators Association., "Guidelines for the management of offshore helideck operations.," no. 5, 2005.

[2] International Organization for Standarisation, "ISO 19901-3:2024(en) Oil and gas industries including lower carbon energy — Specific requirements for offshore structures — Part 3: Topsides structure." International Organization for Standarisation, 2024. [Online]. Available: https://www.iso.org/standard/43812.html

[3] T. Norsok and N. M. Industries, "NORSOK STANDARD Helicopter deck on offshore installations," no. September, 2004.

[4] A. P. I. (API), "PI RP 2L Recommended Practice for Planning, Designing, and Constructing Heliports for Fixed Offshore Platforms." 1996. [Online]. Available: https://standards.globalspec.com/std/1647345/api-rp-2l

[5] A. Emmanuel, I. E. Douglas, and T. Daniel, "Structural Design Analysis of Helideck for an Existing Ocean- Going Vessel American Journal of Engineering Research (AJER)," *Am. J. Eng. Res. 2019*, no. 5, pp. 365–377, 2019.

[6] A. Jose, R. P. Nair, B. Sanoob, and J. Paul, "Structural Optimisation of Helideck Structure Using FEM Finite Element Analysis of Helideck Structure Subjected to Impact Load during Helicopter Landing," no. December 2019, pp. 7–10, 2021, doi: 10.1007/978-3-030-26365-2.

[7] P. JS and S. JK, "Deformation-Based Design Method of Aluminium Helideck for Eurocode 9," *J. Appl. Mech. Eng.*, vol. 06, no. 02, pp. 2–7, 2017, doi: 10.4172/2168-9873.1000258.

[8] B. Kim, C. Kim, and S. H. Ha, "Multiple damage detection of an offshore helideck through the two-step artificial neural network based on the limited mode shape data," *Sensors*, vol. 21, no. 21, 2021, doi: 10.3390/s21217357.

[9] M. Vaghefi, H. Bagheri, and S. R. Mohebpour, "Nonlinear analysis of offshore helidecks due to the helicopter emergency landing loads," *Middle East J. Sci. Res.*, vol. 13, no. 10, pp. 1351–1358, 2013, doi: 10.5829/idosi.mejsr.2013.13.10.1219.

[10] M. Xie, Y. Wang, W. Xiong, J. Zhao, and X. Pei, "A Crack Propagation Method for Pipelines with Interacting Corrosion and Crack Defects," *Sensors*, vol. 22, no. 3, 2022, doi: 10.3390/s22030986.

[11] A. Ismail, A. Zubaydi, B. Piscesa, E. Panangian, R. C. Ariesta, and T. Tuswan, "A strength analysis of conventional and sandwich plate deck using finite element method," *IOP Conf. Ser. Mater. Sci. Eng.*, vol. 1034, no. 1, p. 012026, Feb. 2021, doi: 10.1088/1757-899X/1034/1/012026.

[12] J. Barsom and S. Rolfe, *Fracture and Fatigue Control in Structures: Applications of Fracture Mechanics, Third Edition*. 1999. doi: 10.1520/mnl41-3rd-eb.

IOP Conf. Series: Earth and Environmental Science **1473** (2025) 012049 doi:10.1088/1755-1315/1473/1/012049

The characterization of blue shark (*Prionace glauca*) cartilage potential as nature-derived drug material

T Agustin[1,2], H. Nursyam[3*], M. Firdaus[3], M. Rifa'i[4], I. D. Ekaputri[2], and Mahmiah[2]

[1] Fisheries and Marine Science Graduate Program, Faculty of Fisheries and Marine Science, University of Brawijaya. Jl. Veteran Malang 65149, East Java, Indonesia
[2] Fisheries Study Program, Faculty of Engineering Marine Science, University of Hang Tuah, Jl. Arief Rahman Hakim 150, Surabaya 60111, Indonesia
[3] Department of Aquatic Product Technology, Faculty of Fisheries and Marine Science, University of Brawijaya. Jl. Veteran Malang 65149, East Java, Indonesia
[4] Department of Biology, Faculty of Mathematics and Natural Sciences, University of Brawijaya, Malang, Indonesia

*E-mail: happy_nsy@ub.ac.id

Abstract. Shark's cartilage contains several health-promoting substances, with glucosamine and chondroitin being key components. This study examined the proximate, mineral, amino acid, and chondroitin sulfate content of blue shark (Prionace glauca) cartilage, focusing on how different cartilage diameters affect these characteristics. Using a Complete Randomized Design, the experiment included three treatments based on cartilage diameter: small (A, 1-2 cm), medium (B, 2.1-3 cm), and large (C, 3.1-4 cm), each with four replications. Observed parameters included proximate composition, amino acid profile, mineral content, and chondroitin sulfate levels. Data were analyzed using ANOVA, followed by a Least Significant Difference test. Thehighest protein content (46.20%) was found in the smallest diameter cartilage (A). The cartilagecontained 10 essential and 8 non-essential amino acids, with arginine (6.16%) and glycine (10.91%) being the most abundant in the largest cartilage (C). The highest calcium (184,500 ppm) and phosphorus (89,450 ppm) levels were also found in the largest cartilage. Chondroitinsulfate was most abundant (11.54%) in the medium-sized cartilage (B). ANOVA results showed no significant effect of cartilage diameter on any parameter (P>0.05), suggesting that cartilage size does not determine its potential as a natural health supplement.

1. Introduction

Blue shark (*Prionace glauca*) is one of maritime and pelagic species with adequately wide dispersion in the untamed expanse of tropical and subtropical locales. This species is mostly found in the Indian Ocean from west Sumatera to south Nusa Tenggara in Indonesia's waters [1]. Blue sharks are also known as hiu aer (Baliness), hiu lalaek, and cucut selendang (Javaness) in Bali and Java. They have a high economic value, because products can be made from any body part. Their meat is not widely consumed by fishermen or Indonesian society despite its high protein content [2]. Blue sharks, according to IUCN (International Union for Conservation of Nature), are categorized as nearly threatened species [3]. This species is considered as bycatch in pelagic fisheries, especially tuna fisheries in southern Indian Ocean [4] and [5]. Sharks are fish that live in the deep sea and have complete cartilaginous skeletons. They can live hundreds of millions of years. Cartilage makes up all of the body's organs [6]. Shark ligament contains bioactive mixtures that are useful to people, especially in the injury recuperating process. [7].

Shark cartilage extracts are frequently used medicinally to treat pain and swelling, treat scars, and prevent arthritis. Sharks are also frequently dried and sent out to different countries and the ligament utilized is normally spine [1].

[8] have studied the bioactive compounds of the cartilage and found that the cartilage consists of collagen, amino acid, mineral, and glycosaminoglycan. [9] indicated that the proximate composition and calcium (Ca) content in fresh cartilage are 18,74% carbohydrate, 3,64% protein, 1,54% fat, 74,67%water, 1,41% ash, and 9864,7 ppm Ca. [10] stated that glucosamine and chondroitin content of the cartilage are 28,36% and 6,06%, respectively, with mean rendement of shark cartilage powder of 18,39%.

Glucosamine and chondroitin are glucose amino as major component of the shark cartilage functioning to stimulate the formation of glycosaminoglycan, proteoglycan, and hyaluronic acid. Glucosamine can reduce the symptoms of arthritis and is a good supplement for joint and bone health whereas chondroitin is also an important supplement drug in osteoarthritis treatment [11]. Chondroitin sulphate is a sulphated glycosaminoglycan generally adhered to protein as part of proteoglycan [12]. Thus, shark cartilage is potential as natural drug. Besides, the use of shark cartilage wastes in Indonesia has not also been maximal so the characterization and identification of the shark cartilage-derived active compounds are needed. The present study is expected to be able to develop shark cartilage-based product as natural material-based drug and can increase its economic value as wastes of processed sharks.

2. Materials and Methods

2.1. Materials and Equipment

This study used frozen shark's cartilage obtained from a fish industry in Sidoarjo, All chemicals used in this study were pro-analyst quality. Equipment used included a knife, cutting board, bucket, dryer, grinding machine, analytic balance, porcelain cup, oven, furnace, desiccator, evaporator flask, kjeldahl distillator, stative universal, burette tap, small funnel, erlenmeyer, pipette, soxhlet extractor, beaker glass. Other tools were also used to identify bioactive compounds, namely: HPLC (High-performance liquid chromatography), UPLC (Ultra Performance Liquid Chromatography), LC-MS (Liquid chromatography-mass spectrometry) and ICP-MS (Inductively coupled plasma mass spectrometry).

2.2. Methods

2.2.1.Sample preparation

The frozen shark's cartilage was thawed with running water, and all adhered meat was removed. The cartilage was then washed and separated with diameter size, small (1-2 cm), medium (2.1-3cm), and large (3.1-4cm). These were cut in small and thin pieces, dried at 50°C for 12 h, and ground.

2.2.2. Experimental Design

The experiment applied a Complete Randomized Design with three treatments, A: small diameter (1-2 cm), B: medium-sized diameter (2.1-3 cm), C: large-sized diameter (3.1-4 cm) and four replications. The parameters measured were proximate composition, amino acids profile, mineral content and chondroitin sulphate content.

2.2.3.Proximate composition test

The proximate composition was assessed according to standards [13], which included evaluating moisture content, crude protein, crude fat, and ash. The protein content was calculated using the Kjeldahl method by multiplying the nitrogen content by 6.25, and the moisture content was determined by placing the sample in an oven set at 105°C for 24 hours. The crude fat content of the sample was analyzed using the Soxhlet Extraction Method with petroleum ether at 80°C. The ash content was determined using the dry ashing method. The amount of carbohydrates was then calculated by the difference.

2.2.4. Analysis of amino acids

We used High-Performance Liquid Chromatography (HPLC) to analyze the amino acid content of L-Tryptophan. First, we prepared a standard series of L-Tryptophan in a 10 mL volumetric flask with at least 6 concentrations. Then, we placed a 0.1 g test portion into a 20 mL headspace vial, added 4.2 M NaOH solution, and hydrolyzed it at 110°C for 20 hours before transferring it into a 50 mL beaker glass. Next, we added a citrate buffer solution, adjusted the pH to 4.25 with HCl or NaOH solution, and transferred the test solution to a 50 mL beaker glass. We then added bi-distilled water up to the mark line, homogenized the solution, placed it into a 2 mL tube, and centrifuged it. After centrifuging, the supernatant was filtered using a 0.45 μm 2 mL syringe filter and injected it into the HPLC system. The HPLC system used an RP-18 column and the mobile phase consisted of A = 0.0085 M sodium acetate; B = methanol. The pump system was set to isocratic, with an injection volume of 15 μL, the column temperature was maintained at ambient, and detection was performed using a PDA at a wavelength of 280 nm.

The analysis of amino acid content, cystine, and methionine was conducted using LC-MS (Liquid Chromatography-Mass Spectrometry). The procedure began with the preparation of a series of standards in a 2 mL vial with seven different concentrations. A standard solvent, filtered through a 0.2 μm GHP/RC, was used. The test portion, weighing 0.50 g, was placed in a 20 mL vial kept in a cooling water bath, then transferred to a freezer at -20-0°C for a minimum of 15 minutes. Afterward, it was mixed with an oxidizing solution, and the vial was closed without crimping, shaken gently for homogenization, and stored in the freezer at -20-0°C for 16 hours. Following this, the vial was opened, the sample was thawed, and sodium bisulfite was gradually added from a previously prepared solution and left for at least 3 hours. The oxidized sample was combined with the hydrolysis solution in an acid room, and the headspace vial was covered. It was then heated in an oven at 110°C for 1 hour, capped, heated again at 110°C for 21 hours, removed, and cooled. Next, the solution was transferred into a 100 mL beaker, and the headspace vial was rinsed with bi-distilled water until the volume reached approximately 25 mL. The pH was adjusted to 2.20 using a cooling water bath and then moved into a 50 mL volumetric flask. Distilled water was added up to the mark line, and the solution was homogenized before being transferred into a 2 mL tube, centrifuged, filtered through a 0.2 μm RC membrane filter RC into a 2 mL tube, and injected into the LC-MS/MS system. The LC-MS/MS conditions were as follows: column: MS Imtakt Intrada Amino Acid, mobile phase: A= 0.1% formic acid in acetonitrile and B= 100 mM ammonium formate, flow rate = 0.4 mL min-1, pump system: gradient, column temperature: 37°C, injection volume: 2 μL, and autosampler temperature: 8°C.

Fifteen different kinds of amino acids, other than tryptophane, cysteine, and methionine, were analyzed using Ultra-Performance Liquid Chromatography (UPLC). The following process was followed: An internal standard was used to produce a standard concentration. A 20 mL head space vial containing up to 0.1 g of the test part of solid sample or 1 mL of the test portion of liquid sample was made, hydrolyzed with HCl solution, transferred into a 50 mL volumetric glass, filled with bi-distilled water to the mark line, and homogenized. After passing the test solution through a 0.2 μm syringe filter and collecting the filtrate, the internal standard was added to the mixture. In the derivatization step, the solution was injected into the UPLC apparatus. The UPLC conditions were column: C18, mobile phase: Eluent accq. tag ultra; bi-distilled water, pump system: gradient, column temperature: 49°C, detector: PDA.

2.2.5. Analysis of mineral content

As much as 0.3 g sample was put into an Erlenmeyer, added with 6 mL of concentrated HNO3, inserted into a digestion microwave, and predigested for 10-15 min. at room temperature. The temperature was then set at 150°C - 200°C. After destruction, the erlenmeyer was taken out and cooled down to the room temperature, then the solution was poured into the volumetric flask and diluted to 6 mL and inserted in the sample tube of the ICP-MS for analysis.

2.2.6. Analysis of chondroitin sulphate

Chondroitin sulphate was analyzed as follows: as much as 1.0 g of sample was put in a 25.0 mL flask and added with WFI (Water for Injection) up to the correct mark, homogenized using an ultrasonic homogenizer for 15 min., and filtered through a 0.45 µm nylon membrane into HPLC vial. The HPLC was placed on the sample holder in the HPLC under the following conditions: HPLC Agilent 1220 Infinity LC with autosampler, Column: SeQuant ZIC-HILIC Merck, 4.6 x 150 mm, 5 µm, Detector: Agilent ELSD (Evap = 80°C, Neb = 50°C, Gas Flow Rate = 1.10 SLM, LED Intensity = 100%), Eluent: ACN : Dapar Amonium Format 20 mM pH 4.5 = 77:23, Flow Rate: 1.0 mL min., temperature: 35°C, and injection volume: 10 µL.

2.3. Statistical analysis

One-way ANOVA was used to examine the data, and IBM SPSS Version 25 was then used to perform the Least Significance Difference (LSD) test. The impacts of diameter size on the characterization of shark cartilage were investigated using one-factor analysis.

3. Results and discussion

3.1. Proximate content

Nutritional content analyses covered moisture, fat, ash, protein, and carbohydrate. The proximate content is presented in Table 1.

Table 1. Mean proximate content

Sample code	Mean content (%)				
	Moisture	Ash	Protein	Fat	Carbohydrate
A	6.59±0.89	42.64±2.18	46.20±8.62	0.39±0.33	5.66±3.13
B	7.48±1.25	45.55±0.31	45.79±8.27	0.38±0.36	4.53±1.72
C	6.83±1.05	46.78±2.54	45.41±5.46	0.46±0.25	3.54±1.93
Agustin et al (2012)	5.43	6.20	30.70	1.92	55.74
Jeevithan et al (2014)	-	25.41	8.95	1.69	-

Notes: A = small cartilage (1 – 2 cm), B = medium sized cartilage (2.1 – 3 cm), C = large cartilage (3.1 – 4 cm)

ANOVA showed that different diameter sizes had no significant effects on the shark cartilage powder's moisture properties ($p>0.05$). The highest moisture content was recorded in the treatment of medium-sized cartilage diameter (B), 7.48% (Table 1). It is different from [9] who examine the moisture content of the skark cartilage powder of the backbone and throat cartilage using IRM (*InfraRed Moisture Balance*) method, 5.43%. According to Indonesian National Standard (SNI 01-3158-1992), the bone meal contains a maximum of 8% moisture. Lower moisture content products will have a longer shelf life and the decomposing bacteria can't live on the product.

The ash content of shark cartilage powder was significantly ($p<0.05$) affected by the diameter size where the large diameter cartilage (C) showed a higher ash content 46.78% (Table 1) compare to medium (B) and small diameter cartilage (A) respectively. Based on the LSD test, it is known that the ash content of large diameter (C) is significantly different from that of small diameter (A) but not significantly different from that of medium diameter (B). [10] reported that the backbone shark cartilage powder and throat shark cartilage powder had 21.8% and 9.52% ash content, respectively. [14] found that the ash content of blue shark cartilage powder was more than 25%, whereas the ash content of silver shark (*Carcharhinus albimarginatus*) was 25.41% [15]. High values of ash content are indicative of high mineral content (e.g. macrominerals).

The protein content in shark cartilage powder was not significantly influenced by different diameter size (p>0.05). The highest protein content of the shark cartilage powder was recorded in the treatment of small-sized cartilage diameter (A), 46.20% (Table 1). This protein content is higher than that reported by [9] that the protein content on the backbone shark cartilage powder and the throat shark cartilage powder was 30.74% and 30.35%, respectively. [14] said that the shark cartilage rich in minerals type-apatite, proteins and glycosaminoglycans (GAGs).

The present study found lower fat level (Table 1) than that reported in [9] who indicated the fat content of shark backbone cartilage powder as much as 1.92% and throat cartilage powder as much as 1.94%. Different diameter sizes showed no significant effects on the powder's fat properties (p>0.05). The highest fat content was recorded in the treatment of large-sized cartilage diameter (C), 0.46% (Table 1). The Indonesian National Standard (SNI 01-3158-1992) of fish bone fat meal is 3% for 1st quality and 6% for 2nd quality. Thus, the fat content does not meet the standard quality of the fat content of fish bone meal. Low fat content could result from high water content of the sample. It is in agreement with [16] that water content has inverse relationship with fat and protein content, in which high water content causes fat and protein content of the processed food decrease.

The carbohydrate content ranges from 3.54% to 5.66% (Table 1). It was found that different diameter sizes of the shark cartilage did not significantly affect the carbohydrate content (p>0.05). The carbohydrate content of shark cartilage powder was related to the moisture content, protein, fat and ash values. The highest carbohydrate content (5.66%) was obtained from small diameter (A) treatment. [9] found that the carbohydrate content 55.74% and 57.15% in the backbone shark cartilage powder and the throat shark cartilage powder, rerspectively. According to [17], the shark cartilage composition varies depending on the shark species. The shark cartilage powder from *Squatina oculata* contained 29.34% carbohydrates.

3.2. Amino Acids Profile

Table 2 demonstrates that the present study has identified 18 amino acids in the cartilage of *P. glauca*, 10 essential amino acids and 8 non-essential amino acids, respectively. Total amino acids in this report are higher than that reported in [9], but lower than that reported in [18]. [9] reported only 9 essential amino acids and 6 non-essential amino acids, whereas [18] identified 9 essential amino acids and 8 on-essential amino acids in the cartilage of the blacktip shark (*Carcharhinus limbatus*).

Three amino acids found in large amount are glycine, arginine, and proline (Table 2). Glycine ranged from 8.026 – 10.909% with the highest was recorded in large diameter treatment (C). The highest arginine and proline were also found in the same treatment (C), 6.615% and 5.546%, respectively. Nevertheless, the previous studies reported different amount and types of amino acids in the shark cartilage. [9] found that the highest amino acid was arginine (4.000%) and [18] reported that the highest amino acid was alanine (11.90%). It is supported by [19] that the different nutritional value of fish depends upon the internal and external factors. The former is related with fish species, sex, age, and reproductive stage, whereas the latter is related with the fish habitat, such as environment, food availability, and water quality.

Table 2. Amino acid profiles of blue shark (*Prionace glauca*) cartilage.

No	Amino Acid	Shark cartilage on study			Reference	
		A	B	C	Blue shark cartilage (Agustin et al., 2012)	Blacktip shark (Kittiphattanabawon *et al*, 2010)
1	L-Alanine	2.91	2.67	3.66	1.41	11.90
2	L-Arginine*	4.66	3.59	6.16	4.00	5.40
3	L-Aspartic acid	1.84	1.66	2.14	1.63	4.20
4	Glycine	8.31	8.03	10.91	2.12	31.70
5	L-Glutamic acid	3.93	3.64	4.72	3.10	7.70
6	L-Isoleusine*	1.11	1.06	1.34	0.68	1.90
7	L-cystine	0.14	0.12	0.08	-	0.10
8	L-Leusine*	1.89	1.85	2.24	1.45	2.50
9	L-Lysine*	0.87	0.78	1.46	0.92	2.70
10	Methionine*	0.24	0.25	0.22	0.26	1.40
11	L-Triptofan*	0.07	0.07	0.07	-	-
12	L-Valine*	1.35	1.29	0.73	0.89	2.60
13	L-Fenilalanine*	1.73	1.66	2.30	1.00	1.40
14	L-Proline	4.45	4.21	5.55	-	10.50
15	L-Serine	2.05	1.99	2.51	0.81	3.00
16	L-Treonine*	1.62	1.58	1.97	3.75	2.10
17	L-Tyrosine	0.60	0.462	0.63	0.38	0.30
18	L-Histidine*	1.02	3.466	1.31	0.37	0.80
	Total	**38.82**	**38.388**	**47.98**	**23.39**	**90.2**

*Essential amino acids

ANOVA indicated that the cartilage size diameter did not give significant effect on the glycine level of the cartilage ($p>0.05$). Glycine is one of the dominant amino acid types in collagen extract from the skin of blue shark [20]. It is also the highest amino acid in type 2 collagen of the blue shark cartilage [21]. According to [22] glycine has anti-inflammatory, immunomodulatory, and cytoprotective actions that become a potential therapeutic in mucositis. It belongs to a non-essential amino acid useful as anti-inflammation by reducing cytokine pro inflammation and free fatty acid concentration to increase insulin response [23]. Also, the benefit of glycine for health is to stimulate serotonin hormone production that can bring up good feelings, as a key component of collagen, protein forming the structures of bone, skin, muscle, and connective tissue [24].

3.3. Mineral content

Tabel 3 demonstrates that the highest mineral content in the shark cartilage powder was recorded as much as 184,500.00 ppm calcium and 89,450.00 ppm) in large cartilage diameter (C). [25] stated that the elasmobranchii skeleton is not normally mineralized called as endophytic mass (EPM), growing into the uncalcified cartilage skeleton. EPM consists of a crystal phase enriched with calcium (Ca) and phosphorous (P).

Table 3. Mean mineral content of the shark cartilage meal (ppm)

Minerals	Concentration (%)		
	A (1-2cm)	B (2,1-3cm)	C (3,1-4cm)
Ca*	170,000±14,142.14	173,500±27,577.16	184,500±14,849.24
P*	82,350±6,434.67	86,250±12,232.95	89,450±8,414.57
Zn	74.8±8.77	71.35±10.11	93.1±35.21
Mn	7.74±3.20	6.68±0.12	5.91±0.27
Ba	2.02±0.37	1.78±0.2	1.55±0.31

*Macrominerals

According to [26], the spinal column of the elasmobranchii grows radially and consists of the cartilage calcification that stores calcium phosphate appositionally in the organic matrix. [9] found as much 10,673.59 ppm calcium in the cartilage powder of throat and 10,262.39 ppm in the cartilage powder of backbone. According to [27] the composite (Ca+Zn+P) maps' magenta contrast indicates areas where Ca and P are concentrated but not Zn, while the white contrast denotes high concentrations of all three. The mineral density of the cones is higher than that of the intermedialia, as evidenced by the higher Ca and P intensities in the cones relative to the wedge tissue. Ca and P fluorescence maps clearly show the complex band structure observed in lab microCT. Bands with strong vs. low Ca and P intensity reflect variable bioapatite concentration.

[28] stated that calcium is important for bone formation and maintains the bone strength. Sufficient daily calcium intake can prevent an osteoporosis in future. According to [29], the inadequate consumption of Ca and P intake can influence growth and contribute to stunting. During growth, the need for bone mineralization is high, and low calcium absorption can result in low level of new bone deposition matrix mineralization and osteoblast dysfunction. It is supported by [30] that sufficient calcium and phosphorous is crucial to ensure the bone mineralization formed by the osteoblast. Calcium has also very important roles in physiology and pathology beside its most widely recognized function as a critical structural component of bone [31] and can help prevent and treat osteoporosis [32]). The shark cartilage powder containing high calcium and phosphorous can be used as Ca and P source for health supplement or in food product fortification.

3.4. Chondroitin sulphate quantity

The results revealed that the chondroitin sulphate content of shark cartilage powder reached 8.00 – 11.54 (Table 4).

Table 4. Chondroitin sulphate of the blue shark cartilage powder

Replication	Chondroitin sulphate (%)		
	A (1-2 cm)	B (2.1-3 cm)	C (3.1-4cm)
1	6.96	7.98	2.90
2	10.44	12.17	8.59
3	11.91	10.39	11.18
4	12.99	15.61	9.33
Mean	10.57	11.54	8.00

The highest chondroitin sulphate of the blue shark cartilage in the present study was recorded in sample B (2.1-3 cm), 11.54%, and the lowest in sample C (3.1-4 cm), 8.00%. ANOVA indicated that the blue shark cartilage diameter did not give significant effect on the chondroitin sulphate concentration ($P>0.05$). According to [33], glucosamine and chondroitin sulphate analyses using HPLC method indicated that the shark cartilage extract contains glucosamine and chondroitin sulphate with a retention time of 1.917. [34] found that the absorbent addition to the blue shark cartilage extract yielded a stable extract with 2.62% chondroitin sulphate. According to [17], cartilage contains a crucial element known as chondroitin sulfate (CS). CS consists of a sequence of sulfated and/or unsulfated D-glucuronic acid (GlcA) and N-acetyl-D-galactosamine (GalNAc) residues linked through alternating β(1-3) and β(1-4) bonds.[11] said that the chondroitin sulphate is an important component of the connective tissue, has many pharmacological effects, such as antiinflammation, antioxidant, and protects heart and cerebrovascular system. It also has immunomodulator and anticancer effects. Chondroitin sulphate is mainly used for eye, cardiovascular and cerebrovascular diseases, and osteoarthritis prevention and treatment (OA).

Conclusion

The present study found that the cartilage meal of *P. glauca* contained 46.2% protein, 184,500 ppm Ca, and 89,450 ppm P, respectively. The highest amino acid was glycine, 10.91% and chondroitin sulphate, 11.54%. The use of different cartilage diameters did not give significant effect on all measured parameters.

Acknowledgement

A high appreciation is addressed to the Rector of Hang Tuah University for financial support to my PhD program. We also thank the Testing Service Unit of Airlangga University and the Testing Laboratory of PT. Saraswanti Indo Genetech, Surabaya for their contribution in sample analysis.

References

[1] Fahmi and Darmadi. (2013). Review of the Status of Shark Fisheries and Their Conservation Efforts in Indonesia. First Edition. Directorate of Conservation of Areas and Fish Species. Directorate General of Marine, Coastal and Small Islands. Ministry of Marine Affairs and Fisheries. ISBN : 978-602-913- 09-7.

[2] Nursanto, R. M., Mustofa. A., and Widanti. Y. A. (2019). Shark Nuggets (Carcharhinus amblyrhynchos) with a variation of Oyster Mushroom (Pleurotus sp.). *JITIPARI (Scientific Journal of Food Technology and Industry UNISRI),* **4**. no.1.

[3] Rigby, C.L., Barreto, R., Carlson, J., Fernando, D., Fordham, S., Francis, M.P., Herman, K., Jabado, R.W., Liu, K.M., Marshall, A., Pacoureau, N., Romanov, E., Sherley, R.B. and Winker, H. (2019). *Prionace glauca*. The IUCN Red List of Threatened Species.

[4] Novianto, D. and Nugraha, B. (2014). Composition of bycatch and target fish for tuna rawai fisheries in the Eastern Indian Ocean. *Marine Fisheries*, **5** (2), 119–127.

[5] Novianto, D., Rochman. F., Bahtiar. A., Nugraha. B, and Jatmiko. I. (2015). Blue Shark (*Prionace glauca*) Length Composition From Indonesian Longline Fleet In The Indian Ocean: period 2005-2014. (OTC- 2015-WPEB11-23, 1-12).

[6] Dean, M.N., Laura Ekstrom, Efrat Monsonego-Ornan, Jim Ballantyne, P. Eckhard Witten, Cyrena Riley, Wouter Habraken and Sidney Omelon. (2015). Mineral homeostasis and regulation of mineralization processes in the skeletons of sharks, rays and relatives (Elasmobranchii). Semin Cell Dev Biol (2015), http://dx.doi.org/10.1016/j.semcdb.2015.10.022.

[7] Rakhmiyati, T. Widiyani and A. Budiharjo (2022) Histological Structure of Regenerate Tail in Lizard (Mabouya multifasciata Kuhl) After Treatment Using Shark Cartilage Ointment. *In Proceeding International Conference on Religion, Science and Education*. **1**, pp. 527-536.

[8] Agustin, T.I., W. Sulistyowati, and E. Yatmasari. (2016). Study on the Bioactive Compounds of Shark (Prionace glauca) Cartilage and Its Inflammatory Activity. *International Journal of PharmTech Research*. CODEN (USA): IJPRIF, ISSN: 0974-4304. **9**, No.1, pp 171-178.

[9] Agustin, T.I., M. Febriani. and E. Yatmasari. (2012). Proximate Composition and Amino Acid Profile of Water Shark (Prionace glauca) Cartilage. *Proceedings of the MPHPI* National Seminar (Indonesian Fishery Product Processing Society). Brawijaya University. Hapless.

[10] Sulistyowati, W., Titiek. I.A, A. Arbai, and E. Yatmasari. (2015). Glucosamine and chondroitin sulphate content of shark cartilage (*Prionace glauca*) and its potential as anti-aging supplements. *International Journal of ChemTech Research*, **8**, no.10, pp.163-168.

[11] Ren, Z., Ji, Y., Wang, Y. and Dong, L. (2018). Chondroitin sulphate from *Scophthalmus maximus* for Treating Osteoarthritis. *International journal of biological macromolecules*. **108**, pp.1158-1164.

[12] Sherman, A. L., Ojeda-Correal, G., and Mena, J. (2012). Use Of Glucosamine And Chondroitin In Persons With Osteoarthritis. *PM&R*, **4**, S110-S116.

[13] AOAC (2012) Official Method of Analysis: Association of Official Analytical Chemists. **19th** Edition, Washington DC, 121-130.

[14] Vazquez, J. A., Fraguas, J., González, P., Serra, J. and Valcarcel, J. (2020). Optimal Recovery Of Valuable Biomaterials, Chondroitin sulphate And Bioapatites, From Central Skeleton Wastes Of Blue Shark. *Polymers*, **12**, no.11, pp. 2613.

[15] Jeevithan, E., Bin Bao, Yongshi Bu, Yu Zhou, Qingbo Zhao and Wenhui Wu. (2014): Type II Collagen and Gelatin from Silvertip Shark (Carcharhinus albimarginatus) Cartilage: Isolation, Purification, Physicochemical and Antioxidant Properties. *Mar. Drugs*, **12**, 3852-3873; doi:10.3390/md12073852.

[16] Triasih, D., T.A. Laksanawati and S. Nurlailatul. (2021). Chemical Characteristics of Salami with the Addition of Angkak Extract (Red Mold Rice). *Journal of Animal Husbandry of the Archipelago.* **7** pp 7-10.

[17] Sim, J., A-Rang Im, S.M. Cho, H.J. Jang, J. H. Jo and Y.S. Kim. (2007). Evaluation Of Chondroitin sulphate in Shark Cartilage Powder as A Dietary Supplement: Raw Materials and Finished Products. *Food Chemistry* **101** : 532–539.

[18] Kittiphattanabawon, P., Benjakul, S., Visessanguan, W., and Shahidi, F. (2010). Isolation and characterization of collagen from the cartilages of brownbanded bamboo shark (Chiloscyllium punctatum) and blacktip shark (Carcharhinus limbatus). *LWT-Food Science and Technology*, **43**. No.5, pp.792-800.

[19] Hafiluddin, Y. Perwitasari and S. Budiarto. (2014). Analysis of Nutritional Content and Sludge Odor of Milkfish (*Chanos chanos*) from Two Different Locations. *Journal of the Ocean,* **7**, No.1, pp.33-44

[20] Jeevithan, Elango., Jung W. L., Shujun W., Yves H., José E. M.S.de Val, J. M. Regenstein, S.Y. Lim, Bin Bao and Wenhui Wu (2018) : Evaluation of Differentiated Bone Cells Proliferation by Blue Shark Skin Collagen via Biochemical for Bone Tissue Engineering. *Mar. Drugs* **16**, 350; doi:10.3390/md1610035.

[21] Pan, Z., B. Ge, M. Wei, Jeevithan E. and W. Wu. (2023). Isolation and Biochemical Properties of Type II Collagen from Blue Shark (*Prionace glauca*) Cartilage. *Mar. Drugs.*, **21**, 260. https://doi.org/10.3390/md21050260.

[22] Sa, O.S., N.N.F. Lopes, M.T.S. Alves and E.M.M. Caran. (2018). Effects of Glycine on Collagen, PDGF, and EGF Expression in Model of Oral Mucositis. *Nutrients* **.10.** 1485; doi:10.3390/nu10101485.

[23] Aguayo-Ceron, K.A., F. Sanchez-Munoz, R. A. Gutierrez-Rojas, L. N. Acevedo-Villavicencio, A. V. Flores-Zarate, F. Huang, Abraham Giacoman-Martinez, S. Villafana and R. Romero-Nava. (2023). Glycine: The Smallest Anti-Inflammatory Micronutrient. *Int. J. Mol. Sci.* **24**, 11236. https://doi.org/10.3390/ijms241411236.

[24] Myhre, J. and Dennis S. (2023). Glycine: Benefits, Side Effects, and Risks. Verywellhealth. https://www.verywellhealth.com/glycine-overview-4583816?print

[25] Seidel R., Blumer M., Zaslansky P., Knötel D., Huber D.R., Weaver J.C., Fratzl P., Omelon S., Bertinetti L. and Dean M.N. (2017). Ultrastructural, material and crystallographic description of endophytic masses – a possible damage response in shark and ray tessellated calcified cartilage. *J. Struct Biol*; **198** (1):5-18. doi: 10.1016/j.jsb.2017.03.004. PMID: 28286227.

[26] Mohan J.A., Miller N.R., Herzka S.Z., Sosa-Nishizaki O., Kohin S., Dewar H., Kinney M., Snodgrass O. and Wells R.J.D. (2018) Elements of time and place: manganese and barium in shark vertebrae reflect age and upwelling histories. *Proc. R. Soc. B* 285: 20181760. http://dx.doi.org/10.1098/rspb.2018.1760.

[27] Stock, S. R., Morse, P. E., Stock, M. K., James, K. C., Natanson, L. J., Chen, H. and Park, J. S. (2021). 3D tomography of shark vertebrae via energy dispersive diffraction. SPIE Digitial Library, 11840

[28] Beto, J.A. (2015). The Role of Calcium in Human Aging. Review Article. *Clin Nutr Res* **4**:1-8 pISSN 2287-3732 · eISSN 2287-3740 http://dx.doi.org/10.7762/cnr.2015.4.1.1.

[29] Sari, E. M., M. Juffrie., N. Nurani., and M. N. Sitaresmi. (2016). Protein, calcium and phosphorus intake in stunted and non-stunted children aged 24-59 months. *Indonesian Journal of Clinical Nutrition*, **12**, No. 4, pp.152-159.

[30] Reid, I.R. (2014) Should We Prescribe Calcium Supplements for Osteoporosis Prevention?. *J.Bone Metab* ; **21** (1):21–28.

[31] Reid, I.R, Birstow S.M. and Bolland M.J. (2017). Calcium and Cardiovascular Disease. *Endocrinol Metab.* 2017; **32**(3):339–349.

[32] Li, K., Xia-Fang wang, Ding-You Li, Yuan-Cheng Chen, Lan-Juan Zhao, Xiao G.Liu, Yan-Fang Guo, Jie Shen, Xu Lin, Jeffrey Deng, Rou Zhou and Hong-wen Deng. (2018). The Good, The Bad, and The Ugly of Calcium Supplementation: A Review Of Calcium Intake On Human Health. *Clinical Interventions in Aging*, **13.** 2443–2452.

[33] Rakhmiyati, Tetri Widiyani and Agung Budiharjo (2023): *High Performance Liquid Chromatography* (HPLC) for Detection of Glucosamine and Chondroitin sulphate Compounds. *Biology, Medicine, & Natural Product Chemistry*, **12**, (1) Pages: 5-8 | DOI: 10.14421/biomedich.2023.121.5-8. ISSN 2089-6514 (paper). ISSN 2540-9328 (online).

[34] Agustin, T.I, Risma, R. Sari and D.Setyawan. (2023) The Microstructure and Potential of Chondroitin sulphate in Shark Cartilage Extract. *Squalen Bulletin of Marine and Fisheries Postharvest and Biotechnology*.**18** (2): 74-80. ISSN: 2089-5690, e-ISSN: 2406-9272. https://doi.org/10.15578/squalen.760.

IOP Conf. Series: Earth and Environmental Science **1473** (2025) 012050 doi:10.1088/1755-1315/1473/1/012050

The role of platelet rich fibrin in environmental health

N Nilawati[1,2], I Wardani[1], B Santoso[1]
[1] Faculty of Dentistry, Hang Tuah University, Surabaya, Indonesia
[2] RSUD Haji East Java Province, General Hospital, Indonesia

*E-mail address : ninanilawati.62@gmail.com

Abstract. Platelet Rich Fibrin (PRF), is use and management reflect broader concerns in environmental health. PRF is obtained from the patient's blood, with centrifuging at a certain speed and time. The aim of this study was to determine the centrifugation speed and time that would produce the platelet-derived growth factor (PDGF) and insulin-like growth factor 1 (IGF-1) present in PRF. This was a laboratory study using blood samples from seven individuals who volunteered for the study. 18 ml of blood was taken from each individual and placed into six 3 ml vacuum tubes. Next, it was centrifuged at speeds of 1500rpm, 2000rpm, 2400rpm, and 3000rpm with a centrifugation time of 8 minutes, 10 minutes, and 12 minutes. PDGF and IGF-1 examination uses the dot blot method; then, the dot blot results are quantified using images. Statistical analysis used a parametric one-way ANOVA test with a significance value of $0.000 < 0.05$. Data processing uses the Shapiro Wilk Test. The results of this study showed that blood centrifuged at 2000 rpm for 12 minutes produced the highest amounts of PRF, PDGF, and IGF-1.

1. Introduction

Platelet-rich fibrin (PRF) plays a major role in regenerative medicine and wound healing. PRF connect to environmental health due to PRF is an autologous biomaterial. PRF contains growth factors, cytokines, and leukocytes [1]. A second-generation platelet concentration called PRF has the ability to to increase periodontal regeneration by quicken both soft and hard tissue healing, because PRF releases growth factors including platelet-derived growth factor (PDGF) [2,3]. PRF accelerates wound healing, because the gradual process of fibrinogen turning into fibrin will create a strong, rigid structure that is hard to change shape [4-6]. PRF can speed up the healing of gingiva, periodontal ligament and alveolar bones [5,6]. The three-dimensional structure of fibrin makes it strong and resistant to tearing, making it useful as a barrier for surgical treatment. Based on the nature and content of PRF, it can be a substitute for bone grafts in situations where the periodontal disease needs tissue regeneration during periodontal flap surgery [7]. PRF supports sustainable healthcare practices by utilizing the body's natural healing mechanisms, aligning with broader environmental health goals of reducing chemical and material waste. PRF is fibrin rich in fibroblast growth factor (FGF), basic fibroblast growth factor (b-FGF), vascular endothelial growth factor (VEGF), angiopoietin, and platelet-derived growth factor (PDGF), and transforming growth factors (TGF), hepatocyte growth factor (HGF), epidermal growth factor (EGF), insulin-like growth factors 1 and 2 (IGF-1 and IGF-2), and interleukins, that important to angiogenesis and cell proliferation [8-10]. PRF has a dense fibrin network with leukocytes, cytokines, glycoprotein, and thrombospondin-1, which last up to around 7 days [9-11]. The slow polymerization process of fibrinogen into fibrin, resulting in a physiological form that is useful for the wound healing process [12,13]. PRF was initially created by Chouckron for use in oral surgery [11]. The PRF contains a thick fibrin network with white blood cells, cytokines, glycoprotein, and thrombospondin-1, and it can remain active for about a week [9]. PRF has thrombin, which is made naturally when fibrinogen turns into fibrin through a slow process, and helps wound healing in the body [12,13]. PRF is taken from the patient's blood by spinning it at a specific speed and for a set amount of time. According to Choukroun, the protocol for making PRF is centrifugation of run the machine at 3000 revolutions per minute for duration of 12 minutes. Marenzi et al. (2015) made PRF by By spinning the blood at a high speed of 2700 revolutions per minute for 12 minutes [14]. Farshidfar (2020) used run the machine at 2400 revolutions per minute for 3 minutes [15]. Khaled et al. (2020) used of run the machine at 3000 revolutions per minute for 10 minutes [16]. Even though making PRF is a free and easy protocol, differences in protocol can affect the quality and effectiveness of the material in the healing process. Differences in PRF manufacturing protocols will result in different PRF biological compositions, which will ultimately affect clinical success.

Given the information and issues discussed above, it is necessary to carry out research to find out whether the speed and duration of blood centrifugation would affect the produced PRF. The indicators chosen were PRF weight, PDGF, and IGF-1. Although many similar studies have been carried out, there has been no research to determine the speed and length of centrifugation to obtain PRF with maximum growth factors and PDGF and IGF-1 indicators. The PDGF indicator was chosen because PDGF, as a platelet derivative, it has been commonly used as a framework in cell therapy and tissue engineering. PDGF is a growth factor that is able to regenerate periodontal tissue. PDGF obtained from PRF comes from platelets, which are the most abundant element in PRF [17]. It is very important to look at the PDGF indicator, because of its function in regenerative dentistry, including periodontic cases for infected periodontal defects or periodontal plastic surgery [10]. The indicator IGF-1 was chosen, because a growth factor that helps bone was healing and regeneration, as well as cell differentiation, inhibiting apoptosis in many tissues, increasing proliferation, and playing a major role in stimulating osteoblast function. IGF-1 stimulates osteoblast replication and bone matrix synthesis. In periodontitis, which involves alveolar bone damage, growth factors such as IGF-1, contained in PRF, are needed [18]. This research has to be done because its data and results can help people in the field to understand how long and fast to use PRF to produce high PDGF and IGF-1. This research is a continuation of the previous research with the conclusion that PRF works just as well as a bone graft, so PRF can be used as an alternative for treating chronic periodontitis that requires flap surgery [7].

2. Method

The kind of research was a scientific study conducted in a laboratory, using 21 blood samples from 7 volunteers to study how speed and time variations affect the results. The sampling method is simple random sampling. Sample criteria consist of inclusion and exclusion criteria. Inclusion criteria were blood samples taken from healthy adults, aged 19–59 years, with a body weight 50–70 kg, and not taking medication in the last 2 weeks. Exclusion criteria were blood samples that had been exposed to air for more than 2 minutes. The research variables consist of dependent variables, namely centrifugation speed and centrifugation time. The independent variables are PDGF and IGF-1. This research has received an ethical certificate with number 445/104/KOM.ETIK/2024 from the hospital ethics committee of RSUD Haji East Java Province.

This research was carried out at the Clinical Pathology Laboratory of Haji General Hospital, East Java Province, and the Biochemistry Laboratory of Faculty of Medicine of Brawijaya University. Blood collection was carried out in accordance with blood collection procedures for routine examinations. 18 ml of blood was taken from one volunteer and collected in six vacuum tubes; each containing 3 ml. Centrifugation was carried out with six centrifuges in the same specifications (Rotina 380, Germany).

2.1 Preparation and weight calculation of platelet-rich fibrin (PRF)

The blood samples from the six vacuum tubes were split into two groups. Group A was given centrifugation treatment at a speed of 3000 rpm at different times. Group B was given centrifugation treatment for 12 minutes at different speeds. After the blood sample is centrifuged, three layers are obtained: the top layer is plasma; the middle layer is PRF, and the blood cells at the bottom (Figure1.A). The clear top part is separated then, the middle layer, which is PRF, is taken (Figure 1.B). Next, the PRF is put in a labelled plastic capsule (Figure 1.C). Each capsule containing PRF is weighed use analytic scale (Precisa, Swiss) (Figure 1.D). PRF weight is the weight of the capsule containing PRF minus the weight of the empty capsule.

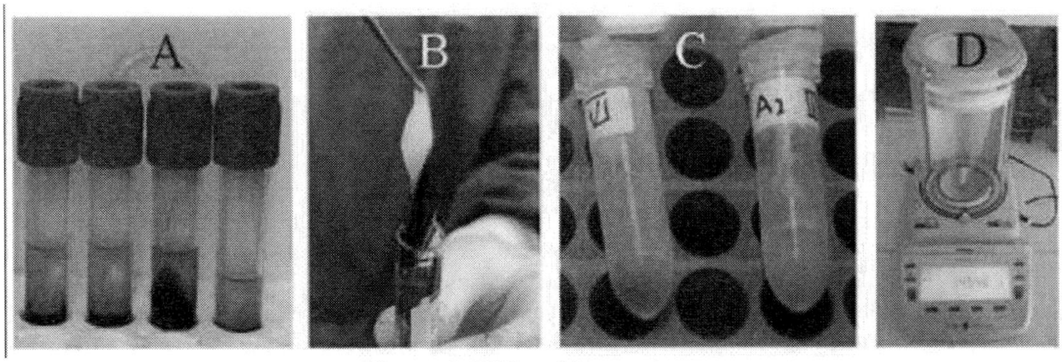

Figure 1. PRF making processes

2.2 Platelet-derived growth factor (PDGF) examination using dot blot assay

Using a fine-bore pipette tip, 2 µl of sample was slowly added to the nitrocellulose membrane at the centre of a 2–4 mm diameter grid. The membrane was allowed to dry at room temperature. Non-specific protein blocking was performed using 5% BSA in TBS-T in a 10 cm petri disk leave at room temperature for 1 hour. Then, it was incubated with the primary antibody (mouse monoclonal anti-PDGF, Santacruz Biotech) at 1:1,000 for 60 minutes at room temperature and washed three times with TBS-T (3 x 5 minutes). Next, this was incubated with secondary antibody, conjugated anti-mouse AP, for 30 min at room temperature and washed three times with TBS-T (1×15 min and 2×5 min) and then once with TBS (5 min). Finally, it was incubated with the western blue reagent for 1 minute, then wash three times with TBS-T (3 × 5 min), and rinsed with H20.The blot results were dark blue and analyzed using image J.

2.3 Examination of insulin-like growth factor 1 (IGF-1) using the dot blot test

The method for examining IGF-1 was same as examining PDGF; however, the primary antibody used was the anti-IGF1 monoclonal antibody, Abcam Cat#ab9572.

3. Statistical Analysis

The obtained data were checked for normality and homogeneity, and a parametric one-way ANOVA test was performed with a significance value of 0.05.

4. Results and discussion

4.1 Platelet-rich fibrin (PRF) weight calculation

Based on the homogeneity of variances test, the data is normal and homogeneous distributed, then, the hypothesis is tested using the one-way ANOVA parametric statistical test, with the result that there is no difference in the average values in all centrifugation treatment groups for 8 minutes, 10 minutes, or 12 minutes with a probability value of sig 0.08 > 0.05. In the group that was given centrifugation treatment for 12 minutes between speeds of 1500 rpm, 2000 rpm, and 2400 rpm, there was the greatest increase in weight when was centrifuged at a speed of 2000 rpm for 12 min compared to blood samples centrifuged at a speed of 2400 rpm or 1500 rpm. The significance value is 0.939 > 0.05.

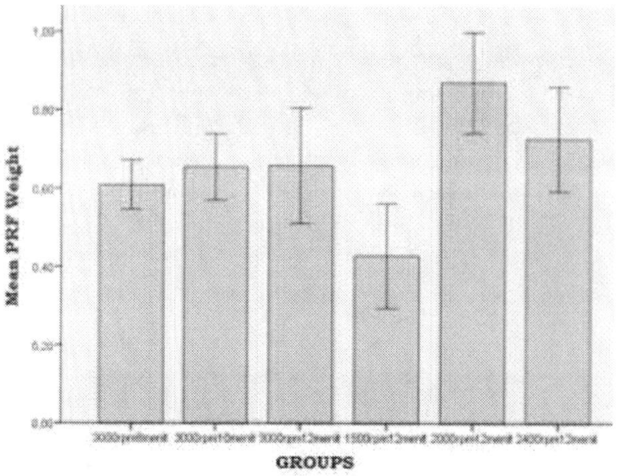

Figure 2. Graph of PRF weight results in each group. The maximum weight gain occurs when centrifuge at 2000 rpm for 12 minutes.

The findings of this research demonstrate that blood centrifuged at different times and speeds produces PRF with varying weights. This aligns with a study by Pancawinata et al. (2018), which states that the amount of time and how fast the blood spin in the centrifuge machine can change how many platelets are in the PRF that you make [19]. However, these findings contradict the research conducted by Zwittnig et al. (2022), which suggests that slower speeds and longer centrifugation times yield the highest PRF production [20]. When the centrifugation speed decreases, the relative separation of the PRF layer is minimized, and the formed PRF agglomerates are also smaller. By reducing the speed of centrifuge, possible platelet contamination fragments of red blood cell, so the possible decrease in platelets when separating blood through centrifugation. Another study indicates that increasing spin speed and longer time, there will be fewer platelets in the substance [21]. The findings of this research show that blood samples that were given centrifugation treatment going quickly at 3000 rpm for 10 or 12 mins produced the same amount of PRF compared to those treated for 8 minutes. Blood samples treated with centrifugation at a speed of 2000 rpm for 12 minutes produced PRF with the highest weight compared to blood samples treated with centrifugation at a speed of 2400 rpm and 1500 rpm for 12 minutes.

4.2 Platelet-derived growth factor (PDGF) examination

According to homogeneity of variance test, it can be deduced that the three groups are from the same population (homogeneous). Therefore, hypothesis testing utilizes, a parametric statistical test called one-way ANOVA test is used. The probability or significance value is 0.400 > 0.05. This indicates that there were no significant differences in mean scores between groups. Specifically, the average values for the groups subjected to centrifugation treatment at a speed of 3000 rpm were similar for the 8-minute, 10-minute, and 12-minute groups.

In the treatment group with speeds of 1500 rpm, 2000 rpm, and 2400 rpm for 12 minutes, the sig probability value was 0.035, which is less than 0.05. This suggests there was a significant difference in scores within the groups that received centrifugation treatment for 12 minutes. Consequently, further testing was conducted to explore more specific differences between time and centrifugation speed on PDGF. The post hoc test method, specifically the Turkey test, was employed as a multiple comparison test. The results revealed that the group with a speed of 2000 rpm for 12 minutes and the group with a speed of 1500 rpm for 12 minutes had a sig value of 0.039, which is less than 0.05.

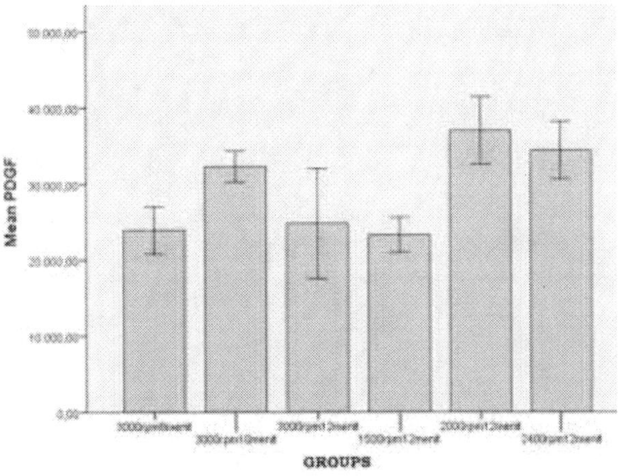

Figure 3. Graph of PDGF results in each group. This indicates that the highest production of PDGF occurred when PRF was rotated at 2000 rpm for 12 minutes.

PDGF is gaining prominence in the medical field, especially in regenerative dentistry, which includes regenerative endodontic, infra bony periodontal abnormalities, periodontal plastic surgery, and oral and maxillofacial surgery [10]. Tissue regeneration has become increasingly important in dentistry, prompting many dentists to implement it. Among the various growth factors, PDGF is the most rapidly growing one [22]. It is widely utilized as a scaffold in tissue engineering and cell-based treatment [8]. In this study, blood samples that were centrifuged for 10 minutes at 3000 rpm revealed increased amounts of PDGF. This finding supports the recommendation by Abdullah et al. (2017) to centrifuge blood at 3000 rpm for 10 minutes [23]. The results of this study support several previous studies that found that PDGF has a high affinity for fibrin, where fibrin binds to several growth factors [16]. This research shows that high PRF results will also produce high PDGF, namely at a speed of 3000 rpm for 10 minutes. Sneha (2022) also found that the direction of tool movement and centrifugation speed influence the release of growth factors [24].

4.3 Insulin-like growth factor-1(IGF-1) examination

According to the test for homogeneity of variances, it may be said that the three groups originate from a homogeneous population, which has the same variance. Therefore, hypothesis testing utilizes a parametric statistical test, namely the one-way ANOVA test, with a probability value of sig $0.262 > 0.05$. The average value did not significantly differ across the groups. Specifically, the average values for the groups subjected to centrifugation treatment at a speed of 3000 rpm were similar for the 8-minute, 10-minute, and 12-minute groups. However, blood samples centrifuged at 2000 rpm for 12 minutes in this study yielded the highest levels of both PDGF and IGF-1. However, in the group that underwent centrifugation treatment for 12 minutes at different speeds (1500 rpm, 2000 rpm, and 2400 rpm), The significance level was found to be $0.035 < 0.05$. This implies that there was a significant difference in the values among the groups that received centrifugation treatment for 12 minutes. The centrifugation treatment for 12 minutes at speeds of 1500 rpm, 2000 rpm, and 2400 rpm had a significant impact on the average value. To further investigate the specific differences between centrifugation time and speed on the production of IGF-1, a post hoc test method, specifically the Turkey test, was conducted as a multiple comparison test.

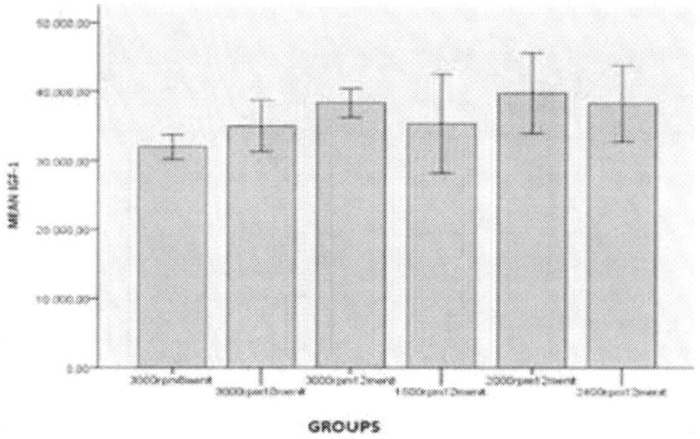

Figure 4. Graph of IGF-1 results in each group. The results showed that the group subjected to centrifugation for 12 minutes at a speed of 2000 rpm produced the highest amount of Insulin like growth factor-1.

IGF-1 produced at 3000 rpm for 12 minutes showed similar results to those produced by centrifugation at 2400 rpm for 12 minutes. IGF-1, is a polypeptide hormone that is widely distributed in the bloodstream and can also be generated after platelet degranulation [25, 26]. Some growth factor is significant in bone formation include IGF-I and PDGF. IGF-I was widely distributed in blood and is released after platelet degranulation. IGF-1 is another substance that is present in platelet α granules and is involved in cell proliferation, differentiation, and migration. Its receptor is located on the surface of platelets, and it may help to amplify platelet responses [27].

5. Conclusion

The speed and duration of blood centrifugation affect the growth factors produced by PRF. From the results of this study, it can be concluded that blood centrifuged at a speed of 2000 rpm for 12 minutes will produce PRF with maximum PDGF and IGF-1. Further research needs to be carried out with other methods and larger samples.

Acknowledgement

The author would like to thank Hang Tuah University, RSUD Haji Provinsi Jawa Timur Hospital, and the people who agreed to participate as subjects of this research

References

[1] Damsaz M, Castagnoli CZ, Eshghpour M, Alamdari DH, Alamdari AH, Noujeim ZEF, et al. Evidence-based clinical efficacy of leukocyte and platelet rich fibrin in maxillary sinus floorlLift, Graft and surgical augmentation procedures. *Front Surg.* 2020; **7** (November):1–13.

[2] Alves R, Grimalt R. A review of platelet rich plasma: History, Biology, Mechanism of Action, and Classification. *Ski Appendage Disord.* 2018; **4** (1):18–24.

[3] Doiphode AM, Hegde P, Mahindra U, Kumar SMS, Tenglikar PD, Tripathi V. Evaluation of the efficacy of platelet-rich plasma and platelet-rich fibrin in alveolar defects after removal of impacted bilateral mandibular third molars. *J Int Soc Prev Community Dent.* 2016; **6** :S47–52.

[4] Pitzurra L, Jansen IDC, de Vries TJ, Hoogenkamp MA, Loos BG. Effects of L-PRF and A-PRF+ on periodontal fibroblasts in in vitro wound healing experiments. *J Periodontal Res.* 2020;55(2):287–95.

[5] Blinstein B, Bojarskas S. Efficacy of autologous platelet rich fibrin in bone augmentation and bone regeneration at extraction socket. *Stomatologija.* 2018; **20** (4):111–8.

[6] Mijiritsky E, Assaf HD, Peleg O, Shacham M, Cerroni L, Mangani L. Use of PRP, PRF and CGF in periodontal regeneration and facial rejuvenation-a narrative review. *Biology (Basel).* 2021;**10** (4):1–23.

[7] Nilawati N, Wedarti YR, Rizal MB. The effectiveness of platelet rich fibrin compared with bone graft for the treatment of chronic periodontitis. *Journal of Dental and Medical Sciences (IOSR-JDMS)* e-ISSN:2023; **22** (9):38–42.

[8] Li D, Huang LT, Zhang CP, Li Q, Wang JH. Insights Into the Role of Platelet-Derived Growth Factors: Implications for Parkinson's Disease Pathogenesis and Treatment. *Front Aging Neurosci.* 2022;**14** (July):1–30.

[9] Bhaumik Thakkar et.al. Comparison of regenerative potential of platelet rich fibrin alone and in combination with bovine bone graft in intraosseous defect by single flap approach: A clinical and radiographic study. *J Int Soc Prevent Communit Dent* 2020;**10** :743-51.

[10] Cano-Durán JA, Peña-Cardelles JF, Ortega-Concepción D, Paredes-Rodríguez VM, García-Riart M, López-Quiles J. The role of leucocyte-rich and platelet-rich fibrin (L-PRF) in the treatment of the medication-related osteonecrosis of the jaws (MRONJ). *J Clin Exp Dent.* 2017; **9** (8):e1051–9.

[11] Ozcay N. Reply to Comment on "Role of platelet-rich fibrin on intestinal anastomosis wound healing in a rat." *Biomed Mater.* 2018; **13** (6):5–7.

[12] Kazemi D, Fakhrjou A. Leukocyte and platelet rich plasma (L-PRP) versus leukocyte and platelet rich fibrin (L-PRF) for articular cartilage repair of the knee: A comparative evaluation in an animal model. *Iran Red Crescent Med J.* 2015;**17** (10).

[13] Clark D, Rajendran Y, Paydar S, Ho S, Cox D, Ryder M, et al. Advanced platelet-rich fibrin and freeze-dried bone allograft for ridge preservation: A randomized controlled clinical trial. *J Periodontol.* 2018; **89** (4):379–87.

[14] Faot F, Deprez S, Vandamme K, Camargos G V., Pinto N, Wouters J, et al. The effect of L-PRF membranes on bone healing in rabbit tibiae bone defects: Micro-CT and biomarker results. *Sci Rep.* 2017; **7** (June 2016):1–10.

[15] Farshidfar N, Amiri MA, Jafarpour D, Hamedani S, Niknezhad SV, Tayebi L. The feasibility of injectable PRF (I-PRF) for bone tissue engineering and its application in oral and maxillofacial reconstruction: From bench to chairside. *Mater Sci Eng C.* 2021; 1–42.

[16] Khaled S, Olgun E, Hendek M. Trombositten zengin fibrin ve diş hekimliğinde kullanımı. *J Med Palliat Care.* 2022; **3** (4):386–93.

[17] Nishimoto S, Fujita K, Sotsuka Y, Kinoshita M, Fujiwara T, Kawai K, et al. Growth factor measurement and histological analysis in platelet rich fibrin: A Pilot Study. *J Maxillofac Oral Surg.* 2015;**14** (4):907–13.

[18] Gao J, Song Y, Huang X, Wang D, Wang H. The expression of platelet-derived growth factor, epidermal growth factor, and insulin-like growth factor-ii in patients with polycystic ovary syndrome and its correlation with pregnancy outcomes. *Ann Palliat Med.* 2021;**10** (5):5671–8.

[19] Pascawinata A, Andriansyah A, Bismanevi R. Effect of speed and length of blood centrifugation time on platelet count in the process of making platelet rich fibrin. *B-Dent J Dental Mask Univ Baiturrahmah.* 2021; **8** (3): 285–92.

[20] Zwittnig K, Kirnbauer B, Jakse N, Schlenke P, Mischak I, Ghanaati S, et al. Growth factor release within liquid and solid PRF. *J Clin Med.* 2022; **11** (17):1–10.

[21] Nagrani T, Kumar S, Haq MA, Dhanasekaran S, Gajjar S, Patel C, et al. Use of injectable platelet rich fibrin accompanied by bone graft in socket endurance: A radiographic and histological study. *Cureus.* 2023; **15** (10).

[22] Nasirzade J, Kargarpour Z, Hasannia S, Strauss FJ, Gruber R. Platelet-rich fibrin elicits an anti-inflammatory response in macrophages in vitro. *J Periodontol.* 2020; **91** (2):244–52.

[23] Abdullah A, Murriky A, Shafik S. Influence of platelet rich fibrin on post-extraction socket healing : A clinical and radiographic study. *Saudi Dent J.* 2017; **29** (4):149–55.

[24] Sneha K, Rani AJ, Kumar P. The G-force conundrum in platelet rich fibrin generation : Management of a problem hidden in plain sight. *Journal of Contemporary Clinical Dentistry.* 2022;150–5.

[25] Rathod A, Jaiswal P, Masurkar D. Advanced platelet rich fibrin in periodontal regeneration. *AIMS Bioengineering,* 2023; **10** (April):175–82.

[26] Pavlovic V, Ciric M, Jovanovic V, Stojanovic P. Platelet rich fibrin : Basics of biological actions and protocol modifications. *Open Access. published by De Gruyter International License.* 2021;446–54.

[27] Baca-Gonzalez L, Serrano ZR, Rancan L, González FT F, Fernández-Tresguerres I, López-Pintor RM, et al. Plasma rich in growth factors (PRGF) and leukocyte platelet rich fibrin (L-PRF): Comparative release of growth factors and biological effect on osteoblasts. *Int J Implant Dent.* 2022; **8** (1).

Mechanical strength comparison of natural resources composite resin with coconut coir (*Cocos nucivera L.*) particulate fiber 70% cellulose filler and RMGIC

F Pangabdian[1], Y Hadinata[1], Y Emilda[2] T A Cevanti[1], D Soesilo[1] and R A Faradila[3]

[1]Conservative Department, Faculty of Dentistry, Universitas Hang Tuah, Surabaya, Indonesia
[2]Pediatric Department, Faculty of Dentistry, Universitas Hang Tuah, Surabaya, Indonesia
[3]Student of Faculty Dentistry Universitas Hang Tuah, Surabaya, Indonesia

* E-mail: fani.pangabdian@hangtuah.ac.id

Abstract. One of the intermediate layer materials that use in dentistry is flowable composite resin. Synthetic fiber usage as filler for composite resin has relatively weak adhesion to the polymer matrix components causing low mechanical strength. Coconut coir filler as natural resources have good interfacial bonding with matrix resin, low toxicity, quite low modulus of elasticity and very high elongation at break. Mechanical tests on the intermediate layer material need to be carried out in order to minimize microleakage and be able to withstand chewing loads in the oral cavity. This research aimed to determine the differences in the mechanical strength of coconut particulate fiber 70% cellulose filler composite resin and RMGIC. Samples divided into two control groups and a treatment group, which contain 6 samples for each group. All samples underwent a flexural strength modulus of elasticity, impact strength and compressive strength test. The results show that the flexural strength value was $p=0.092$ ($p>0.05$), impact strength value was $p=0.065$ ($p>0.05$), compressive strength value was $p=0.850$ ($p>0.05$). There was no differences in mechanical strength between the coconut particulate fiber 70% cellulose filler composite resin and RMGIC. This research supports the development of natural treatments and natural resources.

1. Introduction

A good restorative material is a material that has the ability to adapt in the form of good adhesion between the restoration and the cavity wall, besides that it must also have the property of protecting the cavity from secondary caries [1]. Currently, the use of composite resin restorative materials continues to grow and is the most widely available in dentistry [2]. Research conducted by Farrel *et al*, stated that there are several restoration failures using composite resin, including internal contraction during polymerization which can damage the marginal seal of the composite resin, low compression strength, and shrinkage can occur during polymerization, causing adhesive failure and secondary caries. pulp irritation, post-restoration sensitivity, and microleakage [3]. Several techniques to reduce the main problem, namely microleakage, include the application of an intermediate layer material between the cavity surface and packable composite resin [4]. Coconut coir fiber is a natural material that is very good for use as a composite filler because it has advantages such as being light, heat resistant, cheap, strong, recyclable and easier to obtain naturally compared to other natural fibers. Based on this, coconut coir fiber is starting to be used as an alternative to synthetic fiber to strengthen composite resin as an intermediate layer material [5,6]. The intermediate layer material must have good mechanical strength, including flexural strength, impact strength and compressive strength in order to prevent microleakage of a restoration, so that the composite resin can last a long time in the oral cavity [3,7]. The mechanical properties of a material describe the relationship between load or force and its deformation [8] . In this research, the flowable cellulose coir composite resin consisted of 70% coconut fiber and 30% matrix. Coconut coir was chosen as a filler because it has good interfacial bonding with the matrix resin, low toxicity, quite low modulus of elasticity and very high elongation at break [9,10]. Coir cellulose composite with a fiber ratio of 70% produces particulate filler. Mixing fibers in making cellulose coir composite resin aims to increase mechanical strength, increase the bond between filler and matrix. Coconut coir fiber has good characteristics in the form of a rough surface without any particles so that it has a low elastic modulus and can be used as a flowable composite filler [7,11]. The impact strength of a material can be determined by carrying out an impact test. The impact test is the energy required to break the material with an impact force [5]. A sudden impact can come from trauma due to a sudden blow from either a vertical or horizontal direction. This test can be used as a parameter to see the ability of a material to withstand loads such as impacts [12].

Flexural strength is the ability of a material to withstand compressive and tensile forces or the ability of a material to bend before experiencing fracture [13]. The flexural strength of coconut fiber composites with fiber volume fractions of 0%, 10%, 20%, 30%, 40%, 50%, 60% and 70% which have been treated with alkali (NaOH) varies greatly. At a fiber volume fraction of 70%: 30% matrix, a flexural strength of 54.77934995 MPa is obtained with the minimum flexural strength that a restoration material polymerized by light must have according to ISO 10477 (2020) is 50 MPa. Based on the results of these studies, it can be seen that the more fiber volume in the filler, the greater the flexural strength produced. Therefore, in this study, the concentration used to make coir cellulose composites uses a fiber mass fraction concentration of 70% and 30% matrix because it is based on the viscosity of the flowable composite required and is expected to obtain good flexural strength [11,14].

Compressive strength is a mechanical property of restoration materials that is very important in the chewing process, especially in the posterior region. The composition of the restorative material, the material hardening process, and the oral cavity environment can influence the compressive strength of the restorative material [3,9]. Compressive strength is the mechanical property most needed to minimize the occurrence of fractures due to the chewing load received. The chewing process has a natural pressure that influences the length of time a restoration lasts in the oral cavity [15]. This research aimed to determine the mechanical test of coconut cellulose filler composite resin (*Cocos nucivera L.*) which was formed from granular fiber with a mass ratio of 70% of the fiber matrix. The urgency of this research was to find alternative intermediate layer materials from coconut cellulose filler composite resin formulations (*Cocos nucivera L.*).

2. Method

This type of research was classified as true experimental with a post – test only control group design. The experimental unit used was coconut fiber cellulose filler composite resin (*Cocos Nucivera L.*) fiber particulates form by testing mechanical strength (flexural, elastic modulus, impact and compression). Based on the formula calculation, the minimum number of samples for each research group were 6 samples. Making a natural fiber composite base began with preparing coconut fiber filler material through a delignification process using 60% ethanol. Bleaching was carried out to fade the brown color of the coconut fiber using H_2O_2, NaOH, and demineralized water on a rotary evaporator. Meanwhile, for the synthesis of cellulose fiber, the fiber was dissolved using NaOH and urea, then the dissolved fiber was poured into a 99% ethanol antisolvent solution at a rate of 10ml/min. then let it sit until solid cellulose sedimentation occured. Cellulose sedimentation was then carried out by solvent exchange using demineralized water until the pH was neutral, then the fiber was put into a freezer at a temperature of -25°C. Then the frozen cellulose was freeze dried at -45°C with a pressure of 5 mTorr for 24 hours. The cellulose fiber powder in the glass beaker was then weighed at 1 gram along with a mixture of 25% by weight of BisGMA matrix resin and 5% by weight of TEGDMA, and DGEBA (10% by weight of the matrix). Mixing of fiber and matrix was carried out using a 25°C sonicator for 60 minutes. Then, 1% camphorquinone with the Sigma-Aldrich brand was added and mixed using a magnetic stirrer for around 24 hours.

2.1. Flexural and Modulus Elasticity Test

Making flexural and modulus of elasticity test molds, the molds were made from block-shaped acrylic measuring 2mm x 2mm x 25mm. The coconut fiber cellulose composite resin was inserted into the mold using a syringe every 2 mm and then exposed to visible light cured for 40 seconds. The direction of illumination was perpendicular. After mold setting, the sample was removed from the mold. Samples were stored in an incubator at 37°C for 24 hours. Flexural strength and elastic modulus testing could be carried out using a Universal Testing Machine (UTM) using the three-point bending test method with the sample placed on a support board or UTM bending span with a span of 10 mm and the crosshead directed perpendicular to the width of the sample with a crosshead speed of 1mm/minute constantly then wait until the sample fractures [12,14].

2.2. Impact test

A class I cavity was made in the maxillary premolar tooth then put the premolar tooth in block-shaped acrylic on 1/3 of the tooth root with a size of 5x3x0.5 cm and the tip of the root visible. Maxillary premolars were prepared using a round bur with a bur diameter of ± 1.5 mm to the extent that the round bur could enter the cavity, then the cavity walls were flattened using a fissure bur with a depth of 4 mm and an area of 2x3 mm^2. Application of flowable cellulose coir composite to premolar tooth cavities. Teeth that have been prepared are then divided into 2 groups. The first group used RMGIC and the second group used particulate coir cellulose composite resin. The composite resin was applied as an intermediate layer 2 mm thick using a syringe. Then, light cure was carried out for 40 seconds with the light direction perpendicular to the intermediate layer. The intermediate layer that has been applied was closed using 3M packable composite according to the anatomical shape of the premolar teeth. In this study, impact strength testing was carried out using a KRY type Mini Impact Tester. The sample was placed on the cross section, then the pendulum or load was raised by α (the angle of the pendulum before it swings). Then the pendulum was released so that it hit the specimen and the final angle of the pendulum was recorded. After hitting the specimen, the pendulum would still swing by β (pendulum angle after swinging) [16].

2.3. Compressive Strength

In this research, compressive strength testing was carried out using a Universal Testing Machine. The way this tool worked that a sample in the form of a cylindrical tube was placed right in the middle of the test table. The test eye or indenter was placed in a position directly above the sample so that the tip of the test eye touches the surface of the sample. Then pressure was applied by moving the arrowhead at a speed of 1mm/minute so that the test eye moved down and presses the sample until the sample fractures. Observe fractures in the sample which could be identified by hearing a cracking sound in the sample. If the sample had broken, the pressure application was stopped. Record the number on the digital screen of the measuring instrument when a fracture occurred, which was the force (F) on the Newton scale. Carrying out the same test on samples then converting the data to the Mpa scale with the formula: Compressive strength (MPa) = 4F/□d2 [17,18].

3. Results and Discussion

3.1. Flexural Strength And Modulus Elasticity

Search that has been carried out, namely the flexural strength and modulus of elasticity of coconut coir cellulose composite resin formed from particulate fiber, showed the results in the form of numbers in units of N/mm^2.

Table 1. Mean value and standard deviation

Group	Mean ± Std. Deviation (N/mm^2)
Treatment	39 ± 11
Control	40 ± 12

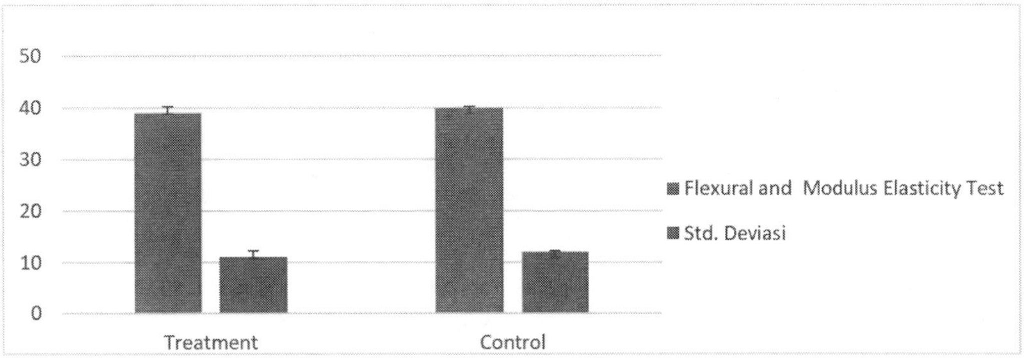

Figure 1. Descriptive test result (Mean values ± standart deviations)

Based on the results of hypothesis testing with the Independent Sample T-Test Unequal Variance, the significance test value was 0.092 ($p>0.05$). So, it could be concluded that there was no significant (real) influence or difference between the 70% particulate fiber-formed coconut fiber cellulose composite resin and RMGIC. This meant that the flexural strength and elastic modulus of coir cellulose composite resin as an intermediate layer were the same as RMGIC

3.2. Impact Test

The research that has been carried out, which were the impact strength of coir cellulose composite resin and RMGIC, showed the results in $Joule/mm^2$ units.

Table 2. Mean value and standard deviation

Group	Mean ± Std. Deviation ($Joule/mm^2$)
Treatment	**0,0283 ± 0,00983**
Control	**0,0271 ± 0,00908**

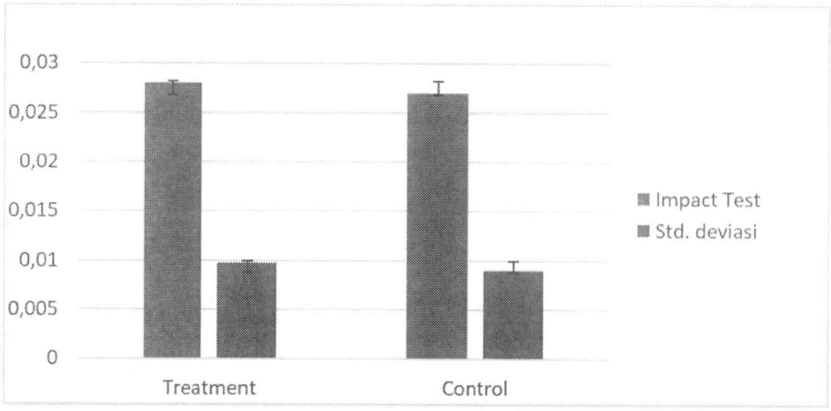

Figure 2. Descriptive test result (Mean values ± standart deviations)

Based on the results of hypothesis testing with the non-parametric Mann-Whitney test obtained a significance value of $p=0.065$ ($p>0.05$). The resulting significance value showed that there was no significant difference between the two groups from the results of the impact strength test. This meant that the impact strength value of coir cellulose composite resin as an intermediate was the same as RMGIC.

3.3. Compressive Strength

Table 3. Mean value and standard deviation

Group	**Mean ± Std. Deviation (MPa)**
Treatment	**0,0283 ± 0,00983**
Control	**0,0271 ± 0,00908**

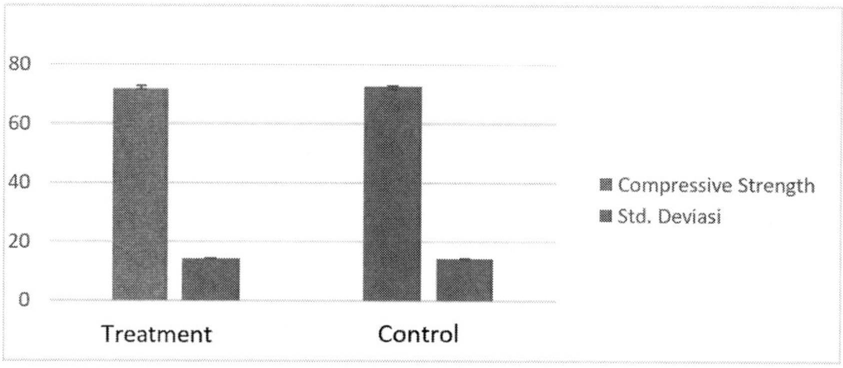

Figure 3. Descriptive test result (Mean values ± standart deviations)

The results of the parametric independent sample t-test have a significance value of $p=0.850$ ($p>0.05$). The results showed that the compressive strength of the control (GIC) and treatment groups (coconut fiber composite filler) was not significantly different. The results of the parametric independent sample t-test, which showed no significant difference in compressive strength between the control (GIC) and treatment groups (coconut fiber composite filler), align with the broader context of intermediate layer materials in dentistry. These materials, such as RMGIC, are developed to prevent microleakage due to fractures after restorative treatment, and their flexural strength and elastic modulus values are crucial for maintaining the integrity of restoration adaptations

The use of intermediate layer materials in the field of dentistry has been widely developed to prevent microleakage due to fractures after restorative treatment with the aim of maintaining the integrity of restoration adaptations [19]. Based on the results of descriptive research, it was found that RMGIC has the same average flexural strength and elastic modulus values as the coconut coir cellulose filler composite formed from particulate fiber. Based on hypothesis testing using the Independent T-test, it was found that there were no significant differences in the flexural values and elastic modulus between the treatment group and the control group. Flexural strength is very important for an intermediate layer material, which is the ability of a material before it breaks or undergoes fracture. The high value of flexural strength is expected to be able to prevent microleakage [20]. The minimum flexural strength value that light-polymerized composite resin must have according to ISO 10477 (2020) is 50 N/mm^2, while the RMGIC is 40-60 N/mm^2.

This research used samples of coconut fiber cellulose filler composite resin formed by fiber particulates and RMGIC. The composition of the coconut coir cellulose filler composite resin particulate fiber consists of 70% fiber and 30% matrix, which were 25% Bis-GMA and 5% TEGDMA. The consistency of the coconut coir cellulose filler composite resin formed from particulate fiber was flowable, so the viscosity of the composite was low, and has high flexibility so it was easy to apply[17,20]. RMGIC was used as a positive control group because RMGIC is a composite resin material consisting of a combination of powdered glass fluoro-alumino silicate particles and liquid methacrylate resin, polyacid, HEMA, water, and initiator [2,19]

The impact test is the energy required to break the material under impact or impact force. The impact test can be used to describe the reaction of a stationary object to a sudden impact with a moving object. This impact can be used to describe trauma from a fall or a blow that hits the dentition area[16,19]. The impact test is used to simulate the material conditions of materials that are often encountered where the load comes suddenly, a material is said to be strong if its ability to accept shock loads is large enough [20]

Based on the results of research and descriptive analysis, it was found that the impact strength value of cellulose coir composite had an average of 0.0283 joules/mm^2 while RMGIC had an average of 0.0183 joules/mm^2. Based on the hypothesis test, the impact strength value using the Mann-Whitney test there was no significant difference between RMGIC and coir cellulose composites. One of the things that can influence the impact strength of a material is whether or not there are pores in the material. If the surface of a material is smooth, it will make the material less brittle and its mechanical properties will be better [12,15]. Apart from that, the factors that influence the impact strength of composite resin are the resin matrix, filler, coupling agent and the length of exposure used [14]

Composite resin restorations require the application of an intermediate layer placed between the restoration and the cavity base to form adhesion that can prevent shrinkage due to polymerization. The use of an intermediate layer with a low elastic modulus causes stress due to polymerization to spread evenly [5,19]. When the impact test was carried out, the first part of the sample that was damaged first was the teeth. The condition of the intermediate layer in the treatment group tended to be more intact than in the control group. In line with research, composite restorations on posterior teeth incorporating RMGIC as an intermediate layer resulted in many fractures [7,11].

Coconut fiber plant fiber has good strength for composite resin. Flowable composites can flow easily, spread evenly and are able to adapt to the shape of the cavity. In addition, flowable composites have great flexibility [8,14]. In this research, coir cellulose composite resin with a filler ratio of 70% is flowable, so that during the application process as an intermediate layer it could flow according to the basic shape of the preparation cavity. Therefore, coir cellulose composite resin can reach marginal gaps which can prevent marginal leakage due to polymerization and thus reduce the occurrence of fractures.

Compressive strength is an important parameter for the success or failure of applying restorative materials to the tooth structure, which can result in the formation of marginal gaps and fractures. The compression test method used in this research was static movement. The tool used was a Universal Testing Machine (UTM) with the Lloyd Instrument model, LRX Plus, Fareham UK combined with the EZ Series and Nexygen Plus programs. The sample specimen in this study was a composite of coconut fiber filler and RMGIC which was molded in a cylindrical shape with a height of 6 mm and a diameter of 4 mm to meet ISO 10477 standards. The descriptive research results showed that the average compressive strength of the treatment group (coconut fiber filler composite) was 64.9 MPa, while the control group (RMGIC) obtained a value of 78.5 MPa. Hypothesis testing shows that the compressive strength of RMGIC compared to coconut fiber filler composite does not show a significant difference. Compressive strength can be influenced by the thickness of the sample and the degree of polymerization, the thicker it is, the more light intensity will be prevented from reaching the tensile base composite [12,16]. The less homogeneous shape, size and smoothness of the sample also influences the reduction in compressive strength test results.

RMGIC has certain strength specifications with the advantage of containing hydroxylmethacrylate (HEMA) so that it can achieve perfect polymerization supported by an acid-base reaction which plays a role in the polymerization reaction on parts that are not exposed to light cure rays and improves mechanical properties. RMGIC as an intermediate layer material has the advantage of being able to bond chemically with the tooth structure, micromechanically with the composite, and release fluorine. Composites with particulate coconut fiber filler which has a compressive strength value that is not much different from RMGIC can be an option because it is environmentally friendly because it is easily decomposed. The strength bases parameter according to Power, et al has a range of 5.5-96 MPa. Based on this and the results of the compression test, the compressive strength of the particulate-formed coconut fiber composite filler can be a base material for restorative treatment [18,20].

4. Conclusion

There was no differences in mechanical strength between coconut particulate fiber 70% cellulose filler composite resin (*Cocos nucivera L.)* and RMGIC. This research supports the development of natural treatments and natural resources.

Acknowledgment

The authors would like to thank Hang Tuah University for supporting this research.

References

[1] Farrel, D. A., Yulianto and Zulfitriyanto (2022). *Indonesian Journal of Social Technology*, **3** (2), pp.1–16.

[2] Kondo, Y. and Arsyad, M. (2018). *Proceedings of the Research Results Seminar (SNP2M),* 2018 **1**, pp. 40–44

[3] Sakaguchi R, Ferracane J, Powers J. Craig's (2018). Elsevier Inc, pp. 1–31 .

[4] Pertiwi OS, Darmawangsa D, Widyawati W (2019). Journal of Dentistry, Baiturrahmah University ; **4** (2), pp. 104–9

[5] Fronza BM., Ayre APA., Pacheco RR (2017). Oper Dent.Jul-Aug:**42** (4), pp. 445- 455.

[6] Zulkifli and Dharmawan, I. B (2019). *Jurnal Polimesin*, **17**, No 1 (2019): Polimesin), pp. 41–46.

[7] Maulida, F., Sari, W. P. and Darmawangsa, D. (2019). *University Journal of Dentistry Padjadjaran*, **31** (1), pp. 43–46.

[8] Cevanti, TA, Sri NSP, Isnaini SI, Rois MF, Setyawan H, Soetojo A, Widjiastuti I. (2021). Journal of International Dental and Medical Research ISSN, **14,** pp. 1401–1406.

[9] Lumintang R, Marpaung JL, Sutrisno A (2017). J Online Poros Tek Mesin **6** (2), pp. 151–62.

[10] Shen C, Rawls HR, Esquivel-Upshaw JF (2022). Phillip's Science of Dental Material. ELSEVIER, pp. 231-278

[11] Setiawan A, Santoso E, Kusuma GE (2018). Semin MASTER PPNS. 2018. **1509,** pp. 95–98.

[12] Garoushi S, Gargoum A, Vallittu PK, Lassila L (2018). J Invest Clin Dent. 1-9.

[12] Younis S, Alaa E (2020). Egypt Dent J. 2020; **66** (3), pp. 1837–43.

[13] McCabe JF, Walls AW (2015). Ninth Edition. ninth. Blackwell. Blackwell Munksgaard, pp. 176–303

[14] Dessy Adhira Nur Safitri, Eko Hadianto, Erdianto Setya Wardhana (2022). 4(March), pp. 59–64.
[15] Nurhapsari, A. and Kusuma, A. R. P. (2018). *ODONTO Dental Journal*, **5** (1), pp. 67–75.
[16] Galvani LD, Bordini EAF, Otarola WGE, Morais JMP, Santos JREV, Dantas AAR, Kuga MC (2020). Dental Press Endod. **10** (1), pp. 32-7.
[17] Martha M, Herda E, Soufyan A (2010). PDGI ; **59** (1). pp. 29–34.
[18] Hadianto E, Syifa LL, Hanafie HF (2018). ODONTO Dent J, **5** (2). pp. 139-179.
[19] Rosenstiel S, Land MF, Fujimoto J (2016). 5th ed. Elsevier. pp: **43**

The Impact of *Pluchea indica L.* Leaf Ethanol Extract on In Vitro Biofilms of Microorganisms that are Associated with Water Pollution

V Yudo[1], D Irawati[1*], and R Budiarti[1,2]

[1] Department of Microbiology, Faculty of Medicine, Hang Tuah University
[2] Head of Microbiology Departement, Faculty of Medicine, Hang Tuah University

[*] Email: desyirawati.fkuht@gmail.com

Abstract. The presence of *Escherichia coli* is one of the microbiological indicators of water pollution. By forming biofilms, these bacteria possess the capacity to tolerate and adapt to very challenging conditions. In terms of water contamination, the formation of *Escherichia coli* biofilms is very important. Co-infection with *Candida albicans*, which can also form biofilms, can worsen the situation. The etanol extract of *Pluchea indica L.* leaves contains flavonoids, which are bioactive substances with antibiofilm potential, along with tannins, saponins, terpenoids, and fenolic acid. The purpose of this study is to evaluate the effect of etanol extract from *Pluchea indica L.* leaves as an antibiofilm agent in *Escherichia coli* and *Candida albicans*. This study was conducted using the Microtitter Plate Biofilm Assay method by reading the optical density (OD) value using a microplate reader with a wavelength of 595 nm. The results of this study indicate that increasing the concentration of extract will cause a decrease in biofilm as indicated by a decrease in OD. With the One-Way ANOVA test, $p = 0.000$. This study concludes that the extract of *Pluchea indica L.* leaves has the capacity to inhibit microbial biofilms of *Escherichia coli* and *Candida albicans*.

1. Introduction

By 2022, it is anticipated that over 1.7 billion people throughout the world would be consuming water that has been contaminated with feces. Contamination of drinking water by microorganisms found in feces is the most significant threat to public health [1]. Coliform groups of bacteria have long served as an indicator organism for microbial contamination in water, shaping the historical understanding of public health security. *Escherichia coli* serves as a key indicator of fecal contamination among coliform bacteria [2].

Peritrichous flagella are often responsible for the motility of *Escherichia coli*, which is a Gram-negative bacteria that does not produce spores. In addition to being the cause of acute enteritis in humans, *Escherichia coli* is also the general cause of "traveler's diarrhoea," which is a condition that is similar to dysentery, and haemorrhagic colitis, which is sometimes referred to as "bloody diarrhoea". This can occur when people drink water contaminated with *Escherichia coli* [3].

The adhering cells would build a matrix of extracellular polysaccharides that would greatly promote adhesion and cohesion. A biofilm is a cellular population that attaches to diverse biotic and abiotic surfaces where it would adhere [4]. Specifically, *Escherichia* coli may quickly grow a biofilm on the surface of both living and non-living carriers, which increases the risk of water contamination

[5]. In addition, every strain has the ability to adhere itself to a surface and replicate in order to build a biofilm structure [6].

Three municipalities in the midwestern Brazil experienced the pollution of drinking groundwater from residence wells in 2022 due to the presence of *Candida spp* [7]. Polymicrobial interkingdom interaction can result in biofilm formation and infection development. Biofilms correlated with polymicrobial illnesses, such as those originated from bacteria and fungus, have recently attracted worldwide attention. [8].

Pluchea indica L. (Less) is a plant species in the Asteraceae family. *Pluchea indica L.* widely grown in Indonesia and and is commonly known as 'beluntas'. The ethanol extract of Beluntas leaf (BL) contains flavonoids, which are bioactive substances with antibiofilm potential, along with tannins, saponins, terpenes, and phenolic acid [9]. Several reports show that ethanol extract of BL has proven to have antibacterial and antibiofilm activity [10].

This study aimed to determine that the ethanol extract of beluntas leaf (BL) can inhibit the formation of biofilms from *Escherichia coli* and *Candida albicans* in vitro.

2. Material and Methode

Beluntas leaf powder comes from Materia Medica Batu, UPT Herbal Laboratory, East Java Provincial Health Service, and is then extracted at the Microbiology Laboratory of the Surabaya Pharmacy Academy. A maceration extract of Beluntas leaf was made by utilizing a solvent that was concentrated using a vacuum rotary evaporator and containing 96% ethanol. This process was repeated until a thick extract was obtained.

The concentration of ethanol extract of BL was 100%, 50%, 25%, and 12.5%. The samples used were *Escherichia coli* ATCC 25922, and *Candida albicans* ATCC 10231. Muller Hinton Agar (MHA) medium was used to cultivate *Escherichia coli* and *Candida albicans*, which were then incubated at 37 degrees Celsius for a period of twenty-four hours. In addition, a suspension of isolates was prepared using 0.5 McFarland standards, which is equivalent to 1.5×10^8 CFU/ml.

Using a Microtiter Plate Biofilm Assay, the inhibition of biofilm formation was tested. Bacteria suspension followed by culture in every microtiter plate (a 96-well polycarbonate tissue culture plate with a flat bottom). The number of wells for the test's replication were adjusted and labeled according to concentration of each extract.

After that, add 80 μL of sample solution with concentration variations of 100%, 50%, 25%, and 12.5% into each well, 80μL of TSB media, and 40 μL of microbial suspension, then the microplate is incubated at 37°C for 24 hours. After 24 hours, the microplate is washed with PBS 3 times to remove the microbial suspension that does not form a biofilm. Next, 200 μL of methanol is added and left for 30 minutes at room temperature. After that, the methanol is discarded and left again for 10 minutes at room temperature. A total of 200 μL of 1% crystal violet is added into the well and then incubated at room temperature for 15 minutes, after which the microplate is washed with sterile water. Observations are made with a microplate reader (λ=595 nm).

3. Result

Visually, it can be seen that the lower concentration of the extract, the more purple the microplate well will appear (Figure 1). This indicates that the biofilm formed is thicker because it absorbs more crystal violet.

Based on the results of reading the microplate (Table 1), it is possible to see that the optical density (OD) of the biofilm containing *Escherichia coli* and *Candida albicans* decreases according to the increase in the amount of the extract.

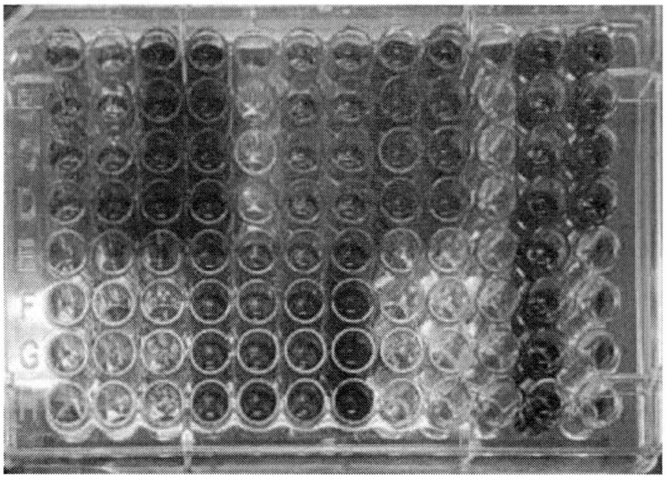

Figure 1. Microtiter Plate Biofilm Assay

The categorization of optical density values in *Escherichia coli* reveals that the biofilm is still strongly adherent at an extract concentration of 12.5%. This is shown by the fact that the biofilm is detected. At an extract concentration of 25%, it can already inhibit biofilm growth. The optical density reduces with increasing concentration of BL extract (Figure 2). And at a concentration of 100%, the ethanol extract of BL can inhibit the growth of *Escherichia coli* biofilm completely. Meanwhile, the OD value in *Candida albicans* indicates that the ethanol extract of BL, at a concentration of 50%, has only just started to inhibit the formation of biofilms.

Table 1. Optical Density Results

Microorganisme	**Concentration of ethanol extract *Pluchea indica* L. leaves**				**Control Group**	
	100 %	**50 %**	**25 %**	**12,5 %**	**(+)**	**(-)**
Escherichia coli						
- R1	0.338	0.366	0.790	1.415	3.900*	0.201
- R2	0.247	0.186	0.783	1.851	3.789	0.238
- R3	0.322	0.309	0.683	1.851	3.891	0.265
- R4	0.248	0.704	0.856	2.293	3.900*	0.256
- Mean	0.288 (N)	0.391 (W)	0.778 (M)	1.852 (S)	3.870	0.239
Candida albicans						
- R1	0.298	0.387	1.990	1.537	3.295	0.283
- R2	0.620	0.513	1.742	1.339	3.589	0.250
- R3	0.300	0.429	1.585	3.133	3.678	0.283
- R4	0.346	0.431	1.721	2.462	3.487	0.214
- Mean	0.391 (W)	0.440 (W)	1.759 (S)	2.117 (S)	3.512	0.257
Escherichia coli + Candida albicans						
- R1	0.338	0.366	0.920	1.415	3.900*	0.238
- R2	0.247	0.186	0.783	1.851	3.900*	0.264
- R3	0.322	0.309	0.823	1.851	3.406	0.268
- R4	0.248	0.407	0.653	2.293	3.476	0.243
- Mean	0.289 (W)	0.317 (W)	0.795 (M)	1.853 (S)	3.671	0.253

Description: * = maximum value of microplate reader, R = replication, (+) = positive control, (-) = negative control, S = strong adherent, M = modetare adherent, W = weak adherent, N = no adherent

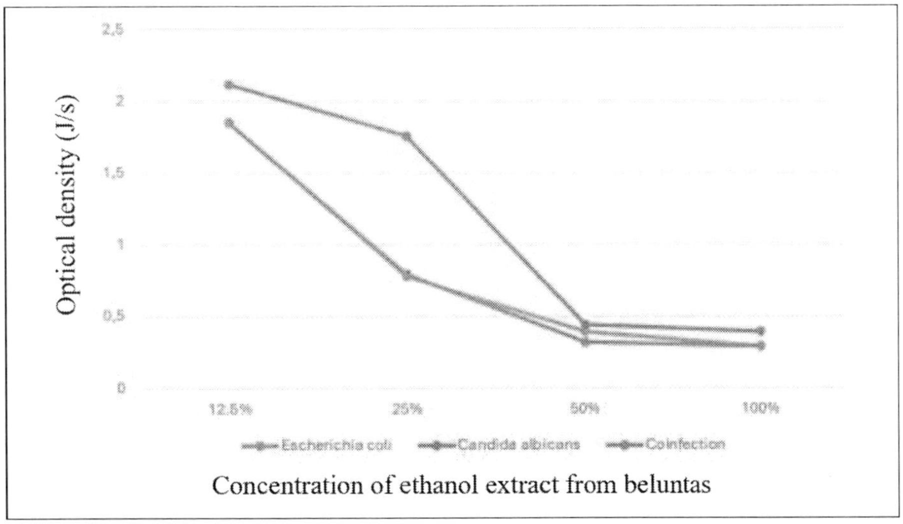

Figure 2. Chart comparing optical density with the concentration of ethanol extract from beluntas leaves

We used the one-way ANOVA statistical test to determine how the OD values vary among treatment groups. The *Escherichia coli* sample showed a significance value (p) = 0.000, meaning (p) <0.05, indicating a significant difference between treatment groups. We then used the Tukey test to determine the real difference between each extract concentration group. Tukey's results for *Escherichia coli* showed a significance value (p) <0.05 between each extract concentration. It can be said that there is a significant difference between each extract concentration when compared to the group without extract (control). This also happened in the *Candida albicans* sample. The *Candida albicans* sample also showed a significance value (p) = 0.000, meaning (p) <0.05, indicating a significant difference between treatment groups. Then, to see the real difference between each extract concentration group, the Tukey test was also used with a significance value (p) <0.05 between each extract concentration. It can be said that there is a significant difference between each extract concentration when compared to the group without extract (control).

4. Discussion

The majority of bacteria that are found in drinking water distribution networks are adherent to the surfaces of pipes, which results in the formation of biofilms. However, in general, only water samples are included in the analysis [11]. Multiple laboratory-scale experiments have shown that *Escherichia coli* can be included in biofilms [12], endure in the biofilm for a maximum of 40 days [13], and remain there in a metabolically active state, even when chlorine levels are measurable [14]. Some writers propose that *Escherichia coli* is capable of thriving in drinking water distribution systems [15] due to its ability to adapt to the oligotrophic environment. In the absence of suitable nutrition, *Escherichia coli* has the capacity to live entirely by using the metabolic byproducts of other bacteria [16].

While doing the microtiter plate experiment, it was observed that both biofilm *Escherichia coli* and *Candida albicans* showed a decrease in the average OD value. The extract of BL contains chemical components that have the capacity to suppress the formation of biofilm.. BL contain flavonoids and polyphenols in their extract.

The BL include flavonoids, which are chemicals that have active properties. Flavone and naringenin are two examples of flavonoids. It has been proven that flavone has the capacity to function as an inhibitor of the development of biofilm. In order to interfere with the quorum sensing signaling pathway, these compounds have the capability of disrupting the contact that occurs between the acyl-homoserine lactone (AHL) and its receptor. These Gram-negative bacteria are responsible for the production of an autoinducer or signaling chemical known as AHL. The mechanism of quorum sensing relies heavily on it as an essential component [17].

Naringenin's function in restricting the quorum sensing system is most likely owing to its capacity to interact with and diminish the amounts of AHL molecules and the Lux-R type transcription factor. This, in turn, causes a reduction in the expression of genes associated with quorum sensing [18]. Quorum sensing is a regulatory mechanism that controls the production of extracellular polymeric substances (EPS), also known as polysaccharides, and is involved in the creation of bacterial biofilms. Inhibiting the quorum sensing pathway leads to the inhibition of EPS production, which in turn inhibits the formation of bacterial biofilms [19].

BL contain not only flavonoids but also a substance known as polyphenol. Polyphenols contain anti-adhesive qualities that enable them to prevent the formation of biofilms. The study findings indicate that bacterial adhesion is a contributing component in both bacterial colonization and biofilm development. Bacteria have the capability to attach to host cells even when adhesin is present. Adhesin and polysaccharide matrix are responsible for attaching the biofilm matrix to the surface of a substrate, allowing it to form three-dimensional structures and mature over time. Polyphenols have the ability to attach themselves to the collagen binding protein, thereby inhibiting the activity of certain genes in bacteria. The rupture of the collagen binding protein will lead to a decrease in the ability of bacteria to bind to extracellular matrix proteins, which will consequently prevent the formation of biofilms [20].

The high levels of optical density (OD) shown in the *Escherichia coli* and *Candida albicans* groups were attributed to minor impairments in biofilm formation. The resistance of *Escherichia coli* and *Candida albicans* to antibacterial or antibiofilm agents may be the cause. Both *Escherichia coli* and *Candida albicans* have the ability to aggregate and create biofilms, which can make them a thousand times more resistant to antibacterial agents. When bacteria create a biofilm, their genetic and metabolic processes undergo a transformation, resulting in the development of a complex matrix that serves as a defense mechanism against antimicrobial agents [21].

5. Conclusion

This study concludes that the extract of *Pluchea indica L.* leaves has the capacity to inhibit microbial biofilms of *Escherichia coli* and *Candida albicans*.

References

[1] World Health Organization 2024 *Guidelines for drinking-water quality: small water supplies* (Geneva: World Health Organization)

[2] Some S, Mondal R, Mitra D, Jain D, Verma D, Das S 2021 *Energy Nexus* **1**

[3] Percival SL and Williams DW 2014 *Escherichia coli Microbiology of Waterborne Diseases: Microbiological Aspects and Risks: Second Edition* (Boston: Elsevier Ltd.) Chapter 6 pp 89–117

[4] Asih DW, Widodo ADW, Setiabudi RJ, Sulistiawati, Tyasningsih W and Wahyunitisari MR 2023 *Journal of Advanced Biotechnology and Experimental Therapeutics* **6** 84–93

[5] Zhou F, Wang D, Hu J, Zhang Y, Tan BK and Lin S 2022 Control Measurements of *Escherichia coli* Biofilm: A Review *Foods* **11**

[6] Abberton CL, Bereschenko L, van der Wielen PWJJ and Smith CJ 2016 *Appl Environ Microbiol* **82** 5320–31

[7] Cupozak-Pinheiro WJ, Araújo de Almeida-Apolonio A, Sasaki MH, Maran NH, Pires de Araújo R, and Beraldo dos Santos Silva D 2022 *Microb Pathog* **169**

[8] Puspitasari F, Widodo ADW, and Purwanta M 2024 *Tropical Journal of Natural Product Research* **8** 7259–63

[9] Chiangnoon R, Samee W, Uttayarat P, Jittachai W, Ruksiriwanich W, and Sommano SR 2022 *Molecules* **27**

[10] Farhamzah, Herli A, and Mursal ILP 2021 *IOP Conf. Series: Materials Science and Engineering* **1071** 012013

[11] Berry D, Xi C, and Raskin L 2006 *Curr Opin Biotechnol* **17** 297–302

[12] Robinson PJ, Walker JT, Keevil CW and Cole J 1995 *FEMS Microbiol Lett* **129** 183–8
[13] Armon R, Starosvetzky J, Arbel T and Green M 1997 *Water Science and Technology* **35** 293–300
[14] Williams MM and Braun-Howland EB 2003 *Appl Environ Microbiol* **69** 5463–71
[15] Camper, AK, Mcfeters GA, Characklis WG and Jones' WL 1991 *Appl Environ Microbiol* **57** (8) 2233-39
[16] Szewzyk U, Manz W, Amann R, Schleifer KH and Stenström TA 1994 *FEMS Microbiol Ecol* **13** 169–76
[17] Manner S, Skogman M, Goeres D, Vuorela P and Fallarero A 2013 *Int J Mol Sci* **14** 19434–51
[18] Vandeputte OM, Kiendrebeogo M, Rasamiravaka T, Stévigny C, Duez P, and Rajaonson S 2011 *Microbiology (N Y)* **157** 2120–32
[19] Vu B, Chen M, Crawford RJ and Ivanova EP 2009 *Molecules* **14** 2535–54.
[20] Borges A, Abreu AC, Malheiro J, Saavedra MJ, and Simões M 2013 *Microbial Pathogen and Strategies for Combating Them: Science, Technology and Education* (Badajoz: Formatex Research Center) pp 32-41
[21] Pargaputri AF, Munadziroh E, and Indrawati R 2017 *Dental Journal* **49** 93

Microstructures and chemical compositions of scallop shells from Probolinggo, East Java

Kiki Syaputri Handayani[1], Victor David Nico Gultom[1], Muhammad Firdaus[1], Sunarti Sinaga[1], Fawzan Bhakti Soffa[1], Asna Adira Finan[2], Rini Suphia Nuryati[2], Isom Hilmi[2*]

[1]Research Center for Marine and Land Bioindustry, Research Organization for Earth Sciences and Maritime, National Research and Innovation Agency (BRIN), Pemenang 83352, Indonesia
[2]Department of Physics, Faculty of Applied Science and Technology (FAST), Universitas Ahmad Dahlan, Yogyakarta 55166, Indonesia

*Email: isom.hilmi@fisika.uad.ac.id

Abstract. Asian moon scallops (*Amusium pleuronectes*) are harvested in Indonesia and South East Asia Region as seafood. This species can be found in Probolinggo, East Java and they are especially abundant during January-March. Apart from its muscle, the shells of the scallops can also potentially be utilized for industrial needs e.g. in the field of biomedical applications. In this paper, the investigations of the microstructure and the chemical compositions of the shells of *A. pleuronectes* with different sizes by means of SEM and EDX, respectively, are reported. The inspection on the microstructures of the surface and cross-sections of the shells using SEM reveals different microstructures of scallop shells of different sizes. According to the EDX measurements, the chemical compositions of the shells consist of predominant Ca contents. A slight difference in chemical compositions of the shells of different scallop's sizes is also observed. This result gives an insight into the physical properties of the local species of scallops and its great potential especially in biomedical applications.

1. Introduction

The maritime areas of Indonesia consist of an extensive coastline. Some areas of the beach have a severe problem with scattered shell trash. The majority of these shells are disposed of in landfills or public waters, bringing about environmental pollution [1]. The mollusc shells are the great solid waste materials released to the environment annually [2], microbial degradation of these shells releases highly poisonous gases, such as NH_3 and H_2S [3]. This condition exacerbates health and sanitation difficulties, harms natural landscapes, and emits unpleasant scents from the breakdown of organics connected to the shells.

In contrast to the common practice of discarding the shell, scallop shell waste is a potential source of calcium carbonate ($CaCo_3$), that is renewable [4] and abundant source of calcium carbonate ($CaCO_3$), constituting 98-99% of the total weight, with the remainder being minor metal oxide impurities [3], [5]. This $CaCO_3$ can be converted into hydroxyapatite (HAp) using phosphoric acid [6]. HAp derived from scallop shells has potential uses in producing inorganic membrane plates for water purification, gas separation, and catalysis [7]. Additionally, HAp and calcium from scallop shells can be applied in biomedical engineering for bone tissue engineering, drug delivery

[8], and biosensors [9]. When $CaCO_3$ from scallop shells is subjected to heat above 700°C, it forms tiny particles. The resulting hydroxyapatite (HA) particle size is less than 45 μm, indicating a highly crystalline phase with irregular particle shape. This is important for the biomedical applications. This particle size affects the bioactivity and integration of HA in biological systems [5], [10], which is useful for use in hard tissue repairs and biomedical [11] with numerous applications [12], [13]. Being popular in Indonesian seafood restaurants with significant shellfish catch and consumption reported in recent years, scallops are rich in protein, essential fatty acids, omega-3, and vitamins [14]. The mineral content of scallop shells is influenced by various factors such as species, habitat, age, size, region, and site-specific environmental conditions [15], [16], [17], [18]. Concentrations of heavy metals in water frequently lead to increased levels in biota bodies. Simple linear regression tests revealed a positive correlation between Cd, Cu, and Zn levels in seawater and metals in mussels in the Indragiri River estuary, but a negative relationship for Pb [19]. According to the study done by Strasser [20], the content of Mn, Sr, and Ba in M.Alaria shellfish is affected by the age of the shellfish, especially in the element Mn:Ca which is influenced by the biological activity of the shellfish, or changes in seawater sources.

The objective of this work is to observe the microstructure and the chemical composition of Asian Moon Scallop's shells (AMS, *Amusium pleuronectes*) that have different sizes using Scanning Electron Microscopy (SEM) and Energy-Dispersive X-ray Spectroscopy (EDX). The AMS is one of 3.000 different species of shells occurring naturally in Indonesia. Its range of occupancies is primarily along the northern coast of Central Java and also along the East Java coastlines. This scallop is part of the *Pectinoidea* family and occurs throughout the Indo-Pacific area [21].

This research provides an insight on the microstructures and chemical compositions of AMS's shells, which could be essential and beneficial for its potential applications. In addition, it can gives insight on the sustainable material cycle of marine ecosystems to improve the quality of environmental stability. Additionally, the activity also supports the Sustainable Development Goals (SDG) especially point 14, namely Maintaining Marine Ecosystems.

2. Materials and Methods

The scallops were collected from two stations around Gili Ketapang, Probolinggo. Six AMS's shells representing three size classes were collected for microstructure and chemical composition analysis. The scallops are categorized into three sizes: small (60-69 mm), medium (70-79 mm), and large (80-90 mm). The SEM and EDX measurements were done to analyse the microstructure and the chemical composition of the shells, i.e. to investigate their correlation in AMS's sizes range to the microstructures and the chemical compositions of the shells. The shell samples of AMS were collected from around Gili Ketapang and subsequently were subjected to the cleaning processes. The left valves of each size class were subsequently fractured in half and cross-sectioned. The interior side of the left valve was investigated by means of SEM. Qualitative energy dispersive X-ray (EDX) spectra of samples were recorded using an Edax (AMATEK Z1 Analyzer) EDX detector mounted on the SEM. This work is basically to investigate the structural and compositional changes of scallop shells in relation to their sizes.

3. Result and Discussion

Firstly, the surface morphologies of the shell were observed using SEM. Figure 1 shows the SEM images of the inner surface of AMS shells of various sizes acquired with the magnification times of 1000×. More specifically, Figures 1a, 1b, and 1c show the SEM images of that of the small, medium, and large scallops, respectively. As seen in Figure 1a, the appearance of the inner-surface topography of the shell of the smallest-sized (youngest in age) AMS looks continuous and smooth. The lined structures are also visible. The surface of the medium-sized AMS's shell looks rather different in which slightly cross-hatches are visible as shown in Figure 1b, forming a bit rougher surface topography. In Figure 1c, showing the SEM image of the inner surface of the largest AMS's shell, it could be discerned that lined structures begin to appear on the surface. Thus, the larger the AMS, the more lined structures are visible on the surface, whereas the smaller the AMS, the smoother the surface. Looking at the line structures at all the three images, it can be noticed that the larger the scallop size, the thinner these layering structures, i.e. the smaller the distances between two adjacent lines. This might also correspond to that the number of the lines is proportional to the size of a scallop. To this point, it can be concluded that the surface topographies of the AMS's inner shells consist of lined-like structures, with different concentrations of micro-sized particulates on the surfaces.

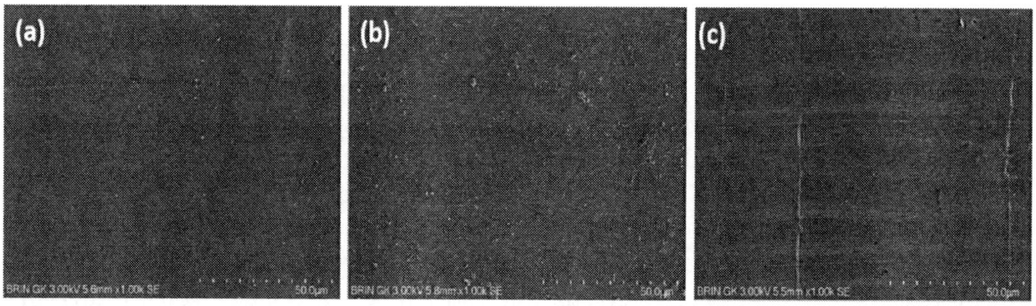

Figure 1. SEM images of the inner surface of Asian Moon Scallops shells from different sizes: (a) small; (b) medium; (c) large.

To comprehend the shell structure's changing tendencies, details on the microstructure can be regarded as the most significant among the relative data [22]. The mechanism behind the ionization (Ca^{2+}) in the environment is poorly understood, ultimately transporting it through the outer surface epithelium (OME) to the growth area of the shell [23], [24]. The ionization of Ca^{2+} has an important role in biomineralization. It refers to the hard tissue formation process of the organism and is characterized as a highly controlled and functional process. Biomineralization can provide a usual metal ion environment related to controlling heavy metals in coastal areas, such as monitoring the crystalline structure of the transition metal ions in the water [25]. The age of the scallop can be estimated by counting the growth rings. Most species form a ring every year, nevertheless, environmental factors may cause variations such as water quality, concentration, and quality of phytoplankton available in their habitat of specific seasons [26].

IOP Conf. Series: Earth and Environmental Science **1473** (2025) 012053 doi:10.1088/1755-1315/1473/1/012053

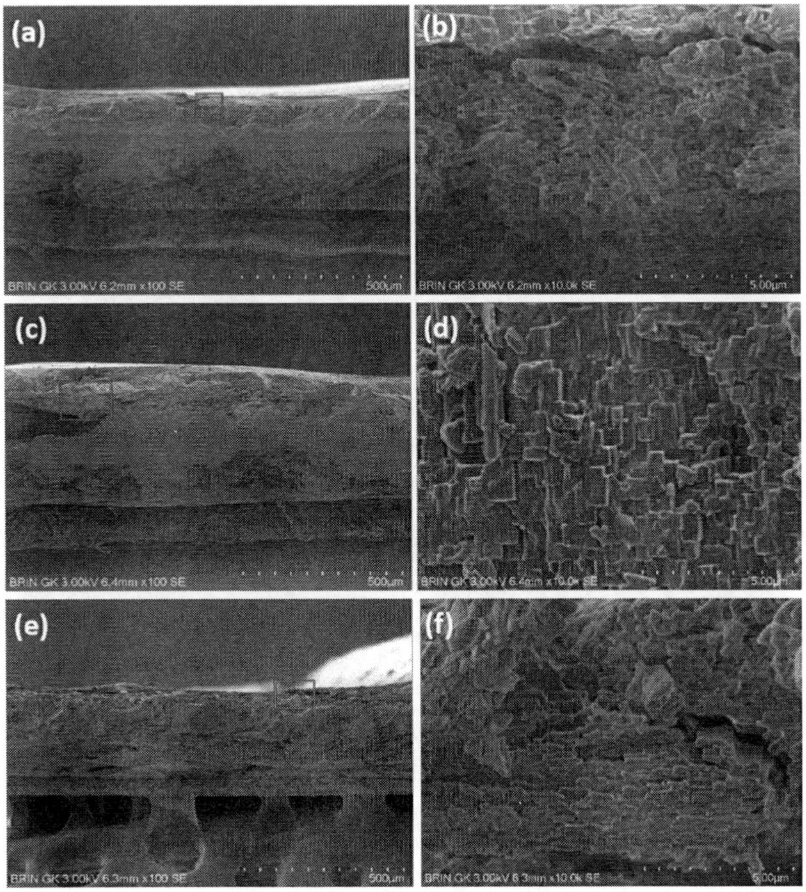

Figure 2. Cross-sectional mode SEM images of the scallop's shells with various scallop sizes.

Furthermore, the microstructures of the shells were also investigated using SEM in cross-sectional mode. Figure 2 shows SEM images of the cross-sections of the AMS shells at the magnification of 100x and 10,000x, as presented at the left and right columns, respectively. The shells have been classified from smallest to largest sizes. Figure 2a shows the SEM images of the scallop's shell with a small height size of 60-69 mm, while Figure 2b shows the magnified image of the area in the rectangular marking in Figure 2a. The area being closely inspected is the area near the surface of the shell because it shows the most pronounced difference between one to another sample. In Figure 2b, an irregular arrangement of overlapping rectangles forming blob-like patterns can be discerned. Similarly, Figures 2c and 2d show that of the medium-sized scallops. In Figure 2d, the pattern in the SEM image appears as an irregular arrangement of thin vertical rectangles glued together. In Figures 2e and 2f, the lower and higher-magnified SEM images of the large-sized scallop are shown. In Figure 2f it looks like a dense arrangement of bricks at the bottom, while there is an irregular arrangement at the top. From the SEM images of the shells of the different-sized scallops, it can be concluded that the microstructures somewhat changed according to the scallop's sizes.

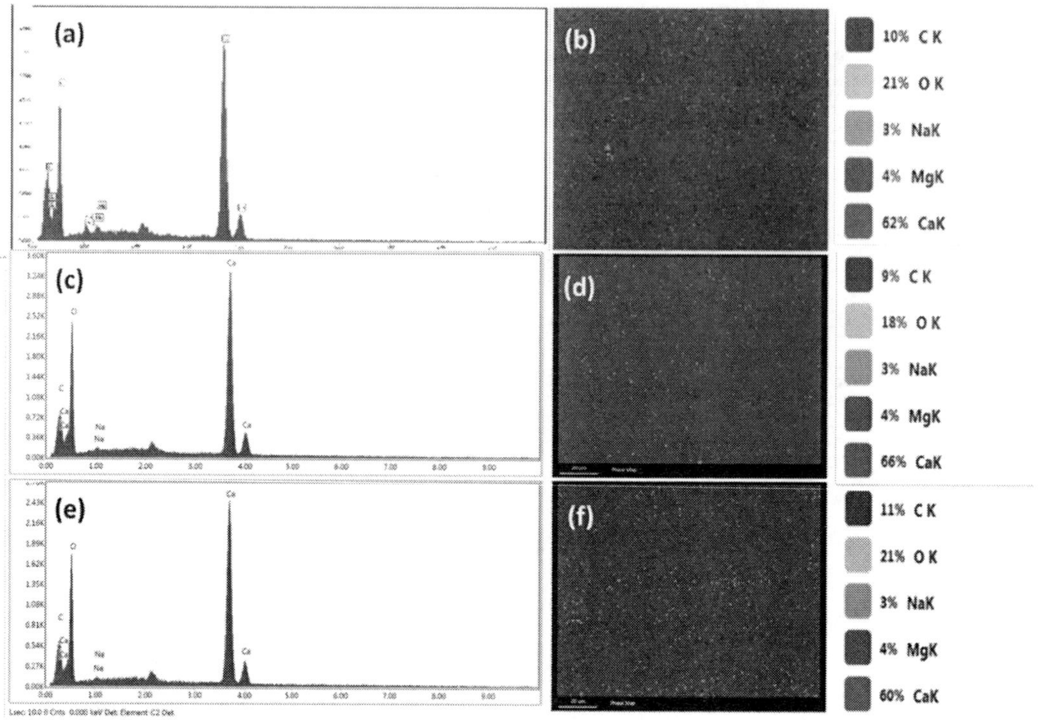

Figure 3. EDX analysis and elemental mapping of the scallop shells at various sizes: (a,b) small; (c,d) medium; (e.f) large.

Figure 3 presents the results of the EDX measurements showing the relative contents of the elements of Ca, O, Na, Mg, and C of the small, medium, and large-sized AMS shells. Figures 3a and 3b show a dominant Ca content of 62%, followed by O (21%), C (10%), Mg (4%), and Na (3%). Figures 3c and 3d show the results of the medium-sized AMS's shell content. Ca is the highest peak with 66%, followed by O (18%), C (9%), Na (3%), and Mg (4%). Figures 3e and 3f show that the large AMS contains the most Ca (60%), O (21%), and C (11%). Comparing the Ca contents of the three samples, it could be noticed that the medium-sized AMS has the highest relative peak intensity of Ca (66%). However, the large scallops have a high frequency in the chemical distribution of the minor elements. To this point, it can be concluded that the medium-sized AMS has the highest Ca content compared to that of the other sizes. These shell chemical discrepancies between small, medium, and large-size AMS might indicate different physiological processes that affect AMS's ability to absorb Ca^{+} from the environment during shell calcification [20]. The appearance of calcium in EDX curves of the scallop shell may be due to calcium carbonate, the primary mineral that makes up the scallop shell [27]. The most common minerals in calcium carbonate are calcite and aragonite [28]. These minerals are considered essential because they play an important role in the formation of tissue and organs [27]. Furthermore, calcium minerals on the shell may be in the form of calcium phosphate, which is widely used in waste remediation due to its adsorbent properties [29], [30]. Because of their composition, which contains a high concentration of $CaCO_3$, they can be used as a low-cost additive or filler in cement production [31].

4. Conclusion

The study observed the presence of calcium carbonate in the shells of these scallops, which showed regular concentrations accompanied by minor elements such as oxygen, carbon, magnesium, and sodium using EDX analysis. SEM and EDX analyses found that as the size of the AMS increased, their surfaces became more layered and showed greater variability in the distribution of the trace elements. The medium-sized AMS's shells contained the highest calcium content of 66%. However, it can be concluded that the shells of the three sizes of AMS contain chemical components, in which they do not show any significant differences between them. These findings highlight the potential of scallops as a sustainable solution to environmental and industrial challenges, serving to renew waste recycling and material recovery.

5. Acknowledgements

The authors would like to thank Riset dan Inovasi Indonesia Maju II, National Research and Innovation Agency (BRIN) and Lembaga Pengelola Dana Pendidikan, Ministry of Finance, Republic Indonesia B-1740/II.7.5/FR/11/2022 and B-15177/III.4/KS.00/11/2022 for supporting this research and innovation. The authors acknowledge the facilities, scientific and technical support from Advanced Characterization Laboratories Yogyakarta, National Research and Innovation Agency through E-Layanan Sains BRIN. IH would like to acknowledge the financial support from Hibah Penelitian Internal UAD 2023 through PDP with contract number of PDP-259/SP3/LPPM-UAD/VIII/2023.

References

[1] K. H. Chung, S. C. Jung, and B. G. Park, "Eco-friendly deicer prepared from waste oyster shells and its deicing properties with metal corrosion," *Environmental Technology (United Kingdom)*, vol. 42, no. 21, pp. 3360–3368, 2021, doi: 10.1080/09593330.2020.1729243.

[2] D. Sahoo, T. Biswal, and S. Acharya, "Synthesis, characterization and biomedical applications of Poly (methyl methacrylate) Chitosan (derived from oyster shell powder)," in *materialstoday:PROCEEDINGS*, ELSEVIER, 2023. Accessed: Aug. 23, 2024. [Online]. Available: https://doi.org/10.1016/j.matpr.2022.10.280

[3] N. Gultekin, "Evaluation of Antimicrobial Activity of Scallop Shell Powder Against Staphylococci Species and Gram Negative Bacteria Isolated From Patients Intensive Care Units," *Biomed J Sci Tech Res*, vol. 13, no. 4, Jan. 2019, doi: 10.26717/bjstr.2019.13.002445.

[4] S. H. Yeom and K. Y. Jung, "Recycling wasted scallop shell as an adsorbent for the removal of phosphate," *Journal of Industrial and Engineering Chemistry*, vol. 15, no. 1, pp. 40–44, Jan. 2009, doi: 10.1016/j.jiec.2008.08.014.

[5] J. Sawai, "Antimicrobial characteristics of heated scallop shell powder and its application (Sawai. 2011)," *Biocontrol Sci*, vol. 16, no. 3, pp. 95–102, 2011.

[6] S. Reni Yenti *et al.*, "Synthesis of hydroxyapatite powder using natural latex particles as pore-creating agent," *Mater Today Proc*, 2023, doi: 10.1016/j.matpr.2023.03.282.

[7] Y. Wibisono *et al.*, "Comparative study of physical properties of hydroxyapatite derived from Asian moon scallop and batik shells using sol-gel techniques for producing inorganic membrane plate," *Chemical Papers*, 2024, doi: 10.1007/s11696-024-03592-2.

[8] A. Szcześ, L. Hołysz, and E. Chibowski, "Synthesis of hydroxyapatite for biomedical applications," Nov. 01, 2017, *Elsevier B.V.* doi: 10.1016/j.cis.2017.04.007.

[9] E. F. Satriawan, C. A. Suryono, I. Widowati, and M. Saputri, "Ecobiology of Asian-moon Scallop Amusium pleuronectes (Linnaeus, 1758) in Kendal Regency Coast, Central Java," *Jurnal Kelautan Tropis Juni*, vol. 27, no. 2, pp. 369–379, 2024, doi: 10.14710/jkt.v27i2.22493.

[10] I. Hilmi, M. Rinastiti, and M. K. Herliansyah, "Synthesis of Hydroxyapatite from Local Bovine Bones for Biomedical Application," in *2011 International Conference on Instrumentation, Communication, Information Technology and Biomedical Engineering*, IEEE, 2011, pp. 1–3.

[11] R. Amalina, R. Ardhani, Y. Yusuf, and H. Susilowati, "Liquid Chitosan-Carbonated Hydroxyapatite http://www.jidmr.com Heni Susilowati and et al Volume • 16 • Number • 2 • 2023 Page 588 Fabrication and Physicochemical Properties of a Novel Gel-Like Liquid Chitosan-Carbonated Hydroxyapatite from Asian Moon Scallop (Amusium Pleuronectes) for Periodontal Application," 2023. [Online]. Available: http://www.jidmr.com

[12] M. Kouzu, A. Kajita, and A. Fujimori, "Catalytic activity of calcined scallop shell for rapeseed oil transesterification to produce biodiesel," *Fuel*, vol. 182, pp. 220–226, Oct. 2016, doi: 10.1016/j.fuel.2016.05.111.

[13] A. Tsuruma, Y. Nomoto, M. Nishio, M. Ishikawa, and J. Sawai, "Efficacy of sorbitol-coated heated scallop-shell powder for the antimicrobial treatment of fresh vegetables," *Food Control*, vol. 110, Apr. 2020, doi: 10.1016/j.foodcont.2019.106972.

[14] A. R. De Setiawan, R. B. Jakaria, I. A. Sari, and T. Sukmono, "Indonesia's Breakthrough in Efficient Shellfish Cleaning Transforms Seafood Industry," *Indonesian Journal of Innovation Studies*, vol. 25, no. 2, Apr. 2024, doi: 10.21070/ijins.v25i2.1133.

[15] H. Yan, J. Chen, and J. Xiao, "A review on bivalve shell, a tool for reconstruction of paleo-climate and paleo-environment," *Chinese Journal of Geochemistry*, vol. 33, no. 3, pp. 310–315, 2014, doi: 10.1007/s11631-014-0692-0.

[16] M. K. Zinchenko, S. I. Zinchenko, M. A. Mazirov, A. O. Ragimov, and A. V. Shitikova, "Biological indicators in the environmental monitoring of gray forest soil of agrosystems," *Caspian Journal of Environmental Sciences*, vol. 19, no. 5, pp. 891–896, Dec. 2021, doi: 10.22124/cjes.2021.5252.

[17] W. M. T. Al-Asadi, A. N. Al-Waheeb, S. A. A. M. Al-Saadi, and S. S. K. Al-Taie, "Effects of Ag and Pb metal accumulation on some biochemical parameters and anatomical characteristics of Sesuvium portulacastrum L. (Aizoaceae) plants," *Caspian Journal of Environmental Sciences*, vol. 20, no. 3, Jul. 2022, doi: 10.22124/CJES.2022.5706.

[18] B. Lakshmanna *et al.*, "Data on Molluscan Shells in parts of Nellore Coast, southeast coast of India," *Data Brief*, vol. 16, pp. 705–712, Feb. 2018, doi: 10.1016/j.dib.2017.11.081.

[19] B. Amin and I. Lestari, "Concentration, distribution and correlation of heavy metals in Asian Moon Scallop (Amusium pleuronectes) and sea water from Indragiri River estuary," in *IOP Conference Series: Earth and Environmental Science*, Institute of Physics Publishing, Nov. 2019. doi: 10.1088/1755-1315/348/1/012001.

[20] C. A. Strasser, L. S. Mullineaux, and B. D. Walther, "Growth rate and age effects on Mya arenaria shell chemistry: Implications for biogeochemical studies," *J Exp Mar Biol Ecol*, vol. 355, no. 2, pp. 153–163, Feb. 2008, doi: 10.1016/j.jembe.2007.12.022.

[21] J. Suprijanto and I. Widowati, "PROSIDING SEMINAR NASIONAL MOLUSKA DALAM PENELITIAN, KONSERVASI DAN EKONOMI Karakteristik Biometrika Kerang Simping Amusium pleuronectes dari Beberapa Daerah di Pantai UtaraJawa Tengah. Jusup Suprijanto dan Ita Widowati," Semarang, 2007.

[22] A. G. Checa, C. Jiménez-López, A. Rodríguez-Navarro, and J. P. Machado, "Precipitation of aragonite by calcitic bivalves in Mg-enriched marine waters," *Mar Biol*, vol. 150, no. 5, pp. 819–827, Feb. 2007, doi: 10.1007/s00227-006-0411-4.

[23] J. Kirsikka Sillanpää, H. Sundh, and K. S. Sundell, "Calcium transfer across the outer mantle epithelium in the Pacific oyster, Crassostrea gigas," *Proceedings of the Royal Society B: Biological Sciences*, vol. 285, no. 1891, Nov. 2018, doi: 10.1098/rspb.2018.1676.

[24] L. Xiang *et al.*, "Amorphous calcium carbonate precipitation by cellular biomineralization in mantle cell cultures of pinctada fucata," *PLoS One*, vol. 9, no. 11, Nov. 2014, doi: 10.1371/journal.pone.0113150.

[25] W. Siriprom, J. Kaewkhao, K. Phachana, and P. Limsuwan, "Crystal structure and morphology dependence of the phase of mollusc shell: A case study of XRD, SEM and ESR," in *Journal of Physics: Conference Series*, Institute of Physics Publishing, 2011. doi: 10.1088/1742-6596/266/1/012124.

[26] A. V. Silina, "Effects of temperature, salinity, and food availability on shell growth rates of the Yesso scallop," *PeerJ*, vol. 11, Feb. 2023, doi: 10.7717/peerj.14886.

[27] O. Kalesaran, C. Lumenta, R. Rompas, and G. Mamuaya, "Mineral composition of the pearl oyster shell (Kalesaran)," *Budidaya Perairan*, vol. 6, no. 1, pp. 25–30, 2018.

[28] P. Southgate and J. Lucas, "The Pearl Oyster," Amsterdam, 2008.

[29] K. S. Vecchio, X. Zhang, J. B. Massie, M. Wang, and C. W. Kim, "Conversion of bulk seashells to biocompatible hydroxyapatite for bone implants," *Acta Biomater*, vol. 3, no. 6, pp. 910–918, 2007, doi: 10.1016/j.actbio.2007.06.003.

[30] S. C. Wu, H. C. Hsu, S. K. Hsu, C. P. Tseng, and W. F. Ho, "Preparation and characterization of hydroxyapatite synthesized from oyster shell powders," *Advanced Powder Technology*, vol. 28, no. 4, pp. 1154–1158, Apr. 2017, doi: 10.1016/j.apt.2017.02.001.

[31] A. Hart, "Mini-review of waste shell-derived materials' applications," May 01, 2020, *SAGE Publications Ltd.* doi: 10.1177/0734242X19897812.

IOP Conf. Series: Earth and Environmental Science **1473** (2025) 012054
doi:10.1088/1755-1315/1473/1/012054

Antibacterial activity of chitosan nanoparticles 4% from blue swimming crab shell waste (*Portunus pelagicus*) against *Staphylococcus aureus*

L Y Yusan[1], Fauliya[1], S Paramita[1], E P Pramesti[1], and K Nurmaningsih[1]

[1]Department of Pharmacy, Hang Tuah University, Surabaya, Indonesia, 60111

*E-mail: liza.yusan@hangtuah.ac.id

Abstract. In this study, chitosan nanoparticles were prepared from waste chitosan of blue swimming crab shell (*Portunus pelagicus*) using ionic gelation method which has the ability to be biodegradable, biocompatible, antimicrobial and non-toxic. The chitosan nanoparticles were characterized by FT-IR spectrophotometry with the deacetylation degree of 84.98%. Chitosan nanoparticles were tested for the purpose of antibacterial action against *Staphylococcus aureus*, which is one of the bacteria that cause gangrene infections in patients with diabetes mellitus. The purpose of this research is to examine the antimicrobial activity of chitosan nanoparticles obtained from blue swimming crab shell waste (*Portunus pelagicus*) against *Staphylococcus aureus*. The method used was antibacterial activity testing with the well diffusion method at concentrations of 4%. The observation results of the antibacterial activity test of chitosan nanoparticles against *Staphylococcus aureus* bacteria were categorized as strong at concentration of 4% with an inhibition zone starting at 10.93 mm. At 4% concentration, the most significant zone of inhibition reached 27.45 mm. At high concentrations, the development of bacteria of *Staphylococcus aureus* was significantly impeded by blue swimming crab shell particles (*Portunus pelagicus*) chitosan nanoparticles, with an inhibition diameter of higher than 20 mm.

1. Introduction

Located in the tropics, Indonesia is a maritime nation with extremely high marine biodiversity, or mega marine biodiversity [1]. The vast area of the ocean and the significant length of the coastline in the country of Indonesia, this makes Indonesia have great potential for marine resources [2]. One of the marine products that are the main stay of Indonesia's exports is blue swimming crab meat (*Portunus pelagicus*) [3]. In the process of taking blue swimming crab meat, blue swimming crab shell waste is produced which reaches 50% to 60% of the crab's total weight [4]. Blue swimming crab shells (*Portunus pelagicus*) contain several useful components, one of which contains chitin (71%) in higher amounts compared to insects and fungi [5]. This chitin can be converted into chitosan through a deacetylation process using a strong base and high temperature [6]. Chitosan is commonly found in *crustacean* exoskeletons, such as crab shells, shrimp shells, and gastropod shells [7]. Research on chitosan from *crustaceans* shows that king crab shells produce the highest percentage of chitosan, which is 87.30%, compared to mud crab shells which have 76.23% and kappa crab shells which contain 79.56% chitosan [8]. This research focuses on the development of chitosan from blue swimming crab shells (*Portunus pelagicus*) because crab is one of the marine products that is widely exported to various countries such as the United States, China, Japan, ASEAN, the European Union, in frozen or canned form. On the east coast of Gresik, crab shell waste reaches 40-60% of the total weight of exported crab [9]. In 2022, the export value of crab increased by 67% compared to 2021, in addition to seaweed [10]. This research is a green economy concept to reduce the risk of environmental damage by handling crab shell waste around the coast by utilizing crab shell waste which contains high chitosan to develop its effectiveness in pharmaceutical technology.

Chitosan processed into nanoparticles can be prepared by the ionic gelation method, which makes it possible to avoid the use of organic solvents and reduce the risk of damage to active ingredients. In crab (*Portunus pelagicus*) shell waste, the nanoparticles produced had the largest particle size of 15.05 nm, which corresponds to the nanoparticle size range between 1-100 nm [11]. Various types of nanoparticles made with different materials and composites have been shown to be effective against various types of bacteria [12,13]. Chitosan has potential as a natural antibacterial because it contains aminopolysaccharide groups and lysozyme enzymes that can inhibit microbial growth. Lysozyme enzyme functions to digest the bacterial cell wall, so that bacteria lose their ability to cause disease because the damaged cell wall results in bacterial cell death [14]. *Staphylococcus aureus* is a pathogenic bacterium that can cause serious infections, such as in patients with diabetic foot ulcers (DFU) [15]. Research by Wada et al. revealed that the most common prevalence of bacterial isolates from DFUs was *Staphylococcus aureus*, with a percentage of 34.34% [16]. Based on these findings, this study aims to test the antibacterial activity of chitosan nanoparticles produced from blue swimming crab (*Portunus pelagicus*) shell waste, focusing on its effectiveness in inhibiting the growth of *Staphylococcus aureus* bacteria that we often find in DFU patients.

2. Method

2.1. Material

Fishermen in Pangkah Kulon Village, Ujungpangkah District, Gresik Regency, East Java, Indonesia, produced waste made from blue swimming crab (*Portunus pelagicus*) shell, where the meat is marketed to export companies in cans. After being cleansed with clean water and prepared at 80°C in the oven, shell samples were ground into a powder using a 60 mesh screen. Determination of *Portunus pelagicus* species was carried out at Airlangga University Surabaya with identification number 20/ULMKILP/UA.FPK/09/2019. The Microbiology Laboratory of the Hang Tuah University Faculty of Medicine and the Pharmaceutical Technology Laboratory of the Hang Tuah University Pharmacy Study Program provided equipment for the research.

2.2. Chitosan preparation

Chitosan was extracted from *Portunus pelagicus* shell waste by means of the deproteination, demineralization, and deacetylation processes [17,18]. Wet crab shells weighing a total of 1000 grams were dried, then pulverized into powder that produced 677 grams. The deproteinization process was carried out with 3% sodium hydroxide solution with a ratio of solute weight per solution volume (w/v), which was heated for 30 minutes at 85°C. This solution was dried for 24 hours at 35°C in an oven and it was neutralized with water to a pH 7. It was then filtered using a *Buchner* filter. At a 1:10 ratio of 1N hydrochloric acid, the procedure of demineralization was carried out for an hour at 75°C with stirring, at the mixture was neutralized to pH 7, filtered, and dried for an additional 24 hours at 35°C. This process produced chitin; remaining calcium carbonate ($CaCO_3$) was eliminated by repeated this step twice. Two forms for deacetylation utilized 40% sodium hydroxide at 100°C for an hour and 5% potassium hydroxide at 150°C for 24 hours. In the deacetylation the process, a 45% sodium hydroxide solution with a 1:20 (w/v) ratio was heated for an hour at 110°C, adjusted to pH 7, filtered, and dried for 24 hours at 80°C in an oven in order to create chitosan [11,17].

2.3. Preparation of nanochitosan

Applying the method of ionic gelation procedure, chitosan nanoparticles were created. First, 0.1 gram of chitosan was dissolved in 100 ml of 1% acetic acid solution, and that mixture was stirring at 1000 rpm until the chitosan dissolved at room temperature. This produced a chitosan solution with a concentration of 0.1%. Subsequently, a 100 ml solution containing 0.1% sodium tripolyphosphate was made and employed as a crosslinking agent. It was stirred until it was dissolved. Drop by drop, the crosslinking solution was slowly added toward the solutions of chitosan in acetic acid by stirring at 1000 rpm for two hours. This process continued until the crosslinking solution was completely consumed. The nanoparticle suspension was then centrifuged at 6000 rpm for 5 minutes. The precipitate formed was washed three times with distilled water, and then dried by freeze drying method for 47 hours [18-20].

2.4. Preparation of bacterial suspension

A single strain of *Staphylococcus aureus* bacteria ATCC 25923 was placed in to a tube with 5 ml of 0.9% sodium chloride solution to create the bacterial suspension, and the method was carried out until the turbidity level reached the *McFarland* standard of 0.5. Using a Shimadzu spectrophotometer, turbidity was identified.

2.5. Antibacterial activity test

Mueller-Hinton Agar (MHA) medium was used in the well diffusion method test for antibacterial activity. First, MHA a medium was put in to a petri dish, Then, 10 µL of the *staphylococcal* bacterial suspension solution was added another petri dish. MHA media that has hardened is then made wells using a 6 mm diameter perforator. Each well was filled with 40 µL of chitosan nanoparticles from blue swimming crab shell waste that had been dissolved in 0.1% DMSO solution. Chitosan nanoparticles were examined at a few concentrations: 100 parts per million (ppm), 200 ppm, 400 ppm, 600 ppm, 800 ppm, and 1000 ppm. Each concentration was carried out 4 times. The positive control used Amoxicillin 500 mg/10 ml water, while the negative control used DMSO 0.1% solution. The inhibitory zone that the formed in the 24-hour incubation duration at 37°C was measured, observed, and characterized at each concentration using Table 1. In addition, this study will be statistically analyzed with ANOVA and Post-Hoc Least Significant Difference (LSD) tests.

Table 1. Classification of bacterial growth inhibition response [21]

Diameter in the inhibition zone	Growth inhibition reaction
≤ 5 (mm)	Mild
6-10 (mm)	Medium
>10-20 (mm)	Powerful
>20-30 (mm)	Extremely Powerful

3. Results and discussion

The drying process of blue swimming crab shell (*Portunus pelagicus*) into fine powder obtained samples of chitosan nanoparticles from waste crab shell. Drying was carried out at the right temperature and time to ensure that the water from the crab shells was released evenly after being washed under running water. The goal is to obtain powder with good texture, dry, and free of lumps. The resulting crab shell powder was then isolated to obtain chitosan through three stages: deproteination, demineralization, and deacetylation [11]. FT-IR spectrophotometry was applied to examine the chitosan nanoparticles derived from crab shell waste. The results showed the highest percentage of deacetylation was 84.98% [11]. This indicates that blue swimming crab shell waste (*Portunus pelagicus*) contains chitosan, because the %DD shows more than 75%. Based on measurements with a Particle Size Analyzer (PSA) in the research of Yusan et al, 15.05 nm is the size of the crab shell waste chitosan nanoparticles with the highest intensity, and its polydispersity index (PDI) value is 0.1140 [11]. This size implies the formation of nanoparticles, each particles occurring between the range 1 nm to 100 nm are categorized as nanoparticles [22]. In addition, the particle's size distribution is shown in the PDI value. A low PDI indicates uniform size distribution and good long-term stability [18].

Chitosan containing lysozyme enzyme and aminopolysaccharide groups can inhibit bacterial growth. Chitosan nanoparticles have more effective antibacterial properties in terms of adsorption due to their large surface and very small size, making it easier to penetrate into bacterial cells. Based on this, an antibacterial activity test was conducted against *Staphylococcus aureus* using the well diffusion method. Table 2 represents the results from the test for the effectiveness of the chitosan nanoparticles derived of blue swimming crabs waste shells (*Portunus pelagicus*) towards *Staphylococcus aureus*, while the average inhibition is shown in Figure 1.

Table 2. Measurement of zone of inhibition of each concentration on chitosan nanoparticles.

Concentration of chitosan nanoparticles (x10^3 ppm)	Inhibition zone replication (mm)				Average±SD	Growth inhibition
	1	2	3	4		
10	11,30	11,10	10,50	10,80	10,93±0,35	Powerful
20	15,25	15,20	15,50	15,60	15,39±0,19	Powerful
40	21,45	21,20	21,10	21,50	21,31±0,19	Extremely powerful
60	24,60	24,25	24,10	24,30	24,31±0,20	Extremely powerful
80	26,30	26,10	26,50	26,25	26,29±0,16	Extremely powerful
100	27,40	27,20	27,50	27,70	27,45±0,20	Extremely powerful
Positive control[a]	20,60	20,40	20,30	20,80	20,53±0,22	Extremely powerful
Negative control[b]	0,00	0,00	0,00	0,00	0,00±0,00	Mild

[a] Positive control using Amoxycillin 500mg/10 ml.
[b] Negative control using DMSO 0.5%.

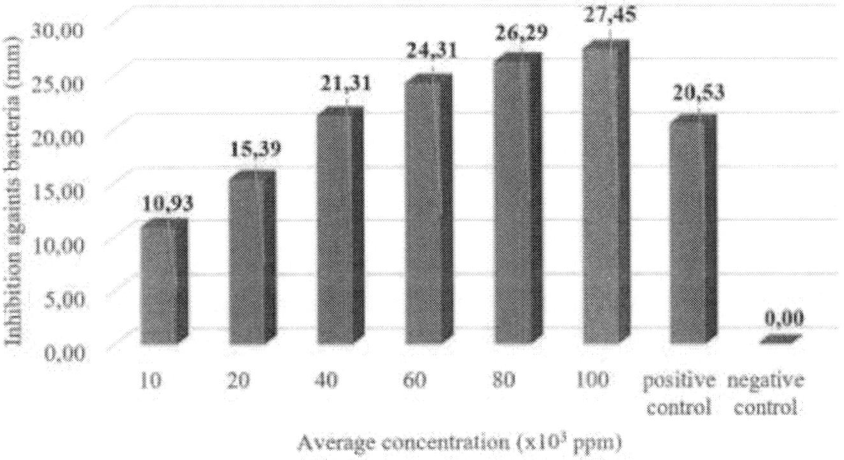

Figure 1. Average inhibition of 4% chitosan nanoparticles of blue swimming crab shell waste (*Portunus pelagicus*).

Figure 1 shows that the highest inhibition of blue swimming crab shell waste chitosan nanoparticles (*Portunus pelagicus*) against *Staphylococcus aureus* at a concentration of 4% chitosan nanoparticles of 27.45±0.20 mm, the growth inhibition response is extremely powerful in inhibiting the growth of *Staphylococcus aureus* bacteria, according to the classification of bacterial growth inhibition response with a clear zone diameter of more than 20–30 mm. Based on the data above, the results obtained for the normality test of the data on the inhibition power of chitosan nanoparticles of blue swimming crab shell waste (*Portunus pelagicus*) against *Staphylococcus aureus* using *Shapiro wilk* resulted in a significant value at a concentration of 10x10^3 ppm of 0.894, 20x10^3 ppm and 40x10^3 ppm of 0.408, 60x10^3 ppm of 0.650, 80x10^3 ppm of 0.900 and for 100x10^3 ppm of 0.995, based on all data the significance is greater than α (0.05), so that normal distributed data is obtained. The results of the homogeneity test of the data on the inhibition of blue swimming crab shell waste chitosan nanoparticles (*Portunus pelagicus*) against *Staphylococcus aureus* using the test of homogeneity of variances resulted in a significance value of 0.045. Based on this value, the data obtained is not homogeneous which is less than α (0.05), so the *Kruskal-Wallis* non-parametric test is carried out with the results of the significance obtained p value = 0.000 which is less than α (0.05) which means that the concentration treatment of the antibacterial activity test on chitosan nanoparticles of blue swimming crab shell waste has a significant difference in the results of inhibition of *Staphylococcus aureus* bacteria, with the highest value at a concentration of $100x10^3$ ppm.

Based on the results of the Post-Hoc Least Significant Difference (LSD) further test, the results of the significance value between concentrations of $10x10^3$ ppm, $20x10^3$ ppm, $40x10^3$ ppm, $60x10^3$ ppm, $80x10^3$ ppm with $100x10^3$ ppm, positive control and negative control obtained a p value of 0.000 which is less than α (0.05), so that the average inhibition zone produced by chitosan nanoparticles of blue swimming crab shell waste against *Staphylococcus aureus* bacteria has a value that is not much different from the six concentrations. This study also shows results where the greater the concentration of chitosan nanoparticles of blue swimming crab shell waste (*Portunus pelagicus*), the greater the inhibition against *Staphylococcus aureus* bacteria. This is in line with the research of Anjugam et al which states that the increasing concentration of nanoparticles will produce greater bacterial inhibition [23]. In other studies found the highest inhibition of crab shell nanokitosan on *Staphylococcus aureus* with a diameter of 30 mm at a concentration of 23 ppm [11,18]. Due to its limitations this study only reported one concentration of chitosan nanoparticles 4% and only examined *Staphylococcus aureus* bacteria, besides that this research is still at the in vitro pre-clinical stage with antibacterial activity testing. This research will still be further developed by testing antioxidant and anti-inflammatory activities and will make a gel formula for wounds where chitosan nanoparticle gel can accelerate cell proliferation so that it can accelerate wound healing.

4. Conclusion

From the analysis of the antibacterial activity test of blue swimming crab shell waste chitosan nanoparticles (*Portunus pelagicus*) against *Staphylococcus aureus* bacteria, the greater the concentration, the greater the inhibition zone response. The result of the this study concluded that chitosan nanoparticles of blue swimming crab shell waste had a substantial inhibitory effect on the bacterial development of *Staphylococcus aureus* at a very strong concentration, with the highest inhibition zone of 27.45±0.20 mm at a concentration of 4% and a significance result of 0.00.

References

[1] Nengsih, N. S., Penerapan indikator pembangunan berkelanjutan di daerah pesisir dalam keanekaragaman hayati laut untuk mensejahterakan masyarakat. Jisipol;**1** (2):151–162. 2020.
[2] Ramadhani, A. A., Potensi Keunggulan Kompetitif Sumber Daya Kelautan Indonesia, Jurnal Ekonomi Sakti (JES); **12** (3):291-296. 2023.
[3] Rejeki, S., Furi, C. A., Ariyanti, R. W., Pengaruh Salinitas Yang Berbeda Terhadap Kelulushidupan Dan Pertumbuhan Rajungan (Portunus pelagicus) Pada Stadia Crab Muda, Pena Akuatika;**18** (1):46-62. 2019.
[4] Amalia, K. P., Ekayani, M., Nurjanah., Pemetaan dan Alternatif Pemanfaatan Limbah Cangkang Rajungan di Indonesia, Jurnal Pengolahan Hasil Perikanan Indonesia; **24** (3):310–318. 2021.
[5] Ningrum EO, Ardiani L, Rohmah NA, Fajar N., Modifikasi Biokomposit Kitosan dari Cangkang Rajungan (Portunus Pelagicus) dan Pektin untuk Aplikasi Edible Fil., Yogyakarta: Proceedings of the National Seminar on Chemical Engineering, UPN Veteran Yogyakarta. ISSN 1693-4393. 2019
[6] Natalia DA, Dharmayanti N, Roswita F., Produksi Kitosan dari Cangkang Rajungan (Portunus sp.) Pada Suhu Ruang, Jurnal Pengelolahan Hasil Perikanan Indonesia; **24** (3):301-9. 2021.

[7] Norhidayah M. Ahyat, F. H. A. Mohamad, Azrilawani Ahmad and Alyza Azzura Azmi., Chitin And Chitosan Extraction From Portunus pelagicus, Journal, Malaysian. 2017.

[8] Ningtyas KR, Muslihudin M, Elsyana V., Isolation and characterization chitosan from varied crab shell, International Conference On Agriculture and Applied Science (ICoAAS). 2020.

[9] Buanasari, Buanasari & Sugiyo, Warlan & Fitriani, Nur & Suryaningsih, Suryaningsih., Potential of Chitosan From Local Crab (Portunus Pelagicus) to Enhance Storability of Musa Paradisiaca L., Jurnal Bahan Alam Terbarukan; **8** :41-46. 2019.

[10] Ministry of Marine Affairs and Fisheries, Marine and Fisheries Data Release. Quarterly I. 2022.

[11] Yusan, L. Y., Nailufa, Y., & Subagio, H., Isolation and Characterization of Chitosan Nanoparticles From Crab Shell Waste (Portunus Pelagicus), International Journal of Applied Pharmaceutics; **16** (2):358–363. 2024.

[12] Ahmed TA, Aljaeid BM., Preparation, characterization, and potential application of chitosan, chitosan derivatives, and chitosan metal nanoparticles in pharmaceutical drug delivery, Drug Des Devel Ther.;**10** :483-507. 2016.

[13] Vishwakarma A, Sriram P, Preetha SP, Tirumurugaan KG, Nagarajan K, Pandian K., Synthesis and characterization of chitosan/TPP encapsulated curcumin nanoparticles and its antibacterial efficacy against colon bacteria, Int J Chem Stud.; **7** :602-6. 2019.

[14] Baharuddin, S., Uji Efektivitas Antijamur Kitosan Cangkang Kepiting Bakau (Scylla sp) Terhadap Pertumbuhan Epidermophyton floccosum dan Candida albicans, Lumbung Farmasi: Jurnal Ilmu Kefarmasian; **2** (2):103-111. 2021.

[15] Du F, Ma J, Gong H, Bista R, Zha P, Ren Y, Gao Y, Chen D, Ran X and Wang C., Microbial Infection and Antibiotic Susceptibility of Diabetic Foot Ulcer in China: Literature review, Front. Endocrinol.;**13** :881659. 2022.

[16] Wada FW, Mekonnen MF, Sawiso ED, Kolato S, Woldegiorgis L, Kera GK, El-Khatib Z, Ashuro AA, Biru M, Boltena MT., Bacterial profile and antimicrobial resistance patterns of infected diabetic foot ulcers in sub-Saharan Africa: a systematic review and meta-analysis, Sci Rep.;**13** (1):14655. 2023.

[17] Ahyat NM, Mohamad F, Ahmad, Ahmad, Azmi AA., Chitin and chitosan extraction from Portunus pelagicus, Malays J Anal Sci.;**21** :770-7. 2017.

[18] Luthfiyana N, Bija S, Nugraeni CD, Lembang MS, Anwar E, Laksmitawati DR., Characteristics and antibacterial activity of chitosan nanoparticles from mangrove crab shell (Scylla sp.) in tarakan waters, Biodiversitas; **23** (8):4018-25. 2022.

[19] Yadav M, Aggarwal P, Yadav D, Singh A., Formulation and evaluation of clobetasol-17-propionate- loaded carboxymethylchitosan nanoparticle, Asian J Pharm Clin Res.;15:88-93. 2022.

[20] Sindhuri GV, Mariappan G, Subramanian S., Formulation and evaluation of epigallocatechin gallate and berberine-loaded chitosan nanoparticles, Int J App Pharm.; **15** :178-89. 2023.

[21] Ernawati & Nur, J., Aktivitas Anti Mikroba Perasan Daun Kirinyuh (Chormolaena odorata L.) Terhadap Candida albicans dan Pseudomonas aeruginosa, Jurnal Kedokteran dan Kesehatan; **17** (2). 2021.

[22] Hosokawa M, Nishino J, Kanno Y., Nanoparticle Technology Handbook, 1st Ed., UK: Elsevier Linacre House. 2007.

[23] Anjugam M, Vaseeharan B, Iswarya A, Divya M, Prabhu NM, Sankaranarayanan K., Biological synthesis of silver nanoparticles using β-1, 3 glucan binding protein and their antibacterial, antibiofilm and cytotoxic potential, Microb Pathog.; **115** :31-40. 2018.

Characterization Comparing the collagen and calcium content in golden sea cucumbers processed using freeze-dry versus grinder method to produce bone supplements

N Prameswari[1*], D Mulawarmanti[2], B Handayani[1], H Setianingsih[3], M Furqon[4], D K Nisa[4], N M Ningsih[4], N P Umbas[1], P S J Wuryanto[1], A I Syahrezi[1], I A Marsha[3], N Z S Tsuroyya[3] and E K Ramadhani[1]

[1] Orthodontic Department, Faculty of Dentistry, Universitas Hang Tuah, Surabaya, Indonesia
[2] Oral Biology Department, Faculty of Dentistry, Universitas Hang Tuah, Surabaya, Indonesia
[3] Department of Anatomy, Faculty of Medicine, Universitas Hang Tuah, Surabaya, Indonesia
[4] PT Nanoherbaltama Internasional, Tangerang, Indonesia

[*]Email: noengki.prameswari@hangtuah.ac.id

Abstract. Background: Osteoporosis, a disease with a multifactorial etiology, affects 10.3% of the population in Indonesia. Bone supplements are essential for preventing osteoporosis. Gold sea cucumber is considered a potential bone supplement because it contains active ingredients such as collagen, calcium, mucopolysaccharides, glucosamine, chondroitin sulfate, and minerals that are beneficial for bone health. Purpose: To compare the collagen and calcium content between golden sea cucumbers processed using the freeze-dry method and grinder method, for producing bone supplements. Methods: The golden sea cucumbers were divided into two groups: V1 (golden sea cucumber processed using the freeze-dry method) and V2 (processed using the grinder method). Both groups were processed into powder form and tested for collagen content using the High Performance Liquid Chromatography (HPLC) method and calcium using AOAC method 985.35 (Association of Official Analytical Chemists) AAS (Atomic Absorption Spectrophometry). Results: Descriptive testing revealed that the collagen content in group V1 was 16.8 ± 0.6, while in group V2 it was 13.0267 ± 0.7024; calcium content in group VI was 0.0600 ± 0.00272 while group V2 it was 0.0532 ± 0.00300 An independent t-test showed a significant difference in collagen content between the two groups, and also calcium with a p-value of $p = 0.000$ and 0,043 ($p < 0.05$). Conclusion: The collagen and calcium content in sea cucumbers processed using the freeze-dry method is higher than that of the grinder method, demonstrating that these method is more optimal for producing collagen and calcium for bone supplements.

1. Introduction

Indonesia, an archipelago with the world's longest coastline of 81,000 km and a marine area of approximately 5.8 million km² (75% of its total land area), is located in the center of the Coral Triangle. This positioning gives the country exceptionally rich marine biodiversity. Indonesia is often referred to as the "Mega Diversity in the World" due to its unparalleled biodiversity [1]. The utilization of food supplements derived from marine organisms is expanding because these organisms possess unique properties not commonly found in terrestrial species [2].

Sea cucumbers are marine organisms with functional materials valuable for biomedical applications [2]. They are typically found in coral reef ecosystems, inhabiting coastal waters at depths ranging from 1 to 40 meters. Sea cucumbers thrive in clean, clear waters with a salinity level of 30-33% and a pH level that supports their growth [3]. Despite their potential, the use of golden sea cucumbers is still limited. Golden sea cucumbers have a protein content of up to 82%, with 80% of that being collagen [4].

Golden sea cucumbers contain several active ingredients, including hyaluronic acid, cell growth factors, chondroitin sulfate, eicosapentaenoic acid, docosahexaenoic acid, and flavonoids, which play important roles in the molecular activity of osteoblasts. For instance, glycosaminoglycans like chondroitin sulfate have been shown to enhance calcium absorption and increase bone regeneration through osteogenesis when administered orally [5]. Additionally, golden sea cucumbers can modulate the inflammatory response, improve the collagen fiber system with shorter treatment durations, and stimulate the modulation and proliferation of fibroblasts [6]. A study also found that the calcium content in golden sea cucumbers is 0.59% [7]. The health benefits of collagen have been extensively documented in previous studies. Collagen is widely recognized for its positive effects on joint health. Reports indicate that collagen can enhance the calcium-binding ability of bones, alleviate arthritis, [8] facilitate joint movement [9], and slow down skin aging [10]. High collagen content is known to improve mineral absorption in the human body. Specifically, collagen can enhance calcium retention, with a retention rate of up to 70% [11].

Calcium and collagen are essential components for bone supplements, which are crucial for maintaining bone health. Without proper maintenance, bones can deteriorate, leading to conditions such as osteopenia, which is a precursor to osteoporosis [12]. In dentistry, calcium and collagen are also beneficial for bone remodeling during orthodontic tooth movement [5]. Nutrition plays a critical role in bone health, influencing the achievement of peak bone density from infancy to around thirty years of age and maintaining bone health throughout adulthood. A balanced diet that meets daily calorie needs and includes adequate calcium and vitamin D intake is essential for reducing bone loss [13]. Therefore, supplements derived from golden sea cucumbers, which are rich in calcium and collagen, can have a significant impact on bone health Indonesia currently faces challenges in developing medicinal or pharmaceutical raw materials. The industrial development for producing pharmaceutical raw materials is progressing slowly, as most industries are focused on the downstream sector, primarily formulating final products rather than producing raw materials [14]. One common preparation method for sea cucumber raw materials is in the form of powder. Powder is a fine, dry material with a very small particle size [15]. Due to its large surface area, powder disperse and dissolves more easily compared to solidified dosage forms [16]. Powder processing can utilize techniques that involve low-temperature heat, such as freeze-drying, to preserve the quality of the resulting product. Freeze-drying, also known as lyophilization, is a process that removes water content from a product by first freezing it and then sublimating the ice to convert it directly into gas, while carefully controlling temperature and pressure during processing [17]. Another food processing technique is grinding, which reduces the particle size of materials from a coarse or large form to a finer or smaller size [18]. The goal of grinding is to produce smaller material particles, which can enhance the separation of valuable minerals from residues, resulting in a more mineral-rich product with lower residue content [15].

Given the rich properties of golden sea cucumbers and their underutilization in bone supplements, it is crucial to conduct research comparing the characterization of calcium and collagen content between the freeze-dry method and the grinding method. This research aims to optimize the use of golden sea cucumbers in the development of new drugs and bone supplements.

2. Material and Methods

This research is a laboratory experimental study. The samples consist of golden sea cucumbers from

Sumenep, East Java, Indonesia. The research design is a posttest-only control group design, featuring two groups: Group V1, which uses golden sea cucumber powder processed by the freeze-dry method, and Group V2, which uses golden sea cucumber powder processed by the grinding method. Replication in the study was determined using the Dahlan formula, resulting in three replications. The tools and materials used include a mask and gloves, Petridish, analytical balance, freeze dryer, grinder, knife, 200-mesh sieve, golden sea cucumber powder, and aquadest.

2.1. Research procedure

2.1.1. Preparation of golden sea cucumber

The preparation of the golden sea cucumber involves several steps. First, the wet golden sea cucumber is cleaned to remove its entrails. Then, it is boiled using sea water. After boiling, the golden sea cucumber is dried in the sun from morning to evening, depending on weather conditions. This drying process continues for several days until the golden sea cucumber appears dry according to the standard. Once dried, the length of the golden sea cucumber is measured using a ruler. Finally, the golden sea cucumbers are weighed to determine their initial weight.

2.1.2. Grinder Method

The preparation continues with the golden sea cucumber being cut into smaller pieces using a knife after measuring its length and weight. The pieces are then dried using an oven at a temperature of 45°C. This drying process is carried out until the golden sea cucumber is thoroughly dried, with its water content reduced to less than 10%. Afterward, the dried golden sea cucumber is placed into a grinder machine to be ground into a fine powder. The resulting powder is then dried again using an oven. Following two rounds of drying, the golden sea cucumber powder is sieved using a 200-mesh sieve to obtain the appropriate particle size.

2.1.3. Freeze Dry Method

The process begins with preparing the golden sea cucumber powder that has been ground. Next, the freeze dryer is turned on by pressing the power ON button. The sample chamber cover tube is opened, and the golden sea cucumber powder is placed inside. The position of the sample container in the chamber is then adjusted. After setting the time on the machine, the drying process is initiated, and the phase changes from Warm-up Vacuum Pump to Main Drying. The drying process continues until the set time is reached, during which the machine automatically adjusts the air pressure in the sample. Once the process is complete, the sample chamber cover is opened, the sample container is removed, and cleaned. Finally, the sample chamber cover is closed again, and the machine is turned off by pressing the power OFF button.

2.1.4. Collagen Assay using HPLC

The powdered golden sea cucumber, after undergoing the smoothing process, is placed into the HPLC machine for collagen content analysis. The computer is then turned on, and a check is performed to ensure that each module on the HPLC device is functioning properly. After verifying that the computer is correctly powered, the HPLC software is opened on the computer. A simple check is conducted to confirm that the module and the computer are connected and able to communicate effectively. Attention is focused on the software, ensuring that the solution used for the mobile phase is available in sufficient quantities. The method is then set up in the HPLC software, and the instrument is operated to obtain a stable baseline. The detector captures the data from the sample and displays it in the software. Finally, the measurement results for collagen content are saved, printed, and recorded, with the same procedure followed for the calcium content test in the golden sea cucumber.

2.1.5. Calcium Assay using AAS spectrometer

Begin by preparing the sample powdered golden sea cucumber. Weigh a specific amount of the sample, usually in milligrams. Next, dissolve the sample in a suitable solvent, often hydrochloric acid (HCl), to extract the calcium into a liquid form. Filter the solution to remove any undissolved particles, and then dilute the sample to a known volume with deionized water or another suitable solvent to ensure that the calcium concentration falls within the working range of the AAS instrument.

Proceed to the calibration step. Prepare a series of calcium standard solutions with known concentrations to create a calibration curve. Run these standard solutions through the AAS to generate a calibration curve, which relates the absorbance measured by the spectrometer to the concentration of calcium in the solution. Next, set up the instrument. Select the AAS to the specific wavelength for calcium detection, typically around 422.7 nm. Use a calcium hollow cathode lamp as the light source. Adjust the settings such as slit width, burner height, and gas flow rates (often using an acetylene-air flame) according to the manufacturer's instructions and the specific requirements of your assay.

For sample analysis, begin with a blank run, usually the solvent without any calcium, to zero the instrument. Aspirate the prepared sample solution into the AAS, and the machine will measure the absorbance, which is directly related to the calcium concentration in the sample. It is recommended to measure each sample multiple times to ensure accuracy and reproducibility. Finally, analyze the data by comparing the absorbance values of your samples against the calibration curve to determine the calcium concentration. Use the dilution factor, if applicable, to calculate the original calcium concentration in the sample. Document the calcium concentration data by saving, printing, and recording the results. Perform quality control to ensure that all results fall within the acceptable range for the assay, and repeat any measurements if necessary.

3. Results and Discussion

The results were then subjected to descriptive tests using SPSS so that the mean values obtained were as shown in Figure 1.

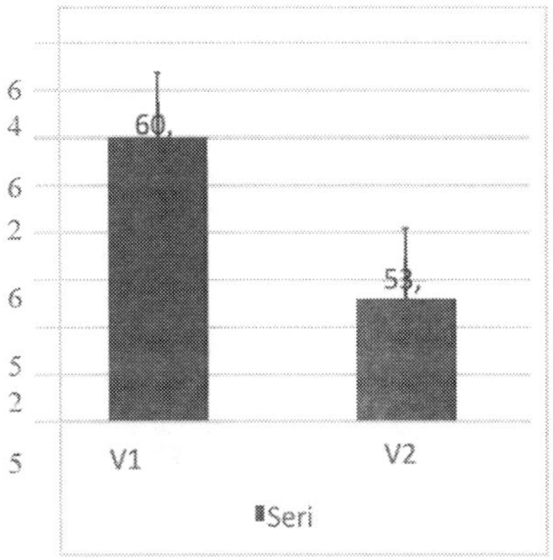

Figure 1. Descriptive Test Results (Mean Values ± Standard deviations) of Calcium Content of Gold Sea Cucumber Powder (Stichopus Hermanii) Using Freeze Dry and Grinder Methods.

The normality test results show that each method obtained Shapiro-Wilk value, Sig. > 0.05, then the calcium content data in the method of making gold sea cucumber powder using freeze dry and grinder in the process of optimizing industrial raw materials for bone health content is normally distributed.

The results of the homogeneity test using Levene's Test show that the significance value or Sig. > 0.05, it can be concluded that the variance of calcium content data in the method of making gold sea cucumber powder using freeze dry and grinder in the process of optimizing industrial raw materials for bone health from the value of this study is homogeneous.

Table 1. The results of the Independent T Test hypothesis test compared the calcium content in the method of making gold sea cucumber powder using freeze dry and grinder.

In dependent Samples T Test	Calcium content
Sig.	0,043*

Based on the results of the Independent T-Test test, the Sig. value of 0.043 <0.05 can be concluded that there is a significant difference in the calcium content test in the method of making gold sea cucumber powder using freeze dry and grinder in the process of optimizing industrial raw materials for bone health.

As for the descriptive test on collagen parameters using SPSS, it shows the mean values as mentioned below:

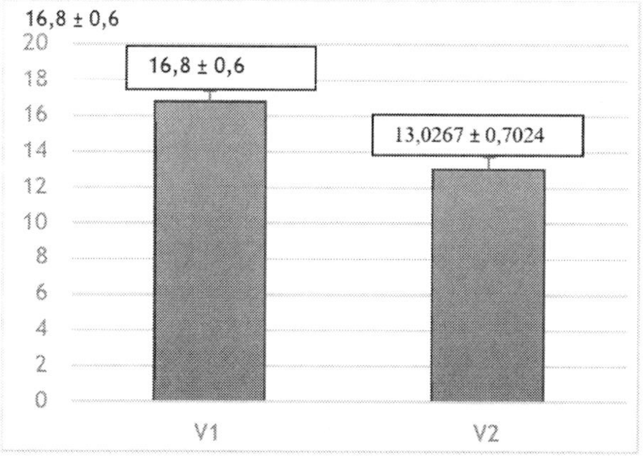

Figure 2. Descriptive Test Results (Mean Values ± Standard deviations) of Collagen Content of Gold Sea Cucumber Powder (Stichopus Hermanii) Using Freeze Dry and Grinder Methods.

Based on the results of the analysis using Saphiro Wilk, the significance results of each group which include collagen levels of the freeze dry treatment group ($p = 0.000$) and the grinder treatment group ($p = 0.843$) show normal data distribution

Based on the results of the Levene Test for collagen content data has a p value = 0.156. It is said to be homogeneous if ($p > 0.05$). This means that the variance of collagen content data is homogeneous, so testing whether there is an effect using parametric analysis, namely by using the Independent Samples T Test.

The results of the Independent Samples Test analysis stated in Table 2.

Table 2. Independent Samples T Test Results of Collagen Levels Between Freeze Dry Treatment Group and Grinder Treatment Group.

In dependent Samples T Test	Collagen content
Sig.	0,000*

Based on the results of the Independent Samples T Test Table 5.4, the p value = 0.000 ($p < 0.05$) means that there is a significant difference in collagen levels between the freeze dry treatment group and the grinder treatment group.

This chapter will discuss the results of laboratory research comparing the calcium and collagen content in gold sea cucumber powder produced using two different methods: freeze-drying and grinding. The research aims to optimize the use of these industrial raw materials for bone health. Gold sea cucumbers were obtained from the sea waters of Madura, specifically in Sumenep Regency, and dried with the help of sunlight. The dried golden sea cucumbers were processed through cutting into small pieces, approximately 5-7 cm in size.

To create raw materials that are efficient and have a relatively long shelf life, powder formulations were chosen in this study for the development of raw materials in the bone health industry to other solid forms. Additionally, powder formulations offer advantages such as extended shelf life and ease of storage [17]. Powder formulations are considered to be faster and more stable in the absorption process within the body compared to other solid formulations [16]. A powder formulation is a combination of two or more homogeneous materials that are pulverized and formed into a dry blend or a chemical substance that is ground until smooth, achieving a particle size of 10,000 to 0.1 micrometers, and then used as a mixed raw material for internal or external use [19].

The golden sea cucumbers undergo various steps before finally becoming a powder preparation. The initial stage is drying, which aims to remove the water content from the golden sea cucumber. Temperature and time adjustments are crucial to achieving high-quality raw materials [20]. Next is the process of cutting the body of the golden sea cucumber and separating the meat from the entrails, ensuring that the final raw materials are clean and free from contamination [21]. The grinding method, in particular, plays a crucial role in maintaining the calcium content in the powder. The freeze-drying method helps preserve the quality of the resulting preparation [22].

This study employed the AAS (Atomic Absorption Spectrophotometry) technique, to determine calcium levels in various samples. The AAS method was selected for this study because it is considered one of the most accurate and reliable methods for measuring calcium levels [23]. In the statistical tests, comparing the calcium content methods in golden sea cucumber. The results were obtained using the freeze-dry method, with an average calcium content of 60.0364 mg per 100 grams, compared to 53.1709 mg per 100 grams using the grinder method. This comparison shows that there is an average difference in the results between the freeze-dry and grinder methods. This result is supported by a t-test, which shows a significant difference in collagen content between the freeze-dry treatment group and the grinder treatment group, as evidenced by a p-value of 0.000 ($p < 0.05$). The test results showed a significant difference, with higher calcium levels obtained from the freeze-dry method. Therefore, the freeze-dry method was chosen in this study as the optimal method for producing powder because the higher calcium content is believed to better support bone health and reduce the risk of osteoporosis [12]. This suggests that one of the factors contributing to differences in calcium is the initial method of pulverization [15]. The freeze-dry method involves a pre-cooling mechanism where the powder is left at a low temperature to freeze. This process aims to reduce the moisture content in the gold sea cucumber powder. Following this, drying is conducted under low pressure and temperature, followed by a sublimation process where the solid turns directly into vapor without passing through the liquid phase. The pressure in the drying chamber is then decreased to increase the sublimation rate, and the temperature is raised to ensure that the final material has a low moisture content, which helps maintain its quality and stability [24]. Understanding the calcium content in this gold sea cucumber powder is important for making informed choices in the development of bone health products, such as drugs and supplements [19].

While this process is effective at preserving the calcium levels, freeze-drying is not recommended for fast production due to several drawbacks. The process takes a relatively long time, incurs high operational costs due to significant energy consumption, and requires large-capacity machinery for large- scale material production. The grinder mechanism begins with an initial grinding process that produces coarse powder. This coarse powder is then reprocessed in the grinder with a finer setting to create smaller particles. The resulting particles are then filtered to ensure a uniform particle size. Although the grinder method produces calcium levels that are not as high as those obtained with the freeze-dry method, this is due to frictional heating between the material particles and the grinder surface, which can lead to a decrease in calcium content. Despite this, the grinder method can be a viable option for producing gold sea cucumber powder when quick production is required, as it is easier to use and the grinding process takes a relatively short amount of time [15].

This study also examining collagen content using HPLC comparing two methods, between grinding and freeze dry. The results of testing and data analysis showed that the average collagen content in the golden sea cucumber group with freeze-dry treatment was 16.8%, while the collagen content in the golden sea cucumber group with grinder treatment was 13.02%. These results indicate that the collagen content in the golden sea cucumber group with freeze-dry treatment is greater than that in the golden sea cucumber group with grinder treatment.

This result is supported by a t-test, which shows a significant difference in collagen content between the freeze-dry treatment group and the grinder treatment group, as evidenced by a p-value of 0.000 ($p < 0.05$). These findings demonstrate that the collagen content in golden sea cucumber powder produced using the freeze- dry method is higher than that produced by the grinder method. Therefore, the freeze-dry method is more optimal for producing collagen, benefiting pharmaceutical and food industry raw materials. The advantage of using the freeze-dry method is that it maintains the quality of ingredients in terms of nutritional value, physical, chemical, and sensory characteristics. However, it has some disadvantages, including relatively high costs and a time-consuming process [17]. The effects of this method on golden sea cucumber include helping to preserve the quality of the extract and ensuring that the physicochemical properties of the golden sea cucumber are well- maintained, resulting in high-quality powder [7].

While the grinder method results in lower collagen content in golden sea cucumbers, this aligns with the findings of other research,15 which indicate that the grinder method is less effective at extracting active compounds and yields lower-quality powder. One disadvantage of this method is that to achieve good- quality powder, the ingredients should ideally be dried first [16]. However, this drying stage can damage the collagen in golden sea cucumbers. Despite this, the grinder method has several advantages, including a simpler process, basic equipment, and suitability for small-scale operations [15].

The freeze-dry method is more effective for obtaining collagen content because collagen is a protein that is highly sensitive to heat. At high temperatures, the triple helix structure of collagen can break down into single chains, leading to an irreversible transition into gelatin. This is consistent with the principles of the freeze- dry method, which begins with freezing the material and then proceeds with drying through sublimation, effectively removing most of the water from the material [24]. Thermal denaturation of collagen occurs under mild conditions, typically when heating collagen at neutral or slightly acidic conditions to around 40°C. Collagen powder and collagen dispersion exhibit different heat stability [13].

The high collagen content in freeze-dried golden sea cucumbers affects the viscosity of the material. Higher collagen content in a product leads to increased viscosity [17]. This is attributed to the physicochemical properties of collagen, which can enhance viscosity. A higher viscosity value indicates greater stability and density of the material [19].

Before being formulated into a preparation, collagen is typically dispersed in weak acid solutions such as acetic acid and citric acid. The isolation process to obtain standardized collagen raw materials can be challenging, time-consuming, and lengthy due to collagen's sensitivity to heat. Therefore, the freeze-dry method is considered suitable because it avoids heat that could damage collagen in golden sea cucumbers. Additionally, freeze-drying helps stabilize the material's structure, inhibits microbial activity, and prevents chemical reactions and enzyme activity that could deteriorate food ingredients [24]. It is also important to determine specific storage conditions for collagen raw materials to maintain their stability before formulation [11].

Collagen is highly valued in the pharmaceutical, food, and cosmetic industries due to its significant economic value. It offers various benefits, such as antioxidant properties that can reduce the risk of several diseases [25]. In addition to its antioxidant effects, collagen also has anti-inflammatory, antidiabetic, and anticancer properties, making it particularly useful in pharmaceutical preparations such as supplements. Collagen possesses favorable physicochemical characteristics, including biocompatibility, biodegradability, low antigenicity, and nontoxicity, which further enhance its advantages. Given its diverse activities in preventing and treating various diseases and the availability of collagen sources, it represents a valuable alternative in modern medicine [26].

The high collagen content in freeze-dried preparations can impact users significantly. Previous research suggested that collagen may enhance calcium retention, which directly influences bone metabolism by promoting growth hormone release and regulating osteoblast function [26],[27]. Additionally, other studies indicate that collagen has positive therapeutic effects on osteoporosis and osteoarthritis. It has the potential to increase bone mineral density, provide protective effects on articular cartilage, and alleviate pain symptoms [28].

In conclusion, the freeze-dry method is more effective in preserving the calcium and collagen content in golden sea cucumbers compared to the grinder method. While both methods can maintain calcium and collagen levels, the freeze-dry process better retains higher levels of calcium and collagen [29]. Therefore, if the goal is to obtain high levels of both calcium and collagen, raw material producers should opt for freeze-drying. However, if time and production costs are constraints, the grinder method may be a more practical choice. Processing golden sea cucumbers using the grinder method can still enhance calcium and collagen content, making it suitable for daily bone supplements. This research supports the development of treatments derived from natural sources, particularly marine biota [30].

4. Conclusion

In conclusion, the freeze-dry method yields higher both calcium and collagen content in golden sea cucumbers compared to the grinder method. While the grinder method offers faster production and lower costs, the freeze-dry method better retains higher levels of these essential nutrients. For producers aiming to maximize calcium and collagen content in their raw materials, the freeze-dry method is recommended. For those constrained by time and production costs, the grinder method remains a practical alternative for enhancing calcium and collagen content in daily bone supplements. This research supports the development of natural treatments, particularly from marine biota, for bone health.

Acknowledgement

Thank you and highest appreciation to the LPPM Hang Tuah University for the 2024 internal research grant.

Refferences

[1] Mulawarmanti, D 2019 Prosiding Seminakel 1–10.
[2] Khatimah, H 2019 Warung Sains Teknologi 1–5.
[3] Ni'mah, U, Pringgenies, D and Santosa, G, 2021 Journal of Marine Research **10** 3 387–94.
[4] Prameswari N, Brahmanta A 2017 Dental Journal (Majalah Kedokteran Gigi) **50** 4 188– 193.
[5] Prameswari, N and Handayani, B 2019 IOP Conference Series: Earth and Environmental Science 217 1.
[6] Prameswari, N Brahmanta, A and Revianti, S 2020 Systematic Reviews in Pharmacy **11** 3 667–73
[7] Damaiyanti, D 2015 Dental Journal (Majalah Kedokteran Gigi) **9** 20 74–81.
[8] Soniwala, S Scinto, K Schott, E Stolarczyk, A Villani, D Dar, Q.-A Grier, A Ketz, J Gill, S Mooney, R.A 2018 Osteoarthr. Cartil. **26** 173–4.
[9] Lin, S Hu, X Li, L Yang, X Chen, S Wu, Y Yang, S 2021 LWT 149 111796.
[10] Bolke, L Schlippe, G Gerß, J Voss, W.A 2019. Nutrients **11** 2494.
[11] Chen, Q Guo, L Du, F Chen, T Hou, H Li, B 2017 Int. J. Food Sci. Technol **52**, 1283–1290.
[12] Sozen, T Ozisik, L & Basaran, N. C A 2017 European Journal of Rheumatology 4 1 46–56.
[13] Xu, N Peng ,X Li, H Liu, J Cheng, J Qi, X Ye, S Gong, H.L Zhao, X Yu, J 2021 Front. Nutr. 8 493.
[14] Kementerian Perindustrian RI 2021 Buku Analisis Pembangunan Industri 1–33.
[15] Marlina, O Grinding 2019 Oki **1**. 1–14.
[16] Latief, R Laga, A & Alang, S 2018 Canrea Journal: Food Technology Nutritions, and Culinary Journal **1** 10 1- 10
[17] Habibi, N Fathia, S and Utami, C 2019 JST (Jurnal Sains Terapan) **5** 2 67–72.
[18] Ardiansyah, D. et al. (2020) 2020 Accurate: Journal of Mechanical Engineering and Science, **1**(2). pp. 13-18.

[19] Tungadi, R 2018 Teknologi Sediaan Solida, Yogyakarta: Pustaka Belajar. Ponorogo: Wade Group.
[20] Sinurat, E and Murniyati, M 2014 Jurnal Pascapanen dan Bioteknologi Kelautan dan Perikanan **92** 133.
[21] Athirafitri, N Indasti, N and Ismayana, A 2021 Jurnal Teknologi Industri Pertanian **31** 3 274– 82.
[22] Rachmat, S and Shovitri, M 2022 Jurnal Sains dan Seni ITS, **10** 2.
[23] Taufik, M Seveline, S and Saputri, E 2018 Agritech **38** 2 187.
[24] Nofrianti, R 2013 Jurnal Aplikasi Teknologi Pangan, **2** 1 1–6.
[25] Furtado M Chen L Chen Z Chen A Cui W 2022 Eng. Reg 2022
[26] Batista, M Fernández Gaspar, F Bronze, M Duarte 2022 Front. Chem. 2022
[27] Liang, F Zuo H 2014 Sci Technol Gelatin **34** 03 109–15.
[28] Akkawi, I & Zmerly, H (2018). Osteoporosis : Current Concepts. 122–127.
[29] Muthukumar, T Sreekumar, G Sastry, T Chamundeeswari M 2018 Adv. Mater. Sci **53** 29–39.
[30] Rahman, A Silva T 2022 Mar. Drugs **20** 170.

IOP Conf. Series: Earth and Environmental Science **1473** (2025) 012056 doi:10.1088/1755-1315/1473/1/012056

Preparation, Characterization and Physical Stability of Salmon Oil Microemulsion with Surfactant and Cosurfactant Ratios

Y A Najih[1*], E Y Rahma[1], D N Rakhma[1] and Y Nailufa[1]

[1]Departement of Pharmaceutics, Faculty of Pharmacy, Universitas hang Tuah, Surabaya, Indonesia.

[*]E-mail: yuli.najih@hangtuah.ac.id

Abstract. Microemulsions are made up of homogeneously dispersed oil and water phases mixed with surfactants and cosurfactants. The purpose of this research is to establish the best surfactant and cosurfactant concentrations for making o/w type salmon oil microemulsions, as well as their physical properties and stability. Salmon oil microemulsion was prepared utilizing a combination of surfactants tween 80 and span 80, as well as sorbitol as a cosurfactant, in the ratio (6:1, 7:1, 8:1). Physical characterization of selected preparations included organoleptic, emulsion type, droplet size, pH, specific gravity, viscosity, and centrifugation. The results showed that the optimum microemulsion formula was F3 with an 8:1 ratio showing a clear and transparent physical appearance of the microemulsion. The type of emulsion formed was oil in water (o/w), with a droplet size of 1.4 μm, a pH value of 7.05 ± 0.2, a specific gravity of 0.0422 ± 0.0003 g/ml, and a viscosity value of 586.6 ± 75.7 cPs. Furthermore, the physical stability test was performed using a cycling test, with 6 cycles at 4°C for 24 hours and then 40°C for 24 hours. The results revealed changes in the organoleptics, pH, specific gravity, and viscosity of the preparation.

1. Introduction

Fish is a commodity that is frequently consumed by the community due to its good taste, availability, ease of access [1], low cost, and high nutritional value [2]. The primary nutrients utilized in the health sector for product development are derived from fish, specifically fish oil. According to [3], fish oil is rich in omega-3 long-chain polyunsaturated fatty acids (PUFA), specifically eicosapentaenoic acid (EPA) and docosahexaenoic acid (DHA). Fish oil offers numerous advantages, including the ability to alleviate symptoms associated with hypertension, stroke, atherosclerosis, myocardial infarction, and inflammation inhibition [4].

Fish utilisation technology is used not only in the food industry but also in the pharmaceutical industry to use fish oil. Salmon oil contains 21.4% EPA and DHA, while shark oil has 20.6% [5]. The unsaturated fatty acids in fish oil are non-polar, which means they are poorly soluble in water [6]. When taken orally, the oil's active components are poorly soluble, resulting in low bioavailability in the body [7,8]. Salmon oil is used in the pharmaceutical sector as emulsion preparations and soft capsules that are administered orally [9]. When fish oil emulsions are given orally, they have low absorption and cause creaming problems during storage, so the preparation must be modified into micro/nano emulsions

[10]. In this study, salmon oil was produced as a microemulsion using span 80 and tween 80 surfactants and sorbitol as a cosurfactant.

Microemulsions are emulsion systems composed of oil, water, and surfactant/cosurfactant phases. These systems typically have droplet sizes ranging from 0.15 to 2 μm and seem cloudy in appearance [11]. The emulsion system exhibits droplets with a size ranging from 10-20 mm and has a milky appearance [12]. The reduced size of doplets enhances the surface area of wetted droplets, thus augmenting the solubility of active components, leading to an improvement in bioavailability [10].

The preparation of a microemulsion system is affected by the composition of the oil phase, water phase, surfactant/cosurfactant, and the speed at which the mixture is stirred [13]. The surfactant/cosurfactant concentration necessary for the preparation of microemulsions typically ranges from 20% to 80% [14]. Prior research has demonstrated the production of arbutin microemulsion using the nonionic surfactant PEG-7 Glyceryl Cocoate, with a surfactant concentration above 30% [15,16] reported that astaxanthin microemulsion can be produced using nonionic surfactants tween 80 and tween 20, with a surfactant concentration of 70-85%. The purpose of this study was to investigate the impact of varying the ratio of surfactant (span 80 and tween 80) and cosurfactant (sorbitol) on the characteristics and physical stability of salmon oil microemulsion.

2. Material and Methods

2.1 Materials
The materials used in this study include distilled water, Golden Bear Laboratory salmon oil produced by Marine Nutraceutical Corp. Pennsylvania 18343 USA, Span 80 (Merck), Tween 80 (Merck), Sorbitol (Merck), BHT, Sodium benzoate, distilled water.

2.2 Preparation of Salmon Fish Oil Microemulsion
Weighing the ingredients according to the concentration determined in table 1. Heating the oil phase, specifically fish oil and BHT, at 55°C and stirring with a hot plate stirrer. Heat the surfactants Tween 80 and Span 80 to 55°C and mix on a hot plate stirrer. Mixing the oil phase and surfactant by gradually incorporating the surfactant into the oil phase while heating to 55°C and stirring at 250 rpm using a hot plate stirrer. After that, gradually add sorbitol cosurfactant to the mixture of oil phase and surfactant while heating at 60°C and stirring at 400 rpm with a hot plate stirrer. Heat the solution containing the oil phase, surfactant, and cosurfactant to 70°C and mix at 570 rpm with a hot plate stirrer. Add the dissolved sodium benzoate preservative solution to the remaining water. Under these conditions, stir for 30 minutes until the mixture becomes clear. The mixture was then homogenized using an ultrasonicator at 40-49°C for 60 minutes.

2.3 Pseudoternary diagram
The concentration of each component of the salmon oil microemulsion was determined using formula optimisation, which involved developing a pseudoternary diagram containing seven formulas. Table 1 and Figure 1 show the concentration optimisation formulas, and one clear and transparent formula is chosen. The chosen formula was then used to create 150 grams of salmon oil microemulsion in three replications using surfactant and cosurfactant ratios of F1 (6:1), F2 (7:1), and F3 (8:1).

2.4 Charaterization of Salmon oil microemulsion
The characterization of microemulsions comprised organoleptic observation, droplet size determination with DelsaTM Nano Submicron Particle Size Analyser, emulsion type determination, specific gravity with pycnometer, pH with Laqua pH meter, viscosity measurement with Viscometer Brookfield, and stability tests using centrifugation with HEALTH centrifuge 8 HOLE+ timer, and cycling with oven Memmert and refrigerator.

2.5 Physical Stability Test

Physical stability tests were conducted using centrifugation and cycling. The centrifuge test was evaluated in terms of organoleptic properties. The evaluation was performed at the start of the microemulsion's production, and after the cycle test, the centrifugation findings revealed that the microemulsion had a distinct odor, a light yellow color, was transparent, and there was no phase separation. As a result, the microemulsion in this situation remained constant.

2.6 Statistical analysis

Data are expressed as mean ± standard deviation from at least three replication experiments. Data processing was carried out by statistical analysis test with one way ANOVA method and continued with Post- Hoc Test Tukey HSD with 95% confidence degree (α=0.05) by comparing the results of physical characteristics of formulation 1, formulation 2 and formulation 3. Then to determine whether there is a significant difference or not from the microemulsion physical characteristics test data before and after centrifugation stability testing and cycling test is the Paired t-test method by comparing the results of physical characteristics such as organoleptic, pH, viscosity and specific gravity.

3. Results

3.1 Pseudoternary diagram

A pseudoternary diagram is made to determine the concentration of each component of salmon oil microemulsion. Tables 1 show concentration optimization emulsion (EM) and microemulsion (ME) formulas and figure 1 shows the pseudoternary diagram.

Table 1. Concentration optimization formula EM and ME.

No	Concentration (wt/v%)			System	appearance
	Oil Phase	Smix (SF-CS)	Water Phase		
1	27,28	12,72	60	EM	Milky white
2	15	25	60	EM	Milky white
3	5	35	60	EM	Milky white
4	5	45	50	ME	Light yellow, cloudy slightly transparent
5	5	50	45	ME	Light yellow, cloudy slightly transparent
6	10	65	25	ME	Light yellow, cloudy slightly transparent
7	15	70	15	ME	Light yellow, clear, transparent

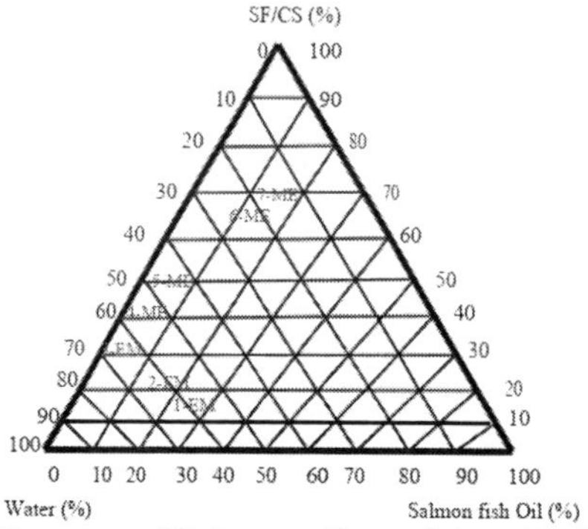

Figure 1. A pseudoternary diagram of Salmon oil emulsion (EM) and microemulsion (ME).

According to Table 1 and Figure 1, the percentage concentration of microemulsion components that can make a salmon oil microemulsion with a clear/transparent appearance is 15% oil, 70% surfactant and cosurfactant (Smix), and 15% water. The formula with the chosen concentration of oil, surfactant, and water ratio (15:70:15) produced a 150 gram salmon oil microemulsion in three replications with surfactant and cosurfactant ratios F1 (6:1), F2 (7:1), and F3 (8:1). Table 2 shows the concentration of microemulsion (ME) formulas. Physical properties and stability of each formula's microemulsion preparations were assessed.

Table 2. Concentration formula ME

Material	Concentration (wt%)		
	F1	F2	F3
Salmon Oil	22,50 g	22,50 g	22,50 g
Tween 80	64,76 g	66,12 g	67,14 g
Span 80	25,23 g	25,76 g	26,16 g
Sorbitol	15 g	13,12 g	11,70 g
BHT	0,15 g	0,15 g	0,15 g
Sodium benzoat	0,15 g	0,15 g	0,15 g
Aquadest	22,21 g	22,21 g	22,21 g

3.2 Charaterization of Salmon oil microemulsion

The organoleptic test or physical appearance of the microemulsion revealed that F3 (8:1) was clearer and more transparent than microemulsion formulae F2 (7:1) and F1 (6:1). The pH test in F3 was higher at 7.05 compared to the pH size in F2 of 6.50 and F1 of 6.66. The pH value found is consistent with the pH of the intestine, which ranges from 6-7 [19]. Figure 2 shows a histogram of the mean pH test of F1, F2, and F3 microemulsions with three replications.

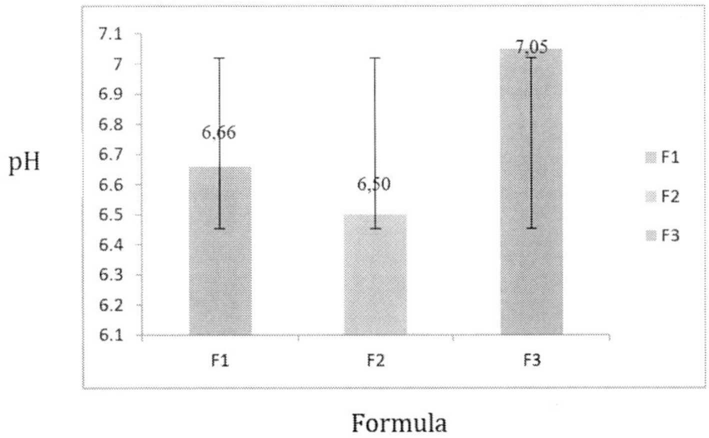

Figure 2. shows a histogram of the mean pH test of F1, F2, and F3 microemulsions with three replications.

The results of the microemulsion specific gravity test can be seen in Figure 3 Histogram of the average specific gravity test of F1, F2 and F3 microemulsions with replication 3 times.

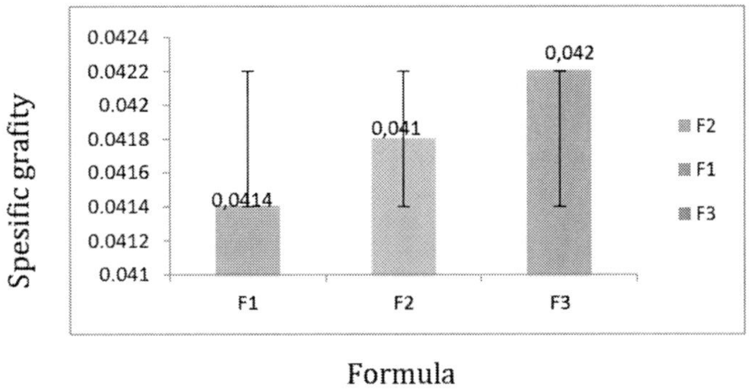

Figure 3. Histogram of the average specific gravity test of F1, F2 and F3 microemulsions with replication 3 times.

Based on Figure 3, it shows that the specific gravity of formula 3 was higher at 0.0422±0.0003 g/ml compared to the specific gravity of F1 at 0.0414±0.0005 g/ml and F2 at 0.0484±0.0001 g/ml.

The microemulsion created is known as Oil in Water (O/W) microemulsion or oil in water. The addition of methylene blue to the O/W type salmon oil microemulsion preparation turns the water phase (dispersion medium) blue and the oil phase (dispersed phase) colourless. Following the microemulsion tests, the droplet size was measured.

The droplet size shows that the droplet size of F3 is smaller at 1436,0nm±108,8 (1.4μm) compared to F1 2067,7 nm±42,5 (2.1μm) and F2 2137,6 nm±101,5 (2.1μm). The uniformity of particle size can be observed with the polydispersity index value. The polydispersity index values for each formula are F1 0.249±0.01; F2 0.382±0.05; and F3 0.431±0.01. The droplet size data can be seen in Figure 4 Histogram of the average droplet size of F1, F2 and F3 microemulsions with replication 3 times. Normality testing shows the droplet size of the obtained significance value of each formula, among others: F1: 0.209; F2: 0.525; F3: 0.658 so that the significance value >0.05, the data is normally distributed. Furthermore, homogeneity testing was carried out which showed that the value of droplet size was homogeneous. Statistical analysis test with one way ANOVA method shows a significance value of 0.000 <0.05 which explains that when given treatment in each formula gives a significant difference in the value of droplet size.

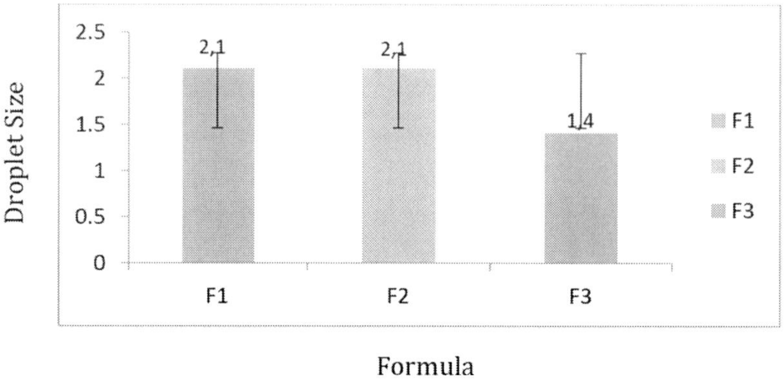

Figure 4. Histogram of the average droplet size of F1, F2 and F3 microemulsions with replication 3 times.

The viscosity characteristics of the initial microemulsion formation showed that the viscosity of the O/W type salmon oil microemulsion with a shear rate of 30 rpm in F3 of 586.6±75.7 cPs was greater than that of 540±52.9 cPs in F1 and 580±41.6 cPs in F2. The viscosity data can be seen in Figure 5 Histogram of the average viscosity of F1, F2 and F3 microemulsions with replication 3 times.

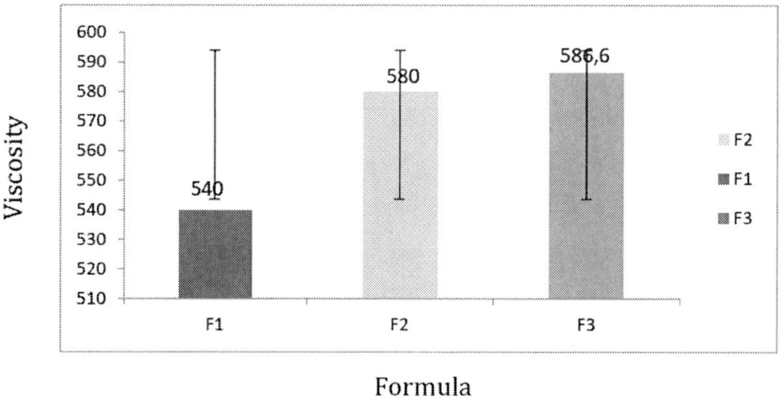

Figure 5. Histogram of the average viscosity of F1, F2 and F3 microemulsions with replication 3 times.

3.3 Physical Stability of salmon oil microemulsion

Physical stability tests were conducted using centrifugation and cycling. The centrifuge test was evaluated in terms of organoleptic properties. The evaluation was performed at the start of the microemulsion production, and after the cycle test, the centrifugation results revealed that the microemulsion had a distinct odor, a light yellow color, and no phase separation. The cycling test lasted 12 days and consisted of 6 cycles of 24 hours at 4°C ± 40°C. Following the cycle test, organoleptic, pH, specific gravity, and viscosity measurements were performed.

The pH evaluation after the cycling test showed that the pH value of F3 was lower at 6.29±0.2 compared to the pH size of F2 at 6.57±0.3 and F1 at 6.72±0.2. Figure 6 shows a histogram of the mean pH test of F1, F2, and F3 microemulsions with three replications.

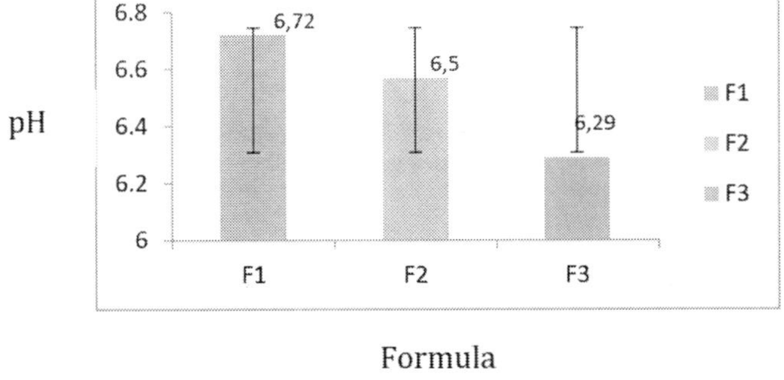

Figure 6. shows a histogram of the mean pH test after cycling test of F1, F2, and F3 microemulsions with replication 3 times.

Evaluation of specific gravity after the cycling test showed that the specific gravity of formula 3 was higher at 0.0419±0.0006 g/ml compared to the specific gravity of F1 at 0.0414±0.0003 g/ml and F2 at 0.0409±0.0003 g/ml.

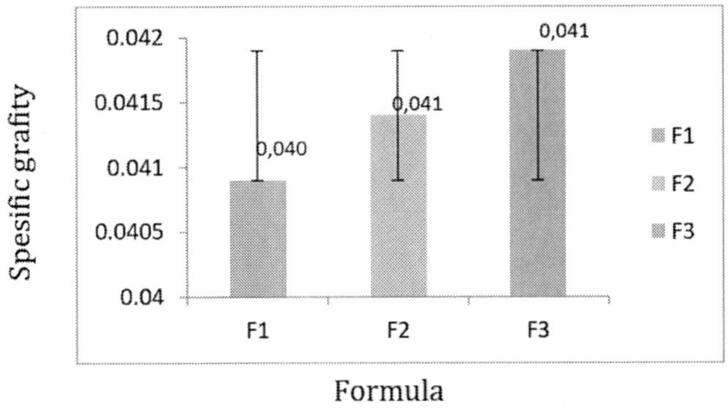

Figure 7. Histogram of the average specific gravity test after cycling test of F1, F2 and F3 microemulsions with replication 3 times.

The specific gravity value was then tested for normality with Saphiro wilk showing the specific gravity values of all formulas were normally distributed because the specific gravity of all formulas showed a significance value >0.05. Furthermore, homogeneity testing was carried out which showed that the specific gravity values were homogeneous. Statistical analysis test with SPSS one-way ANOVA method shows a significance value of 0.107>0.05 which explains that when given treatment in each formula, there is no significant difference in the specific gravity value. Furthermore, a comparison was made between the specific gravity before the cycling test and the specific gravity after the cycling test using the Paired t-test method. A significance value of 0.116>0.05 was obtained, indicating that there was no significant difference between the specific gravity before and after the cycling test.

The viscosity characteristics of the microemulsion with a shear rate of 30 rpm showed that the viscosity of the microemulsion after the cycling test in F3 with an average of 480 ± 124.9 cPs was greater than F1 at 433.3 ± 41.6 cPs and F2 440 ± 52.9 cPs.

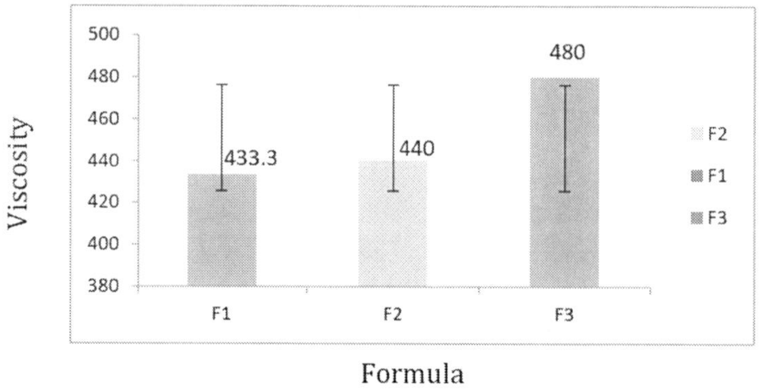

Figure 8. Histogram of the average viscosity after cycling test of F1, F2 and F3 microemulsions with replication 3 times.

The decrease in microemulsion viscosity after the cycling test can be caused by an increase in temperature which causes displacement of oil molecules and a decrease in cohesive tension between molecules. The results of the viscosity normality test using Saphiro wilk showed that the viscosity values of all formulas were normally distributed with a normality value >0.05. Furthermore, homogeneity testing was carried out which showed that the viscosity values of all formulas were homogeneous. Then the statistical analysis test with SPSS using the one-way ANOVA method shows a significance value of $0.762>0.05$ which explains that when given treatment in each formula shows no significant difference in viscosity value. Viscosity before and after the cycling test was compared using the Paired t-test method. A significance value of $0.000<0.05$ was obtained, indicating that there was a significant difference between the viscosity of the microemulsion before and after the cycling test.

4. Discussion

During the final phase of preparing microemulsions, an ultrasonic device is used to reduce the size and produce droplets of varied diameter sizes [18]. The length of 1 hour was chosen since the microemulsion created during the 30 minute sonication procedure did not appear clear, therefore additional time was extended to 60 minutes or 1 hour. The microemulsion system created appears clear in the presence of mist. After 24 hours of storage room temperature, it formed a clear and translucent microemulsion. Storing the microemulsion preparation at room temperature improved the clarity of the produced microemulsion [20]. Temperature and high surfactant concentrations influence cloudy formation in microemulsion systems, causing surfactants to mix and water loss as a result of hydration between molecules [11].

The pseudoternary phase diagram has three components: water, oil, and surfactants - cosurfactants with a 100% concentration [21,22]. At the start of the formula optimization process, the concentration of surfaktan and kosurfaktan was 12.72%. This is consistent with surfactan's 1-10% concentration as an emulsifying agent [23]. According to figure 1 the percentage concentration of microemulsion components that can make a salmon oil microemulsion with a clear/transparent appearance is 15% oil, 70% surfactant and cosurfactant (Smix), and 15% water. [15] found that microemulsions were generated at surfactant concentrations greater than 30%, with 70% surfactant showing clear results in the preparation. [16] found that the surfactant mixture was 70-85%. The microemulsions of salmon oil surfactant and cosurfactant F1 (6:1), F2 (7:1), and F3 (8:1) were synthesized using the formula with the selected concentrations of oil, surfactant, and water ratio (15:70:15). Previous research has shown that microemulsions can be generated and match the characterization criteria when the surfactant and cosurfactant concentration ratios are 5:1 and 9:1 [8,17].

Physical characterization with the organoleptic test or physical appearance of the microemulsion revealed that F3 (8:1) was clearer and more transparent than microemulsion formulae F2 (7:1) and F1 (6:1). The pH test in F3 was higher at 7.05 compared to the pH size in F2 of 6.50 and F1 of 6.66. The pH value found is consistent with the pH of the intestine, which ranges from 6-7 [19]. The pH normality test using Saphiro wilk yielded a normally distributed pH with the following formulas significant values: F1: 0.484; F2: 0.949; F3: 0.529, indicating that the significance value is more than 0.05. Furthermore, homogeneity testing was performed, which revealed that the pH value was uniform with a significance value of 0.567>0.05. Statistical analysis test with SPSS using the one-way ANOVA method on pH value yields a significance value of 0.162>0.05, indicating that there is no significant difference in pH value across treatments.

The specific gravity results obtained were then tested for normality with Saphiro wilk which showed that the specific gravity of F2 <0.05 indicated that it was not normally distributed while F1 and F3>0.05 indicated that the data was normally distributed. So the test was continued with Kruskal wallis, because the normality value of specific gravity in one of the formulas did not meet the requirements of ANOVA testing where the data must be normally distributed. The Kruskal wallis test shows that the specific gravity of F3 is greater than F1 and F2, then the significance result is 0.061>0.05 which indicates that when treatment is carried out in each formula, it does not make a significant difference in the specific gravity value.

According to particle size data, this can happen because when there is a high concentration of surfactant, more surfactant molecules are adsorbed at the oil-water interface, causing surface tension to decrease and smaller droplet sizes to develop [24]. The droplet size results are in accordance with the droplet size range according to [11] which explains that the emulsion-microemulsion droplet size is in the range of 0.15-2μm. Previous investigations found microemulsion droplet sizes ranging from 1.71μm to 7.93μm [8,25] and 9.34μm in research [26]. However, in several studies that have been conducted, the droplet size obtained with units of μm can be said if the preparation is a microemulsion.

The uniformity of droplet size in the microemulsion system can be observed with the polydispersity index value. The polydispersity index must be less than 0.5 to indicate uniform droplet size or monodispersion, whereas greater than 0.5 indicates non-uniform droplet size or polydispersion [27]. The polydispersity index values of the three formulas, F1 0.249; F2 0.382; and F3 0.431, indicate values less than 0.5, thus the droplet size distribution is uniform, forming a monodisperse system. The drawback of this study did not test zeta potential to determine the stability of the microemulsion system developed with zeta potential requirements of ± 30 mV. This study focuses on particle size, polydispersity index and physical appearance, specifically the clear / transparent development of a microemulsion system.

Stirring time, stirring speed, and temperature are further factors that influence the development of microemulsion systems in addition to oil, surfactant-cosurfactant, and water ratios [28]. The contact area between wetted particles and surfactants-cosurfactants can be influenced by the amount of time stirred during microemulsion preparation. Meanwhile, increasing the stirring speed can improve particle uniformity and minimize particle size [13].

The results of the viscosity normality test using Saphiro wilk showed that the significance value of each formula included: F1: 0.363; F2: 1.000; F3: 0.253 so that the significance value is> 0.05. Furthermore, homogeneity testing was carried out which showed that the significance value of viscosity was 0.095>0.05. Statistical analysis test with SPSS using the one-way ANOVA method shows a significance value of 0.559>0.05 which explains that when given treatment in each formula, there is no significant difference in viscosity value. The flow properties of the microemulsion formed are pseudoplastic non-newtonian flow properties.

Physical stability tests were conducted using centrifugation and cycling. The centrifugation test results showed a stable O/W microemulsion with no phase separation or stabil. The physical appearance of the microemulsion O/W after the cycling test reveals that microemulsion formula F3 (8:1) is clearer and more transparent than microemulsion formulas F2 (7:1) and F1 (6:1), as F1 appears cloudy in comparison to F3, which appears clear or transparent. It becaused nonionic surfactants, particularly those with polyoxyethylene groups like Tween 80, are temperature sensitive and so impact the thermodynamic stability of the system. The nonionic surfactant becomes increasingly lipophilic as the temperature rises, as the polyoxyethylene group, which acts as a polar group, dehydrates, This increases the interfacial tension between oil and water, which might alter the appearance of the microemulsion.

The reduction in pH value from before the cycling test can be attributed to the hydrolysis reaction. Normality test with Saphiro wilk showed that the pH values of all formulas were normally distributed because the pH values of all formulas were >0.05. Furthermore, homogeneity testing was carried out which showed that the pH value was homogeneous. Statistical analysis test with one-way ANOVA method shows a significance value of 0.228>0.05 which explains that when given treatment in each formula shows no significant difference in pH value. Comparison between pH before cycling test and pH after cycling test using Paired t-test method. The results obtained a significance value of 0.302>0.05 which indicates that there is no significant difference between pH values before and after the cycling test.

Based on the data above, it shows that the physical characteristics of O/W type microemulsions before and after the cycling stability test have changed, among others, from the pH value where there is an increase in F1 and F2, while in F3 the pH value has decreased. The specific gravity value decreased in all formulas after the cycling test. And the viscosity of the microemulsion also decreased in all formulas.

Acknowledgement
The authors are grateful to the authorities of Department of Pharmaceutics, Faculty of Pharmacy, Universitas Hang Tuah Surabaya, Indonesia for the facilities.

Conflict Of Interest
The authors declare no conflict of interest.

References
[1] Wenaty, A., Mabiki, F., Chove, B., & Mdegela, R. 2018. Fish Consumers Preferences, Quantities Of Fish Consumed And Factors Affecting Fish Eating Habits: A Case Of Lake Victoria In Tanzania. International Journal of Fisheries and Aquatic Studies, **6**(6), 247-252.
[2] Riawa, A.O., Yulinda E., dan Darwis, A.N. 2017. Factors Influencing Community Consumption Of Fresh Fish In Rumbai Pesisir District. Pekanbaru, Riau.
[3] Venugopalan, V. K., Gopakumar, L. R., Kumaran, A. K., Chatterjee, N. S., Soman, V., Peeralil, S., ... & Nagarajarao, R. C. 2021. Encapsulation And Protection Of Omega-3-Rich Fish Oils Using Food-Grade Delivery Systems. Foods, **10**(7), 1566.
[4] Zhang, B., Xiong, K., Cai, J., & Ma, A. 2020. Fish Consumption And Coronary Heart Disease: A Meta-Analysis. Nutrients, **12**(8), 2278.
[5] Damongilala, L. J. 2021. The Nutritional Content Of Fish Food. CV. Patra Media Grafindo, Bandung, pp. 29
[6] Setyawardhani, D. A., Laras, D. S., & Prasetya, K. J. 2020. The Influence of Solvent Variation and Composition on the Extraction of Omega Fatty Acids from Grape Seed Oil through Urea Complexation. Journal of Chemical Process Engineering, **5**(1), 1-8.
[7] Beandrade, M. U. 2018. Formulation And Characterization Of SNEDDS From Black Cumin Extract (Nigella Sativa) With Shark Liver Oil Phase (Centrophorus Sp) And Immunostimulant Activity Testing. Journal of Pharmaceutical Science and Clinical Research, **1**(5), 234-244.
[8] Ratnasari Devi, Harry Noviardi, Baiq Annisa. 2017. The Influence Of Surfactant And Co-Surfactant Ratios On The Characteristics And Stability Of Olive Oil Microemulsions (Olive Oil). Pharmamedika Journal. **2**(2).
[9] Saber, S. 2022. of Thesis: New Trends In The Field Of Pharmaceutical. sciences, **10**(1), 43-52.
[10] Aprilya, A., Rahmadevi, R., & Meirista, I. 2021. Nanoemulsion Formulation With Fish Oil (Oleum Iecoris Aselli) Base Ingredients. Journal of Science and Health, **3**(3), 370-375.

[11] Martin, A. 2011. Physical Pharmacy: Physical Chemical Principles in the Pharmaceutical Sciences. **50th** ed. Maryland: Williams and Wilkins.
[12] Kale, S. N., & Deore, S. L. 2017. Emulsion Micro Emulsion And Nano Emulsion: A Review. Systematic Reviews in Pharmacy, **8**(1), 39.
[13] Nazari, M., Mehrnia, M. A., Jooyandeh, H., & Barzegar, H. 2019. Preparation and Characterization Of Water In Sesame Oil Microemulsion By Spontaneous Method. Journal of Food Process Engineering, **42**(4), e13032.
[14] Swarbrick James. 2007. Encyclopedia of Pharmaceutical Technology Third Edition. USA : Informa Healthcare,Inc. pp. 1548-1561
[15] Najih, Y. A., Rahma, D. N., Nailufa, Y., & Prasetyowati, R. D. 2020. The Effect Of Vegetable Oil Combination And Surfactant PEG 7 Glyceryl Cocoate Ratio On Physical Characteristics And Physical Stability Of Arbutin Microemulsion. Romanian Journal of Pharmaceutical Practice Vol. **13**. 52(3).
[16] Gunarto, C., Ju, Y. H., Putro, J. N., Tran-Nguyen, P. L., Soetaredjo, F. E., Santoso, S. P., ... & Ismadji, S. (2020). Effect Of A Nonionic Surfactant On The Pseudoternary Phase Diagram And Stability Of Microemulsion. Journal of Chemical & Engineering Data, **65**(8), 4024-4033.
[17] Pessoa R.S, Eduardo L.Z, Elton B.R. 2015. Microemulsion of Babassu oil as a Natural Product to Improve Human Immmune System Function. Drug design, Development and Therapy: **9**. Dovepress.
[18] Manickam, S., Sivakumar, K., & Pang, C. H. 2020. Investigations On The Generation Of Oil-In-Water (O/W) Nanoemulsions Through The Combination Of Ultrasound And Microchannel. Ultrasonics Sonochemistry, **69**, 105258.
[19] Jusnita, N., & Nasution, K. 2019. Formulation Of Nanoemulsion From Moringa Leaf Extract (Moringa Oleifera Lamk). Industria: Journal of Technology and Agro-Industrial Management, **8**(3), 165-170.
[20] Barzegar, H., Mehrnia, M. A., Nasehi, B., & Alipour, M. 2018. Fabrication Of Peppermint Essential Oil Nanoemulsions By Spontaneous Method: Effect Of Preparing Conditions On Droplet Size. Flavour and Fragrance Journal, **33**(5), 351-356.
[21] Deshmukh P.D, Salunkhe K.S. et al. 2016. Microemulsions: A Novel Approach For Drug Delivery System. Journal of Advanced Drug Delivery.**3**(2).
[22] Muzzaffar F, Singh UK, Chauhan L. 2013. Review on Microemulsion As Futuristic Drug Delivery. International Journal of Pharmacy and Pharmaceutical Sciences. **5** (3).pp. 39-52.
[23] Rowe, R.C., Sheskey, P.J., and Quinn, M.E., 2009, Handbook of Pharmaceutical Excipients, 6th Ed, Washington D.C: Pharmaceutical Press and American Pharmacists Association., pp. 580-584.
[24] Saberi AH, Fang Y, McClements, DJ. 2013. Fabrication of Vitamin E-enriched Nanoemulsions: Factors Affecting Particle Size Using Spontaneous Emulsification, Journal of Colloid and Interface Science **391**. pp 95-102
[25] Irawati Putri Silvi, Dina Rahmawanty, Mia Fitriani. 2017. Characterization of Patchouli Oil Microemulsion (Pogostemon cablin Beth) Using Virgin Coconut Oil (VCO), Polysorbate 80, and Sorbitol. Journal of Pharmascience. **4**(1) .pp.109-115.
[26] Husniyah Jauharatul, Rahmi Annisa,Burhan Ma'arif. 2018. Formulation and Antibacterial Activity Test of Staphylococcus aureus Microemulsion of Anting-anting (Acalypha Indica) Leaf Extarct Using Isoprophyl Miristat as Oil Phase. J Islamic Pharm **3**(1): 1-7.
[27] Destiyana, O. Y., & Rijai, L. 2018, December. Formulation Of Nanoemulsion Combining Rose Flower Extract (Rosa Damascena Mill.) And Yam Bean Tuber Extract (Pachyrhizus Erosus L.) Using Virgin Coconut Oil (VCO) As The Carrier Oil. In Proceeding of Mulawarman Pharmaceuticals Conferences, **8**, pp. 254-259.
[28] Setyopratiwi, A., & Hanifah, H. T. U. 2022. Formulation and Stability of Oil-in-Water Microemulsions with Virgin Coconut Oil (VCO) as the Oil Phase Using Emulsification Method. In Proceedings of the National Chemistry Seminar. **1**, pp. 108-123.

IOP Conf. Series: Earth and Environmental Science **1473** (2025) 012057 doi:10.1088/1755-1315/1473/1/012057

Hydrolyzed Marine Collagen Loaded Nanoemulsion : Characteristics and Irritation Study Based on Vegetable Oils Combination

D N Rakhma[1*], Y A Najih[1] and Y Andhiarto[1]

[1]Pharmacy Program Study, Faculty of Pharmacy, Universitas Hang Tuah

*E-mail: dita.nurlita@hangtuah.ac.id

Abstract. The exploration of natural sources for collagen to combat skin aging has gained significant attention, particularly through marine sources known as marine collagen. This study focuses on Hydrolyzed Marine Collagen, which, due to its hydrophilic nature, faces challenges in penetration. The nanoemulsion delivery system offers the advantage of overcoming Hydrolyzed marine collagen limitations. The study examined four formulations using different fixed oils— Virgin Coconut Oil (VCO), Olive Oil, and Jojoba Oil—combined with the surfactant PEG-7 Glyceryl Cocoate. The ratios of oil phase to surfactant to water phase were as follows: FN 1 (4:40:56), FN 2 (6:45:49), FN 3 (10:65:25), and FN 4 (15:70:15). All formulations resulted in a clear yellow consistency and were oil-in-water (o/w) emulsions. pH testing revealed that F4 had the highest pH at 6.72 ± 0.03, while the particle size analysis showed F1 had the smallest particles at 20.24 ± 1.89 nm. All formulations had a polydispersity index of less than 1, indicating uniformity. Irritation tests using HET-CAM indicated no irritation from any of the formulations. The study concludes that the ratios of oils, surfactants, and water in the nanoemulsions influence their characteristics while ensuring no irritation.

1. Introduction

Extended exposure to ultraviolet (UV) radiation can cause changes in skin texture, which can accelerate premature skin aging [1]. The manifestation of visible symptoms of premature aging, including dullness, wrinkles, dark spots, and other markers, has a detrimental impact on self-confidence [2,3]. Therefore, it is imperative to use topical therapy to tackle this problem, and counter-aging cosmetics represent a possible solution. With increasing awareness among those struggling with premature aging, there is a corresponding growth in the demand for effective anti-aging cosmetics. This phenomenon has significantly increased the distribution of such products on the market [4].

Collagen is one of the key active components in these cosmetics; this protein is essential for preserving the strength and stability of bodily tissues [5,]. Skin must remain tight and elastic to maintain its natural production of collagen, one of the most common proteins in the body [6]. In cosmetics, collagen is utilized topically and through injections to repair and replace damaged tissue, thereby reducing the signs of aging. Collagen has extensive use in diverse domains, including pharmaceutical administration, wound healing, and tissue regeneration [7]. Collagen can be obtained from both plant and animal origins [8]. Its versatility and effectiveness make it a popular choice in formulating anti-aging products.

Collagen derived from marine biota is called marine collagen. Previous research in vivo tests have shown that collagen extracted from Tilapia fish scales (*Oreochromis niloticus*) can produce wound-healing effects in mice induced by burns [9,10]. Hydrolyzed Marine Collagen is marine collagen with a molecular weight between 0.3 and 8 kDa, which suggests that its low molecular weight could allow for easier skin penetration. However, due to collagen's hydrophilic nature, it faces challenges in penetrating the stratum corneum barrier [11]. To overcome this, an effective delivery system is necessary to enhance the penetration of active ingredients through the stratum corneum [12].

Nanoemulsions have become an up-and-coming delivery technique in the cosmetic industry, specifically for this objective. Nanoemulsion is a formulation of oil-based delivery systems characterized by globules ranging from 50 to 500 nm in size. This system comprises an oil phase, a water phase, surfactants, and cosurfactants to stabilize the combined system [13]. The presence of the oil phase in the nanoemulsion system improves the lipophilic active ingredient's solubility and imparts the specific properties of the nanoemulsion. An emerging trend in the cosmetic industry is using vegetable oil in formulae development [14]. The emollient classification of vegetable oils in cosmetics is attributed to their effective softening and smoothing characteristics. They inhibit water loss by creating a protective layer on the epidermis, softening the stratum corneum, and minimizing skin inflammation because of the fatty acid content in vegetable oil [15]. Using vegetable oil as the oil phase in the nanoemulsion system can reduce the permeation barrier, thereby increasing penetration into the skin [16].

This study used a combination of VCO, Olive oil, and Jojoba oil as oil phases. Virgin Coconut Oil (VCO) is rich in high medium-chain fatty acids designed to enhance skin health by acting as an anti-inflammatory agent [17]. Olive oil and Jojoba oil can enhance skin hydration by preventing trans-epidermal water loss [18,19].PEG 7 Glyceryl Cocoate was used as a surfactant to form a Hydrolyzed Marine Collagen nanoemulsion. The evaluations included characteristics and in vitro irritation to obtain the optimal formula for anti-aging cosmetics.

2. Materials and Methods

2.1. Materials

The materials used in this study are Hydrolyzed Marine Collagen as the active ingredient; Virgin Coconut Oil (VCO), Olive Oil, and Jojoba Oil as the oil phase; the surfactant is PEG-7 Glyceryl Cocoate, and distilled water. Methylene blue is used for emulsion-type coloring. The equipment used in this study is glassware, a Hotplate stirrer, a Particle Size Analyzer, and a pH meter.

2.2. Preparation of Nanoemulsion

Formulation process started by dissolving Hydrolyzed Marine Collagen In 3 mL of water. This solution was then combined with PEG 7 Glyceryl Cocoate and heated in a water bath at 30°C using a porcelain cup. Next, the mixture was blended with the oil phase, which included VCO, olive oil, and jojoba oil, on a hotplate while stirring at a speed of 560 rpm and maintaining a temperature of 30°C. The stirring continued for about 5 minutes until the mixture appeared clear. The completed formulations were transferred into containers for further characteristic and irritation testing (Table 1).

2.3. pH Characteristics Test

The nanoemulsion is poured into a beaker. Then, the electrode of the pH meter is immersed in the nanoemulsion until a stable reading is obtained. The measurements are taken at room temperature, with the desired pH range for this study being 4-6.

2.4. Particle Size and Polydispersity Index Measurement

The test was conducted using a Particle Size Analyzer (PSA) by weighing 1.0 grams of the preparation and diluting it with distilled water to a total volume of 10 ml. The sample was placed in a cuvette, which was then inserted into the sample holder. After powering the device, the particle size option was selected. The average droplet diameter and Polydispersity Index (PI) were recorded. The Polydispersity Index (PI) indicates the variation within the sample, with a PI value below 0.3 signifying that the sample is monodisperse.

2.5. Nanoemulsion Type
One gram of nanoemulsion was applied to a watch glass, adding a drop of blue methylene. Then, visual observation was carried out to determine whether the type of nanoemulsion was oil in water (o/w) or water in oil (w/o).

2.6. HET-CAM Irritation Test
The Hen's Egg Test Chorioallantoic Membrane (HET-CAM) method was employed for the irritant test. The fertilized Leghorn chicken eggs were incubated at 37°C, with the air cavity positioned at the top. The eggs underwent daily rotation for a duration of ten days. On the tenth day, the eggs were ignited to assess fertilization, and any eggs lacking viable embryos were disposed of. The air sac of the viable eggs was delineated, and the exterior eggshell in the designated region was delicately severed with aseptic scissors. In order to facilitate this operation, the shell is softened using a sterile 0.9% NaCl solution.

Table 1. Nanoemulsion Formula.

Formula	Ingredients		
	VCO : Olive Oil : Jojoba Oil (1 :1 : 1) (%)	Water (%)	PEG 7 *Glyceryl cocoate* (%)
FN 1	4	56	40
FN 2	6	49	45
FN 3	10	25	65
FN 4	15	15	70

Following the removal of the outer shell, the egg's outer membrane is soaked with a warm 0.9% NaCl solution and then returned to the incubator for 5-20 minutes to facilitate the removal of the outer membrane. Only eggs that have no damage to the chorioallantoic membrane (CAM) after removing the outer membrane are chosen. A sample weighing 300 mg is applied onto the CAM and allowed to adhere for 20 seconds. After 20 seconds, the CAM is rinsed with sterile 0.9% sodium hypochlorite acid. The observation period of 300 seconds commences immediately after the cleaning of the CAM from the sample. As an irritating control, sodium lauryl sulfate is employed (PC gorup), while water is the negative control (NC group).
The data obtained in the HET-CAM test is calculated using the formula:

$$T = \frac{301-H}{300} \times 5 + \frac{301-L}{300} \times 7 + \frac{301-C}{300} \times 9 \tag{1}$$

Description:
T = Irritation score
H = The time necessary to induce hemorrhage (seconds)
L= The time necessary to induce lysis (seconds)
C = The time necessary to induce coagulation (seconds)

The damage to the hen's egg chorioallantoic membrane was categorized based on the mean as non-irritant (0–0.9), slight irritant (1.0–4.9), moderate irritant (5–8.9), and extreme irritant (9–21).

2.7. Statistical Analysis
An analysis of the quantitative physical attributes testing findings was conducted using SPSS software with the One Way Anova approach, with a significance level of $p < 0.05$. A significant difference is detected, and the post hoc test is further conducted using Tukey's Test.

3. Results and Discussion
The concentration of the oil phase and surfactant used in the nanoemulsion formula will determine the physical characteristics of the system formed (Table 2)[20]. One of them is the pH of each formula. This is affected by differences in the concentration of surfactants used in the formula where the more surfactant concentration used, the pH of the preparation increases, namely FN 1 < FN 2 < FN 3 < FN1. PEG-7 Glyceryl Cocoate surfactant has a pH of 7, so the more surfactant is used, the higher the pH of the preparation will be compared to the value of PEG-7 Glyceryl Cocoate [21]. The pH test results showed that all nanoemulsion formulas met the skin pH range of 4-6. If the pH is too acidic, it can cause skin irritation; if the skin pH is too alkaline, it will cause scaly skin [22].

Table 2. Physcial Characteristics Test Result.

Formula	Visual Characteristic	pH	Particle Size (nm)	Polydispersity Index	Emulsion Type
FN 1	Clear, Yellow	5,11 ± 0,03	20,24±1,89	0,357± 0,011	o/w
FN 2	Clear, Yellow	5,34± 0,02	25,75±1,41	0,383± 0,007	o/w
FN 3	Clear ,Yellow	5,6± 0,02	124,70±0,26	0,224± 0,004	o/w
FN 4	Clear, Yellow	5,72± 0,03	126,40±1,15	0,220± 0,005	o/w

The statistical analysis resulted in a significant difference ($p < 0.05$) in particle size based on the order of formulas with the smallest particle size: F1 (20.24 ± 1.89 nm) < F2 (25.75 ± 1.41 nm) < F3 (124.70 ± 0.46 nm) = F4 (126.40 ± 1.15 nm). These findings suggest that the micelle formation mechanism is responsible for the increase in particle size, which is aligned with the increasing concentration of surfactants. When conditions above the Critical Micelle Concentration (CMC) threshold, surfactants lose their capacity to diminish surface tension, since it has attained saturation, resulting in micelles existing in dynamic equilibrium [23]. The addition of an excessive amount of surfactant will result in the surfactant remaining in the liquid, causing the molecules to aggregate and form a micelle structure. Consequently, the particle size will increase [24]. All formulas have a polydispersity index value close to zero. This value indicates a homogeneous or uniform particle distribution, while an index value exceeding 0.5 suggests that the particles have a high level of heterogeneity [25].

Table 3. HET-CAM Test Result.

	Formula					
	NC	PC	FN 1	FN 2	FN 3	FN 4
Irritation Score	0 ± 0	19,30 ± 0,7	0 ± 0	0 ± 0	0 ± 0	0,3 ± 0,1

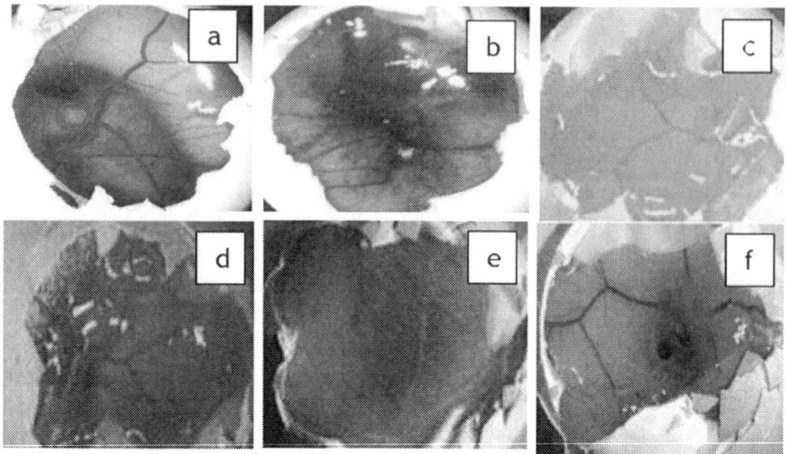

Figure 1. Visual appearance of the eggs after HET-CAM test a). NC (Negative Control), b). PC (Positive Control), c). FN 1, d)FN 2, e) FN 3, f)FN 4

The HET-CAM method carried out an irritation test (Figure 1). In formula FN 1- 3, the irritation score was 0; in formula FN 4, the irritation score showed a value of 0,3. The scoring test results show that all formulas have a score <0,9, which means they are not irritating (table 3). These results indicate that using three combinations of vegetable oil and PEG-7 Glyceryl Cocoate surfactants up to 70% does not risk irritation. Each concentration of PEG 7 Glyceryl Cocoate as a surfactant used in this work was non-irritating and exhibited good physical properties. However, the stability and efficacy of each formula still need to be discovered. Therefore, further investigation is required to gather the necessary data to develop a product suitable for application.

4. Conclusions

According to the study findings, the ratio of oil mixtures to surfactant influences the physical characteristics of Hydrolyzed Marine Collagen nanoemulsions, such as pH, particle size, and polydispersity index. All evaluated formulations exhibited advantageous properties and showed no irritation, rendering them appropriate as anti-aging nanoemulsion formulations. This indicates that the Hydrolyzed Marine Collagen nanoemulsion system may be developed into a skincare product.

References

[1] Krutmann J, Schikowski T, Morita A, Berneburg M. Environmentally-induced (extrinsic) skin aging: Exposomal factors and underlying mechanisms. *Journal of Investigative Dermatology*. 2021 Apr 1;**141**(4):1096-103. https://doi.org/10.1016/j.jid.2020.12.011

[2] Murase D, Kusaka-Kikushima A, Hachiya A, Fullenkamp R, Stepp A, Imai A, Ueno M, Kawabata K, Takahashi Y, Hase T, Ohuchi A. Autophagy declines with premature skin aging resulting in dynamic alterations in skin pigmentation and epidermal differentiation. *International Journal of Molecular Sciences*. 2020 Aug 9;**21**(16):5708. https://doi.org/10.3390/ijms21165708

[3] Passeron T, Krutmann J, Andersen ML, Katta R, Zouboulis CC. Clinical and biological impact of the exposome on the skin. *Journal of the European Academy of Dermatology and Venereology*. 2020 Jul;**34**:4-25. https://doi.org/10.1111/jdv.16614

[4] Sharma A, Kuhad A, Bhandari R. Novel nanotechnological approaches for treatment of skin-aging. *Journal Of Tissue Viability*. 2022 Aug 1;**31**(3):374-86. https://doi.org/10.1016/j.jtv.2022.04.010

[5] Resende DI, Ferreira M, Magalha˜es C, Lobo JS, Sousa E, Almeida IF. Trends in the use of marine ingredients in anti- aging cosmetics. Algal Research. 2021 May 1;**55**:102273. https://doi.org/10.1016/j.algal.2021.102273

[6] Wang H. A review of the effects of collagen treatment in clinical studies. Polymers. 2021 Nov 9;**13**(22):3868. https://doi.org/10.3390/polym13223868

[7] Barzkar N, Sukhikh S, Babich O, Venmathi Maran BA, Tamadoni Jahromi S. Marine Collagen: Purification, Properties And Application. Frontiers In Marine Science. (2023). **28**;10:1245077. https://doi.org/10.3389/fmars.2023.1245077

[8] Reilly DM, Lozano J. Skin Collagen Through The Lifestages: Importance For Skin Health And Beauty. *Plast Aesthet Res*. 2021;8:2. http://dx.doi.org/10.20517/2347-9264.2020.153

[9] Sionkowska A, Adamiak K, Musiał K, Gadomska M. Collagen based materials in cosmetic applications: A review. Materials. (2020). **23**;13(19):4217. https://doi.org/10.3390/ma13194217

[10] Aguirre-Cruz G, Leo´n-Lo´pez A, Cruz-Go´mez V, Jime´nez-Alvarado R, Aguirre-A' lvarez G. Collagen Hydrolysates For Skin Protection: Oral Administration And Topical Formulation. *Antioxidants*. 2020 Feb 22;**9**(2):181. https://doi.org/10.3390/antiox9020181

[11] Felician FF, Xia C, Qi W, Xu H. Collagen from marine biological sources and medical applications. Chemistry & biodiversity. 2018 May;**15**(5):e1700557. https://doi.org/10.1002/cbdv.201700557

[12] Geahchan S, Baharlouei P, Rahman A. Marine Collagen: A Promising Biomaterial For Wound Healing, Skin Anti-Aging, And Bone Regeneration. *Marine Drugs*. 2022 Jan 10;**20**(1):61. https://doi.org/10.3390/md20010061

[13] Kaul S, Gulati N, Verma D, Mukherjee S, Nagaich U. Role Of Nanotechnology In Cosmeceuticals: A Review Of Recent Advances. *Journal of Pharmaceutics*. 2018;**2018**(1):3420204. https://doi.org/10.1155/2018/3420204

[14] Ribeiro AR, Silva SS, Reis RL. Challenges And Opportunities On Vegetable Oils Derived Systems For Biomedical Applications. *Biomaterials Advances*. 2022 Mar 1;**134**:112720. http://dx.doi.org/10.1016/j.msec.2022.112720

[15] Ahmad A, Ahsan H. Lipid-Based Formulations In Cosmeceuticals and Biopharmaceuticals. *Biomedical Dermatology*. 2020 Dec;**4**:1-0. https://doi.org/10.1186/s41702-020-00062-9

[16] Poljsˇak, N., Kreft, S., & Kocˇevar Glavacˇ, N. (2020). Vegetable Butters and Oils in Skin Wound Healing: Scientific Evidence For New Opportunities In Dermatology. *Phytotherapy research*, **34**(2), 254-269. https://doi.org/10.1002/ptr.6524

[17] Wiyani L, Aladin A, Juniar ME. Antioxidant activity of Virgin Coconut Oil and Virgin coconut Oil Emulsion. *Systematic Reviews in Pharmacy*. 2020 Dec 1;**11**(12).

[18] Elmowafy M, Musa A, Alnusaire TS, Shalaby K, Fouda MM, Salama A, Al-Sanea MM, Abdelgawad MA, Gamal M, Fouad SA. Olive Oil/Pluronic Oleogels For Skin Delivery Of Quercetin: In Vitro Characterization And Ex Vivo Skin Permeability. *Polymers*. 2021 May 31;**13**(11):1808. https://doi.org/10.3390/polym13111808

[19] McMullen RL. The Benefits and Challenges of Treating Skin With Natural Oils. *International Journal of Cosmetic Science*. 2024 Aug;46(4):553-65. https://doi.org/10.1111/ics.12960

[20] Teeranachaideekul V, Boribalnukul P, Morakul B, Junyaprasert VB. Influence of Vegetable Oils on In Vitro Performance of Lutein-Loaded Lipid Carriers for Skin Delivery: Nanostructured Lipid Carriers vs. Nanoemulsions. *Pharmaceutics*. 2022 Oct 11;**14**(10):2160. https://doi.org/10.3390/pharmaceutics14102160

[21] Najih, Y. A., RahmA, D. N., NailufA, Y., & Prasetyowati, R. D. he effect of vegetable oil combination and surfactant PEG 7 glyceryl cocoate ratio on physical characteristics and physical stability of arbutin microemulsion. *Romanian Journal of Pharmaceutical Practice* Vol. XIII. 2020;**52**(3). https://doi.org/10.37897/RJPhP.2020.3.5

[22] Proksch E. pH in Nature, Humans and Skin. *The Journal of Dermatology*. 2018 Sep;**45**(9):1044-52. https://doi.org/10.1111/1346-8138.14489

[23] Mustan F, Politova-Brinkova N, Vinarov Z, Rossetti D, Rayment P, Tcholakova S. Interplay Between Bulk Aggregates, Surface Properties and Foam Stability of Nonionic Surfactants. *Advances in Colloid and Interface Science*. 2022 Apr 1;**302**:102618. https://doi.org/10.1016/j.cis.2022.102618

[24] Zhu Y, Free ML. Introduction to Surfactants. In Surfactants in Precision *Cleaning*. 2022 Jan 1 (pp. 1-53). https://doi.org/10.1016/B978-0-12-822216-4.00003-3

[25] Danaei MR, Dehghankhold M, Ataei S, Hasanzadeh Davarani F, Javanmard R, Dokhani A, Khorasani S, Mozafari MR. Impact of particle size and polydispersity index on the clinical applications of lipidic nanocarrier systems. *Pharmaceutics*. 2018 May 18;**10**(2):57. https://doi.org/10.3390/pharmaceutics10020057

Antibacterial activity of chitosan from natural resource: *Penaeus monodon* shell with various concentrations and deacetylation degrees against *Lactobacillus acidophilus*

W Prananingrum[1*], Sularsih[1], F Rahmitasari[1], D G Ramadan[1], P B Prabowo[1], E Radhianto[2] and K Parisihni[3]

[1]Department of Dental Materials, Faculty of Dentistry, Universitas Hang Tuah, Surabaya, Indonesia;
[2]Research Centre of Microbiology, Universitas Airlangga, Surabaya, Indonesia
[3]Department of Oral Biology, Faculty of Dentistry, Universitas Hang Tuah, Surabaya, Indonesia.

*E-mail: widyasri.prananingrum@hangtuah.ac.id

Abstract. *Lactobacillus acidophilus (L. acidophilus)* is commonly found in the deep caries. Calcium hydroxide, a standard for direct pulp capping, has drawbacks such as tunnel defects. This study investigates the antibacterial activity of chitosan synthesized from *Penaeus monodon (P. monodon)* shell at various concentrations and deacetylation degrees (DD) against *L. acidophilus*. Chitosan synthesized from *P. monodon* shell as natural resource from Tarakan, Indonesia with DD 81.21% and 92.96% was prepared, each diluted on concentrations of 3%; 1.5%; 0.75%; 0.375%; 0.188%; 0.094%; 0.046%; and 0.023%. Antibacterial activity was examined by dilution method and colony counting on BHI agar. Data were analyzed using *Kruskal-Wallis* and *Mann-Whitney* tests. Chitosan DD 81.21% showed zero growth at 0.375% to 3%, while 0.023% to 0.188% performed growth (11.25 to 115 CFU/ml). Significant differences were observed, except for the 0.375% to 3% ($p<0.05$). Chitosan 92.96% showed zero growth at 0.188% to 3%, while 0.023% to 0.094% performed growth (10.5 to 86.5 CFU/ml). Significant differences were observed, except for the 0.188% to 3% ($p<0.05$). In conclusion, chitosan DD 81.21% was effective at 0.375% to 3%, while chitosan DD 92.96% exhibited pronounced effectivity at 0.188% to 3%. Increasing concentration and DD of chitosan from *P. monodon* shell enhances antibacterial activity.

1. Introduction

The exposed vital pulp due to mechanical trauma from a bur or deep caries cleaning is treated through direct pulp capping therapy by applying a dental material to the exposed area to promote the development of reparative dentin tissue and preserve pulp vitality. Applying a wound-healing material on the exposed pulp aims to block bacterial entry, initiate soft tissue wound healing, and reparative dentin in the exposed area to prevent progression to irreversible pulpitis, which can ultimately lead to pulp necrosis. Calcium hydroxide ($Ca(OH)_2$) remains the material of standard for direct pulp capping therapy due to its ability to promote reparative dentin formation. Additionally, $Ca(OH)_2$ is highly soluble. $Ca(OH)_2$ has antibacterial properties because $Ca(OH)_2$ ionizes into Ca^{2+} and OH^-, releasing hydroxyl ions which subsequently increase the pH, creating a strong basic or alkaline medium (pH 11 - 12.5) [1]. A basic environment inhibits bacterial growth [2,3]. However, this alkaline property with such a pH can increase the risk of inflammation and/or necrosis in the superficial pulp layers and apical lesions. Meanwhile, this condition stimulated the pulp to form reparative dentin, studies show that 89% of reparative dentin formed with $Ca(OH)_2$ contained tunnel defects. These defects may create dead spaces and result in microleakage which may compromise the long-term effectiveness of the biological seal against bacterial infection [4,5]. In this respect, achieving a well-sealed restoration is crucial to prevent leakage and ensure long-term success. Furthermore, the addition of antibacterial substrates to the direct pulp capping materials is preferred for enhanced advantages [5].

Chitosan is a natural resource widely used in medical fields due to its biocompatibility and biodegradability. Chitosan has antibacterial properties due to its functional $-NH_2$ groups of chitosan. The number of the functional $-NH_2$ groups reflects the extent to which acetyl groups have been removed from chitin in the deacetylation process to produce chitosan. Chitosan is a polysaccharide also known as *β-1,4-2 acetamido-2-dioksi-D-glukosa* synthesized from the deacetylation process of crustacean shells. Indonesia is a country rich in crustacean production, including shrimp. Indonesia estimated increase in shrimp production in 2025 will reach 1.216.086 tonnes and *Penaeus monodon* (*P. monodon*) cultivation is still superior in Tarakan, Kalimantan [6-8]. The quality of chitosan is determined by the source and the method by which it is synthesized [9]. Chitosan from shrimp shells results in higher degrees of deacetylation (DD), solubility, and viscosity than from crab and squilla shells [10]. Similar research also indicates that chitosan from giant tiger prawn *(P. Monodon*) can produce a high degree of deacetylation. Moreover, using shrimp shell waste to produce chitosan will pronounce economic and biological value by approaching the zero-waste process [11]. In global, *P. monodon* is one of the most commercially of shrimp species and about 40% of the head and body carapace shrimp is discarded as waste [12]. The concentration of alkali used in deacetylation is the most critical element for effectively extracting chitosan from shrimp wastes [13]. Various chitosan synthesis methods result in different DD, but only chitosan with DD of more than 70% can be used as an industrial product. Chitosan is classified with 56-69% as a low deacetylation degree, 70-85% as a middle deacetylation degree, and 86-95% as a high deacetylation degree [14]. The antibacterial characteristics of chitosan are affected by various factors, including the source, deacetylation degree, molecular weight, concentration, and bacterial species [15].

Lactobacillus acidophilus (*L. acidophilus*) is one of the most commonly found bacteria in deep caries, with a prevalence of 42.3% in dentin carious lesions [16]. *L. acidophilus* is a gram-positive, facultative anaerobic, rod-shaped bacterium that produces acid from carbohydrates and rapidly decreases the pH. Evaluation of the antibacterial effect of $Ca(OH)_2$ using the diffusion method shows an inhibition zone of 3 mm against *L. acidophilus*. A study on 2.5% chitosan synthesized from haruan fish with a deacetylation degree of 85.25% shows antibacterial properties against *L. acidophilus* with an inhibition zone diameter of 8.6 mm [17]. Meanwhile, 1.5% chitosan synthesized from haruan fish can inhibit the enzyme glucosyltransferase activity of *Streptococcus mutans* by 82.69 units [18].

The antibacterial activity of chitosan synthesized from *P. monodon* shell at various concentrations and deacetylation degrees against *L. acidophilus* has not been examined yet. Nevertheless, Kovacs et al. [19] investigated the MIC value and concentration dependence of commercially available chitosan derivatives with varying molecular weights against *Lactobacillus plantarum*. The MIC values were consistently 0.007% regardless of the molecular weights of the chitosan and its inhibitory effects, which were dependent on the chitosan concentration. The impact of the deacetylation degree of chitosan synthesized from the carapace of *P. monodon* on the antibacterial activity of *Escherichia coli*, *Staphylococcus aureus*, and *Candida albicans* was examined by Hongpattarakere et al [20]. However, this chitosan was produced through a deacetylation process using 50% NaOH at 100°C under vacuum, nitrogen, and regular atmospheres, resulting in chitosan with DD in the range 13.03% to 74.80% [20].

Our study aims to investigate the antibacterial activity of chitosan from *P. monodon* shell at various concentrations and deacetylation degrees against *L. acidophilus*. Thus, we evaluated the colony counting of *L. acidophilus* for chitosan with deacetylation degrees (DD) 81.21% and 92.96%; and each was diluted on concentrations of 3%; 1.5%; 0.75%; 0.375%; 0.188%; 0.094%; 0.046%; and 0.023%.

2. Materials and Methods

Chitosan powder synthesized from *P. monodon* shell from Tarakan, Kalimantan, Indonesia with deacetylation degree (DD) 81.21% and DD 92.96% was prepared. The *P. monodon* shells were first cleaned with water and then sun-dried. Next, the shells were ground and homogenized through a 200 mesh sieve to produce a fine powder (< 74 μm). Chitosan DD 81.21% was synthesized by the 4-step method: deproteinization, demineralization, depigmentation, and deacetylation process. Thus, each step was done at 65°C. Chitosan DD 92.96% was produced by the 4-step method: deproteinization, demineralization, depigmentation, and deacetylation process. Thus, each step was done at 75°C.

Deproteinization stage: *P. monodon* shell powder was mixed with 3.5% NaOH (Merck, USA) in a 1:10 (w/v) ratio. The mixture was stirred for 2 h and then the residue was rinsed with distilled water until it reached a neutral pH. It was then dried for 24 hours. Demineralization stage: chitin powder was treated with 1 M HCl (Merck, USA) in a 1:15 (w/v) ratio. The mixture was stirred for 1 hour at room temperature and then the residue was rinsed with distilled water until the neutral pH was achieved. Depigmentation stage: chitin powder was immersed in acetone (Merck, USA) in a 1:10 (w/v) ratio for 20 hours. The solution was then filtered and rinsed with distilled water until a neutral pH was obtained. The powder was dried for 1 hour. Deacetylation stage: chitin powder from the depigmentation stage was treated with 50% NaOH (Merck, USA) in a 1:10 (w/v) ratio. The mixture was heated, filtered, and rinsed with distilled water until a neutral pH was achieved. Finally, the powder was dried for 1 hour, resulting in chitosan powder. Afterward, chitosan DD 81.21% named the Chi-1 group, and chitosan DD 92.96% named the Chi-2 group. Each was made into a solution by dissolving 1.5 gr chitosan powder in 50% ml of 1% (v/v) acetic acid to obtain a 3% (w/v) chitosan solution. Thereafter, this solution was diluted to prepare 3%; 1.5%; 0.75%; 0.375%; 0.188%; 0.094%; 0.046%; and 0.023% chitosan solution.

L. acidophilus (ATCC 4356) was utilized for this study. The *L. acidophilus* culture procedures adhered to the Research Ethics Committee of the Faculty of Dentistry at Universitas Hang Tuah, with approval number EC/017/KEPK-FKGUHT/II/2024. The *L. acidophilus* was grown on Brain Hearth Infusion Broth (BHIB) medium and incubated at 37°C for 48 h. The *L. acidophilus* cultures were then standardized to 0.5 McFarland, equivalent to 1.5×10^8 CFU/ml. Antibacterial activity was assessed using a dilution method and colonies were counted by spreading on Muller-Hinton agar [21].

Two variations of chitosan solution, each of 3% chitosan with DD 81.21% and DD 92.96% were prepared in eight tubes. Tubes 2 to 8 contained 0.5 ml BHIB medium, while tube 1 had 1 ml of 3% chitosan solution. Sequential dilutions were made by transferring 0.5 ml from tube 1 to tube 2 through 8, resulting in chitosan concentrations of 3%; 1.5%; 0.75%; 0.375%; 0.188%; 0.094%; 0.046%; and 0.023%. Each tube was inoculated with 0.1 ml of *L. acidophilus* suspension at 0.5 Mc Farland standard (1.5×10^8 CFU/ml). Positive control tubes contained BHIB with 0.1 *L. acidophilus* suspension, while negative control contained BHIB without *L. acidophilus*. The tubes were incubated anaerobically at 37°C for 48 h. The *L. acidophilus* subcultures were then prepared by spreading 0.1 ml of *L. acidophilus* from each tube onto Muller-Hinton agar, which was incubated at 37°C for 48 h, and the colonies were counted using the Quebec Colony Counter, reported as CFU/ml.

The data were analyzed using version 20.0 of the Statistical Package for Social Science (SPSS) for Windows, provided by IBM corporation, Ilinois, Chicago, US. The statistical analysis was performed using a non-parametric *Kruskal-Wallis* test followed by *Mann-Whitney* tests for multiple comparisons. Significant difference at $p < 0.05$.

3. Results and Discussion

The bacterial colonies of *L. acidophilus* in this study are presented in Figure 1 (A) and (B). There was zero growth of *L. acidophilus* after being treated with chitosan synthesized from *P. monodon* DD 81.21% at 3%; 1.5%; 0.75%; and 0.375% concentrations. Meanwhile, the growth of *L. acidophilus* started to appear from the concentration of 0.188% and increased at 0.094%; 0.046%; and 0.023% concentrations. Figure 1 (B) showed zero growth of *L. acidophilus* after being treated with chitosan synthesized from *P. monodon* DD 92.96% at 3%; 1.5%; 0.75%; 0.375%; and 0.188% concentrations, while the growth of *L. acidophilus* started from concentration of 0.094% and increased at 0.046%; and 0.023% concentrations.

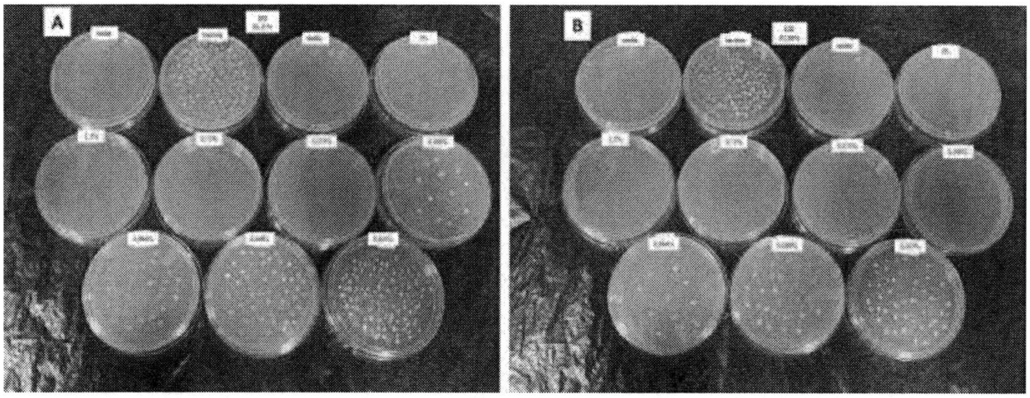

Figure 1. The bacterial colonies of *L. acidophilus* after treated with chitosan synthesized from *P. monodon* DD 81.21% at 3%; 1.5%; 0.75%; 0.375%; 0.188%; 0.094%; 0.046%; and 0.023% concentrations (A). The bacterial colonies of *L. acidophilus* after treated with chitosan synthesized from *Penaeus monodon* DD 92.96% at 3%; 1.5%; 0.75%; 0.375%; 0.188%; 0.094%; 0.046%; and 0.023% concentrations (B).

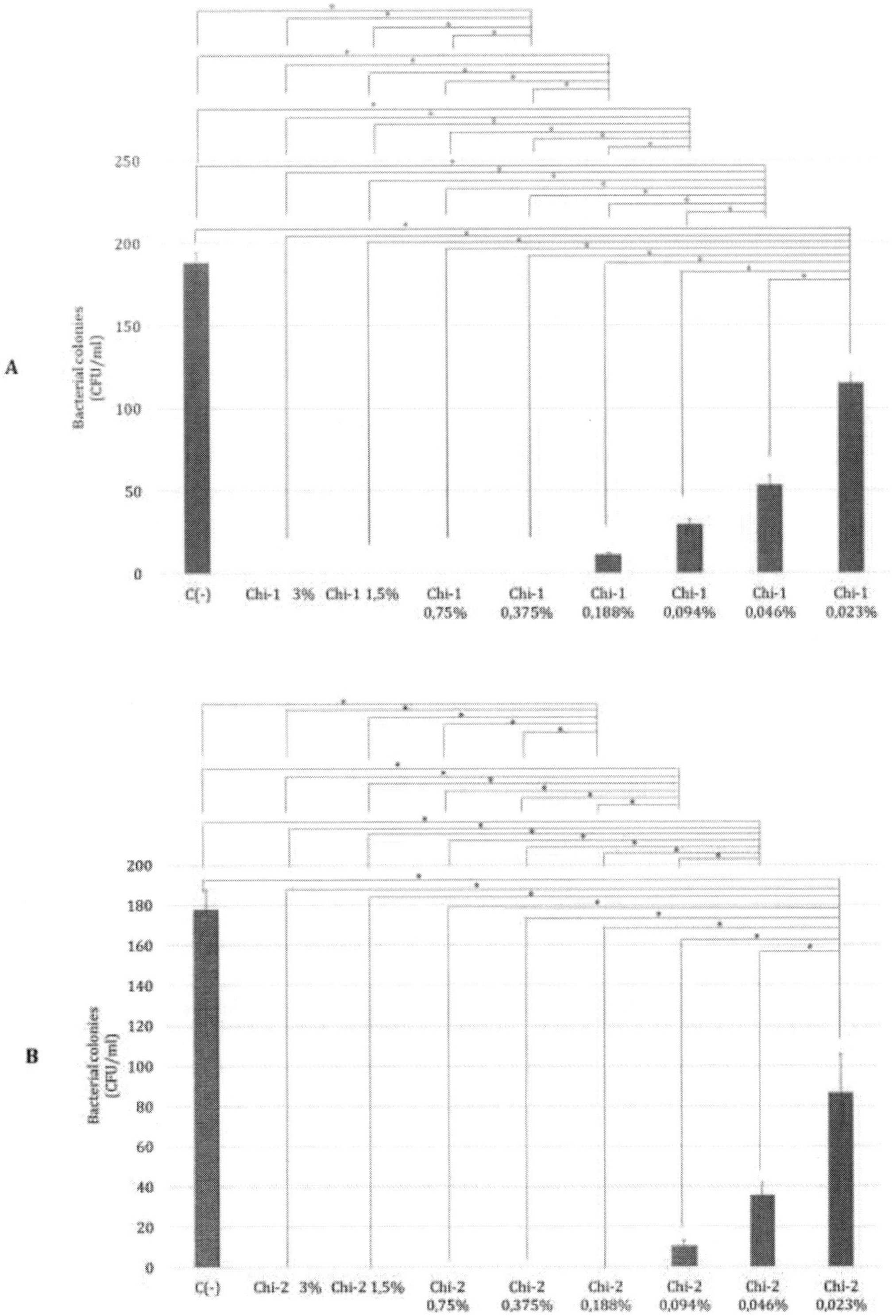

Figure 2. The bacterial colonies of *L. acidophilus* after treated with chitosan synthesized from *P. monodon* DD 81.21% at 3%; 1.5%; 0.75%; 0.375%; 0.188%; 0.094%; 0.046%; and 0.023% concentrations (A). The bacterial colonies of *L. acidophilus* after treated with chitosan synthesized from *Penaeus monodon* DD 92.96% at 3%; 1.5%; 0.75%; 0.375%; 0.188%; 0.094%; 0.046%; and 0.023% concentrations (B). Significant difference at $p < 0.05$ (*).

The mean of bacterial colonies of *L. acidophilus* in this study is presented in Figure 2 (A) and (B). Furthermore, Figure 2 (A) showed the mean of bacterial colonies of *L. acidophilus* after being treated with chitosan synthesized from *P. monodon* DD 81.21% (Chi-1) at 3%; 1.5%; 0.75%; and 0.375% concentrations showed zero, while 0.188%; 0.094%; 0.046%; and 0.023% performed 11.25; 29.75; 53.75; and 115 CFU/ml, respectively. The growth of the bacterial colonies of *L. acidophilus* started from a concentration of 0.188%, i.e. 11.25 CFU/ml. The statistical analysis revealed significant differences among the groups, except groups of 3%; 1.5%; 0.75%; and 0.375% concentrations. Figure 2 (B) showed the mean of bacterial colonies of *L. acidophilus* after being treated with chitosan synthesized from *P. monodon* DD 92.96% (Chi-2) at 3%; 1.5%; 0.75%; 0.375%; and 0.188% concentrations showed zero, while the concentration of 0.094%; 0.046%; and 0.023% performed 10.5; 35.75; and 86.5 CFU/ml, respectively. The growth of the bacterial colonies of *L. acidophilus* started from the concentration of 0.094%, i.e. 10.5 CFU/ml. The statistical analysis revealed significant differences among the groups, except groups of 3%; 1.5%; 0.75%; 0.375%; and 0.188% concentrations.

It seems the chitosan synthesized from *P. monodon* shells with DD 81.21% started a decrease in antibacterial activity against *L. acidophilus* from 0.188% to 0.023% concentration. Likewise, chitosan synthesized from *P. monodon* shells with DD 92.96% also demonstrated a decrease in antibacterial activity against *L. acidophilus* from 0.094% to 0.023% concentration. The results indicated the variations of chitosan concentration for both groups showed varying antibacterial activities. As the concentration of chitosan decreased, the number of bacterial colonies of *L. acidophilus* increased. Conversely, as the concentration of chitosan increased, the number of bacterial colonies of *L. acidophilus* decreased. The chitosan concentration reflects the amount of chitosan in a solution. Chitosan contains amino groups (-NH_2) which have high reactivity towards bacterial cell walls. Some studies suggested that a high concentration of chitosan improves antibacterial activity by enhancing its coverage of bacterial surfaces and decreasing the membrane permeability [19,22]. The interaction between the amino groups (-NH_2) and the bacterial cell wall can lead to the release of cytoplasmic substances, thereby compromising the membrane integrity and causing elevated levels of potassium and phosphate ions in the cytoplasm [19]. A similar study also proved the antibacterial activity of chitosan synthesized from three different sources (*Blaps lethifera*, *Pimelia fernadezlopezi*, and *Musca doemstica*). increased with increasing the concentration of chitosan [23].

Chitosan DD 92.96% is classified as a high deacetylation degree of chitosan, while chitosan DD 81.21% is classified as a middle deacetylation degree of chitosan [9]. This study showed chitosan synthesized from *P. monodon* shells with DD 92.96% exhibits antibacterial activity against *L. acidophilus* at lower concentrations compared to chitosan DD 81.21%. Chitosan DD 92.96% showed antibacterial activity against *L. acidophilus* from the first dilution to the sixth dilution or from 3% to 0.094% concentration. Meanwhile, chitosan DD 81.21% showed antibacterial activity against *L. acidophilus* only up to the fifth dilution or 0.188% concentration. The degree of deacetylation (DD) of chitosan is indicated by the number of free amino groups (-NH_2) present in polysaccharides. It reflects the extent to which acetyl groups have been removed from chitin to produce chitosan [13]. Chitosan DD 92.96% has longer molecular chains due to the higher number of amino groups (-NH_2) compared to chitosan DD 81.21%. The amino groups (-NH_2) of chitosan have bactericidal and bacteriostatic properties [24]. The amino groups (-NH_2) on chitosan have positive charges and high reactivity, leading to differences in electro-negativity when chitosan binds with the negatively charged bacterial cell surface. The amino groups (-NH_2) also have free electron pairs that can attract Mg^{2+} and Ca^{2+}. Then, form covalent bonds where Mg^{2+} and Ca^{2+} act as electron acceptors, disrupting the cytoplasmic membrane and thereby bacterial metabolism and growth. A higher degree of deacetylation (DD) leads to greater positive charge density, resulting in stronger electrostatic interactions [25].

It can be suggested from the findings of this study, that there seems to be a concentration and deacetylation degree-dependent antibacterial activity against *L. acidophilus*. Within the limitation of this study, the results indicated that DD 92.96% with concentrations ranging from 0.188% to 3% exhibits the strongest antibacterial activity against *L. acidophilus*. The optimal balance of concentration, deacetylation degree, and molecular weight of chitosan synthesized from *P. monodon* may be a crucial factor for its antibacterial effectiveness against *L. acidophilus*.

4. Conclusion

In conclusion, the chitosan DD 81.21% demonstrated significant antibacterial activity with concentrations ranging from 0.375% to 3%. Meanwhile, the chitosan DD 92.96% exhibited pronounced antibacterial activity with concentrations ranging from 0.188% to 3%. Increasing the concentration and DD of chitosan synthesized from the *P. monodon* shell enhances its antibacterial activity due to the decrease in the number of bacterial colonies of *L. acidophilus*. The chitosan synthesized from *P. monodon* shell from Tarakan Waters with DD 92.96% and concentrations ranging from 0.188% to 3% can be considered for direct pulp capping materials because the antibacterial properties contribute to direct pulp capping treatment. However, further studies are required, including diverse examinations and in vivo studies.

Acknowledgment

This research was supported by Internal Research Funding, Universitas Hang Tuah for the fiscal year 2024, under appointment number: B/061/UHT.C2/II/2024.

References

[1] Nicholson J, Czarnecka B 2016. *Elsevier Woodhead publishing*.
[2] Da Silva LAB, Leonardo MR, Da Silva RS, Assed S, Guimarães LF 1997. *Int. Endod. J.* **30** 403- 6.
[3] Leonardo MR, Silva LAB, Tanomaru-Filho M, Bonifácio KC, Ito IY 2000. *J. Endod.* **26** 391-4.
[4] Taira Y, Shinkai K, Suzuki M, Kato C, Katoh Y 2011. *Odontology.* **99** 135-147.
[5] Komabayashi T, Zhu Q, Eberhart R, Imai Y 2016. *Dent Mater Journal.* **35** 1-12.
[6] Fisheries M of M A 2015. Jakarta, Indonesia: *The Center for Data, Statistic and Information MMAF.*
[7] Pantjara B, Suwoyo HS, Rusdi I 2021. *IOP Conf. Ser.: Earth Environ. Sci.* **860** 012032.
[8] Pantjara B 2008. *Proceedings of the National Seminar on Fisheries and Marine Affairs BDP FPIK-UNBRA-BBRSEKP-BRKP*. 174-80
[9] Yadav M, Kaushik B, Rao GK, Srivastava CM, Vaya D 2023. *Carbohydrate Polymer Technologies and Applications* **5** n. pag.
[10] Parthiban F, Balasundari S, Gopalakannan A, Rathnakumar K, Felix S 2017. *Curr World Environ.* **12.**
[11] Phuong PTD, Minh NC, Cuong HN, Van Minh N, Han NT, Van Hoa N, Yen HTH, Trung TS 2017. *J Food Sci Technol* **54** 1850-1856.
[12] Wani AK, Akhtar N, Mir T.u.G. et al 2024. *Environ Sci Pollut Res.* **31** 38960–38989.
[13] Hosney A, Ullah S, Barčauskaitė KA 2022. *Mar Drugs.* **20** 675.
[14] Li B, Elango J, Wu W 2020. *Applied Sciences (Switzerland).* **10** 1–24.
[15] Ardean C, Davidescu CM, Nemeş NS, Negrea A, Ciopec M, Duteanu N, Negrea P, Duda-Seiman D, Musta V 2021. *Int J Mol Sci.* **22** 7449.
[16] Rôças I, Alves F, Rachid C, Lima K, Assunção I, Gomes P, Siqueira Jr J 2016. *PLOS ONE.* **10** 1-13.
[17] Wahyuda TW, Putri DKT, Nahzi MYI 2023. *Dentin (J. Ked. Gigi).* **VII** 108 – 113.
[18] Rachfa MAF, Putri DKT, Dewi RK 2021. *Dentin (J. Ked. Gigi).* **V** 87 – 91.

[19] Kovács R, Erdélyi L, Fenyvesi F, Balla N, Kovács F, Vámosi G, Klusóczki Á, Gyöngyösi A, Bácskay I, Vecsernyés M, Váradi J 2022. Pharmaceutics. **15** 18.
[20] Hongpattarakere T, Riyaphan O 2008. *J. Sci. Technol.* **30** 1-9.
[21] Nugraha AP, Triwardhani A, Sitalaksmi RM, *et al.* 2023. *J. Oral Biol Craniofac Res.* **13** 720-726.
[22] Wahjuningrum DA, Setyabudi S, Hadinata D, Amalia IT, Ro'ifa RA, Juniarti DE 2022. *Malaysian Journal of Fundamental and Applied Sciences.* **18** 483-488.
[23] Amor IB, Hemmami H, Laouini SE *et al* 2024. *Biomass Conv. Bioref.* **14** 16245–16255.
[24] Matica MA, Aachmann FL, Tøndervik A, Sletta H, Ostafe V 2019. *Inter J of Mol Sci.* **20**.
[25] Confederat LG, Tuchilus CG, Dragan M, Sha'at M, Dragostin OM 2021. *Molecules.* **26** 3694.

IOP Conf. Series: Earth and Environmental Science **1473** (2025) 012059 doi:10.1088/1755-1315/1473/1/012059

The Coast's Natural Resources Offer a Promising in Silico Approach for Alzheimer's Disease Medication

P Wibowo and R P Putra

Faculty of Medicine, Hang Tuah University, Surabaya, Indonesia

*E-mail: prajogo.wibowo@hangtuah.ac.id

Abstract. Alzheimer's disease (AD) is becoming a significant health issue on an ecumenical scale, with a substantial financial and health impact on families and societies. It is characterized mostly by complete dementia, which includes memory loss, cognitive impairment, executive dysfunction, pearsonality alteration, and behavior changes. Furthermore, the majority of patients experience symptoms of mental disturbance. There are presently no treatments that can prevent or cure AD; however, medicine can provide temporary respite for these symptoms. Several studies have shown a correlation between the development of AD, disturbances in lipid metabolism, and neuroinflammation. We used an in silico strategy to investigate the bioactive flavonoids in the coastal prickly pear cactus cladode. This cactus cladode has four active compounds that bind well to and interact with the inhibitor sites of cyclooxygenase-2 (COX-2), soluble epoxide hydrolase (sEH), and cytosolic 3-hydroxy-3-methylglutaryl-CoA synthase (HMGCS1). The study suggests coastal prickly pear cactus cladode, which contains four bioactive flavonoids, may hold promise as a potential natural AD medication.

1. Introduction

Alzheimer's disease (AD) has become a prominent and increasingly important health issue on an ecumenical level, imposing considerable economic and health challenges on society and families [1,2]. Presently, the global population that lives with dementia exceeds 55 million, with over 60% of them residing in developing nations.

AD is predominant kind of dementia, accounting for approximately 60–70% of cases. Estimates suggest that approximately 27.9% of Indonesia's population aged 65 and over suffers from dementia, including Alzheimer's disease. More than 4.2 million Indonesians are currently living with dementia, emphasizing a significant public health concern for this country [3].

AD is mostly identified by a form of dementia known as complete dementia. This form of dementia includes memory loss, cognitive impairment, executive dysfunction, and changes in behavior and personality. Moreover, symptoms of mental disturbance are experienced by the majority of patients [4]. Although meticulous care and medication can alleviate these symptoms temporarily, no proven therapies exist to prevent or cure AD at this time [5]. To find a new drug to treat Alzheimer's disease, more clinical studies with newer inhibitors are needed. One of these is an active substance from the coastal prickly pear cactus.

Opuntia littoralis is commonly referred to as the coastal prickly cactus. The main habitat of this plant is sandy soil. *Opuntia littoralis* grow in irregular clusters or as shrub-like mounds, which can reach a height of up to 10 feet. The presence of this particular kind of coastal prickly pear cactus is observed on the coast of Samas, located in Bantul Regency, within the Jogjakarta Province. The coastal prickly pear cactus has five functional parts: leaves, blooms, fruits, cladodes, and roots. Cactus plants, like many

IOP Conf. Series: Earth and Environmental Science **1473** (2025) 012059 doi:10.1088/1755-1315/1473/1/012059

other plants, contain various bioactive flavonoids. Various literature reviews have demonstrated the beneficial health effects of several cactus components.

Using an in silico approach, this study investigated the potential of the flavonoids in the coastal cactus as a natural treatment for AD.

2. Methods and experimental design

2.1. Protein target preparation
We have three protein targets for AD. The three-dimensional structures of cyclooxygenase-2 (PDB ID: 5IKT) [6], soluble epoxide hydrolase (PDB ID: 4JNC) [7], and HMG-CoA synthase (PDB ID: 2P8U) [8] can be accessed directly at http://www.rscb.org/pdb.

2.2. Ligand preparation
The three-dimensional structures of the eleven active compounds in the coastal prickly pear cactus cladodes were downloaded from the NCBI PubChem database in SDF format, along with their Canonical SMILE representation, as documented through various research studies (Table 1) [9]. We used the PASS online tool (https://way2drug.com/PASSOnline/index.php) to predict bioactivity [10].

Table 1. List of bioactive compounds in the coastal cactus cladode

No	Bioactive Compound	PubChem CID[a]
1	Gallic acid	370
2	Coumaric	637542
3	3,4-dihydroxybenzoic	72
4	4-hydroxybenzoic	135
5	Ferulic acid	445858
6	Salicylic acid	338
7	Isoquercetin	5280804
8	Isorhamnetin-3-O-glucoside	5318645
9	Nicotiflorin	5318767
10	Rutin	5280805
11	Narcissin	5481663

[a] CID = compound ID number from PubChem

2.3. Docking simulation
Table 2 contains the Molegro Virtual Docker 5.0.0 docking parameters for the coordinates of the center of active sites and the radius of protein targets [11]. The interactions were studied using Discovery Studio 21.1.1, and the docking results were shown in three dimensions using PyMol 2.3.

Table 2. List of the coordinates and radius for the center of active sites

Protein Targets	PDB ID	Grid X	Grid Y	Grid Z	Radius
Cyclooxygenase-2	5IKT	166.29	186.03	192.14	10
soluble Epoxide Hydrolase	4JNC	-12,99	26,62	-12,04	10
HMG-CoA synthase	2P8U	9,5	42,5	74,25	12

IOP Conf. Series: Earth and Environmental Science **1473** (2025) 012059 doi:10.1088/1755-1315/1473/1/012059

3. Results and discussion

3.1. Biological activities of four compounds in the coastal prickly pear cactus cladode

Coastal prickly pear cactus cladode gave us four compounds: rutin, nicotiflorin, isoquercetin, and isorhamnetin. These results were based on eleven active flavonoids in this plant's structural bioactivity predictions [9].

3.2. Docking score analysis

The three protein targets for this study are COX-2, sEH, and HMGCS1 for AD. COX-2 is a specific type of the cyclooxygenase enzyme family. It promotes the synthesis of prostanoids from arachidonic acid (AA), a crucial fatty acid. Prostaglandins, epoxyeicosatrienoic acids (EETs), and leukotrienes are some of the bioactive lipid mediators that AA (a PUFA with 20 carbon atoms) can offer [12]. The evidence indicates that COX-2 levels have risen in the brains of individuals with AD, suggesting a potential role in the disease's development.

Every part of the brain contains the enzyme sEH, which has two functions. Its two terminal domains are responsible for separate enzymatic activities; a hydrolase is located on the C-terminal, while a phosphatase is located on the N-terminal. Some research has connected the sEH's C-terminal part to its anti-inflammatory properties. Phosphorylated lipids, including some that are intermediates in the cholesterol biosynthesis pathway, seem to be the endogenous substrates for the sEH's N-terminal action. It regulates AA metabolism and modulates EET concentrations by transforming them into metabolites with reduced activity [13].

An essential enzyme, HMGCS1, regulates the production of cholesterol in the cytosols of brain cells. This enzyme is part of the mevalonate pathway. It facilitates the alteration of acetyl-CoA into acetoacetyl-CoA, which then transforms into HMG-CoA, a cholesterol building block [14].

Figure 2 displays the docking scores of the chemical complexes with the COX-2, sEH, and HMGCS1 protein. Upon binding to HMGCS1, rutin demonstrates the lowest score of -322.8 kJ/mol. Within the sEH complex, isorhamnetin had the lowest docking score. Furthermore, the anti-inflammatory aim specifically targets the COX-2 protein, which has the lowest binding energy of the nicotiflorin.

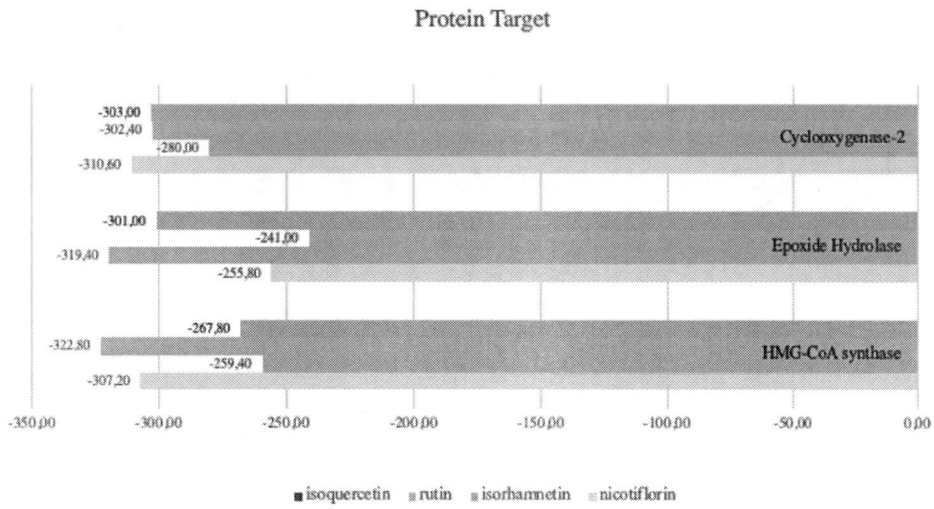

Figure 1. Barchart of MolDock Score

3.3. The four bioactive compounds in the coastal cactus cladode interact with protein targets

3.3.1. *COX-2 interaction.* The four bioactive flavonoids obtained from the cactus cladode display an overwhelming interaction with a shared active site. The study of ligand-protein complexes has shown several interactions, such as electrostatics, unfavorable bonds, hydrogen bonds, and van der Waals interactions (Figure 2). The isoquercetin active sites invariably contain the subsequent residues: ALA527, MET522, LEU352, SER353, ARG120, VAL349, VAL523, LEU531, and SER530 (Figure 3). It has been found that rutin's active site residues are at SER530, MET522, ALA527, LEU352, SER353, VAL523, PHE205, PHE381, TYR385, VAL349, LEU359, LEU531, PHE209, and LEU534. Several COX-2 residues are bound by isorhamnetin, including PHE518, SER353, GLN192, GLY526, HIS90, SER353, LEU384, MET522, TRP387, LEU352, VAL523, LEU352, and ALA527. Nicotiflorin's active site links to the COX-2 enzyme at residues MET522, SER530, ALA527, SER353, VAL523, PHE205, PHE381, TYR385, VAL349, LEU352, LEU531, PHE209, and LEU534. Depending on their active sites, three substances—isoquercetin, rutin, and nicotiflorin—bind to COX2 in the inhibitor region, specifically Ser530 and Tyr385. This suggests that these three compounds can inhibit COX-2 activity and may have anti-inflammatory properties.

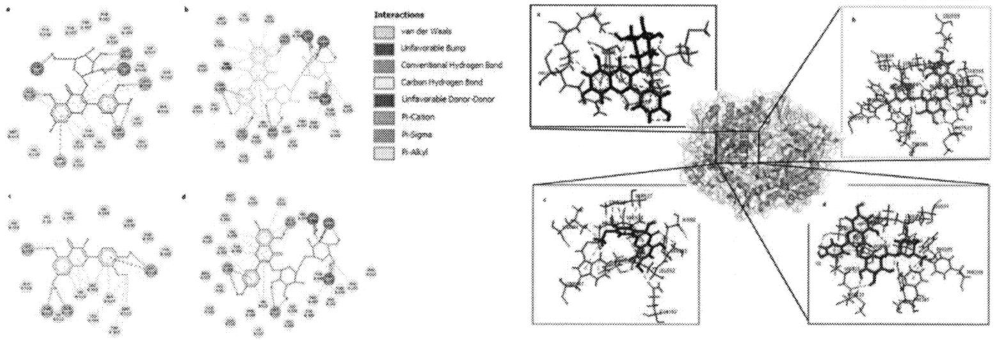

Figure 2. The interaction between cyclooxygenase-2 (COX-2) and bioactive compounds of the coastal prickly pear cactus cladode: (a) isoquercetin, (b) isorhamnetin, (c) nicotiflorin, and (d) rutin.

3.3.2. *sEH interaction.* According to the three-dimensional perspective of the ligand complex with the chemical, all four bioactive chemicals share a common binding site and serve as inhibitors of epoxide hydrolase (Figure 3). The interaction of these four cactus active chemicals induces anti-inflammatory and vasodilatory response.

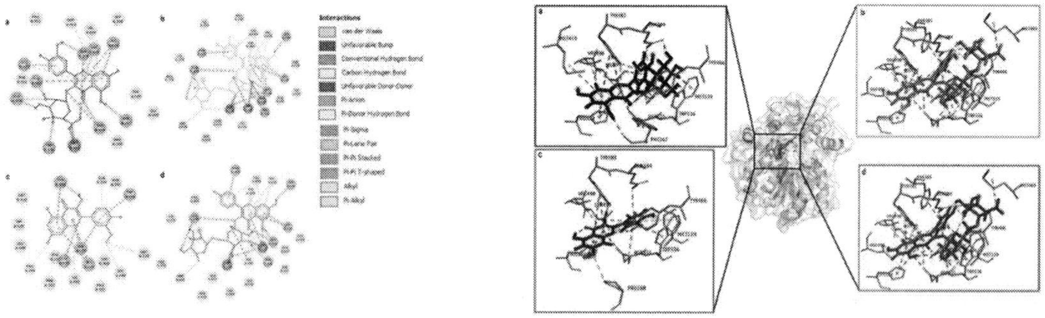

Figure 3. The interaction between soluble epoxide hydrolase (sEH) and bioactive compounds of the coastal prickly pear cactus cladode: (a) isoquercetin, (b) isorhamnetin, (c) nicotiflorin, and (d) rutin.

3.3.3. *HMGCS1 interaction.* Given the three-dimensional structure, all four substances exhibit binding affinity for HMG-CoA synthase in the same region. Notably, the isorhamnetin molecule specifically binds to the catalytic area of HMGCoA synthase, particularly His264, whereas Gly166's standard binding location is the binding region of HMGCoA synthase with the substrate. Isorhamnetin tethers the HMG CoA enzyme, resulting in its instantaneous inactivation, and its activity diminishes when it interacts with rutin (Figure 4).

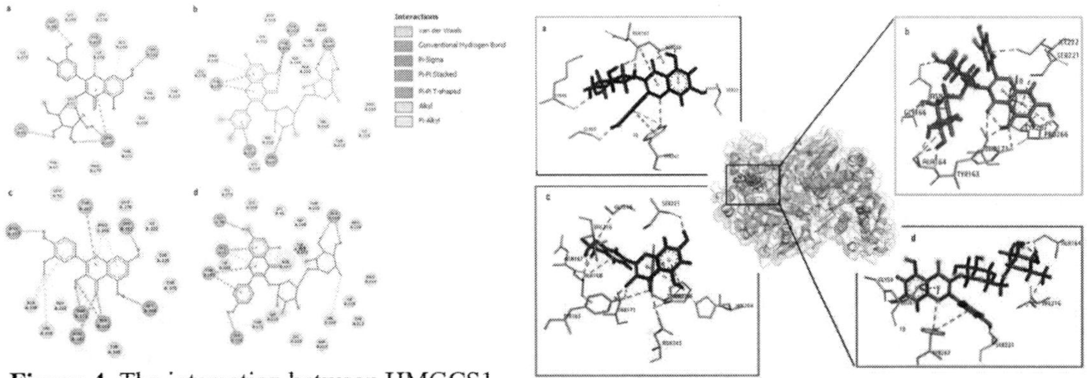

Figure 4. The interaction between HMGCS1 enzyme and bioactive compounds of the coastal prickly pear cactus cladode: (a) isoquercetin, (b) isorhamnetin, (c) nicotiflorin, and (d) rutin.

3.4. *Mechanisms of COX2, sEH, and HMGCS1 inhibitors provide significant benefits.*

Multiple variables contribute to the development of AD. Amyloid plaques, resulting from the buildup of Aβ, abundant tau protein, and intracellular neurofibrillary tangles, are a hallmark of AD. Synaptic degeneration, neuronal death, and astrocyte proliferation are the end outcomes of these processes [13].

Up to 23 percent of the cholesterol in the human body is located in the brain. Some neurodegenerative illnesses, like AD, alter the brain's cholesterol metabolism [15]. An increased risk of AD may be associated with elevated blood cholesterol levels. One possible indicator of how the brain processes cholesterol is the oxysterol 24S-hydroxycholesterol, or 24S-OHC. Deterioration of neurons, as seen in several neurological disorders, might explain why 24S-OHC levels fluctuate. Cholesterol was the primary concern of this portion because of its role in the formation and progression of AD, especially in relation to Aβ [16]. The coastal prickly pear cactus cladode has bioactive compounds that can inhibit HMGCS1 in cholesterol metabolism.

Researchers have discovered that COX-2, an important enzyme in the production of proinflammatory chemicals, significantly influences the development of AD. Neurofibrillary tangles and tau hyperphosphorylation are outcomes of COX-2 overexpression, according to research. The delicate equilibrium between tau phosphorylation and dephosphorylation relies on kinases and phosphatases, which COX-2 is known to regulate. This kinase-phosphatase imbalance is believed to cause tau hyperphosphorylation in large part. The close relationship between COX-2, tauopathies, and neurological inflammation has led researchers to consider inhibiting COX-2 as a potential therapeutic candidate for managing AD [17]. Finding natural substances that can suppress COX-2 is the authors' goal of this research.

The cytochrome P450 (CYP450) enzymes catalyze AA conversion into four regioisomeric epoxyeicosatrienoic acids: 5,6-EET, 8,9-EET, 11,12-EET, and 14,15-EET. The resulting compounds serve as potent signaling molecules and play significant functions in various physiological processes. As EETs affect multiple molecular processes related to neurodegeneration, including inflammatory, apoptotic, angiogenic, and oxidative stress, among others, they have the potential to act as neuroprotective agents [18]. Studies have shown that specific epoxyeicosatrienoic acids (EETs), such as 5,6-EET and 8,9-EET, modulate the function of Transient Receptor Potential Vanilloid 4 (TRPV4) channels. These channels are ion channels located in various tissues, including astrocytes, blood vessels, and the blood-brain barrier. They regulate calcium entry, affecting biological processes like vasodilation. In contrast, 11,12-EET blocks inflammatory stimuli's ability to activate and translocate NF-┤B to the nucleus. This contributes to lowering the inflammatory response by decreasing the expression of genes that promote inflammation [19]. Our findings indicate that isorhamnetin, isoquercetin, nicotiflorin, and rutin are beneficial in treating Alzheimer's disease during its early stages.

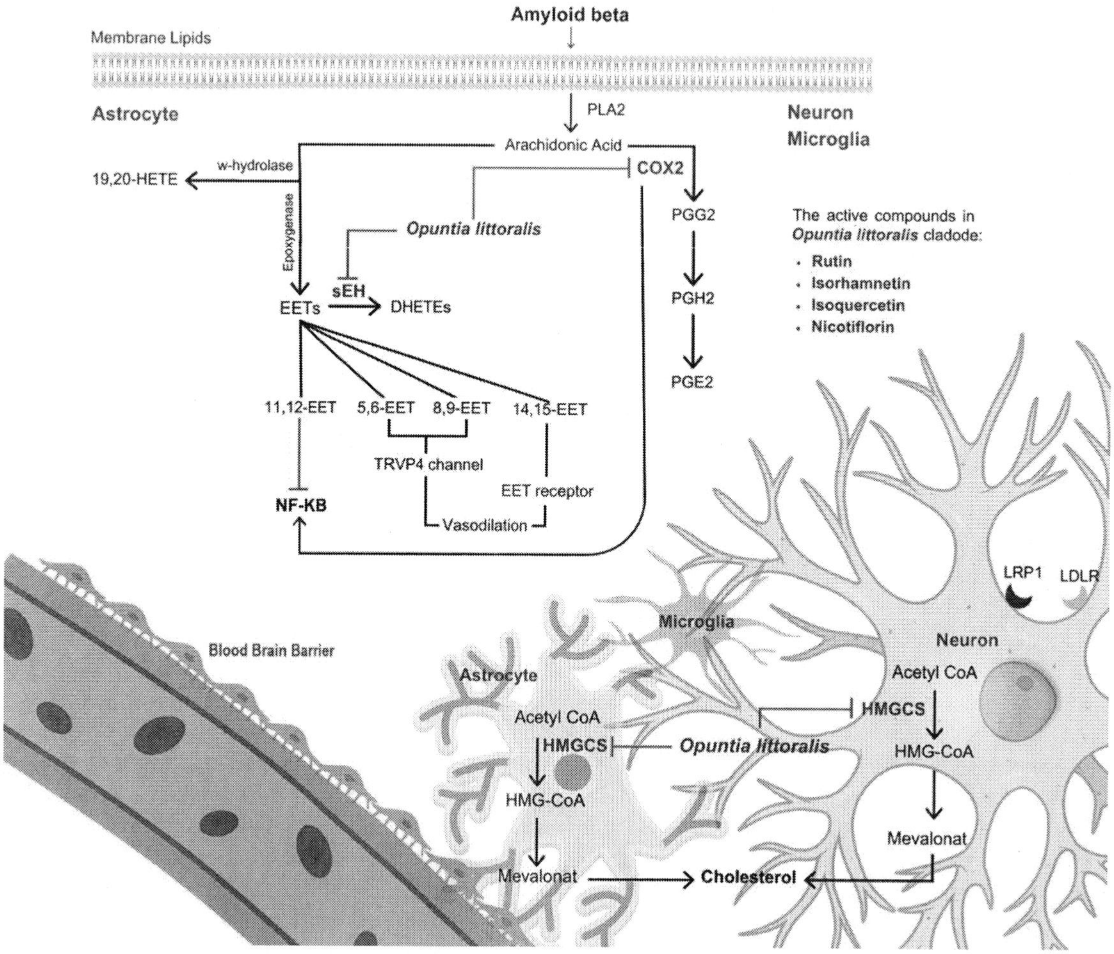

Figure 5. The mechanism of four bioactive compounds in *Opuntia littoralis* cladode as COX2, sEH, and HMGCS inhibitors.

4. Conclusion

This study has identified four bioactive compounds inside the cladode of the coastal prickly pear cactus, exhibiting promise as a natural AD treatment.

References

[1] Nichols E, Szoeke CEI, Vollset SE, Abbasi N, Abd-Allah F, Abdela J, et al 2019 Global, regional, and national burden of Alzheimer's disease and other dementias, 1990–2016: a systematic analysis for the Global Burden of Disease Study 2016. *Lancet Neurol.* **18**. 88–106

[2] Nichols E, Steinmetz JD, Vollset SE, Fukutaki K, Chalek J, Abd-Allah F, et al 2022 Estimation of the global prevalence of dementia in 2019 and forecasted prevalence in 2050: an analysis for the Global Burden of Disease Study 2019. *Lancet Public Health.* **7**. e105–25

[3] Farina N, Roxanne J, Yuda T, Irfani FF, Marguerite S, Imelda T, et al 2023 Comprehensive measurement of the prevalence of dementia in low- and middleincome countries: STRiDE methodology and its application in Indonesia and South Africa. *BJPsych Open.* **9**

[4] Lyketsos CG, Carrillo MC, Ryan JM, Khachaturian AS, Trzepacz P, Amatniek J, Cedarbaum J, Brashear R and Miller D S 2011 *Alzheimer's and Dementia.* **7**. 532–9

[5] Srivastava S, Ahmad R and Khare SK 2021 *Eur J Med Chem.* **216**. 113320

[6] Orlando BJ and Malkowski MG 2016 *Journal of Biological Chemistry.* **291**. 15069–81

[7] Thalji RK, et al 2013 *Bioorg Med Chem Lett.* **23**. 3584–8

[8] Shafqat N, Turnbull A, Zschocke J, Oppermann U and Yue WW 2010 *J Mol Biol.* **398**. 497–506

[9] El-Mostafa K, El Kharrassi Y, Badreddine A, Andreoletti P, Vamecq J, El Kebbaj MS, Latruffe N, Lizard G, Nasser B and Cherkaoui-Malki M 2014 *Molecules.* **19**. 14879–901.

[10] Filimonov DA, Lagunin A, Gloriozova T, Rudik AV, Druzhilovskii DS, Pogodin P and Poroikov V 2014 *Chem Heterocycl Compd (N Y).* **50**. 444–57

[11] Bitencourt-Ferreira G and de Azevedo WF 2019 Molegro Virtual Docker for Docking *Docking Screens for Drug Discovery* ed W F de Azevedo Jr. (New York, NY: Springer New York). 149–67

[12] Tyagi A, Kamal MA and Poddar NK 2020 *Front Neurosci.* **14**

[13] Domingues MF, Callai-Silva N, Piovesan AR and Carlini CR 2020 *Front Mol Neurosci* **12**

[14] Zhang K, Shen F, Lei W, Han Y, Ma X, Lu Y, Hou Y, Liu W, Jiang M, Zhang T and Bai G 2023 *Biomedicine and Pharmacotherapy.* **166**

[15] Dietschy JM 2009 *Biol Chem.* **390**. 287–93

[16] Wood W G, Li L, Müller WE and Eckert GP 2014 *J Neurochem.* **129**. 559–72

[17] Moussa N and Dayoub N 2023 *Saudi Pharmaceutical Journal.* **31**

[18] Griñán-Ferré C, Jarné-Ferrer J, Bellver-Sanchís A, Codony S, Puigoriol-Illamola D, Sanfeliu C, Oh Y, Lee S, Vázquez S and Pallàs M 2024 *CNS Neurosci Ther.* **30**

[19] Imig JD 2019 *Physiol Res.* **68.** 695–704

Analysis of proximate and porosity variations in collagen powder derived from natural resources: *Thunnus albacares* skin with different NaCl concentrations

F Rahmitasari[1*], D Sanjaya[2], TA Cevanti[3], W Prananingrum[1], Sularsih[1]
[1]Department of Dental Materials, Faculty of Dentistry, Universitas Hang Tuah, Surabaya, Indonesia
[2]Student of Faculty of Dentistry, Universitas Hang Tuah, Surabaya, Indonesia
[3]Conservative Dentistry Departement, Faculty of Dentistry, Universitas Hang Tuah, Surabaya, Indonesia

*E-mail: fitria.rahmitasari@hangtuah.ac.id

Abstract. Collagen from natural resources like yellowfin tuna skin (Thunnus albacares) skin is an alternative source of halal collagen that is rich in protein. The characterization of collagen starts with proximate tests like water, ash, protein, and fat content. Collagen powder was synthesized with a variation of NaCl using the PaSC method to produce a higher collagen yield. The requirements for a material to be implemented in the oral cavity are biocompatible and have an ideal porosity size as a biomaterial. This research is to analyze proximate and determine the porosity level of the results of collagen extraction from Thunnus albacares skin with variations in NaCl concentration. The methods used for proximate analysis follow the SNI standards and the porosity was tested using a Scanning Electron Microscope (SEM). Proximate composition was 62,14% moisture content, 0,41% ash content, 1,5% fat content, and 34,9% protein content. The porosity level of collagen powder in groups K, P1, P2, and P3 were 71.04 µm, 79.99 µm, 103.06 µm, 129.41 µm sequentially. Collagen powder with the addition of 1.7 M NaCl has a greater porosity level compared to the groups without NaCl, 0.9 M NaCl, and 1.3 M NaCl.

1. Introduction

Indonesia is an archipelago located between two large oceans, the Pacific Ocean and the Indian Ocean, which has great potential Indonesia's strategic location positions it well to become a leading global producer of tuna. Tuna (Thunnus sp) is a high-value fish and one of Indonesian waters' most highly sought-after species [1]. Tuna is a fishery product that has good economic value as an export commodity and for local consumption. Tuna is a member of the Scombridae family with the genus Thunus, consisting of large and small tuna. Large tuna like Thunnus albacares, Thunnus alalunga, Thunnus obesus, and Thunnus macoyii [2]. Tuna exports have increased by 13.5% every year [3]. Waste from processing fishery products reaches 20-60% of the raw material. Fishery products process waste with raw tuna materials, which consist of the head, skin, viscera, bones, and fins [4]. Collagen is commonly sourced from cattle, chickens, mammals, pigs, and birds. Marine organisms offer a promising alternative for collagen extraction because religious practices do not restrict them and do not pose risks of infectious diseases. With a significant amount of fisheries waste being produced globally, there is a growing emphasis on utilizing this waste to enhance economic value, support biotechnological advancements, and reduce overall waste [5].

In this study, the skin of yellowfin tuna (Thunnus albacares) is one of the alternatives that will be utilized as natural resources of collagen because it contains glycine, glutamic acid, proline, arginine, hydroxyproline, alanine, serine, and aspartic acid in large enough quantities [6]. The use of yellowfin tuna is preferred because the protein contained in yellowfin tuna is higher than that of albacore tuna, which is 20.54 ± 0.26%, and in yellowfin tuna 24.63 ± 0.12%.

In dentistry, collagen is widely used as a material produced in membrane preparations or as a barrier to trigger tissue regeneration. Collagen can be designed into a scaffold suitable for bone regeneration because it is absorbed by the body and offers strong cell adhesion [7]. This makes it a highly effective support structure for bone repair, as it naturally breaks down and facilitates excellent cell attachment [8]. Collagen is also used as a medium for local drug administration in periodontitis treatment, such as tetracycline chips. In addition, type I collagen plays a role in periodontal regeneration [9].

Collagen preparations can be in the form of powder, gel, and scaffold, at the stage of material characterization testing porosity testing to determine the level of microporosity of yellowfin tuna skin collagen material. From the results of collagen, it can be seen the microporosity of collagen using the Scanning Electron microscope (SEM) method. SEM test can be seen in the comparison of pore size with Archimedes' principle. Meanwhile, the porosity level is obtained from the division of the difference between wet and dry weight with the difference between the final volume and the initial volume multiplied by 100% [10]. Observation of collagen through SEM photos can shows the character of collagen tissue in the form of white fibers or fibers, arranged irregularly so that it appears that collagen fibers or fibers are arranged tightly, tenuously, thick, and thin. The characteristics of tuna fish skin collagen fibers produced are thinner, smoother, longer, and many cavities [11].

Collagen extraction methods generally use acids or pepsin enzymes. Still, pepsin enzymes come from pig animals and cannot be used in general because there are restrictions in certain religions. Hence, the collagen extraction method in this study uses papain enzymes derived from the fruit and sap of papaya trees, besides the use of papain enzymes produces higher collagen yields than using acids [12]. Before looking at the porosity through the SEM test, the collagen yield was precipitated using NaCl in the presence of tris (hydroxymethyl) aminomethane at pH 7.5 with the appropriate salt concentration to remove impurities and optimize collagen refinement to the maximum [13]. Previous research on NaCl variations on collagen extraction from tuna fish skin proved that each NaCl concentration of 0.9M, 1.3M, and 1.7M produced different collagen yields. Where the highest yield was obtained at a NaCl concentration of 1.7 M with a yield of 45.06 ± 0.644% and the lowest yield at a NaCl concentration of 0.9 M with a yield of 28.12 ± 1.5% [11]. Based on the above explanation, the researchers wanted to know the analysis of proximate of thunnus albacares skin and size of porosity in the synthesis of collagen powder from yellowfin tuna skin (Thunnus albacares) using the papain soluble collagen method with variations in NaCl concentration.

2. Materials and Methods

This study used a true experimental design, characterized by rigorous control over all external variables that could influence the outcomes. There are three components consisting of replication, randomization, and control or comparison, if the three components are met then the study is called a true experimental (true experimental) [14,15].

The tools used in this study are a cutting board, knife, analytical scales, micropipette, glass jar, sieve, basin, glass stirring rod, erlenmeyer, measuring cup, 6000 rpm centrifugation, 10 ml conical tube, freeze dryer, specimen container, 4 °C refrigerators, magnetic stirrer, conical tube, Scanning Electron Microscope (SEM). The materials needed in this study are yellowfin tuna skin (Thunnus albacares), papain enzyme with activity of 30,000 USP/mg (EC. 3.4.22.2), NaOH, NaCl, CH_3COOH, aquabidest, Whatman filter paper, 14 kDa dialysis bag, flannel cloth, water, pH indicator paper, pH meters. For the proximate analysis, the tools required are analytical scales (with a sensitivity of 0.0001 g), Kjeldahl digestion apparatus, steam distillation equipment, measuring flasks, measuring cups, goblet glasses, stirring rods, muffle furnaces, blenders, desiccants, plastic containers, scissors, porcelain cups, vacuum ovens, and non-vacuum ovens.

2.1 Proximate Analysis

Proximate analysis is used to determine the chemical composition of a material. Proximate analysis of Thunnus albacares skin, was conducted through four tests: water, ash, fat, and protein content. The methods used for this analysis follow the standards SNI 2354.2:2015 (moisture content), SNI 01-2354.1-2006 (ash content), SNI 01-2354.4-2006 (protein content), and SNI 2354-3:2017 (fat content) [16].

2.1.1 Moisture Content

Initially, porcelain cups are heated in an oven at 105°C for 1 hour, then cooled in a desiccator for 15 minutes, and weighed until a consistent weight is achieved. Following this, a 2-gram sample is weighed, placed in a dish, and dried in an oven at 105°C for three hours. The dish is then cooled in a desiccator for 30 minutes and weighed until it reaches a stable weight. The moisture content is then calculated using the formula provided:

$$\text{Moisture content (\%)} = \frac{\text{B} - \text{C}}{\text{B} - \text{A}} \text{x } 100\%$$

A = Weight of empty cup (g)
B = Porcelain cup weight + Initial sample (g)
C = Cup weight + Dry sample

2.1.2 Protein Content

To determine the protein content, 2 grams of the sample, 2 catalyst tablets, and boiling stones are added to a digestion flask. Then, 15 ml of concentrated sulfuric acid (H_2SO_4) and 3 ml of sulfuric acid (H_2SO_4) are carefully added. The mixture is allowed to stand in the acid chamber for 10 minutes before being heated to 410°C until the solution turns clear. After digestion is complete, the solution is cooled, and 50-75 ml of distilled water is added. An Erlenmeyer flask containing 25 ml of a 4% boric acid (H_3BO_4) solution with indicators is prepared to capture the distillate. The digestion flask is then set up for steam distillation, and 50-75 ml of a sodium hydroxide-thiosulfate solution is introduced for the distillation process. The distillate is collected in the Erlenmeyer flask until the total volume reaches 150 ml. Finally, the distillate is titrated with 0.2 N hydrochloric acid (HCl) until a neutral gray color is achieved. The protein content is then calculated using the specified formula:

$$\text{Protein content (\%)} = \frac{(\text{VA} - \text{VB}) \text{ HCl x } 14{,}007 \text{ x } 6{,}25 \text{ x } 100\%}{\text{W x } 100} x\ 100\%$$

VA = ml HCl for sample titration
VB = ml HCl for blank titration
N = Normality of standard HCl used
14,007 = Atomic weight of nitrogen
6,25 = Protein conversion factor for fish
W = Sample weight (g)

2.1.3 Ash Content

Two grams of homogenized samples are placed in porcelain ash cups and then heated in an oven at 100°C for 24 hours. The porcelain cups are then transferred to an ashing furnace, where the temperature is gradually increased to 550°C ± 5°C and maintained until the samples turn into white ash. Afterward, the porcelain cups are cooled in a desiccator for 30 minutes. The ash is then moistened with aquadest, dried on a hot plate, and reheated at 550°C until it reaches a consistent weight. The temperature is then lowered to around 40°C, and the porcelain ash cups are returned to the desiccator for 30 minutes before being weighed once cooled. The ash content is calculated using the following formula:

$$\text{Ash content (\%)} = \frac{B - A}{\text{Sample weight (g)}} \text{x } 100\%$$

A = Porcelain cup weight (g)
B = Weight of the cup with ash (g)

2.1.4 Fat Content
The fat content was measured using the Soxhlet extraction method. First, 3 grams of the sample, 20 ml of HCl, 30 ml water, and boiling stones were added to a 250 mL flask and boiled for 15-20 minutes. After boiling, the mixture was filtered until the pH was neutral, and then it was dried in an oven at 100°C for 10 minutes. Next, the fat was extracted using a Soxhlet extractor. The sample, enclosed in filter paper, was placed in the fat sleeve, and 50 mL of diethyl ether was added to the round bottom flask. The fat sleeve was then inserted in the extractor, and extraction was performed for 3 hours with approximately 5-minute cycles. After extraction, the diethyl ether was evaporated, and the fat was transferred to a round-bottom flask, which was then heated in an oven at 105°C for about 2 hours. The flask was allowed to cool in a desiccator for 30 minutes, and its weight, along with the fat, was recorded until a constant weight was reached.

$$\text{Fat content (\%)} = \frac{(C - B)}{A} \text{x } 100\%$$

A = Sample weight (g)
B = Weight of the empty round base (g)
C = Weight of empty round base and extracted fat (g)

2.2 Collagen Powder Fabrication
The research design carried out is with a post-test-only control design, where there are 2 kinds of groups, each of which is randomly selected. This study focused on the treated group (experimental group) and the untreated group (control group) [17]. Preparation of yellowfin tuna skin collagen powder starts with cleaning the tuna skin, there is no meat attached to the fish skin and cutting the tuna skin into a square shape with a size of 1x1 cm [18-20]. Tuna skin was soaked in 0.1 M NaOH solution for 12 hours at 10°C with a ratio of 1:10 (b/v). Replacement of NaOH solution was done every 2 hours and stirred with a magnetic stirrer for 12 hours. The tuna skin was subsequently rinsed with distilled water until it reached a neutral pH (pH 7-7.5). Afterward, the hydrolysis and extraction process was performed by combining papain enzyme (EC 3.4.22.2) at a concentration of 10,000 U/g with a 0.5 M CH_3COOH solution. This mixture was then allowed to soak for 72 hours (3 days). The next step was filtration using Whatman filter paper to obtain the filtrate [11]. The obtained filtrate was divided into 5 groups and given 400 ml of NaCl solution with concentrations of 0.9M, 1.3M, and 1.7M for 12 hours. Each group of filtrate is put into a conical tube for centrifugation for 1 hour and the results will be obtained in the form of pellets and supernatant. Furthermore, the pellet was dissolved with 0.5 M CH_3COOH solution in a ratio of 1:1 (b/v) and dialyzed with 0.1 M CH_3COOH solution and distilled water for 24 hours. The pellet was put into a 14 kDa dialysis bag. Then the freeze-dry stage was carried out for 72 hours to obtain collagen powder [11,17].

2.3 Scanning Electron Microscope (SEM)
For each group, 2 grams of collagen powder samples are prepared and coated with gold to enhance electrical conductivity. The samples are then attached to a holder with carbon tape and placed into the SEM machine. The SEM is used to observe the pore shape and connectivity displayed on the screen.

IOP Conf. Series: Earth and Environmental Science **1473** (2025) 012060 doi:10.1088/1755-1315/1473/1/012060

Transverse cross-sectional images are analyzed to determine the pore sizes, including pore area and diameter, using ImageJ 1.49V software.

2.4 Data analysis

To assess porosity in this research was conducted using SPSS 25. To perform parametric statistical tests, two conditions must be met: the data must be normally distributed and homogeneous. Therefore, normality and homogeneity tests are required. The Shapiro-Wilk test is used for normality assessment due to the sample size being less than 50. Next, Levene's test is used to check for homogeneity. If the data meet both conditions ($p>0.05$), a parametric statistical test, specifically One-Way ANOVA, is applied since the data scale is the ratio and the study involves more than two independent, non-paired sample groups. If significant differences are found ($p<0.05$), the Post Hoc Least Significant Difference (LSD) test is performed.

3. Results

Chemical composition analysis is conducted to determine the chemical makeup of tuna skins. Understanding the chemical composition of the sample helps assess the suitability of raw materials for collagen production. The analysis involves testing for moisture content, ash content, fat content, and protein content. The results of the study in Table 1 show the percentage value of proximate from yellowfin tuna (Thunnus albacares).

Table 1. Proximate analysis of Thunnus albacares skin (%)

Parameter	Mean (%)	Sd
Moisture content	62,14	±0,27
Ash content	0,41	±0,05
Protein content	34,90	±0,41
Fat content	1,50	±0,03

Based on the research that has been done, it aims to determine the porosity level of collagen powder from yellowfin tuna skin using papain soluble collagen with four test treatments namely without NaCl, 0.9 M, 1.3 M, 1.7 M NaCl. Each treatment unit group consisted of 5 experimental replications (n). The mean and standard deviation of collagen powder of Thunnus albacares using SEM method are shown in Table 2, meanwhile, the results of the porosity test using SEM with 250x magnification can be seen in Figure 1.

Table 2. Mean and standard deviation of collagen powder of Thunnus albacares using SEM method

Group	n	Mean (μm)	Sd
Without NaCl (K)	5	71.04	±17.398
0,9 M NaCl (P1)	5	79.99	±14.054
1,3 M NaCl (P2)	5	103.06	±8.234
1,7 M NaCl (P3)	5	129.41	±18.300

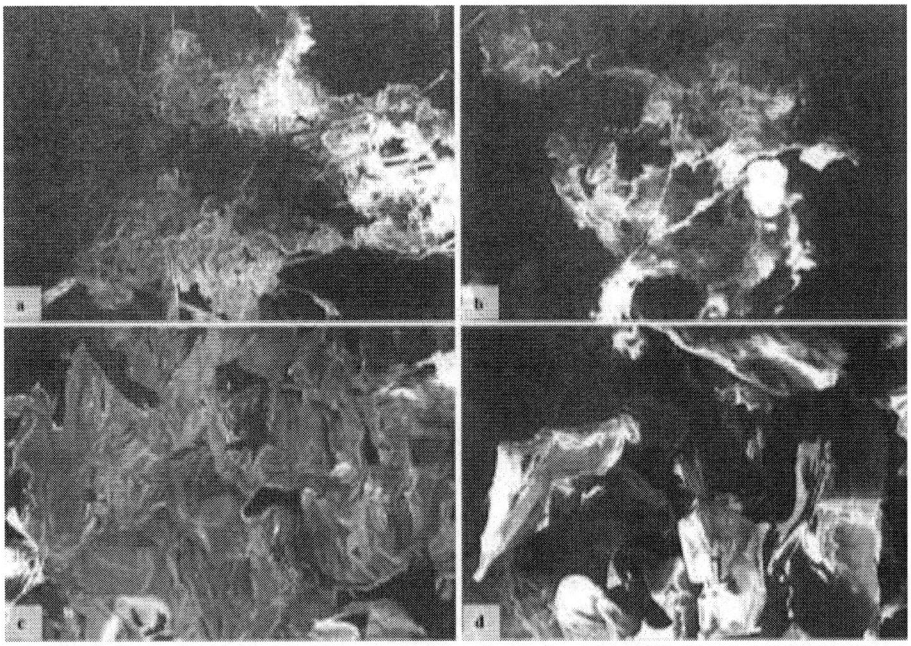

Figure 1. SEM photo results with 250x magnification: a) without NaCl, b) 0.9 M NaCl, c) 1.3 M NaCl, d) 1.7 M NaCl with 250X magnification

Based on the descriptive statistical results of the porosity test of collagen powder, it is known that in group P3 (synthesis of collagen powder with 1.7 M NaCl), the highest porosity value was obtained, then decreased in group P2 (synthesis of collagen powder with 1.3 M NaCl) and further decreased in group P1 (synthesis of collagen powder with 0.9 M NaCl).

One-Way ANOVA test was conducted because the normality test results were normally distributed and the homogeneity test results showed homogeneous data in all groups. A one-way ANOVA test (Table 3) was conducted to determine the difference in porosity level between all groups of yellowfin tuna skin collagen powder (Thunnus albacares) with various NaCl concentrations.

Table 3. The One-Way ANOVA test show the results ($p<0.05$) which can be concluded that there are significant differences between each treatment group.

One-Way ANOVA	*df*	*Asymp.Sig*
	3	0,000

After the One-Way ANOVA test was carried out, it was continued with the post-hoc test (Table 4) which aims to determine significant differences in porosity levels between each group.

Table 4. Post-Hoc Test Results

Group	Without NaCl (K)	0,9 M NaCl (P1)	1,3 M NaCl (P2)	1,7 M NaCl (P3)
Without NaCl (K)	-	0.358	0.003*	0.000*
0,9 M NaCl (P1)	-	-	0.025*	0.000*
1,3 M NaCl (P2)	-	-	-	0.012*
1,7 M NaCl (P3)	-	-	-	-

*($p<0,05$) = There is a significant difference.

4. Discussion

In the results of the Thunnus albacares skin proximate test, it was known that the percentage of moisture content was 62.14%, ash content was 0.41%, protein content was 34.9%, and fat content was 1.5%. The moisture content of tuna skin in the sample of 62.14% has a value above 50%. High moisture content can affect the deterioration of product quality. The higher the moisture content value, the more likely the raw material will experience a process of deterioration in quality. The water content can be affected by the type of species of fish as well as the fishing season [20]. The ash content in this research was 0.41%. The analysis of ash content shows the large amount of minerals contained in a material. The ash content is determined by the demineralization process and the concentration of the solvent used, where the lower the ash content, the higher the quality and purity [21]. In this study, the protein content of tuna skin in this study was 34.9%. Ash content can affect the amount of collagen yield to be produced. Fish skin with high protein levels will have a large amount of collagen obtained. Protein levels can be affected by food, temperature, age of fish, and fish habitat [22]. Fat is one of the main components found in fish skin in addition to protein and water. The skin fat content of yellowfin tuna in this study was 1.5%. Fish with a fat content of more than 4% are classified as high-fat fish, while fish with a fat content of less than 4% are classified as low-fat. High-fat content can affect the pretreatment stage of fish skin [23].

The normality test shows that all treatment groups have a significant value of $p>0.05$, which means that the data is normally distributed and can be continued with the homogeneity test. Based on the results of the homogeneity test using the Levene test, the significance value is 0.215 ($p>0.05$), which means that the porosity level test results of yellowfin tuna skin collagen powder are homogeneous so that it can be continued with the parametric test, namely the One-way ANOVA test. The results of the One-Way ANOVA test show the results ($p <0.05$) which can be concluded that there are significant differences between each treatment group.

Based on the descriptive statistical results of the porosity test of collagen powder, it is known that in group P3, the synthesis of collagen powder with the administration of 1.7 M NaCl obtained the highest average of 129.41, then decreased in group P2: synthesis of collagen powder with the administration of 1.3 M NaCl with an average of 103.06 and further decreased in group P1: synthesis of collagen powder with the administration of 0.9 M NaCl with an average of 79.99 and group K has the lowest average of 71.04.

There is a difference in the size of the porosity of collagen powder in the group without NaCl (K) compared to the P2 and P3 groups, due to the treatment with yellowfin tuna skin collagen powder (Thunnus albacares) given 1.3 M NaCl and 1.7 M NaCl in the P2 and P3 groups so that there is a correction in the three polypeptide chains that affect the porosity level. There was no significant difference between the group without NaCl (K) and the P1 group because the 0.9 M NaCl in the P1 group had a molarity level that was too low so there was no visible effect with the group without NaCl.

There is a difference in the porosity size of collagen powder in group P1 with groups P2 and P3, because of the treatment with yellowfin tuna skin collagen powder (Thunnus albacares) given NaCl 1.3 M and NaCl 1.7 M in groups P2 and P3 so that there is a reaction on three polypeptide chains that affect the porosity level.

There is a difference in the porosity size of collagen powder in group K (71.04) with group P1 (79.99), due to the treatment with yellowfin tuna skin collagen powder (Thunnus albacares) given 0.9 M NaCl in group P1 so that there is a reaction on the three polypeptide chains that affect the porosity level. There is a difference in porosity size There is a difference in the porosity of collagen powder in group K (71.04) with group P2 (103.06), due to the treatment with yellowfin tuna skin collagen powder (Thunnus albacares) with 1.3 M NaCl in group P2 so that there is a reaction in the three polypeptide chains that affect the porosity level. There is a difference in the porosity size of collagen powder in group K (71.04) with group P3 (129.41), due to the treatment with collagen powder of yellowfin tuna skin (Thunnus albacares) with 1.7 M NaCl in group P3 so that the reaction occurs in the three polypeptide chains that affect the porosity level.

There is a significant difference between group K (without NaCl) with group P2 (NaCl 1.3 M) and P3 (NaCl 1.7 M). The result of the significant difference is that group K, which is the control group, was not given any treatment. Other factors that can affect the porosity of collagen powder are external disturbing factors such as temperature, pH, bacteria, and fungi [24]. The collagen powder in groups P1 (NaCl 0.9 M), P2 (NaCl 1.3 M), and P3 (NaCl 1.7 M) were given NaCl during the collagen extraction process so that there was a precipitation process in the form of wet collagen precipitation which influenced the particle size produced. The particle size produced in collagen powder K (without NaCl) is smaller than P1 (NaCl 0.9 M), P2 (NaCl 1.3 M), and P3 (NaCl 1.7 M) so it follows the theory of particle size and porosity, if the particle size is small it can produce a small porosity size and if the particle size is large it will produce a large porosity size [25].

NaCl can provide differences in porosity due to the interaction between NaCl and the base material. In their research, Li, et al (2016) wrote that NaCl can increase the dispersion of elemental fibers, neatly arrange fibers, and form a porous fiber network due to the effects of dehydration and salt precipitation by NaCl. The presence of NaCl in the process increases the porosity of the collagen fiber network. This is due to the effects of dehydration and salt precipitation by NaCl, which positively affect the formation of a porous fiber network. NaCl contributes to the formation of a porous collagen fiber network. The resulting porosity can have a positive impact on mechanical properties [26]. In his research Romadhon, et al, (2019), it is known that the factor that can affect porosity is NaCl concentration, Increasing NaCl concentration can increase the dehydration and salt precipitation effects, which can affect the porosity of collagen fiber networks. Soaking Time, Longer soaking time can provide more time for salt to interact with the base material, thus directly affecting the formation of porous fiber network. Penetration Ability of NaCl, Good penetration can result in even dehydration effect and uniform pore formation. Salt crystallization pattern, Regular and organized crystallization patterns can produce better porous fiber networks [27].

The results in this study showed that the porosity test of the P3 group was higher than the other groups, this could be due to variations in the concentration of Nacl given to the synthesis of yellowfin tuna skin collagen powder, but it could also be due to differences in stirring time between one treatment and another. This can affect the porosity of the collagen powder so the P3 group obtained these results.

Meanwhile, group K obtained the smallest result than the other treatment groups (P1, P2 and P3) due to the different concentrations of NaCl given which can also be caused by differences in stirring time between one treatment and another. Based on the results of this research, the P3 group obtained the largest mean than the other groups which means that the NaCl concentration with a NaCl concentration of 1.7 M greatly affects the three polypeptide chains because the three polypeptide chains are the constituent molecules in collagen so that they can affect the porosity of collagen powder.

5. Conclusion

Based on the research that has been done, it can be concluded that the Thunnus albacares skin proximate test, it was known that the percentage of moisture content was 62.14%, ash content 0.41%, protein content 34.9%, and fat content 1.5%. The porosity of P3 group gets the largest average of 129.41 μm than other groups which means that the NaCl concentration with a NaCl concentration of 1.7 M greatly affects the three polypeptide chains because the three polypeptide chains are the constituent molecules in collagen so that they can affect the porosity of collagen powder.

6. References

[1] Mutmainnah M, Chadijah S, Rustiah WO 2017. *Al-Kimia*. **5(2)**:119–26.
[2] Triharyuni S, Prisantoso BI 2012. *J. Saintek Perikanan*. **8(1)**:52–8.
[3] Kementerian Kelautan dan Perikanan 2018. Jakarta.
[4] Nugroho G, Ekawati AW, Kartikaningsih H 2020. *Res. J. of Life Sci*. **7(2)**:101–7.
[5] Coppola D, Oliviero M, Vitale GA, et al. *Mar Drugs* 2020;**18(214)**:1-23.
[6] Handaratri A, Hudha MI 2021. *Reka Buana J. Ilm Tek Sipil dan Tek Kim*. **6(2)**:104–11.
[7] Rahmitasari F 2016. *J. Mater. Ked. Gigi*. **5(2)**:1-7.
[8] Rahmitasari F, Rahayu RP, Munadziroh E. *Acta Med Philipp* 2022. **56(8)**:43-48.
[9] Rahman VR, Bratadiredja MA, Saptarini NM 2021. *Maj. Farmasetika*. **6(3)**:253–86.
[10] Naibaho E, Wibowo P 2022. *J. Pendidik Kewarganegaraan Undiksha*. **10(1)**:494–500.
[11] Nursyam H 2010. *J. Penelit Perikanan*. **13(1)**:107–13.
[12] Cahyono H, Trilaksani W, Uju 2022. *J. Pengolah Has Perikan Indones*. **25(1)**:1–17.
[13] Cuschieri P 2020. Malta.
[14] Sugiyono 2016. Bandung: PT Alfabet.
[15] Hardani, Auliya NH, Andriani H, Fardani RA, Ustiawaty J, Utami EF, et al 2020. Yogyakarta: Pustaka Ilmu.
[16] Herson NA, Sumual MF, Rumengan IFM, Pongoh J, Mandey LC 2023. *J. Agroekoteknologi Terap*. **4(2)**:428–33.
[17] Nurjanah N, Nurhayati T 2021. *J. Pengolah Has Perikan Indones*. **24(2)**:174-87.
[18] Nurilmala M, Hizbullah HH, Karnia E, Kusumaningtyas E, Ochiai Y 2020. *Mar. Drugs*. **18(2)**:98.
[19] Kolanus JP, Hadinoto S, Idrus S 2019. *Indones J. Ind. Res*. **11(1)**:99–110.

[20] Shon J, Eo J-H, Hwang SJ, Eun J-B 2011. *Food Science and Biotechnology*. **20(1)**: 99-106.
[21] Oktaviani RZI, Perdana F, Nasution AY 2017. *J. of Pharmacy Science*. **1(4)**:1-8.
[22] Azara R 2017. *J. Rekapangan*. **11(1)**: 62-69
[23] Sun DW 2006. Boca Rason: CRC Press Taylor and Francis Group.
[24] Krismariono A, Wiyono N, Prahasanti C 2020. *J. Int Dent Med Res*. **13(2)**:412–6.
[25] Widarti BN, Wardhini WK, Sarwono E 2015. *J. Integr. Proses*. **5(2)**:75–80.
[26] Li X, Wang Y, Li J, Shi B 2016. *JALCA*. **111**:230–7.
[27] Romadhon R, Darmanto YS, Kurniasih RA 2019. *J. Pengolah Has Perikan Indo*. **22(2)**:403-a10.

Coastal Planning Based on Mangrove Conservation with an Integrated Coastal Zone Management Approach: Study Case of Cianjur Regency

A Lestari[1] and A Hadianti*[1]
[1]Department of Architecture and Planning, Universitas Gadjah Mada, Grafika Road Number 2 Yogyakarta, Indonesia

*atrida.h@ugm.ac.id

Abstract. Cianjur Regency on the southern coast of Java Island has prominent coastal resources. In contrast, coastal communities face challenges due to poverty, natural disasters, and environmental degradation, which also threaten the sustainability of the coastal ecosystem. Developing the mangrove ecosystem through the Integrated Coastal Zone Management (ICZM) as a holistic/comprehensive approach to address economic and ecological challenges in the coastal areas to achieve resilience and sustainability. Currently, local communities have not utilized mangroves to their maximum potential. Hence, 17 coastal villages in the southern part of Cianjur Regency were planned to use the Double Diamond Design Process (DDDP) method. The ICZM is used to derive planning elements and synthesis. This method and framework are utilized to address regional-scale problems related to the economic and ecological vulnerability of coastal zones in an integrated manner. The zoning plan contains analysis and detailed planning focusing on mangrove ecosystem development for each village. Optimal spatial planning requires a comprehensive land suitability assessment, which can be achieved if a thorough knowledge of land suitability characteristics supports the process. Meanwhile, integrated coastal spatial planning can be achieved if the ICZM's objectives in institutional integration and human resource management, natural resource conservation, and responsible resource utilization can be met.

1. Introduction

Coastal areas are areas with very complex disaster vulnerabilities and have the potential to cause various disaster risks [1]. Some of the vulnerabilities that cause these risks include economic pressure, rising sea levels, earthquakes and tsunamis, extreme waves and abrasion, and other vulnerabilities that threaten the sustainability of living things. As an archipelagic country where part of the population lives in coastal areas, these vulnerabilities could seriously undermine the achievement of Indonesia's targets with prosperous coastal communities, sustainability of the marine ecosystem, and increased economic growth.

Based on the presentation of the 7th Vice President of the Republic of Indonesia, Ma'ruf Amin, at the Meeting on Handling Extreme Poverty in Coastal Areas, 147 out of 2016 districts/cities, or 69 percent of districts/cities targeted for handling extreme poverty in 2022 are coastal areas [2]. The extreme poverty rate in coastal areas also reached 4.19 percent or 0.19 percent higher than the national extreme poverty average [3]. Coastal areas are areas rich in various potential resources [4]. This implies that coastal areas tend to be higher in terms of economic vulnerability and the economic potential of Indonesia's coast has not been utilized properly [5].

This social condition occurs in small fishermen in the southern region of West Java Province because the coastal communities are below the poverty line compared to other communities [6]. Poverty caused by the underutilization of Indonesia's coastal resources occurs in various regions of West Java, one of which is in Cianjur Regency. Around 90,480 extremely poor people are dominated by the southern region which is a coastal area [7]. Additionally, this can be attributed to the fact that fishermen who experience poverty tend to have never received coaching and training in skills to support the economy [8]. Furthermore, natural conditions/phenomena such as extreme weather, high waves, and lean fishing seasons also affect the income of coastal communities in Cianjur Regency.

Coastal natural disasters' potential hazards can increase the risk of coastal vulnerability in the region. The earthquake and tsunami are two out of four natural disasters potential hazards that are the focus of handling in Cianjur Regency [9]. In addition to these two disasters, rising sea levels due to the climate crisis [10] and coastal erosion due to extreme waves and abrasion [11] are threatening the ecological sustainability of coastal areas. Enhancing regional resilience is important/crucial/essential for mitigating disaster and economic risks.

Increasing economic and disaster resilience in Cianjur Regency can be improved through the development of mangrove ecosystems. Mangroves can provide direct benefits in the form of increasing fishermen's income and developing tourism, to indirect benefits in the form of wave breakers and biodiversity value [12]. The existence of mangrove forests can also provide ecosystem service benefits to enhance regional resilience compared to tropical forests, subtropical forests, temperate forests, boreal forests, and other biomes [13]. These benefits can be achieved if the mangrove ecosystem is managed and developed properly. Strategic policy and directions are needed for the optimal use of mangrove natural resources.

In line with this, the West Java Provincial Government has conducted spatial planning of the Cianjur coast through the West Java Provincial Spatial Plan (RTRW) document which has been integrated with the West Java Provincial Coastal and Small Islands Zoning Plan (RZWP3K) for 2022-2042. This document contains various directions where Cianjur Regency is designated as one of the areas for maintaining local protected areas for coastal boundaries, area for rehabilitation and conservation of coastal ecosystems, protection and improvement of the quality of mangrove ecosystems, improvement of coastal security and adaptation capacity to disasters, development, and arrangement of pond areas, to integrated management space with the development of the silvofishery model (mangroves and fisheries).

However, the regulations in the West Java Provincial RTRW have not been lowered to the district level because there has been no update to the Cianjur Regency RTRW 2011–2031. The spatial plan regulations (i.e. Cianjur Regency RTRW) only include protection of local areas (coastal boundaries), management of the southern coastal area, disaster mitigation studies, development of capture fisheries, and plans for the allocation of green open spaces along the coast. The regulations do not directly state the sustainability of the mangrove ecosystem and silvofishery program. However, in general, the Regency RTRW has objectives that are in line with the Provincial RTRW which has been integrated with the RZWP3K document related to the management of coastal areas to create a resilient and sustainable coast.

Although the 2011–2031 Cianjur Regency RTRW has not been adjusted to the 2022–2042 West Java Provincial RTRW, Cianjur Regency has designated the coastal area as a strategic area from an economic perspective, namely as a strategic area for the arrangement of the southern coastal area. Therefore, conducting a Detailed Spatial Plan (RDTR) and Zoning Regulations for the coastal area of Cianjur Regency is essential. Consequently, it is necessary to prepare a detailed spatial plan in the coastal area that is conceptualized and integrated.

Conceptual and integrated planning is expected to encourage efforts to optimally utilize space in increasing disaster resilience, maintaining the sustainability of the coastal environment to prevent the environment, and improving the coastal community's economy. This expectation is grounded in the capability of spatial planning to effectively incorporate advanced technologies for disaster mitigation facilities while simultaneously facilitating the establishment of activity centers that enhance community capacity [14]. Therefore, there needs to be a clear zoning plan and scheme on how this economic improvement and disaster resilience in Cianjur Regency.

2. Planning Methodology

The approach used in this planning is technocratic, utilizing scientific methods and thinking frameworks to achieve regional development goals and targets [15]. This planning decision-making is based on data and facts found during field surveys as well as the results of non-spatial and spatial data analysis to manage the organization and resources effectively. Therefore, this approach is said to be appropriate for formulating coastal area planning where natural resources tend not to be well managed institutionally down to the local community level. Although technocratic, the author also considers information from the public in the survey process as relevant and accurate information in the planning process. The method used in the analysis and writing of this plan is the Double Diamond Design Process (DDDP) which includes the Spatial Multi Criteria Analysis (SMCA) process, SWOT, and benchmarking.

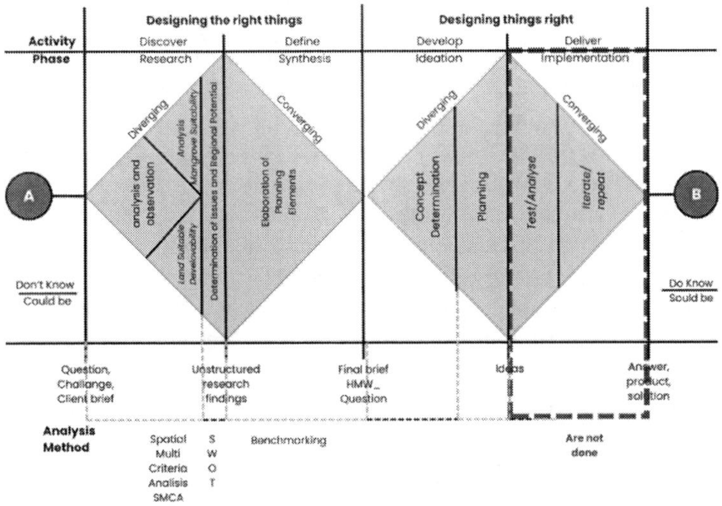

Figure 1. Double Diamond Design Process in Planning

3. Study Area

3.1. Identification of Coastal Characteristics of Cianjur District

The coastal area of Cianjur Regency consists of 3 sub-districts consisting of 17 coastal villages covering an area of 370 km^2 (Figure 1 (a)). The planning area, which is a coastal area of Cianjur, has a distance of more than 100 km from the surrounding urban areas (Figure 1 (b)). Based on data from the Cianjur Central Statistics Agency (BPS), in 2022 there will be 246,08 thousand or 10.55 percent of the total population living below the poverty line in Cianjur Regency. Based on data presented by the Social Service in Cianjur Open Data, the number of recipients of the Program Keluarga Harapan (i.e. national program addressing the poverty issue) in Cianjur Regency in 2022 is around 3 to 7 percent per District. The dominant recipients of aid are the people of southern Cianjur and central Cianjur.

Gross Regional Domestic Product (GRDP) at Constant Prices for Cianjur Regency from 2018 to 2020, the agriculture, forestry, and fisheries sectors consistently accounted for the largest 29 percent contributed to Cianjur GRDP. Economic activities seen in coastal areas are agriculture and fishing in coastal areas. Based on West Java Province BPS data regarding Capture Fisheries and Public Water Production from 2019 to 2020, public water fisheries in Cianjur Regency were the third largest in West Java in 2019. This shows that Cianjur Regency has potential in terms of public water fisheries. Unfortunately, the production value of fisheries and aquaculture in the southern coastal region tends to be lower than those in the northern region. This difference is not due to the amount of fish resources, but rather due to economic conditions related to materials, education, and social status [6]. Based on data from the Fisheries Service in 2016, Cianjur Regency has the highest number of households in fisheries cultivating cages, floating nets, and rice fields.

In addition, to stimulate coastal economic growth, the Ministry of Maritime Affairs and Fisheries (KKP) and Perum Perhutani are collaborating by establishing a pilot sustainable shrimp pond cluster consisting of 15 production ponds in an area of 4 hectares [16]. Furthermore, information was obtained that the profit that could be obtained after the 110-day maintenance period was the first harvest of

Vaname shrimp of approximately 30 tons worth approximately 2 billion. These harvests increase the income of group members up to three times and increase the income of the surrounding community which helps with operational processes.

The condition of the mangroves in Cianjur Regency is starting to deteriorate and only a few spots remain. One of them is at Tipar Beach, Sinarlaut where there has been a decrease in mangrove area of around 0.9 ha from 2012-2022 with the criticality level of the land being 4.6 ha heavily damaged, 1.5 ha damaged, and 3.1 ha not damaged [17]. The existence of mangroves in Cianjur Regency rarely lasts long because they are often affected by tidal floods during the rainy season and there is also a lack of management. An example of ineffective management is the occurrence of sea pine, and mangrove fires caused by human activity in Sinarlaut Village, Agrabinta District on December 31, 2023, around 2 to 3 ha, which occurs almost every year and has occurred around 3 times [18]. The use of mangroves in Cianjur Regency is still limited to preventing disasters. Several efforts have been made, including planting by the central government through the National Mangrove Planting Program with the theme Mangrove for Better Life, which has carried out simultaneous planting in all regions in Indonesia, one of which is in Cianjur Regency, totaling 2,500 trees by 2023.

Based on the 2018-2022 Cianjur Regency Disaster Risk Assessment Document, Cianjur Regency has a high level of disaster vulnerability for forest and land fire disasters and a medium vulnerability class for other disasters [9]. Meanwhile, this document also explains that the disaster capacity in all regions in this Regency is in the medium class with a value of 0.53, which shows that the Cianjur Regency needs to make various efforts to reduce disaster risk. Regarding extreme wave and abrasion disasters as well as tsunamis, Cianjur Regency is exposed in a high-risk category [9]. Apart from tsunamis, the Cianjur coast is also threatened by floods, flash floods, and liquefaction, as well as waves and abrasion.

Based on Sentinel-2 data obtained from the Impact Observatory for Esri, land use data for the coastal area of Cianjur Regency in 2021, land use in Cianjur Regency is dominated by forest areas, at around 76 percent. In the coastal area, the residential area is around 4 percent of the total area of the 3 sub-districts. Based on the suitable developability analysis, which was carried out using the SMCA method where the weighting was carried out as stated in Appendix 4, the analysis showed that 42 percent of the areas were suitable areas for development, 15 percent of the areas were moderate areas, and 43 percent of the areas are unsuitable for development. Areas close to the coast are areas that are dominated by areas suitable for development. Based on land suitability analysis of the 100-meter area of the administrative boundary of Cianjur Regency which is assumed to be the coastal boundary with analysis was carried out using the SMCA method and obtained the suitability values for mangrove land was very unsuitable 1.45 km^2, unsuitable 1.69 km^2, moderate 2.06 km^2, suitable 2.78 km^2, and very suitable 3.40 km^2.

IOP Conf. Series: Earth and Environmental Science 1473 (2025) 012061 doi:10.1088/1755-1315/1473/1/012061

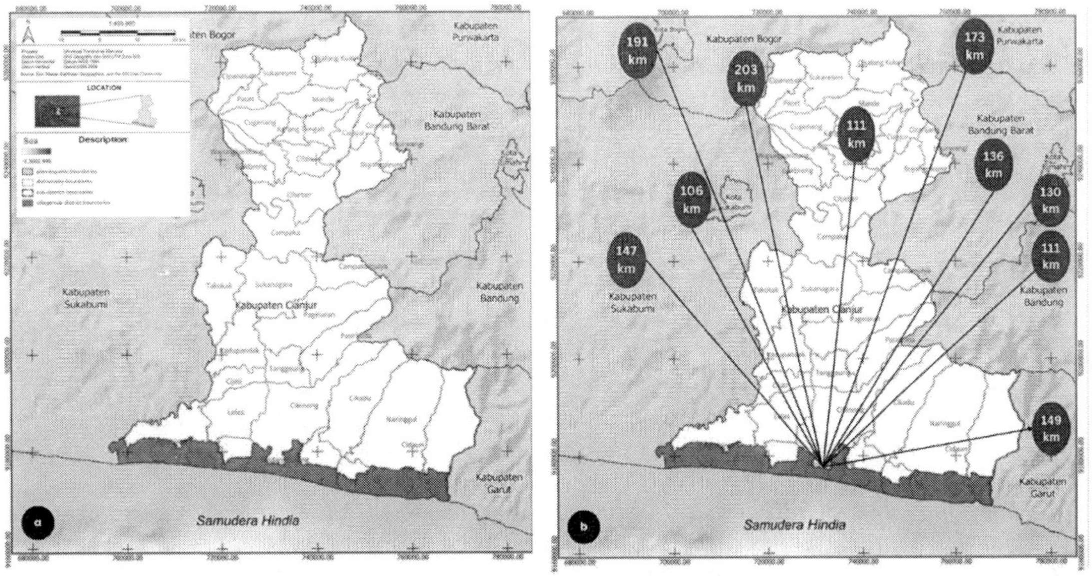

Figure 2. Study area and spatial distribution of various types of data used in this study: (a) Planning Area Boundaries, and (b) Constellation with Surrounding Areas

3.2. SWOT Analysis of Coastal Characteristics of Cianjur District

Based on the analysis carried out using the SWOT approach, the following analysis of the coastal characteristics of Cianjur Regency is described as follows.

Table 1. SWOT Analysis of Coastal Characteristics of Cianjur District

No.	Aspects	Elements
1.	Strengths	• The collector road connecting the west coast to the east and nearby districts is well-maintained • The community is active and knowledgeable in capture fisheries and aquaculture development • The community demonstrates creativity in utilizing natural resources • Conservation and rehabilitation efforts are underway
2.	Weaknesses	• There is no center for community development and training focused on economic and ecological advancement • Accessibility to the western coast (Agrabinta District) remains limited • Economic empowerment for coastal communities lacks diversity and is concentrated in the fisheries sector. • No collaboration exists among stakeholders to enhance human resources, economic growth, and disaster resilience • There are no ecosystem rehabilitation and conservation centers • No program prioritizes ecosystem-based decision-making • There is no initiative to accelerate conservation, long-term management, or the development of structures to boost ecosystem resilience • No socialization on ecosystem services and conservation has been conducted within the community • Education on sustainable resource use is insufficient, and environmentally unfriendly products dominate coastal area consumption • Community activities offer social benefits, like cooperation in tourism and Fish Auction Place areas, but lack new behaviors that drive economic growth
3.	Opportunities	• Seventeen villages have low-density populations, offering potential for development • The provincial government has multiple programs focused on forestry, fisheries, and agriculture • Private institutions and NGOs run programs for ecosystem rehabilitation, conservation, and fisheries • Advancements in technology support development • Provincial policies guide coastal area development in Cianjur
4.	Threats	• Coastal management by private institutions tends to conflict with regulations and is profit-based • Lack of synchronization between stakeholders • Environmental degradation and natural disasters

4. Result

4.1. Zoning Planning Concept

Cianjur Regency coastal area planning is district-scale planning which is implemented in all coastal villages of Cianjur Regency and is branded "One Village One Mangrove". This village-level implementation is carried out to facilitate administrative management and develop the unique characteristics of each village. This planning is carried out through the ICZM approach to manage and conserve coastal areas and natural resources to minimize resource and human losses [19]. ICZM is also defined as an effort aimed at conserving coastal resources, ecological functions, and values by implementing adequate planning and coordination in political, social, institutional, and economic contexts, to achieve sustainable development of coastal areas [20]. This approach is used so that Cianjur Regency can increase its capacity in terms of disaster and economics without ignoring 3 operational objectives of ICZM based on the World Bank through efforts: (1) improve sectoral management through increasing training capacity, establishing laws, and improving the quality of human resources; (2) defense and protection of the productivity and diversity of coastal ecosystems from habitat destruction and over-exploitation; and (3) guidelines/direction of wise development in the sustainable use of coastal resources.

The application of ICZM in spatial planning has been conducted in several countries and integrated with national priorities. For example, in Montenegro, the implementation of ICZM into spatial planning was carried out through the elaboration of the ecosystem approach (EcAp) [21] which is a solution to the overlap of geographical aspects between Marine Spatial Planning (MSP) and ICZM Directive 2014/89/EU (EC 2014). The key factors in this regulation are the analysis prioritize ecological considerations, including conservation status, environmental pressure, biodiversity, habitat and population, and others. Meanwhile, Cambodia implements ICZM with several principles that focus on socio-economic development Including strategies to improve the quality of life of coastal residents and poverty alleviation, striving for sustainable human resource management, and several other aspects related to communities and institutions [22]. This causes the aspects that are considered in spatial analysis in Cambodia to be land use and human activities. As an effort to realize this goal, the coastal area conservation of Cianjur Regency is carried out through the Nature-Based Solution (NBs) concept by utilizing supporting mangrove ecosystem services which can support provisioning and regulating ecosystem services. In general, mangrove forests provide numerous benefits for ecosystem services and their inhabitants [13]. The Nature-Based Solution (NBs) concept is implemented through mangrove conservation efforts which are implemented through protecting mangroves from damage and preserving mangroves by planting. In this planning process, mangrove spaces and surrounding areas are divided/segmented by their functions/activities, namely: 1) Protection areas, which are areas formed in sensitive spaces as zones. preservation and protection from disasters. This area is prioritized in coastal areas that are not too close to settlements and focuses on protecting beaches from abrasion, coastline erosion, and other disasters; 2) the Silvofishery development area, is an area formed to support economic activity. This area is prioritized in areas that are quite close to residential areas and existing community ponds; 3) Ecological development area is an area formed to increase the diversity of organisms and their environment. This area is prioritized in areas close to river mouths and settlements.

4.2. Zoning Planning Detail

After analyzing based on the direction of the Provincial RTRW, Cianjur Regency RTRW, the potential and issues that exist in the coastal area of Cianjur Regency, the objective of The Cianjur Regency Coastal Area Plan, namely, "Creating an Active, Adaptive and Productive coastal area of Cianjur Regency in creating prosperous coastal communities in 2045." This goal is achieved through a mangrove conservation zoning plan for each coastal village in Cianjur Regency. Program implementation planning in each village is carried out to create local wisdom and culture, increase community awareness, and facilitate management in institutions, administration, and budgets at the village level. The mangrove development activities have 3 focus functions, namely, mangrove management, utilization of silvofishery ponds, and processing of small and medium industrial fishery products. This categorization is adjusted to the area of mangrove function plans and the socio-economic conditions of the local community. In Agrabinta District, management is focused on mangrove management and processing of fishery products. In Sindangbarang District, management focuses on 3 uses, but in the urban center, it focuses on the small and medium fisheries industry. Meanwhile, in Cidaun District the focus is on mangrove management and the use of silvofishery ponds.

The general service center system in coastal areas consists of the Local Activity Center (Pusat Kegiatan Lokal/PKL) in Sindangbarang District and Cidaun District, as well as the Neighborhood Service Center (Pusat Pelayanan Lingkungan/PPL) in Agrabinta District as depicted in (Figure 3 (a)). This Service Center Plan is adapted to the service center plan in the 2011-2031 Cianjur Regency RTRW for Sindangbarang District and Agrabinta District but it is recommended that service functions be increased in Cidaun District. Meanwhile, the planned spatial structure in 2045 is to develop the main service center for mangrove development and management in Saganten Village, Sindangbarang District, and 17 mangrove development and management centers in every coastal village in Agrabinta District to Cidaun District (Figure 3(b)).

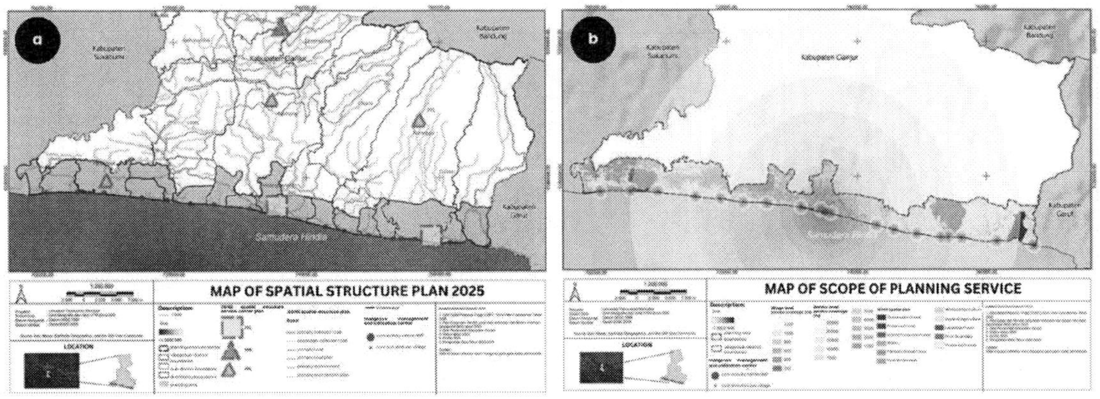

Figure 3. (a) Spatial Structure Map and (b) Scope of Planning Services

The land use plan of the Cianjur coastal area is made at 17 activity center points depicted on a 1:5.000 scale map. This pattern planning is adjusted to suit the provincial spatial pattern plan, district spatial pattern plan, suitable development, and the suitability of mangrove land. This spatial pattern plan focuses on mangrove development, resulting in an additional 299.3 ha of mangrove area which is divided into 3 sub-districts, namely 65.5 ha in Agrabinta District, 125.4 ha in Sindangbarang District, and 108.3 ha in 95 Agrabinta Districts. In more detail, the spatial pattern plan for the southern part of Cianjur Regency is planned as depicted in Figure 4.

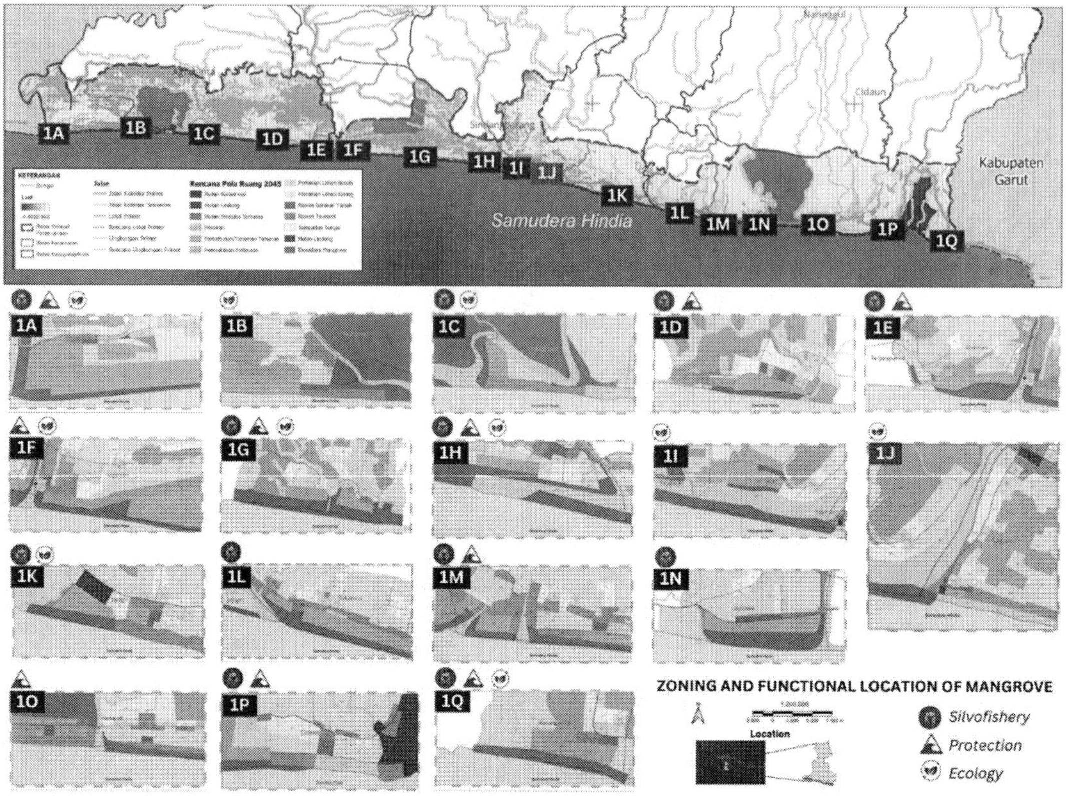

Figure 4. Mangrove Planning and Function Zone Map

The mangrove zone is located in these 3 sub-districts and the mangroves are planted using a spaced clump technique and equipped with hybrid wave breaker engineering, this plan aims to mitigate the risk of erosion to mangroves caused by the intense waves in the coastal area. In this zoning plan, three zoning categories focus on the main functions of mangroves, namely the function of increasing ecological diversity, coastal protection, and improving the economy through the development of silvofishery. Three zoning areas focused on the function of mangrove utilization were made through several considerations, namely proximity to river mouths, proximity to settlements, and proximity to existing ponds and plans. The determination of mangrove area uses consideration of mangrove suitability, suitable development, and existing properties such as settlement orientation and existing land use conditions. In terms of total area, mangroves that serve an ecological function are 106.1 ha, 68.8 ha for protection, and 124.3 ha for silvofishery as explained in the Table 2.

Table 2. Mangrove Area and Its Utilization in Each Village

Area Zoning Code	Utilization Area (ha)				Type of Fisheries Developed
	Mangrove Ecological Function	Mangrove Protection Function	Mangrove Silvofishery Function	Pond Silvofishery Function	
		Agrabinta Subdistrict			
1A (Wangunjaya Village)	5,1	8,7	9,8	19,4	Asian tiger shrimp
1B (Sinarlaut Village)	12	-	-	-	-
1C (Mekarsari Village)	20,4	-	4,2	6,4	Vaname shrimp
1D (Tanjungsari Village)	-	1,1	3,8	8,8	Crabs
		Sindangbarang Subdistrict			
1E (Wanasari Village)	-	1	10,3	8,9	Crabs
1F (Hegarsari Village)	10,1	4,8	-	-	-
1G (Kertasari Village)	7,3	4	12,8	10,6	Asian tiger shrimp

Area Zoning Code	Utilization Area (ha)				Type of Fisheries Developed
	Mangrove Ecological Function	Mangrove Protection Function	Mangrove Silvofishery Function	Pond Silvofishery Function	
1H (Talagasari Village)	5,8	16,1	5,4	7	Crabs
1I (Sirnagalih Village)	30,7	-	-	-	-
1J (Saganten Village)	1,4	-	-	-	-
1K (Jayagiri Village)	7,2	-	8,4	10,5	Vaname shrimp
Cidaun Subdistrict					
1L (Sukapura Village)	-	-	29,9	32,9	Crabs
1M (Cisalak Village)	-	8,5	9,4	10,8	Vaname shrimp and Asian tiger shrimp
1N (Jayapura Village)	-	-	20,4	16,1	Crabs
1O (Kertajadi Village)	-	9,1	-	-	-
1P (Cidamar Village)	-	9,6	4,3	5,3	Udang Vaname
1Q (Karangwangi Village)	5,6	5,8	5,7	9	Crabs

Each functional focus has several zones based on certain types of species. A detailed plan for the location of zoning, spatial patterns, and mangrove uses as depicted in Figure 5. In this figure, the image with solid line corners is the part that borders another directed zone, while the part that is not clear is the part that borders the river mouth or other function like agriculture. On the zoning map, there is a white line which is the boundary between one function and another.

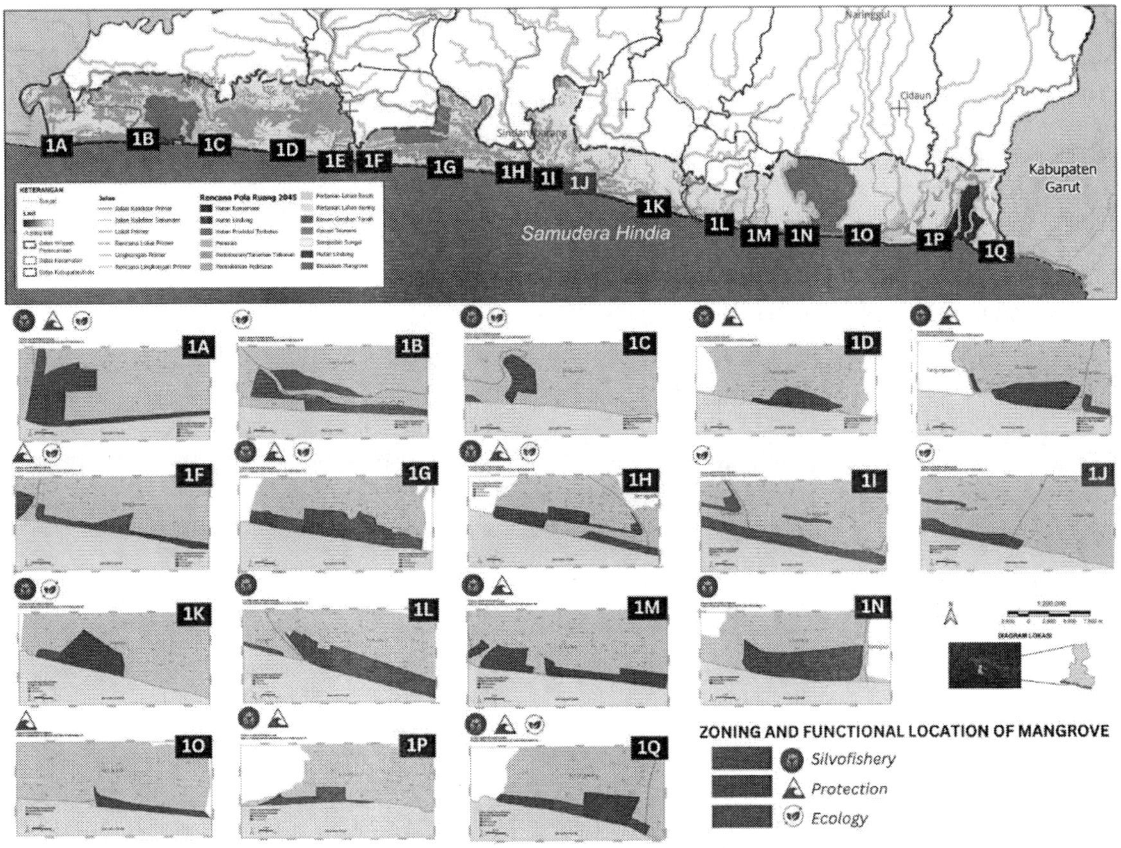

Figure 5. Mangrove Functional Zoning Map

The coastal communities of Cianjur Regency have diverse socio-economic activities. However, the most dominant livelihoods are in the primary sector, such as farmers and fishermen whose harvests or fish catches are sold directly to collectors or at fish auction places without being processed first. This causes the added value of agricultural and fishery products to tend to be low. Meanwhile, the coastal communities are also vulnerable to the threats of environmental degradation and coastal disasters. Therefore, community development initiatives are essential to stimulate economic transformation and enhance the resilience of the mangrove ecosystem by fostering new livelihood opportunities. The community development activities include managing mangrove forests, developing tourism and research areas, developing silvofishery pond cultivation, and developing processed fishery products. The concept of developing processed products is carried out to increase community productivity in producing unique and quality local wisdom products, utilizing local resources, and producing sustainably.

The government (e.g. the central government, West Java Provincial government, and Cianjur Regency government) is responsible for realizing and managing activities. Meanwhile, the community, which is hosted by community institutions in terms of citizen groups, communities, and non-government organizations, plays a role in managing and carrying out various activities. Furthermore, the private sector such as investors and private companies are encouraged to collaborate in realizing the expected plans.

The Cianjur Regency Coastal Area Plan will be carried out within 20 years from 2025 to 2045. Total financing is IDR. 2,648,766,400,355 which was issued in stages over 20 years of development. This budget comes from five sources, namely the Regional Revenue and Expenditure Budget (APBD), the Environmental Fund Management Agency (BPDLH), Corporate Social Responsivity (CSR), transfer funds in the form of Village Funds, and others in the form of fisheries income.

4.3. Logical Framework

Aligned with the Cianjur Regency Coastal Area Plan objectives of enhancing environmental quality, strengthening coastal resilience, and promoting community development, it will be measured by Logical Framework. This framework concludes inputs, outputs, outcomes, and impacts can be achieved through activities as explained in the Table 3.

Table 3. Logical Framework

	Narrative	Indicators	Verification Data Source	Assumption
	Ecological Benefits in the Form of Improving the Quality of the Environment			
Impact/Goals	Improving the quality of the living environment	Increasing the Environmental Quality Index (IKLH) of the Cianjur Regency to 67,64 (new formula) in 2025 from 67,17 in 2023	IKLH (Ministry of Environment and Forestry) measurement result	
Outcome	Increasing the quality of land cover in Cianjur Regency	Land Cover Quality Index (IKTL) 69,25 in 2045 from 67,74 in 2023	IKTL (Ministry of Environment and Forestry) measurement results	Other land cover indicators have increased and/or have not experienced a decrease in number or change in function; Other indicator data is available to measure; There is consistency of measurement and monitoring every year
Output	Effective mangrove rehabilitation	The minimum mangrove area will be 299.3 ha in 3 sub-districts by 2045	Ministry of Agrarian and Spatial Planning Land Use/Land Cover Data; Land Use/Land Cover Data Public Works Public Housing	Data from the West Java Province Maritime and Fisheries Service There is no conversion of mangrove land to other functions
Input	Implementation of the Cianjur Regency Coastal Area Zoning Planning Program Based on Mangrove Conservation, especially ecological zoning	The number of mangroves for ecological zoning is 106.10 ha in 2036	Data on the mangrove area of the West Java Province Maritime and Fisheries Service accessed from the West Java Province Open Data;	Mangrove Management Institutional Data is managed effectively and transparently Mangroves were successfully planted according to the agenda year and managed well

Ecological Benefits in the Form of Disaster Resilience				
Impact/Goals	Increasing regional capacity for disasters	Regional capacity index to 0.6 in 2045 from 0.53 in 2017	Regional Capacity Index data measured by the Regional Disaster Management Agency	
Outcome	Improving the quality of disaster mitigation	The effectiveness value of disaster prevention and mitigation will reach 0.6 in 2045, which was initially around 0.5 in 2017.	Results of the assessment of the effectiveness of disaster prevention and mitigation by the Regional Planning Agency	There was an increase in the quality and quantity of twelve indicators of the effectiveness of disaster prevention and mitigation; 7 priority indicators of regional resilience experienced and did not increase decline
Output	Mangrove rehabilitation is running effectively	The mangrove area will be 299.3 ha in 3 sub-districts in 2045	Land Use/Land Cover Data for Public Works, Public Housing; Data from the West Java Province Maritime and Fisheries Service	There is no conversion of mangrove land to other functions; Mangroves grow well
Input	Implementation of the Cianjur Regency Coastal Area Zoning Planning Program Based on Mangrove Conservation, especially protection zoning	The number of mangroves for protection zoning covering an area of 8.8 ha was successfully planted and grew in 2039 Mangrove area data	West Java Province Maritime and Fisheries Service accessed from West Java Province Open Data; Mangrove Management Institutional Data	Mangroves were successfully planted according to the agenda year and managed well
Economic Benefits in the Form of Economic Resilience				
Impact/Goals	Increasing the welfare of coastal communities	Per Capita expenditure in three coastal sub-districts (Agrabinta, Sindangbarang, and Cidaun sub-districts) will reach IDR 5,100,000.00* in 2045 per month from IDR 867,547.00 in 2022 *projection of average expenditure for West Java Province per month 2045 Central Statistics Agency data	Research result data; Central Bureau of Statistics	
Outcome	Increasing the number of business fields in the form of pond managers and micro and small industries in coastal areas	• Increase in the number of pond managers to more than 500 people • The increase in the number of micro and small industries in Agrabinta, Sindangbarang, and Cidaun sub-districts	Data from the Department of Cooperatives, Small, Medium Enterprises and Industry	There is no reduction in the number of food industries, whether processing fishery products or not; the industry runs sustainably; The income of coastal communities due to industry increases
Output	Increasing the production of aquaculture fishermen	Increased fishery production (shrimp) in silvofishery ponds by 21,911 tons and crabs by 12,043.6 tons by 2045 *Based on projected production results of silvofishery ponds with a pond area of 90 percent of the area allocated for silvofishery ponds	Aquaculture Production Data provided by the West Java Province Maritime and Fisheries Service which is accessed from West Java Province Open Data	Effective and efficient production and distribution management; there is community creativity in processing fishery products, especially shrimp so that the micro and small food industry grows
Input	Implementation of the Cianjur Regency Coastal Area Zoning Planning Program Based on Mangrove Conservation, especially zoning of mangroves and silvofishery ponds	The number of mangroves for silvofishery covering an area of 124.3 ha and ponds covering an area of 124.3 ha was successfully grown and implemented in 2034	Data on Cultivation Types in Aquaculture Production data provided by the West Java Province Maritime and Fisheries Service which is accessed from West Java Province Open Data; Mangrove Management Institutional Data	Mangroves are used by silvofishery; there is training for the community; Silvofishery pond land is developed and well-utilized

5. Discussion

In the process of implementing this plan, several challenges need to be anticipated by each stakeholder involved, namely:

1) Establishing a sustainable mangrove ecosystem is a time-intensive process. After planting, it takes several years for the mangrove ecosystem to reach maturity and provide significant environmental benefits. During this period, diligent care and monitoring are crucial to ensure the optimization of the growth and quality of the ecosystem Pollution in coastal areas either by rubbish or waste can inhibit the growth and development of mangroves. Plastic waste and water pollution by industrial or domestic waste can clog root systems, inhibit access to nutrients, and hurt the health and growth of mangroves. Toxic substances in rubbish and waste can damage mangrove roots and tissue, disrupt the reproductive system, and reduce biodiversity in the surrounding ecosystem. Therefore, efforts to manage waste and control water pollution are very important to ensure that mangroves can grow optimally.
2) The lack of compliance with spatial planning regulations has led to land use conflicts among stakeholders, including local communities, pond farmers, and local governments on mangrove areas and ponds. A potential solution to address those circumstances by to overcome this conflict is through the application of Zoning Regulation Techniques (TPZ) which can provide options for handling and utilizing land according to the characteristics, development objectives, and problems faced in certain zones so that control of space utilization can be achieved effectively.
3) Coastal areas are faced with diverse weather and climate conditions. In certain seasons, coastal areas are faced with rising sea levels, extreme waves, and extreme weather. This condition can inhibit the growth of mangroves, in the early period of mangrove growth. Therefore, before planting, there needs to be further analysis regarding weather forecasts and potential various threats to mangrove growth and development.
4) The limitation of human resource capacity and skills can hamper the targets of planned planting, management, and various transformation programs. Training, education, and community empowerment are key steps that need to be taken to optimize human resource capacity in coastal areas. These initiatives will empower communities to actively participate through training, education, and empowerment, communities can actively participate in decision-making and implementation of environmental programs. Communities will be 'Agents of Change' in preserving and managing coastal ecosystems.

6. Conclusions

The coastal area of Cianjur Regency requires planning that can utilize the potential to resolve environmental degradation problems, the threat of disaster risks due to suboptimal disaster capacity, and low economic resilience. The ICZM approach involves comprehensive planning to integrate economic and ecological elements. The embodiment of this planning is the development of mangrove ecosystems which are poured into an integrated spatial plan. This spatial plan, a mangrove utilization plan, a community socio-economic activity plan, an institutional scheme, and planning financing. In this planning process, there is a Logical Framework evaluation scheme that can be integrated with the evaluation of the regional development long-term policy/strategic plan to evaluate in the next 20 years.

References

[1] Diposaptono S. Mitigasi bencana alam di wilayah pesisir dalam kerangka pengelolaan wilayah pesisir di Indonesia. Agency for the Assessment and Application of Technology. 2003;**8**.

[2] Setkab. Pemerintah Targetkan Hapus Kemiskinan Ekstrem di 147 Kabupaten/Kota Wilayah Pesisir di Tahun 2022. Pemerintah Targetkan Hapus Kemiskinan Ekstrem Di 147 Kabupaten/Kota Wilayah Pesisir Di Tahun 2022. 2021. https://setkab.go.id/pemerintah-targetkan-hapus-kemiskinan-ekstrem-di-147-kabupaten-kota-wilayah-pesisir-di-tahun-2022/ (accessed September 26, 2023).

[3] Indraswari DL. Ironi kemiskinan wilayah pesisir yang kaya potensi ekonomi kelautan. kompas.id. 2023. https://www.kompas.id/baca/riset/2023/01/25/ironi-kemiskinan-wilayah-pesisir-yang-kaya-potensi-ekonomi-kelautan (accessed September 26, 2023).

[4] Kementerian PPN/Bappenas. Indonesia Blue Economy Roadmap. 2023.

[5] Rukin R. Pembangunan perekonomian masyarakat desa pesisir sebagai upaya peningkatan kesejahteraan. JSH. 2020;**11**:1. https://doi.org/10.30997/jsh.v11i1.1998.

[6] Suryana AAH, Nurhayati A, Dewanti LP, Andhikawati A, Dewi RR. Pemetaan daya saing subsektor perikanan tangkap di Provinsi Jawa Barat. BIMSEKP. 2021;**7**:115. https://doi.org/10.15578/marina.v7i2.8881.

[7] Antara News. Memberantas kemiskinan ekstrem di Cianjur. Antara News. 2021. https://www.antaranews.com/berita/2497417/memberantas-kemiskinan-ekstrem-di-cianjur (accessed September 27, 2023).

[8] Kompas.com. Ironi, kemiskinan di Lumbung Padi Jawa Brat. KOMPAS.com. 2021. https://regional.kompas.com/read/2021/11/03/112711778/ironi-kemiskinan-di-lumbung-padi-jawa-barat (accessed September 27, 2023).

[9] BNPB Kabupaten Cianjur. Dokumen kajian risiko bencana Kabupaten Cianjur 2018-2022. 2017.

[10] Azuga NA. Kajian kerentanan kawasan pesisir terhadap bencana kenaikan muka air Laut (sea level rise) di Indonesia. J-Tropimar. 2021;**3**. https://doi.org/10.30649/jrkt.v3i2.41.

[11] Septian ART. Analisis ancaman bencana gelombang ekstrem dan abrasi di wilayah pesisir Gresik Jawa Timur. JOMCER: Journal of Disaster Mitigation and Civil Engineering Research. 2022;**01**.

[12] Lugina M, Indartik I, Aulia Pribadi M. Economic valuation of mangrove ecosystem and their contribution to household income: Case studies at Pemogan, Tuban, and Kutawaru Villages. j.penelit.sosial.ekonomi.kehutanan. 2019;**16**:197–210. https://doi.org/10.20886/jpsek.2019.16.3.197-210.

[13] Taye FA, Folkersen MV, Fleming CM, Buckwell A, Mackey B, Diwakar KC, et al. The economic values of global forest ecosystem services: A meta-analysis. Ecological Economics. 2021;**189**:107145. https://doi.org/10.1016/j.ecolecon.2021.107145.

[14] Hadianti A, Ulinnuha H, Heliani LS, Sarwadi A, Kurniawan A, Budisetiawan J, et al. Integrating Multi-Hazard Risk Analysis into Spatial Planning for Small Island: Study Case of Sangihe Island. IOP Conf Ser: Earth Environ Science. 2021;**799**:012008. https://doi.org/10.1088/1755-1315/799/1/012008.

[15] Sidik S. Implementasi perencanaan pembangunan daerah (Studi tentang Musrenbang di Kabupaten Sumenep). JPAP. 2016;**2**. https://doi.org/10.30996/jpap.v2i01.710.

[16] Nursyamsi M. Tambak udang berkelanjutan penggerak ekonomi pesisir Cianjur. Perhutani 2021. https://www.perhutani.co.id/tambak-udang-berkelanjutan-penggerak-ekonomi-pesisir-cianjur/ (accessed March 16, 2024).

[17] Darmayanti HP. Analisis dinamika perubahan tutupan lahan dan tingkat kekritisan lahan mangrove di kawasan Pantai Tipar Kabupaten Cianjur, Jawab Barat. Skripsi. Institut Teknologi Bandung, 2023.

[18] Nurmuslim B. Libur Tahun Baru, Lahan Cemara Laut dan Mangrove Terbakar di Agrabinta. www.logikanews.co 2023. https://logikanews.co/libur-tahun-baru-lahan-cemara-laut-dan-mangrove-terbakar-di-agrabinta/ (accessed May 9, 2024).

[19] World Bank. ICZM guidelines. The International Bank; 1996.

[20] Hossain MS, Failler P, M. G. Hussain, M.K. Alam. Integrated Coastal Zone Management (ICZM) Marine Spatial Planning (MSP). Bangladesh: Maritime Affairs Unit, Ministry of Foreign Affairs, Government of Bangladesh with the support of the European Union in association with the University of Chittagong. 2018.

[21] UNEP/MAP/PAP. Guidelines for the preparation of National ICZM Strategies required by the Integrated Coastal Zone Management (ICZM) Protocol for the Mediterranean. Split: Priority Action Programme. 2015.

[22] Lyngby JE, Jeppesen G, Vann M. Integrated coastal zone management and planning principle in Cambodia. 2017.

Maritime Society Involvement In Environment Management By The Government

Kuncowati[1]*, S T Wiyono[1] and Mudiyanto[1]

[1] Ship Operations Engineering Technology Study Program, Faculty of Shipping Vocational, Hang Tuah University, Surabaya, Indonesia

* E-mail: kuncowati@hangtuah.ac.id

Abstract. Environmental management is an integral part of efforts to maintain the sustainability of global ecosystems. The role of the government in managing and protecting the environment requires active involvement from society as one of the main pillars of policy implementation. This study aims to analyze the influence of maritime society involvement, which consists of the level of environmental awareness and cooperation between the government and the society that affects the level of involvement in environmental management by the government. The research method used is a quantitative method with differential statistical analysis techniques with Smart-PLS to describe the involvement of maritime communities in various environmental management contexts. The novelty of this study is that it aims to increase society's involvement in environmental management. It is hoped that this research will contribute to improving environmental management by the government by encouraging the active role of maritime communities in preserving natural resources for future generations.

1. Introduction

The management of the maritime environment by the government is very important in reducing the impact of maritime activities on marine ecosystems. Research shows that ocean transport contributes significantly to ocean pollution, thus requiring effective intervention strategies to minimize marine pollution. [1]The shipping industry contributes significantly to the degradation of the marine environment, which requires sustainable management. [2]The complexity of shipping's impact on the environment and the monitoring and enforcement of regulatory compliance are factors that affect the government's management of the maritime environment. [3].

Maritime environmental management involves complex frameworks that require coordination at different levels and sectors; current governance arrangements often need to be improved due to poor interaction between stakeholders, which can exacerbate environmental problems. [4] Traditional stakeholder groups, including marine convention experts, resource users, and the general public, also participate in management's decision-making process [5], [6].

The involvement of maritime communities in environmental management is increasingly recognized as essential for the sustainable management of coastal and marine resources. In the management of waste on ships, the perception of ship passengers has a positive and significant effect [7]. The culture of safety and occupational health of the crew is the strongest path that positively and significantly affects the separation of solid waste as an effort to manage the marine environment [8]. In an effort to prevent marine pollution, the crew must understand the regulations and make efforts to prevent marine pollution [9]. Crew awareness of safety and understanding of their role as crew members in waste handling affects waste handling on board [10].

Collaborative governance is needed to manage the maritime environment. This involves integrating society's voices in decision-making and increasing society's capacity in ecosystem management. [11]In addition, the study highlights the importance of aligning society and government efforts to ensure the effective management of marine ecosystems through a society-participation approach. [4]. However, maintaining society engagement remains a challenge as social and economic factors significantly affect engagement. [12]. Society's involvement in environmental management, such as solid waste treatment, effective involvement is hampered by a lack of structured education and communication. [13]. Thus, encouraging collaboration between society and the government is essential to achieve the sustainability of the maritime environment. [14]. Conservation messages should emphasize aspects of the marine environment that directly impact societies to increase awareness of conservation strategies and initiatives [15].

The level of maritime society's awareness of environmental management is influenced by knowledge of environmental issues and understanding of policies to reduce marine litter. [16]. Regulatory socialization plays an important role in raising awareness among stakeholders. [17]. Meanwhile, the level of societal cooperation in the management of the maritime environment is influenced by effective collaboration between society and the government and government entities related to policy alignment in the management of marine ecosystems. [18] and effective collaboration between stakeholders [19].

Based on the results of the study [16], [10] It shows the influence of public awareness on the prevention of marine pollution. Still, the influence of awareness and cooperation of the maritime society on their involvement in environmental management by the government has yet to be shown. In addition, the results of the research [17] Society's cooperation with the government has not indicated the influence of social, educational, and social communication factors in collaboration for environmental management. So, this study aims to get a model of society's involvement in environmental management by the government by analyzing awareness, cooperation, and societal involvement.

The acquisition of a waste management model by the government can provide input to the government and the maritime society in managing the coastal and marine environment more effectively and sustainably. To minimize the gap in society, actions to protect the marine ecosystem should be encouraged, namely by encouraging actions and lifestyles that support the marine environment. [20] [21].

2. Method

A quantitative research method with the Structural Equation Modeling-Partial Least Square (SEM PLS) statistical analysis technique was used in this study to analyze the involvement of the maritime society in environmental management by the government. SEM-PLS analysis techniques can be used to determine causal relationships in a theoretically defined model.

The maritime society selected as a sample for this study is 100 crew members operating in Indonesian waters. The probability sampling technique is simple because it involves randomly sampling from the population without regard to the strata in the population [22]. The number of samples is determined by the maximum Similarity estimate, which is between 100 and 200 [7].

The questionnaire is based on the indicators of research variables, which include exogenous variables of the level of awareness of the maritime society with indicators of knowledge about environmental issues. [16], policy understanding[16] and socialization of policy regulations [17]. Exogenous variables of the level of cooperation between the maritime society and indicators of effective collaboration between the society and the government and government entities related to policy alignment in marine ecosystem management [18] and effective collaboration between stakeholders [19]. Endogenous variables of society's involvement with indicators of social and economic levels [12], education and communication skills [13]The government links variables to environmental management, using indicators of the impact of sea transportation on the environment. [1]–[3], enforcement of regulatory compliance [3], environmental governance [3].

Furthermore, questionnaires were distributed to 100 crew members. The respondents' answers were carried out by an external model test, which included a validity and reliability test, an internal model test testing the direct influence of research variables in construction, and a quadratic R test was carried out. *R square* has a value of 0.50-0.75, then the model is moderate, and if R *Field* is greater than 0.75, then it is a powerful model. [22].

3. Results

SEM-PLS analysis was carried out by testing the outer model and the inner model. External model tests are conducted to test validity and reliability. From the results of the validity test, the external loading value is obtained, as shown in Table 1

Table 1.Research on external loading on environmental management by the government

Indicators of the researcher's variables	Outer loading
Maritime society awareness (X1)	
Knowledge of environmental issues (X1.1)	0,974
Policy understanding (X1.2	0,979
Regulatory socialization (X1.3)	0,985
Maritime society cooperation (X2)	
Collaboration between the society and the government (X2.1)	0,982
Government entities in policy alignment (X2.2)	0,945
Effective proof between stakeholders (X2.3)	0,965
Maritime society involvement (Y1)	
Social (Y1.1)	0,962
Education (Y1.2)	0,968
Communication skills (Y1.3)	0,802
Environmental management by the government (Y2)	
Impact of marine transportation on the environment (Y2.1)	0,889
Enforcement of regulatory compliance (Y2.2)	0,895
Environmental Governance (Y2.3)	0,840

Source: SmartPLS Research Report, 2024

Based on Table 1, it can be explained that all the indicators of the research variables of maritime society awareness, maritime society cooperation, maritime society involvement, and environmental management by the government are more than 0.7, so all indicators are declared valid for further SEM-PLS analysis. In addition to conducting validity tests for all research variables, reliability tests are carried out. The reliability test can be seen from the results of the Average Variance Extracted (AVE) value, composite reliability value, and Cronbach alpha shown in the following table 2.

Table 2. Results of construction validity and reliability tests

Variable	AVE	Alpha Cronbach	Composite Reliability
Maritime society awareness	0,959	0,978	0,986
Maritime society cooperation	0,930	0,962	0,975
Maritime society involvement	0,835	0,899	0,938
Environmental management by the government	0,765	0,857	0,907

Source: SmartPLS Research Report, 2024

Based on Table 2 of the validity and reliability test, it can be seen that the AVE value of all research variables in environmental management by the government is more than 0.5; this shows that the correlation in the research construction is good or valid. Furthermore, from Table 2, it can be seen that the total composite value of the research variables is more than 0.7, so it can be stated that all variables in the research on society's involvement in environmental management by the government are reliable and can be used for further analysis. The reliability of this research variable is also strengthened by the Cronbach alpha results, which show that all research variables have a Cronbach alpha value of more than 0.6, so it can be stated that all research variables have a high level of reliability.

The next SEM-PLS analysis is bootstrap, which is used to obtain the results of the internal model test, both the direct influence and the direct influence of each research variable. The results of the test of the direct influence of maritime society awareness, maritime society cooperation, and maritime society involvement in environmental management by the government are shown in Table 3—the direct influence of the variables of society's involvement in environmental management by the government.

Table 3. Direct influence of society involvement variables in environmental management by the government

	Original sample	Average Sample	Standard deviation	P Value
Maritime Society awareness -> Maritime Society Involvement	0,531	0,526	0,129	0,000
Maritime Society Cooperation -> Maritime Society Involvement	0,453	0,459	0,130	0,001
Maritime society involvement -> environmental management by the government	0,877	0,882	0,013	0,000

Source: SmartPLS Research Report, 2024

Table 3 shows the direct influence of each research variable on maritime society's involvement in environmental management by the government. The research variable was stated to have a positive and significant influence if it had a value of less than 0.005. The results of the hypothesis test show that maritime society awareness has a positive and significant direct influence on maritime society involvement of 0.531. Maritime society cooperation has a positive and significant effect on the involvement of the maritime society by 0.453, and the involvement of the maritime society has a positive and significant effect on environmental management by the government by 0.877. Furthermore, based on the results of the test of the indirect influence of the variables of Maritime Society Awareness and Maritime Society through the variables between Maritime Society Involvement on the variables of Environmental Management by the Government, the research model of society involvement in environmental management by the government is shown in the following table 4.

Table 4. Indirect influence of research variables

	Original sample	Average Sample	Standard deviation	P Value
Maritime society awareness -> Maritime society involvement -> Environmental management by the government	0,465	0,464	0,114	0,000
Maritime society cooperation -> Maritime society involvement -> Environmental management by the government	0,397	0,0,405	0,115	0,001

Source: SmartPLS Research Report, 2024

From the results of the indirect influence test in Table 4, it can be explained that maritime society awareness has a positive and significant influence on environmental management variables by the government through the variable between maritime society involvement of 0.465 units. Maritime society cooperation has a positive and significant effect on environmental management variables by the government indirectly through variables including the involvement of the maritime society of 0.397 units. Meanwhile, the results of the R-squared test on the model of society involvement in environmental management by the government are shown in the following table 5.

Table 5. R square

	R square
Maritime society involvement	0,947
Environmental management by the government	0,769

Source: SmartPLS Research Report, 2024

Table 5 shows that the R Square results of the maritime society awareness and cooperation variables greatly affect the maritime society involvement variable in the structural model because the R Square is 0.947. The results of the R Square analysis of the variables of maritime society awareness, maritime society cooperation, and maritime society involvement greatly affect the variables of environmental management by the government because the R Square value was obtained at 0.769.

4. Discussion

Based on the results of the SEM-PLS outer model test, it can be seen that all indicators of the research variables of society involvement in maritime environmental management by the government are valid because they have a value of 0.7. The results of this outer model test also showed that the level of awareness of the maritime society was influenced by the factors of public knowledge about the environment and policy understanding and was most strongly influenced by regulatory socialization with a value of 0.985. The results of this study can add to the research findings [17]. Public awareness is not only influenced by the socialization of regulations but also by knowledge and understanding of maritime society. To increase awareness of maritime society, efforts are needed to increase public knowledge about the environment, understanding of policies, and especially the socialization of national and international regulations in environmental management.

From the results of the outer model test, it can be explained that the level of cooperation of the maritime society is influenced by the factors of government entities in policy alignment, effective cooperation between stakeholders, and the highest is influenced by the factor of collaboration between the society and the government. This is in line with the results of the study. [18] Which states that effective cooperation between the society and the government is needed in the management of the marine environment. The results of this study can add to the findings of the research. [18] Good cooperation in environmental management is not only influenced by the collaboration factor between society and the government. Still, it must also consider the effectiveness factor of cooperation and government entities in policy alignment.

The results of the outer model also show that social factors, education, and communication skills of the maritime society influence the involvement of the maritime society. The results of this study provide additional findings. [18] This states that society's involvement requires collaboration between society and the government. Therefore, to increase maritime society's involvement, social factors, education, and communication skills need to be improved.

The results of the study show that government environmental management is most influenced by the regulatory compliance factor of 0.895. This strengthens the results of the study [3] This states that the complexity of shipping's impact on the environment and monitoring and enforcement of compliance with regulations are factors that affect the government's management of the maritime environment. Therefore, improving the factor of enforcing society's compliance with applicable regulations is very important.

The results of the inner model test of society involvement in environmental management by the government showed the direct influence of the maritime society awareness variable on the positive and significant involvement of the maritime society of 0.531. This is in line with the research presented by [15] The conservation message should focus on elements of the marine environment that directly resonate with society to enhance awareness of conservation strategies and initiatives. The results of this study mean that to increase society's involvement, maritime society's awareness must be increased, especially by increasing the socialization of environmental management regulations that society must comply with.

The results of the SEM-PLS analysis also showed that maritime society cooperation had a positive and significant effect on maritime society involvement by 0.453. This is in line with research that states that maritime environmental management involves complex frameworks that require coordination at various levels and sectors; current governance arrangements often experience difficulties due to poor interactions between stakeholders that can exacerbate environmental problems [4]To increase society's involvement, cooperation between society and the government must be increased, and the main way to do this is to increase government entities' policy alignment.

Environmental management by the government is directly and significantly influenced by the involvement of the maritime society by 0.877 units. This is in line with research that states that the involvement of maritime societies in environmental management is important for sustainable management of coastal and marine resources, such as collaborative governance by integrating society voices in decision-making and increasing society capacity in ecosystem management. [23]This study found that maritime society's involvement is the strongest variable that directly affects the government's management of the environment compared to the variable of awareness and cooperation of the maritime society.

Meanwhile, the results of the indirect influence test of the research variables show that the strongest path that affects environmental management by the government is the maritime society awareness variable, which indirectly affects environmental management by the government through the variable between the involvement of the maritime society, which is 0.645. The relationship between variables in the model of society involvement in environmental management by the government from the results of SEM-PLS analysis is shown in Figure 1.

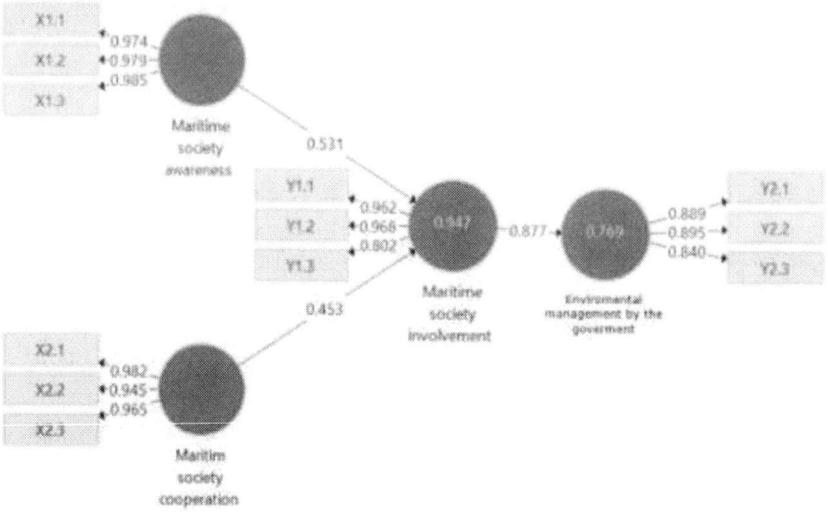

Figure 1. Model of maritime society involvement in environmental management by the government
Source: Results of research analysis with SEM-PLS, 2024

Based on Figure 1, the model of society's involvement in environmental management by the government, the results of the inner model analysis with SEM-PLS show that the maritime society's awareness (X1) is 0.531 units (53.1%), cooperation is (X2) 0.453 (45.3%), and the society's involvement (Y1) is 0.877 (87%), which has a positive and significant effect on environmental management by the government (Y2).

Based on Figure 1, it is evident that the results of the R-square test indicate a strong combined influence of the variables of maritime society awareness and maritime society cooperation on the variable of maritime society involvement, with an R-square value of 0.947. The variables of awareness, cooperation, and involvement of the maritime society together affect environmental management by the government strongly, with the R Square analysis value of 0.769. This model is a model in the strong category because it has an R square value of more than 0.750. This means that there is a strong relationship between variables in the research construction.

The model of public involvement in environmental management by the government is a novelty in this study because it is based on previous research [16, [10] It has yet to show the influence of awareness and cooperation of the maritime society on its involvement in environmental management by the government. In addition, the results of the research on Society's cooperation with the government have yet to indicate the influence of social, educational, and social communication factors in collaboration for environmental management.

5. Conclusion

The novelty of this research is the acquisition of a model of maritime society involvement in environmental management by the government, which shows a strong relationship between maritime society awareness, cooperation, and society involvement in environmental management by the government. The strongest path in this model that directly affects environmental management by the government is societal involvement. Meanwhile, from the results of the indirect influence test, the research variable shows that the strongest path affecting environmental management by the government is the maritime society awareness variable, which indirectly affects environmental management by the government through the variable between the involvement of the maritime society.

Based on the research conclusions regarding society's involvement in environmental management by the government, it is recommended that the government, particularly the Directorate General of Sea Transportation and the Ministry of Environment, enhance public participation in maritime environmental management. This can be achieved by raising awareness within maritime communities through education, training, and the dissemination of regulations related to maritime environmental management, with careful consideration of the community's social background, education level, and communication skills.

Additionally, it is recommended that maritime communities, especially crew members, improve their compliance with maritime environmental management regulations. Since the scope of this study was limited to crew members, future research should also include other groups, such as the fishing community, coastal residents, and other maritime environment users, to broaden the scope of involvement.

Reference

[1] W. C. Lin, "Maritime Environment Assessment And Management Using Through Balanced Scorecard By Using Dematel And Anp Technique.," *Public Health*, **19**, p. 2873, 2022, doi: Https://Doi.Org/10.3390/ Ijerph19052873.

[2] P. Q. and F. M. Wang L, Yao J, Zhang H, "A Sustainable Shipping Management Framework In The Marine Environment: Institutional Pressure, Ecodesign, And Cross-Functional Perspectives," *Mar. Sci*, 2023, doi: 10.3389/Fmars.2022.1070078.

[3] R. J. K. Byrnes, T. A., & Dunn, "Boating- And Shipping-Related Environmental Impacts And Example Management Measures," *J. Mar. Sci. Eng.*, **8**, no. 11, p. 908, 2020, doi: 10.3390/Jmse8110908.

[4] A. Ross, H., Adhuri, D. S., Abdurrahim, A. Y., & Phelan, "Opportunities In Community-Government Cooperation To Maintain Marine Ecosystem Services In The Asia-Pacific And Oceania," *Ecosyst. Serv.*, 2019, doi: 10.1016/J.Ecoser.2019.100969.

[5] C. Flécher, "Le Travail En Mer Depuis L'introduction Du Code Ism. In Revue D'histoire Maritime. Hal.Archives-Ouvertes," 2014.

[6] L. Brennan, J., Fitzsimmons, C., Gray, T., & Raggatt, "Eu Marine Strategy Framework Directive (Msfd) And Marine Spatial Planning (Msp): Which Is The More Dominant And Practicable Contributor To Maritime Policy In The Uk?," *Mar. Policy*, vol. 43, pp. 359–366, 2014, doi: 10.1016/J.Marpol.2013.07.011.

[7] Kuncowati, "Passenger Attitudes And Waste Management On Ships, International Journal Of Sustainable Development And Planning.," *J. Homepage*, **19,** no. 17, pp. 2683–2689, 2024.

[8] Kuncowati, "Solid Waste Separation Model on Ships based on Occupational Health Safety Culture and Work Environment," *ournal Environ. Sci.*, **22,** no. 4, pp. 1829–8907, 2024.

[9] K. Roni Ashiddiqi, Ogik Laotuju, Mudiyanto, "Ship Waste Management as an Effort to Prevent Marine Pollution in KM. Nusantara Belt 115," *J. Shipp. Port Appl.*, **14,** no. 1, pp. 36–41, 2023, doi: 10.30649/japk.v14i1.101.

[10] G. C. & K. Kuncowati, Qomariyatus Sholihah, "Handling Waste on Ships: A Model Based on Safety and Awareness, LOGI," *Sci. J. Transp. Logist.*, **13,** no. 1, 2022, doi: 10.2478/logi-2022-0020.

[11] J.-L. Chen, "Engaging Local People In Conserving The Socio-Ecological Production Landscape And Seascape By Practicing Collaborative Governance In Mao'ao Bay, Chinese Taipei," pp. 203–225, 2023.

[12] A. T. S. Emma Legaspi Ballad., Yoshinori Morooka., "Ensuring Sustainability Of Community Participation In Locally-Managed Marine Protected Area In North-Western Cagayan," *Philipp. Marit.*, 2022.

[13] O. C. Uzoaru, "Involvement Of Community Members In Solid Waste Management And Environmental Protection In Rivers State: Implications For Sustainable Cities," *Int. J. Res. Granthaalayah*, 2021, doi: Doi.Org/10.29121/Granthaalayah.V10.I1.2022.4433.

[14] S. H. Erga Yuhandra, Iman Jalaludin Rifa'i, "Local Government Policies In The Management Of Coastal And Marine Resources Through Community Participation, Unification," *J. Law*, 2023, doi: Org/10.25134/Unifikasi.**10i1.7539**.

[15] E. S. Easman, K. E. Abernethy, and B. J. Godley, "Assessing public awareness of marine environmental threats and conservation efforts," *Mar. Policy*, **87,** no. June 2017, pp. 234–240, 2018, doi: 10.1016/j.marpol.2017.10.030.

[16] U. K. Pranyoto, Prapti Utamib, "The Effect of Knowledge and Awareness of Human Resources on Attitudes in Efforts to Control Pollution in Waters and Ports Maritime Dynamics," *Marit. Dyn.*, **3,** no. 1, pp. 44–51, 2022.

[17] C. Y. T. Salmah Sharon, Yuyun Karystin Meilisa Suadel, "Socialization Of Regional Regulations Regarding Waste Management To Increase Awareness Of Environmental Cleanliness For Msmes," *J. Community Serv. Merdeka Univ. Malang*, **8,** no. 1, pp. 188–198, 2023.

[18] S. La Nalefo, Bahtiar, "The Study Of Dynamic Of Coastal Resource Community-Based Management In Wabula Village Of Buton Regency," *ournal Sustain. Dev.*, **11,** no. 6, 2018.

[19] A. L. L. Mungcal., "Department Of Anthropology, University Of The Philippines-Diliman," *J. Wetl. Environ. Manag.* **1**, no. 1, pp. 5–15, 2013, doi: Org/10.20527/Jwem.01.01.02.

[20] M. J. A. R.L. Jefferson, I. Bailey, D. d'A. Laffoley, J.P. Richards, "Public per- ceptions of the UK marine environment," *Mar. Policy*, **43,** pp. 327–337, 2014, doi: http://dx. doi.org/10.1016/j.marpol.2013.07.004.

[21] A. C. J. Vincent, "Saving the shallows: focusing marine conservation where people might care, Aquat. Conserv.," *Mar. Freshw. Ecosyst.*, **21,** pp. 495–499, 2011, doi: http://dx. doi.org/10.1002/acq.2220.

[22] I. Ghozali, *Application of Multivariate Analysis with IBM SPSS 23 Program (8th Edition).* Diponegoro University Publishing Board, 2016.

[23] J. Brennan, C. Fitzsimmons, T. Gray, and L. Raggatt, "EU marine strategy framework directive (MSFD) and marine spatial planning (MSP): Which is the more dominant and practicable contributor to maritime policy in the UK?," *Mar. Policy*, **43**, pp. 359–366, 2014, doi: 10.1016/j.marpol.2013.07.011.

IOP Conf. Series: Earth and Environmental Science **1473** (2025) 012063 doi:10.1088/1755-1315/1473/1/012063

Strengthening National Legal Frameworks for Coral Reef Protection: Toward a Sustainable Blue Economy

Chomariyah[1*]

[1] International Law Department of Hang Tuah University, 150 Arif Rahman Hakim Street, Surabaya, 60111, Indonesia

*Email: chomariah@hangtuah.ac.id

Abstract. The worldwide decline of coral reefs presents major risks to marine ecosystems, coastal protection, and the well-being of millions of people who depend on marine resources. The Sustainable Blue Economy concept is centres on utilizing ocean resources for economic development while maintaining the health of ocean ecosystems. The primary issues currently endangering coral reefs include coral bleaching and elevated sea temperatures. These conditions cause corals to expel the symbiotic algae residing in their tissues, resulting in a loss of colour. Prolonged bleaching can ultimately lead to the death of the corals. The study aims at analysing, how national legal frameworks can be leveraged to enhance coral reef conservation, thereby contributing to the development of a sustainable blue economy. Method, using the comparative analysis with identify successful strategies and best practice adopted by various states (Indonesia, Australia and Malaysia) and case studies. The result of this research show that national legal frameworks are critical in driving effective coral reef conservation. The comparative analysis highlights that successful strategies often involve a combination of legal protections, community engagement, and adaptive management. By adopting best practices from different countries, a more resilient and sustainable approach to coral reef conservation can be developed, supporting the broader goals of a sustainable blue economy.

1. Introduction

In 2024, the occurrence of coral bleaching worldwide has increased due to extreme sea temperatures. This event has caused significant damage, including mass bleaching in Australia's Great Barrier Reef. Research indicates that climate change is threatening the survival of coral reefs, including some that are hundreds of years old coral [1]. Although coral reefs occupy less than 1% of the ocean's seabed, they are the most diverse ecosystems in the marine world, supporting at least 25% of marine species. Spanning over 100 countries and territories, these ecosystems boast the highest global biodiversity and are among the planet's most valuable. They play a vital role in ensuring safety, coastal protection, well-being, food security, and economic stability for hundreds of millions of people, especially in developing nations, transitional economies, and small island developing states. The goods and services provided by coral reefs, including tourism, fisheries, coastal protection, and medicinal resources, are valued at approximately $2.7 trillion annually. Despite their critical importance, coral reefs are facing serious threats from human activities that contribute to climate change and, as a result, coral reef bleaching.

Max-Planck-Gesellschaft 2024 report [2], global coral bleaching events are projected to become increasingly common in the coming decades due to climate change. The report predicts that if carbon emissions are not significantly reduced, all of the world's reefs could experience bleaching by the end of the century. The severe coral bleaching is anticipated to occur annually, endangering the reefs' ability to recover unless they can adapt to rising temperatures. Coral reefs offer essential ecological, economic, and social benefits, including support for marine biodiversity, billions in revenue from fisheries and tourism, and protection for coastal communities against natural disasters. Nevertheless, coral reefs are under substantial threat from climate change, pollution, overfishing, and harmful fishing practices. Climate change causes bleaching and ocean acidification, which weaken corals, while pollution deteriorates water quality. Overfishing and destructive fishing techniques further upset the ecological balance and cause physical damage to the reefs. This information is highlighted in reports from the National Oceanic and Atmospheric Administration (NOAA) and The International Coral Reef Initiative (ICRI) [3] Coral reefs worldwide are undergoing a mass bleaching event driven by the climate crisis and unprecedented ocean heat, with experts cautioning that this could become the most severe bleaching period ever recorded. Over the past year, more than 54% of coral reef areas globally have been affected, impacting at least 53 countries and territories across extensive regions of the Atlantic, Pacific, and Indian Oceans.

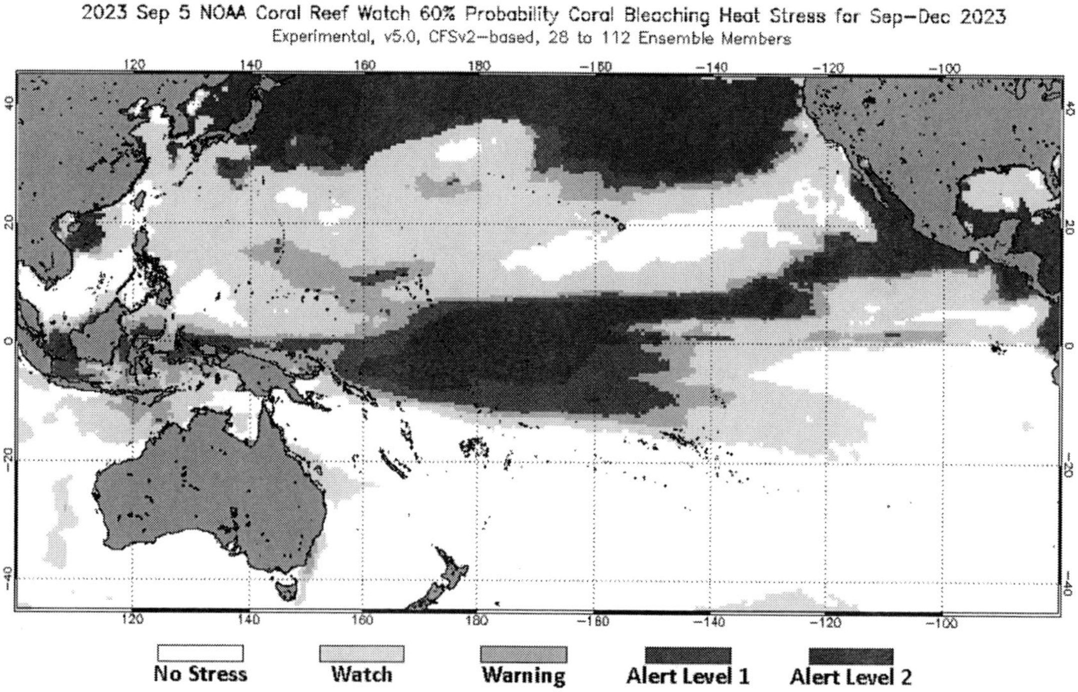

Figure 1. NOAA Coral Reef Watch's , coral bleaching alert area.
Source from coral reef watch, NOAA.gov

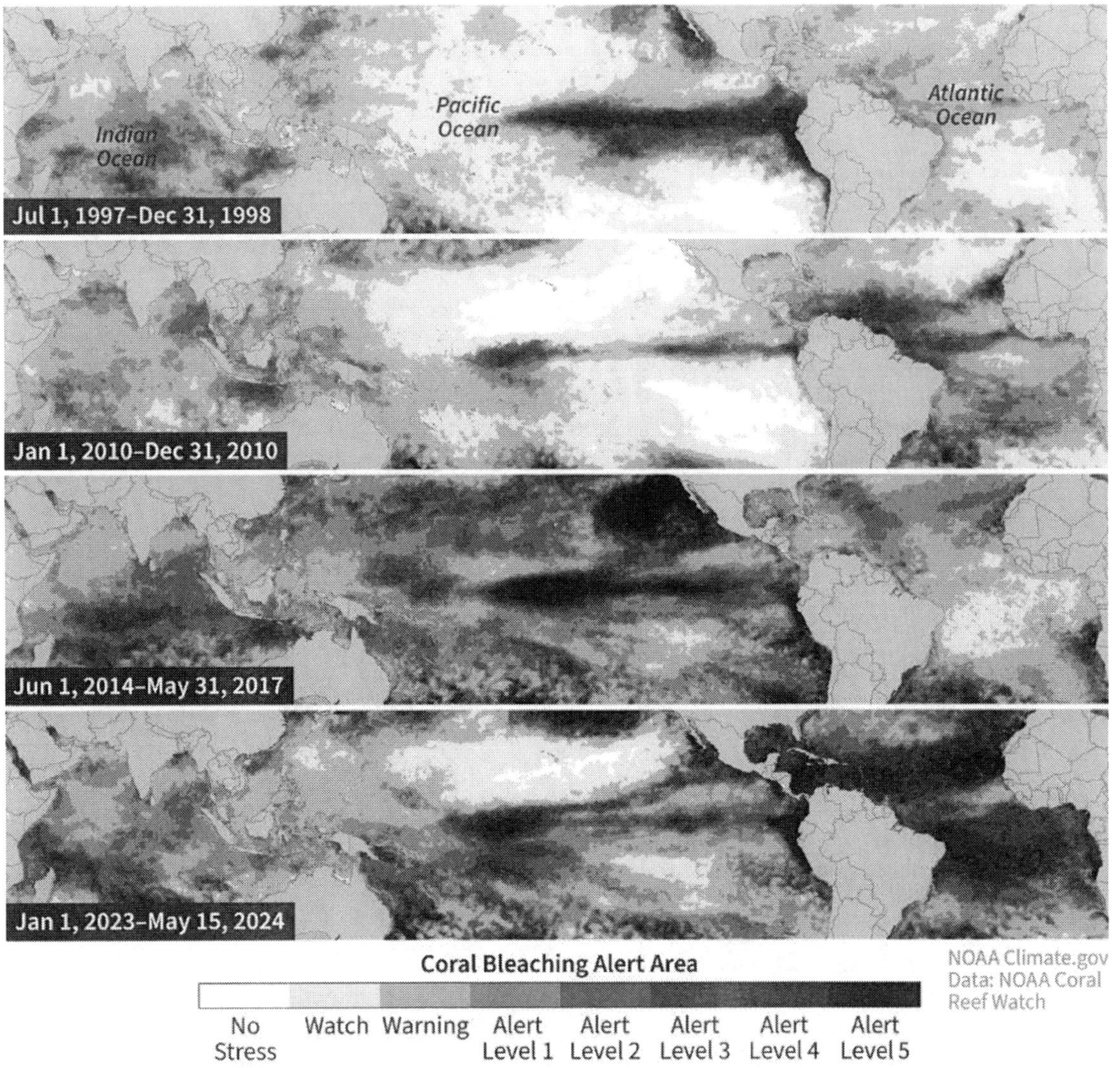

Figure 2. Comparing Global Coral Bleaching Events.
Source: https://www.climate.gov/media/16084

That figure shows that the leading cause of coral bleaching is climate change. In ocean temperature triggering corals to expel algae. Other factors like low tides, pollution, and excessive sunlight can also cause bleaching. Marine heatwaves have caused four major coral bleaching on the Great Barrier Reef, resulting in a reduction of shallow water corals by as much as 50%. Although coral reefs have the potential to recover, this is contingent upon a decrease in temperatures and the restoration of normal conditions.

2. Legal Framework, Methodology and Literature Review

2.1. Legal Framework

This research involved legal analysis through the examination of documents and regulations. Legal research refers to the process of identifying and gathering authoritative sources relevant to a legal issue, and it is the field of study focused on effectively organizing and utilizing these authorities to address questions of law [4]. From literature review stated [5] Indonesia's policy has motivated various practitioners to engage in reef restoration, but many projects lack coordination with broader restoration networks or scientists. Notably, only 16% of these projects include post-installation monitoring frameworks. To improve restoration results, it is crucial to set clear objectives and implement long-term monitoring during the planning phases, as well as to focus on sharing knowledge and collaborating with the international scientific community. These steps will enable Indonesia to realize its significant potential as a global leader in coral reef restoration. And from another research stated [6], Bangladesh needs to strengthen the enforcement of environmental laws, particularly those related to marine protection, by revising its pollution prevention policies, developing a policy on prosecution, and adopting a cooperative approach to implementing its legal framework for coral reef protection. The country could benefit from studying and incorporating lessons from global laws in these areas.

2.2. Methodology

The research methodology involves several key steps:

2.2.1. Literature Review

This includes collecting and analysing relevant international legal documents, such as UNCLOS 1982 , CBD, the Ramsar Convention, and other treaties related to marine conservation. Additionally, the review will cover academic articles, legal commentaries, and reports from international organizations (e.g., UNEP, IUCN) that explore the legal dimensions of coral bleaching and marine environmental protection.

2.2.2. Case Studie

Identify and examine case studies from specific regions or countries where legal frameworks for coral reef protection have been implemented, with a focus on examples like Australia's Great Barrier Reef Marine Park.

2.2.3. International Legal Instruments

Investigate international legal instruments addressing coral bleaching, with particular attention to their scope, objectives, and whether they are binding or non-binding.

2.2.4. Regional Legal Frameworks

Assess regional legal frameworks, such as the Coral Triangle Initiative, and evaluate how they align with and complement global instruments.

2.2.5. National Implementation

Although the research has a global focus, it will also briefly examine how international frameworks are implemented at the national level in key countries with significant coral reefs, such as Australia, Indonesia, and Malaysia.

2.3. Literature Review

2.3.1. Relationship between Coral Reef Conservation and the Sustainable Bue Economy

United Nation Environment Programme was launched report about Coral Reef Restoration [7]. The objective is to assist practitioners, managers, and decision-makers in assessing whether and how coral reef restoration can be employed as a strategy to safeguard coral reefs at local, regional, and global scales. The report recommends that coral reef restoration strategies adhere to four key principles: 1) planning and evaluating based on specific goals and objectives, 2) identifying adaptive approaches to manage risks, 3) involving local stakeholders and communities throughout the restoration process, and 4) establishing long-term monitoring plans to support adaptive management and enhance understanding of restoration success for specific objectives.

The concept of a Sustainable Blue Economy [8] offers crucial advantages to both present and future generations by restoring, protecting, and preserving diverse, productive, and resilient ecosystems. It is dependent on the use of eco-friendly technologies, renewable energy resources, and the circulation of materials within a closed-loop system. It relies on sustainable technologies, green energy sources, and the continuous reuse of materials in a circular economy model. Healthy coral reefs play a crucial role in this economy, supporting fishing and tourism, which provide millions of jobs and contribute significantly to global economies. Additionally, coral reef organisms are sources of important pharmaceuticals, including treatments for cancer, arthritis, and viruses. However, corals face significant threats from pollution and climate change.

The relationship between coral reef conservation and the sustainable blue economy is both complex and mutually beneficial. Coral reefs are essential to the health of marine ecosystems and are foundational to the blue economy, particularly in sectors such as fisheries, tourism, and coastal protection. Examples, First, coral reefs provide essential habitats and breeding grounds for numerous marine species, many of which are crucial for both commercial and subsistence fishing. Globally, reef-associated fisheries are vital for food security and livelihoods, especially in coastal communities. Reefs also attract millions of tourists annually, generating significant revenue through activities like snorkelling, diving, and eco-tourism. This tourism is a major economic driver for many island and coastal nations. Secondly, reefs function as natural barriers that shield shorelines from storm surges and coastal erosion, thereby decreasing the reliance on expensive artificial infrastructure. This protective role is especially important in areas that are susceptible to natural disasters.

2.3.2. International Legal Framework for Coral Reefs Protection

The United Nations Convention on the Law of the Sea (UNCLOS) 1982 Part XII, addresses the Protection and Preservation of the Marine Environment. This section requires states to implement measures to safeguard and maintain the marine environment, including coral reefs. Countries must prevent, reduce, and control marine pollution from various sources such as land-based activities, seabed operations, and vessel-related pollution. Article 194 specifically mandates that states must take all necessary actions to prevent, reduce, and control pollution from any source, ensuring that their activities do not harm other states or the marine environment. This is particularly relevant to coral bleaching, as pollution and other human activities increase stress on coral reefs. Additionally, under UNCLOS 1982, coastal states have sovereign rights over natural resources within their Exclusive Economic Zone (EEZ), extending up to 200 nautical miles from their coast. This includes the management and conservation of coral reefs within their EEZ. Countries like Indonesia have the right and responsibility to manage and protect coral reefs within their EEZ, including implementing measures to mitigate coral bleaching. Continental Shelf: States have rights over the continental shelf, which often includes coral reefs. These rights are to be exercised with due regard for environmental protection, particularly when exploiting resources like oil and gas, which could impact coral reefs. Article 212 (Pollution from or through the Atmosphere): This provision relates to atmospheric pollution, including greenhouse gases that contribute to climate change—a primary driver of coral bleaching. While UNCLOS 1982 does not explicitly mention climate change, the obligations under this article can be interpreted as requiring states to take action to reduce emissions that contribute to global warming and, consequently, coral bleaching. Article 61 (Conservation of the Living Resources): States must ensure that the living resources within their jurisdiction are not threatened by excessive exploitation. This involves managing fisheries in a manner that does not damage coral reef ecosystems, which are vital for the well-being of numerous marine species. Article 62 (Utilization of the Living Resources) stipulates that while states have the right to exploit marine resources, they must do so sustainably, taking into account the effects on the marine environment, including coral reefs. This is critical in the context of coral bleaching, as overfishing and destructive fishing practices can exacerbate the vulnerability of coral reefs to bleaching.

3. Result and Discussion

3.1. Regional Regulation on Coral Reef Conservation

One of the key regional organizations is the Coral Triangle Initiative on Coral Reefs, Fisheries, and Food Security (CTI-CFF), a multilateral partnership formed by six countries: Indonesia, Malaysia, the Philippines, Papua New Guinea, Solomon Islands, and Timor-Leste. The CTI-CFF aims to address threats to marine and coastal resources while promoting sustainable management practices [9]. The CTI-CFF Strategic Plan for 2025-2035 focuses on several key goals and objectives, Sustainable Marine and Coastal Management, Identification and effective management of Priority Seascapes, Adoption of an Ecosystem Approach to Fisheries Management (EAFM), Establishment and management of Marine Protected Areas (MPAs). For Climate Change Adaptation, the plan begins with developing and implementing a comprehensive regional Early Action Plan for climate adaptation. It also includes the creation of interconnected National Centres of Excellence focused on this issue. In terms of Biodiversity Conservation, the strategy aims to improve the conservation status of endangered species such as sharks, sea turtles, seabirds, marine mammals, corals, seagrass, and mangroves.

Table 1. Annual Programs and Activities of CTI-CFF [9].

	Programs	Activities
2025	Conservation 1	Launch of investment plans for "Priority Seascapes." Strengthening policy frameworks for EAFM.
2026	Conservation 2	Expansion of the Coral Triangle MPA System (CTMPAS). Initiation of community-based climate adaptation projects.
2027-2028	Conservation of Fisheries	Enhanced measures for sustainable tuna stock exploitation. Development of sustainable coastal fisheries and poverty reduction initiatives
2029-2030	Action Plan of Climate Change	Complete execution of the Early Action Plan for Climate Change Adaptation. Establishment of operational Climate Change Adaptation Centres of Excellence.
2031-2035	Monitoring Improvement	Continuous monitoring and improvement of biodiversity conservation efforts. Strengthening public-private partnerships for sustainable marine resource management.

IOP Conf. Series: Earth and Environmental Science **1473** (2025) 012063 doi:10.1088/1755-1315/1473/1/012063

Coral reef bleaching is a significant environmental issue, particularly in the Indian Ocean, where reefs are vital to marine biodiversity, coastal protection, and local economies. Bleaching happens when corals, under stress from alterations in conditions like temperature, light, or nutrient levels, release the symbiotic algae (zooxanthellae) that reside within their tissues. This results in the corals turning white and becoming more susceptible to disease and death. Preventive actions in regional areas, including those around the Indian Ocean, are crucial to protecting these ecosystems. Here's an analysis of the preventive measures with Monitoring and Early Warning Systems ensure to establishing comprehensive monitoring systems that track sea surface temperatures, coral health, and other environmental parameters can help detect early signs of stress. This enables timely interventions. And programs like the Indian Ocean Coral Bleaching Alert System (IOCBAS) provide real-time data to predict bleaching events, allowing for preventive measures to be activated [9]. The preventive measures (precautionary approach) [10] however, emphasizes the need to take action earlier in the process. It serves as a method for addressing situations where there is potential harm but scientific uncertainty, ensuring that management measures are not delayed.

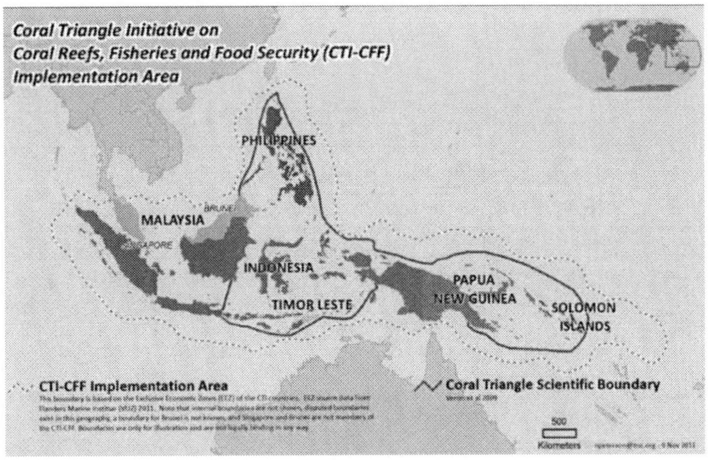

Figure 3. Coral Triangle Scientific Boundary [9].

CTI-CFF Establishing and enforcing Marine Protected Areas MPAs can help reduce local stressors on coral reefs, such as overfishing, coastal development, and pollution. MPAs create refuges where corals can recover and adapt to changing conditions. Reducing pollution from agricultural runoff, sewage, and industrial activities can improve water quality and reduce stress on coral reefs. Nutrient-rich waters promote algal blooms that can smother corals, while pollutants can weaken corals' resistance to bleaching. And implementing and enforcing sustainable fishing practices can help maintain the balance of reef ecosystems, preventing overfishing and the destruction of coral habitats by destructive fishing methods like blast fishing or cyanide fishing. Indian Ocean countries participating in international climate agreements, like the Paris Agreement, centripetal to global efforts to mitigate climate change and, consequently, coral bleaching. Supporting research into coral resilience, such as identifying and cultivating more heat-resistant coral species, can provide long-term solutions to prevent bleaching. Creating artificial structures that mimic natural reefs can provide new habitats for marine life and support coral growth in areas where natural reefs have been degraded.

IOP Conf. Series: Earth and Environmental Science **1473** (2025) 012063 doi:10.1088/1755-1315/1473/1/012063

3.2. Best Practice from National Legal Frameworks

3.2.1. Australia, Indonesia and Malaysia

Since coral bleaching in Australia, the critical measures Australia must take to combat coral bleaching, primarily driven by rising sea temperatures due to global climate change. Key strategies include reducing greenhouse gas emissions through investments in renewable energy, stricter carbon emission regulations, and active participation in international climate agreements. Enhancing carbon sequestration via reforestation and soil carbon projects is also emphasized. The abstract emphasizes the significance of safeguarding marine areas, creating no-take zones, minimizing agricultural and industrial runoff, and encouraging sustainable tourism practices to alleviate pressure on coral reefs and aid in their recovery.

Table 2. Comparing Regulation and policies between Indonesia, Malaysia and Australia.

State / Subject	Indonesia	Malaysia	Australia
Conservation	*Act No. 5 of 1990 regarding the Protection of Biodiversity and Ecosystems.*	*Malaysia National Biodiversity Policy 1998*	*Environment, Protection and Biodiversity Conservation Act 1999 (EPBC Act)*
Management	*Act No. 31/2004 on Fisheries as amended by Act No. 45/2009*	*Fisheries Act 1985*	
Management and Protection	*Act No. 32 of 2009 on Environmental Protection and Management.*	*Environmental Quality Act 1974, Act 127*	
Management	*Act No. 27 of 2007 on the Management of Coastal Areas and Small Islands.*		*The Great Barrier Reef Marine Park 1975. The Act provides the legal framework for managing and conserving the reef's biodiversity.*
Restoration	*Ministerial Regulation No. 28/2021 on the Implementation of Marine Spatial Regulation in Indonesia,*	*Integrated Coastal Zone Management (ICZM): A framework for the sustainable management of coastal areas, balancing development and conservation needs*	*Reef 2050 This is the overarching strategy for protecting and managing the Great Barrier Reef until 2050. It integrates actions across various sectors to improve reef health*
Management and Protection	*Ministerial Regulation No. 31/PERMEN-KP/2020 on the*	*National Coastal Zone Physical Plan (2012):*	*Reef 2050 Water Quality Improvement Plan 2017-2022*

	Management of Coral Reef Ecosystems and the Protection of Coral Species	*Provides guidelines for the sustainable development of Malaysia's coastal zones, with specific measures for coral reef protection.*	
	Indonesia has established marine protected areas (MPAs) and implemented zoning regulations to protect coral reefs. Laws such as the Fisheries Law and Coastal and Small Islands Management Law provide legal backing for these protection	*Malaysia's National Policy on Biological Diversity and the Marine Parks Malaysia Act play crucial roles in coral reef conservation. The country has designated several marine parks to protect coral ecosystems.*	*Australia's Great Barrier Reef Marine Park Act and the Environment Protection and Biodiversity Conservation Act are central to reef protection efforts. The country has also established a comprehensive reef management plan*

Source: regulation from Australia, Indonesia and Malaysia

3.2.2. Case studies in Indonesia, Malaysia and Australia

3.2.2.1. There are some case studies highlighting the coral reefs bleaching in Indonesia.

Coral reef bleaching in Indonesia has been extensively studied due to the country's vast and diverse coral ecosystems. Here are some notable case studies that highlight the impact of coral bleaching and the subsequent conservation efforts in different regions of Indonesia: (1) 2016 Coral Bleaching Event in the Raja Ampat Archipelago, Raja Ampat, West Papua, Raja Ampat is known for its rich biodiversity, often referred to as the marine biodiversity core. In 2016, the region experienced a significant coral bleaching event triggered by elevated sea temperatures associated with the El Niño phenomenon. The bleaching event affected a large portion of the coral reefs, with reports indicating that up to 85% of corals were affected in some areas. This led to a significant loss of coral cover and a decline in reef fish populations. Post-bleaching, there were concerted efforts to monitor and rehabilitate the affected reefs. The involvement of local communities in reef monitoring and the implementation of marine protected areas (MPAs) helped to mitigate further damage. Restoration efforts included coral transplantation and the promotion of sustainable tourism practices to reduce human impact. (2) 2009-2010 Coral Bleaching in the Wakatobi National Park, Southeast Sulawesi. Wakatobi National Park is one of Indonesia's most important marine protected areas, home to a diverse array of coral species. During 2009-2010, Wakatobi experienced widespread coral bleaching due to prolonged high sea temperatures. The bleaching event caused significant coral mortality, particularly among sensitive species such as Acropora. The decline in coral health also impacted the local fishing community, which relies heavily on reef fish. The management of Wakatobi National Park responded by increasing the enforcement of fishing regulations and expanding no-take zones within the park. Additionally, community-based coral restoration projects were initiated, focusing on coral gardening and the establishment of coral nurseries to aid in reef recovery. (3) In 2010, the reefs in this area were significantly impacted by a bleaching event due to increased sea temperatures. The bleaching led to widespread coral mortality, with some reefs experiencing up to 50% coral loss. The decline in coral cover resulted in reduced habitat complexity, which negatively impacted the biodiversity of the reef. (4) The 2016-2017 global bleaching event, exacerbated by rising sea temperatures, impacted the park's coral reefs. The bleaching affected approximately 40-60% of the coral cover in certain areas of the park. This had significant ecological repercussions, including a decline in fish diversity and abundance.

3.2.2.2. There are some case studies highlighting to coral reefs bleaching in Malaysia

Coral Bleaching Event in Malaysia, (1) The 2010 coral bleaching event was one of the most severe in Malaysia's history, triggered by elevated sea temperatures linked to the global El Niño phenomenon. Around 60% of coral species in Pulau Redang experienced bleaching, with a notable percentage suffering from partial or total mortality. The hard corals, especially those of the Acropora species, were the most severely impacted. (2) 2016 Coral Bleaching Event in Tioman Island, Tioman Island experienced significant coral bleaching during the global coral bleaching event of 2016, driven by elevated sea temperatures. Up to 30-40% of the coral reefs around Tioman Island were bleached, with some reefs experiencing up to 90% bleaching in certain areas. The most affected species included branching corals and table corals. (3) 2019-2020 Bleaching Event in Sipadan Island, Sipadan Island, known for its rich marine biodiversity, experienced a coral bleaching event from late 2019 to early 2020, linked to unusually high sea temperatures. The bleaching affected a range of coral species, with some dive sites reporting up to 50% of corals bleached. This event raised concerns about the long-term health of Sipadan's coral reefs, which are vital to the local tourism economy. (4) Coral Bleaching in Tun Sakaran Marine Park (2020) Tun Sakaran Marine Park, Sabah, Malaysia, Marine Park, located in the Semporna Islands, experienced a coral bleaching event in 2020 due to rising sea temperatures and a prolonged period of calm, clear weather. The bleaching affected a large percentage of coral species, particularly in the shallow reef areas. The damage was most severe in corals exposed to direct sunlight and in areas with less water circulation.

3.2.2.3. There are some case studies highlighting to coral reefs bleaching in Australia

Significant coral bleaching events, primarily impacting the Great Barrier Reef. These events have further underscored the vulnerability of coral reefs to climate change and the ongoing challenges in reef conservation. (1) In early 2020, the Great Barrier Reef underwent its third major bleaching event in only five years, caused by increasing sea temperatures. This was largely attributed to ongoing global warming and local heatwaves. Approximately 60% of the reef suffered from varying degrees of bleaching. The event was particularly severe in the central and southern parts of the reef, areas that had been less affected in previous events. Coral mortality varied, with some areas experiencing significant losses, while others saw partial recovery. The Australian government and Great Barrier Reef Marine Park Authority (GBRMPA) conducted extensive surveys and monitoring to assess the damage. (2) In 2022, another significant bleaching event occurred, marking the first time in recorded history that the reef had bleached during a La Niña year, a period typically associated with cooler ocean temperatures. The 2022 bleaching affected around 91% of the reefs surveyed, with some areas experiencing severe bleaching. This event was widespread, impacting areas that had previously been resilient. (3) In 2023 and 2024, while no mass bleaching event of the scale seen in 2020 or 2022 occurred, ongoing marine heatwaves and localized bleaching incidents were reported. The persistent threat of high sea temperatures due to global warming continues to loom over the Great Barrier Reef and other coral systems in Australia. Continued stress on coral reefs has led to a gradual decline in coral cover in some areas, with certain species more vulnerable to repeated bleaching.

Coral reef conservation is integral to the sustainable blue economy, providing both direct and indirect economic benefits while preserving vital ecological functions. By focusing on the protection and sustainable management of coral reefs, we can ensure that these ecosystems remain vital in supporting livelihoods, safeguarding coastlines, and contributing to global biodiversity. This, in turn, supports a resilient and sustainable blue economy, balancing economic growth with environmental stewardship.

4. Conclusion

National legal frameworks are vital for the effective governance of coral reefs and the promotion of a sustainable blue economy. By establishing clear guidelines for conservation, regulating harmful activities, and promoting sustainable practices, legal frameworks can ensure the long-term health of coral reefs and the economic benefits they provide. Collaborative efforts at national, regional, and global levels are essential to address the complex challenges facing coral reefs and secure their future.

Despite the progress made in some regions, many challenges remain in implementing effective national legal frameworks for coral reef governance. These include limited enforcement capacity, lack of funding, and insufficient stakeholder engagement. To address these challenges, this research recommends strengthening legal frameworks, enhancing enforcement mechanisms, increasing funding for conservation initiatives, and fostering community involvement.

Acknowledgments
The authors sincerely express their gratitude to Hang Tuah University for providing financial support for this research.

References

[1] BBC. 2024. "400-year record heat threat to Great Barrier Reef'. Available online: www.bbc.com/news/articles/cy0ngx130kxo.
[2] Max-Planck-Gesellschaft. 2024. The fourth mass coral bleaching. Available online: https://www.mpg.de/21887931/coral-bleaching
[3] UNEP. 2024. *Ocean heat is driving a global coral bleaching event and it could be the worst record,* available online: https://www.unep.org/news-and-stories/story/worlds-corals-are-bleaching-heres-why-and-what-it-means-oceans-future
[4] Ian McLeod,. 2020. *Legal Method*, New Edition, Macmillan, Macmillan Law Masters. p.106-11
[5] Tries B. Razak, Lisa Boström-Einarsson, Cut Aja Gita Alisa, Rindah Talitha Vida, Timothy A.C. Lamont. 2022. *Coral reef restoration in Indonesia: A review of policies and projects*, Marine Policy, **137**, 104940, ISSN 0308-597X, https://doi.org/10.1016/j.marpol.2021.104940.
[6] Shovon, K. M. A., Hannan, M. A. and Rahman, M. R. 2021. Legal Aspects of Coral Reef Management and Conservation in Bangladesh. *Open Access Library Journal*, 8, e7775.: http://dx.doi.org/10.4236/oalib.1107775.
[7] United Nations Environment Programme 3,. 2021. "*Coral Reef Restoration: A guide to coral restoration method*". available online: https://www.unep.org/resources/report/coral-reef-restoration-guide-coral-restoration-method.
[8] United Nations Environment Programme 4. 2023. "*Sustainable Blue Economy*". available online: https://www.unep.org/topics/ocean-seas-and-coasts/ecosystem-based-approaches/sustainable-blue-economy.
[9] Coral Triangle Initiative, "On Coral Reefs, Fisheries and Food Security, History of CTI-CFF", avalaible online: https://www.coraltriangleinitiative.org/about.
[10] Chomariyah, 2015. "*Oceans Governance: Implementation of the Precautionary Approach to Anticipate in Fisheries Crisis*". Procedia Earth Planet. Sci., **14**, 94–100. https://doi.org/10.1016/j.proeps.2015.07.089.
[11] Dubbing, Kelly Heber, 2018. "*Managing Coral Reef*: An Ecological and Institutional analysis of Ecosystem in Southeast Asia", Anthem Press Publishing, London, UK
[12] Laurence H. De Clippele, Laura Alonso Díaz, Dominic A. et all., 2023, " *Evaluating annual severe coral bleaching risk for marine protected areas across Indonesia*", Marine Policy, **148**, 105428, https://doi.org/10.1016/j.marpol.2022.105428.
[13] Unsworth, R. K. F., Clifton, J., Smith, D. J. , 2010, "*Marine Research and Conservation in the Coral Triangle: The Wakatobi National Park*". Nova Science, USA.
[14] Razak, F., Lua, W. Y., Abd Rasid, N. H., Aziz, N., et. all., 2024, "Adopting Marine Spatial Planning (MSP) using coral health assessment as indicator: A case study in Pulau Redang Marine Park, Malaysia. *Ocean and Coastal Management*", **248**. https://doi.org/10.1016/j.ocecoaman.2023.106943
[15] Islam, M. Rezaul , et. all., edited., 2021, "*Field Guide for Research in Community Settings, Tools, Methods, Challenges and Strategies*", Edward Elgar Publishing, Cheltenham, UK.
[16] Precht, William F., 2006, "*Coral Reef Restoration Handbook*", Taylor & Francis Group, US.
[17] Isabelle M. Cote, 2006, "*Coral Reef Conservation*", Cambridge University Press, UK.

[18] R. V. Salm, John R. Clark, 2000, *"Marine and Coastal Protected Areas : A Guide for Planners and Managers"*, Third Edition, IUCN International Union for Conservation of Nature, Washington DC, xxi+371pp.
[19] Grimsditch, Gabriel D and Salm, Rodney V. 2006. *"Coral Reef Resilience and Resistance to Bleaching"*. IUCN, Gland, Switzerland. 52pp.
[20] Bruce Prideaux and Anja Pabel. 2018. *"Coral Reefs, Tourism, Conservation and Management"*, Earthscan from Routledge, UK.
[21] United Nations Convention on Law of the Sea. 1982. (Came into Force on 16 November 1994, (1833 UNTS 397)). Available online: https://www.un.org/depts/los/convention_agreements/texts/unclos/unclos_e.pdf. (accessed on 14 August 2024).
[22] Convention on Biological Diversity. 1992. (Came into Force on 29 December 1993, (1760 UNTS 79)). Available online: https://www.cbd.int/doc/legal/cbd-en.pdf. (accessed on 14 August 2024).
[23] United Nations Framework Convention on Climate Change. 1992. (1771 UNTS 107). Available online: https://unfccc.int/ (accessed on 14 August 2024).
[24] Pedro Fidelman, Chris McGrath, Maxine Newlands, Kirstin Dobbs, Belinda Jago, Karen Hussey, *"Regulatory implications of coral reef restoration and adaptation under a changing climate"*, Environmental Science & Policy, **100**, 2019, Pages 221-229, https://doi.org/10.1016/j.envsci.2019.04.016.

Challenges and Policy Framework of Hazardous and Poisonous Waste Management in Indonesia: Policy Implementation and Public Impact

M R L Ekowanti[1], M H Tamrin[1*] and R N Madaniyah[1]

[1] Department of Public Administration, Universitas Hang Tuah, Surabaya, Indonesia

*E-mail: m.husnitamrin@hangtuah.ac.id

Abstract. This study explores the challenges and policy frame in the operation of hazardous and poisonous waste (B3) in Indonesia, focusing on policy implementation and its impact on society. Despite comprehensive regulations such as PermenLHK No. 6 of 2021 governing B3 waste management, field implementation still faces significant issues, including regulatory violations and illegal waste dumping. Policy changes, particularly those resulting from the Omnibus Law (Job Creation Law) that abolished Article 102 of the Environmental Protection and Management Law, have exacerbated the situation by reducing criminal penalties and increasing the risk of violations. The study employs a qualitative system with a case study approach to analyze the impacts of communication, resources, and bureaucratic structure on B3 waste management. The novelty in this research lies in its in-depth analysis of the effects of policy changes on waste violations and illegal dumping, alongside specific strategic recommendations for improving management effectiveness. Findings indicate that key challenges include insufficient regulatory understanding, a shortage of environmental inspectors, and inadequate infrastructure. The study recommends enhancing communication, strengthening human resources, and upgrading infrastructure and technology to improve B3 waste management effectiveness and mitigate negative environmental and public health impacts.

1. Introduction

Hazardous and toxic waste (B3) management is a pressing global challenge, especially as industrial activity and urbanization increase. B3 waste, which includes hazardous chemicals, industrial waste, and consumer products containing hazardous substances, if not correctly handled, will cause severe effects on human health and the environment [1-3]. Various countries have developed policies and technologies to address the problem of B3 waste, although their implementation and impact often face various obstacles [4,5].

At the international level, countries similar as the United States, Japan, and the European Union have implemented an extensive B3 waste operation programs and practices. The United States, through the Resource Conservation and Recovery Act (RCRA), establishes a frame for waste operation with a focus on waste prevention, reduction, and handling [6]. The European Union regulates the operation of electronic waste and dangerous materials through directives similar as Waste Electrical and Electronic Equipment (WEEE) and Restriction of Hazardous Substances [7]. Japan is known for its efficient waste management system, supported by cutting-edge technology and active community participation in recycling programs. [8,9].

However, in many lower to lower-middle income countries, including Indonesia, B3 waste management faces significant challenges. Lower to lower-middle income countries often suffer from a lack of infrastructure and trained human resources, as well as ineffective oversight [10-12]. Inadequate regulation and low public awareness of the importance of B3 waste management exacerbate this problem [10,13]. In Indonesia, although there are regulations such as Law No. 32 of 2009 and Government Regulation No. 101 of 2014, their implementation still faces various obstacles such as lack of infrastructure, resources, and weak supervision [14]. The reduction in criminal sanctions in the Job Creation Law reduces law enforcement and facilitates illegal activities, adding to the complexity of B3 waste management problems [15].

Regulations for hazardous and poisonous waste (B3) in Indonesia are regulated through the Regulation of the Minister of Environment and Forestry (PermenLHK) 6 of 2021, covering various aspects from storage to landfill [16]. However, regulatory violations such as unlicensed operations and improper management often occur, causing environmental pollution [17]. Cases of illegal dumping of B3 waste in several areas show the need for education and increased public awareness about the dangers of B3 waste, as well as the importance of stricter law enforcement. [18,19].

The literature review shows that although many studies have discussed B3 waste management policies from various perspectives, there is still a lack of thorough examination of policy application and its consequences [14,15,18]. This study aims to fill this gap by identifying challenges in implementing B3 waste management policies in Indonesia and evaluating their impacts on society. With this approach, it is hoped that it can provide new insights for more effective and sustainable B3 waste management, as well as contribute to formulating better policy strategies to overcome existing challenges and protect health and the environment [20-22].

Public education also plays an important role in B3 waste management, increasing public awareness about the dangers of B3 waste and how to manage it can significantly influence waste management practices [23,24]. Effective education programs and awareness campaigns can help reduce illegal dumping and increase community participation in waste management programs [23,25]

Technological innovation in B3 waste management is also an important factor. The development of innovative, greener technologies can address issues related to B3 waste management [26]. However, the adoption of this technology for waste is often hampered by budget constraints and a lack of adequate policy support [27].

In order to address these challenges, it is important for the government and stakeholders to work together to formulate better strategies for B3 waste management. Collaboration between government and business sector, institusional capacity bulding and strengthening of monitoring systems can help improve policy effectiveness and minimize adverse effects on health and the environment [10,28,29].

By understanding the challenges and constraints in enforcing policies for B3 waste management policies, as well as comparing effective international practices, it is hoped that this research can provide significant contributions in formulating better and more sustainable policy strategies. This is important to create a healthier and safer environment for the community, as well as protecting the ecosystem from the negative impacts of B3 waste.

2. Method

This research uses a qualitative approve by utilizing a case study method to investigate concerns related to hazardous waste management in Indonesia [30]. Case studies were selected to provide an in - depth understanding of hazardous waste management practices and challenges in several locations that experience significant problems [31]. The study will be conducted in areas in East Java that are known for hazardous waste dumping cases. Data were collected through in - depth interviews with government officers, waste managers, company representatives, and local communities; direct observation of hazardous waste management practices; and analysis of documents related to regulations and reports of violations[32,33].

Thematic analysis techniques will be used to identify and group key themes, and to compile interpretations regarding policy implementation and its impacts. To ensure the accuracy and reliability of the data, this study will apply data triangulation and member checking [32]. This study will also comply with ethical principles, including obtaining permission from relevant parties and maintaining confidentiality of information. The research flow is summarized in the following diagram, which illustrates the process from data collection to analysis and interpretation of findings.

3. Result and Discussion

3.1. Communication

Communication in B3 waste management in Indonesia, especially in East Java, plays an important role in ensuring that policies set by the central government can be implemented properly in the field. Information related to B3 waste management policies is conveyed from the central government to local governments and implementers through various media such as technical guidance, socialization, interactive dialogues, talk shows, and coaching clinics held at various events. In East Java, activities such as technical guidance and socialization are often carried out periodically to ensure comprehensive understanding at the regional level. This socialization does not only involve implementers in the field but also business actors and related agencies, with the aim of preventing misunderstandings in policy implementation. For example, every socialization and technical guidance held by the Ministry of Environment and Forestry (KLHK) is often broadcast live via the YouTube platform, such as on the KLHK Directorate General of PSLB3 channel, so that it can be accessed by anyone, anytime and anywhere. However, communication obstacles remain, especially when there are regulatory changes, such as after the enactment of the Job Creation Law (UUCK), which caused confusion during the transition period, although various efforts have been made by the Ministry of Environment and Forestry, including in East Java, such as through national coordination meetings and functional environmental extension workers. Violations in the operation of hazardouse waste are still set up, most of which are caused by a lack of understanding by business actors or even deliberate violations by business actors because of the expensive cost (seeking as much profit as possible). In addition, cases of B3 waste hoarding that have occurred in various regions in East Java, such as Jombang, Mojokerto, and Gresik, show that communication and understanding problems in the field still require serious attention.

The lack of clarity in communication regarding regulations is one of the causes of violations in the management of B3 waste in East Java. Although regulations such as PermenLHK No. 6 of 2021 have explained the duties and responsibilities of the central government, regional governments, and business actors in managing B3 waste, implementation in the field is not always as expected. In some cases, implementers in the field do not fully understand the regulation, which can be caused by a lack of effective socialization or the complexity of the regulation itself. For example, regarding the determination of the status of B3 waste as regulated in Appendix IX of PP 22 of 2021, there is still waste produced by business actors whose status is not clear whether it is B3 waste or not, causing confusion between business actors and field officers. This situation is often exacerbated by rapid regulatory changes and a lack of in-depth training, especially at the regional level such as East Java. In addition, the elimination of criminal sanctions in Article 102 of the Law on Environmental Protection and Management by the Job Creation Law also makes things worse. This facilitates illegal activities in managing B3 waste without a permit, which is very risky for environmental quality and public health. In East Java, cases of illegal stockpiling or dumping of B3 waste, such as what happened recently in the Bangkalan area, Madura, are clear evidence that weak law enforcement can have serious impacts. Strict law enforcement and more effective communication are needed to prevent further violations. Public education must also be improved, considering the many cases of B3 waste stockpiling carried out without awareness of the dangers caused. This awareness is important, especially in areas that are often used as B3 waste stockpiles, so that the public can play an active role in monitoring and reporting illegal activities related to B3 waste operation.

In the context of George Edward III's theory about public policy implementation, the communication factor is one of the key elements that influences the effectiveness of policy implementation in the field. According to Edward III, clarity and consistency of communication between policy makers and implementers in the field greatly determine the success of policy implementation [34]. When information and policy guidelines are not conveyed properly or change, as happened in B3 waste management in East Java, this can cause confusion and ineffectiveness in policy implementation. These communication limitations, as seen in the handling of B3 waste, have the potential to hinder proper understanding by business actors and field officers, resulting in the emergence of violations and illegal practices that are detrimental to the environment and society.

3.2. Resources

From a resource perspective, the analysis of hazardouse waste management in East Java shows challenges in terms of the availability and utilization of human resources, facilities, and access to information and technology.

Human Resources

In terms of human resources, regions should have sufficient manpower to implement B3 waste management policies. Based on the Law on Environmental Protection and Management, the central and regional governments are responsible for determining and implementing policies regarding B3 waste, providing guidance and supervision of business actors' compliance with B3 waste management. These tasks are carried out by the Minister, Governor, and Mayor/Regent and can be delegated to officers below them. In addition, there are special functional positions such as environmental impact controllers, environmental extension workers (LH), and LH supervisors who have their respective main duties and functions (tupoksi) in implementing B3 waste management.

However, in practice, there is a shortage in the number of environmental supervisors (PPLH) in several districts/cities in East Java. Based on the approval of the determination of formation needs by the Minister of State Apparatus Empowerment and Bureaucratic Reform, each region should have at least three PPLH personnel, but in reality there are still many regions that do not have adequate officers. The following is the number of Provinces/Cities/Regencies in Indonesia that do not have PPLH in 2023, as follows:

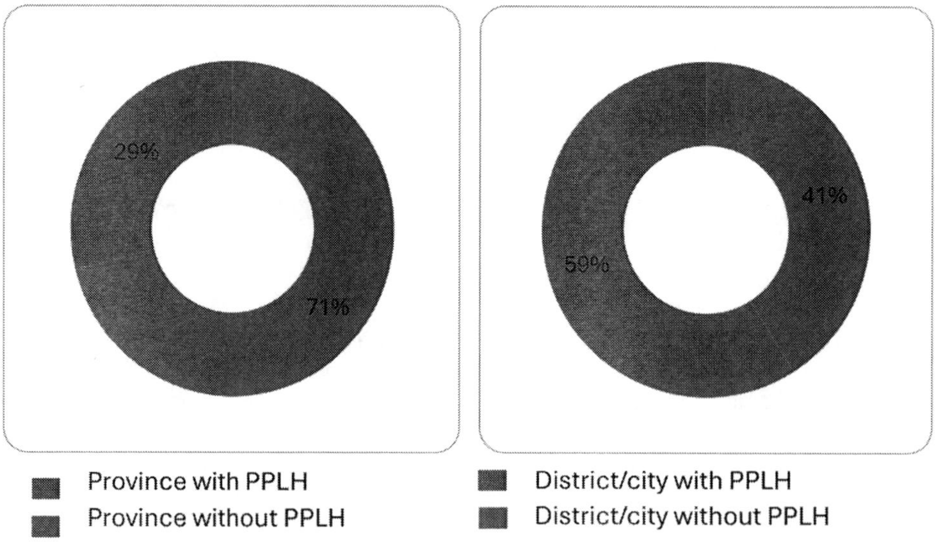

Figure 1. Data on environmental extension workers and supervisors in regions throughout Indonesia.

In addition, business actors who produce B3 waste are required to have special officers who are certified to handle B3 waste. Despite this obligation, data collection on all businesses to ensure the ownership of competent officers seems to be lacking. However, in the management of B3 waste management permits, this competency certificate must be attached, indicating that there are minimum standards that must be met.

3.2.1. Facilities and Infrastructure
Facilities and infrastructure for B3 waste management are also a concern. Not all local governments have adequate facilities for storing and processing B3 waste, because this obligation is more borne by business actors who produce B3 waste. Currently, there are many business actors who provide B3 waste management services with various facilities and technologies. However, the adequacy of these facilities is usually verified through a B3 waste management permit, where field verification is carried out before the permit is issued. This shows that even though facilities are available, standardization and fulfillment of appropriate criteria need to be continuously monitored.

3.2.2. Access to Information and Technology
Access to new information and technology to reduce B3 waste generation and process B3 waste is still a challenge. The central government, through the Ministry of Environment and Forestry and the Directorate General of PSLB3, routinely holds technical guidance and socialization related to innovations in B3 waste management. For example, in August 2024, technical guidance and talk shows were held involving successful B3 waste managers. However, the distribution of this information and training needs to be more even, especially to reach areas that may not yet have adequate access. More intensive training and more accessible information for implementers in the field are needed to improve their capacity to manage B3 waste effectively.

In the context of George Edward III's theory about public policy implementation, the availability of resources, including human resources, facilities, and access to information, is a crucial determinant of the success of policy implementation [34]. Edward III emphasized that policies will not be effective if the necessary resources are inadequate or not well distributed. In the case of hazardous waste management in East Java, the limited number of environmental supervisors (PPLH) and the lack of adequate facilities at the regional level hinder effective policy implementation. In addition, unequal access to information and new technology worsens the situation, as implementers in the field may not have the knowledge and skills needed to manage B3 waste according to established standards. Therefore, to achieve optimal policy implementation, there needs to be an increase in the allocation and utilization of resources at all levels of government and related sectors.

3.3. Disposition
From the perspective of policy implementation, the attitude and readiness of policy implementers towards B3 waste management have a significant impact in the success of the policy implementation. Strong support from policy implementers at the regional level is reflected in the existence of various regulations governing B3 waste management. The existence of these regulations not only shows a commitment to B3 waste management, but also provides a clear legal framework for implementation in the field.

However, policy implementation often faces challenges, especially related to incentives and support for policy implementers. Programs such as PROPER by the Ministry of Environment and Forestry, which provides awards for business actors who comply with B3 waste management, as well as rewards for exemplary environmental supervisors, play an important role in motivating policy implementers. The ADIPURA program also provides additional encouragement for local governments that demonstrate good environmental management performance. These incentives are expected to increase the enthusiasm and dedication of policy implementers, but their effectiveness depends on consistent and transparent implementation.

However, challenges in implementing B3 waste management policies remain. The procurement of functionalities such as Environmental Supervisors, Environmental Impact Controllers and Environmental Extension Workers which are not evenly distributed in several regions and the concerns of local governments regarding the potential negative impact on the flight of investment investors indicate bureaucratic obstacles and local economic concerns. To overcome these obstacles, it is important to strengthen the capacity of regulatory and law enforcement institutions, as well as improve synergy between local governments, policy implementers, and the community. Periodic evaluation and adjustment of implementation strategies are also needed to ensure that B3 waste management policies can be effectively put into practice and provide maxium advantages for both the environment and the community. In the framework of George Edward III about public policy implementation, the attitude and readiness of policy implementers are key elements that influence successful implementation. According to Edward III, the commitment and motivation of policy implementers, as well as effective incentive support, greatly influence how policies are implemented in the field. In East Java, although there are incentive programs such as PROPER and ADIPURA, bureaucratic challenges and local economic concerns can hinder the commitment of policy implementers. To overcome this, a more holistic approach is needed, which focuses not only on providing incentives but also on increasing capacity and structural support. This is in line with Edward III's view that the success of policy implementation depends on the alignment between policy design, resource availability, and the motivation of implementers at all levels, so that B3 waste management policies can be implemented effectively and sustainably.

3.4. Bureaucratic Structure and Implementation of Hazardous Waste Management Policy

Analysis of the bureaucratic structure in supporting or hindering the implementation of *Hazardous* waste management policy shows that standard operating procedures have been regulated in government regulations and regulations of the Minister of Environment and Forestry. In addition, B3 waste management procedures are stated in the B3 waste management permit held by each business actor. Although these guidelines are available on the website of the Directorate General of PSLB3, the flexibility of their implementation in the field can vary depending on the compliance of business actors and the level of supervision by both central and regional governments.

The division of tasks and responsibilities between the central, provincial, and district/city governments in B3 waste management is regulated in PP 22 of 2021 and PP 5 of 2021. Ideally, this division of tasks is designed to avoid duplication and ensure effective coordination. However, after the enactment of the Job Creation Law, several clauses have triggered multiple interpretations, which can disrupt coordination. Although there is a circular from the center to address this, the problem of coordination remains a challenge.

Policy changes due to the Job Creation Law have an impact on the bureaucratic structure, especially in terms of the division of authority between the central, regional, and provincial governments. Although there are changes, their impact on the way the bureaucracy works may not be significant overall. The main changes include the new classification of waste types, namely registered Non-B3 waste, but technically, the provisions related to B3 waste management have not undergone major changes. The need for adaptation in the bureaucratic structure remains to ensure effective policy implementation and in accordance with the latest regulations.

The role of the bureaucratic structure in enforcing B3 waste management policies lies not only in the implementation of SOPs for laws and regulations and licensing for each business actor, but also in the effectiveness of coordination between institutions. A good bureaucratic structure must be able to integrate various levels of government and ensure that all parties understand and comply with applicable regulations. Clear regulations regarding the division of tasks between the central, provincial, and district/city governments are very important to avoid implementation errors and ensure that every aspects of hazardous waste management are handled comprehensively.

However, policy implementation is often hampered by problems of coordination and interpretation of regulations. Although the existence of a circular from the center can help overcome multiple interpretations, problems in implementation in the field are still possible. For example, policy changes from the Job Creation Law can affect the way the bureaucracy works and change priorities and responsibilities in B3 waste management. A bureaucratic structure that is not flexible or not ready to face these changes can hinder effective policy enforcement.

To improve the effectiveness of policy implementation, periodic evaluation and adjustment of the bureaucratic structure is needed. Strengthening the capacity and training of policy implementers, as well as improving communication between institutions, can help overcome existing challenges. With an adaptive and collaborative approach, it is hoped that the bureaucratic structure can support the implementation of hazardous waste management policies more efficiently and effectively, and meet the goals of environmental protection and public health.

In the context of George Edward III's about public policy implementation, the bureaucratic structure is one of the important factors influencing the effectiveness of policy implementation. Edward III emphasized that a clear and well-organized bureaucratic structure is key to ensuring that policies can be implemented effectively in the field. In hazardous waste management in East Java, although the bureaucratic structure has been regulated through various regulations, coordination problems and multiple interpretations of regulations due to policy changes such as the Job Creation Law show that an inflexible bureaucratic structure can be an obstacle to policy implementation. Therefore, to achieve optimal implementation, it is important for the bureaucratic structure to be more adaptive and responsive to regulatory changes, as well as strengthen coordination between government institutions at various levels. Thus, Edward III's theory emphasizes the need for alignment between bureaucratic structures and existing policies so that environmental protection and public health goals can be achieved effectively.

4. Conclusion

The analysis of B3 waste management policy implementation in East Java reveals significant challenges related to communication, resource allocation, disposition of implementers, and bureaucratic structure. Effective communication, as emphasized by George Edward III's theory of public policy implementation, is crucial for ensuring clarity and consistency in the field. However, communication gaps, particularly during regulatory changes, have led to confusion and non-compliance among business actors and field officers. Similarly, the availability and utilization of resources, including human resources, facilities, and access to technology, are found to be inadequate, hindering the effective implementation of B3 waste management policies.

Moreover, the disposition and readiness of policy implementers, influenced by incentives and bureaucratic challenges, have a significant impact in the success of policy enforcement. The bureaucratic structure, though regulated, faces issues of coordination and interpretation, especially after the introduction of the Job Creation Law, which has impacted the division of authority and priorities in B3 waste management. To achieve effective policy implementation, it is essential for the bureaucratic structure to be adaptive and responsive to changes, with strengthened coordination and periodic evaluation to ensure that environmental protection and public health objectives are met.

References

[1] Kumar A A critical review on sustainable hazardous waste management strategies: a step towards a circular economy. *Environmental Science and Pollution Research*, **30**(48). 105030–105055 2023

[2] Blagojevic M, Buchal S, Chute S, Shareefdeen Z and Elkamel A Hazardous Waste Transport in the Environment in *Hazardous Waste Management: Advances in Chemical and Industrial Waste Treatment and Technologies*, Z. Shareefdeen, Ed., Cham: Springer International Publishing, **2022**. 273–295

[3] Neksumi M, Zishan M, Sushmita B and Manzoor U The Global Menace of Hazardous Waste: Challenges and Management," in *Handbook of Solid Waste Management: Sustainability through*

Circular Economy, C. Baskar, S. Ramakrishna, S. Baskar, R. Sharma, A. Chinnappan, and R. Sehrawat, Eds., Singapore: Springer Nature Singapore, **2022**. 1715–1742

[4] Banerjee T, Nair AP and S. MS Chapter 21 - Hazardous waste management: lessons from developed countries in *Waste Management and Resource Recycling in the Developing World*, P. Singh, P. Verma, R. Singh, A. Ahamad, and A. C. S. Batalhão, Eds., Elsevier, **2023**. 487–504

[5] Agamuthu A and Babel S Waste management developments in the last five decades: Asian perspective. *Waste Management & Research*, **41**(12). 1699–1716, Oct. 2023

[6] M. K. Awasthi *et al.*, "Chapter 6 - Sustainable Management of Solid Waste in *Sustainable Resource Recovery and Zero Waste Approaches*, M. J. Taherzadeh, K. Bolton, J. Wong, and A. Pandey, Eds., Elsevier. **2019**. 79–99

[7] Chandrappa R and Das DB Waste from Electrical and Electronic Equipment in *Solid Waste Management: Principles and Practice*, Cham: Springer International Publishing, **2024**, 343–367

[8] Hu JL, Honma S and Chang TM Recycling Efficiencies in Japan's Administrative Regions: Findings from Network Data Envelopment Analysis. *Circular Economy and Sustainability*, **4**(1). 1–21. 2024

[9] Agamuthu P and Babel S Waste management developments in the last five decades: Asian perspective. *Waste Management & Research*, **41**(12). 1699–1716, Oct. 2023

[10] Al-Worafi YM Healthcare Facilities in Developing Countries: Infrastructure in *Handbook of Medical and Health Sciences in Developing Countries: Education, Practice, and Research*. Cham: Springer International Publishing, **2023**.1–21.

[11] Al-Worafi YM Quality of Healthcare Systems in Developing Countries: Status and Future Recommendations in *Handbook of Medical and Health Sciences in Developing Countries: Education, Practice, and Research*. Cham: Springer International Publishing, **2023**. 1–28

[12] Gaudin S and Yazbeck AS Identifying Major Health-System Challenges in Developing Countries Using PERs: Equity is the Elephant in the Room. *Health Syst Reform*. **7**(2)

[13] Salvia G The wicked problem of waste management: An attention-based analysis of stakeholder behaviours. *J Clean Prod*, **326**. 129200

[14] Ashraf M, Harvelian A and Kartika T Regulation and Enforcement of Environmental Law in Handling Forest Fires Based on Law Number 32 of 2009. *Humaniorum*. **1**. 99–104, Jan. 2024

[15] Jiwanti A Criminal Law Policy in Efforts to Overcome Environmental Crimes in the Job Creation Law. *Justisi*, **9**. 158–174, May 2023

[16] KemenLHK Regulation of the Minister of Environment and Forestry of the Republic of Indonesia Number 6 of 2021 concerning Procedures and Requirements for Hazardous and Toxic Waste Management. **2021**. 1–301

[17] Patil RA and Ramakrishna S A comprehensive analysis of e-waste legislation worldwide. *Environmental Science and Pollution Research*, **27**(13). 14412–14431, 2020

[18] Junaidi, Simbolon Y, Siahaan P and Batu D Criminal Liability for Perpetrators of Environmental Pollution as a Result of B3 Waste (Case Study of Decision Number 1482/Pid.Sus-LH/2021/PT MDN). *Nusantara Journal of Social Science*. **10**. 4301–4314, Oct. 2023

[19] Santoso WY Legal Aspects in Management of Hazardous and Toxic Waste. *Legal Image*. **29**(2). 335–345, Jun. 2017

[20] Arisman and Fatimah YA Waste Management in Indonesia: Strategies and Implementation of Sustainable Development Goals (SDGs) and Circular Economy in *Circular Economy Adoption: Catalysing Decarbonisation Through Policy Instruments*. Singapore: Springer Nature Singapore, **2023**. 131–157.

[21] Das S, Lee SH, Kumar P, Kim KH, Lee SS and Bhattacharya SS Solid waste management: Scope and the challenge of sustainability," *J Clean Prod*. **228**. 658–678, 2019

[22] Treacy J Policy Implementation on Waste Management and Achievement of Related SDGs in *SDGs in the European Region*. Cham: Springer International Publishing, **2022**. 1–29

[23] Yasmeen R, Sarfraz M, Shah WUH, Ivascu L and Cifuentes J The impact of public awareness, infrastructure, and technological development with economic growth on solid waste

management of European countries: does governance quality matters. *Environmental Science and Pollution Research*. **30**(53). 113442–113456. 2023

[24] Franchetti MJ Solid Waste Public Involvement and Education in *Recovery of Materials and Energy from Urban Wastes: A Volume in the Encyclopedia of Sustainability Science and Technology, Second Edition*. New York, NY: Springer New York, **2019**. 197–220

[25] Sewak A, Deshpande S, Rundle Thiele S, Zhao F, and Anibaldi R Community perspectives and engagement in sustainable solid waste management (SWM) in Fiji: A socioecological thematic analysis. *J Environ Manage*, **298**. 113455, 2021

[26] Govani J, Singh E, Kumar A, Zacharia M, Rena and Kumar S Chapter 4 - New generation technologies for solid waste management in *Current Developments in Biotechnology and Bioengineering*, S. Kumar, R. Kumar, and A. Pandey, Eds., Elsevier. **2021**. 77–106

[27] Tanveer M, Khan SAR, Umar M, Yu Z, Sajid MJ and Haq IU Waste management and green technology: future trends in circular economy leading towards environmental sustainability," *Environmental Science and Pollution Research*, **29**(53). 80161–80178, 2022

[28] Tamrin MH and Lubis L Stakeholder Collaboration in The Development of Essential Ecosystem Area. *Journal of Social and Development*. **39**(1). 35–43. 2023

[29] Tamrin MH and Raharja WT Collaboration To Build The Country, **1**. Jakarta: PT. Akses Prima Cendekia, 2021

[30] Creswell JW and Creswell JD Research Design Qualitative, Quantitative, and Mixed Methods Approaches. 2018

[31] Yin RK Case Study Research and Applications Sixth Edition. 2018

[32] Miles MB, Huberman AM and Saldana J, Qualitative Data Analysis. 2014.

[33] Lacey A and Luff D Qualitative Data Analysis. **1**. Nottingham, 2009.

[34] Setyawan D, Priantono A and Firdausi F George Edward III Model," *Publicio: Scientific Journal of Politics, Policy and Society*. **3**. 9–19. Jul 2021

Conservation Efforts for Sustainable Razor Clam (*Solen sp.*) Resources after Enacting the Job Creation Law

Kamarudin[1*] and N Trisyani[2]

[1]Faculty of Law, Universitas Hang Tuah, Surabaya, Indonesia
[2]Study Program of Fisheries, Universitas Hang Tuah, Surabaya, Indonesia

[*]Email: kamarudin@hangtuah.ac.id

Abstract. *Pamekasan* is one of the regions producing razor clam resources in Indonesia, but the population growth rate is lower than other regions. Local community is responsible for keeping sustainable fish resources. Local government need to make various efforts to conserve razor clam resources. However, since the Job Creation Law has enacted, central government determines norms, standards, procedures, and criteria for fisheries businesses, including razor clams, the local government shall hold them. This study aims to examine razor clam conservation efforts in *Pamekasan* in regulatory mechanism. The research constitutes normative legal research on the principles, norms, and legal system in Indonesia. Results of the study indicate that efforts to look after, preserve and make use of razor clam resources ecosystems, species and genetics can be realized by regulating them in a local regulation and proposing coastal areas producing razor clams to be designated as a regional fishery conservation area.

1. Introduction

Razor clam, also known as knife clam, bamboo clam or *Solen sp.*, is one of the aquatic species that is not found in all water areas in the world and Indonesia. In *Pamekasan* Regency Indonesia, there are centres producing razor shells, among others, in *Pademawu* District, *Galis* District and *Larangan* District. East *Pademawu* Village, *Pademawu* Sub-district, has four hamlets whose residents work as razor clam seekers, namely *Mangunan* Hamlet, *Sawahan* Hamlet, West *Mungging* Hamlet and East *Mungging*. In these hamlets, fishermen's wives work to find razor shells. Meanwhile, in Galis Sub-district, there are two villages that have the potential of razor clam fisheries, namely *Polagan* Village and *Galis* Village [1]. In *Larangan* sub-district, the presence of razor clams can be found on the *Talang Siring* beach in *Montok* village. Although there are centres of razor clam production in *Pamekasan* Regency, the level of pollination growth is low compared to other regions. In addition, the fishing techniques used tend to be classified as overexploitation with dredges or hoes [2]. Till now, those centres of razor clam production are legally not stipulated yet as waters conservation areas. Therefore, there is a real need to limit the fishing time during the spawning period to provide opportunities to reproduce and increase stocks in nature. Efforts to cultivate razor mussels should also be initiated in an effort to develop this species into a commercial species. Efforts to conserve razor clam resources, especially in *Pamekasan* Regency, shall be carried out in various approaches, including legal approaches.

The elucidation of Article 7 paragraph (5) of Act Number 31 of 2004 on Fisheries states that shellfish (*mullusca*) is one of the fish species. In addition to regulating the management, this law also regulates fish resource conservation through ecosystem, species and genetic conservations. Efforts to conserve fish resources can not be basically separated with efforts to manage fish resources and their environments. In addition, the Government i.e. the Minister could stipulate conservation areas relating to fisheries. After enacting the Job Creation Law, Act Number 6 of 2023 replacing Act Number 11 of 2020, stipulating conservation areas is asserted to be the central government's authority, while every person and local government shall obey all stipulated provisions. Nevertheless, local governments and communities still have duty to protect and preserve razor clam, so that its population in *Pamekasan* can be sustainable and beneficial [3]. Because the characteristics of fish resources and their environment have a high sensitivity to be influenced by global and seasonal climates and ecosystem linkages between local, regional and global water areas which probably across country's sovereignty boundaries, efforts to develop and manage fish resource conservation shall be based on the precautionary principle with the support of scientific evidence [4].

According to Gifford Pinchot [5], conservation, the optimal utilization of natural resources that can be carried out for a long period of time into the future, needs to be done in order to maintain and/or improve the quality of human life. Anthropocentrism is one model of conservation ethics where humans are at the centre of the natural system. Humans and their interests are considered the most decisive component in the ecosystem order and the most important in determining policies related to nature. Humans with an anthropocentric perspective are considered to be prone to exploitative, destructive and uncaring behaviour towards the limited capabilities, sustainability and balance of nature.

The conservation of razor clams, especially in *Pamekasan* Regency, requires joint efforts between the local government and the local community, so that conservation can be realized. Efforts to conserve razor clam resources in *Pamekasan* Regency have been researched and written by previous researchers. However, conservation efforts through regulatory aspect, local government and community can carry out in optimising conservation achievements, have not been widely researched and written yet. Therefore, the paper aims to analysing regulatory mechanism to conserve razor clam resources after enacting the Job Creation Law, so that it is expected to protect sustainable razor clam population and to eventually bring sustainable prosperity to the community.

2. Methods

The research was normative legal research which the used data were secondary data. They were obtained from legal sources and library sources, i.e. a) primary legal sources that are laws and regulations containing rules on the conservation of marine resources and other relevant regulations; b) secondary legal sources that are relevant books, journals, research results and papers to the research problem; and c) tertiary legal sources, in the form of legal sources that provide guidance and explanation of primary and secondary legal sources, such as dictionaries and encyclopaedias. The secondary data were obtained through reading and studying and analysing legal sources, such as laws and regulations, books, documents and others.

The research applied descriptive analysis by describing regulatory mechanism as the conservation efforts of razor clam resources in *Pamekasan* Regency Indonesia. The analysis was carried out qualitatively on the collected and processed data, in order to formulate the conclusions of the research. The analysis was carried out by describing the facts studied, connected and legally analysed by using an analytical knife, namely the concept of the legal system.

3. Results and Discussion

3.1. Regulating Fish Resources Conservation After Enacting the Job Creation Law

According to the provisions of Article 27 number 1 of Law Number 6 Year 2023 on Job Creation, conserving the fish resources including razor clams is an attempt to keep, maintain, and operate fish resources, including ecosystems, species, and genetics for ensuring their existence, availability, and sustainability, and for maintaining and improving the quality value and diversity of fish resources. After the enactment of the Job Creation Law, provisions regarding the conservation of fish resources are regulated in several laws and regulations, namely: 1) Law Number 6 of 2023 on Job Creation, 2) Law Number 31 of 2004 on Fisheries, 3) Government Regulation Number 27 of 2021 on the Implementation of Marine and Fisheries Sector, and 4) Government Regulation Number 60 of 2007 on Conservation of Fish Resources.

Based on Article 1 numbers 2, 3, and 4 of Government Regulation Number 60 of 2007 on Conservation of Fish Resources, the potential of all fish species shall be conserved and its quality and diversity shall be maintained and improved. This effort is the conservation of fish resources which includes ecosystem, species and genetic conservation. Ecosystem conservation means efforts to protect, preserve and utilize ecosystem functions as a habitat for aquatic biota life support at present and in the future. Fish species conservation is defined as an effort to protect, conserve, and utilize fish resources, to ensure the existence, availability, and sustainability of fish species for present and future generations. Meanwhile, fish genetic conservation is an effort to protect, conserve and utilize fish resources, to ensure the existence, availability and sustainability of fish genetic resources for current and future generations.

Ecosystem conservation is carried out in all types of ecosystems related to fish resources. Ecosystem types of fish resources consist of marine, seagrass, coral reef, mangrove, estuary, beach, swamp, river, lake, reservoir, pond, artificial water ecosystem. Conservation of these ecosystems is carried out through protecting and rehabilitating fish habitats and populations, researching and developing, utilizing fish resources and environmental services, developing community socio-economic aspects, supervising and controlling, and monitoring and evaluation. One or more ecosystem types may be designated as a marine protected area by the Central Government. Water areas are protected and managed with a zoning system in order to realize sustainable management of fish resources and the environment. It has been suggested that methods for protecting fish resources should be in line with ecosystem conservation techniques, since these can poster the development of biotic resources, and assist the preservation of natural habitats and ecosystems. Regulations governing conservation management should cover all major areas in order to successfully conserve aquatic biodiversity [6].

Species conservation is carried out through classifying fish species, determining fish species protection status, maintaining, breeding, and researching and development. Fish species are classified as protected and unprotected species. Criteria for protected fish species include endangered, rare, limited distribution areas (endemic), drastic decline in the number of fish populations in nature, and/or low levels of reproductive ability. Therefore, more study is required, particularly on the biology and ecology of fish species. Additionally, coordinated efforts are needed to formulate regulations that assure the sustainable usage of fish resources. There is need for more development, investigation, and study into the possibility of using breeding and culture of endangered species to create "seed banks" for those species. To increase the natural populations, the fingerlings generated by these seed banks can subsequently be replenished into the lake [7].

Conservation of fish genetic resources is carried out through efforts of maintenance, breeding, research, and preservation of gametes. Gamete preservation of fish genetic resources is carried out in frozen condition. Maintenance is carried out on protected and unprotected fish species through the collection of live fish in a controlled medium as an artificial habitat. Maintenance of fish species can be done by taking fish from natural habitats or from the results of breeding. Maintenance of protected fish species and unprotected fish species in artificial habitats is carried out after fulfilling the following requirements, i.e. fish health standards, a large enough, safe, and comfortable place, and employing experts in the medical and fish maintenance fields. Maintenance of protected fish species and unprotected fish species can be carried out by individuals, community groups, Indonesian legal entities, research institutions, and/or universities. Successful conservation programs for fish genetic resources frequently necessitate the cooperation of numerous stakeholders [8]. The conservation of wild fish genetic resources is essentially nature conservation, with little overlap or similarity to the conservation of terrestrial farm animal genetic resources, as noted by Pullin [9], who has proposed a number of options and approaches for the conservation of farmed fish genetic resources. The Food and Agriculture Organization recently produced technical guidelines and recommendations in response to an expert solicitation on the condition of aquatic genetic resources worldwide [10]. These guidelines and recommendations offer guidance on important matters pertaining to the management and conservation of fish genetic resources, such as risk assessment, cryo-based gene banking, and genetic management of cultured populations [11]. The sustainable regulation and management of fish genetic resources requires institutional and political assistance, particularly in underdeveloped nations where regulatory systems are frequently [12].

Breeding is conducted for protected fish species and unprotected fish species through hatcheries in a controlled environment, hatching of eggs, rearing of wild-caught juveniles or transplantation. Breeding of protected fish species and unprotected fish species is carried out by maintaining the genetic purity of the fish. Breeding the protected and unprotected fish species shall fulfil the qualification standards for breeding fish species. It can be carried out by individuals, community groups, Indonesian legal entities, research institutions, and/or universities. For the purpose of controlling the breeding activities of protected and unprotected fish species, marking of broodstock and breeding fish may be carried out. Research and development is conducted on protected and unprotected fish species [13].

Utilization of fish resource conservation includes utilizing water conservation areas and fish species and genetics. Utilizing the water conservation areas is carried out through fishing, fish farming, aquatic nature tourism, or research and education. Meanwhile, utilizing the fish species and genetics is carried out through research and development, breeding, trade, aquaria, exchange, and maintenance for pleasure [14]. Utilizing the marine protected areas for fishing and for fish farming is carried out in sustainable fisheries zones. Fishing and fish farming activities in the fisheries management area of the Republic of Indonesia that are not for commercial purposes shall obtain approval from the Central Government or Regional Government in accordance with their authority based on norms, standards, procedures, and criteria set by the Central Government. Utilizing them for nature tourism can be carried out in utilization zones and sustainable fisheries zones. Water nature tourism in water conservation areas can be conducted through water nature tourism activities, and/or water nature tourism business. The utilization for research and education activities can be conducted in the core zone, sustainable fisheries zone, utilization zone, or other zones. Foreigners and foreign legal entities that will conduct research activities in marine protected areas may be granted permits after fulfilling research licensing requirements based on the provisions of laws and regulations.

Fish species and genetics utilization is carried out on protected and unprotected fish species. Utilization of fish species and fish genetics can be done through taking fish from nature. The collection of protected and unprotected fish species from nature shall obtain a permit from the Minister or a designated official, after receiving a recommendation from the scientific authority. Any protected fish species taken from the wild for breeding and aquaria activities are declared as state-owned fish. Every person who takes fish from the wild is obliged to pay fisheries levies in accordance with the provisions of laws and regulations.

Utilizing the fish species and genetics for research and development activities can be carried out on protected and unprotected fish species. Research and development can be conducted by individuals, universities, non-governmental organisations, and research and development institutions. Utilizing them for research and development activities shall obtain a utilization permit from the Minister or a designated official. The results of research and development shall be submitted to the Minister. Research and development on protected fish species and unprotected fish species carried out by foreigners and foreign legal entities in Indonesia shall be carried out in accordance with the provisions of laws and regulations. Utilizing them for breeding can be carried out on protected and unprotected fish species. Breeding can be carried out by individuals, community groups, Indonesian legal entities, research institutions, and universities. Breeding activities shall obtain a permit from the Minister or a designated official after the applicant fulfils technical and administrative requirements.

Fish species utilization for trade includes protected fish species of breeding result, unprotected fish species, and tradable fish based on international law provisions. The Minister determines the quota amount for taking unprotected fish from the wild for trade after receiving recommendations from scientific authorities. Utilizing the fish species for trade can be carried out by individuals and corporations. In trading, individuals and corporations shall obtain a permit from the Minister or a designated official, after fulfilling technical requirements. Utilizing them for trade can be done for export, import, or re-export. Fish species utilization for export, import, or re-export shall at least be equipped with a letter of delivery from and to overseas, shipping or transport documents, a letter of acquisition of trade quota, a certificate of origin, and a certificate of breeding results. The exported, imported, or re-exported fish species shall be quarantined in accordance with laws and regulations.

Utilizing fish species for aquaria can be carried out on protected and unprotected fish species. Aquaria can be conducted by Indonesian legal entities, research institutions, or universities. Indonesian legal entities, research institutions, or universities conducting fish aquaria shall obtain a licence from the Minister or a designated official. Indonesian legal entities, research institutions, or universities conducting fish aquaria shall be responsible for the fish health, safety, and security. Aquaria are carried out through collecting live fish in a controlled medium as an artificial habitat, collecting dead fish including their parts, and demonstrating live fish attractions.

The utilization for exchange can be carried out on protected and unprotected fish species. Exchange of fish species can be carried out by government, local government, Indonesian legal entity, or university. Species exchange shall hold a permit from the Minister or a designated official. The exchange of protected and unprotected fish species shall be conducted on an equal basis. Finally, the utilization for pleasure maintenance can be carried out on protected and unprotected fish species. Maintenance can be carried out by individuals. The protected fish species for pleasure can only be maintained from breeding results. Individuals conducting the maintenance of protected fish species shall obtain a licence from the Minister or a designated official. Individuals who keep protected fish species for pleasure shall maintain fish species health, comfort, safety and security, and provide a place and facilities that meet the standards of fish species maintenance. Therefore, a delicate balance between resource utilization and conservation becomes increasingly crucial. Balancing acts as compliance with fisheries regulations and environmental laws. These ensure that fishing activities align with legal requirements, preventing illegal practices that can further harm ecosystems [15].

In order that fish resources conservation can reach expected goals, supervision should be carried out through guarding and patrolling water conservation areas and monitoring the utilization of protected fish species and genetics. Supervision is carried out by fisheries supervisors, consisting of Fisheries Civil Servant Investigators and Non-Fisheries Civil Servant Investigators.

Violating utilization of fish resource conservation areas and utilization of protected fish species and genetics are subject to administrative sanctions [16]. Administrative sanctions are imposed by the utilization licensor in accordance with its authority in the form of written warning, license suspension, license revocation, and fines. Administrative fines are non-tax state revenues deposited into the State Treasury.

Administrative sanctions, written warnings, are given to permit holders who do not fulfil their obligations a maximum of three times consecutively within a maximum period of fourteen calendar days for each warning. Administrative sanctions, licence suspension, are imposed on any licence holder who until the expiry of the third written warning period does not carry out its obligations. Administrative sanctions in the form of licence suspension are imposed for six months since the sanction is imposed. Administrative sanctions in the form of licence revocation are imposed in the event that the period of licence suspension has expired, the licence holder does not carry out its obligations. Violating permit holders are subject to administrative sanctions in the form of permit suspension and administrative sanctions in the form of a minimum of ten times and a maximum of fifteen times of the fisheries levies that are their obligations. The licence suspension is imposed for six months since the sanction is imposed. In case of the freezing period has expired, the permit holder still does not carry out its obligations, the permit for the utilization of fish species and genetic fish is revoked. Revocation of the licence does not remove the administrative fine sanction.

To increase public awareness about the importance of fish resource conservation [17], community development is carried out by the Government, local governments, the business world, non-governmental organisations, or community groups. In the context of coaching, the community may be given awards for efforts to manage ecosystem, fish species and genetics conservation. Awards can be given by the

Government or local governments to individuals or those who are meritorious in the field of fish resource conservation.

3.2. Regulating Conservation of Razor Clam (Solen sp.) Resources in Pamekasan Indonesia

The enactment of the Job Creation Law has caused problems for the conservation of razor clam resources in *Pamekasan* Regency. Steps the local government and local communities can carry out in the legal aspect for realizing the conservation of razor clam resources include regulating conservation through local regulations, local government supervision, and increasing public legal awareness.

3.3.Regulating by Local Regulations

Local regulations of *Pamekasan* regency are laws and regulations, the Local People's Representative Council of *Pamekasan* establish with the joint approval of the *Pamekasan* regent. They contain the regency authority, the authority which location is within the regency, the authority which users are within the regency, the authority which benefits or negative impacts are only within the regency, and the authority which use of resources is more efficient if it is carried out by the regency. They may contain the imposition of the cost of coercive enforcement/implementation of the local regulations in whole or in part to the violator in accordance with the regulation provisions. In addition, they may contain a maximum imprisonment of six months or a maximum fine of fifty million, imprisonment or a fine, and sanctions restoring the original situation and administrative sanctions.

Drafting Regional Regulations [18] within the *Pamekasan* government is carried out with procedures i.e. the regent instructs the initiating regional apparatus to prepare a regency local regulation draft based on the regency regulations establishing programmes. In drafting the regency local regulation, the regent forms a drafting team for the regency local regulation which is stipulated by a decree of the regent. The drafting team members consist of regent, regency local secretary, initiator local apparatus, local apparatus of legal affairs, related local apparatus, and legislative drafter. The Regent may include related vertical agencies and/or academics as members of the drafting team. The drafting team is led by a chairman appointed by initiator local apparatus. In case of that the team leader is another appointed official, the head of the initiator local apparatus is responsible for the contents of local regulation draft.

In preparing the local regulation draft, the drafting team may invite researchers and/or experts from universities or community organisations as needed [19]. The drafting team leader reports the encountered development and/or problems in drafting the regency local regulation to the regency secretary to obtain direction or decision. The leader and the initiator local apparatus sign the drafted local regulation. The leader submits the drafted regency regulation to the regent through the regency secretary to be harmonized, finalized, and conceptually strengthen. The regency secretary assigns the head of the regency apparatus of legal affairs to coordinate harmonizing, finalizing, and strengthening the conception of the regency regulation draft. In coordinating them, the head of the regency apparatus of legal affairs can include ministry vertical agencies governing legal affairs. The regency secretary submits the results of harmonizing, finalizing, and strengthening the conception to the proponent and the head of the relevant regency apparatus in order to obtain approval signatures on each page of the regency regulation draft. The regency secretary submits the approved and signed draft to the regent. Each final draft to be submitted to the Local People's Representative Council shall be presented by the team leader to the regent.

Pamekasan regency regulation draft initiated by the Local People's Representative Council can be submitted by members, commissions, joint commissions of the Local People's Representative Council, or Agency for Establishing Regency Regulations based on the regency regulations establishing programmes. The submitted regency regulation paper is delivered to the head of the Local People's Representative Council with explanations or information and/or academic papers. The explanation or statement contains the main ideas and regulated content, a list of names, and the proponent signature. The reviewed and harmonised academic paper contains the drafting background and purposes, objectives to be realized; subject matter, scope, or object to be regulated, and scope and direction of regulation. The secretariat of the Local People's Representative Council put on the submitted regency local regulation draft a main number.

The draft regulating regency budgets, regency local regulation revocation or limited amendments of regency local regulations is submitted by accompanying the explanation or statement containing the main thoughts and regulated content material. The chairman of the Local People's Representative Council submits the draft to Agency for Establishing Regency Regulations to be studied for harmonising, rounding up, and stabilising the draft conception. The agency submits the assessment results of the draft to the chairman. The leader of the Council submits the results of the agency's review in a plenary meeting of the Council. The leader of the Council submits the draft to members of the Council within a maximum period of seven days before the plenary session of the Council. In the plenary meeting of the Council, the proponent provides an explanation, while the fractions and other members of the Council provide views and the proponent provides answers to their views. The plenary meeting of the Council decides the proposed draft in the form of approval, approval with amendments or rejection. In case of approval with amendments, the head of the Council assigns the commission, joint commission, the Agency, or special committee to improve the draft. The draft refinement is then resubmitted to the leaders of the Council. The draft, the Council prepares is submitted by the leaders of the Council to the regent for discussion. If in one session period, the Council and the regent submit a draft on the same material, the draft submitted by the Council shall be discussed and the draft submitted by the regent shall be used as material for comparison.

3.4. Submitting to Be a Water Conservation Area

There are three categories of water conservation areas i.e. parks, sanctuaries and maritime conservation areas [20]. The fish resources e.g. razor clam conservation includes a fishery sanctuary which serve to maintain and improve the quality of fish resources. The sanctuary category itself has the criteria of having a typical, unique, rare, endemic, and/or endangered fish species in its habitat that requires protection and preservation efforts in order to ensure the continuity of its natural development, having a large habitat of targeted species that supports the continuity of its life cycle, living and breeding place for a certain fish species that needs to be protected and preserved, having one ecosystem type as a habitat for certain fish species that is relatively natural, and the surrounding environmental conditions that support the sustainable fisheries developing efforts. *Pamekasan* regency can apply for the status of local conservation area for the of razor clam resources conservation.

The conservation area of fishery sanctuary is determined through the following stages i.e. initiatively submitting the prospective conservation area, identifying and taking inventory, reserving, and determining. The initiative proposal for prospective conservation area can be submitted by the regional government, customary law communities managing stipulated area, or any person. Submitting the initiative proposal of conservation area candidate shall be delivered to minister and a copy be submitted to the regent and the relevant regent, and be delivered to the regent and a copy be submitted to the Minister and the relevant regent. Submitting the candidate is an initial study and location map that includes conservation targets. Conservation targets are biodiversity, fish resources and their habitats or traditional cultural sites. Location map includes district, regency/city, island, beach, bay, cape, strait, sea, and/or ocean. If the candidate constitutes an area, the customary law communities manage, part or all of the waters area managed by customary law communities is handed over to the government or local government which is stated in the handover minutes. The area managed by the customary law communities includes an area that has been determined in the zoning plan document. The handed over area is managed by the government or local government by involving the customary law communities through partnerships.

Based on submitting the initiative proposal for conservation area candidate, the minister or regent in accordance with their authority shall conduct identification and inventory, and assessment of the identification and inventory results. Identification and taking inventory are conducted through survey, technical consultancy and public consultation. In carrying out identification and inventory, the minister or regent in accordance with his authority establishes a working group. In term that the submission of the initiative proposal for conservation area candidates is addressed to the minister, the working group is chaired by the director general. In case that the submission of an initiative proposal for conservation area candidate is addressed to the regent, the working group is chaired by the head of the regency apparatus organisation. The working group is in charge of preparing preliminary and final documents that should be completed within a maximum period of twelve months since the working group is appointed. Surveys are conducted to obtain data and information that at least contain important habitats and biota distribution, spawning areas, potential threats to conservation targets, condition of resource utilization status, stakeholder mapping, existence of potential traditional cultural sites, and important habitats representation, repeatability, and connectivity.

Data and information are used as analytical material for preparing the preliminary document that contains at least the establishing purpose of the conservation area, the conservation area size, conservation target priorities, conservation area zoning, provisions on conservation area utilization, conservation area categories, work plans, and conservation area maps with a minimum accuracy level of 1:50,000 (one to fifty thousand) scale. The working group submits a technical consultancy request on the preliminary document to the director general or head of the regency apparatus organisation in accordance with their authority for obtaining a response during a maximum period of fifteen working days since the complete technical consultancy application is received. The response to the technical consultancy results is suggestions for improving the preliminary document, and technical approval to proceed to the public consultation stage. If the director general or head of the regency apparatus organisation does not provide it during fifteen working days, the working group can conduct a public consultation. It is conducted to obtain input, responses, or suggestions for improvement from stakeholders or related parties, and to agree the initial document. The results of the consultation are written in public consultation minutes and location map signed by the working group chairman and stakeholder or related party representatives. The public consultation minutes are used as materials for preparing the final document on conservation area backup, that is the result of improving the initial document after the public consultation.

The final document is used by the minister or regent in accordance with their authority as the basis for backing up conservation areas that are then stipulated by the minister or regent. Determining the conservation area reservation shall at least contain the establishing purpose of the conservation area, the conservation area size, conservation target priorities, conservation area zoning, provisions on conservation area utilization, conservation area categories, work plans, and conservation area maps with a minimum accuracy level of 1:50,000 (one to fifty thousand) scale containing the outer boundaries and the conservation area zoning. The determined conservation area reservation is integrated into the the zoning plan document.

The conservation area Reservation, the minister or the regent has stipulated, is proposed to the minister to be designated as a conservation area. The regent submits a proposal for designing conservation area to the minister if the conservation area reservation is stipulated by the regent. The proposal is submitted to the minister during a maximum period of six months after the conservation area reservation is designated. The proposal shall attach at least the regent's decision on the conservation area reservation, the final document, and the plan for establishing the Management Organisation Unit. If the proposal for designing the conservation area is not submitted by the regent, the minister shall designate the reserved conservation area. In the framework of designation process as a conservation area, the minister shall evaluate the designated conservation area, and the proposal for designing the conservation area from the regent. The evaluation is carried out by the director general during a maximum period of three months after the minister stipulates the conservation area reservation and/or since the complete proposal for conservation area designation from the regent is received. The director general submits the evaluation results to the minister as the basis for the conservation area designation. Based on the evaluation results, the minister can designate conservation areas. The designated conservation areas are integrated into the zoning plan document.

The allocation of marine space or spatial pattern for conservation areas that has been included in the zoning plan document is equalised as a conservation area reservation. Based on the conservation area reservation, the final document is prepared. Space allocation for conservation areas that have been included in the zoning plan document for coastal areas and small islands is proposed by the regent to the minister to be designated as a conservation area during a maximum period of two years after the zoning plan document for coastal areas and small islands is stipulated. The marine spatial pattern for conservation areas is processed to be designated as a conservation area during a maximum period of one year after the zoning plan document of the marine area is stipulated. If the space allocation for included conservation areas in the zoning plan document for coastal areas and small islands is not proposed by the regent to the minister, the minister shall stipulate it as the conservation areas. The designated conservation area is managed by the regent in accordance with his/her authority. The conservation area stipulated by the minister is followed up with boundary arrangement which includes publishing the conservation area in the Indonesian sea map issued by the agency handling government affairs in the field of hydrography and oceanography, socialisation, and installation of boundary markers.

4. Conclusions

Conservation of razor clam resources can be realized through regulation in a Local Regulation on the protection, preservation and utilization of razor clam ecosystems, species and genetics. Thus, supervision and law enforcement efforts, and community development efforts by the local government can be carried out more optimally. In addition, the conservation of razor clam resources can be realized through the proposal of razor clam-producing coastal areas in the regency to the central government to be designated as local fisheries sanctuary conservation areas. The local government is expected to conduct a comprehensive study to prepare an academic paper and draft of local regulation on the conservation of razor clam resources or fish resources in general. The study is also conducted to prepare a proposal for the razor clam producing areas in the regency to be designated as local waters conservation areas.

Acknowledgments

Authors thank Prof. Ninis Trisyani for helping to collect *solen sp.* data, Prof. Chomariyah for providing his invaluable expert knowledge on the law of fisheries and Prof. M. Khoirul Huda for proof reading the manuscript. This project has received funding from Universitas Hang Tuah, Surabaya trough Center for Research and Community Service.

References

[1] Fajar, Baharudin Zain Surya. 2019. "Hubungan Panjang Berat Kerang Bambu (Solen spp) di Perairan Pesisir Pamekasan Pulau Madura." PhD diss., University of Muhammadiyah Malang, 2019.

[2] Trisyani N. Fishing technique and environmental factors affecting the size of razor clam Solen sp. in Indonesia coast. Aquaculture, Aquarium, Conservation & Legislation. 2018 Feb 1;**11**(1):29-36.

[3] Trisyani N. and Kamarudin. Population growth and fishery status of the Lorjuk shellfish (Solen sp.) on Pamekasan beaches, Indonesia. IOP Conference Series: Earth and Environmental Science 2020 Feb 1 (**441**, No. 1, p. 012074). IOP Publishing.

[4] Trisyani N. Keanekaragaman Hayati dan Konservasi (Study pada spesies Lorjuk, Solen sp.)*. Seminar Nasional Kelautan XII. 2017.

[5] Pinchot G. Attractive, Durable Roofing Made From Recycled Plastic and Wood Fiber. Forest Products Laboratory, Technology Marketing Unit. 2005.

[6] Irfan, S. and Alatawi, A.M.M. (2019) Aquatic Ecosystem and Biodiversity: A Review. *Open Journal of Ecology*, **9**, 1-13. https://doi.org/10.4236/oje.2019.91001

[7] Outa NO, Yongo EO, Keyombe JLA, Ogello EO, Namwaya Wanjala D. A review on the status of some major fish species in Lake Victoria and possible conservation strategies. *Lakes & Reserv*. 2020; **25**: 105–111. https://doi.org/10.1111/lre.12299

[8] Lind, C., Ponzoni, R., Nguyen, N. and Khaw, H. (2012), Selective Breeding in Fish and Conservation of Genetic Resources for Aquaculture. Reproduction in Domestic Animals, **47**: 255-263. https://doi.org/10.1111/j.1439-0531.2012.02084.x

[9] Pullin RS. Conservation of farmed fish genetic resources: broad options, strategies, availability of methods, and comparisons with those for conservation of the genetic resources of other farmed animals. InInternational Workshop "Options and Strategies for the Conservation of Farm Animal Genetic Resources" 2005 (p. 87).

[10] Bartley DM, Harvey BJ, Pullin RS, editors. Workshop on Status and Trends in Aquatic Genetic Resources: A Basis for International Policy: 8-10 May 2006, Victoria, British Columbia, Canada. Food & Agriculture Org.; 2007.

[11] Food and Agriculture Organization, 2008: *Aquaculture development. 5. Genetic resource management.* Vol. FAO Technical Guidelines for Responsible Fishereis No. 5, Suppl. 3 FAO, Rome, p. 125.

[12] Gjedrem, T. and Robinson, N. (2014) Advances by Selective Breeding for Aquatic Species: A Review. Agricultural Sciences, **5**, 1152-1158. http://dx.doi.org/10.4236/as.2014.512125.

[13] M. Blanco, C.G. Sotelo, M.J. Chapela, R.I. Pérez-Martín, Towards sustainable and efficient use of fishery resources: present and future trends, Trends in Food Science & Technology, **18**, Issue 1, 2007, Pages 29-36, https://doi.org/10.1016/j.tifs.2006.07.015.

[14] Daris L, Yusuf M. Conflict dynamics of fishery resources utilization in Maros District, South Sulawesi Province, Indonesia. Aquaculture, Aquarium, Conservation & Legislation. 2019 Jun 1;**12**(3):786-91.

[15] Oachiş A, Zoican Ş, Vornicu L, Şmuleac L, Paşcalău R. Fisheries and Environmental Issues Balancing Resource Utilization and Conservation. Research Journal of Agricultural Science. 2023 Oct 1;**55**(4).

[16] Kastro E, Mardiyati S, Hendri S. Law Enforcement of Dangerous Fishing Methods to the Conservation of Inland Fisheries Resources in Public Waters South Sumatera Province. International Journal of Social Science Research and Review. 2024 Jul 11;**7**(7):36-45.

[17] Easman ES, Abernethy KE, Godley BJ. Assessing public awareness of marine environmental threats and conservation efforts. Marine Policy. 2018 Jan 1;**87**:234-40.

[18] Handayani IG. Legal Drafting Local Regulations in Local Parliament. International Journal of Business. Economics and Law. 2013 Jun;**2**(3).

[19] Asnawi E. Legal Politics of Legislation of Regional Regulations in Indonesia. Legal Politics of Legislation of Regional Regulations in Indonesia. 2021 Sep 25;**3**(2):1-7.

[20] Government Regulation Number 60 of 2007 on Conservation of Fish Resources

Pro Marine Environmental Behavior On College Students

D Mustami'ah[1*] And N S Widanti[1]

[1] Psychology, Psychology Faculty University of Hang Tuah, Surabaya, Indonesia

*E-mail: dewi.mustamiah@hangtuah.ac.id

Abstract. Pro-marine environmental behavior is important to be done by everyone, including college students as the nation's next generation. This is so important that every individual should participate in preserving the marine environment. This study aims to determine the factors that play the most important role in shaping marine environmental behavior. Respondents of this research were 322 students of the University of Hang Tuah. The Likert scale was used to collect data research. The research results show that students are still less motivated to take part in preserving the marine environment. Likewise, the willingness to reduce the use of plastic materials also tends to be lacking. The research results also show that students are willing to participate in activities to preserve the marine environment, for example cleaning the beach and planting mangroves. Students are also interested in participating in marine environmental observer group activities, but recycling plastic waste and bringing their reusable bags when shopping is still lacking. The study results show that the correlation between attitudes toward the environment and marine environmental behavior is strong. Subjective norms are strongly related to marine environmental behavior and perceived behavioral control is also strongly related to marine environmental behavior.

1. Introduction

Humans are very dependent on the environment around them, both biotic and abiotic. It is all because humans and nature are interconnected. Human behavior towards the environment affects the human life. One of the important environments for human life is the sea. The sea has the function of stabilizing the climate, producing oxygen, storing carbon, as a source of food reserves, energy, minerals, recreation, and also for transportation.

Considering the important role of the sea, preventive measures against damage to the marine environment due to human activities must be taken. Nowadays, marine and coastal ecosystems around the world face major threats due to irresponsible human actions. Plastic waste is increasing day by day and it is estimated that by 2050 the weight of plastic waste in the sea will exceed the weight of fishes. This is caused by millions of tons of plastic waste that is not managed properly and then thrown into the sea. This causes more than 100 million marine animals to be lost every year and the marine ecosystem is destroyed [1]. This condition is certainly not following the objectives of Sustainable Development Goal 14 (SDG 14), which is to conserve and sustainably utilize marine and ocean resources for sustainable development. This aims for humans to participate in preserving and utilizing oceans, seas, and marine resources properly so that they remain sustainable. This is because sustainable oceans are important for humans and other life on Earth.

Efforts to maintain the sustainability of the marine environment are very important so as not to harm human life. Research on the marine environment has been widely conducted by previous researchers, including [1] related to the importance of students having awareness of the marine environment. Marine environmental awareness is the ability to understand marine environmental problems and commit to behaving well in everyday life. Several studies have developed tools to measure

marine environmental awareness [2]. Furthermore, several studies aimed to develop marine environmental awareness in students based on the Theory of Planned Behavior [3]. Research to prevent damage to the marine environment relates to awareness, management, and restoration of marine ecosystems[4][5]. Pro-marine environment behavior is influenced by marine environmental awareness[6][7][2][8]. Marine environmental awareness is important for every human being because it is the foundation for environmental behavior [9].
This research is important considering Hang Tuah University which has a maritime and marine vision and mission.

Students need to have the knowledge to prevent damage to the marine environment. Prevention of damage to the marine environment can be seen from how a person behaves pro-marine environment or not. Pro-environmental behavior is behavior that can prevent damage to the marine environment[10], behavior to protect against environmental damage[11], conservation lifestyle behavior, supporting proper land management, and being involved in pro-environmental policies.

There are many versions of measuring pro-environmental behavior. There are 4 categories in measuring pro-environmental behavior, namely the social environment, which is efforts to educate or socialize environmental issues to others; land use which is related to actions for environmental conservation such as cleaning up trash in public areas; lifestyles that support conservation such as recycling, managing trash, saving water and energy; and the last aspect related to citizen concern for the environment such as behavior to participate in making pro-environmental policies [12]. There is also another opinion that states that measuring pro-environmental behavior has 6 aspects, energy saving, fueled vehicles, managing waste, recycling, consumerism, and behaviors that aim to conserve nature[13]. This research on pro-marine environmental behavior refers to the opinion of Larson et al [12] that students can be said to behave pro-marine environment if they can socialize the marine environmental issues; are involved in marine environmental management through actions like cleaning the marine environment; have the lifestyle that supports marine environmental conservation such as recycling, managing waste; and behave in a way that shows concern for the marine environment such as participating in making pro-marine environmental policies.

Factors influencing pro-environmental behavior are complex[14]. Other researchers state that knowledge, attitudes, personal norms, perception of locus of control, gender, race, socioeconomic status, extrinsic motivation, and place of residence can influence pro-environmental behavior[15][16]. Positive attitude influences pro-environmental behavior [6];[8][17]. Positive attitude influences pro-environmental behavior[18][19]. Marine environmental attitudes are feelings toward the marine environment[20], characterized by positive or negative attitudes of individuals regarding the fragile and vulnerable balance of nature; the possibility that nature will experience a crisis; assessing the reality of limited growth of nature; individual belief in the rights held by nature as equal to human rights; and realizing the limitations of human ability to improve nature. Students' positive attitudes affect protecting the environment.

Subjective norms influence pro-environmental behavior [21]. Subjective norms are an individual's perception of the judgment of the social environment and how strongly the individual complies [22]. This norm is related to the individual's perception of whether or not others will support the realization of the action [23]. Subjective norms are influenced by social pressures that require individuals to behave in a certain way [24]. Subjective norms are related to the perceptions of individuals around them, for example, family members, friends, or colleagues support their behavior or not[25].

Pro-environmental behavior is influenced by behavioral control, which is an assessment of a person's ability to do something, whether there are obstacles or not in realizing behavior. [26]. Individuals who have the characteristics of perceived behavioral control can be identified from control beliefs, sources and opportunities needed to generate behavior, power of factors, and individual perceptions of the availability of sources needed to generate or prevent behavior from occurring [27]. Perceived behavioral control can influence ecological behavior [28].

Given the importance of pro-environmental behavior, research is important, because students are the successors of the nation who will manage Indonesia's marine wealth in the future.

2. Methods

The research was conducted in a quantitative research setting. Primary data is used in this study, namely, data obtained directly from the main source which is the respondent's response to a series of questions asked with a psychological scale. The sampling method is non-probability sampling. The respondents are students of University of Hang Tuah who are willing to fill out the Google form. Based on the data obtained, the total number of respondents is 322, consisting of 86 students from the health sciences field, 152 students from the social sciences field, and 84 students from engineering sciences fields. The research tool used to collect data is the pro-marine environmental behavior scale which consists of 5 alternative answers from those stating agree to disagree, starting with: Strongly Agree (SS), Agree (S), Neutral (N), Disagree (TS), Strongly disagree (STS).

3. Results and Discussion

3.1. Population

Based on descriptive data, it is known that the research subjects consist of :

Table 1. Respondents description

	Characteristics	Amount	Percent
Gender	Male	126	39
	Female	196	61
sciences field	health sciences field	86	26
	Social Sciences field	152	48
	Engineering sciences field	84	26

N = 322

From Table 1 it is known that the research respondents were 126 male students or 39%, while the number of female students was 196, or 61%. Also, know that the respondents were more dominant female students. The subjects in the study consisted of 3 study program groups: health sciences field 85 students (26%), Engineering sciences fields students (48%), and engineering sciences field program 85 students (26%). Based on this data the most dominant respondents were students from social study programs which is 48%.

3.2 Measuring

The measuring instrument is declared valid if the items are arranged based on the indicators from each aspect. In addition, they are measuring to meet the requirements of the corrected item-total correlation score that is more than 0,3. The description of each item can be seen in Table 2.

Table 2. Pro marine environmental behavior Items

Number	Items	Corrected item-total correlation	Cronbach's Alpha if item deleted
1	I am motivated to take part in preserving the marine environment.	0.479	0.788
2	I reduce the use of plastic materials	0.515	0.786

3	I will remind anyone who throws plastic waste carelessly	0.353	0.798
4	I limit the use of products that can pollute seawater.	0.507	0.786
5	I am willing to plant mangroves to prevent seawater intrusion	0.530	0.782
6	I am willing to join in cleaning the beach	0.573	0.777
7	I am willing to join the environmental club	0.411	0.794
8	I am ready to be a volunteer to save marine life	0.426	0.792
9	I recycle plastic waste	0.379	0.797
10	I bring reusable bags when shopping at the market or supermarket	0.348	0.999
11	I buy environmentally friendly detergent even though it is more expensive.	0.399	0.796
12	I am willing to spend money to participate in the marine environmental sustainability campaign.	0.504	0.785

This result shows that all items meet the requirements for measuring pro-marine environmental behavior.Furthermore, the results of this study also describe respondents' awareness of marine environmental sustainability. The results of the description of marine environmental sustainability awareness from research respondents are as follows in Table 3:

Table 3. Distribution score of marine environmental awareness scale

No	Indicators	Min	Max	Mean	Std.Dev	Variance
1.	Socializing marine environmental issues.	1	3	2.25	.371	.138
2.	Management of the marine environment.	1	3	2.71	.422	.178
3.	Lifestyle that supports conservation.	1	3	2.40	.553	.306
4.	Care for the marine environment.	1	3	2.15	.474	.224

Based on Table 3, it is known that the item socializing marine environmental issues has a mean value of 2.25 or 22.5%. The minimum value is 1 and the maximum is 3 with a standard deviation of 0.371. Management of the marine environment has a mean value of 2.71 or 27.1%, with a standard deviation of .422. A lifestyle that supports conservation has a mean value of 2.40 or 24%. The minimum value is 1 and the maximum is 3 with a standard deviation of 0.553. Student concern for the marine environment has a mean value of 2.15 or 21.5%. The minimum value is 1 and the maximum is 3 with a standard deviation of 0.474.

Based on the results of the description of each indicator above, it is known that the indicator has the highest value in forming respondents to study managing marine environment with a mean value of 27.1%. Willingness act pro-marine environment in a real and easy way, including by holding beach cleaning activities from garbage or activities related to environmental conservation. Marine environmental management is related to actions for environmental conservation, maintaining the coastal environment, and keeping the sea free of rubbish which will strengthen individuals to always behave in a pro-environmental manner [29] [30]. Pro-environmental behavior itself is an effort to prevent the negative impacts of individual behavior and prevent environmental damage [11]. This is supported by the willingness of subjects to volunteer to save marine biota, where 63.6% of subjects agreed, 32.4% sometimes and 4% did not agree.

Furthermore, the indicator of the lifestyle that supports conservation has a mean value of 24%. Pro-environment forms such as supporting conservation, for example, recycling, managing waste, and saving water and energy as a manifestation of environmental behavior. If students have the habit of managing waste by getting used to throwing away waste in its place, then the individual is considered to be behaving pro-marine environment [11]. Of subjects who always recycle plastic waste 15% sometimes 58.6%, and 26.5% stated that they never recycle. This means that students still need to improve their lifestyle to support conservation.

Students also stated that they were willing to conduct socialization on issues related to the marine environment, which was 22.5%. This means that students are willing to make efforts to introduce marine environmental issues to friends or people around them. Students can do this if they have sufficient knowledge regarding the marine environment. someone who knows the intention to protect the marine environment and will then behave in a pro-marine environment manner [20], [31], [32]. Based on this, students who have taken the MARINE SCIENCE AND TECHNOLOGY course should have the will to share their knowledge with friends or people around them so that they can help preserve the marine environment by reminding them when someone throws plastic waste carelessly. The results of this study show that 56.3% of subjects answered that they always reminded, 42.2% sometimes reminded and 1.5% never reminded.

The indicator concern for marine has a mean value of 21.5%, which means it is the lowest compared to other indicators. Caring for the marine environment is an attitude and action to prevent damage to the marine environment and attempts to repair such damage. This can be seen from the answers of most respondents who state that they never buy environmentally friendly detergent because it is expensive, that is 50.3%. Respondents who sometimes buy are 28.1% and 21.6% state always buy. As many as 31.6% of respondents answer that they are never willing to spend money to participate in environmental sustainability campaigns, as many as 45.3% answer that they sometimes spend money on campaigns and only 23.1% are always willing to spend money on campaigns.

Based on the analysis results, it is also known that there are differences between the three Faculties, which are health sciences fields, social sciences fields, and engineering sciences fields. For more details, it can be seen in Table 4.

Table 4. Pro-marine environmental behavior based on science fields

Science Fields	Mean
health sciences field	84,74
Social Sciences field	85,67
Engineering sciences field	86,329

Based on Table 4, it is known that students who are included in the Engineering sciences field are in the study programs of Naval Engineering, Naval Systems Engineering, Oceanography and Fisheries, and Naval Systems. The average score of pro-marine environmental behavior among students in this science field is 86.329. Meanwhile, students in Social Sciences have an average score of 85.67, while health students have an average score of 84.74.

The results of this study reflect that students in the engineering sciences field are likely to gain more knowledge about the ocean than in other science fields, so they have a better understanding of the causes of damage to the marine environment. In addition, students of the engineering sciences field understand the impacts and ways to overcome environmental damage. The students' learning activities are also more frequent at sea compared to students in other sciences field. However, these results still need to be deepened because other studies have emphasized that a definite relationship between academic field groups and pro-environmental behavior has not been found. Previous research findings that discussed pro-environmental behavior in China found that highly educated, more urban, younger, female, and environmentally oriented people showed more pro-environmental behavior than their other counterparts [1].

4. Conclusion

Students who were respondents tended to behave pro-marine environment. These students have the will to participate in managing and preserving the marine environment, especially having the will to act in a pro-marine environment manner. Activities that can support this include holding beach cleaning activities from garbage or activities related to environmental conservation. Students that have the highest tendency to act more pro marine environment behavior are those that included in the engineering sciences field which are the study programs of Naval Engineering, Naval Systems Engineering, Oceanography and Fisheries, and Naval Systems. Their educational background which related to may make those higher in awareness compared to students from other sciences fields.

Acknowledgments

The researcher would like to thank the Head of Hang Tuah University and the research members so that the research is by the objectives achieved.

Refernces

[1] Chen C L , and Tsai C H, 2015 Marine environmental awareness among university students in Taiwan: a potential signal for the sustainability of the oceans. **22**, 7 p. 958–977.
[2] Umuhire M L and Fang Q, 2016 Method and application of ocean environmental awareness measurement: Lessons learned from university students of China *Pollution Bulletin* **102**, 2 p. 289–294.
[3] Mustami'ah D Harahap N Marjono and Tarno H, 2023 Development of Planned Behavior Theory in Increasing Awareness of the Marine Environment in Students *International Journal on Engineering Applications* **11**, 3 p. 143–151.
[4] Gkargkavouzi A Paraskevopoulos S and Matsiori S, Jan. 2020 Public perceptions of the marine environment and behavioral intentions to preserve it: The case of three coastal cities in Greece *Mar Policy* **111**.
[5] Yoon A Jeong D and Chon J, Jun. 2021 The impact of the risk perception of ocean microplastics on tourists' pro-environmental behavior intention *Science of the Total Environment* **774**.
[6] Liu P Teng M and Han C, Aug. 2020 How does environmental knowledge translate into pro-environmental behaviors?: The mediating role of environmental attitudes and behavioral intentions *Science of the Total Environment* **728**.
[7] Soares J Miguel I Venâncio C Lopes I and Oliveira M, Jun. 2021 Public views on plastic pollution: Knowledge, perceived impacts, and pro-environmental behaviours *J Hazard Mater* **412**.
[8] Yang M X Tang X Cheung M L and Zhang Y, Jan. 2021 An institutional perspective on consumers' environmental awareness and pro-environmental behavioral intention: Evidence from 39 countries *Bus Strategy Environ* **30**, 1 p. 566–575.
[9] Holt D and Barkemeyer R, 2012 Media Coverage of Sustainable Development Issues – Attention Cycles or Punctuated Equilibrium? Sustainable Development **20**, 1 p. 1–17.
[10] Schultz P W and Kaiser F G, 2012, Promoting Pro-Environmental Behavior, in *The Oxford Handbook of Environmental and Conservation Psychology*, (Oxford University Press), p. 556–580.
[11] Blok V Wesselink R Studynka O and Kemp R, 2015 Encouraging sustainability in the workplace: A survey on the pro-environmental behaviour of university employees in *Journal of Cleaner Production* **106** p. 55–67.
[12] Larson L R Stedman R C Cooper C B and Decker D J, Sep. 2015 Understanding the multi-dimensional structure of pro-environmental behavior *J Environ Psychol* **43** p. 112–124.
[13] Kaiser F G Oerke B and Bogner F X, Sep. 2007 Behavior-based environmental attitude: Development of an instrument for adolescents *J Environ Psychol* **27**, 3 p. 242–251.

[14] O'Brien K, 2013 Global environmental change III: Closing the gap between knowledge and action *Prog Hum Geogr* **37**, 4 p. 587–596.
[15] Steg L and Vlek C, Sep. 2009 Encouraging pro-environmental behaviour: An integrative review and research agenda *J Environ Psychol* **29**, 3 p. 309–317.
[16] Scannell L and Gifford R, Sep. 2010 The relations between natural and civic place attachment and pro-environmental behavior *J Environ Psychol* **30**, 3 p. 289–297.
[17] Vicente-Molina M A Fernández-Sainz A and Izagirre-Olaizola J, 2018 Does Gender Make A Difference In Pro-Environmental Behavior? The Case Of The Basque Country University Students *J. Clean. Prod* **176** p. 89–98.
[18] Maichum K Parichatnon S and Peng K C, 2016 Application of the extended theory of planned behavior model to investigate purchase intention of green products among Thai consumers *Sustainability (Switzerland)* **8**, 10.
[19] Yadav R and Pathak G S, Apr. 2017 Determinants of Consumers' Green Purchase Behavior in a Developing Nation: Applying and Extending the Theory of Planned Behavior *Ecological Economics* **134** p. 114–122.
[20] Chen C-L and Tsai C-H, Oct. 2016 Marine environmental awareness among university students in Taiwan: a potential signal for sustainability of the oceans *Environ Educ Res* **22**, 7 p. 958–977.
[21] Yadav R and Pathak G S, 2016 Young Consumers' Intention Towards Buying Green Products In A Developing Nation: Extending The Theory Of Planned Behavior *J. Clean. Prod* **135** p. 732–739.
[22] Cordano M Welcomer S Scherer R F Pradenas L and Parada V, Sep. 2011 A cross-cultural assessment of three theories of pro-environmental behavior: A comparison between business students of chile and the united states *Environ Behav* **43**, 5 p. 634–657.
[23] Baron R A, 2012 *Social psychology / Robert A. Baron, Nyla R. Branscombe | OPAC Perpustakaan Nasional RI.* 13th ed. Boston: Pearson Education.
[24] Ham M Jeger M and Frajman Ivković A, Jan. 2015 The role of subjective norms in forming the intention to purchase green food *Economic Research-Ekonomska Istraživanja* **28**, 1 p. 738–748.
[25] Jaén I Moriano J A and Liñán F, 2013, Personal values and entrepreneurial intentions: An empirical study., in *Conceptual Richness and Methodological Diversity in Entrepreneurship Research*, p. 15–31.
[26] Zhang J Cheng M Wei X Gong X and Zhang S, 2019 Internet Use And The Satisfaction With Governmental Environmental Protection: Evidence From China *J. Clean. Prod* **212** p. 1025–1035.
[27] Fishbein M and Ajzen I, Jan. 2005 Theory-based Behavior Change Interventions: Comments on Hobbis and Sutton *J Health Psychol* **10**, 1 p. 27–31.
[28] Lee T H Jan F H and Huang G W, 2015 The Influence Of Recreation Experiences On Environmentally Responsible Behavior: The Case Of Liuqiu Island, Taiwan. *J. Sustain. Tour.* **23**, 6 p. 947–967.
[29] Gosling E and Williams K J H, Sep. 2010 Connectedness to nature, place attachment and conservation behaviour: Testing connectedness theory among farmers *J Environ Psychol* **30**, 3 p. 298–304.
[30] Cooper C Larson L Dayer A Stedman R and Decker D, Apr. 2015 Are wildlife recreationists conservationists? Linking hunting, birdwatching, and pro-environmental behavior *Journal of Wildlife Management* **79**, 3 p. 446–457.
[31] Boubonari T Markos A and Kevrekidis T, Jan. 2013 Greek pre-service teachers' knowledge, attitudes, and environmental behavior toward marine pollution *Journal of Environmental Education* **44**, 4 p. 232–251.
[32] Rannikko P, 2016 Local Environmental Conflicts and the Change in Environmental Consciousness **39**, 1 p. 57–72.

Village Governance Model Based on Local Wisdom

S J Utama[1*] and D Arieffiani[2]

[1] Public Administration, Faculty of Social Sciences, Hang Tuah University, Surabaya Indonesia
[2] Public Administration, Faculty of Social Sciences, Hang Tuah University, Surabaya Indonesia

*E-mail: sasmito.jatiutama@hangtuah.ac.id

Abstract. This study examines the model of village governance based on local wisdom. The locus in this paper is Sekapuk Village, Ujungpangkah District, Gresik Regency. This study is an exploratory study that combines qualitative and quantitative research approaches. Village governance based on local wisdom in the study location can run optimally, this is indicated by the preparation of village development planning and budget planning that have utilized local knowledge, local values, local skills, local resources and local decision-making mechanisms. Local wisdom in the study location has effectively utilized village government resources that are managed independently, participatory and aspirational.

1. Introduction

A village is a defined community with its own boundaries that holds the authority to govern and oversee local affairs based on traditional customs and inherited rights that are acknowledged within the Indonesian legal framework. The Law Number 6 of 2014 on Villages aims to empower village communities by granting them more autonomy in managing their governance, with the goal of fostering fair development and enhancing the well-being of village inhabitants. This inclusive approach is intended to address issues like regional disparities, poverty, and socio-cultural challenges, ultimately leading to a more equitable and prosperous village life.

The law focuses on empowering villages and recognizing their inherent rights. Article 19 specifically mentions the concept of "village origin rights" and grants villages with "village-scale local authority". This gives villages the autonomy to govern and administer themselves without being overly regulated by higher levels of government. Moreover, villages are also tasked with executing certain authorities delegated by the government, whether it be at the national, provincial, or district/city level[1].

Defining the concept of "village governance" poses a challenge. According to the UNDP (2005)[2], governance can be best described as a sophisticated structure composed of mechanisms, procedures, connections, and organizations that enable different parties to express their interests, fulfill their obligations, and address any disagreements that may arise. Another related definition by Chhotray and Stoker (2009) [3] portrays governance as the rules governing collective decision-making in environments with multiple actors or organizations, where formal control systems do not dictate the terms of relationships between these entities[4].

The concept of village government governance encompasses the principles, methods, and strategies employed in the organization and administration of government functions at the village level. It aims to establish efficient and effective governance practices to ensure that village governments can successfully deliver public services, manage resources, and address the needs of the local community. The key to effective village governance lies in the community's active participation in the decision-making process., transparency in public information disclosure, regular monitoring of government performance, a fair and transparent mechanism for resolving disputes, and alignment with community goals and aspirations [5].

Villages have a close relationship with local wisdom, especially how the village runs its government and the dynamics of development that are implemented. Local wisdom takes various forms,

IOP Conf. Series: Earth and Environmental Science **1473** (2025) 012067 doi:10.1088/1755-1315/1473/1/012067

serving as a melting pot of diversity in social interactions grounded in cultural insights. These forms mirror the essence of societal norms, encompassing values, guidelines, morals, beliefs, traditions, rituals, and distinct behavioral standards. Within the realm of local wisdom, noble virtues surface prominently, including reverence for the Divine, moral integrity, accountability, self-discipline, autonomy, integrity, reverence, compassion, empathy, reliability, innovation, diligence, resilience, fairness, leadership aptitude, philanthropy, modesty, open-mindedness, peace advocacy, and solidarity [6].

According to Mitchell (2019)[7], local wisdom encompasses six key dimensions.[6–11] Firstly, the Dimensions of Local Knowledge highlight each community has the capability to adjust and thrive in its surroundings by utilizing their indigenous knowledge to manage natural elements. This includes understanding climate change and other environmental phenomena. Secondly, the Local Value Dimensions underscore the unique norms and values that govern the behavior of individuals within a community. These values may vary across different communities depending on their societal progress. For example, behaviors like tattooing and body piercing may be accepted in one community, such as among the Dayak tribe, but not in others.

Thirdly, the Dimensions of Local Skills emphasize the essential abilities every society possesses to sustain their livelihoods. This includes traditional methods like hunting, gathering, farming, and cottage industries to fulfill their basic needs and maintain their survival economy. Fourthly, the Dimensions of Local Resources stress the importance of utilizing local resources responsibly according to the community's needs. It is crucial for communities to avoid exploiting resources on a large scale or for commercial gain to maintain a healthy balance with nature and prevent harmful impacts on the environment. The fifth dimension refers to the concept of decentralized decision-making, where individual communities are empowered to govern themselves through local institutions.. This institutional framework ensures that citizens adhere to agreed-upon rules and regulations. In case of any violations, sanctions are imposed through a process of decision-making led by the tribal leader. The sixth dimension is local group solidarity, which highlights the importance of social relations and cooperation between individuals in a community including mutual support and collaboration in various tasks, including environmental preservation and protection.

The existing capabilities and knowledge of the community still need to be used as a basis or foundation in formulating or planning regional management policies as well as natural resources. [12] The emergence of local wisdom is expected from traditional villages or village autonomy, as a legal community unit that has the authority to manage and do self-manage. It means that the village has autonomous rights that are different from the formal autonomy belonging to the provincial, district and city governments [13].

The challenge of overseeing villages lies in effectively managing the resources and knowledge of the local community. By harnessing the wisdom and values unique to each village, governance can be strengthened to promote equal prosperity for all residents.[14–16]. In the midst of evolving regional economic conditions and a more democratic political atmosphere, the advancement of villages has become notably arduous, encountering numerous obstacles and challenges. The insufficient focus on village development has been highlighted in various research studies [17–20].

The hampered progress in village development is often attributed to the community's inability to adapt to rapid and dynamic changes. Hence, the obstacles become more demanding in a growing transparent regional market and democratic political climate. Various studies have highlighted the lack of focus on village development as a significant issue.[21–24] Previous research shows that various obstacles occur in implementing local wisdom in village government. Mulda (2020) from the results of his research in Mekkatta Village, Majene Regency showed that the Village Government management process was relatively ineffective, this can be seen from the planning process to the evaluation carried out on each village government program. This condition is caused by minimal human resources and minimal financial support for village development [25].

However, there have been many successful village governance practices involving the role of local communities and local wisdom. Sasi for example which is a mutual agreement between indigenous leaders, respected individuals, and local communities in protecting and maintaining the environment and natural resources through the wisdom inherited from these communities.. [26,27]. It improves resource management and implements more successful development strategies by directly involving villagers in governance processes [28,29].

Sekapuk Village, as the location of this research, is located in Ujungpangkah District, Gresik Regency.. Which is of concern because of its success in developing the community's economy. Sekapuk Village was previously a village faced with various problems that required special solutions and attention. Some of the main problems of the village are low levels of community income, limited access to resources, waste problems, and the lack of optimal use of tourism potential. Sekapuk was able to change by advancing its local potential to become a more advanced village [30].

The transformation of Sekapuk Village is quite significant. so that it has received various awards. The title of Village Without Corruption, Setigi Destination which was named the Best Artificial Tourism in the 2021 Gresik Tourism Award, the Best Brilliant Village 2020 for competition organized by Bank Rakyat Indonesia (BRI).[30] This achievement was followed by other awards such as the Head of Sekapuk Village receiving the Indonesia Leaders Awards 2022 category of Most Promissing Leaders Innovation & Performance 2022 by Seven Media Asia in 2022 and the Asia Leaders Awards 2023 Category of Visionary Leaders Indonesia in 2023. Sekapuk Village was selected as 1 of 9 villages representing Indonesia in the Association of Southeast Asian Nations Villages Network (AVN).[31]. Due to this achievement, Sekapuk Village has become a role model for many villages from all over Indonesia.

In relation to this, this study examines how the development model is made, as well as the factors that form it and those that inhibit it. Practically, this study is useful in the field of village development, and specifically focuses on the aspect of local wisdom in village governance.

2. Research Methods

This research is exploratory research, with a qualitative approach that is also complemented by quantitative. Data is compiled through discussions and field surveys, which allowed researchers to obtain relevant and contextual information. The grounded theory stage is used to analyze data to help compile information systematically, resulting in valid and accountable information.

3. Results and Discussion

3.1. Location Overview

Sekapuk Village with an area of 297.33 ha has a height of 20 meters, this village is surrounded by views of white limestone mountains that add to its natural beauty. The climate conditions in Sekapuk Village are quite supportive for agriculture, with an average rainfall of 1.4 thousand mm per year and temperatures reaching 32°C. The existing agricultural potential, coupled with natural resources in the mining sector, make this village have great opportunities for economic development [32].

The accessibility of the village is also quite good, because the main gate is near the Deandles highway, which connects Gresik and Tuban. This strategic geographical location allows Sekapuk Village to interact with the surrounding areas and utilize the existing potential for the advancement of society. Sekapuk Village has an area of village treasury land of 2.50 ha, village hall complex of .78 ha, cemetery land of 1.50 ha, field land of 0,80 ha, dry fields of 147.70 ha, residents' yards of 11.60 ha, other land of 38.226 ha. While the area of other public infrastructure is 2.47 ha, offices of .26 ha, park area of .02 ha, plantation/forest area of 2 ha [32].

Sekapuk Village has a vision of "Creating a Civilized, Just and Prosperous Society". Meanwhile, the implemented Missions encompass various key areas such as enhancing physical infrastructure and fostering aesthetically pleasing and contemporary urban planning. They also focus on streamlining an efficient and accountable government system to better serve the populace. Additionally, efforts are being made to promote educational development and nurture creativity, revive local culture and traditions, bolster village economies, and empower the community financially. Furthermore, the emphasis is on safeguarding the rich diversity that exists within society, ensuring equal opportunities for all individuals, and protecting the rights of the community as a whole [32].

3.2. Authority Based on Original Rights and Local Authority on a Village Scale

Table 1. Local Wisdom Based on the Authority of Original Rights in Sekapuk Village.

Local Wisdom Criteria	Total Value of Relatedness of Each Authority	Percentage Level of Relatedness
Local Knowledge	10	83.33%
Local Values	12	100%
Local Skills	0	0%
Local Resources	4	33.33%
Local Decision Making	10	83.33%

Gresik Regency has established guidelines for the arrangement of village authority, namely Number 6 of 2016 as a regulation stipulated by the regent. The follow-up in Sekapuk Village is Regulation Number 3 of 2018 as a village regulation that regulates this authority.[33] Local wisdom can be seen in village regulation number 3. Which contains 12 authorities based on original rights and 184 local authorities on a village scale. The assessment of each authority is carried out by giving a weight of 1 for each authority that is related to the criteria of local wisdom.[6–11] While those that have no relevance to the criteria are given a weight of 0. The assessment of the 12 authorities based on original rights in Sekapuk Village is in table 1 below.

The relationship between the 12 authorities based on original rights and local wisdom shows that all of these authorities are relevant to the application of local values. The forms of authority include: arrangement of the village apparatus organizational system; appointment of the caretaker of the tomb; management of village treasury land, which comes from rewards and other similar land; facilitate the recording of land rights in villages; facilitation of village level land dispute resolution; land use planning and mapping; preservation of the culture of mutual cooperation: community service, community service, social service and/or other terms; restoration of ancestral graves; development of village community organization systems: development of community associations, development of community discussions; preservation of traditional arts groups; preservation of customs.

The percentage of local value implementation reached 100%. This confirms a number of factors for the sustainability of local wisdom in society are largely driven by the empirical experiences that individuals have gone through in applying such wisdom. These experiences often manifest in the form of practices and prohibitions, both of which are known to have a positive influence on the community.[34]. In local culture, there is often unique wisdom and values that have important meaning in society, so that there is independence in interpreting and realizing this local wisdom [35].

Meanwhile, the relevance of the authority of 184 authorities shows the dominant criteria as in table 2 below.

Table 2. Local Wisdom Based on Village-Scale Authority in Sekapuk Village

Local Wisdom Criteria	Total Value of Relatedness of Each Authority	Percentage Level of Relatedness
Local Knowledge	170	92.39%
Local Values	60	32.61%
Local Skills	128	69.57%
Local Resources	91	49.46%
Local Decision Making	164	89.13%

The percentage value of the relevance of the application of local knowledge reached 92.39% and Local Decision Making reached 89.13%. Therefore, local wisdom is important in maintaining and preserving a culture while safeguarding it from negative external influences. It encompasses the beliefs, values, and practices passed down through generations within a community, guiding them in navigating challenges and fulfilling their necessities [8].

The criteria for local wisdom that are fairly consistent in having a high percentage level of connection is local decision-making. It is only natural that the government is responsible for leading, providing direction and fostering an environment that effectively encourages progress and growth in all aspects.[18,36,37] This includes policy development, village administration, and efforts to reduce poverty [38,39].

3.3. Local Wisdom in the Use of Resources

Identification of resource wisdom is related to the portrait of the types of resources available in the village and the existence of resource management in the village. In terms of natural resources in the Sekapuk Village area, they include: (a) 148 Ha of agricultural/dry land whose productivity can still be increased because it has not been worked on optimally, (b) the existence of brick and dolomite mining which can be used as building materials, (c) the existence of former mining sites which can be used as village tourism areas, (d) the existence of bengkok land which can be used for managing plantation and livestock land as potential Agro Business Tourism, (e) the large amount of cow and goat manure, makes it possible to develop an organic fertilizer manufacturing business, and (f) the existence of peanut, corn, cassava and other harvests which have not been managed properly.[40] Sekapuk has potential development resources, with many facilities and infrastructure, both self-help and assistance. As for institutions, Sekapuk has groups in the village such as Village-Owned Enterprises, Karang Taruna, farmer groups, and religious groups.

3.3.1 Natural Resource Management

The first is the Limestone Mine. Sekapuk Village is known for its limestone mountains. This limestone mountain is managed by a company, namely PT. Polowijo Gosari. However, after the development of BUMDes, the village government requested that part of the limestone mountain be managed by the Village-Owned Enterprise with a profit-sharing system. The goal is to overcome unemployment and increase village income. There are around 80 plots of land on the limestone mountain that are managed by the Sekapuk Village-Owned Enterprise by renting them to workers. Later, for every white brick sold for an average of around IDR 2 thousand per piece, the Village-Owned Enterprise gets a profit share of IDR 75 per piece. This mining business unit is the second largest contributor to income after tourism. In

2020, the Village-Owned Enterprise's turnover from this mining unit reached IDR 1.86 billion and in 2021 it was IDR 3.62 billion and became a contributor to the Village-Owned Enterprise's income that year. Although in 2022 the income from this mining unit fell to IDR 1.68 billion due to the pandemic.

The second is village tourism. Setigi Tourism is the first village tourist spot built in 2019 and premiered in January 2020. This tourism has capital from bank loans and contributions from Sekapuk residents, Setigi tourism which was formerly a former mining excavation has now been transformed into a tourist spot that attracts a million charms. When it was first launched, this tourist spot which covers an area of 5 hectares was crowded with tourists. Then on February 2, 2022, this Village Tourism unit launched the second tourist spot, namely the "Pak Inggih" (KPI) Garden. The launch of this KPI deliberately took the 2nd of the month of 2 because it was the second tourist spot after Setigi. KPI is a former village land area of 2.5 hectares which is used as an asset of the Village-Owned Enterprise. In this garden, 17 cottages were built for accommodation and there are thousands of trees of various types. The construction of this KPI cost around IDR 7.8 billion in the initial stage and is still being improved. Most of the capital comes from the Village-Owned Enterprises, as well as Sekapuk residents through share ownership. This Tourism Unit is the largest contributor to the income of the Sekapuk Village-Owned Enterprises. In 2020, the village tourism unit contributed a turnover of 5.07 billion. Then in 2021, the village tourism income fell to IDR 3.26 billion and became the second largest contributor after the mining unit. Then in 2020 when the pandemic arrived, this village tourism unit actually became the largest contributor to village income compared to the other five units, namely IDR 2.84 billion. Although the value is smaller, this Village Tourism unit is quite effective in maintaining income. This is inseparable from the hard work of the Sekapuk village government which continues to promote their village tourism even in the midst of restrictions imposed by the government.

3.3.2 Development Resource Management

The first development resource management is Community Water Management (PAM). This PAM unit serves the clean water needs of Sekapuk residents. The Sekapuk Village-Owned Enterprise charges a rate of IDR 2.5 thousand per cubic meter of water. This PAM unit contributed a turnover of IDR 340.26 million in 2020 and increased to IDR 393.17 million in 2021 and jumped again to IDR 433.22 million in 2022. This PAM unit is one of the business units that has consistently recorded revenue growth in the last three years after the Multi Jasa unit for the Sekapuk Village-Owned Enterprise.

Second, is the Village Cleaning Unit (UKD). This business unit is specifically designed to manage community waste which is one of the main problems in Sekapuk Village. After the formation of this UKD, waste can finally be managed well and even provide additional income for the local community. Through this UKD, waste is collected, sorted and sold. The Village Government also promoted the slogan "waste is exchanged for gold" in the hope that the community will be motivated to process waste. Through this program, UKD buys waste from the community, where the payment is in gold. The community does not directly get money from the sale of their waste, but it is saved until the value of the savings is equivalent to the price of one gram of gold. This UKD has employees who collect waste from the community every day and take it to the shelter. The community is asked to pay IDR 15 thousand per month for the transportation of the waste. But for non-organic waste, UKD buys it from residents who are coordinated by 32 neighborhood associations. Residents are also encouraged to save so that they can only get the proceeds from the sale after the value is enough to buy one gram of gold. That is why this program is called waste exchanged for gold. This UKD generated a turnover of IDR 81.65 million in 2020 and increased significantly to IDR 119.39 million in 2021, although it fell again to IDR 91.59 million in 2022.

The third development resource management is Integrated Sports Infrastructure (Usport). This land and sports facilities is also expected to improve Human Resources, achievements and health. Initially, this land was not managed properly even though it had been fenced around and had stands. After being taken over by the Sekapuk Village-Owned Enterprise, this land was again used to support the Village economy by renting out this sports facility. Usport has not yet provided significant income to the Sekapuk Village-Owned Enterprise. In 2021, Usport began to provide income of IDR 2.06 million

IOP Conf. Series: Earth and Environmental Science **1473** (2025) 012067 doi:10.1088/1755-1315/1473/1/012067

and increased in 2022 to IDR 7.76 million. The Sekapuk Village-Owned Enterprise does not simply seek profit from managing this facility, but also encourages the public to maintain their health by utilizing this sports facility. Moreover, this facility is free for Sekapuk residents.

Fourth is the Multi-Service Sector. This business provides loans for micro-businesses, banking services including savings and loans and tax payments. This Multi-Service Unit is in the same office as the Village-Owned Enterprise and sees a number of residents carrying out transaction and payment activities served by two employees. In this business unit, the people of Sekapuk Village can transfer money to the bank, withdraw cash, send documents and letters, pay for credit, pay taxes, and save. This multi-service unit records a turnover of up to IDR 800 million per month. This Multi-Service Unit recorded a turnover of IDR 136.29 million in 2020 and increased to IDR 169.20 million in 2021 and reached 177.98 million in 2022.

The management situation in Sekapuk village is in accordance with the development of an entrepreneurial style government system which is an urgent need for today's government, especially for village governments and communities who have long been left behind in development [41].

3.3.3 Social Resource Management

Social resource management is self-funding. The Sekapuk community saves in the Village-Owned Enterprise without any administrative fee deductions, starting from IDR 8 thousand per day. The tourism project that has been designed in Sekapuk has also begun to be implemented with a budget of IDR 2.4 billion to build Setigi tourism. Community participation in building Setigi is by becoming shareholders. A total of 1 thousand shares were issued with a nominal value of IDR 2.4 million per share. This is the support of residents who absorb 400 shares. While the other 600 shares are owned by the Village-Owned Enterprise as the controller. The community receives a certificate of proof of ownership of Setigi shares, with a promise of a return in the form of a dividend of 10% per year. The local wisdom in utilizing the results of the analysis is shown in table 3 below. Local wisdom from the results of the study shows that it is running optimally, with the existence of forms of management of natural resources, humans, development and social which provide economic benefits for local residents.. Empowerment of community institutions as partners of village government at the level of community participation. This right is supported by the role of local institutions as facilitators, planners, implementers, and controllers at every level of government [42].

Table 3. Local Wisdom in the Use of Resources.

Local Wisdom Criteria	Indicator	Findings
Local Knowledge	Utilization of resources is related to local knowledge such as geography, demographics, and sociography.	There are natural, human, development and social resources that are understood by the community and summarized in planning documents including the Village Medium-Term Development Plan
Local Values	The use of resources is related to local values in society such as prohibitions and local rules set by the village government	Use of resources determined based on village government regulations
Local Skills	The use of resources that encourage and accommodate local skills, one of which is the potential for home industry	The potential of home industries, namely Village-Owned Enterprises, Mining Units, Dapur Mbok Inggih with processed food products
Local Resources	Utilization of resources that can optimize the existence of biological resources, water and housing	Business units are developing and management institutions such as community groups and Drinking Water Users Associations are growing
Local Decision Making	That encourages a democratic climate through locally regulated decision-making mechanisms that are implemented jointly and in a hierarchical manner	Institutional arrangements and inter-institutional relations.

3.4. Local Wisdom in Supervision of Village Government Implementation

Identification of village government supervision policies in Sekapuk Village related to the evaluation of village government performance. The results of the field study show that all development activities and target achievements will always be evaluated routinely and involve the community.. The monitoring, evaluation and accountability are carried out using approaches including (a) evaluating the process of implementing activities both physically, costly and administratively, (b) evaluating the achievement of activities physically (volume and quality), (c) evaluating the achievement of targets and impacts and (d) evaluating the preservation and sustainability of activities.

The form of monitoring and evaluation that can be applied later is joint monitoring by the community and the Village Consultative Institusion (BPD). This is a unique phenomenon and contradicts Putri's (2022) opinion that Local wisdom can impede the implementation of reporting violations, potentially leading to a decrease in the quality of information provided to stakeholders. This can affect accountability and transparency efforts in striving for good governance [43].

Table 4. Local Wisdom in Supervision of Village Government Implementation.

Supervision Aspects	Findings
Application of Local Knowledge	Village Consultative Institusion as the organizer of supervision
Application of Local Values	The community through the Neighborhood Association and Citizens Association in monthly meeting forums evaluates the performance of village government administration.

All development activities and target achievements will always be evaluated routinely and involve the community (participatory). The existence of this community involvement is shown by the activities of the Sekapuk Village Government routinely holding joint deliberation meetings with the Neighborhood Association, Citizens' Association, Village-Owned Enterprises and the Karang Taruna Youth Group, in these activities the aim is to collect community aspirations, accommodate all problems in the village and at the same time find joint solutions, as in the following visualization in figure 1.

Figure 1. Monthly Community Meeting with the Sekapuk Village Government.

Submission and evaluation of the performance of community representatives, namely each RT, RW, and Village Government through the forum. so that if there are obstacles, effective solutions can be determined in a participatory and communal manner. Furthermore, the results of the analysis regarding local wisdom in supervising the implementation of the Village Government can be presented in the following table 4.

The wisdom of supervision in Sekapuk Village is not only in formal institutions that are formed with a supervisory function (BPD), but participatory supervision is also given space through forums so that the climate of transparency in the performance of government implementation can be conducive and develop. This is also driven by democratization in the village. As Patterman argues that participating in democracy not only results in well-informed and engaged citizens, but also contributes to building stronger communities and improving governance overall.[44] Grassroots democracy is based on the idea that political power and the process of making decisions should be distributed and available to regular individuals. This allows them to play a direct role in determining how their communities operate [45].

3.5. Influential factors

The first driving factor in Sekapuk Village is the leadership aspect. The existence of a vision and mission that is applied, local citizen figures, good communication patterns with residents, strong commitment, a spirit of sacrifice, and innovative development ideas are inherent in the leadership of the Head of Sekapuk Village. The second is natural resources by the existence of supporting natural resources. The existence of limestone mines and extensive village land support the management of village assets that contribute to the local economy.

The third is the aspect of development resources. The existence of infrastructure through clean water facilities that support the formation of the Drinking Water Users Association, waste facilities for the Village Cleaning Unit, the institution of the Village-Owned Enterprise with 6 business units further strengthens the governance of village governance that is oriented towards improving community welfare. While the last is the aspect is participations.. The Sekapuk community saves in the Village-Owned Enterprise without any deduction of administration fees. Community participation in building Setigi is by becoming shareholders.

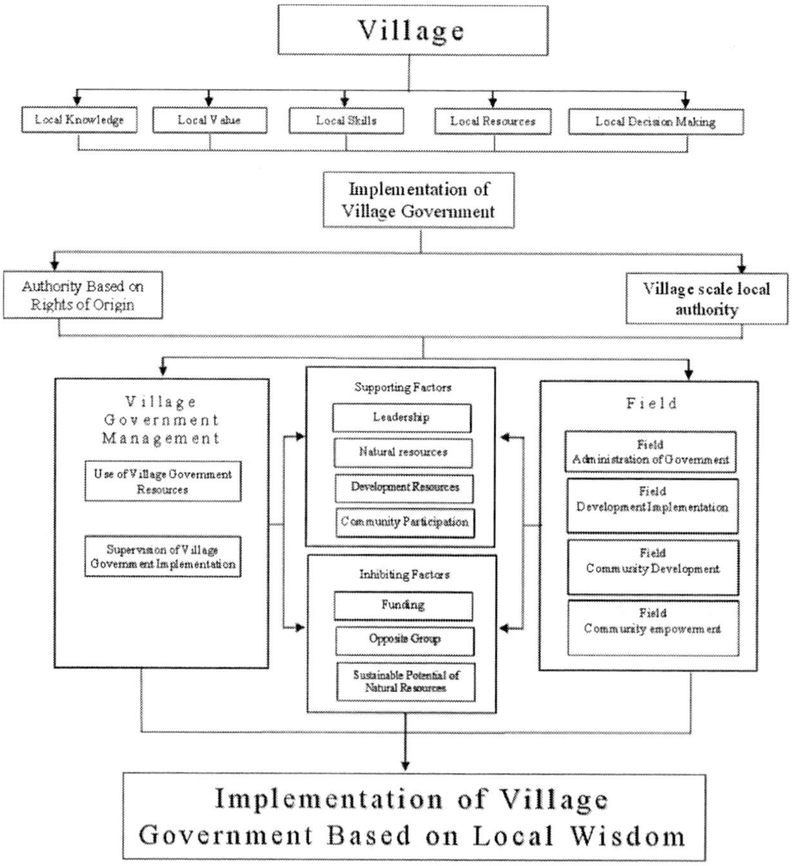

Figure 2. Village Government Model Based on Local Wisdom

Meanwhile, the first inhibiting factor is funding. Efforts to realize development programs are faced with large budget needs. Therefore, funding innovation is needed to support the development of village government programs, especially in economic development and tourism. Second is the opposing group. There is a small group of residents who are opposed to the village government, and have the potential to trigger conflict. Lime mines that have the potential to run out in the future. The management of village government which begins with planning, organizing the use of resources and its supervision is carried out transparently and participatory by the local community. The visualization model is as follows in figure 2.

This study confirms the great role of village head leadership. In addition, the existence of resources, both natural resources and development resources, also determines what is supported by community participation in priority and superior development programs in their areas.

4. Conclusion

The findings of this research suggest that the implementation of village governance based on local wisdom in the study area is able to operate efficiently. This is evident from the careful development of village development plans, Village Medium-Term Development Plans, Village Government Work Plans, and revenue and expenditure budgets that incorporate local knowledge, values, skills, resources, and decision-making processes. Furthermore, the village institutions, their roles and responsibilities, and the patterns of institutional relationships have been established according to regulations that authenticate their presence. The local wisdom in the study area has effectively utilized the village government resources in a manner that is independent, participatory, and forward-thinking.

References

[1] Silubun AJ, Kalalo JJJ, Inggit AB, Kalalo CN, Rahail EB. Village authority and position in realizing village autonomy. IOP Conf Ser Earth Environ Sci, **473**, Institute of Physics Publishing; 2020. https://doi.org/10.1088/1755-1315/473/1/012032.

[2] United Nations Development Programme B for DP. UNDP project on Pro-Poor and Gender Sensitive Governance Indicators for Policy Reform Framework for Piloting. **2005**.

[3] Chhotray V, Stoker G. Governance in Development Studies. In: Chhotray V, Stoker G, editors. Governance Theory and Practice: A Cross-Disciplinary Approach, London: Palgrave Macmillan UK; **2009**, p. 97–119. https://doi.org/10.1057/9780230583344_5.

[4] Tagivakatini S, Ferreira C, Matos F. Village governance in relation to coastal resource management in Fiji: a case study of Namada and Navukailagi villages. GOT - Journal of Geography and Spatial Planning **2016**:355–79. https://doi.org/10.17127/got/2016.10.017.

[5] Febriyanti DP, Ramadhan IN, Fitriyah L, Kusuma AD, Vianisa Y, Mudhoffar DR. The Implementation of Standard Operational Procedures for Village-Business Entity (BUMDes) in Hippa Rice Fields Irrigation to Support SDGS 2030, **2023**, p. 345–54. https://doi.org/10.2991/978-2-38476-152-4_32.

[6] Tri Haryanto J. Local Wisdom Supporting Religious Harmony in Tengger Community, Malang, East Java, Indonesia. Jurnal "Analisa" 2014;**21**:201–13.

[7] Mitchell B. Resource and Environmental Management: Third EditionThird Edition. **2019**. https://doi.org/10.1093/oso/9780190885816.001.0001.

[8] Putra TB, Thamrin, Saam Z, Husein S. Management of Bioculture Potential with Environmental Perspective Based on Local Wisdom. Journal of Environmental Management and Tourism 2023;**14**:2345–53. https://doi.org/https://doi.org/10.14505/jemt.

[9] Arismayanti NK. Community Local Wisdom and Efforts to Create Quality Marine Tourism. Webology 2021;**18**:243–60. https://doi.org/10.14704/web/v18i2/web18319.

[10] Luriawati Naryatmojo D. Internalization the Concept of Local Wisdom for Students in the Listening Class. Arab World English Journal 2019;**10**:382–94. https://doi.org/10.24093/awej/vol10no1.31.

[11] Nofiyanti F, Zulyanti Nasution D, Octarina D, Agie Pradhipta RMW. Local Wisdom for Sustainable Rural Tourism: The Case Study of North Tugu Village, West Java Indonesia. E3S Web of Conferences, vol. 232, EDP Sciences; **2021**. https://doi.org/10.1051/e3sconf/202123202031.

[12] Kamal MA, Utomo MR, Lukman HAKIM M, Dwi QURBANI I, Dzil IKRAM A. Peatland Management Based on Local Wisdom Through Rural Governance Improvement and Agroindustry. **4**. 2023.

[13] Malau H, Supriyono B, Hakim A, Muluk MRK. The Implementation of Customary Values on The Administration of Governance : A Studi on Village Governance Base on Local Wisdom at Nagari Kapau Agam Regency of West Sumatra Indonesia. Russ J Agric Socioecon Sci 2020;**98**:49–58. https://doi.org/10.18551/rjoas.2020-02.07.

[14] Arifin R, Muhtada D, Sri Utari I. Preserving Local Wisdoms for Village Good Governance Management in Industry 4.0 Era (Study of North Coastal Area, Central Java, Indonesia). Proceedings of First International Conference on Culture, Education, Linguistics and Literature, CELL **2019**, 5-6 August, Purwokerto, Central Java, Indonesia 2019. https://doi.org/10.4108/eai.5-8-2019.2289816.

[15] Rodiyah W&. Acceleration Of Acquisition Of Copyright Through Village-Owned Enterprises (BUMDES) In Indonesia (Study In Kedungkelor Village, Warureja, Tegal Regency, Central Java). South East Asia Journal of Contemporary Business, Economics and Law n.d.;**18**.

[16] Rodiyah R. The Accelerated Model of The Right to Work Through The Harmonization of Man Power Policies (A Case Study of Outsourcing Manpower after the Constitutional Court Decision Number 27/PUU-IX/2011 about the Review of Law Number 13 of 2003 Concerning Substantive Justice Attainment). South East Asia Journal of Contemporary Business, Economics and Law 2014;**4**.

[17] Helliwell JF, Huang H, Grover S, Wang S. Empirical linkages between good governance and national well-being. J Comp Econ 2018;**46**:1332–46. https://doi.org/10.1016/J.JCE.2018.01.004.

[18] Anugerah Ash-shidiqqi E, Wibisono Ellectrananda Anugerah Ash-shidiqqi H, Wibisono H. Corruption and Village: Accountability of Village Fund Management on Preventing Corruption (Problems and Challenges). Journal of Indonesian Legal Studies 2018;**3**:2548–1584.

[19] Anggun MM. BPD Empowerment Program in Optimizing Village Financial Implementation (Perspective Law No. 6 of 2014). Journal of Indonesian Legal Studies 2018;**3**.

[20] Arifin R. Democracy on Indonesian Legal Reform: How Can People Participate on Laws and Regulations Establishment Process. Journal of Indonesian Legal Studies 2017;**2**,Isuue 02:155–8.

[21] Kurniawan JA. Contested land, contesting laws. A context of legal pluralism and industrialization in Indonesia. Sortuz: Oñati Journal of Emergent Socio-Legal Studies 2014;**6**:93–106.

[22] Doli Muhammad Ja B, Dalimunthe far, Muda I, Muhammad Ja D, Dalimunthe α far, σ F, et al. Study on the Impact of Government Complexity and Regional Government's Size on Human Development Index in North Sumatera, Indonesia. Global Journal of Management and Business Research: A Administration and Management 2014;**14**.

[23] Wismadi A, Brussel M, Zuidgeest M, Sutomo H, Nugroho LE, van Maarseveen M. Effect of neighbouring village conditions and infrastructure interdependency on economic opportunity: A case study of the Yogyakarta region, Indonesia. Comput Environ Urban Syst 2012;**36**:371–85. https://doi.org/10.1016/J.COMPENVURBSYS.2012.02.001.

[24] Widodo I. Politik Indonesia Indonesian Political Science Review Dana Desa dan Demokrasi dalam Perspektif Desentralisasi Fiskal. Indonesian Political Science Review 2017;**2**:65–85. https://doi.org/https://doi.org/10.15294/jpi.v2i1.8487.

[25] Mulda R. MANAJEMEN PEMERINTAHAN DESA: STUDI KASUS PENYELENGARAAN PEMERINTAHAN DESA MEKKATTA, KABUPATEN MAJENE. Jurnal Trias Politika 2020;**4**:214–28.

[26] Alvayedo MB, Erliyana A. Tinjauan Hukum Kedudukan Dan Keterlibatan Kearifan Lokal Masyarakat Adat Maluku Berupa Sasi Dalam Pengelolaan Lingkungan Hidup. Jurnal Ilmu Sosial Dan Pendidikan (JISIP) **2022**. https://doi.org/10.36312/jisip.v6i3.3220/http://ejournal.mandalanursa.org/index.php/JISIP/index.

[27] Rahman Haulussy R, Idris R, Dody May Putra Agustang A. The Sustainability Of The Sasi Lola Tradition And Customary Law (Case Study In Masawoy Maluku, Indonesia). INTERNATIONAL JOURNAL OF SCIENTIFIC & TECHNOLOGY RESEARCH 2020;**9**:2.

[28] Putra F. Village Governance and Public Participation in Indonesia. Journal of Interdisciplinary Socio-Economic and Community Study 2023;**3**:55–64. https://doi.org/10.21776/jiscos.03.2.02.

[29] De Nisa G. Community Participation in Village Development Planning: A Comparative Study in Bakambat and Cindai Alus Villages. Journal of Transformative Governance and Social Justice J-TRAGOS 2024;**2**:25–41. https://doi.org/10.26905/jtragos.v2i1.11240.

[30] Wahyu D. Dulu Miskin dan Tertinggal, Begini Cara Desa Sekapuk Kembangkan Ekonomi Masyarakat Hingga Jadi Desa Miliarder - UKMINDONESIA.ID. UkmindonesiaId 2023. https://ukmindonesia.id/baca-deskripsi-posts/dulu-miskin-dan-tertinggal-begini-cara-desa-sekapuk-kembangkan-ekonomi-masyarakat-hingga-jadi-desa-miliarder (accessed August 25, 2024).

[31] Laoli N. Mengenal Desa dengan Pergaulan Tingkat ASEAN. Noverius Laoli 2023. https://jelajahekonomi.kontan.co.id/ekonomidesa/news/mengenal-desa-dengan-pergaulan-tingkat-asean (accessed August 25, 2024).

[32] Halim A. Laporan Penyelenggaraan Pemerintahan Desa Akhir Masa Jabatan Kepala Desa Sekapuk Periode 2018-2023 **2023**. https://drive.google.com/file/d/1ctgg_HF5aZAU3ck6ZEqHqdU_EWWjn-Qn/view (accessed August 25, 2024).

[33] Arieffiani D, Jati Utama S. Implementation of Village Government Based on Local Wisdom in Planning and Organization Perspective. International Journal of Scientific and Management Research 2023;**06**:01–15. https://doi.org/10.37502/ijsmr.2023.6901.

[34] Hutagalung SS, Indrajat H. Adoption Of Local Wisdom In Disaster Management In Indonesia. International Journal Of Scientific & Technology Research 2020;**9**:48–52.

[35] Parameswara A, Wulandari A. Sustaining Local Communities through Cultural Industries Based on Local Wisdom in Tigawasa Village. J Sustain Dev 2020;**13**:139. https://doi.org/10.5539/jsd.v13n6p139.

[36] Lewis BD. Decentralising to Villages in Indonesia: Money (and Other) Mistakes. Public Administration and Development 2015;**35**:347–59. https://doi.org/10.1002/PAD.1741.

[37] Antlöv H, Wetterberg A, Dharmawan L. Village Governance, Community Life, and the 2014 Village Law in Indonesia. Bull Indones Econ Stud 2016;**52**:161–83. https://doi.org/10.1080/00074918.2015.1129047.

[38] Setyawan Jhody P, Puguh Setyawan Jhody R. Poverty Reduction in Perspective of Public Service Reform: A Study on Legal and Social Analysis (Case of Sragen, Indonesia). Journal of Indonesian Legal Studies 2017;**2**:131–44.

[39] Miranti R, Duncan A, Cassells R. Revisiting the Impact of Consumption Growth and Inequality on Poverty in Indonesia during Decentralisation. Bull Indones Econ Stud 2014;**50**:461–82. https://doi.org/10.1080/00074918.2014.980377.

[40] Anonymous. Peraturan Kepala Desa Sekapuk No. 02 Tahun 2016 Tentang Rencana Pembangunan Jangka Menengah Desa (RPJM Desa) Desa Sekapuk Tahun 2018 - 2023 **2018**. https://www.scribd.com/document/390351007/2-RPJMDes (accessed August 25, 2024).

[41] Yayat Rukayat. KOMBINASI RESIPROKAL ASAS REKOGNISI DAN SUBSIDIARITAS PEMERINTAHAN DESA DENGAN KONSEP PEMERINTAHAN BERGAYA WIRAUSAHA. Jurnal Academia Praja 2021;**4**:180–200.

[42] Angelia N, Batubara BM, Zulyadi R, Hidayat TW, Hariani RR. Analysis of Community Institution Empowerment as a Village Government Partner in the Participative Development Process. Budapest International Research and Critics Institute (BIRCI-Journal): Humanities and Social Sciences 2020;**3**:1352–9. https://doi.org/10.33258/birci.v3i2.991.

[43] Putri PPS. Local Wisdom: A Lesson Learned in Village Governance. International Journal of Religious and Cultural Studies 2022;**4**. https://doi.org/10.34199/ijracs.2022.04.05.

[44] Pateman C. Participatory Democracy Revisited. Perspectives on Politics 2012;**10**:7–19. https://doi.org/10.1017/S1537592711004877.

[45] Nguyen KNV. The Role of Implementing Grassroot Democracy in Vietnam Currently. Social Science and Humanities Journal 2024;**8**:34748–54. https://doi.org/10.18535/sshj.v8i03.990.

IOP Conf. Series: Earth and Environmental Science **1473** (2025) 012068 doi:10.1088/1755-1315/1473/1/012068

Career Aspiration Among Undergraduate Students Majoring in Marine and Maritime Sciences

W Sulistiani[1*] and D W Untari[1]

[1] Department of Psychology, Universitas Hang Tuah, Surabaya, Indonesia

*E-mail: wiwik.sulistiani@hangtuah.ac.id

Abstract. Career aspiration is a primary career development task for individuals to find a career that matches their self-concept. Therefore, students need to set career goals according to their ideal situation. This study aimed to provide an overview of career aspirations, examining the comparison of career aspirations between male and female, level education undergraduate students. The participants in this study were 192 undergraduate students majoring in fisheries and marine scattered in various universities in East Java. Data were collected using a career aspirations scale. The findings in this study indicate that students have the following career aspirations: 71.9% were in the high category, 27.6% were moderate and 0.5% were low. Male students have higher career aspirations compared to female students. Vocational students have higher career aspirations compared to undergraduate students. The research findings have implications for educators to consider contextual variables in designing career interventions for university students.

1. Introduction

Indonesia, as the largest archipelagic country in the world, has enormous marine and maritime potential. The economic value of Indonesia's marine resources is estimated to be $8.22 trillion per year [1], making it a crucial sector in the nation's economic development. The country's vast waters and wealth of marine natural resources have made the marine and maritime sector one of the mainstays of its national economy. To strengthen its maritime identity and global role, the government has implemented policies such as the global maritime Fulcrum [2]. The Indonesian government has also designated this sector as a priority in its sustainable development efforts, with the goal of making Indonesia the world's maritime axis. In light of this, it is essential to have a workforce with high competence and a strong career orientation in the marine and maritime sector.

However, in reality, the appeal of maritime careers is still relatively low, and promotional efforts and the design of educational programs in this field need to be improved [3]. Likewise, undergraduate students who have taken up this field have career aspirations in maritime and marine fields that are still less than optimal. Many students do not have a clear vision of their career prospects after graduation, which can affect their motivation to learn and their involvement in relevant academic and non-academic activities. Understanding student motivation in marine and maritime education is important [3]. The study found that aligning educational programs with students' career expectations can increase their aspirations and commitment to maritime and marine fields. High and focused career aspirations are fundamental for students to maximize their potential and contribute significantly to the future development of the maritime and marine sectors.

Career aspiration plays an important role in building future careers. Previous studies have shown that clear career aspirations can guide students in achieving academic achievement, improving employability, and adaptability in the workplace. Students with clear career aspirations tend to have a clear vision of their goals, thus helping them make informed decisions regarding their education and career paths [4]. Career aspirations are very important because one's career achievement is greatly influenced by aspirations in pursuit of the desired career path [5]. In addition, career aspirations are closely related to higher levels of motivation, which lead to increased academic performance and a stronger determination to succeed in their chosen fields. Research conducted by Creed et al. also supports this: individuals with high career aspirations tend to be more skilled at planning their careers and exploring various aspects related to their careers and abilities [6].

Career aspirations and individuals are influenced by various factors, including gender, parental educational background, socioeconomic status, parental occupation, educational level, parental expectations, and other aspects [7]. Based on the description above, research on career aspirations in students studying marine and maritime studies is important to improve the quality of human resources in the marine and maritime sector in Indonesia. Currently, this study aims to describe the career aspirations of students majoring in marine and maritime studies in Indonesia.

2. Methods

2.1. Participants

The participants in this study were 192 students studying marine and maritime science, including study programs in shipbuilding engineering, oceanography, fisheries, fisheries cultivation engineering, and fisheries agribusiness, in several cities in East Java (50.5% male and 49.5% female, average age 20.3 years, 60 undergraduate students and 132 vocational students). Convenience sampling was used in this study.

2.2. Measures

This study used the Career Aspirations Scale developed by O'Brien [8] and adapted by Sawitri et al. [9]. This scale reveals the career goals of individuals, which consist of career achievement, namely the individual's strong desire to achieve a higher career in the career field being pursued; leadership, namely the strong desire to become a leader, including the ideal of becoming a leader and carrying out the role of a leader in the career field being pursued; and further education, namely the individual's interest in continuing studies that will support their career field in the future. Participants were asked to answer 10 career aspiration statements using a 6-point Likert scale, ranging from 1 (strongly disagree) to 6 (strongly agree). An example item is "When I am established in my career, I want to provide training to others". Previous research [9] has reported a reliability coefficient of 0.76.

2.3. Procedure

The data in this study were collected using a scale that was filled in by the participants through a link. Before filling out the scale, participants were given information about the purpose of the study and asked to provide informed consent stating their willingness to participate in this study. Completing the scale took approximately 15 minutes.

2.4. Statistical Analysis

Statistical analysis was performed using descriptive statistics to calculate the mean, frequency, percentage, and standard deviation. The category of career aspirations of students was calculated by comparing hypothetical and empirical means. The hypothetical mean is the midpoint in the measurement scale, whereas the empirical mean is the mean score obtained from the research sample [10]. Categorization based on the hypothetical mean assesses the participants' actual conditions and may result in proportions that do not follow a normal distribution [11]. Frequency measurements were used to determine how many students were in the high, medium, and low categories. The authors ranked each indicator of career aspiration in students based on calculations using the comparison of total item scores and total items on each indicator. Meanwhile, to see differences in career aspiration by gender and level of education, the analysis used the Mann-Whitney U test.

3. Result and Discussion

Table 1. Score category.

	Category	Frequency	Percent	Valid Percent	Cumulative Percent
Career Aspiration	Low	1	.5	.5	.5
	Moderate	53	27.5	27.6	28.1
	High	138	71.9	71.9	100
Leadership	Low	3	1.6	1.6	1.6
	Moderate	54	28.1	28.1	29.7
	High	135	70.3	70.3	100
Career Achievement	Low	0	0	0	0
	Moderate	85	44.3	44.3	44.3
	High	107	55.7	55.7	100
Educational Aspiration	Low	2	1	1	1
	Moderate	77	40.1	40.1	41.1
	High	113	58.9	58.9	100

Based on the results of the analysis, the participants in this study mostly had career aspirations in the medium and high categories. In detail, 71.9% of the participants had career aspirations in the high category, 27.5% in the medium category, and 0.5% in the low category. This finding shows that not all students who take marine and maritime education have career aspirations in the higher category. When viewed from the career aspiration dimension, the results for the leadership dimension were 70.3% of students in the high category, 28.1% of students in the medium category and 1.6% of students in the low category. In the career achievement dimension, 55.7% of students were in the high category, 44.3% of students were in the medium category, and no students had career achievements in the low category. In the further education dimension, 58.9% of students have a high desire to continue their studies to a higher level, 40.1% have a desire for further study in the medium category, and 1% have a desire for further study in the low category.

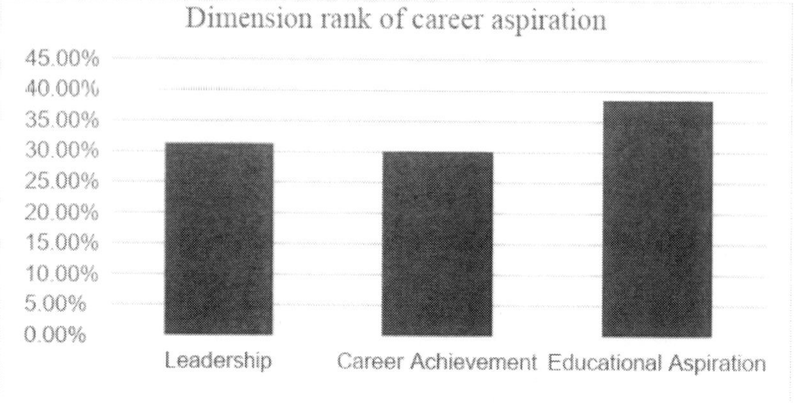

Figure 1. Dimension rank of career aspiration.

Based on figure 2, it can be seen the rankings of the career aspiration dimensions. The dimension with the highest ranking is educational aspiration, at 38.56%. The second-ranked dimension is leadership, at 31.35%, and the career achievement dimension is ranked third. This indicates that the majority of subjects feel that enhancing skills through further education or certified courses is important.

Table 2. Results of comparison between the Empirical Mean and Hypothetical Mean

Career Aspiration	Empirical Mean				Hypothetical Mean			
	Min	Max	Mean	SD	Min	Max	Mean	SD
	27.00	60.00	47.7083	7.94539	10	60	35	8.3333

According to the comparison of the empirical and hypothetical means shown in Table 2, the empirical mean has a greater value than the hypothetical mean. This finding shows that career aspirations of students who take marine and maritime fields are high. The results of the comparison of standard deviation values obtained a higher hypothetical standard deviation value than the empirical standard deviation. This implies that the career aspirations of students in the marine and maritime fields have low variation. Simply put, the participants have the same level of career aspiration.

The results showed that most students had career aspirations in the medium-to-high category. Similarly, in each dimension of career aspirations. These results indicate that participants in this study had a "strong" level of career aspirations towards marine and maritime fields. Students are seriously concerned about their current career choices. Students feel proud and have a positive view of their chosen career. Previous studies have found that individuals with high career aspirations tend to have clearer career goals; therefore, they are more motivated to act, adapt, and engage in activities that support these career aspirations. This condition directs individuals' self-perceived employability of individuals [9]. Individuals with strong career aspirations know what they will do in the future [12]. This will make individuals more adaptive to their careers so that they will feel more life satisfaction [13]. When viewed from the existence of the country, Indonesia is an archipelago with around 13,000 islands. Indonesia has a coastline of approximately 81,000 km, making it the third-largest coastline in the world. With such an area, Indonesia has great potential to develop maritime resources. Marine fisheries, inland waters, and the maritime capture and cultivation sector still have great opportunities to continue to be developed [14]. Opportunities in the fish processing industry include the following: 1) the market for processed aquatic products, both domestic and international, is still diversified; 2) The Indonesian government strongly supports sustainable development in the fish processing industry sector; 3) consumer demand for ready-to-eat processed foods continues to increase; 4) there is a very large potential supply of raw materials; and 5) abundant labour is available [15].

Table 3. Wilcoxon signed-rank test results based on participants' gender

Group	N	Mean Rank	Sum of Ranks	z	P
Male	97	109.85	10655.50	-3.373	.001
Female	95	82.87	7872.50		

Based on Table 3, the test results of differences in career aspirations in terms of gender obtained a p-value of 0.001 <0.05, meaning that there are differences in the career aspirations of students taking marine and maritime education based on gender. The mean rank value of male is greater than the mean rank value of females (109.85> 82.87), this means that male have higher career aspirations than female.

This study showed significant differences in career aspirations between male and female students. The results showed that males had higher career aspirations than females. These results are in line with previous studies that reported differences in career aspirations based on gender [16] [17]. Barrett's research results [17] found that gender plays a major role in career aspirations. Boys tend to be interested in careers related to mechanics (engineering, construction, and mechanics), whereas female prefer jobs related to humanity (teachers, psychologists, and nurses). In addition, there tended to be no difference between boys and girls in careers related to science.

The results of the study show that students who take marine and maritime fields have higher career aspirations, which could be because shipping and maritime majors are sciences that tend to be more towards mechanics and science. Some researchers argue that even though there are families who try to create gender equality, where husbands and wives work together to earn a living, gender inequality still exists [18]. After marriage, male tend to spend more time taking care of their houses and their children [18]. Additionally, in Indonesian culture, male have generally more dominant in interpreting gender roles in everyday life.

Table 4. Wilcoxon signed-rank test results based on participants' level of education.

Group	N	Mean Rank	Sum of Ranks	z	P
Undergraduate Students	60	84.59	5075.50	-2.007	.045
Vocational Students	132	101.91	13452.50		

Based on Table 4, the results of the test for differences in career aspirations in terms of education level obtained a p-value of 0.001 <0.05, meaning that there are differences in the career aspirations of students taking marine and maritime education based on education level. The mean rank value of vocational students is greater than the mean rank value of undergraduate students (101.91 > 84.59), this means that vocational students have higher career aspiration than undergraduate students.

This study also found significant differences in career aspirations based on education level. Vocational students have higher career aspirations than undergraduate students. This may be due to vocational students having clear career goals. Vocational students have a strong vocational self-concept that increases their career readiness [19].

The results of open-ended questions related to factors that support and inhibit career aspirations in students in the marine and maritime fields obtained results including education and skills in the marine and maritime fields, parents who have jobs in the same field, family economic status, friends, and boyfriends. The results of this study revealed factors that support and inhibit career aspirations in students in the marine and maritime fields, including education and skills in the marine and maritime fields, parents who have jobs in the same field, family economy, friends, and boyfriends. These results are in accordance with previous studies that stated that career aspirations are influenced by gender and parental income [20].

4. Conclusion

This study uses descriptive analysis. This study found that the career aspirations of students who take marine and maritime education are in the medium and high categories, although there are still some students who are in the low category. The results of the analysis per dimension show that in the leadership dimension, most students are in the high and medium categories. In the career achievement dimension, most students are in the high and medium categories. Furthermore, in the further education dimension, the results are not much different, most students have a high desire to continue their studies. The results of the study also found that male students have higher career aspirations than female students. Vocational students have higher career aspirations than undergraduate students. Future researchers can use the results of this study as a basis and reference for future research involving career aspirations in students who take marine and maritime education.

Acknowledgment

This research was supported by the Faculty of Psychology and LPPM Hang Tuah University. We want to thank the Dean of the Faculty of Psychology, Ka. LPPM Hang Tuah University and Rector of Hang Tuah University. We also thank the research subjects who have greatly helped us.

References

[1] H. Hastuti, A. Muhidu, R. Rastin, and E. Agus Mokodompit, "Indonesia's Marine Economic Potential As A Maritime Country: Marine Economy," *International Journal of Science, Technology & Management*, **4**, no. 4, pp. 813–825, Jul. 2023, doi: 10.46729/ijstm.v4i4.897.

[2] Briliantina Putri, "Approaches in the Indonesian Maritime identity Construction," *JGSS*, **3**, no. 2, Dec. 2023, doi: 10.36859/jgss.v3i2.1795.

[3] Y.-Y. Lau, C. Dragomir, Y.-M. Tang, and A. K. Y. Ng, "Maritime Undergraduate Students: Career Expectations and Choices," *Sustainability*, **13**, no. 8, 2021, doi: 10.3390/su13084297.

[4] Bhanu Pratap Singh and Bharti Dogra, "A Study of Career Aspirations of College Going Students Pursuing Academic and Professional Undergraduate Courses," pp. 37–44, 2024, doi: https://doi.org/10.69758/GIMRJ2406IIV12P005.

[5] R. D. Febriani, A. M. Yusuf, and M. Iswari, "Perbedaan aspirasi karier siswa ditinjau dari jenis kelamin, jurusan, dan tingkat pendidikan orangtua serta implikasinya terhadap pelayanan bimbingan dan konseling," *Konselor*, **5**, no. 3, pp. 160–171, 2016.

[6] P. Creed, N. Buys, C. Tilbury, and M. Crawford, "The relationship between goal orientation and career striving in young adolescents," *Journal of Applied Social Psychology*, **43**, no. 7, pp. 1480–1490, 2013.

[7] M. A. Al-Bahrani, S. M. Allawati, Y. A. Abu Shindi, and B. S. Bakkar, "Career aspiration and related contextual variables," *International Journal of Adolescence and Youth*, **25**, no. 1, pp. 703–711, Dec. 2020, doi: 10.1080/02673843.2020.1730201.

[8] K. M. O'Brien, "The influence of psychological separation and parental attachment on the career development of adolescent women," *Journal of Vocational Behavior*, **48**, no. 3, pp. 257–274, 1996, Accessed: Sep. 13, 2024. [Online]. Available: https://www.sciencedirect.com/science/article/pii/S000187919690024X

[9] D. R. Sawitri and K. S. Dewi, "Aspirasi karir, regulasi diri, dan self-perceived employability pada mahasiswa," *Jurnal Psikologi*, **17**, no. 1, pp. 68–76, 2018, Accessed: Sep. 13, 2024. [Online]. Available: http://download.garuda.kemdikbud.go.id/article.php?article=1405621&val=1286&title=ASPIRASI%20KARIR%20REGULASI%20DIRI%20DAN%20SELF-PERCEIVED%20EMPLOYABILITY%20PADA%20MAHASISWA

[10] Y. N. Sufren, *Belajar Otodidak SPSS Pasti Bisa*. Elex Media Komputindo, 2014. Accessed: Sep. 13, 2024. [Online]. Available: https://books.google.com/books?hl=id&lr=&id=meBMDwAAQBAJ&oi=fnd&pg=PP1&dq=Sufren,+%26+Natanael,+Y.+(2014).+Belajar+otodidak+SPSS+Pasti+Bisa.+Jakarta:+PT+Elex+Media+Komputindo.&ots=RPu5cgsuho&sig=bE9VBmTJ6P6xbRAG9H3VTZRI1As

[11] W. Widhiarso, "Membuat kategori skor hasil pengukuran dari skala, " *Fakultas Psikologi. Universitas Gadjah Mada. Yogyakarta*, **2010**.

[12] Q. Oracion Jr and I. L. S. Abina, "The Mediating Effect of Students' Attitude to Student Career Aspiration and Mathematics Achievement," *Journal of Research and Advances in Mathematics Education*, **6**, no. 3, pp. 158–173, 2021, Accessed: Sep. 13, 2024. [Online]. Available: https://eric.ed.gov/?id=EJ1322868

[13] W. Sulistiani and W. R. Syanti, "Does Career Adaptability Serve as a Mediator between Optimism and Life Satisfaction?," *Journal of Educational, Health & Community Psychology (JEHCP)*, **12**, no. 4, 2023, Accessed: Sep. 13, 2024. [Online]. Available: https://core.ac.uk/download/pdf/598490847.pdf

[14] A. Agus, "Marine/Fisheries Resource Using (Case Study Ternate Manucipality, North Molucca)," *Torani Journal of Fisheries and Marine Science*, pp. 93–103, 2018, Accessed: Sep. 13, 2024. [Online]. Available: http://journal.unhas.ac.id/index.php/torani/article/view/4511

[15] A. Talib, "Peluang dan tantangan industri teknologi pengolahan hasil perikanan dalam mendukung terwujudnya lumbung ikan nasional (LIN) di Maluku Utara," *Agrikan: Jurnal Agribisnis Perikanan*, **11**, no. 1, pp. 19–27, 2018, Accessed: Sep. 13, 2024. [Online]. Available: https://ejournal.stipwunaraha.ac.id/AGRIKAN/article/view/197

[16] E. Rahmiyati, "Tingkat Aspirasi Karir Siswa di Tinjau dari Jenis Kelamin," *G-Couns: Jurnal Bimbingan dan Konseling,* [G-Couns: Journal of Guidance and Counseling], **8**, no. 3, pp. 1743–1751, 2024, Accessed: Sep. 13, 2024. [Online]. Available: http://journal.upy.ac.id/index.php/bk/article/view/5280

[17] E. Barrett, "Career aspirations of teenagers and the future of gender equality in occupations," *Journal of Education and Work*, **34**, no. 2, pp. 110–127, Feb. 2021, doi: 10.1080/13639080.2021.1887829.

[18] J. W. Santrock, *Life-span development 14th ed. New York (14 ed)*. McGraw-Hill Education, 2013.

[19] S. S. Ahmid, C. C. Tiew, and M. N. L. Y. Abdullah, "The influence of innovative characteristics, work readiness, and vocational self-concept on employability of vocational college students," *International Journal for Research in Vocational Education and Training (IJRVET)*, **10**, no. 3, pp. 288–317, 2023, Accessed: Sep. 13, 2024. [Online]. Available: https://www.econstor.eu/handle/10419/280347

[20] N. Widiastuti, "Aspirasi karier siswa SMA berdasarkan status sosial ekonomi dan gender," *Indonesian Journal of Educational Counseling*, **1**, no. 2, pp. 109–128, 2017, Accessed: Sep. 13, 2024. [Online]. Available: http://ijec.ejournal.id/index.php/counseling/article/view/9

IOP Conf. Series: Earth and Environmental Science **1473** (2025) 012069 doi:10.1088/1755-1315/1473/1/012069

Design of an Automated Paint Spray Machine to Support the Implementation of a Green Shipyard

D Tjipto[1], R S Widjaja [1*], J Prananda[2], D B Purwanto[1]

[1] Department of Naval Architecture, Institute Teknologi Sepuluh Nopember, Surabaya.
[2] Departement of Marine Engineering, Institut Teknologi Sepuluh Nopember Surabaya
[*] Corresponding author, email: sjarief@na.its.ac.id

Abstract. Conventional painting methods for vessel plates have shortcomings in terms of quality, human failure, and production waste that pollute the environment. Although painting technology is constantly evolving, specific applications for ship plates are still very limited. The research developed the Eco-Automated Painting Spray Machine to support green shipyard applications by reducing environmental pollution from production waste, especially waste that flows freely in the air. This experimental study examines the design and building of machines with automatic control systems that allow for precise adjustment in paint applications. To determine machine efficiency, performance assessments are performed in a variety of operating situations. The results reveal that this automatic sprayer design is able to eliminate paint residue that pollutes the shipyard environment, which is in line with the philosophy of environmentally friendly shipyards. The implementation of the Eco-Automated Painting Spray Machine is projected to be a reference for the development of automated painting technology.

1. Introduction

The development of science and technology in the shipping sector is presently continuing to be developed to optimize work processes in the area. The rapid and dynamic development of the ship production process has an significant function in the level of effectiveness of the shipbuilding sector. The increasing need for sea transportation continues to encourage the advancement of efficient technology as the main supporting mode for ship production [9].

Automatic spray painting technology in chambers has developed as an innovation in the shipping industry and vehicle production, making it possible to achieve better results and as a supporting role for a shipyard industry that is free of environmental pollution. This research aims to develop and optimize automatic primer painting technology which has been developed to help increase the efficiency of the production process and reduce the negative impact of painting residues on the environment [2]. In the conventional painting process, painting is done manually which can result in less than optimal results and is very dependent on the skills and qualifications of the painter. The use of automatic painting technology is needed to help improve the quality of plate painting results and reduce the possibility of environmental pollution due to the spread of dangerous chemicals from paint residues that are scattered in the air during the painting process. The use of automation technology can help increase production efficiency and reduce operational costs, thereby supporting the implementation of green shipyards.

This research is related to previous research which examined the Development of Automated Spray-Painting System for Anti-Static Coating Process which was researched by Rudzuan M.N. [7]. In general, the aim of this research is to overcome inconsistencies in paint thickness as well as overcome environmental and health problems from manual painting on workers who use automated tools using a paint booth of 7000mm x 5500mm. Important factors considered in this research include automatic primer painting capabilities, primer paint thickness parameters, and paint sprayer movement patterns. The type of primer used will be adjusted to the paint application material. The use of painting criteria and specifications will be adjusted to the needs in the field. By using an automated paint spray machine in the chamber, it is hoped that there will be no need to consume excessive amounts of paint due to wind blowing because the air in the chamber moves statically.

2. Methodology

Painting is a type of coating on metal and non-metal objects with the aim of beautifying their appearance or to protect them from direct contact with the surrounding environment. The durability of a painting is influenced by the ability of the coating to stick to the material to be painted. If the adhesive power is not strong, apart from poor paint results, there is also a very large possibility of moisture entering between the material and the paint. Spray thickness and evenness of the coating layer are the main problems in the painting process, many studies have been carried out in the analysis of spray path, spray overlap, spray gun orientation and spray flow rate as reported by Feng et al. [4].

The thickness of the surface protective paint is usually between 200-300 microns in dry film thickness or only 3-5% of the surface thickness of the material with the normal thickness of each paint layer being around 30-50 microns or stated in the paint product specifications [5]. The success of the painting process depends on the material's shape, cleanliness, and paint adhesion. Factors like sandblasting results and unclean material can also influence the outcome. HR skills' consequences are related to the angle, distance, and pressure used on the material. Increasing the spray angle reduces surface roughness until a minimum level is reached at a 90° spray angle, as sand energy decreases [8].

2.1 *Painting Indicator*

Dry film thickness (DFT) and wet film thickness (WFT) need to be considered in achieving painting success. To achieve a coating system with effective material protection, consider the film thickness of the painting. Less thickness can cause failure and repeating paint, while excessive thickness can cause cracking and loss of adhesion. The recommended film thickness in the coating or paint technical data can guide the paint, while the DFT calculation can determine the final thickness [1]. The general formula used to determine film thickness is shown in formula 1 below.

$$WFT = \frac{DFT}{\%solids\ by\ Volume} \quad \text{..................} \quad (1)$$

2.2 *Painting Process*

The painting process is carried out based on the results of the preparation of the ship's sheet steel plates which have been sand blasted and then painted using a spray gun. Because the painting area is large or has an uneven shape, slight defects will often occur. Defects are usually only discovered after the final painting process, making the polishing process very troublesome. The use of paint spray angles generally uses a vertical angle perpendicular to the material which the tool will adjust to an optimum pressure of 0.8 – 1 bar. According to Rasyid et al., 2019 [6], there are four important things in spray gun control, namely: (1) spray gun distance, (2) spray gun angle, (3) swing stroke speed, (4) overlapping pattern, as follows:

1) Spray Gun Distance
   The painting distance or distance between the spray gun and the painted area depends on the processes and the object to be painted. If it is too close, the paint will melt, and if this happens to metal paint, it will result in uneven paint results caused by metal particles collecting. If it is too far, the surface will become rough. If the spray distance is irregular, the results will be mottled and not shiny. Spray guns usually measure 15-20 cm, acrylic lacquer 10-20 cm, and enamel 15-25 cm.

IOP Conf. Series: Earth and Environmental Science **1473** (2025) 012069 doi:10.1088/1755-1315/1473/1/012069

2) Spray Gun Angle
Paint spraying must be positioned parallel to the work-piece and follow the shape of the work-piece, whether horizontal or curved to get effective painting results. The spraying direction must form a 90° angle with the work plane. In addition, painting is done from top to bottom to avoid fatigue in work and ensure better results.
3) Painting Swing Step Speed
The speed of movement of the sprayer should be stable, both horizontally and vertically. If it is too slow, the paint will melt, if it is too fast then the painting will be uneven. If the speed is less stable then the painting results will be uneven and less shiny. The spray gun swing stroke speed must be constant, which is recommended to be approximately 1.2 feet/second. As for the use of the formulation for painting time, use the following formula 2.

$$painting\ time, t\ [\mathrm{min}] = \frac{solution\ vol\ [\mathrm{cc}]}{flow\ rate\ [^{\mathrm{cc}}/_{\mathrm{min}}]} \quad \text{......................} \quad (2)$$

4) Overlapping Patterns (Over lapping)
Over lapping is a painting technique used on the surface of a work-piece to ensure that the first and subsequent sprays can connect perfectly. When over lapping, it is very important to ensure that the spray pattern has an even thickness, so that the resulting layer is even and invisible. The appropriate over lapping width is around 1/2 to 2/3 of the spray pattern, so as to ensure that the resulting layer is even and not visible. By using overlapping, you can ensure that the resulting layer is even and invisible, thereby improving the quality of the painting and ensuring better results. The illustration shown in Figure 1 is a front view illustration that shows the distance of the nozzle to the work-piece and the area of the work-piece that is exposed to paint using a 90° angle of the spray gun to the plane.

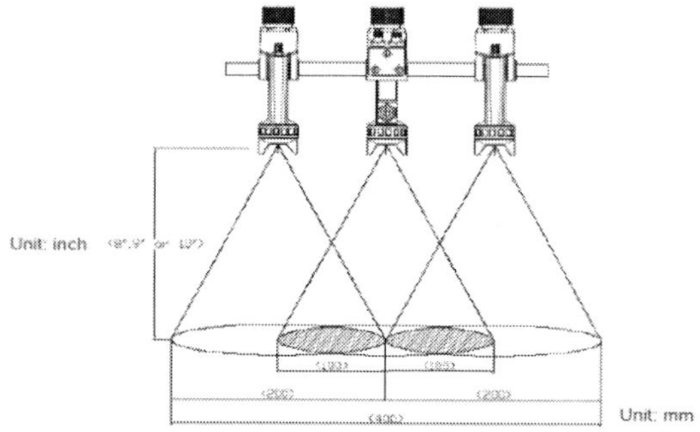

Figure 1. Illustration of Overlapping Painting Patterns

The overlapping painting pattern shown in Figure 2 below is the width of the overlap with a distance of 2/3 and 1/2 which is applied to the vertical plane.

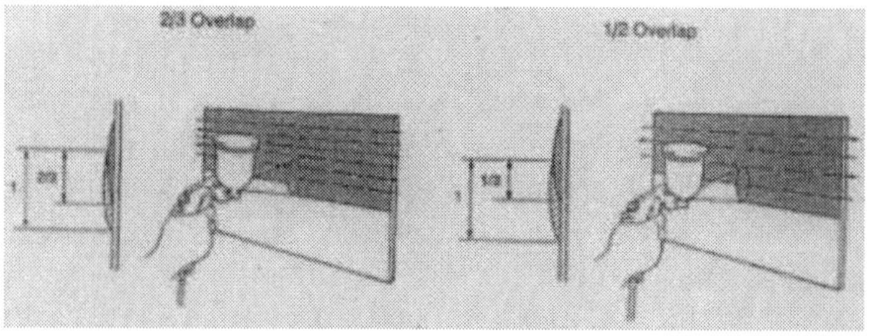

Figure 2. Overlapping Painting Pattern

2.3 *Painting Mechanism*

The automatic airless paint spray gun system depicted on Figure 3, the diagram consists of several key components that work together to facilitate the precise application of paint. Starting with the compressor, it supplies the necessary compressed air to the system. This air is then regulated by air transformers, which condition and adjust the pressure for optimal operation. The regulated air is split into two streams: one stream, known as operating control air, is directed through a three-way solenoid valve to control the triggering and operation of the spray gun. The other stream is used as atomizing air, which is essential for breaking up the paint into fine droplets, allowing for even coverage during spraying. The paint itself is stored in a paint tank and is pumped into the system by a diaphragm paint pump. This pump ensures a consistent and controlled flow of paint from the tank to the spray gun. The paint then travels through the fluid nipple, entering the spray gun where it meets the atomizing air.

The spray gun has two key markings: CAP marked for the atomizing air entry and CYL marked for the air that controls the gun's cylinder, which likely influences the spray pattern or trigger mechanism. Together, these components ensure that the paint is finely atomized and sprayed evenly, resulting in a smooth and efficient application process.

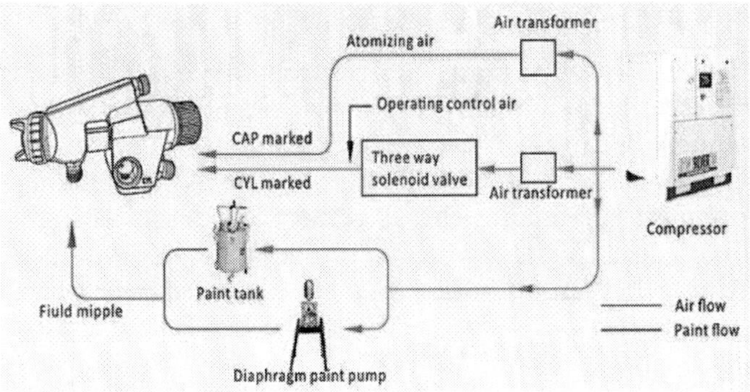

Figure 3. Automatic Paint Spray Diagram

2.4 *Types of Painting Method*

The main function of painting on ship steel plates is generally used to protect against corrosion, but to date the development of the painting function has increased a lot. Based on its function, painting is categorized into three layers, namely primer coating, anti-corrosion (AC) coating, and anti-fouling (AF) coating. Painting effectively combats corrosion by creating a protective layer between the surface and environment, shielding iron through a galvanic effect. The paint contains metallic zinc, aluminum, and zinc phosphate, which ensures suitable protection by interacting with the iron [3]. The barrier effect is produced by painting with a thick layer between 200 microns and 350 microns. The added effect of flakes oriented parallel to the surface of the material means that a better barrier effect can be achieved. The flake pigment that is arranged will have an effect on the water entry process so that it has a longer path to reach the surface of the material. The barrier effect pattern is shown in Figure 4.

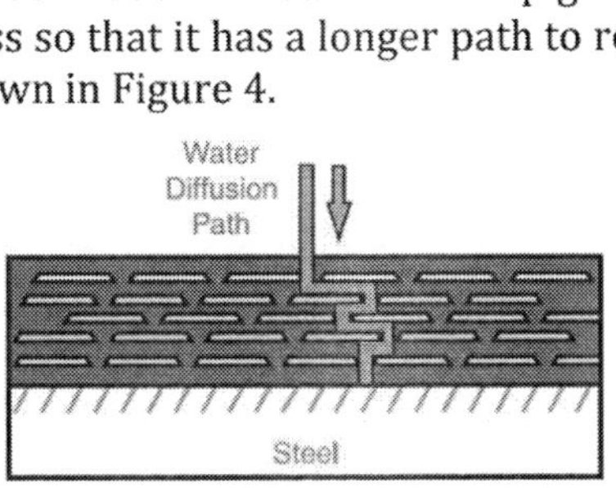

Figure 4. Water in Barrier Effect

IOP Conf. Series: Earth and Environmental Science **1473** (2025) 012069 doi:10.1088/1755-1315/1473/1/012069

A Programmable Logic Controller (PLC) is a digital system that controls machines through sequential control, timing, counting, and arithmetic operations. Originally designed to replace relay panels, PLCs make more informed decisions than traditional relays. Initially used in automotive parts, they are crucial in repetitive tasks like drilling machines and conveyor systems. PLCs use a ladder diagram and are connected to a Video Display Unit for control and monitoring.

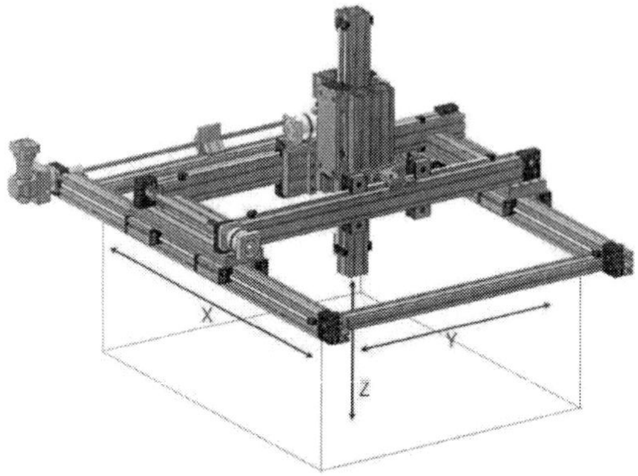

Figure 5. Paint Spray Machine Framework Model

The system uses a gantry with a spray gun, controlled by an explosion-proof servo motor and height sensor, and a PLC for optimal process management, providing data on workpiece type, pressure, paint volume, XYZ spray direction, and position [7]. PLCs are categorized into micro-PLC (less than 32 terminals), mini PLC (32-128 terminals), and large PLC (more than 128 terminals). Each category has unique characteristics and is suited for different applications. Micro-PLCs have limited instruction types, while larger PLCs offer wider and more flexible capabilities. PLCs with fewer inputs/outputs have more limited instruction types.

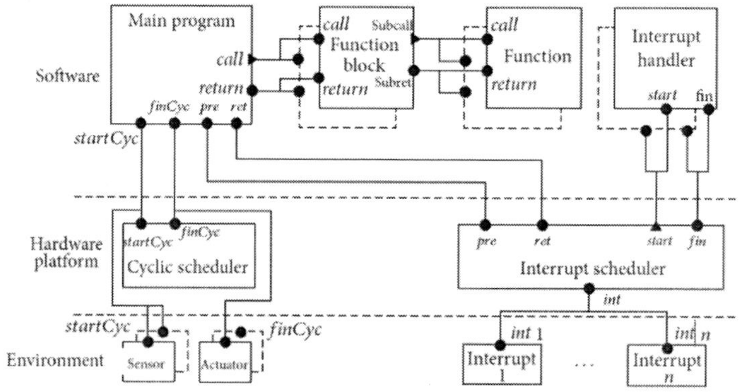

Figure 6. PLC System

The PLC system architecture comprises three layers, as shown in Figure 6: software, hardware platform, and environment, designed to model complex software-hardware systems, emphasizing the interrelation between software execution, hardware platform, and environment [10]. The hardware platform model is independent of the software application and can be translated automatically for existing PLC software, facilitating simulation and verification using the BIP tool. The three layers are detailed as follows:

- Software Layer: Includes all application program organization, with components that can call functions or blocks of functions.
- Hardware Platform Layer: An abstract model describing PLC features like cyclic execution mode and interrupt handling.

- Environment Layer: A model of a controlled device where sensors collect data, send it to the PLC, and produce control signals. It includes components like communication, alarm, and clock interrupts, and data is sent via startCyc port.

In general, a PLC consists of two main components, namely, the Central Processing Unit (CPU) and the input/output interface system which is shown in Figure 7 (a). The CPU has a function to regulate all processes that occur in the PLC. There are three main components that make up a CPU, namely: Processor, Memory, Power supply. The interaction relationship or working diagram in the CPU is shown in Figure 7 (b) below.

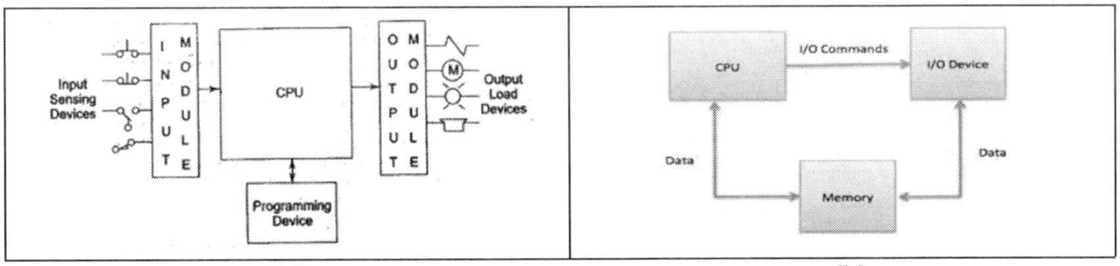

(a) (b)

Figure 7. (a) Basic Component of PLC (b) Work Diagram CPU of PLC

3. RESULT AND DISCUSSION

The acquisition of conventional painting data was carried out through field conditions surveys with field practitioners. There are some important points related to the use of equipment and materials. The data obtained showed that 18 liters of paint were enough to paint 8 to 9 sheets of 6-foot plates. (untuk pengecatan satu sisi pelat). The main equipment used in this process is an airless pump used to regulate air pressure (5–35 psi) and spray paint with a resulting pressure ranging from 500–1500 psi so that the application of paint obtains an optimal pressure. The size of the gap used is based on the pump input requirements (size 6.35 mm), and the painting distance between the nozzle and the surface of the plate adjusted to the worker ranges from 300 to 400 mm.

(a) (b) (c)

Figure 8. (a) Air Compressor, (b) Paint Suction Pump Operator, (c) Painting Operation

A conventional worker takes 1-2 minutes to complete one side of a plate size 6 feet by 20 feet at an average speed of 0.8–1 second for every 1 time the plate width. Although this method is easy and fairly fast done by the workers, it is still found to be deficient in terms of the efficiency of the use of paint affected by the wind blowing conditions that affect the paint stretch. Automatic painting machines in closed containers can significantly reduce the paint cracks that often occur in painting in open spaces, thereby increasing the efficiency of the use of the amount of paint and reducing the environmental impact of the paint wasted. The design of the Automatic Painting Machine container is depicted in the Figure 9.

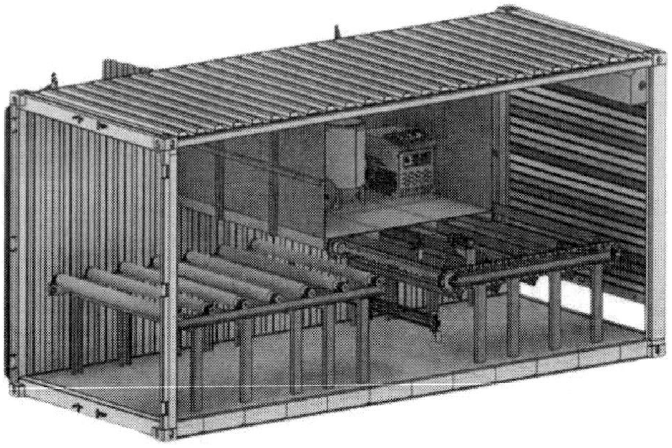

Figure 9. Paint Chamber 3D Ilustration

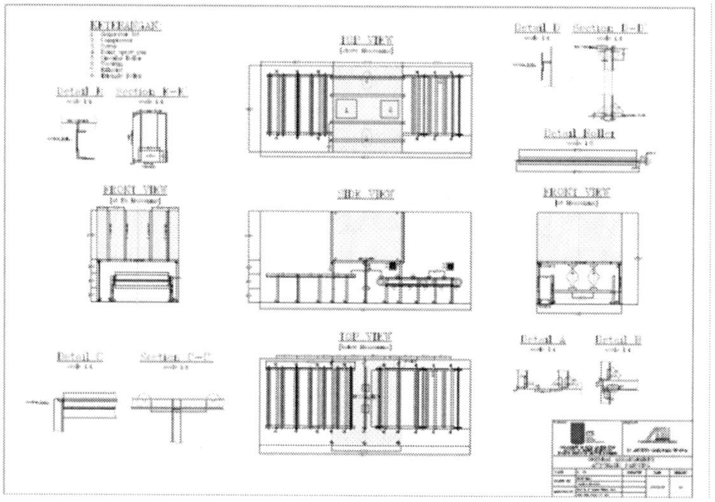

Figure 10. Paint Chamber Drawing

The nozzle employed in this system features a diameter specification ranging from 1 to 1.5 mm, and its discharge rate varies between 180 and 330 milliliters per minute. The resulting paint radius is contingent upon the nozzle opening and the applied air pressure. Specifically, the maximum paint radius spans from 220 to 290 millimeters, while the optimal distance between the nozzle and the plate is approximately 200 to 250 millimeters.

The automatic painting method approach demonstrates a substantial boost in efficiency. Based on the design specification calculations, the suggested conveyor speed and the recommended nozzle movement speed are approximately 2.28 meters per minute, enabling the two-sided coating of a single sheet measuring 6096 x 1828 mm to be completed in approximately 2.674 minutes. In a single day, this automated system can coat up to 50 sheets of plates on both sides, necessitating a total paint volume of 66.85 liters (approximately 1.337 liters per two-sided plate). In contrast, the conventional method is limited to coating 50 sides in one day and requires approximately 4.22 liters of paint for two sides of one sheet. Thus, when this automated painting design is applied as a techniques, there will be a 100% gain in productivity and a 68.3% decrease in paint use, making it a more efficient alternatice in terms of time and material usage. Under optimal operating circumstances, including machine downtime, an automated approach with a conveyor speed of 2.28 meters per minute can coat approximately 141 two-sided plates within a standard working day of 7 hours. In addition, this design generates much reduced air pollution emissions, principally owing to its enclosed environment and superior control systems intended to decrease VOC emissions.

4. CONCLUSION

Based on the research method and results of the design of the automated paint-spray machine, it can be concluded as follows:

- An automated paint-spray machine has played significant roles to ensure that the painting process complied with the standard requirement of paint surface at steel plate, and also completed by the suction blower and drying mechanism, so that all plates can be used directly.
- The process of automated pain-spray machines can be isolated, it can not harm the environment and damage to the worker health.
- Due to isolated process, so that the paint consumption could be minimized.

REFERENCES

[1] American Bureau of Shipping, A 2017 Guidance notes on the inspection, maintenance and application of marine coating systems. *Journal of Chemical Information and Modeling, January*, 192.

[2] Ariany, Z. 2014 Kajian Reparasi Pengecatan Pada Lambung Kapal (Studi Kasus Km. Kirana 3). *Teknik, 35*(1), 27–32. https://doi.org/10.14710/teknik.v35i1.6822

[3] Dwilaksana, D., Jatisukamto, G., & Teknik Mesin, J. 2017 Analisis kekasaran permukaan proses sand blasting dengan variasi tekanan, waktu dan sudut menggunakan metode taguchi. *J-Proteksion, 2*(1), 27–30.

[4] Feng, X., Bin, Z., Zhaoyi, Y., Jiahao, Z., & Huafeng, D. 2023 *Design and implementation of a 7-DOF cable-driven serial spray-painting robot with motion-decoupling mechanisms.*

[5] Luangkularb, S., Prombanpong, S., & Tangwarodomnukun, V. 2014 Material consumption and dry film thickness in spray coating process. *Procedia CIRP, 17*, 789–794. https://doi.org/10.1016/j.procir.2014.02.046

[6] Rasyid, A. H. A., Santoso, D. I., & Utama, F. Y. 2019 Pemilihan Parameter Pengecatan Untuk Mendapatkan Ketebalan Lapisan Cat Yang Tepat Untuk Permukaan Tidak Merata. *Otopro, 12*(2), 82. https://doi.org/10.26740/otopro.v12n2.p82-87

[7] Rudzuan, M. N., Khairunizam, W., Zunaidi, I., Razlan, Z. M., Shahriman, A. B., Rozman, A. R., & Shaharizal, A. 2019 Development of Automated Spray-Painting System for Anti-Static Coating Process. *IOP Conference Series: Materials Science and Engineering, 557*(1). https://doi.org/10.1088/1757-899X/557/1/012001

[8] Sulistyo, E., & Setyarini, P. H. 2011 Pengaruh Waktu Dan Sudut Penyemprotan Pada Proses Sand Blasting Terhadap Laju Korosi Hasil Pengecatan Baja AISI 430. *Jurnal Rekayasa Mesin, 2*(3), 205–208.

[9] Utomo, E. 2015 *Robot Application in Shipbuilding Industry*. 31–38.

[10] Wang, R., Guan, Y., Liming, L., Li, X., & Zhang, J. 2013 Journal of Applied Mathematics - 2013 - Wang - Component-Based Formal Modeling of PLC Systems.pdf. *Journal of Applied Mathematics, 2013*, 1–8. https://doi.org/http://dx.doi.org/10.1155/2013/721624

IOP Conf. Series: Earth and Environmental Science **1473** (2025) 012070 doi:10.1088/1755-1315/1473/1/012070

Design of an Automated Dust Collector for Sandblasting Machine to Support the Implementationof a Green Shipyard

Faris Taruna Nugraha[1], Raden Sjarief Widjaja[1*], Yuda Apri Hermawan[1], Totok Yulianto[1], Sri Rejeki[1]

[1] Department of Naval Architecture, Institute Teknologi Sepuluh Nopember, Surabaya.
[1*] corresponding author : email: sjarief@na.its.ac.id

Abstract. The background of this research is based on the serious problems posed by silica dust exposure due to sandblasting activities in shipyards. It focuses on the development of a "Dust Collector" as part of an "Eco-Automated Sandblasting Chamber", with the main objective of reducing the negative impact of silica dust in shipyard areas. Based on various literature studies exist on dust collectors and pollution control, this research innovates by designing a more efficient and eco-friendly system, featuring a vacuum exhaust within a portable container and a vibrating sieving device for reusable dust separation. This research involves a design approach. Key stages include literature review, data collection, concept design, technical calculations, detailed engineering design (DED) on "Eco-Automated Sandblasting Chamber". This research aims to make an innovative design that can significantly contribute to pollution control technology in the shipyard industry and become an innovative solution, widely adopted. The conclusion is to present an innovative design and the effectiveness of the system in reducing dust and waste emissions, improving work safety, and protecting the environment in the shipyard and its surroundings.
Sandblasting process, abrassive material, sorting and re-cycle mechanism, dust collector machine.

1. Introduction

In shipbuilding and repair, the hull's surface must undergo rigorous cleaning before painting to prevent corrosion and ensure the longevity of the ship. One of the most widely used methods for surface preparation is sandblasting, which efficiently removes rust, dirt, and other contaminants from steel plates. However, the sandblasting process also generates significant amounts of silica dust, a byproduct that poses serious environmental and health risks. Silica dust is categorized as hazardous (B3 waste) under Indonesian regulations, and prolonged exposure can lead to respiratory diseases such as silicosis. Shipyards that rely on open sandblasting methods face several challenges, including the release of airborne dust particles that contaminate the surrounding environment and endanger workers' health. Despite its efficiency and adaptability to complex surfaces, sandblasting, particularly with silica sand, is under increasing scrutiny due to its negative environmental impact. Therefore, there is an urgent need for innovative solutions that mitigate these risks while maintaining the effectiveness of the process.

To address this issue, the design of an automated dust collector system for sandblasting machines is proposed. This system aims to capture and filter silica dust at the source, reducing both the environmental footprint and health hazards associated with the process. The dust collector is integrated with an eco-friendly sandblasting chamber, designed to recycle and reuse abrasive materials, thereby contributing to the broader objective of a green shipyard. By implementing such technology, shipyards can comply with environmental regulations, protect workers, and enhance operational sustainability. This paper focuses on the design and technical

IOP Conf. Series: Earth and Environmental Science **1473** (2025) 012070 doi:10.1088/1755-1315/1473/1/012070

development of this automated dust collector. The following sections will detail the methodology used, including literature review, technical calculations, and design development, ultimately demonstrating the system's potential to reduce silica dust pollution in shipyards.

2. Methodology

2.1 Sandblasting

Shipyards use sandblasting, also known as abrasive blast cleaning, to remove rust and dirt from ship surfaces and create a rough profile on metal to improve the adhesion of protective materials like paint. However, Indonesian regulations classify the silica sand waste during sandblasting as hazardous and toxic (B3 waste). Inhalation of silica dust can lead to silicosis, making it crucial to find innovative uses for the excess silica sand waste to minimize health risks and environmental impact [2].

Figure 1. Ship hull sandblasting process
[Example figure taken from https://eltamacoating.com/]

Sandblasting is a surface cleaning method that involves shooting sand particles onto a material to remove contaminants like rust, paint, salt, oil and also to create a rough surface for better paint adhesion. It offers speed and flexibility, especially for complex shapes. Key factors affecting sandblasting results include the operator's skill, air pressure, sand size, spraying time, and distance. The hardness of the workpiece surface and the erosion rate are crucial for ensuring effective coating adhesion [3].

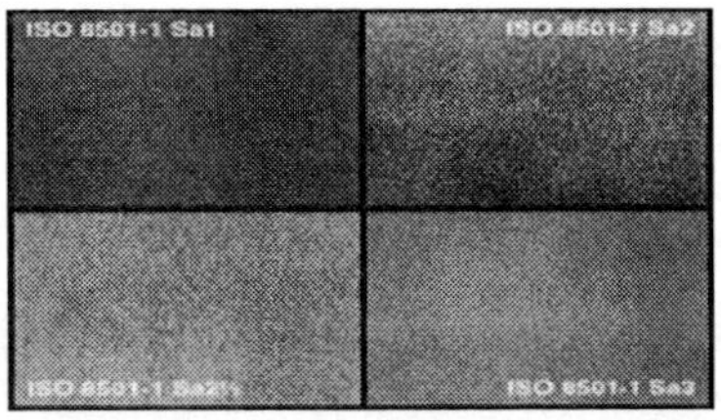

Figure 2. Class preparation standards
[Example figure taken from http://id.worldironsteel.com/]

Sandblasting roughness standards include SA 2, SA 2.5, and SA 3. SA 2 (Thorough Blast Cleaning) requires a 65% clean surface, free of visible grease, dirt, loose scale, rust, paint, and foreign matter, with any remaining contamination firmly adhered. SA 2.5 (Very Thorough Blast Cleaning) ensures an 85% clean surface with a grayish color, free from visible contaminants, allowing only small spots or lines of contamination. SA 3 (Blast Cleaning to Pure Metal) achieves a 100% clean, uniform metallic surface, completely free of all visible grease, dirt, scale, rust, paint, and foreign matter [4]. Depending on the conditions of the object, we can classify sandblasting into two types: Dry sandblasting is used for non-flammable metal or iron objects like car frames, ship bodies, and piles, while wet sandblasting is applied to metal or iron objects that are prone to burning or in fire-prone locations, such as fuel tanks, offshore oil refineries, and gas stations. Wet sandblasting involves mixing silica sand with anti-rust chemicals to minimize sparks. The equipment remains the same for both methods, including the compressor, air filter tube, sand collector tube, hose, nozzle, and the operator's special helmet [1].

2.2 Abrasive Material Type

a) Silica Sand b) Steel Shot c) Steel Grit

Figure 3. Types of Abbrasive Materials

Silica sand, an abrasive material found in granite, quartz, and other rocks, typically has a hexagonal prism shape with a hexagonal pyramid tip and contains about 90% free silica. It is effective for cleaning surfaces but less suitable for creating deep profiles. However, silica sand can be hazardous to humans, as the high levels of crystalline silica dust it produces can lead to serious health issues such as silicosis and cancer when inhaled.

Steel shot, a round-shaped steel product, contains 1% silica. It is effective for cleaning surfaces but less effective for creating profiles, so it is often mixed with steel grit. Steel shot is reusable and ideal for use in enclosed areas. Steel grit, which is also made of steel but has a sharp shape, contains 1% silica. Both steel shot and steel grit can rust and contaminate surfaces, so they must be rust-free before use. They can be reused several times and are generally used for shop blasting [5].

2.3 Air Pollution Through Dust

Various natural and mechanical processes, such as breaking, grinding, sieving, beating or blasting, cutting, and crushing, produce dust, a solid particle. These dust particles are inhaled through the air when we breathe, and most of the dust will stay in the lungs. In general, the size of dust particles varies from very small to quite large, from invisible to the naked eye to visible. Dust, or Suspended Particulate Matter (SPM), ranges from 1 to 500 microns in size, with particles over 50 microns visible to the naked eye. It includes inorganic (metals, non-metals) and organic (plants, animals, synthetic) materials. Dust can remain airborne for extended periods, posing health risks when inhaled, and may combine with other substances through chemical reactions.

2.4 Pollution Monitoring

The human body has a defense mechanism to remove most inhaled dust, but some particles can remain in the lungs depending on their size. The accumulation of dust is influenced by airway anatomy, airflow velocity, and dust size. According to WHO (1996), dust particles dangerous to humans range from 0.1 to 5 microns, or up to 10 microns. The Indonesian Ministry of Health identifies hazardous dust as particles between 0.1 and 10 microns. These particles can be inhaled in significant volumes, with 100μ Thoraric, 10μ respirable, and $4\mu^3$ particles being particularly concerning. Dust levels in an area can be calculated using specific formulas:

$$V = \frac{(F1 + F2)}{2} \times t \times \frac{Pa}{Ta} \times \frac{298}{101,3} \qquad \text{.........................(1)}$$

$$TSP = \frac{(W2 - W1) \times 1000}{V} \qquad \text{.........................(2)}$$

Where:

TSP	= Dust content in the air (mg/M^3)
V	= Volume of air inhaled (M^3)
W1	= Filter weight before sampling (mg)
W2	= Filter weight after sampling (mg)
F1	= Initial flow rate (M^3/Minute)
F2	= Final flow rate (M^3/Minute)
t	= Time for taking test samples (Minutes)
Pa	= Average barometric pressure during test sampling (kPa)
Ta	= Average temperature during test sampling (K)
298	= Temperature under normal conditions 298 K
101.3	= Pressure under normal conditions 1 atm (kPa) 1000 = Concentration (g)

Filling dry chemicals (such as talc, cement, or fertilizer), sandblasting, drilling, using smoothing machines, and rust removal can release dust into the air. The breakdown or disturbance of materials during these processes generates dust, resulting in widespread dust exposure in the workplace [6].

2.5 Vacuum Machines

Depending on the hazard level (light, medium, or high), we must appropriately classify industrial vacuums for hazardous dust as Class L, M, or H. They are designed to safely capture and store dust, unlike standard household or commercial vacuums, which are inadequate for protecting against harmful dust and should not be used in industrial settings. Class M and H vacuums have safety features that include:

- Safe dust bag removal
- Alarm indicator if air flow falls below 20 cubic meters per second
- Prevention of accidental opening and release of dust when not in use.

In industries like engineered stone countertops, Class H vacuums are essential for managing highly hazardous dust, such as respirable crystalline silica (RCS) and asbestos. They are equipped with safety features to prevent harmful dust from being released back into the environment. Class L vacuums manage light dust, Class M vacuums handle medium dust, and Class H vacuums specifically design for the most dangerous dust, ensuring worker protection and environmental safety.

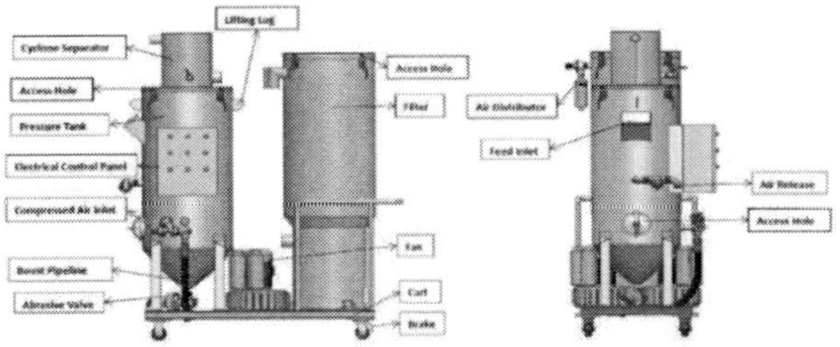

Figure 4. Vacuum blast machine equipment
[Example figure taken from www.tsblasting.com]

2.6 Dust Collector

A dust collector machine, similar to a vacuum cleaner, is used to maintain cleanliness by sucking up dust particles from production machines, including silica sand dust and paint. Designed to

IOP Conf. Series: Earth and Environmental Science **1473** (2025) 012070 doi:10.1088/1755-1315/1473/1/012070

work with sandblasting machines, it functions as a ventilation system that reduces air pollution by separating dust from clean air. Effective dust collection is crucial, especially when airborne dust contains silica, requiring adequate airflow to remove contaminants from workers' breathing zones. The recommended airflow rate is 50-100 cubic feet per minute per square foot of facial area (0.25 m^3/s per square meter) based on the specific application.

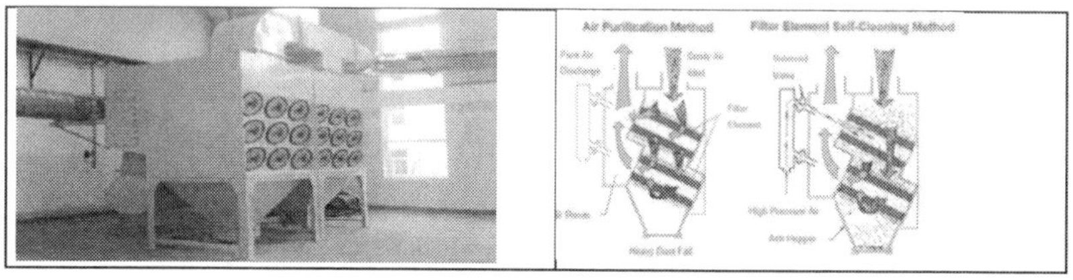

Figure 5. Dust collector machine
[Example figure taken from www.tsblasting.com]

Several parts are contained in the Dust Collector System: Dust Collector System Parts:

- Hood: Captures contaminants, taking into consideration the shape, speed and direction of contaminant release.
- Ducts: Channel contaminants that have been captured by the hood to a dust collector.
- Air Cleaner: Filters pollutants from the inhaled air before being discharged.
- Blower: An electrically powered device for sucking up contaminants, using a fan with a wind speed that can be changed according to needs.

2.7 Airport Draft Curtain

Airport draft curtains are an important barrier in the operation of standard conveyors, which act as barrier doors in wall openings at airports. Airports place these barriers in wall gaps where baggage conveyors transition from public to private areas. The metal frame holds two rows of overlapping opaque (usually black) vinyl strips at the top of the wall opening. The ball-like double thick vertical edges that most curtains have help the strips hang straighter, insulate better, and last longer. This configuration facilitates baggage movement but obscures the exchange of airflow between the two sectors and preserves private areas.

Figure 6. Airport Draft Curtain
[Example figure taken from www.asgco.com]

2.8 Silica Sand Sieving Machine

The silica sand sieve machine was created because of the demand for silica sand with a certain and uniform size, such as 0.8-1.4 millimeters (mesh 14x20). This aims to avoid mixtures with coarser or lower silica sand sizes. Mesh sizes are used to describe various sizes of silica sand. Sieving is a mechanical separation process used industrially to sort silica sand by particle size. It produces two results: oversize (particles larger than the sieve holes) and undersize (particles smaller than the sieve holes).

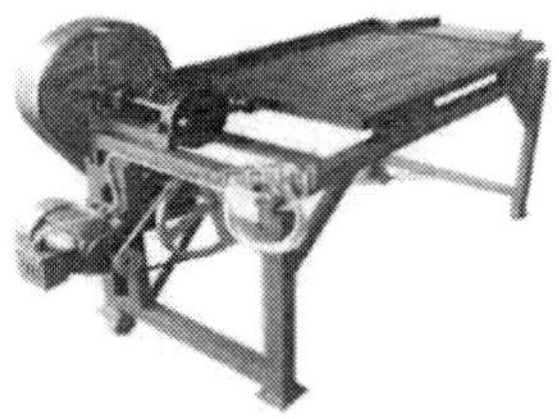

Figure 7. Silica Sand Sifter Machine
[Example figure taken from www.grahamesin.com]

Mechanical sieving, also known as vibratory, shaking, or shake sieving, is performed using a machine that usually has a series of sieves with different hole sizes. During the sieving process, solid materials are thrown or dropped onto the surface of the sieve. Smaller particles (undersize) or fines can pass through the sieve holes, while larger particles (oversize) or tails cannot pass through the sieve holes. Dry sieving is more common.

2.9 Silica Mesh

Figure 8. Silica Mesh Sand 12-24
[Example figure taken from https://image.amapurewater.com]

A sieve consists of a frame and a mesh network, which denotes the number of holes per square inch. For instance, a 40 mesh sieve has 40 holes per inch (1,600 holes per square inch), while a 100 mesh sieve has 100 holes per inch (10,000 holes per square inch). In some sieves, additional equipment is included, such as a rocking mechanism connected to a moving wheel. The mesh size indicates the size of the holes: larger mesh numbers correspond to smaller holes, resulting in finer material after sieving [7].

3. RESULT AND DISCUSSION

To conduct this research, it is necessary to survey the field conditions where the equipment will operate. At Adiluhung Saranasegara Shipyard in Bangkalan, Madura, sandblasting is done in an open area with silica sand, which is used only once and then discarded. The residual dust is allowed to disperse into the air, and the fallen dust is used for landfilling. Despite its potential for reuse or recycling, the high cost of recycling leads to the waste being sent to landfills.

Figure 9. (a) Air Compressor, (b) Paint Suction Pump Operator, (c) Painting Operation

To evaluate the effectiveness of the dust collector, calculations were performed based on the expected dust generation and the capacity of the dust collector system. Assumed to be 10 kg/hour of silica sand with a dust concentration of 5%. Required airflow rate for effective dust capture was calculated using the formula for dust concentration.

$$Air\ Flow\ Rate = \frac{Dust\ Content\ \times Volume\ of\ Air}{Filter\ Efficiency} \quad \text{.......................................(3)}$$

For a dust content of 50 mg/m^3 and a volume of air at 1 m^3/min, the required airflow was calculated to be 0.25 m^3/s. For the efficiency, the dust collector was designed to achieve a minimum efficiency of 95% in capturing airborne dust, based on the results from similar systems and literature.

In general, the designed dust collector consists of a reciprocating motion sand sieve and two separate dust collection bins. The first bin collects and disposes of fine sand with a mesh size smaller than the sieve mesh, while the second bin collects silica sand particles larger than the sieve mesh. This dust collector is positioned directly beneath the mezzanine of the sandblasting room with dimensions of L x W x H (2000 x 1700 x 400 mm).

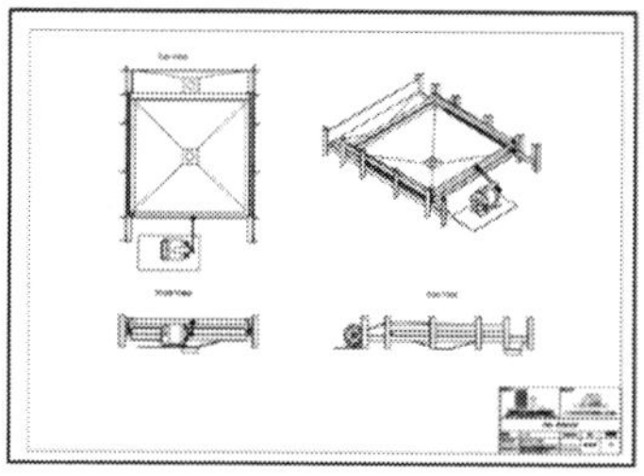

Figure 10. Design of Dust Collector

The reciprocating sand sieve measures 1690 x 1592 mm and is built with 75 x 75 x 7 mm L-shaped profiles on three sides. It features an 60 mesh wire mesh bolted to the frame and is supported by a 50 mm diameter caster wheel on 4 points under sand sieve, creating a 2-3 degree slope for efficient particle separation. A 3 kW, 1500 rpm electric motor drives the sieve through a connecting rod that connect on rotating flange coupling. The motor's rotation generates the necessary oscilation for the sieving process. If the flange radius used is 24 mm, the oscillation stroke of the sand sieve is 2 times the flange radius which is 48 mm. The connecting rod used is 462 mm long. In 1 minute, the resulting oscillatory motion is equivalent to 25 Hz vibration frequency. Assuming that the sand sieve's mass is 80 kg, the force on the connecting rod is 47,669 N.

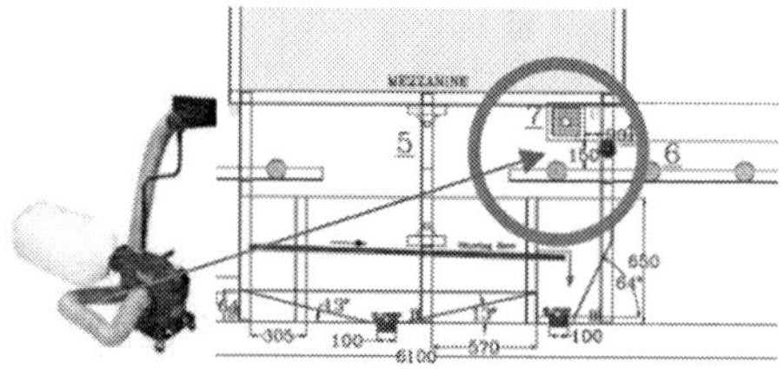

Figure 11. Exhaust Vacuum of Dust Collector
[Example figure taken from https://www.tokopedia.com]

In the dust collection system, difficult-to-settle dust is suctioned by an exhaust blower positioned on the mezzanine. The 16-inch diameter exhaust hose is placed 150 mm above the conveyor, and 90 mm behind the steel brush. The dust is directed into bags on both sides of the container. The exhaust blower operates at 220 volts with an air volume of 2010 CMH.

The dust collector captures a large amount of reusable silica sand, which is separated from fine dust using a reciprocating sieve. Usable silica is returned to the sandblasting process, while finer silica, reduced to a mesh size below 60, can be used in concrete production. This approach reduces the need for landfilling and minimizes environmental impact from silica waste.

4. CONCLUSION

Based on the research and design results, the automated dust collector significantly improves dust control and worker safety in shipyards. It effectively reduces harmful silica dust, mitigating environmental and health risks while enabling abrasive material recycling, which lowers reliance on landfills and reduces waste management costs. By minimizing hazardous waste and supporting sustainable practices, the system enhances operational efficiency and compliance with environmental regulations. Future improvements, such as increased automation and real-time monitoring, could further optimize its performance in creating greener shipyards.

REFERENCES

[1] Dwi Kurniawan, W. (2023). Proses Sandblasting dan Coating Pada Kapal di PT. Dok Perkapalan Surabaya *Otopro* **13** 44-53
[2] Hendrawan, A., Aprilian, R. (2020). SANDBLASTING PADA KAPAL MV. BERLIAN INDAH *Jurnal Saintara* **4** 25-32
[3] Setyarini, P. H., & Sulistyo, E. (2011). Optimasi Proses Sand Blasting Terhadap Laju Korosi Hasil Pengecatan Baja Aisi 430. *Jurnal Rekayasa Mesin* **2** 106-109
[4] Sugiantoro (2018). *ANALISIS KEKASARAN PERMUKAAN PROSES SAND BLASTING DENGAN VARIASI TEKANAN, WAKTU DAN SUDUT MENGGUNAKAN METODE TAGUCHI*. Thesis (UNIVERSITAS JEMBER)
[5] Harmoko, P., Hakim, M., & Dzikrullah Faza, M. (2023). Penerapan Usaha Sandblasting Untuk Meningkatkan Daya Saing Bengkel Motor Di Desa Bugangan *JURNAL EKONOMI, MANAJEMEN, BISNIS DAN SOSIAL (EMBISS)* **3** 237-252
[6] Risna (2013). *STRATEGI PENGELOLAAN LINGKUNGAN PT.INDUSTRI KAPAL INDONESIA MAKASSAR DALAM MENGENDALIKAN PENCEMARAN AIR DAN UDARA.* Thesis (UNIVERSITAS HASANUDDIN MAKASSAR)
[7] Prasetya, H. *PENGARUH UKURAN MESH PASIR SILIKA PADA PROSES SAND BLASTING TERHADAP WETTABILITY PERMUKAAN STAINLESS STEEL AISI 316L.* Thesis (UNIVERSITAS BRAWIJAYA)

IOP Conf. Series: Earth and Environmental Science **1473** (2025) 012071 doi:10.1088/1755-1315/1473/1/012071

Offshore Wind Farm As Blue Economic Pillar And Support For Energy Sustainability

I Isnurwanto
Faculty of Law, Hang Tuah University, Indonesia, 60111

E-mail: bhima216@gmail.com

Abstract. Energy is not only limited to life, but is also a crucial factor of global economy. As a basic need of all aspects of life, current energy is still fossil-generated, causing greenhouse gas emission that contributes to climate change and air pollution, ecosystem damages, and health risks over humans. Therefore, it is essential to find out optional renewable energy resources as blue energy that is environmentally-friendlier. Furthermore, a descriptively qualitative method was used by means of qualitative analysis of objects or phenomena being researched based on the actual condition when the research was carried out. Findings had revealed that Indonesia has been utilizing land breeze as an energy resource to generate energy by means of onshore wind farm (PLTB), instead of offshore wind farm (PLTA). Therefore, putting offshore wind farm as a priority for power plant to generate clean, environmentally-friendly, and sustainable energy is urgent. As blue energy, offshore wind farm plays important roles in support of blue economy and holistic energy sustainability. Also, as a pillar of blue economy, offshore wind farm brings about impacts to the economic sector and contributes to energy resource diversification and local economic growth.

1. Introduction

Amidst this current modern life with all its advances, energy remains the thing we cannot leave behind. In every aspect of life, energy has become an unseen foundation, but very urgent. Not only as a basic need, energy has also been the backbone for all aspects of human's life. Simply put, lamps we turn on and industrial machines we use every day do need energy, which indicates how significant energy is in taking up human's life, to make it balanced and advanced. Thus, it is easily imaginable about how humans would bear with life when no energy is available. Availability of energy allows us to live our life comfortably and efficiently, pampering us with ease to carry out any activities.

Energy causes impacts, not only limited to life, but also global economy, such as industry, transportation, technology, and others in need of energy. However, on the other hand, use of energy also may cause massive consequences upon environment. Utilization of fossil-generated energy resources, such as coal and oil, will result in greenhouse gas emission that contributes to climate change and air pollution. Body of International Energy (IEA) proclaims that fossil fuels can contribute to 44% out of the total CO2 global emission[1]. The effects can not only ruin the ecosystem, but also cause health risks to humans, such as respiratory illnesses and other health issues.

Establishing sustainability and energy availability is of importance. Nonetheless, if fossil resources are used constantly, environmental degradation will occur massively. Meanwhile, depending on energy import will result in weighty challenges for Indonesia as it may influence the energy sovereignty and economic stability. Thus, evolving blue energy as renewable energy needs to be the main priority. Utilizing blue energy, in addition, will not only be answers for technological problems, but also reinforce the energy sovereignty through the use of abundantly available local resources.

Through transition to new and renewable energy (NRE), solar, wind, and water energy offers cleaner and more sustainable energy. Some of excellences of renewable energy include: low cost and easy access to it, wasteless, less impacts to global temperature, resistance to increase of fuel cost [2]. In addition, blue energy comes up as a promising solution, offering environmentally-friendlier and highly-efficient options. Diverting to using this kind of energy might deflate potentials of environmental damages. Therefore, it is not only about protecting our life, but also ensuring the sustainability of energy resources for the future generation. Clean and affordable energy is seen to cause positive impacts to social welfare. In-depth understanding and sustainable effort to divert to environmentally-friendlier options help us ensure that energy will remain the first pillar in support of life and to preserve the environmental stability for the future. Given continual increase of population and global economic growth, needs of energy will become bigger as well.

Blue energy generated by wind power for power plant has been quite massively developed in Indonesia, yet still by means of onshore wind farm. To July 2024, Indonesia has been making remarkable advancement in the development of wind power plant. In fact, some onshore wind farms in Indonesia have been running for operation, namely [3]:

1. PLTB Sidrap in Sidenreng Rappang Regency, South Sulawesi;
2. PLTB Selat Panjang, Kepulauan Meranti Regency, Riau;
3. PLTB Mamuju, Mamuju Regency, West Sulawesi;
4. PLTB Rote Ndao, Rote Ndao Regency, East Nusa Tenggara;
5. PLTB Belitung, Belitung Islands, Bangka Belitung.

With great wind potentials and support from dynamic policies, it is expected that wind power sector keeps on going in support of transition to clean and sustainable energy in Indonesia. However, the development of onshore wind farms in Indonesia, to date, has not been coincided with that of offshore wind farms. In fact, technology used for both is of close likelihood. It fully needs notices that marine energy, especially offshore wind, has been identified as potential renewable energy, aiming at decarbonizing and deflating greenhouse gas emission [4].

Why is it important to build up offshore wind farm in Indonesia? It is needed to know that Indonesia is the biggest archipelagic country in the world, not only because of its wonder but also abundantly potential marine energy. Its marine spans from Pacific to Hindia Oceans, surrounded by shallow marine rich of energy. In addition, its coastal line spans for 81,290 kilometers, covering many as 17,504 islands [5]. Indonesia is in possession of offshore wind energy that is usable for blue energy. Indonesian marine, especially at Selat Sunda [6] and North Papua marine offers potential of offshore wind potential with constant speed. By the speed, turbines or wind power plant can supply stable and reliable electricity power. On the other hand, however, the hardest challenge lies on finding out the most strategic marine area with literally eligible wind power for establishing offshore wind farm. Thus, knowledge and understanding of which location gives the most potential wind power are fundamental [7].

Utilizing offshore wind energy will not only be a significant move in the diversification of energy resources in Indonesia, but also minimize dependence on fossil fuels that has been dominating the energy sector. By the use of existing potential offshore wind, Indonesia can develop renewable energy that is sustainable and environmentally-friendly in support of the existing onshore wind farms. Development of offshore wind farms to sustain onshore ones becomes a strategic effort by Indonesia in its contribution to the global move in response to climate change. Applying precise strategies and investments at technology and infrastructure is so crucial since Indonesia is potential enough to be the originator for global blue energy development. It ensures, in addition, that Indonesia does not only exploit its natural wonder, but also optimize existing natural resources for greener and more sustainable future. Further, blue energy, through use of onshore and offshore wind and by means of apt technology, is coincidently very important for blue economic growth and holistic sustainability of energy.

2. Research Problems

In order to overtake new and renewable energy, blue energy offers environmentally-friendly and efficient alternatives, which seems to be quite effective at decreasing negative impacts and minimizing environmental damages. Indonesia has been using and developing wind power by means of onshore wind, called onshore wind farm, and not offshore one. In fact, in some advanced countries, offshore wind farms have been quite popular. Basically, onshore and offshore wind farm shares noticeable similarities, especially in terms of sources, work principles, environmental benefits, and infrastructural needs. However, if we take a closer look at the location and topography, offshore wind will provide greater, more stable, and more consistent energy than the onshore will due to flat marine areas and no hindrances that will block the wind. For that reason, in the future, establishing offshore wind farm needs to be prioritized by the government as a strategic global move to abolish the dependence on fossil fuels that is not so environmentally-friendly. There must be numerous challenges and issues in its development and construction. However, both of wind farm belong to new and renewable energy, which is clean, sustainable, renewable, and needed in Indonesia in order to complete needs of environmentally-friendly energy. Thus, if offshore wind farm, with the onshore as well, is massively developed in Indonesia, we are of great optimism that it can be the main pillar of blue economy for energy sustainability.

3. Research Method

Literature review was carried out for the current research, which means that data were collected through review over relevant literature. Data of the research were secondary data obtained from books, scientific journals, and articles. Meanwhile, for the research method, a descriptively qualitative method was used in which it attempted to describe, illustrate, and elaborate certain objects or phenomena by data in accordance with the real conditions where research was carried out [8]. Qualitative research is basically descriptive, with analysis as the core of research since the main goal of this design is to describe and illustrate findings [9]. Ulfatin asserts that qualitative research is used to figure out things under investigation. It also requires researchers to identify the naturalness of certain phenomena, deeply understand issues phenomenologically, symbolic interaction, ethnographic, and case studies, and define qualitative features [10]. Furthermore, it is stated that all of the types of qualitative research are descriptive since its goal is to describe and/or illustrate characteristics of certain phenomena. Thus, the present research involved analysis of literature upon existing references, books, and other literatures in regards to offshore wind farm as a part of blue economy and support for energy sustainability. Findings, thus, are constructed narratively, with logical illustration based on analysis of existing phenomena.

4. Literature Study

Kurnia, *et* al (2021) [11] in an article entitled Feasibility Study of Offshore Wind Farms in the Indonesian Archipelago had explained eligibility of offshore wind farm development in Indonesia's archipelago. The research, in addition, is expected to provide readers with priceless knowledge of how Indonesia could make use of potentials of offshore wind power to facilitate transition to renewable energy. Simply put, the research main goal is to study eligibility of offshore wind farm projects in Indonesia, including evaluation of offshore wind energy, technology, and infrastructure as well as analysis of economy, finance, social and environment. In discussion, it has been found that there are so many locations in Indonesian archipelago with very fine wind for offshore wind farm development. Meanwhile, the real challenge is the initial cost that is relatively high just for technological development for necessary turbine and other infrastructure. However, with systematic planning and rules, the cost remains manageable. There are, in fact, some impacts to marine ecosystem, which needs further and holistic study. In addition, involvement of local people to succeed in the project is also crucial. Regardless of the benefits of the project, the development is still in need of wary attention to environmental issues before gaining social support for the sake of success.

Moreover, Yang, *et al* (2023) [12] in an article entitled *Techno-Economic Feasibility of Offshore Wind Farms: A Review* reviewed technical and economic eligibilities of offshore wind farms in order to explore numerous aspects to influence feasibility of these projects. Foci lied on technology, cost, and benefits. In addition, the research aimed at presenting new advancement at turbine technology, analyzing cost and economic advantages, discussing environmental impacts and the wind farm sustainability, and also scrutinizing opportunities and challenges for further development. Results of the discussion had shown that turbine technology has been massively developed, which leads to high efficiency and cost dropping for each of unit energy. Regardless of the necessity for big investments, offshore wind farms offer relatively low-cost operation with highly potential energy production. Environmental impacts caused by the offshore wind farms are also less dangerous than those by fossil fuels that may need high-cost installation and maintenance and specific locations open for extra innovation and technical solution.

In regards to above mentioned studies, it is clear that authors of both articles focused on discussing technology, challenges, and possible issues in the development of offshore wind farms, as well as how offshore energy could be a crucial part of global renewable energy. On the other hand, current research is focused, not only on describing technology, challenges, and issues in developing offshore wind farms in Indonesia, but also explaining how offshore wind farms serve as blue economic pillar and support for energy sustainability, indicating core differences from previous studies and remarking the novelty of the article.

5. Findings and Discussions

a. *Blue Energy and Blue Economy: The Concept*

Blue energy uses marine wealth, and our marine areas are innovative solutions amidst global challenges for the next transition to cleaner energy resources, annulling greenhouse gas emission and sustaining less-polluted air. Regardless of the fact that blue energy is still partially used as renewable energy, it comes up with huge potential in support of energy sustainability and development of blue economy.

Moreover, blue energy, in fact, includes some technological features designed in order to transform energy from aquatic resources to sources of electricity. There are several types of blue energy that can be very promising and generated by marine, such as wave energy, tidal energy, marine current energy, marine thermal energy, and offshore wind energy. Every each of those aforementioned kinds uses different elements from marine features to generate power [13].

1) Wave energy uses marine wave that is continually rolling. To capture the energy, moving float is technologically designed to follow the move of wave. It can be also by means of wave converter system that converts marine wave into energy. By the speed and consistency that are predictable and generated by the marine, wave power offers stable and sustainable energy resources. There are some countries having adopted this sort of technology, such as England, Portugal, Australia dan Scotland.
2) Tidal energy uses changes of marine water volume due to ebb and flow. By means of barrage or tidal turbine, energy generated by water move will be convertible to electricity. The main excellence of this energy lies on its ability to provide reliable and measurable energy after constant ebb and flow. This energy has been popular in such countries as France, South Korea, England, and Canada.
3) Marine current energy uses constant and strong marine current, with under-water turbines similar to wind turbine. Turbines are going to capture kinetic energy from currents. With nearly never-ending potentials, marine current energy might be very stable and efficient energy resources. The energy has been vastly used in England, South Korea, Scotland, and Canada.
4) Marine thermal energy explores thermal differences between marine surface with warm temperature and deep ocean with colder one. Technology of Ocean Thermal Energy Conversion (OTEC) converts the thermal differences into energy. It offers huge potentials for sub-tropical and tropical areas where thermal differences are significantly clear. Some countries have adopted this sort of energy, such as Japan, Solomon Islands, and France.

5) Offshore wind energy is one of many innovations that is so promising in the field of renewable energy. It is generated by wind power breezing over marine surface. Numerous countries have used this technology, such as Germany, England, USA, Denmark, and the Netherlands.

By using huge potentials in blue energy, we will not only able to minimize dependence on fossil fuels, but also to support blue economic growth. Further, blue economy focuses on sustainable utilization of marine resources, and it actually can be significantly encouraged by investments and technological development of blue economy. As a matter of the fact, it unlocks new opportunities for economic development and sustainability, like infrastructure development, job vacancy, and technological innovations for further and better industrial expansion. Blue economy constitutes a policy of marine resource utilization in regards to environmental principles to sustain economic, welfare, and job opportunity growth, as well as marine ecosystem preservation [14].

There are some challenges that need to be faced in order to succeed in the technology development; some of which are extremely high cost needed and complexity to design kinds of technology in response to dynamics of marine situation. Regardless of sustainability of investments and technological advancement, blue energy is potential to hold a very crucial role in the transition to cleaner and more sustainable energy. It does not only provide us with solutions for energy in the future, but also contributes to more far-reaching blue economic growth, making the energy a key factor to global energy sustainability and economic welfare in the future.

Blue economy is sustainable marine economy for economic and social advantages, in addition to ensuring environmental preservation for long-term utilization on the resources [15]. Blue economy signifies a result of global consensus of Conference Rio+20 in 2012 based upon a premise that healthy and sustained marine ecosystem will be more productive, and it is strongly recommended that marine economy be implemented [16]. It, in addition, includes multiple sectors with a purpose of elevating people's welfare and economic growth without damaging marine ecosystems. As a matter of the fact, an economic approach model derived from blue economy will no longer count on destructive development, or over-exploitation. In other words, new understanding will be generated, provoking mindset transformation and development structure through environmental utilization that typifies a breakthrough to keep up with social economic betterment [17]. In this context, the term 'blue' is not only referred to marine color, but also a philosophy of sustainability and innovations leading people to using massive potentials from the marine by means of environmentally-friendly ways.

Blue economy appears as an innovative approach integrating economic growth and marine ecosystem preservation. Blue economy, in addition, proposes a new way of making use of marine wealth, while preserving the sustainability of this planet to go towards cleaner, greener, and more sustainable future. There are some principles of blue economy, including:

1) Social involvement, where local people and stakeholders are involved for decision making and marine resource management in order to ensure that all policies and regulations are fitted to their necessities and aspirations.
2) Integrated management, where multiple sectors allied to marine, such as fisheries, tourism, and energy are integrated into a coherent management framework for the sake of better results and conflict-free situations between one to another sector.
3) Sustainability, in which marine and coastal resources are managed sustainably for ensuring that economic advantages obtained do not cause damages or threats to the marine ecosystem sustainability.
4) Innovation, where new technological advancements and methods are developed in order to maintain the efficiency and effectiveness of marine resource management, including innovations for renewable energy and fishery management sectors.

Blue economy, in addition, will not only cause economic advantages, but also ensure that marine and coastal areas are always protected and productive. Seeing marine as valuable resources that need smart management in order to ensure long-term advantages for life is a holistic approach. Through the

IOP Conf. Series: Earth and Environmental Science **1473** (2025) 012071 doi:10.1088/1755-1315/1473/1/012071

adoption of principles of blue economy, we can create better future, where economic growth and environmental sustainability go hand in hand. The vision helps us go towards sustainable utilization of marine wealth, ensuring that marine stays as sources of life for the next generation.

Blue energy and economy are mutual concepts in regards to sustainable utilization of marine resources. Through smart utilization of blue energy, economy will be meaningfully influenced. It can be ensured that people and future generation can reach the top of economic welfare and have good quality life.

b. Offshore Wind Farm (PLTA): Challenges, Issues, and Opportunities

1) Offshore Wind Farms

When the wind blows through blades of turbines, it causes them to rotate. This rotation is converted into electricity in regard to the principle of electromagnetism, allowing magnets to rotate inside a coil of conductive wire. Then, generated electrical energy is distributed onshore through cables, which is functioning for the needs of homes and businesses [18].

Then, how do offshore wind farms work? Here is to illustrate [19].

a) Electromagnetism: The Basics
   i. Electricity can be generated from magnetic rotation inside a coil of conductive wire.
   ii. Inside the wind turbines, rotation occurs through clean and natural wind power.

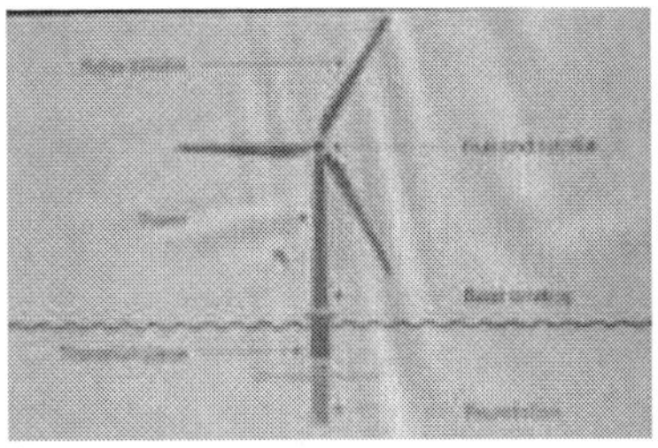

Figure 1.1

b) Inside the wind turbines
   i. To capture wind power, turbines' upper part rotates, facing off the wind. Three blades are set up at precise angles, and the air moving through the blades causes them to rotate.
   ii. In a compartment placed at nacelle, a non-moving part of turbines, rotation of the blades is sent through the drive shaft, but oftentimes through the gear box, in order to rotate magnets in the coil and generate alternating current.

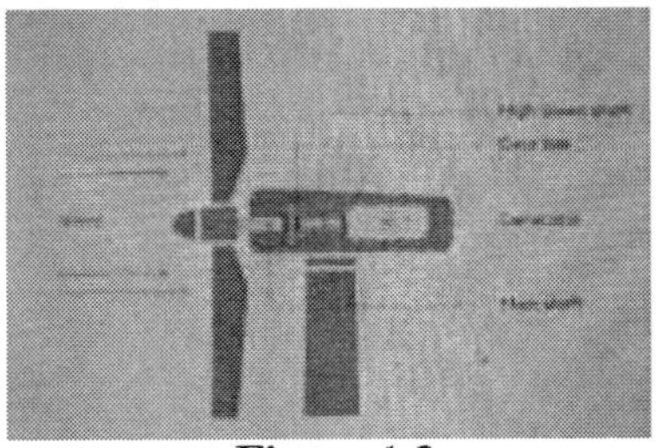

Figure 1.2

c) Wind farm as a power plant
   i. A wind turbine can result in an amount of megawatt (MW) power, which is so big in amount, but relatively smaller than that of steam turbine for conventional power plant.
   ii. Wind turbines are collected in a wind farm as a giant power plant, without any emissions after electricity is generated.

iii. An offshore wind farm consists of several turbines spread over in far-reaching sea. Each of turbines is tightly set on the foundation in the bottom of the sea, with a tower soaring to the sky, allowing blades to rotate at higher speed.

d) Transferring power onshore

i. Each turbine transfers power via wire, heading to the bottom of towers located at the sea bottom where the offshore main shelter stands. At this point, energy is alleviated into higher voltage, which is ready to be sent off onshore via high-voltage wire. Higher voltage indicates small portion of energy loss during the transmission.

ii. Onshore, voltage of the main shelter is readjusted so electricity is transferrable to the electric network and distributable via electric channel to facilities in need of electricity.

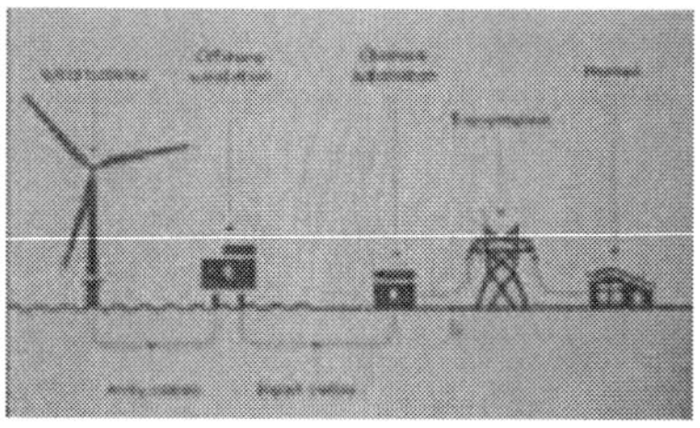

Figure 1.3

In comparison to onshore wind farms generated by land breeze, work principles of offshore ones as abovesaid share very close similarities. It needs an extra infrastructure that connects off- and onshore shelters. Work principles of onshore wind farms can also be found in the wind farms operating in Indonesia. According to Ali, *et. al.,* in a book entitled *Energy Conversion Technique* (2022), characteristics of onshore wind farms are as follows [20].

a. Work principles of onshore wind farms are dependent on turbines or windmills as means of converting wind into electricity.
b. Process starts when wind exists and rotates the turbines.
c. Wind blows blades and automatically rotates rotors in the generator. Generator converts kinetic energy from the wind to electricity due to electromagnetic field.
d. In shafts located in the generator, there are permanent ferromagnetic materials. When turbines rotate, the shafts do as well and cause magnetic flux change at the stator, a loop-shaped wire.
e. At this moment, electric voltage and current are generated. Electricity is, further, stored in a battery, and is ready to use.

Figure 1.4 PLTB Sidrap
in Sidenreng Rappang Regency, South Sulawesi [21]

Other than onshore wind farms, why are offshore ones also very important to build, and why can it be the top priority for development in Indonesia? Nowadays, if the world is trying to bear with climate change and to set free from dependence on using fossil fuels, Indonesia has been in the right and promising track to make use of renewable energy. One of the most possible options that needs alert attention is to get equipped with offshore wind farms. With due considerations of some important factors, offshore wind farm development appears as a main target to carry out strategies for Indonesia's renewable energy expansion.

1. Potentials and space availability – Indonesia as the biggest archipelagic country with far-reaching coastal line is so potential for offshore wind farm development. Some spaces potential to build wind farms are Javanese, Sumatran, and Kalimantan oceans, with all of which offering stable and strong wind power.
2. Technological development – Technology of offshore turbines has significantly been advanced, and is proved reliable in some countries across the universe. The countries like Denmark, Germany, and England have succeeded in advancement of offshore wind farms. It literally indicates that this sort of technology is effectively applicable. Indonesia, in some extents, can learn from experiences and innovation exemplified by those countries to accelerate technological implementation that has been proved successful. Success of these international projects has been a real proof that this sort of technology is relatively adaptable in Indonesia, with less technical risks, to alleviate the development efficiency.
3. Economic growth – Offshore wind farm development has offered intense economic benefits. In addition to generating renewable energy, this project can be a determinant factor of multiple-sector growth, like construction, logistics, and maintenance services. Investments for offshore wind farms will create new job opportunities, elevate skills of workforces', and keep sustaining local and national economic growths. By running the projects, industrial ecosystem for renewable energy in Indonesia can be reinforced, and national competitiveness in global market for clean energy is enhanced.
4. Governmental commitment and support policies – Indonesian government has shown determined commitment to supporting the development of renewable energy through policies and incentives that are positive with blue energy. For now, onshore wind farms are already massive, so offshore ones are just in need of technological development of the existing technology. This regulation opens up opportunities for investments for and development of renewable energy projects. Support policies are urgent to ensure the success of offshore wind farm and long-term sustainability of the investments for renewable energy.

2) Comparing Onshore and Offshore Wind Farms

In regards to global strategies for climate change and transformation to cleaner energy resources, a number of technologies for renewable energy have offered multiple solutions. Amidst the existing, offshore wind farm is showy and one of the most promising methods, together with onshore wind farm and others. However, if we need to choose one that is more preferable between onshore and offshore wind farms, some factors need to be considered to fit in the aptest solution based on the actual situation. Thus, how do the onshore and offshore wind farms differ one from another?

a) Energy Potentials and Efficiency

i. Offshore wind farm offers enormous potential energy. Marine areas obviously possess stronger and more consistent wind power than those of on land. It means that offshore turbines are operated by means of high-efficient energy, making it more effective in resulting in greater amount of energy and, of course, more stable.

ii. On the other hand, onshore wind farm varies depending on locations and topographical characteristics. That is why onshore turbines possess equitably lower efficiency than those of the offshore in some wide-ranging conditions.

b) Cost

i. Cost is a key factor to every decision making in the development of renewable energy. To build up offshore wind farm, early investments in huge sum of money are needed, especially for the turbine installment offshore and support infrastructure construction, like foundation, submarine cables, and operational facilities. In addition, maintenance cost for offshore turbines is costlier due to more difficult access and weather issues in the oceans.

ii. Onshore wind farm, meanwhile, is relatively cheaper in terms of initial investments as it needs no special marine infrastructure. Unlike the offshore, maintenance cost for it is much more affordable due to easier access and friendlier environmental conditions.

c) Environmental impacts

i. Regarding environmental impacts, offshore wind farm has lower risks of errors due to lack of disturbances because of land usage. Offshore turbines are located in the oceans, so it causes no conflicts of land usage and results in insignificant visual impacts. On the other hand, considering potential impacts that might harm marine ecosystem, like fish and bird migration, and environmental impacts after installment and operation is crucial.

ii. Onshore wind farm results in easily manageable and controllable impacts, but causes disturbance upon land usage and wild life animals. In addition, installment of onshore turbines may worsen natural views and other needs of land usage.

d) Operational Challenges

i. Offshore wind farms are built with challenges, especially inaccessibility and extreme weather, which can influence operational and maintenance performances. However, the excellence of offshore wind farms lies on offshore wind speed and consistency. However, greater energy and more stable energy production after the installment can pay off those challenges.

ii. Onshore wind farms have better accessibility and are more affordable for maintenance and operation. However, irregular wind speed and limit to land usage can be an issue to achieve optimum energy and its efficiency.

Simply put, off- and onshore wind farms have their own excellences and weaknesses. The former offers greater potentials and more efficient energy, but with costly investments and maintenance, while the latter needs more affordable investments in the beginning and easier maintenance procedures, yet with less significant potentials and less efficient energy due to varied wind situations.

3) Challenges, issues and opportunities in establishing offshore wind farms

a) Challenges in establishing offshore wind farms

i. High initial cost – Initial cost needed to build up offshore wind farms is the biggest challenge as it must be very costly. Cost includes advanced technology of wind turbines, robust support structure, and transmissive infrastructure for submarine energy. Also, research and development (RnD) for new and more efficient technology also costs extra price.

ii. Marine condition – Offshore wind farms might be operated in certain marine areas with high waves, strong wind, and possibility of proneness to corrosion because of the water. Consequently, special designs of technology might be needed in order to defend from such a situation, which must require high cost.

iii. Environmental impacts – Offshore wind farm is environmentally-friendlier than other fuels. However, it is still potential to bring about effects to marine ecosystem, including habitats inside. In addition, noise due to construction and operation activity may affect marine life.

iv. Regulations and Licensing – Both may retard progress of development since obtaining licenses and obeying rigid regulations in regards to environment, security, and usage of marine space take time and energy-draining. In addition, harmonizing projects towards users' interest is oftentimes problematic due to complicated process of negotiation. In addition, legal status for establishing offshore wind farms over marine usage to plant in the turbines is in need of certain regulations due to the fact that marine space usage will be limited in order to ensure the security and local people interests.

b) Issues in establishing offshore wind farms

i. High Cost – Offshore wind farms are in need of advanced and special technology and infrastructure, in addition to development, and it is necessary to support its sustainability by provision of equipment that also cost big amount of money for its investment and research needs.

ii. Integrated power – Energy resulted by marine must be forwarded to land through submarine cables, which also needs very high cost.

iii. Natural conditions – Though offshore wind power is more consistent that of onshore, seasonal variances and extreme weather change influence energy generation. Further, environmental features, such as marine temperature and salinity can affect performance and durability of used equipment.

c) Opportunities in establishing offshore wind farms

i. With challenges and issues, offshore wind farm development offers one most promising opportunity, huge potential energy. In addition, offshore wind farms possess capacity to generate great amount of energy, especially at the spaces with strong and consistent wind power.

ii. Reduction of carbon emission – Offshore wind farms generate electricity with no carbon emission, which is positive with global strategies to diminish effects of climate change. By limitless wind resources, offshore wind farm is an alternative for sustainable energy in the future.

iii. Unlocking job opportunities and better economy – Offshore wind farm development and operation unlock more job opportunities in multiple sectors, such as construction, engineering, and maintenance. Investments for technology will also stimulate economic growth around the coastal areas by means of infrastructural improvement and industrial activities.

iv. Energy resource diversification – Offshore wind farms will enlarge diversity of energy resources, help reduce risks due to dependence on one specific type of energy resource, and strengthen defensive system of energy. By integration of multiple renewable energy resources, more stable and efficient energy system can be created.

c. Offshore Wind Farm as Blue Economic Pillar

Amidst the global challenges regarding climate change and urgent needs in transition to renewable energy resources, offshore wind farm comes up with its excellence as one of blue economic pillars. Blue economy, in addition, is focused on the sustainable use of marine resources in support of economic growth, unlocking more job opportunities, and protection of marine ecosystem, which requires strong partnership for its technological advancement. With ability to make use of offshore wind power, offshore wind farms offer several benefits positive with blue economic principles.

1) Economic potentials

a) Investment effects to economy – Investment for offshore wind farms does not only alter point of view of energy usage, but also significant effects to the economy. Establishing offshore wind farms in the future will influence environmental aspects and energy, unlocking significant opportunities for wide-ranging economic sectors. Investments as mentioned include:

i. Unlocking job opportunities – Job opportunity will be one of live effects of investments for offshore wind farm establishment. During construction, hundreds of job opportunities are created in order to fill in such professional demands as designers, engineers, technicians, and construction workers.

ii. Infrastructural development – During the infrastructural establishment of offshore wind farms, special port facilities to transport and assemble the wind turbines are highly needed to elevate transportational infrastructure to carry out fulfilment of other needs. Further, investments are not only meant to sustain infrastructure, but also hoist economic activities around the ports.

iii. Income and tax – In this mega project, extra incomes to local government can be obviously obtained from property, company, and other contribution taxes. In addition, success of this project can attract more investments for renewable energy and other relevant sectors.

iv. Life quality of local community – With the infrastructural establishment of offshore wind farms, life quality of local community will automatically be hoisted positively. Corporate Social Responsibility (CSR) serves to be support for establishment projects, in charge of compensating and taking care of local community activities, such as education and social programs. Economic diversification due to offshore wind farm establishment helps diminish dependence on traditional industries, like fishery and agriculture, which multiple times is considered as 'backbone' of local economy.

b) Projection of sectoral growth and contribution to gross domestic product (GDP) – Investments for offshore wind farm development is not only significant for energy resource diversification, but also for local economic growth. Existing offshore wind farms will play an important role at economic development in the nearest future. By support to auxiliary technology and policies, offshore wind farms will always be the main support for economic growth and energy sustainability in the future.

2) Unlocking Job Opportunities

a) Offshore wind farm industry will be significantly progressing, unlocking the gate to more promising job opportunities. Job opportunity starts from planning to operational phases. Creating job opportunity is not only to support development of energy, but also to significantly contribute to sustainable economy. Establishment involves planning, designing construction, and operational maintenance, which are essential to ensure the success of projects in support of local economic in addition to accelerating transition to cleaner and more sustainable energy resources.

b) Offshore wind farm development project will not only cause advancement at renewable energy, but also is so significant for job skill development and training. Providing opportunities for special training reinforces education and innovations in developing training method. Consequently, it does not only strengthen individual skill, but also local and national economic growth, indeed. By means of integrated and sustainable approach, offshore wind farms contribute to developing skillful and readier workforces in the future with all possible challenges in renewable energy industries.

3) Infrastructural Development

a) In the next future, offshore wind farm project will be more advanced as it is projected as solution for sustainable renewable energy, which needs integrated infrastructure to support all aspects, from establishment to operational development. It includes special ports, ships, storage facilities, and other support elements that are helpful to ensure the success of project from the beginning the wind farms are being operated. If investments are held precisely for this infrastructure, offshore, together with onshore wind farms can collaboratively be operated efficiently, which helps ensure that renewable energy is provided and used stably and sustainably so that it helps contribute to global strategies in diminishing dependence on fossil fuels usage.

b) Existence of offshore wind farms will be key to transformation of renewable energy, with the onshore that has been being operated until this time. However, the effects are not only limited to the energy sector. Massive contribution of the sector can also vastly affect relevant industries, especially transportation and logistics.

By infrastructural reinforcement for some dimensions needed, transportation, logistic management, job opportunity provision, and technological innovation will be in need of energy sector's roles. The effects, in addition, will not only be the cause of offshore wind farm success, but also help advance industrial capability holistically.

4) Economic Diversification

With the discourse of offshore wind farm development, it is obvious that offshore wind farms will be main support in the transition to clean energy, vastly contributing to minimizing global dependence on fossil resources. Offshore wind farm, in addition to onshore, offers an effective and sustainable solution to the extent of urgent climate change and needs of energy resource diversification. By providing clean energy, elevating energy resource diversification, generating large-scale energy, upturning climate change effects, and encouraging sustainable investments, offshore wind farms can accelerate transition to cleaner and more sustainable energy. It constitutes a vital strategy in order to reduce environmental consequences due to energy, and creates environmentally-friendlier solution for the next generation.

d. Offshore Wind Farms as Support for Sustainable Energy

In a certain epoch in the middle of global awareness of climate change and importance of sustainable energy, offshore wind farm appears as a solution to ensure the future of clean and sustainable energy. In fact, this sort of technology does not only offer environmentally-friendlier solution than fossil fuels, but also has played an important role in support of global energy sustainability.

1) Clean and Renewable Energy
   a) In the middle of global move against the climate change and greenhouse gas emission, offshore wind farm appears as one of the most possible solutions by the use of wind power in order to produce clean and sustainable energy.
   b) Unlike other power plants with fossil fuels that generate hazardous smoke and carbon-dioxide, offshore wind farms generate electricity without greenhouse gas emission. In addition, offshore wind farms will not only generate energy, but also directly contribute to declining air pollution and betterment upon environmental damages.
   c) Offshore wind farms will also be predicted to be able to contribute to creating more stable and sustainable energy network since it is able to generate constant electricity. These wind farms also help stabilize electric network and reduce any energy resources that can negatively affect the environment.
   d) Offshore wind farms, in addition, are not only going to remark technological advancement, but also a concrete strategy in the transition to greener and more sustainable future. Further, the wind farms serve to be the main pillar to decline greenhouse gas emission and to protect our planet for the next generation and for cleaner and healthier life.

2) Energy Stability
   a) The main excellence of offshore wind farms lies on stability and consistency at producing energy due to absence of obstructive factors, e.g. mountain and big buildings, in the middle of the wide-ranging sea. This condition will create an ideal environment for turbines to rotate, allowing more efficient usage of stronger wind speed.
   b) Basically speaking, wind speed is relatively strong, which allows the turbines offshore to be able to generate more energy with higher efficiency level. In addition, marine will not

only provide strong and speedy wind, but also sustain the consistency of energy production every day and year. That is the reason why offshore wind farms can generate clean and stable energy.

3) Less Environmental Damages
   a) It has been explained previously that during its operation, offshore wind farms will not result in greenhouse gas emission, which means that it will not be the cause of global warming, but generate clean and renewable energy and decline air pollution instead.
   b) Possibility of environmental damages that need the most thoughtful attention occurs during establishment and infrastructural installment that will likely cause defect on marine ecosystem [22]. In addition, vibration and noise due to usage of heavy equipment may also cause audio pollution and disrupt marine species.
   c) Offshore wind farms offer much cleaner and environmentally-friendlier option with less significant impacts, especially in regard to air quality and ecosystemic health, than fossil energy, like coal, oil, and gas. Fossil-generated combustion is the reason why greenhouse gas emission exists, which lively contributes to climate change, air and water pollutions, and wide-ranging devastation on the natural habitats.
   d) Other possible excellence post-establishment of offshore wind farms is that the submarine buildings can be functioned as artificial reefs that can enrich local diversity in the local marine areas.

6. Conclusion

Energy as the essential need for every aspect of human life is mostly fossil-generated. It results in greenhouse gas emission that contributes to climate change and air pollution. Blue energy as renewable energy must be the main priority, so in this case, offshore wind is the blue energy as meant to be used as power plant to generate clean energy that is friendlier to environment and sustainable though onshore wind farms have been operated initially. In addition, offshore wind farms by the use of offshore wind plays an important role to support blue economic growth as well as holistic energy sustainability.

a. Offshore Wind Farms as Blue Economic Pillar
   1) Investments for offshore wind farms will not only transform the point of view over energy usage, but also economic sector.
   2) Investments for offshore wind farms will not only contribute to diversification at energy resource, but also local economic growth.
   3) Investments for offshore wind farms will significantly contribute to global dependence on fossil-generated resources.

b. Offshore Wind Farms as Support for Energy Sustainability
   1) Offshore Wind Farms are generated by offshore wind, which generates clean and sustainable energy in response to climate change and in order to decline global warming effects and greenhouse gas emission,
   2) Offshore Wind Farms have excellences in regards to stability and consistency for energy production.
   3) Offshore Wind Farms have played an important role at generating more stable and sustainable energy due to its consistent ability for electricity generation.

c. Challenges and Issues in the Establishment of Offshore Wind Farms
   1) High initial development cost
   2) Dynamic marine condition that influences infrastructural development
   3) Difficult and rigid regulations and licensing due to complexity of marine issue, security, and marine space usage
   4) More complicated technology and infrastructure needed than that of onshore.

d. Opportunity in the Establishment of Offshore Wind Farms
   1) Offshore wind farms possess capacity to generate large-scale energy, mainly at certain locations with robust and consistent wind power.
   2) Offshore wind farms generate electricity with no carbon emission, which appears to be a strategic solution for sustainable energy in the future.
   3) Offshore wind farms unlock multiple-sector job opportunities, and can stimulate economic growth in coastal areas.
   4) Offshore wind farms enrich energy resource diversity, and help decline risks of dependence on one single energy source only.

Acknowledgment
The authors would like to thank Hang Tuah University for supporting this research. The author also would like to thank the BRIN Energy Conversion and Conservation Research Center, especially the wind energy sector, which has supported and provided enlightenment in writing this research.

References
[1] Econusa. Eco-defender, a renewable energy generator. [Online Article]. Accessed from https://econusa.id/id/ecodefender/pembangkit-listrik-energi-terbarukan/
[2] Daryanto, Y. Wind Potential Study for Wind Power Plant. Yogyakarta: PPTAGGUPT-LAGG. 2007.
[3] Katadata. Green energy and new energy: A cluster of the biggest wind farms in ASEAN. [Online Article]. Accessed from https://katadata.co.id/ekonomi-hijau/energi-baru/66352602e89ac/sederet-lokasi-pembangkit-tenaga-angin-ri-ada-yang-terbesar-di-asean
[4] International Energy Agency. Offshore wind outlook in reviewing the ecological impacts of offshore wind farms. [Online Article]. Accessed from https://iea.blob.core.windows.net assets/495ab264-4ddf-4b68-b9c0-514295ff40a7/Offshore_Wind_Outlook_2019
[5] Ministry of Energy and Natural Resources. Geo-marine surveyor ship as the solution. [Online Article]. Accessed from https://www.esdm.go.id/id/media-center/arsip-berita/kapal-survei-geomarin-iii-sebagai-sebuah-jawaban
[6] Academia Edu. Study of wind tidal wave and current potential in Sunda Strait as an alternative energy. [Online Article]. Accessed from https://www.academia.edu/64297479/Study_of_Wind_Tidal_Wave_and_Current_Potential_in_Sunda_Strait _as_an_Alternative_Energy
[7] Rizkiani, I., et. al. Analysis and Simulation of Energy Conversion from Wind to Electricity by means of Feedback Linearization Control Method. Journal Science and Art ITS. 2012.
[8] Sugiyono. (2017). Educational Research Methodology: Approach to Quantitative, Qualitative, and R&D Research. Bandung: Alfabeta.
[9] Ramdhan, M. Methods of Research and Media Creation for the Nation (CMN). 2021.
[10] Ulfatin, N. Qualitative Method for Education Field: Theory and Its Application. Media Nusantara Creative.
[11] Kurnia, Aisyah., et al. Feasibility study of offshore wind farms in the Indonesian archipelago. Journal of Renewable and Sustainable Energy. 2021.
[12] Yang, Minxian, et. al. Techno-economic feasibility of offshore wind farms: A review. Journal Renewable and Sustainable Energy Reviews. 2023.
[13] Lestari Kompas. Kinds of marine energies that can be used for electricity resources. [Online article]. Accessed from https://lestari.kompas.com/read/2023/06/21/140000986/4-jenis-energi-laut-yang-bisa-dijadikan-listrik?page=all#google_vignette
[14] Coordinator Ministry of Economy of the Republic of Indonesia. (September 2, 2024). About blue energy, Minister Airlangga claimed that energy transition should be based on benefits and fair distribution. [Online Press Conference]. Accessed from

https://www.ekon.go.id/publikasi/detail/4504/singgung-energi-biru-menko-airlangga-sampaikan-transisi-energi-harus-dengan-manfaat-dan-biaya-yang-terdistribusi-secara-adil.
[15] Nasution, M. Potential and challenges of blue economy in support of economy in Indonesia: A literature study. Journal Budget **7**, 2nd Edition: Centre of Budgeting Study. 2022.
[16] Damanik, Riza., et. al. Strategical Projects of Blue Economy towards Advanced Country 2045. Laboratory of Indonesia 2045 (LAB 45). 2023.
[17] Prayuda, R. and Sary, D. V. Indonesia's strategies in implementing blue economic concept towards coastal people empowerment in ASEAN economic era. Indonesian Journal of International Relations, **3**, No. 2., pp 46-64. 2023.
[18] https://us.orsted.com/renewable-energy-solutions/offshore-wind/what-is-offshore-wind-power/how-do-offshore-wind-turbines-work, diakses tanggal 29 Juli 2024 jam 11.16 WIB
[19] https://us.orsted.com/renewable-energy-solutions/offshore-wind/what-is-offshore-wind-power/how-do-offshore-wind-turbines-work, diakses tanggal 29 Juli 2024 jam 11.16 WIB
[20] Ali, M. T., et. al. Energy Conversion Engineering. Accessed from https://books.google.co.id/books/about/Teknik_Konversi_Energi.html?id=LaaBEAAAQBAJ&redir_esc=y. Rizmedia Pustaka Indonesia. 2002.
[21] Makassar Kompas. Sidrap offshore wind farms, the giant windmill to harvest energy for lightning. [Online Article]. Accessed from https://makassar.kompas.com/read/2022/01/20/215429378/pltb-sidrap-kincir-angin-raksasa-yang-memanen-energi-untuk-menerangi
[22] Hall, R., João, E., & Knapp, CEnvironmental impacts of decommissioning: Onshore versus offshore wind farms. Environment Impact Assessment Rev. **83**, 106404 in reviewing the ecological impacts of offshore wind farms. Available at https://www.nature.com/articles/s44183-022-00003-5#ref-CR1. Accessed on August 20, 2024, at 07:13 a.m. 2020.

IOP Conf. Series: Earth and Environmental Science **1473** (2025) 012072 doi:10.1088/1755-1315/1473/1/012072

Characteristics Surface Current for Sustainable Fishing on Map FMA-718

R Tambun[1], Supartono[1], V D Prasita[1], S Widiyanto[1], Poundra[2].

[1]Magister of Marine Engineering, Hang Tuah University, Surabaya 60111, Indonesia.
[2]Department of Naval Arcitechture, Hang Tuah University, Surabaya 60111, Indonesia.

E-mail: Robert.tambun@hangtuah.ac.id

Abstract. This research aims to investigate surface currents in Fish Management Area 718, located between 128°-141° East Longitude and 4°-11° South Latitude. The study seeks to create a thematic map and identify FMA-718 based on four seasons in specific Indonesian waters. It focuses on determining the spatiotemporal distribution of current direction and speed, derived from the Pushidrosal sea map and OFS data, and their relation to Fishing area characteristics. Data were processed at the Computer Laboratory of Marine Engineering, Faculty of Engineering and Marine Science, Hang Tuah University, using primary data from the Meteorology, Climatology, and Geophysics Agency's Ocean Forecast System (OFS) for 2021-2023, and secondary data from the Sea Map of Hidros digitizer no.151 Papua and Surrounding Islands. Analysis and map creation were done using MapInfo Pro 15 software. The study aims to identify temporal differences in fishing spots by examining the spatiotemporal distribution of current data, useful for fishing business actors in planning operations. Findings indicate that the Aru Sea, Arafura Sea, and Timor Sea are ideal for fishing due to favorable surface current conditions, characterized by weak flows year-round, supporting fish resource survival. However, during the east season, westward currents may reduce fertility in fishing areas.

1. Introduction

Fish Management Area 718 (FMA) waters, which encompass a geographic region between 127°-141° East Longitude and 4°-11° South Latitude, can be divided into three distinct areas, each possessing unique characteristics. The northern and eastern waters face the Papua mainland, while the western waters open up to the Banda Sea, and the southern waters connect to the Arafura Sea. The southern region comprises the Indonesian Exclusive Economic Zones, which borders the waters of the Australian Continent, and the Arafura Sea.

The prevailing current pattern in FMA waters moves horizontally in both the east and west directions and vertically due to variations in water depth (DISHIDROS, 2015). The sea current pattern in Indonesian waters is generally influenced by changes in monsoon winds and tidal influences with significant amplitudes, especially in the surface layer. During the east monsoon, which occurs from June to August, water masses from the Banda Sea are pushed towards the Flores Sea, then into the Java Sea and the Makassar Strait, driven by winds coming from the west across the Flores Sea towards the Banda Sea [1].

Current flow involves the movement of water masses either vertically or horizontally. Vertical currents or seafloor currents are influenced by the surface or bottom relief of the seafloor and move towards the surface, where the current originates from the South Pole towards the Equator [2]. Horizontal currents in the surface layer are influenced by the wind, which moves towards balance due to the magnetism of the Earth-Moon-Sun solar system [3]. In light of changes in wind direction movements, research is needed to determine surface current movement patterns according to the season at FMA 718.

The objective of this study is to identify the spatiotemporal distribution of current direction and speed sourced from Pushidrosal nautical charts and Ocean Forecast System (OFS) data in relation to fishing area characteristics. This information is expected to help fishing business actors in place time and location of fishing operations at FMA 718 by considering the temporal distribution of fishing spots.

2. Research Methods

The research conducted for this study utilizes data from the Pushidrosal map No. 151 Papua and Surrounding Islands [4]. The geographical boundaries for this study were determined by reference points from the 66 FMA 718 map. The surface current data used in this study was obtained from the BMKG website for the period December 2021 to November 2023. The spot location was chosen based on its position between land, proximity to contour lines, and its representation of the entire research location for surface current data. The data collected from all observation spots is used to display surface current patterns according to the seasons at FMA 718.The analysis of surface current information was conducted in two stages. The first stage utilized the GIS software MapInfo Pro version 15.2.0 (64 Bit) Release Build 218 [5] to determine surface current patterns using thematic mapping type Bar Charts and visual dimensions to analyze information from BMKG maps and Pushidrosal nautical charts. The second stage divided the Pushidrosal map no. 151 at a geographic position of 127°- 141° W and 4°-11° S into several grids using grid maker tools, providing an index code (A1-N7) on the map [6].

3. Results and discussion

The detection of current direction and speed can be determined at various depths on the BMKG Maritime Meteorology Center - Ocean Forecast System (OFS) site, starting from the surface to 10m, 25m, 50m, 100m, and 250m. [7] The maximum current speed occurs at the surface of the water and decreases with increasing depth. The movement of water masses from one place to another is known as ocean currents [8]. Surface currents are influenced by wind and tides [9]. Information about currents is crucial for various purposes in marine and fisheries matters.

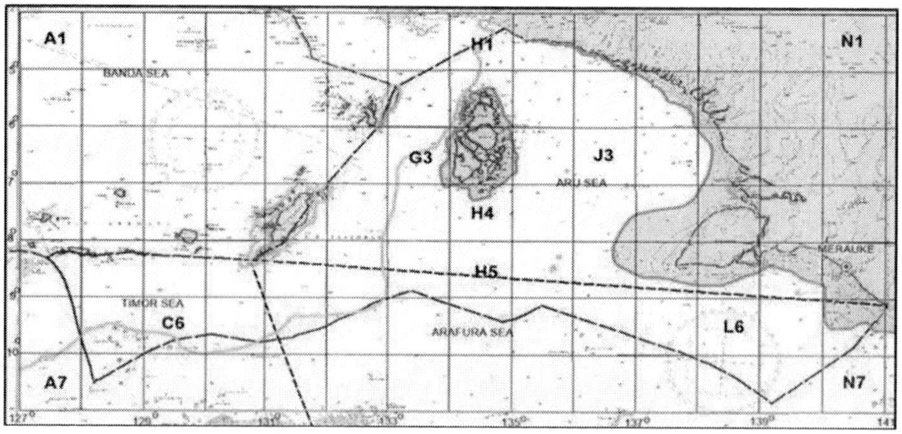

Figure 1. Fish Management Area 718 map layout.

In FMA 718, two types of current circulation exist: the monsoon current system and the Indonesian Cross Ocean Current (ARLINDO). The latter transfers water from the Pacific to the Indian Ocean, with seasonal variations in the Arafura Sea [10]. The distribution and abundance of aquatic resources are closely linked to oceanographic conditions and variations. Comprehensive and accurate information about these characteristics is essential for sustainable aquatic resource management [11]. Surface current patterns are stable during the West Season (WS) and East Season (ES) when currents align with the wind. During transitional seasons (T1 and T2), currents are affected by sea water movement between land masses, with constantly changing wind directions. [12] Who noted that currents from the west, such as from the Southeast Indian Ocean, South Java Current, and Flores Sea, flow eastward towards the Timor Sea and Sawu Sea, following the Northwest monsoon wind.

The relationship between vertical and horizontal currents and the life cycle of marine biota in FMA 718 is a complex and intricate one. The characteristics of the waters in the western part are such that they are deep and open, while those in the eastern part are shallow and land-covered. [13] Fish eggs and larvae typically remain within a radius of 160 km from their spawning origins. However, under certain conditions, such as the presence of currents, they may migrate up to a radius of 300 km. It is not uncommon for fish species to experience changes in spawning behavior as a means of avoiding the mortality of eggs and larvae offshore, which is often caused by seasonal changes in current direction and speed. These currents can have a significant impact on the spatial distribution of fish eggs and larvae by carrying them to nursery areas [14].

3.1. Aru Sea Area (H1, G3, J3 and H4 Zones)

The observation location in the waters of the Aru Sea in the H1 zone is situated at a geographic position ranging from 134-135° East Longitude and 4-5° South Latitude. This position is located between the Aru Islands and the mainland of Papua, to the east of the Kai Islands or the Banda Sea. The depth of the water in H1 zone ranges from 14-3375 meters. This depth falls within the bathypelagic zone, which has a depth of 1000-4000 meters.

The variability of current direction and speed in the H1 zone is influenced by the monsoon season. The patterns of current movement based on seasonal differences in the H1 zone are depicted in Figure 2. During the west season (WS), currents tend to move towards the east (E) and southeast (SE). The current moving towards the east (E) has speeds ranging from 20-30 cm/s, 10-20 cm/s, and 30-45 cm/s, while the current moving towards the southeast (SE) has a speed of 10-20 cm/s (Figure 2a). This is consistent with the findings of Nontji (1993), who stated that when the MB flows towards the east, the currents tend to move in that direction.

During the West-East transition season (T1), current movement patterns tend to point towards the east (E) and northwest (NW). The currents moving towards the east are dominated by speeds ranging from 30-45 cm/s and 20-30 cm/s, while those moving towards the northwest (NW) are dominated by speeds ranging from 20-45 cm/s (Figure 2b). The variability in current direction is influenced by wind direction, which tends to fluctuate during the transition season. This is consistent which suggest that wind plays an important role in generating the direction and speed of currents

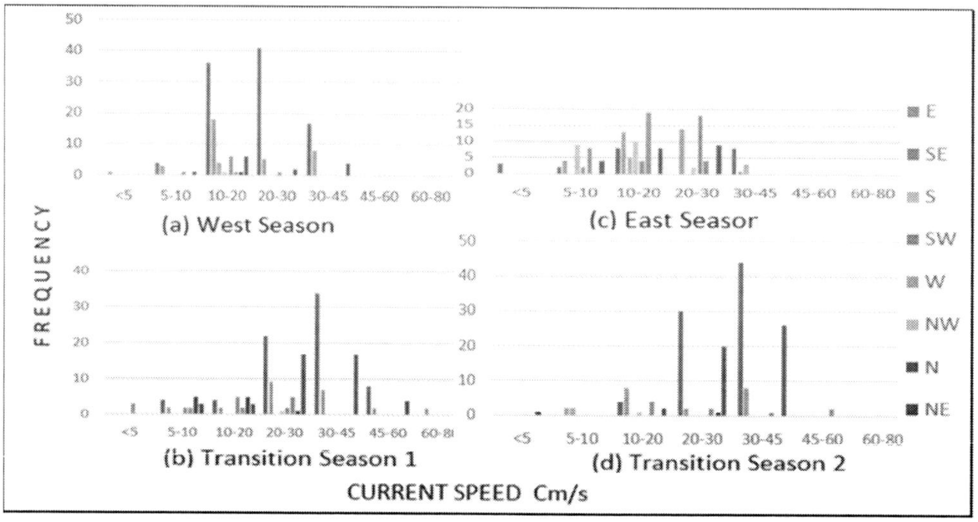

Figure 2. Direction and Speed of Flow According to Season in H1 Zone.

The location for observing the water currents of the Aru Sea in the G3 zone is situated between 133-134° East Longitude and 6-7° South Latitude, between the Kai Islands and the Aru Islands, and it is bordered by open waters to the northwest, southwest, and south. The characteristic water depth in the G3 zone ranges from 9-3435 meters, as reported in 2015 by DISHIDROS. This depth is within the bathypelagic zone, which ranges from 1000-4000 meters.

The pattern of current movement based on seasonal differences is displayed in Figure 3. During the West Season (WS), currents tend to move towards the northeast and east. The current flowing to the northeast is predominantly at a speed of 10-20 cm/s, while the current flowing towards the east has a speed of 5-15 cm/s, as shown in Figure 3a. This is consistent with Nontji's findings in 1993, who stated that the direction of the current is influenced by the direction of the wind, which flows to the east during the WS. In the West-East transition season (T1), the current movement pattern tends to be southwest (SW) and southeast (SE). The currents moving towards the southwest (SW) are primarily at speeds of 10-30 cm/s and 25-40 cm/s, while those moving towards the southeast are primarily at speeds of 20-30 cm/s, as shown in Figure 3b.

During the East Season (ES), the currents typically move in a southwesterly (SW) and northwesterly (NW) direction at speeds of 20-30 cm/s, 30-45 cm/s, and 10-20 cm/s, respectively (as depicted in Figure 3c). Additionally, during the East-West transition season (T2), the current movement pattern is somewhat similar to ES, which is dominant towards the West. The current speed heading west during the T2 season is 20-30 cm/s, 30-45 cm/s, and 10-20 cm/s, as shown in Figure 3d. [15] noted that Eddy currents can occur in the middle of the T2 season, when the current flows towards the southeast and causes a decline in chlorophyll-a to a depth of 400 meters, where Tuna habitats are often found. Similarly, Supartono et al. (2019) and Tambun et al. (2018) reported that Tuna fishing areas are typically located on contour lines greater than 200 meters.

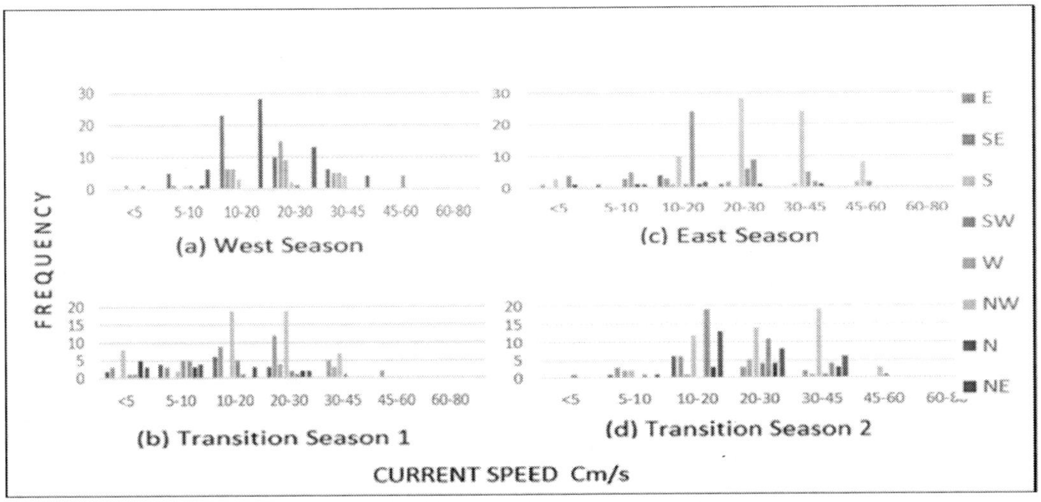

Figure 3. Direction and Speed of Current According to Season in G3 Zone.

The current observation site in the Aru Sea is located in J3 zone, between 136-137°E and 6-7°S, bordered by the Aru Islands to the west and mainland Papua to the north, east, and southeast. Water depths in J3 zone range from 27-43 meters. [16] This area falls within the neritic zone, defined as less than 200 meters deep. The neritic zone, part of the euphotic zone, supports abundant phytoplankton and seaweed due to high oxygen and sunlight levels. Fertile fishing grounds around the Aru Islands are influenced by extensive mangrove forests and large river estuaries on the Papua coast, which enhance primary productivity and serve as buffer zones for fish resources. These areas are crucial nursery grounds for juvenile fish, particularly species that migrate to deeper waters as adults.

The current direction and speed in the J3 zone are influenced by the monsoon season. Figure 4 illustrates the pattern of current movement based on seasonal differences. During the West Season (WS), currents tend to move towards the east, northeast, and southeast at speeds of 10-20 cm/s (Figure 4a). In the West-East transition season (T1), the pattern of current movement in the J3 zone varies predominantly southwest (SW) at a speed of <5 cm/s and northeast at a speed of 10-20 cm/s (Figure 4b). During the East Season (ES), the direction and speed of the J3 zone vary significantly, with a dominant northwest direction (NW) at a speed of 10-20 cm/s and a southwest direction (SW) at a speed of 5-10 cm/s (Figure 4c). In contrast, the East-West transition season (T2) experiences a dominant southwest direction at a speed of 10-20 cm/s, with a current heading south at a speed of 10-20 cm/s and southeast (SE) at a speed of 10-20 cm/s (Figure 4d). Noted that eddy currents can occur in the middle of the T2 season, when the current flows southeast and carries algae blooms as a substitute for chlorophyll-a at the 20-200 m contour line. This area is ideal as a habitat for shrimp, as reported by Tambun et al. [17]

The observation location in the Aru Sea's waters, in the H4 zone, is situated between 134-135° East Longitude and 7-8° South Latitude. This location lies south of the Aru Islands, with open water to the eastern, southern, and western. The water depth in the H4 zone ranges from 11-52 m, as reported by DISHIDROS in 2015. This depth falls within the neretic zone, which has a depth of <200 m.

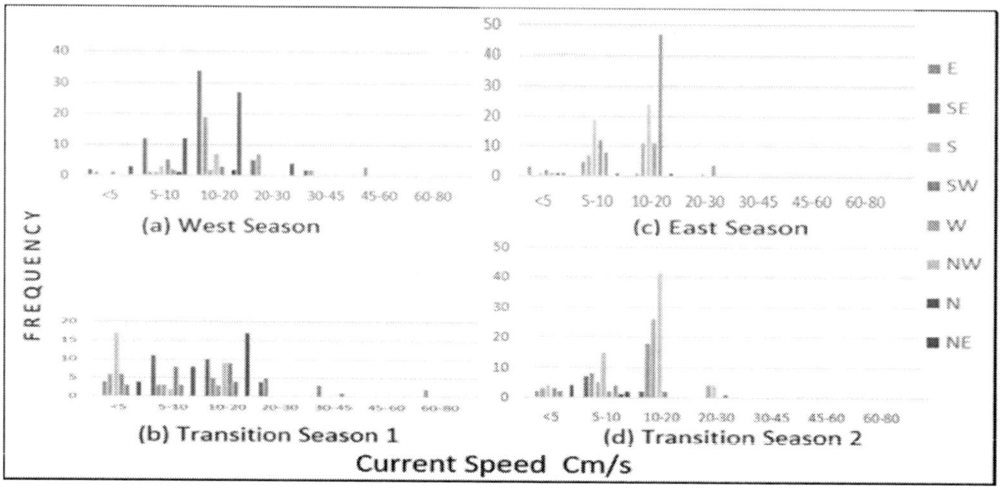

Figure 4. Direction and Speed of Current According to Season in J3 Zone.

The variability of current direction and speed in the H4 zone is influenced by the monsoon season. The current movement patterns based on different seasons can be observed in Figure 5. During the West Season (WS), currents tend to move towards the eastern (E), southeast (SE), and northeast (NE). The dominant current speed moving towards the east is 5-10 cm/s, while the current moving towards the southeast has a speed of 45-60 cm/s and ≤ 5 cm/s, as depicted in Figure 5a. The direction and speed of this current are, of course, influenced by the direction and speed of the wind.

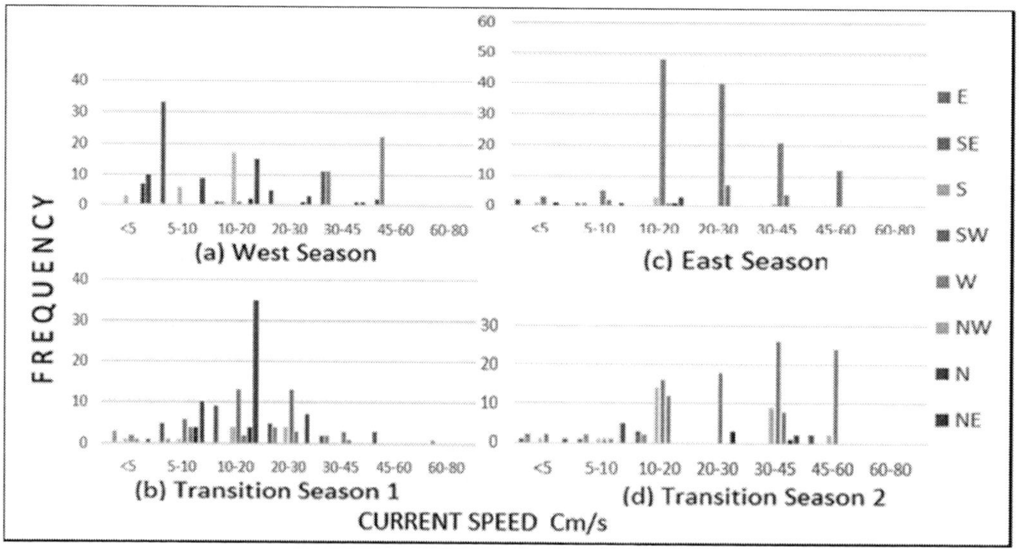

Figure 5. Direction and Speed of Current According to Season in H4 Zone.

In the West-East transition season (T1), the movement pattern of current directions is more varied than west season (WS), with a dominant direction towards the northeast (NE) and west (W). The current moving towards the northeast is dominated by a speed of 10-20 cm/s, while the speed of the current moving towards the west (W) decreases, with a dominance of 10-30 cm/s, 20- 30 cm/s, 30-45 cm/s, and ≤ 5 cm/s, as shown in Figure 5b. These observations align with the opinion of Pranowo et al. (2012), which states that during the East Monsoon or Transition 1 (April), east winds generate currents that move westward. During the east season (ES), the current primarily flows westward (W) at a speed of 10-20 cm/s, 20-30 cm/s, and 30-45 cm/s (as shown in Figure 5c). Additionally, in the East-West transition season (T2) for the westward (W) direction, the speed increases compared to the east season, reaching 30-45 cm/s, 45-60 cm/s, and 20-30 cm/s. Furthermore, the current speed in the transition season is more variable than in the east season, with speeds ranging from 30-45 cm/s and 5-10 cm/s (as illustrated in Figure 5d).

3.2. Arafura Sea Region (H 5 and L6 Zones)

The waters of the Arafura Sea are connected to the Timor Sea and the Banda Sea, so there will be mixing (mixing) between masses of fresh water originating from Papua mainland and the Arafura Sea. The periodic fertilization process occurs due to the process of upwelling and continuous refreshing from the Pacific Ocean through the Indonesian Cross Flow mechanism. These three natural processes that occur regularly cause the fertility of the Arafura Sea to be undoubtedly accompanied by high primary productivity, so that it is a potential fishing area, especially shrimp [18] also added that the Arafura Sea has relatively moderate to strong currents, which allows the spread of planktonic microorganisms such as foraminifera.

The Arafura Sea, including the Sahul shelf, stretches along the eastern longitude from 132° to 141°, spanning a distance of approximately 999 kilometers. Its water depth is ≤ 420 meters, with a bottom layer consisting of thick mud and a small amount of sand covering almost 70% of its water area (Nontji 1993; DISHIDROS, 2015). Given its position between the Aru Sea and Timor Sea, there is mixing between the masses of fresh water originating from Papua mainland and the Timor Sea. The Arafura sea's fertility is sustained by the periodic upwelling of fresh water from the Pacific and Indian Oceans through the Indonesian Cross Current and the Timor Trench and Aru Basin. These natural processes occur regularly, resulting in high levels of primary productivity in the Arafura sea [19].

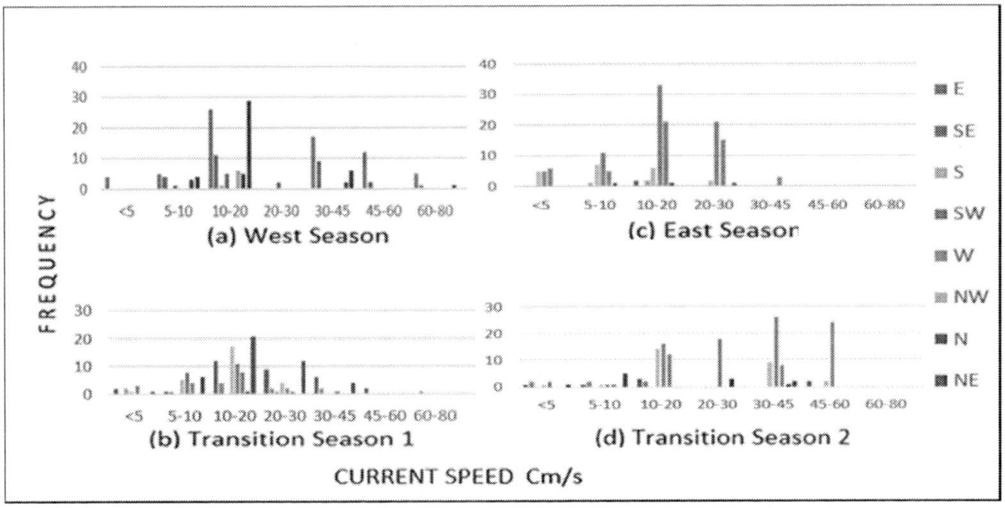

Figure 6. Direction and Speed of Current According to Season in H5 Zone.

The observation site in the Arafura sea's H5 zone, located between 134-135° East Longitude and 8-9° South Latitude within Indonesia's Exclusive Economic Zone, features open waters with depths ranging from 36-50 meters. This zone falls within the neritic region, characterized by depths less than 200 meters (Molles and Manuel, 2013). Current direction and speed in the H5 zone vary with the Monsoon season, as illustrated in Figure 6. During the West Season (WS), currents generally flow northeast (NE) and eastern (E), with speeds of approximately 10-20 cm/s and 30-60 cm/s (Figure 6a). This current direction corresponds with Nontji's 1993 findings, which indicate an eastward flow during the WS. During the West-East transition season (T1), the movement pattern predominantly heads northeast (NE) at velocities of 10-20 cm/s, 5-10 cm/s, and 20-30 cm/s. Conversely, the westward current maintains a more uniform speed of 10-20 cm/s, 20-30 cm/s, and 5- 10 cm/s (Figure 6b). In the East Season (ES), currents generally move western (W) and northwest (NW) at speeds of 10-20 cm/s, 20-30 cm/s, 5-10 cm/s, and ≤5 cm/s, with westward velocities of 10-20cm/s, 20-30 cm/s, and 5-10 cm/s (Figure 6c). During the East-West transition season (T2), currents primarily flow west and southwest at 10-20 cm/s, and northeast at 20-30 cm/s (Figure 6d).

The observation location in the Arafura sea's waters in L6 zone is situated between 138-139° E and 9-10° S within Indonesia's EEZ, characterized by open water. The water depth in L6 zone ranges from 111-118 m falling within the Neritic zone (<200 m). The maximum depth of the Aru Sea, Arafura Sea, and Timor Sea is 5000 m.

The variability of current direction and speed in L6 zone is influenced by the monsoon season. The pattern of current movement based on seasonal differences can be seen in Figure 7. In the West Season (WS), the steady current moves eastern (E), to the northeast (NE). The speed of the current moving to the east is dominated by speeds of 10-20 cm/s and 20-30 cm/s, while the direction of the current moving to the northeast has relatively less frequency and a speed of 10-20 cm/s, 20-30 cm/s (Figure 7a). The direction of the current in this study is in line with the findings of Nontji (1993) who said that the direction of the current corresponds to the direction of the wind, namely when the West Season (WS) the current is heading east.

During the West-East transition season (T1), currents predominantly move westward, eastern, and northeast. Westward currents feature speeds of 10-20 cm/s and 5-10 cm/s, while northwest currents have speeds of 5-10 cm/s. The westward currents are stronger than those moving eastern and northeast, which are primarily 10-20 cm/s (Figure 7b), indicating the West Season's continued dominance in T1. In the East Season (ES), currents generally flow west at 10-20 cm/s and northwest at 20- 30 cm/s, with speeds reaching 45-60 cm/s and 10-20 cm/s (Figure 7c). During the East-West transition season (T2), currents usually move west at 10-20 cm/s, 20-30 cm/s, and 5-10 cm/s, and northwest at 10-20 cm/s (Figure 7d).

3.3. Eastern Timor Sea Region (C6 Zone)

The Timor Sea is an extension of the Indian Ocean located between the island of Timor and the Northern Territory of Australia. The Timor Sea is divided between Indonesia for the eastern sea and Timor Leste for the western sea. The current observation location in the waters of the eastern Timor Sea in C6 zone is located in a geographic position between 129-130° Eastern Longitude and 9-10° South Latitude. This location is located in the EEZ of Indonesia, which is open water. The water depth in C6 zone ranges from 18-412 m. This depth is in the vertical structure of ocean depths, including the mesopelagic zone which has a depth of ≤ 1000 m. [20]

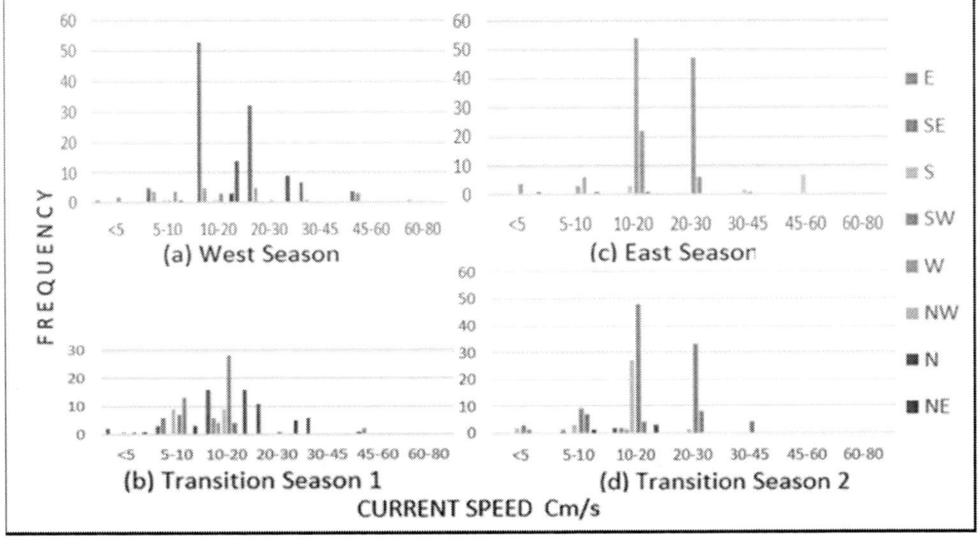

Figure 7. Direction and Speed of Flow According to Season in L6 Zone.

The variability of current direction and speed in the C6 zone is influenced by the Monsoon Season. The pattern of current movement based on seasonal differences can be seen in Figure 8. In the West Season (WS), currents tend to move towards the southeast (SE), northeast (NE) and eastern (E), current variability is dominated at a speed of 10-20 cm/s currents moving to the southeast (SE) are dominated by speeds of 10-20 cm/s, then 5-10 cm/s, 20-30 cm/s and are homogeneous at 30-80 cm/s, while currents moving to the northeast (NE) and eastern (E) have speeds of 10-20 cm/s and ≤5 cm/s (Figure 8a). In the West-East transition season (T1) the current speed towards the west is higher with a current speed of 10-60 cm/s compared to the eastern with a current speed <5-30 cm/s. The current moving to the west (W) is dominated by a speed of 30-45 cm/s and is homogeneous at 45-60 cm/s, 20-30 cm/s and 10-20 cm/s. Currents move to the northwest (NW) and western (W) with a speed of 20-60 cm/s, while those moving to the eastern (E), northeast (NE) and southeast (SE) are dominated by a speed of 10 -20 cm/s (Figure 8b).

In the East Season (ES), currents more often move to the southwest (SW) at a speed of 30-45 cm/s, 10-30 cm/s and <5 cm/s and to the west (W) at a speed of 20-30 cm/s, 10-20 cm/s and <5 cm/s and the dominant flow direction is southwest with a speed of 30-45 cm/s (Figure 8c). Furthermore, in the East-West transition season (T2) the current movement pattern is somewhat similar to the East Season which is dominant towards the southwest (SW) and western (W) with a speed of 20- 30 cm/s, 10-20 cm/s and 30-45 cm/s, and the dominant speed is 20-30 cm/s (Figure 8d).

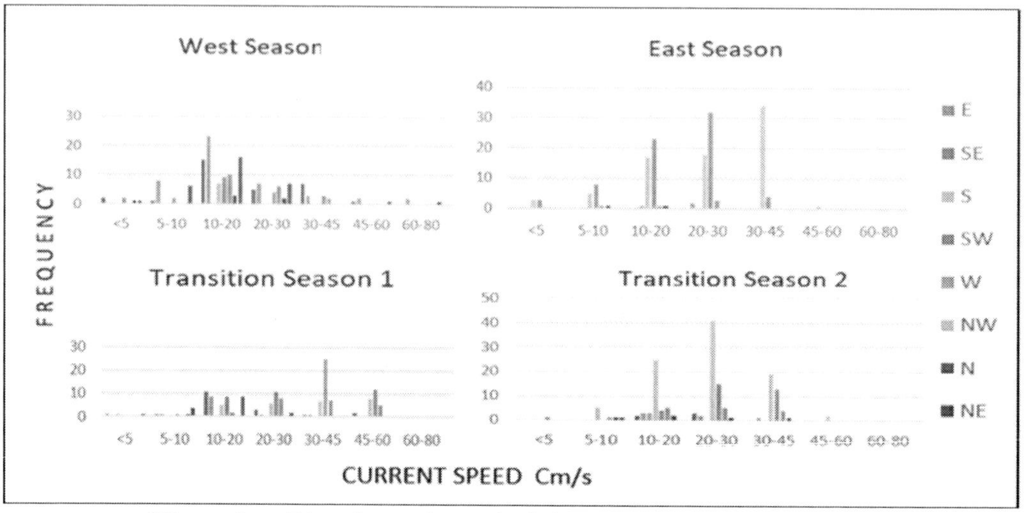

Figure 8. Direction and Speed of Flow According to Season in C6 Zone.

4. Conclusion

The Aru Sea, Arafura Sea and Timor Sea are highly suitable for fishing due to their ideal characteristics of surface currents. These surface currents are predominantly weak and consistent throughout the year, which supports the abundance of fish resources in the area. However, during the east season (ES) and the East-West transition season (T2), the currents heading western can cause a lack of fertility in the fishing areas. The Aru Sea exhibits a specific pattern in the movement of currents across its H1, G3, J3 and H4 zones during each season. In the West Season (WS), the current movement tends towards the eastern and northeast, while in the West-East transition season (T1), the current tends to move towards the northeast, southeast, and eastern. During the east season (ES), the current moves westward, and in the East-West transition season (T2), the current tends to move western and southwest. The current speed in the Aru Sea ranges from 5 to 60 cm/s.

The Arafura Sea's spatiotemporal distribution of current directions also exhibits specific patterns across its H5 and L6 zones during each season. In the West Season (WS), the current movement pattern tends towards the southeast and east, while in the West-East transition season (T1), the current tends to move towards the west and northeast. During the East Season, the current moves west and southwest, and in the East-West transition season (T2), the current tends to move southwest. The current speed in the Arafura Sea ranges from 5 to 80 cm/s. Finally, the Timor Sea's spatiotemporal distribution of current directions also exhibits specific patterns across its C6 zone, during each season. In the West Season (WS), the current movement pattern tends towards the southeast and eastern, while in the West-East transition season (T1), the current tends to move western and southwest. During the East Season (ES), the current moves western and southwest, and in the East-West transition season (T2), the current tends to move southwest. The current speed in the Timor Sea ranges from 5 to 80 cm/s.

Acknowledgements

We extend our gratitude to BMKG-OFS, PUSHIDROSAL and Hang Tuah University for their support of this system. We also thank the Faculty of Engineering and Marine Science (FTIK) at Hang Tuah University for funding the experiment. A special thanks goes to the Magister of Ocean Engineering at Hang Tuah University for their comprehensive support throughout the experiment.

References

[1] Grote B, Ekau W, Hagen W, Huggett JA, Verheye HM. Early life-history strategy of Cape Hake in the Benguela Upwelling Region. Fish Res. 2007; **86**(2-3):179-87.

[2] Hassanudin M. Arus Lintas Indonesia. Oseana. 1998; **23**(2).

[3] HIDROS (Hidrografi dan Oseanografi). Peta Laut Indonesia No. 151 PAPUA dan Pulau-Pulau Sekitarnya. 2015. Type: JPG file; Size: 8.18 MB.

[4] Koch DH. Earth from Space [Internet]. National Geographic; 2014. Available from: https://www.youtube.com/watch?v=kpFryXQbVEA&t=4029s.

[5] MapInfo. Software GIS [Internet]. 2015. Available from: http://web.pb.com/minpro.

[6] Masiyah S. Biodiversitas Mangrove di Kabupaten Merauke Propinsi Papua. J Ilmiah Agribisnis Perikanan 2016; **9**(1):1-7 Available from: https://doi.org/10.29239/j.agrikan.10.1.1-7.

[7] Mengjie Z, Jing T, Fuquan X, Rui W. Point grid map: A new type of thematic map for statistical data associated with geographic points. Cartogr Geogr Inf Sci. 2016; **43**(5):364-75. Available from: https://doi.org/10.1080/15230406.2016.1160797.

[8] Molles C, Manuel Jr. Ecology: Concepts and Applications. 6th ed. New York: McGraw-Hill; 2013.

[9] Mullineaux LS, Mills SW, Sweetman AK, Beaudreau AH, Metaxas A, Hunt HL. Vertical, lateral and temporal structure in larval distributions at hydrothermal vents. Mar Ecol Prog Ser. 2005; **293**:1-16.

[10] OFS. Pusat Meteorologi Maritim BMKG (Badan Meteorologi Klimatologi Geofisika). Ocean Forecast System (OFS) [Internet]. https://peta-maritim.bmkg.go.id/ofs- static.

[11] Permen KP. Nomor: 18/PERMEN-KP/2014 tentang Wilayah Pengelolaan Perikanan Negara Republik Indonesia. 2014.

[12] Pranowo WS, Philips H, Wiffels S. Upwelling event 2003 along South Java Sea and Lesser Sunda Islands. J Segara. 2005;**1**(3):119-26.

[13] Pratiwi ANW, Luthfi OM, Ibrahim F, Putri GA. Studi pola arus perairan Selatan Daerah Istimewa Yogyakarta dengan Menggunakan Metode Penginderaan Jauh. J Ilmiah Rinjani. 2018; **6**(1).

[14] Rosana N, Tambun R. Aplikasi Sistem Informasi Geografis untuk Penyajian Posisi dan Data Pemeriksaan Kapal Penangkap Ikan di Wilayah Timur Indonesia. In: Seminar Nasional Perikanan Tangkap IV; 2011; Institut Pertanian Bogor. Bogor.

[15] Rosana N, Djanat PV, Tambun R. Model based spatial for monitoring surveillance of fisheries to ward illegal fishing in waters of eastern Indonesia. Int J Eng Sci. 2014; **3**(3):1-7.

[16] Rosana N, Djanat PV, Tambun R. GIS for monitoring the operation on inspection and termination of fishing vessels in the eastern Indonesian waters. Int J Eng Sci.2014; **3**(3):20-28

[17] Supartono, Tambun R, Sutisna DH, Mualim R, Krisnafi Y. Forecasting potential fishing areas in Fisheries Management Areas 718 in Indonesia using GIS and Monte Carlo simulation. AACL Bioflux. 2019; **12**(1):1-7. Available from: http://www.bioflux.com.ro/aacl.

[18] Tambun R, Simbolon D, Wahju R, Supartono. Model based spatial for monitoring surveillance of fisheries ships to fishing area at Aru Sea, Arafura Sea and East Timor Sea in FMA-RI 718. Int J Eng Sci. 2018;**7**(10):1-7.

[19] Tambun R, Simbolon D, Wahju R, Supartono. Zona Potensial Penangkapan Ikan Berdasarkan Musim di WPPNRI 718. J Ilmu Teknol Kelautan Tropis. 2018; **10**(3):757-68.

[20] Van der Molen J, Rogers SI, Ellis JR, Fox CJ, McCloghrie P. Dispersal patterns of the eggs and larvae ofspring-spawning fish in the Irish Sea. J Sea Res. 2007; **58**(4):313-30.

AUTHOR INDEX